10/98

WITHDRAWN

ENCYCLOPEDIA OF ACOUSTICS

EDITORIAL BOARD

ENCYCLOPEDIA OF ACOUSTICS

Volume One

MALCOLM J. CROCKER, *Editor-in-Chief*
Auburn University

A Wiley-Interscience Publication

JOHN WILEY & SONS, INC.

New York • **Chichester** • **Weinheim** • **Brisbane** • **Singapore** • **Toronto**

Copyright © 1997 by John Wiley & Sons, Inc.

Library of Congress Cataloging in Publication Data:
Encyclopedia of acoustics / [edited by] Malcolm J. Crocker
      p.   cm.
   Includes index.
   ISBN: 0-471-17767-9 (Volume One)
   ISBN: 0-471-80465-7 (set)
   1. Acoustics—Encyclopedias.   I. Crocker, Malcolm J.
QC221.5.E53   1997
534—dc21   96-37424

Printed in the United States of America

10 9 8 7 6 5 4 3

CONTENTS

Volume One

Volume Three

Volume Four

FOREWORD

Both the Editor-in-Chief Malcolm Crocker, and his publisher, John Wiley & Sons, are to be warmly congratulated on their successes in the arduous tasks of compilation, and ultimate completion, of this comprehensive *Encyclopedia of Acoustics*. They are volumes that will superbly introduce readers to advanced knowledge and skills covering all areas of acoustics as they have developed in the second half of the twentieth century.

This, of course, was a period during which great strides forward were being made in the fundamental mechanics and physics of gases, liquids, and solids—and when such vigorous advances created a strong base for massively increased understanding of acoustics and vibration in their mechanical, physical, and engineering aspects. Simultaneously, huge developments in the modern human and environmental sciences were applied to deepen understanding of the physiology and psychology of human hearing, as well as of its responses to coherent stimuli (including speech and music) and to potentially damaging noise; and to facilitate architectural and engineering improvements aimed at enhancing reception of desired signals and at the limitation and control of noise levels. In the meantime, developments throughout the life sciences stimulated a flowering of bioacoustics, just as the data-processing revolution was transforming acoustic modeling, signal processing, and instrumentation.

An encyclopedia designed to cover such an enormous field needs to be organized with consummate skill, so that readers are continually being helped to find their way through material which, necessarily, is extremely wide-ranging. With this aim in view the *Encyclopedia of Acoustics* has been carefully divided into some 18 parts covering different clearly defined areas of the subject, with the coherence of each part being ensured through an explanatory "Introduction" that appears as the part's first chapter. Yet not one out of the 18 parts is self-sufficient; on the contrary, links to the subject matter of other parts are continually emphasized through cross-references. Indeed a remarkably coherent picture of the entire discipline emerges from this organization of the encyclopedia into 18 parts, each carefully introduced to give internal coherence, while all of them are tightly bound together by such comprehensive cross-linking.

At the same time, none of the devotion shown by all those preparing Introductions to its major parts would have succeeded in making this encyclopedia so outstandingly valuable to readers had it not been for the very special qualities of scientific authority and expository excellence shown by those contributing each of the specialized chapters. The Editor-in-Chief and his Editorial Board are to be congratulated, then, not only on the admirable organization of the four volumes as a whole but also on their brilliant selection of authors for every one of the work's 166 chapters. Indeed, the encyclopedia's exceptional importance stems not only from its refined large-scale design, but also from its in-depth treatment of every significant aspect of contemporary acoustics.

Acoustical scientists entered the twentieth century stimulated by Lord Rayleigh's two great volumes entitled *The Theory of Sound* (2nd Ed., 1896). Today, of course, no work by just one author could perform a similar function. Nevertheless, devotees of acoustics are, as the twenty-first century approaches, being offered an encyclopedia which, once again, may help them to see as a whole their marvelously wide-ranging science.

JAMES LIGHTHILL

PREFACE

In the foreword, Sir James Lighthill generously compares the importance of this book on acoustics with Lord Rayleigh's *Theory of Sound*. Whether this comparison is justified or not will be judged by the readers. At first sight they appear to be very different books. But on closer examination there are similarities. The first edition of Lord Rayleigh's *Theory of Sound* appeared in two volumes in 1876 and 1877, just 120 years ago. The second edition (revised and enlarged) was printed in two volumes in 1894 and 1896. Rayleigh stated in the preface to the first edition of his book that "at the present time many of the most valuable contributions to science are to be found only in scattered periodicals and transactions of societies, published in various parts of the world and in several languages, and are often practically inaccessible to those who do not happen to live in the neighbourhood of large public libraries." Rayleigh also wrote in his preface that it was the purpose of his book to "lay before the reader a connected exposition of the theory of sound, which should include the more important of the advances made in modern times by Mathematicians and Physicists." Few would deny that Rayleigh succeeded magnificently in his "endeavour" at the end of the nineteenth century and that his *Theory of Sound* is a *tour de force*.

When one considers the situation in the field of acoustics one hundred years after Rayleigh's wonderful book, the situation seems surprisingly similar. Although there are several excellent textbooks on different aspects of acoustics theory or practice and some several excellent textbooks on different aspects of acoustics theory or practice and some handbooks on specialized topics in acoustics and vibration, no one book gives a comprehensive treatment. Up-to-date discussions on all topics of acoustics are certainly to be found in journals, periodicals, books, and conference proceedings. But the serious reader who seeks a good understanding of all the important topics in acoustics must consult at least fifteen to twenty journals and five to ten books. Considering the vast expansion of scientific knowledge and the fact that many journal articles are written at a very high level for just a few specialists, the situation is much more difficult for the reader now at the end of the twentieth century than a hundred years ago. It is immediately apparent that no one author can really do justice to the whole field of acoustics and that a multiple-author book is necessary.

Discussions with colleagues Tony Embleton, Manfred Heckl, William Lang, Bruce Lindsay, Richard Lyon, and Ted Schultz in the early 1980s soon convinced me that there was a need for a comprehensive treatment of acoustics, and my planning of a detailed outline for a multiple-author book with 160 chapters began. Because of other commitments it was difficult to make much progress on the book until the mid- to late 1980s. At that time I was fortunate to obtain the help of additional colleagues and form the Editorial Board shown at the beginning of this book. Sadly some of the Editorial Board members have since passed away.

The Editorial Board members provided invaluable help in suggesting the best possible author to write each chapter. In the late 1980s and early 1990s frequent meetings of the Editorial Board were held at meetings of the Acoustical Society of America and some international conferences overseas to coordinate the project and exchange information.

Before the authors began to write their chapters in the early 1990s, I prepared a detailed contents of the whole book which (with the help of the Editorial Board and some authors) included a brief outline of each chapter. The PACS* classification of subjects used by the Journal of the Acoustical Society of America was very useful in planning for the book contents but it was by no means the only comprehensive subject classification system consulted. Classification schemes used in various other journals also proved useful. The tentative book contents and chapter outlines were shared with each author before they began to write their chapters. By this method, it has been possible to reduce extensive overlapping in treatment of topics and to ensure that there is adequate cross-referencing to related material in the book. After each chapter was written it was evaluated by several reviewers and their anonymous comments were shared with the authors who updated their chapters and provided their final versions late in 1995 and early 1996. Each author was asked to prepare a chapter on their topic that would be understandable to general readers not just useful for specialists in their field. At the same time, authors were asked to write chapters that were at a reasonably high scientific level. Also they were asked to provide suitable, up-to-date, readily accessible references for the reader who wished to study a topic in more depth. To prevent the book becoming unreasonably long, each author was given a page allocation. To meet all of these objectives in a limited space was a difficult assignment, but my impression is that most authors have responded admirably.

The book is divided into 18 main parts. A member of the Editorial Board has been associated with each part and in most cases has written the first chapter to serve as an introduction to that part. Coverage in this book includes linear and nonlinear acoustics and cavitation, aeroacoustics and atmospheric sound propagation, underwater acoustics, ultrasonics, mechanical

vibrations and shock, statistical methods, noise control, architectural acoustics, signal processing, physiological acoustics, psychological acoustics, speech communication, musical acoustics, bioacoustics, animal bioacoustics, measurements and instrumentation, and transducer design. With coverage of so many topics, the book is expected to be of interest to engineers, scientists, architects, musicians, physicians, psychologists, and all those whose life acoustics touches in one way or another.

That this book would not have been possible without the dedicated efforts of its 180 authors is obvious and my thanks is extended to each and every one. However, I am also heavily indebted to the almost 500 reviewers who donated their efforts to provide helpful comments to the authors. About 350 of these have allowed their names to be printed in this book. The remainder have wished to remain anonymous. I have already mentioned the members of the Editorial Board who have all been helpful; but some of whom provided great help, beyond the call of duty. I should like to thank all members. I was supported by very able staff at my university especially Yana Sokolova, who possessed excellent organizational skills and who provided really splendid assistance with this project and also by Lisa Beckett in the final stages. I would also like to thank Xinche Yan for providing considerable help in one of the early proofing stages of this book. The staff at Wiley must also be thanked, in particular Bob Hilbert whose scientific knowledge, patience, and dedication to this project made him a joy to work with. Last and not least I would like to thank my wife, Ruth, and daughters, Anne and Elizabeth, for their support, patience, and understanding and to my mother, Alice, for her continual encouragement with this and all of my projects.

MALCOLM J. CROCKER

*Section 4.3 "Acoustics", of the Physics and Astronomy classification Scheme – 1995 (PACS) of the American Institute of Physics.

February, 1997

CONTRIBUTORS

Paul J. Abbas, University of Iowa, Iowa City, IA 52242

Robert E. Apfel, Department of Mechanical Engineering, Yale University, New Haven, CT 06520-8286

Bishnu S. Atal, Bell Laboratories, Lucent Technologies, 600 Mountain Ave., Murray Hill, NJ 07974

Anthony A. Atchley, Naval Postgraduate School, Monterey, CA 93943

Jeffrey C. Bamber, Joint Department of Physics, The Institute of Cancer Research and The Royal Marsden Hospital, Sutton, Surrey SM2 5PT U.K.

James E. Barger, BBN Systems and Technologies, 70 Fawcett St., Cambridge, MA 01238

Henry E. Bass, National Center for Physical Acoustics, University of Mississippi, University, MS 38677

James W. Beauchamp, University of Illinois at Urbana-Champaign, Urbana, IL 61801

Elliott H. Berger, E-A-R/Aearo Company, 7911 Zionsville Rd., Indianapolis, IN 46268-1657

A. J. Berkhout, Laboratory of Seismics and Acoustics, Delft University of Technology, The Netherlands

David A. Bies, Department of Mechanical Engineering, University of Adelaide, South Australia, 5005, Australia

David T. Blackstock, Mechanical Engineering Department and Applied Research Laboratories, The University of Texas at Austin, Austin, TX 78713-8029

William K. Blake, David Taylor Model Basin, Bethesda, MD 20084-5000

Joseph E. Blue, Naval Undersea Warfare Center, Orlando, FL 32856

O. Z. Bluy, Defence Research Establishment Atlantic, Dartmouth, Nova Scotia, Canada

J. M. Bowsher, Formerly with Department of Physics, University of Surrey, Guildford, Surrey GU2 5XH U.K.

Mack A. Breazeale, National Center for Physical Acoustics, University of Mississippi, University, MS 38677

David G. Browning, Browning Biotech, Kingston, RI 02881

Robert D. Bruce, Collaboration in Science and Technology Inc., 15835 Park Ten Place, Houston, TX 77084-5131

P. V. Brüel, Brüel Acoustics, Holte, Denmark

John C. Burgess, Department of Mechanical Engineering, University of Hawaii, Honolulu, HI 96822

Ilene J. Busch-Vishniac, Mechanical Engineering Department, The University of Texas at Austin, Austin, TX 78712

Søren Buus, Communication and Digital Signal Processing Center, Department of Electrical and Computer Engineering, Northeastern University, Boston, MA 02115

Marvin Camras, Illinois Institute of Technology, Chicago, IL 60616

John H. Cantrell, NASA Langley Research Center, Hampton, VA 23681-0001

R. R. Capranica, 1115 West Maximilian Way, Tucson, AZ 85704-3028

Edwin L. Carstensen, The University of Rochester, Rochester, NY 14627

John G. Casali, Industrial and Systems Engineering, Virginia Tech, Blacksburg, VA 24061

Josko A. Catipovic, Woods Hole Oceanographic Institution, Woods Hole, MA 02543

William J. Cavanaugh, Cavanaugh Tocci Associates, Inc., Sudbury, MA 01776

R. P. Chapman, Defence Research Establishment Pacific, Victoria, British Columbia, Canada

Brian L. Clarkson, University of Wales, Swansea, U.K.

Guy Cloutier, Biomedical Engineering Laboratory, Institute of Research in Clinical Medicine, 110 Avenue des Pins Ouest, Montreal, Québec H2W 1R7 Canada

Robert D. Collier, Consultant in Acoustics, rcollier@tpk.net

Alan B. Coppens, P.O. Box 335, Black Mountain, NC 28711-0335

A. Craggs, Department of Mechanical Engineering, University of Alberta, Edmonton, Alberta, Canada

D. G. Crighton, Department of Applied Mathematics & Theoretical Physics, University of Cambridge, Cambridge CB3 9EW U.K.

Malcolm J. Crocker, Department of Mechanical Engineering, Auburn University, Auburn, AL 36849

Lawrence A. Crum, Applied Physics Laboratory, University of Washington, Seattle, WA 98105

Gilles A. Daigle, National Research Council, Ottawa, Ontario K1A 0R6 Canada

P. O. A. L. Davies, Institute of Sound and Vibration Research, The University of Southampton, Southampton SO17 1BJ U.K.

LeRoy M. Dorman, Scripps Institution of Oceanography, University of California, San Diego, La Jolla, CA 92093-0215

Earl H. Dowell, School of Engineering, Duke University, Durham, NC 27708-0271

A. P. Dowling, Engineering Department, Cambridge University, Cambridge, U.K.

Louis R. Dragonette, Applied Acoustics Section, Naval Research Laboratory, Washington, DC 20375

Floyd Dunn, Bioacoustics Research Laboratory, University of Illinois, Urbana, IL 61801

Ira Dyer, Massachusetts Institute of Technology, Cambridge, MA 02139

Peggy L. Edds-Walton, Marine Biological Laboratory, Woods Hole, MA 02543

Kenneth McK. Eldred, Ken Eldred Engineering, East Boothbay, ME 04544

S. J. Elliott, Institute of Sound and Vibration Research, University of Southampton, Southampton SO17 1BJ U.K.

E. Carr Everbach, Department of Engineering, Swarthmore College, Swarthmore, PA 19081-1397

F. J. Fahy, Institute of Sound and Vibration Research, University of Southampton, Southampton SO17 1BJ U.K.

Gunnar Fant, Department of Speech Music and Hearing, KTH, Box 70014, Stockholm S-10044 Sweden

Wolfgang Fasold, Friedenstrasse 2, D-15566 Schöneiche, Germany

Richard R. Fay, Department of Psychology, Parmly Hearing Institute, Loyola University Chicago, Chicago, IL 60626

David Feit, David Taylor Model Basin, Bethesda, MD 20084-5000

Leopold B. Felsen, Department of Aerospace and Mechanical Engineering, and Department of Electrical and Computer Engineering, Boston University, Boston, MA 02215

J. E. Ffowcs Williams, Department of Engineering, University of Cambridge, Cambridge CB2 1PZ U.K.

Sanford Fidell, BBN Systems and Technologies, 21120 Vanowen Street, Canoga Park, CA 91303-2853

Frederick H. Fisher, Marine Physical Laboratory, Scripps Institution of Oceanography, University of California, San Diego, La Jolla, CA 92093-0701

J. L. Flanagan, Rutgers University, Piscataway, NJ 08855-1390

Neville H. Fletcher, Research School of Physical Sciences and Engineering, Australian National University, Canberra 0200 Australia

Kenneth G. Foote, Institute of Marine Research, N-5024 Bergen, Norway

T. G. Forrest, Department of Biology, University of North Carolina at Asheville, Asheville, NC 28804

Shimshon Frankenthal, Tel Aviv Institute of Technology, Ramat Aviv, Israel

C. R. Fuller, Vibration and Acoustics Laboratories, Virginia Polytechnic Institute and State University, Blacksburg, VA 24061

Thomas B. Gabrielson, Applied Research Laboratory, The Pennsylvania State University, State College, PA 16804

Charles F. Gaumond, Naval Research Laboratory, Washington, DC 20375

Thomas L. Geers, Center for Acoustics, Mechanics and Materials, Department of Mechanical Engineering, University of Colorado, Boulder, CO 80309-0427

David M. Green, 399 Federal Point Road, East Palatka, FL 32131

Steven Greenberg, International Computer Science Institute, Berkeley, CA 94704 and University of California, Berkeley, CA 94720

James F. Greenleaf, Ultrasound Research, Mayo Clinic, Rochester, MN 55905

Ervin R. Hafter, University of California, Berkeley, CA 94720

Mark F. Hamilton, Department of Mechanical Engineering, The University of Texas at Austin, Austin, Texas 78712-1063

Colin H. Hansen, Department of Mechanical Engineering, University of Adelaide, South Australia 5005, Australia

William Morris Hartmann, Department of Physics and Astronomy, Michigan State University, East Lansing, MI 48824

Marc D. Hauser, Department of Anthropology and Psychology, Harvard University, Cambridge, MA 02138

Manfred A. Heckl, Institut für Technische Akustik, Technische Universität, Berlin, Germany

Robert Hickling, Sonometrics Inc., 8306 Huntington Rd., Huntington Woods, MI 48070

Joseph E. Hind, University of Wisconsin, Madison, WI 53706

P. C. Hines, Defence Research Establishment Atlantic, Dartmouth, Nova Scotia, Canada

Elmer L. Hixson, Electrical and Computer Department, The University of Texas at Austin, Austin, TX 78712

R. M. Hoover, Hoover & Keith, Inc., 11381 Meadowglen, Houston, TX 77082

Larry E. Humes, Department of Speech and Hearing Sciences, Indiana University, Bloomington, IN 47405

Finn Jacobsen, Department of Acoustic Technology, Technical University of Denmark, Lyngby DK-2800 Denmark

Valery K. Kedrinskii, Lavrentyev Institute of Hydrodynamics, Russian Academy of Sciences, Novosibirsk, Russia

R. H. Keith, Hoover & Keith, Inc., 11381 Meadowglen, Houston, TX 77082

Gary R. Kidd, Department of Speech and Hearing Sciences, Indiana University, Bloomington, IN 47405

Mead C. Killion, Etymotic Research, 700 Perric Dr., #462, Elk Grove, Village, IL 60307

D. O. Kim, Division of Otolaryngology, Department of Surgery, University of Connecticut Health Center, Farmington, CT 06030-1110

Larry S. King, Consultant, 675 West End Avenue, New York, NY 10025

Howard F. Kingsbury, Acoustical Consultant

David Lloyd Klepper, Hebrew University, Givat Ramand, Mt. Scopus, Jerusalem, Israel

William L. Konrad, 54 Laurel Hill Drive, Niantic, CT 06357

Govindappa Krishnappa, National Research Council, Vancouver, British Columbia V6T 1W5 Canada

Robert W. Krug, Cirrus Research, Inc., Waukausha, WI 53188

William A. Kuperman, Marine Physical Laboratory, University of California, San Diego, La Jolla, CA 92037

Ulrich J. Kurze, Müller-BBM GmbH, Robert-Koch-Str. 11, D-82152 Planegg, Germany

K. Heinrich Kuttruff, Aachen University of Technology, D–52056 Aachen, Germany

William W. Lang, Vassar College, Poughkeepsie, NY 12601

Werner Lauterborn, Drittes Physikalisches Institut, Universität Göttingen, D-37073 Göttingen, Germany

Moisés Levy, Department of Physics, University of Wisconsin–Milwaukee, Milwaukee, WI 53201

James Lighthill, Department of Mathematics, University College London, London WC1F 6BT U.K.

Richard H. Lyon, RH Lyon Corp, 691 Concord Ave., Cambridge, MA 02138

Robert C. Maher, University of Nebraska at Lincoln, Lincoln, NE 68588

George C. Maling, Jr., Noise Control Foundation, 62 Timberline Drive, Poughkeepsie, NY 12603

Jerome E. Manning, Cambridge Collaborative, Inc., 689 Concord Avenue, Cambridge, MA 02138

Philip L. Marston, Department of Physics, Washington State University, Pullman, WA 99164-2814

A. Harold Marshall, Marshall Day Associates, Wellesley St., Auckland 01 Australia

J. D. Maynard, Department of Physics, The Pennsylvania State University, University, Park, PA 16802

Dennis McFadden, Institute for Neuroscience and Department of Psychology, The University of Texas at Austin, Austin, TX 78712

Robert H. Mellen, Kildare Corp., New London, CT 06320

Jürgen Meyer, Physikalisch-Technische Bundesanstalt, D-38116 Braunschweig, Germany

Harry B. Miller, Naval Undersea Warfare Center, New London, CT

Joanne L. Miller, Department of Psychology, Northeastern University, Boston, MA 02115

Mark B. Moffett, Naval Undersea Warfare Center, Newport, RI 02841

Brian C. J. Moore, Department of Experimental Psychology, University of Cambridge, Cambridge CB2 3EB U.K.

Charles T. Moritz, Collaboration in Science and Technology Inc., 15835 Park Ten Place, Houston, TX 77084-5131

Michael Möser, Institut für Technische Akustik, Technische Universität, Berlin, Germany

Victor Nedzelnitsky, National Institute of Standards and Technology, Gaithersburg, MD 20899

P. A. Nelson, Institute of Sound and Vibration Research, University of Southampton, Southampton SO17 1BJ U.K.

David E. Newland, Department of Engineering, University of Cambridge, Cambridge CB2 1PZ U.K.

Stephen Nowicki, Department of Zoology, Duke University, Durham, NC 27706

Douglas L. Oliver, Department of Anatomy, University of Connecticut Health Center, Farmington, CT 06032

Kanji Ono, Department of Materials Science and Engineering, University of California, Los Angeles, CA 90095-1595

Emmanuel P. Papadakis, Quality Systems Concepts, Inc., 379 Diem Woods Drive, New Holland, PA 17557-8800

K. Parham, Division of Otolaryngology, Department of Surgery, University of Connecticut Health Center, Farmington, CT 06030-1110

G. Pavic, Acoustics Department, C.E.T.I.M., Senlis 60300 France

William T. Peake, Research Laboratory of Electronics, Massachusetts Institute of Technology, Cambridge, MA 02139 and Eaton-Peabody Laboratory, Massachusetts Eye and Ear Infirmary, Boston, MA 02114

Karl S. Pearsons, BBN Systems and Technologies, 21120 Vanowen Street, Canoga Park, CA 91303-2853

Allan D. Pierce, Department of Aerospace and Mechanical Engineering, Boston University, Boston, MA 02215

Allan G. Piersol, Piersol Engineering Company, 23021 Brenford Street, Woodland Hills, CA 91364

J. Pope, Pope Engineering Company, Newton Centre, MA 02159

Arthur N. Popper, Department of Zoology, University of Maryland, College Park, MD 20742

Alan Powell, Department of Mechanical Engineering, University of Houston, Houston, TX 77204-4792

M. G. Prasad, Noise and Vibration Control Laboratory, Department of Mechanical Engineering, Stevens Institute of Technology, Hoboken, New Jersey 07030

Lawrence R. Rabiner, AT&T Labs-Research, 600 Mountain Ave., Murray Hill, NJ 07974

Richard Raspet, Department of Physics and Astronomy, University of Mississippi, University, MS 38677

Paul J. Remington, BBN Systems and Technologies, Cambridge, MA 02138

Bernard E. Richardson, Department of Physics and Astronomy, University of Wales Cardiff, CF1 3TH U.K.

N. Riley, School of Mathematics, University of East Anglia, Norwich NR4 6XG U.K.

John J. Rosowski, Department of Otology and Laryngology, Harvard Medical School and Eaton-Peabody Laboratory, Massachusetts Eye and Ear Infirmary, Boston, MA 02114

Thomas D. Rossing, Physics Department, Northern Illinois University, Dekalb, IL 60115

Mario A. Ruggero, Department of Communication Sciences and Disorders, Northwestern University, Evanston, IL 60208-3550

Vincent Salmon, Acoustical Consultant, Menlo Park, CA

Joseph Santos-Sacchi, Department of Surgery/Section of Otolaryngology, School of Medicine, Yale University, New Haven, CT 06510-2757

Bertram Scharf, Northeastern University, Boston, MA 02115 and CNRS Centre de Recherche en Neurosciences Cognitives, Marseille, France

Hans G. Schneider, FWG, D-24148 Kiel, Germany

Susan C. Schneider, Department of Electrical and Computer Engineering, Marquette University, Milwaukee, WI 53233

William F. Sewell, Department of Otolaryngology and the Program in Neuroscience, Harvard Medical School, Boston, MA

A. F. Seybert, Department of Mechanical Engineering, University of Kentucky, Lexington, KY 40506

Edgar A. G. Shaw, Institute for Microstructural Sciences, National Research Council, Ottawa, Ontario K1A OR6 Canada

F. Douglas Shields, National Center for Physical Acoustics, University of Mississippi, University, MS 38677

U. S. Shirahatti, University of Chicago, Chicago, IL 60637

K. Kirk Shung, Bioengineering Program, The Pennsylvania State University, University Park, PA 16802

James A. Simmons, Department of Neuroscience, Brown University, Providence, RI 02912

Norma B. Slepecky, Department of Bioengineering and Neuroscience, Institute for Sensory Research, Syracuse University, Syracuse, NY 13244-5290

Philip H. Smith, Department of Anatomy, University of Wisconsin-Madison, Madison, WI 53706

Robert C. Spindel, Applied Physics Laboratory, University of Washington, Seattle, WA 98105

Bradley M. Starobin, Polk Audio, 5601 Metro Drive, Baltimore, MD 21215

Peter Stepanishen, Department of Ocean Engineering, University of Rhode Island, Narragansett, RI 02882-1197

Kenneth N. Stevens, Research Laboratory of Electronics and Department of Electrical Engineering and Computer Science, Massachusetts Institute of Technology, Cambridge, MA 02139

Nobuo Suga, Department of Biology, Washington University, St. Louis, MO 63130

Johan Sundberg, Speech Music Hearing, KTH, Stockholm, Sweden

Kenneth S. Suslick, Department of Chemistry, University of Illinois, Urbana, IL 61801

Louis C. Sutherland, Consultant in Acoustics, 27803 Longhill Dr., Rancho Palos Verdes, CA 90275

Gregory W. Swift, Condensed Matter and Thermal Physics Group, Los Alamos National Laboratory, Los Alamos, NM 87545

Gregory C. Tocci, Cavanaugh Tocci Associates Inc., Sudbury, MA 01776

Mikio Tohyama, Kogakuin University, Hachioji-shi, Tokyo 192 Japan

Ivan Tolstoy, Knockrennie, Castle Douglas DG7 3PA SW Scotland, U.K.

Constanine Trahiotis, University of Connecticut Health Center, Farmington, CT 06032

Herbert Überall, Department of Physics, Catholic University of America, Washington, DC 20064

Eric E. Ungar, BBN Corporation and Acentech Incorporated, 125 Cambridge Park Drive, Cambridge, MA 02140

Arnie Lee Van Buren, Naval Undersea Warfare Center, Newport, RI

Lawrence N. Virgin, School of Engineering, Duke University, Durham, NC 27708-0300

Henning E. von Gierke, Biodynamics and Bioengineering Division, Armstrong Laboratory, Wright-Patterson AFB OH 45433-7901 and School of Medicine, Wright State University, Dayton, OH 45401-0927

W. Dixon Ward, Department of Otolaryngology, University of Minnesota, Minneapolis, MN 55455

A. C. C. Warnock, Institute for Research in Construction, National Research Council, Ottawa Ontario K1A OR6 Canada

Charles S. Watson, Department of Speech and Hearing Sciences, Indiana University, Bloomington, IN 47405

Gabriel Weinreich, University of Michigan, Ann Arbor, MI 48109-1120

D. E. Weston, BAeSEMA Ltd., Apex Tower, 7 High Street, New Malden, Surrey KT3 4LH U.K.

Ewart A. Wetherill, Paoletti Associates, 40 Gold Street, San Francisco, CA 94133

Gary A. Williams, Department of Physics and Astronomy, University of California, Los Angeles, CA 90095

J. Woodhouse, Engineering Department, Cambridge University, Cambridge, U.K.

Peter F. Worcester, Scripps Institution of Oceanography, University of California, San Diego, La Jolla, CA 92093-0225

T. W. Wu, Department of Mechanical Engineering, University of Kentucky, Lexington, KY 40506

Tom C. T. Yin, Department of Neurophysiology, University of Wisconsin-Madison, Madison, WI 53706

William A. Yost, Parmly Hearing Institute, Loyola University, Chicago, IL 60026

William T. Yost, NASA Langley Research Center, Hampton, VA 23681-0001

H. K. Zaveri, Brüel & Kjœr, Nœrum, Denmark

Randy Zelick, Department of Biology, Portland State University, Portland, OR 97207

William E. Zorumski, NASA Langley Research Center, Hampton, Virginia 23681

Allan J. Zuckerwar, NASA Langley Research Center, Hampton, VA 23681

PART I

GENERAL LINEAR ACOUSTICS

1

INTRODUCTION

M‍ALCOLM J. C‍ROCKER

1 INTRODUCTION

The fluid mechanics equations, from which the acoustics equations and results may be derived, are quite complicated. However, because most acoustic phenomena involve very small perturbations, it is possible to make significant simplifications to these fluid equations and to linearize them. The results are the equations of linear acoustics. The most important equation, the wave equation, is presented in this chapter together with some of its solutions. Such solutions give the sound pressure explicitly as functions of time and space, and the general approach may be termed the wave acoustics approach. This chapter presents some of the useful results of this approach but also briefly discusses some of the other alternative approaches, sometimes termed ray acoustics and energy acoustics, that are used when the wave acoustics approach becomes too complicated.

The first purpose of this chapter is to present some of the most important acoustics formulas and definitions, without derivation, which are used in the chapters following in Part I and in many of the other chapers. The second purpose is to make some helpful comments about the chapters that follow in Part I and about other chapters as it seems appropriate.

2 WAVE MOTION

Some of the basic concepts of acoustics and sound wave propagation used in Part I and also throughout the rest of this book are discussed here. For a more advanced

Encyclopedia of Acoustics, edited by Malcolm J. Crocker
ISBN 0-471-80465-7 © 1997 John Wiley & Sons, Inc.

mathematical treatment the reader is referred to Chapter 2 and later chapters.

Wave motion is easily observed in the waves on stretched strings and as the ripples on the surface of water. Waves on strings and surface water waves are very similar to sound waves in air (which we cannot see), but there are some differences that are useful to discuss.

If we throw a stone into a calm lake, we observe that the water waves (*ripples*) travel out from the point where the stone enters the water. The ripples spread out circularly from the source at the wave speed which is independent of the wave height. Somewhat like the water ripples, sound waves in air travel at a constant speed, almost independent of their strength. Like the water ripples, sound waves in air propagate by transferring momentum and energy between air particles. There is no net flow of air away from a source of sound, just as there is no net flow of water away from the source of water waves. Of course, the waves on the surface of a lake are circular or two dimensional, while in air, sound waves in general are spherical or three dimensional.

As water waves move away from a source, their curvature decreases, and the *wavefronts* may be regarded almost as straight lines. Such waves are observed in practice as *breakers* on the seashore. A similar situation occurs with sound waves in the atmosphere. At large distances from a source of sound, the spherical wavefront curvature decreases, and the wavefronts may be regarded almost as plane surfaces.

Plane sound waves may be defined as waves that have the same acoustic properties at any position on a plane surface drawn perpendicular to the direction of propagation of the wave. Such plane sound waves can exist and propagate along a long straight tube or duct (such as an air conditioning duct). In such a case, the waves propagate in a direction along the duct and the plane waves are perpendicular to this direction (and are represented by duct cross sections). Such waves in a duct are one dimensional, like the waves on a long string or rope under tension (or like the ocean breakers described above).

Although there are many similarities between one-dimensional sound waves in air, waves on strings, and surface water waves, there are some minor differences. In a fluid such as air, the fluid particles vibrate back and forth in the same direction as the direction of wave propagation; such waves are known as *longitudinal*, *compressional*, or *sound waves*. On a stretched string, the particles vibrate at right angles to the direction of wave propagation; such waves are usually known as *transverse waves*. The surface water waves described are also partly transverse partly longitudinal waves, with the complication that the water particles move up and down and back and forth horizontally. (This movement describes elliptical paths in shallow water and circular paths in deep water. The vertical particle motion is much greater than the horizontal motion for shallow water, but the two motions are equal for deep water.) The water wave direction is, of course, horizontal.

Surface water waves are not compressional (like sound waves) and are normally termed *surface gravity waves*. Unlike sound waves, where the wave speed is independent of frequency, long wavelength surface water waves travel faster than short wavelength waves, and thus water wave motion is said to be *dispersive*.[1] Bending waves on beams, plates, cylinders, and other engineering structures are also dispersive (see Chapter 11). There are several other types of waves that can be of interest in acoustics: shear waves, torsional waves, and boundary waves (see Chapter 12), but the discussion here will concentrate on sound wave propagation in fluids.

3 PLANE SOUND WAVES

If a disturbance in a thin element of fluid in a duct is considered, a mathematical description of the motion may be obtained by assuming that (1) the amount of fluid in the element is conserved, (2) the net longitudinal force is balanced by the inertia of the fluid in the element, (3) the process in the element is adiabatic (i.e., there is no flow of heat in or out of the element), and (4) the undisturbed fluid is stationary (there is no fluid flow). Then the following equation of motion may be derived.[1-6]

$$\frac{\partial^2 p}{\partial x^2} - \frac{1}{c^2}\frac{\partial^2 p}{\partial t^2} = 0. \tag{1}$$

This equation is known as the one-dimensional equation of motion, or *acoustic wave equation*, and it relates the second rate of change of the sound pressure with the coordinate x with the second rate of change of the sound pressure with time t through the square of the speed of sound c. Identical wave equations may be written if the sound pressure p in Eq. (1) is replaced with the particle displacement ξ, the particle velocity u, fluctuating density ρ, or the fluctuating temperature T. However, the wave equation in terms of the sound pressure in Eq. (1) is perhaps most useful, since the sound pressure is the easiest acoustic quantity to measure (using a microphone) and is the acoustic perturbation we sense with our ears. The sound pressure p is the acoustic pressure perturbation or fluctuation about the time-averaged, or undisturbed, pressure p_0.

The *speed of sound waves* c is given for a perfect gas by

$$c = (\gamma R T)^{1/2}. \tag{2}$$

The speed of sound is proportional to the square root of the absolute temperature T. The ratio of specific heats γ and the gas constant R are constants for any particular gas. Thus Eq. (2) may be written as

$$c = c_0 + 0.6 T_C, \tag{3}$$

where, for air, $c_0 = 331.6$ m/s, the speed of sound at $0°$C, and T_C is the temperature in degrees Celsius. Note that Eq. (3) is an approximate formula valid for T_C near room temperature. The speed of sound in air does not depend on the atmospheric pressure. For a complete discussion of the speed of sound in fluids, see Chapter 6.

A solution to (1) is

$$p = f_1(ct - x) + f_2(ct + x), \tag{4}$$

where f_1 and f_2 are arbitrary functions such as sine, cosine, exponential, log, and so on. It is easy to show that Eq. (4) is a solution to the wave equation (1) by differentiation and substitution into Eq. (1). Varying x and t in Eq. (4) demonstrates that $f_1(ct - x)$ represents a wave traveling in the positive x-direction with wave speed c, while $f_2(ct + x)$ represents a wave traveling in the negative x-direction with wave speed c (see Fig. 1).

The solution given in Eq. (4) is usually known as the *general solution*, since, in principle, any type of sound wave form is possible. In practice, sound waves are usually classified as impulsive or steady in time. One particular case of a steady wave is of considerable importance. Waves created by sources vibrating sinusoidally in

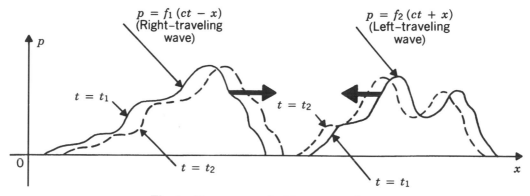

Fig. 1 Plane waves of arbitrary wave form.

time (e.g., a loudspeaker, a piston, or a more complicated structure vibrating with a discrete angular frequency ω) vary both in time t and space x in a sinusoidal manner (see Fig. 2):

$$p - p_1 \sin(\omega t - kx + \phi_1) + p_2 \sin(\omega t + kx + \phi_2). \quad (5)$$

At any point in space, x, the sound pressure p is simple harmonic in time. The first expression on the right of Eq. (5) represents a wave of amplitude p_1 traveling in the positive x-direction with speed c, while the second expression represents a wave of amplitude p_2 traveling in the negative x-direction. The symbols ϕ_1 and ϕ_2 are *phase angles*, and k is the *acoustic wavenumber*. It is observed that the wavenumber $k = \omega/c$ by studying the ratio of x and t in Eqs. (4) and (5). At some instant t the sound pressure pattern is sinusoidal in space, and it repeats itself each time kx is increased by 2π. Such a repetition is called a wavelength λ. Hence, $k\lambda = 2\pi$ or $k = 2\pi/\lambda$. This gives $\omega/c = 2\pi f/c = 2\pi/\lambda$, or

$$\lambda = \frac{c}{f}. \quad (6)$$

The wavelength of sound becomes smaller as the frequency is increased. At 100 Hz, $\lambda \approx 3.5$ m ≈ 10 ft. At 1000 Hz, $\lambda \approx 0.35$ m ≈ 1 ft. At 10,000 Hz, $\lambda \approx 0.035$ m ≈ 0.1 ft ≈ 1 in.

At some point x in space, the sound pressure is sinusoidal in time and goes through one complete cycle when ω increases by 2π. The time for a cycle is called the period T. Thus, $\omega T = 2\pi$, $T = 2\pi/\omega$, and

$$T = \frac{1}{f}. \quad (7)$$

4 IMPEDANCE AND SOUND INTENSITY

We see that for the one-dimensional propagation considered the sound wave disturbances travel with a constant

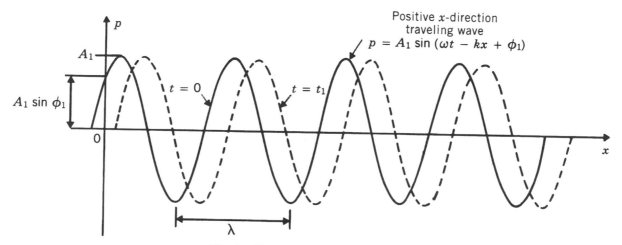

Fig. 2 Simple harmonic plane waves.

wave speed c, although there is no net, time-averaged movement of the air particles. The air particles oscillate back and forth in the direction of wave propagation (x-axis) with velocity u. We may show that for any plane wave traveling in the positive x-direction at any instant

$$\frac{p}{u} = \rho c, \tag{8}$$

and for any plane wave traveling in the negative x-direction

$$\frac{p}{u} = -\rho c. \tag{9}$$

The quantity ρc is called the characteristic impedance of the fluid, and for air, $\rho c = 428$ kg/m^{2}s at 0°C and 415 kg/m^{2}s at 20°C.

The *sound intensity* is the rate at which the sound wave does work on an imaginary surface of unit area in a direction perpendicular to the surface. Thus, it can be shown that the *instantaneous* sound intensity in the x-direction, I, is obtained by multiplying the *instantaneous* sound pressure p by the *instantaneous* particle velocity in the x-direction, u. Therefore

$$I = pu, \tag{10}$$

and for a plane wave traveling in the positive x-direction this becomes

$$I = \frac{p^2}{\rho c}. \tag{11}$$

The time-averaged sound intensity for a plane wave traveling in the positive x-direction, $\langle I \rangle_t$, is given as

$$\langle I \rangle_t = \frac{\langle p^2 \rangle_t}{\rho c}, \tag{12}$$

and for the special case of a sinusoidal (pure-tone) wave

$$\langle I \rangle_t = \frac{\langle p^2 \rangle_t}{\rho c} = \frac{p_1^2}{2\rho c}, \tag{13}$$

where p_1 is the pressure amplitude, and the mean-square pressure is thus $\langle p^2 \rangle_t = p_{rms}^2 = \frac{1}{2}p_1^2$.

We note, in general, for sound propagation in three dimensions, that the instantaneous sound intensity $\mathbf{I}$ is a vector quantity equal to the product of the sound pressure and the instantaneous particle velocity $\mathbf{u}$. Thus $\mathbf{I}$ has magnitude and direction. The vector intesity $\mathbf{I}$ may be resolved into components $\mathbf{I}_x$, $\mathbf{I}_y$, and $\mathbf{I}_z$. For a more complete discussion of sound intensity and its measurement see Chapter 156 and Ref. 7.

5   THREE-DIMENSIONAL WAVE EQUATION

In most sound fields, sound propagation occurs in two or three dimensions. The three-dimensional version of Eq. (1) in Cartesian coordinates is

$$\frac{\partial^2 p}{\partial x^2} + \frac{\partial^2 p}{\partial y^2} + \frac{\partial^2 p}{\partial z^2} - \frac{1}{c^2}\frac{\partial^2 p}{\partial t^2} = 0. \tag{14}$$

This equation is useful if sound wave propagation in rectangular spaces such as rooms is being considered. However, it is helpful to recast Eq. (14) in spherical coordinates if sound propagation from sources of sound in free space is being considered. It is a simple mathematical procedure to transform Eq. (14) into spherical coordinates, although the resulting equation is quite complicated. However, for propagation of sound waves from a spherically symmetric source (such as the idealized case of a pulsating spherical balloon known as an omnidirectional or monopole source) (Table 1), the equation becomes quite simple (since there is no angular dependence):

$$\frac{\partial^2(rp)}{\partial r^2} - \frac{1}{c^2}\frac{\partial^2(rp)}{\partial t^2} = 0. \tag{15}$$

Here, r is the distance from the origin and p is the sound pressure at that distance.

Equation (15) is identical in form to Eq. (1) with p replaced by rp and x by r. The general and simple harmonic solutions to Eq. (15) are thus the same as Eqs. (4) and (5) with p replaced by rp and x with r. The general solution is

$$rp = f_1(ct - r) + f_2(ct + r), \tag{16}$$

or

$$p = \frac{1}{r}f_1(ct - r) + \frac{1}{r}f_2(ct + r), \tag{17}$$

where f_1 and f_2 are arbitrary functions. The first term on the right of Eq. (17) represents a wave traveling outward from the origin; the sound pressure p is seen to be inversely proportional to the distance r. If the distance r is doubled, the sound pressure level [Eq. (29)] decreases by $20 \log_{10}(2) = 20(0.301) = 6$ dB. This is known as the *inverse-square law*. The second term in Eq.

TABLE 1   Models of Idealized Spherical Sources: Monopole, Dipole, and Quadrupole[a]

Monopole distribution representation	Velocity distribution on spherical surface	Oscillating sphere representation	Oscillating force model
Monopole			
Dipole			
Quadrupole (Lateral quadrupole shown)			(Longitudinal quadrupole) (Lateral quadrupole)

[a]For simple harmonic forces, after one half-period the velocity changes direction; positive sources become negative and vice versa.

(17) represents sound waves traveling inward toward the origin, and in most practical cases these can be ignored (if reflecting surfaces are absent).

The simple harmonic (pure-tone) solution of Eq. (15) is

$$p = \frac{p_1}{r} \sin(\omega t - kr + \phi_1) + \frac{p_2}{r} \sin(\omega t + kr + \phi_2), \quad (18)$$

where p_1 and p_2 are the sound pressure amplitudes.

6   SOURCES OF SOUND

The second term on the right of Eq. (18), as before, represents sound waves traveling inward to the origin and is of little practical interest. However, the first term represents simple harmonic waves of angular frequency ω traveling outward from the origin, and this may be rewritten as[5]

$$p = \frac{\rho c k Q}{4\pi r} \sin(\omega t - kr + \phi_1), \quad (19)$$

where Q is termed the *strength of an omnidirectional (monopole) source* situated at the origin, and $Q = 4\pi p_1/\rho c k$. The mean-square pressure p_{rms}^2 may be found[5] by time-averaging the square of Eq. (19) over a period T:

$$p_{\text{rms}}^2 = \frac{(\rho c k)^2 Q^2}{32\pi^2 r^2}. \quad (20)$$

From Eq. (20), the mean-square pressure is seen to vary with the inverse square of the distance r from the origin of the source for such an idealized omnidirectional point source everywhere in the sound field. Again, this is known as the inverse-square law. If the source is idealized as a sphere of radius a pulsating with a simple harmonic velocity amplitude U, we may show that Q has units of volume flow rate (cubic metres per second). If the source radius is small in wavelengths so that $a \ll \lambda$ or $ka \ll 1$, then we can show that the strength $Q = 4\pi a^2 U$.

Many sources of sound are not like the simple omnidirectional monopole source just described. For example, an unbaffled loudspeaker produces sound both from the back and front of the loudspeaker. The sound from the front and the back can be considered as two sources that are 180° out of phase with each other. This system can be modeled[5,8] as two out-of-phase monopoles of source strength Q separated by a distance l. The sound pressure produced by such a dipole system is

$$p = \frac{\rho c k Q l \cos \theta}{4\pi r}$$
$$\cdot \left[\frac{1}{r} \sin(\omega t - kr + \phi) + k \cos(\omega t - kr + \phi) \right], \quad (21)$$

where θ is the angle measured from the axis joining the two sources (the loudspeaker axis in the practical case). Unlike the monopole, the dipole field is not omnidirectional. The sound pressure field is directional. It is, however, symmetric and shaped like a figure-eight with its lobes on the dipole axis.

For a dipole source the sound pressure has a near-field and a far-field behavior similar to the particle velocity of a monopole. Close to the source (the near field), for some fixed angle θ, the sound pressure falls off rapidly, $p \propto 1/r^2$, while far from the source (the far field $kr \gg 1$), the pressure falls off more slowly, $p \propto 1/r$. In the near field the sound pressure level decreases by 12 dB for each doubling of distance r. In the far field the decrease in sound pressure level is only 6 dB for doubling of r (like a monopole). The phase of the sound pressure also changes with distance r, since close to the source the sine term dominates and far from the source the cosine term dominates. The particle velocity may be obtained from the sound pressure [Eq. (21)] and use of Euler's equation [Eq. (22)]. It has an even more complicated behavior with distance r than the sound pressure, having three distinct regions.

As discussed in Chapter 9, an oscillating force applied at a point in space gives rise to results identical to Eq. (21), and hence there are many real sources of sound that behave like the idealized dipole source described above,

for example, pure-tone fan noise, vibrating beams, unbaffled loudspeakers, and even wires and branches (which sing in the wind due to alternate vortex shedding).

The next higher order source is the quadrupole. It is thought that the sound produced by the mixing process in an air jet gives rise to stresses that are quadrupole in nature. See Chapters 9, 27, and 28. Quadrupoles may be considered to consist of two opposing point forces (two opposing dipoles) or equivalently four monopoles. (See Table 1.) We note that some authors use slightly different but equivalent definitions for the source strength of monopoles, dipoles, and quadrupoles. The definitions used in Sections 6 and 8 of this chapter are the same as in Refs. 5 and 8 and result in expressions for sound pressure, sound intensity, and sound power, which although equivalent are different in form from those in Chapter 9, for example.

The expression for the sound pressure for a quadrupole is even more complicated than for a dipole. Close to the source, in the near field, the sound pressure $p \propto 1/r^3$. Farther from the source, $p \propto 1/r^2$; while in the far field, $p \propto 1/r$.

Sound sources experienced in practice are normally even more complicated than dipoles or quadrupoles. The sound radiation from a vibrating piston is described in Chapters 9 and 11, the sound radiation from vibrating cylinders in Chapter 9.

The discussion in Chapter 9 considers steady-state radiation. However, there are many sources in nature and created by people that are transient. As shown in Chapter 10, harmonic analysis of these cases is often not suitable and time-domain methods have given better results and understanding of the phenomena. These are the approaches adopted in Chapter 10.

7   SOUND INTENSITY AND DIRECTIVITY

The radial particle velocity in a spherically spreading sound field is given by Euler's equation as

$$u = -\frac{1}{\rho} \int \frac{\partial p}{\partial r} \, dt, \quad (22)$$

and substituting Eqs. (19) and (22) into (10) then using Eq. (20) and time-averaging gives the magnitude of the radial intensity in such a field as

$$\langle I \rangle_t = \frac{p_{\text{rms}}^2}{\rho c}, \quad (23)$$

the same result as for a plane wave [see Eq. (12)].

The sound intensity decreases with the inverse square of the distance r. Simple omnidirectional monopole sources radiate equally well in all directions. More complicated idealized sources such as dipoles, quadrupoles, and vibrating piston sources create sound fields that are directional (see Fig. 3). Of course, real sources such as machines produce even more complicated sound fields than these idealized sources. (For a more complete discussion of the sound fields created by idealized sources, see Chapter 9.) However, the same result as Eq. (23) is found to be true for any source of sound as long as the measurements are made sufficiently far from the source. The intensity is not given by the simple result of Eq. (23) close to sources such as dipoles, quadrupoles, or more complicated sources of sound. Close to such sources Eq. (10) must be used for the instantaneous radial intensity, or

$$\langle I \rangle_t = \langle pu \rangle_t \tag{24}$$

for the time-averaged radial intensity.

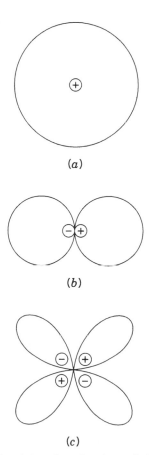

(a)

(b)

(c)

Fig. 3 Polar directivity plots for the radial sound intensity in the far field of (a) monopole, (b) dipole, and (c) (lateral) quadrupole.

The time-averaged radial sound intensity for a dipole is given by[5,8]

$$\langle I \rangle_t = \frac{\rho c k^4 (Ql)^2 \cos^2 \theta}{32\pi^2 r^2} . \tag{25}$$

The sound intensity radiated by a dipole is seen to depend on $\cos^2 \theta$. In general, a directivity factor $D_{\theta,\phi}$ may be defined as the ratio of the radial intensity $\langle I_{\theta,\phi} \rangle_t$ (at angles θ and ϕ and distance r from the source) to the radial intensity $\langle I_s \rangle_t$ at the same distance r from an omnidirectional source of the same total power. Thus

$$D_{\theta,\phi} = \frac{\langle I_{\theta,\phi} \rangle_t}{\langle I_s \rangle_t} . \tag{26}$$

A directivity index $\mathrm{DI}_{\theta,\phi}$ may be defined, where

$$\mathrm{DI}_{\theta,\phi} = 10 \log D_{\theta,\phi}. \tag{27}$$

8 SOUND POWER

The *sound power P* of a source is given by integrating the intensity over any closed surface S around the source (see Fig. 4).

$$P = \int_s \langle I_n \rangle_t \, dS. \tag{28}$$

The normal component of the intensity I_n must be measured in a direction perpendicular to the elemental area dS. If a spherical surface is chosen, then the sound power of an omnidirectional (monopole) source is

$$P_m = \langle I_r \rangle_t 4\pi r^2, \tag{29}$$

$$P_m = \frac{p_{\mathrm{rms}}^2}{\rho c} 4\pi r^2, \tag{30}$$

and from Eq. (20) the sound power of a monopole is[5,8]

$$P_m = \frac{\rho c k^2 Q^2}{8\pi} . \tag{31}$$

It is apparent from Eq. (31) that the sound power of an idealized (monopole) source is independent of the distance r from the origin, where the source is located. This is the result required by conservation of energy and also to be expected for all sound sources.

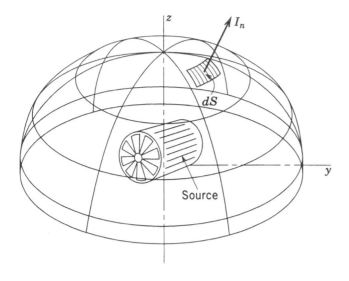

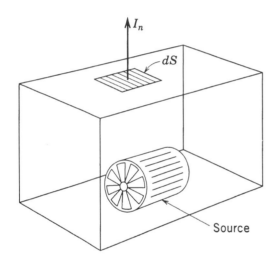

Fig. 4   Sound intensity I_n being measured on a segment dS of a hemispherical enclosure surface surrounding a source having a sound power P.

Equation (30) shows that for an omnidirectional source (in the absence of reflections) the sound power can be determined from measurements of the mean-square pressure made with a single microphone. Of course, for such a source, measurements should really be made with a *reflection-free (anechoic) environment* or very close to the source where reflections are presumably less important.

The sound power of a dipole source is obtained by integrating the intensity given by Eq. (25) over a sphere around the source. The result for the sound power is[5,8]

$$P_d = \frac{\rho c k^4 (Ql)^2}{24\pi}.$$ (32)

The dipole is obviously a much less efficient radiator than a monopole, particularly at low frequency.

In practical situations with real directional sound sources and where reflections are important, use of Eq. (30) becomes difficult and less accurate, and then the sound power is more conveniently determined from Eq. (28) with a sound intensity measurement system. See Chapter 156.

It is important to note that the sound power radiated by a source can be significantly affected by its environment. For example, if a monopole source with a high internal impedance (whose strength Q will be unaffected by the environment) is placed on a floor, its sound power will be doubled (and its sound power level increased by 3 dB). If it is placed at a floor–wall intersection, its sound power will be increased by four times (6 dB); and if it is placed in a room corner, its power is increased by eight times (9 dB).

9   DECIBELS AND LEVELS

The range of sound pressure magnitude and sound power experienced in practice is very large (see Figs. 5 and 6). Thus, logarighmic rather than linear measures are often used for sound pressure and power. The most common is the *decibel*. The decibel represents a relative measurement or ratio. Each quantity in decibels is expressed as a ratio relative to a *reference sound pressure, power,* or *intensity*. Whenever a quantity is expressed in decibels, the result is known as a *level*.

The decibel (dB) is the ratio R_1 given by

$$\log_{10} R_1 = 0.1, \qquad 10 \log_{10} R_1 = 1 \text{ dB}.$$ (33)

Thus, $R_1 = 10^{0.1} = 1.26$. The decibel is seen to represent the ratio 1.26. A larger ratio, the *bel* is sometimes used. The bel is the ratio R_2 given by $\log_{10} R_2 = 1$. Thus, $R_2 = 10^1 = 10$. The bel represents the ratio 10.

The *sound pressure level* L_p is given by

$$L_p = 10 \log_{10} \left(\frac{\langle p^2 \rangle_t}{p_{\text{ref}}^2} \right) = 10 \log_{10} \left(\frac{p_{\text{rms}}^2}{p_{\text{ref}}^2} \right)$$

$$= 20 \log_{10} \left(\frac{p_{\text{rms}}}{p_{\text{ref}}} \right) \qquad \text{dB},$$ (34)

where p_{ref} is the reference pressure $p_{\text{ref}} = 20\ \mu\text{Pa} = 0.00002\ \text{N/m}^2\ (= 0.0002\ \mu\text{bar})$. This reference pressure was originally chosen to correspond to the quietest sound (at 1000 Hz) that the average young person can hear.

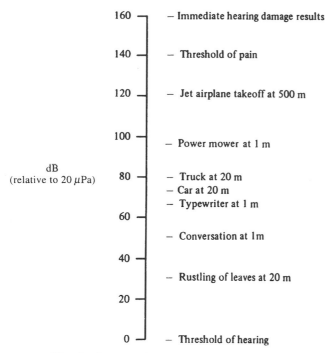

Fig. 5   Some typical sound pressure levels.

The sound power level L_W is given by

$$L_W = 10 \log_{10} \left(\frac{W}{W_{\text{ref}}} \right) \quad \text{dB} \quad (35)$$

where W is the sound power of a source and $W_{\text{ref}} = 10^{-12}$ W is the reference sound power.

The sound intensity level L_i is given by

$$L_i = 10 \log_{10} \left(\frac{I}{I_{\text{ref}}} \right) \quad \text{dB}, \quad (36)$$

where I is the component of the sound intensity in a given direction and $I_{\text{ref}} = 10^{-12}$ W/m^2 is the reference sound intensity.

Some typical sound pressure and sound power levels are given in Figs. 5 and 6.

If two sound sources radiate independently (uncorrelated sources), then because of the interfering effects of the source pressures time-average to zero, the mean

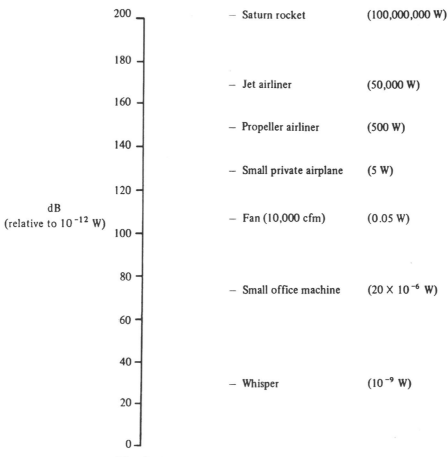

Fig. 6   Some typical sound power levels.

$L_{tot} - L_1$
Decibels to be Added to Higher Level

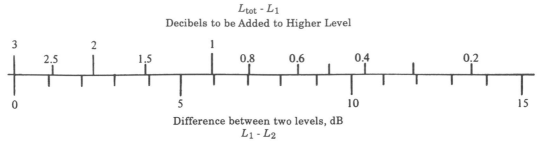

Difference between two levels, dB
$L_1 - L_2$

Fig. 7 Chart for combination levels of decibels (for uncorrelated sources).

square pressures are additive and the total sound pressure level at some point in space, or the total sound power level, may be determined using Fig. 7. *First example:* If two independent sound sources each create sound pressure levels operating on their own of 80 dB, at a certain point, what is the total level? *Answer:* The difference in levels is 0 dB; thus the total sound pressure level is 80 + 3 = 83 dB. *Second example:* If two independent sound sources have sound power levels of 70 and 73 dB, what is the total level? *Answer:* The difference in levels is 3 dB; thus the total sound power level is 73 + 1.8 = 74.8 dB.

Figure 7 and these two examples do *not* apply to the case of two pure tones of the same frequency. For such a case in the first example instead of 83 dB, the total sound pressure level can range anywhere between 86 dB (for in-phase sound pressures), to $-\infty$ dB (for out-of-phase sound pressures). For the second example the total sound power radiated by the two pure-tone sources depends on the phasing and separation distance. As discussed in Section 7, two out-of-phase closely spaced pure-tone sound sources are termed a *dipole*.

10 REFLECTION, REFRACTION, SCATTERING, AND DIFFRACTION

For a homogeneous plane sound wave at normal incidence on a fluid medium of different characteristic impedance ρc, both reflected and transmitted waves are formed (see Fig 8).

From energy considerations (provided no losses occur at the boundary) the sum of the reflected intensity I_r and transmitted intensity I_t equals the incident intensity I_i,

$$I_i = I_r + I_t, \tag{37}$$

and dividing throughout by I_i,

$$\frac{I_r}{I_i} + \frac{I_t}{I_i} = R + T = 1, \tag{38}$$

where R is the *reflection coefficient* and T is the *transmission coefficient*. For plane waves at normal incidence on a plane boundary between two fluids (see Fig. 8)

$$R = \frac{(\rho_1 c_1 - \rho_2 c_2)^2}{(\rho_1 c_1 + \rho_2 c_2)^2}, \tag{39}$$

and

$$T = \frac{4\rho_1 c_1 \rho_2 c_2}{(\rho_1 c_1 + \rho_2 c_2)^2}. \tag{40}$$

Some interesting facts can be deduced from Eqs. (39) and (40). Both the reflection and transmission coefficients are independent of the direction of the wave, since interchanging $\rho_1 c_1$ and $\rho_2 c_2$ does not affect the values of R and T. For example, for sound waves traveling from air to water or water to air, almost complete reflection occurs, independent of direction, and the reflection coefficients are the same and the transmission coefficients are the same for the two different directions.

As discussed before, when the characteristic impedance ρc of a fluid medium changes, incident sound is both reflected and transmitted. It can be shown that if a

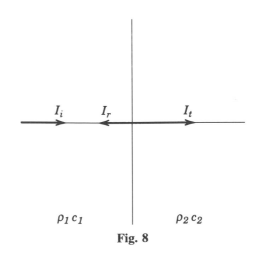

Fig. 8

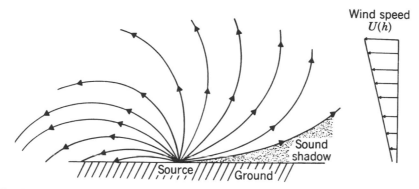

Fig. 9   Refraction of sound in air with wind speed $U(h)$ increasing with altitude h.

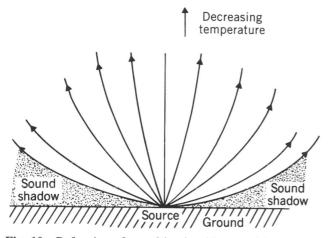

Fig. 10   Refraction of sound in air with normal temperature lapse (temperature decreases with altitude).

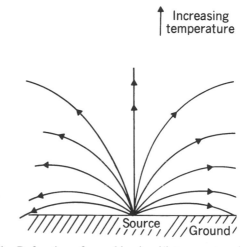

Fig. 11   Refraction of sound in air with temperature inversion.

both reflected and transmitted. It can be shown that if a plane sound wave is incident at an oblique angle on a plane boundary between two fluids then the wave transmitted into the changed medium changes direction. This effect is called *refraction*. Temperature changes and wind speed changes in the atmosphere are important causes of refraction.

Wind speed normally increases with altitude, and Fig. 9 shows the refraction effects to be expected for an idealized wind speed profile. Atmospheric temperature changes alter the speed of sound c, and temperature gradients can also produce sound shadow and focusing effects, as seen in Figs. 10 and 11.

When a sound wave meets an obstacle, some of the sound wave is deflected. The *scattered* wave is defined to be the difference between the resulting wave with the obstacle and the undisturbed wave without the presence of the obstacle. The scattered wave spreads out in all directions interfering with the undisturbed wave. If the obstacle is very small compared with the wavelength, no sharp-edged sound shadow is created behind the obstacle. If the obstacle is large compared with the wavelength, it is normal to say that the sound wave

is reflected (in front) and *diffracted* (behind) the obstacle (rather than *scattered*). In this case a strong sound shadow is caused in which the wave pressure amplitude is very small. In the zone between the sound shadow and the region fully "illuminated" by the source the sound wave pressure amplitude oscillates. These oscillations are maximum near the shadow boundary and minimum well inside the shadow. These oscillations in amplitude are normally termed *diffraction bands*. One of the most common examples of diffraction caused by a body is the diffraction of sound over the sharp edge of a *barrier* or *screen*. For a plane homogeneous sound wave it is found that a strong shadow is caused by high frequency waves where $h/\lambda \gg 1$ and a weak shadow where $h/\lambda \ll 1$, where h is the barrier height and λ is the wavelength. For intermediate cases where $h/\lambda \approx 1$ a variety of interference and diffraction effects are caused by the barrier. Scattering is caused not only by obstacles placed in the wave field, but also by fluid regions where the properties of the medium such as its density or compressibility change their values from the rest of the medium. Scattering is also caused by turbulence (see Chapter 32) and from rain or fog particles in the atmosphere and bubbles in water and by rough or absorbent areas on wall surfaces.

11   RAY ACOUSTICS

There are three main modeling approaches in acoustics, which may be termed wave acoustics, ray acoustics, and energy acoustics. So far in this chapter we have mostly used the wave acoustics approach in which the acoustic quantities are completely defined as functions of space and time. This approach is practical in certain cases where the fluid medium is bounded and in cases where the fluid is unbounded as long as the fluid is homogenous. However, if the fluid properties vary in space due to variations in temperature or due to wind gradients, then the wave approach becomes more difficult and other simplified approaches such as the ray acoustics approach described here and in Chapter 3 are useful. This approach can also be extended to propagation in fluid-submerged elastic structures, as described in Chapter 4. The energy approach is described in Section 12.

In the ray acoustics approach, rays are obtained that are solutions to the simplified eikonal equation [Eq. (41)].

$$\left(\frac{\partial S}{\partial x}\right)^2 + \left(\frac{\partial S}{\partial y}\right)^2 + \left(\frac{\partial S}{\partial z}\right)^2 - \frac{1}{c^2} = 0. \quad (41)$$

The ray solutions can provide good approximations to more exact acoustic solutions. In certain cases they also satisfy the wave equation.[6] The eikonal $S(x, y, z)$ represents a surface of constant phase (or wavefront) that propagates at the speed of sound c. It can be shown that Eq. (41) is consistent with the wave equation only in the case when the frequency is very high.[6] However, in practice, it is useful, provided the changes in the speed of sound c are small when measured over distances comparable with the wavelength. In the case where the fluid is homogeneous (constant sound speed c_0 and density ρ throughout) S is a constant and represents a plane surface given by $S = (\alpha x + \beta y + \gamma z)/c_0$, where α, β, and γ are the direction cosines of a straight line (a ray) that is perpendicular to the wavefront (surface S). If the fluid can no longer be assumed to be homogeneous and the speed of sound $c(x, y, z)$ varies with position, the approach becomes approximate only. In this case some parts of the wavefront move faster than others, and the rays bend and are no longer straight lines. In cases where the fluid is not stationary, the rays are no longer quite parallel to the normal to the wavefront. This ray approach is described in more detail in several books and in Chapter

3 (where in this chapter the main example is from underwater acoustics). The ray approach is also useful for the study of propagation in the atmosphere and is a method to obtain the results given in Figs. 9–11. It is observed in these figures that the rays always bend in a direction toward the region where the sound speed is less. The effects of wind gradients are somewhat different since in that case the refraction of the sound rays depends on the relative directions of the sound rays and the wind in each fluid region. Chapter 4 presents an extension of the ray approach to fluid-submerged structures such as elastic shells. The same high-frequency assumption is made that λ/L must be small, where L is a characteristic dimension in the fluid field (environment or structural dimension).

12   ENERGY ACOUSTICS

In enclosed spaces the wave acoustics approach is useful, particularly if the enclosed volume is small and simple in shape and the boundary conditions are well-defined. In the case of rigid walls of simple geometry, the wave equation is used, and after the applicable boundary conditions are applied, the solutions for the natural (eigen) frequencies for the modes (standing waves) are found. See Chapters 7 and 92 for more details. However, for large rooms with irregular shape and absorbing boundaries, the wave approach becomes impracticable and other approaches must be sought. The ray acoustics approach together with the multiple-image-source concept is useful in some room problems, particularly in auditorium design or in factory spaces where barriers are involved. However, in many cases a statistical approach where the energy in the sound field is considered is the most useful. See Chapters 76–78 and 91–92 for more detailed discussion of this approach. Some of the fundamental concepts are briefly described here.

For a plane wave progressing in one direction in a duct of unit cross section, all of the sound energy in a column of fluid c metres in length must pass through the cross section in 1 s. Since the intensity $\langle I \rangle_t$ is given by $p_{\text{rms}}^2/\rho c$, then the total sound energy in the fluid column c metres long must also be equal to $\langle I \rangle_t$. The energy per unit volume ϵ (joules per cubic metre) is thus

$$\epsilon = \frac{\langle I \rangle_t}{c} \quad (42)$$

or

$$\epsilon = \frac{p_{\text{rms}}^2}{\rho c^2} \quad (43)$$

The energy density ϵ may be derived by alternative

means and is found to be the same as that given in Eq. (42) in most acoustic fields, except very close to sources of sound and in standing wave fields. In a room with negligibly small absorption in the air or at the boundaries, the sound field created by a source producing broadband sound will become very reverberant (the sound waves will reach a point with equal probability from any direction). In addition, for such a case the sound energy may be said to be diffuse if the energy density is the same anywhere in the room. For these conditions the time-averaged intensity incident on the walls (or on an imaginary surface from one side) is

$$\langle I \rangle_t = \tfrac{1}{4}\,\epsilon c, \qquad (44)$$

or

$$\langle I \rangle_t = \frac{p_{\text{rms}}^2}{4\rho c}. \qquad (45)$$

In any real room the walls will absorb some sound energy (and convert it into heat). The *absorption coefficient* $\alpha(f)$ of the wall material may be defined as the fraction of the incident sound intensity that is absorbed by the wall surface material:

$$\alpha(f) = \frac{\text{sound intensity absorbed}}{\text{sound intensity incident}}. \qquad (46)$$

The absorption coefficient is a function of frequency and can have a value between 0 and 1. The *noise reduction coefficient* (NRC) is found by averaging the absorption coefficient of the material at the frequencies 250, 500, 1000, and 2000 Hz (and rounding off the result to the nearest multiple of 0.05). See Chapter 92 for more detailed discussion on the absorption of sound in enclosures.

If we consider the sound field in a room with a uniform energy density ϵ created by a sound source that is suddenly stopped, then the sound pressure level in the room will decrease. We define a reverberation time in such a room as the time that the sound pressure level takes to drop by 60 dB. We may show that the reverberation time T_R is given as

$$T_R = \frac{0.161V}{S\overline{\alpha}}, \qquad (47)$$

where V is the room volume in cubic metres, S is the wall surface area in square metres, and $\overline{\alpha}$ is the average absorption coefficient of the wall surfaces.

By considering the sound energy radiated into a room

by a broadband noise source of sound power W, we may sum together the mean squares of the sound pressure contributions caused by the direct and reverberant fields and after taking logarithms obtain the sound pressure level in the room:

$$L_p = L_w + 10\log\left(\frac{D_{\theta,\phi}}{4\pi r^2} + \frac{4}{R}\right), \qquad (48)$$

where $D_{\theta,\phi}$ is the directivity factor of the source (see Section 7) and R is the so-called room constant

$$R = \frac{S\overline{\alpha}}{1 - \overline{\alpha}}. \qquad (49)$$

A plot of the sound pressure level against distance from the source is given for various room constants in Fig. 12. It is seen that there are several different regions. The near and far fields depend on the type of source (see Section 11 and Chapter 9) and the free field and reverberant field. The free field is the region where the direct term $D/4\pi r^2$ dominates, and the reverberant field is the region where the reverberant term $4/R$ in Eq. (48) dominates. The so-called critical distance $r_c = (D_{\theta,\phi}R/16\pi)^{1/2}$ occurs where the two terms are equal.

13 SOUND RADIATION FROM IDEALIZED STRUCTURES

The sound radiation from plates and cylinders in bending (flexural) vibration is discussed in Chapter 11. There are interesting phenomena observed with free-bending waves. Unlike sound waves, these are dispersive and travel faster at higher frequency. The bending-wave speed is $c_b = (\omega \kappa c_l)^{1/2}$, where κ is the radius of gyration $h/(12)^{1/2}$, h is the thickness, and c_l is the longitudinal wave speed $\{E/[\rho(1-\sigma^2)]\}^{1/2}$, where E is Young's modulus of elasticity. When the bending-wave speed equals the speed of sound in air, the frequency is called the critical frequency (see Fig. 13). The critical frequency is

$$f_c = \frac{c^2}{2\pi\kappa c_l}. \qquad (50)$$

Above this frequency f_c the coincidence effect is observed because the bending wavelength λ_b is greater than the wavelength in air λ (Fig. 14) and trace wave matching always occurs for the sound waves in air at some angle of incidence. See Fig. 15. This has important consequences for the sound radiation from structures

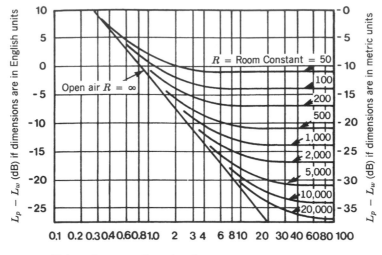

Fig. 12 Sound pressure level in a room (relative to sound power level) as a function of distance (r).

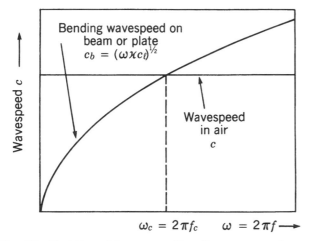

Fig. 13 Variation of frequency of bending wave speed c_b on a beam or panel and wave speed in air c.

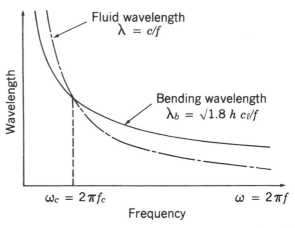

Fig. 14 Variation with frequency of bending wavelength λ_b on a beam or panel and wavelength in air λ.

and also for the sound transmitted through the structures from one air space to the other.

For free-bending waves on infinite plates above the critical frequency the plate radiates efficiently, while below this frequency (theoretically) the plate cannot radiate any sound energy at all. (See Chapter 11.) For finite plates, reflection of the bending waves at the edges of the plates causes standing waves that allow radiation (although inefficient) from the plate corners or edges even below the critical frequency. In the plate center, radiation from adjacent quarter-wave areas cancels. But radiation from the plate corners and edges, which are normally separated sufficiently in acoustic wavelengths, does not cancel. At very low frequency, sound is radiated mostly by corner modes, then up to the critical frequency mostly by edge modes. Above the critical fre-

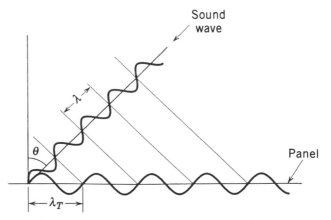

Fig. 15 Diagram showing trace wave matching between waves in air of wavelength λ and waves in panel of trace wavelength λ_T.

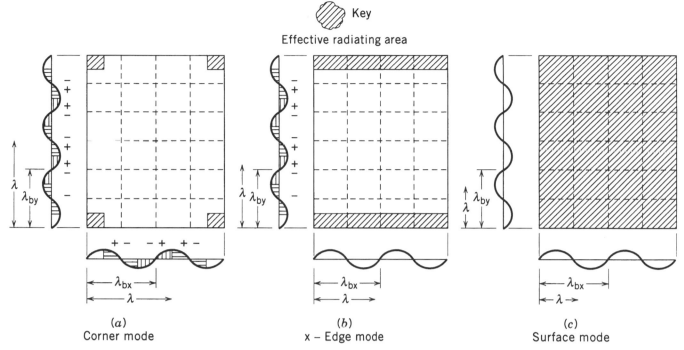

Fig. 16 Wavelength relations and effective radiating areas for corner, edge, and surface modes. The acoustic wavelength is λ, and λ_{bx} and λ_{by} are the bending wavelengths in the x- and y-directions, respectively.

quency the radiation is caused by surface modes with which the whole plate radiates efficiently (see Fig. 16). Radiation from bending waves in plates and cylinders is discussed in detail in Chapter 11. Sound transmission through structures is discussed in Chapters 83, 93, and 94.

14 BOUNDARY WAVES IN ACOUSTICS

Boundary waves in acoustic systems and elastic structures are observed on all scales—from the submillimetre (acoustic delay lines) to planetary size (Rayleigh modes in seismology). As their existence is essentially independent of that of the "body" or "volume" waves of acoustics or elasticity, they constitute a distinct and separate class of modes. The essential mechanism is one of storage at a boundary; the role of the boundary is *intrinsic*. In contrast, body or volume waves are in no way governed by the boundary, whose role is thus *extrinsic*. The energy density of boundary waves decreases exponentially or at a higher rate with increasing distance from the boundary. Reciprocally the excitation of such modes decreases exponentially with the distance of the source from the boundary. Homogeneous plane waves incident on a plane boundary do not generate true surface modes. The theory and practice of acoustics and elasticity offer

many examples of such modes: surface waves traveling along a rigid periodic structure of corrugations on a hard wall bounding a fluid half space, Stoneley waves along plane elastic–elastic interfaces, Scholte waves along plane fluid–elastic interfaces, boundary roughness modes, Rayleigh waves on the surface of an elastic solid, and so on. Boundary waves exhibit the remarkable property of channelling spherically propagating energy produced by a point source into two-dimensional propagating configurations, that is, from an r^{-2} geometric energy spreading loss to an r^{-1} law. That is why Rayleigh waves of shallow-focus earthquakes dominate seismograms at middle or long ranges—an effect also observed in the laboratory with boundary roughness modes in acoustics. Boundary waves are discussed in detail in Chapter 12.

15 STANDING WAVES

Standing-wave phenomena are observed in many situations in acoustics and the vibration of strings and elastic structures. Thus they are of interest with almost all musical instruments (both wind and stringed) (see Chapters 131–137); in architectural spaces such as auditoria and reverberation rooms; in volumes such as automobile and aircraft cabins; and in numerous cases of vibrating structures, from tuning forks, xylophone bars, bells and cymbals, to windows, wall panels and innumerable other engineering systems including aircraft, vehicle, and ship structural members. With each standing wave is associated an eigen (or natural) frequency and a mode shape (or shape of vibration). Some of these systems can be idealized to simple one-, two-, or three-dimensional systems. For example with a simple wind instrument such as a whistle, Eq. (1) above together with the appropriate spatial boundary conditions can be used to predict the predominant frequency of the sound produced. Similarly, the vibration of a string on a violin can be predicted with an equation identical to Eq. (1) but with the variable p replaced by the lateral string displacement. With such a string, solutions can be obtained for the fundamental and higher natural frequencies (overtones) and the associated standing-wave mode shapes (normally sine shapes). In such a case for a string with fixed ends, the so-called overtones are just integer multiples (2, 3, 4, 5, ...) of the fundamental frequency. The standing wave with the whistle and string can be considered mathematically to be composed of two waves of equal amplitude traveling in opposite directions.

A similar situation occurs for bending waves on bars, but because the equation of motion is different (dispersive), the higher natural frequencies are not related by simple integers. However, for the case of a beam with simply supported ends, the higher natural frequencies are given by 2^2, 3^2, 4^2, 5^2, ... or 4, 9, 16, 25, ... and the mode shapes are sine shapes again.

The standing waves on two-dimensional systems (such as bending vibrations of plates) may be considered mathematically to be composed of four opposite traveling waves. For simply supported rectangular plates the mode shapes are sine shapes in each direction. For three-dimensional systems such as the air volumes of rectangular rooms, the standing waves may be considered to be made up of eight traveling waves. For a hard-walled room, the sound pressure has a cosine mode shape with the maximum pressure at the walls, and the particle velocity has a sine mode shape with zero normal particle velocity at the walls. See Chapter 7 for the natural frequencies and mode shapes for a large number of acoustic and structural systems.

16   WAVEGUIDES

Waveguides can occur naturally where sound waves are channeled by reflections at boundaries and by refraction. The ocean can be considered to be an acoustic waveguide that is bounded above by the air–sea interface and below by the ocean bottom (see Chapter 36). Similar channeling effects are also sometimes observed in the atmosphere. (See Chapters 3 and 32.) Waveguides are also encountered in musical instruments and engineering applications. Wind instruments may be regarded as waveguides. In addition, waveguides comprised of pipes, tubes, and ducts are frequently used in engineering systems, for example, air conditioning ducts and the ductwork in turbines and turbofan engines. The sound propagation in such waveguides is similar to the three-dimensional situation discussed in Section 15 but with some differences. Although rectangular ducts are used in air conditioning systems, circular ducts are also frequently used, and theory for these must be considered as well. In real waveguides, air flow is often present and complications due to a mean fluid flow must be included in the theory.

For low-frequency excitation only plane waves can propagate along the waveguide (in which the sound pressure is uniform across the duct cross section). However, as the frequency is increased, the so-called first cut-on frequency is reached above which there is a standing wave across the duct cross section caused by the first higher mode of propagation.

For excitation just above this cut-on frequency, besides the plane-wave propagation, propagation in this higher order mode can also exist. The higher mode propagation in each direction in a rectangular duct can be considered to be composed of four traveling waves each with a vector (ray) almost perpendicular to the duct walls and with a phase speed along the duct that is almost infinite. As the frequency is increased, these vectors move increasingly toward the duct axis, and the phase speed along the duct decreases until at very high frequency it is only just above the speed of sound c. However, for this mode, the sound pressure distribution across the duct cross section remains unchanged. As the frequency increases above the first cut-on frequency, the cut-on frequency for the second higher order mode is reached, and so on. For rectangular ducts, the solution for the sound pressure distribution for the higher modes in the duct consists of cosine terms with a pressure maximum at the duct walls, while for circular ducts, the solution involves Bessel functions. Chapter 8 explains how sound is propagated in both rectangular and circular waveguides and includes discussion on the complications created by a mean flow, dissipation, discontinuities, and terminations. Chapter 162 discusses the propagation of sound in another class of waveguides: horns.

17   ACOUSTIC LUMPED ELEMENTS

When the wavelength of sound is large compared with the physical dimensions of the acoustic system under consideration, then the lumped-element approach is useful. In this approach it is assumed that the fluid mass, stiffness, and dissipation distributions can be "lumped" together to act at a point, significantly simplifying the analysis of the problem. The most common example of this approach is its use with the well-known Helmholtz resonator in which the mass of air in the neck of the resonator vibrates at its natural frequency against the stiffness of its volume. A similar approach can be used in the design of loudspeaker enclosures and the concentric resonators in automobile mufflers in which the mass of the gas in the resonator louvres (orifices) vibrates against the stiffness of the resonator (which may not necessarily be regarded completely as a lumped element). Dissipation in the resonator louvers may also be taken into account. Chapter 13 reviews the lumped-element approach in some detail.

18 NUMERICAL APPROACHES: FINITE ELEMENTS AND BOUNDARY ELEMENTS

In cases where the geometry of the acoustic space is complicated and where the lumped element approach cannot be used, then it is necessary to use numerical approaches. In the late 1960s, with the advent of powerful computers, the acoustic finite element method (FEM) became feasible. In this approach the fluid volume is divided into a number of small fluid elements (usually rectangular or triangular), and the equations of motion are solved for the elements, ensuring that the sound pressure and volume velocity are continuous at the node points where the elements are joined. The FEM has been widely used to study the acoustic performance of elements in automobile mufflers and cabins.

The boundary element method (BEM) was developed a little later than the FEM. In the BEM approach the elements are described on the boundary surface only, which reduces the computational dimension of the problem by 1. This correspondingly produces a smaller system of equations than the FEM and thus saves computational time considerably because of the use of a surface mesh rather than a volume mesh. For sound propagation problems involving the radiation of sound to infinity, the BEM is more suitable because the radiation condition at infinity can be easily satisfied with the BEM, unlike with the FEM. However, the FEM is better suited than the BEM for the determination of the natural frequencies and mode shapes of cavities.

Recently FEM and BEM commercial software has become widely available. The FEM and BEM are described in Chapters 14 and 15.

19 ACOUSTIC MODELING USING EQUIVALENT CIRCUITS

Electrical analogies have often been found useful in the modeling of acoustic systems. There are two alternatives. The sound pressure can be represented by voltage and the volume velocity by current, or alternatively the sound pressure is replaced by current and the volume velocity by voltage. Use of electrical analogies is discussed in Chapter 13. They have been widely used in loudspeaker design and are in fact perhaps most useful in the understanding and design of transducers such as microphones where acoustic, mechanical, and electrical systems are present together and where an overall equivalent electrical circuit can be formulated (see Chapters 161, 162, and 164). Beranek makes considerable use of electrical analogies in his books.[9, 10] In Chapter 16, their use in the design of automobile mufflers is described.

REFERENCES

1. M. J. Lighthill, *Waves in Fluids*, Cambridge University Press, Cambridge, 1978.

2. P. M. Morse and K. U. Ingard, *Theoretical Acoustics*, Princeton University Press, Princeton, NJ, 1987.

3. L. E. Kinsler, A. R. Frey, A. B. Coppens, and J. V. Sanders, *Fundamentals of Acoustics*, Wiley, New York, 1982.

4. A. D. Pierce, *Acoustics: An Introduction to Its Physical Principles and Applications*, McGraw-Hill, New York, 1981.

5. M. J. Crocker and A. J. Price, *Noise and Noise Control*, Vol. 1, CRC Press, Cleveland, 1975.

6. R. G. White and J. G. Walker (Eds.), *Noise and Vibration*, Halstead Press, Wiley, New York, 1982.

7. F. J. Fahy, *Sound Intensity*, Second Edition, E&FN Spon, Chapman & Hall, London, 1995.

8. E. Skudrzyk, *The Foundations of Acoustics*, Springer Verlag, New York, 1971.

9. L. L. Beranek, *Acoustical Measurements*, rev. ed., Acoustical Society of America, New York, 1988.

10. L. L. Beranek, *Acoustics*, Acoustical Society of America, New York, 1986 (reprinted with changes).

2

MATHEMATICAL THEORY OF WAVE PROPAGATION

ALLAN D. PIERCE

1 INTRODUCTION

This chapter gives a more detailed discussion of some of the basic concepts referred to in Chapter 1. Of particular concern here is the mathematical embodiment of those concepts that underly the wave description of acoustics. This mathematical theory[1] began with Mersenne, Galileo, and Newton and developed into its more familiar form during the time[2,3] of Euler and Lagrange. Prominent contributors during the nineteenth century include Poisson, Laplace, Cauchy, Green, Stokes, Helmholtz, Kirchhoff, and Rayleigh.

The basic mathematical principles underlying sound propagation, whether through fluids or solids, include the principles of continuum mechanics, which include a law accounting for the conservation of mass, a law accounting for the changes in momenta brought about by forces, and a law accounting for changes in energy brought about by work and the transfer of heat. The subject also draws upon thermodynamics, upon symmetry considerations, and upon various known properties of the substances through which sound can propagate. The theory is inherently approximate, and there are many different versions, which differ slightly or greatly from each other in just what idealizations are made at the outset. The concern here is primarily with the simpler mathematical idealizations that have proven useful in acoustics, and in particular with those associated with linear acoustics, but the discussion begins with the more nearly exact nonlinear equations and explains how the linear equations of acoustics follow from them.

Encyclopedia of Acoustics, edited by Malcolm J. Crocker
ISBN 0-471-80465-7 © 1997 John Wiley & Sons, Inc.

2 CONSERVATION OF MASS

The matter through which an acoustic wave travels is characterized by its density ρ, which represents the local spatial average of the mass per unit volume in a macroscopically small volume. This density varies in general with position, generically denoted here by the vector $\mathbf{x}$, and with time t. The material velocity $\mathbf{v}$ with which the matter moves is defined so that $\rho\mathbf{v}$ is the mass flux vector within the fluid. The significance of the latter is that, were one to conceive of a hypothetical stationary surface within the material with a local unit-normal vector $\mathbf{n}(\mathbf{x})$ pointing from one side to the other, then $\rho\mathbf{v} \cdot \mathbf{n}$ gives the mass crossing this surface per unit time and per unit area of the surface. Here, also, a local average is understood; the material velocity $\mathbf{v}(\mathbf{x}, t)$ associated with a point and time may be considerably different from the instantaneous velocity of a molecule in the vicinity of that point.

The concept of a mass flux vector leads to the prediction that the net rate of flow of mass out through any hypothetical closed surfae S within the material is the area integral of $\rho\mathbf{v} \cdot \mathbf{n}$. (The volume V enclosed by this surface is sometimes referred to as a control volume.[4]) The conservation of mass requires that this area integral be the same as the rate at which the volume integral of ρ decreases with time, so that

$$\int_S \rho\mathbf{v} \cdot \mathbf{n}\, dS = -\frac{d}{dt} \int_V \rho\, dV. \qquad (1)$$

The partial differential equation expressing the conservation of mass results, after application of Gauss's theorem[5] to the surface integral, with the recognition that the volume of integration is arbitrary, so that

$$\frac{\partial \rho}{\partial t} + \nabla \cdot (\rho \mathbf{v}) = 0, \tag{2}$$

where the partial derivative with respect to time implies that the position $\mathbf{x}$ is held fixed in the differentiation.

An alternate form of the above partial differential equation is

$$\frac{D\rho}{Dt} + \rho \nabla \cdot \mathbf{v} = 0 \tag{3}$$

with the Stokes abbreviation[6]

$$\frac{D}{Dt} = \frac{\partial}{\partial t} + \mathbf{v} \cdot \nabla \tag{4}$$

for the total time derivative operator. The two terms in the latter correspond to (i) the time derivative as would be seen by an observer at rest and (ii) the convective time derivative.

If the material is incompressible, then the density viewed by someone moving with the flow should remain constant, so $D\rho/Dt$ should be zero, and consequently the conservation of mass requires that $\nabla \cdot \mathbf{v} = 0$. Although much of the analysis of fluid mechanics is based on this idealization, it is ordinarily inappropriate for acoustics, because the compressibility of the material plays a major role in the propagation of sound.

3 THE EULER EQUATION

The local rate of material acceleration is identified as the total time derivative $D\mathbf{v}/Dt$ of the material velocity. Consequently, the generalization of Newton's second law to a continuum requires that $\rho D\mathbf{v}/Dt$ equal the force per unit mass exerted on the material. The force on any small identifiable aggregate of material consists of a sum of a body force $\mathbf{F}_B$ and a surface force $\mathbf{F}_S$. Possible body forces are gravitational or electromagnetic; the former is customarily expressed as

$$\mathbf{F}_B = \int_V \rho \mathbf{g} \, dV, \tag{5}$$

where $\mathbf{g}$ is the acceleration associated with gravity. The surface force is expressible as

$$\mathbf{F}_S = \int_S \sum_{i,j=1}^{3} \sigma_{ij} \mathbf{e}_i n_j \, dS, \tag{6}$$

where the σ_{ij} are the Cartesian components of the stress tensor.[7] Here the quantities $\mathbf{e}_i$ are the unit vectors appropriate for the corresponding Cartesian coordinate system, and the n_j are the Cartesian components of the unit vector pointing outward to the enclosing surface S. The stress tensor's Cartesian component σ_{ij} can be regarded as the ith component of the surface force per unit area on a segment of the surface of a small element of the continuum, when the unit outward normal to the surface is in the jth direction. This surface force is caused by interactions with the neighboring particles just outside the surface[8] or by the transfer of momentum due to diffusion of molecules across the surface. The condition that the net force per unit volume be finite in the limit of macroscopically infinitesimal volumes requires[9] that the stress components σ_{ij} be independent of the shape and orientation of the hypothetical surface S, so the stress is a tensor field associated with the material that depends in general on position and time. Considerations of the net torque[10] that such surface forces would exert on a small element and the requirement that the angular acceleration of the element be finite in the limit of very small element dimensions lead to the conclusion $\sigma_{ij} = \sigma_{ji}$, so the stress tensor is symmetric.

With the above identifications for the surface and body forces and a subsequent application of Gauss's theorem to transform the surface integral to a volume integral, the application of Newton's second law yields Cauchy's equation of motion, which is written in Cartesian coordinates as

$$\rho \frac{D\mathbf{v}}{Dt} = \sum_{ij} \mathbf{e}_i \frac{\partial \sigma_{ij}}{\partial x_j} + \mathbf{g}\rho. \tag{7}$$

Here the Eulerian description is used, with each field variable regarded as a function of actual spatial position coordinates and time.

For disturbances in fluids, such as air and water, and with the neglect of viscous shear forces, the surface force associated with stress can only be normal[11] to the surface. Since such would have to hold for all possible orientations of the surface, the stress tensor has to have the form

$$\sigma_{ij} = -p\delta_{ij}, \tag{8}$$

where p is identified as the pressure and where δ_{ij}, equal to unity when the indices are equal and zero otherwise, is the Kronecker delta. The pressure p is understood to be such that $-p\mathbf{n}$ is the force per unit area exerted on the material on the interior side of a surface when the outward unit-normal vector is $\mathbf{n}$. The insertion of the above idealization for the stress into Cauchy's equation of motion (7) yields Euler's equation of motion for a fluid:

$$\rho \frac{D\mathbf{v}}{Dt} = -\nabla p + \mathbf{g}\rho. \tag{9}$$

Although viscosity is often important adjacent to solid boundaries[12] and can lead to the attenuation of sound for propagation over very large distances, it is often an excellent first approximation to use the Euler equation as the embodiment of Newton's second law for a fluid in the analysis of acoustic processes. Also, gravity[13] has a minor influence on sound and can often be ignored in analytical studies.

The Euler equation can be alternately written, via a vector identity, as

$$\frac{\partial \mathbf{v}}{\partial t} - \mathbf{v} \times (\nabla \times \mathbf{v}) + \nabla \left(\frac{\mathbf{v}^2}{2} \right) = -\frac{1}{\rho} \nabla p + \mathbf{g}, \tag{10}$$

and the curl of this yields the corollary

$$\frac{D}{Dt} (\nabla \times \mathbf{v}) + (\nabla \times \mathbf{v})\nabla \cdot \mathbf{v} - [(\nabla \times \mathbf{v}) \cdot \nabla]\mathbf{v}$$
$$= \frac{1}{\rho^2} \nabla \rho \times \nabla p. \tag{11}$$

This equation governs the dynamics of the vorticity $(\nabla \times \mathbf{v})$ and provides a means for assessing the common idealization that the vorticity is identically zero.

4 THERMODYNAMIC PRINCIPLES IN ACOUSTICS

The quasi-static theory of thermodynamics[14] presumes that the local state of the material at any given time can be completely described in terms of a relatively small number of variables. One such state variable is the density ρ; another is the internal energy u per unit mass (specific internal energy). For a fluid in equilibrium, these two variables are sufficient to specify the state. Other quantities such as the pressure p and the temperature T can be determined from equations of state that give these in terms of ρ and u. An important additional state

variable, although one not easily measurable, is the equilibrium entropy $s(\mathbf{x}, t)$ per unit mass (specific entropy), the changes of which can be related to heat transfer, but which also can be regarded as a function of u and ρ. The physical interpretation of s requires that there be no heat transfer when s is constant, and conservation of energy requires that work done by pressure forces on a surface of a fluid element during a quasi-static process result in a corresponding increase in internal energy $\rho V \, du$. For a sufficiently small element, this work is p times the net decrease, $-dV$, in volume. Since mass is conserved, $\rho \, dV = -V \, d\rho$, and the work is consequently $p\rho^{-1}V \, d\rho$. This yields $\rho V \, du = p\rho^{-1}V \, d\rho$ when $ds = 0$, and leads to the prediction that, to first order in differentials, ds must be proportional to $du - p\rho^{-2} \, d\rho$, where the proportionality factor must be a function of u and ρ. There is an inherent arbitrariness in the actual form of the function $s(u, \rho)$, but the theory of thermodynamics requires that it can be chosen so that the reciprocal of the proportionality factor has all the properties commonly associated with absolute temperature. Such a choice yields

$$T \, ds = du - p\rho^{-2} \, d\rho, \tag{12}$$

with the identification of $T \, ds$ as the increment of "heat energy" that has been transferred per unit mass during a quasi-static process.

For an ideal gas with temperature-independent specific heats, which is a common idealization for air, the function $s(u, \rho)$ is given by

$$s = \frac{R_0}{M} \ln(u^{1/(\gamma - 1)}\rho^{-1}) + s_0. \tag{13}$$

Here the constant s_0 is independent of u and ρ^{-1}, while M is the average molecular weight (average molecular mass in atomic mass units), and R_0 is the universal gas constant (equal to Boltzmann's constant divided by the mass in an atomic mass unit), equal to 8314 J/kg·K. The quantity γ is the specific heat ratio, equal to approximately $\frac{7}{5}$ for diatomic gases, $\frac{5}{3}$ for monatomic gases, and $\frac{9}{7}$ for polyatomic gases whose molecules are not collinear. For air, γ is 1.4 and M is 29.0. (The expression given here for entropy neglects the contribution of internal vibrations of diatomic molecules, which cause the specific heats and γ to depend slightly on temperature. In the explanation of the absorption of sound in air, a nonequilibrium entropy[15] is used that depends on the fractions of O_2 and N_2 molecules that are in their first excited vibrational states in addition to the quantities u and ρ^{-1}). With the aid of the differential relation (14), the corresponding expressions for temperature and pressure appear as

$$T = (\gamma - 1) \frac{M}{R_0} u, \qquad p = (\gamma - 1)\rho u, \qquad (14)$$

and from these two relations emerges the ideal-gas equation

$$p = \rho \frac{R_0}{M} T. \qquad (15)$$

For other substances, a knowledge of the first and second derivatives of s, evaluated at some representative thermodynamic state, is sufficient for most linear acoustic applications.

An immediate consequence of the above line of reasoning is that, for a fluid in equilibrium, any two state variables are sufficient to determine the others. The pressure, for example, can be expressed as

$$p = p(\rho, s), \qquad (16)$$

where the actual dependence on ρ and s is intrinsic to the material. For an ideal gas, for example, this function has the form

$$p = K(s)\rho^{\gamma}, \qquad (17)$$

where

$$K(s) = (\gamma - 1)p_0\rho_0^{-\gamma}\exp\left((\gamma - 1)\frac{M}{R_0}(s - s_0)\right) \qquad (18)$$

is a function of specific entropy only.

The formulation of acoustics that is most frequently used, and with which the present chapter is concerned, assumes that the processes of sound propagation are quasi-static in the thermodynamic sense. This implies that the p given by the equation of state (16) is the same as that which appears in the Euler equation (9). Also, Eq. (12) must hold for any given possibly moving element of fluid, so that the differential relation yields the partial differential equation

$$T \frac{Ds}{Dt} = \frac{Du}{Dt} - p\rho^{-2}\frac{D\rho}{Dt}. \qquad (19)$$

With the neglect of heat transfer mechanisms within the fluid, the conservation of energy[16] requires that

$$\rho \frac{D}{Dt}\left\{\frac{v^2}{2} + u\right\} = -\nabla \cdot \{p\mathbf{v}\} + \rho\mathbf{g} \cdot \mathbf{v}, \qquad (20)$$

where the quantity within braces on the left is the energy (kinetic plus potential) per unit mass and the terms on the right correspond to the work done per unit volume by pressure and gravitational forces, respectively. The Euler equation (9) yields the derivable relation

$$\rho \frac{D}{Dt}\left\{\frac{v^2}{2}\right\} = -\mathbf{v} \cdot \nabla p + \rho\mathbf{g} \cdot \mathbf{v}, \qquad (21)$$

and the subtraction of this from Eq. (20) yields

$$\rho \frac{Du}{Dt} = -p\nabla \cdot \mathbf{v} = p\rho^{-1}\frac{D\rho}{Dt}. \qquad (22)$$

The latter equality results from the version (3) of the conservation-of-mass equation. Therefore, the conservation-of-energy relation (20) requires that the right side of Eq. (19) be identically zero, so

$$\frac{Ds}{Dt} = 0, \qquad (23)$$

which implies that the entropy per unit mass of any given fluid particle remains constant. A disturbance satisfying this relation is said to be isentropic.

The total time derivative of Eq. (16) with the replacement of Ds/Dt by 0 yields the equation

$$\frac{Dp}{Dt} = c^2 \frac{D\rho}{Dt}, \qquad (24)$$

where c^2 abbreviates the quantity

$$c^2 = \left(\frac{\partial p}{\partial \rho}\right)_s. \qquad (25)$$

Here the subscript s implies that the differentiation is carried out at constant entropy. Thermodynamic considerations based on the requirement that the total entropy increases during an irreversible process requires in turn that this derivative be positive. Its square root c is identified as the speed of sound, for reasons discussed in a subsequent portion of this chapter.

The relation between the total time derivatives of pressure and density that appears in Eq. (24) can be combined with the conservation-of-mass relation (3) to yield

$$\frac{Dp}{Dt} + \rho c^2\nabla \cdot \mathbf{v} = 0, \qquad (26)$$

which is often of greater convenience in acoustics than

the mass conservation relation because the interest is typically in pressure fluctuations rather than in density fluctuations.

For an ideal gas, the differentiation of the expression $p(\rho, s)$ in Eq. (17) yields

$$c^2 = \gamma K(s)\rho^{\gamma-1}, \qquad (27)$$

which can equivalently be written[17]

$$c^2 = \frac{\gamma p}{\rho} = \gamma \, \frac{R_0}{M} \, T. \qquad (28)$$

For air, R_0/M is 287 J/kg·K and γ is 1.4, so a temperature of 293.16 K (20°C) and a pressure of 10^5 Pa yield a sound speed of 343 m/s and a density ρ of 1.19 kg/m³.

Acoustic properties of materials are discussed in depth elsewhere in this handbook, but some simplified formulas for water are given here as an illustration of the thermodynamic dependences of such properties. For pure water, the sound speed[18] is approximately given in metres-kilograms-seconds (MKS) units by

$$c = 1447 + 4.0\Delta T + 1.6 \times 10^{-6}p. \qquad (29)$$

Here c is in meters per second, ΔT is the temperature relative to 283.16 K (10°C), and p is the absolute pressure in pascals. The pressure and temperature dependence of the density is approximately given by

$$\rho \approx 999.7 + 0.048 \times 10^{-5}p - 0.088\Delta T - 0.007(\Delta T)^2. \qquad (30)$$

The values for sea water[19] are somewhat different because of the presence of dissolved salts. An approximate expression for the speed of sound in sea water is given by

$$c \approx 1490 + 3.6\Delta T + 1.6 \times 10^{-6}p + 1.3\Delta S. \qquad (31)$$

Here ΔS is the deviation of the salinity in parts per thousand from a nominal value of 35.

5 THE LINEARIZATION PROCESS

Sound results from a time-varying perturbation of the dynamic and thermodynamic variables that describe the medium. The quantities appropriate to the ambient medium are customarily represented[20] by the subscript 0, and the perturbations are represented by a prime on the corresponding symbol. The total pressure is $p =$

$p_0 + p'$, and there are corresponding expressions for fluctuations in specific entropy, fluid velocity, and density. The linear equations that govern acoustic disturbances are then determined by the first-order terms in the expansion of the governing nonlinear equations in the primed variables. The zeroth-order terms cancel out completely because the ambient variables should themselves correspond to a valid state of motion of the medium. Thus, for example, the linearized version of the conservation-of-mass relation in Eq. (2) is

$$\frac{\partial \rho'}{\partial t} + \mathbf{v}_0 \cdot \nabla \rho' + \rho'\nabla \cdot \mathbf{v}_0 + \mathbf{v}' \cdot \nabla \rho_0 + \rho_0 \nabla \cdot \mathbf{v}' = 0. \quad (32)$$

The possible forms of the linear equations differ in complexity according to what is assumed about the medium's ambient state and according to what dissipative terms are included in the original governing equations. In the establishment of a rationale for using simplified models, it is helpful[21] to think in terms of the order of magnitudes and characteristic scales of the coefficients in more nearly comprehensive models. If the spatial region of interest has bounding dimensions that are smaller than any scale length over which the ambient variables vary by a respectable fraction, then it may be appropriate to idealize the coefficients in the governing equations as if the ambient medium were spatially uniform or homogeneous. Similarly, if the characteristic wave periods or propagation times are much less than characteristic time scales for the ambient medium, then it may be appropriate to idealize the coefficients as being time independent. Examination of orders of magnitudes of terms suggests that the ambient velocity may be neglected if it is much less than the sound speed c. The first-order perturbation to the gravitational force term can be ordinarily neglected if the previously stated conditions are met and if the quantity g/c is sufficiently less than any characteristic frequency of the disturbance.

A discussion of the restrictions on using linear equations and of neglecting second- and higher order terms in the primed variables is outside the scope of the present chapter, but it should be noted that one regards p' as small if it is substantially less than $\rho_0 c^2$ and $|\mathbf{v}'|$ as small if it is much less than c, where c is the sound speed defined via Eq. (25). It is not necessary that p' be much less than p_0, and it is certainly not necessary that $|\mathbf{v}'|$ be less than $|\mathbf{v}_0|$.

6 ACOUSTIC FIELD EQUATIONS

The customary equations for linear acoustics for a medium where there is no ambient flow can be derived from the Euler equation (9) with the neglect of the gravity term and from Eq. (26), which combines the conservation of mass with thermodynamic principles. The ambient medium is taken as time independent, so the ambient quantities depend at most only on position $\mathbf{x}$. Since the zeroth-order equations must hold when there is no disturbance, the Euler equation requires that the ambient pressure be independent of position. It is not precluded, however, that the ambient density ρ_0 and sound speed c_0 vary with position. Thus, the linearized version of Eq. (26) becomes

$$\frac{\partial p'}{\partial t} = c_0^2 \left(\frac{\partial \rho'}{\partial t} + \mathbf{v}' \cdot \nabla \rho_0 \right). \quad (33)$$

The inference that

$$p' = c_0^2 \rho' \quad (34)$$

applies only if the ambient entropy is constant, which for the present circumstances (no ambient flow) requires ρ_0 to be constant. If $\nabla \rho_0$ were nonzero, the appropriate alternate would be

$$p' = c_0^2 (\rho' + \boldsymbol{\xi} \cdot \nabla \rho_0), \quad (35)$$

where $\boldsymbol{\xi}$ is the displacement vector (position shift) of the fluid particle nominally at $\mathbf{x}$ in its ambient position.

To allow for the possibility that the ambient medium is inhomogeneous, it is more convenient to deal with Eq. (26), and thereby take advantage of $\nabla p_0 = 0$, rather than to deal with the mass conservation equation. Equation (26) for the circumstances just described leads after linearization to

$$\frac{\partial p'}{\partial t} + \rho_0 c_0^2 \nabla \cdot \mathbf{v}' = 0, \quad (36)$$

and the Euler equation of Eq. (9) leads to

$$\rho_0 \frac{\partial \mathbf{v}'}{\partial t} = -\nabla p'. \quad (37)$$

The two equations just stated are a complete set of linear acoustic equations, insofar as there are as many partial differential equations as there are unknowns.

An alternate set of equations of comparable simplicity results when there is an ambient irrotational homoentropic flow. The term *homoentropic* means that the entropy is uniform (homogeneous) throughout the ambient medium. Because the total time derivative of the

entropy is constant for the idealization considered here in which there is no heat transfer, a fluid that is homoentropic at any given instant will remain so for all time, so s will be independent of both position and time. This means that the pressure can be regarded as a function of density only, so that

$$\frac{\partial p}{\partial t} = c^2 \frac{\partial \rho}{\partial t}, \qquad \nabla p = c^2 \nabla \rho, \qquad p' = c_0^2 \rho', \quad (38)$$

where the latter results from the linearization of $p = p(p, s_0)$, even though c_0^2 is not necessarily constant.

When the fluid is homoentropic, it follows from Eq. (38) that $\nabla \rho \times \nabla p = 0$, so the vorticity dynamics equation (11) reduces to

$$\frac{D}{Dt} (\nabla \times \mathbf{v}) + (\nabla \times \mathbf{v}) \nabla \cdot \mathbf{v} - [(\nabla \times \mathbf{v}) \cdot \nabla]\mathbf{v} = 0. \quad (39)$$

This has the implication that, if the vorticity $(\nabla \times \mathbf{v})$ is initially everywhere zero, then it will stay everywhere zero for all time. This allows one to conceive of an ambient medium that is both homoentropic and irrotational, so that $\nabla \times \mathbf{v}_0 = 0$.

For these circumstances, with a homoentropic and irrotational ambient medium, the linearized equations, after a modest amount of manipulation, take the form

$$\rho_0 \left[\frac{\partial}{\partial t} + \mathbf{v}_0 \cdot \nabla \right] \left(\frac{p'}{\rho_0 c_0^2} \right) + \nabla \cdot (\mathbf{v}' \rho_0)$$
$$= 0, \quad (40)$$

$$\frac{\partial \mathbf{v}'}{\partial t} - \mathbf{v}_0 \times (\nabla \times \mathbf{v}')$$
$$= -\nabla \left[\frac{p'}{\rho_0} + \mathbf{v}' \cdot \mathbf{v}_0 \right]. \quad (41)$$

In these equations, the ambient quantities $\mathbf{v}_0$, ρ_0, and c_0 may depend on both time and position, although they are constrained by the zeroth-order fluid dynamic equations. A lengthier[22] analysis suggests that these equations remain a good approximation for arbitrary inhomogeneous time-dependent media with ambient flow, provided that the characteristic wavelengths of the sound disturbance are sufficiently short. Furthermore, they remain good approximations for all but extremely low infrasonic frequencies when one incorporates gravity into the derivation, with account taken of the associated variation of density with height (as for the air in the atmosphere).

In the limit when $\mathbf{v}_0$ is zero and when ρ_0 and c_0 are constant, Eqs. (36) and (40) are equivalent, while Eqs.

(37) and (41) are also equivalent. Because neither of the two sets of equations just given involve the ambient pressure p_0 and the density perturbation ρ', it is customary to delete the prime on p' and the subscript 0 on ρ_0 and c_0 in discussions of linear acoustics. A further simplification is to delete the prime on $\mathbf{v}'$ but to retain the subscript on $\mathbf{v}_0$, so that the total fluid velocity is $\mathbf{v}_0 + \mathbf{v}$. In the remainder of this chapter, p is the acoustic part of the total pressure, $\mathbf{v}$ is the acoustic part of the fluid velocity, and ρ is the ambient density, unless stated otherwise.

7 WAVE EQUATIONS

For a medium that is at rest but possibly inhomogeneous, the field equations (36) and (37), upon elimination of $\mathbf{v}$, lead to the partial differential equation

$$\rho c^2 \nabla \cdot \left(\frac{1}{\rho} \nabla p \right) - \frac{\partial^2 p}{\partial t^2} = 0. \tag{42}$$

A very good approximation to this when the ambient density ρ is slowly varying with position (but not necessarily constant) is

$$c^2 \nabla^2 \left(\frac{p}{\sqrt{\rho}} \right) - \frac{\partial^2}{\partial t^2} \left(\frac{p}{\sqrt{\rho}} \right) = 0, \tag{43}$$

but when the ambient density ρ is constant, one can use the ordinary wave equation

$$\nabla^2 p - \frac{1}{c^2} \frac{\partial^2 p}{\partial t^2} = 0, \tag{44}$$

where the dependent field variable is simply p.

The wave equation is sometimes rewritten in a more compact form,

$$\Box^2 p = 0, \tag{45}$$

where the operator

$$\Box^2 = \nabla^2 - c^{-2} \frac{\partial^2}{\partial t^2} \tag{46}$$

is called the d'Alembertian, because d'Alembert was the first (in 1747) to derive the one-dimensional version of Eq. (44), although his derivation was for the case of a vibrating string. The discovery of the wave equation for sound in fluids is due primarily to Lagrange and Euler.

For the linear acoustic equations (40) and (41), which apply when there is an ambient inhomogeneous flow, an analogous wave equation[23] also results. Equation (41) admits the possibility that $\nabla \times \mathbf{v} = 0$. Acoustic disturbances in fluids typically adhere to this and consequently can be described in terms of a velocity potential Φ such that

$$\mathbf{v} = \nabla \Phi. \tag{47}$$

Equation (41) is then satisfied identically with

$$p = -\rho D_t \Phi, \qquad D_t = \frac{\partial}{\partial t} + \mathbf{v}_0 \cdot \nabla, \tag{48}$$

where the operator D_t represents the total time derivative operator following the ambient flow. The substitution of Eqs. (47) and (48) into Eq. (40) then yields the wave equation

$$\frac{1}{\rho} \nabla \cdot (\rho \nabla \Phi) - D_t \left(\frac{1}{c^2} D_t \Phi \right) = 0. \tag{49}$$

This reduces to the wave equation of Eq. (44) when ρ and c are constant and there is no ambient flow.

8 ENERGY CONSERVATION COROLLARY

A consequence of the linear acoustic equations (36) and (37) (no ambient flow, medium possibly inhomogeneous, but time independent) is the sound energy conservation corollary[24]

$$\frac{\partial w}{\partial t} + \nabla \cdot \mathbf{I} = 0, \tag{50}$$

with

$$w = \frac{1}{2} \rho v^2 + \frac{1}{2} \frac{1}{\rho c^2} p^2, \tag{51}$$

$$\mathbf{I} - p\mathbf{v} \tag{52}$$

identified as the sound energy density and sound intensity (energy flux vector), respectively. The two terms in the energy density can be interpreted as the kinetic energy per unit volume and the potential energy per unit volume, respectively. The intensity corresponds to sound power flow per unit area.

This interpretation of an equation such as Eq. (50) as a conservation law follows after integration of both sides over an arbitrary fixed volume V within the fluid and after replacement of the volume integral of the divergence of $\mathbf{I}$ with a surface integral by means of Gauss's theorem. Doing this yields

$$\frac{d}{dt} \int_V w \, dV + \int_S \mathbf{I} \cdot \mathbf{n} \, dS = 0, \tag{53}$$

where $\mathbf{n}$ is the unit-normal vector pointing out of the surface S enclosing V. This relation states that the net rate of increase of "sound energy" within the volume must equal the "sound power" flowing into the volume across its confining surface. If dissipation terms such as those associated with viscosity or relaxation processes are included within the linear acoustic equations, then the zero on the right side is replaced[25] by a term equal to the negative of the energy that is being dissipated per unit time within the volume.

No equation of simplicity comparable to Eq. (50) applies with generality to a medium with inhomogeneous ambient flow. However, if c, ρ, and $\mathbf{v}_0$ are constant, then Eqs. (40) and (41) yield

$$\frac{\partial w}{\partial t} + \nabla \cdot (\mathbf{I} + \mathbf{v}_0 w) = 0, \tag{54}$$

where the quantities w and $\mathbf{I}$ are the same as given above. The interpretation that results from this is that the w given by Eq. (51) is the sound energy density regardless of whether there is an ambient flow. The quantity $\mathbf{I}$ is the energy flux vector relative to the medium itself. If the medium is flowing with an ambient velocity, then the energy flux vector seen by someone in a stationary coordinate system must also include the energy convected by the flow, which is just $\mathbf{v}_0 w$, so the energy flux vector perceived by someone at rest is $\mathbf{I} + \mathbf{v}_0 w$.

For acoustic disturbances in inhomogeneous moving flows, it is customary to use w and $\mathbf{I} + \mathbf{v}_0 w$ as identification for energy density and intensity, but without the expectation that sound energy is conserved, as there is a possibility that energy can be interchanged between the ambient flow and the acoustic field. (In the geometric acoustics[26] limit, where the disturbance everywhere locally resembles a plane wave, a conservation law does apply, but it is not interpretable as the conservation of sound energy. Instead, what is conserved is a quantity termed *wave action*.)

9 EQUATIONS OF ELASTICITY

Sound in solids is governed by analogous equations to those that govern sound in fluids. Cauchy's equation of motion [Eq. (7)] applies to solids as well as to fluids, but because a solid can support shear stresses, it is inappropriate to take the stress tensor as diagonal, with diagonal components equal to the negative of a pressure, as is done in Eq. (8). Instead, the idealization[27] of a linear elastic homogeneous solid is often used, and the Cauchy equation is linearized at the outset, with ρ set to the ambient density and with $D\mathbf{v}/Dt$ taken as the second derivative with respect to time of a particle displacement vector, with components (ξ_1, ξ_2, ξ_3), where $\xi_i(\mathbf{x}, t)$ is the ith Cartesian component of the displacement of the particle nominally at $\mathbf{x}$ from its ambient position. For any given Cartesian coordinate system, the Cauchy equation of motion then reduces to

$$\rho \frac{\partial^2 \xi_i}{\partial t^2} = \sum_{j=1}^{3} \frac{\partial \sigma_{ij}}{\partial x_j}, \tag{55}$$

with the normally insignificant gravitational term deleted. The generalization of Hooke's law to a linear elastic solid causes the stress tensor components σ_{ij} to be linearly related to the strain tensor components

$$\epsilon_{ij} = \frac{1}{2} \left(\frac{\partial \xi_i}{\partial x_j} + \frac{\partial \xi_j}{\partial x_i} \right). \tag{56}$$

Given the usual idealization that the material is isotropic, the stress–strain relations take the form

$$\sigma_{ij} = 2\mu \epsilon_{ij} + \lambda \delta_{ij} \sum_{k=1}^{3} \epsilon_{kk}. \tag{57}$$

The Lamé constants λ and μ (the latter being the same as the shear modulus G) are related to the elastic modulus E and Poisson's ratio ν by the relations

$$\lambda = \frac{\nu E}{(1 + \nu)(1 - 2\nu)}, \qquad \mu = G = \frac{E}{2(1 + \nu)}. \tag{58}$$

Alternative quantities that are convenient to use are

$$c_1^2 = \frac{\lambda + 2\mu}{\rho}, \qquad c_2^2 = \frac{\mu}{\rho}, \tag{59}$$

The quantities c_1 and c_2 are referred to as the dilatational and shear wave speeds, respectively.

When λ and μ are both constant, the substitution of the stress–strain relation into Cauchy's equation of motion yields

$$\frac{\partial^2 \xi}{\partial t^2} = (c_1^2 - c_2^2)\nabla(\nabla \cdot \xi) + c_2^2 \nabla^2 \xi. \qquad (60)$$

This leads to the ordinary wave equation (44) for two special circumstances. If the displacement field is irrotational such that $\nabla \times \xi = 0$ (as is so for sound in fluids), then one can set $\xi = \nabla \Phi$, where the displacement scalar potential Φ is constant outside the acoustically perturbed region. In this circumstance, Φ and the components ξ_i satisfy

$$\nabla^2 \Phi - \frac{1}{c_1^2}\frac{\partial^2 \Phi}{\partial t^2} = 0. \qquad (61)$$

The other circumstance is when the displacement field is solenoidal, such that $\nabla \cdot \xi = 0$. Then it is possible to set $\xi = \nabla \times \Psi$, where the components of the vector potential Ψ satisfy the wave equation

$$\nabla^2 \Psi - \frac{1}{c_2^2}\frac{\partial^2 \Psi}{\partial t^2} = 0. \qquad (62)$$

For relatively general circumstances, any displacement field governed by the elastodynamic equations is decomposable to

$$\xi = \nabla \Phi + \nabla \times \Psi, \qquad (63)$$

where Φ satisfies Eq. (61) and the components of Ψ satisfy Eq. (62).

An energy conservation corollary of the form (5) also holds for sound in solids. The appropriate identifications for the energy density w and the components I_i of the intensity are

$$w = \frac{1}{2}\rho \sum_i \left(\frac{\partial \xi_i}{\partial t}\right)^2 + \frac{1}{2}\sum_{i,j} \epsilon_{ij}\sigma_{ij}, \qquad (64)$$

$$I_i = -\sum_j \sigma_{ij}\frac{\partial \xi_j}{\partial t}. \qquad (65)$$

10   PLANE WAVES IN FLUIDS

A solution of the wave equation that plays a central role in many acoustic concepts is that of a plane wave, which is such that all acoustic field quantities vary with time and with one Cartesian coordinate, taken here as x, but are independent of y and z. The Laplacian ∇^2 reduces thus to $\partial^2/\partial x^2$, and the d'Alembertian can be expressed as the product of two first-order operators, so that the wave equation takes the form

$$\left(\frac{\partial}{\partial x} + \frac{1}{c}\frac{\partial}{\partial t}\right)\left(\frac{\partial}{\partial x} - \frac{1}{c}\frac{\partial}{\partial t}\right)p = 0. \qquad (66)$$

The general solution of

$$\left(\frac{\partial}{\partial x} + \frac{1}{c}\frac{\partial}{\partial t}\right)f = 0 \qquad (67)$$

is any function $f(x - ct)$ that depends on x and t only in the combination $x - ct$. Similarly, the general solution of

$$\left(\frac{\partial}{\partial x} - \frac{1}{c}\frac{\partial}{\partial t}\right)g = 0 \qquad (68)$$

is any function $g(x + ct)$ that depends on x and t only in the combination $x + ct$. Consequently, the general solution of Eq. (66) is given by

$$p(x, t) = f(x - ct) + g(x + ct), \qquad (69)$$

where f and g are two arbitrary functions.

The quantity $f(x - ct)$ represents a plane wave traveling forward in the $+x$-direction at a speed c, while $g(x + ct)$ represents a plane wave traveling backward in the $-x$-direction, also at a speed c. The appropriateness of these identifications is established by the setting of x to the position $x_P(t)$ of a moving sensor. The function $f(x_P - ct)$ will appear to this sensor to be constant in time if $dx_P/dt = c$. Thus, crests, troughs, zero crossings, and other characteristic waveform features, at which the wave amplitude is some fixed value, appear to be moving with the speed c. An individual term, such as $f(x - ct)$ or $g(x + ct)$, which represents a wave traveling in just one direction, is referred to as a traveling wave. For a traveling plane wave, in contrast to traveling spherical waves (discussed in a subsequent section), not only the shape but also the amplitude is conserved during propagation.

The above description can be generalized to propagation in an arbitrary direction $\mathbf{n}$, the acoustic part of the pressure then being given by

$$p = f(\mathbf{n} \cdot \mathbf{x} - ct) \qquad (70)$$

for some generic function $f(\psi)$. Furthermore, p is the solution of the progressive wave equation

$$\frac{\partial p}{\partial t} + c\mathbf{n} \cdot \nabla p = 0, \qquad (71)$$

which is the generalization of Eq. (67).

The expression (70) for a plane traveling wave may be regarded as also being of the form

$$p = F(t - \tau), \qquad (72)$$

where τ, referred to as the eikonal, is a function of position $\mathbf{x}$ only. If the traveling wave is propagating in the direction $\mathbf{n}$, then

$$\tau = \frac{1}{c}\, \mathbf{n} \cdot \mathbf{x}, \qquad \nabla \tau = \frac{\mathbf{n}}{c}, \qquad (73)$$

and this establishes that, regardless of the direction of $\mathbf{n}$, the function τ must be a solution of the eikonal equation

$$(\nabla \tau)^2 = \frac{1}{c^2}. \qquad (74)$$

This particular equation[28] is used to describe the long-term evolution of waves that locally resemble plane waves but for which the wavefronts are possibly curved and for circumstances where the ambient sound speed may possibly vary with position. (Wavefronts are surfaces in space at which similar waveform features, such as crests, are being simultaneously received.)

If there is an ambient flow, then the idealization that $\mathbf{v}_0$, ρ, and c are constant allows Eq. (49) to apply directly to the pressure, so that

$$\nabla^2 p - \frac{1}{c^2}\, D_t^2 p = 0. \qquad (75)$$

Plane traveling waves for such circumstances exist of the generic form

$$p = F(t - \tau) = F\left(t - \frac{1}{v_{\mathrm{ph}}}\, \mathbf{n} \cdot \mathbf{x}\right), \qquad (76)$$

where the phase speed v_{ph} is the solution of

$$\left(\frac{1}{v_{\mathrm{ph}}}\right)^2 = \frac{1}{c^2}\left(1 - \frac{\mathbf{v}_0 \cdot \mathbf{n}}{v_{\mathrm{ph}}}\right)^2, \qquad (77)$$

or, equivalently, where the eikonal τ is the solution of the eikonal equation

$$(\nabla \tau)^2 = \frac{1}{c^2}\, (1 - \mathbf{v}_0 \cdot \nabla \tau)^2, \qquad (78)$$

with the correspondence $\nabla \tau = \mathbf{n}/v_{\mathrm{ph}}$. The latter is consequently referred to as the wave slowness vector.

The appropriate solution of Eq. (77) for the phase speed, being that which reduces to the sound speed c

when there is no ambient flow, is

$$v_{\mathrm{ph}} = c + \mathbf{v}_0 \cdot \mathbf{n}, \qquad (79)$$

which indicates the sound speed is augmented by the ambient velocity component in the direction normal to the wavefront.

The wave equation (75) for homogeneous moving media can also be obtained from the ordinary wave equation (44) for homogeneous nonmoving media by means of a Galilean transformation, to a new coordinate system where the coordinate axes are moving with velocity $-\mathbf{v}_0$ relative to the original coordinate axes. Such causes the substitutions

$$t \rightarrow t, \qquad \mathbf{x} \rightarrow \mathbf{x} - \mathbf{v}_0 t, \qquad (80)$$

$$\frac{\partial}{\partial t} \rightarrow \frac{\partial}{\partial t} + \mathbf{v}_0 \cdot \nabla, \qquad \nabla \rightarrow \nabla. \qquad (81)$$

The plane-wave solution (70) consequently transforms to

$$p = f(\mathbf{n} \cdot \mathbf{x} - [\mathbf{v}_0 \cdot \mathbf{n} + c]t). \qquad (82)$$

which is of the general form of Eq. (76), with the identification (79) for the phase speed v_{ph}. This can be alternately written as

$$p = f(\mathbf{n} \cdot [\mathbf{x} - \mathbf{v}_{\mathrm{gr}} t]), \qquad (83)$$

where the group velocity[29] $\mathbf{v}_{\mathrm{gr}}$ is given by

$$\mathbf{v}_{\mathrm{gr}} = c\mathbf{n} + \mathbf{v}_0. \qquad (84)$$

The term *group velocity* is used here because this is the velocity with which a wave disturbance of finite spatial extent, although locally resembling a plane wave, would appear to be propagating through space. The general rule represented by these above relations is that acoustic waves that locally resemble plane waves always propagate relative to the ambient fluid itself in directions perpendicular to their wavefronts with the sound speed. If the ambient medium is not moving, then the disturbance has a velocity $c\mathbf{n}$. However, if the ambient medium is moving with a velocity $\mathbf{v}_0$, then an observer at rest sees a velocity that is the vector sum of the velocity of the fluid and the velocity of the wave relative to the fluid.

Unlike the case for electromagnetic waves, the wave equation for sound does not appear the same for two different coordinate systems, one of which is moving at constant velocity relative to the other. (That such is the case for the wave equation governing electromagnetic waves in a vacuum is a principal tenet of Einstein's theory of relativity.)

11   RELATIONS BETWEEN ACOUSTIC VARIABLES IN A PLANE SOUND WAVE

For the special case of a plane sound wave moving in the +x-direction through a nonmoving homogeneous medium, the acoustic part of the pressure is given by Eq. (85), and Euler's equation yields

$$v_x = \frac{1}{\rho c} f(x - ct) - \frac{1}{\rho c} g(x + ct). \qquad (85)$$

Thus, in the expression for the acoustic part of the pressure, the term $f(x - ct)$, which corresponds to propagation in the same direction as that in which the fluid flow component v_x is directed, is multiplied by a factor $1/\rho c$, with a positive sign. The other term $g(x + ct)$, which corresponds to propagation in the direction opposite to that in which the fluid flow component v_x is directed, is multiplied by a factor $-1/\rho c$, with a negative sign.

The general rule that emerges from the one-dimensional example just cited is that, for a traveling plane wave propagating in the direction corresponding to unit vector $\mathbf{n}$, where, as in Eq. (70), the acoustic part of the pressure is given by $p = f(\mathbf{n} \cdot \mathbf{x} - ct)$ for some generic function $f(\psi)$, the acoustically induced fluid velocity is

$$\mathbf{v} = \frac{\mathbf{n}}{\rho c} p. \qquad (86)$$

Because the fluid velocity is in the same direction as that of the wave propagation, such waves are said to be longitudinal. (Electromagnetic plane waves in free space, on the other hand, are transverse. Shear waves in solids, as are discussed further below, are also transverse.)

The relation (86) also holds for a plane wave propagating through a medium with an ambient fluid velocity $\mathbf{v}_0$. This can be derived from the linear acoustic equations (40) and (41), but it also follows directly because the above relation is invariant under Galilean transformations. The unit vector $\mathbf{n}$ appears the same to an observer moving with speed $-\mathbf{v}_0$ as it does to one at rest. The acoustic portion $\mathbf{v}$ of the total fluid velocity is the fluid velocity relative to the ambient flow velocity. Although the ambient flow velocity may appear different to observers moving relative to each other, the acoustic portion of the fluid velocity appears the same. The pressure and the density both appear unchanged when viewed by observers moving at constant speed. The sound speed c, because it is a thermodynamic property

of the medium, as stated in Eq. (25), is also a Galilean invariant.

Both the density and the temperature fluctuate in an acoustic disturbance. Given that the medium is homogeneous, with possibly a uniform flow, and regardless of whether or not the disturbance is a plane wave, the density fluctuation is related to the pressure fluctuation by Eq. (34), $p' = c'\rho'$. The temperature fluctuation T' can be derived from the equation of state that expresses absolute temperature in terms of total pressure and total density, so that

$$T' = \left\{ \left(\frac{\partial T}{\partial p} \right)_{\rho, 0} + \frac{1}{c^2} \left(\frac{\partial T}{\partial \rho} \right)_{p, 0} \right\} p'. \qquad (87)$$

which by thermodynamic identities reduces to

$$T' = \left(\frac{T\beta}{\rho c_p} \right)_0 p', \qquad (88)$$

where

$$\beta = \rho \left(\frac{\partial (1/\rho)}{\partial T} \right)_{p, 0}, \qquad c_p = T_0 \left(\frac{\partial s}{\partial T} \right)_{p, 0} \qquad (89)$$

denote the coefficient of thermal expansion and the specific heat at constant volume. If the fluid is an ideal gas, where $\beta = 1/T_0$ and $c_p = (R_0/M)\gamma/(\gamma - 1)$, the temperature fluctuation reduces to

$$T' = (\gamma - 1) \left(\frac{T}{\rho c^2} \right)_0 p'. \qquad (90)$$

The prediction that the temperature fluctuation is nonzero is in accord with the requirement that sound propagation is more appropriately idealized as an adiabatic (no entropy fluctuation) process rather than as an isothermal process.

Relations such as Eqs. (34), (86), and (88), which relate different field components in a traveling wave, are categorically referred to as polarization relations, in analogy with those relations with the same name that characterize the ratios of the electric and magnetic field components in a propagating electromagnetic wave.

For a traveling plane acoustic wave in a homogeneous nonmoving medium, the energy conservation corollary of Eq. (50) applies. It follows from Eqs. (51) and (86) that the kinetic and potential energies are the same (Rayleigh's principle[30] for progressive waves) and that the energy density is given by

$$w = \frac{1}{\rho c^2} \, p^2. \tag{91}$$

The intensity becomes

$$\mathbf{I} = \mathbf{n} \, \frac{p^2}{\rho c}. \tag{92}$$

In regard to propagation through a moving medium, the intensity (perceived by an observer at rest when a plane wave is passing through) is, in accord with Eq. (54),

$$\mathbf{I} + \mathbf{v}_0 w = (c\mathbf{n} + \mathbf{v}_0)w. \tag{93}$$

This yields the interpretation that the energy in a sound wave is moving relative to the fluid with the sound speed in the direction normal to the wavefront. The velocity with which the energy is moving as seen by an observer at rest is consequently $c\mathbf{n} + \mathbf{v}_0$ and includes as an additive term the ambient velocity of the fluid. The appearance of the factor $c\mathbf{n} + \mathbf{v}_0$ in Eq. (93) supports the identification given in the previous section of this chapter that such a quantity is a group velocity (energy velocity).

12   PLANE WAVES IN SOLIDS

Plane acoustic waves in isotropic elastic solids have properties similar to those of waves in fluids. Dilatational (or longitudinal) plane waves are such that the curl of the displacement field vanishes, so the displacement vector must be parallel to the direction of propagation. A comparison of Eq. (61) with the wave equation of Eq. (44) indicates that such a wave must propagate with the speed c_1 determined by Eq. (59). Thus, a wave propagating in the $+x$-direction has no y- and z-components of displacement and has an x-component described by

$$\xi_x = F(x - c_1 t), \tag{94}$$

where F is an arbitrary function. The stress components can be deduced from Eqs. (56)–(59). These equations as well as symmetry considerations require, for a dilatational wave propagating in the x-direction, that the off-diagonal elements of the stress tensor vanish. The diagonal elements are given by

$$\sigma_{xx} = \rho c_1^2 F'(x - c_1 t), \tag{95}$$
$$\sigma_{yy} = \sigma_{zz} = \rho(c_1^2 - 2c_2^2)F'(x - c_1 t). \tag{96}$$

(Here the prime denotes a derivative with respect to the total argument.)

The divergence of the displacement field in a shear wave is zero, so a plane shear wave must cause a displacement perpendicular to the direction of propagation. Shear waves are therefore transverse waves. Equation (62), when considered in a manner similar to that described above for the wave equation for waves in fluids, leads to the conclusion that plane shear waves must propagate with speed c_2. A plane shear wave polarized in the y-direction and propagating in the x-direction will have only a y-component of displacement, given by

$$\xi_y = F(x - c_2 t). \tag{97}$$

The only nonzero stress components are the shear stresses

$$\sigma_{yx} = \rho c_2^2 F'(x - c_2 t) = \sigma_{xy}. \tag{98}$$

13   EQUATIONS GOVERNING WAVES OF CONSTANT FREQUENCY

Insofar as the governing equations are linear with coefficients independent of time, disturbances that vary sinusoidally with time can propagate without change of frequency. Such sinusoidally varying disturbances of constant frequency have the same repetition period T (reciprocal of frequency f) at every point, but the phase will in general vary from point to point. In mathematical descriptions of waves with fixed frequency, it is customary to use an angular frequency $\omega = 2\pi f$, which has units of radians per second.

For a plane wave of fixed angular frequency ω traveling in the $+x$-direction at the sound speed c, such that the sound pressure is a function of $t - (x/c)$, as in the first term of Eq. (69), one can in general write

$$p = |P| \cos\left[\omega\left(t - \frac{x}{c}\right) + \phi_0\right] = |P| \cos(\omega t - kx + \phi_0), \tag{99}$$

where $|P|$ is the amplitude of the disturbance, ϕ_0 is a phase constant, and $k = \omega/c$ is termed the wavenumber. The wavelength λ is the increment in propagation distance x required to change the argument of the trigonometric function by 2π radians, so $k = 2\pi/\lambda$. Also, the increment in t required to change the argument by 2π is the period T, which is the reciprocal of the frequency f; this observation yields the simple rule $\lambda = c/f$, relating wavelength, sound speed, and frequency.

In general, for a disturbance of fixed frequency or for one frequency component of a multifrequency disturbance, it is convenient to use a complex-number representation, such that each field amplitude is written[31]

$$p = \text{Re}\{\hat{p}e^{-i\omega t}\}. \tag{100}$$

Here $\hat{p}$ is called the complex amplitude of the sound pressure and in general varies with position. Two conventions are in common use for such a complex-number representation; the second uses $e^{+i\omega t}$ instead of $e^{-i\omega t}$ in expressions such as that of Eq. (100). In the latter case the identification of $\hat{p}$ would be the complex conjugate of what is used here. The $e^{-i\omega t}$ convention is predominant in literature on wave propagation as such, while the latter is predominant in literature on vibrations. If the reader wishes to translate from the first convention to the latter in the equations that appear in the remainder of this chapter, it is only necessary to replace i by $-i$.

Insertion of an expression such as that of Eq. (100) into a homogeneous linear ordinary or partial differential equation with real time-independent coefficients yields a result that can always be written in the form $\text{Re}\{\Phi e^{-i\omega t}\} = 0$, where the quantity Φ is an expression depending on the complex amplitudes and their spatial derivatives, but not depending on time. The requirement that the real part of $\Phi e^{-i\omega t}$ should be zero for all values of time consequently can be satisfied if and only if $\Phi = 0$. Moreover, the form of the expression Φ can be readily obtained from the original equation with a simple prescription: Replace all field variables by their complex amplitudes and replace all time derivatives using the substitution

$$\frac{\partial}{\partial t} \rightarrow -i\omega. \tag{101}$$

Thus, for example, the linear acoustic equations given by Eqs. (36) and (37) reduce to

$$-i\omega\hat{p} + \rho c^2 \nabla \cdot \hat{\mathbf{v}} = 0, \tag{102}$$
$$-i\omega\rho\hat{\mathbf{v}} = -\nabla\hat{p}. \tag{103}$$

The wave equation in Eq. (44) reduces to

$$\nabla^2\hat{p} + k^2\hat{p} = 0, \tag{104}$$

which is the Helmholtz equation[32] for the complex pressure amplitude.

14   SPHERICAL WAVES

Another wave type of fundamental importance is that of a spherically symmetric wave spreading out radially from a source in an unbounded medium. The symmetry implies that the acoustic field variables be functions of only the radial coordinate r and time t. The Laplacian reduces then to

$$\nabla^2 p = \frac{\partial^2 p}{\partial r^2} + \frac{2}{r}\frac{\partial p}{\partial r} = \frac{1}{r}\frac{\partial^2(rp)}{\partial r^2}, \tag{105}$$

so the wave equation of Eq. (44) becomes

$$\frac{\partial^2(rp)}{\partial r^2} - \frac{1}{c^2}\frac{\partial^2(rp)}{\partial t^2} = 0, \tag{106}$$

with the general solution

$$p(r,t) = \frac{f(r-ct)}{r} + \frac{g(r+ct)}{r}. \tag{107}$$

Causality considerations (no sound before source is turned on) lead to the conclusion that the second term on the right side of Eq. (107) is not an appropriate solution of the wave equation when the source is concentrated near the origin. The expression

$$p(r,t) = \frac{f(r-ct)}{r}, \tag{108}$$

which describes the sound pressure in an outgoing spherically symmetric wave, has the property that listeners at different radii will receive (with a time shift corresponding to the propagation time) waveforms of the same shape but of different amplitudes. The factor of $1/r$ is characteristic of spherical spreading and implies that the peak waveform amplitudes in a spherical wave decrease with radial distance as $1/r$.

The fluid velocity associated with an outgoing spherical wave is purely radial and has the form

$$v_r = \frac{1}{\rho c}[-r^{-2}F(r-ct) + r^{-1}f(r-ct)]. \tag{109}$$

Here the function F is such that its derivative is the function f that appears in Eq. (108). Because the first term (a near-field term) decreases as the square rather than the first power of the reciprocal of the radial distance, the fluid velocity v_r asymptotically approaches $p/\rho c$, which is the same as the plane-wave relation of Eq. (86).

For outgoing spherical waves of constant frequency,

the complex amplitudes of the pressure and fluid velocity are

$$\hat{p} = A \, \frac{e^{ikr}}{r}, \qquad \hat{v}_r = \frac{1}{\rho c} \left[1 - \frac{1}{ikr} \right] \hat{p}, \qquad (110)$$

where A is a constant. The expression for $\hat{p}$ here is a solution of the Helmholtz equation (104), with the Laplacian given by Eq. (105).

15   CYLINDRICAL WAVES

For cylindrically symmetric waves, there is no dependence on the azimuthal angle or on the axial coordinate, so the Laplacian in cylindrical coordinates reduces to

$$\nabla^2 = \frac{1}{r} \, \frac{\partial}{\partial r} \left(r \frac{\partial}{\partial r} \right), \qquad (111)$$

where r is here the radial distance from the symmetry axis. Consequently, the wave equation of Eq. (44) takes the form

$$\frac{\partial^2(\sqrt{r}p)}{\partial r^2} - \frac{1}{c^2} \frac{\partial^2(\sqrt{r}p)}{\partial t^2} + \frac{\sqrt{r}p}{4r^2} = 0. \qquad (112)$$

and the Helmholtz equation of Eq. (104) can be written in either of the forms

$$\frac{d^2\hat{p}}{dr^2} + \frac{1}{r} \frac{d\hat{p}}{dr} + k^2\hat{p} = 0,$$

$$\frac{d^2(\sqrt{r}p)}{dr^2} + \left[k^2 + \frac{1}{4r^2} \right] \sqrt{r}p = 0. \qquad (113)$$

The solution of the latter that corresponds to an outgoing wave is

$$\hat{p} = A H_0^{(1)}(kr), \qquad (114)$$

where the indicated function is a Hankel function[33] of the first kind, which asymptotically approaches the limit

$$\lim_{kr \to \infty} H_0^{(1)}(kr) = \left(\frac{2}{\pi kr} \right)^{1/2} e^{-i\pi/4} e^{ikr}. \qquad (115)$$

For cylindrical waves that are not of constant fre-

quency, an outgoing solution can be taken as

$$p = \int_{-\infty}^{\infty} R^{-1} F(t - c^{-1}R) \, dz_0, \qquad (116)$$

where $R = [r^2 + z_0^2]^{1/2}$ and F is an arbitrary function. Waveform shapes of outward propagating cylindrical waves tend to distort with increasing propagation distance, especially so at small r. However, at larger values of r, it is often a good approximation to neglect the last term in Eq. (112), resulting in an approximate solution of the generic form, $p(r, t) \approx f(r - ct)/\sqrt{r}$, which is similar to the expression of Eq. (108) for an outgoing spherical wave, only here the amplitude drops off with r as $1/\sqrt{r}$. The latter approximate expression is consistent with the constant-frequency solution given by Eq. (114) when the Hankel function is replaced by its asymptotic limit (115).

The fluid velocity induced by outgoing cylindrical waves is not as simply related to the corresponding sound pressure as that induced by a plane wave, although symmetry directs that the velocity must be in the appropriate radial direction when the propagation is cylindrically symmetric. For the constant-frequency case, an expression may be determined from the radial component of (103) such that the velocity is expressed in terms of the Hankel function of first order. However, a simple approximate result emerges in the limit of large radial distance r, this being the plane-wave relation of Eq. (86). (Here large r implies large compared to a characteristic wavelength, or compared to c divided by a characteristic angular frequency.)

16   SOUND SOURCES AND SOUND POWER

Many sound fields can be idealized as being steady, such that long-term time averages are insensitive to the duration and the center time of the averaging interval. Constant-frequency sounds and continuous noises fall into this category. For such sounds, the time derivative of the sound energy density will average out to zero over a sufficiently long time period, so the acoustic energy corollary of Eq. (50), in the absence of dissipation, yields the time-averaged relation

$$\nabla \cdot \mathbf{I}_{av} = 0. \qquad (117)$$

This implies that the time-averaged vector intensity field is solenoidal in regions that do not contain sound sources. This same relation holds for any frequency component of the acoustic field or for the net acoustic contribution to the field from any given frequency band. The above yields the integral relation

$$\int_S \mathbf{I}_{av} \cdot \mathbf{n}\, dS = 0, \tag{118}$$

which is interpreted as a statement that the net sound power flowing out of any region not containing sources must be zero on the time average and for any given frequency band.

For a closed surface that encloses one or more sources such that the governing linear acoustic equations do not apply at every point within the volume, the reasoning above allows definition of the time-averaged net sound power of these sources as

$$\mathscr{P}_{av} = \int_S \mathbf{I}_{av} \cdot \mathbf{n}\, dS, \tag{119}$$

where the surface S encloses the sources. If follows from Eq. (118) that the sound power of a source computed in such a manner will be the same for any two choices for the surface S, provided that both surfaces enclose the same source and no other sources. The value of the integral is independent of the size and of the shape of S. This result is of great practical importance, as it allows considerable lattitude in the measurement of source power.

In the computation of the time-averaged intensity for a field of fixed frequency, a useful mathematical relation is

$$\{a(t)b(t)\}_{av} = \tfrac{1}{2} \operatorname{Re}\{\hat{a}\hat{b}^*\}, \tag{120}$$

where $a(t)$ and $b(t)$ are any two quantities oscillating with the same angular frequency and $\hat{a}$ and $\hat{b}$ are their complex amplitudes, in the sense of Eq. (100). Thus the time-averaged intensity becomes

$$\mathbf{I}_{av} = \tfrac{1}{2} \operatorname{Re}\{\hat{\mathbf{v}}\hat{p}^*\}. \tag{121}$$

17 BOUNDARY CONDITIONS AT INTERFACES

For the model of a fluid without viscosity or thermal conduction, such as is governed by Eqs. (36) and (37), the appropriate boundary conditions at an interface are that the normal component of the fluid velocity be continuous and that the pressure be continuous. At a rigid nonmoving surface, the normal component must consequently vanish, but no restrictions are placed on the tangential component of the velocity. The model also places

no requirements on the value of the temperature at a solid surface.

In many cases involving boundary surfaces, it is helpful to make use of the concept of specific acoustic impedance[34] or unit area acoustic impedance $Z_S(\omega)$, which is defined as

$$Z_S(\omega) = \frac{\hat{p}}{\hat{v}_{in}}, \tag{122}$$

where $\hat{v}_{in}$ is the component of the fluid velocity directed into the surface under consideration. Typically, the specific acoustic impedance, often referred to briefly as impedance without any adjective, is used to describe the acoustic properties of materials. In many cases, surfaces of materials abutting fluids can be characterized as locally reacting, so that Z_S is independent of the detailed nature of the acoustic pressure field. In particular, the locally reacting hypothesis implies that the velocity of the material at the surface is unaffected by pressures other than in the immediate vicinity of the point of interest. At a nominally motionless and passively responding surface, and when the hypothesis is valid, the appropriate boundary condition on the complex amplitude $\hat{p}$ that satisfies the Helmholtz equation (104) is

$$i\omega\rho\hat{p} = -Z_S \nabla\hat{p} \cdot \mathbf{n}, \tag{123}$$

where $\mathbf{n}$ is the unit-normal vector pointing out of the material into the fluid. A surface that is perfectly rigid has $|Z_S| = \infty$. The other extreme, where $Z_S = 0$, corresponds to the ideal case of a pressure-release surface. This is, for example, what is normally assumed for the upper surface of the ocean in underwater sound. Since a passive surface absorbs energy from the sound field, the time-averaged intensity component into the surface should be positive or zero. This observation leads to the requirement that the real part (specific acoustic resistance) of the impedance should always be nonnegative. The imaginary part (specific acoustic reactance) may be either positive or negative.

18 REFLECTION AT PLANE SURFACES AND INTERFACES

When a plane wave reflects at a surface with finite specific acoustic impedance Z_S, a reflected wave is formed such that the angle of incidence θ_I equals the angle of reflection (law of mirrors). Here both angles are reckoned from the line normal to the surface and correspond to the directions of the two waves. If one takes the y-axis as pointing out of the surface and the surface as coinciding with the $y = 0$ plane, then an incident plane wave propagating obliquely in the +x-direction will have a complex pressure amplitude

$$\hat{p}_{\text{in}} = \hat{f} e^{ik_x x} e^{-ik_y y}, \tag{124}$$

where $\hat{f}$ is a constant. (For transient reflection, the quantity $\hat{f}$ can be taken as the Fourier transform of the incident pressure pulse at the origin.) The two indicated wavenumber components are $k_x = k \sin \theta_{\text{I}}$ and $k_y = k \cos \theta_{\text{I}}$. The reflected wave has a complex pressure amplitude given by

$$\hat{p}_{\text{refl}} = \mathcal{R}(\theta_{\text{I}}, \omega) \hat{f} e^{ik_y y} e^{ik_y y}, \tag{125}$$

where the quantity $\mathcal{R}(\theta_{\text{I}}, \omega)$ is the pressure amplitude reflection coefficient.

Analysis that makes use of the boundary condition of Eq. (123) leads to the identification

$$\mathcal{R}(\theta_{\text{I}}, \omega) = \frac{\xi(\omega) \cos \theta_{\text{I}} - 1}{\xi(\omega) \cos \theta_{\text{I}} + 1} \tag{126}$$

for the reflection coefficient, with the abbreviation $\xi(\omega) = Z_S/\rho c$, which represents the ratio of the specific acoustic impedance of the surface to the characteristic impedance of the medium.

The above relations also apply, with an appropriate identification of the quantity Z_S, to sound reflection[35] at an interface between two fluids with different sound speeds and densities. Translational symmetry requires that the disturbance in the second fluid have the same apparent phase velocity (ω/k_x) (trace velocity) along the x-axis as does the disturbance in the first fluid. This requirement is known as the trace velocity matching principle and leads to the observation that k_x is the same in both fluids.

If the trace velocity is higher than the sound speed c_2, then $c_2 < c_1/\sin \theta_{\text{I}}$ and a propagating plane wave (transmitted wave) is excited in the second fluid, with complex pressure amplitude

$$\hat{p}_{\text{trans}} = \mathcal{T}(\omega, \theta_{\text{I}}) \hat{f} e^{ik_x x} e^{ik_2 y \cos \theta_{\text{II}}}, \tag{127}$$

where $k_2 = \omega/c_2$ is the wavenumber in the second fluid and θ_{II} (angle of refraction) is the angle at which the transmitted wave is propagating. The trace velocity matching principle leads to Snell's law[36]:

$$\frac{\sin \theta_{\text{I}}}{c_1} = \frac{\sin \theta_{\text{II}}}{c_2}. \tag{128}$$

The change in propagation direction from θ_{I} to θ_{II} is the phenomenon of refraction.

The requirements that the pressure and normal component of the fluid velocity be continuous across the interface yield the relation

$$1 + \mathcal{R} = \mathcal{T}, \tag{129}$$

and this in conjunction with the continuity of the normal component of the fluid velocity yields the reflection coefficient

$$\mathcal{R} = \frac{Z_{\text{II}} - Z_{\text{I}}}{Z_{\text{II}} + Z_{\text{I}}}, \tag{130}$$

which involves the two impedances $Z_{\text{I}} = \rho_1 c_1 / \cos \theta_{\text{I}}$ and $Z_{\text{II}} = \rho_2 c_2 / \cos \theta_{\text{II}}$.

The other possibility, that the trace velocity is lower than the sound speed c_2, can only occur when $c_2 > c_1$ and, moreover, only if θ_{I} is greater than the critical angle $\theta_{\text{cr}} = \arcsin(c_1/c_2)$. In this circumstance, an inhomogeneous plane wave propagating in the x-direction but dying out exponentially in the +y-direction is excited in the second medium. Instead of Eq. (140), the transmitted pressure is given by

$$\hat{p}_{\text{trans}} = \mathcal{T}(\omega, \theta_{\text{I}}) \hat{f} e^{ik_x x} e^{-\beta k_2 y}, \tag{131}$$

with

$$\beta = \left[\left(\frac{c_2}{c_1} \right)^2 \sin^2 \theta_{\text{I}} - 1 \right]^{1/2}. \tag{132}$$

The previously stated equations governing the reflection and transmission coefficients are still applicable, but with the replacement of $\cos \theta_{\text{II}}$ by $i\beta$. This causes the magnitude of the reflection coefficient $\mathcal{R}$ to become unity, so the time-averaged incident energy is totally reflected. Sound energy is present in the second fluid, but its time average over a wave period stays constant once the steady state is reached.

REFERENCES

1. R. B. Lindsay, *Acoustics: Historical and Philosophical Development*, Dowden, Hutchinson, and Ross, Stroudsburg, 1972. (Many of the older articles cited in this chapter are reprinted in English translation in this book.)

2. C. Truesdell, "The Theory of Aerial Sound, 1687–1788," *Leonhardi Euleri Opera Omnia*, Ser. 2, Vol. 13, Orell Füssli, Lausanne, 1955, pp. 19–72.

3. C. Truesdell, "Rational Fluid Mechanics, 1687–1765," *Leonhardi Euleri Opera Omnia*, Ser. 2, Vol. 12, Orell Füssli, Lausanne, 1954, pp. 9–125.

4. G Batchelor, *An Introduction to Fluid Dynamics*, Cambridge University Press, Cambridge, 1967.

5. O. D. Kellogg, *Foundations of Potential Theory*, Dover, New York, 1953.

6. G. G. Stokes, "On the Theories of the Internal Friction of Fluids in Motion, and of the Equilibrium and Motion of Elastic Fluids," *Trans. Camb. Phil. Soc.*, Vol. 8, 1845, pp. 74–102.

7. P. A. Thompson, *Compressible-Fluid Dynamics*, McGraw-Hill, New York, 1972.

8. Kirkwood, J. G., "The Statistical Mechanical Theory of Transport Processes, I: General Theory," *J. Chem. Phys.*, Vol. 14, 1946, pp. 180–201.

9. Y. C. Fung, *Foundations of Solid Mechanics*, Prentice-Hall, Englewood Cliffs, NJ, 1965.

10. C. S. Yih, *Fluid Mechanics*, McGraw-Hill, New York, 1969.

11. H. Lamb, *Hydrodynamics*, Dover, New York, 1945.

12. L. Cremer, "On the Acoustic Boundary Layer Outside a Rigid Wall," *Arch. Elektr. Übertrag.*, Vol. 2, 1948, pp. 136–139.

13. P. G. Bergmann, "The Wave Equation in a Medium with a Variable Index of Refraction," *J. Acoust. Soc. Am.*, Vol. 17, 1946, pp. 329–333.

14. A. H. Wilson, *Thermodynamics and Statistical Mechanics*, Cambridge University Press, London, 1957.

15. A. D. Pierce, "Aeroacoustic Fluid Dynamic Equations and Their Energy Corollary with O_2 and N_2 Relaxation Effects Included," *J. Sound Vibr.*, Vol. 58, 1978, pp. 189–200.

16. G. R. Kirchhoff, "On the Influence of Heat Conduction in a Gas on Sound Propagation," *Ann. Phys. Chem.*, Vol. 134, 1868, pp. 177–193.

17. P. S. Laplace, "On the Velocity of Sound through Air and through Water," *Ann. Chim. Phys.*, Ser. 2, Vol. 3, 1816, pp. 238–241.

18. W. D. Wilson, "Speed of Sound in Distilled Water as a Function of Temperature and Pressure," *J. Acoust. Soc. Am.*, Vol. 31, 1959, pp. 1067–1072.

19. W. D. Wilson, "Equation for the Speed of Sound in Sea Water," *J. Acoust. Soc. Am.*, Vol. 32, 1960, pp. 641–644, 1357.

20. A. D. Pierce, *Acoustics: An Introduction to Its Physical Principles and Applications*, McGraw-Hill, New York, 1981.

21. C. Eckart, "Vortices and Streams Caused by Sound Waves," *Phys. Rev.*, Vol. 7, 1948, pp. 68–76.

22. A. D. Pierce, "Wave Equation for Sound in Fluids with Unsteady Inhomogeneous Flow," *J. Acoust. Soc. Am.*, Vol. 87, 1990, pp. 2292–2299.

23. D. I. Blokhintzev, "The Propagation of Sound in an Inhomogeneous and Moving Medium I," *J. Acoust. Soc. Am.*, Vol. 18, 1946, pp. 322–328.

24. G. R. Kirchhoff, *Vorlesungen über mathematische Physik: Mechanik*, Teubner, Leipzig, 1877.

25. C. Eckart, "The Thermodynamics of Irreversible Processes," *Phys. Rev.*, Vol. 58, 1940, pp. 267–269.

26. W. D. Hayes, "Energy Invariant for Geometric Acoustics in a Moving Medium," *Phys. Fluids*, Vol. 11, 1968, pp. 1654–1656.

27. S. H. Crandall, N. C. Dahl, and T. J. Lardner, *An Introduction to the Mechanics of Solids*, McGraw-Hill, New York, 1978.

28. P. G. Frank, P. G. Bergmann, and A. Yaspan, "Ray Acoustics," in *Physics of Sound in the Sea*, Department of the Navy, Headquarters Naval Material Command, Publication NAVMAT P-9675, 1969, pp. 41–68.

29. M. J. Lighthill, *Waves in Fluids*, Cambridge University Press, Cambridge, 1978.

30. J. W. S. Rayleigh, "On Progressive Waves," *Proc. Lond. Math. Soc.*, Vol. 9, 1877, pp. 21–26.

31. C. J. Bouwkamp, "Contributions to the Theory of Acoustical Radiation," *Phillips Res. Rep.*, Vol. 1, 1946, pp. 251–277.

32. H. Helmholtz, "Theory of Air Oscillations in Tubes with Open Ends," *J. Reine Angew. Math.*, Vol. 57, 1860, pp. 1–72.

33. M. Abramowitz and I. A. Stegun, *Handbook of Mathematical Functions*, Dover, New York, 1965.

34. P. M. Morse and R. H. Bolt, "Sound Waves in Rooms," *Rev. Modern Phys.*, Vol. 16, 1944, pp. 69–150.

35. G. Green, "On the Reflexion and Refraction of Sound," *Trans. Camb. Phil. Soc.*, Vol. 6, 1838, pp. 403–412.

36. W. B. Joyce and A. Joyce, "Descartes, Newton, and Snell's Law," *J. Opt. Soc. Am.*, Vol. 66, 1976, pp. 1–8.

3

RAY ACOUSTICS FOR FLUIDS

D. E. WESTON

1 INTRODUCTION

Ray acoustics is also called geometric acoustics and is probably the easiest and best known way of thinking about sound propagation problems. In a homogeneous medium the sound may be pictured as traveling in a straight line directly from the source to the receiver, and at high frequencies this is virtually what does happen, just as in ray optics.

In general, there is a three-dimensional distribution of sound speed c, and the ray approach proceeds in two stages. First, it finds the ray path or paths by considering the wavefront or the equivalent acoustic phase, which may vary quite rapidly with position. Second, it models the acoustic amplitude, which is assumed to vary only slowly. Reflections from the continuous changes in speed and density are ignored, so the result is usually an approximation, though often a good one.

Technically the approach stems from Huygens' principle, in which each point on a wavefront is imagined to be the source of a spherically spreading wavelet, as shown in Fig. 1. After a short time interval the envelope of the wavelets is taken to define the new wavefront, and then the process repeats indefinitely. The direction of ray travel is normal to the wavefront, except that there can be complications for moving fluids and anisotropic solids.

This chapter presents the basic equations, discusses applications (one is dramatized in Fig. 2, to be examined more fully later, gives a very brief account of expansions and extensions to ray theory, and finally comments on validity.

Encyclopedia of Acoustics, edited by Malcolm J. Crocker
ISBN 0-471-80465-7 © 1997 John Wiley & Sons, Inc.

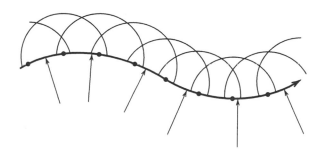

Fig. 1 Wavefront, Huygens wavelets, and ray directions.

2 BASIC EQUATIONS

It is necessary to start by introducing the eikonal $\tau(x, y, z)$, named from the Greek for image but actually specifying the travel time to a point on the acoustic wavefront. Thus any constant value of the eikonal τ defines a surface that is a wavefront. For example, with a point source in a homogeneous medium each τ value corresponds to a different spherical surface. For an extended source or an inhomogeneous medium the τ value may depend upon position in a complicated manner, with wavefronts that are also complicated, as in Fig. 1.

Defining the eikonal as a time has advantages both in simplicity and in reaching an equation that does not involve frequency, since the ray concept is independent of frequency. But it has sometimes been defined as the length $c_0\tau$, where c_0 is a reference sound speed, and sometimes as the phase $\omega\tau$, where ω is the angular frequency. Huygens' principle leads to the eikonal equation,[2-4] expressed in Cartesian coordinates as

$$\left(\frac{\partial\tau}{\partial x}\right)^2 + \left(\frac{\partial\tau}{\partial y}\right)^2 + \left(\frac{\partial\tau}{\partial z}\right)^2 = \frac{1}{c^2}. \tag{1}$$

Sound velocity (m s^{-1})

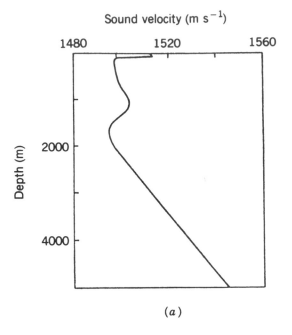

(a)

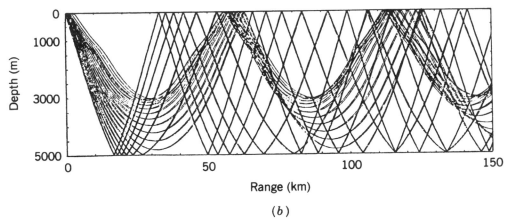

(b)

Fig. 2 (a) Typical deep-water profile of sound velocity vs. depth, Northeast Atlantic, October. (b) Computer-generated ray trace for profile (a), source at 38 m depth, ray angles at source every 1° between ±17°. Vertical scale is exaggerated by a factor of 10. From Ref. 1, © Crown Copyright, reproduced with the permission of the Controller of HMSO.

The point of Eq. (1) is that it enables the development in time of the wavefront to be followed. If the fluid is allowed to move, there are necessary modifications, as discussed by Pierce.[2]

Another form of Eq. (1) uses the local wavenumber k, really a propagation vector, together with the wavenumber components k_x, k_y, k_z resolved along the Cartesian axes,

$$k_x^2 + k_y^2 + k_z^2 = k^2. \qquad (2)$$

The ray direction normal to the wavefront is specified by the local direction cosines k_x/k, k_y/k, and k_z/k, so that Eqs. (1) or (2) also allow the tracing of rays.

In the most straightforward case k_x, k_y, and k_z are all real, and we have an ordinary ray or wave in which the amplitude variations are very slow. But if, for example, k_x^2 is negative and k_x imaginary, it is still possible to find a physical interpretation. It is that of a so-called inhomogeneous plane wave[5] in which the amplitude changes exponentially across the wavefront as $\exp(ik_x x)$: a good example is its occurrence in a second medium when the waves from the first suffer total internal reflection.

The eikonal equation is the basic one in ray acoustics but takes us on to another important result, Fermat's principle. This states that the travel time along a ray path from a to b is at a local extremum with respect to any small change δ in the path s,

$$\delta\left(\int_a^b c^{-1}\,ds\right) = 0. \qquad (3)$$

In practice this usually turns out to be a path of least time.

To model amplitudes or intensities, narrow ray tubes or beams must be considered. Energy or power is conserved along the tube, and a transport equation or flux approach shows that intensity must vary *inversely as the tube cross-sectional area*. Thus, in a homogeneous medium the intensity will vary as the familiar r^{-2} law, where r is the range or distance from a point source.

3 APPLICATIONS

For any acoustic propagation problem a ray theory treatment should at least be considered. Problems run from those with solid structures in the laboratory and elsewhere; to those with fluids in the laboratory, industry, and even physiology; to noise in factories and music in concert halls (Chapters 91 and 92)[6]; and to long-range noise in the atmosphere (Chapter 32),[7] in the ocean (Chapter 36),[1,4] and in the earth's crust. This chapter concentrates on fluids; however, in scattering from targets and from irregularities (Chapters 4 and 43) ray theory may be applied not only within structures but also for surface-guided or creeping waves on the outside. Here the applications are considered less for their own sake, since they are covered elsewhere, and more as illustrations of particular ray effects.

Perhaps the classic ray acoustic effects are those for high frequencies in the laboratory, where many ray optic phenomena may be reproduced, including that of focusing with a suitable lens.

Figure 3 introduces multiple paths and images for a homogeneous medium with range-independent geometry. This could represent a laboratory arrangement, a first approximation to room acoustics, or a coastal water experiment. In fact, the acoustics of rooms, or of concert halls, involves the extreme form of this type of complication since the enclosures are three dimensional and the multiple arrivals determine the critical time history on reception. Pure ray theory breaks down on each reflection at a sharp boundary, and the connection to a continuing ray has to be made by solving a boundary condition problem (Chapter 2). In a room most boundaries are lossy and are not simple plane surfaces, so the process

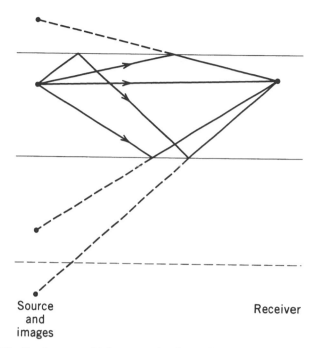

Source and images   Receiver

Fig. 3 Some multiple ray paths due to boundary reflections, illustrating the concept of image sources.

is one of scattering or diffraction. In practice, for room acoustics the geometry is usually so complicated that ray acoustics is to be preferred over wave acoustics, detailed modeling with the latter being too difficult.

In any application with multiple paths it is necessary to combine the various contributions. With continuous sources the simplest way is to add intensities, or, strictly, since intensity is a vector, to add the mean-square values of one of the field quantities, such as acoustic pressure. This incoherent addition may be satisfactory if source and receiver are well away from boundaries and if an average result is acceptable. Otherwise a great improvement is possible by coherent addition of amplitudes, that is, taking account of phase. In all these propagation complexities the principle of reciprocity tells us that the transmission is unchanged if the source and receiver are interchanged, however complicated the environment, provided the transducers and the medium remain still.

There remain the major applications to atmospheric acoustics and to underwater acoustics, and the discussions on these will be intertwined since each involves a layered medium and there are basic similarities. In effect, long-range underwater acoustics is an upside-down version of long-range atmospheric acoustics. The height of the isothermal atmosphere is of the same order as the depth of the deep ocean. But sound speed in water is more than 4 times that in air, and comparisons are better made for the same wavelength rather than the same frequency. Also, the absorption coefficients for air are about two orders of magnitude greater than those for water at the same frequency. Thus at around 100 Hz any acoustic signal in air disappears beyond a few 100 km, whereas in water it can extend halfway round the world. (An alternative basis for comparison of air and water is to choose the respective frequencies so that the absorptions are the same.) For both media the ranges can greatly exceed the heights or depths, and it is not surprising that the grazing angles for the rays are typically very small. Also, for both media, ray methods have to compete with a variety of wave approaches.

The scene is really set by the profiles of sound speed c, whereas the density variation, especially in a ray treatment, is unimportant. The sound speed (Chapter 6) in air is mainly a function of the temperature and the wind, although gravity plays an indirect part at very low frequencies. Sound speed in water depends mainly on temperature, salinity, and ambient pressure or depth, with current as a minor influence. A common assumption in modeling is that the environment, specifically the sound speed profile, is independent of range.

In a general range-independent layered medium the connection between the grazing angle θ of a ray (i.e., measured with respect to the horizontal) at one height or one depth and that at another is given by Snell's law, expressed as the invariant

$$c \sec \theta = \text{const.} \quad (4)$$

For example, if the velocity gradient c' is constant, this produces ray paths that are arcs of circles of radius c/c', c being measured where the ray is horizontal.

For layering with central or circular symmetry there is Bouguer's law:

$$R^{-1} c \sec \theta = \text{const.} \quad (5)$$

where R is the range from the center of symmetry and θ is measured with respect to the local layering. This may be used to allow for the sphericity of the earth, by equating R with the earth radius, but the resulting correction is small.

One special aspect of atmospheric acoustics is the magnitude of the refraction differences between day and night. In the daytime the warmer air, with higher sound speed, usually lies near the ground and tends to curve the rays upward, away from the surface. At night the cooler air is usually near the ground and tends to curve the rays downward. Another aspect is the magnitude of the dependence on wind direction; for sound propagating upwind the rays tend to curve upward, and for propagation downwind the rays tend to curve downward. Further atmospheric ray problems concern the influence of topography and propagation within city streets or within forests.

In underwater acoustics the propagation literature is very extensive, reflecting the great variability in the environmental conditions. There is the deep ocean, the Continental Slope of intermediate depth, and the shallow coastal waters, all with a variety of sound speed profiles and ray diagrams. In addition, underwater landscape is as varied as that above water.

A detailed example of ray behavior is shown in Figure 2, with a fan of rays from the source. Although the illustration could have come from either air or water, this diagram is for range-independent layered water. There are shadow zones where no rays penetrate; note especially the clear near-surface region around the 20-km range. The closer range boundary of a shadow is typically formed by one ray, the limit ray. The longer range boundary of a shadow is typically formed by the envelope of a series of rays that form a caustic, nominally with an infinite concentration of rays. Of course, the presence of an infinity indicates that ray theory is not working properly at these positions. But high levels are certainly encountered in the neighborhood of a caustic; note in particular the high concentration of rays at the convergence zones near the surface, around both 60 and 120 km in range. Convergence zones can occur at comparable ranges in the atmosphere.

The profile in Fig. 2a shows sound speed minima at three depths that identify very efficient sound channels; the surface sound channel, the shallow sound channel, and the main, or SOFAR, channel. For a source within a channel, some downgoing rays are refracted upward above the channel bottom, then either refracted shallow or reflected at the surface, and continue cycling along in this manner. Channeling also occurs in atmospheric acoustics.

It is possible to cope with range-dependent profiles, but it is common to assume cylindrical symmetry about the source. An intensity formula follows from the ratio of ray tube areas that also covers the range-independent case:

$$\frac{I_b}{I_a} = \left| \frac{\cos \theta_a}{x \sin \theta_b} \frac{\partial \theta_a}{\partial x} \right|_{\text{const } z} = \left| \frac{\cos \theta_a}{x \cos \theta_b} \frac{\partial \theta_a}{\partial z} \right|_{\text{const } x}. \quad (6)$$

Here a and b refer to source and receiver, I_b is the desired intensity, I_a is the intensity at unit range (1 m), x is horizontal range, and z is measured vertically. The result might be quoted as a transmission loss $10 \log(I_a/I_b)$ in decibels. In practice, I_b should be corrected for the attenuation along the path as well as for losses on reflection, where appropriate.

One special application of Eq. (6) is to the location of caustics, actually surfaces in three dimensions, found by putting $\partial\theta_a/\partial x$ or $\partial\theta_a/\partial z$ equal to infinity. Quite complicated caustic formations can occur.

4 WENTZEL–KRAMERS–BRILLOUIN APPROXIMATION FOR MODES

The ray concept comes in three varieties.[8] First, there are the rays that join a specific source and a specific receiver, as in Figure 3. Second are the rays broadcast in a fan from a source, defining a ray field as in Fig. 2b. Third is the infinite-distance concept with neither specific source nor specific receiver, the so-called plane-wave ray, for which in a homogeneous medium the wavefront is planar. This last type brings us towards the concept of normal modes (Chapter 36).

Consider a layered medium in which refraction or reflection at both upper and lower points leads to channeling. The rate of phase change can be integrated over a complete loop or cycle of the plane-wave ray, with the phase changes ϕ_1 and ϕ_2 at top and bottom included. The condition for reinforcement provides an estimate of the mode eigenvalues, that is,

$$2 \int k_z \, dz + \phi_1 + \phi_2 = 2n\pi. \tag{7}$$

Reinforcement occurs if n is integral, so that after one loop the ray has the same phase as when it started, and n is then known as the mode number. This is the Wentzel–Kramers–Brillouin (WKB) approximation.[4] By adding together the downgoing and upgoing rays, it is possible to model the model intensity as a function of depth. Although based on ray theory the WKB approach avoids problems such as those with caustics.

There is in general a dichotomy between the description of propagation using rays and using modes,[4] either exact modes or WKB modes. Ray descriptions can be transformed into WKB descriptions and vice versa, whether the angular interval considered is complete or restricted. For example, a single ray may be replaced by a summation over modes for which the corresponding plane-wave angle is close to the ray angle or even by an approximate integration over these modes. This last idea may also be used in propagation modeling, especially as in the RAYMODE program.[9]

5 IMPROVEMENTS

Much work has gone into modifying or improving ray theory, often by patching in a local wave theory solution in the neighborhood where ray theory breaks down.

The prime example is due to Brekhovskikh,[5] who fits an Airy function to the wave field near a caustic. Among other things this shows that the geometric calculation of phase for a ray that has passed through a simple caustic must be corrected by a $\frac{1}{2}\pi$ phase advance; physically the ray may be thought of as sensing the region of higher speed beyond the caustic. The same phenomenon occurs in WKB theory at refraction turning points, where the ray is horizontal.

Another improvement concerns reflection at a real medium, as opposed to an idealized sound-hard or sound-soft boundary, and this medium will be assumed to have a higher sound speed. The rays are assumed to be relatively shallow in angle and of the plane-wave type. They will undergo total internal reflection, nominally without loss in amplitude but with a phase change ϕ that depends on the angle. This phase change is equivalent to the reflection having taken place at an imaginary sound-soft boundary (i.e., a free or pressure-release surface) situated a distance δ_v beyond the real boundary, equivalent again to a "horizontal" displacement δ_h along the boundary. These wave shifts or displacements are given as

$$\delta_v = \frac{\pi+\phi}{2k_z}, \qquad \delta_h = \frac{k_x(\pi+\phi)}{k_z^2}. \tag{8}$$

They can be useful in the calculation of modal phase velocities and eigenvalues; for example, simplified version of Eq. (7) may be possible.

However, if the concern is with the amplitude distribution for a beam having a small range of angles, a different pair of displacements must be used:[5,8]

$$\Delta_v = \frac{1}{2}\frac{\partial\phi}{\partial k_z}, \qquad \Delta_h = -\frac{\partial\phi}{\partial k_x}. \tag{9}$$

These are the better known beam displacements, and Δ_h is the same as the Goos–Hänchen shift in optics. It can be used to improve ray theory calculations of group or energy velocity and of modal cycle distance, the latter being illustrated in Fig. 4.

Note that as grazing angle θ approaches zero, δ_v and

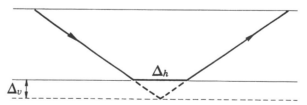

Fig. 4 Beam displacement on reflection, with effect on cycle distance.

Δ_v approach the same constant value. The beam displacement idea is related to that of the lateral ray,[5] and in underwater refraction shooting this is the ground wave arrival. In effect, it travels along the upper boundary of the seabed at a speed equal to that of the seabed material and for an indefinite distance. It is not a trapped or channeled interface wave. Note that Fermat's principle still applies to special ray types such as this one.

6 RANGE-DEPENDENT ENVIRONMENT

The general case is a three-dimensional medium with substantial and complicated sound speed changes as a function of all three coordinates. Ray acoustics may work in theory, but in practice, any calculations will be difficult to organize and perhaps also difficult to interpret. However, a conventional ray approach can cope well with a range-dependent layered medium, especially if the layering changes only slowly and, as already noted, there is cylindrical symmetry about the source that obviates cross-track complications. For sufficiently slow changes there is a ray invariant[10] given by an integration of ray angle with respect to a vertical line, so that the ray paths at one locality can be found from those at another locality without the need to know what has happened in between:

$$\int \frac{\sin\theta \, dz}{c} = \frac{1}{\omega} \int k_z \, dz = \text{const.} \qquad (10)$$

This can be useful for long ranges in both atmospheric and underwater acoustics. For a homogeneous ("isospeed") medium this result states that $H\sin\theta$ is constant, where H is channel width. For example, for shallow coastal water, with propagation downslope, H increases, and the successive reflections at the sloping bottom cause $\sin\theta$ to decrease in compensation.

The result in Eq. (10) is obviously closely related to the WKB eigenvalue equation (7). It is equivalent to assuming that all the energy in a given normal mode at one locality will transfer into the corresponding normal mode at another locality, where the layering and mode characteristics may be quite different, provided again that the layering changes sufficiently slowly. This assumption, known as the adiabatic approximation, ignores coupling between different modes.

Development of the idea of either ray invariants or adiabatic modes gives a measure of the average intensity as a function of range r and of vertical coordinate (height or depth).[10] In effect, this is a range averaging over a distance interval comparable to cycle distance X; it gets rid of fluctuations due to interference, and the result $\overline{I_b}$ is a useful quantity. Keeping the values of grazing angle

small yields

$$\frac{\overline{I_b}}{I_a} = \frac{4}{r} \int \frac{d\theta_a}{\theta_b X_b} . \qquad (11)$$

For the isospeed constant-depth case this becomes simply $2\theta_m/rH$. Here θ_m is the maximum allowed angle, perhaps set by the critical angle for total internal reflection.

If the environment does vary across the track, the rays or modes may be deflected sideways, and this horizontal refraction of the projected ray path can be treated using another application of ray theory, in two dimensions.[1,8] The effect is usually small—but not always!

7 VALIDITY OF RAY APPROACH

The derivation of the eikonal equation (1) using Huygens' principle already suggests the approximate nature of the result. An alternative derivation from the more firmly based wave equation (Chapter 2) gives the validity condition[3]

$$\left| \frac{A''}{A(\omega\tau')^2} \right| \ll 1. \qquad (12)$$

Here A is the amplitude of a field quantity such as acoustic pressure, the prime refers to spatial differentiation, and the inequality must hold for any direction. One consequence of this is the condition[4]

$$\left| \frac{1}{k_z} \frac{d}{dz} (\ln k_z) \right| \ll 1, \qquad (13)$$

where the resulting inequality again must hold for all directions, the choice of z being merely illustrative. This result is more useful than inequality (12) since it controls reflectivity. Thus a ray normally incident on a discontinuity with a sharp change in sound speed will have a significant reflected component, whereas it will pass virtually straight through if there is a gradual change of the same magnitude. Inequality (13) is also applicable and also important for the validity of the WKB mode calculation. But it does not cover changes in density.

A weaker and simpler version of inequality (13) is given below followed by two equivalents of this weaker version, these three calling attention to different aspects of validity:

$|\lambda c'/c| \ll 1$, or the fractional velocity change over a wavelength should be very small.

$|\lambda A'/A| \ll 1$, or the fractional amplitude change over a wavelength should be very small.

$|\lambda c\tau''| \ll 1$, or the ray radius of curvature should be much larger than a wavelength.

Generally it appears that the wavelength λ should be smaller than any other pertinent length, such as shadow edge thickness (including caustics), focal size, obstacle size, roughness scale, range from source, and channel width H.

Concerning height of atmosphere or water depth, a different type of criterion is obtained by calculating the approximate ratio of the number of effective ray arrivals to effective mode arrivals. Conceptually and for convenience it can pay to use the ray approach if this ratio is less than unity. For source and receiver both well away from the boundaries of a channel, this produces[1]

$$\frac{r\lambda}{2H^2} < 1. \tag{14}$$

Comprehensive advice on when to use ray theory is difficult to formulate, as it depends on which version of the formulas is to be used, and it is often best just to try ray acoustics. It is likely to do well in comparison with other methods at high frequency, at short range, for great channel width, and for range-dependent environments. It can give exact results in some simple situations.

8 CONCLUDING REMARKS

Ray acoustics is a very useful tool for a wide variety of applications; it should not be misused, but it can some-

times be rewarding to walk the tightrope at the limit between ray and wave approaches.

REFERENCES

1. D. E. Weston and P. B. Rowlands, "Guided Acoustic Waves in the Ocean," *Rep. Prog. Phys.*, Vol. 42, 1979, pp. 347–387.

2. A. D. Pierce, *Acoustics: An Introduction to Its Physical Principles and Applications*, McGraw-Hill, New York, 1981; revised edition, Acoustical Society of America, New York, 1989.

3. P. G. Frank and A. Yaspan, "Ray Acoustics," in *Physics of Sound in the Sea*, Chapter 3, Department of the Navy, Washington, DC, 1969; originally issued 1946.

4. I. Tolstoy, and C. S. Clay, *Ocean Acoustics: Theory and Experiment in Underwater Sound*, McGraw-Hill, New York, 1966; paperback edition, Acoustical Society of America, New York, 1987.

5. L. M. Brekhovskikh, *Waves in Layered Media*, Academic, New York, 1960; second edition, 1980.

6. H. Kuttruff, *Room Acoustics*, 3rd ed., Applied Science, London, 1991.

7. J. E. Piercy, T. F. W. Embleton, and L. C. Sutherland, "Review of Noise Propagation in the Atmosphere," *J. Acoust. Soc. Am.*, Vol. 61, 1977, pp. 1403–1418.

8. D. E. Weston, "Rays, Modes and Flux, in L. B. Felsen (Ed.), *Hybrid Formulation of Wave Propagation and Scattering*, Nijhoff, Dordrecht, 1984, pp. 47–60.

9. P. C. Etter, *Underwater Acoustic Modeling*, Elsevier, New York, 1991.

10. D. E. Weston, "Acoustic Flux Formulas for Range-dependent Ocean Ducts," *J. Acoust. Soc. Am.*, Vol. 68, 1980, pp. 269–281.

4

RAY ACOUSTICS FOR STRUCTURES

Leopold B. Felsen

1 INTRODUCTION

Under high-frequency or short-wavelength conditions, wave fields described by linear wave equations behave in a particularly simple form due to a phenomenon known as *localization*. The high-frequency or short-wavelength regime is characterized by the condition $\lambda/L \ll 1$, where λ is the wavelength and L is a characteristic length that governs the environment wherein the field propagates. Thus, L may be a large observation distance, the size of a scattering object, the scale over which an inhomogeneous propagation medium varies appreciably, and so on. Ray methods refer to a discipline that organizes high-frequency wave phenomena excited by sources in the presence of globally complex environments into an interactive sequence of simpler *local wave* problems. The local wave fields (in magnitude and phase) are transported, and carry the local wave energy, along trajectories called *rays*, and the local wave fields are correspondingly referred to as *ray fields*. When a particular ray field encounters obstacles, discontinuities, and other perturbing features as it progresses through the environment, such encounters give rise to new species of reflected, transmitted, and scattered ray fields whose initial amplitude, phase, and direction are determined by solving the relevant local wave problems in the vicinity of the point of impact. In this way, the high-frequency wave field at an observation point P is given by the sum of all ray fields passing through P, provided that these ray fields are *individually distinct* in their magnitude, phase, and propagation direction. The latter condition defines the limitations of simple ray theory. The condition is violated when two or more ray fields reaching P are nearly codirectional and therefore interfere such that their local independence is affected. A more sophisticated "uniform" ray theory that refines the simple ray theory must now be invoked to account for the more complicated wave behavior engendered by the self-consistent interaction of the simple ray fields. The confined affected space domains where this occurs are referred to as "transition regions." In this manner, by blending simple ray theory, which works "almost everywhere," with uniform ray theory in transition regions, one has available a versatile *quantitative* algorithm that is matched directly to high-frequency wave phenomenology. The governing rules are established by a variety of high-frequency asymptotic techniques. Early pioneering contributions are due to J. B. Keller,[1] who formalized the classification and tracking of various species of ray fields into the geometric theory of diffraction (GTD)[1–7] (see also Chapter 42).

When the multiplicity of source-excited progressing (traveling) ray fields at P is generated by systematic successive reflections or scatterings (e.g., between layer boundaries or scattering centers), their collective effect can often be summed into oscillatory (standing-wave) *modal ray fields*. Modal ray fields characterize the high-frequency behavior of guided (e.g., trapped or leaky) modes that satisfy the relevant *source-free* wave equations plus boundary conditions (see Chapters 7 and 8).

The present chapter aims at a tutorial introduction to ray theory, especially its spectral foundation (see Section 2), defining terminology and relying primarily on physical concepts with simple examples. The reader may consult the cited references for mathematical details and a wide range of applications to fluid–structure interaction (see also Chapters 11 and 93). It is to be noted that ray theory applies to high-frequency linear wave phenomena in *any* discipline (acoustic, elastic, electromagnetic, etc.) provided that its underlying assumptions based on

Encyclopedia of Acoustics, edited by Malcolm J. Crocker
ISBN 0-471-80465-7 © 1997 John Wiley & Sons, Inc.

locality are satisfied. Throughout the years, the method has been extended, refined, and tested[8-12] so that it can now be regarded as one of the leading tools for quantitative calculation of high-frequency linear wave fields under environmental conditions of broad generality.

2 RAY THEORY (SPECTRAL FOUNDATIONS)

Source-excited time-harmonic acoustic wave fields $u(x, y, z)$ defined by a linear wave equation in an unbounded homogeneous fluid can generally be decomposed into continuous spectra of plane waves. The spectral variables are the wavenumbers (k_x, k_y) corresponding to a conveniently chosen $(x, y, 0)$ plane, in which the decomposition is performed via (x, y) Fourier transforms with kernel $\exp[i(k_x x + k_y y)]$ applied to the wave equation. By spectral synthesis, the wave field is represented as[13-16]

$$u(\mathbf{r}) = \frac{1}{2\pi} \iint_{-\infty}^{\infty} dk_x \, dk_y \, U(k_x, k_y) \exp(i\mathbf{k} \cdot \mathbf{r}) \quad (1)$$

where $U(k_x, k_y)$ is the excitation amplitude of each spectral plane wave, $\mathbf{r} = (x, y, z)$ and (k_x, k_y, k_z) are the position vector and wave vector, respectively, $k_z = k_z(k_x, k_y) = [k^2 - (k_x^2 + k_y^2)]^{1/2}$ is the wavenumber along z, $k = \omega/c = 2\pi/\lambda$ is the acoustic wavenumber in the fluid, ω is the radian frequency, λ is the wavelength, and c is the acoustic propagation speed in the fluid. Since $-\infty < (k_x, k_y) < \infty$, k_z is real when $(k_x^2 + k_y^2)^{1/2} < k$ but imaginary when $(k_x^2 + k_y^2)^{1/2} > k$. The former condition describes plane waves that propagate away from the $z = 0$ plane into the region $z > 0$ along real directions $\theta_{x,y,z}$ defined by $k_{x,y} = k \sin \theta_{x,y}$, $k_z = k \cos \theta_z$. The latter condition describes evanescent wave fields that decay exponentially away from the $z = 0$ plane into the region $z > 0$; this implies that $k_z = \pm i|k_z|$, corresponding to time dependence $\exp(\pm i\omega t)$, respectively, with $\theta_{x,y,z}$ complex. These formal considerations apply to wave fields excited by (a) actual sources (e.g., transducers) and (b) virtual sources that can be employed to model induced vibrations on elastic scatterers insonified by incident excitation; in case (b), the resulting wave field is the scattered field. For details, see Chapter 5.

When $\mathbf{k} \cdot \mathbf{r}$ is large with k_z real, the exponential in the integrand of Eq. (1) fluctuates rapidly with respect to the spectral amplitude U almost everywhere in the (k_x, k_y) range. Since the value of an integral equals the area under the curve traced out by the integrand, and since positive and negative portions during each rapid fluctuation cycle cancel approximately, the major contribution to the integral arises from those (k_x, k_y) intervals where the phase

varies slowly; that is, the intervals around the *stationary phase* points $(k_{xs}(\mathbf{r}), k_{ys}(\mathbf{r}))$, where $(\partial/\partial x)(\mathbf{k}_s \cdot \mathbf{r}) = (\partial/\partial y)(\mathbf{k}_s \cdot \mathbf{r}) = 0$. Stationary phase evaluation of the integral in Eq. (1) yields the approximate wave field (Ref. 16, p. 428)

$$u(\mathbf{r}) \sim A_s(\mathbf{r}) \exp(i\mathbf{k}_s \cdot \mathbf{r}), \qquad \mathbf{k}_s \cdot \mathbf{r} \gg 1, \quad (2)$$

where $\sim$ means "asymptotically equal to", $\mathbf{k}_s = (k_{xs}, k_{ys}, k_{zs})_\mathbf{r}$, the subscript $\mathbf{r}$ implies that $k_{xs}, \ldots$ are $\mathbf{r}$ dependent, and

$$A_s(\mathbf{r}) = \frac{\exp(i\pi\sigma/4)}{|\det[(\partial^2/\partial k_x \, \partial k_y)(\mathbf{k}_s \cdot \mathbf{r})]|^{1/2}} U(k_{xs}, k_{ys})_\mathbf{r}. \quad (2a)$$

Here, $\sigma = (\text{sgn } d_1 + \text{sgn } d_2)$, where $\text{sgn } d = \pm 1$ for $d \gtrless 0$ and $d_{1,2}$ are the eigenvalues of the 2×2 matrix with elements $(\partial^2/\partial k_i \, \partial k_j)(\mathbf{k}_s \cdot \mathbf{r})$, $i, j = x, y$. It is assumed that $|\det(\cdot)|$, which is proportional to the ray tube cross section $dS(\mathbf{r})$ in Fig. 1, does not vanish.

The wave object in Eq. (2) with its $\mathbf{r}$-dependent amplitude A and wave vector $\mathbf{k}_s$ is referred to as a *local plane wave*, in contrast with the spectral *true plane waves* in the integrand of Eq. (1) with their $\mathbf{r}$-independent amplitudes U and wave vectors $\mathbf{k}$ (see Fig. 1). The local plane wave in Eq. (2) can be interpreted as tracking the constructive interference maxima of the true spectral plane wave bundle that surrounds $\mathbf{k}_s(\mathbf{r})$ in the integrand of Eq. (1). The trajectories of these maxima are the *rays*, and the local plane-wave fields transported along these trajectories are correspondingly the *ray fields* (Fig. 1). The surface $\psi_s(\mathbf{r}) = (\mathbf{k}_s \cdot \mathbf{r})/k = \text{const}$, which is perpendicular to the wave vector $\mathbf{k}_s$, defines the wavefront, and the family of rays associated with this wavefront forms a *ray congruence*. This establishes the spectral foundation of ray theory. Ray theory can also be developed directly from the wave equations that describe the wave field in question by inserting the *ansatz* (i.e., the assumed form) $u(\mathbf{r}) \sim A(\mathbf{r}) \exp[ik\psi(\mathbf{r})]$ into the wave equation and deriving expressions for the normalized phase $\psi(\mathbf{r})$ and the amplitude $A(r)$ [Ref. 16, Chapter 1]; the asymptotic ordering is facilitated by exhibiting the assumed large wavenumber k explicitly in the phase. The result is consistent with the local plane wave in Eq. (2), but the direct derivation does not highlight the connection with spectral theory.

The above methods are effective when all fields reaching an observer can be computed with sufficient accuracy. This may become difficult when multiple scatterings proliferate unless these traveling-wave (progressing) multiples can be converted collectively into standing-wave (oscillatory) modal forms that constitute source-

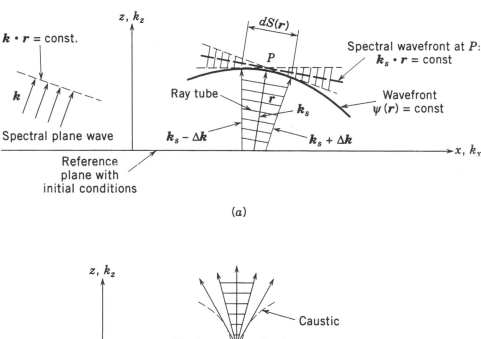

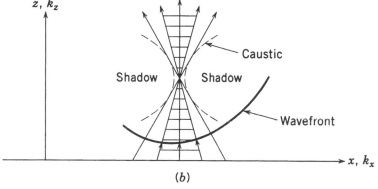

Fig. 1 Spectral (true) plane waves and local plane waves, with ray field modeling as in Eq. (2); two-dimensional. The spectral wave vector bundle (ray tube) with spread $\pm\Delta\mathbf{k}$ surrounding $\mathbf{k}_s$ is shaded horizontally, and the corresponding wavefront bundle is shaded vertically. The global wavefront $\psi(\mathbf{r})$ = const is intersected perpendicularly by the local rays, and the local plane wavefronts are tangent to the global wavefront at the ray intersection points. The family of rays corresponding to the wavefront $\psi(\mathbf{r})$ = const forms a *ray congruence*. The above sketches apply to homogeneous media where the rays are straight lines; in inhomogeneous media, the rays are curved. (a) Convex wavefront portion: diverging-ray congruence [$dS(\mathbf{r})$ expands; defocusing initial conditions]. (b) Concave wavefront portion: converging-ray congruence [($dS(\mathbf{r})$) contracts; focusing initial conditions]. The envelope of the converging congruence, called "caustic" (dashed), separates regions that are illuminated or dark (shadowed) with respect to rays. Because $dS(\mathbf{r}) \to 0$ on the caustic, simple ray theory fails in its vicinity, which therefore defines a transition region.

free solutions of the relevant wave equations. Such "modal ray" fields can be synthesized in special environments with appropriate symmetries (see Section 3).

3 MODAL RAY THEORY

The synthesis, by ray methods, of modal fields in "waveguides" requires self-consistent superposition of opposite traveling-ray field congruences so as to satisfy the waveguide boundary conditions in the transverse cross-sectional domain[17] (*waveguide* here refers broadly to any channel that transversely confines propagating wave fields by reflection or refraction). Self-consistency can be implemented by following a ray field in one of the congruences through one complete reflection cycle, after which it must be phase coherent with the same phase front from which it was launched. This implies that the total round-trip phase accumulation is an integer multiple of 2π. Guided modes can be defined globally in configurations that render the relevant wave equations coordinate separable (see Chapter 86).

3.1 Plane-Parallel Waveguide

The process for a homogeneous plane-parallel guiding region filled with a fluid medium is schematized in Fig. 2. Points 1 and 4 are on the same wavefront so that the phase accumulation along path 1–2–3–4–1 is $Q = k(l_{12} + l_{23} + l_{34}) + \tau_1 + \tau_2$, where l_{ij} is the path length between points i and j and $\tau_{1,2}$ are phase changes (if any) introduced by (local plane-wave) reflection at the boundaries. Self-consistency is enforced by $Q(\theta_m) = 2m\pi$, $m =$ integer, which selects the set of angles θ_m, the inclination angles of the modal plane wave congruences. The self-consistency condition can be converted into a *transverse resonance condition* by projecting the phase increments along ray paths into corresponding increments along a waveguide cross section $z = $ const. With $k_{xm} = k \sin \theta_m$ as the transverse wavenumber, the resonance condition becomes

$$2ak_{xm} + \tau_1(k_{xm}) + \tau_2(k_{xm}) = 2m\pi \qquad (3)$$

This form of enforcing closure, being localized on a waveguide cross section, is actually preferable under more general conditions where the modal congruences are not as easily determined as in this example (because the local plane waves in the parallel-plane geometry are true plane waves, the ray-synthesized modal eigenvalues θ_m or k_{xm} and the corresponding modal fields are exact).

3.2 Wedge Waveguide

A less trivial example is furnished by a fluid wedge waveguide with perfectly reflecting boundaries at $\phi = \phi_{1,2}$, which still renders the wave equation coordinate separable (in cylindrical (ρ, ϕ) coordinates) but has modal congruences with nonparallel rays and therefore with nonplanar phase fronts. These congruences are generated by tangential shedding from circular *caustics* with phase progression $\exp(\pm i\mu\phi)$ on constant coordinate surfaces $\rho = \rho_\mu$ (Fig. 3a), and they define local plane-wave fields. On the concave side $\rho < \rho_\mu$, there are no real rays, and the fields are evanescent.

For source-free solutions in the presence of the wedge boundaries (Fig. 3b), the caustic radius ρ_m and phasing μ_m must be determined as before to satisfy closure of a ray field departing from and returning to the same wavefront after a complete round trip. By analogy with Eq. (3), the condition is enforced most directly by a round-trip excursion along a $\rho = $ const surface, conveniently taken along the caustic; because the ray fields are tangential to the caustic, the wavenumber k_ϕ along the caustic equals k. Assuming reflection coefficient phases $\tau_{1,2} = 0$ and referring to Eq. (3), one finds the guided mode parameters

$$\mu_m = \frac{m\pi}{\alpha}, \qquad \rho_m = \frac{\mu_m}{k}, \qquad \alpha = \phi_2 - \phi_1, \qquad (4)$$

which completely characterize the modal ray field in the wedge waveguide. Note that the caustic $\rho = \rho_m$ separates the propagating regime $\rho > \rho_m$ from the evanescent regime $\rho < \rho_m$ and thereby defines the ray field cutoff transition for mode m.

If the waveguide boundaries lie along the circles $\rho = \rho_{1,2}$ (Fig. 3c), modal ray fields may be generated by circular caustics lying either interior to the inner boundary ($\rho_\mu < \rho_1$) or between the outer and inner boundaries $\rho_1 < \rho_\mu < \rho_2$. In the former case, the azimuthally propagating modal ray field fills the entire cross section, whereas in the latter case, the modal ray field clings to the outer (concave) boundary and is referred to as a "whispering gallery mode." Whispering gallery modes can exist just on the *concave* boundary alone because the modal caustic provides the necessary confinement.

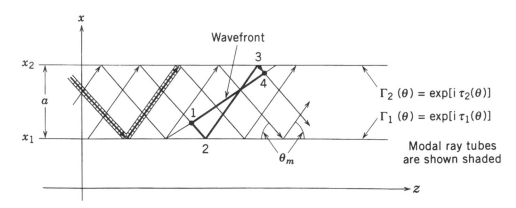

Fig. 2 Modal rays, congruences, and ray tracing in plane-parallel waveguide. Ray tubes are shaded.

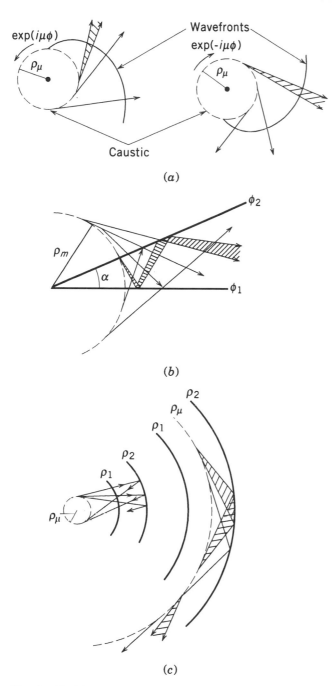

Fig. 3 Modal rays, congruences, and ray tracing in wedge and annular waveguides. Ray tubes are shaded. (*a*) Counterclockwise and clockwise progression. (*b*) Congruences for wedge waveguide. (*c*) Congruences for annular and whispering gallery waveguide.

As before, synthesis by self-consistent closure and ray field tracking for either the exterior or interior caustic yields fields that agree with the high-frequency asymptotic approximation of the exact solution. Mode fields in other curvilinear geometries may be synthesized by similar considerations.[17]

4 HYBRID RAY–MODE THEORY

In general propagation and scattering scenarios, it may be advantageous to treat certain portions of the problem by tracking progressive local ray fields and other portions by tracking collectively restructured oscillatory modal ray fields. This can be done by recourse to the hybrid ray–mode algorithm, which combines ray fields and mode fields self-consistently. The ray–mode partitioning can be arbitrary, but for practical reasons, one seeks to effect the decomposition so as to take advantage of the "best" features of each. Theory and applications are discussed elsewhere.[18–24]

An example for submerged elastic structures is discussed next.

5 FLUID–STRUCTURE-COUPLED RAY THEORY: SUBMERGED, INTERNALLY, LOADED ELASTIC CYLINDRICAL SHELL

Acoustic scattering by submerged elastic structures generally comprises contributions due to external phenomena and contributions due to internal phenomena. The latter are excited by fluid–structure coupling and leave their imprint on the scattered field in the fluid by structure–fluid coupling. Under conditions that validate ray theory internally, the tracking of the ray fields into and out of the structure can be accomplished by GTD generalized to accommodate elastic media. When the structure can support guided modes (either at the fluid–structure interface or due to internal layering), their excitation from the fluid and their shedding into the fluid takes place under conditions of phase matching (on the fluid–structure boundary) between the wave vectors of the external ray fields and the internal mode fields.[25–33]

Figure 4 schematizes typical ray species and the associated leakage mechanisms for a two-dimensional thin shell. Far-field plane-wave backscattering from an internally loaded, submerged, thin elastic cylindrical shell is treated quantitatively in Fig. 5. The physical configuration with relevant ray trajectories is sketched in Fig. 5*a*, the shell being loaded internally by a simple mass–spring system. The results in Figs. 5*b*, and *c* are taken from Ref. 28. The plots depict the backscatter form function $f(0)$, which is related to the far-zone scattered pressure as

$$p(r, \theta) = \left[p_0 \left(\frac{a}{2r} \right)^{1/2} \exp(ikr) \right] f(\theta), \qquad (5)$$

where p_0 is the amplitude of the normally incident plane wave and k is the acoustic wavenumber in the fluid. In

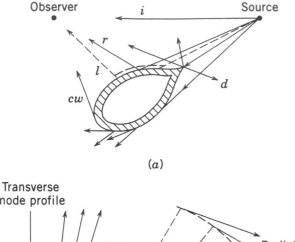

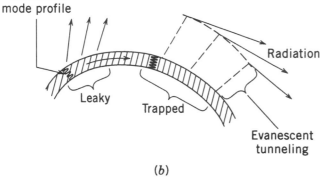

Fig. 4 Ray species. (*a*) Ray fields in the fluid; all categories incorporated from the GTD: incident (*i*) and reflected (*r*), edge diffracted (*d*), creeping (*cw*) (along smooth convex portions), and leaky (*l*) (phase matched to interior). (*b*) Modal ray fields in plane or smoothly curved shell sections: leaky (supersonic; direct coupling to fluid by phase-matched radiation) and trapped (evanescent into the fluid but can radiate from curved sections after evanescent tunneling).

the frequency range $5 < ka < 20$, the shell material supports a compressional supersonic leaky mode [see (*l*) in Fig. 5*a*] that sheds radiation tangentially at the phase-matching angle $\theta_l = \sin^{-1}(v/v_l)$ with respect to the surface normal, where v is the wave speed in the fluid and $v_l > v$ is the azimuthal phase speed of the compressional mode in the shell. Excitation of this leaky mode by the incident field, in turn, arises from those incident-wave components that locally satisfy the phase-matching condition $\theta_{inc} = \theta_l$. The shell also supports a subsonic (trapped) flexural mode that is, however, not excited by the incident plane wave.

The ray constituents (see Fig. 5*a*) comprise the following:

1. The specularly reflected field (see *r* in Fig. 5*a*), with $\theta = 0$ for backscatter.

2. The compressional leaky modes in the shell, launched on the shell at those angular locations where $\theta_{inc} = \theta_l$. There are two such locations,

launching clockwise and counterclockwise traveling modes, respectively. For far-field backscattering, these leaky wave fields detach into the fluid from the same locations on the shell.

3. The fields excited by the induced forces F_0 and F_π at the load attachment points $\theta = 0$, π, respectively, on the shell. These forces, which are induced by the internal load, excite direct radiation into the fluid (labeled *e*) as well as shell-guided trapped (*t*) and leaky modes; contributions from the latter must be added to the leaky-mode fields in 2. Since F_π is not visible from the observation angle $\theta = 0$, only F_0 contributes direct radiation.

These individual contributions are shown separately in Fig. 5*b*, and their phase-coherent superposition is seen to reconstruct with good accuracy the intricate pattern of maxima and minima in the angular harmonics reference solution in Fig. 5*c*. This suggests that the hybrid GTD and modal ray fields are well matched to the wave physics. It may be noted that the leaky-mode amplitude in Fig. 5*b* exhibits periodicities characteristic of the leaky compressional circumferential-mode resonant maxima whereas the amplitude variation of the induced source field carries the corresponding resonant signature of the trapped flexural modes, which are excited and radiated via diffraction at the load attachment points.

6 SUMMARY

Ray acoustics is one of the important disciplines for predictive modeling of mid to high-frequency fluid–structure-coupled wave interactions, both in the near and far zones of a submerged elastic object. Combining GTD-type progressing ray fields with modal ray fields provides algorithms of broad versatility that are matched to the relevant wave physics. This is true especially for pulsed excitation where space–time-resolved wave packets essentially propagate along ray trajectories. Ray theory also plays a role in the propagation and scattering of beam-type fields. Applicability is restricted to the localizing conditions stated in Section 1. How these conditions pertain to various scenarios and how well the algorithms can be implemented is best learned by looking at applications reported in the cited references.

Broadly speaking, ray methods work well over "regular" portions of a scattering object. Portions with complexity that either mitigates against or invalidates ray modeling can often be treated numerically. Future developments can be expected to deal with hybrid (numerical) (ray–mode) schemes that connect the numerically and ray acoustically treated problem portions.

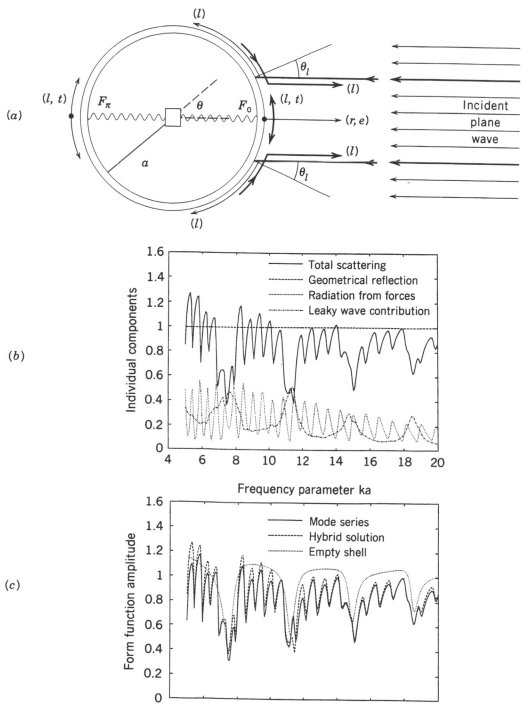

Fig. 5 Plane-wave backscattering from a submerged cylindrical shell with a centrally loaded mass–spring system. Relevant ray trajectories are shown as well. (*a*) Physical configuration. (*b*) Amplitudes of individual contributing constituents and their combined effect. Solid curve: total field. Dashed curve: geometric reflection. Dotted curve: direct radiation from force at attachment point $\theta = 0$ illuminated by the incident wave; the attachment point $\theta = \pi$ is in the shadow region. Dash–dot curve: contribution from compressional leaky waves, which includes three components—one due to the incident field and two from the induced forces at the two attachment points. (*c*) Backscatter form function amplitude in the frequency range $5 < ka < 20$. Comparison between exact harmonic series reference solution (solid curve) and results from hybrid ray–mode synthesis (dashed curve). Results for the empty shell are shown for comparison (dotted curve).

REFERENCES

1. J. B. Keller, "Geometrical Theory of Diffraction," *J. Opt. Soc. Am.*, Vol. 52, 1962, pp. 116–130.

2. R. C. Hansen (Ed.), *Geometric Theory of Diffraction*, IEEE Press, New York, 1981.

3. G. L. James, *Geometrical Theory of Diffraction for Electromagnetic Waves*, Peter Peregrinus, London, 1976.

4. L. B. Felsen and G. A. Deschamps (Eds.), "Rays and Beams," *Proc. IEEE*, Vol. 62, 1974 (special issue).

5. R. Stone (Ed.), "Radar Cross Sections of Complex Objects," *Proc. IEEE* Vol. 77, 1989 (special issue).

6. L. B. Felsen, "Target Strength: Some Recent Theoretical Developments," *IEEE J. Ocean-Eng.*, Vol. OE-12, 1987, pp. 443–452.

7. W. G. Neubauer, "Acoustic Reflection from Surfaces and Shapes," Naval Research Lab., Washington, D.C.

8. V. A. Borovikov and N. D. Veksler, "Scattering of Sound Waves by Smooth Convex Elastic Cylindrical Shells," *Wave Motion*, Vol. 7, 1985, pp. 143–152.

9. N. D. Veksler and V. M. Korunskii, "Analysis and Synthesis of Backscattering from a Circular Cylindrical Shell," *J. Acoust. Soc. Am.*, Vol. 83, 1990, pp. 943–962.

10. P. L. Marston, "GTD for Backscattering from Elastic Spheres and Cylinders in Water and Coupling of Surface Waves with the Acoustic Field," *J. Acoust. Soc. Am.*, Vol. 83, 1988, pp. 25–37.

11. S. G. Kargl and P. L. Marston, "Ray Synthesis of Lamb Wave Contributions to the Total Scattering Cross Section for an Elastic Spherical Shell," *J. Acoust. Soc. Am.*, Vol. 88, 1990, pp. 1103–1113.

12. L. G. Zhang, N. H. Sun, and P. L. Marston, "Midfrequency Enhancement of the Backscattering of Tone Bursts by Thin Spherical Shells," *J. Acoust. Soc. Am.*, Vol. 91, 1992, pp. 1862–1874.

13. L. M. Brekhovskikh, "Waves in Layered Media," Academic, New York, 1980.

14. L. M. Brekhovskikh and O. A. Godin, "Acoustics of Layered Media I," *Wave Phenomena Series*, Springer, New York, 1990.

15. L. M. Brekhovskikh and O. A. Godin, "Acoustics of Layered Media II," *Wave Phenomena Series*, Springer, New York, 1992.

16. L. B. Felsen and N. Marcuvitz, *Radiation and Scattering of Waves*, Prentice-Hall, Englewood Cliffs, NJ, 1973; classic reissue by IEEE Press, Piscataway, NJ, 1994.

17. L. B. Felsen, "Ray Methods for Modal Fields in General Closed and Open Propagation Environments," *Proc. IEEE*, Vol. 79, 1991, pp. 1391–1400.

18. A. Kamel and L. B. Felsen, "Hybrid Green's Function for SH Motion in a Low Velocity Layer," *Wave Motion*, Vol. 5, 1983, pp. 83–97.

19. L. B. Felsen, "Progressing and Oscillatory Waves for Hybrid Synthesis of Source Excited Propagation and Diffraction," *IEEE Trans. Antennas Propagat.*, Vol. AP-32, 1984, pp. 775–796.

20. E. Heyman and L. B. Felsen, "A Wavefront Interpretation of the Singularity Expansion Method," *IEEE Trans. Antennas Propagat.*, Vol. AP-33, 1985, pp. 706–718.

21. L. T. Lu and L. B. Felsen, "Ray, Mode and Hybrid Options for Source Excited Propagation in an Elastic Plate," *J. Acoust. Soc. Am.*, Vol. 78, 1985, pp. 710–714.

22. L. B. Felsen and H. Shirai, "Rays, Modes and Beams for Plane Wave Coupling into a Wide Open-ended Parallel Waveguide," *Wave Motion*, Vol. 9, 1987, pp. 301–317.

23. E. Heyman, G. Friedlander, and L. B. Felsen, "Ray-Mode Analysis of Complex Resonances of an Open Cavity," *IEEE Proc. Special Issue on Radar Cross Sections of Complex Objects*, Vol. 77, 1989, pp. 780–787.

24. I. T. Lu, L. B. Felsen, and J. M. Klosner, "Observables Due to Beam-to-Mode Conversion of a High-Frequency Gaussian *P*-Wave Input in an Aluminum Plate in Vacuum," *J. Acoust. Soc. Am.*, Vol. 87, 1990, pp. 42–53.

25. L. B. Felsen, J. M. Ho, and I. T. Lu, "Three-Dimensional Green's Function for Fluid-loaded Thin Elastic Cylindrical Shell: Formulation and Solution," *J. Acoust. Soc. Am.*, Vol. 87, 1990, pp. 543–553.

26. L. B. Felsen, J. M. Ho, and I. T. Lu, "Three-Dimensional Green's Function for Fluid-loaded Thin Elastic Cylindrical Shell: Alternative Representations and Ray-Acoustic Forms," *J. Acoust. Soc. Am.*, Vol. 87, 1990, pp. 554–569.

27. J. M. Ho and L. B. Felsen, "Nonconventional Traveling Wave Formulations and Ray-Acoustic Reductions for Source-excited Fluid-Loaded Thin Elastic Spherical Shells," *J. Acoust. Soc. Am.*, Vol. 88, 1990, pp. 2389–2414.

28. L. B. Felsen and Y. P. Guo, "Hybrid Ray-Mode Parametrization of Acoustic Scattering from Submerged Thin Elastic Shells with Interior Loading," *J. Acoust. Soc. Am.*, Vol. 94, 1993, pp. 888–895.

29. T. Kapoor and L. B. Felsen, "Hybrid Ray-Mode Analysis of Frequency and Time Domain Scattering by a Finite Submerged Elastic Plate," *Wave Motion*, Vol. 22, 1995, pp. 109–131.

30. A. D. Pierce, "Waves on Fluid-loaded Inhomogeneous Elastic Shells of Arbitrary Shape," *ASME J. Vib. Acoust.*, 1992, Vol. 115, pp. 1809–1829.

31. A. N. Norris and D. A. Rebinsky, "Acoustic and Membrane Wave Interaction at Plate Junctions," *J. Acoust. Soc. Am.*, Vol. 97, 1995, pp. 2063–2073.

32. J. M. Ho, "Acoustic Scattering by Submerged Elastic Cylindrical Shells: Uniform Ray Asymptotics," *J. Acoust. Soc. Am.*, Vol. 95, 1993, pp. 2936–2946.

33. L. B. Felsen and A. N. Norris (guest editors), "Mid-to High Frequency Acoustic Scattering and Radiation from Fluid-Loaded Structures: Asymptotic Techniques," *Wave Motion* (special issue), Vol. 22, 1995, pp. 1–131.

5

INTERFERENCE AND STEADY-STATE SCATTERING OF SOUND WAVES

HERBERT ÜBERALL

1 INTRODUCTION

Although the science of acoustics reaches back into antiquity, the father of modern acoustics is without any doubt the English physicist John William Strutt, Lord Rayleigh (1842–1919), whose textbook *Theory of Sound* is still relevant today.[1] Among his many achievements is an elegant mathematical expression describing the scattering of sound waves by material objects that is termed the *Rayleigh series*.

From these beginnings, acoustics has seen a rapid and extensive development that today covers many different applications such as nondestructive materials testing, medical acoustics, ocean acoustics and the detection and recognition of submerged targets, and so on. Apart from the understanding of acoustic propagation required here, many of these applications are based on the subject of acoustic *scattering*, which represents an important ingredient in their design and interpretation. In the following, acoustic scattering will be discussed regarding its history and modern developments. These, starting from Rayleigh's normal-mode series, have proceeded to a separation of the scattering amplitude into "geometric" parts (specular reflection, transmission, and internal reflection) and parts that correspond to diffraction via the effects of circumferential (surface) waves on the scattering object. The latter are shown to give rise to prominent resonance effects for submerged elastic scatterers, which were clarified theoretically by the *resonance scattering theory* (RST) of Flax, Dragonette, and Überall and experimentally by the *method of isolation and identification of resonance* (MIIR) of Maze and Ripoche. The

Encyclopedia of Acoustics, edited by Malcolm J. Crocker
ISBN 0-471-80465-7 © 1997 John Wiley & Sons, Inc.

modifications of these phenomena for objects in air and for multiple scatterers are also discussed.

2 RECENT HISTORY: THE CONCEPT OF NORMAL MODES AND SURFACE WAVES

The World War II effort gave a great impetus to acoustics research. The subsequent fundamental work of Faran[2] should be stressed. He derived the acoustic field scattered from submersed elastic objects having the "canonical" shapes of cylinders and spheres, in response to an incident plane wave, mathematically representing the fields in the form of Rayleigh series. He evaluated the field expressions numerically and carried out experiments that verified the correctness of the theory. More interesting from the physics viewpoint, he showed that resonances of the scattering objects were excited when the frequency of the incident sound wave coincided with one of the "natural" or eigenfrequencies of vibration of the submerged object. Using incident sound pulses, he observed that a "ringing" of the resonances (i.e., a continuing vibration of the object after the traveling pulse had passed on) took place when the carrier frequency of the pulse was equal to the eigenfrequency.

Pulse scattering was subsequently studied experimentally by Barnard and McKinney,[3] who observed periodic echoes returned by finite-size smooth metallic objects in water in response to a single incident pulse. Diercks et al.[4] interpreted this physically as the effects of circumferential pulses generated by the incident pulse that encircle the scatterer and radiate off shock waves as they propagate along, the shock waves from successive revolutions of the circumnavigating pulse reaching the observer in a sequence. This picture is based on the "creeping waves" discussed earlier by Franz[5] for the case of impenetrable objects. The difference is, however, that on impenetrable (rigid or soft) objects the circumferential (peripheral, or surface) waves propagate outside the object in the ambient fluid with a speed below the sound speed in the fluid, while for a (penetrable) elastic body the strongly dominant peripheral waves are of an elastic nature, propagating on the interior side of the scatterer's surface with speeds that are close to the higher wave speeds in the elastic material (although the slower Franz waves are present here also, but with smaller amplitudes). Both wave types radiate into the fluid: the fast elastic-type surface waves emit "shock" waves at a grazing angle ϑ with the scatterer's surface, given by

$$\cos \vartheta = \frac{c}{c_l}, \qquad (1)$$

where c is the speed of sound in the ambient fluid and c_l the phase velocity of the lth-type surface wave. For the Franz-type (creeping) waves where $c_l < c$, the radiated wave is emitted tangentially and may be called a "slip" wave.[6] This situation is shown in Fig. 1. Note that the locus on the scatterer at which the surface waves are generated is also determined by Eq. (1). The field radiated by the creeping waves may be identified with the phenomenon of diffraction.

For the case of steady-state scattering, where one does not have an incident pulse but has a "long" pulse that has been approaching for an infinite time and continues to do so, all these wave types are superimposed on each other, together with a geometrically (or specularly) reflected

wave as well as "transmitted" waves refracted into the inside of the elastic scattering object. (For an impenetrable, i.e., rigid or soft, object, which is a mathematical idealization not found in nature, only the reflected and the creeping waves would be present).

As to the internally transmitted waves, Fig. 1 only shows their first refraction. They continue being retransmitted to the exterior of the scatterer or reflected back into it (Fig. 2). At every interaction with the boundary, the interior waves, which are of the two types that an elastic medium can support (compressional, or P, and shear, or S), further split into these two types again (*mode conversion*). These retransmitted waves, which are of dominant importance at very large frequencies ($ka \ggg 1$, where $k \equiv \omega/c = 2\pi/\lambda$ is the wavenumber in the fluid and a is a typical dimension of the scatterer) have been experimentally demonstrated by Quentin et al.[7] The scattering at lower frequencies ($ka \sim 1$, but mainly $10 \lesssim ka \lesssim 100$) is dominated by the effects of the elastic surface waves (at least for the common case of metal objects in water), where the circumferential (surface) waves at certain frequencies can match their phases when closing into themselves after repeated circumnavigations, forming standing waves around the object and thus generating highly noticeable resonance effects in the scattered amplitude when plotted versus frequency. By themselves, the surface waves would thus give rise to a series of isolated peaks on such a plot, which can be referred to as the *resonance spectrum* of a scattering object characteristic for its shape and composition. In practice, however, one always has the reflected wave present in addi-

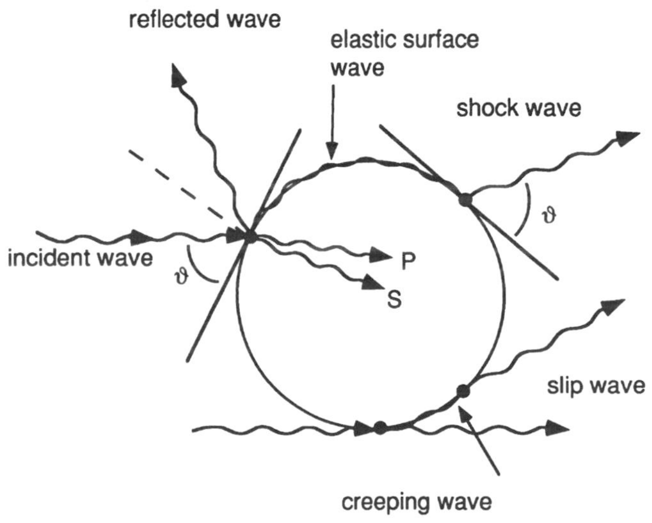

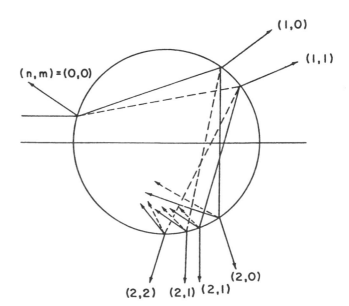

Fig. 1 Peripheral wave types and the waves radiated from them; also, reflected and transmitted (refracted) waves of compressional (P) and shear type (S).

Fig. 2 Geometric-optical picture of refracted waves inside an elastic object; n indicates the total number of internal segments of a ray and m the number of segments of shear (S) type.

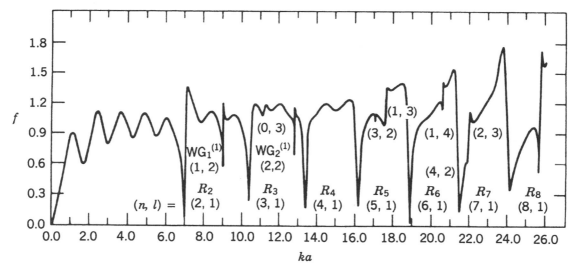

Fig. 3 Modulus of the form function f (backscattering amplitude less geometric spreading factor) plotted vs. ka for a WC sphere in water. (From Ref. 8.)

tion (as well as the retransmitted waves wherever they arc of importance), and when the scattering amplitude is plotted versus frequency, the reflected (plus transmitted) wave(s) give rise to a smooth background spectrum with which the isolated resonances interfere, for the case of metal objects in water mainly destructively, so that they manifest themselves as a series of isolated dips. This is shown in Fig. 3 for a solid tungsten carbide (WC) sphere in water.[8] The resonances are labeled by two numbers (n, l), where l is the type of circumferential wave that causes them (on a solid body, $l - 1$ is referred to as a *Rayleigh wave*, causing the succession of deep dips labeled R_n, and $l \geq 2$ as "*Whispering Gallery*"

waves, causing narrow dips; on an airfilled shell, the various wave types l are referred to as *Lamb waves*) and n is the number of wavelengths of the resonant standing wave closing into itself over the circumference of the scatterer. The quantity plotted here is the modulus of the *form function*, defined as the far-field backscattering amplitude with its geometric spreading factor ($1/r^{1/2}$ for two-dimensional scattering, e.g., from cylinders, $1/r$ for three-dimensional scattering) removed.

If the "background" of the reflected wave is subtracted out from the total scattering amplitude, then indeed only the pure resonances remain, as shown in Fig. 4 for the resonance spectrum of a WC sphere in water,[8] indicat-

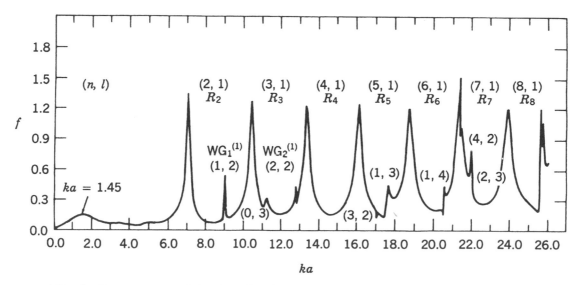

Fig. 4 Resonant part of the form function (rigid background subtracted), or resonance spectrum, of a WC sphere plotted vs. ka. (From Ref. 8.)

ing broad Rayleigh wave and narrow Whispering Gallery wave resonances. The RST of Flax et al.,[9, 10] to be discussed below, describes the isolated resonances by a resonance formula, and a prerequisite for its application is the separation of the "geometric" (reflected and refracted/transmitted) background amplitude (known in nuclear scattering theory as the *potential scattering term*[11]) from the resonance terms in the scattered amplitude. Experimentally, this separation has been achieved by the long-pulse MIIR of Maze and Ripoche,[12] and a long series of investigations has been performed by these acousticians at the University of Le Havre, France, on all aspects of resonance scattering.[13–15] However, a short-pulse method by Numrich et al.[16] and de Billy[17] has been devised for the same purpose. On the theoretical side, scattering from objects of noncanonical shape (mainly spheroidal) has been calculated by Werby[18,19] employing the so-called *T*-matrix method of Waterman[20]; he also has derived an accurate background formula for the specular reflection from air-filled shells[21] that correctly goes over into a soft background for very thin shells and into a rigid background for thick shells (including solid bodies). Related work is given in Refs. 22 and 23.

The form function, or rather the normalized far-field scattering amplitude at an observer's angular position ϑ, for a sphere can be written in the form of a Rayleigh series as[9]

$$f(\vartheta) = \frac{2}{ka} \sum_{n=0}^{\infty} (2n + 1)e^{i\delta_n} \sin \delta_n P_n(\cos \vartheta). \quad (2)$$

Here, $P_n(\cos \vartheta)$ is the Legendre polynomial. The individual terms in the series are referred to as *normal modes* or *partial waves*. We introduced the scattering function

$$S_n = e^{2i\delta_n}, \quad (3)$$

where δ_n is the scattering phase shift. For a rigid sphere, for example, one has

$$S_n^{(r)} = -\frac{h_n^{(2)'}(x)}{h_n^{(1)'}(x)} \equiv e^{2i\xi_n}, \quad (4)$$

where $x = ka$ is a normalized frequency variable, a is the sphere radius, k the wavenumber in the ambient fluid, and $h_n^{(1)'}(x)$ the derivative of the spherical Hankel function. For elastic spheres, S_n is known from the boundary conditions. As in Eq. (4), it has a quotient form with denominator $D_n(x)$, which is a function of two variables: the frequency x and the mode number n.

The equation

$$D_n(x) = 0 \quad (5)$$

can be solved in either variable if the other one is held (real and) constant. Keeping the frequency x real and constant leads to a solution

$$n \to \nu_l(x), \qquad l = 1, 2, \ldots, \quad (6)$$

which introduces poles of the scattering amplitude in the complex mode number plane (known as *Regge poles* in quantum physics[24]). They lead to the dominant contributions (as shown by the so-called Watson transformation[5]) expressed via the asymptotic forms of P_n:

$$P_{\nu_l}(\cos \vartheta) \cong \frac{2}{\pi(\nu_l + \frac{1}{2}) \sin \vartheta} \frac{1}{2} \sum_{\epsilon = \pm 1} e^{i\epsilon(\nu_l + 1/2)\vartheta - i\epsilon\pi/4}, \quad (7)$$

which quite obviously describe circumferential waves that encircle the sphere in the angular direction ϑ with propagation constant $\nu_l + \frac{1}{2}$ in both senses ($\epsilon = \pm 1$).

Alternately, keeping n constant and equal to real integers, Eq. (5) has the solution

$$x = x_{nl}, \qquad l = 1, 2, \ldots, \quad (8)$$

where x_{nl} is a complex-pole position in the complex-frequency plane, as shown for a WC sphere[19] in Fig. 5. In this representation, the poles are called the *singularity expansion poles*, from a corresponding electromagnetic scattering theory.[25] Figure 5 shows that the poles fall into distinct families, corresponding to the Rayleigh wave ($l = 1$) and to the Whispering Gallery waves ($l = 2, 3, \ldots$).

Scattering theory may be extended to noncanonical objects (e.g., spheroids) by applying the *T*-matrix (null field, or extended boundary condition) method of Waterman.[20] In acoustic scattering, this has been extensively carried out by Werby.[18,19,26] (The method requires time-consuming computation at higher frequencies and may even show stability and convergence problems there.) One may introduce a spherical basis

$$\psi_n(\mathbf{r}) = \gamma_{mn}^{1/2} h_n^{(1)}(kr) Y_{mn}^{\sigma}(\hat{r}),$$
$$\hat{\psi}_n(\mathbf{r}) = \gamma_{mn}^{1/2} j_n(kr) Y_{mn}^{\sigma}(\hat{r}), \quad (9)$$

with γ_{mn} the normalization constants and Y_{mn}^{σ} the spherical harmonics. The incident wave is expanded as[8]

$$p_{\text{inc}} = \sum_n a_n \hat{\psi}_n(\mathbf{r}), \quad (10)$$

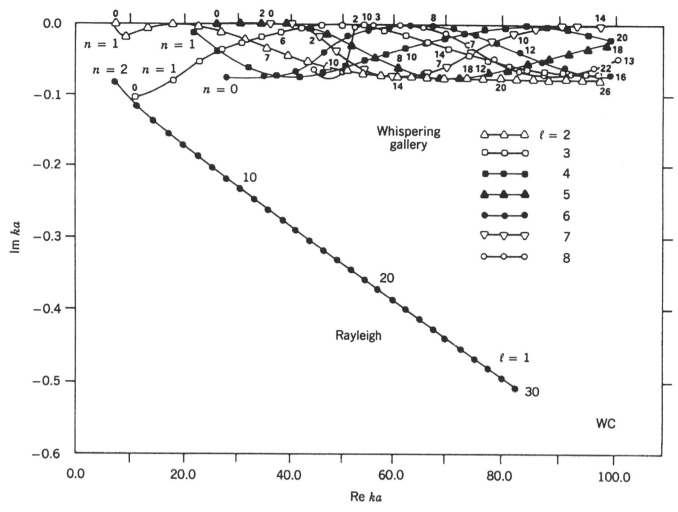

Fig. 5   Complex-frequency poles of the scattering amplitude for a WC sphere of radius a immersed in water (units ka, where $k = \omega/c$, c = sound speed in water). (From Ref. 19.)

with known coefficients a_n, the scattered wave as

$$p_{sc} - \sum_n c_n \psi_n(\mathbf{r}), \tag{11}$$

and the interior field (assuming a fluid body) as

$$p_{int} = \sum_n d_n \hat{\psi}_n^{(0)}(\mathbf{r}), \tag{12}$$

where, in $\hat{\psi}_n^{(0)}$, $k \equiv \omega/c$ is replaced by $k_0 \equiv \omega/c_0$ (c_0 being the sound velocity in the interior, assumed fluid). The unknown expansion coefficients c_n and d_n are expressed by a_n by satisfying the appropriate boundary condition on the scatterer's surface S. The resulting relation

$$c_i = T_{ij}a_j \tag{13}$$

determines the scattered wave via the T matrix, which is found as

$$T = -Q^{-1}\hat{Q} \tag{14}$$

(a generalization of the mentioned former quotient), where, for example,

$$Q_{pq} = k \int_S \left(\frac{\rho_0}{\rho} \hat{\psi}_p^{(0)} \frac{\partial \psi_q}{\partial n} - \psi_q \frac{\partial \hat{\psi}_p^{(0)}}{\partial n} \right) dS, \tag{15}$$

ρ_0 being the density of the interior fluid.

Calculating form functions and resonance spectra in

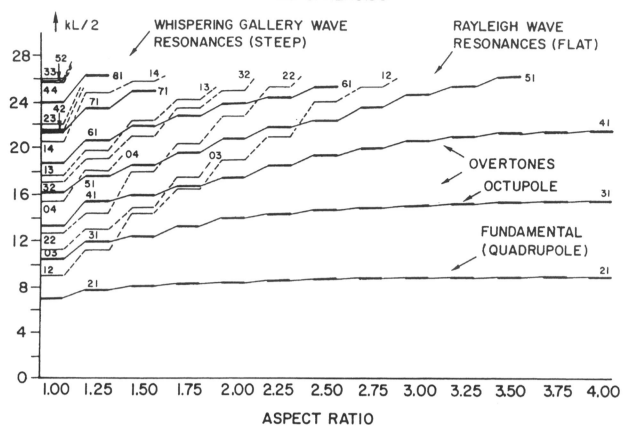

Fig. 6 Spectroscopic level diagram of the real resonance frequencies of prolate WC spheroids, as obtained from a T-matrix calculation for excitation by an axially incident plane wave as a function of aspect ratio. (For aspect ratio 1, levels are the real parts of pole positions in Fig. 5). (From Ref. 18.)

this way, the resonance frequencies have been obtained for a series of WC spheroids of various aspect ratios (up to $4:1$) in water, subject to an axially incident plane wave. This is presented in Fig. 6, showing that the Rayleigh wave resonances move slowly upward with increasing aspect ratio, while the Whispering Gallery wave resonances move rapidly and cross over the Rayleigh resonances, resembling the "level crossing" phenomenon of deformed nuclei in nuclear physics.

3 RESONANCE SCATTERING THEORY AND PHASE MATCHING OF SURFACE WAVES

Resonance scattering theory was formulated in acoustics by Flax et al.[9,10] in 1977–1978. Patterned after quantum mechanical resonance scattering, it showed that the form function of Eq. (2) could be written as

$$f(\vartheta) = \frac{2}{ka} \sum_n (2n+1)e^{2i\xi_n}$$

$$\cdot \left[e^{-i\xi_n} \sin \xi_n + \sum_l \frac{\frac{1}{2}\Gamma_{nl}}{x_{nl} - x - \frac{1}{2}i\Gamma_{nl}} \right]$$

$$\cdot P_n(\cos \vartheta), \tag{16}$$

which makes obvious its superposition character of a smooth specular background (first term in braces) and a series of resonances (second term) with width Γ_{nl}. This expression demonstrates that the singularity expansion poles are located at

$$x = x_{nl} - (i/2)\Gamma_{nl}, \tag{17}$$

that is, in the lower complex x-plane, as shown in Fig. 5. These x-poles may be obtained from the Regge poles by solving Eq. (6):

$$\nu_l(x_{nl}) = n, \tag{18}$$

which is essentially the resonance condition.

This discussion has so far referred to spheres, and it may similarly be applied to cylinders. For finite objects of more general shape, it can be shown that Eq. (18) can be written in the form[27, 28] (surface wave phase-matching condition)

$$\oint \frac{ds}{\lambda_l} = n + \tfrac{1}{2}, \tag{19}$$

the wavelength of the lth surface wave, of speed $c_l(k)$, being

$$\lambda_l(k) = \frac{(2\pi/k)c_l(k)}{c}. \tag{20}$$

Equation (19) is integrated over the shortest path (i.e., a geodesic) over the object's surface that closes into itself, following Fermat's principle. The term $\tfrac{1}{2}$ in Eq. (19) stems from a $\lambda/4$ phase jump at each of two caustics that the geodesics touch. It is absent for cases with no caustics. If the λ_l or $c_l(k)$ are known or are modeled, Eq. (19) may serve to obtain the complex resonance frequencies in this case. At the resonance frequencies, the phase and group velocities may alternately be found as

$$\frac{c_l(k)}{c} = \left.\frac{ka}{n + \tfrac{1}{2}}\right|_{\text{at res}}, \tag{21}$$

$$\frac{c_l^{gp}(k)}{c} = \left.\frac{d(ka)}{dn}\right|_{\text{at res}}. \tag{22}$$

In this way, the resonance frequencies, for example, for spheroids, similar to those given in Fig. 5 for a sphere, may be obtained from the phase-matching principle. The latter can be applied even for more complex objects where the T-matrix method fails. Numerous examples of resonance scattering have been discussed in Ref. 29 and of surface waves in Ref. 30.

4 METHOD OF ISOLATION AND IDENTIFICATION OF RESONANCES

This experimental method was devised by Maze and Ripoche[12] and is based on the scattering of long pulses of a spatial extent much larger than the size of the scatterer. A modulated square pulse is incident, as in Fig. 7a. If its carrier frequency does not coincide with a resonance frequency, its shape will be unchanged after scattering. At the resonance frequency, an initial transient appears, as well as a ringing tail, as shown in Fig. 7b for an aluminum cylinder in water. The ringing amplitude (free transient) is a sensitive function of frequency, and when plotted versus ka (bottom part of Fig. 8), it perfectly furnishes the resonance spectrum of the cylinder.[31] Measuring the amplitude of Fig. 7b a little ahead of the start of the ringing (where the incident pulse has not yet cut off, i.e., a quasi-stationary "forced" regime has become established) furnishes the form function, however (top part of Fig. 8; a detector bias here reduced its value at both ends).

The shape of Fig. 7b is explained[6, 32] as shown in Fig. 9. The observed specular wave train and the successive circumferential wave trains overlap (in the case of Fig. 7, the wave train wraps around the cylinder 60 times!). Off resonance, the phases are random and the circumferential waves annihilate each other, leaving only the specular return. At resonance, they are in phase through Eq. (18) or (19); all overlapping waves add up, but in the example of Fig. 7, they interfere destructively with the specular wave, thus causing the central constriction in Fig. 7b.

The MIIR also identifies the resonances by assigning their order number n. The cosine of $(n + \tfrac{1}{2})\vartheta$, combining the two exponentials in Eq. (7) at resonance, shows that the number of lobes in the angular diagram of $f(\vartheta)$ for a sphere is determined by the value of n. A similar expression holds for a cylinder (with n instead of $n + \tfrac{1}{2}$ since there is no caustic), or approximately also for more generally shaped bodies.[8] The measured "daisy pattern" of Fig. 10 for the scattered-wave angular distribution[33] shows eight lobes (note that the incision at $\vartheta = 180°$, i.e., backscattering, stems from the source and receiver blocking each other), thus indicating $n = 4$ for the 330-kHz resonance of the object (a hemispherically capped cylindrical brass shell in water).

5 SCATTERING FROM OBJECTS IN AIR

When metallic scattering objects are contained in air, rather than submersed in water as was considered above, their large impedance contrast with air will lead to a negligible excitation of elastic transmitted waves as well as of (subsurface) elastic-type circumferential waves by an incident airborne acoustic signal; that is, they will appear as if they were rigid objects. Consequently, only the airborne incident and the Franz-type exterior circumferential ("creeping") waves will be present. Their complex resonance frequencies (poles of the scattering amplitude)

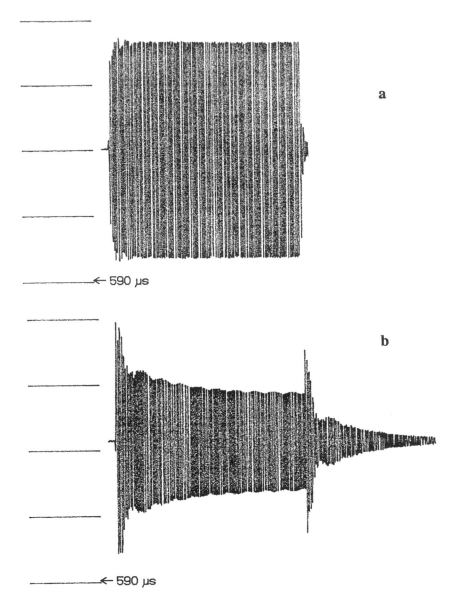

Fig. 7 Reflected pulse for an aluminum cylinder in water (*a*) off resonance and (*b*) at resonance. (From Refs. 12 and 14.)

would appear in the WC sphere plot of Fig. 5 considerably below[34] those of the Rayleigh poles, due to their larger imaginary parts (attenuations), but close to those of the Franz poles of the actual WC sphere (not shown in Fig. 5) and with a downward-sloping pattern similar to that of the Rayleigh poles. Measurements of sound scattering by duraluminium cylinders and spheres in air[35] could therefore be interpreted on the basis of creeping waves on a rigid cylinder alone. These measurements included a verification of their theoretical dispersions and absorptions as a function of frequency. Figure 11 shows

the normalized *target strength* σ (the absolute-squared backscattering amplitude with spreading factor removed) versus frequency for a rigid cylinder of radius *a* and a normally incident plane wave with the following theoretical results:

1. *Dashed Curve.* Exact rigid-cylinder scattering theory based on the cylinder analog of Eq. (2)

2. *Solid Curve.* Creeping-wave theory based on the Watson transformation and retention of only the lowest order creeping wave

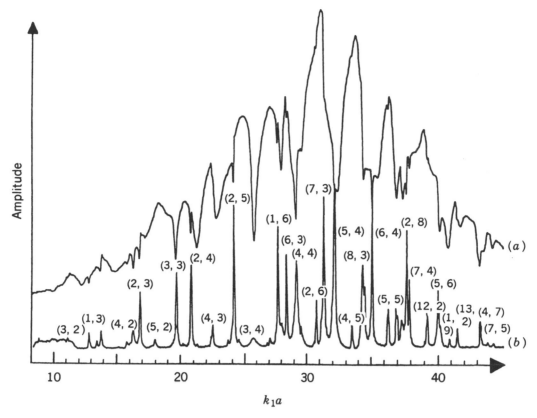

Fig. 8 Backscattering response for an aluminum cylinder in water: (*a*) form function (forced regime); (*b*) resonance spectrum (free transient). (From Refs. 14 and 31.)

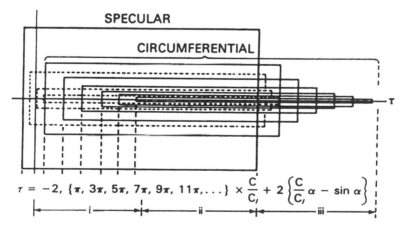

$$\tau = -2, \{\pi, 3\pi, 5\pi, 7\pi, 9\pi, 11\pi, \ldots\} \times \frac{C}{C_l} + 2 \left\{ \frac{C}{C_l} \alpha - \sin \alpha \right\}$$

Fig. 9 Schematic view of superposition (without any coherent or incoherent additions) of specular (large solid rectangle), penetrating and multiply internally reflected (dotted rectangles), and circumferential wave trains (small solid rectangles), showing initial transient region (i), quasi-steady-state region (ii), and final transient region, or ringing (iii).

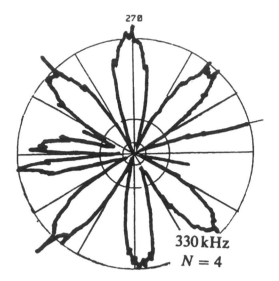

Fig. 10 Angular scattering diagram for a hemispherically capped cylindrical brass shell in water. (From Ref. 33.)

3. *Dot–Dashed Curve.* Use of the "physical optics" or Kirchhoff approximation[36] (which is obviously inapplicable here)

The dots are the experimental results, confirming the physical interpretation based on creeping waves. The undulatory character of the curves is due to the interference of the creeping wave with the (flat) specular background. The wide peaks are not interpretable as creeping-wave resonances since the large imaginary parts of the creeping-wave poles render these excessively broad. Rather, the interference minima can be shown to arise mathematically from zeros in the resonant amplitude that lie near the real axis.

For scattering objects of a density much below that of metals, appreciable penetration of sound into their interior will occur. Experimental results showing this, employing a spherical scatterer of polyurethane foam in air, are available together with their theoretical interpretation.[37] This analysis showed that it was not possible to model the scatterer as an absorbing fluid sphere. Rather, the foam material had to be described by the theory of Biot[38] for porous solids, in which in addition to the shear wave a fast and a slow dilatational wave are present, coupled by the entrained fluid mass in the pores and its friction. The experiment employed an incident one-cycle sine pulse; it measured both transmitted fields inside the sphere and scattered fields in the exte-

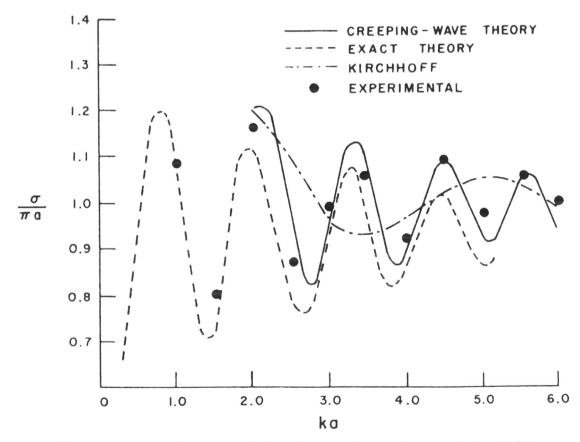

Fig. 11 Comparison of experimental data with calculations for the normalized scattering intensity of a rigid circular cylinder as a function of frequency. (From Ref. 35)

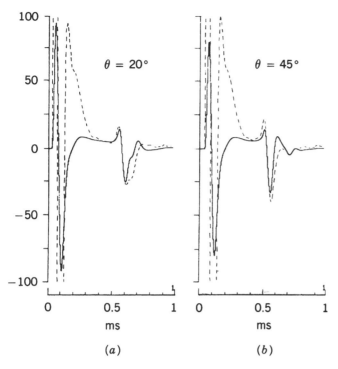

Fig. 12 Pulse scattering amplitude vs. time, at 20° or 45° from the backscattering direction, for a viscoelastic polyurethane foam sphere in air. (From Ref. 37.)

rior. The transmitted pulses exhibit a time delay due to the dispersion of the interior waves. The calculated scattered pulses, which are shown as solid curves in Fig. 12 as observed at $\theta = 20°$ and 45° from the backscattering direction and which closely agree with the measured pulses, consist of the specularly reflected pulse and the later arriving first creeping pulse. They are compared with calculated pulse shapes for a rigid sphere (dashed curves). The difference is due to the absorptivity of the porous material, leading to an attenuation of the observed specular (and to a lesser degree, creeping) pulse that increases with θ.

6 SCATTERING FROM MULTIPLE OBJECTS

The theory of scattering from a manifold of objects goes back to Foldy, Lax, and later Twersky,[39] who assumed the objects to be an arbitrary configuration of parallel cylinders. This assumption is sufficient to highlight the physics involved in this process: The resulting field consists of the incident wave plus a sum of various orders of scattering. The first order, which may be considered as the *single-scattering approximation* results from the excitation of each object by only the incident wave, or primary excitation. The second order of scattering results from the excitation of each object by the first order of

scattering from the remaining objects, and so on, to an infinite order of scattering. The first order thus consists of waves scattered by one object, the second order of waves scattered by two objects, and so on. Mathematically, this is treated by a technique of *multiple scattering* in which a hierarchy of coupled multiple-scattering equations is developed, usually to be truncated at a given stage in order to facilitate the solution of the problem.

This approach was applied to randomly distributed cylinders of arbitrary cross section[40] and to finite arbitrary gratings[41] and periodic gratings[42] of compliant tubes. Scattering from two parallel rigid cylinders[43] or two cylindrical shells[44] was considered as a special case.

As an illustration, Fig. 13 shows the insertion loss (defined as $20 \log_{10} p_i/p_t$, where p_i is the incident pressure and p_t the pressure transmitted through a grating), both measured (solid or dash–dotted curves) and calculated (dashed or dotted curves), for two different gratings of parallel steel tubes of highly eccentric rectangular cross section in water. The geometry of the two gratings is indicated in the figure. The first dip position is normalized to unity by dividing by f_1, the single-tube (without water loading) fundamental resonance frequency, and the resonance frequency of the tubes in the second grating, f_2, is indicated. In the case of randomly distributed scatterers,[40] one may consider the imbedding medium (the "matrix") together with the included scatterers as a composite "effective medium," for which the wave phase velocities and attenuations may be obtained from the multiple-scattering theory.

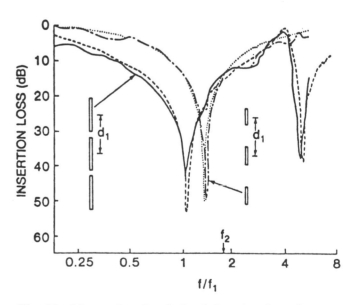

Fig. 13 Measured and calculated insertion loss for two single gratings of cylindrical steel tubes with highly eccentric rectangular cross sections in water. (From Ref. 42.)

It is hoped that this chapter has conveyed to the reader a picture of the present status of acoustic scattering theory and experiment, showing that modern methods of both measurement and analysis have brought us a long way in understanding the scattering phenomenon since the pioneering days of Faran.[2]

REFERENCES

1. Rayleigh, Lord (John William Strutt), *Theory of Sound*, 1877; reprinted by Dover, New York, 1937.

2. J. J. Faran, "Sound Scattering by Solid Cylinders and Spheres," *J. Acoust. Soc. Am.*, Vol. 23, 1951, p. 405.

3. G. R. Barnard and C. M. McKinney, "Scattering of Acoustic Energy by Solid and Air-filled Cylinders in Water," *J. Acoust. Soc. Am.*, Vol. 33, 1961, p. 226.

4. K. J. Diercks, T. G. Goldsberry, and W. W. Horton, "Circumferential Waves in Thin-walled Air-filled Cylinders in Water," *J. Acoust. Soc. Am.*, Vol. 35, 1963, p. 59.

5. W. Franz, "Über die Greenschen Funktionen des Zylinders und der Kugel," *Z. Naturf.*, Vol. A9, 1954, p. 705.

6. N. Veksler, *Acoustic Resonance Spectroscopy*, Springer, Berlin and Heidelberg, 1993.

7. G. Quentin, M. de Billy, and A. Hayman, "Comparison of Backscattering of Short Pulses by Solid Spheres and Cylinders at Large *ka*," *J. Acoust. Soc. Am.*, Vol. 70, 1981, p. 870.

8. M. F. Werby, H. Überall, A. Nagl, S. H. Brown, and J. W. Dickey, "Bistatic Scattering and Identification of the Resonances of Elastic Spheroids," *J. Acoust. Soc. Am.*, Vol. 84, 1988, p. 1425.

9. L. Flax, L. R. Dragonette, and H. Überall, "Theory of Elastic Resonance Excitation by Sound Scattering," *J. Acoust. Soc. Am.*, Vol. 63, 1978, p. 723.

10. H. Überall, "Modal and Surface Waves Resonances in Acoustic-Wave Scattering from Elastic Objects and in Elastic-Wave Scattering from Cavities," in J. Miklowitz and J. Achenbach, Eds., *Proceedings of the International Union Theoretical and Applied Mechanics (IUTAM) Symposium: Modern problems in elastic wave propagation*, Wiley, New York, 1978, pp. 239–263.

11. A. M. Lane and R. G. Thomas, "*R*-Matrix Theory of Nuclear Reactions," *Rev. Mod. Phys.*, Vol. 30, 1958, p. 257.

12. G. Maze and J. Ripoche, "Méthode d'isolement et d'identification des resonances (MIIR) de cylindres et de tubes soumis à une onde acoustique plane dans l'eau," *Rev. Phys. Appl.*, Vol. 18, 1983, p. 319.

13. B. Poirée, Ed., N. GESPA, *La Diffusion Acoustique*, CEDOCAR, Paris, 1987.

14. H. Überall, Ed., *Acoustic Resonance Scattering*, Gordon and Breach, New York, 1991.

15. H. Überall, G. Maze, and J. Ripoche, "Experiments and Analysis of Sound Induced Structural Vibrations," in A. Guran and D. J. Inman, Eds., *Stability, Vibration, and Control of Structures*, World Scientific, River Edge, NY, and Singapore, 1994.

16. S. K. Numrich, N. H. Dale, and L. R. Dragonette, "Generation of Plate Waves in Submerged Air-filled Shells," in G. C. Everstine and M. K. Au-Yang, Eds., *Advances in Fluid–Structure Interaction 1984*, American Society of Mechanical Engineers, New York, PVP vol. 78/AMD vol. 64, 1984, pp. 59–74.

17. M. de Billy, "Determination of the Resonance Spectrum of Elastic Bodies via the Use of Short Pulses and Fourier Transform Theory," *J. Acoust. Soc. Am.*, Vol. 79, 1986, p. 219.

18. M. F. Werby, J. J. Castillo, A. Nagl, R. D. Miller, J. M. D'Archangelo, J. W. Dickey, and H. Überall, "Acoustic Resonance Spectroscopy for Elastic Spheroids of Varying Aspect Ratios, and the Level Crossing Phenomenon," *J. Acoust. Soc. Am.*, Vol. 88, 1990, p. 2822.

19. M. F. Werby, Y. J. Stoyanov, J. W. Dickey, M. Keskin, J. M. D'Archangelo, A. Nagl, N. J. Stoyanov, J. J. Castillo, and H. Überall, "Resonant Acoustic Scattering from Elastic Spheroids," *J. d'Acoust.*, Vol. 3, 1990, p. 201.

20. P. C. Waterman, "New Formulation of Acoustic Scattering," *J. Acoust. Soc. Am.*, Vol. 45, 1969, p. 1417.

21. M. F. Werby, "Recent Developments in Scattering from Submerged Elastic and Rigid Targets," in H. Überall, Ed., *Acoustic Resonance Scattering*, Gordon and Breach, New York, 1991; "The Acoustical Background for Submerged Elastic Shells," *J. Acoust. Soc. Am.* Vol. 90, 1991, p. 3279; "The Isolation of Resonances and the Ideal Acoustical Background for Submerged Elastic Shells," *Acoust. Lett.*, Vol. 15, 1991, p. 65.

22. S. G. Kargl and P. L. Marston, "Longitudinal Resonances in the Form Function for Backscattering from a Spherical Shell: Fluid Shell Case," *J. Acoust. Soc. Am.*, Vol. 88, 1990, p. 1114.

23. N. D. Veksler, "Intermediate Background in Problems of Sound Wave Scattering by Elastic Shells," *Acustica*, Vol. 76, 1992, p. 1.

24. S. C. Frautschi, *Regge Poles and S-Matrix Theory*, W. A. Benjamin, New York, 1963.

25. C. E. Baum, "The Singularity Expansion Method," in L. B. Felsen, Ed., *Transient Electromagnetic Fields*, Springer, Berlin and Heidelberg, 1976, pp. 129–179.

26. M. F. Werby and G. J. Tango, "Numerical Study of Material Properties of Submerged Elastic Objects Using Resonance Response," *J. Acoust. Soc. Am.*, Vol. 79, 1986, p. 1260.

27. H. Überall, Y. J. Stoyanov, A. Nagl, M. F. Werby, S. H. Brown, J. W. Dickey, S. K. Numrich, and J. M. D'Archangelo, "Resonance Spectra of Elongated Objects," *J. Acoust. Soc. Am.*, Vol. 81, 1987, p. 312.

28. D. G. Vasil'ev, "The Frequency Distribution Function of a Shell of Revolution Immersed in a Liquid," *Dokl. Akad.*

Nauk SSSR, Vol. 248, 1979, p. 325 (in Russian); English translation *Sov. Phys. Dokl.*, Vol. 24, 1979, p. 720.

29. L. Flax, G. C. Gaunaurd, and H. Überall, "Theory of Resonance Scattering," in W. P. Mason and R. N. Thurston, Eds., *Physical Acoustics*, Vol. 15, Academic, New York, 1981, pp. 191–294.

30. H. Überall, "Surface Waves in Acoustics," in W. P. Mason and R. N. Thurston, Eds., *Physical Acoustics*, Vol. 10, Academic, New York, 1973, pp. 1–60.

31. G. Maze, "Diffusion d'une onde acoustique plane par des cylindres et des tubes immergés dans l'eau: Isolement et identification des résonances," Ph.D. Dissertation, University of Rouen, 1984.

32. W. E. Howell, S. K. Numrich, and H. Überall, "Complex Frequency Poles of the Acoustic Scattering Amplitude and Their Ringing," *IEEE Trans. Ultrason. Ferroelec. Freq. Control*, Vol. UFFC-34, 1987, p. 22.

33. S. K. Numrich and H. Überall, "Scattering of Sound Pulses and the Ringing of Target Resonances," in A. D. Pierce, Ed., *Physical Acoustics*, Vol. 21, Academic, New York, 1991, pp. 235–318.

34. H. Überall, G. C. Gaunaurd, and J. D. Murphy, "Acoustic Surface Wave Pulses and the Ringing of Resonances," *J. Acoust. Soc. Am.*, Vol. 72, 1982, p. 1014.

35. M. L. Harbold and D. N. Steinberg, "Direct Experimental Verification of Creeping Waves," *J. Acoust. Soc. Am.*, Vol. 45, 1969, p. 592.

36. P. M. Morse and H. Feshbach, *Methods of Theoretical Physics*, McGraw-Hill, New York, 1953.

37. G. Deprez, "Sphère absorbante: diffraction aérienne en régime impulsionnel," in B. Poirée, Ed., N. GESPA, *La Diffusion Acoustique*, CEDOCAR, Paris, 1987.

38. M. A. Biot, "Theory of Propagation of Elastic Waves in Fluid-saturated Porous Solids," *J. Acoust. Soc. Am.*, Vol. 28, 1956, p. 168.

39. V. Twersky, "Multiple Scattering of Radiation by an Arbitrary Configuration of Parallel Cylinders," *J. Acoust. Soc. Am.*, Vol. 42, 1952, p. 42.

40. V. K. Varadan, V. V. Varadan, and Y. H. Pao, "Multiple Scattering of Elastic Waves by Cylinders of Arbitrary Cross Section," *J. Acoust. Soc. Am.*, Vol. 63, 1978, p. 1310.

41. C. Audoly and G. Dumery, "Modeling of Compliant Tube Underwater Reflectors," *J. Acoust. Soc. Am.*, Vol. 87, 1990, p. 1841.

42. R. P. Radlinski, "Scattering by Multiple Gratings of Compliant Tubes," *J. Acoust. Soc. Am.*, Vol. 72, 1982, p. 607.

43. J. W. Young and J. C. Bertrand, "Multiple Scattering by Two Cylinders," *J. Acoust. Soc. Am.*, Vol. 58, 1975, p. 1190.

44. J. P. Sessarego and J. Sageloli, "Diffusion acoustique par deux coques cylindriques parallèles," *J. d'Acoust.*, Vol. 5, 1992, p. 463.

6

SPEED OF SOUND IN FLUIDS

ALLAN J. ZUCKERWAR

1 INTRODUCTION

Sound is a physical phenomenon identified with the propagation of a mechanical disturbance through a medium. The *speed of sound* is defined as the distance traversed per unit time by a given point on the disturbance, provided the disturbance does not change its shape (as would be the case in an ideal medium). If the disturbance is resolved into spatially harmonic components, an observed point on any one component may be designated by a phase angle, and if directional information is included, then the distance traversed per unit time is called the *phase velocity*. If the phase velocity is the same at all frequencies, the medium is said to be *nondispersive*; otherwise it is *dispersive*.

The phase velocity appears in the familiar wave equation, which governs the propagation of sound through the medium. The wave equation is derived from an equation of motion, a constitutive equation or equation of state, and a continuity equation. In an unbounded homogeneous medium the magnitude of the phase velocity, that is, the speed of sound, depends on the adiabatic bulk modulus and the density of the undisturbed medium. The speed of sound is sensitive to such properties as temperature, pressure, composition, and absorption. These are considered in this chapter for gases and liquids. Additional effects as the sensitivity to sound pressure amplitude, to specific absorption processes, and to medium boundaries are discussed in Chapters 7, 8, 9, and 56.

Encyclopedia of Acoustics, edited by Malcolm J. Crocker
ISBN 0-471-80465-7 © 1997 John Wiley & Sons, Inc.

2 SPEED OF SOUND IN GASES

2.1 Molecular Foundation

Ideal Gas   Consider a gas having a pressure P, volume V, density ρ, temperature T, entropy S, and molecular mass M. Kinetic analysis yields, for the speed of sound,

$$c = \sqrt{\frac{M_s}{\rho_0}}, \tag{1}$$

where ρ_0 is density of the undisturbed medium and M_s the adiabatic bulk modulus defined by

$$M_s = \rho \left(\frac{\partial P}{\partial \rho} \right)_s. \tag{2}$$

The equation of state of an ideal gas,

$$P = \frac{\rho R T}{M}, \tag{3}$$

and the first two laws of thermodynamics,

$$T\, dS = C_v\, dT - \frac{P}{\rho^2}\, d\rho = 0, \tag{4}$$

yield the adiabatic bulk modulus $M_s = \gamma P$ and, from this, the speed of sound in terms of the properties of the gas:

$$c = \sqrt{\frac{\gamma P}{\rho_0}} \tag{5}$$

$$= \sqrt{\frac{\gamma R T}{M}}, \tag{6}$$

where γ is the specific heat ratio and R the ideal-gas constant. Equation (6), based on the ideal-gas law, is slightly less accurate than Eq. (5), which is amenable to real-gas values. The speed of sound increases with temperature, decreases with molecular mass, and for a gas obeying Eq. (3) precisely is independent of pressure.

The specific heat ratio $\gamma = C_p / C_v$ can be written as

$$\gamma = \frac{C_v + R}{C_v} \tag{7}$$

since the specific heat at constant pressure $C_p = C_v + R$ for an ideal gas. The specific heat C_v at constant volume contains contributions from the molecular translational, vibrational, and rotational degrees of freedom (electronic contribution being negligible):

$$C_v = \frac{3R}{2} + R \sum_i \Phi_v \left(\frac{\Theta_{vi}}{T} \right) + \frac{R}{2} \sum_i \Phi_R \left(\frac{\Theta_{Ri}}{T} \right). \tag{8}$$

The translational contribution is simply $\frac{1}{2}R$ for each of the three degrees of freedom, as given by first term on the right.

The vibrational contribution depends upon the Planck–Einstein function

$$\Phi_v(x) = \frac{x^2 e^x}{(e^x - 1)^2}, \tag{9}$$

where $x = \Theta_{vi}/T$ and

$$\Theta_{vi} = \frac{h \nu_i}{k_B}. \tag{10}$$

In Eq. (10), h is Planck's constant, k_B Boltzmann's constant, Θ_{vi} the *characteristic temperature*, and ν_i the frequency of the ith vibrational mode. If the characteristic temperature is low ($x \ll 1$), then $\Phi_v(x) \to 1$, the classical value; if it is high ($x \gg 1$), then $\Phi_v(x) \approx x^2 e^{-x}$ is small and the mode is said to be "frozen in." In a molecule consisting of N atoms, the number of vibrational modes is $3N - 5$ if the molecule is structurally linear and $3N - 6$ if it is nonlinear. Each vibrational mode, containing both kinetic and potential energy, is responsible for two degrees of freedom. The characteristic vibrational temperatures of most molecules are so high that at standard temperature (273.15 K) the vibrational degrees of freedom make only a small contribution to the specific heat.

The characteristic temperatures Θ_{Ri} for rotation, in contrast to those for vibration, are generally so small that the rotational degrees of freedom are fully excited at room temperature and the rotational function $\Phi_R \approx 1$ (H_2 being a notable exception). Since there are three axes of rotation, the rotational contribution to the specific heat is generally $\frac{3}{2}R$; but if a molecule is linear, rotation about the collinear axis is frozen in and makes no measurable contribution. Therefore, for nearly all gases at standard temperature Eq. (8) can be written as

$$C_v = \frac{R}{2} (3 + F) + R \sum_i \Phi_v \left(\frac{\Theta_{vi}}{T} \right) \tag{11}$$

where F is 0 for a monatomic molecule, 2 for a diatomic or linear molecule, and 3 for a nonlinear molecule; the sum is taken over all the vibrational modes.

If the vibrational contribution is neglected, then Eqs. (7) and (11) give the following values of γ:

$$\gamma = \begin{cases} \frac{5}{3} & \text{for a monatomic gas} \\ \frac{7}{5} & \text{for a diatomic gas or linear polyatomic gas,} \\ \frac{4}{3} & \text{for a nonlinear polyatomic gas.} \end{cases}$$

Example Compute the speed of sound in CO_2 at standard temperature and pressure (STP) ($T = 273.15$ K and $P = 1$ atm $= 1.0133 \times 10^5$ Pa) from molecular data: $M = 0.0441$ kg/mol, $\Theta_{v1} = 1997.5$ K (symmetric stretch), $\Theta_{v2} = 960.1$ K (doubly degenerate bending), $\Theta_{v3} = 3380.2$ K (antisymmetric stretch), $F = 2$ (linear).

From Eqs. (11), (9), (7), and (6) one finds

$$\Phi_v \left(\frac{\Theta_{v1}}{T} \right) = 0.0357, \qquad \Phi_v \left(\frac{\Theta_{v2}}{T} \right) = 0.390,$$

$$\Phi_v \left(\frac{\Theta_{v3}}{T} \right) = 0.00065,$$

$$C_v = (3+2)(\tfrac{1}{2}R) + (0.0357 + 2 \times 0.390$$

$$+ 0.00065)R = 3.316R,$$

$$\gamma = \frac{(3.316 + 1)R}{3.316R} = 1.302,$$

$$c = \sqrt{\frac{1.302 \times 8.3145 \times 273.15}{0.0441}} = 258.9 \text{ m/s}$$

[from Eq. (6)],

which is somewhat higher than the experimental determination[1] $c = 257.5$ m/s. Equation (6) yields too high a sound speed because Eq. (3) underestimates the density.

Knowledge of the sound speed of a gas permits one to infer an important microscopic property. Comparison of Eq. (6) with the well-known expression for the root-mean-square (rms) thermal speed of the gas molecules $v_{rms} = \sqrt{(3RT/M)}$ yields the relationship

$$v_{rms} = \sqrt{\frac{3}{\gamma}} \, c. \tag{12}$$

Thus CO_2 molecules move at an rms speed of $\sqrt{(3/1.302)} \times 257.5 = 390.8$ m/s at STP.

The validity of Eqs. (5) and (6) is based on three fundamental assumptions: namely, that the medium is isentropic, linear, and continuous. The consequences of departure from these assumptions are considered in Sections 2.2–2.5.

2.2 Isothermal Sound Speed

In situations where sound propagation is confined to a narrow duct or other small enclosure, rapid heat exchange between the solid boundaries and the gas may dictate the use of the isothermal sound speed,

$$c_T = \frac{c}{\sqrt{\gamma}}, \tag{13}$$

rather than the adiabatic sound speed. Quantitatively, the use of Eq. (13) is appropriate when the thickness of the thermal boundary layer

$$d_h = \sqrt{\frac{2K}{\rho \omega C_p}} \tag{14}$$

exceeds the lateral dimensions of the enclosure.[2] In Eq. (14), K is the thermal conductivity of the gas and ω the acoustic angular frequency, related to frequency f by $\omega = 2\pi f$.

Example The air gap between the membrane and backplate of a condenser microphone is typically $h = 25$ μm. Does sound propagate within at the adiabatic or isothermal sound speed?

Since there is a boundary layer at each boundary, the criterion for choosing the isothermal sound speed becomes $h < 2d_h$. Substitution of known values for air at 20°C into Eq. (14) leads to the result

$$f < (2 \times 0.0025/25 \times 10^{-6})^2 = 40 \text{ kHz}$$

Consequently sound in the air gap propagates at the isothermal sound speed at frequencies of up to 40 kHz.

2.3 Speed of Sound at High Gas Pressures

With increasing pressure, deviations from the ideal-gas law become more pronounced and affect the sound speed through modifications to Eqs. (3) and (7). Based on the virial equation of state,[3] these equations are replaced by

$$P = \frac{\rho RT}{M} (1 + B_p P + \cdots), \tag{15}$$

$$\gamma = \gamma^0 \left\{ 1 + \frac{PR}{C_v^0} \left[\frac{RT}{C_p^0} \frac{\partial^2 (TB_p)}{\partial T^2} + 2 \frac{\partial (TB_p)}{\partial T} + \cdots \right] \right\}. \tag{16}$$

Substitution of Eqs. (15) and (16) into Eq. (5) leads to the sound speed of a real gas:

$$c = c^0 \left\{ 1 + 2P \left[B_p + \frac{R}{C_v^0} \frac{\partial (TB_p)}{\partial T} \right. \right.$$

$$\left. \left. + \frac{1}{2} \frac{R^2 T}{C_v^0 C_p^0} \frac{\partial^2 (TB_p)}{\partial T^2} \right] + \cdots \right\}^{1/2}, \tag{17}$$

where the superscript 0 designates ideal-gas values [Eqs. (3), (6), and (7)]. Here B_p is the second virial coefficient for a power series in pressure; it is related to B_v, the corresponding coefficient for a power series in volume, by $B_p = B_v/RT$.

Example Correct the sound speed of CO_2 at STP for the nonideality of the gas. Repeat for $P = 10$ atm.

Reference 3 gives the second virial coefficient of CO_2 as a function of temperature (converted here to TB_p):

$$TB_p = 0.335 - 8092T^{-1.5} - 3.489 \times 10^{10}T^{-4.5} \text{atm}^{-1}.$$

After inserting into Eq. (17) one obtains

$$c = 258.9[1 + 2 \times 1 \times (-0.00672 + 0.00485 - 0.00206)]^{1/2}$$
$$= 257.9 \text{ m/s},$$

which now brings the sound speed close to the experimental value. Note that the magnitudes of all the correction terms in Eq. (17) are comparable. Thus the effect of pressure on the specific heat ratio [Eq. (16)] is substantial, and it is not sufficient to account for the bulk modulus [Eq. (15)] alone. At $P = 10$ atm, Eq. (17) yields $c = 248.5$ m/s, but no precise measured value is available for comparison.

2.4 Speed of Sound at Low Gas Pressures

A decrease in gas pressure to sufficiently low levels introduces some interesting physical effects, which have a profound impact upon the sound speed. The prominence of any of these effects depends upon the relative magnitudes of three characteristic lengths: the propagation path s (i.e., enclosure dimension), the acoustic wavelength λ, and the molecular mean free path l. Accordingly, these effects are organized into the following dispersion regimes: *classical, $s \gg l$, $\lambda \gg l$; Burnett, $s \gg l$, $\lambda \approx l$; transition, $s \approx l$; and near, $s \ll l$.*

In the classical regime collisions among molecules are far more frequent than their collisions with the walls of the enclosure; furthermore, the transport properties (like viscosity and thermal conductivity) attendant to such collisions respond with such rapidity, relative to the period of the perturbing sound wave, that sound propagation remains nearly adiabatic and conforms to the relationships given in the previous sections. In the Burnett regime the intermolecular collisions are still dominating, but the lengthened response time of the transport properties, now comparable to the period of the sound wave, leads to significant "classical absorption" and accompanying dispersion of the sound wave. In the near regime intermolecular collisions concede to molecule–wall collisions; the constitutive relationships of continuum mechanics are no longer valid. Both the transport properties and sound speed are functions of frequency and geometry of the enclosure. In the transition regime the intermolecular collision rate is comparable to that with the walls.

Classical theory is based on linear relationships among the transport properties. In a nonrelaxing gas (see Chapter 56) the conservation laws and constitutive equations yield a quadratic equation for the square of the complex wavenumber. Greenspan[4] has shown that the solution for a propagating wave can be approximated by a remarkably simple equation:

$$\left(\frac{\alpha\lambda}{2\pi} + i\frac{c^0}{c}\right)^2 = -\left(1 + i\frac{\gamma_F}{r}\right)^{-1}, \quad (18)$$

where γ_F is an effective specific heat ratio ($\frac{7}{5}$ for a monatomic gas, $\frac{4}{3}$ for a polyatomic gas having a Prandtl number equal to $\frac{3}{4}$, α the attenuation due to viscous and thermal transport, c^0 the sound speed in an ideal gas, and r a viscous Reynolds number,

$$r = \frac{P}{2\pi f\mu}, \quad (19)$$

dependent upon μ the absolute viscosity. Note that the γ_F do not correspond to the γ given in Section 2.1. Equation (18) can be used to find the sound speed over the classical and Burnett regimes.

The physical idea behind the theory of sound propagation in the near regime is that the distribution function of molecular velocities is dominated by the velocity imparted by the sound transmitter rather than by the thermal velocities. The resulting expression for the sound speed involves an integral that cannot be evaluated in closed form; but it may be concluded that the sound speed is independent of pressure and falls very slowly with the product $2\pi f s$.

The plot in Fig. 1 of normalized sound speed versus the frequency–pressure ratio shows typical boundaries of demarcation for the classical, Burnett, transition, and near regimes in air. The first two are based on the real part of Eq. (18), which shows good agreement with experiment. The last is estimated from the data of Ref. 4 and corresponds to specific experimental conditions; scaling laws for other experimental conditions are cited in this reference.

2.5 Sound Dispersion in a Relaxing Gas

The theory of molecular relaxation in a gas appears in a later chapter (Chapter 56). For the present purpose it is important only to note that a relaxation process is characterized by two parameters: a relaxation time τ and a

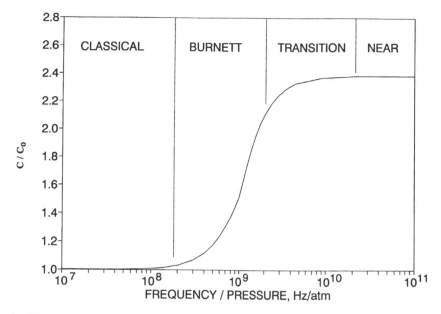

Fig. 1 Dispersion of sound in air. The sound speed c, normalized to the low-frequency limiting value c_0, is plotted against frequency–pressure ratio. The boundaries represent a gradual transition from one regime to another.

relaxation strength ϵ. The phenomenologic impact of a relaxation process on sound propagation is that it causes sound absorption and dispersion, revealed as a *relaxation peak* and *dispersion step*, respectively. The relationship between the squared sound speed and relaxation parameters is given in the equation[5]

$$\frac{c^2}{c_0^2} = \left(1 + \frac{\epsilon}{1 - \epsilon} \frac{\omega^2 \tau^2}{1 + \omega^2 \tau^2}\right). \qquad (20)$$

This function is plotted in Fig. 2. Here it is seen that the "center" of the dispersion step occurs at a frequency

$$f_R = (2\pi\tau)^{-1}, \qquad (21)$$

the *relaxation frequency*, and that its "height" has a magnitude

$$\frac{c_\infty^2 - c_0^2}{c_0^2} = \frac{\epsilon}{1 - \epsilon}. \qquad (22)$$

At low frequencies ($\omega\tau \ll 1$) $c = c_0$, the "relaxed" sound speed, while at high frequencies ($\omega\tau \gg 1$) $c = c_\infty$, the "unrelaxed" sound speed. The relaxation strength of a relaxing degree of freedom in an ideal gas is found to be

$$\epsilon = \frac{RC_i}{C_v(C_p - C_i)}, \qquad (23)$$

where C_v and C_p are low-frequency specific heats and C_i the specific heat of the relaxing degree of freedom, taken from Eq. (8).

Example For CO_2 at STP find the relaxation strength ϵ, the unrelaxed sound speed c_∞, and, given the vibrational relaxation time $\tau = 4.7 \times 10^{-6}$ s, the relaxation frequency f_R and the sound speed at $f = 1$ kHz and at $f = 100$ kHz.

Assume that the relaxation time is that of the doubly degenerate bending mode. The data of Section 2.1 shows $C_v = 3.316R$, $C_p = 4.316R$, and $C_i = 0.780R$. Equations (20), (22), and (23) yield

$$\epsilon = \frac{R \times 0.708R}{3.316R(4.316R - 0.780R)} = 0.0645,$$

$$c_\infty = c_0(1 - \epsilon)^{-1/2} = 257.5(1 - 0.0645)^{-1/2}$$

$$= 266.2 \text{ m/s},$$

$$f_R = (2\pi\tau)^{-1} = (2\pi \times 4.7 \times 10^{-6})^{-1} = 34 \text{ kHz},$$

$$c(1 \text{ kHz}) = 257.5 \left(1 + \frac{0.0645}{1 - 0.0645} \frac{0.0295^2}{1 + 0.0295^2}\right)^{1/2}$$

$$= 257.5 \text{ m/s},$$

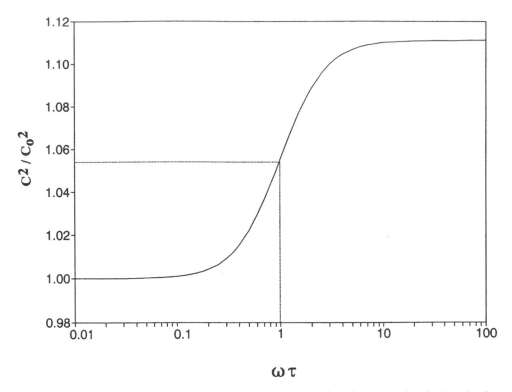

Fig. 2   Dispersion of sound in a relaxing gas. Here the relaxation strength ϵ is 0.1. At the relaxation frequency, corresponding to the condition $\omega\tau = 1$, the square of the sound speed is halfway between the low-frequency and high-frequency limits.

$$c(100 \text{ kHz}) = 257.5 \left(1 + \frac{0.0645}{1 - 0.0645} \frac{2.95^2}{1 + 2.95^2}\right)^{1/2}$$

$$= 265.4 \text{ m/s},$$

since $\omega\tau = 0.0295$ at $f = 1$ kHz ($0.0294 f_R$) and 2.95 at $f = 100$ kHz ($2.94 f_R$). This example illustrates the effect of frequency on the sound speed of a relaxing gas.

2.6   Speed of Sound in Gas Mixtures

Equation (6) holds for a mixture of ideal gases having mole fractions X_1, X_2, ... if one substitutes

$$\gamma = 1 + \frac{R}{C_v} = 1 + R(X_1 C_{v1} + X_2 C_{v2} + \cdots)^{-1}, \quad (24)$$

$$M = X_1 M_1 + X_2 M_2 + \cdots. \quad (25)$$

Example   Find the speed of sound in air at STP for relative humidities of 0 and 100%.

First consider dry air as an ideal gas. The data are compiled in Table 1. Equations (24), (25), and (8) yield

$M = 28.9644$ kg/kmol, $\gamma^0 = 1.4005$, and $c^0 = 331.39$ m/s, all of which agree with Greenspan.[6] Now consider dry air as a real gas. Greenspan cites data for B_v, converted here to data for B_p:

$$B_p = -6.023 \times 10^{-4} \text{ atm}^{-1},$$

$$\frac{d}{dT}(TB_p) = 2.9446 \times 10^{-3} \text{ atm}^{-1},$$

$$\frac{d^2}{dT^2}(TB_p) = -2.532 \times 10^{-5} \text{ atm}^{-1} \cdot \text{K}^{-1}.$$

Then Eq. (17) yields

$$c = 331.39 \left[1 + 2 \times 1 \left(-6.023 + \frac{29.446}{2.4967}\right.\right.$$

$$\left.\left. - \frac{273.15 \times 0.2532}{2 \times 3.4967 \times 2.4967}\right) \times 10^{-4}\right]^{1/2} = 331.45 \text{ m/s}.$$

In humid air containing a mole fraction X_h of water vapor, the mole fractions of all the other constituents are

TABLE 1 Properties of the Constituents of Air at $T = 0°C$

Constituent	M_j, kg/kmol[a]	$X_j{}^a$	C_{vj}°/R^b
N_2	28.0134	0.78084	2.5019
O_2	31.9988	0.209476	2.5208
Ar	39.948	0.00934	1.5001
CO_2	44.00995	0.000314	3.3255
CO	28.01	0.19×10^{-6}	2.503
H_2	2.01594	0.5×10^{-6}	2.4418
Ne	20.183	18.18×10^{-6}	1.5000
Kr	83.80	1.14×10^{-6}	1.4995
CH_4	16.04303	2×10^{-6}	3.2822
He	4.0026	5.24×10^{-6}	1.5000
N_2O	44.0128	0.27×10^{-6}	3.6165
Xe	131.30	0.087×10^{-6}	1.4998
Air (dry)	28.9644	—	2.4967
H_2O (100% RH^c)	18.01534	0.006026	3.01584^d
Air (100% RH)	28.8984	—	2.4998

^{a}From Ref. 7.

^{b}For N_2–H_2, $C_v^\circ = C_p^\circ - R$, where C_p° is taken from Ref. 8 at 0.01 atm. For Ne–Xe, $C_v^\circ \approx C_p' - R$, where C_p' is the real gas value taken from Ref. 7; the contributions from these gases scarcely affect the fifth significant figure.

^{c}Relative humidity.

^{d}Based on Eq. (11) using data from Ref. 9.

reduced to $X_j' = X_j(1 - X_h)$. At 273.15 K the vapor pressure of water is 4.58 Torr, corresponding to a mole fraction $X_h = 4.58/760 = 0.006026$ (100% relative humidity). When the X_j are substituted into Eqs. (24) and (25), B_p is assumed unchanged, and H_2O is included, then Eq. (17) yields $c^0 = 331.71$ m/s and $c = 331.77$ m/s.

3 SPEED OF SOUND IN LIQUIDS

3.1 Liquid Structure and Its Relation to Sound Speed

From an acoustical point of view the liquid is a state intermediate between a gas and a solid. It is gaslike in that, in the absence of losses, it does not offer resistance to a shear stress and solidlike in that its bulk modulus is determined by intermolecular bonding forces and not by external forces (e.g., gravity) or constraints (enclosure walls), as in a gas.

This state of affairs is best explained by a modern picture of liquid structure.[10] The radial distribution function (molecular number density vs. radial distance) obtained from x-ray and neutron diffraction patterns clearly shows regions of highly ordered structure as found in a crystalline solid. These regions, called *aggregates* or some-

times *clusters*, are of limited spatial extent, ranging from a few molecular diameters to macroscopic dimensions. Liquids showing little tendency toward aggregation are called *unassociated*; those containing a substantial distribution of small-sized aggregates are called *associated*; and those containing long chains of aggregates are called *polymerized*. The prevailing amorphous matter contains an abundance of microcavities, called *holes*, of varying size and distribution, which are responsible for the fluidity of the liquid. The major impact of the holes is an enhancement of the *free volume*, which plays a key role in the emergence of many equilibrium and transport properties of the liquid.

The derivation of the sound speed of a liquid from its microscopic properties requires realistic expressions for the equation of state and the internal energy, leading in turn to the bulk modulus and specific heat. Successful theoretical developments in these areas have been limited. Among the difficulties are the incorporation of the free volume into the equation of state and the tabulation of contributions to the free energy, which is not as straightforward as in a gas. The most successful efforts are, for the most part, at least partially empirical.

3.2 The Equation of State

The general form of an equation of state

$$P = \rho f(\rho, T, \ldots) \qquad (26)$$

yields a bulk modulus

$$M_S = \gamma \rho \left(\frac{\partial P}{\partial \rho}\right)_T = \gamma \left[P + \rho \left(\frac{\partial f}{\partial \rho}\right)_T\right]. \qquad (27)$$

As a gas condenses to a liquid, the second term becomes decisive owing to the large increase in density.

The development of a successful equation of state of a liquid must address the issues of molecular bonding, association and polymerization, and free volume in both the ordered and disordered structure. The principal approaches are as follows:

1. *Semirigorous Derivations Based on a Liquid Model.* These include the celebrated Lennard-Jones and Devonshire cell model, the Eyring hole model, and correlation models to determine the radial distribution function. Although these models have had some success in predicting many physical properties of liquids, they have been less successful in predicting sound speed.

2. *Semiempirical Modifications to Existing Equations of State.* A noteworthy example is the theory of Schaaffs,[11] who modified the van der Waals equation to include an improved expression for molecular volume as well as the specifics of bonds between various types of atoms or molecular complexes. Schaaffs's treatment has enjoyed remarkable success in predicting the sound speeds of unassociated organic liquids.

3. *Empirical Expressions Fitted to Experimental Data.* A common procedure is to expand one state variable in a series (not necessarily a power series) of the others. In applications to sound speed, however, the specific heat at one reference point remains an unknown parameter. Therefore it is just as useful to expand the sound speed itself in terms of state variables, an approach taken by van Itterbeek and co-workers[12] for cryogenic liquids.

3.3   Speed of Sound in Unassociated Liquids: Schaaffs's Method

Schaaffs's formulas[11] to compute the sound speed are

$$c = c_R \left(\frac{\rho B}{M} - \frac{\beta}{\beta + B} \right), \qquad (28)$$

$$M = \sum_i z_i M_i, \qquad (29a)$$

$$B = \sum_i z_i B_i, \qquad (29b)$$

$$\beta = \sum_i z_i \beta_i, \qquad (29c)$$

where M is the molecular mass, B the *external addendum* (molecular volume per mole), β the *internal addendum* (heavy atom corrective volume), z_i the number and M_i the mass of the ith species, $c_R = 4450$ m/s, and B_i and β_i contributions from the ith species obtained from a table. Some excerpts are listed in Table 2.

Example   Compute the sound speed of carbon tetrachloride (CCl_4), chloroform ($CHCl_3$), benzene (C_2H_6), and methanol (CH_3OH) at 20°C. Values of the respective densities ρ are 1.5939, 1.487, 0.878, and 0.7913 g/cm³. The atomic weights of H, C, Cl, and O are 1.008, 12.01, 35.457, and 16.00 g/mol.

Inserting the data into Eqs. (28)–(29) yields, for CCl_4,

$$M = 1 \times 12.01 + 4 \times 35.457 = 153.838 \text{ g/mol},$$

TABLE 2   Parameters Entering Schaaffs' Formulas (28)–(29) for Selected Atomic and Molecular Bonds

Bond Type	External Addendum, B_i
H (all bonds)	1.06
C (4 monovalent)	3.06
C (1 divalent)	3.36
C–Cl (monovalent)	6.92

Bond Type	Internal Addendum, β_i
C–Cl (monovalent)	0.25, 0.52, 0.75, 0.87 for 1, 2, 3, 4 atoms present
C–OH (monovalent)	0.6

$$B = 1 \times 3.06 + 4 \times 6.92 = 30.74 \text{ cm}^3,$$

$$\beta = 4 \times 0.87 = 3.48 \text{ cm}^3,$$

$$c = 4450 \left(1.5939 \times \frac{30.74}{153.838} - \frac{3.48}{3.48 + 30.74} \right)$$

$$= 964.8 \text{ m/s}.$$

The results for the above liquids are summarized in Table 3. The agreement between theory and experiment is remarkable, even for CH_3OH, which is considered an associated liquid.

3.4   Speed of Sound in Cryogenic Liquids

Despite the simplicity of cryogenic liquid structure, which consists of monatomic molecules bonded by central forces, Schaaffs-like expressions for the sound speed are not available. Over the years van Itterbeek and co-workers have presented expressions for the sound speed of cryogenics versus temperature and pressure based on polynomial fits to a wide range of data. For example, their expressions for liquid nitrogen (LN_2) and liquid oxygen (LO_2) at the saturation temperatures (77.395 K and 90.19 K at 1 atm), after interpolation and unit conversion, are

$$c(LN_2) = 854.1 + 0.8370P - 0.9072 \times 10^{-3}P^2 + 0.9697$$
$$\times 10^{-6}P^3 - 0.4904 \times 10^{-9}P^4, \qquad (30a)$$

$$c(LO_2) = 907.7 + 0.5641P - 0.4389 \times 10^{-3}P^2 + 0.3632$$
$$\times 10^{-6}P^3 - 0.1429 \times 10^{-9}P^4, \qquad (30b)$$

where c is in metres per second and P is in atmospheres. At $P = 1$ atm, Eqs. (30a) and (30b) yield 854.9 and 908.3 m/s, compared to experimental values[13] of 852.9 and 907.0 m/s.

TABLE 3   Sound Speed in Organic Liquids after Schaaffs

Liquid	CCl_4	CH_3Cl	C_6H_6	CH_3OH
M (g/mol)	153.838	119.389	78.108	32.042
B (cm^3)	30.74	24.88	26.52	10.77
β (cm^3)	3.48	2.25	0	0.16
Theoretical c (m/s)	964.8	1009.9	1326.5	1118.4
Experimental c (m/s)	935	1001	1326	1121

3.5   Speed of Sound in Associated Liquids (Water)

A successful treatment of many physical properties of an associated liquid must account for the role of structural complexes. For illustration, Hall's model of water[3]—considered an exemplary associated liquid—assumes the liquid composition to be a mix of the open *tridymite* (icelike) structure and a more closely packed quartzlike structure. Eucken's model[3] assumes the structure to contain a distribution of polymeric units, namely the monomer, dimer, tetramer, and octamer, the last of which has the open tridymite structure. As the temperature increases toward the boiling point, the closely packed constituents gain at the expense of the vanishing tridymite. The competition between structural redistribution and thermal expansion explains such exceptional properties of water as the density maximum at 4°C and the sound speed maximum at 74°C.

A precise expression for the sound speed in saturated water versus temperature, promising an error not to exceed 0.05%, is given by Chavez, Sosa, and Tsumura[14]:

$$c = \left(1 - \frac{T}{T_c}\right)^a \sum_{k=0}^{5} b_k T^k \qquad \text{m/s}, \qquad (31)$$

where T_c, a, and the coefficients are given in Table 4. Equation (31) is plotted in Fig. 3. At 20°C, Eq. (31) yields $c = 1481.8$ m/s. An increase in pressure increases the sound speed.[15]

3.6   Speed of Sound in Electrolytic Solutions

The presence of ions in a polar solvent such as water completely reorganizes the structure due to the orientation of solvent dipoles about the ions. The considerable electric fields of the ions compress the volume of the solvent, thereby increasing the bulk modulus and speed of sound.

Although theoretical treatments based on microscopic properties are lacking, Weissler and Del Grosso[16] made an interesting observation concerning the contribution of the solutes to the sound speed of sea water. If the ith solute at a molarity X_i individually produces a sound speed $c_i(X_i)$, then the sound speed of the composite solution is

$$c = c_0 + \sum_i [c_i(X_i) - c_0], \qquad (32)$$

where c_0 is the sound speed of pure water (implying negligible ionic interaction). Values in metres per second for the seven leading salts in sea water at 30°C are listed in Table 5. When the speed of pure water $c_0 = 1510.0$ m/s and total increment of 36.2 m/s are inserted into Eq. (32), the resulting sound speed is 1546.2 m/s, compared to a measured value of 1545.8 m/s.

3.7   Temperature Dependence of the Sound Speed in Liquids: Rao's Constant

Rao[17] established an empirical relation between the temperature coefficient of the sound speed and the coefficient of thermal expansion for an unassociated liquid:

$$\frac{1}{c}\left(\frac{\partial c}{\partial T}\right) \bigg/ \frac{1}{V}\left(\frac{\partial V}{\partial T}\right)_P = -k, \qquad (33)$$

TABLE 4   Parameters Entering Chavez, Sosa, and Tsumura's[a] formula (31) for Temperature Dependence of Sound Speed in Pure Water

k	b_k
0	−19214.88484
1	230.1609318
2	−1.028803876
3	0.002414336487
4	$-2.902395566 \times 10^{-6}$
5	$1.430493449 \times 10^{-9}$
a	0.75
T_c (K)	647.067

^{a}Reference 14.

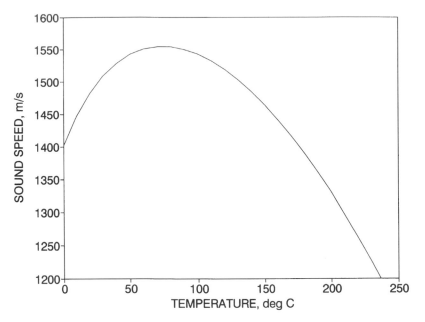

Fig. 3 Speed of sound in saturated water vs. temperature.

where the constant k is nearly independent of temperature and has a nominal value of 3. Although Rao proposed Eq. (33) as a strictly empirical relation, Schaaffs provides a derivation based on the semiempirical concepts leading to Eq. (28). Deviations from Eq. (33) are presumed to indicate association of the liquid. Integration of Eq. (33) yields

$$Vc^{1/k} = \mathcal{R}, \qquad (34)$$

where the constant $\mathcal{R}$ is of theoretical interest because it is related to the repulsive exponent in the Lennard-Jones potential, and it serves as the basis for an additivity relationship in liquid mixtures. The value of Rao's constant k is listed in Table 6 for several liquids.

4 SPEED OF SOUND IN TWO-PHASE FLUIDS

The speed of sound in two-phase fluid mixtures, such as bubbly liquids or foggy gases, is strongly influenced by interfacial heat, mass, and momentum transfer. In the idealized mixture (absence of temperature differentials, phase transitions, and slip between phases), the density ρ and sound speed c are weighted combinations of those of the individual phases[18]:

$$\rho = \rho_1(1 - \phi) + \rho_2\phi, \qquad (35)$$

$$\frac{1}{c^2} = \rho\left(\frac{1 - \phi}{\rho_1 c_1^2} + \frac{\phi}{\rho_2 c_2^2}\right), \qquad (36)$$

where subscripts 1 and 2 refer to the liquid and gas,

TABLE 5   Effect of Solutes on Sound Speed in Sea Water at 30°C

Solute	Molarity, X_i	Sound Speed, $c_i(X_i)$ (m/s)	Increment, $c_i(X_i) - c_0$ (m/s)
NaCl	0.4649	1538.2	28.2
MgSO$_4$	0.0281	1513.4	3.4
MgCl$_2$	0.0263	1512.9	2.9
CaCl$_2$	0.0105	1510.9	0.9
KCl	0.00997	1510.6	0.6
NaCHO$_3$	0.00246	1510.2	0.2
NaBr	0.00083	1510.0	0.0
Total			36.2

TABLE 6   Rao's Constant k for Selected Liquids

Liquid	Temperature Range, K	k
Nitrogen	70–76	2.4
Oxygen	80–90	2.3
Potassium	337–433	0.97
Tin	505–653	3.10
Glycerine	293–313	0.78
CCl$_4$	288–318	2.8
Octane	298–318	3.0
Acetone	293–313	2.6
Benzene	288–318	3.0
Toluol	293–313	3.1

respectively. The quantity ϕ is the *void fraction*, defined as

$$\phi = v_2 \rho, \qquad (37)$$

where v_2 is the volume of the gas per unit mass of the mixture. According to Eq. (36), the sound speed of the mixture is less than that of the individual phases.

If ϕ is small but sufficiently large that $\phi \rho_1 c_1^2 \gg (1 - \phi)\rho_2 c_2^2$, as in a moderately bubbly liquid, then the density is essentially that of the liquid, but the compressibility is dominated by the gas. The sound speed increases strongly with pressure but is still bounded by that of the liquid.

If the gas is a condensing vapor on the gas–liquid phase boundary, as in the presence of fog in the atmosphere, then ϕ can vary due to a phase change induced by acoustic pressures. When the condensation/evaporation entropy is taken into account, then the sound speed can be approximated by

$$c^2 = \frac{P^2 L^2 M_2^2}{C_{p1} R^2 \rho_1^2 T^3}, \qquad (38)$$

where L is the enthalpy of evaporation, M_2 the molecular weight of the vapor, and C_{p1} the specific heat at constant pressure of the liquid.

Wei and Wu[19] modeled sound propagation through atmospheric fog as a relaxation process by including the effects of droplet growth; their low-frequency limit of the sound speed is somewhat more complex than Eq. (38).

REFERENCES

1. F. W. Giacobbe, "Precision Measurement of Acoustic Velocities in Pure Gases and Gas Mixtures," *J. Acoust. Soc. Am.*, Vol. 94, 1993, pp. 1200–1210.

2. P. M. Morse and K. U. Ingard, *Theoretical Acoustics*, McGraw-Hill, New York, 1968, p. 290.

3. K. F. Herzfeld and T. A. Litovitz, *Absorption and Dispersion of Ultrasonic Waves*, Academic, New York and London, 1959.

4. M. Greenspan, "Sound Waves in Gases at Low Pressures," in W. P. Mason, Ed., *Physical Acoustics*, Vol. IIA, Academic, New York and London, 1965.

5. H. J. Bauer, "Theory of Relaxation Phenomena in Gases," in W. P. Mason, Ed., *Physical Acoustics*, Vol. IIA, Academic, New York and London, 1965.

6. M. Greenspan, "Comments on 'Speed of Sound in Standard Air,'" *J. Acoust. Soc. Am.*, Vol. 80, 1987, pp. 370–372.

7. G. S. K. Wong and T. F. W. Embleton, "Variation of Specific Heats and of Specific Heat Ratio in Air with Humidity," *J. Acoust. Soc. Am.*, Vol. 76, 1984, pp. 555–559.

8. J. Hilsenrath et al., *Tables of Thermal Properties of Gases*, National Bureau of Standards Circular 564, U.S. Department of Commerce, Washington, DC, 1955.

9. A. J. Zuckerwar and R. W. Meredith, "Low-Frequency Absorption of Sound in Air," *J. Acoust. Soc. Am.*, Vol. 78, 1985, pp. 946–955.

10. A. Muenster, "Theory of the Liquid State," in A. van Itterbeek, Ed., *Physics of High Pressures and the Condensed Phase*, North-Holland, Amsterdam, 1965.

11. W. Schaaffs, *Molekularakustik*, Springer-Verlag, Berlin, 1963.

12. A. Van Itterbeek and W. Van Dael, "Velocity of Sound in Liquid Oxygen and Liquid Nitrogen as a Function of Temperature and Pressure," *Physica*, Vol. 28, 1962, pp. 861–870.

13. A. J. Zuckerwar and D. S. Mazel, "Sound Speed Measurements in Liquid Oxygen–Liquid Nitrogen Mixtures," *NASA TP 2464*, National Technical Information Service, Springfield, VA, 1985.

14. M. Chavez, V. Sosa, and R. Tsumura, "Speed of Sound in Pure Water," *J. Acoust. Soc. Am.*, Vol. 77, 1985, pp. 420–423.

15. C. C. Chen and F. J. Millero, "Reevaluation of Wilson's Sound-Speed Measurements for Pure Water," *J. Acoust. Am.*, Vol. 60, 1976, pp. 1270–1273.

16. A. Weissler and V. A. Del Grosso, "The Velocity of Sound in Sea Water," *J. Acoust. Soc. Am.*, Vol. 23, 1951, pp. 219–223.

17. M. R. Rao, "A Relation between Velocity of Sound in Liquids and Molecular Volume," *Indian J. Physics*, Vol. 23, 1940, pp. 109–116.

18. V. E. Nakoryakov, B. G. Pokusaev, and I. R. Shreiber, *Wave Propagation in Gas–Liquid Media*, CRC Press, Boca Raton, FL, 1993.

19. R. Wei and J. Wu, "Absorption of Sound in Water Fog," *J. Acoust. Soc. Am.*, Vol. 70, 1981, pp. 1213–1219.

7

STANDING WAVES

U. S. SHIRAHATTI AND MALCOLM J. CROCKER

1 INTRODUCTION

The main focus of this chapter is on standing waves or modes of vibration. Engineers tend to think of vibrations in terms of modes and sound in terms of waves, and quite often it is forgotten that the two are simply different ways of looking at the same physical phenomenon. When considering the interactions between sound waves and the vibration of structures, it is important to have a working knowledge of both physical models.

A mode of vibration on a taut, fixed string can be interpreted as being composed of two waves of equal amplitude and wavelength traveling in opposite directions between the bounded ends. Alternatively, it can be interpreted as a "standing wave," that is, the string oscillates with a spatially varying amplitude within the confines of a specific stationary waveform.

Central to this concept of wave–mode duality is the finite or limited extent of the medium. In fact, it is not possible to have standing waves or modes in a medium of infinite or unlimited extent. A further point to be remembered is that an infinite or unlimited medium can vibrate freely at any frequency. In contrast, a finite- or limited-extent medium can vibrate freely only at specific frequencies known as the natural frequencies.[1]

2 ONE-DIMENSIONAL STANDING WAVES

In Chapter 2, the derivation of the wave equation for a fluid medium is presented [see Eq. (44)]. This form of wave equation can also be used to study standing waves

or modes in strings, bars, and acoustic waveguides. The only thing that is different is the wave speed c, which depends on the nature of the medium. The wave equation may be interpreted as

Acceleration of any particle at any instant
= square of wave speed
× curvature of waveform at that point
and instant.

The expressions for the speeds of the several other important kinds of waves will not be derived here. The results of such derivations may be summarized by stating that the *wave speed c* is always given by some simple fraction under the square-root sign. The numerator of the fraction is always that particular elastic coefficient that defines the elastic property of the medium that is responsible for the wave motion. The denominator of the fraction is always a term denoting the mass, linear density, or total density (i.e., mass per unit length, area, or volume) of the medium transmitting the wave. In general,

$$c = \sqrt{\frac{E}{\rho}}. \tag{1}$$

Table 1 lists expressions for wave speeds of several types of waves.

The speed of almost all waves that exist in mechanical media such as strings, plates, liquids, or solids is usually too great for the eye to be able to follow the waveform as it travels. For small wave speeds, the elastic constant E should be small and the inertia ρ large.

Encyclopedia of Acoustics, edited by Malcolm J. Crocker
ISBN 0-471-80465-7 © 1997 John Wiley & Sons, Inc.

TABLE 1   Values of Wave Speed $c = \sqrt{E/\rho}$

Type of Wave	E Represents	ρ Represents
Transverse wave in a String	Tension T	Linear density
Longitudinal (compressional) waves in a liquid	Bulk modulus K	Density
Longitudinal waves in a rigid rod	Young's modulus E	Linear density
Transverse waves in a rigid rod	Rigidity modulus G	Linear density
Torsional waves in a rigid rod	Moment of torsion J_0	Linear density
Compressional waves in a gas	Atmospheric pressure p_0	Density

2.1   Standing Waves in Strings

Free Vibrations   Since all physical media that carry waves are bounded, that is, limited in extent, the simpler results that follow from this fact should be described, at least. In the case of waves on a real string, which is never very long, the idea of a wave that travels on and on indefinitely is certainly unrealistic.

D'Alembert has shown that the solution of the wave equation consists of the superposition of two terms, the functions $f_1(ct-x)$ and $f_2(ct+x)$. The function $f_1(ct-x)$ denotes a wave moving to the right (or positive x-direction) and the function $f_2(ct+x)$ denotes a wave moving to the left. There are several equivalent expressions for the string displacement $y = f(ct - x)$, such as

$$y = A \sin \frac{2\pi}{\lambda} (ct - x), \qquad (2)$$

$$y = A \sin 2\pi \left(ft - \frac{x}{\lambda} \right), \qquad (3)$$

$$y = A \sin 2\pi(\omega t - kx), \qquad (4)$$

where $k = 2\pi/\lambda = \omega/c$ is the wavenumber, f is the frequency in hertz, and λ is the wavelength in meters. It is also possible to represent y using complex exponential notation as

$$y = Ae^{i(\omega t - kx)}, \qquad i = \sqrt{-1}. \qquad (5)$$

For example, let us consider a string of length l with its ends rigidly clamped. The displacement y on the string at any point is given by

$$y = Ae^{i(\omega t - kx)} + Be^{i(\omega t + kx)}, \qquad (6)$$

where A and B are complex constants to be determined from the boundary conditions at $x = 0$ and $x = l$. These boundary conditions for a clamped string are $y = 0$ at $x = 0$ and $x = l$. The condition $y = 0$ at $x = 0$ yields $A = -B$ from Eq. (5); physically this means that the wave

is reflected at either end with a $180°$ phase change. Hence

$$y = Ae^{i\omega t}(e^{-ikx} - e^{ikx}) \qquad (7)$$

$$= (-2i)Ae^{i\omega t} \sin kx. \qquad (8)$$

The condition $y = 0$ at $x = l$ for all times t leads to $\sin kl = 0$, or $kl = n\pi$ ($n = 1, 2, 3$), or $k = n\pi/l$, which in terms of angular frequency becomes $\omega_n = n\pi c/l$.

These frequencies are called *natural frequencies* or *eigenfrequencies*. When the string is vibrating at a particular natural frequency, it has an associated deflected shape, known as its *mode of vibration*. It may be noticed that a string can theoretically vibrate in an infinite number of *modes*. Returning now to Eq. (5), we see that the *displacement* of the nth mode of vibration is given by

$$y_n = (A_n \cos \omega_n t + B_n \sin \omega_n t) \sin\left(\frac{\omega_n x}{c}\right), \qquad (9)$$

where the *amplitude* of the nth mode is given by

$$\sqrt{A_n^2 + B_n^2} = 2A. \qquad (10)$$

Forced Vibrations   Consider a very long string such that when it is set into oscillation with an oscillator, waves can be seen moving away with a speed c, but no waves are seen returning. At $x = 0$ the force applied on the string by the oscillator is $F_y = F_0 e^{i\omega t}$ and it can be shown that, at $x = 0$, the ratio of the force to the velocity v becomes

$$\frac{F_0 e^{i\omega t}}{v} = \frac{T}{c} = Z_0, \qquad (11)$$

where Z_0 is the characteristic mechanical impedance of the lossless infinite string. Since from Table 1 $c = \sqrt{T/\rho}$, we may also write the input impedance as $Z_0 = \rho c$, which is a real number and is not complex. Every lossless medium in which a wave propagates has a characteristic impedance of this type.

If the string is of finite length l and it is driven at $x = 0$ as before but is supported at $x = l$, the *input impedance is quite different* because of the reflected waves. The outgoing and the reflected waves add at all times to give a motion contained within the wave envelope. The string at any point x moves up and down through its equilibrium position $y = 0$ with a frequency ω. The amplitude of the so-called standing-wave pattern (mode of vibration) is different at various points x.

Energy of a Vibrating String

A vibrating string possesses both kinetic and potential energy. The kinetic energy of an element of length dx and linear density ρ is given by $\frac{1}{2}\rho\,dx(\partial y/\partial t)^2$ and the totak kinetic energy is the integral of this term along the length of the string. Hence

$$E_k = \frac{1}{2}\,\rho \int_0^L \left(\frac{\partial y}{\partial t}\right)^2 dx. \qquad (12)$$

The potential energy is the work done by the tension T in extending an element dx to a new length ds when the string is vibrating. Hence

$$E_p = \int_0^L T(ds - dx)$$

$$= T\int_0^L \left\{\left[1 + \left(\frac{\partial y}{\partial x}\right)^2\right]^2 - 1\right\} dx, \qquad (13)$$

or using Taylor's approximation and neglecting higher order terms

$$E_p = \frac{1}{2}\,T\int_0^L \left(\frac{\partial y}{\partial x}\right)^2 dx. \qquad (14)$$

In a standing wave the total energy is a constant. However, energy is continually being exchanged between kinetic and potential forms. Further, when the displacement is maximum, all the energy is potential, while with zero displacement, the velocity is maximum and all the energy is kinetic. It may be noted also that maximum values of both potential and kinetic energies are identical.

TABLE 2   Standing-Wave Patterns for a Vibrating String[a]

Situation	Boundary Conditions	Natural Frequency Equation	Mode Shape
Unequal, spring loaded; k_1 and k_2 are spring stiffness, T is tension in string	$\left(\dfrac{\partial y}{\partial x}\right)_{x=0} = \dfrac{K_1}{T}\,y(0),$ $\left(\dfrac{\partial y}{\partial x}\right)_{x=l} = \dfrac{K_2}{T}\,y(l),$ $n = 1, 2, 3, \dots$	$\omega_n^2 = \dfrac{\Omega_n^2 C^2}{l^2},$ $\tan\Omega_n = \dfrac{\Omega_n(\alpha_1 - \alpha_2)}{\alpha_1\alpha_2 + \Omega_n^2}$	For $\alpha_1 \neq \infty$: $y_n(x) = \dfrac{\alpha_1}{\Omega_n}\sin\left(\dfrac{\Omega_n x}{l}\right) + \cos\left(\dfrac{\Omega_n x}{l}\right),$
		for $\alpha_1 \neq 0$: $\alpha_1 = \dfrac{lk_1}{\rho c^2}, \alpha_2 = \dfrac{lk_2}{\rho c^2}$	$y_n(x) = \sin\left(\dfrac{\Omega_n x}{l}\right) + \dfrac{\Omega_n}{\alpha_1}\cos\left(\dfrac{\Omega_n x}{l}\right)$
Free–free	$k_1 = k_2 = 0$	$\omega_n = \left(\dfrac{n\pi c}{l}\right)$	$y_n(x) = \cos\left(\dfrac{n\pi x}{l}\right)$
Clamped–clamped	$k_1 = k_2 = \infty$	$\omega_n = \left(\dfrac{n\pi c}{l}\right)$	$y_n(x) = \sin\left(\dfrac{n\pi x}{l}\right)$
Equal, spring loaded	$k_1 = k_2$	$\omega_n = \left(\dfrac{n\pi c}{l}\right),$ or $y_n(x) = \sin\left(\dfrac{n\pi x}{l}\right) + \dfrac{\alpha_1}{n\pi}\cos\left(\dfrac{n\pi x}{l}\right),$ $n = 1, 2, 3, \dots$	$y_n(x) = \dfrac{\alpha_1}{n\pi}\sin\left(\dfrac{n\pi x}{l}\right) + \cos\left(\dfrac{n\pi x}{l}\right)$

[a]From Ref. 2 with permission.

Natural Frequencies and Mode Shapes of a String for Different Boundary Conditions The natural frequencies and mode shapes or standing-wave patterns for a vibrating string with different boundary conditions are listed (see Table 2).[2]

2.2 Standing Waves in a Duct with Various Boundary Conditions

Free Vibrations Standing waves are readily formed in a tube of length l with various boundary conditions at the two ends. Table 3 lists some typical boundary conditions along with the resonance frequencies. Table 3 also gives expressions for sound pressure, particle velocity, and intensity inside the duct.

Forced Vibrations Consider plane sound waves traveling along a hard-walled tube such that only a single frequency is present. These plane waves are excited by the oscillations of a piston located at $x = 0$. Now, at $x = l$, it is possible to impose three types of terminations.

- *A Perfect Sound Absorber.* There is no reflection of sound at the termination of the tube; therefore no standing waves are formed.
- *A Perfectly Rigid Termination.* The waves are perfectly reflected at the termination. The reflected waves traveling back to the left have the same amplitude as incident waves traveling to the right. These two traveling waves combine to give rise to perfect standing waves.
- *A Partially Absorptive Termination.* There is a partial reflection at the termination so that a weaker wave returns from the right to the left. The two opposite traveling waves add together to give rise to a weaker form of standing waves.

In Table 4, expressions are given for sound pressure, particle velocity, intensity, and natural frequencies for various boundary conditions.

From the expressions in Table 4 note the following:

1. If $Z = \rho c$ (*anechoic termination*), then

TABLE 3   Standing Waves in a Duct: Free Vibrations

Case	Boundary Condition	Field Variables
Closed at both ends $x=0$ $\qquad$ $x=l$	$\zeta(0) = 0,$ $\zeta(l) = 0$	$\zeta = B \sin kx \sin \omega t$ $p = -\rho c \omega B \cos kx \sin \omega t$ $u = \omega B \sin kx \cos \omega t$ $I = -\frac{1}{4}\rho c \omega^2 B^2 \sin(2kx)\sin(2\omega t)$ $\omega_n = \left(\dfrac{n\pi c}{l}\right)$
Open at both ends $x=0$ $\qquad$ $x=l$	$\left(\dfrac{\partial \zeta}{\partial x}\right)_{x=0} = 0,$ $\left(\dfrac{\partial \zeta}{\partial x}\right)_{x=l} = 0$	$\zeta = A \sin kx \sin \omega t$ $p = \rho c \omega A \sin kx \sin \omega t$ $u = \omega A \cos kx \cos \omega t$ $I = \frac{1}{4}\rho c \omega^2 A^2 \sin(2kx)\sin(2\omega t)$ $\omega_n = \left(\dfrac{n\pi c}{l_a}\right),$ where $l_a = 0.6a + l + 0.6a$
Open at $x = l$ and Closed at $x = 0$ $x=0$ $\qquad$ $x=l$	$\zeta(0) = 0,$ $\left(\dfrac{\partial \zeta}{\partial x}\right)_{x=l} = 0$	$\zeta = B \sin kx \sin \omega t$ $p = \rho c \omega B \cos kx \sin \omega t$ $u = \omega B \sin kx \cos \omega t$ $I = -\frac{1}{4}\rho c \omega^2 B^2 \sin(2kx)\sin(2\omega t)$ $\omega_n = \left(\dfrac{n\pi c}{l_a}\right),$ where $l_a = 0.6a + l + 0.6a$

Note: ζ = particle displacement, p = sound pressure, u = sound velocity, I = instantaneous intensity, a = radius of duct, l = length of duct; A and B are constants.

TABLE 4   Standing Waves in a Duct: Forced Vibrations

Case	Boundary Conditions	Field Variables
1. *Piston and closed end* $x=0$ $x=l$	$\zeta(0) = r \sin \omega t$ $\zeta(l) = 0$	$\zeta = r \dfrac{\sin[k(l-x)]}{\sin kl} \cos \omega t$ $u = \omega r \dfrac{\sin[k(l-x)]}{\sin kl} \cos \omega t$ $p - \rho c \omega r \dfrac{\cos[k(l-x)]}{\sin kl} \sin \omega t$ $I = \frac{1}{4} \rho c (\omega r)^2 \dfrac{\sin[2k(l-x)]}{\sin^2 kl} \sin(2\omega t)$ $\langle I \rangle = 0$, where $\langle \cdot \rangle$ signifies time average
2. *Piston and open end* $x=0$ $x=l$	$\zeta(0) = r \sin \omega t$ $\left(\dfrac{\partial \zeta}{\partial x} \right)_{x=l} = 0$	$\zeta = r \dfrac{\cos[k(l-x)]}{\cos kl} \sin \omega t$ $u = \omega r \dfrac{\cos[k(l-x)]}{\cos kl} \cos \omega t$ $p = -\rho c \omega r \dfrac{\sin[k(l-x)]}{\cos kl} \sin \omega t$ $I = -\frac{1}{4} \rho c (\omega r)^2 \dfrac{\sin[2k(l-x)]}{\cos^2 kl} \sin(2\omega t)$ $\langle I \rangle = 0$, where $\langle \cdot \rangle$ signifies time average
3. *Piston and Termination* Impedance $Z = R + jX$ $x=0$ $x=l$	$u(0) = \omega r e^{j\omega t}$ $\dfrac{p(l)}{u(l)} = Z$	$p = \rho c \omega r \left(\dfrac{\cos[k(l-x)] + j(\rho c/Z)\sin[k(l-x)]}{(\rho c/Z)\cos kl + j \sin kl} \right) e^{j\omega t}$ $u = \omega r \left(\dfrac{(\rho c/Z)\cos[k(l-x)] + j \sin[k(l-x)]}{(\rho c/Z)\cos kl + j \sin kl} \right) e^{j\omega t}$ $I = \rho c \omega r^2 \left(\dfrac{(\rho c/Z)\cos^2[k(l-x)] + j(\rho c/Z + 1)\cos[k(l-x)]\sin[k(l-x)]}{(\rho c/Z)^2 \cos^2(kl) - \sin^2 kl + 2j(\rho c/Z)\cos kl \sin kl} \right) e^{2j\omega t}$ $\langle I \rangle = \dfrac{(\frac{1}{2})(\rho c \omega r)^2 (2R/ZZ^*)}{(\rho c/ZZ^*)\rho c \cos^2 kl + \sin^2 kl - (\rho c/ZZ^*)\cos kl \sin kl (2X)}$

Note: Asterisk denotes complex conjugate

TABLE 5   Standing Waves in Membranes of Finite Area

Case	Boundary Conditions	Mode Shape	Natural Frequency
Clamped rectangular membrane	$W(x,0,t) = 0,$ $W(x,d,t) = 0,$ $W(0,y,t) = 0,$ $W(b,y,t) = 0,$	$W(x,y) = \sin\left(\dfrac{m\pi x}{b}\right)\sin\left(\dfrac{n\pi y}{d}\right),$ $m, n = 1, 2, 3, \ldots$	$\omega_{mn}^2 = (m\pi)^2 + \left(\dfrac{n\pi b}{d}\right)^2$ $m, n = 1, 2, 3, \ldots$
Clamped solid circular membrane	$W(b, \theta) = 0$	$W_{mn,j}(r,\theta) = J_m\left(\dfrac{\omega_{mn} r}{b}\right)\psi_j(m\theta),$ $j, m = 0, 1, 2, \ldots, n = 1, 2, \ldots,$ $\psi_j(m\theta) = \sin(m\theta + j\pi/2),$ $m = 0, 1, 2, \ldots, j = 0, 1$	$J_m(\omega_{mn}) = 0,$ $m = 0, 1, 2, \ldots,$ $n = 1, 2, 3, \ldots$

$$ZZ^* = (\rho c)^2, \qquad R = \rho c, \qquad (15)$$

and the time-averaged intensity at any point in the duct $\langle I \rangle = \frac{1}{2}\rho c(\omega r)^2$, which indicates that there are no standing waves.

2. If $Z \to \infty$ (*rigid termination*), then $\langle I \rangle = 0$, which indicates perfect standing waves.

3. If

$$Z = \frac{c}{\pi a^2}\left(\frac{k^2 a^2}{4} + j0.6ka\right),$$

TABLE 6   Roots of $J_m(\omega_{mn}) = 0$

m	$n = 1$	$n = 2$	$n = 3$	$n = 4$	$n = 5$
0	2.408	5.5201	8.6537	11.7915	14.9309
1	3.8317	7.0156	10.1735	13.3237	16.4706
2	5.1356	8.4172	11.6198	14.7960	17.9598
3	6.3802	9.7610	13.0152	16.2235	19.4094
4	7.5883	11.0647	14.3752	17.6610	20.8269
5	8.7714	12.3386	15.7002	18.9801	22.2178
6	9.9361	13.5893	17.0038	20.3208	23.5861
7	11.8064	14.8213	18.2876	21.6416	24.9349

TABLE 7   **Acoustical Modes and Natural Frequencies**

Description	Figure	Natural Frequency f_{ijk} (Hz)	Mode Shape Φ_{ijk}	
Slender tube, both ends closed	$D \ll L$; $\frac{ic}{2L}$ $D \ll \lambda$, where $\lambda = c/f$	$\cos\dfrac{i\pi x}{L}$ $i = 0,1,2,\dots$		
Slender tube, one end closed, one end open	$\dfrac{ic}{4L}$ $D \ll \lambda$, where $\lambda = c/f$	$\cos\dfrac{i\pi x}{2L}$ $i = 1,3,5,\dots$		
Slender tube, both ends open	$\dfrac{ic}{2L}$ $D \ll \lambda$, where $\lambda = c/f$	$\sin\dfrac{i\pi x}{L}$ $i = 1,2,3,\dots$		
Closed rectangular volume	$\dfrac{c}{2}\left(\dfrac{i^2}{L_z^2} + \dfrac{j^2}{L_y^2} + \dfrac{k^2}{L_z^2}\right)^{1/2}$	$\cos\dfrac{i\pi x}{L_z}\cos\dfrac{j\pi y}{L_y}\cos\dfrac{k\pi x}{L_z}$ $\begin{aligned} i&=0,1,2,\dots\\ j&=0,1,2,\dots\\ k&=0,1,2,\dots\end{aligned}$		
Closed cylindrical volume	$\dfrac{c}{2\pi}\left(\dfrac{\lambda_{jk}^2}{R^2} + \dfrac{i^2\pi^2}{L^2}\right)^{1/2}$ λ_{jk} from Table (a) below	$J_j\left(\lambda_{jk}\dfrac{r}{R}\right)\cos\dfrac{i\pi x}{L}\begin{cases}\sin j\theta_i & i = 0,1,2,\dots\\ or; & j = 0,1,2,\dots\\ \cos j\theta & k = 0,1,2,\dots\end{cases}$ $J_j = j$th-order Bessel function		
Closed spherical volume	$\dfrac{c\lambda_i}{2\pi R}$ λ_i from Table (b) below	Modes symmetric about center $\dfrac{R}{\lambda_i r}\sin\dfrac{\lambda_i r}{R}$ $i = 0,1,2,\dots$		
Arbitrary closed volume	$\dfrac{c}{2L}$	L—Maximum linear dimension fundamental natural frequency (approximate): Finite element analysis		

	Table (a)							Table (b)				
λ_{jk}			j					i	0	1	2	3	4
k	0	1	2	3	4	5	6	λ_i	0	4.4934	7.7253	10.9041	14.0662
0	0.	1.8412	3.0542	4.2012	5.3176	6.4156	7.5013	$\lambda_i = \pi(i + 1/2)$ for $i \geq 4$ $\tan\lambda_i = \lambda_i$					
1	3.8317	5.3314	6.7061	8.0152	9.2824	10.5199	11.7349						
2	7.0156	8.5363	9.9695	11.3459	12.6819	13.9872	15.2682						
3	10.173	11.7060	13.1704	14.5859	15.9641	17.3128	18.6374						

$$\lambda_{j=0,k} = \pi(k + 1/4) \quad \text{for} \quad k \geq 3 \quad (J_j'(\lambda_{jk}) = 0)$$

Reprinted, with permission, from Ref. 3: R. D. Blevins, "Fluid Systems," in *Formulas for Natural Frequency and Mode Shape*, Van Nostrand Reinhold, New York, 1984 (and Ref. 4).

where $ka < 0.5$ is the radiation impedance for a pipe with an infinite flange radiating out into the atmosphere, the results from case 3 in Table 4 must be used and not those from case 2. In this case a weaker form of standing waves exists.

3 TWO-DIMENSIONAL STANDING WAVES

To illustrate two-dimensional standing waves, consider membranes. The two-dimensional wave equation is represented by[2]

$$\frac{\partial^2 W}{\partial t^2} = c^2 \nabla^2 W, \tag{16}$$

where ∇^2 is the two-dimensional Laplace operator and W is the membrane displacement. In Table 5, the boundary conditions, mode shapes, or standing-wave patterns along with the natural frequencies for rectangular and circular membranes are given.

TABLE 8 Summary of Equation for Lateral Vibration of Single-Span Uniform

End Conditions	Equations[a]
	(1) $u(0,t) = u'(0,t) = u(l,t) = u'(l,t) = 0$ (2) $\cos \beta l \cosh \beta l = 1$ (3) $\phi(x) = A(\cos \beta x - \cosh \beta x) + (\sin \beta x - \sinh \beta x)$ $A = -\dfrac{\sin \beta l - \sinh \beta l}{\cos \beta l - \cosh \beta l} = \dfrac{\cos \beta l - \cosh \beta l}{\sin \beta l + \sinh \beta l}$
	(1) $u''(0,t) = u'''(0,t) = u''(l,t) = u'''(l,t) = 0$ (2) $\cos \beta l \cosh \beta l = 1^*$ (3) $\phi(x) = A(\cos \beta x + \cosh \beta x) + (\sin \beta x + \sinh \beta x)$ $A = -\dfrac{\sin \beta l - \sinh \beta l}{\cos \beta l - \cosh \beta l} = \dfrac{\cos \beta l - \cosh \beta l}{\sin \beta l + \sinh \beta l}$
	(1) $u(0,t) = u'(0,t) = u(l,t) = u''(l,t) = 0$ (2) $\tan \beta l = \tanh \beta l$ (3) $\phi(x) = A(\cos \beta x - \cosh \beta x) + (\sin \beta x \quad \sinh \beta x)$ $A = -\dfrac{\sin \beta l - \sinh \beta l}{\cos \beta l - \cosh \beta l} = -\dfrac{\sin \beta l + \sinh \beta l}{\cos \beta l + \cosh \beta l}$
	(1) $u(0,t) = u''(0,t) = u''(l,t) = u'''(l,t) = 0$ (2) $\tan \beta l = \tanh \beta l^*$ (3) $\phi(x) = A \sin \beta x + \sinh \beta x$ $A = \dfrac{\sinh \beta l}{\sin \beta l} = \dfrac{\cosh \beta l}{\cos \beta l}$
	(1) $u(0,t) = u''(0,t) = u(l,t) = u''(l,t) = 0$ (2) $\sin \beta l = 0$ (3) $\phi(x) = A \sin \beta x$
	(1) $u(0,t) = u'(0,t) = u''(l,t) = u'''(l,t) = 0$ (2) $\cos \beta l \cosh \beta l = -1$ (3) $\phi(x) - A(\cos \beta x - \cosh \beta x) + (\sin \beta x - \sinh \beta x)$ $A = -\dfrac{\sin \beta l + \sinh \beta l}{\cos \beta l + \cosh \beta l} = \dfrac{\cos \beta l + \cosh \beta l}{\sin \beta l - \sinh \beta l}$

Lateral deflection $\sum \phi(x)q(t)$ m = length mass ω = frequency (rad/s)

Beam equation: $\dfrac{\partial^2 u}{\partial t^2} + a^2 \dfrac{\partial^4 u}{\partial x^4} = 0$ $\dfrac{d^4 \phi}{dx^4} - \beta^4 \phi = 0$ $\dfrac{d^2 q}{dt^2} + \omega^2 q = 0$

$$\beta^4 = \frac{\omega^2}{a^2} = \frac{m\omega^2}{EI}$$

[a](1) End condition, (2) frequency equation; (3) eigenfunction, * semidefinite. Reprinted, with permission, from Ref. 5: F. S. Tse, I. E. Morse, and R. T. Hinkle. *Mechanical Vibrations.* Allyn and Bacon. Boston. 1978.

4 THREE-DIMENSIONAL STANDING WAVES

The analysis of standing waves in a tube can be readily extended to a three-dimensional cavity in an enclosure. For simple geometries and boundary conditions the acoustic modes can be expressed analytically as shown in Table 6.[3,4] If the geometry is complicated, it becomes essential to use a numerical method such as the finite element method, which is discussed in Chapter 14.

4.1 Standing Waves in Beams and Plates

Beams In Chapter 64, Section A.2.1, the equation of motion for bending waves in beams (the Bernoulli–Euler equation) has been given. We set here for convenience

TABLE 9   Frequency Coefficients and Different Mode Shapes for Plates

Boundary Conditions	Deflection Function or Mode Shape	N	K
	$\left(\cos\dfrac{2\pi x}{a} - 1\right)\left(\cos\dfrac{2\pi y}{b} - 1\right)$	2.25	$12 + 8\left(\dfrac{a}{b}\right)^2 + 12\left(\dfrac{a}{b}\right)^4$
	$\left(\cos\dfrac{3\pi x}{2a} - \cos\dfrac{\pi x}{2a}\right)\left(\cos\dfrac{2\pi y}{b} - 1\right)$	1.50	$3.85 + 5\left(\dfrac{a}{b}\right)^2 + 8\left(\dfrac{a}{b}\right)^4$
	$\left(1 - \cos\dfrac{\pi x}{2a}\right)\left(\cos\dfrac{2\pi y}{b} - 1\right)$	0.340	$0.0468 + 0.340\left(\dfrac{a}{b}\right)^2 + 1.814\left(\dfrac{a}{b}\right)^4$
	$\left(\cos\dfrac{2\pi x}{a} - 1\right)\sin\dfrac{\pi y}{b}$	0.75	$4 + 2\left(\dfrac{a}{b}\right)^2 + 0.75\left(\dfrac{a}{b}\right)^4$
	$\left(\cos\dfrac{2\pi x}{a} - 1\right)\dfrac{y}{b}$	0.50	$2.67 + 0.304\left(\dfrac{a}{b}\right)^2$
	$\cos\dfrac{2\pi x}{a} - 1$	1.50	8
	$\left(\cos\dfrac{3\pi x}{2a} - \cos\dfrac{\pi x}{2a}\right)\left(\cos\dfrac{3\pi y}{2b} - \cos\dfrac{\pi y}{2b}\right)$	1.00	$2.56 + 3.12\left(\dfrac{a}{b}\right)^2 + 2.56\left(\dfrac{a}{b}\right)^4$
	$\left(\cos\dfrac{3\pi x}{2a} - \cos\dfrac{\pi x}{2a}\right)\left(1 - \cos\dfrac{\pi y}{2b}\right)$	0.227	$0.581 + 0.213\left(\dfrac{a}{b}\right)^2 + 0.031\left(\dfrac{a}{b}\right)^4$
	$\left(1 - \cos\dfrac{\pi x}{2a}\right)\left(1 - \cos\dfrac{\pi y}{2b}\right)$	0.0514	$0.0071 + 0.024\left(\dfrac{a}{b}\right)^2 + 0.0071\left(\dfrac{a}{b}\right)^4$
	$\left(\cos\dfrac{3\pi x}{2a} - \cos\dfrac{\pi x}{2a}\right)\sin\dfrac{\pi y}{b}$	0.50	$1.28 + 1.25\left(\dfrac{a}{b}\right)^2 + 0.50\left(\dfrac{a}{b}\right)^4$
	$\left(\cos\dfrac{3\pi x}{2a} - \cos\dfrac{\pi x}{2a}\right)\dfrac{y}{b}$	0.333	$0.853 + 0.190\left(\dfrac{a}{b}\right)^2$
	$\cos\dfrac{3\pi x}{2a} - \cos\dfrac{\pi x}{2a}$	1.00	2.56
	$\left(1 - \cos\dfrac{\pi x}{2a}\right)\dfrac{\pi^2}{b^2}\sin\dfrac{\pi y}{b}$	0.1134	$0.0156 + 0.0852\left(\dfrac{a}{b}\right)^2 + 0.1134\left(\dfrac{a}{b}\right)^4$
	$\left(1 - \cos\dfrac{\pi x}{2a}\right)\dfrac{y}{b}$	0.0756	$0.0104 + 0.0190\left(\dfrac{a}{b}\right)^2$
	$1 - \cos\dfrac{\pi x}{2a}$	0.2268	0.0313
	$\sin\dfrac{\pi x}{a}\sin\dfrac{\pi y}{b}$	0.25	$0.25 + 0.50\left(\dfrac{a}{b}\right)^2 + 0.25\left(\dfrac{a}{b}\right)^4$
	$\left(\sin\dfrac{\pi x}{a}\right)\dfrac{y}{b}$	0.1667	$0.1667 + 0.0760\left(\dfrac{a}{b}\right)^2$
	$\sin\dfrac{\pi x}{a}$	0.50	0.50

Reprinted, with permission, from Ref. 6: A. Leissa, *Vibration of Plates*, Acoustical Society of America, New York, 1993.

$\kappa^2 S = I$, the moment of inertia of the beam, and $\rho = m/S$, is the mass per unit length of the beam.

In Table 7 the boundary conditions and the natural frequency equation along with the mode shape or eigenfunction are presented for several beam end conditions.

Plates The governing equation for bending waves in orthrotropic plates using the Euler–Bernoulli theory is given in Chapter 64, Section A.5. In the case of an isotropic plate

$$D_1 = D_3 = 2(D_\mu + 2D_G) = \frac{Eh^3}{12(1 - \mu^2)} = D, \qquad (17)$$

and the governing differential equation becomes

$$D \nabla^4 \zeta_2 + \rho h \frac{\partial^2 \zeta_2}{\partial t^2} = 0. \qquad (18)$$

Note that E is Young's modulus, h is the thickness of the plate, μ is Poisson's ratio, ζ_2 is the displacement perpendicular to the plane of the plate, and ρ is the mass per unit volume of the plate. One can rewrite this equation in terms of x- and y-coordinates and $\tilde{\rho}$, the density per unit area, as

$$D \nabla^4 w + \tilde{\rho} \frac{\partial^2 w}{\partial t^2} = 0, \qquad (19)$$

where we have set $\zeta_2 = w$.

Altogether there are 21 combinations of simple boundary conditions [that is, clamped (C), simply supported (SS), or free (F)] for rectangular plates. In Table 8 the boundary conditions, mode shape, and factors for computing the natural frequency using $\omega^2 = (\pi^4 D/a^4\rho)(k/N)$ are presented. The results presented here were computed using Rayleigh's method and hence yield only the upper bound on the fundamental frequencies.[6]

REFERENCES

1. M. P. Norton, *Fundamentals of Noise and Vibration Analysis for Engineers*, Cambridge University Press, Cambridge, 1989.

2. E. B. Magrab, *Vibrations of Elastic Structural Members*, Sijthoff & Nordhoff, MD, 1979.

3. R. D. Blevins, "Fluid Systems," in *Formulas for Natural Frequency and Mode Shape*, Van Nostrand Reinhold, New York, 1984.

4. L. Beranek and I. Ver, *Noise and Vibration Control Engineering*, Wiley, New York 1992.

5. F. S. Tse, I. E. Morse, and R. T. Hinkle, *Mechanical Vibrations*, Allyn and Bacon, Boston, 1978.

6. A. Leissa, *Vibration of Plates*, Acoustical Society of America, New York, 1993.

8

WAVEGUIDES

P. O. A. L. DAVIES

1 INTRODUCTION

This chapter is concerned with sound propagation in pipes and ducts, where, in contrast to free space, it is always strongly influenced or controlled by the presence of the confining boundaries. Pipe and duct networks feature widely in the engineering infrastructure of the process, power generation, automotive, and transport industries, of domestic services, and so on. In many such practical applications, the frequencies of interest remain sufficiently low, so that the wavelength remains a fraction of the transverse dimensions of the pipe or waveguide. Then the propagating waves remain plane, or one dimensional. When such is not the case, wave propagation may occur in one or more higher order modes. Both possibilities are addressed in this chapter.

This account of methods for the description and calculation of sound propagation in waveguides, or ducts, begins with an outline of the classical approach to establishing the guiding physical principles and analytical techniques. This is essentially a search for solutions of the appropriate form of the wave equation that satisfy the boundary conditions imposed by the duct walls.

2 CLASSICAL THEORY AND ASSUMPTIONS

The usual assumptions of classical theory are as follows:

1. The acoustic medium is a frictionless, homogeneous (ideal) fluid.
2. The processes associated with the wave motion are isentropic.

3. Fluctuating pressure amplitudes are sufficiently small that the linearizing acoustic assumptions remain valid.
4. The wave propagation remains wholly axial and directed along x.
5. The duct walls are rigid (acoustically hard), continuous, and of infinite axial extent.
6. The duct is rectangular or cylindrical with constant dimensions.

The classical account of waveguide acoustic behavior continues to the end of Section 5, with the application to lined ducts in Section 6.

The account continues with a sequence of selected examples, since in practical application due consideration must also be given to one or more of the particular features of the problem, such as the following:

1. Uniform sections of duct are of finite length L.
2. Boundary conditions will then include wave reflection and transmission at terminations and other discontinuities.
3. There are significant effects arising from the presence of a mean flow.
4. Thermal and viscous effects may introduce significant damping.
5. The temperature and flow conditions may vary axially or transversely within the duct.
6. Relevant geometric and other features such as axial taper or irregularities exist in the wall shape or properties.

Excluded here, however, is any discussion of the sources of excitation (see, e.g., Chapter 9). These selected examples are followed in Section 11 by a discussion of para-

Encyclopedia of Acoustics, edited by Malcolm J. Crocker
ISBN 0-471-80465-7 © 1997 John Wiley & Sons, Inc.

metric models of waveguide performance, and then a short summary in Section 12 completes the chapter.

2.1 The Wave Equation

The generation of propagating sound waves in a duct containing fluid by a source in the duct is subject to certain conditions. First the fluctuating acoustic pressure, density, and velocities must satisfy an appropriate wave equation, which itself must satisfy the conservation of momentum, mass, and energy in the bulk of the fluid. The second condition is associated with the physical boundary conditions at the duct walls, where the fluctuating pressure and the components of the displacement of the fluid near the duct wall must equal those on the duct wall. It is customary to apply the continuity of velocity rather than displacement, since for simple harmonic motion the two conditions are identical except when there is relative motion between the fluid close to the wall and the wall. The third condition is the boundary at the source, where the pressure and displacement fields on the source side and on the fluid side must be continuous. Waves reflecting from the walls interfere with each other, giving rise to interference patterns over the cross section of the duct. An infinite number of such modes is possible, but at a given frequency only a finite number can propagate down the duct.

The sound field in the waveguide must satisfy the wave equation

$$\frac{D^2 p}{Dt^2} = c^2 \, \nabla^2 p, \tag{1}$$

where $p = p(x_i, t)$ is the fluctuating acoustic pressure at the coordinate position x_i, c is the adiabatic speed of sound, and the symbols D/Dt denote the material derivative. In classical acoustics it is normal to assume that there is no mean flow, so that Eq. (1) becomes

$$\frac{\partial^2 p}{\partial t^2} = c^2 \, \nabla^2 p, \tag{2}$$

when assumptions 1–6 above are included. With the existence of a mean flow having a uniform time-averaged axial velocity u_0 so that $Dp/Dt = \partial p/\partial t + u_0 \, \partial p/\partial x$, while $D^2 p/Dt^2 = (\partial/\partial t + u_0 \, \partial/\partial x) \, Dp/Dt$, the term on the left side of Eq. (1) becomes

$$\frac{D^2 p}{Dt^2} = \frac{\partial^2 p}{\partial t^2} + 2u_0 \frac{\partial^2 p}{\partial x \, \partial t} + u_0^2 \frac{\partial^2 p}{\partial x^2}, \tag{3}$$

which is relevant to situations of practical interest con-

sidered later. When significant additional mean velocity gradients are present, as may be the case with bends or area and other discontinuities, it will be necessary to include the associated factors in the expansion of $D^2 p/Dt^2$; see, for example Section 16.5 of Ref. 1 and Ref. 2. Finally, it is worthwhile noting that a wave equation expressing the distribution in space and time of the scalar fluctuating velocity potential φ may be more readily derived and more easily and conveniently solved than the corresponding form with the acoustic pressure p. See, for example, Section 9.

2.2 Application of the Wave Equation

To apply the wave equation to a specific problem, it is appropriate to select the coordinate system that conveniently matches the wall boundaries when expressing the Laplacian $\nabla^2 p$ occurring on the right side of Eq. (1) or (2). Thus with rectangular ducts the rectangular coordinates x, y, z are appropriate, so that $\nabla^2 = \partial^2/\partial x^2 + \partial^2/\partial y^2 + \partial^2/\partial z^2$. Similarly, with straight cylindrical pipes the selection of cylindrical polar coordinates x, r, θ gives

$$\nabla^2 = \frac{\partial^2}{\partial x^2} + \frac{1}{r} \frac{\partial}{\partial r} \left\{ r \frac{\partial}{\partial r} \right\} + \frac{1}{r^2} \frac{\partial^2}{\partial \theta^2}. \tag{4}$$

With conical expansions and contractions spherical polar coordinates are appropriate, and so on. Similar considerations apply to the selection of the appropriate finite elements when these are used to obtain numerical solutions, for example as explained in Chapter 16 of Ref. 1 and Chapter 7 of Ref. 3, both of which have extensive sets of relevant references, such as the sequence by Astley and Eversman. An alternative numerical scheme may be based on boundary elements.[4]

A review of appropriate mathematical methods of solution, with listed references illustrating their application, can be found in Section 5 of Ref. 5. The separation-of-variables method was chosen for what follows here, since insight is thus maintained into the distribution of the fluctuating pressure and velocity associated with the modes that can exist in any specific case.

3 MODES OF A RECTANGULAR WAVEGUIDE

Consider a duct extending to infinity on the x-axis but bounded at $y = 0$ and l_y and $z = 0$ and l_z by acoustically hard (rigid) duct walls. Note that it is convenient here to have the axis run along one corner. It is assumed that the medium inside the duct is homogeneous and has no time-averaged motion so that the sound field must satisfy Eq. (2). Since this equation is linear in $p(x, y, z, t)$

with coefficients that are functions only of the selected spatial coordinates x, y, z, no generality is lost if it is Fourier transformed on the time t. Then let p be represented by the product of four functions each depending on one variable only, so that

$$p(x, y, z, t) = X(x)Y(y)Z(z)\exp(i\omega t), \qquad (5)$$

where ω is the radian frequency of excitation. From Eqs. (2) and (5) one obtains

$$\frac{1}{X}\frac{d^2X}{dx^2} + \frac{1}{Y}\frac{d^2Y}{dy^2} + \frac{1}{Z}\frac{d^2Z}{dz^2} + k^2 = 0, \qquad (6)$$

where the first factor is a function of x only, the second only of y, and the third only of z, and $k^2 = (\omega/c)^2$. One can investigate each spatial dependency in turn, with the corresponding constants denoted by $-k_x^2$, $-k_y^2$, and $-k_z^2$. One then has $(1/X)(d^2X/dx^2) = -k_x^2$, and similarly, for the other two space functions, and so for Eq. (6) to be satisfied, one must have

$$k^2 = k_x^2 + k_y^2 + k_z^2. \qquad (7a)$$

Corresponding to this, one can define a vector wavenumber $\mathbf{k}$ with the x, y, z components k_x, k_y, k_z.

As mentioned, with $X(x)$ now to be determined as the solution of the ordinary differential equation $d^2X/dx^2 + k_x^2X = 0$, one can take

$$X(x) = A^+ \exp(-ik_x x) + A^- \exp(ik_x x), \qquad (7b)$$

where A^+ and A^- correspond, respectively, to the complex amplitudes of the component waves traveling along the waveguide in the direction of positive and negative x, respectively.

In the same way, one finds that the spatial dependency in the y- and z-directions is described respectively by

$$Y(y) = C_1 \cos(k_y y) + C_2 \sin(k_y y), \qquad (7c)$$

$$Z(y) = D_1 \cos(k_z z) + D_2 \sin(k_z z). \qquad (7d)$$

The forms of solution in Eq. (7b) and in Eqs. (7c) and (7d) are equivalent, but the latter form is more convenient when acoustically hard boundary conditions at the wall are applied.

3.1 Boundary Conditions

The spatial distribution in the y- and z-directions must also satisfy the boundary conditions imposed by the walls. Thus with hard walls, the particle displacement

and hence, for simple harmonic fluctuations, the fluctuating particle velocity normal to them must be zero. This can be related to the fluctuating pressure by satisfying dynamic equilibrium (Euler's equation) at the wall, which is expressed for the y-direction and associated particle velocity v by $\rho\, \partial v/\partial t = -\partial p/\partial y$, giving $v = (-1/i\omega\rho)(\partial p/\partial y)$, with a similar expression for the velocity w in the z-direction. The hard-wall boundary conditions are satisfied for $\partial p/\partial y = 0$ when $y = 0$, l_y and also for $\partial p/\partial z = 0$ when $z = 0$, l_z. Substitution of these conditions into Eqs. (7c) and (7d) leads to $C_2 = D_2 = 0$, $k_y = m\pi/l_y$, $k_z = n\pi/l_z$, m, $n = 0, 1, 2, \ldots$. The integers m and n indicate the number of fluctuating pressure amplitude minima (nodes) in the spatial standing-wave distribution across the duct in the y- and z-directions, respectively. Thus $m = n = 0$ corresponds to plane-wave propagation with the fluctuating pressure in phase with the velocity and at constant phase over the whole cross section.

With the inclusion of the wall boundary conditions and substitution of all three spatial dependencies defined by Eqs. (7b)–(7d) into Eq. (5), acoustic propagation in a hard-walled rectangular waveguide in the direction of positive x is expressed by

$$p(x, y, z, t) = A_{mn}^+ \cos\left(\frac{m\pi y}{l_y}\right) \cos\left(\frac{n\pi z}{l_z}\right)$$
$$\cdot \exp[i(\omega t - k_{x,mn}x)], \qquad (8)$$

with $k_{x,mn}^2 = k^2 - (m\pi/l_y)^2 - (n\pi/l_z)^2$ and A_{mn}^+ the amplitude distribution in the (m, n) duct mode. Wave propagation in the opposite direction (negative x) is described by replacing the factors $A^+ \exp[i(\omega t - k_x x)]$ by $A^- \exp[i(\omega t + k_x x)]$ in Eq. (8). Both components normally occur together in ducts of finite length as sets of interfering axially propagating waves.

4 MODES OF A CYLINDRICAL WAVEGUIDE

Consider a hard-walled cylindrical duct of radius a with its axis aligned along x. Substituting the appropriate expression [Eq. (4)] for the Laplacian $\nabla^2 p$ into Eq. (3) leads to the wave equation

$$\frac{\partial^2 p}{\partial x^2} + \frac{1}{r}\frac{\partial p}{\partial r} + \frac{\partial^2 p}{\partial r^2} + \frac{1}{r^2}\frac{\partial^2 p}{\partial \theta^2} - \frac{1}{c^2}\frac{\partial^2 p}{\partial t^2} = 0. \qquad (9)$$

Again using the method of separation of variables described in Section 3, one writes

$$p(x, r, \theta, t) = X(x)R(r)\Theta(\theta)\exp(i\omega t). \qquad (10)$$

Substituting Eq. (10) into Eq. (9) leads to

$$\frac{1}{X}\frac{d^2X}{dx^2} + \frac{1}{R}\left(\frac{d^2R}{dr^2} + \frac{1}{r}\frac{dR}{dr}\right)$$
$$+ \frac{1}{r^2\Theta}\frac{d^2\Theta}{d\theta^2} + k^2 = 0, \qquad (11)$$

where $k^2 = (\omega/c)^2$. Proceeding initially as before with rectangular ducts, letting $-k_x^2 = (1/X)(d^2X/dx^2)$, one finds that the resulting axial distribution function $X(x)$ is again given by Eq. (7b). Selecting next the circumferential distribution function Θ, one notes that the periodicity in the coordinate θ together with the physical limitation on the sound field that it be single valued at any position θ imply that

$$\Theta(\theta) = \exp(\pm im\theta), \qquad m = 0, 1, 2, \ldots, \qquad (12a)$$

so that $p(x, r, \theta, t)$ is represented by a Fourier series in θ, with coefficients depending on r, x, t. The integer m is the number of circumferential modes (or radial nodal lines) in the harmonic complex pressure amplitude. The plus-or-minus sign in Eq. (12a) corresponds to interference patterns rotating in the negative/positive θ directions, respectively.

The radial distribution function $R(r)$ must then satisfy

$$\frac{d^2R}{dr^2} + \frac{1}{r}\frac{dR}{dr} + \left\{k_r^2 - \left\{\frac{m}{r}\right\}^2\right\}R = 0, \qquad (11a)$$

where $k_r^2 = k^2 - k_x^2$, with the radial wavenumber k_r constant. A transformation of variable from r to $k_r r$ converts this to a standard form of Bessel's equation, for which the appropriate solution for a cylindrical duct is

$$R(k_r r) = J_m(k_r r), \qquad k_r \neq 0, \qquad (12b)$$

where $J_m(k_r r)$ is the Bessel function of the first kind of order m. The complete solution to Eq. (11a) may also include a corresponding Bessel function of the second kind, the Neumann functions N_m, which have values that are infinite for zero argument. Thus the latter set are omitted for cylindrical ducts but should be included[6] in the solution for annular ducts.

Some general features of Bessel functions $J_m(\psi)$, $\psi = k_r r$ in terms of their order m and argument ψ, are of practical interest. That of zero order ($m = 0$), which is plotted in Fig. 1, differs from all those of higher order ($m > 0$),

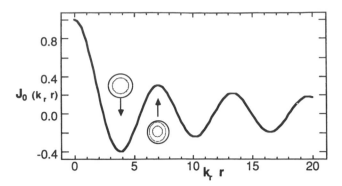

Fig. 1 $J_0(k_r r)$, defining radial pressure distribution in a cylindrical waveguide with no spinning modes ($m = 0$). The two insets give the positions of the nodal lines for $n = 1$ and $n = 2$ with rigid walls.

being unity at $\psi = 0$ and oscillating as ψ increases in the general manner of a damped cosine wave. Those with $m > 0$ are all zero at $\psi = 0$, increasing in value with ψ to a peak at approximately $\psi = m + 1$ and then again continuing to oscillate in the manner of a damped cosine wave, as can be seen for $m = 9$ in Fig. 2. Finally, the slope $J_m'(\psi) = d[J_m(\psi)]/d\psi$ is zero at $\psi = 0$ in all cases except when $m = 1$, when it is 0.5.

Finally, it remains to specify the values of $k_r r$ that satisfy the boundary conditions imposed by the duct walls. With hard walls, as for rectangular ducts, we require $\partial p/\partial r = 0$ at the duct wall at radius a corresponding to the values of $k_r r$ where the gradient $d[J_m(\psi)]/d\psi$ is zero. These are designated as the physically tenable values $k_r = k_{mn}$, where $n = 0, 1, 2, 3, \ldots$ and k_{mn} represents the nth root of the corresponding eigenequation. Collecting the solutions (7b), (12a), and (12b) and substituting them in Eq. (10) leads to the pressure pattern

$$p(x, r, \theta, t) = A_{mn}^{\pm} J_m(k_{mn}r)\exp\, i(\omega t \mp m\theta \mp k_x x), \qquad (13)$$

where $A^{\pm}$ and the subsequent minus-or-plus signs relate respectively to the positively and negatively spinning- and propagating-wave patterns [see also the discussion following Eqs. (8) and (12a)].

The influence of the presence of spinning modes on the radial distribution of fluctuating pressure is demonstrated by comparing Fig. 2, where there are nine, present with Fig. 1, where there are none. The relative positions of the nodal lines associated respectively with one and two radial modal circles in the two figures demonstrate the progressive concentration of acoustic energy toward the duct periphery that is found to occur as the number of spinning modes increases. A comprehensive study of the sound field in rigid-walled annular ducts can be found in Ref. 6.

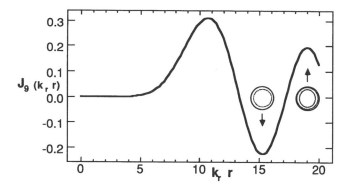

Fig. 2 $J_q(k_r r)$, defining radial pressure distribution in a cylindrical waveguide with nine spinning modes ($m = 0$). The two insets give the position of the nodal lines for $n = 1$ and $n = 2$ with rigid walls.

5 AXIALLY PROPAGATING MODES, CUTOFF MODES

Although an infinite number of modes are physically possible, only those whose axial wavenumbers k_x remain wholly real can propagate indefinitely, while the remainder will decay exponentially. The modes that propagate can be identified by reference to the appropriate condition for rigid-walled rectangular ducts, which is [see Eq. (8)]

$$k_{x,mn}^2 = \left(\frac{\omega}{c}\right)^2 - \left(\frac{m\pi}{l_y}\right)^2 - \left(\frac{n\pi}{l_z}\right)^2 > 0. \quad (14)$$

Similarly, for cylindrical ducts,

$$k_{x,mn}^2 = \left(\frac{\omega}{c}\right)^2 - (k_{r,mn})^2 > 0. \quad (15)$$

These equations demonstrate that the plane-wave mode ($m = n = 0$) always propagates, while higher order modes will not propagate unless condition (14) or (15) is satisfied.

At frequencies for which the condition is not satisfied, one has $k_{x,mn} = \mp i|k_{x,mn}|$, with minus or plus respectively, in the $\pm x$-direction. Thus the modes decay axially at an exponential rate with distance away from a source or reflection location. The axially decaying modes are called cutoff modes, with the frequency at which this occurs for each one called the cutoff frequency. Axial propagation will take place in modes with frequencies of excitation above the cutoff frequency, and these modes are then termed cut-on.

With rectangular ducts the condition for propagation becomes $\omega/c > \pi/l_y$ when l_y is the larger duct dimension, that is, when the free-field wavelength $\lambda_f < 2l_y$.

TABLE 1   Values of $k_{mn}a$ for $J'_m(k_{mn}a) = 0$ Corresponding to Cut-on of Axial Propagation in Rigid Cylindrical Waveguide of Radius a

J_m	$n =$				
	0	1	2	3	4
J_0	0	3.83	7.02	10.17	13.32
J_1	1.84	5.33	8.54	11.71	14.86
J_2	3.05	6.71	9.97	13.17	16.35
J_3	4.20	8.02	11.35	14.59	17.79
J_4	5.32	9.28	12.68	15.96	19.20
J_9	10.71	15.29	19.00	22.50	25.89

The corresponding conditions for cylindrical ducts, with radius a, can be established by reference to Table 1. This lists the first few values of $k_{mn}a$, for which $J'_m(k_{mn}a)$ is zero, thus defining the cut-on frequencies for the relevant modes. Noting that $k = \omega/c$, one sees from the table that the first purely spinning mode ($m = 1, n = 0$) propagates so long as the reduced frequency $\omega a/c > 1.84$, the second ($m = 2, n = 0$) when $\omega a/c > 3.05$, and so on.

Alternatively, inspection of Eq. (13) shows that at any fixed axial position x, the angular velocity of spin $\Omega_p = \partial\theta/\partial t = \omega/m$, while the corresponding peripheral velocity at the wall is $\omega a/m$ with Mach number $M_p = \omega a/mc$. Since such spinning modes propagate when $\omega a/c > \phi(m)$, where the values of $\phi(m)$ are given in the table, an alternative condition that such modes should propagate is $M_p > \phi(m)/m$. Noting too that $\phi(m) \approx (m+1)$ with $n = 0$, an approximate condition is that the spinning modes will propagate so long as $M_p > (m+1)/m$. In general, one can show that the cutoff condition represents a very rapid decay rate for such modes, which diminishes as $M_p^2 \to 1$.

6 LINED DUCTS

When the walls are not rigid and are lined with an absorbing material, the boundary conditions at the wall or lining surface require that the fluctuating pressures and normal fluid displacement (or velocity) at the boundary surface must equal those on the surface. The absorbent properties of the boundary are usually characterized by the normal acoustic impedance Z_b, which is defined as the ratio of acoustic pressure to normal velocity at the surface.

6.1 Lined Rectangular Waveguides with Absorbing Boundaries

If the walls are symmetrically lined with respect to the duct axis, it is more convenient to have a coordinate system based on the axis. The acoustic modes will be symmetric or antisymmetric about the axis, giving $\partial p/\partial y = 0$ and $\partial p/\partial z = 0$ for symmetric and $p = 0$ for antisymmetric modes for rectangular cross-sectional ducts at $y = z = 0$. The boundary conditions at $y = \pm\frac{1}{2}l_y$ become for

$$Z_{by} = \rho c \; \frac{k}{k_y} \; i \cot(k_y l_y/2), \quad \text{for symmetric modes,}$$

$$(16a)$$

$$-Z_{by} = \rho c \; \frac{k}{k_y} \; i \tan(k_y l_y/2) \quad \text{for antisymmetric ones,}$$

$$(16b)$$

with similar expressions for Z_{bz}. These transcendental equations are known as eigenequations, having an infinite number of solutions yielding k_{ym} and k_{zm}, which have complex values when the boundaries are absorbing. The axial wavenumber will then be

$$k_{mn}^2 = k^2 - k_{ym}^2 - k_{zn}^2 = (\beta_{mn} - i\alpha_{mn})^2, \qquad (17)$$

which represents an axially decaying wave motion due to energy absorption at the walls. The pressure distribution can be obtained by the appropriate substitution in Eq. (8). A sharp cutoff frequency does not exist for this case as all modes decay to some extent.

6.2   Lined Cylindrical Waveguides

The corresponding eigenequation for a lined cylindrical duct with a boundary impedance Z_a, which is normally complex, is given by

$$ikJ_m(k_r a) + \frac{Z_a}{\rho c} \; k_r J'_m(k_r a) = 0, \qquad (18)$$

which can be solved for k_r, which is in general complex. The prime denotes the derivative with respect to the argument. The axial wavenumber components k_{mn} of the wave vector then follow from Eq. (15), where these are in general also complex. Their imaginary parts then determine the axial decay rates for the cut-on modes, as with Eq. (17), with again no sharp cutoff frequency.

7   DUCTS WITH UNIFORM AXIAL FLOW

Sound propagation through flow ducts is described by the appropriate form of the convected wave equation (1). Solutions of the wave equation for a cylindrical wave-

guide with a steady uniform axial flow with velocity u_0 are presented here to illustrate the significant effects on axial wave propagation of the existence of a mean flow. The wave equation (1) is then expressed by

$$\left[\frac{D^2}{Dt^2} - c^2 \left(\frac{\partial^2}{\partial x^2} + \frac{\partial^2}{\partial r^2} + \frac{1}{r} \frac{\partial}{\partial r} + \frac{1}{r^2} \frac{\partial^2}{\partial \theta^2} \right) \right] p = 0,$$

$$(19)$$

with $D^2 p/Dt^2$ expressed by Eq. (3) where the velocity u_0 is explicitly included. Sound propagation in rectangular ducts is described by substituting the appropriate expression for the Laplacian $\nabla^2 p$ in Eq. (1), while solutions for more general cases can be found among the references cited.

7.1   Axial Wavenumber k_x

For rigid boundaries in a cylindrical duct, substitution of Eq. (10) into Eq. (19) followed by separation of variables leads to the general eigenequation for X,

$$(1 - M^2) \frac{\partial^2 X}{\partial x^2} - 2ikM \frac{\partial X}{\partial x} + (k^2 - k_r^2)X = 0,$$

where the flow Mach number $M = u_0/c$ and k_r^2 and k^2 are both constants. This has a solution $X(x) = A \exp(-ik_x x)$, with k_x expressed as

$$k_x = \frac{-kM \pm [k^2 - (1 - M^2)k_r^2]^{0.5}}{1 - M^2}, \qquad (19a)$$

which, with $M = 0$, is the result previously expressed by Eq. (15) found with rigid boundaries. In this case the pressure pattern in the duct will be described by Eq. (13) with the appropriate substitution for k_x. We note, from Eq. (19a), that k_x is wholly real when $k^2 > (1 - M^2)k_r^2$. Thus the frequency at which modes become cut on is reduced by the factor $(1 - M^2)^{0.5}$. Also, by analogy with the results in Sections 2 and 3, the axial wavenumber k_x for a rigid rectangular duct is given by the substitution of $k_y^2 + k_z^2$ for k_r^2 in Eq. (19a), with a consequent reduction to the frequency of cut-on by the same factor for propagating modes described by Eq. (14).

With plane-wave excitation and propagation one sets $k_r = 0$, so that $k_x = k(M \pm 1)/(1 - M^2)$. Thus with propagation in the direction of positive x, $k_x = k/(1 + M)$, and in the direction of negative x, $k_x = -k/(1 - M)$. Thus the wavelength is stretched relative to that in the free field for waves traveling in the direction of flow and contracted for waves traveling in the reverse direction, due to the convection by the mean flow.

7.2 Radial Wavenumber k_r in Lined Ducts with Mean Flow

To investigate the convective effects of the mean flow on the values of the radial wavenumber k_r (and thus on k_x), one must first adopt the appropriate matching conditions at the interface. One cannot match particle velocity in this case since, with particle displacement ζ one finds in general that $u = D\zeta/Dt = \partial\zeta/\partial t + u_0\,\partial\zeta/\partial x$ in the flow while in the lining u_b is simply $\partial\zeta/\partial t$, without a convective term. Expressing conservation of radial momentum (Euler's equation) in terms of ζ in the lining, one has $(-1/\rho)(\partial p/\partial r) = \partial u_0/\partial t = \partial^2\zeta/\partial t^2$, with $u_b = d\zeta/dt = i\omega\zeta$ and $\zeta = u_b/i\omega = p/i\omega Z_b$, where Z_b is the surface impedance of the lining. In the flow, conservation of radial momentum is expressed by $(-1/\rho)\,\partial p/\partial r = D^2\zeta/\partial t^2$. Hence, matching radial displacements ζ at the interface $r = a_i$, we find the boundary condition to be satisfied is expressed by $(1/\rho)(\partial p/\partial r)+(1/i\omega Z_b)\,D^2p/Dt^2 = 0$.

With this boundary condition one finds that the radial wavenumber k_r must satisfy

$$\frac{Z_b}{\rho c}\,\frac{\partial^2 p}{\partial r\,\partial t} + \frac{1}{c}\left(\frac{\partial}{\partial t} + u_0\,\frac{\partial}{\partial x}\right)^2 p = 0, \qquad r = a_i.$$

The corresponding values of k_r are then given by

$$\frac{Z_b}{\rho c}\,k_r J'_m(k_r a_i) + ik\left(\frac{1-Mk_x}{k}\right)^2 J_m(k_r a_i) = 0. \quad (20)$$

The implications of this result show that when the wall impedance is high, so that the radial displacements remain relatively insignificant at the wall, the modal patterns approach those for rigid boundary since $J'_m(k_r a_i) \to 0$. However, the radial wave numbers found with soft walls and hence the radial pressure patterns are modified by the convective effects of mean flow, while the circumferential modes are not. A similar result is found for lined rectangular ducts.[3] The combined effects of soft walls and sheared flow have been described by Ko.[7] Similar studies[8] have been made for annular flow ducts.

7.3 Sound Power Flux

The usefulness of sound power stems from the fact that the total acoustic energy, the sum of its kinetic and potential energy components, is conserved if the acoustic medium is lossless (i.e., for an inviscid non-heat-conducting fluid). In a stationary acoustic medium the sound power W is defined as $\int_S I_i\,dS_i$, where $I_i = \langle pu_i\rangle$ are the components of the sound intensity, which is the time-averaged sound energy flux, with $\langle\cdot\rangle$ denoting the time averaging. In a moving medium, however, the sound energy flux includes transport of the mean energy density and intensity by the mean flow. Provided viscous and thermal diffusion effects are negligible and the flow is steady, irrotational, and uniform, the sound power flux per unit area normal to the flow, or the sound intensity,[9] is given by

$$I = (1 + M^2)\langle pu\rangle + M\left(\frac{\langle p^2\rangle}{\rho c} + \rho c\langle u^2\rangle\right). \quad (21)$$

With propagation in several modes simultaneously, care may be necessary when performing the averages or in interpreting overall sound power measurements. Intensity probes do not appear to give reliable measurements in this environment so that other methods of measurement should be adopted.[10]

7.4 Sound Propagation with Vibrating Walls

When the duct walls are not rigid but react elastically to the fluctuating internal pressure field, the cross section will vary with time according to $S(t) = \pi[a(t)]^2$ for a cylindrical tube with radius a. With mean flow neglected, the phase speed c_v will be defined[11] by

$$\left(\frac{1}{c_v}\right)^2 = \left(\frac{1}{c_0}\right)^2 + \left(\frac{1}{c_w}\right)^2 = \left(\frac{1}{S_0}\right)\left|\frac{d(\rho S)}{dp}\right|_{p=p_0},$$

where p_0 and S_0 are, respectively, the undisturbed or time-averaged pressure and area, c_0 is the usual isentropic sound speed, and c_w is a perturbation resulting from the wall vibration. Adopting the linear approximation $|dS/dp|_{p=p_0} = 2\pi a_0|da/dp|_{p=p_0}$ and a locally reacting wall with (per unit area) mass m, resistance R, and stiffness K, the perturbation in phase velocity becomes

$$c_w^2 = \frac{a_0}{2\rho_0}\,[m(\omega_0^2 - \omega^2) + iR\omega],$$

where $\omega_0 = (K/m)^{0.5}$. The corresponding wavenumber modified by wall vibration is, to first order,

$$k_v = \left(\frac{\omega}{c_0}\right)\left(1 + \frac{\rho_0 c_0^2/a_0}{m(\omega_0^2 - \omega^2) + iR\omega}\right). \quad (22)$$

One should note that this approximation is not strictly valid when the excitation frequency $\omega = \omega_0$, the resonance frequency of the wall, since radiation damping of the vibration may then make a significant contribution to R.

8 TERMINATIONS AND SOURCES

In practice, a waveguide comprises one or more sections of uniform or regularly shaped duct, each of length L_i, say, $i = 1, 2, \ldots$, but in general of different cross section, connected in sequence between the primary source of excitation and the final termination. Axial sound propagation in each individual section is described by relations such as Eqs. (8) and (13) once the modal distribution has been related to the shape and physical characteristics of the walls as well as to the frequency and to the properties of the acoustic medium and where relevant to the mean flow. The values of the two sets of arbitrary constants $A_{mn}^{\pm}$ that these equations include depend on the boundary conditions at the ends of each section. Thus they depend on the factors controlling excitation and wave reflection at the source end, those controlling reflection and wave transfer at the junctions between the component elements, and those controlling wave reflection and transmission at the final termination. The acoustic response of any uniform duct of finite length to any specific form of excitation is strongly influenced by the boundary conditions at this termination as well as its length. (See Section 11.) Thus it is logical to begin acoustic analysis of the system at its final termination,[10] where conditions are normally well defined and invariant, and then proceed along the system toward the source.

The termination may be characterized by its termination impedance $Z_{mn} = p_{mn}/u_{mn}$, by its radiation resistance, by its pressure reflection coefficient A_{mn}^-/A_{mn}^+, or by the corresponding power reflection coefficient for each mode. In the absence of reflections (i.e., $A_{mn}^- = 0$), the termination impedance is equal to $\rho ck/k_{mn}$. Specification of the reflection and transfer characteristics at the termination and at any junction between elements requires that the complete set of expressions describing the sound field are matched at the transfer plane for continuity of pressure and displacement[10, 12–15] (or alternatively when appropriate, the velocity). Inclusion of the relevant boundary conditions demonstrates that, in general, the matching at such discontinuities involves the addition of further modes to the reflected and transmitted waves, though these may decay rapidly and thus not propagate. An interesting result[13] is that any well-cut-on mode of higher order suffers little reflection but carries on past the termination as if it were continuing down an infinite duct of the same cross section.

A primary sound source may be represented by a distribution of fluctuating velocity or pressure or of both together. Similar matching conditions apply at the source plane, so that, for example, the sum of the component velocities over all modes must equal the source velocity. The sound power transferred from the source to the waveguide in each mode will depend on the complex load impedance offered by the duct at the source plane with the corresponding internal impedance of the source (see Section 11.2). Sound may also be generated or amplified at other positions along flow ducts by various mechanisms,[16, 17] as a result of the transfer of energy from the mean motion by the action of shear layers, with their associated vorticity. Those sources associated with turbulent mixing, such as those of boundary layer noise, tend to be broadband while those associated with resonant or reactive behavior are tonal in character. Such secondary sources may be situated at specific locations or distributed along elements of the system.

9 AXIAL SOUND PROPAGATION THROUGH PRACTICAL WAVEGUIDES

The classical approach to the analysis of sound propagation in realistic waveguides with complex geometry normally neglects mean flow and thus ignores features that may be significant in practice. In some instances, the boundary conditions have also been oversimplified, as for example with applications of an analogy with lumped electrical networks. A more physically valid approach is considered here, illustrated by reference to specific examples. For instance, flow duct elements, or the junctions between them, may concern rapid or distributed axial changes in the duct geometry or in the magnitude and direction of the wave propagation vector. Acoustic transfer can be described by satisfying conservation of mass, energy, and momentum both within and at the bounding surfaces of an appropriate control volume.[10] Successful predictions that correlate well with observation have generally been confined to plane waves or where significant energy transfer between the modes does not occur.[18] More recent examples are described in Ref. 19, which includes some 20 contributed papers, and others are summarized in a survey by Cummings[20] with many more relevant references.

As noted in Section 7, flow ducts may be represented as a sequence of uniform or geometrically regular elements with discontinuities at their junctions. For simplicity, the discussion presented here is restricted to axial acoustic propagation along the elements and across the junctions between them. The pressure density and velocity will all be functions of axial displacement $\mathbf{x}$ and time t and expressed as the sum of mean time-averaged and fluctuating parts, with pressure expressed as $p_0(\mathbf{x}) +$

$p(\mathbf{x}, t)$, and so on. For potential mean flow along the elements with velocity potential $\phi_0(\mathbf{x}) + \phi(\mathbf{x}, t)$, one has $u_0(\mathbf{x}) + u(\mathbf{x}, t) = -\partial\phi_0(\mathbf{x})/\partial\mathbf{x} - \partial\phi(\mathbf{x}, t)/\partial\mathbf{x}$, the first term representing the irrotational mean flow velocity $u_0(\mathbf{x})$ and the second with zero time average, the fluctuating particle velocity. As before, the pressures, velocity, and so on, will be Fourier transformed on the time t and the discussion will refer to spectral components of frequency $\omega = 2\pi f$.

Doak[21] has shown that given any three-dimensional geometric region (regular element) in which there is a low-Mach-number mean potential flow for which the velocity potential is $\phi_0(\mathbf{x})$, then the Fourier spectral density $\Phi(\mathbf{x}, \omega)$ of any acoustic velocity potential $\phi(\mathbf{x}, t)$, due only to sources on the terminations of the region is given to first order in the mean flow Mach number $-\nabla\phi_0(\mathbf{x})/c_0$ by

$$\Phi(\mathbf{x}, \omega) = \exp\left[-i\left(\frac{k}{c_0}\right)\phi_0(\mathbf{x})\right]\varphi_0(\mathbf{x}, \omega), \qquad (23)$$

where $\varphi_0(\mathbf{x}, \omega)$ satisfies the (no-flow) scalar Helmholtz equation $(\nabla^2 + k^2)\varphi_0(\mathbf{x}, \omega) = 0$. In addition, the approximate spectral density $\Phi(\mathbf{x}, \omega)$ of the acoustic velocity potential [and hence $\varphi_0(\mathbf{x}, \omega)$] must satisfy the appropriate boundary conditions at both end junctions of the element (region). For example, with a uniform flow duct of constant cross section, Eq. (23) becomes[10]

$$p(x, \omega) = p^+(0, \omega)\exp(ik^*Mx)[\exp(-ik^*x) + r_0$$
$$\cdot \exp(ik^*x)]\exp(i\omega t), \qquad (24)$$

where $p^+(0, \omega)$ is the complex amplitude of the component wave incident at the downstream termination $x = 0$, where the reflection coefficient is r_0 and $k^* = (\omega/c)/(1 - M^2)$. This describes the superposition of a standing wave and a traveling wave.

9.1 Wave Transfer across Junctions and Other Discontinuities

Where the junctions between elements represent an area or any other discontinuity, a part of the incident-wave energy is reflected, with the remainder transmitted or dissipated, though generally the associated processes remain almost isentropic so that the resulting entropy fluctuations may be negligibly small. Following the procedure outlined above, one first defines an appropriate control volume V with surface S at the junction, including any relevant fixed boundaries. With axial wave motion, the integral relations describing conservation of mass and momentum over the surface S are expressed as

$$\int_S \rho'u'\, dS = 0, \qquad (25a)$$

$$\int_S [\rho'(u')^2 + p']\, dS = \sum F, \qquad (25b)$$

where $\rho' = \rho_0 + \rho$, $u' = u_0 + u$, $p' = p_0 + p$, and $\sum F$ represents the externally applied boundary forces, which are zero for rigid surfaces. With isentropic processes the acoustic pressure and density fluctuations are related by $c_0^2\rho = p$. When the entropy fluctuations are significant, though small, one can represent the fluctuating entropy[2] by $\varepsilon = (-1/\rho_0 T_0)\delta/(\gamma - 1)$ and then $c_0^2\rho = (p + \delta)$. Conservation of energy is expressed as

$$\int_S [h' + 0.5(u')^2]\, dS = 0, \qquad (25c)$$

$$h' = h_0 + \frac{p}{\rho_0} + T_0\varepsilon, \qquad (25d)$$

where T_0 is the ambient absolute temperature.

To evaluate the acoustic characteristics of the junction, one first subtracts the terms relating to mean motion from Eqs. (25a)–(25d) and then discards all fluctuating terms of order higher than the first (the acoustic approximation). Having expressed fluctuating velocity u, density ρ, and pressure p in terms of the equivalent progressive component wave amplitudes p^+ and p^-, one Fourier transforms the result on the time t. These three equations, with any necessary equations defining the boundary conditions, are then solved for each spectral component $p_1^\pm$ on one side in terms of $p_2^\pm$ on the other and, where relevant, δ. Where the shape of the wavefront changes across the junction, for example, from spherical to plane,[22] V should represent a finite volume within which the readjustment takes place. When the propagating waves remain plane on both sides, V may also be reduced to a plane, with an appropriate end correction included at the junction,[10] or alternatively, an appropriate set of higher order modes[18, 23] may be introduced there. Such additions are necessary to account for the presence of evanescent waves required to satisfy the boundary conditions, with any corresponding adjustments to the wavefronts from one side to the other.

The transmission of plane waves at an abrupt area expansion with outflow provides a common example of such wave transfer. An analytic solution[18] that includes mean flow but with the restrictive assumption that reflected waves are absent beyond the junction shows fair agreement with corresponding observations.[24] Another analytic solution[23] that avoids this assumption but in which mean flow is neglected shows good agreement with measurements without flow. An alternative approach[10, 24] that also provides predictions that agree well with all these and further sets of measurements, including those with mean flow, represents the evanescent waves and the associated fluctuating mass at the junction by an appropriate end correction. Similar comparisons[25, 10] can be made concerning transfer at an expansion with reversal of the direction of propagation or with the addition of a tuned side branch[10, 23] at the discontinuity.

Examples of extended control volumes include the ends of conical expansions with the associated transfers between spherical and plane-wave motion,[22] the analysis for which remains valid provided the flow does not separate from the walls. This has been shown to provide predictions that compare well with observation[19] for Mach numbers up to 0.2. An earlier analysis of sound propagation in a flow duct with perforated walls backed by an enclosed annular cavity provides predictions that also closely match observation.[26, 27] Here the acoustic field within the lined duct has been matched to that within the surrounding annulus by specifying the impedance[26, 28] of the perforated wall with grazing flow. Other examples (e.g., describing the acoustic performance with various arrangements of absorbing linings) can be found in Ref. 19 as contributed papers with extensive lists of relevant references.

10 VISCOUS AND THERMAL INFLUENCES

The propagation of sound in ducts subject to viscous and thermal effects was first described by Kirchhoff in 1868. A comprehensive review of the problem with a numerical solution for tubes of Kirchhoff's equation has been presented by Tijdeman[29] and supplemented by further comment by Kergomard.[30] The main parameters governing sound propagation in rigid cylindrical tubes are found to be the shear wave, or Stokes, number $S_n = a(\omega/\nu)^{0.5}$, where a is the tube radius and ν is the kinematic viscosity of the medium, as well as the reduced frequency, or Helmholtz, number $\omega a/c$, together with γ, the ratio of the specific heats and $\sigma = 1/\sqrt{P_r}$, with P_r the Prandtl number for the medium. The solution with Kirchhoff's wide-tube approximation,[31] which is shown to be appropriate for $S_n > 10$, replaces the free-space wave propagation constant $k = \omega/c$ by $\beta = k + \alpha(1 - i)$, where α is given by

$$\frac{\alpha}{k} = \frac{(1/S_n)[1 + (\gamma - 1)\sigma]}{\sqrt{2}}. \tag{26}$$

For air the factor $[1 + (\gamma - 1)\sigma]/\sqrt{2}$ is close to unity. The resulting wave dispersion (i.e., the dependence of wave speed on frequency) represented by the factor α/k is inversely proportional to the Stokes number and thus to the square root of frequency. Thus these effects become relatively less significant as the frequency of excitation increases although α increases in proportion to the Stokes number (but see Ref. 30 for very high frequencies). An extension to tubes of arbitrary cross section is provided by Stinson.[32] With narrow tubes, the viscous effects predominate, and both dispersion and attenuation become dominant. In the presence of a turbulent mean flow, the damping of acoustic waves is influenced to some extent by the action of the turbulent stresses near the wall.[33] With smooth walls, so that a laminar sublayer is present, the influence depends on the value of the dimensionless acoustic boundary layer thickness $\delta^+ = (2u_\tau^2/\nu\omega)^{1/2}$, where u_τ is the wall friction velocity $(\tau_0/\rho_0)^{1/2}$ and τ_0 the wall shear stress. With $\delta^+ < 12.5$, say, observations indicate that the influence of wall shear stress may be neglected.[33] Thus with the exception of smooth walls with low frequencies of excitation and perhaps high-mean-flow Mach numbers or for very narrow tubes, the influence of mean pipe friction losses on sound propagation remains small. When such is the case, a useful first approximation to plane-wave propagation along flow ducts is expressed by Eq. (24) when k is replaced by β evaluated from Eq. (26).[10, 19] Note that the Froude friction factor term given in Ref. 3, and first introduced in Eq. 1.99 there, is based on a quasi-static assumption that applies only at very low frequencies.

Flow ducts with axial temperature gradients provide a common practical problem in which both the flow and physical properties vary with gas temperature and hence with axial distance x along the duct. Thus, associated with the flow temperature $T(x)$ there will be a corresponding axial velocity $u(x, t)$, density $\rho(x, t)$, pressure $p(x, t)$, and entropy $s(x, t)$, with corresponding changes in gas properties. Relations expressing conservation of mass, momentum, and energy for a general case are respectively

$$\frac{1}{\rho} \frac{D\rho}{Dt} + \frac{\partial u_i}{\partial x_i} = 0, \tag{27a}$$

$$\frac{Du_i}{Dt} + \frac{1}{\rho} \frac{\partial p}{\partial x_i} - \frac{1}{\rho} \frac{\partial e_{ij}}{\partial x_j} = 0, \tag{27b}$$

$$\frac{\rho Ds}{Dt} - \frac{\partial}{\partial x_j} \left(\frac{k_T}{T} \frac{\partial T}{\partial x_j} \right)$$

$$= \frac{1}{T} \left[\frac{k_T}{T} \left(\frac{\partial T}{\partial x_j} \right)^2 + e_{ij} \frac{\partial u_i}{\partial x_j} \right], \tag{27c}$$

$$\rho = \rho(s, p), \qquad (27d)$$

where k_T is the thermal conductivity of the gas and e_{ij} is the viscous stress tensor. If one assumes small viscosity and neglects diffusion, conservation of energy simplifies to $\rho(Ds/Dt) = 0$, with corresponding changes to Eq. (27b). With one-dimensional flow in the x-direction $D/Dt = \partial/\partial t + u_0 \partial/\partial x$, as for Eq. (3). Solutions[34] for this problem with plane waves and moderate temperature gradients, so that $T_x = T_0(1 - \alpha x)$, $\alpha x < 0.1$, justify subdividing the duct into appropriately short axial segments, with flow conditions and gas properties in each segment represented by their averaged properties. An approximate solution for plane waves with an exponential axial temperature distribution has been derived by Cummings.[35]

Wave propagation where the fluctuating pressure amplitude is sufficiently large to introduce significant changes in the gas properties and speed of sound represents a nonlinear problem that is best approached in the time domain. Though the general literature on this subject is extensive, successful predictions of continuous sound propagation under these conditions appear to be lacking so far, except for some rather special circumstances.[36,37]

11 CALCULATION AND SPECIFICATION OF WAVEGUIDE PERFORMANCE

If is often convenient for acoustic performance assessment (e.g., during design optimization for noise control) to employ compact descriptions summarizing the acoustic behavior of the waveguide or each of its constituent elements. For simplicity, the discussion will be confined to plane-wave propagation, since the principles and approach can readily be extended to higher order modes. Acoustic plane-wave transmission across an individual element or sequence of elements is illustrated in Fig. 3. Here S represents the source of excitation for the element or sequence T, for which Z represents the acoustic impedance at its termination. With isentropic plane waves the complex amplitudes p^+ and p^- are related to the corresponding acoustic pressure p and particle velocity u by

$$p = p^+ + p^-, \qquad (28a)$$
$$\rho c u = p^+ - p^-, \qquad (28b)$$

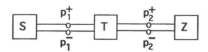

Fig. 3 Plane acoustic wave transfer across a waveguide or one of its elements.

Equation (28a) is valid[10] irrespective of the presence of a steady mean flow u_0, while Eq. (28b) is similarly valid so long as viscothermal effects remain negligible.

The wave components $p_1^\pm$ on the source side of T can be related to $p_2^\pm$ on the load side by

$$p_1^+ = T_{11}p_2^+ + T_{12}p_2^-, \qquad (29a)$$
$$p_1^- = T_{21}p_2^+ + T_{22}p_2^-, \qquad (29b)$$

where T_{11}, $T_{12} T_{21}$, and T_{22} are the four elements of the scattering transmission matrix $[T]$ defining the transfer. The complex values of these four elements are functions of geometry, of the undisturbed values of ρ and c, of the frequency of excitation, of the mean flow Mach number M, but remain independent of the value of Z. Obviously Eqs. (28a) and (28b) imply that the fluctuating acoustic pressure p_1 and velocity u_1 on the source side of T can be related to p_2 and u_2 on the load side in terms of the corresponding impedance transfer matrix $[t]$ by expressions analogous to Eqs. (29a) and (29b). The relative difficulty of measuring acoustic particle velocity u, compared with measurements of p combined with the restricted validity[10] of Eq. (28b), suggests that the scattering matrix $[T]$ provides a more robust description of the measured wave transfer than does the impedance matrix $[t]$.

To return to Fig. 3, we see that acoustic wave transmission across T can also be expressed in terms of the transmission coefficients

$$T_i = \frac{p_1^+}{p_2^+}, \qquad (30a)$$
$$T_r = \frac{p_1^-}{p_2^-}, \qquad (30b)$$

together with the reflection coefficient r, where

$$r = \frac{p_2^-}{p_2^+} \qquad (31a)$$

or

$$r = \frac{\zeta - 1}{\zeta + 1}, \qquad (31b)$$

and $\zeta = Z/\rho c$, also noting $p_1^-/p_1^+ = r T_r/T_i$. The transmission coefficients are related to the elements of the scattering matrix $[T]$ by

$$T_i = T_{11} + r T_{12}, \qquad (32a)$$
$$T_r = \frac{T_{21}}{r} + T_{22}, \qquad (32b)$$

and it is evident that T_i and T_r are both functions of r and hence of Z as well as of the element's geometry, ρ, c, f, and M. To evaluate the elements of $[T]$, one can easily show[38] that calculation or measurement of T_i and T_r with two different loads Za, Zb are necessary unless $[T]$ possesses reciprocal properties. This cannot be true when mean flow is present or when viscothermal or flow-associated losses are significant and, as it turns out, when the element is not geometrically reciprocal. Thus transfer coefficients might be preferred to transfer matrix representations, despite the apparent convenience of the latter's independence of the load Z. Finally, one should note that both representations remain independent of the source of excitation, so long as it remains external to the element.

11.1   Single-Parameter Descriptions of Acoustic Performance

Single-parameter descriptions have several practical uses. Normally they define the acoustic attenuation of the system or element T or of its transmission loss index TL or insertion loss index IL, all expressed in decibels. The transmission loss of an acoustic element is usually defined as the difference in power flux in free space between that incident on and that transmitted across it and is then an invariant property of the element. With reference to the element T in Fig. 3, after establishing anechoic conditions by setting $Z = \rho c$ or $r = 0$, then

$$TL = 10 \log_{10} \frac{S_1 I_1}{S_2 I_2}, \qquad (33)$$

where the intensity or power flux may be found from Eq. (21) and S_i is the cross-sectional area of the duct.

The insertion loss is the measured change in power flux at a specified receiver when the acoustic transmission path between it and the source is modified by the insertion of the element, provided that the impedances of both source and receiver remain invariant. If W_1 is the power at the receiver before the insertion of the element and W_2 that measured or calculated afterward, then the insertion loss index is defined by

$$IL = 10 \log_{10} \frac{W_1}{W_2}. \qquad (34)$$

In practice, it is normally necessary to provide a reference duct system of known performance to measure or calculate W_1 and then repeat this procedure with the element or system of interest. This also implies that the source impedance should be known and invariant; see Eqs. (39) and (40). Since the impedance presented at the source by the reference system will normally differ from

that under test, due compensation to the results may be necessary to account for any corresponding changes to the source when its impedance is not effectively infinite.

A simple index of performance, which remains independent of an external source but is dependent only on the acoustic characteristics of the waveguide with its operating load Z, is the attenuation index AL, expressed as

$$AL = 20 \log_{10} |T_i|, \qquad (35)$$

which can provide a convenient guide when the source impedance is unknown.

11.2   Waveguide Excitation

The transfer of acoustic energy or sound power from the source to the system is strongly influenced by the acoustic impedance or load that the system presents to the source. One recalls too that the acoustic characteristics of both system and source are functions of the frequency, so the discussion that follows here applies strictly to each spectral component, although it also applies in general to all of them. The acoustic circuit corresponding to Fig. 3 has been extended to include two representations of the source in Fig. 4.

In the acoustic circuit in Fig. 4a the source is now represented by a fluctuating volume velocity with amplitude V_s and has an internal effective shunt impedance Z_e, while the system impedance at the source plane $Z_1 = p_1/u_1$. This representation corresponds to a fluctuating mass injection with an effective volume velocity $V_s = u_s S_s$. One notes that at the source plane the circuit model represents continuity of pressure with the source strength equated to a discontinuity of volume velocity. Thus this is equivalent to an acoustic monopole source in free space. The power output of this source is given by

$$W_r = 0.5 \,\mathrm{Re}(p_1^* V_s) = \frac{0.5 |V_s^2| \,\mathrm{Re}[Z_1/(1 + Z_1/Z_e)]}{S_s}, \qquad (36a,b)$$

where the asterisk represents the complex conjugate. Alternatively, when the source is associated with fluctuating aerodynamic forces, the circuit model in Fig. 4b now represents continuity of velocity u_S across the source plane with the source strength f_s equated to the discontinuity of pressure $(P_s - p_1)$ acting over area S_s, or $(P_s - p_1)S_s$. This is equivalent to an acoustic dipole source in free space. The power output is given by

$$W_f = 0.5 \,\mathrm{Re}(f_s^* u_s) = 0.5 |f_s^2| \,\mathrm{Re}(M_I), \qquad (37a,b)$$

Fig. 4 Acoustic circuits for system excitation: (*a*) by fluctuating volume velocity; (*b*) by fluctuating aerodynamic forces.

where the *input mobility* $M_1 - u_s/f_s$, which also equals $1/Z_e S_s$ for the circuit in Fig. 4*b*. Analogous expressions can be derived for the sources of excitation by mixing noise associated with turbulent shear layers.

In practice, the source of excitation may include a combination of factors related to fluctuating mass, fluctuating force, and flow turbulence. In all such cases the associated acoustic power delivered by the source to the system depends on the magnitude of the source impedance, among other factors, while the acoustic emission from the open termination to the surroundings depends also on the efficiency of power transfer through it. To illustrate the influence of the source and system impedance on such emissions, consider again the acoustic circuit in Fig. 4*a*. At the source plane the mass flux balance is given by

$$u_1 = u_s - \frac{p_1}{Z_e} \tag{38a}$$

or

$$Z_e u_s = p_1 + \rho_0 c_0 u_1 \zeta_e, \tag{38b}$$

where $\zeta_e = Z_e/\rho_0 c_0$. Substitution for p_1 and $\rho_0 c_0 u_1$ in terms of the corresponding wave component amplitudes $p_1^{\pm}$ from Eqs. (28a) and (28b) into Eq. (38b) and then making use of Eqs. (30a) and (30b) and Eq. (31a), one obtains

$$p_2^+ = \frac{Z_e u_s}{(1 + \zeta_e)T_i + (1 - \zeta_e)rT_r]}. \tag{39}$$

If W_1 is the observed reference power and W_2 is the observed power with system T_2, when a reference system T_1 is replaced by another system T_2, provided the termination impedance Z and thus r remain constant and also similarly for Z_e and u_s, one finds that

$$\frac{W_1}{W_2} = \left(\frac{(p_2^+)_1}{(p_2^+)_2} \right)^2, \tag{40}$$

which can be substituted for W_1/W_2 in Eq. (34).

12 CONCLUSIONS

Any practical waveguide consists of a sequence of elements having uniform and regularly shaped boundaries connected by discontinuities with various configurations including expansions, branches, resonators, changes in bounding wall impedance, and so on. Wave energy is transmitted along the duct from a source or distributed sources of excitation to its final termination normally open to the surroundings. Provided that the fluctuating pressure amplitude remains a sufficiently small fraction of the ambient pressure in the medium, wave propagation is well described by adopting the acoustic approximation and formulated with linear models. Provided also that the transverse dimensions remain a small fraction of the acoustic wavelength, a further valid approximation is that wave propagation along each element remains effectively one dimensional.

A physically realistic analysis of sound propagation along waveguides must begin with a complete analytic model of the associated fluid motion, that is, the equations of mass, linear momentum, and energy transport and the fluid's constitutive equations, taken together with the geometric and surface properties of the boundaries. In many cases, useful and sufficiently valid descriptions are obtained by assuming that transfer processes remain isentropic and the system is time invariant. With the acoustic approximation, sound propagation along the regular elements can then be described for small mean flow Mach number by a linearized wave equation that may normally be Fourier transformed on the time t. The resulting solutions will then describe the spectral characteristics of the acoustic wave motion in such elements.

Similar spectral descriptions of wave transfer across discontinuities may also be derived by matching the associated fluid motion over appropriate surfaces on either side of the discontinuity, which together with any relevant fixed boundaries form a control volume. The equations expressing conservation of mass, energy, and momentum are then expressed in integral form. The processes associated with the boundary conditions may include losses as well as generation of nonpropagating (evanescent) higher order modes, whose influence, normally reactive, must be included. One classical method of doing so is by the addition of an appropriate end correction at such junctions. Many applications cited in the literature reduce the control volume to a plane, with continuity of pressure and particle velocity across it, the latter replacing continuity of particle displacement, though this simplification may not always be realistic.

ACKNOWLEDGMENT

The author expresses his thanks to M. J. Fisher for his helpful suggestions and discussion during the preparation of this chapter.

REFERENCES

1. R. G. White and J. G. Walker, *Noise and Vibration*, Ellis Horwood, Chichester, 1982.

2. P. Munger and G. M. L. Gladwell, "Acoustic Wave Propagation Is a Sheared Fluid Contained in a Duct," *J. Sound. Vib.*, Vol. 9, 1969, pp. 28–58.

3. M. L. Munjal, *Acoustics of Ducts and Mufflers*, Wiley, New York, 1987.

4. R. D. Ciskowski and C. A. Brebia (Eds.), *Boundary Element Methods in Acoustics*, Elsevier, London, 1991.

5. W. Rostafinski, "Monograph on Propagation of Sound Waves in Curved Ducts," NASA reference publication 1248, 1991.

6. C. L. Morfey, "Rotating Pressure Patterns in Ducts: Their Generation and Transmission," *J. Sound Vib.*, Vol. 1, 1964, pp. 60–87.

7. S. H. Ko, "Sound Attenuation in Acoustically Lined Circular Ducts in the Presence of Uniform Flow and Shear Flow," *J. Sound Vib.*, Vol. 22, 1972, pp. 193–210.

8. S. H. Ko, "Theoretical Prediction of Sound Attenuation in Acoustically Lined Annular Ducts in the Presence of Uniform Flow and Shear Flow," *J. Acoust. Soc. Am.*, Vol. 54, 1974, pp. 1592–1606.

9. C. L. Morfey, "Acoustic Energy in Non Uniform Flows," *J. Sound Vib.*, Vol. 14, 1971, pp. 159–170.

10. P. O. A. L. Davies, "Practical Flow Duct Acoustics," *J. Sound Vib.*, Vol. 124, 1988, pp. 91–115.

11. M. J. Lighthill, *Waves in Fluids*, Cambridge University Press, Cambridge, 1978.

12. P. E. Doak, "Excitation, Transmission and Radiation of Sound from Ducts," *J. Sound Vib.*, Vol. 31, 1973, I, pp. 1–72, II, pp. 137–174.

13. C. L. Morfey, "A Note on Radiation Efficiency of Acoustic Duct Modes," *J. Sound Vib.*, Vol. 9, 1969, pp. 367–372.

14. P. Munger, H. E. Plumblee, and P. E. Doak, "Analysis of Acoustic Radiation in a Jet Flow Environment," *J. Sound Vib.*, Vol. 36, 1974, pp. 21–52.

15. R. H. Munt, "Acoustic Transmission Properties of a Jet Pipe with Subsonic Flow. I. The Cold Jet Reflection Coefficient," *J. Sound Vib.*, Vol. 142, 1990, pp. 413–436.

16. E. A. Müller (Ed.), *The Mechanics of Sound Generation in Flows*, Springer, Berlin, 1979.

17. P. O. A. L. Davies, "Flow Acoustic Coupling in Ducts," *J. Sound Vib.*, Vol. 77, 1981, pp. 191–209.

18. A. Cummings, "Sound Transmission at a Sudden Area Expansion with Mean Flow," *J. Sound Vib.*, Vol. 38, 1975, pp. 149–155.

19. H. G. Jonasson (Ed.), *Science for Silence*, Vol. 1: *Proc Internoise 90*, Göteburg, Sweden, Noise Control Foundation, NY, 1990, pp. 517–702.

20. A. Cummings, "Prediction Methods for the Performance of Flow Duct Silencers," in H. G. Jonasson (Ed.), *Science for Silence*, Göteburg, Sweden, Noise Control Foundation, NY, 1990, pp. 17–38.

21. P. E. Doak, "Acoustic Wave Propagation in a Homentropic Irrotational Low Mach Number Mean Flow," *J. Sound Vib.*, Vol. 155, 1992, pp. 545–548.

22. P. O. A. L. Davies and P. E. Doak, "Wave Transfer to and from Conical Diffusers with Mean Flow," *J. Sound Vib.*, Vol. 138, 1990, pp. 345–350.

23. K. S. Peat, "The Acoustical Impedance at the Junction of an Extended Inlet or Outlet Duct," *J. Sound Vib.*, Vol. 150, 1991, pp. 101–110.

24. D. Ronneberger, "Experimentelle Untersuchungen zum akustiskhen Reflexionsfaktor von unstetigen Querschnittsänderungen in einem luftdurchströmten Rohr," *Acoustica*, Vol. 19, 1967/68, pp. 222–235.

25. A. Cummings, "Sound Transmission in a Folded Annular Duct," *J. Sound Vib.*, Vol. 41, 1975, pp. 375–379.

26. J. W. Sullivan, "A Method of Modelling Perforate Tube Muffler Components," *J. Acoust. Soc. Am.*, Vol. 66, 1979, I, pp. 772–778, II, pp. 779–788.

27. J. L. Bento Coelho, "Acoustic Characteristics of Perforate Liners in Expansion Chambers," Ph.D. Dissertation, University of Southampton, 1983.

28. J. L. Bento Coelho and P. O. A. L. Davies, "Prediction of Acoustical Performance of Cavity Backed Perforate Liners in Flow Ducts," *Proc 11 Institute of Acoustics Meeting, UK*, 1983, pp. 317–321.

29. H. Tijdemann, "On the Propagation of Sound Waves in Cylindrical Tubes," *J. Sound Vib.*, Vol. 39, 1975, pp. 1–33.

30. J. Kergomard, "Comments on Wall Effects on Sound Propagation in Tubes," *J. Sound Vib.*, Vol. 98, 1985, pp. 149–152.

31. Lord Rayleigh, *Theory of Sound*, Macmillan, London 1896.

32. M. Stinson, "The Propagation of Plane Sound Waves in Narrow and Wide Circular Tubes and Generalization to Uniform Tubes of Arbitrary Cross-Sectional Shape," *J. Acoust. Soc. Am.*, Vol. 89, 1991, pp. 550–558.

33. M. C. A. M. Peters, A. Hirshberg, and A. P. J. Wijnands, "The Aero-acoustic Behaviour of an Open Pipe Exit," *Proc. Noise-93*, Vol. 6, 1993, pp. 191–196.

34. P. O. A. L. Davies, "Plane Acoustic Wave Propagation

in Hot Gas Flows," *J. Sound Vib.*, Vol. 122, 1988, pp. 389–392.

35. A. Cummings, "Ducts with Axial Temperature Gradients," *J. Sound Vib.*, Vol. 51, 1977, pp. 55–68.

36. P. O. A. L. Davies and G. Jiajin, "Finite Amplitude Wave Reflection at an Open Exhaust," *J. Sound Vib.*, Vol. 141, 1990, pp. 165–166.

37. N. Sugimoto, "Burgers Equations with Fractional Derivatives; Hereditary Effects on Nonlinear Acoustic Waves," *J. Fluid Mech.*, Vol. 225, 1991, pp. 631–653.

38. P. O. A. L. Davies, "Transmission Matrix Representation of Exhaust System Acoustic Characteristics," *J. Sound Vib.*, Vol. 151, 1991, pp. 333–338.

9

STEADY-STATE RADIATION FROM SOURCES

A. P. DOWLING

1 INTRODUCTION

Insight into the behavior of many practical sound sources, such as vibrating surfaces, jet flows, and combustion, can be obtained by considering elementary sources. This chapter begins by discussing sources that are so small in comparison with the wavelength of the sound they produce that the sources can be considered as concentrated at a single point.

The simple monopole point source was introduced in Chapter 2. It is an acoustic source that leads to an omnidirectional sound field and causes an unsteady volume outflow of fluid from any surface enclosing the source. A small pulsating sphere is one practical realization of a point monopole source. The characteristics of the sound field generated by a monopole source are discussed in more detail in this chapter, with particular emphasis on the case where the source is periodic in time. Dipole and quadrupole point sources are introduced. These are less effective than monopoles at generating a distant sound field, a characteristic that has important implications in low-Mach-number aeroacoustics (see Part III).

Many sources of sound involve vibrating surfaces, and analytical results for some simple geometries are reviewed. These include pulsating and vibrating spheres and cylinders and baffled pistons. The different forms of the sound field radiated from these surfaces at low and at high frequencies provide physical insight into practical acoustic sources involving vibrating surfaces, such as loudspeakers and active sonar systems.

The chapter concludes with a discussion of the effects of source motion on the radiated sound field.

Encyclopedia of Acoustics, edited by Malcolm J. Crocker
ISBN 0-471-80465-7 © 1997 John Wiley & Sons, Inc.

2 MONOPOLE POINT SOURCE

2.1 Sound Pressure Field

A source that is concentrated at a point and produces an omnidirectional sound field is called a *simple source* or a *monopole point source*. It generates a pressure field consisting of spherical wavefronts traveling out from the source with the sound speed c, with amplitude decaying inversely with distance from the source. The pressure at position $\mathbf{x}$, a distance r away from a monopole point source of *strength* $Q(t)$, is[1-6] given as

$$p_M(\mathbf{x}, t) = \frac{Q(t - r/c)}{4\pi r} \quad (1)$$

at time t; the pressure perturbation a distance r away from the source is related to the source strength at the earlier time $t - r/c$.

If $Q(t)$ is simple harmonic with frequency ω, it is convenient to write the function $Q(t)$ in terms of its complex amplitude $\hat{Q}$. Then $Q(t) - \mathrm{Re}(\hat{Q}e^{i\omega t})$ and Eq. (1) reduces to

$$p_M(\mathbf{x}, t) = \frac{\hat{Q}}{4\pi r} e^{i\omega(t - r/c)}. \quad (2)$$

The real part of $\hat{Q}e^{i\omega t}$ represents the actual source strength, and the real part of the complex pressure in Eq. (2) describes the actual pressure.

2.2 Velocity Field

The particle velocity $\mathbf{u}_M(\mathbf{x}, t)$ produced by a monopole point source is radially outward from the source and can be readily calculated from the radial momentum equation $\rho\, \partial u_{Mr}/\partial t = -\partial p_M/\partial r$ to give[1-6]

$$u_{Mr}(\mathbf{x}, t) = \frac{1}{\rho c}\left(1 + \frac{1}{ikr}\right)\frac{\hat{Q}}{4\pi r}\, e^{i\omega(t - r/c)}$$

$$= \frac{1}{\rho c}\left(1 + \frac{1}{ikr}\right) p_M(\mathbf{x}, t), \qquad (3)$$

where $k = \omega/c$ is the wavenumber and ρ is the mean density.

In the *near field*, the region $kr \ll 1$, the velocity field varies with the inverse square of distance from the source and lags the pressure perturbation by $90°$.

In the *far field*, $kr \gg 1$, $u_{Mr} = p_M/\rho c$, recovering the unidirectional plane-wave result.

2.3   Equivalence to Injection of Fluid

Integration of the radial velocity over the surface of a small sphere centered on the source shows that there is a net outward volume flow rate from the source of $\hat{Q}e^{i\omega t}/i\omega\rho$. A point monopole is equivalent to the injection of fluid at the point, the monopole source strength being equal to the product of the density and the rate of change of the volume flux.

This equivalence highlights the main characteristic of any physical source that approximates to a point monopole. It must produce an unsteady volume outflow from a region that is very small in comparison with the wavelength. Pulsating bubbles, sirens, and unsteady combustion are all examples of monopole sources, as are small bodies vibrating so that they undergo a change in volume.

2.4   Intensity[1-6]

The mean intensity I_M is entirely radial and decays with the inverse square of distance from the source. It is given by

$$I_{Mr} = \overline{p_M u_{Mr}} = \frac{\overline{p_M^2}}{\rho c} = \frac{|\hat{Q}|^2}{32\pi^2 \rho c r^2}. \qquad (4)$$

2.5   Sound Power[1-6]

Since the sound field is omnidirectional, the radiated sound power $P_M = 4\pi r^2 I_{Mr}$, that is,

$$P_M = \frac{|\hat{Q}|^2}{8\pi \rho c}. \qquad (5)$$

2.6   Inhomogeneous Wave Equation

The pressure field due to a monopole point source of strength $Q(t)$ at the origin satisfies the inhomogeneous wave equation[3-5]

$$\left(\frac{1}{c^2}\frac{\partial^2}{\partial t^2} - \nabla^2\right) p_M = Q(t)\delta(\mathbf{x}). \qquad (6)$$

In this equation ∇^2 is the Laplacian operator $\partial^2/\partial x^2 + \partial^2/\partial y^2 + \partial^2/\partial z^2$. The function $\delta(\mathbf{x})$ is zero for $\mathbf{x} \neq \mathbf{0}$, indicating that the source is concentrated at $\mathbf{x} = \mathbf{0}$. Away from the origin the pressure perturbations satisfy the usual homogeneous acoustic wave equation (see Chapters 1 and 2).

3   SUPERPOSITION OF MONOPOLE SOURCES

Since sound is a linear motion, the sound pressure due to two sources of strengths $Q_1(t)$ and $Q_2(t)$ at $\mathbf{y}_1$ and $\mathbf{y}_2$, respectively, can be found by adding their individual pressure fields:

$$p(\mathbf{x}, t) = \frac{Q_1(t - r_1/c)}{4\pi r_1} + \frac{Q_2(t - r_2/c)}{r\pi r_2}, \qquad (7)$$

where r_1 is the distance of $\mathbf{x}$ from the source at $\mathbf{y}_1$, $r_1 = |\mathbf{x} - \mathbf{y}_1|$, and r_2 is the distance of $\mathbf{x}$ from $\mathbf{y}_2$, $r_2 = |\mathbf{x} - \mathbf{y}_2|$.

Similarly the pressure field due to a distributed source can be determined by integration of the field of a single source.

3.1   Line Source

A line source concentrated on the 3-axis of uniform strength $q_l(t)$ per unit length produces a pressure perturbation

$$p_l(\mathbf{x}, t) = \int_{-\infty}^{\infty} \frac{q_l(t - |\mathbf{x} - \mathbf{y}|/c)}{4\pi|\mathbf{x} - \mathbf{y}|}\, dy_3, \qquad (8)$$

where $\mathbf{y} = (0, 0, y_3)$.

If $q_l(t)$ is simple harmonic, $q_l(t) = \hat{q}_l e^{i\omega t}$, Eq. (8) simplifies to

$$p_l(\mathbf{x}, t) = \frac{\hat{q}_l}{4i} H_0^{(2)}(k\sigma)e^{i\omega t}. \qquad (9)$$

Here, $H_n^{(2)}(z)$ is a Hankel function[7] of order n and σ is the radial distance of $\mathbf{x}$ from the 3-axis, $\sigma = (x_1^2 + x_2^2)^{1/2}$. The line source produces a volume outflow rate of $\hat{q}_l e^{i\omega t}/i\omega\rho$ per unit length of 3-axis.

In the near field, $k\sigma \ll 1$,

$$p_l(\mathbf{x}, t) \simeq -\frac{\hat{q}_l}{2\pi} \ln(k\sigma) e^{i\omega t}. \qquad (10)$$

In the far field, $k\sigma \gg 1$.

$$p_l(\mathbf{x}, t) \simeq \sqrt{\frac{1}{8\pi k\sigma}} \, \hat{q}_l e^{i\omega(t - \sigma/c) - i\pi/4}. \qquad (11)$$

Waves travel radially outward with speed c, their amplitude decaying with the inverse square root of distance from the line source.

Intensity and Power   The mean intensity I_l is entirely radial and decays with the inverse of the radial distance from the 3-axis:

$$I_l = \frac{|\hat{q}_l|^2}{16\pi\rho\omega\sigma}. \qquad (12)$$

The radiated sound power per unit length of the 3-axis is

$$P_l = \frac{|\hat{q}_l|^2}{8\rho\omega}. \qquad (13)$$

A small length l of the line source radiates a sound power $l|\hat{q}_l|^2/8\rho\omega$. In contrast, Eq. (5) shows that the power radiated by a point source with the same volume outflow rate, $l\hat{q}_l e^{i\omega t}/i\omega\rho$, is $l^2|\hat{q}_l|^2/8\pi\rho c$. The power output of a small source of length l is enhanced by the factor $\pi(kl)^{-1}$ when it acts alongside other sources to produce a two-dimensional sound field.

3.2 Surface Source

A surface source concentrated in the $x_1 = 0$ plane of uniform strength $q_a(t)$ per unit area produces a pressure perturbation

$$p_a(\mathbf{x}, t) = \int_{-\infty}^{\infty} \int_{-\infty}^{\infty} \frac{q_a(t - |\mathbf{x} - \mathbf{y}|/c)}{4\pi|\mathbf{x} - \mathbf{y}|} \, dy_2 \, dy_3, \quad (14)$$

where $\mathbf{y} = (0, y_2, y_3)$.

If $q_a(t)$ is simple harmonic, $q_a(t) = \hat{q}_a e^{i\omega t}$, Eq. (14) simplifies to

$$p_a(\mathbf{x}, t) = \frac{c\hat{q}_a}{2i\omega} \, e^{i\omega(t - |x_1|/c)}; \qquad (15)$$

one-dimensional waves travel away from the $x_1 = 0$ plane with speed c and unchanging amplitude. This source produces a volume outflow rate of $\hat{q}_a e^{i\omega t}/i\omega\rho$ per unit area.

Intensity and Power[2]   The mean intensity I_a is in the 1-direction and its amplitude is independent of x_1:

$$I_a = \text{sgn}\,(x_1) \frac{c|\hat{q}_a|^2}{8\rho\omega^2}. \qquad (16)$$

The radiated sound power per unit area is given by

$$P_a = \frac{c|\hat{q}_a|^2}{4\rho\omega^2}. \qquad (17)$$

A small area A of the surface source radiates a sound power $Ac|\hat{q}_a|^2/4\rho\omega^2$. A point source with the same volume outflow rate of $A\hat{q}_a e^{i\omega t}/i\omega\rho$ only generates a sound $A^2|\hat{q}_a|^2/8\pi\rho c$. The power output of the source of small area A is enhanced by the factor $2\pi/k^2 A$ when it acts alongside other surface sources to produce a one-dimensional sound field. The power output of a small loudspeaker is enhanced by a horn, which constrains the radiation to be one dimensional near the source.

3.3 Two Monopole Point Sources with Different Phase

Consider two monopole point sources of frequency ω and amplitude $\hat{Q}$ positioned a distance l apart, as shown in Fig. 1. One source leads the other by β, that is, $Q_1(t) = \hat{Q}e^{i(\omega t + \beta)}$ and $\hat{Q}_2(t) = \hat{Q}e^{i\omega t}$. These sources generate the pressure field

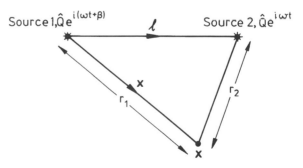

Fig. 1 Two monopole point sources a distance l apart.

$$p(\mathbf{x}, t) = \frac{\hat{Q}}{4\pi r_1} e^{i[\omega(t - r_1/c) + \beta]} + \frac{\hat{Q}}{4\pi r_2} e^{i\omega(t - r_2/c)}, \quad (18)$$

where r_1 is the distance from $\mathbf{x}$ to the source at the origin, $r_1 = |\mathbf{x}|$, and r_2 is the distance from $\mathbf{x}$ to the source at l, $r_2 = |\mathbf{x} - \mathbf{l}|$.

The sound power emitted by each individual source can be found by integrating the intensity over a small sphere centered on the source. This gives

$P_1 =$ sound power emitted by source 1

$$= \frac{|\hat{Q}|^2}{8\pi\rho c} \left(1 + \frac{1}{kl} \sin(kl + \beta)\right), \quad (19a)$$

$P_2 =$ sound power emitted by source 2

$$= \frac{|\hat{Q}|^2}{8\pi\rho c} \left(1 + \frac{1}{kl} \sin(kl - \beta)\right), \quad (19b)$$

$P_\infty =$ sound power radiated to infinity

$$= P_1 + P_2 = \frac{|\hat{Q}|^2}{4\pi\rho c} \left(1 + \frac{1}{kl} \sin(kl) \cos \beta\right). \quad (19c)$$

Comparison with the single-source result in Eq. (5) shows that the sound power emitted by a source is altered when there is a coherent source nearby.

Special Cases

(i) $\beta = 0$, $P_1 = P_2 = \dfrac{P_\infty}{2} = \dfrac{|\hat{Q}|^2}{8\pi\rho c} \left(1 + \dfrac{1}{kl} \sin(kl)\right).$

$$(20)$$

The power output of a monopole source is increased when there is a second source of the same phase within half a wavelength, that is when $l < \frac{1}{2}\lambda$, where the wavelength $\lambda = 2\pi/k$:

(ii) $\beta = \frac{1}{2}\pi$, $P_1 = \dfrac{|\hat{Q}|^2}{8\pi\rho c} \left(1 + \dfrac{1}{kl} \cos(kl)\right),$

$$(21a)$$

$$P_2 = \frac{|\hat{Q}|^2}{8\pi\rho c} \left(1 - \frac{1}{kl} \cos(kl)\right), \quad (21b)$$

$$P_\infty = \frac{|\hat{Q}|^2}{4\pi\rho c}. \quad (21c)$$

There is now an exchange of energy between the two sources. Both sources emit power $|\hat{Q}|^2/8\pi\rho c$, which is radiated to infinity, but source 1 emits $\cos(kl)|\hat{Q}|^2/8\pi\rho ckl$ additional power, which is absorbed by source 2:

(iii) $\beta = \pi$, $P_1 = P_2 = \dfrac{P_\infty}{2} = \dfrac{|\hat{Q}|^2}{8\pi\rho c} \left(1 - \dfrac{1}{kl} \sin(kl)\right).$

$$(22)$$

The power output of a monopole source is decreased when there is a second source out of phase within half a wavelength, that is when $l < \lambda/2$. In particular, for $kl \ll 1$, Eq. (22) becomes

$$P_1 = P_2 = \frac{P_\infty}{2} = \frac{(|\hat{Q}|kl)^2}{48\pi\rho c}. \quad (23)$$

3.4 Line Array of Monopole Sources[3]

Consider a line array consisting of N harmonic monopole point sources equally spaced along the 1-axis at distance d apart, as shown in Fig. 2.

If all these sources have the same monopole strength $\hat{Q}e^{i\omega t}$, they generate the pressure field

$$p(\mathbf{x}, t) = \sum_{n=0}^{N-1} \frac{\hat{Q}}{4\pi r_n} e^{i\omega(t - r_n/c)}, \quad (24)$$

where $r_n = |\mathbf{x} - nd\mathbf{e}_1|$. When $\mathbf{x}$ is sufficiently far from the line array so that $r = |\mathbf{x}| \gg L$, where $L = (N - 1)d$ is the length of the array, $r_n \simeq r - nd\cos\theta$, with $\cos\theta = x_1/r$. The sum in Eq. (24) can then be evaluated to give

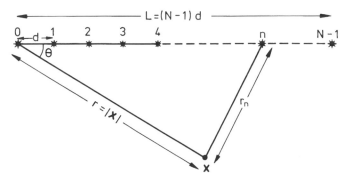

Fig. 2 The N sources forming a line array.

$$p(\mathbf{x}, t) = \frac{N\hat{Q}}{4\pi r} \frac{\sin(\frac{1}{2}Nkd\cos\theta)}{N\sin(\frac{1}{2}kd\cos\theta)}$$

$$\times e^{i[\omega t - kr - (k/2)L\cos\theta]}. \qquad (25)$$

At any angle θ, the mean intensity vector is radial, with

$$I_r(r, \theta) = \frac{N^2|\hat{Q}|^2}{32\pi^2\rho cr^2} \left| \frac{\sin(\frac{1}{2}Nkd\cos\theta)}{N\sin(\frac{1}{2}kd\cos\theta)} \right|^2. \qquad (26)$$

In a direction normal to the array ($\theta = 90°$)

$$I_r(r, \tfrac{1}{2}\pi) = \frac{N^2|\hat{Q}|^2}{32\pi^2\rho cr^2}. \qquad (27)$$

Hence at a general angle θ to the array,

$$I_r(r, \theta) = I_r(r, \tfrac{1}{2}\pi) \left| \frac{\sin(\frac{1}{2}Nkd\cos\theta)}{N\sin(\frac{1}{2}kd\cos\theta)} \right|^2. \qquad (28)$$

The function $D(\theta) = |\sin(\frac{1}{2}Nkd\cos\theta)/[N\sin(\frac{1}{2}kd\cos\theta)]|^2$ describes how the acoustic intensity varies with angle. It is usually called the *directional factor* and is plotted as a function of polar angle θ in Fig. 3 for $N = 5$ and various values of kd.

Low-Frequency Limit   A source whose size L is small in comparison with k^{-1} is said to be *compact*. For $Nkd \ll 1$, the array length is compact, and $D(\theta)$ is approximately equal to unity for all values of θ. The pressure field is omnidirectional (see Fig. 3a) and its strength is N times that produced by a single monopole point source.

High-Frequency Limit   The term $D(\theta)$ has a maximum value of unity that it attains in directions normal to the array ($\theta = 90°$). The pressure field vanishes whenever

$$\cos\theta = \frac{2\pi n}{Nkd} \quad \text{for integers } n \text{ that are not multiples of } N \qquad (29)$$

The region $\pi - \cos^{-1}(2\pi/Nkd) > \theta > \cos^{-1}(2\pi/Nkd)$ is called the *major lobe*.

If kd is large enough, the pressure field can have a

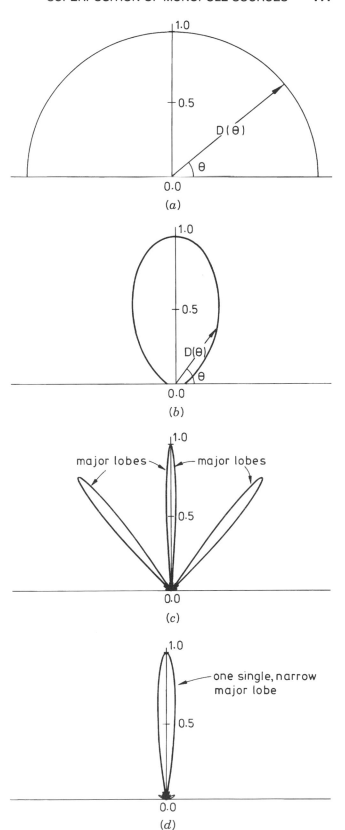

Fig. 3   Directional factor $D(\theta)$ for a line array with $N = 5$: (a) $kd = 0.1$; (b) $kd = 1.0$; (c) $kd = 10.0$; (d) $kd = 2\pi(N - 1)/N = 1.6\pi$.

second major lobe, since $D(\theta)$ is also equal to unity at $\theta = \cos^{-1}(2\pi/kd)$ for $kd > 2\pi$. In general, a major lobe is obtained whenever $\theta = \cos^{-1}(2\pi M/kd)$ for some integer $M < kd/2\pi$.

In some applications a single major lobe as narrow as possible is required. A good approximation to this can be achieved by choosing $kd = 2\pi(N - 1)/N$. This ensures that there is no second major lobe (see Fig. 3d). The primary major lobe is confined to angles $\pi - \cos^{-1}[1/(N-1)] > \theta > \cos^{-1}[1/(N - 1)]$, and the width of this lobe can be decreased by increasing the number of elements N.

Steering   Sometimes it is required to transmit a beam at a specific angle to the array. This can be achieved by inserting a time delay $n\tau$ in the source strength of the nth element. Then

$$p(\mathbf{x}, t) = \sum_{n=0}^{N-1} \frac{\hat{Q}}{4\pi r_n} e^{i\omega(t - n\tau - r_n/c)}. \qquad (30)$$

This procedure is called *steering* and, for $r \gg L$, leads to the directional factor

$$D(\theta) = \left| \frac{\sin[\frac{1}{2}Nk(d\cos\theta - c\tau)]}{N\sin[\frac{1}{2}k(d\cos\theta - c\tau)]} \right|^2. \qquad (31)$$

The primary major lobe is now transmitted in a direction $\theta_s = \cos^{-1}(c\tau/d)$ to the line array.

4   DIPOLE POINT SOURCE

4.1   Sound Pressure Field[1-6]

A *dipole point source* at the origin of *vector strength* $\mathbf{F}(t)$ produces a pressure field

$$p_D(\mathbf{x}, t) = -\mathrm{div}\left(\frac{\mathbf{F}(t - r/c)}{4\pi r} \right)$$

$$= -\sum_{i=1}^{3} \frac{\partial}{\partial x_i}\left(\frac{F_i(t - r/c)}{4\pi r} \right), \qquad (32)$$

where $r = |\mathbf{x}|$. The derivative can be evaluated to show that

$$p_D(\mathbf{x}, t) = \frac{1}{4\pi}\left(\frac{\mathbf{x} \cdot \mathbf{F}(t - r/c)}{r^3} + \frac{\mathbf{x}}{r^2 c} \cdot \frac{\partial \mathbf{F}(t - r/c)}{\partial t} \right). \qquad (33)$$

When the direction of the dipole is constant, this simplifies to

$$p_D(\mathbf{x}, t) = \frac{\cos\theta}{4\pi}\left(\frac{1}{cr}\frac{\partial F(t - r/c)}{\partial t} + \frac{F(t - r/c)}{r^2} \right), \qquad (34)$$

where $F = |\mathbf{F}|$ and θ is the angle between $\mathbf{F}$, the direction of the dipole axis, and $\mathbf{x}$, the radius vector to the observation point.

If $F(t)$ is simple harmonic of frequency ω, $F(t) = \mathrm{Re}(\hat{F}e^{i\omega t})$, Eq. (34) reduces to

$$p_D(\mathbf{x}, t) = \frac{ik\cos\theta}{4\pi r}\left(1 + \frac{1}{ikr} \right)\hat{F}e^{i\omega(t - r/c)}. \qquad (35)$$

The pressure field is directional and proportional to $\cos\theta$. It has maximum amplitude in line with the dipole axis and decreases to zero at $\theta = 90°$. In the far field, $kr \gg 1$, the pressure varies inversely with distance from the source, while in the near field, $kr \ll 1$, it is proportional to the inverse square of the distance from the source.

The ratio of the pressure perturbation at $r = R \gg k^{-1}$ to that at $r = \epsilon \ll k^{-1}$ in the same direction θ is $k\epsilon^2/R$. This is a factor $k\epsilon(\ll 1)$ smaller than the corresponding result for the pressure generated by a monopole point source. A point dipole is therefore less efficient than a monopole at converting the near-field pressures into far-field sound.

4.2   Velocity Field[1,2,5,6]

The particle velocity follows directly from Eq. (35) and the momentum equation. In terms of spherical polar coordinates (r, θ, ϕ), centered on the point dipole,

$$u_{Dr} = \frac{ik\cos\theta}{4\pi\rho cr}\left(1 + \frac{2}{ikr} - \frac{2}{k^2r^2} \right)\hat{F}e^{i\omega(t - r/c)}, \qquad (36a)$$

$$u_{D\theta} = \frac{\sin\theta}{4\pi\rho cr^2}\left(1 + \frac{1}{ikr} \right)\hat{F}e^{i\omega(t - r/c)}, \qquad (36b)$$

$$u_{D\phi} = 0. \qquad (36c)$$

In the far field the particle velocity is radial with $u_{Dr} = p_D/\rho c$, recovering the plane-wave result. The near acoustic field is more complicated than that generated by a monopole point source, and the particle velocity is not entirely radial.

4.3 Equivalence to a Point Force

The dipole source produces no net volume outflow. However, it exerts a force $\mathbf{F}(t)$ on the fluid. A vibrating body, which is very small in comparison with the wavelength and whose volume is fixed, approximates to a point dipole (apart from the special case when the body has the same density as the surrounding medium; then the net dipole strength is zero).[2]

4.4 Intensity[1,2,5,6]

The mean intensity vector $\mathbf{I}_D = \overline{p_D \mathbf{u}_D}$ is entirely radial and decays with the inverse square of the distance from the source:

$$I_{Dr} = \frac{k^2 \cos^2 \theta |\hat{F}|^2}{32\pi^2 \rho c r^2}. \qquad (37)$$

4.5 Sound Power[1–6]

The sound power radiated to infinity is given as

$$P_D = \frac{k^2 |\hat{F}|^2}{24\pi \rho c}. \qquad (38)$$

4.6 Inhomogeneous Wave Equation[4,5]

The pressure field due to a dipole point source at the origin of vector strength $\mathbf{F}(t)$ satisfies the inhomogeneous wave equation

$$\left(\frac{1}{c^2} \frac{\partial^2}{\partial t^2} - \nabla^2 \right) p_D = -\text{div}(\mathbf{F}(t)\delta(\mathbf{x})). \qquad (39)$$

4.7 Dipole as Superposition of Monopoles[1,6]

For two simple sources close together, equal in magnitude, and of opposite sign (as shown in Fig. 1 with $\beta = \pi$), Eq. (18) can be expanded as a Taylor series in l. This shows that

$$p(\mathbf{x}, t) = -\text{div}\left(\frac{\hat{Q}\mathbf{l}}{4\pi r} e^{i\omega(t - r/c)} \right) \qquad (40)$$

for $l = |\mathbf{l}| \ll r$ and k^{-1}. This is equal to the dipole sound field described by Eq. (32) provided $\mathbf{F} = \hat{Q}\mathbf{l}e^{i\omega t}$. Therefore a point dipole is equivalent to two equal and opposite nearby point monopoles. The vector dipole strength is equal to the product of the monopole source strength and their distance apart in magnitude and is in the direction of the separation between the monopoles.

5 QUADRUPOLE POINT SOURCE

5.1 Sound Pressure Field[4,5]

A *quadrupole point source* at the origin of strength $T_{ij}(t)$, $i = 1, 2, 3, j = 1, 2, 3$, produces a pressure field

$$p_Q(\mathbf{x}, t) = \sum_{i=1}^{3} \sum_{j=1}^{3} \frac{\partial^2}{\partial x_i \, \partial x_j} \frac{T_{ij}(t - r/c)}{4\pi r}. \qquad (41)$$

The quadrupole source strength can be conveniently written as a matrix:

$$(T_{ij}) = \begin{pmatrix} T_{11} & T_{12} & T_{13} \\ T_{21} & T_{22} & T_{23} \\ T_{31} & T_{32} & T_{33} \end{pmatrix}. \qquad (42)$$

The strength of a point quadrupole source is specified by the functional form of its components.

A quadrupole point source can be considered as the superposition of two nearby dipoles of equal magnitude and opposite strengths (see Section 5.3). One practical example of a quadrupole source is a tuning fork, whose prongs vibrate in antiphase. Quadrupole sources are important because the momentum flux in air flows is a quadrupole source and is the main source mechanism in jet noise (see Chapters 27 and 28).

If $T_{ij}(t)$ is harmonic of frequency ω, $T_{ij}(t) = \text{Re}(\hat{T}_{ij}e^{i\omega t})$, the derivatives in Eq. (41) can be evaluated to give

$$p_Q(\mathbf{x}, t) = -\sum_{i=1}^{3} \sum_{j=1}^{3} \left[\frac{x_i x_j}{r^2} \left(1 + \frac{3}{ikr} - \frac{3}{k^2 r^2} \right) \right.$$
$$\left. - \delta_{ij} \left(\frac{1}{ikr} - \frac{1}{k^2 r^2} \right) \right] \frac{k^2 \hat{T}_{ij}}{4\pi r} e^{i\omega(t - r/c)}. \qquad (43)$$

In the far field this simplifies to

$$p_Q(\mathbf{x}, t) = -\frac{k^2 \hat{T}_{rr}}{4\pi r} e^{i\omega(t - r/c)}, \qquad (44)$$

where $\hat{T}_{rr} = (x_i x_j / r^2)\hat{T}_{ij}$ is the component of $\hat{T}_{ij}$ in the observer's direction.

Simple expressions for the pressure field can be determined if only one component of T_{ij} is nonzero. If this is one of the diagonal terms, the quadrupole .is called *longitudinal*.[4] If T_{11} is the only nonzero component, then

$$p_Q(\mathbf{x}, t) = -\frac{k^2 \hat{T}_{11}}{4\pi r} \left[\cos^2 \theta \left(1 + \frac{3}{ikr} - \frac{3}{k^2 r^2} \right) \right.$$
$$\left. - \left(\frac{1}{ikr} - \frac{1}{k^2 r^2} \right) \right] e^{i\omega(t - r/c)}. \qquad (45)$$

where $\cos \theta = x_1/r$. The far-field pressure of a longitudinal quadrupole has directivity $\cos^2 \theta$ and decays inversely with distance from the source. In the near field the pressure varies with the inverse cube of distance from the source and has directivity $1 - 3\cos^2 \theta$.

If one of the off-diagonal terms of T_{ij} is the only nonzero component, the quadrupole is called *lateral*.[4] If T_{12} is the only nonzero component, then

$$p_Q(\mathbf{x}, t) = -\frac{k^2 \hat{T}_{12}}{4\pi r} \cos \theta \sin \theta \cos \phi$$
$$\cdot \left(1 + \frac{3}{ikr} - \frac{3}{k^2 r^2} \right) e^{i\omega(t - r/c)}, \qquad (46)$$

where (r, θ, ϕ) are spherical polar coordinates with $\cos \theta = x_1/r$, $\sin \theta \cos \phi = x_2/r$. A lateral quadrupole has the same directional dependence in the pressure field at all distances from the source. The far-field pressure decays inversely with distance from the source, while the near-field pressure decays with the inverse cube of distance from the source.

A quadrupole source that has $T_{11} = T_{22} = T_{33} \neq 0$ and all its off-diagonal components $T_{12} = T_{13} = T_{23} = 0$ is called an *isotropic quadrupole source*. It produces a pressure field

$$p_Q(\mathbf{x}, t) = -\frac{k^2 \hat{T}_{11}}{4\pi r} e^{i\omega(t - r/c)}. \qquad (47)$$

The pressure field of an isotropic quadrupole is identical to that of a monopole of strength $-k^2 \hat{T}_{11} e^{i\omega t}$. It is omnidirectional and everywhere decays inversely with distance from the source.

5.2 Sound Power[4]

A longitudinal quadrupole $\hat{T}_{11} e^{i\omega t}$ emits a sound power $k^4 |\hat{T}_{11}|^2 / 40\pi\rho c$. A lateral quadrupole $\hat{T}_{12} e^{i\omega t}$ emits a sound power $k^4 |\hat{T}_{12}|^2 / 120\pi\rho c$.

5.3 Quadrupole as a Superposition of Dipoles[4]

A quadrupole point source can be considered as two nearby dipoles of equal but opposite strengths. Equation

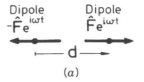

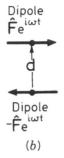

Fig. 4 Quadrupoles as a superposition of dipoles: (*a*) a longitudinal quadrupole. (*b*) a lateral quadrupole.

(32) shows that the pressure field generated by a point dipole of strength $\hat{\mathbf{F}} e^{i\omega t}$ can be written in the form

$$p_D(\mathbf{x}, t) = -\text{div} \left(\frac{\hat{\mathbf{F}} e^{i\omega(t - r/c)}}{4\pi r} \right).$$

The combined pressure field generated by a dipole of strength $\hat{\mathbf{F}} e^{i\omega t}$ at $\mathbf{x} = \mathbf{d}$ and $-\hat{\mathbf{F}} e^{i\omega t}$ at $\mathbf{x} = \mathbf{0}$ can be written as

$$p(\mathbf{x}, t) = \sum_{i=1}^{3} \sum_{j=1}^{3} d_i \hat{F}_j \frac{\partial^2}{\partial x_i \, \partial x_j} \left(\frac{e^{i\omega(t - r/c)}}{4\pi r} \right) \qquad (48)$$

for $|\mathbf{d}| \ll k^{-1}$ and r. This is equal to the quadrupole sound field described by Eq. (41) provided $T_{ij}(t) = d_i \hat{F}_j e^{i\omega t}$. A point quadrupole is equivalent to two equal and opposite nearby point dipoles.

For a longitudinal quadrupole the separation between the dipoles is in the direction of their axis, as shown in Fig. 4a. For a lateral quadrupole the separation is perpendicular to the dipole axis (Fig. 4b).

6 RECIPROCITY OF SOURCE AND FIELD POINTS

6.1 Reciprocity in an Unbounded Region

Denote the sound field generated at $\mathbf{x}$ by a harmonic monopole point source of unit strength at $\mathbf{x}_0$ by $G(\mathbf{x}|\mathbf{x}_0)$.

It is evident from Eq. (2) that

$$G(\mathbf{x}|\mathbf{x}_0) = \frac{e^{i\omega(t - |\mathbf{x} - \mathbf{x}_0|/c)}}{4\pi|\mathbf{x} - \mathbf{x}_0|}. \quad (49)$$

This is symmetric in $\mathbf{x}$ and $\mathbf{x}_0$, so that

$$G(\mathbf{x}_0|\mathbf{x}) = G(\mathbf{x}|\mathbf{x}_0); \quad (50)$$

from the definition of G. The function $G(\mathbf{x}_0|\mathbf{x})$ denotes the pressure at $\mathbf{x}_0$ due to a unit point monopole at $\mathbf{x}$. Equation (50) states that this is identical to the pressure at $\mathbf{x}$ due to a unit point monopole at $\mathbf{x}_0$. The pressure field is unchanged when the positions of the source and the listener are exchanged. This is referred to as *reciprocity* and can be generalized to other sources.

The pressure field $p_Q(\mathbf{x})$ due to a point monopole of strength $\hat{Q}e^{i\omega t}$ at $\mathbf{x}_0$ is given by

$$p_Q(\mathbf{x}) = \hat{Q}G(\mathbf{x}_0|\mathbf{x}). \quad (51)$$

That is, the pressure at $\mathbf{x}$ due to a monopole at $\mathbf{x}_0$ is equal to the product of the monopole source strength and the field at $\mathbf{x}_0$ due to a unit monopole at $\mathbf{x}$.

The pressure field $p_F(\mathbf{x})$ due to a point dipole of strength $\hat{\mathbf{F}}e^{i\omega t}$ at $\mathbf{x}_0$ is given by

$$p_F(\mathbf{x}) = \sum_{i=1}^{3} \hat{F}_i \frac{\partial}{\partial x_{0i}} G(\mathbf{x}_0|\mathbf{x})$$

$$= -\sum_{i=1}^{3} \hat{F}_i \frac{\partial}{\partial x_i} G(\mathbf{x}_0|\mathbf{x}). \quad (52)$$

That is, the pressure field at $\mathbf{x}$ due to a dipole at $\mathbf{x}_0$ is equal to the product of the dipole source strength and the gradient of the field at $\mathbf{x}_0$ due to a unit monopole at $\mathbf{x}$.

The pressure field $p_T(\mathbf{x})$ due to a point quadrupole of strength $\hat{T}_{ij}e^{i\omega t}$ at $\mathbf{x}_0$ is given by

$$p_T(\mathbf{x}) = \sum_{i=1}^{3} \sum_{j=1}^{3} \hat{T}_{ij} \frac{\partial^2}{\partial x_{0i} \partial x_{0j}} G(\mathbf{x}_0|\mathbf{x})$$

$$= \sum_{i=1}^{3} \sum_{j=1}^{3} \hat{T}_{ij} \frac{\partial^2}{\partial x_i \partial x_j} G(\mathbf{x}_0|\mathbf{x}). \quad (53)$$

That is, the pressure field at $\mathbf{x}$ due to a quadrupole at $\mathbf{x}_0$ is equal to the product of the quadrupole source strength and the double gradient of the field at $\mathbf{x}_0$ due to a unit monopole at $\mathbf{x}$.

6.2 Reciprocity in a Bounded Region

The reciprocal theorems expressed in Eqs. (50)–(53) were derived here for an unbounded region. They remain true[4] in the presence of a wide variety of different types of surfaces. These include hard and soft surfaces, elastic surfaces such as membranes, shells and plates,[8] and locally reacting surfaces,[5] whose boundary conditions can be expressed in the form

$$\frac{\partial p}{\partial n} = K(\mathbf{x}, \omega)p \quad \text{for some } K. \quad (54)$$

Here, $K(\mathbf{x}, \omega)$ may be a function of both position and frequency.

7 MULTIPOLE-SOURCE EXPANSION[2,4,5]

Consider a source distributed over a volume V with strength $\hat{q}(\mathbf{x})e^{i\omega t}$/unit volume. This source radiates a pressure field

$$p(\mathbf{x}, t) = \frac{e^{i\omega t}}{4\pi} \int \hat{q}(\mathbf{y}) \frac{e^{-i\omega|\mathbf{x} - \mathbf{y}|/c}}{|\mathbf{x} - \mathbf{y}|} \, dV. \quad (55)$$

Let the origin be within V and the maximum radial distance of points within V from the origin be l. If kl is small, the source region is said to be *compact*.

For a compact source region, when $r = |\mathbf{x}|$ is large in comparison with l, $e^{-i\omega|\mathbf{x} - \mathbf{y}|/c}/|\mathbf{x} - \mathbf{y}|$ can be expanded as a Taylor series in $\mathbf{y}$:

$$\frac{e^{-ik|\mathbf{x} - \mathbf{y}|}}{|\mathbf{x} - \mathbf{y}|} = \frac{e^{-ikr}}{r} - (\mathbf{y} \cdot \nabla)\left(\frac{e^{-ikr}}{r}\right)$$

$$+ \frac{1}{2} (\mathbf{y} \cdot \nabla)(\mathbf{y} \cdot \nabla)\left(\frac{e^{-ikr}}{r}\right) + \cdots .$$

$$(56)$$

where the operator $\nabla = (\partial/\partial x_1, \partial/\partial x_2, \partial/\partial x_3)$.

Substitution of this expansion into Eq. (55) leads to

$$p(\mathbf{x}, t) = \frac{\hat{Q}e^{i\omega(t - r/c)}}{4\pi r} - \hat{\mathbf{F}} \cdot \nabla\left(\frac{e^{i\omega(t - r/c)}}{4\pi r}\right)$$

$$+ \sum_{i=1}^{3} \sum_{j=1}^{3} \hat{T}_{ij} \frac{\partial^2}{\partial x_i \partial x_j}\left(\frac{e^{i\omega(t - r/c)}}{4\pi r}\right)$$

$$+ \cdots \quad \text{for} \quad r \text{ and } k^{-1} \gg l, \quad (57)$$

where

$$\hat{Q} = \int_V \hat{q}(\mathbf{y}) \, dV, \tag{58a}$$

$$\hat{\mathbf{F}} = \int_V \mathbf{y}\hat{q}(\mathbf{y}) \, dV, \tag{58b}$$

$$\hat{T}_{ij} = \int_V \frac{1}{2} y_i y_j \hat{q}(\mathbf{y}) \, dV. \tag{58c}$$

At distances large compared with the source size, the acoustic field generated by a source distributed over a compact region is equivalent to a multipole expansion of point sources at the origin.

7.1 Compact Source Region with Monopole Far Field

If $\hat{Q} \neq 0$, $\hat{F}_i \sim O(l\hat{Q})$, $\hat{T}_{ij} \sim O(l^2\hat{Q})$, and so on, and

$$p(\mathbf{x}, t) \simeq \frac{\hat{Q}e^{i\omega(t - r/c)}}{4\pi r} \quad \text{for} \quad r \text{ and } k^{-1} \gg l. \tag{59}$$

When the total monopole source strength is nonzero, a distributed source generates a distant sound field equivalent to that of a monopole point source[2] at the origin of strength $\int \hat{q}(\mathbf{y}) \, dV$.

7.2 Compact Source Region with Dipole Far Field

If $\hat{Q} = 0$ and $\hat{\mathbf{F}} \neq \mathbf{0}$, then

$$p(\mathbf{x}, t) \simeq -\hat{\mathbf{F}} \cdot \nabla \left(\frac{e^{i\omega(t - r/c)}}{4\pi r} \right) \quad \text{for} \quad r \text{ and } k^{-1} \gg l. \tag{60}$$

The sound field is equivalent to that of a dipole point source at the origin,[2] the strength of the equivalent dipole being the first moment of $\hat{q}(\mathbf{y})$, and $\hat{\mathbf{F}} = \int \mathbf{y}\hat{q}(\mathbf{y}) \, dV$.

7.3 Compact Source Region with Quadrupole Far Field

If $\hat{Q} = 0$, $\hat{\mathbf{F}} = \mathbf{0}$ and $\hat{T}_{ij} \neq 0$, then

$$p(\mathbf{x}, t) = \sum_{i=1}^{3} \sum_{j=1}^{3} \hat{T}_{ij} \frac{\partial^2}{\partial x_i \, \partial x_j} \left(\frac{e^{i\omega(t - r/c)}}{4\pi r} \right). \tag{61}$$

The sound field is equivalent to that of a quadrupole point source at the origin, the strength of the equivalent quadrupole being dependent on the second-order moments of $\hat{q}(\mathbf{y})$, $\hat{T}_{ij} = \int \frac{1}{2} y_i y_j \hat{q}(\mathbf{y}) \, dV$.

8 HARMONICALLY PULSATING AND VIBRATING SPHERES

8.1 Pulsating Sphere

Consider an expanding and contracting sphere whose radius at time t is $a + \hat{\epsilon}e^{i\omega t}$, where $\hat{\epsilon}$ is small in comparison with both a and k^{-1}. Such a sphere generates a pressure field[2-6]

$$p(\mathbf{x}, t) = p(r, t) = \frac{i\omega a^2 \rho \hat{v}_s}{r(1 + ika)} e^{i\omega[t - (r - a)/c]}, \tag{62}$$

where $\hat{v}_s e^{i\omega t}$ is the surface velocity on $r = a$, $\hat{v}_s = i\omega\hat{\epsilon}$. This pressure field is omnidirectional and has the same dependence on radial distance r as that produced by a monopole point source [see Eq. (2)]. The *specific acoustic impedance* on $r = a$, Z_s, is defined by

$$Z_s = \frac{p(a, t)}{\hat{v}_s e^{i\omega t}} = \frac{ika}{1 + ika} \rho c. \tag{63}$$

The sound power radiated to infinity can be determined by integrating the energy flux through a large sphere,

$$P_\infty = 2\pi \frac{(ka)^2}{1 + (ka)^2} \rho c a^2 |\hat{v}_s|^2. \tag{64}$$

High-Frequency Limit[2,4,5]

$$p(\mathbf{x}, t) = \frac{a}{r} \rho c \hat{v}_s e^{i\omega[t - (r - a)/c]} \quad \text{for} \quad ka \gg 1. \tag{65}$$

The specific acoustic impedance on $r = a$ is ρc, recovering the plane-wave (or geometric acoustics) result. The factor a/r in Eq. (65) leads to a factor $(a/r)^2$ in intensity, and accounts for the fact that energy fed over the surface area $4\pi a^2$ of the sphere spreads out over the larger area $4\pi r^2$ at radius r. The time delay $(r - a)/c$ in Eq. (65) is the time taken for sound to reach $\mathbf{x}$ from the nearest point on the sphere. In the geometric acoustics limit, the pressure perturbation at $\mathbf{x}$ is produced by sound traveling along a ray, which takes the shortest path from the body surface to $\mathbf{x}$.

Since the surface pressure and velocity are virtually in phase, the sphere does a significant amount of work on the fluid, which is radiated away as sound energy, and

$$P_\infty = 2\pi\rho c a^2 |\hat{v}_s|^2 \quad \text{for} \quad ka \gg 1. \tag{66}$$

Low-Frequency Limit[2–6]

$$p(\mathbf{x}, t) = \frac{\hat{Q}}{4\pi r} \, e^{i\omega(t - r/c)} \quad \text{for} \quad ka \ll 1, \quad (67)$$

with $\hat{Q} = i\omega\rho 4\pi a^2 \hat{v}_s$. A compact pulsating sphere is equivalent to a monopole point source of strength $\hat{Q}e^{i\omega t}$, where $\hat{Q}e^{i\omega t}$ is the product of the density and the rate of change of volume enclosed by the sphere (cf. Section 2.1). This remains true even if the change in radius is not small in comparison with a, provided that $\hat{Q}e^{i\omega t}$ is the time derivative of the actual volume enclosed by the sphere, the sphere remains compact, and the speed of the surface motion is much smaller than the speed of sound.[4,9] On a compact sphere, the surface pressure and velocity are virtually 90° out of phase. This means that the sphere does little work on the surrounding fluid. The surface motion is therefore inefficient at radiating sound power. Indeed, $P_\infty = 2\pi(ka)^2 \rho c a^2 |\hat{v}_s|^2$, a factor $(ka)^2$ smaller than the power radiated by the large sphere in Eq. (66).

8.2 Vibrating Sphere

A rigid sphere vibrating with small amplitude about the origin in the 1-direction with speed $\hat{U}e^{i\omega t}$ has a normal surface velocity $\hat{U}\cos\theta e^{i\omega t}$, where $\cos\theta = x_1/r$. This leads to a pressure field[2,4–6]

$$p(\mathbf{x}, t) = p(r, \theta, t) = -\frac{(ka)^2 \rho c a \cos\theta}{2(1 + ika) - (ka)^2} \left(1 + \frac{1}{ikr}\right)$$

$$\times \frac{\hat{U}}{r} \, e^{i\omega[t - (r - a)/c]}, \quad (68)$$

where $k = \omega/c$. The pressure field has the same directivity and dependence on r as that produced by a dipole point source with its axis in the 1-direction [see Eq. (35)]. The specific acoustic impedance on $r = a$ is given by

$$Z_s = \frac{ika(1 + ika)}{2(1 + ika) - (ka)^2} \, \rho c. \quad (69)$$

The force $\mathbf{f}(t)$ exerted on the fluid by the sphere is in the 1-direction, with

$$f_1(t) = \int_0^\pi p(a, \theta, t) \cos\theta \, 2\pi a^2 \sin\theta \, d\theta$$

$$= \frac{2\pi}{3} \frac{ika(1 + ika)}{1 + ika - \frac{1}{2}(ka)^2} \, \rho c a^2 \hat{U} e^{i\omega t}. \quad (70)$$

The *mechanical impedance* of the surface of the sphere, Z_m, can be found by dividing $f_1(t)$ by the complex velocity in the 1-direction:

$$Z_m = \frac{2\pi}{3} \frac{ika(1 + ika)}{1 + ika - \frac{1}{2}(ka)^2} \, \rho c a^2. \quad (71)$$

The real part of Z_m is a measure of the work done by the sphere on the fluid and is commonly called the *resistance*. The imaginary part of Z_m is called the *reactance*.

The sound power radiated to infinity is equal to

$$P_\infty = \frac{1}{2} \, \text{Re} \, Z_m |\hat{U}|^2 = \frac{\pi}{6} \frac{(ka)^4}{1 + \frac{1}{4}(ka)^4} \, \rho c a^2 |\hat{U}|^2. \quad (72)$$

High-Frequency Limit[2,4,5]

$$p(\mathbf{x}, t) = \frac{a}{r} \, \rho c \hat{U} \cos\theta e^{i\omega[t - (r - a)/c]} \quad \text{for} \quad ka \gg 1, \quad (73)$$

and the specific acoustic impedance is ρc, the geometric acoustics result. The force exerted on the fluid by the sphere is in phase with the velocity:

$$f_1(t) = \frac{4\pi}{3} \, \rho c a^2 \hat{U} e^{i\omega t} \quad \text{for} \quad ka \gg 1. \quad (74)$$

The mean rate at which this force does work on the fluid is $(2\pi/3)\rho c a^2 |\hat{U}|^2$, and this power is radiated to infinity:

$$P_\infty = \frac{2\pi}{3} \, \rho c a^2 |\hat{U}|^2 \quad \text{for} \quad ka \gg 1. \quad (75)$$

Low-Frequency Limit[2,4,5]

$$p(\mathbf{x}, t) = -\frac{1}{2} \, (ka)^2 \rho c \hat{U} \cos\theta \frac{a}{r} \left(1 + \frac{1}{ikr}\right) e^{i\omega(t - r/c)}$$

$$\text{for} \quad ka \ll 1. \quad (76)$$

It is evident from a comparison with Eq. (35) that a compact vibrating sphere produces the same pressure field as a point dipole: The strength of the equivalent point dipole is $2\pi i\omega\rho a^3 \hat{U} e^{i\omega t}$.

On $r = a$, $Z_s \simeq \frac{1}{2} ika\rho c$. The surface pressure leads the normal velocity by approximately $90°$. The largest term in the force $\mathbf{f}$ is therefore in phase with the acceleration and describes the force the sphere must exert on the fluid to overcome its inertia. There is also a very much smaller component of $\mathbf{f}$ in phase with the velocity. That in-phase component of $\mathbf{f}$ is $\frac{1}{3}\pi(ka)^4\rho ca^2\hat{U}e^{i\omega t}$. The rate at which the sphere does work on the fluid is therefore $\frac{1}{6}\pi(ka)^4\rho ca^2|\hat{U}|^2$. This power is radiated to infinity,

$$P_\infty = \frac{\pi}{6}(ka)^4\rho ca^2|\hat{U}|^2 \quad \text{for} \quad ka \ll 1. \quad (77)$$

This is a factor $(ka)^4/4$ smaller than the power radiated by the large vibrating sphere in Eq. (75) and is a measure of the very small amount of power that is radiated into the sound field by the rigid motion of a compact sphere.

9 HARMONICALLY PULSATING AND VIBRATING CYLINDERS

9.1 Pulsating Cylinder[1]

The sound field generated by an expanding and contracting cylinder whose radius at time t is $a + \hat{\epsilon}e^{i\omega t}$ (where $|\hat{\epsilon}| \ll a$ and k^{-1}) is given by

$$p(\mathbf{x}, t) = i\rho c \frac{H_0^{(2)}(k\sigma)}{H_1^{(2)}(ka)}\hat{v}_s e^{i\omega t}. \quad (78)$$

Here, σ is the radial distance of $\mathbf{x}$ from the cylinder axis and $H_n^{(2)}(z)$ is a Hankel function[7] of order n; $\hat{v}_s e^{i\omega t} = i\omega\hat{\epsilon}e^{i\omega t}$ is the radial surface velocity. This cylindrically symmetric pressure field has the same dependence on σ as that due to the line source in Section 3.1.

The specific acoustic impedance on $\sigma = a$,

$$Z_s = i\frac{H_0^{(2)}(ka)}{H_1^{(2)}(ka)}\rho c. \quad (79)$$

The sound power radiated to infinity per unit length of cylinder can be determined by integrating the energy flux through a large cylinder:

$$P_\infty = \frac{2\rho c}{k}\frac{|\hat{v}_s|^2}{|H_1^{(2)}(ka)|^2} \quad \text{per unit length of cylinder.} \quad (80)$$

High-Frequency Limit

$$p(\mathbf{x}, t) = \left(\frac{a}{\sigma}\right)^{1/2}\rho c\hat{v}_s e^{i\omega[t - (\sigma - a)/c]} \quad \text{for} \quad ka \gg 1. \quad (81)$$

Equation (81) describes the geometric acoustics result. The pressure and particle velocity are related by the plane-wave formula $p = \rho cu_\sigma$. The factor $(a/\sigma)^{1/2}$ in Eq. (81) leads to an intensity proportional to a/σ. This accounts for the fact that energy fed in over a surface area $2\pi al$ of the cylinder spreads out over the larger area $2\pi\sigma l$ at radius σ. Since the surface pressure and velocity are virtually in phase, the cylinder does a significant amount of work on the fluid, which is radiated away as sound energy, and

$$P_\infty = \pi\rho ca|\hat{v}_s|^2 \quad \text{per unit length of cylinder } ka \gg 1. \quad (82)$$

Low-Frequency Limit[4]

$$p(\mathbf{x}, t) = \frac{\hat{q}_l}{4i}H_0^{(2)}(k\sigma)e^{i\omega t} \quad \text{for} \quad ka \ll 1, \quad (83)$$

with $\hat{q}_l = i\omega\rho 2\pi a\hat{v}_s$. A pulsating cylinder of compact radius is equivalent to a line source (see Section 3.1), where the strength of the line source is the product of the density and the rate of change of volume enclosed by the cylinder. The pressure field has a different form in the near and far fields, as discussed in Section 3.1.

The specific acoustic impedance on $\sigma = a$ is given by

$$Z_s \simeq -ika\ln(ka)\rho c \quad \text{for} \quad ka \ll 1. \quad (84)$$

Since the surface pressure and velocity are virtually $90°$ out of phase, the cylinder does little work on the surrounding fluid. Indeed, $P_\infty = \frac{1}{2}\pi^2(ka)\rho ca|\hat{v}_s|^2$ per unit length of cylinder, a factor $\frac{1}{2}\pi ka$ smaller than the power radiated per unit length by a cylinder of large radius in Eq. (82).

9.2 Vibrating Cylinder[1]

A rigid cylinder with axis in the 3-direction and vibrating with small velocity $\hat{U}e^{i\omega t}$ in the 1-direction about the origin produces a pressure field

$$p(\mathbf{x}, t) = p(\sigma, \theta, t) = -i\rho c \, \frac{H_1^{(2)}(k\sigma)}{H_1^{(2)\prime}(ka)} \, \hat{U} \cos\theta e^{i\omega t}, \quad (85)$$

where $\sigma = (x_1^2 + x_2^2)^{1/2}$, $\cos\theta = x_1/\sigma$, and the prime denotes differentiation with respect to the argument.

The specific acoustic impedance on $\sigma = a$ is given by

$$Z_s = -i \, \frac{H_1^{(2)}(ka)}{H_1^{(2)\prime}(ka)} \, \rho c. \quad (86)$$

The force $\mathbf{f}(t)$ exerted on the fluid by unit length of the cylinder is in the 1-direction, with

$$f_1(t) = \int_0^{2\pi} p(a, \theta, t) \cos\theta \, a \, d\theta$$

$$= -i\pi \, \frac{H_1^{(2)}(ka)}{H_1^{(2)\prime}(ka)} \, \rho c a \hat{U} e^{i\omega t}. \quad (87)$$

The mechanical impedance per unit length of the cylinder is given as

$$Z_m = \frac{f_1(t)}{\hat{U} e^{i\omega t}} = -i\pi \, \frac{H_1^{(2)}(ka)}{H_1^{(2)\prime}(ka)} \, \rho c a. \quad (88)$$

The sound power radiated to infinity is calculated as

$$P_\infty = \frac{\rho c |\hat{U}|^2}{k |H_1^{(2)\prime}(ka)|^2} \quad \text{per unit length of cylinder.}$$

$$(89)$$

High-Frequency Limit

$$p(\mathbf{x}, t) = \left(\frac{a}{\sigma}\right)^{1/2} \rho c \hat{U} \cos\theta e^{i\omega[t - (\sigma - a)/c]} \quad \text{for} \quad ka \gg 1,$$

$$(90)$$

and the specific acoustic impedance is ρc, the geometric acoustics result. The force exerted on the fluid by the cylinder is in phase with the velocity:

$$f_1(t) = \pi \rho c a \hat{U} e^{i\omega t} \quad \text{for} \quad ka \gg 1. \quad (91)$$

The mean rate at which the cylinder does work on the fluid is $\frac{1}{2}\pi\rho c a |\hat{U}|^2$ per unit length of cylinder. This power is radiated to infinity:

$$P_\infty = \tfrac{1}{2}\pi\rho c a |\hat{U}|^2 \quad \text{per unit length of cylinder}$$
$$\text{for } ka \gg 1. \quad (92)$$

Low-Frequency Limit[1]

$$p(\mathbf{x}, t) = \tfrac{1}{2}\pi(ka)^2 \rho c \hat{U} \cos\theta \, H_1^{(2)}(k\sigma) e^{i\omega t} \quad \text{for} \quad ka \ll 1.$$

$$(93)$$

In the near field the pressure is inversely proportional to σ, whereas in the far field it decays as $\sigma^{-1/2}$. On $\sigma = a$, $Z_s \simeq ika\rho c$. The surface pressure leads the normal velocity by approximately 90°. The largest term in $\mathbf{f}$, the force per unit length, is therefore in phase with the acceleration,

$$f_1(t) \simeq \pi i \omega \rho a^2 \hat{U} e^{i\omega t} \quad \text{for} \quad ka \ll 1. \quad (94)$$

There is also a very much smaller component of $\mathbf{f}$ in phase with the velocity. That in-phase component of $\mathbf{f}$ is $\frac{1}{2}\pi^2(ka)^3 \rho c a \hat{U} e^{i\omega t}$. The rate at which the cylinder does work on the fluid is therefore

$$P_\infty = \frac{\pi^2}{4}(ka)^3 \rho c a |\hat{U}|^2 \quad \text{per unit length of cylinder.}$$

$$(95)$$

This is a factor $\frac{1}{2}\pi(ka)^3$ smaller than the power radiated by the unit length of a cylinder of large diameter in Eq. (92).

10   BAFFLED CIRCULAR PISTON[1-5]

A circular piston of radius a mounted flush in an infinite hard wall, $x_1 = 0$, and vibrating with normal velocity $\hat{U} e^{i\omega t}$ generates a pressure field

$$p(\mathbf{x}, t) = \frac{i\omega\rho \hat{U} e^{i\omega t}}{2\pi} \int_0^a \int_0^{2\pi} \frac{e^{-i\omega R/c}}{R} \, \sigma \, d\sigma \, d\mu, \quad (96)$$

where the origin has been chosen to be at the center of the piston. Here, R is the distance from the observer at x to an element of the piston face with cylindrical polar coordinates $(0, \sigma, \mu)$:

$$R = [x_1^2 + (x_2 - \sigma \cos \mu)^2 + (x_3 - \sigma \sin \mu)^2]^{1/2}. \quad (97)$$

10.1 Distant Sound Field

For $r = |\mathbf{x}| \gg a^2 k$ and a, the integrals in Eq. (96) may be evaluated analytically to give

$$p(\mathbf{x}, t) = p(r, \theta, t) = \frac{i\omega\rho a^2 \hat{U}}{2r} \frac{2J_1(ka \sin \theta)}{ka \sin \theta} e^{i\omega(t - r/c)}, \quad (98)$$

where $k = \omega/c$, $\cos \theta = x_1/r$, and $J_1(z)$ denotes a Bessel function.[7] The function $2J_1(z)/z$ is plotted in Fig. 5.

At any angle θ, the mean sound intensity vector is radial, with

$$I_r(r, \theta) = \frac{(ka)^2 \rho c a^2 |\hat{U}|^2}{8r^2} \left| \frac{2J_1(ka \sin \theta)}{ka \sin \theta} \right|^2. \quad (99)$$

On the axis

$$I_r(r, 0) = \frac{(ka)^2 \rho c a^2 |\hat{U}|^2}{8r^2}. \quad (100)$$

Hence, at an angle θ to the axis,

$$I_r(r, \theta) = I_r(r, 0) \left| \frac{2J_1(ka \sin \theta)}{ka \sin \theta} \right|^2. \quad (101)$$

The directional factor $D(\theta) = |2J_1(ka \sin \theta)/(ka \sin \theta)|^2$

describes how the sound intensity varies with angle and is plotted as a function of polar angle θ in Fig. 6 for two different values of ka.

Low-Frequency Limit When the piston is compact, $2J_1(ka/\sin \theta)/(ka \sin \theta)$ is approximately equal to unity for all values of θ. The pressure field is therefore omnidirectional and can be written in the form

$$p(\mathbf{x}, t) = \frac{\hat{Q}}{2\pi r} e^{i\omega(t - r/c)} \quad \text{for} \quad ka \ll 1, \quad (102)$$

with $\hat{Q} = i\omega\rho\pi a^2 \hat{U}$. The amplitude of the pressure field is twice that produced in unbounded space by a monopole point source with a volume outflow of $\pi a^2 \hat{U} e^{i\omega t}$. The factor 2 describes the effect of the rigid baffle. An image of the piston in the baffle surface leads to a doubling of the pressure field.

High-Frequency Limit The term $J_1(z)$ has zeros at $z = 3.83, 7.02, 10.17, \ldots$. For $ka \geq 3.83$, $D(\theta)$ therefore vanishes at an angle θ_1 such that $\sin \theta_1 = 3.83/ka$. The major lobe occupies the region $|\theta| < \theta_1$. If the piston is sufficiently noncompact that $ka \geq 7.02$, $D(\theta)$ has a second zero at an angle θ_2 such that $\sin \theta_2 = 7.02/ka$. The secondary lobe is detected at angles between θ_1 and θ_2. It is clear from the shape of the $2J_1(z)/z$ plot in Fig. 5 that the maximum value of the intensity in the secondary lobe is smaller than that in the major lobe (see Fig. 6b). If the piston is grossly noncompact, a large number of side lobes may be present, each with a smaller peak intensity than the preceding lobe.

10.2 Pressure Field on the Symmetry Axis[3,4]

On the symmetry axis, $x_2 = x_3 = 0$, the integrals in Eq. (96) can be evaluated exactly for any value of x_1 to yield

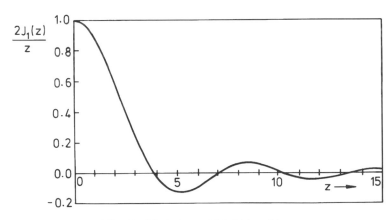

Fig. 5 Variation of $2J_1(z)/z$ with z.

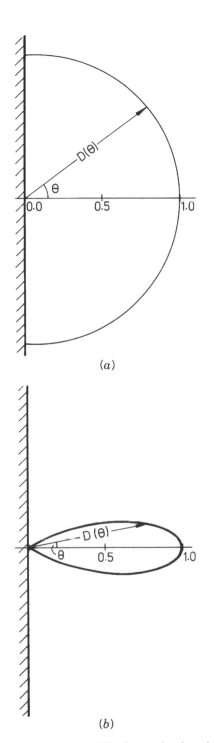

(a)

(b)

Fig. 6 Directional factor $D(\theta)$ for a circular piston in an otherwise rigid wall: (a) $ka = 0.5$; (b) $ka = 5.0$.

$$p(x_1, 0, t) = 2i\rho c\hat{U} \exp i\{\omega t - \tfrac{1}{2}k[x_1 + (x_1^2 + a^2)^{1/2}]\}$$
$$\times \sin \{\tfrac{1}{2}k[(x_1^2 + a^2)^{1/2} - x_1]\}. \qquad (103)$$

The parameter $|\hat{p}|$ is plotted as a function of x_1 for $ka = 11\pi$ in Fig. 7.

The pressure vanishes whenever $k(x_1^2 + a^2)^{1/2} - kx_1$ is a multiple of 2π, that is, when

$$kx_1 = \frac{(ka)^2 - (2n\pi)^2}{4n\pi}, \qquad (104)$$

where n is a positive integer less than $ka/2\pi$. Thus when ka is 11π, as in Fig. 7, there are five pressure nodes along the x_1-axis.

In the distant field, $x_1 \gg ka^2$, $x_1 + (x_1^2 + a^2)^{1/2} \simeq 2x_1$ and $(x_1^2 + a^2)^{1/2} - x_1 \simeq \tfrac{1}{2}a^2/x_1$, and Eq. (103) reduces to

$$p(x_1, 0, t) = \frac{i(ka)\rho ca\hat{U}}{2x_1} e^{i\omega(t - x_1/c)}, \qquad (105)$$

in agreement with the form for the distant-field axial pressure obtained by putting $\theta = 0$ in Eq. (98).

10.3 Mechanical Impedance[3,4]

The force exerted by the piston on the fluid is in the 1-direction, and its strength is equal to the integral of the pressure over the surfaces of the piston:

$$f_1(t) = \int_S p(0, x_2, x_3, t) \, dx_2 \, dx_3. \qquad (106)$$

After substitution for $p(\mathbf{x}, t)$ from Eq. (96), the integrals can be evaluated to give

$$f_1(t) = \pi\rho ca^2 \hat{U} e^{i\omega t}[R_1(2ka) + iX_1(2ka)], \qquad (107)$$

where $R_1(z) = 1 - 2J_1(z)/z$, $X_1(z) = 2\mathbf{H}_1(z)/z$, and $\mathbf{H}_1(z)$ is Struve's[7] function. The parameters $R_1(z)$ and $X_1(z)$ are plotted in Fig. 8.

The mechanical impedance Z_m is obtained by dividing this force by the complex velocity of the piston:

$$Z_m = \frac{f_1(t)}{\hat{U} e^{i\omega t}} = \pi\rho ca^2[R_1(2ka) + iX_1(2ka)]. \qquad (108)$$

The resistance Re Z_m describes the rate at which work is done by the piston on the fluid. This power is radiated to infinity,

$$P_\infty = \tfrac{1}{2}\pi\rho ca^2 R_1(2ka)|\hat{U}|^2. \qquad (109)$$

High-Frequency Limit For $k_a \gg 1$ the functions $J_1(2ka)$ and $\mathbf{H}_1(2ka)$ can be approximated by their large-argument asymptotic forms[7]:

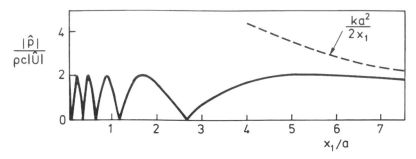

Fig. 7 Variation of $|\hat{p}|$ with x_1 on symmetry axis of a circular piston, $ka = 11\pi$. (Reprinted with permission, from Ref. 4, Fig. 5–10.)

$$R_1(2ka) \simeq 1 - \frac{\cos(2ka - 3\pi/4)}{\pi^{1/2}(ka)^{3/2}}, \qquad (110a)$$

$$X_1(2ka) \simeq \frac{2}{\pi ka} + \frac{\sin(2ka - 3\pi/4)}{\pi^{1/2}(ka)^{3/2}}. \qquad (110b)$$

The limiting values of 1 and $2/\pi ka$ are approached in an oscillatory manner, as shown in Fig. 8. To leading order,

$$Z_m = \rho c \pi a^2, \qquad (111)$$

the geometric acoustics result. The force is in phase with the piston velocity and is equal to the product of the surface pressure $\rho c \hat{U} e^{i\omega t}$ and the piston surface area. The power radiated to infinity is simply given by the product of the mean intensity on the piston $\frac{1}{2}\rho c |\hat{U}|^2$ and the piston surface area:

$$P_\infty = \frac{1}{2}\pi \rho c a^2 |\hat{U}|^2. \qquad (112)$$

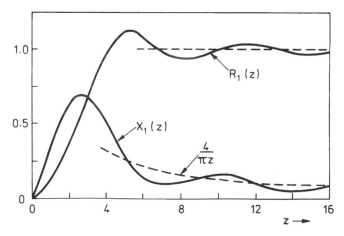

Fig. 8 Piston impedance functions $R_1(z)$ and $X_1(z)$ for a circular piston.

Low-Frequency Limit For small values of z, $J_1(z) \sim \frac{1}{2}z - \frac{1}{16}z^3$, and $\mathbf{H}_1(z) \simeq 2z^2/3\pi$, so that

$$Z_m = \pi \rho c a^2 \left[\frac{(ka)^2}{2} + i \frac{8ka}{3\pi} \right]. \qquad (113)$$

The largest term in Z_m is its imaginary part, the reactance. This describes the force the piston must exert on the fluid to overcome its inertia. That force is equal to $\frac{8}{3}i\omega\rho a^3 \hat{U} e^{i\omega t}$, so that an effective volume $\frac{8}{3}a^3$ of fluid is accelerated.

The real part of Z_m describes the rate at which work is done on the fluid by the piston and hence the radiated sound power:

$$P_\infty = \frac{1}{2} \text{Re} \, Z_m |\hat{U}|^2 = \frac{1}{4}\pi(ka)^2 \rho c a^2 |\hat{U}|^2, \qquad (114)$$

which is a factor $\frac{1}{2}(ka)^2$ smaller than the high-frequency result in Eq. (112).

10.4 Generalization to Other Geometries

A baffled piston of arbitrary cross-sectional area mounted flush in an infinite hard wall, $x_1 = 0$, and vibrating with normal velocity $\hat{U} e^{i\omega t}$ generates a pressure field

$$p(\mathbf{x}, t) = \frac{i\omega\rho \hat{U} e^{i\omega t}}{2\pi} \int_S \frac{e^{-i\omega R/c}}{R} \, dy_2 \, dy_3, \qquad (115)$$

where $R = |\mathbf{x} - \mathbf{y}|$ and the integral is over the surface area S of the piston. At very large distances this simplifies to

$$p(\mathbf{x}, t) = \frac{i\omega\rho \hat{U} e^{i\omega(t - r/c)}}{2\pi r} \int_S e^{i\omega(x_2 y_2 + x_3 y_3)/rc} \, dy_2 \, dy_3$$
$$(116)$$

for $r \gg a$ and ka^2, where $r = |\mathbf{x}|$ and a denotes the piston size. In Section 9.1, this integral was evaluated for a circular piston. It can also be evaluated analytically when the piston is rectangular.[5] For an arbitrarily shaped piston, analytical results can only be derived in the limiting cases of low and high frequencies.

Low-Frequency Limit When the piston is compact, the pressure field does not depend on the details of the piston geometry. Then

$$p(\mathbf{x}, t) = \frac{\hat{Q}}{2\pi r}\, e^{i\omega(t - r/c)}, \qquad (117)$$

where $\hat{Q} = i\omega\rho S\hat{U}$ and S is the surface area of the piston. This results in a radiated sound power

$$P_\infty = \frac{\rho(\omega S|\hat{U}|)^2}{4\pi c}. \qquad (118)$$

High-Frequency Limit On the face of the piston, $p \simeq \rho c \hat{U} e^{i\omega t}$. Hence the force exerted on the fluid by the piston is approximately $\rho c S \hat{U} e^{i\omega t}$, and the radiated sound power is $\frac{1}{2}\rho c S|\hat{U}|^2$.

Here the baffle is an infinite-plane hard surface. Analytical results can also be obtained for a plane impedance surface and for a circular piston set in a rigid sphere.[1] Unbaffled radiators can be treated numerically by the boundary element method discussed in Chapter 15.

11 SOURCES IN UNIFORM MOTION

11.1 Monopole Point Source[1,5]

The sound field produced by a monopole point source of frequency ω in uniform motion with speed U_1 in the 1-direction satisfies the inhomogeneous wave equation

$$\left(\frac{1}{c^2} \frac{\partial^2}{\partial t^2} - \nabla^2 \right) p = \frac{\partial}{\partial t}\, (\delta(\mathbf{x} - \mathbf{U}t)\hat{Q} e^{i\omega t}), \qquad (119)$$

where $\mathbf{U} = (U_1, 0, 0)$ and $\mathbf{U}t$ is the source position at time t. This equation has solution

$$p(\mathbf{x}, t) = \frac{\partial}{\partial t} \left(\frac{\hat{Q} e^{i\omega \tau}}{4\pi R|1 - M \cos \theta|} \right). \qquad (120)$$

The Mach number M is equal to U_1/c; R and θ are the distance and direction of the listener at $\mathbf{x}$ from the source at emission time τ, as illustrated in Fig. 9:

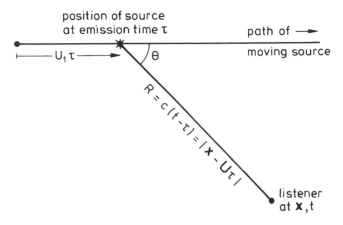

Fig. 9 Emission from a source with velocity $\mathbf{U}$.

$$R = |\mathbf{x} - \mathbf{U}\tau| \quad \text{and} \quad \cos \theta = (x_1 - U_1\tau)/R. \qquad (121)$$

The emission time τ satisfies

$$c(t - \tau) = |\mathbf{x} - \mathbf{U}\tau|. \qquad (122)$$

If U_1 is subsonic, Eq. (122) has one root. When U_1 is supersonic, it has no real roots outside the Mach cone, $U_1 t - x_1 = (M^2 - 1)^{1/2}(x_2^2 + x_3^2)^{1/2}$, and two real roots within it. The pressure is then the sum of contributions from these two emission times.

Differentiation of Eq. (122) shows that

$$\left. \frac{\partial \tau}{\partial t} \right|_{\mathbf{x}} = \frac{1}{1 - M \cos \theta}, \qquad (123)$$

which may be used to evaluate the derivative in Eq. (120):

$$p(\mathbf{x}, t) = \frac{i\omega \hat{Q} e^{i\omega \tau}}{4\pi R(1 - M \cos \theta)|1 - M \cos \theta|}$$
$$+ \frac{U_1(\cos \theta - M)\hat{Q} e^{i\omega \tau}}{4\pi R^2 |1 - M \cos \theta|^3}. \qquad (124)$$

In the far field the first term on the right-hand side of Eq. (124) is the largest. Then there is no effect of source motion in the direction $\theta = 90°$. The pressure directly in front of the moving source ($\theta = 0$) is larger than the pressure the same distance behind the source ($\theta = \pi$) by a factor $(1 + M)^2/(1 - M)^2$.

The frequency ω_l of the sound heard by a listener at $\mathbf{x}$ can be defined as the time derivative of the phase $\omega\tau$:

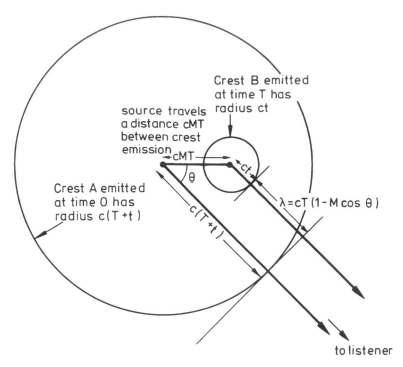

Fig. 10 Sound field at time $T + t$. Moving source emits crests at intervals $T = 2\pi/\omega$. In a direction θ, crest A leads crest B by a wavelength $cT(1 - M\cos\theta)$.

$$\omega_l = \omega \left.\frac{\partial\tau}{\partial t}\right|_\mathbf{x} = \frac{\omega}{1 - M\cos\theta}. \qquad (125)$$

The sound radiated by a moving source of frequency ω is heard at $\mathbf{x}$ at the *Doppler-shifted* frequency $\omega/(1 - M\cos\theta)$. The frequency shift occurs because source motion causes a contraction $1 - M\cos\theta$ in the wavelength of sound traveling in the direction θ (see Fig. 10).

11.2 Dipole Point Source[1,5]

The distant sound field produced by a dipole point source of frequency ω in uniform motion with speed U_1 in the 1-direction satisfies the inhomogeneous wave equation

$$\left(\frac{1}{c^2}\frac{\partial^2}{\partial t^2} - \nabla^2\right)p = -\frac{\partial}{\partial x_i}\left(\delta(\mathbf{x} - \mathbf{U}t)\hat{F}_i e^{i\omega t}\right). \quad (126)$$

This has solution

$$p(\mathbf{x}, t) = -\frac{\partial}{\partial x_i}\left(\frac{\hat{F}_i e^{i\omega\tau}}{4\pi R|1 - M\cos\theta|}\right) \qquad (127)$$

$$= \frac{i\omega\hat{F}_R e^{i\omega\tau}}{4\pi Rc(1 - M\cos\theta)|1 - M\cos\theta|}$$

$$+ \frac{[(1 - M^2)\hat{F}_R - M(1 - M\cos\theta)\hat{F}_1]e^{i\omega\tau}}{4\pi R^2|1 - M\cos\theta|^3},$$

$$(128)$$

where $\hat{F}_R = \hat{\mathbf{F}}\cdot(\mathbf{x} - \mathbf{U}\tau)/R$ is the component of $\hat{\mathbf{F}}$ in the direction of the listener at emission time.

Dipole Axis in the Direction of Motion If $\hat{\mathbf{F}}$ is in the 1-direction, $\hat{\mathbf{F}} = (\hat{F}_1, 0, 0)$, Eq. (128) simplifies to

$$p(\mathbf{x}, t) = \frac{i\omega\cos\theta\,\hat{F}_1 e^{i\omega\tau}}{4\pi Rc(1 - M\cos\theta)|1 - M\cos\theta|}$$

$$+ \frac{(\cos\theta - M)\hat{F}_1 e^{i\omega\tau}}{4\pi R^2|1 - M\cos\theta|^3}. \qquad (129)$$

Dipole Axis Normal to Direction of Motion If $\hat{\mathbf{F}}$ is in the 2-direction, $\hat{\mathbf{F}} = (0, \hat{F}_2, 0)$, Eq. (128) simplifies to

$$p(\mathbf{x}, t) = \frac{i\omega x_2 \hat{F}_2 e^{i\omega\tau}}{4\pi R^2 c(1 - M\cos\theta)|1 - M\cos\theta|}$$

$$+ \frac{x_2(1 - M^2)\hat{F}_2 e^{i\omega\tau}}{4\pi R^3 |1 - M\cos\theta|^3}. \qquad (130)$$

Far-Field Pressure   When the listener is far from the source, the first term of the right-hand sides of Eqs. (128)–(130) describes the major contribution to the sound field.

11.3   Pulsating Sphere[5,10]

In Section 8.1, a stationary, compact pulsating sphere was found to produce the same sound field as a point monopole. That does not remain true when the sphere is in motion, even if the flow is considered to be irrotational. The far-field pressure generated by a compact sphere of variable radius $a + \hat{\epsilon}e^{i\omega t}$ and velocity $\mathbf{U} = (U_1, 0, 0)$ is given as

$$p(\mathbf{x}, t) = -\frac{\omega^2 \rho a^2 \hat{\epsilon} e^{i\omega\tau}}{R(1 - M\cos\theta)^{7/2}} \quad \text{for} \quad M^2 \ll 1. \quad (131)$$

The effect of motion on a pulsating body is different from the effect of motion on a monopole point source. This is because a moving pulsating body producing a mass flux necessarily has a momentum flux associated with it. Hence the sound field produced is that of a moving monopole and a coupled dipole. Moreover, the pressure field due to the dipole is smaller by only a factor of the order of the Mach number than that due to the monopole and so affects the Doppler amplification factor.

11.4   Vibrating Sphere[10]

In Section 8.2, a stationary, compact vibrating sphere was found to produce the same sound field as a point dipole. However, its pressure field is affected differently by source motion from that of a point dipole. In irrotational flow, a rigid sphere with velocity $\mathbf{U} + \hat{\mathbf{U}}e^{i\omega t}$ generates a pressure field

$$p(\mathbf{x}, t) = -\frac{\omega^2 \rho a^3}{2Rc} \left(\frac{\hat{U}_R}{(1 - M\cos\theta)^4} - \tfrac{1}{3}M\hat{U}_1 \right) e^{i\omega\tau},$$

$$M^2 \ll 1, \qquad (132)$$

where $\mathbf{U} = (\mathbf{U}_1, 0, 0)$ and $M = U_1/c$; $\hat{U}_R$ is the component of $\hat{U}$ in the direction of the listener at emission time. If the sphere vibrates in the same direction as its mean velocity, the last term in Eq. (132) is nonzero. It represents an omnidirectional field, which produce amplification even at $90°$ to the motion. In practice, such a sphere would shed vorticity, leading to additional effects of flow.

REFERENCES

1. P. M. Morse and K. U. Ingard, *Theoretical Acoustics*, McGraw-Hill, New York, 1968; reissued Princeton University Press, 1987.

2. M. J. Lighthill, *Waves in Fluids*, Cambridge University Press, Cambridge, 1978.

3. L. E. Kinsler, A. R. Frey, A. B. Coppens, and J. V. Sanders, *Fundamentals of Acoustics*, Wiley, New York 1982.

4. A. D. Pierce, *Acoustics: An Introduction to Its Physical Principles and Applications*, McGraw-Hill, New York, 1981; revised edition, Acoustical Society of America, New York, 1989.

5. A. P. Dowling and J. E. Ffowcs Williams, *Sound and Sources of Sound*, Ellis Horwood, Chichester, 1983.

6. E. Skudrzyk, *Simple and Complex Vibratory Systems*, Pennsylvania State University Press, 1968.

7. M. Abramowitz and I. A. Stegun, *Handbook of Mathematical Functions*, Dover, New York, 1965.

8. L. M. Lyamshev, "A Question in Connection with the Principle of Reciprocity in Acoustics," *Sov. Phys. Dokl.*, Vol. 4, 1959, pp. 405–409.

9. P. A. Frost and E. Y. Harper, "Acoustic Radiation from Surfaces Oscillating at Large Amplitude and Small Mach Number," *J. Acoust. Soc. Am.*, Vol. 58, 1975, pp. 318–325.

10. A. Dowling, "Convective Amplification of Real Simple Sources," *J. Fluid Mech.*, Vol. 74, 1976, pp. 529–546.

10

TRANSIENT RADIATION

PETER STEPANISHEN

1 INTRODUCTION

Acoustic signals of interest are transient in nature since the signals exist over a limited time duration. Common acoustic transients include speech, impact noise such as a baseball striking a bat, musical sounds, and numerous other biologic and/or machinery-induced acoustic transients. In addition to these more well known sources of acoustic energy, acoustic transients are generated by a wide variety of devices or transducers for applications that include the remote location of underwater objects or subbottom oil and gas deposits and biomedical applications that involve noninvasive diagnostic testing and imaging using ultrasonic pulse echo systems.

This chapter briefly addresses the subject of linear acoustic transient radiation from a variety of sources, whereas the inclusion of nonlinear effects and the propagation of finite-amplitude waves in fluids that are of importance in transient explosive and large-amplitude problems are addressed in Chapter 19. The recent use of time-domain methods has led to an improved physical understanding of acoustic transient phenomena and will thus be emphasized here.

A statement of the general linear acoustic transient radiation problem is first presented. The solution of the initial-value problem in which the acoustic field is specified at a reference time in a fluid with no boundaries or sources is then presented, and the introduction of sources into the fluid is then addressed. Finally, acoustic transient radiation from sources on boundaries and impact noise are addressed.

Encyclopedia of Acoustics, edited by Malcolm J. Crocker
ISBN 0-471-80465-7 © 1997 John Wiley & Sons, Inc.

2 STATEMENT OF THE GENERAL PROBLEM

Consider now the acoustic transient radiation and scattering problem of interest as illustrated in Fig. 1. A specified acoustic field is assumed to exist in the fluid at time t_0. A general distribution of space- and time-dependent acoustic sources is also present in the fluid and is specified here to be zero for time $t < t_0$. In addition, an elastic structure of arbitrary shape is present in the fluid.

The basic problem of interest is to determine the acoustic pressure field in the fluid that results from the above-noted initial conditions, sources in the fluid, and/or the vibration of the structure. From a mathematical viewpoint the acoustic field can be represented as the solution of an initial-boundary-value problem.[1] For the sake of generality normalized space and time coordinates $(\mathbf{x}, t)$ are introduced here, as shown in Table 1.

The general initial boundary-value problem of interest can now be stated in normalized coordinates as

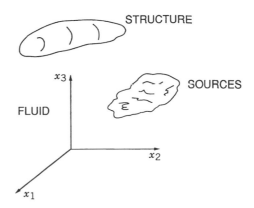

Fig. 1 Acoustic sources and a structure in a fluid.

TABLE 1   Normalization Factors

Variable	Normalization Factor
Length	L
Velocity	c
Density	ρ
Time	L/c
Pressure	ρc^2

Note: L = characteristic length; c = acoustic wave speed of fluid; ρ = density of fluid.

$$\left(\frac{\partial}{\partial t^2} - \nabla^2\right)\phi(\mathbf{x}, t) = q(\mathbf{x}, t) \quad \text{where } t_0 < t < \infty, \quad \mathbf{x} \text{ in } V;$$

$$\phi(\mathbf{x}, t_0) \equiv f_1(\mathbf{x});$$

$$\frac{\partial\phi(\mathbf{x}, t)}{\partial t} \equiv f_2(\mathbf{x}), \qquad t = t_0;$$

$$\hat{n} \cdot \nabla\phi = u(\mathbf{x}, t), \quad \mathbf{x} \text{ on } \sigma;$$

$$p(\mathbf{x}, t) = \frac{\partial\phi(\mathbf{x}, t)}{\partial t}, \tag{1}$$

where $\phi(\mathbf{x}, t)$ is the velocity potential, $p(\mathbf{x}, t)$ is the pressure, $u(\mathbf{x}, t)$ is the normal velocity of the structure in the normal direction $\hat{n}$, and $f_1(\mathbf{x})$ and $f_2(\mathbf{x})$ are initial conditions. The distribution $q(\mathbf{x}, t)$ represents the sources in the fluid with volume V.

Rather than present a general solution to the above initial-boundary-value problem, a series of reduced acoustic transient problems will be addressed in the following sections. The solutions of these individual problems more clearly illustrate the general phenomena of interest associated with intitial conditions, sources in the fluid, and sources on structures. Furthermore, the individual solutions can be superimposed to address the more general problem since the basic problem as formulated in Eq. (1) is a linear one. The transient scattering and diffraction problem is addressed in Chapter 42.

3   INITIAL-VALUE PROBLEM

The simplified case of a source-free fluid with no structure and a specified set of finite-valued initial conditions in the fluid is first addressed using Eq. (1). After Laplace transforming Eq. (1) with respect to time and then solving the resulting inhomogeneous equation using a Green's function method, a straightforward inverse Laplace transform leads to the following solution[2] for the time-dependent velocity potential:

$$\phi(\mathbf{x}, t) = \int_V dV_0\, g(\mathbf{x}, t | \mathbf{x}_0, t_0) f_2(\mathbf{x}_0)$$

$$+ \frac{\partial}{\partial t}\left[\int_V dV_0\, g(\mathbf{x}, t | \mathbf{x}_0, t_0) f_1(\mathbf{x}_0)\right], \tag{2}$$

where the indicated integrals are volume integrals and $g(\mathbf{x}, t | \mathbf{x}_0, t_0)$ denotes the time-dependent free-space Green's function,[1,2] that is,

$$g(\mathbf{x}, t | \mathbf{x}_0, t_0) = \frac{\delta(t - t_0 - |\mathbf{x} - \mathbf{x}_0|)}{4\pi|\mathbf{x} - \mathbf{x}_0|},$$

$$-\infty < t, t_0 < \infty, \qquad \mathbf{x}, \mathbf{x}_0 \text{ in } V, \tag{3}$$

and δ denotes the Dirac delta function.

After positioning a spherical coordinate system at the spatial point of interest and using the sifting property of the Dirac delta function, it is easily shown that the volume integral in Eq. (2) can be reduced to the following surface integral solution, which was originally developed by Poisson (1819):

$$\phi(\mathbf{x}, t) = \frac{1}{4\pi}(t - t_0)\int_\Omega d\Omega\, f_2(\mathbf{x} + (t - t_0)\hat{\mathbf{a}}_x)$$

$$+ \frac{1}{4\pi}\frac{\partial}{\partial t}(t - t_0)\int_\Omega d\Omega\, f_1(\mathbf{x} + (t - t_0)\hat{\mathbf{a}}_x)$$

$$\text{for } t \geq t_0, \tag{4}$$

where Ω indicates a solid angle and $\hat{\mathbf{a}}_x$ represents a unit vector in the radial direction centered at $\mathbf{x}$.

The solution of Eq. (4) has a simple and intuitively obvious interpretation. In brief, the two terms in Eq. (4) correspond to an initial acoustic pressure and a particle velocity field, respectively, that are distributed throughout space at $t = t_0$. As time progresses, the resultant field at any point is simply related to the spatial average of the initial field over a spherical surface centered at the spatial point of interest and with a radius equivalent to the elapsed time $t - t_0$. Such a wave interpretation is consistent with the use of Huygens's principle[3,4] to develop the solution to the initial-value problem.

4   SPACE- AND TIME-DEPENDENT VOLUMETRIC SOURCES

The next simplest type of transient field problem to address is the case of a field with a space- and time-dependent source distribution in the fluid but with no initial conditions or internal structure in the fluid. Mathematically the statement of the problem reduces to the solution of an inhomogeneous wave equation with a radiation condition on the field.[1,2] For a specified acoustic source distribution $q(\mathbf{x}, t)$ the use of a standard Green's function development leads to the following solution for the time-dependent velocity potential:

$$\phi(\mathbf{x}, t) = \int_{t_0}^{t} dt_0 \int_V d\mathbf{x}_0 \, g(\mathbf{x}, t | \mathbf{x}_0, t_0) q(\mathbf{x}_0, t_0). \quad (5)$$

Transient radiation from point source time-dependent multipoles is of fundamental importance in understanding transient radiation from sources since they can be used as building blocks for addressing radiation from complex sources. Steady-state radiation from harmonic multipoles is addressed in Chapter 9. The most useful of these time-dependent multipoles for present purposes are the monopole and dipole. Monopole sources are associated with time-varying volume changes within a fluid and as such are of fundamental importance in sound generation by bubble oscillations,[5,6] combustion noise,[5] parametric sources,[7] and laser-induced sound.[8] Dipole sources are associated with forces acting on a fluid and are of fundamental importance in sound generation by structures or boundaries in a fluid.[9] Quadrupole sources are of fundamental importance in noise generation by turbulence, which is discussed in Chapter 27.

A simple illustration of a monopole source is a uniformly pulsating sphere with a volume flow rate denoted by $Q(t)$. Such a source generates an omnidirectional time-dependent field. In contrast to the monopole, the net volume flow rate for a dipole source is zero and the dipole source can be used to represent the radiation from a small translating sphere.[10] The pressure field for the monopole can be expressed in several forms as

$$p(\mathbf{x}, t) = \frac{\partial}{\partial t} \frac{Q(\mathbf{x}_0, t - r)}{4\pi r}$$

$$= \frac{\partial}{\partial t} [g(\mathbf{x}, t | \mathbf{x}_0, 0) \otimes Q(\mathbf{x}_0, t)], \quad (6)$$

where $r = |\mathbf{x} - \mathbf{x}_0|$, $Q(\mathbf{x}_0, t)$ denotes the volume flow rate of the source at $\mathbf{x}_0$, and the $\otimes$ denotes the convolution operator in time,[11] that is,

$$a(\mathbf{x}, t) \otimes b(\mathbf{x}, t) = \int a(\mathbf{x}, t - \tau) b(\mathbf{x}, \tau) \, d\tau. \quad (7)$$

The pressure field for the dipole can also be expressed in several forms as

$$p(\mathbf{x}, t) = \mathbf{\nabla}_0 \cdot \frac{\mathbf{f}(\mathbf{x}_0, t - r)}{r}$$

$$= g_d(r, t) \otimes f(t) \quad \text{where } \mathbf{f}(\mathbf{x}, t) = f(t) \hat{r}_f,$$

$$\cos \theta = \hat{r}_f \cdot \hat{r}_r,$$

$$g_d(r, t) = \cos \theta \, \frac{\partial g(r, t)}{\partial r}$$

$$= \frac{\cos \theta}{4\pi} \left(\frac{\delta(t - r)}{r^2} + \frac{1}{r} \delta'(t - r) \right). \quad (8)$$

In contrast to the field from the monopole source, which exhibits the same time history at all field points, apart from the inverse range dependence, the field from the dipole source exhibits a near-field r^{-2} and a far-field r^{-1} range dependence along with a directionality effect indicated by the $\cos \theta$ dependence for the pressure field.

In general, the transient radiated pressure field from a monopole source distribution in a fluid can be determined via the principle of superposition, which leads to the equation

$$p(\mathbf{x}, t) = \frac{1}{4\pi} \frac{\partial}{\partial t} \int_V \frac{q(\mathbf{x}_0, t - |\mathbf{x} - \mathbf{x}_0|)}{|\mathbf{x} - \mathbf{x}_0|} \, dV_0, \quad (9a)$$

which in the far field can be expressed via the usual approximations as

$$p(\mathbf{x}, t) = \frac{1}{4\pi |\mathbf{x}|} \frac{\partial}{\partial t} \int_V q(\mathbf{x}_0, t - |\mathbf{x} - \mathbf{x}_0|) \, dV_0, \quad (9b)$$

where $q(\mathbf{x}, t)$ is used to denote a volumetric source strength density. In a similar manner the transient radiated pressure field from a dipole source distribution in a fluid can be determined via the principle of superposition, which leads to the equation

$$p(\mathbf{x}, t) = -\frac{1}{4\pi} \int_V \frac{\mathbf{\nabla}_0 \cdot \mathbf{f}(\mathbf{x}_0, t - |\mathbf{x} - \mathbf{x}_0|)}{|\mathbf{x} - \mathbf{x}_0|} \, dV_0, \quad (10)$$

where $\mathbf{f}(\mathbf{x}, t)$ is used here to denote a force density. Equations (9) and (10) form the basis for evaluating the acoustic transient radiation from an elastic structure in a fluid, as noted in the following section.

To illustrate some interesting transient phenomena, consider a thin finite-length line source with a uniform source strength density $q(t)$. The volume integral in Eq. (9) can then be reduced to a line integral. In the far field it can be shown that the pressure resulting from an arbitrary pulsed excitation $q(t)$ can be represented as a sum of two space-dependent pulses where each pulse is a scaled and time-delayed replica of $q(t)$ which appears to originate at an end of the line distribution. If the pulse duration of $q(t)$ is less than the travel time over the source, then two distinct pulses will be observed in regions of the field.

As a second example, consider the case of a uniform monopole source density distribution distributed over a thin finite planar surface. Once again the pressure field can be simply obtained from Eq. (9), with the volume integral now reduced to a surface integral over the aperture. It can be shown that transient radiation from a finite-sized uniform planar source distribution can thus be expressed as a sum of a plane-wave contribution and an edge-wave contribution that originates from the boundary of the distribution. Such a result is to be expected in light of the importance of endpoint contributions for the line source problem as noted above.

5   TRANSIENT RADIATION FROM STRUCTURAL BOUNDARIES

Consider now the acoustic transient radiation from a structure in a fluid, as illustrated in Fig. 1. The initial conditions in the fluid are assumed to be zero and the medium is homogeneous and source free. It can be readily shown by a standard Green's function development[1,2] that the pressure field can be expressed as the following integral over the surface of the structure:

$$p(\mathbf{x}, t) = \frac{1}{4\pi} \int_\sigma \left(\frac{1}{r} [a] + \frac{1}{r^2} \frac{\partial r}{\partial n} [p] \right.$$
$$\left. + \frac{1}{r} \frac{\partial r}{\partial n} \left[\frac{\partial p}{\partial t} \right] \right) dS_0, \qquad (11)$$

where $r = |\mathbf{x} - \mathbf{x}_0|$ and $[\cdot]$ denotes the retarded time of the quantity in the brackets, that is, $[a] = a(\mathbf{x}_0, t - r)$, where a is acceleration. It is thus apparent that the normal acceleration, pressure, and its time derivative over the surface are in general required to determine the pressure in the fluid via the use of Eq. (11), which is the well-known Kirchhoff retarded-potential solution of the acoustic transient-field problem.

An alternative form of the surface integral solution for the pressure field can be obtained via the sifting property of the Dirac delta function, which then leads to the expression

$$p(\mathbf{x}, t) = \int_\sigma \left[\frac{\partial u(\mathbf{x}_0, t)}{\partial t} \otimes g(\mathbf{x}, t | \mathbf{x}_0, 0) + p(\mathbf{x}_0, t) \right.$$
$$\left. \otimes g_d(\mathbf{x}, t | \mathbf{x}_0, 0) \right] dS_0, \qquad (12)$$

which is a convolution integral form of the solution in which the dependence of the external field upon the surface acceleration and pressure is explicitly noted. It is immediately apparent from Eqs. (6) and (8) that the pressure field can now be interpreted as arising from a monopole source distribution associated with the surface acceleration and a dipole source distribution associated with the surface pressure.

If the structure of interest is compact for the time-dependent signals of interest, that is, if the smallest time scale of interest is much greater than the maximum acoustic travel time between any two points on the structure, Eq. (12) reduces to the field associated with a point monopole and a point dipole. The associated monopole and dipole strengths correspond to surface-integrated values. This low-frequency result is useful if the time-dependent velocity and pressure are known or can be estimated.

Acoustic transient radiation from a structure of arbitrary shape with a specified normal velocity or acceleration is now discussed as a result of its relative simplicity. Via a standard Green's function development[1,2] the field resulting from the specified normal velocity of the structural boundary surface shown in Fig. 1 can be expressed as

$$\phi(\mathbf{x}, t) = \int_t \int_\sigma G(\mathbf{x}, t | \mathbf{x}_0, t_0) u(\mathbf{x}_0, t_0) \, dS_0 \, dt_0,$$
$$p(\mathbf{x}, t) = \frac{\partial \phi(\mathbf{x}, t)}{\partial t}, \qquad (13)$$

where $G(\mathbf{x}, t | \mathbf{x}_0, t_0)$ is the Green's function for the Neumann boundary-value problem of interest. Since the normal velocity of the structure, that is, $u(\mathbf{x}, t)$, can be expressed as

$$u(\mathbf{x}, t) = \sum_n u_n(t) \psi_n(\mathbf{x}) \qquad (14)$$

where the $\psi_n(\mathbf{x})$ are a suitably chosen set of basis functions and $u_n(t)$ are the associated modal velocities, it follows after some simple algebra that the solution to the associated boundary-value problem can be expressed as

$$\phi(\mathbf{x}, t) = \sum_n \int_\sigma \int_t G(\mathbf{x}, t | \mathbf{x}_0, t_0) u_n(t_0) \psi_n(\mathbf{x}_0) \, dS_0 \, dt.$$

(15)

An impulse response representation of the solution then results from integrating Eq. (15) over space to obtain

$$\phi(\mathbf{x}, t) = \sum_n h_n(\mathbf{x}, t) \otimes u_n(t),$$

(16)

where the $h_n(\mathbf{x}, t)$ are space- and mode-dependent impulse response functions defined as

$$h_n(\mathbf{x}, t) = \int_0 G(\mathbf{x}, t | \mathbf{x}_0, 0) \psi(\mathbf{x}_0) \, dS_0.$$

(17)

The case of a planar structure with a specified space- and time-dependent velocity is perhaps the simplest case. Although the Green's function for the planar problem is well known, that is, $G = 2g$, in cases of practical interest involving a finite-sized planar structure, the normal velocity is not known over the entire plane. It is a common assumption to assume that the structure is set in an infinite rigid planar baffle surrounding the radiator. The normal velocity over the entire plane of the structure reduces to that shown in Eq. (14), and the Green's function is known.

The impulse response method[12-16] for planar vibrators is well suited to investigate the spatial and temporal properties of transient acoustic fields from planar structures including ultrasonic transducers. General properties of the impulse response functions have been determined. For the case of planar radiators of finite size, the impulse response functions are nonzero only over a finite time duration, and the correspondence to finite impulse response filters[11] is noted here. Closed-form expressions for $h_n(\mathbf{x}, t)$ have been obtained for circular[13,14] and rectangular vibrators, and relatively simple expressions are available for $h_n(\mathbf{x}, t)$ with $\mathbf{x}$ in the far-field region of rectangular[15] and circular piston[16] sources.

To illustrate typical acoustic transient phenomena of interest for planar sources, consider a simple example of a circular piston or ultrasonic transducer with a specified gated sinusoidal velocity in which the characteristic length for the problem is the radius of the piston. The velocity of the piston is specified to consist of a single cycle with a normalized pulse duration T. In the far field, the on-axis pressure is simply a scaled replica of the acceleration, that is, a single-cycle cosine pulse, for all values of T. The on-axis far-field pressure is thus related to the volumetric acceleration of the source as previously indicated in Eqs. (6) and (9).

When $T \gg 1$, the pressure field from the piston source is essentially omnidirectional and the on-axis pressure is representative of the pressure field at all polar angles. This is to be expected since the piston is omnidirectional for the predominantly low frequency components in the energy spectrum of the velocity. As T decreases, there is an upward shift in the principal frequency components in the energy spectrum of the velocity, and the pressure field thus exhibits more directional effects. In addition a multipulse behavior occurs for $T \ll 1$. The multipulse structure is associated with the edge of the piston in a manner similar to that previously noted for the finite-line and planar source distributions. This behavior, which has been experimentally verified, is to be expected from the extensive work by Freedman[17-19] on related problems.

Acoustic transient radiation from a nonplanar structure of arbitrary shape can also be investigated using the impulse response method. It is noted that the normal velocity of such a structure can be represented as an eigenfunction expansion of the form shown in Eq. (14) in which the basis functions are of course dependent on the shape of the structure. Equations (16) and (17) are thus applicable to the more general nonplanar case when the appropriate Green's function G is used to evaluate the space- and mode-dependent impulse response functions. It is apparent from some related solutions of transient problems for specific geometries[20-22] that the impulse responses for nonplanar structures will in general exhibit decaying oscillatory responses characteristic of infinite impulse response filters, that is, the responses are not time limited or of finite time duration.

As a specific example of a nonplanar vibrator consider the case of a spherical vibrator.[20] Since the Green's function can be obtained in closed form using inverse Fourier transform methods, the impulse response functions can also be so determined. Hence, acoustic transient radiation from a spherical source with an arbitrary space- and time-dependent normal velocity can be determined using an impulse response method in the manner indicated earlier. Due to space constraints, the general case is not addressed here; however, some comments regarding the fields for sphere vibrating with a spherically symmetric time-dependent radial motion and a sphere vibrating as a rigid body with a rectilinear motion are noted here. The velocity of the sphere is again specified to consist of a single cycle of a sinusoidal signal with a normalized pulse duration T in which the characteristic length for the problem is now the radius of the sphere.

The pressure field resulting from a spherically symmetric time-dependent radial motion is first discussed. It can be shown that the pressure for $T \gg 1$ at each point is a cosine pulse that exhibits an inverse range dependence. The pressure in this regime is thus proportional to the surface acceleration, which is to be expected for this low-frequency case for the reasons noted earlier. The pressure in the high-frequency region ($T \ll 1$) corresponds to the velocity as would be expected. Results for the intermediate or resonance region T require more detailed study.

The pressure field for the case of a sphere vibrating as a rigid body with a rectilinear motion for a gated sinusoidal velocity is now discussed. The time history of the pressure field for this case changes with both range and pulse duration T. For the low-frequency case where $T \ll 1$ the spatial pressures exhibit a time dependence similar to that of the velocity. Spatial pressures in the high-frequency region ($T \ll 1$) generally exhibit an early time-impulsive-type behavior followed in time by an exponentially decaying negative tail. Such results illustrate the complex nature of the acoustic transient field from even a simple vibrator.

In light of the preceding example, it is apparent that rigid bodies in nonuniform motion can radiate acoustic transients. A relatively simple example is the pressure field generated by a rigid sphere with an impulsive acceleration[21-23]; that is, the velocity of the sphere is a step, or Heaviside, function. An exact solution for the pressure field can be obtained using either time-domain methods or the inverse Fourier transform method. The results for the pressure field indicate an exponentially decaying oscillatory pressure at each point in the field. This oscillatory field is associated with oscillatory surface pressure waves that traverse the surface of the sphere as a result of the spatially nonuniform normal velocity, which results in tangential pressure gradients and associated tangential accelerations. As such, the period of the oscillations is associated with the travel time for an acoustic disturbance to circumnavigate the sphere.

An interesting discussion of the energy exchange process between the source and the acoustic field[22] for the rigid-body sphere shows that acoustic energy is of course propagated to the far field during the acceleration of the body of a uniform velocity; however, an equivalent amount of energy is "entrained," or trapped, in the near field and is associated with the hydrodynamic mass. This equipartition of energy is clearly illustrated in the work of Junger,[21-23] who further illustrates that the total energy required to accelerate the body is radiated to the far field if the body is ultimately deaccelerated to rest or zero velocity. The kinetic energy associated with the acoustic near field is thus converted to the far field during the deacceleration process.

It is apparent from the preceding example that rigid bodies of arbitrary shape undergoing arbitrary transient motions will in general radiate oscillatory acoustic transients. As noted by Junger and Thompson,[22] these oscillations are to be expected as a result of the spatially nonuniform normal velocity and associated tangential accelerations that result in oscillatory surface pressure fields. In contrast to the case of the rigid sphere, more complex bodies undergoing spatially nonuniform time-dependent normal velocities can be expected to exhibit complex oscillatory pressure fields in which the periods of oscillations are linearly related to the size of the body. Although such a result can be inferred from the residue solution of the sphere, it is equally apparent from the solution of the Kirchhoff retarded-potential solution in Eq. (12). Such results are important in addressing the general subject of impact noise as noted in the following section.

6 IMPACT NOISE

Impact noise is an important topic involving acoustic transient phenomena. Such noise can arise in a vast number of different mechanical processes, as noted by Richards et al.[24-26] These processes include stamping, rivetting, forging, firing of reciprocating engines, and numerous other hammer-type operations. In addition to being of importance in the immediate workplace, the transmission of such impact noise through floors and other structures into other areas is also of concern. The subject of footfall noise is a specific case of interest, and related work has been recently summarized by Beranek and Ver,[27] who addressed periodic impact noise in buildings.

Impact noise can be broadly subdivided into two major areas: noise generated by the initially high surface accelerations and deaccelerations of impacting bodies during the time of contact and noise arising from the free vibration of the bodies following impact. The acceleration/deacceleration component of the process controls the early time response of the time-dependent field pressures, whereas the long time response is dominated by the free vibrations of the impacting bodies. Peak pressures are a function of the contact time of the excitation and the dynamic response of the impacting structures. The results of the preceding section clearly illustrate the complexity of the general problem.

The acceleration component of impact noise was addressed by Richards et al.[24] using a model of two impacting spheres. The importance of impact duration on the noise radiation process was clearly observed in an associated experimental study that led to curves for the peak sound pressure level as a function of contact time between the impacting bodies. For compact bodies incapable of flexural motion the study concludes that acceleration noise energy is of the same order of magnitude as that due to ringing and is less than 0.015% of the kinetic energy of the impacting bodies at the time of impact.

Subsequent studies by Richards and co-workers[25] focused on the importance of ringing noise following impact. These studies clearly illustrate the importance of flexural waves in the structures under impact in determining the relative contribution of acceleration versus ringing-noise energy. For noncompact bodies undergoing flexural motions following impact, the ringing-noise energy can be several orders of magnitude greater than the acceleration component. Although significant progress has been made in this area, it is clear that considerable work remains to be completed.

7 CONCLUDING COMMENTS

A brief overview of acoustic transient phenomena has been presented. Due to space limitations, a number of important related topics have been omitted. Some of these topics of interest are the following: acoustic transients in structures with internal fluid, for example, air conditioning and fluid piping systems exhibiting water hammer effects; acoustic transient loading effects in internally and externally fluid loaded structures; and acoustic transient radiation from fluid-loaded structures.

To address the acoustic transient radiation problem from structures subject to transient loading in air or a heavy fluid such as water, it is necessary to first address the transient structural vibration problem to determine the space- and time-dependent velocity or acceleration. Classical methods of addressing such transient vibration problems are well known.[28] A brief review of the vibration problem for bodies undergoing impact is also available.[22] Transient acoustic fluid-loaded vibration problems are a subject of present research interest. In this latter area the interested reader is directed to the texts of Junger and Feit[21] and Fahy,[30] which provide the basis for addressing the analogous harmonic problems. An understanding of the fundamentals related to the harmonic problem is considered essential prior to addressing the analogous transient problem.

REFERENCES

1. I. Stakgold, *Boundary Value Problems of Mathematical Physics*, Vol. 2, Macmillan, New York, 1968.
2. P. Morse and H. Feshbach, *Methods of Theoretical Physics*, McGraw-Hill, New York, 1953.
3. B. B. Baker and C. T. Copson, *The Mathematical Theory of Huygens Principle*, Clarendon, Oxford, 1953.
4. S. Hanish, "Review of World Contributions from 1945 to 1965 to the Theory of Acoustic Radiation," U.S. Naval Research Laboratory Mem. Rep. 1688, Washington, DC, 1966, Chapter 3.
5. A. P. Dowling and J. E. Ffowcs Williams, *Sound and Sources of Sound*, Halsted, Wiley, New York, 1983.
6. D. Ross, *Mechanics of Underwater Noise*, Pergamon, New York, 1976.
7. P. J. Westervelt, "Parametric Acoustic Array," *J. Acoust. Soc. Am.*, Vol. 35, 1963, pp. 535–537.
8. P. J. Westervelt and R. S. Larson, "Laser Excited Broadside Array," *J. Acoust. Soc. Am.*, Vol. 54, 1973, pp. 121–122.
9. S. Tempkin, *Elements of Acoustics*, Wiley, New York, 1981.
10. P. M. Morse and K. U. Ingard, *Theoretical Acoustics*, McGraw-Hill, New York, 1968.
11. A. Papoulis, *Signal Analysis*, McGraw-Hill, New York, 1977.
12. P. R. Stepanishen, "Transient Radiation from Pistons in an Infinite Planar Baffle," *J. Acoust. Soc. Am.*, Vol. 49, 1971, pp. 1629–1638.
13. P. R. Stepanishen, "Acoustic Transients from Planar Axisymmetric Vibrators Using the Impulse Response Approach," *J. Acoust. Soc. Am.*, Vol. 70, 1981, pp. 1176–1181.
14. P. R. Stepanishen, "Transient Radiation and Scattering from Fluid Loaded Oscillators, Membranes, and Plates," *J. Acoust. Soc. Am.*, Vol. 88, 1990, pp. 374–385.
15. P. R. Stepanishen, "Comments on Farfield of Pulsed Rectangular Radiator," *J. Acoust. Soc. Am.*, Vol. 52, 1972, p. 434.
16. P. R. Stepanishen, "Acoustic Transients in the Far Field of a Baffled Circular Piston Using the Impulse Response Approach," *J. Sound Vib.*, Vol. 32, 1974, pp. 295–310.
17. A. Freedman, "Transient Fields of Acoustic Radiators," *J. Acoust. Soc. Am.*, Vol. 48, 1970, pp. 135–138.
18. A. Freedman, "Farfield of Pulsed Rectangular Acoustic Radiator," *J. Acoust. Soc. Am.*, Vol. 49, 1971, pp. 738–748.
19. A. Freedman, "Sound Field of a Pulsed, Planar, Straight-Edged Radiator," *J. Acoust. Soc. Am.*, Vol. 51, 1972, pp. 1624–1639.
20. J. Brillouin, "Rayonnement transitoire des sources sonores et problemes conneses," *Ann. Telecommun.*, Vol. 5, 1950, pp. 160–172, 179–194.
21. M. C. Junger and D. Feit, *Sound Structures and Their Interaction*, MIT Press, Cambridge, MA, 1972.
22. M. C. Junger and W. Thompson, "Oscillatory Acoustic Transients Radiated by Impulsively Accelerated Bodies," *J. Acoust. Soc. Am.*, Vol. 38, 1965, pp. 978–986.
23. M. C. Junger, "Energy Exchange between Incompressible Near and Acoustic Far Field for Transient Sources," *J. Acoust. Soc. Am.*, Vol. 40, 1966, pp. 1025–1030.
24. E. J. Richards, M. E. Westcott, and R. K. Jeypalan, "On

the Prediction of Impact Noise, I: Acceleration Noise," *J. Sound Vib.*, Vol. 62, No. 4, 1979, pp. 547–575.

25. E. J. Richards, M. E. Westcott, and R. K. Jeypalan, "On the Prediction of Impact Noise, II: Ringing Noise," *J. Sound Vib.*, Vol. 65, No. 3, 1979, pp. 419–451.

26. J. M. Cushieri and E. J. Richards, "On the Prediction of Impact Noise IV: Estimation of Noise Energy Radiated by the Impact Excitation of a Structure," *J. Sound Vib.*, Vol. 86, No. 3, 1983, pp. 319–342.

27. I. L. Ver, "Interaction of Sound Waves with Solid Struc-tures," in L. L. Beranek and I. L. Ver (Eds.), *Noise and Vibration Control Engineering*, Wiley, New York, 1992.

28. R. C. Ayre, in C. M. Harris and C. E. Crede (Eds.), *Shock and Vibration Handbook*, McGraw-Hill, New York, 1976.

29. W. C. Hoppmann, in C. M. Harris and C. E. Crede (Eds.), *Shock and Vibration Handbook*, McGraw-Hill, New York, 1976.

30. F. Fahy, *Sound and Structural Vibration: Radiation Trans-mission and Response*, Academic, New York, 1985.

11

ACOUSTIC INTERACTION BETWEEN STRUCTURES AND FLUIDS

F. J. FAHY

1 INTRODUCTION

Nearly all solid structures exist in surface contact with one or more fluid media, of which the most common are air and water. A few man-made exceptions now exist in space. Vibration generated in a solid structure is communicated to a fluid with which it is in contact via normal motion of the media interface. Everyday examples include the generation of audible sound in the air by vibrating structures such as machines, building components, and stringed instruments: This mechanism is one of the most common sources of acoustic noise. A related phenomenon of less obvious, but nonetheless considerable, practical importance is that of the excitation of vibration in solid structures by sound generated by sources in a contiguous fluid. In its more dramatic manifestations it can be responsible for severe damage to structures and connected components and even failure; a more commonly experienced example is the transmission of airborne sound through party walls between dwellings. Acoustic coupling between structural and fluid systems can significantly alter the free and forced vibration behavior of the coupled components from their uncoupled forms; as evidenced, for example, by the propagation of pressure waves in flexible pipes containing liquids, such as blood vessels: The speed of propagation is very much less than that in the fluid itself.

The general aims of this chapter are to explain the fundamental mechanism of the interaction phenomenon, to make the reader aware of situations and conditions in which fluid–structure interaction has to be considered as a major feature of system behavior, to present relevant and useful basic expressions and relationships, and to offer guidance on sources of more comprehensive information relating to the subject (including material contained in this book). The scope of the chapter does not extend to problems involving the interaction of vibrating structures with two-phase fluid media, such as the generation of surface waves in the sea by the motion of marine structures or the vibration of heat exchanger pipes filled with a mixture of vaporized and liquid water. The frequency range of concern is the so-called audio range of 20 Hz–20 kHz: Ultrasonic phenomena are excluded.

2 NATURE OF FLUID–STRUCTURE INTERACTION

When a solid structure is caused to vibrate, it produces vibrational disturbances in any fluid with which it is in contact. Whatever the frequency of vibration, the resulting internal fluid forces (pressures) and motions are governed by the same equation, known as the *acoustic wave equation*: The disturbances constitute a *sound field*. This sound field is uniquely determined by (i) the properties of the fluid, (ii) the geometry of the vibrating surface(s), (iii) the acoustic properties and geometric distribution of any other "passive" surfaces bounding the fluid, and (iv) the spatial distribution of the component of vibrational acceleration normal to the surface of the vibrating structure(s). The agent that couples the fluid to the structure is the fluid pressure at the interface, and that which couples the structure to the fluid is the surface acceleration. The relationship is expressed mathematically by the Kirchhoff–Helmholtz integral equation (see Chapters 13 and 15).

Encyclopedia of Acoustics, edited by Malcolm J. Crocker
ISBN 0-471-80465-7 © 1997 John Wiley & Sons, Inc.

It is useful to make a distinction between two basic geometric configurations: a structure may completely, or largely, enclose a fluid with which it interacts or it may be immersed in a fluid volume of which it does not itself form the outer boundaries and that may extend to very large distances in terms of structural dimensions. Examples are, respectively, the air within an aircraft fuselage and the sea in which a submarine is immersed. The reason for making this distinction is that the nature of the fluid reaction to boundary vibration takes significantly different forms in the two cases. An enclosed sound field exhibits the phenomenon of *standing waves* or *acoustic modes*, which possess characteristic frequencies; the fluid can consequently resonate and can also store vibrational energy. The result is that the fluid reaction to boundary motion can vary very strongly with frequency and energy can oscillate between the fluid and bounding elastic structures. An "unenclosed" fluid volume does not exhibit resonant behavior, and sound energy generated by boundary vibration flows away, not to return (although some energy flux oscillation does occur very close to a vibrating boundary). The fluid-loading effects in this latter case are generally less strong and less variable with frequency than those in the enclosed-fluid configuration.

Whether the independent sources of vibration operate directly on the fluid or on the structure, fluid–structure interaction couples the two components to form an integral vibrating system. The influence of such coupling on the vibrational behavior of structures and fluids varies greatly from system to system. The strength of fluid–structure interaction is proportional to mean fluid density; consequently, the fluid-loading effect of liquids and highly compressed gases is generally very much greater than that of gases at near atmospheric pressure. However, the high-frequency vibration of stiff, lightweight structures, such as aerospace honeycomb sandwich constructions, can also be significantly affected by air loading.

Vibrations of structural and fluid systems are wave phenomena. Consequently, fluid–structure acoustic interaction is, by definition, an "extended" reaction; this means that the field at any one point on the interface between the media is, in principle, influenced by conditions at all other points on the interface. This makes analysis and evaluation rather complicated in most practical situations for which simple analytical models are not generally available. In many engineering systems, the geometric configurations and material characteristics of the components are so complicated that only numerical solutions to the equations governing coupled fluid–structure vibration are viable (see Chapters 15 and 41). However, the general nature of the interaction process may be qualitatively revealed by idealized examples, as illustrated throughout this chapter.

3   INTERACTION OF PLANE STRUCTURES WITH SEMI-INFINITE FLUID VOLUMES

3.1   Simple Boundary Source Model: The Rayleigh Integral

Vibration of a surface structure displaces fluid volume at the interface. The pressure field generated in an infinitely extended fluid at a distance r from a small piston element of area δS set in an otherwise *rigid plane* surface, and in time-harmonic oscillation at circular frequency ω, is given by $p(r, t) = (i\omega\rho_0 V_n \delta S/2\pi r) \exp[i(\omega t - kr)]$, where V_n is the surface normal velocity, ρ_0 is the mean fluid density, and k is the wavenumber ($=\omega/c$), where c is the fluid sound speed. The pressure is seen to decrease linearly with distance from the source. A continuous nonuniform distribution of surface normal velocity may be represented by a distribution of such piston elements, and according to the principle of linear superposition, the total sound pressure generated by such vibration may be represented by a summation, due account being taken of distance and phase through r and kr. The expression in integral form is

$$p(r, t) = \frac{i\omega\rho_0}{2\pi} e^{i\omega t} \int_S \frac{V_n(r_s)e^{-ikR}}{R} \, dS, \qquad (1)$$

where $R = |r - r_s|$ is the distance between the observation point and the position of the surface element at r_s.

This is known as Rayleigh's first integral. Note carefully that it strictly applies only to infinitely extended plane surfaces on which the surface normal velocity V_n is *known at all points* on the plane; consequently, it cannot be directly applied to a plane vibrating structure of limited extent, because the normal velocity on the plane beyond the limits of the structure is not known a priori. The fluid pressure (sound) field is determined by the distribution of *normal acceleration* $i\omega V_n$ over the plane surface. (Rayleigh's second integral expresses the radiated pressure in terms of the distribution of pressure over the whole surface.)

3.2   Fluid Loading on Oscillating Rigid Circular Piston

Application of the Rayleigh integral to a system consisting of a rigid, plane, circular piston of radius a that is surrounded by an in-plane infinite rigid baffle and vibrates harmonically in a direction normal to the plane yields the following expression for the ratio of the spatial average fluid pressure on the piston to the volume velocity ($V_n \pi a^2$) of the piston, defined as the radiation impedance of the piston:

$$Z_{rad} = R_{rad} + iX_{rad}, \qquad (2)$$

where

$$R_{rad} = \frac{\rho_0 c}{\pi a^2} \left[1 - \frac{2J_1(2ka)}{2ka} \right],$$

$$X_{rad} = \frac{\rho_0 c}{\pi a^2} \left[\frac{2H_1(2ka)}{2ka} \right],$$

where J_1 is the Bessel function of order 1 and H_1 is the Struve function of order 1. The real (in-phase) and the quadrature (phase $\pi/2$) components of the nondimensional impedance are known respectively as the acoustic resistance ratio and reactance ratio; they are plotted in Fig. 1. The piston circumference equals the acoustic wavelength when $ka = 1$. Note that the asymptotic value ($2ka \to \infty$) of the resistance ratio is unity (i.e., the same as a plane wave), whereas the reactance decreases to negligible values. By contrast, the reactance greatly exceeds the resistance when $ka < 2$. When $ka \ll 1$, the expressions for radiation resistance and reactance are well approximated by $R_{rad} \approx (\rho_0 c/\pi a^2)(ka)^2/2$ and $X_{rad} \approx (\rho_0 c/\pi a^2)(8/3\pi)(ka)$.

The resistive and reactive components of the radiation impedance have distinct physical interpretations. The former represents the ability of the piston to do net work on the fluid over one cycle of oscillation, that is, to radiate sound power into the fluid, given by $W = \frac{1}{2} \int_S \text{Re}\{pV_n^*\} dS$, where the quantities in brackets are complex amplitudes.

The latter represents the generation of reactive power by which kinetic energy is put into the fluid close to the piston during half a cycle, and equal energy is returned to the piston during the other half. Positive reactance indicates that the fluid applies an inertial-type load to the piston, which, when $ka \ll 1$, can be expressed in terms of an added mass of magnitude $M_a = \frac{8}{3}\rho_0 a^3$. The sound power radiation acts like a viscous damper of coefficient given by $B_a = \frac{1}{2}[\rho_0 c\pi a^2(ka)^2]$. If such a piston of mass M is mounted on a damped spring suspension of stiffness S and damping coefficient B, the mechanical impedance of the coupled system will be given by $Z_c = i\omega(M + M_a - S/\omega^2) + (B + B_a)$, and the resonance frequency of the piston will be lowered by the fluid loading.

An analysis of *unbaffled* piston radiation is presented in Ref. 1. Figure 1 compares baffled and unbaffled radiation impedances.

The nondimensional radiation impedance $Z_{rad}(\pi a^2/\rho_0 c)$ of a rigid piston radiating into an anechoically terminated tube of the same radius, when $ka < 1$, is equal

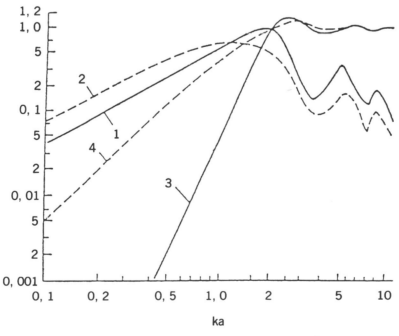

Fig. 1 Radiation impedance of an oscillating piston: (1) $-X_{rad}\pi a^2/\rho_0 c$ for an unbaffled piston; (2) $-X_{rad}\pi a^2/\rho_0 c$ for a baffled piston; (3) $R_{rad}\pi a^2/\rho_0 c$ for an unbaffled piston; (4) $R_{rad}\pi a^2/\rho_0 c$ for a baffled piston.[1]

to unity; the reactance is zero. The physical difference between these cases is that, close to a baffled piston, the fluid elements are free to move parallel to the piston surface and along elliptical trajectories, by which little strain is produced in the fluid at greater distances from the surface, whereas in a tube they are all forced to move in the direction normal to the piston surface, by which the strain is effectively passed on to fluid further along the tube.

Suppose a small ($ka \ll 1$) baffled piston is encircled by an annular piston of the same area that oscillates in opposite phase; the net volume velocity of the source is zero. The resistive component of pressure acting on each piston is negligible, but the reactive component acting on the central piston is altered only little from its value in the absence of the annulus. This simple example illustrates that the reactive component of fluid loading at a point on a vibrating surface depends mainly on the surface normal acceleration in the immediate vicinity of the point; however, the resistive component, and therefore the radiated power, generally depends upon the distribution of surface acceleration over a much larger area.

3.3   Sound Field Generated by Plane Surface Waves on Plane Boundary

Many structures of practical interest, such as room enclosures, machine casings, passenger compartment envelopes, and the hulls and bulkheads of ships, may, for the purpose of vibration and acoustic analysis, be represented by flat plates. Consequently, it is important to understand the nature of the acoustic interaction of vibrational waves traveling in plates and a contiguous fluid. Analysis of the sound field generated in a semi-infinite volume of fluid bounded by an *infinitely extended* plane surface carrying a harmonic plane transverse wave of wavenumber k_x yields the following expressions for the ratio of the complex amplitudes of surface pressure to surface normal velocity (specific acoustic impedance):

$$k_x < k: \qquad z = \frac{\rho_0 c}{[1 - (k_x/k)^2]^{1/2}}, \qquad (3a)$$

$$k_x > k \qquad z = \frac{i(\rho_0 \omega / k_x)}{[1 - (k/k_x)^2]^{1/2}}. \qquad (3b)$$

This form of impedance, which is associated with a given surface wavenumber, may be termed a *wave impedance*. The physical interpretation of these expressions is that when the surface wavenumber exceeds the acoustic wavenumber, the impedance presented by the fluid is purely reactive and inertial in character, and no sound energy is radiated away into the far field: The sound

field decays exponentially with distance perpendicular to the plane. The inertial loading corresponds to that of a layer of fluid of thickness $(k_x^2 - k^2)^{-1/2}$ moving with the surface. When the surface wavenumber is less than the acoustic wavenumber, the impedance is purely resistive, and sound energy is radiated. In the limit $k/k_x \to 0$, the impedance approaches the characteristic impedance $\rho_0 c$ of the fluid. The fluid loading increases without limit as k_x approaches k. This is a feature unique to the infinitely extended surface wave field. The failure of the surface to radiate sound energy when $k_x > k$ may be qualitatively explained by analogy with the two-piston example described in Section 3.2 above: Regions of opposite phase motion are within a distance of less than half an acoustic wavelength of each other, and the resulting pressure phase is at 90° to that of the surface normal velocity. Clearly, k_x/k is a very important parameter, having a critical value of unity that separates two quite different regimes of fluid behavior.

If the plane-surface vibration is produced by *free* flexural propagation in a uniform plane elastic structure, the critical condition corresponds to equality of k and the free bending wavenumber $k_b = \omega^{1/2}(m/D)^{1/4}$ which occurs at a particular *critical frequency* given by $f_c = (c^2/2\pi)(m/D)^{1/2}$, where c is the sound speed in the fluid and D and m are, respectively, the bending stiffness per unit width and mass per unit area of the structure (see Fig. 2). For homogeneous isotropic plates of thickness h, f_c is uniquely related to h by a constant that depends only on the material properties and the fluid sound speed. Table 1 lists values for common materials in air; For water the values should be multiplied by 19. The acoustic wave impedance acts in series with the in vacuo plate wave impedance, which is given by $z_p = -(i/\omega)[Dk_x^4 - \omega^2 m]$. Free-wave propagation corresponds to a condi-

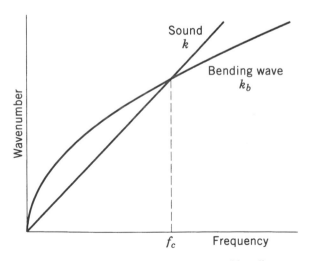

Fig. 2   Dispersion curves for sound waves and bending waves in a flat plate.

TABLE 1   Product of Thickness and Critical Frequency for Uniform Homogeneous Flat Plates in Air[a] at 20°C

Material	hf_c (m/s)
Steel	12.4
Aluminium	12.0
Brass	17.8
Copper	16.3
Glass	12.7
Perspex	27.7
Chipboard	23[b]
Plywood	20[b]
Asbestos cement	17[b]
Concrete	
Dense	19[b]
Porous	33[b]
Light	34[b]

[a]To obtain values in water, multiply by 18.9.
[b]Variations of up to ±10% possible.

tion of zero external loading, and therefore, zero total wave impedance; free-wave propagation cannot occur at $k_b = k$, when the fluid wave impedance is infinite. The most dramatic effects of plate loading by an unenclosed of fluid are produced by liquids at low frequencies. At frequencies for which $k_b \gg k$, the fluid-loaded free structural wavenumber is given approximately by $k_b' = (\omega^2 \rho_0/D)^{1/5}$, which corresponds to a phase speed given by $c_{ph} = (\omega^3 D/\rho_0)^{1/5}$. At such frequencies the inertial loading by the fluid greatly outweighs the plate inertia, and the latter does not control the wave speed. At higher frequencies for which $k_b > k$, the inertial loading corresponds approximately to an additional mass per unit area given by $m' = \rho_0/k_b$. This loading can significantly lower structural natural frequencies in the ratio $[m/(m + m')]^{1/2}$. At frequencies close to and above the plate critical frequency, an undamped flexural wave can travel in a fluid-loaded plate at a phase speed slightly less than that of sound in the fluid, but this wave is very difficult to excite mechanically. In addition, above the plate critical frequency there exists a wave with a supersonic phase velocity. This is the so-called leaky wave, which radiates energy away at the Mach angle; the amplitude of the wave decays exponentially with distance along the plate as energy is lost. In cases of heavy fluid loading, this decay rate is very high, so that the wave is not present in the plate at large distances from the excitation.

3.4   Sound Radiation by Baffled Rectangular Flat Plates

As indicated above, the rectangular, uniform, flat-plate model is a useful approximation to many structures of

interest. The vibration modes take a particularly simple form if simply supported boundary conditions are assumed. Many analyses of modal sound radiation have been published, for example, Refs. 2–7. Examples from Wallace are shown in Fig. 3, in which the radiation efficiency (ratio) is defined by $\sigma = W_{rad}/\rho_0 c S \langle v^2 \rangle$, where W_{rad} is the radiated sound power and $\langle \overline{v^2} \rangle$ is the space-averaged mean-square normal velocity of the surface of area S. The efficiency is asymptotic to unity when $k \gg k_b$. In most cases, the fundamental mode is the most efficient at frequencies below f_c. Notice that frequency does not appear in Fig. 3. Below f_c, adjacent cells of uniform vibration phase tend to short circuit each other, as in the two-piston case above. As a result, radiated sound originates principally from the uncanceled volume velocity at the boundaries; modes that radiate in this manner are termed *edge modes*. Consequently, the radiation efficiency of *unbaffled* panels is so low below f_c that, in practice, it can be neglected; above f_c it is unity. Perforated plates have, as expected, very low radiation efficiency.

A radiation efficiency curve for *resonant, multimodal* baffled plate radiation is presented in Fig. 4.[8] Note that low-frequency efficiency increases with decrease of plate size. (This curve is *not applicable* to acoustically excited plain panels or locally excited, nonuniform, nonreverberant panels.) Below f_c, incident plane waves couple most strongly with plate modes well above their resonance frequencies, and the associated transmission, which exceeds resonant transmission (except in cases of very lightly damped and/or small plates), is therefore mass controlled. The radiation loss factor is related to σ by $\eta_{rad} = (\rho_0/\rho_s)(1/kh)\sigma$ where ρ_s is the plate density.

Practical platelike structures are rarely uniform, which reduces the short-circuiting effect below f_c; σ may be considerably increased by the presence of irregularities such as localized stiffeners and concentrated masses. The addition of stiffeners or masses for the purposes of noise control influences both σ and the vibration level induced by mechanical sources. It is therefore impossible to present generalized data on this subject; however, it is advised that added stiffeners should also increase structural damping levels, wherever possible. Where plates are excited by localized force inputs, the region in the immediate vicinity of excitation radiates sound power, given by $W_f = \rho_0 c k^2 F^2/4\pi m^2 \omega^2 = \rho_0 F^2/4\pi c m^2$, where F is the harmonic force amplitude and m is the plate mass per unit area. The ratio of sound powers radiated by the force and the surface is given by $W_f/W_s = (4\pi)(\omega/\omega_c)(\eta/\sigma)$. Excitation by a harmonic point velocity v_0 gives $W_v = 16\rho_0 v_0^2 D/\pi c m = 16\rho_0 v_0^2 c^3/\pi \omega_c^2$. The sound power generated per unit length of uniform line force excitation is $W_f' = \rho_0 c k \omega^2 F^2/4D^2 k_b^8 = \rho_0 F^2/4\omega m^2$, the ratio of pow-

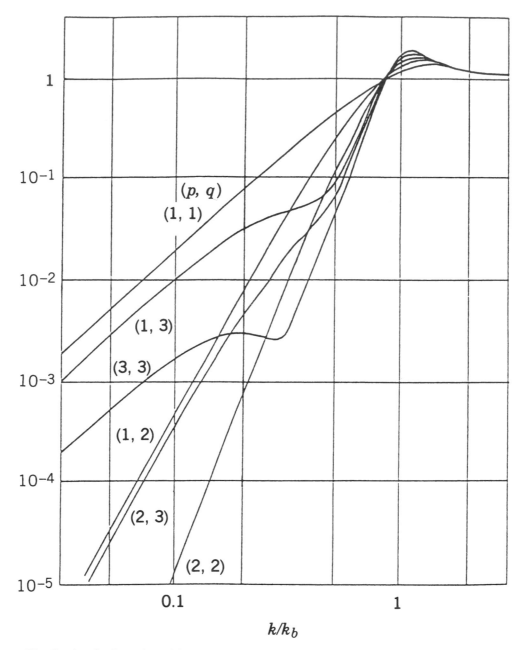

Fig. 3 A selection of modal radiation ratios for a baffled rectangular, simply supported, square plate (p, q = mode orders).[2]

ers radiated from the excitation region and the plate are given by $W_f'/W_s = 2(\omega/\omega_c)^{1/2}(\eta/\sigma)$, and the power radiated per unit length of line velocity excitation is given by $W_v' = 2\rho_0 D^{1/2} v_0^2/m^{1/2} = 2\rho_0 c^2 v_0^2/\omega_c$.

3.5 Fluid-Loading Effects on Flat Plates

The acoustic forces imposed on engineering structures by atmospheric air are generally rather small compared with the internal structural forces, although the stiffness and resonant reaction of air in a small enclosed cavity, such as that between wall partitions, can significantly influence structural motion.[9] Lightweight aerospace structures such as honeycomb panels are exceptions; acoustic damping often exceeds mechanical dissipation damping, especially in the frequency range close to f_c.[10] Loading by liquid significantly lowers the natural frequencies of flat plates, the effect decreasing with increasing mode

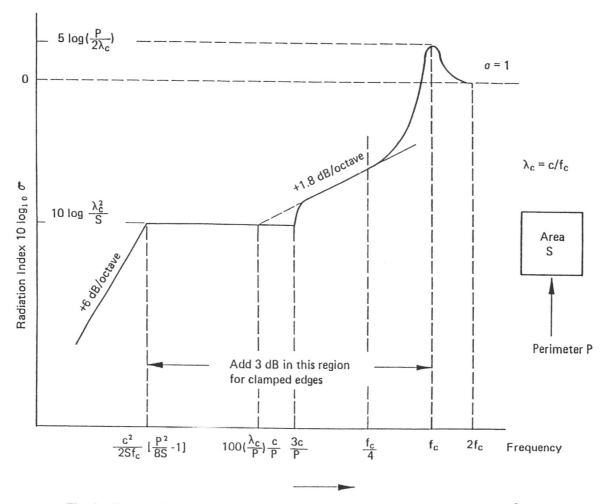

Fig. 4 Theoretical model-averaged radiation efficiency of a baffled rectangular panel.[8]

order. The ratio of fluid-loaded to in vacuo natural frequency is approximately given by $f_l/f \approx (1+\rho_0/mk_b)^{-1/2}$ where k_b is the free plate wavenumber.[11] Fluid loading also couples the in vacuo modes of structures,[12] but this coupling may often be neglected for practical purposes.

The critical frequencies for most plates in water are so high that uniform plate mode radiation efficiencies are very low; consequently, sound is radiated mainly from the sites of mechanical excitation and structural irregularities, such as stiffeners. For $\rho_0 c/\omega m \gg 1$ and at frequencies well below f_c, the sound radiated from plate structures underwater can be estimated rather accurately by assuming that twice the applied forces and plate–stiffener reaction forces are applied *directly* to the water. The expressions for sound powers radiated by point and line forces, respectively, are $W_f = k^2F^2/12\pi\rho_0 c$ and $W'_f = \pi k F^2/16\rho_0 c$. Note, these contain no plate parameters. A more detailed account of fluid-loading effects is contained in Chapter 41.

4 INTERACTION OF CIRCULAR CYLINDRICAL STRUCTURES WITH INFINITE FLUID VOLUMES

4.1 Vibration of Circular Cylindrical Shells

Many engineering structures have the basic form of circular cylindrical shells; examples include fluid-conducting pipes and tubes, aircraft fuselages, fluid storage tanks, and electric motor and generator casings. Vibrational waves in circular shells are very complex in nature because of coupling between the displacements in the axial, circumferential, and radial directions and because stresses arise from both flexural and median surface (membrane) strains.[13] Three "families" of wave types can propagate in the axial direction. Except at very low frequencies, each family contains waves in which displacements in one of the principal directions is dominant: They are consequently referred to as *longitudinal*, *torsional*, and *flexural* waves. The last mentioned is

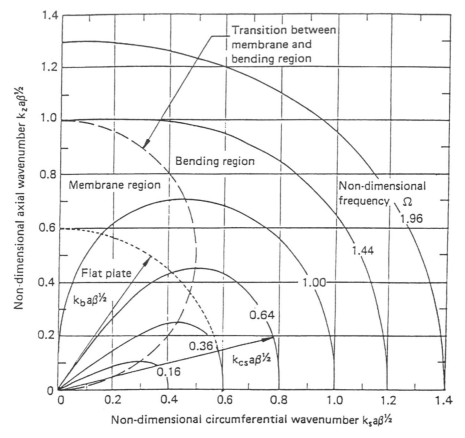

Fig. 5   Universal constant-frequency loci for flexural waves in thin-walled cylindrical shells.

the most involved in fluid–structure interaction because radial displacements are dominant.

4.2 Sound Radiation from Circular Cylindrical Shells

The acoustic interaction with surrounding fluids is considerably different in character from that of flat plates because shell curvature and the resulting membrane effects lead to the existence of surface vibrational wavenumbers less than that of sound at frequencies below the critical frequency of the flat plate of the same thickness as the shell wall. This behavior may be understood by reference to a shell wavenumber diagram, as shown in Fig. 5, using the coordinate system shown in Fig. 6. The constant-frequency loci for the equivalent flat

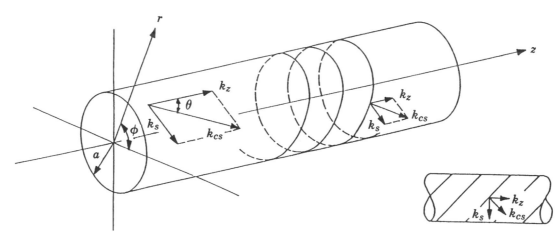

Fig. 6   Cylindrical shell coordinates and wavenumbers.

TABLE 2 Cutoff Frequencies of Flexural Modes of Thin-Walled Circular Cylindrical Shells[14]

Circumferential Mode, Order n	$\Omega_n/\beta n^2$	Ω_n/β
2	0.67	2.68
3	0.85	7.65
4	0.91	14.56
5	0.95	23.75
6	0.96	34.56
7	0.97	47.10

Note: Ω_n = frequency/ring frequency; β = thickness parameter = $h/(12)^{1/2}a$; h = shell wall thickness; a = shell radius.

plate are quarter circles of radius k_b, as shown: membrane effects "pull" the loci toward the origin, reducing the shell wavenumbers and increasing the wave phase speed. The nondimensional frequency Ω is the ratio of the frequency to the ring frequency, which is given by $f_r = c_l'/2\pi a$, where a is the shell radius and c_l' is the phase speed of longitudinal waves in a plate. The wall thickness h is nondimensionalized in the parameter $\beta = h/\sqrt{12}a$. The circumferential wavenumber k_s can take only discrete values n/a (integer n). Flexural wave modes of circumferential order n may propagate axially only at frequencies greater than their cutoff frequencies. These are given in Table 2.

Analysis of the sound field radiated by a flexural shell wave of given circumferential wavenumber n/a traveling along an infinite uniform cylinder with axial wavenumber k_z produces resistive and reactive components of the nondimensional surface specific acoustic impedance of the form shown in Figs. 7a, b.[13] (*Note:* The curves in Fig. 7 relate to a specific ratio of axial wavelength to cylin-

der radius.) The impedance is purely reactive if $k_z > k$, irrespective of k_s: No sound power is radiated. The resistive impedance peaks when $k^2 = k_s^2 + k_z^2$, as expected and then asymptotes to unity when $k^2 \gg k_s^2 + k_z^2$. Some useful approximate expressions for the reactive and resistive components of cylinder radiation impedance are presented in Table 3.[13] The radiation characteristics of finite cylinders are difficult to evaluate and to describe in general terms because of the diffraction of the sound by the ends of the cylinder; numerical methods are generally employed for quantitative analysis. In some cases, it may be reasonable to assume that a cylinder is extended beyond its vibrating region by rigid cylindrical baffles. In this case the theoretical results for infinite cylinders may be applied by means of Fourier synthesis.

When the cylinder circumference considerably exceeds the acoustic wavelength at frequencies of interest, for example, in cases of large aircraft fuselages at frequencies in excess of 100 Hz, the curvature of the surface has little geometric influence on sound radiation. Consequently, the forms of analysis of modal radiation introduced in Section 3.4 for flat plates may be applied. Modes of which the principal axial wavenumber exceeds k and for which $n < ka$ exhibit the acoustic short-circuiting phenomenon along their lengths, leaving a ring of uncanceled sources at each end; these correspond to plate edge modes. Statistical analysis of resonant, multimode radiation by large-diameter, thin-walled shells yield the set of curves shown in Fig. 8, in which the parameter is the ratio of ring to critical frequencies.[15]

In many cases of practical interest, the acoustic wavelength greatly exceeds the cylinder circumference at frequencies of interest ($ka \ll 1$), for example, sound radiation from typical industrial pipes at frequencies below 300 Hz. Sound radiation is then dominated by the low-

TABLE 3 Asymptotic Values of Specific Radiation Impedance for Infinite Cylinders Carrying a Standing Wave of Axial Wavenumber k_z

$$k_z > k$$

$$z_n = i\omega m_n$$

where

$$m_n \approx \begin{cases} -\rho_0 a \ln(|k_z^2 - k^2|^{1/2}a) & n = 0 \\ \rho_0 a/n & n \geq 1 \\ \rho_0/(k_z^2 - k^2)^{1/2} \end{cases} \quad \begin{aligned} & (k_z^2 - k^2)a^2 \ll 2n + 1 \\ & \\ & (k_z^2 - k^2)^{1/2}a \gg n^2 + 1 \end{aligned}$$

$$k_z < k$$

$$z_n = i\omega m_n + r_n$$

where

$$m_n \approx \rho_0 a/2(k^2 - k_z^2)a^2 \qquad (k^2 - k_z^2)^{1/2}a \gg n^2 + 1$$

$$r_n \approx \begin{cases} \rho_0 c\pi ka(k^2 - k_z^2)^n a^{2n}/(n!)^2 2^{|2n-1|} & 0 < (k^2 - k_z^2)a^2 \ll 2n + 1 \\ \rho_0 c & (k^2 - k_z^2)^{1/2}a \gg n^2 + 1 \end{cases}$$

From Ref. 13.

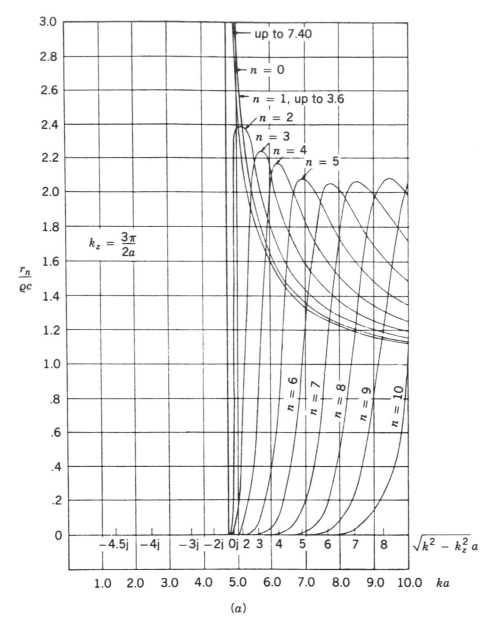

Fig. 7 (a) Specific acoustic resistance associated with cylindrical wave harmonics ($k_z = 3\pi/2a$).[13] (b) Specific acoustic reactance associated with cylindrical wave harmonics ($k_z = 3\pi/2a$).[13]

n modes (0, 1, 2). In practice, the "breathing" mode ($n = 0$) is not easily excited at low frequencies far below the ring frequency because it is very stiff. The "bending" mode ($n = 1$) is readily excited, even by plane sound waves in a contained fluid, through the presence of structural nonuniformities in the pipe, pipe bends, and pipe supports. The radiation efficiency of the bending mode is very low unless the axial structural wavelength exceeds the acoustic wavelength.[14] The $n = 2$ structural (ovalling) mode has a cutoff frequency given by $f/f_r = 2.68\beta$, above which it is often observed to

dominate pipe vibration. In industrial pipes the axial wavelength associated with these modes usually greatly exceeds the acoustic wavelength, and the $n = 0$, 1, 2 modes radiate in a similar manner to line monopoles, dipoles, and quadrupoles, respectively. The radiation efficiency of a uniformly vibrating long, thin cylinder is shown in Fig. 9.[14] The addition of stiffening rings to thin cylindrical shells does not greatly alter the sound power generated by localized excitation of the shell wall because it is dominated by radiation from modes of low circumferential order, of which the strain energy is pre-

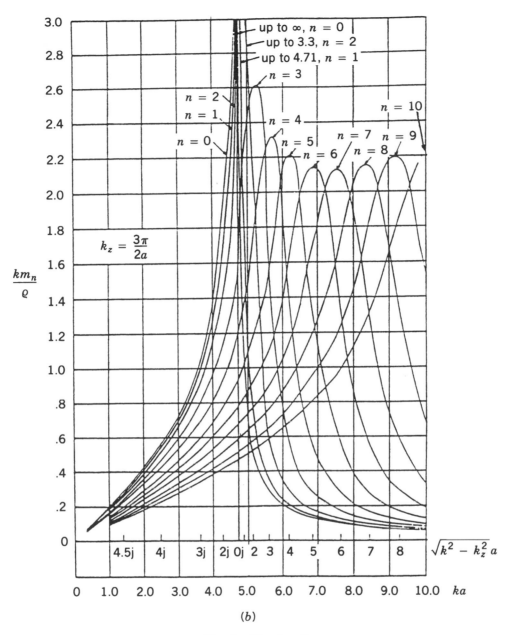

Fig. 7 (*Continued*)

4.3 Fluid-Loading Effects on Circular Cylindrical Shells

Sound pressures on a cylinder vibrating in a surrounding fluid produce two distinct effects: (1) the component in phase with surface normal velocity drains away vibrational energy and therefore damps the motion and (2) the component at 90° to the velocity mass loads the struc-

dominantly associated with membrane (midplane) strain, and not with flexural deformation: Rings influence the former far less than the latter.

ture, thereby reducing flexural wave speeds and modal natural frequencies. Air loading of cylinders generally has little effect, except in cases of large-diameter, honeycomb sandwich shells, in which the critical and ring frequencies may be comparable and the radiation loss factor may exceed the structural dissipation loss factor. One result is that damping treatment may be ineffective in reducing sound radiation. The most marked effects of fluid loading occur on structures submerged in a liquid, most commonly water.

Inertial loading can greatly reduce modal natural frequencies, especially those of low circumferential order,

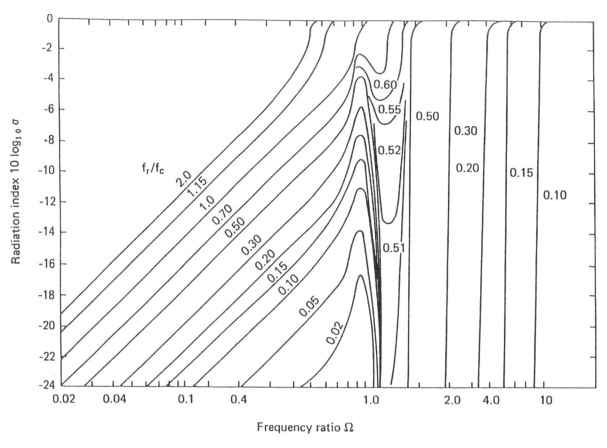

Fig. 8 Modal-averaged radiation efficiency of thin-walled, large-diameter cylindrical shells.[15]

as indicated in Table 3 and illustrated by Fig. 10. A method for estimating the reactive (inertial) effect of fluid loading on shell natural frequencies is presented in Ref. 16. The inertial loading experienced by thin-walled, pipelike structures vibrating in bending at frequencies for which $ka \ll 1$ is equivalent to an additional mass per unit length equal to that of a cylinder of fluid of radius a. Resistive fluid loading greatly influences the relative contributions to sound radiation by the various modes of submerged cylinders excited by vibrational forces. Those that have radiation loss factors close in value to their internal loss factors are the most effective. A remarkable result is that the sound power radiated by a point-force-driven cylindrical shell into water may be rather similar in magnitude to that which it radiates into air, although the modes principally responsible for power radiation are quite different.[17] Consequently, measures taken to increase mechanical damping can significantly alter the relative modal contributions to radiation.

Heavy fluid loading greatly affects the distribution of sound intensity radiated by a uniform plate or cylinder excited by a localized force or moment, tending to confine the principal region of radiation to the immediate vicinity of the point of excitation. As with the plate, the radiated sound power may be rather accurately estimated

by assuming that twice the force (moment) is applied directly to the fluid.

5 INTERACTION OF STRUCTURES WITH CONTAINED FLUIDS

Many structures take the forms of fluid containment vessels; examples include steam pipes and car passenger compartments. Acoustic interaction is important in the first example because noise generated inside the pipe can cause structural damage as well as unacceptably high levels of radiated noise. In the second example, it is the noise levels created inside the compartment by various external noise and vibration sources that are of concern.

As explained in Section 2, interaction between fluids and the structures that contain them can be stronger and more frequency dependent than for surrounding fluids due to the resonant behavior of the fluid volume. In most cases of air containment, the effects of fluid loading on structural vibration are negligible, so that fully coupled vibration analysis may be avoided. However, if the resonance frequencies of a structural mode and an acoustic mode of the (assumed) rigidly bounded fluid volume are very close together, coupled modes having

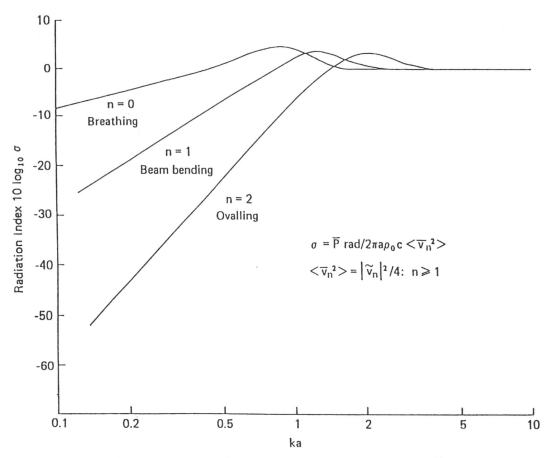

Fig. 9 Radiation efficiencies of uniformly vibrating cylinders.[14]

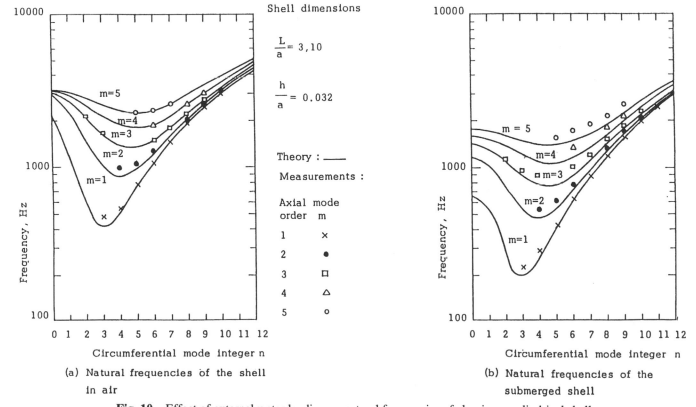

(a) Natural frequencies of the shell in air

(b) Natural frequencies of the submerged shell

Fig. 10 Effect of external water loading on natural frequencies of aluminum cylindrical shell with closed ends.

frequencies slightly altered from the uncoupled values will occur.[14] Such behavior may be observed in vehicle compartments.

At low frequencies, a volume of fluid acts as an elastic spring, which can significantly affect the natural frequency of a coupled lightweight structure. For example, the air contained within a loudspeaker cabinet is usually stiffer than the mechanical suspension and controls the low-frequency performance. A shallow cavity can exert a strong effect on a large panel that bounds it, as exemplified by the mass–air–mass resonance in lightweight double partitions, which limits the maximum achievable low-frequency transmission loss of such structures (see Chapter 93).

Acoustic interaction between structures and the fluids they contain is most marked in liquid containment vessels, including pipes. Modes involve both fluid and structural motion and may not rigorously be described as "fluid modes" or "structural modes"; however, in most cases, the vibrational energy of a mode resides principally in either the fluid or solid component. One of the most common examples of this form of interaction is observed in water-filled hoses in which the flexibility of the walls significantly alters the speed of propagation of acoustic disturbances along the pipe. Well below the ring frequency of the pipe wall, the elastic reaction of the wall to expansion of the cross section greatly exceeds the inertial reaction. A simple model of the interaction, which accounts only for the elastic reaction of the pipe wall and neglects inertial effects, yields the following expression for wave speed[13]: $c_0/c = [1 + (2a\rho_0 c^2/h\rho_s c_p^2)]^{-1/2}$, where ρ_s and c_p are, respectively, the density and speed

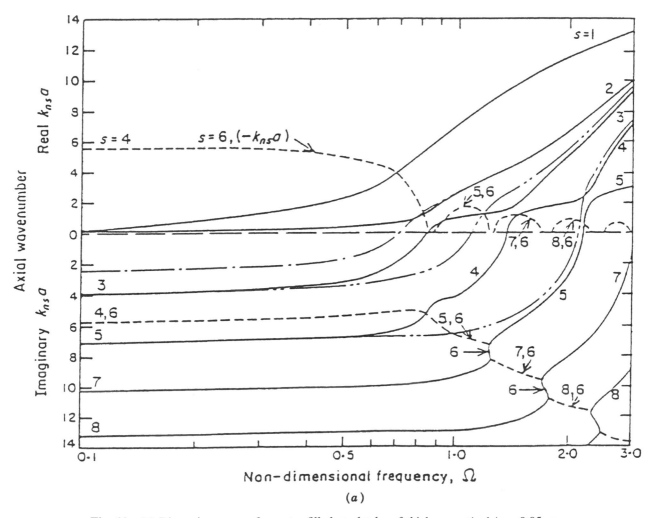

Fig. 11 (*a*) Dispersion curves for water-filled steel tube of thickness ratio $h/a = 0.05$, $n = 0$: (——) purely real and imaginary $k_{ns}a$; (---) real and imaginary parts of complex $k_{ns}a$; (— · —) pressure release duct solution; (— · · —) rigid walled duct solution. (*b*) Dispersion curves for water-filled steel tube of thickness ratio $h/a = 0.05$, $n = 1$: (——) purely real and imaginary $k_{ns}a$; (— · —) real and imaginary parts of complex $k_{ns}a$; (— · · —) rigid walled duct solution.

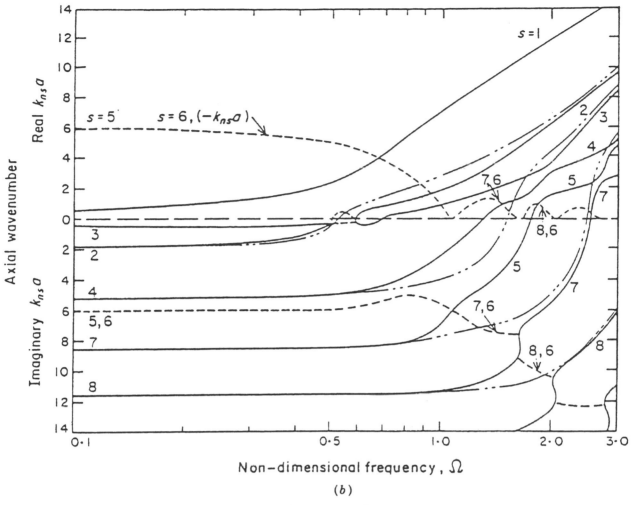

Fig. 11 (*Continued*)

of sound in the pipe material. Well above the pipe ring frequency the wall inertia dominates, and a corresponding model that neglects wall elasticity but accounts for wall mass yields the following expression for the axial wavenumber[13]: $\gamma = k[1 - (2\rho_0/\rho_s k^2 ha)]^{1/2}$ when $\rho_0 a/\rho_s h \ll 1$. At frequencies less than a cutoff frequency given by $f_c = (c/\pi)(\rho_0/2\rho_s ha)^{1/2}$, wave propagation is not possible in the absence of wall stiffness. At f_c, the fluid compressibility is balanced by the wall inertia. Blood vessels are so flexible that the speed of propagation of pressure disturbances is of the order of only 10–15 ms^{-1}. A general analysis of fluid-filled cylinders yields axial dispersion curves of which the form depends very much on the nondimensional parameter $K = \rho_0 a/\rho_s h$, where ρ_0/ρ_s is the ratio of fluid to solid densities and a/h is the ratio of pipe radius to wall thickness. When K is large (large-diameter, thin-walled cylinders containing liquid), one set of branches of the dispersion curves corresponds closely to the acoustic dispersion curve for a cylinder of fluid with a zero-pressure boundary. The "shell wave" branches are signifi-

cantly altered from their in vacuo forms by fluid loading. The natural frequencies of purely circumferential, low-order (low-n) "shell" modes of cylinders are significantly reduced by fluid loading. The ratio of in vacuo (see Table 2) to fluid-filled natural frequencies for $n \geq 2$ is given approximately by $(1 + K/n)^{1/2}$, provided that the in vacuo frequency is below the ring frequency as well as below the lowest transverse acoustic mode cutoff frequency given by $f = 1.84c/2\pi a$.[18] Examples of the dispersion curves of a steel pipe filled with water for $n = 0$ and $n = 1$ are shown in Figs. 11*a,b*.[19]

6 ACOUSTICALLY INDUCED VIBRATION OF STRUCTURES

6.1 Practical Aspects

The process of airborne sound transmission through partitions involves the vibrational response of the structure to sound incident on one side, together with the consequent radiation from the other. The influence on partition motion of the sound pressures it generates is related to the nondimensional fluid-loading parameter $\rho_0 c/\omega m$. In most cases of practical interest in air, this parameter is much smaller than unity, in which case the influence of the air loading on the vibration of the partition is extremely small, except for very lightweight partitions such as plastic films and membranes and very low frequencies. In water, however, fluid-loading forces often dominate, and a partition has little influence on sound transmission. This topic is treated in Chapters 41 and 93.

High levels of sound ($>$140 dB) can cause fatigue damage to engineering structures and produce malfunctions in mechanical and electrical equipment, either by direct excitation or, more commonly, via vibration induced in supporting structures. Acoustically induced fatigue is of serious concern to aircraft designers, and the noise from the rockets of launch vehicles can adversely affect the operational integrity of electronic component packages onboard payloads. Noise has also been known to threaten industrial plant components such as steam and gas control valves and associated piping. Gas-cooled nuclear reactor structures may also be at risk from damage by gas circulator noise. The vibrational response of ship structures to incident sound and the consequent reradiation of sound constitute sources of interference to sonar receivers. The optimization of panel (membrane) sound absorbers for studio sound control requires understanding of the principle of matching of internal and radiation resistances.

6.2 General Principles

The sound field that results from the incidence of a sound wave upon a linear elastic structure may be expressed exactly as the sum of three components: (i) the unobstructed incident wave; (ii) the field produced by the interaction of the incident field with the body when considered to be completely rigid; and (iii) the field radiated by the vibrational surface motion produced by the incident sound. Superposition of components (i) and (ii) gives the "blocked" field. Field component (ii) may be thought of as the result of radiation by the body when vibrating in such a manner that the surface normal acceleration is equal in magnitude and opposite in sign to the corresponding component of the unobstructed incident field at the body surface. A structure vibrates in response to the total sound pressure on its surface, which is the sum of all three components. The structure is coupled to the fluid via the radiated field component. The general equation of structural normal displacement may be expressed as $L(w) + m\ddot{w} = p_{bl} + p_{rad}$, where L is a differential operator. Since p_{rad} is a function of $\ddot{w}$, through the Kirchhoff–Helmholtz integral equation (see Chapter

15), this relationship has the form of an integrodifferential equation. The physical effect of radiation loading is to increase the mass and damping (and occasionally stiffness) of the structure, and hence the term p_{rad} may be taken across to the left-hand side of the equation of motion, which in this modified form represents the response of the fluid-loaded structure to the blocked pressure. This form is particularly convenient when the in vacuo modes of a structure are known, since radiation loading generally alters the mode shapes rather little, and hence the generalized blocked force may be evaluated by rigid-body scattering analysis or modal reciprocity analysis, as indicated below.

6.3 Relationship between Acoustically Induced Structural Response and Sound Radiation

Point Reciprocity Lyamshev explicitly expresses Rayleigh's reciprocity relationship between sound pressure at a field point generated by a point-force-excited linear elastic structure and the response of that structure to sound generated by a point monopole source at the same field point[20] (Fig. 12). This relationship is of great practical value because it obviates the need to simulate vibrational forces in operating systems, for the purpose of determining the resulting sound, by replacing the direct measurement with an observation of the vibrational response to a point source placed at the sound field observation position.[21, 22]

Modal Reciprocity The response of a structural mode to acoustic excitation by a point source in a free field may be expressed in terms of its sound radiation characteristics.[23, 24] The ratio of the modal velocity amplitude V_m to the amplitude of the incident sound pressure at the position of, but in the absence of, the structure at the modal resonance frequency ω_m is given by

$$\frac{(V)_m^2}{P_0^2} = \frac{4\pi D(\theta, \phi)}{\rho_0 c k^2} \frac{R_{rad}^m}{(R_{int}^m + R_{rad}^m)^2}, \qquad (4)$$

where $D(\theta, \phi)$ is the directivity factor of modal radiation in the direction of the position of the source and R_{rad}^m and R_{int}^m are the modal radiation and internal resistances, respectively. If the source is assumed to be very distant in terms of the maximum dimensions of the structure, this expression gives the response to plane-wave excitation from direction (θ, ϕ). Maximum response occurs when internal and radiation damping are equal. The expression for broadband excitation is

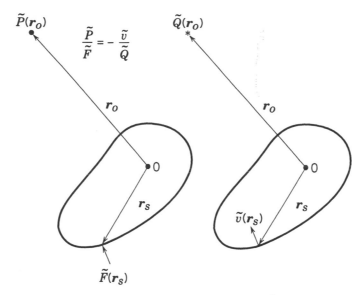

Fig. 12 Lyamshev's vibroacoustic reciprocity relationship: $\tilde{Q}$ = point monopole volume velocity (source strength); $\tilde{F}$ = point force; $\tilde{v}$ = vibration velocity; $\mathbf{r}$ = position vector; $\tilde{P}$ = sound pressure.

$$\frac{(V_m^2)}{G_{p0}(\omega)} = \frac{2\pi^2 cD(\theta,\phi)}{\rho_0 \omega_m^2 M_m} \frac{R_{\text{rad}}^m}{R_{\text{int}}^m + R_{\text{rad}}^m}, \tag{5}$$

where $G_{p0}(\omega)$ is the uniform spectral density of the sound pressure and M_m is the modal mass. The corresponding expressions for diffuse field excitation are

$$\frac{V_m^2}{P_0^2} = \frac{4\pi}{\rho_0 c k^2} \frac{R_{\text{rad}}^m}{(R_{\text{int}}^m + R_{\text{rad}}^m)^2} \tag{6}$$

and

$$\frac{V_m^2}{G_{p0}(\omega)} = \frac{2\pi^2 c}{\rho_0 \omega_m^2 M_m} \frac{R_{\text{rad}}^m}{R_{\text{int}}^m + R_{\text{rad}}^m}. \tag{7}$$

If a number of modes have close natural frequencies, the total energy E_s of vibrational response to broadband diffuse sound is given by

$$\frac{\overline{E_s}}{P_0^2} = \frac{2\pi^2 c n_s(\omega)}{\rho_0 \omega_c^2} \left\langle \frac{R_{\text{rad}}^m}{R_{\text{int}}^m + R_{\text{rad}}^m} \right\rangle_m, \tag{8}$$

where $n_s(\omega)$ is the modal density (see Chapter 156) and the angle brackets denote modal average. The practical interpretation of these relationships is that the response is controlled by internal (dissipation) damping only if it considerably exceeds acoustic damping. When the latter greatly exceeds the former, the response to broad-

band excitation reaches an upper limit independent of damping, but the pure tone response *decreases* as R_{rad} increases.

7 NUMERICAL ANALYSES OF STRUCTURE–FLUID INTERACTION

The majority of systems of practical engineering interest exhibit such complexity of geometry and structural detail that their vibrational and acoustic behavior cannot be accurately predicted by analytical mathematical methods. Modern, computer-based techniques of numerical analysis are available; these are either finite element methods or boundary element methods or combinations thereof. These techniques are described in Chapter 15. The reader should be warned that the *broadband* noise radiation from complex vibrating structures it is generally rather difficult to predict with great precision for the following reasons: (i) the amplitude and phase distribution of surface vibration is not normally known with sufficient accuracy, or in sufficient detail, to provide a suitable input to the computational procedure; (ii) analysis is performed one frequency at a time, thereby requiring vast computing times for broadband noise problems (however, new techniques of frequency interpolation can significantly speed up the calculation); and (iii) most practical sources are highly irregular in their geometry, thereby requiring a large and complex array of surface elements. Another major impediment to the application of boundary element methods to practical problems at the present time is that they cannot routinely deal with the forms of sound insulation construction commonly installed in vehicles, pipe lagging, and machinery covers, where the sound-absorbent material may exhibit nonlocal reaction. Finite element analysis is best suited to the computation of sound fields in enclosed volumes, but external radiation fields may be represented by the *wave envelope* element method, which is available in several finite element software packages.

Alternative methods to the boundary element method for sound radiation calculation are currently under development, for example, the multipole representation by which an array of monopole sources is distributed within the source volume so as to produce a sound field that matches, as well as possible, the actual distribution of normal acceleration on the source surface.[25, 26]

REFERENCES

1. E. L. Shenderov, "Sound Radiation by an Unbaffled Oscillating Disk (by a Disk in an Acoustically Compliant Baffle)," *Sov. Phys. Acoust.*, Vol. 34, 1988, pp. 191–198.

2. C. E. Wallace, "Radiation Resistance of a Rectangular Panel," *J. Acoust. Soc. Am.*, Vol. 51, 1972, pp. 946–952.

3. G. Maidanik, "Response of Ribbed Panels to Reverberant Sound Fields," *J. Acoust. Soc. Am.*, Vol. 34, 1962, pp. 809–826.

4. F. G. Leppington, E. G. Broadbent, and K. H. Heron, "The Acoustic Radiation Efficiency of Rectangular Panels," *Proc. Roy. Soc. Lond. Ser. A*, Vol. 382, 1982, pp. 245–271.

5. F. G. Leppington, E. G. Broadbent, and K. H. Heron, "Acoustic Radiation from Rectangular Panels with Constrained Edges," *Proc. Roy. Soc. Lond. Ser. A*, Vol. 392, 1984, pp. 67–84.

6. R. Timmel, "The Radiation Efficiency of Rectangular Thin Homogeneous Plates in an Infinite Baffle," *Acustica*, Vol. 73, 1991, pp. 1–11 (in German).

7. R. Timmel, "Investigations of the Effect of Edge Boundary Conditions for Flexurally Vibrating Rectangular Panels on the Radiation Efficiency as Exemplified by Clamped and Simply-Supported Panels," *Acustica*, Vol. 73, 1991, pp. 12–20 (in German).

8. L. L. Beranek and I. Vér (Eds.), *Noise and Vibration Control Engineering*, Wiley, New York, 1992.

9. N. Kiesewetter, "Impedance and Resonances of a Plate Before an Enclosed Volume of Air," *Acustica*, Vol. 61, 1986, pp. 213–217.

10. C. E. Wallace, "The Acoustic Radiation Damping of the Modes of a Rectangular Panel," *J. Acoust. Soc. Am.*, Vol. 81, 1987, pp. 1787–1794.

11. N. S. Lomas and S. I. Hayek, "Vibration and Acoustic Radiation of Elastically Supported Rectangular Plates," *J. Sound. Vib.*, Vol. 52, 1977, pp. 1–25.

12. H. G. Davies, "Low Frequency Random Excitation of Water-loaded Rectangular Plates," *J. Sound Vib.*, Vol. 15, 1971, pp. 107–126.

13. M. C. Junger and D. Feit, *Sound, Structures and Their Interaction*, 2nd ed., MIT Press, Cambridge, MA, 1986.

14. F. J. Fahy, *Sound and Structural Vibration*, Academic Press, London, 1987.

15. E. Szechenyi, "Modal Densities and Radiation Efficiencies of Unstiffenend Cylinders Using Statistical Methods," *J. Sound Vib.*, Vol. 19, 1971, pp. 65–82.

16. M. K. Au-Yang, "Natural Frequencies of Cylindrical Shells and Panels in Vacuum and in a Fluid, *J. Sound Vib.*, Vol. 57, 1978, pp. 341–355.

17. B. Laulagnet and J-L. Guyader, "Modal Analysis of a Shell's Acoustic Radiation in Light and Heavy Fluids," *J. Sound Vib.*, Vol. 131, 1989, pp. 397–416.

18. P. G. Bentley and D. Firth, "Acoustically Excited Vibrations in a Liquid-filled Cylindrical Tank," *J. Sound Vib.*, Vol. 19, 1971, pp. 179–191.

19. C. R. Fuller and F. J. Fahy, "Characteristics of Wave Propagation and Energy Distribution in Cylindrical Shells Filled with Fluid, *J. Sound Vib.*, Vol. 81, 1981, pp. 501–518.

20. L. M. Liamshev, "Theory of Sound Radiation by Thin Elastic Shells and Plates," *Sov. Phys. Acoust.*, Vol. 5, 1960, pp. 431–438.

21. T. Ten Wolde, "On the Validity and Application of Reciprocity in Acoustical, Mechano-acoustical and Other Dynamical Systems," *Acustica*, Vol. 28, 1973, pp. 23–32.

22. F. J. Fahy, "The Vibroacoustic Reciprocity Principle and Applications to Noise Control," *Acustica*, Vol. 81, 1995, pp. 544–558.

23. P. W. Smith, "Response and Radiation of Structural Modes Excited by Sound," *J. Acoust. Soc. Am.*, Vol. 34, 1962, pp. 640–647.

24. G. Chertock, "General Reciprocity Relation," *J. Acoust. Soc. Am.*, Vol. 34, 1962, p. 989.

25. G. H. Koopmann, L. Song, and J. B. Fahnline, "A Method of Computing Acoustic Fields Based on the Principle of Wave Superposition," *J. Acoust. Soc. Am.*, Vol. 86, 1989, pp. 2433–2438.

26. M. Ochmann and F. Wellner, "Calculation of the Three-dimensional Sound Radiation from a Vibrating Structure Using a Boundary Element–Multigrid Method," *Acustica*, Vol. 73, 1991, pp. 177–190 (in German).

12

BOUNDARY WAVES

IVAN TOLSTOY

1 INTRODUCTION

Boundary waves form a class of phenomena separate and distinct from "volume" or "body" waves. They are characterized by energy storage in the neighborhood of an interface and exhibit hybrid properties of propagation and vibration. Their existence is *not* dependent upon that of volume (body) waves, although in the presence of such, given a source of energy, for example, a point source, boundary modes will effectively capture part of the spherically spreading acoustic energy and convert it into cylindrically diverging waves with amplitudes falling off with range r like $r^{-1/2}$.

Elementary models of boundary waves are *surface tension* and *gravity waves* on the free surface of a fluid. Other classic examples are *Rayleigh waves* on the free surface of an elastic solid (the dominant arrival and primary source of destruction from shallow-focus earthquakes) and *Stoneley waves* at the interface of two elastic solids or at a liquid–solid interface (sometimes called *Scholte waves*). *Boundary roughness modes* are a good example of purely acoustic boundary waves, propagating along the interface between a compressible half-space and a rough rigid wall (or between two half-spaces). On the other hand, contrary to a common misconception, Love waves are *not* boundary waves; they are simply shear waves trapped by total reflection in a low-velocity surface layer.

2 BASIC FORMULATION

The principles involved are illustrated in their simplest form in two dimensions x, z, assuming plane harmonic

Encyclopedia of Acoustics, edited by Malcolm J. Crocker
ISBN 0-471-80465-7 © 1997 John Wiley & Sons, Inc.

waves and neglecting attenuation. An ordinary acoustic or elastic volume (body) wave incident upon a plane interface generates reflected and transmitted waves: the effect of the boundary is here *extrinsic*—it modifies the total field, but its presence or absence does not affect the incident volume wave. On the other hand, the energy of a boundary wave is concentrated exponentially at the interface, whose role is now *intrinsic*: Without the interface there can be no boundary wave.

Let φ be a displacement potential corresponding to the field displacement **d** and pressure perturbation p:

$$\mathbf{d} = \mathbf{i}\xi + \mathbf{k}\zeta = \nabla\varphi, \tag{1}$$

$$p = -\rho\varphi_{tt}, \tag{2}$$

where **i**, **k** are the x, z unit vectors, ρ is the fluid density and subscripts indicate differentiation with respect to time t. The acoustic wave equation is

$$\nabla^2\varphi = c^{-2}\varphi_{tt}, \tag{3}$$

c being the speed of sound.

Let the fluid be homogeneous: $c = \text{const}$. Assume a harmonic wave of angular frequency ω propagating in two dimensions x, z. Separation of variables yields possible solutions:

$$\varphi = qe^{\pm i\alpha x}e^{\pm i\gamma z}e^{\pm i\omega t}, \tag{4}$$

$$\alpha^2 + \gamma^2 = k^2 = \omega^2 c^{-2}. \tag{5}$$

If the separation constants α, γ are real, Eq. (4) represents plane waves of x, z wavenumbers α, γ. The corresponding field displacements ξ, ζ are given as

$$\xi = \varphi_x = \pm i\alpha\varphi, \tag{6}$$

$$\zeta = \varphi_z = \pm i\gamma\varphi. \tag{7}$$

The particular combination

$$\varphi = q e^{i\gamma z} e^{i\alpha x} e^{-i\omega t} \tag{8}$$

is a plane wave traveling in the direction of increasing x, z (Fig. 1), with angle of incidence θ with respect to the xy plane:

$$\alpha = k \sin \theta, \tag{9}$$

$$\gamma = k \cos \theta, \tag{10}$$

Mathematically legitimate solutions are also obtained by making one or more of the separation constants imaginary, for example, $\gamma = \pm ig$, and

$$\varphi = q e^{\pm gz} e^{i(\alpha x - \omega t)}, \tag{11}$$

$$g = (\gamma^2 - k^2)^{1/2}, \qquad \alpha \geq k, \qquad g \geq 0. \tag{12}$$

When an acoustic wave represented by Eq. (8), traveling in a half-space $z > 0$ of sound velocity c_1, is incident upon the plane surface $z = 0$ of a half-space $z < 0$ of sound velocity $c_2 \geq c_1$, if the angle of incidence exceeds the critical angle $\theta_c = \sin^{-1} c_1/c_2$, then the wave is totally reflected and the field in the half-space $z < 0$ is represented by Eqs. (11) and (12). In this example the energy is not concentrated near the $z = 0$ interface, since it is distributed sinusoidally in the region $z > 0$. The exponential field in $z < 0$ is not an independent mode in its own right: that is, it is not a boundary wave mode.

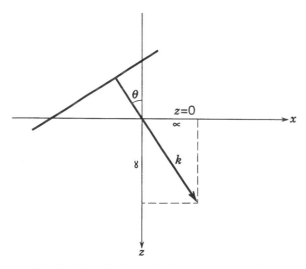

Fig. 1   Coordinates and plane-wave geometry.

3   NECESSARY CONDITIONS FOR BOUNDARY MODES; ENERGETICS

The trapping of energy near a $z = 0$ boundary imposes definite characteristics for the energy distribution and flow; for Eqs. (11) and (12) to describe bona fide boundary modes, certain conditions must be satisfied.

First, it is clear that the energy density cannot be allowed to oscillate sinusoidally or to increase indefinitely away from the interface $z = 0$. This is known as the boundary condition at infinity. It means, for example, that in the half-space $z > 0$ we must choose the negative exponent in Eq. (11):

$$\varphi_1 = q_1 e^{-g_1 z} e^{i(\alpha x - \omega t)}, \qquad g_1 > 0, \qquad z > 0. \tag{13}$$

Second, this solution must obey all relevant boundary conditions at $z = 0$. In the general case of an interface between two media there will be two or more potential function solutions, and the boundary conditions yield a system of linear equations between the field functions and their derivatives. Substitution of the field functions into these gives a homogeneous system of n linear equations in n unknowns q_j. The condition for this system to have nontrivial solutions is that the determinant of the coefficients of the q_j must vanish. The real roots of this equation give the characteristic wavenumbers and phase velocities of the undamped boundary modes, if any.

The simplest problem of this type would correspond to a homogeneous acoustic half-space $z > 0$, with zero energy density (no disturbance) in $z < 0$. The boundary wave will then be described by Eq. (13). A nontrivial solution of this kind occurs if, for instance, the boundary condition is of the kind

$$\varphi_z = -\eta\varphi \quad \text{at } z = 0, \qquad \eta > 0. \tag{14}$$

There exist specific boundaries, for example, slightly rough interfaces, yielding this kind of *impedance* boundary condition. Equations (13) and (14) give then

$$g = \eta, \tag{15}$$

where in general η is a function of the frequency and of the physical parameters describing the interface (see Section 4). This is the characteristic (dispersion) equation which, by virtue of Eq. (12), defines a relation between α and ω and thus gives the phase velocity v and the group velocity U along the boundary:

$$v = \frac{\omega}{\alpha}, \tag{16}$$

$$U = \frac{d\omega}{d\alpha}. \tag{17}$$

The group velocity gives the rate of energy transport. If $\omega = \alpha c$, where c is a constant, $v = c$ is not a function of frequency; then $v = U$ and propagation is not dispersive. In the dispersive case v is a function of ω and $v \neq U$.[1]

The most obvious property of such a boundary wave is the exponential concentration of energy near the boundary; the energetics are also special in other ways.

The classic expressions for the kinetic and potential energy densities are

$$T = \tfrac{1}{2}\rho(\xi_t^2 + \zeta_t^2), \qquad (18)$$

$$V = \tfrac{1}{2}\rho c^2(\xi_x + \zeta_z)^2, \qquad (19)$$

For the traveling acoustic volume wave of Eq. (8)

$$T = V = \tfrac{1}{2}\rho c^2 k^4 \varphi^2. \qquad (20)$$

For the case of Eqs. (6)–(8), the contributions of each pair (ξ_t, ξ_x) or (ζ_t, ζ_z) to the energy are either in phase or out of phase by a multiple of π. In *volume waves*, therefore, transport of energy consists of equal parts of *kinetic and potential energy in phase* with each other, traveling in half-wave packets with the speed of sound.

In *boundary waves*, on the other hand, Eqs. (6), (7), and (13) show that ζ_t and ζ_z in Eqs. (18) and (19) are in quadrature, as are the pressure p and the vertical velocity ζ_t. Insofar as the vertical displacement field is concerned, then, energy is being exchanged periodically between the kinetic and potential forms: There is *no mean energy flow normal to the boundary*. This is typical of linear oscillators, in which energy is shuttled periodically back and forth between two forms. This illustrates the energy storage mechanism at the interface: A boundary wave has the hybrid characteristics of a traveling wave (x-coordinate, along the interface) and an oscillator (z-direction, normal to the interface).

From the fact that energy is trapped by the interface, we may deduce that boundary waves travel in two dimensions; that is, they spread cylindrically. Thus, given a point source at or near the interface, the far-field amplitude of the boundary wave will fall off with range r like $r^{-1/2}$, whereas the acoustic volume amplitude falls off like r^{-1}. Under certain conditions (low attenuation) the boundary wave will then dominate the far field.

4 MODES OF MEMBRANE OVER AN ACOUSTIC HALF-SPACE

A thin membrane under tension on the surface of a compressible half-space offers a simple model of an inter-

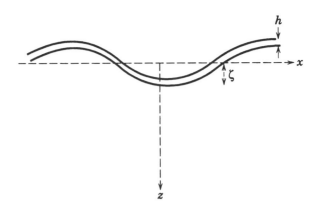

Fig. 2 Thin membrane on surface of acoustic half-space $z > 0$.

face capable of storing potential energy and thus carrying boundary waves (Fig. 2).

Assume a membrane of thickness $h(kh \ll 1)$ and specific mass m under tension T. Let it be stretched on the surface $z = 0$ of a compressible half-space $z > 0$ of density ρ_f and sound velocity c. Using the usual small displacement and slope approximations, the downward elastic force per unit area on the membrane is $-T\zeta_{xx}$. In the absence of fluid this would be balanced by a force of inertia $m\zeta_{tt}$, where $m = \rho_m h$ is the mass per unit area of the membrane. In the presence of an underlying fluid we add the pressure term $p = -\rho_f c^2 \, \nabla^2 \varphi$ to the balance of forces to obtain the boundary condition

$$-T\zeta_{xx} + \rho_m h\zeta_{tt} = -\rho_f c^2 \, \nabla^2 \varphi \qquad z = 0. \qquad (21)$$

Dropping the subscript on g, Eq. (13) yields the *characteristic equation*

$$\rho_f \omega^2 + \rho_m h\omega^2 g - T\alpha^2 g = 0 \qquad (22)$$

or

$$\omega^2 = T\alpha^2 \rho_f^{-1}(\alpha^2 - k^2)^{1/2}\left(1 + \frac{hg\rho_m}{\rho_f}\right)^{-1}. \qquad (23)$$

(It is easy to introduce the acceleration of gravity G into the picture: Following the discussion of Ref. 1 (p. 132), we simply add $G\alpha$ to the numerator.)

For these boundary waves to exist, it is clearly necessary to have the following:

1. $T \neq 0$: Potential energy storage at the surface is the essential mechanism here.

2. $\alpha > k$, that is, the phase velocity of these waves must be less than the speed of acoustic waves

(otherwise we merely have a supersonic corrugation generating, or generated by, a train of acoustic volume waves).

3. For $h \to 0$, T becomes the surface tension of the fluid. In the incompressible case $k = 0$, $g = \alpha$, and Eq. (23) gives the usual "capillary wave" result:

$$\omega^2 = T\rho_f^{-1}\alpha^3. \tag{24}$$

This underlies the feature that the role of compressibility is inessential: The existence of boundary waves is independent of that of acoustic volume waves in the half-space.

A similar type of boundary wave corresponds to the flexure modes of a thin plate over a fluid and may be used to model the low-frequency behavior of ice sheets.

5 ACOUSTIC BOUNDARY WAVES NEAR ROUGH WALL

At certain kinds of rough boundaries, the scattered field obeys a boundary condition of the type of Eq. (14).[2–4]

Consider a slightly rough, perfectly hard wall. The roughness elements may be seen as individual scatterers rigidly attached to the average plane $z = 0$ of the wall. Let a be the mean radius of the scatterers and l their average separation. For an equation such as (14) to hold, the following conditions must be met:

1. The roughness elements must be *compact*:

$$ka \leq 1. \tag{25}$$

However, for the hard surface case, it can be shown[2b] that this condition may be relaxed to something like $ka < 2$.

2. The roughness elements on the wall must *not be too sparse*, that is,

$$kl \leq 1. \tag{26}$$

If these conditions are met and Eq. (14) is obeyed, boundary waves will exist either for regularly spaced roughness elements (of any shape) or for stochastic roughness of mean height a and correlation distance l.[3–5]

The numerical value of η in Eqs. (14) and (15) may be derived from first principles using standard compact scatterer approximations that, in the case of an incident acoustic field, replace the scattered field by that of two point sources: a monopole of strength Q and a dipole of moment $\mathbf{M}$.[5,6] These quantities can be calculated for

scatterers of *any* shape,[5] and the scattered field of a single scatterer is

$$\varphi_{sc} = \frac{Qe^{ikr}}{r} + \mathbf{M} \cdot \nabla\left(\frac{e^{ikr}}{r}\right) \tag{27}$$

If a distribution of such scatterers is attached to the wall, the sum of their fields may be represented by a surface integral. This gives the total scattered field. A standard theorem from potential theory then allows one to express the normal derivative of the coherent scattered field φ_s at $z = 0$ in terms of the total field φ. For a hard wall $\partial\varphi_s/\partial z = \partial\varphi/\partial z$, and one obtains

$$\frac{\partial\varphi}{\partial z} = -\eta\varphi, \qquad z = 0, \tag{28}$$

where some simple algebra yields

$$\eta = -2\pi N\left(Q + \frac{\nabla \cdot \mathbf{M}}{\nu}\right), \tag{29}$$

where for a rigid wall $Q < 0$, $\nabla \cdot \mathbf{M} > 0$, N is the number of scatterers per unit area, ν is a factor introduced to take account of near-field interaction between dipoles[5] (for tightly packed hemispheres $\nu \simeq 1.3$ and for contiguous circular hemicylinders $\nu \simeq 1.8$).

The parameters Q, $\mathbf{M}$ are easily calculated for compact roughness elements of any shape. Particularly simple results are obtained for hemispherical bosses or cylindrical corrugations. Thus, in the former case, for a point source and receiver at z_0 and z, it may be shown that the ratio of the amplitude p_B of the boundary wave to that of the direct spherically spreading acoustic wave p_D is[7,8] given as

$$\frac{p_B}{p_D} = \epsilon(2\pi r)^{1/2}k^{3/2}e^{-r\delta}e^{-g(z+z_0)}. \tag{30}$$

For both source and receiver on the wall $z = z_0 = 0$:

$$\frac{p_B}{p_D} = \epsilon(2\pi r)^{1/2}k^{3/2}e^{-r\delta}, \tag{31}$$

where, for hexagonal packing,

$$\epsilon = \frac{2\pi}{3}Na^3\left(\frac{1}{2} - \frac{\pi^2a^3}{4l^3}\right)\left(1 + \frac{\pi^2a^3}{4l^3}\right)^{-1}. \tag{32}$$

The attenuation coefficient δ is obtained by calculating the energy reradiated by the scatterers in all directions away from the wall[8]:

$$\delta = \frac{7}{6} a^3 \epsilon^2 k^6. \tag{33}$$

Thus for δ not too large and k not too small the boundary wave may, for increasing r, become larger than the direct, spherically spreading acoustic arrival. But since δ is proportional to k,[6] the effect of increasing frequency appears abruptly, effectively giving a high-frequency cutoff. Experiments by Medwin and colleagues[9,10] have demonstrated these phenomena very clearly (Fig. 3).

Since a plane wave corresponds to a source at infinity, Eq. (30) gives a vanishing result ($z_0 \to \infty$): *plane waves do not excite boundary modes* (a generally valid result).

In the case of stochastically rough surfaces perturbation theory gives similar results: All these equations remain valid, provided one replaces relevant quantities by their proper statistical averages.[3,4,11]

Equations (13)–(15) show that we must have $\eta > 0$. In view of Eq. (29) this means that the dipole effect, which scatters energy parallel to the wall, is responsible for the

existence of the boundary wave; this leads to a concentration of energy near the interface.

6 BOUNDARY WAVES ON ELASTIC SOLID BOUNDARIES

The classic example of boundary waves in acoustics and elasticity is that of Rayleigh waves on the free surface of an ideal, homogeneous elastic half-space. They provide the basic mechanism for the largest and most destructive waves generated by shallow-focus earthquakes.[12]

To describe the propagation of small disturbances in a homogeneous elastic solid, it is customary to introduce two potential functions: a *scalar* displacement potential Φ and a *vector* potential $\mathbf{\Psi}$. The displacement field is then

$$\mathbf{d} = \nabla\Phi + \mathbf{curl}\,\mathbf{\Psi}. \tag{34}$$

The potentials Φ, $\mathbf{\Psi}$ obey the wave equations

$$\nabla^2\Phi = c_P^{-2}\,\frac{\partial^2\Phi}{\partial t^2}, \tag{35}$$

$$\nabla^2\mathbf{\Psi} = c_S^{-2}\,\frac{\partial^2\mathbf{\Psi}}{\partial t^2}, \tag{36}$$

where c_P, c_S are the compressional (P) and shear (S) wave velocities

$$c_P^2 = (\lambda + 2\mu)\rho_s^{-1}, \qquad c_S^2 = \mu\rho_s^{-1}, \tag{37}$$

ρ_s being the solid density, μ the rigidity, and λ the other Lamé constant.

In two dimensions, Rayleigh waves correspond to solutions of the type

$$\Phi = Ae^{-g_P z}e^{i(\alpha x - \omega t)}, \qquad g_P = (\alpha^2 - \omega^2 c_P^{-2})^{1/2}, \qquad z > 0, \tag{38}$$

$$\mathbf{\Psi} = \mathbf{B}e^{-g_S z}e^{i(\alpha x - \omega t)}, \qquad g_S = (\alpha^2 - \omega^2 c_S^{-2})^{1/2}, \qquad z > 0, \tag{39}$$

$\mathbf{B}$ being a vector pointing along the y axis. The boundary conditions (vanishing normal and tangential stresses at $z = 0$) can be expressed in terms of Φ and $\mathbf{\Psi}$; a little

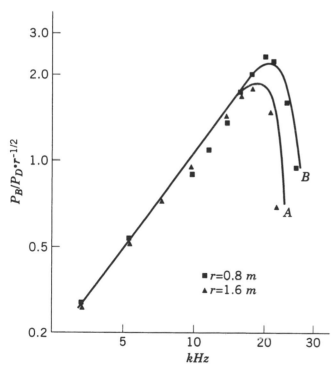

Fig. 3 Ratio of boundary wave pressure amplitude p_B to direct acoustic wave amplitude p_D as function of frequency for point source and receiver on rough surface $z = 0$ for two different ranges. Model used consisted of 2-mm-diameter spheres of lead shot ($a = 1$ mm) on metal plate in air. Spheres are tightly packed ($l = 2$ mm). Solid curves are calculated from Eqs. (31)–(33). The data were obtained by D'Spain, Childs, and Medwin.[10] Vertical axis units are (metres).$^{-1/2}$

algebra then shows that these conditions are satisfied by Eqs. (38) and (39) if[1, 12]

$$4(1 - v^2 c_P^{-2})^{1/2}(1 - v^2 c_S^{-2})^{1/2} = (2 - v^2 c_S^{-2})^2, \quad (40)$$

v being the phase velocity; this is the characteristic equation for Rayleigh waves on the free surface of a homogeneous elastic half-space. It has one real root $v = v_R$ in the domain $v_R < c_S < c_P$. Equation (40) does not contain the frequency, and Rayleigh waves on the surface of a homogeneous isotropic elastic half-space are *not* dispersive.

The numerical value of v_R depends on the dimensionless Poisson's ratio σ and varies from $v_R = 0.85c_S$ ($\sigma = 0$, perfectly compressible solid) to $v_R = 0.955c_S$ ($\sigma = 0.5$, incompressible solid). For $\sigma = 0.25$, $v_R = 0.9194c_S$ (the commonly quoted value).

For shallow-focus earthquakes as much as 70% of the energy goes into the Rayleigh modes. Since these spread cylindrically (in contrast to the spherically spreading P and S waves), they tend to be the most prominent arrivals from shallow shocks. They are also the most destructive (especially in the presence of a resonating surface layer). In practice Earth is layered and the theory of Rayleigh waves for realistic models is more complicated. Layering introduces a length scale and propagation becomes sensitive to wavelength, that is, it is dispersive; the observed dispersion allows one to study the internal structure of the planet.[1, 12, 13] Figure 4 shows a typical comparison between theory and experiment.

It is also possible to demonstrate the possibility of boundary waves at a smooth interface between two elastic half-spaces; they are called Stoneley waves in honor of their discoverer.[14] The algebra is a little more complicated than for Rayleigh waves, but the mechanism is the same: Energy is trapped near the interface by elastic stresses and travels parallel to the boundary. When the half-spaces are homogeneous and isotropic, there is no length scale and propagation is *not* dispersive.

Stoneley-type modes also exist at the interface of an elastic solid and a fluid; in this case they are sometimes referred to as Scholte[15] or Biot[16, 17] waves. Here, again, in the case of homogeneous and isotropic half-spaces, there is no length scale and propagation is not dispersive.

In the solid ($z > 0$) the elastic fields are described by Eqs. (38) and (39). In the fluid half-space the field is given by the usual scalar displacement potential

$$\varphi = q e^{gz} e^{i(\alpha x - \omega t)}, \qquad z < 0, \quad (41)$$

and g is defined by Eq. (12). Equations (38), (39), and (41) must satisfy three boundary conditions at $z = 0$:

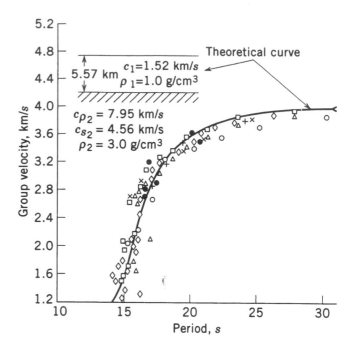

Fig. 4 Observed dispersion of fundamental Rayleigh wave mode for variety of oceanic paths (group velocity vs. period) compared to calculation (solid curve) based on averaged acoustic model of inset (after Ewing et al.[13]).

vanishing of tangential stress in the solid, continuity of normal stress and pressure, and continuity of the normal component of displacement. This yields the characteristic equation[18]

$$4(1 - v^2 c_P^{-2})^{1/2}(1 - v^2 c_S^{-2})^{1/2} - (2 - v^2 c_S^{-2})^2$$
$$= \frac{\rho_f}{\rho_s} v^4 c_S^{-4}(1 - v^2 c_P^{-2})^{1/2}(1 - v^2 c^{-2})^{-1/2}, \quad (42)$$

where ρ_f is the density of the fluid, c its sound velocity. This equation always has a real root $v = v_{ST}$ smaller than the smallest of the two quantities c, c_S. For most models (e.g., water over hard rock or metal) $c_S > c$ and v_{ST} is slightly less than c. In the case of weak solids (low-rigidity sediments) in contact with water, the problem may be complicated by the effects of gravity and buoyancy, and the wave can be very slow.

It was pointed out by Biot[16] that Eq. (42) has a real root in the limit of a massless solid ($\rho_S \to 0$, $c_S \to \infty$, $c_P \to \infty$, $c_S^2 \rho_S \to \mu \neq 0$) and an incompressible fluid ($c \to \infty$, $\rho_f \neq 0$):

$$v = \mu^{1/2} \rho_f^{-1/2}(1 - \sigma)^{-1/2}. \quad (43)$$

This illustrates the energetics of the boundary wave: Kinetic energy is stored in the incompressible fluid above the interface and is exchanged periodically with potential energy stored in the massless solid below the boundary.

Neither of the two half-spaces can sustain body wave modes. This underlines the special nature of boundary waves. It has been suggested by Junger[17] that Stoneley waves of this type be renamed after Biot.

7 OCCURRENCE AND IMPORTANCE OF BOUNDARY WAVES

Acoustic and elastic boundary wave phenomena occur in nature as destructive, earthquake-generated Rayleigh waves, as flexure modes of ice sheets on water surfaces, and as Stoneley-type (Scholte, Biot) arrivals at water–sediment interfaces in acoustical experiments at sea.[15] Acoustic rough-surface boundary waves have been observed in the laboratory[9, 10]; Rayleigh modes have been used in acoustic delay lines.[19] Acoustoelastic boundary waves are thus quite common. This chapter has emphasized that they constitute a class of propagation phenomena distinct and separate from the usual body, or volume, waves of acoustics and elasticity and are characterized by different energetics. The energy of boundary waves is channeled by an interface; they exist as modes in their own right, and they are independent of the existence of volume waves. These attributes distinguish them fundamentally from other, superficially similar phenomena such as volume wave trapping in waveguides (Love waves) or deflection by diffraction (creeping waves).

REFERENCES

1. I. Tolstoy, *Wave Propagation*, McGraw-Hill, New York 1973.

2. M. A. Biot (a) "Generalized Boundary Conditions for Multiple Scatter in Acoustic Reflection," *J. Acoust. Soc. Am.*, Vol. 44, 1968, pp. 1616–1622; (b) "Lagrangian Analysis of Multiple Scatter in Acoustic and Electromagnetic Reflection," *Acad. Roy. Belg. Bull. Cl. Sci.*, Vol. 59, 1973, pp. 153–199.

3. F. G. Bass and I. M. Fuks, *Wave Scattering from Statistically Rough Surfaces*, Pergamon, Oxford, 1979.

4. A. R. Wenzel, "Smoothed Boundary Conditions for Randomly Rough Surfaces," *J. Math. Phys.*, Vol. 15, 1974, pp. 317–323.

5. I. Tolstoy, "Smoothed Boundary Conditions, Coherent Low-Frequency Scatter, and Boundary Modes," *J. Acoust. Soc. Am.*, Vol. 75, 1984, pp. 1–22.

6. M. J. Lightill, *Waves in Fluids*, Cambridge University Press, Cambridge, 1978.

7. I. Tolstoy, "The Scattering of Spherical Pulses by Slightly Rough Surface," *J. Acoust. Soc. Am.*, Vol. 66, 1979, pp. 1135–1144.

8. I. Tolstoy, "Long-Wavelength Scatter from Rough Surfaces," in D. Sette (Ed.), *Frontiers in Physical Acoustics*, North-Holland, Amsterdam, 1986.

9. H. Medwin, J. Bailey, J. Bremhorst, B. J. Savage, and I. Tolstoy, "The Scattered Acoustic Boundary Wave Generated by Grazing Incidence at a Slightly Rough Surface," *J. Acoust. Soc. Am.*, Vol. 66, 1979, pp. 1131–1134.

10. G. L. D'Spain, E. Childs, and H. Medwin, "Low Frequency Grazing Propagation over Steep-sloped Rigid Roughness Elements," *J. Acoust. Soc. Am.*, Vol. 76, 1984, pp. 1774–1790.

11. A. Tolstoy, D. Berman, O. Diachok, and I. Tolstoy, "An Assessment of Second-Order Perturbation Theory for Scattering of Sound by Hard, Statistically Rough Surfaces," *J. Acoust. Soc. Am.*, Vol. 77, 1985, pp. 2074–2080.

12. K. E. Bullen, *An Introduction to the Theory of Seismology*, Cambridge, 1963.

13. W. M. Ewing, W. S. Jardetzky, and F. Press, *Elastic Waves in Layered Media*, McGraw-Hill, New York, 1957.

14. R. Stoneley, "Elastic Waves at the Surface of Separation of Two Solids," *Proc. Roy. Soc. A*, Vol. 106, 1924, pp. 416–428.

15. D. Rauch, "On the Role of Bottom Interface Waves in Ocean Seismo-acoustics: A Review," in T. Akal and J. M. Berkson (Eds.), *Ocean Seismo-Acoustics*, Plenum, New York, 1986.

16. M. A. Biot, "Interaction of Rayleigh and Stoneley Waves in the Ocean Bottom," *Bull. Seism. Soc. Am.*, Vol. 42, 1952, pp. 81–93.

17. M. C. Junger, "A Proposal to Honor the Memory of Maurice A. Biot," *J. Acoust. Soc. Am.*, Vol. 80, 1986, p. 1530.

18. I. Tolstoy and C. S. Clay, *Ocean Acoustics*, American Institute of Physics, New York, 1987.

19. Y. V. Gulayev and A. V. Medved, "Acoustic Surface Wave Devices and Their Applications in Radioelectronics (review)," *Radiophys. Quant. Electron.*, Vol. 26, 1983, pp. 911–948.

13

ACOUSTIC LUMPED ELEMENTS FROM FIRST PRINCIPLES

Robert E. Apfel

1 INTRODUCTION

The term *acoustical lumped elements* refers to the significant simplification in the behavior of sound that occurs when sound interacts with a physical structure that is much smaller than an acoustic wavelength. These elements are analogous to electrical resistors (R), capacitors (C), and inductors (L); therefore, all the mathematical tools developed for these electrical elements can be adopted for acoustical and mechanical elements. This approach has found broad application in areas ranging from loudspeaker and muffler design,[1,2] to noise in the ocean from resonating bubbles,[1] to musical instruments based on Helmholtz resonators (e.g., when one blows over an opening of a wine bottle).

The direct analogy between electrical circuits and mechanical elements is made apparent by considering the equations for an R–L–C series circuit driven by an alternating voltage source and for a spring–mass–dashpot system driven by an alternating forcing function.

2 ACOUSTIC SYSTEM AND DEFINITION OF SYMBOLS

There is a fairly general familiarity among researchers in the electrical and mechanical sciences and engineering with lumped elements and their equivalent circuits. A comparison between electrical and mechanical, one-degree-of-freedom systems is given in Fig. 1.

$$L\ddot{Q} + R\dot{Q} + \frac{1}{C}\,Q = E_0 e^{i\omega t} \qquad M\ddot{X} + c\dot{X} + kX = F_0 e^{i\omega t}$$

Encyclopedia of Acoustics, edited by Malcolm J. Crocker
ISBN 0-471-80465-7 © 1997 John Wiley & Sons, Inc.

For simple harmonic motion:

$$Q = Q_0 e^{i\omega t} \qquad X = X_0 e^{i\omega t}$$

$$I = \dot{Q} = i\omega Q \qquad V = \dot{X} = i\omega X$$

$$Z_E = \frac{E}{I} = R + i\omega L + \frac{1}{i\omega C} \qquad Z_M = \frac{F}{V} = c + i\omega M + \frac{k}{i\omega}$$

The electrical impedance Z_E and mechanical impedance Z_M are defined roughly as the ratios of "cause" to "effect" [electomotive force (EMF) ÷ current for electrical impedance and force ÷ velocity for mechanical impedance].

For simple harmonic motion ($e^{i\omega t}$) in either of these systems, the differential equations can be simplified, leading to an impedance that has three terms: a real term (R or c), an imaginary term proportional to $i\omega$ (L or M), and an imaginary term proportional to $1/\omega$ ($1/C$ or k).

For distributed acoustic systems, the cause–effect relationship is between the change in pressure (disturbance) and the resulting particle velocity of the medium. The definitions of the relevant quantities are as follows:

Acoustic System Definitions

p	Acoustic pressure ($= P - P_0$), instantaneous pressure change from its ambient value P_0 in a sound field
v	Particle velocity; resulting from the passing of a sound wave, it is superimposed upon the thermally established and random molecular velocity
C_s	Speed at which a sound wave propagates (not to be confused with v)

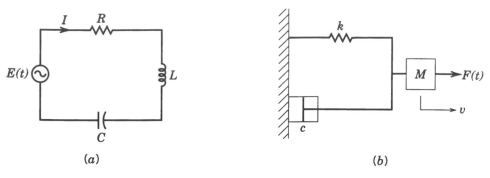

Fig. 1 (*a*) Electrical and (*b*) mechanical one-degree-of freedom systems: L = inductance, R = resistance, $1/C$ = inverse capacitance, E = exciting voltage, Q = charge, I = dQ/dt = current, M = mass, c = damping constant, k = spring constant, F = exciting force, X = displacement, $V = dX/dt$ = velocity.

$p/v \equiv Z$ Specific acoustic impedance, a measure of how much fluid motion is produced for a given pressure disturbance

3 ACOUSTIC ELEMENTS

As a result of a mechanical disturbance in a medium, a sound wave is launched. The motions that arise behave Newton's laws of motion as applied to distributed (continuous) systems. A longitudinal pressure disturbance propagates with a wave velocity C_s. For a simple harmonic disturbance of frequency f, the wave will propagate a distance λ (the wavelength, $= C_s/f$) in one period of oscillation ($1/f$). If the wavelength of sound is long compared to the relevant physical length scales of the problem (as defined in the examples that follow), then acoustic lumped elements can be defined. It should be noted that this wavelength restriction reduces the problem from one of sound propagation to one of incompressible fluid hydrodynamics. Nevertheless, it should be realized that for the propagation of plane waves, the ratio of pressure to particle velocity is just the characteristic impedance $Z_c = \rho_0 C_s$, where ρ_0 is the fluid density (1.2 kg/m^3 for air at 20°C) and C_s is the sound velocity (344 m/s for air at 20°C), and Z equals 416 acoustic ohms (SI units). This impedance is real and thus represents a resistive impedance term, reflecting the fact that energy is lost to the source as acoustic radiation.

3.1 Inductance: Mass Element for a Tube

Consider a tube of length l and cross-sectional area A. A simple harmonic pressure disturbance across the ends caused by an acoustic wave will accelerate the mass of gas according to Newton's law, $F = Ap = M_{\text{gas}}\, dv/dt$,

where dv/dt for harmonic motion can be written as $i\omega v$. The specific acoustic impedance $Z = p/v$ becomes, therefore, $Z = i\omega[M_{\text{gas}}/A]$.

The term in brackets is an inductance-like normalized mass of fluid in the tube. If this distributed mass is to move like a single slug, the pressure must be communicated throughout the tube in a time short compared to the acoustic period, τ, which is equivalent to requiring that the tube length be small compared to the acoustic wavelength. As an example, for a frequency of 100 Hz and an air-filled tube, the requirement for the lumped-element impedance expression to be valid would be that $l \ll 3.4$ m, or $l < 34$ cm.

3.2 Inductance: Mass Element for Vibrating Sphere

When a sphere vibrates, it moves the fluid surrounding it. By calculating the total kinetic energy of the surrounding fluid, with the assumption that the sphere is small compared to the acoustic wavelength, one finds

$$\text{KE} = \tfrac{1}{2}M_{\text{eff}}V_R^2,$$

where $M_{\text{eff}} = 4\pi R^3\rho = 3 \times (\tfrac{4}{3}\pi R^3\rho)$ and V_R is the velocity of the sphere at the outer diameter. The effective mass of fluid "felt" by the pulsating sphere is seen to be three times the mass of outer fluid that could fill the sphere. The force reaction on the sphere by the medium can, therefore, be written as

$$M_{\text{eff}}\,\frac{dV_R}{dt} = i\omega M_{\text{eff}}V_R \quad \text{(harmonic assumption).}$$

The reaction impedance $Z = p/V_R = (F/A)/V_R = i\omega[M_{\text{eff}}/A]$, which is an inductance-type lumped element.

3.3 Inductance: Mass Element for Vibrating Piston

The mathematics is complex, but the results are simple for a uniformly vibrating circular piston of radius a:

$$M_{\text{eff}} = \rho \pi a^2 \Delta l,$$

where

$$\Delta l \doteq \begin{cases} \frac{8}{3}\pi a, & \text{piston in a large rigid baffle,} \\ 0.6a, & \text{no baffle.} \end{cases}$$

In both cases Δl represents the height of a "pill box" of fluid extending out from the vibrating piston. This pill box encloses the effective mass "felt" by the vibrating piston.

3.4 Resistive Elements

Viscous Resistive Elements   When flow occurs through a slit of width w or a circular tube of radius a, energy is lost because of viscous dissipation (viscosity μ). When laminar flow conditions exist and the acoustic wavelength is long compared to the tube or slit length l, then the impedance is given by

$$Z_{\text{slit}} = \frac{12\mu l}{w^2}, \qquad Z_{\text{tube}} = \frac{8\mu l}{a^2}.$$

Radiative Acoustic Elements   Whereas the resistance associated with plane sound waves is $\rho_0 C_s$, the acoustic resistance is less for low frequencies (where the waves are not planar). For a baffled circular tube of radius a, the radiation resistance in cases where the wavelength is much larger than the tube diameter is given by

$$Z_R = \rho_0 C_s \frac{(ka)^2}{2},$$

where k is the wavenumber ($= 2\pi f/C_s$).

3.5 Capacitance: Compliance Element

Consider a sealed loudspeaker box with a speaker diaphragm of area A and peak displacement d. By using the relationship between volume and pressure for an adiabatic change in a gas, one can write an expression relating the pressure change inside the loudspeaker to the diaphragm displacement d: $p = \gamma P_0 A d / V_0$, where γ is the ratio of the specific heats at constant pressure and volume, and it has been assumed that the volume change produced by the diaphragm displacement is small compared to the speaker's volume. For simple harmonic oscillatory motion with diaphragm velocity v, we find that $d = v/i\omega$, and the impedance can be computed as

$$Z = \frac{p}{v} = \frac{1}{i\omega C_M A},$$

where

$$C_M = \frac{V_0}{A^2 \gamma P_0} \qquad \text{(mechanical compliance)}.$$

Once again our analysis assumes that the wavelength of sound is large compared to the longest diagonal dimension of the loudspeaker (which assures that the pressure is essentially uniform within the box). Then the shape of the box is not a factor—only the volume.

4 RESONANCE BETWEEN MASS AND SPRING ELEMENTS

Consider now a sealed enclosure (as in Section 3.5) penetrated by a vibrating piston with a diaphragm of mass $M_{\text{diaphragm}}$ (Section 3.3) and area A. If resonance is defined by the vanishing of the imaginary part of the impedance, then for two acoustic elements in series (which implies that the same fluid volume oscillates in both elements), the resonance frequency f_0 is found by

$$Z = i\omega \frac{M_{\text{eff}}}{A} + \frac{1}{i\omega C_M A} = 0, \qquad \omega = \omega_0,$$

or

$$f_0 = \frac{\omega_0}{2\pi} = \frac{1}{2\pi} \sqrt{\frac{1}{M_{\text{eff}} C_M}}$$

$$= \frac{1}{2\pi} \sqrt{\frac{A\gamma P_0}{[M_{\text{diaphragm}} + \rho A(l + \Delta l)]}},$$

where Δl is given in Section 3.3 and is usually closer to $0.6a$ for most practical cases. This expression can be used for the *Helmholtz resonance*, as is found, for example, when one blows over the top of a partially filled, narrow-mouth bottle of neck length l. (Then, $M_{\text{diaphragm}} = 0$.)

The mechanical Q is a measure of the sharpness of the resonance as one varies the frequency, $Q = f_0/(f_2 - f_1)$, where f_2 and f_1 are the frequencies above and below the resonance frequency f_0 at which the resonator radiates half the power. For a diaphragm-free resonator with tube neck of length l and radius a,

$$Q \cong \frac{(l + 0.6a)C_s}{f_0 \pi a^2}.$$

REFERENCES

1. L. E. Kinsler, A. R. Frey, A. B. Coppens, and J. V. Sanders, *Fundamentals of Acoustics*, 3rd ed., Wiley, New York, 1982.

2. L. Beranek, *Acoustics*, American Institute of Physics, New York, 1986.

14

ACOUSTIC MODELING: FINITE ELEMENT METHOD

A. CRAGGS

1 INTRODUCTION

The problem being considered here is the following: Given an acoustic space with specified conditions on the bounding surfaces, either determine its response to a given source distribution or find its natural frequencies and normal modes. Now there are very few acoustic volumes amenable to an exact analytical solution as is possible with uniform tubes and rectangular and cylindrical cavities. An enclosure that has a complex geometry like the interior passenger space of an automobile or the piping configuration of the exhaust system has to be studied by an approximate numerical analysis, and the finite element method is one of several procedures available.

Essentially, a complex volume can be formed by assembling together a number of small elements that have a simpler geometry and smaller number of degrees of freedom. The more complex the geometry is, the greater the number of elements required. The number of elements is also influenced by the frequency range of interest, as usually the minimum dimension of an element should not be less than half a wavelength. If the frequency is very high, then perhaps the finite element procedure is not the best method to use.

There are a number of ways for formulating approximate numerical models for a continuous system. Some methods approximate the governing partial differential equation, which in this case is the wave equation, with finite differences; finite element equations can also be formed by the method of weighted residuals, the method of Galerkin being well established in this respect. Here, however, a variational procedure is outlined that is sim-

ilar to that of the energy formulations used for elastic structures.

2 THEORY

The procedure is outlined in many standard texts on finite element methods. It is based on a scalar functional that usually contains terms in the kinetic and potential energies and the boundary work.[1–5] The first variation of the functional leads to the exact governing differential equation and the boundary conditions. The belief is that the first variation of an approximate functional will give the optimum approximate equations.

2.1 One-Dimensional Elements

As they are more tractable, one-dimensional pipe finite elements will be considered first, and a further simplification is introduced by considering only harmonic motion that eliminates the time-dependent term in the wave equation. However, allowances are made for a varying pipe geometry. All elements are formulated in terms of the acoustic pressure.

The element is based on the scalar functional

$$\pi = \frac{1}{2} \int_0^l \left(\frac{dp}{dx} \right)^2 A \, dx - \left(\frac{\omega}{c} \right)^2 \frac{1}{2} \int_0^l p^2 A \, dx, \quad (1)$$

where p is the acoustic pressure, A is the cross-sectional area, c is the speed of sound, ω is the radian frequency, l is the length of the element, and x is the distance along the pipe. The first term in the functional is related to the kinetic energy and the second to the strain energy (but they are not equal). In this form the first variation leads directly to the classical form of the wave equation

Encyclopedia of Acoustics, edited by Malcolm J. Crocker
ISBN 0-471-80465-7 © 1997 John Wiley & Sons, Inc.

expressed in standard texts on acoustics and results in an eigenvalue problem in ω^2, rather than its reciprocal.

The stationary values of Eq. (1) lead to the well-known Webster horn equation

$$A\,\frac{d^2p}{dx^2} + \frac{dp}{dx}\,\frac{dA}{dx} + \left(\frac{\omega}{c}\right)^2 Ap = 0 \qquad (2)$$

together with the boundary conditions $A(dp/dx) = 0$ at $x = 0$ and $x = l$. If a number of these elements are to be connected together, then at the element interfaces (see Fig. 2) the boundary conditions are

$$A\left(\frac{dp}{dx}\right)_2 - A\left(\frac{dp}{dx}\right)_3 = 0, \qquad (3)$$

$$p_2 = p_3. \qquad (4)$$

Equation (3) is a form of the continuity equation, while Eq. (4) simply states that on connection the pressures are compatible.

If the problem is restricted to hard (i.e., immobile) boundaries, then the functional form (1) is all that is required. Certainly for many problems in room acoustics the hard boundary condition gives a very good approximation to the truth, and models based on this assumption give reliable predictions of the natural frequencies. However, if the boundary does have significant flexibility, extra terms are needed in the functional. For example, if one of the boundaries, say at $x = l$, has an impedance Z, then a suitable functional is

$$\pi = \frac{1}{2}\int_0^l \left(\frac{dp}{dx}\right)^2 A\;dx - \left(\frac{\omega}{c}\right)^2 \frac{1}{2}\int_0^l p^2 A\;dx$$
$$- \frac{j}{2}\,\rho\omega\,\frac{p^2}{Z}\,A_l, \qquad (5)$$

where j is $\sqrt{-1}$, the complex operator.

This modified functional should only be applied to the terminal element or elements at the boundaries of the global system. Also, the form (5) is only suitable for conservative boundary conditions that involve no energy dissipation; the boundary is then either stiffness or mass controlled.

If the boundary has stiffness k per unit area, then $Z = k/j\omega$; if it is mass controlled with mass m per unit area, $Z = j\omega m$. In both of these simple cases the complex operator j disappears from the functional in Eq. (5) and retains a real form, with the extra term being added to either the first term of the functional (mass-controlled

boundary) or the second term, the stiffness-controlled boundary. In each case, however, the terms disappear when either the stiffness or mass values become large.

When the boundaries are nonconservative, the functional has to be modified in the manner suggested by Morse and Ingard;[6] see also Craggs.[8]

The functional (1) will now be used to set up the approximate equations for a pipe element that allows a linear variation of acoustic pressure throughout the length. The cross-sectional area A will also be given a linear variation; refer to Fig. 1. Thus,

$$p = p_1\left(1 - \frac{x}{l}\right) + p_2\left(\frac{x}{l}\right), \qquad (6)$$

and the cross-sectional area $A(x)$ can be written in terms of the end values A_1 and A_2:

$$A = A_1 + (A_2 - A_1)\left(\frac{x}{l}\right). \qquad (7)$$

To proceed further, Eq. (3) is written in a matrix form:

$$p = \{f(x)\}^{\mathrm{T}}\{P_e\} = \{P_e\}^{\mathrm{T}}\{f(x)\}, \qquad (8)$$

where $\{P_e\}$ is the vector $\{P_1/P_2\}$ containing the nodal acoustic pressures and $\{f(x)\}$ is a vector containing the linear polynomials $1 - x/l$ and x/l. The superscript T denotes that the matrix has been transposed so that a column vector is transposed to a row matrix.

If Eq. (5) is differentiated with respect to x, then

$$\frac{dp}{dx} = \left\{\frac{df}{dx}\right\}^{\mathrm{T}}\{P_e\} = \{P_e\}^{\mathrm{T}}\left\{\frac{df}{dx}\right\}. \qquad (9)$$

Further progress in establishing the approximate equations is obtained by recognizing that

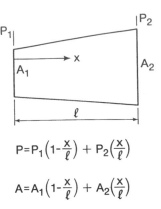

$$P = P_1\left(1 - \frac{x}{\ell}\right) + P_2\left(\frac{x}{\ell}\right)$$

$$A = A_1\left(1 - \frac{x}{\ell}\right) + A_2\left(\frac{x}{\ell}\right)$$

Fig. 1   Pipe acoustic finite element having linear variation of pressure and cross-sectional area.

$$p^2 = \{P_e\}^{\text{T}}\{f\}\{f\}^{\text{T}}\{P_e\},$$

$$\left(\frac{dp}{dx}\right)^2 = \{P_e\}^{\text{T}}\left\{\frac{df}{dx}\right\}\left\{\frac{df}{dx}\right\}^{\text{T}}\{P_e\}; \quad (10)$$

and after substitution, an approximate functional for the element is

$$\pi = \frac{1}{2}\{P_e\}^{\text{T}}\left(\int_0^l A_1 \left\{\frac{df}{dx}\right\}\left\{\frac{df}{dx}\right\}^{\text{T}} dx\right.$$

$$+ \frac{A_2 - A_1}{l}\int_0^l x\left\{\frac{df}{dx}\right\}\left\{\frac{df}{dx}\right\}^{\text{T}} dx\right)\{P_e\}$$

$$- \frac{1}{2}\left(\frac{\omega}{c}\right)^2\{P_e\}^{\text{T}}\left(\int_0^l A_1\{f\}\{f\}^{\text{T}} dx\right.$$

$$+ \frac{A_2 - A_1}{l}\int_0^l x\{f\}\{f\}^{\text{T}} dx\right)\{P_e\}. \quad (11)$$

Carrying out the integrations, the result is

$$\pi = \frac{1}{2}\left(\{P_e\}^{\text{T}}[S]\{P_e\} - \left(\frac{\omega}{c}\right)^2\{P_e\}^{\text{T}}[R]\{P_e\}\right), \quad (12)$$

where

$$[S] = \frac{A_1}{2l}\begin{bmatrix} 1 & 1 \\ -1 & 1 \end{bmatrix} + \frac{A_2}{2l}\begin{bmatrix} 1 & -1 \\ -1 & 1 \end{bmatrix} \quad (13)$$

and

$$[R] = \frac{A_1 l}{12}\begin{bmatrix} 3 & 1 \\ 1 & 1 \end{bmatrix} + \frac{A_2 l}{12}\begin{bmatrix} 1 & 1 \\ 1 & 3 \end{bmatrix}, \quad (14)$$

and setting the first variation to zero, that is, $\delta\pi = 0$, gives the approximate equation for one finite element,

$$[S]\{P_e\} - \frac{\omega^2}{c^2}[R]\{P_e\} = \{0\}. \quad (15)$$

In the above it was assumed that there are no volume source terms, that is, the boundaries are immobile.

Equation (15) is for a single finite element; the assembly of two or more elements can be illustrated by considering the system shown in Fig. 2. Unconnected the elements have four unknowns, p_1, p_2, p_3, and p_4. If the elements are joined at node 2, then Eqs. (3) and (4) need

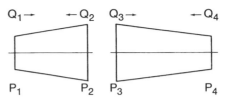

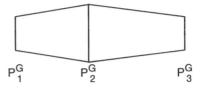

Fig. 2 Assembly of two linear pipe finite elements.

to be satisfied. As a result of these constraints, the four-degree-of-freedom system is reduced to a global system having three degrees of freedom:

$$\begin{bmatrix} S_{11}^A & S_{12}^A & 0 \\ S_{21}^A & S_{22}^A + S_{11}^B & S_{12}^B \\ 0 & S_{21}^B & S_{22}^B \end{bmatrix}\begin{Bmatrix} P_1^G \\ P_2^G \\ P_3^G \end{Bmatrix}$$

$$- \frac{\omega^2}{c^2}\begin{bmatrix} R_{11}^A & R_{12}^A & 0 \\ R_{21}^A & R_{22}^A + R_{11}^B & R_{12}^B \\ 0 & R_{21}^B & R_{22}^B \end{bmatrix}\begin{Bmatrix} P_1^G \\ P_2^G \\ P_3^G \end{Bmatrix}$$

$$= \begin{Bmatrix} 0 \\ 0 \\ 0 \end{Bmatrix}. \quad (16)$$

The individual element matrices are then assembled by overlaying them at the connecting nodes. In the same manner many elements can be linked together. While it is common that they are connected in a chainlike fashion to form a one-dimensional system, it is also possible to form branched systems.

An element that is a little more accurate and allows a quadratic variation in the cross-sectional area and pressure has three degrees of freedom. Thus, if (refer to Fig. 3)

$$p = a_1 + b_1\left(\frac{x}{e}\right) + e_1\left(\frac{x}{e}\right)^2$$

and

$$A = a_2 + b_2\left(\frac{x}{e}\right) + c_2\left(\frac{x}{e}\right)^2,$$

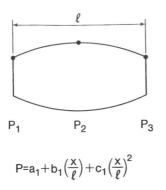

$$P = a_1 + b_1\left(\frac{x}{\ell}\right) + c_1\left(\frac{x}{\ell}\right)^2$$

Fig. 3 The quadratic pipe element.

then the resulting matrices are

$$[S] = \alpha_1 \begin{bmatrix} 7 & -8 & 1 \\ -8 & 16 & -8 \\ 1 & -8 & 7 \end{bmatrix} + \alpha_2 \begin{bmatrix} 3 & -4 & 1 \\ -4 & 16 & -12 \\ 1 & -12 & 11 \end{bmatrix}$$

$$+ \alpha_3 \begin{bmatrix} 3 & -6 & 3 \\ -6 & 32 & -26 \\ 3 & -26 & 23 \end{bmatrix}, \tag{17}$$

where $\alpha_1 = A_1/3l$, $\alpha_2 = (4A_2 - 3A_1 - A_3)/6l$, and $\alpha_3 = (2A_1 - 4A_2 + 2A_3)/15l$, and

$$[R] = \beta_1 \begin{bmatrix} 4 & 2 & -1 \\ 2 & 16 & 2 \\ -1 & 2 & 4 \end{bmatrix} + \beta_2 \begin{bmatrix} 1 & 0 & -1 \\ 0 & 16 & 4 \\ -1 & 4 & 7 \end{bmatrix}$$

$$+ \beta_3 \begin{bmatrix} 2 & -4 & -5 \\ -4 & 64 & 24 \\ -5 & 24 & 44 \end{bmatrix}, \tag{18}$$

where $\beta_1 = A_1 l/30$, $\beta_2 = (4A_2 - 3A_1 - A_3)l/60$, and $\beta_3 = (2A_1 - 4A_2 + 2A_3)l/420$. If the cross section is uniform throughout, then $\alpha_2 = \alpha_3 = \beta_2 = \beta_3 = 0$ and only the first terms remain.

This latter element has been used to determine the natural frequencies of the wine bottle shown in Fig. 4 (i), the interior dimensions being height 20.32 cm, base diameter 7.62 cm, and neck outlet diameter 1.52 cm. The model consists of eight of the quadratic finite elements that when assembled give a global matrix with 17 degrees of freedom. However, this was reduced by applying the boundary condition $p = 0$ at the open end. Figure 4 shows the four lowest modes and their natural frequencies.

The quadratic element was also used to model a one-dimensional version of the branched system shown in Fig. 5. Here the one-dimensional element results were compared with those from three-dimensional study (refer

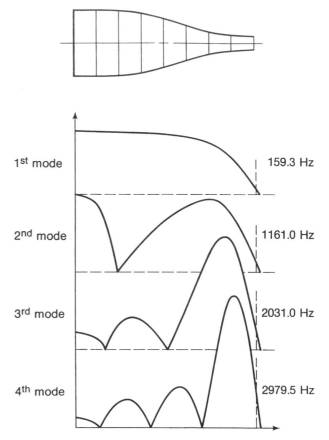

Fig. 4 (Top) Finite element model of a wine bottle from an assembly of quadratic elements. (Bottom) Lowest four normal modes and natural frequencies.

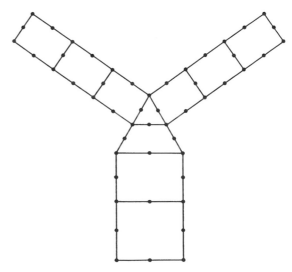

Fig. 5 Finite element model of a Y-branch. This was modeled with one-dimensional elements and three-dimensional elements (shown).

TABLE 1 Comparison of First 10 Modes of Y-System for Pipe Element and HX20

Mode	Pipe Element	HX20
1	6.854	6.772
2	9.874	10.03
3	39.60	40.78
4	62.15	61.50
5	90.16	93.97
6	164.4	164.7
7	179.4	177.4
8	300.0	305.6
9	403.5	400.2

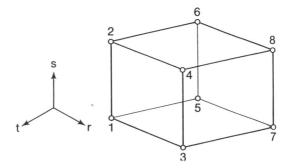

Fig. 6 Eight-node rectangular acoustic finite element.

to the next section). The results from the models are compared in Table 1, and there is close agreement for the first nine modes.

The pipe element is only valid as long as the acoustic wavelength is large compared with the cross-sectional dimensions. If this is not the case and the acoustic pressure is not uniform over the cross section, then it is necessary to allow for this with three-dimensional elements. These will be considered in the next section.

2.2 Three-Dimensional Elements

The steps in forming the finite element matrices $[S]$ and $[R]$ are the same as those given in the previous section. However, because of the extra two dimensions and the implied extra degrees of freedom, it is not a practical proposition to evaluate all of the terms analytically, as was done in the one-dimensional case. It is essential to use numerical procedures throughout, and this in turn results in a greater flexibility of approach.

The procedure is illustrated by developing a hexahedral element. In the first instance this will have a rectangular section, but later the geometry will be allowed to vary to produce elements with curved boundaries that will give the element a wider application:

The basic rectangular element is shown in Fig. 6, and neglecting the source effects, the functional has the form

$$\pi = \frac{1}{2} \iiint \left(\frac{\partial r}{\partial r}\right)^2 + \left(\frac{\partial p}{\partial s}\right)^2 + \left(\frac{\partial p}{\partial t}\right)^2 dr\, ds\, dt$$

$$- \frac{\omega^2}{c^2} \frac{1}{2} \iiint p^2\, dr\, dr\, dt, \qquad (19)$$

and the stationary values of this lead to the Helmholtz equation

$$\frac{\partial^2 p}{\partial r^2} + \frac{\partial^2 p}{\partial s^2} + \frac{\partial^2 p}{\partial t^2} + \frac{\omega^2}{c^2} p = 0, \qquad (20)$$

with the boundary condition $\partial p / \partial \bar{n} = 0$ on the surface, where $\bar{n}$ is the direction of the outgoing normal vector. The formation of the element matrices will be illustrated with the eight-node element, which has a node at each corner and allows a linear variation of the acoustic pressure:

$$p = a_1 + a_2 r + a_3 s + a_4 t + a_5 rs + a_6 rt + a_7 st + a_8 rst,$$

$$p = \{F\}^T \{\alpha\},$$

where

$$\{F\} = \begin{Bmatrix} 1 \\ r \\ s \\ t \\ rs \\ rt \\ st \\ rst \end{Bmatrix}, \qquad \{\alpha\} = \begin{Bmatrix} a_1 \\ a_2 \\ a_3 \\ a_4 \\ a_5 \\ a_6 \\ a_7 \\ a_8 \end{Bmatrix}. \qquad (21)$$

At this stage the pressure is expressed in terms of the generalized coefficients $a_1, \ldots, a_n$ that govern the contribution of each of the simple polynomial terms. It is more useful to relate these to the acoustic pressures at the individual node points. This can be achieved by placing the boundary values c_i, s_i, and t_i for each nodal pressure p_i, which in this case gives eight equations in eight unknowns that can be expressed in the matrix form

$$\{p\} = [T]\{\alpha\}, \qquad (22)$$

$$\{\alpha\} = [M]\{p\}, \quad \text{where } [M] = [T^{-1}], \qquad (23)$$

Using this result, the pressure at any point r, s, t within the element can be written in terms of the nodal pressures

$$p = \{F\}^T [M]\{p\}. \qquad (24)$$

Differentiating this equation, first with respect to r, then with respect to s and t, gives the derivative equations, which in matrix form are

$$\left\{ \begin{array}{c} \dfrac{\partial p}{\partial r} \\[2mm] \dfrac{\partial p}{\partial s} \\[2mm] \dfrac{\partial p}{\partial r} \end{array} \right\} = [G][M]\{P\} \qquad (25)$$

The matrix $[G]$ is a 3×8 matrix, the rows containing the derivatives $\{\partial F/\partial r\}^\mathrm{T}$, $\{\partial F/\partial s\}^\mathrm{T}$, and $\{\partial F/\partial t\}^\mathrm{T}$. Substituting the results (2) and (3) into Eq. (1) gives the approximate functional for the element

$$\pi = \frac{1}{2}\{P\}^\mathrm{T}[S]\{P\} - \frac{1}{2}\left(\frac{\omega}{c}\right)^2 \{P\}^\mathrm{T}[R]\{P\}, \qquad (26)$$

where

$$[S] = \iiint [M]^\mathrm{T}[G]^\mathrm{T}[G][M] \, dr \, ds \, dt, \qquad (27)$$

$$[R] = \iiint [M]^\mathrm{T}\{F\}^\mathrm{T}\{F\}^\mathrm{T}[M] \, dr \, ds \, dt. \qquad (28)$$

Setting the first variation $\delta\pi = 0$ in Eq. (4) gives the approximate form for the three-dimensional Helmholtz equation

$$\left([S] - \frac{\omega^2}{c^2}[R]\right)\{P\} = 0, \qquad (29)$$

which is a linear eigenvalue problem. The limits of integration will depend upon the dimensions of the element. Since the volume is a regular parallelepiped, the integration is not difficult, but there are so many terms that it is best carried out numerically. For example, the eight-node element discussed above will require 64 integrations for each of the $[S]$ and $[R]$ matrices. The higher order element with 20 nodes will require 400 integrations and the 32-node element will require 1024.

The regular element discussed above will have a significant but limited application to rectangular sectioned rooms and ducts. The application can be greatly extended if the geometry also is allowed to be distorted, and it is possible to do this by using the same polynomials that govern the variation in the pressure to govern the variation in geometry. To achieve this, the regular geometry in the r, s, t system is transformed to an irregular geometry in x, y, z (refer to Fig. 7) by the transformations

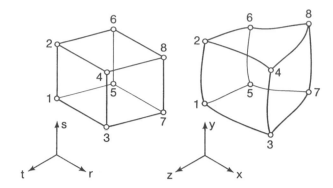

Fig. 7   Transformation from a regular rectangular element in r, s, t to an irregular element in x, y, z.

$$x = \{F\}^\mathrm{T}[M]\{X\}, \qquad y = \{F\}^\mathrm{T}[M]\{Y_i\},$$
$$z = \{F\}^\mathrm{T}[M]\{Z_i\}, \qquad (30)$$

where $\{X\}$, $\{Y\}$, and $\{Z\}$ are the vectors containing the x-, y-, and z-coordinates at the node points. The functional for the element in x, y, z is

$$\pi = \frac{1}{2}\iiint \left(\frac{\partial p}{\partial x}\right)^2 + \left(\frac{\partial p}{\partial y}\right)^2 + \left(\frac{\partial p}{\partial z}\right)^2 dx \, dy \, dz$$
$$- \frac{\omega^2}{c^2}\frac{1}{2}\iiint p^2 \, dx \, dy \, dz. \qquad (31)$$

However, by making use of the results,

$$\left\{ \begin{array}{c} \dfrac{\partial p}{\partial x} \\[2mm] \dfrac{\partial p}{\partial y} \\[2mm] \dfrac{\partial p}{\partial z} \end{array} \right\} = [J]^{-1} \left\{ \begin{array}{c} \dfrac{\partial p}{\partial r} \\[2mm] \dfrac{\partial p}{\partial s} \\[2mm] \dfrac{\partial p}{\partial f} \end{array} \right\} \qquad (32)$$

and $dx \, dy \, dz = \|J\| \, dr \, ds \, dt$, where $[J]$ is the Jacobian matrix and $\|\cdot\|$ denotes the absolute value of the determinant of $[J]$.

The matrices $[S]$ and $[R]$ are given by

$$[S] = \int_{-1}^{+1}\int_{-1}^{+1}\int_{-1}^{+1} [M]^\mathrm{T}[G]^\mathrm{T}[J^{-1}]^\mathrm{T}[J^{-1}][G][M]$$
$$\cdot \|J\| \, dr \, ds \, dt, \qquad (33)$$

$$[R] = \int_{-1}^{+1}\int_{-1}^{+1}\int_{-1}^{+1} [M]^\mathrm{T}\{F\}\{F\}^\mathrm{T}[M]\|J\| \, dr \, ds \, dt. \qquad (34)$$

It is worth noting that the integrals have been expressed in the r, s, and t domain where there is a regular geometry. Further, the limits of integration are -1 to $+1$, which conform with the standard limits of tabulated numerical integration schemes. Equations (24) and (25) are those used in several commercial codes, though some codes prefer using serendipity functions instead of the simple polynomials, as these allow the pressures and its derivatives to be written down directly in terms of the nodal pressures and avoid forming the inverse matrix $[M]$. However, this matrix need only be formed once and its elements stored in memory.

Although the results have been derived for the 8-node element, the same procedure can be used to form the element matrices with 20 or 32 nodes. These more powerful elements, however, need extra polynomials with higher order terms. While they give more accurate results on a degree-of-freedom basis, there is a penalty due to the greater number of integration points needed and extra complexity in the input data. They have their place in specialized programs. The 20-node element has probably the most common usage. An indication of the accuracy of the HEX8, HEX20, and HEX32 elements is given in Table 2, where the results for the lowest modes of a unit cube are given. The highest order element is seen to give the greatest accuracy; although it is not shown here, this gives a more rapid convergence to the solution as the number of elements is increased. A cavity representing a model of a car interior (Fig. 8) has been simulated with the 32-node element. Thirty of these elements were used, resulting in a system with 378 degrees of freedom. A comparison of the predicted and measured frequencies are given in Table 3.

While the three-dimensional isoparametric element is very versatile, there are situations in which the number of degrees of freedom can be reduced if the geometry has some form of symmetry. If there is axial symmetry, then a suitable finite element can be derived from the stationary values of the functional

$$\pi = \frac{1}{2} \iint \left(\frac{\partial p}{\partial r}\right)^2 + \left(\frac{\partial p}{\partial z}\right)^2 2\pi r \, dr \, dz$$

$$- \frac{\omega^2}{c^2} \iint p^2 2\pi r \, dr \, dz, \qquad (35)$$

TABLE 2   Frequency of Lowest Mode (0, 01), (0, 10) (1, 00) for a Unit Cube

Exact	HEX8	HEX20	HEX32
$\pi^2 = 9.8696$	12.000	11.595	9.8751

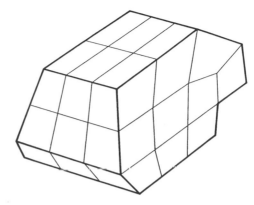

Fig. 8   Acoustic finite element grid for a model car enclosure. Assembly is of the 32-node isoparametric elements.

where r and z are the radial and axial dimensions, respectively. A very effective element is the four-node quadrilateral element. This element is very easy to form, it has only four degrees of freedom, it has pressures at each node and allows a linear variation in pressure and geometry, and yet it gives very acceptable results. It is particularly useful in evaluating axisymmetric mufflers.

2.3   Acoustostructural Problems

In an earlier section a very simple method for dealing with boundary flexibility was considered that is suitable for nondissipative reactive boundaries. However, if the bounding structure also has resonances in the same frequency range as the enclosed acoustic space, then a finite element model of the structure is also needed. The formulation of the mass matrix $[M]$ and the stiffness matrix $[K]$ for a structure is given in the text by Petyt.[10]

Recognizing that the structural equations are formu-

TABLE 3   Natural Frequencies of Irregular Cavity

Mode	Finite Element Simulation	Measured Frequency
100	140.8	154.6
001	160.3	168.2
010	208.9	220.2
200	243.3	251.0
101	267.8	281.8
201	288.8	301.8
110	299.9	312.2
110	299.9	312.2
002	315.4	323.8
102	353.2	355.4
111	347.0	366.6
020	371.3	378.2

lated in terms of displacements $\{w\}$ and the acoustics in terms of pressure $\{p\}$, the coupled equations governing harmonic motion have the form

$$
\begin{bmatrix} S & \vdots & 0 \\ \hline \theta^T & \vdots & K \end{bmatrix} \begin{Bmatrix} p \\ w \end{Bmatrix} - \omega^2 \begin{bmatrix} \dfrac{1}{c^2} & \vdots & \rho\theta \\ \hline 0 & \vdots & M \end{bmatrix} \begin{Bmatrix} p \\ w \end{Bmatrix}
$$
$$
= \begin{Bmatrix} q \\ Q \end{Bmatrix}. \tag{36}
$$

The matrix $[\theta]$ is a rectangular matrix found by integrating the product of the allowed pressure and displacement polynomials over the coupling surface area[7]; $\{q\}$ is an acoustic volume source term and $\{Q\}$ is a force vector representing distributed forces of excitation applied on the structure.

The technique given in Eq. (36) has formed the basis for determining the response of rooms to window excitation by sonic booms,[7] making sound transmission studies in cars,[8] and finding the effect of flexible walls on the transmission loss of silencers.[9]

3 FINAL COMMENTS

The finite element method is a good way to model the acoustics of an enclosed space, and even low-order elements with a linear variation of pressure can give acceptable results. In its simplest form, the resulting equations can involve huge matrices that are costly to solve, especially as they need to be solved over a spectrum of frequencies. Commercial computer codes can make work much easier in that they incorporate automatic grid generators and pre- and postprocessing facilities. Codes like ANSYS and NASTRAN have ready-made acoustic software built into a fully comprehensive engineering package. SYSNOISE is a program that deals solely with acoustic problems.

REFERENCES

1. G. M. L. Gladwell, "A Finite Element Method for Acoustics," paper presented at Fifth International Conference on Acoustics, Liege, 1965, paper L33.

2. V. Mason, "On the Use of Rectangular Finite Elements," Institute of Sound and Vibration Report No. 161, University of Southampton, 1967.

3. A. Craggs, "The Use of Simple Three Dimensional Acoustic Finite Elements for Determining the Natural Modes and Frequencies of Complex Shaped Enclosures," *J. Sound Vib.*, Vol. 23, 1972, pp. 331–339.

4. M. Petyt, J. Leas, and G. H. Koopman, "A Finite Element Method for Determining the Acoustic Modes of Irregular Shaped Cavities," *J. Sound Vib.*, Vol. 45, 1976, pp. 495–502.

5. D. Nefske and L. J. Howell, "Automobile Interior Noise Reduction Using Finite Element Methods," *Trans. SAE*, Vol. 87, 1978, pp. 1727–1737.

6. P. M. Morse and K. U. Ingard, *Theoretical Acoustics*, New York; McGraw-Hill, New York, 1968, pp. 250–256.

7. A. Craggs, "The Transient Response of a Coupled Plate-Acoustic System Using Plate and Acoustic Finite Elements," *J. Sound Vib.*, Vol. 15, 1971, 509–529.

8. A. Craggs, "An Acoustic Finite Element Approach for Studying Boundary Flexibility and Sound Transmission between Irregular Enclosures," *J. Sound Vib.*, Vol. 30, No. 3, pp. 343–357.

9. C. I. J. Young and M. J. Crocker, "Finite Element Acoustical Analysis of Complex Muffler Systems with and without Wall Vibrations," *Noise Control Eng.*, Vol. 9, No. 2, 1977, pp. 86–93.

10. M. Petyt, *Introduction to Finite Element Vibration Analysis*, Cambridge University Press, 1990.

15

ACOUSTIC MODELING: BOUNDARY ELEMENT METHODS

A. F. SEYBERT AND T. W. WU

1 INTRODUCTION

The boundary element method (BEM) is a numerical technique for calculating the sound radiated by a vibrating body or for predicting the sound field inside of a cavity such as a vehicle interior. The BEM may also be used to determine the sound scattered by an object such as a microphone or for predicting the performance of silencers or mufflers. The BEM is becoming a popular numerical technique for acoustical modeling in industry. The major advantage of this method is that only the boundary surface (e.g., the exterior of the vibrating body) needs to be modeled with a mesh of elements. For infinite-domain problems, such as radiation from a vibrating structure, the so-called Sommerfield radiation condition is automatically fulfilled. In other words, there is no need to create a mesh to approximate this radiation condition. A typical BEM input file consists of a surface mesh, a normal velocity profile on the surface, the fluid density, speed of sound, and frequency. The output of the BEM includes the sound pressure distribution on the surface of the body and at other points in the field, the sound intensity, and the sound power. In this chapter, an overview of the BEM is presented with emphasis on its application to industrial noise control problems. Although the BEM may also be formulated in the time domain, the focus here will be on the frequency domain only.

1.1 The Boundary Element Mesh

Figure 1 shows a BEM for the calculation of the sound radiated by a tire. This is an example of an *exterior prob-*

lem in acoustics, so named because the acoustic domain is infinite or, in the case of Fig. 1, semi-infinite due to the reflecting ground surface. The geometry of the tire is represented by a boundary element mesh that is a series of points called *nodes* on the surface of the body that are connected together to form *elements* of either quadrilateral or triangular shape. Note in the case of the tire in Fig. 1 and many other structures there exists a symmetry plane so that only one-half of the body needs to be modeled.

Building the boundary element mesh or model is the first step in solving a problem with the BEM. The size of the elements must be chosen small enough to obtain an acceptable solution but not so small as to result in excessive computer time. In general, a boundary element mesh must meet three requirements to obtain an acceptable solution. First, it must model the geometry of the body accurately. This means that all major surfaces, as well as edges and corners, must be accurately represented. (However, it is not necessary to model regions of the body for which it can be concluded beforehand are not important acoustically, such as surface irregularities that are small compared to the acoustic wavelength.) Second, the mesh must be fine enough to represent the distribution of vibration on the surface of the body. This can be done by using a sufficient number of elements per *structural wavelength*. Finally, the mesh must be fine enough to represent the sound pressure distribution on the surface of the body. This can be done in most cases by selecting the element size to be no larger than a certain fraction of the *acoustic wavelength* for the highest frequency of interest. These last two requirements usually conflict, so one should select the largest element size that satisfies both requirements. The actual element size depends on the type of element used, as discussed in Section 2.3.

Encyclopedia of Acoustics, edited by Malcolm J. Crocker
ISBN 0-471-80465-7 © 1997 John Wiley & Sons, Inc.

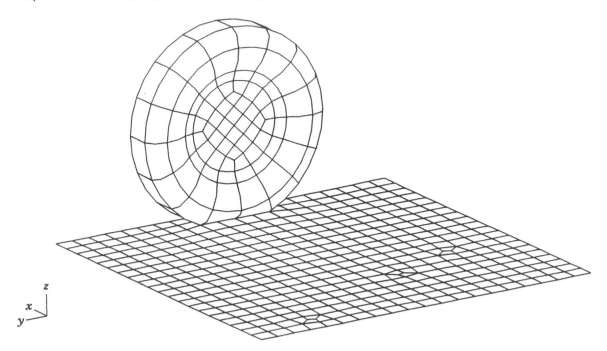

Fig. 1 Boundary element mesh for tire radiation problem.

Most commercially available pre- and postprocessing programs developed for the finite element method (FEM) may be used as well for constructing boundary element meshes. One uses shell or plate elements where the physical and material properties of the elements are immaterial since boundary elements have no thickness or properties. These same programs may be used to visualize the results of the BEM using their standard contouring capabilities.

1.2 Input to the Boundary Element Model

The boundary element mesh covers the entire surface of the radiating body. One needs to know some information (the so-called *boundary conditions*) about the problem at every node point of the mesh. Thus, for example, for most radiation problems one knows the vibration velocity normal to the surface at every point. Both the magnitude and phase of the velocity must be known. If a portion of the surface is covered by a sound-absorbing material, one must know the acoustic impedance of the material at each of the nodes that lie thereon.

1.3 Output of the Boundary Element Model

As described in Section 2, the BEM calculates the sound pressure distribution on the surface of the body from the geometry of the body and the distribution of vibration velocity or other boundary conditions provided by

the user. First, once the sound pressure and the vibration velocity are known on the surface, the sound intensity, sound power, and sound radiation efficiency may be found. Second, the sound pressure, particle velocity, and sound intensity can be calculated at so-called *field points*, that is, points in the acoustic domain that are not on the surface of the body. These field points can even lie on a reflecting plane such as the ground plane shown in Fig. 1.

Figure 2 shows the contour of sound pressure level (SPL) radiated by the tire in Fig. 1 at 1000 Hz. Note that the BEM will determine the SPL at all points in the acoustic domain, both in the near and the far fields. The BEM will also calculate the SPL in the "shadow zone" behind a body due to radiation from the other sides. All of this is done without any additional approximations or restrictions than discussed previously, that is, the need for an adequate mesh and the requirement to know the boundary conditions on the surface of the body.

1.4 Application of the Boundary Element Mode to Interior Problems

A second class of problems in which the BEM may be used is the so-called *interior problem* in acoustics. In contrast to exterior problems in which the acoustic domain extends to infinity, the acoustic domain for interior problems is finite. Typical examples of interior problems include the prediction of the sound field inside a

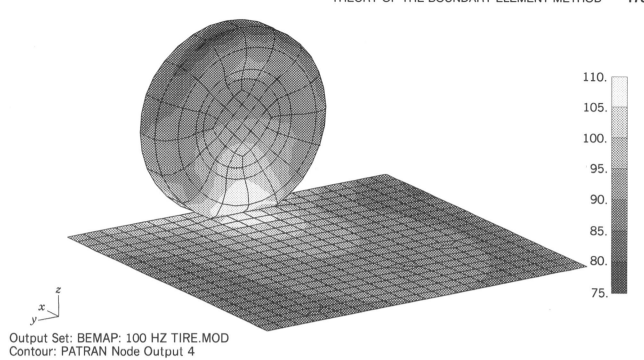

Output Set: BEMAP: 100 HZ TIRE.MOD
Contour: PATRAN Node Output 4

Fig. 2   The SPL contour plot of tire radiation at 1000 Hz.

vehicle or the calculation of the transmission loss of silencers.

From the user's standpoint, there is no difference between the application of the BEM to interior or exterior problems. First, the user creates a mesh with the same requirements discussed in Section 1.1. Second, boundary condition information at every node is supplied to the BEM. The BEM calculates the sound pressure on the inside surface of the cavity as well as the sound pressure level at field points inside the cavity.

Figure 3 shows the boundary element mesh for a simple expansion chamber muffler. Figure 4 shows the transmission loss (TL) of the muffler predicted by the BEM along with experimental results.[1] The BEM results were obtained by specifying the velocity on all nodes to be zero (since the muffler casing was assumed to be rigid) except those nodes at the inlet and exit of the muffler. At the inlet cross section, the velocity was specified as unity, and at the outlet the boundary condition was that the acoustic impedance was equal to the *characteristic impedance* of the medium.

The BEM can reveal a considerable amount of detail about the behavior of the system. For example, Fig. 5 shows the SPL contour inside the muffler at 2900 Hz, whereas the TL is approximately zero in Fig. 4. From Fig. 5 it may be seen that a cross mode in the chamber is excited at this frequency, resulting in a deterioration of the muffler's performance.

2   THEORY OF THE BOUNDARY ELEMENT METHOD

Although the mathematical foundation for the BEM is much older, it was not until the 1960s that the method could be used for problems of even modest size on the computers available at that time.[2–6] This early work provides a good basis for the BEM theory reviewed in this section.

2.1   The Boundary Element Method for Exterior Problems

Figure 6 shows a body denoted by B with surface S that radiates sound into an unbounded acoustic domain called B' having mean density ρ_0 and speed of sound c. The governing equation of linear acoustics is the well-known Helmholtz equation

$$\nabla^2 p + k^2 p = 0, \tag{1}$$

where p is the sound pressure at any point in the domain B' and $k = \omega/c$ is the wavenumber for harmonic waves of frequency ω. Boundary conditions on S are of the general form

$$\alpha p + \beta v_n = \gamma, \tag{2}$$

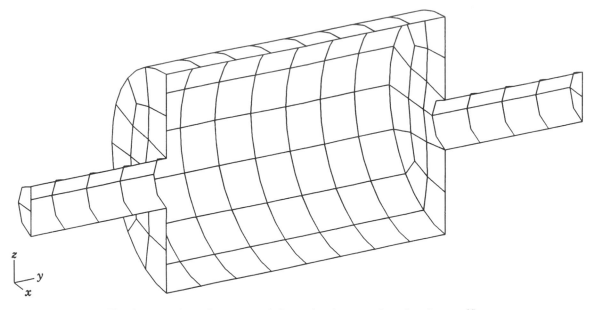

Fig. 3 Boundary element mesh for a simple expansion chamber muffler.

where v_n is the velocity (magnitude and phase) of any point on S in the normal direction, defined by the unit-normal vector n of S directed away from the acoustic domain (i.e., into the body B), and α, β, and γ are complex constants. A number of practical boundary conditions are included in Eq. (2). For example, Eq. (2) may be used to specify the normal velocity at a point by setting $\alpha = 0$, $\beta = 1$, and γ equal to the known velocity. Equation (2) can be used to represent a *locally reacting* normalized acoustic impedance z_n by setting $\alpha = 1$, $\beta = -z_n$, and $\gamma = 0$.

By using the direct formulation (via either the Green's second identity or the weighted residual formulation) and the Sommerfeld radiation condition, Eq. (1) is reformulated into a *boundary integral equation* as follows[7–11]:

$$C(P)p(P) = -\int_S [p(Q)G'(Q,P) + ikz_0 v_n(Q)G(Q,P)] \, dS(Q). \quad (3)$$

In the above equation, $p(P)$ is the sound pressure at a

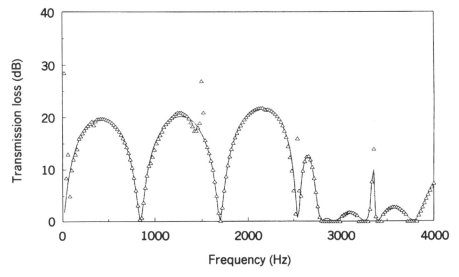

Fig. 4 The TL for the simple expansion chamber muffler: solid line, BEM; symbols, experiment.

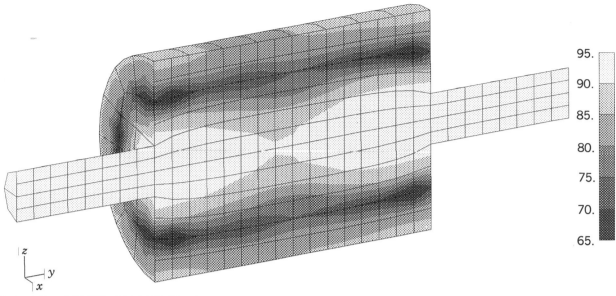

Output Set: BEMAP: 2900 HZ EC_MD
Contour: PATRAN Node Output 4

Fig. 5   The SPL contour plot for the expansion chamber muffler at 2900 Hz.

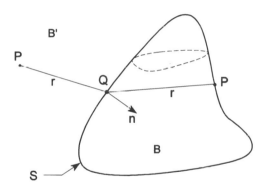

Fig. 6   Nomenclature for the BEM acoustic radiation problem.

point P, $p(Q)$ and $v_n(Q)$ are the sound pressure and normal velocity distributions on the surface of the body, $z_0 = \rho_0 c$ is the characteristic impedance of the medium, and $C(P)$ is a constant whose value depends on the location of the point P. Equation (3) is often referred to as the Helmholtz integral equation. In Eq. (3), G is the free-space Green's function

$$G(Q, P) = \frac{e^{-ikr}}{r},\qquad (4)$$

where $r = |Q - P|$ is the distance between points Q and P and G' is the derivative of G in the direction normal to the body. From a physical standpoint, Eq. (3) states

that the sound pressure at any point P can be found by integrating (summing) the contributions of a dipole distribution [first term in Eq. (3)] and a monopole distribution [second term in Eq. (3)] over the boundary S to the point P in question.

The leading coefficient $C(P)$ in Eq. (3) is 4π for P in the acoustic domain B' and zero for P inside of the body B. When P is located on the surface S, the value of $C(P)$ is determined by[7]

$$C(P) = 4\pi - \int_S \frac{\partial}{\partial n}\left(\frac{1}{r}\right)\, dS(Q),\qquad (5)$$

which has the physical interpretation of the exterior solid angle of S at point P. The value of $C(P)$ when P is on S will be 2π at points where the surface has a unique tangent plane. For corners and edges, $C(P)$ must be evaluated from Eq. (5).

It should be noted that when P coincides with Q on S, both integrals in Eq. (3) are singular because $r = 0$. The singularity of both terms is $O(1/r)$, which can be removed by a coordinate transformation before the integration is attempted.

2.2   The Boundary Element Method for Interior Problems

For interior problems such as the muffler shown in Fig. 3, the BEM formulation[12-14]

$$C^0(P)p(P) = -\int_S [\, p(Q)G'(Q,P)$$
$$+ ikz_0 v_n(Q)G(Q,P)]\, dS(Q) \qquad (6)$$

is similar to that described in the previous section with several exceptions. First, the Sommerfeld radiation condition is not used to derive Eq. (6) because for interior problems the domain is finite. Second, the normal n that is directed away from the acoustic domain is opposite to the normal n for exterior problems. Finally, the value of the leading coefficient $C^0(P) = 4\pi - C(P)$, where $C(P)$ is given by Eq. (5).

2.3 Numerical Implementation of the Boundary Element Method

One (and only one) of the acoustic variables p or v_n may be specified at each point Q on the surface S. (If the impedance $z_n = p/v_n$ is known at a point Q, this is equivalent to knowing either p or v_n, although not explicitly.) A numerical solution to the boundary integral equation [Eq. (3) or (6)] is necessary to determine the unknown variable at each node. A numerical solution can be achieved by discretizing the boundary surface S into a number of surface elements and nodes. In most of the early work[2–6] only piecewise shape functions or planar element discretizations of the surface were used. As element technology developed, the BEM has been extended[7] to adapt *isoparametric elements* such as those shown in Figs. 1 and 3. With isoparametric elements, the same polynomials (*shape functions*) are used to approximate both the surface shape and the variation of the acoustic variables over the element.

Elements may be classified according to order as well as shape. *Linear* elements are elements in which the geometry and the acoustic variables (sound pressure and vibration on the surface of the body) are represented by a linear or first-order approximation. *Quadratic* elements provide a second-order or quadratic approximation (i.e., they have curvature). Compared to linear elements, fewer quadratic elements are needed to obtain a result of given accuracy. All of the elements in Fig. 1 are quadratic elements.

The arrangement of the nodes for quadratic and linear isoparametric elements is shown in Fig. 7. Each quadratic element, whether quadrilateral or triangular, consists of a node at each vertex and one on each side. It is the side nodes that permit the element to assume a curved shape. It is not necessary to locate the side nodes at the midpoint of the segment, but accuracy will decrease if the side nodes are located too close to the vertex nodes. Likewise, accuracy will decrease if the element is severely distorted. Good modeling technique is

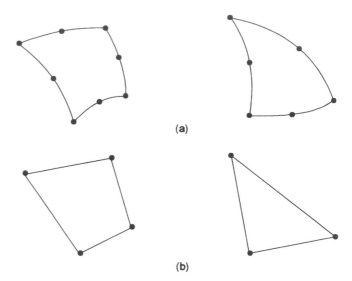

Fig. 7 Isoparametric boundary elements: (*a*) quadratic; (*b*) linear.

achieved by following simple guidelines and common sense with respect to element size and shape.

As mentioned in Section 1.1, the size of the boundary elements is determined by the geometry of the structure being modeled and the structural or acoustic wavelength, whichever is smaller. For linear elements, at least four elements per wavelength are required; for quadratic elements, at least two elements per wavelength are required.

A discretization of the surface S yields a discrete form of either Eq. (3) or (6). By placing a singular point P at each of the N nodes successively, N linearly independent equations are obtained. These equations can be arranged in matrix form to yield

$$[g(\omega)]\{p\} = [h(\omega)]\{v_n\}, \qquad (7)$$

where the quantities in curly brackets are $N \times N$ matrices formed by numerical integration of the G or G' and the element shape functions.[7] After inserting the boundary conditions, Eq. (7) may be solved to determine the unknown acoustic variable at each node. Once both p and v_n are known at every node, it is a straightforward calculation to determine the sound intensity, sound power, and radiation efficiency from these data. Then, using a discretized form of the boundary integral equation, one may determine the sound pressure, particle velocity, and sound intensity at any point P in the acoustic domain B' by simple numerical integration.

It may be seen from Eq. (7) that the frequency of vibration ω is contained in the g and h matrices. This means that these matrices must be recalculated for each frequency for which a solution is desired. It is sometimes advantageous to use the frequency interpolation technique[15,16] to reduce the central processing unit (CPU) time necessary for formation of the g and h matrices. However, this method does not substantially reduce *total* CPU time when the matrix is large because most of the CPU time is expended in solving rather than forming the matrix equation.

2.4 The Multidomain Boundary Element Method

As mentioned in Section 1, the traditional BEM can only be applied to homogeneous acoustic domains [i.e., as shown in Eq. (3) or (6), only one value of the characteristic impedance z_0 may be used]. However, by dividing the acoustic domain into several smaller domains in which z_0 may be different in each and applying the boundary integral equation to each smaller domain, it is possible to use the BEM to solve problems having a nonhomogeneous domain.[17, 18] Where the various domains are joined, continuity of pressure and particle velocity, or other special boundary conditions, are enforced to effect a solution. If one of the domains is an interior one and one an exterior one, the multidomain BEM may be used to solve problems such as radiation of sound from openings in enclosures and sound radiated from open ducts.[19]

The muffler in Fig. 8 is an example where the multidomain BEM is ideally suited. Each of the volumes in the muffler has a different average temperature, which means that z_0 will be different. In addition, the first volume is filled with a *bulk-reacting* absorbing material that, in general, has a complex characteristic impedance (and

cannot be modeled by a normal impedance z_n). With the multidomain BEM it is also possible to model perforations, as in the muffler inlet tube in Fig. 8.

Figure 9 shows the contour of the SPL inside the muffler in Fig. 8 at 700 Hz. For this problem the sound pressure at the inlet was selected to be approximately 85 dB. The progressive reduction of sound by each volume and tube in the muffler is clearly seen from the data in Fig. 9. The multidomain BEM also allows the BEM to be extended to thin bodies. The conventional single-domain BEM has numerical problems when the body is a thin structure, such as a plate or the tube connecting volumes 2 and 3 in Fig. 8.

2.5 Application of the Boundary Element Method to Half-Space Radiation Problems

The BEM may be applied to problems in which a body radiates sound near an infinite reflecting surface by modifying the Green's function G in Eq. (3). The modified Green's function

$$G_H = \frac{e^{-ikr}}{r} + R\,\frac{e^{-ikr_1}}{r_1}, \qquad (8)$$

where $r_1 = |Q - P_1|$, the distance from a point on the body to an image point P_1 of the point P relative to the reflecting surface, and R is the reflection coefficient of the surface. The tire radiation problem in Figs. 1 and 2 shows an application of the BEM to half-space radiation problems.[20]

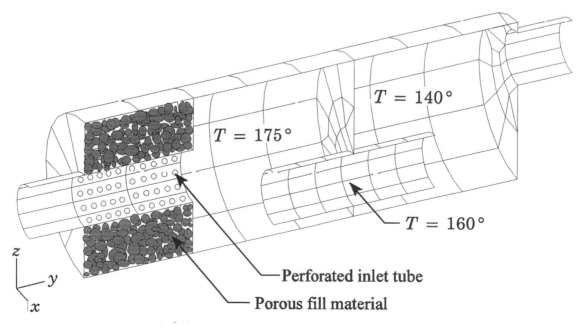

$T = 175°$

$T = 140°$

$T = 160°$

z y x

Perforated inlet tube

Porous fill material

Fig. 8 Muffler modeled using the multidomain BEM.

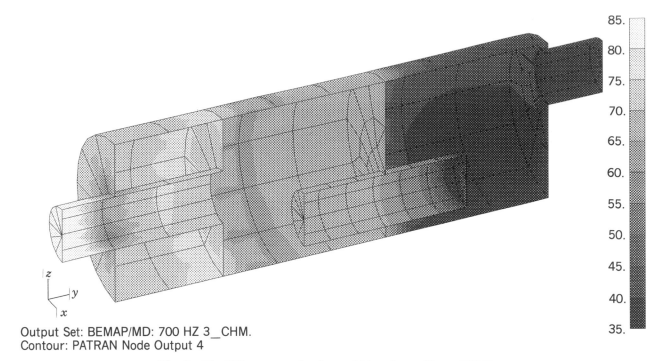

Output Set: BEMAP/MD: 700 HZ 3_CHM.
Contour: PATRAN Node Output 4

Fig. 9 The SPL contour plot for multidomain muffler at 700 Hz.

2.6 Nonuniqueness of the Boundary Element Method Solution

The exterior boundary integral equation [Eq. (3)] does not have a unique solution at certain characteristic frequencies.[21–23] This nonuniqueness is a purely mathematical problem arising from the integral formulation rather than from the nature of the physical problem. The interior boundary integral equation [Eq. (6)] does not have a nonuniqueness problem. Several methods have been developed to overcome the nonuniqueness problem, including CHIEF.[6] In the CHIEF method, Eq. (3) is collocated at one or more interior points of the body where $C(P) = 0$; these equations are combined with the original system of equations in Eq. (7) to obtain a solution by a least-squares approach. In the majority of radiation problems the CHIEF method provides a unique solution.[24] However, the selection of the interior points for collocation may not yield a unique solution if the points fall on the nodal surfaces of a related interior problem. More robust methods have been developed to remedy this problem.[25–31] In addition, the accuracy of the BEM solution may be confirmed by using a so-called interior point checking procedure[24] or an estimated matrix condition number.[32]

2.7 Other Boundary Element Method Formulations

In addition to the *direct* formulation [i.e., Eqs. (3) and (6)] there are various *indirect* integral equation formu-

lations of the sound radiation problem.[33–35] The formulation is called indirect because an unknown density distribution on the boundary must be solved before any physical solutions can be found. Similar to the direct formulation, the indirect formulation also breaks down at a set of characteristic frequencies. The failure in fact is more catastrophic[21] than its counterpart in the direct formulation because it is a nonexistence problem instead of a nonuniqueness problem. Nevertheless, methods to overcome the nonexistence problem do exist.[21, 23]

When the body in Fig. 6 is an axisymmetric body, Eq. (3), or its interior counterpart, Eq. (6), may be separated into an integral over the generator of the body and a second integral over the angle of revolution.[36–39] The second integral is an elliptic integral that may be evaluated analytically. The integral over the generator may be evaluated using one-dimensional (i.e., line) boundary elements.

3 THE BOUNDARY ELEMENT METHOD AS AN ENGINEERING ANALYSIS TOOL

The BEM is a relatively new tool compared to other acoustic analysis techniques such as the FEM. Much progress has been made in enhancing and tailoring the BEM for acoustics. Similar to the FEM, however, it is computationally and memory intensive, perhaps more so for certain applications.

3.1 Comparison between the Boundary Element and the Finite Element Methods

The BEM and the more familiar FEM share a number of features such as common underlying assumptions, the use of element technology to effect a solution, and the use of pre- and postprocessing to manage the information obtained from each method. In some sense, however, the FEM is the more general of the two methods. The FEM may be used for nonlinear problems such as when the sound pressure is extremely high or when the acoustic domain is *nonhomogeneous*, as for example when there are temperature gradients. In its usual form, the BEM is restricted to linear, homogeneous problems. However, a modification of the BEM, referred to as the multidomain BEM, discussed in Section 2.4, overcomes this last limitation for many problems of practical interest.

For most radiation problems, the BEM is preferred over the FEM. With the FEM, one would need to extend the surface mesh into the acoustic domain using three-dimensional volume elements. This is not recommended, as it results in a large number of elements, but more important, it yields an error since no matter how far the mesh is extended, it is not possible to represent an infinite domain with a finite model using the FEM. This illustrates the fundamental difference between the BEM and the FEM: With the BEM one discretizes only the surface of the radiating body while with the FEM the entire acoustic domain must be discretized. Because with the BEM all numerical approximations are confined to the surface, a coarser mesh can be used as compared to the FEM for the same accuracy.

Most commercially available FEM programs are written for structural applications and, therefore, do not have many of the features that are important for acoustic modeling. For example, most FEM programs do not calculate sound intensity or sound power, nor is it possible with most FEM programs to use an impedance boundary condition (for modeling absorbing surfaces). However, it is sometimes useful to perform an eigenvalue search of an acoustic domain. For this objective, the FEM is ideally suited. However, it is possible with the BEM to use a forced-response calculation to determine the resonance frequencies of the acoustic domain.

A primary advantage of the BEM over the FEM is its simplicity, which allows the new user to become proficient in a much shorter time. Because the boundary element mesh is only a surface mesh, it is easy to construct and does not have as many potential variations as one has with the FEM.

3.2 Characteristics of Boundary Element Method Software

Regardless of the BEM formulation (direct or indirect), there are features that are important to most users for a wide variety of applications. The following list is not inclusive since there are certainly other features (such as structural/acoustic coupling or acoustic sensitivities) that are important in some problems. However, a software program that has the following features will be useful for a majority of problems in acoustics:

1. Capable of solving interior, exterior, and multidomain problems
2. Can handle a reflecting surface with an arbitrary reflection coefficient
3. Allows the definition of a plane of symmetry for modeling symmetric bodies
4. Accepts a range of boundary conditions, including the impedance boundary condition
5. Allows the input vibration to be discontinuous at nodes to permit relative motion between adjacent components and to model edges and corners without special attention by the user
6. Incorporates interfaces to popular commercial pre- and postprocessing programs
7. Calculates the sound radiation from surfaces using the Rayleigh integral (for preliminary predictions of sound radiation)
8. Calculates the sound intensity, sound power, and sound radiation efficiency
9. Can include perforates for modeling mufflers and silencers
10. Can include point sources for preliminary studies of cavities and resonance frequencies
11. Includes a robust method to overcome the nonuniqueness problem
12. Incorporates a frequency interpolation method to increase the speed of multifrequency runs
13. Automatically maintains the accuracy of a BEM calculation without increasing the computer time excessively
14. Has an element library that includes both linear and quadratic boundary elements

There are a number of commercially available BEM software programs that include most or all of these features as well as more specialized enhancements. Several such programs are listed in Table 1.

TABLE 1   Partial List of BEM Acoustics Software

Software Product Name	Company	Location
SYSNOISE	LMS	Belgium
COMET	Automated Analysis	Ann Arbor, MI
BEMAP	Spectronics	Lexington, KY

REFERENCES

1. A. R. Mohanty, "Experimental and Numerical Investigation of Reactive and Dissipative Mufflers," Ph.D. Dissertation, University of Kentucky, Lexington, KY, 1993.

2. L. H. Chen and D. G. Schweikert, "Sound Radiation from an Arbitrary Body," *J. Acoust. Soc. Am.*, Vol. 35, 1963, pp. 1626–1632.

3. G. Chertock, "Sound Radiation from Vibrating Surfaces," *J. Acoust. Soc. Am.*, Vol. 36, 1964, pp. 1305–1313.

4. L. G. Copley, "Integral Equation Method for Radiation from Vibrating Bodies," *J. Acoust. Soc. Am.*, Vol. 41, 1967, pp. 807–816.

5. L. G. Copley, "Fundamental Results Concerning Integral Representations in Acoustic Radiation," *J. Acoust. Soc. Am.*, Vol. 44, 1968, pp. 28–32.

6. H. A. Schenck, "Improved Integral Formulation for Acoustic Radiation Problems," *J. Acoust. Soc. Am.*, Vol. 44, 1968, pp. 41–58.

7. A. F. Seybert, B. Soenarko, F. J. Rizzo, and D. J. Shippy, "An Advanced Computational Method for Radiation and Scattering of Acoustic Waves in Three Dimensions," *J. Acoust. Soc. Am.*, Vol. 77, 1985, pp. 362–368.

8. W. Tobacman, "Calculation of Acoustic Wave Scattering by Means of the Helmholtz Integral Equation, I," *J. Acoust. Soc. Am.*, Vol. 76, 1984, pp. 599–607.

9. W. Tobacman, "Calculation of Acoustic Wave Scattering by Means of the Helmholtz Integral Equation, II," *J. Acoust. Soc. Am.*, Vol. 76, 1984, pp. 1549–1554.

10. K. A. Cunefare, G. H. Koopmann, and K. Brod, "A Boundary Element Method for Acoustic Radiation Valid at All Wavenumbers," *J. Acoust. Soc. Am.*, Vol. 85, 1989, pp. 39–48.

11. T. Terai, "On the Calculation of Sound Fields Around Three-Dimensional Objects by Integral Equation Methods," *J. Sound Vib.*, Vol. 69, 1980, pp. 71–100.

12. A. F. Seybert and C. Y. R. Cheng, "Applications of the Boundary Element Method to Acoustic Cavity Response and Muffler Analysis," *ASME Trans. J. Vib. Acoust. Stress Rel. Design*, Vol. 109, 1987, pp. 15–21.

13. R. J. Bernhard, B. K. Gardner, and C. G. Mollo, "Prediction of Sound Fields in Cavities Using Boundary Element Methods," *AIAA J.*, Vol. 25, 1987, pp. 1176–1183.

14. S. Suzuki, S. Maruyama, and H. Ido, "Boundary Element Analysis of Cavity Noise Problems with Complicated Boundary Conditions," *J. Sound Vib.*, Vol. 130, 1989, pp. 79–91.

15. H. A. Schenck and G. W. Benthien, "The Application of a Coupled Finite-Element Boundary-Element Technique to Large-Scale Structural Acoustic Problems," in C. A. Brebbia and J. J. Connor (Eds.), *Advances in Boundary Elements*, Vol. 2, Computational Mechanics Publications, Southampton, 1989, pp. 309–319.

16. A. F. Seybert and T. W. Wu, "Applications in Industrial Noise Control," in R. D. Ciskowski and C. A. Brebbia (Eds.), *Boundary Element Methods in Acoustics*, Computational Mechanics Publications, Southampton, 1991, Chapter 9.

17. C. Y. R. Cheng, A. F. Seybert, and T. W. Wu, "A Multidomain Boundary Element Solution for Silencer and Muffler Performance Prediction," *J. Sound Vib.*, Vol. 151, 1991, pp. 119–129.

18. T. Tanaka, T. Fujikawa, T. Abe, and H. Utsuno, "A Method for the Analytical Prediction of Insertion Loss of a Two-Dimensional Muffler Model Based on the Transfer Matrix Derived from the Boundary Element Method," *ASME Trans. J. Vib. Acoust. Stress Rel. Design*, Vol. 107, 1985, pp. 86–91.

19. A. F. Seybert, C. Y. R. Cheng, and T. W. Wu, "The Solution of Coupled Interior/Exterior Acoustic Problems Using the Boundary Element Method," *J. Acoust. Soc. Am.*, Vol. 88, 1990, pp. 1612–1618.

20. A. F. Seybert and B. Soenarko, "Radiation and Scattering of Acoustic Waves from Bodies of Arbitrary Shape in a Three-Dimensional Half Space," *ASME Trans. J. Vib. Acoust. Stress Rel. Design*, Vol. 110, 1988, pp. 112–117.

21. A. J. Burton and G. F. Miller, "The Application of Integral Equation Methods to the Numerical Solutions of Some Exterior Boundary Value Problems," *Proc. Roy. Soc. Lond. A* Vol. 323, 1971, pp. 201–210.

22. R. E. Kleinman and G. F. Roach, "Boundary Integral Equations for the Three-Dimensional Helmholtz Equation," *SIAM Rev.*, Vol. 16, 1974, pp. 214–236.

23. A. J. Burton, "The Solution of Helmholtz' Equation in Exterior Domains Using Integral Equations," National Physical Laboratory, Report No. NAC 30, Teddington, Middlesex, United Kingdom, 1973.

24. A. F. Seybert and T. K. Rengarajan, "The Use of CHIEF to Obtain Solutions for Acoustic Radiation Using Boundary Integral Equations," *J. Acoust. Soc. Am.*, Vol. 81, 1987, pp. 1299–1306.

25. T. W. Wu and A. F. Seybert, "Acoustic Radiation and Scattering," in R. D. Ciskowski and C. A. Brebbia (Eds.), *Boundary Element Methods in Acoustics*, Computational Mechanics Publications, Southampton, 1991, Chapter 3.

26. T. W. Wu and A. F. Seybert, "A Weighted Residual Formulation for the CHIEF Method in Acoustics," *J. Acoust. Soc. Am.*, Vol. 90, 1991, pp. 1608–1614.

27. G. Krishnasamy, L. W. Schmerr, T. J. Rudolphi, and F. J. Rizzo, "Hypersingular Boundary Integral Equations: Some Applications in Acoustic and Elastic Wave Scattering," *ASME J. Appl. Mech.*, Vol. 57, 1990, pp. 404–414.

28. T. W. Wu, A. F. Seybert, and G. C. Wan, "On the Numerical Implementation of a Cauchy Principal Value Integral to Insure a Unique Solution for Acoustic Radiation and Scattering," *J. Acoust. Soc. Am.*, Vol. 90, 1991, pp. 554–560.

29. D. J. Segalman and D. W. Lobitz, "A Method to Overcome Computational Difficulties in the Exterior Acoustic Problem," *J. Acoust. Soc. Am.*, Vol. 91, 1992, pp. 1855–61.

30. X. F. Wu, A. D. Pierce, and J. H. Ginsberg, "Variational Method for Computing Surface Acoustic Pressure on Vibrating Bodies, Applied to Transversely Oscillating Disks," *IEEE J. Oceanic Eng.*, Vol. OE-12, 1987, pp. 412–418.

31. J. G. Mariem and M. A. Hamdi, "A New Boundary Finite Element Method for Fluid–Structure Interaction Problems," *Intl. J. Numer. Meth. Engr.*, Vol. 24, 1987, pp. 1251–1267.

32. T. W. Wu, "On Computational Aspects of the Boundary Element Method for Acoustic Radiation and Scattering in a Perfect Waveguide," *J. Acoust. Soc. Am.*, Vol. 96, 1994, pp. 3733–3743.

33. C. R. Kipp and R. J. Bernhard, "Prediction of Acoustical Behavior in Cavities Using an Indirect Boundary Element Method," *ASME J. Vib. Acoust. Stress Reliabil. Design*, Vol. 109, 1987, pp. 15–21.

34. C. A. Brebbia and R. Butterfield, "The Formal Equivalence of the Direct and Indirect Boundary Element Methods," *Applied Mathematical Modeling*, Vol. 2, No. 2, 1978.

35. G. Chertock, "Integral Equation Methods in Sound Radiation and Scattering from Arbitrary Surfaces," Naval Ship Research and Development Center, Report 3538, June 1971.

36. A. F. Seybert, B. Soenarko, R. J. Rizzo, and D. J. Shippy, "A Special Integral Equation Formulation for Acoustic Radiation and Scattering for Axisymmetric Bodies and Boundary Conditions," *J. Acoust. Soc. Am.*, Vol. 80, 1986, pp. 1241–1247.

37. B. Soenarko, "A Boundary Element Formulation for Radiation of Acoustic Waves from Axisymmetric Bodies with Arbitrary Boundary Conditions," *J. Acoust. Soc. Am.*, Vol. 93, 1993, pp. 631–639.

38. P. Juhl, "An Axisymmetric Integral Equation Formulation for Free Space Non-Axisymmetric Radiation and Scattering of a Known Incident Wave," *J. Sound Vib.*, Vol. 163, 1993, pp. 397–406.

39. W. L. Meyer, W. A. Bell, M. P. Stallybrass, and B. T. Zinn, "Prediction of Sound Fields Radiated from Axisymmetric Surfaces," *J. Acoust. Soc. Am.*, Vol. 63, 1979, pp. 631–638.

16

ACOUSTIC MODELING (DUCTED-SOURCE SYSTEMS)

M. G. PRASAD AND MALCOLM J. CROCKER

1 INTRODUCTION

Ducted sources are commonly found in mechanical systems. Typical ducted-source systems include engines and mufflers and air-moving devices (flow ducts and fluid machines and associated piping). In these systems, the source is the active component and the load is the path, which consists of elements such as mufflers, ducts, and end terminations. The acoustic performance of the system depends on the source–load interactions. This chapter presents a system model based on electrical analogies that has been found useful in predicting the acoustic performance of systems. The various methods for determining the impedance of a ducted source are given. The acoustic performance of a system with a muffler as a path element is described in terms of the insertion loss and radiated sound pressure.

2 SYSTEM MODEL

The basic source–load representation of a ducted system is shown in Fig. 1.

The equations for the source–load system based on pressure and velocity (complex) source representations are given by

$$\tilde{p}_L = \frac{\tilde{p}_s \tilde{Z}_L}{(\tilde{Z}_s + \tilde{Z}_L)} \tag{1}$$

and

$$\tilde{V}_L = \frac{\tilde{V}_s \tilde{Z}_s}{(\tilde{Z}_s + \tilde{Z}_L)}, \tag{2}$$

Encyclopedia of Acoustics, edited by Malcolm J. Crocker
ISBN 0-471-80465-7 © 1997 John Wiley & Sons, Inc.

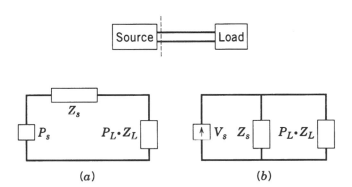

Fig. 1 Electrical analog of a ducted-source-load system: (*a*) pressure source; (*b*) volume velocity source.

where $\tilde{p}_s$ and $\tilde{V}_s$ are the source pressure and volume velocity, respectively; $\tilde{p}_i$ and $\tilde{V}_L$ are the pressure and volume velocity response of the source-load system, respectively; and $\tilde{Z}_s$ and $\tilde{Z}_L$ are the complex source and load impedances, respectively.[1,2]

3 SOURCE CHARACTERISTICS

The source at one end of a duct system represents one of the boundary conditions. It is more difficult to characterize the source compared to the termination because of the dynamic nature of the source.[3]

Although studies have been carried out both in the time and frequency domains, the transmission matrix approach in the frequency domain has been found to be effective in modeling and performance evaluations of ducted systems. Prior to the development of the direct and indirect methods for the measurement of source impedance, frequency-independent values of zero, characteristic (line) or infinite impedances were often assumed to characterize the source in the system model. However, such assumptions for source impedance do not normally yield good predictions.

3.1 Direct Methods

Direct methods for the measurement of source impedance are based on either[3] the standing-wave technique or the transfer function technique. These techniques have been successfully used for active test sources such as engines and pneumatic and electroacoustic drivers given a sufficient signal-to-noise ratio.[4,5] In this context, *signal* refers to the sound pressure level of the secondary measurement source and *noise* refers to the sound pressure level of the test source in operation. A minimum signal-to-noise ratio of 10 dB is required. In direct methods, the microphones are placed inside the duct. A modified transfer function method even for a low signal-to-noise ratio has been developed that only requires an additional measurement of a calibration transfer function.[6]

The complex measured transfer function $\tilde{H}_{12}$ is used to obtain the complex reflection $\tilde{R}$ from which the normalized complex impedance $\tilde{Z}_n$ looking at the test source is calculated:

$$\tilde{R} = e^{j(k_i + k_r)l} \left[\frac{\tilde{H}_{12} - e^{-jk_is}}{e^{-jk_rs} - \tilde{H}_{12}} \right], \tag{3}$$

$$\tilde{Z}_n = \frac{1 + \tilde{R}}{1 - \tilde{R}}, \tag{4}$$

where s is the spacing between microphones, l is the distance between the test source and the microphone farthest from the source, and k_i and k_r are forward and backward wavenumbers in the presence of a mean flow. The transfer function method using white-noise excitation for the secondary source is more effective and also faster to use than the standing-wave method.

3.2 Indirect Methods

Indirect methods are based on the use of different loads and their corresponding responses.[7,8] In Eq. (1), the pressure response $\tilde{P}_L$ is measured for a known load of impedance $\tilde{Z}_L$. Assuming that the source pressure is invariant, the system of equations for two, three, four, or more load impedances is solved to calculate the source impedance. With two and three loads the complex pressure response needs to be used, whereas with four loads the sound pressure level response can be used.

The advantages of indirect methods are that no secondary source is required as the response is measured from the test source and the microphone can be placed outside the duct. However, the methods are sensitive even to slight errors in the measured response, which strongly influences the numerical aspects.

The direct and indirect methods are essentially experimental methods based on frequency-domain analysis. However, the analytical modeling efforts have been carried out mainly in the time domain based on the method of characteristics. There have been some studies based on the modeling of the geometry of sources.[3]

4 PATH ELEMENT AND FOUR-POLE MATRIX

A duct system can be seen as composed of three elements, namely, the source, path, and termination. The path element is the duct system between the predetermined source and termination junctions. The path element can be made up of several types of area discontinuities from a simple expansion to a complicated muffler geometry, as shown in Fig. 2.

The four-pole matrix approach is an effective way of cascading the different subelements of the path element, based on transmission line theory using variables, namely the sound pressure and volume velocity. Any two junctions 1 and 2 in the path can be related using

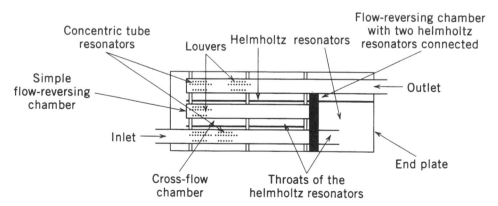

Fig. 2   Cross section of a common U.S. automobile muffler with different parts indicated.

$$
\begin{bmatrix} \tilde{p}_1 \\ \tilde{V}_1 \end{bmatrix} = \begin{bmatrix} \tilde{A} & \tilde{B} \\ \tilde{C} & \tilde{D} \end{bmatrix} \begin{bmatrix} \tilde{p}_2 \\ \tilde{V}_2 \end{bmatrix}, \tag{5}
$$

where $\tilde{p}$ and $\tilde{V}$ are the complex sound pressure and volume velocity and $\tilde{A}$, $\tilde{B}$, $\tilde{C}$, and $\tilde{D}$ are the complex four-pole parameters that describe the spectral response of the element. The four-pole parameters can be obtained for any geometric design and in the presence of mean flow and temperature gradients. The four-pole parameters can be obtained using both classical as well as numerical methods.[2] The four-pole parameters for a straight duct in the presence of a mean flow are given by

$$
\begin{bmatrix} \tilde{p}_1 \\ \tilde{V}_1 \end{bmatrix} = e^{-Mk_c l} \begin{bmatrix} \cos k_c l & j\dfrac{\rho c}{S}\sin k_c l \\ j\dfrac{S}{\rho c}\sin k_c l & \cos k_c l \end{bmatrix} \begin{bmatrix} \tilde{p}_2 \\ \tilde{V}_2 \end{bmatrix}, \tag{6}
$$

where M is the Mach number, $k_c = k/(1 - M^2)$ is the convected wavenumber, and l is the length of the duct element. The four-pole matrices for various types of duct elements are available in the literature.[2, 10] It is seen that the four-pole matrix formulation is very convenient, particularly for use with a digital computer.

5 TERMINATION IMPEDANCE

The termination impedance $\tilde{Z}_r$ offered by the type of termination in a duct system influences the system performance. There are several types of terminations used, namely characteristic, flanged and unflanged open end. The characteristic impedance refers to an infinitely long duct at the termination resulting in no reflections. An unflanged open end is commonly found in exhaust systems and is termed the radiation impedance $\tilde{Z}_r$, which can be obtained by Eq. (3) using the end reflection coefficient, $\tilde{R}$, given by[2]

$$
\begin{aligned}
\tilde{R} &= |\tilde{R}|e^{j(\pi - 2k_0\delta)}, \\
|\tilde{R}| &= 1 + 0.01336k_0 r_0 - 0.59079(k_0 r_0)^2 \\
&\quad + 0.33576(k_0 r_0)^3 - 0.06432(k_0 r_0)^4 \\
&\qquad 0 < k_0 r_0 < 1.5, \\
\dfrac{\delta}{r_0} &= 0.6133 - 0.1168(k_0 r_0)^2, \qquad k_0 r_0 < 0.5, \\
\dfrac{\delta}{r_0} &= 0.6393 - 0.1104(k_0 r_0), \qquad 0.5 < k_0 r_0 < 2, \tag{7}
\end{aligned}
$$

where k_0 is the wavenumber, r_0 is the radius of the duct open end, and δ is the end correction factor.

6 SYSTEM PERFORMANCE

The acoustic performance of a ducted-source system depends on the impedances of the source and load and the four-pole parameters of the path. This complete description of the system performance is termed *insertion loss* (IL). The insertion loss is the difference between sound pressure levels measured at the same reference point (from the termination) without and with the path element, such as a muffler, in place. Figure 3 shows various descriptions:

$$
\text{IL} = (L_{p2} - L_{p1}) \qquad \text{dB}. \tag{8}
$$

Another useful description of the path element is given by the *transmission loss* (TL) as

$$
\text{TL} = 10\log\frac{S_i I_i}{S_t I_t} \qquad \text{dB}, \tag{9}
$$

where I_i is the intensity incident on the path element and I_t is the intensity transmitted through the path element; S_i and S_t are duct cross-sectional areas at incidence and transmission, respectively.

Noise reduction (NR) is another descriptor used to measure the effect of the path element and is given by

$$
\text{NR} = (L_{p1} - L_{p2}) \qquad \text{dB}, \tag{10}
$$

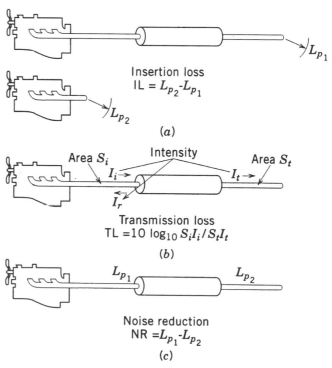

Fig. 3 Definitions of muffler performance.

where L_{p1} and L_{p2} are the measured sound pressure levels upstream and downstream of the path element (such as the muffler), respectively.

The three main descriptions of system performance are given by

$$\text{IL} = 20 \log_{10} \left| \frac{\tilde{A}\tilde{Z}_r + \tilde{B} + \tilde{C}\tilde{Z}_s\tilde{Z}_r + \tilde{D}\tilde{Z}_s}{\tilde{A}'\tilde{Z}_r + \tilde{B}' + \tilde{C}'\tilde{Z}_s\tilde{Z}_r + \tilde{D}'\tilde{Z}_s} \right| \quad \text{dB,} \quad (11)$$

$$\text{TL} = 20 \log_{10} \left| \frac{1}{2} \left(\tilde{A} + \frac{\tilde{B}S}{\rho c} + \frac{\tilde{C}\rho c}{S} + \tilde{D} \right) \right| \quad \text{dB,} \quad (12)$$

$$\text{NR} = 20 \log_{10} \left| \left(\tilde{A} + \frac{\tilde{B}}{\tilde{Z}_r} \right) \right| \quad \text{dB.} \quad (13)$$

The primed four-pole parameters in Eq. (11) refer to the system without the noise-reducing path elements, such as mufflers, in place. Environmental factors such as mean flow and temperature gradient effects can be accounted for in the four-pole parameters.[4]

As seen in Eqs. (8), (9), and (10), the insertion loss is the most useful description for the user, as it gives the net performance of the path element (muffler), including the interaction of the source and termination impedances. Equation (11) shows that the insertion loss depends on the source impedance. It is easier to measure insertion loss than to predict it because the characteristics of most sources are not known. A difficult question to be addressed in system modeling is how the performance (insertion loss) of a muffler varies when used with different sources.

The transmission loss is easier to predict than to measure. The transmission loss is defined so that it depends only on the path geometry and not on the source and termination impedances. The transmission loss is a special case of insertion loss in which the complex source and termination impedances are both characteristic ($Z_s = Z_r = \rho c/S$).

The transmission loss is very useful for the acoustic design of the path geometry. However, it is difficult to measure as it requires two microphones to separate the incident and transmitted intensities. The description of the system performance in terms of noise reduction requires a knowledge of both the path element and the termination in terms of four-pole matrix and impedance respectively.

Although, the above descriptions provide system model performance, an even more important quantity is the sound pressure radiated from the system. This description requires a knowledge of the source strength in terms of the volume velocity $\tilde{V}_s$. The complex radiated

sound pressure can be expressed as

$$|\tilde{p}_r| = |\tilde{Z}_{rs}||\tilde{V}_s|, \quad (14)$$

where $\tilde{Z}_{rs}$ is the transfer impedance of the source–path–termination system:

$$|\tilde{Z}_{rs}| = \frac{|\tilde{Z}_s| \; |\tilde{Z}_r + \rho_r c_r/S_r|}{2|\tilde{A}\tilde{Z}_r + \tilde{B} + \tilde{C}\tilde{Z}_s\tilde{Z}_r + \tilde{D}\tilde{Z}_s|}$$
$$\cdot \sqrt{\frac{S_r}{4\pi r^2} \frac{\rho_0 c_0}{\rho_r c_r}[(1 + M)^2 - (1 - M)^2 R^2]}, \quad (15)$$

$$|\tilde{V}_s| = \frac{|\tilde{p}_r'|}{|\tilde{Z}_s||\tilde{Z}_r|}|\tilde{A}'\tilde{Z}_r + \tilde{B}' + \tilde{C}'\tilde{Z}_s\tilde{Z}_r + \tilde{D}'\tilde{Z}_s| \quad (16)$$

where the primed quantities refer to a straight-duct system. The subscript "r" refers to tail pipe exit.

The radiated sound pressure usually refers to an open system with a finite-length tailpipe with an open end. However, the downstream sound pressure can be evaluated at a downstream field point inside the duct system.

7   APPLICATIONS

Although system modeling studies have been carried out on various ducted sources such as engines, fans, blowers, pumps, and so on, the most extensive work has been carried out on engine–muffler–tailpipe systems.[1–4] This is because unmuffled engine exhausts are usually the dominant noise sources in vehicles. System modeling studies are useful for application of the interaction of the source and load and the location of the muffler relative to the engine. Also, studies have useful application in the optimum design of the geometry of the source. Linear system modeling has been found to be effective and useful. Generally, nonlinear modeling would be challenging.[9]

Source impedances have been measured using the two microphone random-excitation method. In Fig. 4 the predicted insertion loss of a simple expansion chamber on an eight-cylinder engine is compared with that measured. The insertion loss of a simple expansion chamber is predicted using Eq. (11) using the measured engine (source) impedance. The system shown in Fig. 4 is an eight-cylinder engine with a simple expansion chamber. The studies have accounted both flow and temperature gradient effects.

Figure 5 shows comparisons of the sound pressure radiated made both from prediction and from measurements on the same engine–muffler system.[4] The applications here are presented only for an engine–muffler

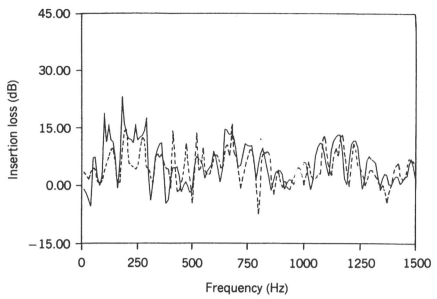

Fig. 4   Insertion loss of an expansion chamber on the engine operating at 2000 rpm, 10 in. Hg: (——) predicted (using measured engine impedance); (------) measured.

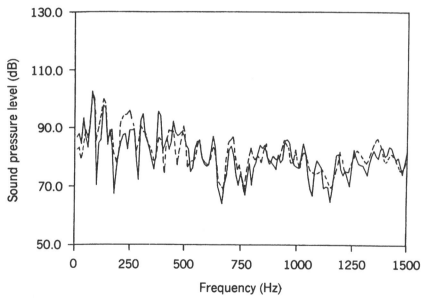

Fig. 5   Radiated sound pressure level with the expansion chamber on the engine operating at 2000 rpm, 10 in. Hg: (——) predicted (using measured engine impedance); (------) measured.

system. However, the modeling of ducted source–load systems is also important for many other ducted systems.[12–15]

REFERENCES

1. M. J. Crocker, "Internal Combustion Engine Exhaust Muffling," *Proceedings of NOISE-CON 77*, 1977, pp. 331–358.

2. M. L. Munjal, *Acoustics of Ducts and Mufflers*, Wiley, New York, 1987.

3. M. G. Prasad, "Characterization of Acoustical Sources in Duct Systems—Progress and Future Trends," in *Proceedings of Noise-Control*, Tarrytown, NY, 1991, pp. 213–220.

4. M. G. Prasad and M. J. Crocker, "Acoustical Source Characterization Studies on a Multi-Cylinder Engine Exhaust System" and "Studies of Acoustical Performance of a Multi-Cylinder Engine Exhaust Muffler System," *J. Sound Vib.*, Vol. 90(4), 1983, pp. 479–508.

5. A. G. Doige and H. S. Alves, "Experimental Characterization of Noise Sources for Duct Acoustics," *Trans. ASME: J. Vib. Acoust.*, Vol. 111, 1989, pp. 108–114.

6. W. Kim and M. G. Prasad, "A Modified Transfer Function Method for Acoustic Source Characterization Duct Systems," presented at the ASME Winter Annual Meeting, Paper 93-WA/NCA-5, 1993.

7. M. G. Prasad, "A Four Load Method for Evaluation of Acoustical Source Impedance in a Duct," *J. Sound Vib.*, Vol. 114, No. 2, 1987, pp. 347–355.

8. H. Boden, "On Multi-Load Methods for Determination of the Source Data of Acoustic One-Port Sources," *J. Sound Vib.*, Vol. 180, No. 5, 1995, pp. 725–743.

9. A. D. Jones, W. K. Van Moorhem, and R. T. Voland, "Is a Full Non-Linear Method Necessary for the Prediction of Radiated Exhaust Noise?" *Noise Control Eng. J.*, March-April 1986, pp. 74–80.

10. A. G. Galaitsis and I. L. Ver, "Passive Silencers and Lined Ducts," in L. L. Beranek and I. L. Ver (Eds.), *Noise and Vibration Control Engineering*, Wiley, New York, 1992, Chapter 10.

11. L. J. Eriksson, "Silencers," in D. E. Baxa (Ed.), *Noise in Internal Combustion Engines*, Wiley, New York, 1982.

12. S. Skaistis, *Noise Control of Hydraulic Machinery*, Marcel Dekker, New York, 1988, pp. 164–169.

13. J. Lavrentjev, M. Abom, and H. Boden, "A Measurement Method for Determining the Source Data of Acoustic Two-Port Sources," *J. Sound Vib.*, Vol. 183, No. 3, 1995, pp. 517–531.

14. M. Abom and H. Boden, "A Note on the Aeroacoustic Source Character of In-Duct Axial Fans," *J. Sound Vib.*, Vol. 186, No. 4, 1995, pp. 589–598.

15. H. Boden and M. Abom, "Maximum Sound Power from In-Duct Sources with Applications to Fans," *J. Sound Vib.*, Vol. 187, No. 3, 1995, pp. 543–550.

PART II

NONLINEAR ACOUSTICS AND CAVITATION

17

NONLINEAR ACOUSTICS AND CAVITATION

David T. Blackstock and Malcolm J. Crocker

1 INTRODUCTION

In Part I and most of the rest of this encyclopedia, the wave behavior of sound, for example, propagation, reflection, transmission, refraction, and diffraction, is described in terms of the linear wave equation (see Chapters 1 and 2). If the wave amplitude becomes high enough, nonlinear effects occur and the linear wave equation becomes inadequate. In nonlinear acoustics, novel phenomena unknown in linear acoustics are observed, for example, waveform distortion, formation of shock waves, increased absorption, nonlinear interaction (as opposed to superposition) when two sound waves are mixed, amplitude-dependent directivity of acoustic beams, cavitation, and sonoluminescence. This chapter begins with a simple description of waveform deformation and then reviews, briefly, the chapters on nonlinear acoustics which follow in Part II.

2 WAVEFORM DISTORTION

This section is devoted to simple physical arguments to explain waveform distortion, which, more than any other phenomenon, defines nonlinear acoustics. Consider a plane wave propagating, say, in the x-direction in a lossless fluid. If it is a small signal (infinitesimal amplitude), the wave phenomena can be described in terms of the linear wave equation

$$\nabla^2 \phi - \frac{1}{c_0^2} \frac{\partial^2 \phi}{\partial t^2} = 0,$$

Encyclopedia of Acoustics, edited by Malcolm J. Crocker
ISBN 0-471-80465-7 © 1997 John Wiley & Sons, Inc.

where ϕ is the velocity potential, t is time, and c_0 is the small-signal sound speed (the value found in tables).

For such cases of small (infinitesimal amplitude) disturbances, the wave propagates without change of shape, i.e., without distortion because all points on the waveform travel with the same speed, namely, $dx/dt = c_0$. If the amplitude of the wave is finite, however, the propagation speed varies from point to point on the waveform. The variation has two causes. First, by its very existence, a propagating wave sets up a longitudinal velocity field u in the fluid through which it travels. Since the moving fluid helps carry, i.e., convects the sound wave along, the propagation speed with respect to a fixed observer is therefore

$$\frac{dx}{dt} = c + u, \tag{1}$$

where c is the sound speed with respect to the moving fluid. Second, c is not quite the same as c_0. To see this, let the fluid be a gas, for which the sound speed varies as $\sqrt{T}$, where T is the absolute temperature. The sound speed is a little higher where the acoustic pressure p is positive (because compression raises the temperature) and a little lower where the p is negative (expansion lowers the temperature). The mathematical description of this relationship is

$$c = c_0 + \frac{\gamma - 1}{2} u, \tag{2}$$

where γ is the ratio of specific heats of the gas. Although explained above in terms of the effect of temperature, the deviation of c from c_0 may be traced to nonlinearity of the pressure–density relation of the fluid. Combination of Eq. (2) with Eq. (1) yields

$$\frac{dx}{dt} = c_0 + \beta u, \qquad (3)$$

where β, called the coefficient of nonlinearity, is given by

$$\beta = \frac{\gamma + 1}{2}. \qquad (4)$$

The dependence of propagation speed on particle velocity, given by Eq. (3) and due physically to the convection and nonlinearity of the pressure–density relation, is what produces distortion of the traveling wave. A complete mathematical description of nonlinear acoustical phenomena requires a nonlinear wave equation. Model nonlinear equations for various propagation, standing wave, and diffraction problems are discussed in Chapter 18. Chapter 19 provides more detail, particularly on various one-dimensional propagation problems.

Examples of waveform distortion are shown in Chapter 19 (Fig. 2). Although even in nonlinear acoustics u is normally much smaller than c_0, the effect of the varying propagation speed is cumulative and eventually causes noticeable distortion. Moreover, the stronger the wave, the more quickly the distortion develops.

The example given above for propagation in a gas is easily extended to other media. Convection, which is responsible for the presence of the factor u in Eq. (1), is the same for compressional waves in all media. Nonlinearity of the pressure–density relation (or its equivalent for solids) is a great deal larger, however, for liquids and solids (compressional waves) than for gases. In the case of liquids, the (normalized) coefficient of the first nonlinear term in the pressure–density relation is known as $B/2A$, and B/A is called the parameter of nonlinearity. In this case Eq. (4) is replaced by

$$\beta = 1 + \frac{B}{2A}. \qquad (5)$$

Chapter 20 is devoted to a discussion about B/A and a tabulation of its value for a large number of liquids. Nonlinear behavior of compressional waves in solids may be expressed in a similar fashion; see Chapter 21, which includes a table of values for β for a variety of solids.

3   REVIEW OF NONLINEAR ACOUSTICS TOPICS

There are two main classes of problems in nonlinear acoustics: (1) *source* problems, where the time variation of an acoustical field variable (e.g., pressure or particle velocity) is specified at a single spatial location and (2) *initial-value* problems, where the spatial variation of the field variable is specified at a particular time. Chapter 18 concentrates on the first class: *source* problems. Sound waves propagating from intense sound sources become distorted and almost always result in the formation of shock waves, which are discussed in Chapters 18, 23, and also in Chapter 31.

Chapter 19 is concerned with the second class: *initial-value* problems. Even when the particle velocity is much less than the speed of sound and the sound pressure is much smaller than $\rho_0 c_0^2$ (where ρ_0 is the static density), over large distances and long times waveform distortion occurs because of the cumulative effects of nonlinearity. The chapter begins with the Riemann analysis for plane waves without dissipation. Then diffusive mechanisms (viscosity and heat conduction) are included in the wave propagation formulation. The formation, propagation, interaction, and structure of shock waves are analyzed as well. Underwater acoustics and the effects of bubble concentrations are also discussed and analyzed in that chapter.

The nonlinearity in acoustic media contributes to the distortion in the wave shape during wave propagation. For fluids isentropic acoustic pressure fluctuations may be related to the density fluctuation through a Taylor series. The parameter of nonlinearity B/A is related to coefficients of the second and first terms in the Taylor series. The coefficient of nonlinearity is $\beta = 1 + B/2A$. Chapter 20 explains how these parameters can be derived from the equation of state of the material (if known) or else measured. Values of B/A are given in Tables 1 and 2 of that chapter for a variety of nonbiological and biological materials.

Chapter 21 discusses nonlinear propagation in solids and shows how the same basic approach can be used as for fluids. Tables 3–5 of that chapter provide values of B/A and β for gases, fluids, and some solids (crystals). The chapter also provides details of how these quantities are measured.

Finite-amplitude standing sound waves are experienced in a number of practical problems including mufflers for internal combustion engines, and, more recently, thermoacoustic heat engines and acoustic compressors. Several different approaches have been used to study such problems including Riemann-invariant analysis, the method of characteristics, and finite-difference analysis. Cavity walls cause phase speed dispersion and losses due to viscous and thermal boundary layers. The dispersion perturbs the resonance frequencies of rigid-wall cavities and influences the harmonic spectrum of the standing waves. In Chapter 22 a simplified nonlinear acoustic wave equation is solved. This approach accounts for shear viscosity and can be used to include additional losses. The equations are solved by a perturbation method.

Many applications of intense sound, such as those in aerospace rocket exhausts, sonar, medicine, and nondestructive evaluation, involve radiation from directive sources. Linear effects of diffraction determine length scales corresponding to nearfields, focal lengths, and beamwidths. However, nonlinear effects can be substantially different, and even dominant. The importance of the nonlinear effects depends on how the diffraction lengths compare with length scales corresponding to absorption and shock formation. Chapter 23 describes these competing effects in detail, beginning with analytic solutions for harmonic generation in weakly nonlinear Gaussian beams, both unfocused and focused. Asymptotic and numerical solutions are used to illustrate shock formation, nonlinear side lobe generation, and self-demodulation of pulses in sound beams radiated by circular pistons.

The lumped-element assumption, in which the sound wavelength is much greater than a typical element dimension, is discussed in Chapter 13. In many practical cases, such as in exhaust mufflers and in jets, intense sound waves exist, and the related nonlinear phenomena must be considered with lumped-element models. Elements in mufflers, for example, exhibit nonlinear behavior that must be included in the acoustic modeling. Chap-

ter 24 discusses both how the nonlinear behavior may be modeled and measured experimentally.

Cavitation can occur in fluids because of a variety of causes, one of which is acoustic excitation. If sufficiently intense sound fields are created in a liquid, then cavitation can occur during the low-pressure phase of the pressure fluctuations when bubbles are produced. Strong noise occurs during acoustic cavitation. It is caused by the cavities or bubbles which are generated and set into oscillation in the sound field. The violent motion of the cavities in the sound field can cause destructive effects not only with soft materials but even with metallic materials and can lead to failure. Chapter 25 discusses all these phenomena in detail.

The last topics in Part II are sonochemistry and sonoluminescence. In the 1930s these phenomena were discovered in Germany when liquids containing bubbles were irradiated with intense sound waves. In such a case the bubbles oscillate, then collapse and omit a flash of light. Very high temperatures, high pressures, and high heating and cooling rates can occur. Chemical reactions can be speeded up immensely during this process. Some have even suggested that, in principle, temperatures within such microscopic bubbles might be raised high enough to cause cold fusion. Chapter 26 reviews all of the important phenomena associated with sonoluminescence.

18

SOME MODEL EQUATIONS OF NONLINEAR ACOUSTICS

David T. Blackstock

1 PROPAGATION: MODEL EQUATIONS FOR PROGRESSIVE WAVES

Almost all model equations for propagation of finite-amplitude waves take advantage of the simplification that results when the fluid flow is so-called simple wave, that is, when the wave field is progressive. To illustrate, we consider small-signal plane waves. Let u stand for particle velocity, x distance, t time, and c_0 small-signal sound speed. If only forward-traveling waves are present, the (second-order) wave equation

$$\frac{\partial^2 u}{\partial x^2} - \frac{1}{c_0^2}\frac{\partial^2 u}{\partial t^2} = 0$$

may be integrated once to yield a first-order wave equation

$$\frac{\partial u}{\partial x} - \frac{1}{c_0}\frac{\partial u}{\partial t} = 0.$$

[*Proof:* The solution of the first-order equation is $u = f(t - x/c_0)$, where f is any function.] The first-order equation is further simplified by transforming from coordinates x, t to coordinates x, τ, where $\tau = t - x/c_0$ is retarded time. The result is

$$\frac{\partial u}{\partial x} = 0, \tag{1}$$

the solution of which is $u = f(\tau)$. Equation (1) is the

building block on which most model equations for more complicated progressive waves are based.

In what follows we frequently distinguish between *source* problems [$u(t)$ specified at a single spatial location, often $x = 0$] and *initial-value* problems [$u(x)$ specified at a single time, usually $t = 0$]. Some of the literature of nonlinear acoustics is devoted to initial-value problems. Since the most common acoustic problems involve radiation, however, we stress source problems in this chapter.

1.1 Lossless Propagation

Plane Waves The model equation suitable for source-generated plane waves of finite amplitudes in a lossless fluid is[1]

$$\frac{\partial u}{\partial x} - \frac{\beta}{c_0^2}\, u\, \frac{\partial u}{\partial \tau} = 0 \tag{2}$$

[notice that linearization yields Eq. (1)], where β is the coefficient of nonlinearity (Chapter 17, Section 2). This equation, its sister equation appropriate for initial-value problems [$\partial u/\partial t + \beta u(\partial u/\partial x')$, where $x' = x - c_0 t$], and some solutions are discussed in Chapter 19, Sections 2–4. For sinusoidal source excitation, that is, $u = u_0 \sin \omega t$ at $x = 0$, Eq. (2) is satisfied by the Fubini solution[2]

$$u = u_0 \sum \frac{2}{n\sigma}\, J_n(n\sigma)\sin n\omega\tau, \tag{3}$$

where $\sigma = \beta\epsilon k x = x/\bar{x}$, $\bar{x} = (\beta\epsilon k x)^{-1}$ is the shock formation distance, $\epsilon = u_0/c_0$, and $k = \omega/c_0$ is the wavenumber. The size of σ, sometimes called the dimensionless

Encyclopedia of Acoustics, edited by Malcolm J. Crocker
ISBN 0-471-80465-7 © 1997 John Wiley & Sons, Inc.

distortion range variable, is a measure of the amount of distortion that has occurred. For example, $\sigma = 1$ indicates shock formation. Equation (15) in Chapter 19 gives a Fubini-like solution for the initial-value problem, $u = u_0 \sin kx$ at $t = 0$.

Other One-Dimensional Waves, Including Ray Theory

For one-dimensional progressive waves that are not plane, geometric spreading, which is described mathematically by an extra term in the model equation, slows down the rate of distortion (see also Chapter 31, Section 4). If the waves are spherical or cylindrical and $kr \gg 1$, the model equation is[3]

$$\frac{\partial u}{\partial r} + \frac{a}{r} - \frac{\beta}{c_0^2} u \frac{\partial u}{\partial \tau} = 0, \qquad (4)$$

where r is the radial distance coordinate, the retarded time is now $\tau = t - (r - r_0)/c_0$, r_0 is a reference distance (e.g., the radius of the source), and a has the value 1 for spherical waves, $\frac{1}{2}$ for cylindrical waves, and 0 for plane waves. Introduction of the coordinate stretching function

$$z = r_0 \ln \frac{r}{r_0} \qquad \text{(spherical waves)}, \qquad (5)$$

$$= 2(\sqrt{rr_0} - r_0) \quad \text{(cylindrical waves)} \qquad (6)$$

and the spreading compensation function

$$w = \left(\frac{r}{r_0}\right)^a u \qquad (7)$$

reduces Eq. (4) to plane-wave form,

$$\frac{\partial w}{\partial z} = \frac{\beta}{c_0^2} w \frac{\partial w}{\partial \tau} = 0. \qquad (8)$$

Any plane-wave solution may therefore be extended to apply to spherical and cylindrical waves simply by replacing u and x with w and z, respectively. For example, the Fubini solution for spherical waves has the same form as Eq. (3). The expression for the distortion range variable for this case, $\sigma = \beta \epsilon k r_0 \ln r/r_0$, shows that the rate of distortion is much slower for spherical waves than for plane waves.

One-dimensional propagation also occurs in ducts of slowly varying cross section, such as horns or, in geometrical acoustics, ray tubes. Given the cross-sectional area $A(s)$ of the horn or ray tube, where s is the distance along the horn or ray tube axis, the coordinate stretching function is

$$z = \int_{s_0}^{s} \left(\frac{A_0}{A}\right)^{1/2} ds', \qquad (9)$$

where s_0 is a reference distance similar to r_0 and $A_0 = A(s_0)$. The spreading compensation function is

$$w = \left(\frac{A}{A_0}\right)^{1/2} u. \qquad (10)$$

In this way Eq. (8) is generalized to apply to finite-amplitude propagation in nonuniform ducts. The generalization may also be extended to cover cases in which the fluid is inhomogeneous.[4,5]

See Chapter 19, Section 10, for some applications of one-dimensional, nonplanar, finite-amplitude waves. Notice that although geometric spreading has been emphasized here, converging waves, which are often important in medical ultrasonics, may be treated in the same way.[1]

1.2 Dissipative Propagation

Although Eq. (8) is useful for a wide variety of finite-amplitude propagation problems, distortion almost always leads to formation of shocks, which are inherently dissipative. Losses must therefore be taken into account. Treated in this section are (1) the Burgers equation, the first really successful model that includes losses; (2) generalizations of the Burgers equation; and (3) weak-shock theory.

Classical Burgers Equation: Plane Waves in Thermoviscous Fluids

Originally proposed as model for turbulence,[6] the Burgers equation was found to be an excellent approximation of the conservation equations for plane progressive waves of finite amplitude in a thermoviscous fluid.[7] In the form suitable for source problems, the Burgers equation is[8]

$$\frac{\partial u}{\partial x} - \frac{\beta}{c_0^2} u \frac{\partial u}{\partial \tau} = \frac{\delta}{2c_0^3} \frac{\partial^2 u}{\partial \tau^2}, \qquad (11)$$

where $\delta = \nu[\mathcal{V} + (\gamma - 1)/\text{Pr}]$ is the diffusivity of sound,[7] $\nu = \mu/\rho_0$ is the kinematic viscosity, μ is the shear viscosity coefficient, ρ_0 is the static density, $\mathcal{V} = \frac{4}{3} + \mu_B/\mu$ is the viscosity number, μ_B is the bulk viscosity coefficient, and Pr is the Prandtl number. Notice that the only difference between the Burgers equation and Eq. (2) is that the latter has no dissipation term. For this reason Eq. (2) is sometimes called the *lossless Burgers equation*.

The fact that the Burgers equation is exactly integrable, a property discovered by Cole[9] and Hopf,[10] makes it a very attractive model. In Chapter 19, Section 5, the form of the Burgers equation appropriate for initial-value problems is given, the Hopf–Cole transformation is presented, and several different applications are discussed.

The most popular acoustical problem to which the Burgers equation has been applied is that for sinusoidal source excitation. Even though exact, the solution is quite complicated.[8,11] For distances larger than $3\bar{x}$ ($\sigma > 3$), however, it reduces to the well-known Fay solution[12]

$$u = u_0 \frac{2}{\Gamma} \sum \frac{\sin n\omega\tau}{\sinh n(1+\sigma)/\Gamma}, \qquad (12)$$

where $\Gamma = \beta\epsilon k/\alpha = 1/\alpha\bar{x}$ characterizes the importance of nonlinear distortion relative to absorption, and $\alpha = \delta\omega^2/2c_0^3$ is the small-signal absorption coefficient at the source frequency.

Generalized Forms of the Burgers Equation

The Burgers equation has been generalized in two ways. First, it has been extended to apply to other one-dimensional waves in thermoviscous fluids. For example, the Burgers equation for cylindrical and spherical waves is the same as Eq. (4) but with the zero on the right-hand side replaced by the right-hand side of Eq. (11). The equation for horns and ray tubes (stratified fluids) is similar. As shown in Chapter 19, Section 10, however, the generalized equation is not exactly integrable. The advantage of a known exact solution therefore stops with plane waves in homogeneous fluids.

In the second form of generalization,[13] plane waves are retained but other forms of dissipation are considered, for example, that due to relaxation effects[14–16] (Chapter 19, Section 7) or thermoviscous boundary layers[1] (Chapter 19, Section 8). Solutions of limited validity for special cases are known, but nothing comparable to the general solution of the classical Burgers equation.

Of course, the two forms of generalization may be combined, for instance in application to nonlinear geometric acoustics in the ocean or atmosphere.[17]

The chief merit of the various generalized forms of the Burgers equation is their relative simplicity (as compared to the full-fledged conservation equations for fluids). They offer a good starting point for numerical solution.[16,17]

Weak-Shock Theory

Weak-shock theory[18–21] (see also Chapter 31, Section 4) is a very effective alternative to the Burgers equation for cases in which dissipation is primarily due to shocks in the traveling wave. Unlike the Burgers equation, weak-shock theory is as effective when generalized to apply to nonplanar waves and/or inhomogeneous media as it is for ordinary plane waves in homogeneous media. The lossless Burgers equation (8) is used for continuous sections of the waveform between shocks, and a low-amplitude approximation of the Rankine–Hugoniot shock relations, sometimes cast as the *equal-area rule*,[20] is used to calculate the position and amplitude of each shock. In this way the continuous sections of the waveform are connected to each other. Well-known applications arc to N-waves and sawtooth waves[1,20,21]; see also Chapter 27 and Sections 6 and 7 in Chapter 31. Weak-shock theory has also been used to show the connection between the Fubini solution [Eq. (3)] and the Fay solution [Eq. (12)].[21]

Weak-shock theory loses its accuracy when the wave becomes so weak that its shocks are dispersed and no longer dominant centers of dissipation. At great distances, therefore, where the wave amplitude becomes very small, asymptotic results based on weak-shock theory may be unsatisfactory.[1]

A computer algorithm based on weak-shock theory, called the Pestorius algorithm,[22,23] has been developed to predict the propagation of complicated signals, such as finite-amplitude noise. The limitation described in the previous paragraph is overcome by periodically correcting the waveform for effects of ordinary dissipation.

2 STANDING WAVES, REFLECTION, AND REFRACTION

Standing waves, reflection, and refraction are problems of compound flow: The sound field is composed of overlapping forward- and backward-traveling waves. Since nonlinearity precludes superposition, compound flow cannot be described simply by adding the progressive wave expressions for the two waves. In other words the two waves interact with each other as well as propagate. One- and two-dimensional problems are considered separately here.

2.1 One-Dimensional Fields

The general nonlinear wave equation for planar flow in lossless gases is

$$c_0^2 \frac{\partial^2\phi}{\partial x^2} - \frac{\partial^2\phi}{\partial t^2} = \frac{\partial}{\partial t}\left(\frac{\partial\phi}{\partial x}\right)^2 + (\gamma-1)\frac{\partial\phi}{\partial t}\frac{\partial^2\phi}{\partial x^2}$$

$$+ \frac{\partial\phi}{\partial x}\left[\frac{\gamma-1}{2}\frac{\partial\phi}{\partial x}\frac{\partial^2\phi}{\partial x^2}\right.$$

$$\left. + \frac{1}{2}\frac{\partial}{\partial x}\left(\frac{\partial\phi}{\partial x}\right)^2\right]. \qquad (13)$$

[When only forward-traveling waves are present, this equation reduces to simple-wave form, Eq. (6) in Chapter 19 or, with minor approximation, Eq. (2).] An alternative representation in terms of Riemann invariants is given in Chapter 19, Section 1. In general, compound-flow problems are much more difficult to solve than simple-wave problems.

The problem of standing waves in a closed-end resonance tube was solved by Chester[24]; for experimental verification, see Cruikshank.[25] Jimenez[26] (theory) and Sturtevant[27] (experiments) considered a range of different end conditions, from closed to open. Chapter 22 deals with closed-end tubes and also with the three-dimensional extension, rectangular cavities with rigid walls.

Reflection of a normally incident plane wave at an interface between two semi-infinite fluids is a related problem involving overlapping wave fields. First let the interface be a rigid wall. Pfriem[28] solved this problem for waves of continuous waveform, that is, a shock-free field. His result shows that pressure doubling occurs only for small signals.[1] Although the pressure amplification produced by the wall can be quite high for strong waves, the shock-free assumption severely limits the applicability of the result. Solution of the corresponding shock reflection problem[29] shows a much smaller wall amplification factor for shock waves, but still greater than 2. For a pressure-release surface the result for small signals, that the particle velocity doubles at the interface, continues to hold even for finite-amplitude waves.[1]

2.2 Two-Dimensional Fields

When plane waves are obliquely incident on the interface, refraction occurs as well as reflection. Equation (13) generalizes to

$$c_0^2 \nabla^2 \phi - \phi_{tt} = [(\nabla \phi)^2]_t + (\gamma - 1)\phi_t \nabla^2 \phi$$
$$+ \nabla \phi [\tfrac{1}{2}(\gamma - 1)\nabla \phi \nabla^2 \phi + \tfrac{1}{2}\nabla(\nabla \phi)^2]. \quad (14)$$

The reflection–refraction problem is very difficult, and only a few investigators have attempted it.[30,31] The question of whether the law of specular reflection and Snell's law hold for continuous finite-amplitude waves is still open. Some results are known for shock waves.[32]

3 MODEL EQUATIONS FOR TWO- AND THREE-DIMENSIONAL FIELDS

The model equations discussed in Section 1 are too simple for anything but one-dimensional propagation. The exact wave equations in Section 2 are more complicated than need be for most two- and three-dimensional problems in nonlinear acoustics (the third-order terms have little impact) and yet do not include the important effects of dissipation. Several more useful two- and three-dimensional model equations have been developed, usually in response to specific needs. For example, the Westervelt equation[33] was an almost incidental product of Westervelt's discovery of the parametric array (see Chapter 53); a desire to include the effect of diffraction in propagation of intense sound beams led Zabolotskaya and Khokhlov to what is now called the KZK equation.[34–36] Naze Tjøtta and Tjøtta and their co-workers have broadened the approach and obtained a family of very general model equations,[37–39] which may be used for a very wide variety of problems in nonlinear acoustics. All previously mentioned model equations are special cases. Here is one version of the Tjøttas' equation for thermoviscous fluids:

$$\Box^2 p + \frac{\delta}{c_0^4} \frac{\partial^3 p}{\partial t^3} = -\frac{\beta}{\rho_0 c_0^4} \frac{\partial^2 p^2}{\partial t^2} - \left(\nabla^2 + \frac{1}{c_0^2} \frac{\partial^2}{\partial t^2}\right) \mathcal{L}, \quad (15)$$

where $\Box^2$ is the d'Alembertian operator

$$\Box^2 = \nabla^2 - \frac{1}{c_0^2} \frac{\partial^2}{\partial t^2} \quad (16)$$

and $\mathcal{L}$ is the Lagrangian density

$$\mathcal{L} = \frac{\rho_0 u^2}{2} - \frac{p^2}{2\rho_0 c_0^2}. \quad (17)$$

For example, for unidirectional sound beams $\mathcal{L} = 0$, and Eq. (15) reduces to the Westervelt equation (in Westervelt's original development the dissipation term was not included).

4 DIFFRACTION

Beginning with the pioneering work of Zabolotskaya and Khokhlov,[34] progress on effects of diffraction in nonlinear acoustics has been substantial. Most of the investigations have been on sound beams, and Chapter 23 is devoted to this topic. The applications are various, for example, parametric arrays (Chapter 53), harmonic distortion, self-demodulation, and focused beams. Pulses as well as time-harmonic signals have been studied, and the work has been both experimental and theoretical. The model equation used for almost all the analytical work in this area is the KZK equation, which, for an axisymmetric beam, is

$$\frac{\partial^2 p}{\partial z \, \partial \tau} = \frac{c_0}{2} \left(\frac{\partial^2 p}{\partial r^2} + \frac{1}{r} \frac{\partial p}{\partial r} \right) + \frac{\delta}{2c_0^3} \frac{\partial^3 p}{\partial \tau^3}$$
$$+ \frac{\beta}{2\rho_0 c_0^3} \frac{\partial^2 p^2}{\partial \tau^2}. \tag{18}$$

The coordinate system is cylindrical: z is the axial coordinate and r is the radial coordinate.

Much computational work has been done on nonlinear effects in beams, most based on the KZK equation (see Chapter 23), some on other models.[40,41]

REFERENCES

1. D. T. Blackstock, "Nonlinear Acoustics (Theoretical)," in D. E. Gray (Ed.), *American Institute of Physics Handbook*, McGraw-Hill, New York, 1972, pp. 3-183 to 3-205.

2. E. Fubini, "Anomalies in the Propagation of Acoustic Waves of Great Amplitude," *Alta Frequenza*, Vol. 4, 1935, pp. 530–581 (in Italian).

3. D. T. Blackstock, "On Plane, Spherical, and Cylindrical Sound Waves of Finite Amplitude in Lossless Fluids," *J. Acoust. Soc. Am.*, Vol. 36, 1964, pp. 217–219.

4. C. L. Morfey, "Nonlinear Propagation in a Depth-Dependent Ocean," Tech. Rep. ARL-TR-84-11, Applied Research Laboratories, University of Texas at Austin, May 1, 1984 (ADA 145 079).

5. W. D. Hayes, R. C. Haefeli, and H. E. Karlsrud, "Sonic Boom Propagation in a Stratified Atmosphere, with Computer Program," Aeronautical Research Associates of Princeton, NASA CR-1299, April 1969.

6. J. M. Burgers, "A Mathematical Model Illustrating the Theory of Turbulence," in R. von Mises and T. von Kármán (Eds.), *Advances in Applied Mechanics*, Vol. I, Academic, New York, 1948, pp. 171–191.

7. M. J. Lighthill, "Viscosity Effects in Sound Waves of Finite Amplitude," in G. K. Batchelor and R. M. Davies (Eds.), *Surveys in Mechanics*, Cambridge University Press, Cambridge, 1956, pp. 250–351.

8. D. T. Blackstock, "Thermoviscous Attenuation of Plane, Periodic, Finite-Amplitude Sound Waves," *J. Acoust. Soc. Am.*, Vol. 36, 1964, pp. 534–542.

9. J. D. Cole, "On a Quasi-Linear Parabolic Equation Occurring in Aerodynamics," *Q. Appl. Math.*, Vol. 9, 1951, pp. 225–236.

10. E. Hopf, "The Partial Differential Equation $u_t + uu_x = \mu u_{xx}$," *Commun. Pure Appl. Math.*, Vol. 3, 1950, pp. 201–230.

11. S. I. Soluyan and R. V. Khokhlov, "The Propagation of Acoustic Waves of Finite Amplitude in a Dissipative Medium," *Vestn. Mosk. Univ., Fiz. Astron.*, Ser. III, Vol. 3, 1961, pp. 52–61 (in Russian).

12. R. D. Fay, "Plane Sound Waves of Finite Amplitude," *J. Acoust. Soc. Am.*, Vol. 3, 1931, pp. 222–241.

13. D. T. Blackstock, "Generalized Burgers Equation for Plane Waves," *J. Acoust. Soc. Am.*, Vol. 77, 1985, pp. 2050–2053.

14. O. V. Rudenko and S. I. Soluyan, *Theoretical Foundations of Nonlinear Acoustics*, Consultants Bureau, Plenum, New York, 1977.

15. A. D. Pierce, *Acoustics: An Introduction to Its Physical Principles and Applications*, McGraw-Hill, New York, 1981, Chap. 11.

16. R. O. Cleveland, M. F. Hamilton, and D. T. Blackstock, "Time-Domain Modeling of Finite-Amplitude Sound in Relaxing Fluids," *J. Acoust. Soc. Am.*, Vol. 99, 1996, pp. 3312–3318.

17. R. O. Cleveland, J. P. Chambers, H. E. Bass, R. Raspet, D. T. Blackstock, and M. F. Hamilton, "Comparison of Computer Codes for the Propagation of Sonic Booms through the Atmosphere," *J. Acoust. Soc. Am.*, Vol. 100, 1996, pp. 3017–3027.

18. K. O. Friedrichs, "Formation and Decay of Shock Waves," *Commun. Pure Appl. Math.*, Vol. 1, 1948, pp. 211–245.

19. G. B. Whitham, "The Flow Pattern of a Supersonic Projectile," *Commun. Pure Appl. Math.*, Vol. 5, 1952, pp. 301–348.

20. L. D. Landau and E. M. Lifshitz, *Fluid Mechanics* (translated from the Russian by J. B. Sykes and W. H. Reid), Addison-Wesley, Reading, MA, 1959, Art. 95.

21. D. T. Blackstock, "Connection between the Fay and Fubini Solutions for Plane Sound Waves of Finite Amplitude," *J. Acoust. Soc. Am.*, Vol. 39, 1966, pp. 1019–1026.

22. F. M. Pestorius, "Propagation of Plane Acoustic Noise of Finite Amplitude," Tech. Rep. ARL-TR-73-23, Applied Research Laboratories, University of Texas at Austin, August 1973 (AD 778 868).

23. F. M. Pestorius and D. T. Blackstock, "Propagation of Finite-Amplitude Noise," in L. Bjørnø (Ed.), *Finite-Amplitude Wave Effects in Fluids*, Proceedings of 1973 Symposium in Copenhagen, IPC Press, Guildford, 1974, pp. 24–29.

24. W. Chester, "Resonant Oscillations in Closed Tubes," *J. Fluid Mech.*, Vol. 18, 1964, pp. 44–64.

25. D. B. Cruikshank, Jr., "Experimental Investigation of Finite-Amplitude Acoustic Oscillations in a Closed Tube," *J. Acoust. Soc. Am.*, Vol. 52, 1972, pp. 1024–1036.

26. J. Jimenez, "Nonlinear Gas Oscillations in Pipes. Part I. Theory," *J. Fluid Mech.*, Vol. 59, 1973, pp. 23–46.

27. B. Sturtevant, "Nonlinear Gas Oscillations in Pipes. Part II. Experiment," *J. Fluid Mech.*, Vol. 63, 1974, pp. 97–120.

28. H. Pfriem, "Reflexionsgesetze für ebene Druckwellen

grosser Schwingungsweite," *Forsch. Gebeite Ingenieurw.*, Vol. B12, 1941, pp. 244–256.

29. R. Courant and K. O. Friedrichs, *Supersonic Flow and Shock Waves*, Interscience, New York, 1948.

30. F. D. Cotaras, "Reflection and Refraction of Finite Amplitude Acoustic Waves at a Fluid–Fluid Interface," Tech. Rep. ARL-TR-89-1, Applied Research Laboratories, University of Texas at Austin, January 3, 1989 (ADA 209 800).

31. K. T. Shu and J. H. Ginsberg, "Oblique Reflection of a Nonlinear P Wave from the Boundary of an Elastic Half Space," *J. Acoust. Soc. Am.*, Vol. 89, 1991, pp. 2652–2662.

32. R. G. Jahn, "Transition Processes in Shock Wave Interactions," *J. Fluid Mech.*, Vol. 2, 1957, pp. 33–48.

33. P. J. Westervelt, "Parametric Acoustic Array," *J. Acoust. Soc. Am.*, Vol. 35, 1963, pp. 535–537.

34. E. A. Zabolotskaya and R. V. Khokhlov, "Quasi-Plane Waves in the Nonlinear Acoustics of Confined Beams," *Sov. Phys. Acoust.*, Vol. 15, 1969, pp. 35–40.

35. V. P. Kuznetsov, "Equations of Nonlinear Acoustics," *Sov. Phys. Acoust.*, Vol. 16, 1971, pp. 467–470.

36. N. S. Bakhvalov, Ya. M. Zhileikin, and E. A. Zabolotskaya, *Nonlinear Theory of Sound Beams*, American Institute of Physics, New York, 1987.

37. S. I. Aanonsen, T. Barkve, J. Naze Tjøtta, and S. Tjøtta, "Distortion and Harmonic Generation in the Nearfield of a Finite Amplitude Sound Beam," *J. Acoust. Soc. Am.*, Vol. 75, 1984, pp. 749–768.

38. J. Naze Tjøtta, and S. Tjøtta, "Interaction of Sound Waves. Part I: Basic Equations and Plane Waves," *J. Acoust. Soc. Am.*, Vol. 82, 1987, pp. 1425–1428.

39. J. Naze Tjøtta and S. Tjøtta, "Nonlinear Equations of Acoustics," in M. F. Hamilton and D. T. Blackstock (Eds.), *Frontiers of Nonlinear Acoustics—12th ISNA*, Elsevier Applied Science, London, 1990, pp. 80–97.

40. P. T. Christopher and K. J. Parker, "New Approaches to the Linear Propagation of Acoustic Fields," *J. Acoust. Soc. Am.*, Vol. 90, 1991, pp. 507–521.

41. P. T. Christopher and K. J. Parker, "New Approaches to Nonlinear Diffractive Field Propagation," *J. Acoust. Soc. Am.*, Vol. 90, 1991, pp. 488–499.

19

PROPAGATION OF FINITE-AMPLITUDE WAVES IN FLUIDS

D. G. CRIGHTON

1 INTRODUCTION

Sound waves propagate linearly when *both* their amplitude is very small *and* the times and distances over which they are observed are not too great. If either of these conditions is violated, one may have to take account of nonlinear effects. When disturbance amplitudes are large, as in explosions or close to a high-speed jet engine exhaust, then there are large local nonlinear effects, and the process cannot in any sense be thought of as acoustic. But even when disturbance amplitudes are locally small (i.e., particle velocity small compared with the speed of sound, pressure fluctuations small compared with the background pressure), a sound wave will still suffer severe waveform distortion, through the cumulative action of small nonlinearities, in its propagation over a sufficiently large time or distance. And no matter how small the initial disturbance, its long-time evolution will necessarily require nonlinear equations (unless the wave is heavily damped by dissipation).

We are concerned in this chapter with those weak nonlinear effects that become important in long-range propagation of a signal that locally satisfies the "infinitesimal-amplitude" criterion for linear acoustics; and if the amplitudes are small but larger than infinitesimal, then the important finite-amplitude effects will simply occur over shorter times and distances. These are the situations covered by the field of *nonlinear acoustics*.

The effects of nonlinearity (or finite amplitude) on sound waves are cumulative and lead eventually to very steep waves. As a consequence of the steep gradients induced in this way, various linear mechanisms become important, even though they were insignifi-

cant in the initial undistorted wave. In particular, diffusive mechanisms (viscosity, heat conduction) and frequency dispersion (minute in pure air or water but large in bubbly liquid, for example) are called increasingly into play as a sound wave steepens under its nonlinear self-distortion. What nonlinear acoustics is really about is this competition, first analyzed in these terms by Lighthill,[1] between nonlinear waveform-steepening mechanisms and linear waveform-easing mechanisms. In some cases these mechanisms come into balance, and then they determine waveforms of some characteristic shape—shock waves when diffusion or relaxation mechanisms balance nonlinearity, soliton pulses when (as in bubbly liquid) frequency dispersion balances nonlinearity. Recall, by contrast, that a linear wave can have any shape, and once that shape is set (by a source), then it remains unchanged in linear (nondissipative, nondispersive) propagation.

This chapter will give a summary of important results illustrating the basic issues in nonlinear acoustics. We start with the Riemann analysis for plane waves with no dissipation, which is important because it shows exactly what the effects of nonlinearity, however small or large, are. Then we show how to include diffusivity for weak waves and analyze the formation, propagation, interaction and structure of shock waves when steepening is resisted both by diffusion and by relaxation processes. In applications to underwater acoustics, the effects of small bubble concentrations are important for linear and nonlinear propagation of sound, and we discuss these effects along with, in later sections, those of divergence or convergence of the rays along which the waves can be regarded as propagating, of the effects of dissipation in tube wall boundary layers, and of wavefront turning and focusing associated with multidimensional nonlinear acoustics. The treatment is necessarily in mathematical

Encyclopedia of Acoustics, edited by Malcolm J. Crocker
ISBN 0-471-80465-7 © 1997 John Wiley & Sons, Inc.

terms, but the results are simply quoted, with emphasis throughout on interpretations in clear physical terms.

2   RIEMANN ANALYSIS FOR PLANE WAVES

For motions of an ideal fluid starting from a uniform rest state (suffix 0) the entropy S remains uniform, and this implies a definite functional relationship between pressure p and density ρ, $p = p(\rho)$. Then one can define the function of state, with dimensions of velocity,

$$P = \int_{\rho_0}^{\rho} \frac{c(\rho)}{\rho} \, d\rho = \int_{p_0}^{p} \frac{dp}{\rho c}, \tag{1}$$

where $c^2 = dp/d\rho$ defines the local sound speed c, with $P = 0$ in the undisturbed medium. Riemann rearranged the mass and momentum conservation equations in the form

$$\left(\frac{\partial}{\partial t} + (u \pm c) \frac{\partial}{\partial x} \right) (u \pm P) = 0, \tag{2}$$

where $u(x, t)$ is the fluid velocity at point x at time t in a one-dimensional (plane) wave. These can be regarded as ordinary differential equations along *characteristic curves* $C_\pm$ in the (x, t) plane:

$$\frac{dR_\pm}{dt} = 0 \quad \text{on } C_\pm : \frac{dx}{dt} = u \pm c, \qquad R_\pm = u \pm P, \tag{3}$$

so that the *Riemann invariant* $R_+ = u + P$ is constant along each C_+ and $R_- = u - P$ is constant along each C_-.[1-3]

If neither of $R_\pm$ has the same value everywhere, we have what is called a *compound-wave* motion, with waves traveling in both directions relative to the local flow. A flow with R_+ (or R_-) constant everywhere is called a left- (or right-) running *simple wave*.

For a perfect gas with specific heat ratio (adiabatic index) $\gamma = c_p/c_v$, we have $p/p_0 = (\rho/\rho_0)^\gamma$ and

$$c = \left(\frac{\gamma p}{\rho} \right)^{1/2}, \qquad R_\pm = u \pm \frac{2c}{\gamma - 1}. \tag{4}$$

3   CAUCHY PROBLEM

Figure 1 shows the (x, t) plane for a Cauchy (initial-value) problem in which $u(x, 0)$ and (say) $(p - p_0)(x, 0)$

are specified and vanish identically for $x < 0$, $x > L$. Imagine that the fluid occupies a long tube, along the x-axis, and is in a uniform state of rest at $t = 0$, except for the segment $0 < x < L$ in which a velocity and pressure perturbation are specified. Regions I, II, and III are undisturbed ($u = 0$, $c = c_0$); region IV is one of compound flow; regions V and VI are simple-wave regions with R_-, R_+ identically zero, respectively. The solution for compound flow must be obtained numerically through step-by-step construction of piecewise linear approximations to the $C_\pm$. For $t > t_1$ we have two independent simple-wave problems; we shall analyze here the right-running wave (region V). This has one initial condition, say for u, specified on the segment $O'L'$ and vanishing elsewhere, with time now measured from t_1. We write this initial condition as $u = f(x)$ at $t = 0$.

Riemann invariant $R_- = 0$ everywhere in V (and also in I and III) gives $c = c_0 + \frac{1}{2}(\gamma - 1)u$, and so the R_+ part of Eq. (3) gives

$$\frac{\partial u}{\partial t} + \left(c_0 + \frac{\gamma + 1}{2} u \right) \frac{\partial u}{\partial x} = 0, \tag{5}$$

or

$$\frac{du}{dt} = 0 \quad \text{on } C_+ : \qquad \frac{dx}{dt} = c_0 + \frac{\gamma + 1}{2} u. \tag{6}$$

Thus the C_+ characteristics are straight in a simple wave with constant R_-, and if a C_+ is labeled by the point $x = \xi$ through which it passes at $t = 0$, then that C_+ carries the constant signal $u = f(\xi)$. The characteristic solution

$$u = f(\xi), \qquad x = \xi + \left[c_0 + \frac{\gamma + 1}{2} f(\xi) \right] t \tag{7}$$

is therefore an implicit solution for u, given (x, t), in terms of the distribution $u = f(x)$ at $t = 0$ ($t = 0$ is the start of the simple-wave flow). We shall explain what the physical content of Eq. (7) is in Section 5.

4   WEAK NONLINEARITY: MULTIPLE SCALES

The results above are exact and hold for waves of any amplitude. By specializing to the "weakly nonlinear" approximation that characterizes nonlinear acoustics, it is however possible to develop the theory greatly through the inclusion of many further effects and mechanisms that preclude exact analysis of the Riemann type. The idea is that *locally* (i.e., over a few wave periods or wavelengths) the nonlinear term in Eq. (5) is smaller

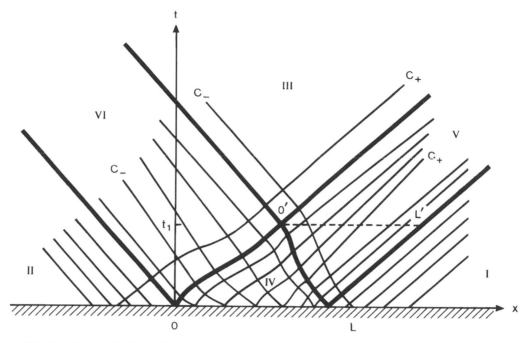

Fig. 1 The (x, t) plane for an initial-value problem in which the segment OL is initially disturbed. In regions I, II, and III the undisturbed state prevails. In region IV there is compound flow. Region V is one of simple-wave propagation (C_+ characteristics straight) to the right, region VI of simple-wave propagation (C_- characteristics straight) to the left. At time t_1 the simple-wave propagation begins with, for region V, an initially disturbed segment $O'L'$.

than the linear ones by a factor of order u/c_0, and therefore if $u/c_0 \ll 1$, the essential balance is between $\partial u/\partial t$ and $c_0 \, \partial u/\partial x$. Accordingly, in any "small" term we may replace $\partial u/\partial t$ by $-c_0 \, \partial u/\partial x$, or vice versa. Thus in nonlinear acoustics Eq. (5) may equally well be taken in the approximate form

$$\frac{\partial u}{\partial t} + c_0 \frac{\partial u}{\partial x} - \frac{\gamma + 1}{2c_0} u \frac{\partial u}{\partial t} = 0, \qquad (8)$$

or, in terms of range x and retarded time (or phase) $\tau = t - x/c_0$ as independent variables, by

$$\frac{\partial u}{\partial x} - \frac{\gamma + 1}{2c_0^2} u \frac{\partial u}{\partial \tau} = 0. \qquad (9)$$

This form is particularly suitable for *signaling problems*, which are more natural in acoustic contexts than Cauchy problems, and for such problems the simple-wave representation often applies throughout the flow with none of the complications of compound waves. If u is specified at $x = 0$, as $g(t)$ say, then we have a signaling problem, and the solution of Eq. (9) is

$$u = g(\phi), \qquad \tau = \phi - \frac{\gamma + 1}{2c_0^2} x g(\phi). \qquad (10)$$

This would give the field u in a plane-wave mode propagating in a duct $0 < x < \infty$, with $g(t)$ representing measured data at $x = 0$ or the velocity of a piston, $g(t) = \dot{X}_p(t)$. [Strictly $u = g(t)$ should be imposed at the piston face $x = X_p(t)$, not at $x = 0$; however, this particular nonlinearity is a local one of absolutely no global consequence and can be ignored if $\dot{X}_p/c_0 \ll 1$.]

The nonlinear acoustics requirement $u/c_0 \ll 1$ is well satisfied even in very strong acoustic waves. A plane wave in a duct at a sound pressure level of 150 dB has $u/c_0 \simeq 10^{-2}$, and even in this rather extreme case the local linear balance $\partial u/\partial t + c_0 \, \partial u/\partial x \simeq 0$ holds with a relative error of only about 1%.

It is natural to try to exploit the smallness of u/c_0 in perturbation theory.[4] A straightforward perturbation expansion applied to the mass and momentum equations, together with $(p/p_0) = (\rho/\rho_0)^\gamma$, is easily carried out to give

$$\frac{u(x, t)}{c_0} = \epsilon F(x - c_0 t)$$

$$- \frac{\gamma + 1}{2} \epsilon^2 c_0 t F(x - c_0 t) F'(x - c_0 t) + \cdots,$$

$$(11)$$

where we consider only waves traveling to the right and

where ϵ = (velocity amplitude at $t = 0$)/(sound speed c_0), $u(x,0) = \epsilon c_0 F(x)$. We see that the weakly nonlinear corrections to the linear prediction contain *secular terms*, proportional to the elapsed time, which for large t make the expansion invalid (and which indicate, therefore, that *linear acoustics itself fails at large t*). The ratio of successive terms in Eq. (11) is of order $\epsilon \omega t$, where ω is a typical frequency, showing that the motion cannot be a weakly nonlinear perturbation of linear acoustics for $\omega t \sim \epsilon^{-1}$ and larger.

To find a description valid at such long times, we introduce *multiple scales*, the *fast scales* x, t and a *slow scale* $T = \epsilon t$, and seek a solution of the form

$$u = \epsilon u_0(x, t, T) + \epsilon^2 u_1(x, t, T) + \cdots,$$

with similar expansions of the other variables. At lowest order we find that u_0 depends on the "fast" variables only in the combination $x \pm c_0 t$ (local linear acoustics; the relations between the leading-order velocity, pressure, and density fluctuations are also simply those of linear theory). Then inspection of the equation for u_1 shows that secular terms will not arise—and therefore $u \simeq \epsilon u_0$ will remain a good approximation even at large times, $\omega t = O(\epsilon^{-1})$—provided the "slow," or long-time, evolution is governed by

$$\frac{\partial u_0}{\partial T} + \frac{\gamma + 1}{2} u_0 \frac{\partial u_0}{\partial \chi} = 0, \qquad (12)$$

where $\chi = x - c_0 t$ and we consider only waves running to the right. This is exactly Eq. (5) of the Riemann analysis. Alternatively, taking $X = \epsilon x$ as a slow space variable and setting $u = \epsilon u_0(x, t, X) + \epsilon^2 u_1(x, t, X) + \cdots$, we find that we can have (x, t) dependence only through $\tau = t - x/c_0$ (for right-running waves) and that

$$\frac{\partial u_0}{\partial X} - \frac{\gamma + 1}{2c_0^2} u_0 \frac{\partial u_0}{\partial \tau} = 0, \qquad (13)$$

which is precisely the (approximate) equation (9). All such equations are formally equivalent for $(x, t) = O(1)$ up to $O(\epsilon^{-1})$. They show that local linear acoustics holds over the fast scales but that *on the slow scales acoustic propagation is inherently nonlinear*.

Multiple-scales analysis of this kind has been used many times recently to derive certain canonical nonlinear wave equations (Burgers, Korteweg–de Vries, nonlinear Schrödinger, and many others) on a formal basis.[4] The underlying idea here is simply that there is *local* linear behavior, that is, $\partial u/\partial t \pm c_0 \, \partial u/\partial x = 0$ and $p - p_0 = \pm \rho_0 c_0 u$ for right- (left-) running waves, together with some nonlinear modulation over the long or slow scales.

5 WAVE DISTORTION: SHOCK FORMATION

The process described by Eq. (5) is one of *pure distortion* of the initial waveform f (aside from trivial translation at speed c_0); the evolved waveform $u = f(\xi)$ contains precisely those "wavelets" (values of the signal u) that were initially present, but at time t a given wavelet has reached not the position $x = \xi + c_0 t$ of linear theory, but the position $x = \xi + c_0 t + \frac{1}{2}(\gamma + 1)ut$. Thus the waveform is subject to uniform shearing[1] at a rate $\frac{1}{2}(\gamma + 1)$ [i.e., the "excess wavelet speed" is $\frac{1}{2}(\gamma + 1)u$] of which a contribution 1 comes from convective nonlinearity, and a contribution $\frac{1}{2}(\gamma - 1)$ from the nonlinearity of the equation of state; the former dominates for air ($\gamma = 1.4$), the latter for water, where the constant equivalent to γ has a value around 7. The waveform distortion as described by Eqs. (12) and (13) is illustrated in Fig. 2. It is evident that the nonlinear processes are cumulative, and over long space and time intervals, $\omega x/c_0 = O(c_0/u_0)$, $\omega t = O(c_0/u_0)$, there is substantial distortion of the wave profile, though there is no change of the values of the signal carried by the wave.

Such waveform distortion can also be seen as transfer of energy to the higher Fourier components produced as the wave steepens. A famous solution exhibiting this explicitly is that due to Fubini; if for Eq. (5) we have a pure sinusoid $u = u_0 \sin kx$ at $t = 0$, then[2,5]

$$u(x, t) = 2u_0 \sum_{n=1}^{\infty} (-1)^{n-1} \frac{J_n(n\beta k u_0 t)}{n\beta k u_0 t}$$

$$\cdot \sin nk(x - c_0 t), \qquad (14)$$

where $\beta = (\gamma + 1)/2$. This comes from direct evaluation of the Fourier coefficients by transformation of the integration over x to one over ξ, a maneuver that requires $t < t_* = 1/\beta k u_0$. Figure 3 shows the initial variation of the Fourier amplitudes according to Eq. (14). Energy is pumped, at least initially, from the low harmonics to the higher ones. It can be shown from Eq. (14) that energy is conserved in this process, provided $t < t_*$, that is,

$$\sum_{n=1}^{\infty} \frac{J_n^2(n\beta k u_0 t)}{(n\beta k u_0 t)^2} = \text{const.}$$

This also follows directly in any continuous simple wave; thus from Eq. (5), for example, we have

$$\frac{d}{dt} \int_A^B u^2 \, dx = 0,$$

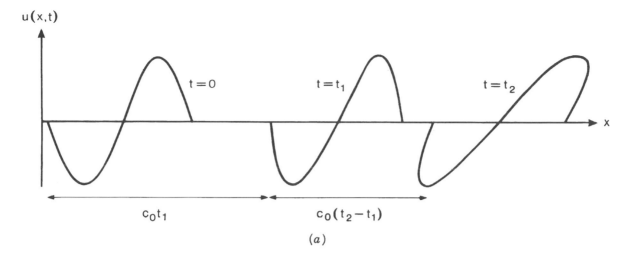

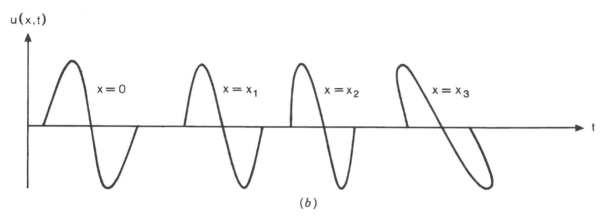

Fig. 2   Cumulative distortion by nonlinear convection, as seen in (*a*) initial-value problems, *u* as a function of *x* at increasing times t_i, and (*b*) signaling problems, *u* as a function of *t* at increasing ranges x_i from the signaling location.

where *A*, *B* correspond to points at which $u = 0$ or that differ by a wavelength of a periodic wave.

Now in all waves except those of pure expansion [pure expansion means $\partial u(x,0)/\partial x > 0$] it is evident that the wave profile will develop an infinite gradient at a finite time $t = t_*$, and that, for $t > t_*$, $u(x,t)$ will be triple valued for some range of *x*. The "shock formation" time t_* can be found by integrating the equation $dS/dt + \beta S^2 = 0$ for $S = \partial u/\partial x$ along the characteristics; or by finding the time at which the Jacobian $\partial x/\partial \xi|_t$ of the transformation to characteristic coordinates first becomes singular; or directly by considering the waveform distortion; it is

$$t_* = 1/\max_{f' < 0} |\beta f'(\xi)|. \qquad (15)$$

The triple-valued waveform for $t > t_*$ is physi-

cally unacceptable and is prevented in reality by one or more linear mechanisms [the *nonlinear* theory of Eq. (5) is exact] that may have been initially negligible but increase in importance as the wave steepens under nonlinear effects. In a medium with significant dispersion, short-wave components generated by steepening may disperse away fast enough to prevent the catastrophe at $t = t_*$, but in acoustic media the dispersion is not normally sufficient, and instead it is dissipation that precludes wave overturning. This leads (see later) to a description in which, for $t > t_*$ and small dissipation, the wave is well described by branches of the "lossless" simple-wave solution, separated by thin regions that, on the overall wave scale, can be treated as *shock discontinuities*; see Fig. 4. To locate the discontinuity in the waveform, it can be argued[1-3] that the creation and propagation of it cannot change the mass or momentum perturbations carried by the wave, and since in the weakly

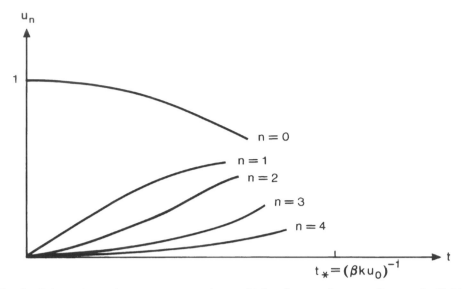

Fig. 3 Schematic to show energy pumping to higher frequencies according to the Fubini solution [Eq. (14)]. The ordinate is $u_n = 2J_n(nt/t_*)/(nt/t_*)$, with $t_* = (\beta k u_0)^{-1}$ the shock formation time.

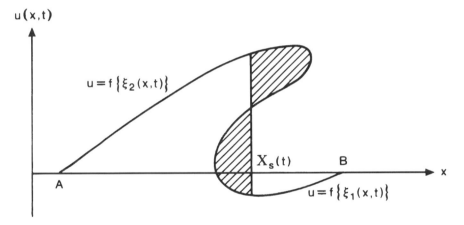

Fig. 4 Avoidance of triple-valued signals by insertion of a shock discontinuity satisfying the equal-areas rule; $u = f\{\xi_1(x,t)\}$, $u = f\{\xi_2(x,t)\}$ are branches of the simple-wave solution ahead of and behind the shock at $x = X_s(t)$.

nonlinear approximation each is proportional to $\int_A^B u\,dx$ (where $u = 0$ at A, B, or else $[A, B]$ contains an integral number of wavelengths of a periodic wave), the shock must cut off lobes of equal area[3] in the graph of u against x, as in Fig. 4 (*equal-areas* rule). The *energy* in segment $[A, B]$, proportional to $\int_A^B u^2\,dx$, is *not preserved* and in fact decreases at a rate that is then fixed regardless of the details of the dissipation mechanism. Those details simply provide the fine structure of the shock, though not all possible dissipation mechanisms are in fact able to provide a smooth transition between arbitrary signals, supplied by simple-wave theory, outside the shock (see Section 8).

Analytically, the equal-areas rule is expressed by

$$\dot{X}_s(t) = c_0 + \frac{\gamma + 1}{4}\,[f(\xi_{1s}) + f(\xi_{2s})], \qquad (16)$$

that is, as the mean of the wavelet speeds $c_0 + \frac{1}{2}(\gamma + 1)u$ on the two sides of the shock. Here $x = X_s(t)$ is the shock path, $\xi_1(x, t)$, $\xi_2(x, t)$ are branches of the characteristic relation obtained by continuous variation from the single-valued regions ahead of and behind the shock, respectively, and the suffix s indicates their limiting values as $x \to X_s(t) \pm 0$.

In general, shock formation leads to wave reflection (and compound flow), and dissipation leads to entropy production. However, the entropy change can be found[1] as

$$S_2 - S_1 \simeq c_v \left(\frac{\gamma^2 - 1}{12\gamma^2} \right) \left(\frac{p_2 - p_1}{p_1} \right)^3, \qquad (17)$$

that is, of the third order in the shock strength $(p_2 - p_1)/p_1$, whereas changes in other linear mechanical and thermodynamic variables are of first order. Thus in nonlinear acoustics one can continue to use simple-wave theory and the uniform-entropy approximation even after shock formation.

Note that for normal fluids waves steepen forward, as in Fig. 2a, and any shock formed by dynamic evolution is necessarily *compressive* ($p_2 > p_1$); a rarefaction shock created somehow would immediately become continuous. There are, however, cases in gasdynamics in which the fundamental derivative $(\partial^2 \rho^{-1}/\partial p^2)_S$ [which is $(\gamma+1)/\gamma^2 p^2 \rho$ for a perfect gas] is negative; van der Waals gases at very high pressure have this property, and actual examples include certain hydro- and fluorocarbons. Then waves steepen *backward*, forming rarefaction shocks and compression expansions, though the entropy of course always increases through any shock [because the factor analogous to $\gamma + 1$ in Eq. (17) is then negative]. Fluids in which this occurs are often called "retrograde," with negative nonlinearity.[6] It is also possible for a gas to have a vanishing coefficient of quadratic nonlinearity on some manifold in the space of thermodynamic variables, and then in the neighborhood of that manifold one needs to account for quadratic and cubic nonlinearity together, with a model equation involving "mixed nonlinearity," of the form

$$\frac{\partial u}{\partial t} + (c_0 + \epsilon_m u + \alpha_m u^2) \frac{\partial u}{\partial x} = 0,$$

where ϵ_m is small but α_m is of order unity in an appropriate sense. Then one may have compression or expansion shocks, or, indeed, shocks that change their character from one to the other as the phase of the wave changes.[6]

The procedure—whatever the type of nonlinearity—in which one uses a combination of simple-wave theory together with shock discontinuities located according to an appropriate conservation principle is called *weak-shock theory*.[3] It holds for moderate times after shock formation, though not necessarily to very large times and distances (see Section 7 below).

6 BURGERS EQUATION

If one includes molecular diffusion of heat and momentum (*thermoviscous* diffusion) as a small effect, nomi-

nally of the same order of magnitude as nonlinearity over a few wave periods, then by a variety of approaches one comes for a simple wave to the Burgers equation[1-5]

$$\frac{\partial u}{\partial t} + \left(c_0 + \frac{\gamma + 1}{2} u \right) \frac{\partial u}{\partial x} = \frac{1}{2} \delta \frac{\partial^2 u}{\partial x^2}, \qquad (18)$$

where δ was called the "diffusivity of sound" by Lighthill[1]; $\delta = \nu[\frac{4}{3} + (\gamma - 1)/\sigma]$ in terms of kinematic viscosity ν and Prandtl number σ. This equation, proposed by Burgers as a model for turbulence, is a fundamental equation of wave theory. It is *exactly integrable* by the Hopf–Cole transformation of 1950–1951 (actually known at least as early as 1906 and now seen in a wider context as a Bäcklund transformation[7] between solutions of two different partial differential equations). In Eq. (18) put

$$u = -\delta \frac{\partial}{\partial X} \ln \psi, \qquad X = x - c_0 t, \qquad (19)$$

and then we find that ψ satisfies the linear diffusion equation

$$\frac{\partial \psi}{\partial t} = \frac{1}{2} \delta \frac{\partial^2 \psi}{\partial X^2}. \qquad (20)$$

Given $u(x, 0)$ we have

$$\psi(X, 0) = \exp \left(-\frac{1}{\delta} \int^X u(X'', 0) \, dX'' \right)$$

$$= \exp \left(-\frac{1}{\delta} \int^X f(X'') \, dX'' \right),$$

in terms of which the solution of Eq. (20) is known to be

$$\psi(X, t) = \frac{1}{(2\pi \delta t)^{1/2}}$$

$$\cdot \int_{-\infty}^{+\infty} \exp \left(-\frac{|X - X'|^2}{2\delta t} \right) \psi(X', 0) \, dX',$$

$$(21)$$

and then $u(x, t)$ follows from Eq. (19) as single valued for all (x, t) for all $\delta > 0$.

The limit $\delta \to 0+$ can be examined quite generally[1-3] by standard techniques for asymptotics of integrals *pro-*

vided (x, t) are fixed as the limit is taken. Then one finds $u = f(\xi_1)$ or $u = f(\xi_2)$ [with branches of $\xi(x, t)$ defined from Eq. (7) by continuous variation from $t = 0$] with jumps between these branches at locations $X_s(t)$ satisfying the equal-areas rule, or the differential equation (16). Thus one recovers weak-shock theory [though not necessarily if x, t become large, e.g., $(x, t) = O(\delta^{-1})$ as $\delta \to 0$], and in addition one can obtain a description of the internal shock structure over a region of thickness $O(\delta)$ around $x = X_s(t)$. This can be written,[1] again with $\beta = \frac{1}{2}(\gamma + 1)$, as

$$
u = \frac{U_2 + U_1}{2} - \frac{U_2 - U_1}{2}
$$

$$
\tanh\left\{\frac{\beta}{2\delta}(U_2 - U_1)(x - X_s(t))\right\},
$$

$$
\dot{X}_s = c_0 + \beta\left(\frac{U_2 + U_1}{2}\right), \qquad (22)
$$

and for normal fluids describes a smooth symmetric compressive transition from $U_1 = f(\xi_{1s})$ ahead to $U_2 = f(\xi_{2s})$ behind, with $U_2 > U_1$; note that U_1, U_2, $\dot{X}_s$ are not necessarily constant but can be taken so over the space and time scales $O(\delta)$ associated with passage of the shock. In the case where they *are* constant—a single shock wave separating semi-infinite media with uniform velocities U_1 and U_2—Eq. (22) is the exact traveling-wave solution of the Burgers equation first found by G. I. Taylor in 1910.

The solution, Eqs. (19)–(21), of the Burgers equation can also be used to examine shock–shock interaction.[1,3] In the Hopf–Cole variable ψ the N-shock solution is given by

$$
\psi = \sum_{i=1}^{N+1} \exp\left\{-\frac{U_i}{\delta}(X - \overline{X}_i) + \frac{1}{2}\frac{U_i^2}{\delta}t\right\}, \qquad (23)
$$

with the ordering $U_1 < U_2 < \cdots < U_{N+1}$. At $t = 0$ the signal $u(x, 0)$ comprises N shock transitions, the one from U_j to U_{j+1} being located at $(U_{j+1}\overline{X}_{j+1} - U_j\overline{X}_j)/(U_{j+1} - U_j), j = 1, \ldots, N$, each of which then propagates at its individual speed $c_0 + \frac{1}{4}(\gamma + 1)(U_j + U_{j+1})$. If the jth and $(j + 1)$th shocks meet at t_j, then for $t > t_j$ there is [from Eq. (23)] no range of x for which u is close to U_{j+1}, and instead there is a single shock from U_j to U_{j+2}. After a finite time all possible *shock mergings* of this kind have taken place, and there remains only a single shock, from U_1 to U_{N+1}. All traces of the previous history of the wave are destroyed (with errors exponentially small, $O(e^{-1/\delta})$). As before, the levels U_j

need not be constant, and if they are not, the shock paths are curved. The shock merging represents a spectral cascade both to lower wavenumbers (the waveform simplification means energy transfer to larger scales) and to higher ones (the few shocks remaining have increased strength and the smallest scale present is reduced to $O[\delta/(U_{N+1} - U_1)]$).

Propagation of the individual shocks and calculation of the various merging times and subsequent propagation of the new shocks can all be carried out in the (x, t) plane using just Eq. (16). If the shock structure is needed, it can be taken as given by Eq. (22) unless the times involved are very long, $O(\delta^{-1})$.

7 SIMPLE EXAMPLES: LONG-TIME BEHAVIOR

1. *Monopolar Pulse.* Suppose $f(x)$ is one signed, vanishing for $|x| > L$, and with $\int_{-\infty}^{+\infty} f(x)\, dx = UL$, say. The weak-shock solution depends on details of f but, for large t, has the same form as if f were initially triangular[3]; thus $u = 0$ for $x < c_0 t - L$, $u = (x - c_0 t + L)/\beta t$ for $c_0 t - L < x < X_s(t)$, and $u = 0$ for $x > X_s(t)$, where $X_s(t) \sim c_0 t + (2UL\beta t)^{1/2}$ and again $\beta = (\gamma + 1)/2$. Transition between $U_1 = 0$ and $U_2 = (2UL/\beta t)^{1/2}$ is given by Eq. (22), and examination of the exact Hopf–Cole solution shows that the weak-shock description continues to hold for arbitrarily large x, t. The reason is that the *Reynolds number* (ratio of nonlinear terms to diffusive terms on the overall wave scale) has a typical value $(\int_{-\infty}^{+\infty} u\, dx)/\delta = \text{const} = UL/\delta$, assumed large. Therefore the wave never becomes linear; the shock width is proportional to $\delta t^{1/2}$ and is small and fixed on the overall scale $t^{1/2}$. The ramp remains described by a nonlinear simple-wave solution, the shock by the (nonlinear and diffusive parts of) the Burgers equation.

2. *N-Wave.* The model case has the ideal form of Fig. 5. Weak-shock theory indicates[1] independent propagation of the positive and negative parts of the wave as in part 1 above, and the Hopf–Cole solution confirms this, for fixed (x, t) as $\delta \to 0$. However, the long-time behavior is very different. Diffusion of mass and momentum across the node $u = 0$ takes place, leading to a reduction of area of the positive or negative parts separately, hence of the wave Reynolds number and thus of the importance of nonlinearity. Ultimately all trace of a narrow shock disappears, and the wave becomes smooth and dies in *old age* under linear mechanisms,[1,8,9]

$$
u \propto \left(\frac{X}{t^{1/2}}\right)\exp\left(-\frac{X^2}{2\delta t}\right), \qquad X = x - c_0 t. \qquad (24)
$$

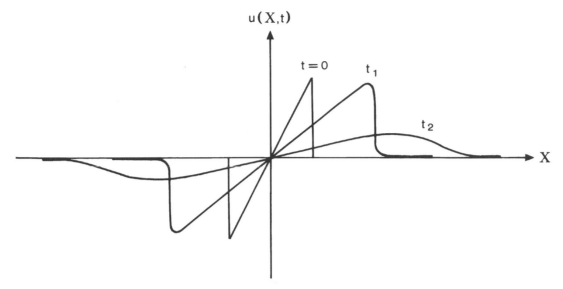

Fig. 5   Propagation of an N-wave, with $X = x - c_0 t$; at $t = t_1$, the wave is fully nonlinear, with overall length scale increasing as $t^{1/2}$ and with thin Taylor shocks of thickness proportional to $\delta t^{1/2}$; at $t = t_2$ the wave has decayed into old age, with propagation governed by linear mechanisms alone.

Weak-shock theory thus fails in this case for very large t (actually because the shocks get increasingly retarded by diffusive effects[1]—"shock displacement due to diffusion"—while keeping their weak-shock theory strength, thus reducing the Reynolds number as the propagation continues until it is so low that diffusion is important over the whole wave profile).

3. *Periodic Wave.* The familiar example is $u = u_0 \sin \omega x / c_0$ at $t = 0$. Shocks form at $\frac{1}{2}(\gamma + 1)\omega t_* = c_0 / u_0$ and by symmetry are permanently centered on $\omega X / c_0 = (2n + 1)\pi$. For $t_* \ll t \ll c_0^2 / \omega^2 \delta$ weak-shock holds, and the wave quickly develops a sawtooth form,[5] with Fourier representation

$$u = \frac{2c_0}{\beta \omega t} \sum_{n=1}^{\infty} \frac{(-1)^{n-1}}{n} \sin\left[n \frac{\omega}{c_0} (x - c_0 t) \right], \quad (25)$$

or

$$u = \frac{X}{\beta t} = \frac{2}{\gamma + 1} \left(\frac{x}{t} - c_0 \right), \qquad -\pi < \frac{\omega X}{c_0} < +\pi. \tag{26}$$

Significant in these is the fact that u is independent of u_0, a characteristic nonlinear effect known as *amplitude saturation*. The shocks have Taylor structure, with $\partial u / \partial x = O(\delta^{-1})$ in them; as a result the energy dissipation rate, proportional to $\delta(\partial u / \partial x)^2$ per unit length of

$0x$, is concentrated at $O(\delta^{-1})$ in the shocks and when integrated over the shocks gives a total dissipation rate independent of δ and precisely equal to the rate of loss of energy implicit in the weak-shock solution. Analytically, if $E = \langle u^2 \rangle$, averaged over a cycle in x, then it follows from Eq. (18) that $dE/dt - -\delta \langle (\partial u / \partial x)^2 \rangle$. Evaluating this for $\delta \to 0$ from the shock contribution alone gives $dE/dt = -(2\pi^2/3\beta^2)(c_0^2/\omega^2 t^3)$, identical to what is obtained from the weak-shock theory, which from Eq. (26) gives $E = (\pi^2/3\beta^2)(c_0^2/\omega^2 t^2)$.

The shock thickness is $O(\delta/(U_2 - U_1)) = O(\delta t / u_0)$ and for $t = O(\delta^{-1})$ is comparable with the wavelength. At this phase of the evolution the effective Reynolds number is unity, and nonlinearity and diffusion are comparable everywhere (and comparable with $\partial u / \partial t$). From the Hopf–Cole solution one can then obtain an asymptotic description valid over the whole wave:

$$u(x, t) \sim \frac{\omega \delta}{\beta c_0} \sum_{n=1}^{\infty} \frac{(-1)^{n-1} \sin\left[n(\omega/c_0)(x - c_0 t) \right]}{\sinh\left(n\omega^2 \delta t / 2c_0^2 \right)}. \tag{27}$$

This is associated with the name of Fay[5] (who worked directly with approximations to the governing equations and not with the Burgers equation itself); amazingly it turns out to be an exact solution of the Burgers equation, though it does not satisfy the initial condition posed here. For $\omega t \gg c_0^2 / \omega \delta$ only the first harmonic survives, and

$$u \sim \frac{4}{\gamma + 1}\left(\frac{\omega\delta}{c_0}\right)\exp\left(-\frac{\delta\omega^2}{2c_0^2}t\right)\sin\left[\frac{\omega}{c_0}(x - c_0t)\right],$$

(28)

showing linear old-age decay with amplitude saturation. Note that the time to old age is just the time at which diffusive decay would have taken over in the absence of nonlinear effects; under the conditions implied here, wave steepening, shock formation, and shock thickening all take place before that time and scramble all trace of the initial amplitude u_0. Weak-shock theory again fails in this case [now for $\omega t = O(c_0^2/\omega\delta)$], the essential mechanism being shock thickening.

8   RELAXATION AND DISPERSION

In many common situations, thermoviscous effects are often dominated over the initial wave scales by dissipation and dispersion associated with relaxation processes. In single-phase polyatomic gases the relaxation process may be the inability of one or more of the vibrational or rotational modes to instantaneously accept its full energy (as specified by the equipartition theorem); in a multiphase medium, such as a gas containing fine dust particles, the processes of relaxation involve the transfer of macroscopic heat and momentum between phases. Such a relaxing medium (with one relaxing mode for simplicity) may be characterized by a relaxation time T (with an assumed exponential relaxation rate $\exp(-t/T)$) and two sound speeds: c_0, the low-frequency (or equilibrium) speed, and $c_\infty > c_0$, the infinite-frequency (frozen) speed. Linear plane waves propagate in such a medium as $\exp[-\alpha x - i\omega(t - x/c)]$, where the attenuation coefficient $\alpha(\omega)$ and phase speed $c(\omega)$ are given by[2,9]

$$\alpha(\omega) = \frac{\omega^2T(c_\infty - c_0)}{c_0^2 + c_\infty^2\omega^2T^2}, \qquad c(\omega) = \frac{c_0^2 + c_\infty^2\omega^2T^2}{c_0 + c_\infty\omega^2T^2};$$

(29)

c increases monotonically from c_0 to c_∞, α increases monotonically from 0 to $(c_\infty - c_0)/c_\infty^2T$, as ω increases from 0 to ∞.

The significant case for nonlinear acoustics is that in which the dispersion $(c_\infty - c_0)/c_0$ is small and comparable with the nonlinearity (Mach) parameter u_0/c_0; however, we want to deal with *finite rate* processes, so no restriction is placed on the typical frequency parameter ωT. Then with inclusion also of thermoviscosity, one can derive the model equation[2,9,10]

$$\left(1 + T\frac{\partial}{\partial t}\right)\left[\frac{\partial u}{\partial t} + \left(c_0 + \frac{\gamma + 1}{2}u\right)\frac{\partial u}{\partial x} - \frac{1}{2}\delta\frac{\partial^2 u}{\partial x^2}\right]$$

$$= (c_\infty - c_0)c_0T\frac{\partial^2 u}{\partial x^2}.$$

(30)

For waves with $\omega T \ll 1$ this reduces to the Burgers equation with a diffusivity of sound $\Delta = \delta + \delta_B$, $\delta_B = 2(c_\infty - c_0)c_0T$; relaxation processes on low-frequency waves are equivalent to a *bulk diffusivity* δ_B (often much larger than the molecular diffusivity δ). All the results for the Burgers equation then go through; in particular, relaxation processes prevent wave overturning, and the "shocks" produced have thickness $O(\delta_B/(U_2 - U_1))$, much larger than those structured by thermoviscous diffusion. This is important for sonic boom propagation in a humid atmosphere. For $\omega T \gg 1$ it can be shown from Eq. (30) that

$$\frac{\partial u}{\partial t} + \left(c_\infty + \frac{\gamma + 1}{2}u\right)\frac{\partial u}{\partial x} + \lambda u - \frac{1}{2}\delta\frac{\partial^2 u}{\partial x^2} = 0, \quad (31)$$

with $\lambda = (c_\infty - c_0)/c_0T$. This equation, sometimes known as the Varley–Rogers equation when $\delta = 0$, is not integrable and no exact solutions are known except when $\delta = 0$. Then, by characteristics, one has

$$u = f(\xi)e^{-\lambda t}, \qquad x = \xi + c_\infty t + \frac{\gamma + 1}{2}f(\xi)\left(\frac{1 - e^{-\lambda t}}{\lambda}\right),$$

(32)

and in the absence of genuine molecular diffusion, wave overturning *will* occur in finite time—despite the relaxation processes—provided $|f'(\xi)| > 2\lambda/(\gamma + 1)$ for some ξ with $f'(\xi) < 0$. Whether, for general ωT and $\delta = 0$, wave overturning will occur in finite time (necessitating the reintroduction of δ) can only be settled by numerical work, though that is frustrated in direct finite-difference or spectral schemes by "numerical viscosity." An intrinsic coordinate approach that gets around this has recently been developed.[11]

The steady traveling waves $u = F(\chi)$, $\chi = x - Vt$, are the only solutions to Eq. (30) that are known (even then we need $\delta = 0$). If we take $u = 0$ ahead of the wave, $u = U$ behind, we find $V = c_0 + \frac{1}{4}(\gamma + 1)U$ (the shock speed for a medium in equilibrium). Then the integration gives

$$\frac{W}{U}\ln F + \left(1 - \frac{W}{U}\right)\ln(U - F) = \frac{\chi - \chi_0}{2VT}.$$

(33)

Here W is defined exactly as $\frac{1}{2}U - 2c_0(c_\infty - c_0)/(\gamma + 1)V$, and for small dispersion we can put $V = c_0$ in the second term, so that $W = 2(V - c_\infty)/(\gamma + 1)$. If $V < c_\infty$, that is, $W < 0$, Eq. (33) gives a smooth transition between 0 and U, described as a *fully dispersed* relaxing shock. If $V \ll c_\infty$, the solution becomes the Taylor solution (22) with $U_1 = 0$, $U_2 = U$, and the diffusivity δ replaced by the bulk diffusivity δ_B. For stronger waves relative to the dispersion $(c_\infty - c_0)/c_0$ the wave profile is markedly asymmetric, and has a kink on the low-pressure side when $W = 0$. For $W > 0$ the curve starting from U doubles back on itself, there is no continuous transition structured by relaxation processes alone, and one must insert a discontinuity (or, equivalently, a thin Taylor shock structured by molecular diffusivity) within the outer relaxing shock.[1,2] Equation (33) is used from $u = U$ forward to the point X^* at which $V - u^* = c_\infty$, and at this point the discontinuity (or Taylor shock) takes the velocity down to zero. The Taylor shock speed of course turns out to be given by Eq. (16) again. In such a case we have a *partly dispersed* relaxing shock,[1] with an embedded diffusive shock; the reason for the discontinuity is that otherwise the lowest (and therefore linear) wavelets would have a propagation speed, into fluid at rest, greater than the highest available sound speed c_∞. Partly dispersed shocks, with a double structure, are common in air, though it is difficult to conduct repeatable experiments in atmospheric air because the relaxation time T (and even more the bulk diffusivity $\delta_B \sim T^2$) is extraordinarily sensitive to the humidity.

9 NONLINEAR WAVES IN TUBES

In tubes of finite length waves propagate in both directions, independently in the linear limit, except for coupling at the ends. For weakly nonlinear waves it is also found that the simple waves propagating in each direction do not interact in the body of the fluid, but only through the boundary condition coupling at the ends. It is the *self-distortion* of a simple wave that leads to cumulative long-time effects, not interaction with a wave traveling in the opposite direction. Reflection at a closed end just requires $u = 0$ there; at an open end one may assume $p = p_0$ (though there is often a need to model vortex-shedding processes that take place when shocks reflect from an open end[12]). Analytically these reflection processes are described by functional equations for the simple-wave shapes. There is an extensive literature on such functional mappings, and it is known, for example, that in some parameter ranges very complicated behavior can develop.

Dissipation by diffusive effects in the Stokes bound-ary layers on the tube walls is usually much greater than the thermoviscous dissipation in the uniform central core. The model equation for a right-running simple wave in a circular tube, with both effects included, was derived by Chester[13] in the form

$$\frac{\partial u}{\partial t} + \left(c_0 + \frac{\gamma + 1}{2}\, u\right)\frac{\partial u}{\partial x} - \frac{1}{2}\,\delta\,\frac{\partial^2 u}{\partial x^2}$$
$$= c_0 \kappa \int_0^\infty \frac{\partial u}{\partial x}(x, t - \xi)\xi^{-1/2}\, d\xi, \qquad (34)$$

where u is velocity averaged over a section and $\kappa = (1/R)(\nu/\pi)^{1/2}[1 + (\gamma - 1)/\sigma^{1/2}]$ for a tube of radius R. As with relaxation, although the tube wall effects are generally larger than the mainstream diffusion, neglect of the latter often leads to partly dispersed shocks of the relaxing-media kind, a thin inner "mainstream diffusion" subshock of Taylor type being embedded in a highly asymmetric outer shock whose much larger thickness is controlled by κ. Limited analytical information is available on solutions of Eq. (34), together with asymptotic and numerical studies.[14,15]

Detailed confirmation of the applicability of Eq. (34) for waves in a very long tube was obtained by Pestorius and Blackstock.[16] They took a complicated signal u, as specified at one end in an experiment, and allowed it to propagate a small distance according to nonlinear lossless theory. The evolved signal was then Fourier analyzed, and phase shifts and amplitude decay factors were applied to the spectral components to account for linear attenuation and dispersion effects in the boundary layer over the small time step. Then the modified Fourier components were synthesized into a time waveform, which was then allowed to propagate a further time step with the same split procedure. Extremely close agreement with experiment was obtained, even over very long ranges. Such methods are probably the best that will ever be available. Equations such as (34) have been shown to be nonintegrable, and there is almost no chance of general analytical progress.

10 NONLINEAR PROPAGATION IN BUBBLY LIQUID

If α_0 is the volume concentration of small gas bubbles in liquid, then it is well known[17] that the low-frequency sound speed in the bubbly suspension is given by $c_0^2 = \gamma p_0/\rho_0 \alpha_0(1 - \alpha_0)$ (except for α_0 very small or very close to 1), with $p/\rho^\gamma = $ const the equation of state for the air in the bubbles and ρ_0 the mixture density. Even for quite small α_0 the speed c_0 may be not only less than the water sound speed c_w but also significantly less than the air sound speed; values as low as 30 ms^{-1} were measured and compared quite favorably with this prediction as long ago as 1954. This implies that a given velocity u_0, generated by a piston source say, is associated with nonlinear effects greater by a factor c_w/c_0 when the motion is generated in bubbly liquid than in pure water (provided frequencies are sufficiently low) and shocks form much earlier, at shorter ranges from a source. Detailed analysis[17] shows that the lossless simple-wave equation takes the form

$$\frac{\partial u}{\partial t} + \left(c_0 + \frac{1}{\alpha_0}u\right)\frac{\partial u}{\partial x} = 0. \tag{35}$$

Shock formation then takes place at time $t'_* = \beta\alpha_0 t_*$, where t_* is the pure-water time (for the same initial f) given in Eq. (15).(Again α_0 must not be too small.) For $t > t'_*$ the motion might be assumed to proceed as in a single-phase medium, with the emergence of thin shocks that might in the end thicken and lead to old-age decay.

The microstructure of the bubbly medium does, how-ever, have a well-defined intrinsic time scale—the bubble volume oscillation period—when there is a sharp clustering of bubble sizes about a single value R_0. This leads to strong frequency dispersion. If $\omega_0 = (3p_0/\rho_w R_0^2)^{1/2}$ is the (Minnaert) isolated bubble resonance frequency, then the phase speed $c_p(\omega)$ of linear waves of frequency ω is given by[17]

$$\frac{1}{c_p^2(\omega)} = \frac{(1-\alpha_0)^2}{c_w^2} + \frac{1}{c_0^2[(1-\omega^2/\omega_0^2) - 2i\Delta(\omega/\omega_0)]}, \tag{36}$$

where we include an ad hoc damping factor Δ in the bubble response. Figure 6 shows Re c_p as a function of ω. If $\Delta = 0$ there is a forbidden frequency band in which the motion is purely reactive, with an infinite sound speed as one approaches the upper limit of this band from above. Dissipation (and variation of bubble size) blurs this picture, but all the essentials have been seen in experiment, including decrease of c_p from the (low) value c_0 almost to zero at $\omega = \omega_0$ and very large values of $c_p(> 3000$ms$^{-1})$ at higher frequencies, with approach to $c_w/(1-\alpha_0)$ at the highest frequencies. In general, the dis-

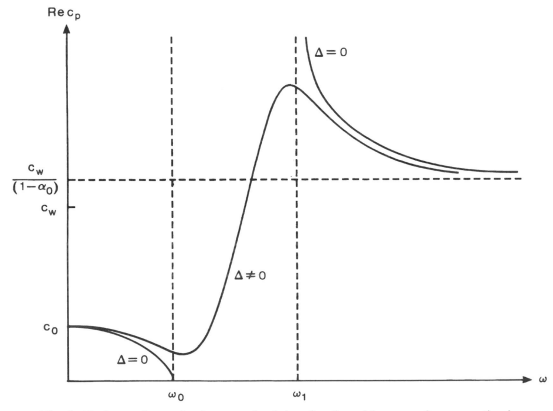

Fig. 6 Real part of acoustic phase speed $c_p(\omega)$ as function of frequency for propagation in a bubbly liquid (schematic only); Δ is the dissipation coefficient. If $\Delta = 0$, then Re $c_p = 0$ throughout the forbidden frequency band $\omega_0 < \omega < \omega_1$.

person is only weak (which is necessary for strong nonlinear interaction between all the harmonics generated by nonlinearity) at low and high frequencies. At low frequencies the dispersion is cubic, $\omega = c_0 k - c_0^3 k^3 / 2\omega_0^2$, and the combined effects of nonlinearity and dispersion are represented by the Korteweg–de Vries (KdV) equation[17]

$$\frac{\partial u}{\partial t} + \left(c_0 + \frac{1}{\alpha_0} u \right) \frac{\partial u}{\partial x} + \frac{c_0^3}{2\omega_0^2} \frac{\partial^3 u}{\partial x^3} = 0. \quad (37)$$

This famous equation is the prototype for integrable equations with *soliton* solutions. These are traveling-wave solutions of sech^2 profile (bell shaped), necessarily representing positive-pressure pulses, traveling at an excess speed proportional to amplitude and with width inversely proportional to (amplitude)$^{1/2}$. A new analytical method (inverse spectral, or scattering, transform) has been devised for such integrable equations and through a sequence of three linear problems provides the solution for $u(x, t)$ given $u(x, 0)$. Analysis shows that in general $u(x, t)$ will contain a definite number [related to properties of $u(x, 0)$] of solitons that propagate with fixed amplitude and shape, together with a decaying oscillatory wave train ("radiation"). The solitons preserve their identity after interaction with each other and with other waves and thus represent permanent large-scale features that dominate the wave field at large time. For a simple introduction to soliton theory, see Drazin and Johnson.[18]

Predictions of soliton theory have been compared quite favorably with experiments on pressure pulses in bubbly liquid. Dissipation cannot always be ignored and typically contributes a Burgers term $-\frac{1}{2}\delta \, \partial^2 u/\partial x^2$ to the left side of Eq. (37). Then one cannot have pulses in which the pressure returns behind the wave to its value ahead but rather shock waves in which the pressure is increased behind. If the dissipation coefficient δ is appropriately large relative to the dispersion coefficient $c_0^3/2\omega_0^2$, the shock transition is monotonic; if not, the signal approaches the value ahead monotonically but the value behind through a decaying oscillation.[19]

At frequencies far above the bubble resonance the dispersion takes the form $\omega = c_w k + (\omega_0^2 c_w/2c_0^2)k^{-1}$, and the appropriate nonlinear wave equation is the nonlinear Klein–Gordon equation

$$\frac{\partial}{\partial x} \left[\frac{\partial u}{\partial t} + \left(c_w + \frac{\Gamma + 1}{2} u \frac{\partial u}{\partial x} \right) \right] - \frac{\omega_0^2 c_w}{2c_0^2} u = 0, \quad (38)$$

where Γ plays the role for water of γ for air. Here the dispersion is not generally strong enough to prevent wave overturning, and shocks form in finite time (no wave

overturning can occur for KdV). Little is known about the subsequent motion (except in the case of periodic waves[19]; the only bounded traveling-wave solutions of Eq. (38) are periodic), but the nonlinear and dispersive effects in this case are much less dramatic than in the low-frequency case.

11 GEOMETRIC AND MATERIAL NONUNIFORMITY

Consider now the propagation of quasi-plane waves in narrow horns (no higher order modes). It is natural in such cases (which include freely diverging or converging cylindrical and spherical waves) to formulate signaling, rather than initial-value, problems. Let s be a (range) coordinate along the horn, $\tau = t - (s - s_0)/c_0$ retarded time relative to a signaling location s_0 for a simple wave propagating "outward," $u(s, \tau)$ velocity averaged across a section of area $A(s)$. Then if the medium parameters ρ, c are uniform, energy conservation in linear nondissipative theory requires $\partial(A^{1/2}u)/\partial s = 0$, or $\partial u/\partial s + \frac{1}{2}u(d/ds) \ln A(s) = 0$. This is also the "transport equation" for high-frequency waves propagating along a ray tube; $\omega L/c_0 \gg 1$ is required, where L is the length scale for variations in $A(s)$. Now if nonlinearity and diffusion are included as of the same nominal magnitudes $[O(\omega L/c_0)^{-1}]$ as the geometric area variations, one has

$$\frac{\partial u}{\partial s} - \frac{\gamma + 1}{2c_0^2} u \frac{\partial u}{\partial \tau} + \frac{1}{2} u \frac{d}{ds} \ln A = \frac{\delta}{2c_0^3} \frac{\partial^2 u}{\partial \tau^2}, \quad (39)$$

called a *generalized Burgers* equation. It has been proved[8] that none of these can be exactly integrated, except when $\delta = 0$ or when $A'(s) = 0$ (Burgers equation), and few exact solutions are known. The transformations

$$v = \left[\frac{A(s)}{A(s_0)} \right]^{1/2} u, \qquad \zeta = \int_{s_0}^{s} \left[\frac{A(s_0)}{A(s)} \right]^{1/2} ds \quad (40)$$

put Eq. (39) in the form of a Burgers equation with range-dependent diffusivity,

$$\frac{\partial v}{\partial \zeta} - \frac{\gamma + 1}{2c_0^2} v \frac{\partial v}{\partial \tau} = \frac{\delta}{2c_0^3} G(\zeta) \frac{\partial^2 v}{\partial \tau^2}, \quad (41)$$

where $G(\zeta) = A^{1/2}[s(\zeta)]/A^{1/2}(s_0)$.

Included here are the cases of freely diverging cylindrical and spherical waves, for which $s = r$, and

$$A(r) \sim r, \qquad\qquad A(r) \sim r^2,$$

$$\zeta = 2r_0^{1/2}(r^{1/2} - r_0^{1/2}), \qquad \zeta = r_0 \ln \frac{r}{r_0}, \qquad (42)$$

$$G(\zeta) = 1 + \frac{\zeta}{2r_0}, \qquad\qquad \exp \frac{\zeta}{r_0},$$

respectively.

For $\delta = 0$, Eq. (41) is integrable by characteristics and the question of shock formation is covered by the plane-wave criteria and the range transformation. For example, if $u = u_0 \sin \omega t$ at $r = r_0$, then shocks form at ranges

$$r_* = r_0 + \frac{2c_0^2}{(\gamma + 1)u_0\omega}, \qquad r_0 \left(1 + \frac{c_0^2}{(r+1)u_0\omega r_0} \right)^2,$$

$$r_0 \exp \left[\frac{2c_0^2}{(\gamma + 1)u_0\omega r_0} \right]$$

in the plane, cylindrical, and spherical cases, respectively.

The transformations also give well-known decay laws for practically important problems. For example, if the signal is an N-wave, then the velocity jump Δu at the shocks decays as $r^{-1/2}$ in plane flow (cf. Section 7), and therefore as $r^{-3/4}$ and $r^{-1}(\ln r)^{-1/2}$ for cylindrical and spherical N-waves, while for the time duration $T(r)$ of the whole N-wave at any range r one has $T(r) \sim r^{1/2}, r^{1/4}, (\ln r)^{1/2}$ for plane, cylindrical, and spherical N-waves. Cylindrical geometry is relevant to the sonic boom of a supersonic aircraft; the flight axis is timelike and the propagation is along ray paths on the surface of a cone, with $A(r) \sim r$.

In the weak-shock theory treatment of Eq. (39), shock discontinuities are inserted, as before, at equal-areas locations in the graph of u (or v) against τ, and the analysis of the lossless simple-wave parts is as for plane flow, after transformation via Eq. (40). For the diffusive structure of the shocks, this is effectively given by the Taylor solution [Eq. (22)] in which we simply replace δ by the equivalent diffusivity $\delta G(\zeta)$ at range r, at least if r is not large as $\delta \to 0$. As with plane waves, however, weak-shock theory often fails at large ranges; indeed, for the N-wave and sinusoidal signals, weak-shock theory holds indefinitely *only* for the exponentially converging horn.[8] For all others, one of a number of nonuniformities takes place at large range, indicating a change of the wave properties and dynamics. Shock thickening is one and is typical for horns diverging no more rapidly than cylindrically. For more rapid divergence, there is typically first a *local* nonuniformity, a breakdown in Taylor shock structure. Conditions outside the shocks are changing so rapidly (because of the rapid area variations) that

the local nonlinear steepening–diffusion balance can no longer be preserved and the shocks become evolutionary [i.e., their structure can only be described by solutions of Eq. (39) in its full form, without neglect of $\partial u/\partial s$]. Then they decay to an error function form that thickens with increasing range (but more slowly than the Taylor shock would). Then a second *global* nonuniformity follows, in which the wave becomes smooth everywhere and dies in old age.[9] Spherically diverging waves typically follow this scenario, but there are several other possible scenarios[8] for the long-range behavior after weak-shock theory has ceased to be valid. The elucidation of these scenarios requires asymptotic and numerical study[20] in the absence of analytical solutions.

If the medium parameters are not uniform, then one must first solve the corresponding linear problem by ray methods to define the horn or ray tube geometry. Take, as a simple example, the case of an isothermal atmosphere, $c(\mathbf{x}) = c_0$, constant, but with density stratification $\rho(z) = \rho_0 \exp(-z/H)$ in the vertical direction z. Then the rays are straight and for a spherically symmetric excitation are just radial lines through the origin, $\theta = $ const ($\theta = $ angle from vertical). Energy conservation (in the absence of winds) implies $\partial(\rho^{1/2}A^{1/2}u)/\partial r = 0$, where $A \sim r^2$, $z = r \cos \theta$, and the retarded time is simply $\tau = t - (r - r_0)/c_0$. Generalizing this by the inclusion of nonlinear terms, we have, as the transport equation of *nonlinear geometrical acoustics*,

$$\frac{\partial u}{\partial r} - \frac{\gamma + 1}{2c_0^2} u \frac{\partial u}{\partial \tau} + \frac{u}{r} + \frac{u \cos \theta}{2H} = 0. \qquad (43)$$

There is a different nonlinear problem to be solved along each ray. Shock formation [easily examined by transformation of Eq. (43) to the form $\partial v/\partial \zeta - v \, \partial v/\partial \tau = 0$] is facilitated by the decreasing density for waves propagating along rays above the horizontal and hindered for waves propagating in directions below the horizontal to the extent that shocks may not form in finite range. If $c(\mathbf{x})$ is not uniform, an equation corresponding to Eq. (43) can be derived by consideration not of just the invariance of energy but also of Snell's law for the phase (see Rogers and Gardner[21] for an application to the propagation of N-waves in the lower atmosphere [$c(\mathbf{x}) = $ const] and the upper atmosphere [$c(\mathbf{x}) \sim z$, giving rays that are arcs of circles]).

12   NONLINEARITY AND DIFFRACTION

The formal basis for nonlinear geometric acoustics (Section 11) rests on the idea that if the geometric (or material) nonuniformity parameter $(\omega L/c_0)^{-1}$ and the nonlinearity parameter u_0/c_0 are comparable, then the field has a phase structure given by local linear ray theory, with modulation along the rays expressed by a generalized Burgers equation. However, if a relatively rapid variation *along* the wavefronts is imposed at a boundary or allowed to develop by diffraction in two or three dimensions, then one-dimensional propagation along rays fails. In the simplest case, think of a beam propagating along the x-axis, with (modulation distance)/(wavelength) $= \epsilon^{-1}$, say, and suppose (scale for lateral y variations)/(wavelength) $= O(\epsilon^{-1/7})$. Here ϵ is a small parameter, comparable with the particle Mach number u_0/c_0. Then the balance between nonlinear modulation and lateral diffraction turns out to be governed by the dZK (dissipative Zabolotskaya–Khokhlov) equation[2,22]

$$\frac{\partial}{\partial x}\left[\frac{\partial u}{\partial t} + \left(c_0 + \frac{\gamma+1}{2}u\right)\frac{\partial u}{\partial x} - \frac{1}{2}\delta\frac{\partial^2 u}{\partial x^2}\right]$$
$$+ \frac{c_0}{2}\frac{\partial^2 u}{\partial y^2} = 0, \tag{44}$$

where we have two space dimensions and u is the x-velocity.

A great deal of work, mainly numerical, has been done on this equation, much of it by Russian scientists; see especially the monographs.[2,10,22,23] The essential feature of Eq. (44) is that its solutions display *nonlinear wavefront turning*, a property seen in numerical simulations but necessarily absent from nonlinear geometrical acoustics. A transformation between solutions of dZK has recently been found,[24] and the infinitesimal version of it can be used to demonstrate the wavefront turning due to amplitude variations along wavefronts.

A key question is whether, in the nondissipative limit $\delta = 0$, diffraction is able to prevent the caustic singularity formation produced in nonlinear geometrical acoustics by convergence of rays to a point. The answer is that it cannot in all cases. Classes of exact solutions to ZK [$\delta = 0$ in Eq. (44)] have now been found from the transformation mentioned above,[24] and these can be used to show that nonlinear caustic formation may indeed occur in ZK theory. The amplitude at the focus will actually be limited either by dissipative effects or by full-wave-mechanics effects (dZK itself may not be applicable in the caustic region) in a way not yet understood.

13 FURTHER COMMENTS

Several items should at least be mentioned here. First, an important practical and scientific field within nonlinear acoustics is concerned with *statistical problems*; for example, suppose the propagation is governed by a generalized Burgers equation, and the input signal $u(s = s_0, t) = u_0(t)$ can only be described statistically. Then one would like to know the statistics of the evolved signal [e.g., the mean-square value $\langle u^2(s)\rangle$ as a function of range s and the frequency spectrum $\Pi(\omega, s)$ of $\langle u^2\rangle$ at each s] in terms of the input statistics. Even in the weak-shock theory description this is a formidable problem, because one has to compute the statistics with allowance for shock formation and multiple-shock coalescence in each realization of the random field. Applications would include, for example, the propagation of high-intensity noise from jet aircraft engines. Other important problems include the propagation of an input pure-tone signal contaminated by broadband random noise and the possibility of enhancement or suppression of one of these components by the other. There is again much Russian work in this field, and the reader is referred to an important monograph[25] devoted entirely to statistical nonlinear acoustics.

Second, not all nonlinearity in acoustic media is of the quadratic-convective kind. For transverse and torsional waves in perfect elastic solids the nonlinearity is cubic-convective, $u^2\,\partial u/\partial x$, while for elastic media with microstructure (pores, voids, cracks) the stress–strain relation may be linear but with a different Young's modulus for compression and for expansion. For chemically active media with Arrhenius rate laws, transcendental nonlinearity $\exp u$ arises. Naturally, quite new phenomena arise in all cases. For cubic nonlinearity (the so-called modified Burgers equation) the striking feature is the formation in finite time, from general initial conditions, of *sonic shocks*.[26] These have a propagation speed precisely equal to the wavelet or characteristic speed ahead of the shock, and the signals on either side of the shock are related in the ratio 2 to -1. Such sonic shocks cannot occur in ordinary gas dynamics but are generic for cubic and all higher nonlinearity. For media with bilinear stress–strain laws, the weak-shock description can be analyzed and shows that a sinusoidal wave would develop shocks whose strength vanishes identically (and therefore the whole signal vanishes identically!) in finite time.[10]

For propagation of pressure waves in reacting media the dominant nonlinearity is in the chemistry initially and leads to explosive singularities in finite time unless fuel depletion is accounted for. Then a very complicated interaction takes place between the thermal field and the gasdynamic velocity field, leading to the launching of shock, deflagration, and ultimately detonation waves.

REFERENCES

1. M. J. Lighthill "Viscosity Effects in Sound Waves of Finite Amplitude," in G. K. Batchelor and R. M. Davies (Eds.), *Surveys in Mechanics*, Cambridge University Press, U.K., 1956, pp. 249–350.

2. O. V. Rudenko and S. I. Soluyan, *Theoretical Foundations of Nonlinear Acoustics*, Consultants Bureau, Plenum, New York, 1977.

3. G. B. Whitham, *Linear and Nonlinear Waves*, Wiley, New York, 1974.

4. J. K. Engelbrecht, V. E. Fridman, and E. N. Pelinovsky, *Nonlinear Evolution Equations*, Longman/Wiley, New York, 1988.

5. R. T. Beyer (Ed.), *Nonlinear Acoustics in Fluids*, Van Nostrand Reinhold, New York, 1984, (a collection of fundamental papers in nonlinear acoustics).

6. A. Kluwick, "Small-Amplitude Finite-Rate Waves in Fluids Having Both Positive and Negative Nonlinearity," in A. Kluwick (Ed.), *Nonlinear Waves in Real Fluids*, Springer-Verlag, Berlin, 1991, pp. 1–43.

7. C. Rogers and W. F. Shadwick, *Bäcklund Transformations and Their Applications*, Academic, New York, 1982.

8. J. J. C. Nimmo and D. G. Crighton, "Geometrical and Diffusive Effects in Nonlinear Acoustic Propagation over Long Ranges," *Phil. Trans. Roy. Soc. Lond.* Vol. A, 320, 1986, pp. 1–35.

9. D. G. Crighton and J. F. Scott, "Asymptotic Solutions of Model Equations in Nonlinear Acoustics," *Phil. Trans. Roy. Soc. Lond. A*, Vol. 292, 1979, pp. 101–134.

10. K. A. Naugol'nykh and L. A. Ostrovsky, *Nonlinear Wave Processes in Acoustics*, Cambridge University Press, U.K., 1996.

11. P. W. Hammerton and D. G. Crighton, "Overturning of Nonlinear Acoustic Waves. Part 1. A General Method. Part 2. Relaxing Gas Dynamics," *J. Fluid Mech.*, Vol. 252, 1979, pp. 585–599, 601–615.

12. J. H. M. Disselhorst and L. van Wijngaarden, "Flow in the Exit of Open Pipes During Acoustic Resonance," *J. Fluid Mech.*, Vol. 99, 1980, pp. 293–319.

13. W. Chester, "Resonant Oscillations in Closed Tubes," *J. Fluid Mech.*, Vol. 18, 1964, pp. 44–64.

14. N. Sugimoto, "Burgers Equation with a Fractional Derivative; Hereditary Effects on Nonlinear Acoustic Waves," *J. Fluid Mech.*, Vol. 225, 1991, pp. 631–653.

15. J. J. Keller, "Resonant Oscillations in Closed Tubes: The Solution of Chester's Equation," *J. Fluid Mech.*, Vol. 77, 1976, pp. 279–304.

16. F. M. Pestorius and D. T. Blackstock, "Propagation of Finite-Amplitude Noise," in L. Bjørnø (Ed.), *Finite-Amplitude Wave Effects in Fluids*, IPC Press, Guildford, 1974, pp. 24–29.

17. L. van Wijngaarden, "One-Dimensional Flow of Liquids Containing Small Gas Bubbles," *Ann. Rev. Fluid Mech.*, Vol. 4, 1972, pp. 369–396.

18. P. G. Drazin and R. S. Johnson, *Solitons: An Introduction*, Cambridge University Press, U.K. 1989.

19. D. G. Crighton, "Nonlinear Acoustics of Bubbly Liquids," in A. Kluwick (Ed.), *Nonlinear Waves in Real Fluids*, Springer-Verlag, Berlin, 1991, pp. 45–68.

20. P. W. Hammerton and D. G. Crighton, "Old-Age Behaviour of Cylindrical and Spherical Nonlinear Waves; Numerical and Asymptotic Results," *Proc. Roy. Soc. Lond. A*, Vol. 422, 1989, pp. 387–405.

21. P. H. Rogers and J. H. Gardner, "Propagation of Sonic Booms in the Thermosphere," *J. Acoust. Soc. Am.*, Vol. 67, 1980, pp. 78–91.

22. N. S. Bakhvalov, Ya. M. Zhileikin, and E. A. Zabolotskaya, *Nonlinear Theory of Sound Beams*, American Institute of Physics, New York, 1987.

23. B. K. Novikov, O. V. Rudenko, and V. I. Timoshenko, *Nonlinear Underwater Acoustics*, American Institute of Physics, Acoustical Society of America, New York, 1987.

24. A. T. Cates and D. G. Crighton, "Nonlinear Diffraction and Caustic Formation," *Proc. Roy. Soc. Lond. A*, Vol. 430, 1990, pp. 69–88.

25. S. N. Gurbatov, A. N. Malakhov, and A. I. Saichev, *Nonlinear Random Waves and Turbulence in Nondispersive Media: Waves, Rays, Particles*, Manchester University Press, U.K., 1991.

26. I. P. Lee-Bapty and D. G. Crighton, "Nonlinear Wave Motion Governed by the Modified Burgers Equation," *Phil. Trans. Roy. Soc. Lond. A*, Vol. 323, 1987, pp. 173–209.

20

PARAMETERS OF NONLINEARITY OF ACOUSTIC MEDIA

E. CARR EVERBACH

1 INTRODUCTION

Nonlinearity in acoustic media contributes to distortion of the wave shape during propagation and can be described by various parameters. Nonlinearity parameters include B/A, the ratio of the second to the first term in a Taylor series expansion of the pressure as a function of density, the coefficient of nonlinearity β, which takes self-convection into account, and the acoustic Mach number M, which differentiates the linear from the nonlinear regimes. During acoustic wave propagation, energy initially carried at the fundamental frequency is moved into the higher harmonics as the wave steepens due to the cumulative effects of nonlinearity. Additional heating of the medium and loss of amplitude at the fundamental frequency are negative consequences, but nonlinearity can also be used to infer properties of the medium or generate useful acoustic sources. Nonlinearity parameters may be measured by quantifying the relationship among thermodynamic properties of a medium or by measuring the waveform distortion and inferring the underlying nonlinearities. Mixture laws have been developed to aid in the calculation of nonlinearity parameters of mixtures of media whose component nonlinearities are known.

2 ACOUSTIC MACH NUMBER

The acoustic Mach number is defined via[1] $M = u_{max}/c_0$, where u_{max} and c_0 are the maximum particle velocity and the small-signal acoustic phase speed in the propagation medium, respectively. For plane-wave propagation in a gas, $M = p_{max}/\gamma P_0$, where p_{max} is the maximum deviation of pressure from the equilibrium pressure P_0 and γ is the ratio of specific heats C_p/C_v for the gas. For linear acoustics, $M \ll 1$.

3 ACOUSTIC NONLINEARITY PARAMETER B/A

For small deviations from P_0, the pressure P may be expanded in a Taylor series in terms of the density ρ and specific entropy s. For changes that induce temperature gradients small enough so that appreciable heat flow does not occur during a fraction of an acoustic period, the system is said to be adiabatic. With the further requirement that the changes are thermodynamically reversible, the entropy does not change from its equilibrium value s_0, the system is said to be isentropic, and the Taylor series becomes

$$p = P - P_0 = A \left[\frac{\rho - \rho_0}{\rho_0} \right] + \frac{B}{2!} \left[\frac{\rho - \rho_0}{\rho_0} \right]^2$$
$$+ \frac{C}{3!} \left[\frac{\rho - \rho_0}{\rho_0} \right]^3 + \cdots, \qquad (1)$$

where p is the excess or acoustic pressure and $(\rho - \rho_0)/\rho_0$ is the condensation or fractional change in density of the propagation medium. Also, for small deviations from equilibrium[2]

Encyclopedia of Acoustics, edited by Malcolm J. Crocker
ISBN 0-471-80465-7 © 1997 John Wiley & Sons, Inc.

$$A = \rho_0 \left[\left(\frac{\partial P}{\partial \rho} \right)_s \right]_{\rho = \rho_0} = \rho_0 c_0^2,$$

$$B = \rho_0^2 \left[\left(\frac{\partial^2 P}{\partial \rho^2} \right)_s \right]_{\rho = \rho_0},$$

$$C = \rho_0^3 \left[\left(\frac{\partial^3 P}{\partial \rho^3} \right)_s \right]_{\rho = \rho_0}, \tag{2}$$

and so on, to higher orders. When the condensation is infinitesimal, the higher order terms are negligible and the acoustic waves will propagate at the (constant) speed c_0. For finite-amplitude waves, the higher order terms B, C, ... become increasingly more important as the amplitude increases. Alternatively, for waves of a given amplitude, materials with larger values of B, C, ... will cause the local sound speed to differ increasingly from c_0. The acoustic nonlinearity parameter B/A is the ratio of the quadratic coefficient in this expression to the linear coefficient[3] and therefore provides a measure of the degree to which the local sound speed deviates from the small-signal (linear) case. Further manipulations yield equivalent expressions:

$$\frac{B}{A} = 2\rho_0 c_0 \left(\frac{\partial c}{\partial p} \right)_s \tag{3}$$

or

$$\frac{B}{A} = 2\rho_0 c_0 \left[\left(\frac{\partial c}{\partial p} \right)_T \right]_{\rho = \rho_0} + \frac{2 c_0 T \kappa_e}{C_p} \left[\left(\frac{\partial c}{\partial T} \right)_p \right]_{\rho = \rho_0} \tag{4}$$

or

$$\frac{B}{A} = \left[\frac{\partial (1/k)}{\partial P} \right]_s - 1, \tag{5}$$

where c is the local sound speed; T is the temperature in Kelvin; and κ_e, C_p, and k are the volume coefficient of thermal expansion, the specific heat at constant pressure, and the adiabatic compressibility (the reciprocal of stiffness) for the propagation medium, respectively.

Equation (3) shows that B/A is proportional to the change of sound speed for a change of pressure, provided that the pressure change is so rapid and smooth that isentropic conditions continue to hold. Equation (4) shows that B/A may be written as the sum of isothermal

and isobaric components[4]; however, the isothermal component of B/A for most materials typically dominates.[5] Equation (5) suggests that materials whose stiffness changes greatly with changes in pressure, such as bubbly liquids,[6,7] will have a large value of the acoustic nonlinearity parameter. Other physical characteristics described by B/A include the geometric packing of molecules[8] and the form of the potential energy function that governs the forces between adjacent molecules in a material.[9] The connection between molecular structure and B/A has been exploited by several investigators[10,11] to determine the relative concentrations of "bound water" and "free water" in water–alcohol mixtures or the contribution of molecular subgroups to the overall nonlinearity of larger molecules in solution.[12] Other authors[13] have suggested a connection between the level of macroscopic structure in soft-organ tissue and its B/A value. Fluids in which B/A can be zero or negative are called "retrograde" fluids and have unusual thermodynamic properties.[14]

4 COEFFICIENT OF NONLINEARITY β

The acoustic nonlinearity parameter describes the steepening of acoustic waves as they propagate through a material.[15] A given point on a traveling acoustic waveform will propagate at the local sound speed dx/dt, given by

$$\frac{dx}{dt} = c_0 + \beta u, \tag{6}$$

where the so-called coefficient of nonlinearity $\beta \equiv 1 + B/2A$. Equation (6) shows that wave steepening has two distinct causes that are assumed to act independently: nonlinearities inherent in the material's properties, which are described by B/A, and those due to convection. The convective term arises from purely kinematic considerations[16] and would therefore exist even if no material nonlinearities were present, that is, for the case of $B/A = 0$. Equation (6) also shows that nonlinear effects are cumulative. The amount of distortion also depends upon the time or distance the wave travels.

5 SPECIFIC INCREMENT OF B/A

Because the increase in B/A value with solute concentration χ of solutions of biologic materials is often linear, the specific increment of B/A with concentration, $\Delta(B/A)/\chi$, can be used to determine the relative contribution of the solute to the total nonlinearity of the solution.[17] The specific increment is defined by differentiating Eq. (4):

$$\frac{\Delta(B/A)}{\chi} = 2\rho_0 c_0 \left\{ \frac{1}{\chi}\Delta\left(\frac{\partial c}{\partial P}\right)_{T_0} \right.$$

$$+ ([c]+[\rho])\left(\frac{\partial c}{\partial P}\right)_{T_0}$$

$$+ \frac{\kappa_e T}{\rho_0 C_p}\left\{ \frac{1}{\chi}\Delta\left(\frac{\partial c}{\partial T}\right)_{P_0} \right.$$

$$\left.\left. + ([c]+[\kappa_e]-[C_p])\left(\frac{\partial c}{\partial T}\right)_{P_0}\right\}\right\}, \quad (7)$$

where $[c] = \Delta c/\chi c_0$, $[\rho] = \Delta\rho/\chi\rho_0$, $[\kappa_e] = \Delta\kappa_e/\chi\kappa_{e0}$, $[C_p] = C_{p0}/\chi C_p$, the subscript 0 denotes the solvent, and Δ means the difference between solution and solvent for the corresponding parameter.

6 HIGHER ORDER NONLINEARITY PARAMETERS

The higher order parameters C/A, D/A,... become important at large acoustic pressure amplitudes or in extremely nonlinear media.[18] From Eqs. (2) and (3) the third-order constant C/A may be written[19]

$$\frac{C}{A} = \frac{3}{2}\left(\frac{B}{A}\right)^2 + \frac{1}{k}\frac{\partial[B/A]}{\partial P}. \quad (8)$$

For most materials, the first term in Eq. (8) exceeds the second term by several orders of magnitude.[20,21]

7 MIXTURE LAWS FOR NONLINEARITY PARAMETERS

If the nonlinearity parameters of each of an n-component mixture of mutually immiscible materials are known, it is possible to derive[22] expressions for the corresponding nonlinear parameters of the mixture as a whole:

$$k^2\beta = \sum_{i=1}^{n}\frac{\rho Y_i k_i^2 \beta_i}{\rho_i} = \sum_{i=1}^{n} X_i k_i^2 \beta_i, \quad (9)$$

where Y_i and X_i are the mass and volume fractions of component i, respectively. In Eq. (9), k is the adiabatic

compressibility and ρ the density of the mixture as a whole, given by

$$k = \sum_{i=1}^{n} k_i X_i \quad \text{and} \quad \rho = \sum_{i=1}^{n} \rho_i X_i,$$

respectively, where the subscripted quantities refer to the component properties. Similar expressions may be developed[23] for C/A, D/A,....

8 METHODS OF MEASURING NONLINEARITY PARAMETERS

Since the coefficients B and A are thermodynamic properties of the propagation medium, they can either be derived from the equation of state of the material or be measured empirically if the equation of state is not known. The B/A value for a gas obeying the perfect-gas law $P/P_0 = (\rho/\rho_0)^\gamma$ is $B/A = \gamma - 1$, and for a perfect gas, $C/A = (\gamma - 1)(\gamma - 2)$.

For materials for which no analytical equation of state exists, B/A must be measured.[24] The finite-amplitude method[25] is a technique in which the distortion of the wave as it propagates is measured via the growth of the second harmonic, and the B/A value is inferred from Eq. (6). Another technique is the thermodynamic method,[26] which relies upon changes in sound speed that accompany changes in ambient pressure and temperature via Eq. (4). The isentropic phase method makes use of Eq. (3): Sound speed is measured during a sufficiently rapid and smooth pressure change that the system is considered thermodynamically reversible.[27–31] This method has the advantage that a detailed knowledge of the thermodynamic properties of the propagation material and of the acoustic field of the source transducer is unnecessary. The precision with which B/A can be measured is about 10% for the finite-amplitude method and 5% for the thermodynamic method. Early isentropic phase methods were capable of measurement precisions of 4%, but methodological improvements[32,33] have increased these to within 1%. These more precise techniques have been necessitated by the use of the acoustic nonlinearity parameter in predictive models of tissue composition[34,35] and nonlinear acoustic propagation in biologic tissues. Other methods for measuring B/A include optical methods,[36] parametric arrays,[37,38] cavity resonance systems,[39–41] and methods involving the measurement of volumetric effects.[42]

Tables 1 and 2 show values of B/A for various materials reported in the literature.

TABLE 1 Published B/A Values of Nonbiologic Materials at 1 atm

Substance	T (°C)	B/A	Reference
Distilled water	0	4.2	43
	20	5.0	43
		4.985 ± 0.063	23
	25	5.11 ± 0.20	29
	26	5.1	31
	30	5.31	5
		5.18 ± 0.033	32
		5.280 ± 0.021	23
	40	5.4	43
	60	5.7	43
	80	6.1	43
	100 (liquid)	6.1	43
Sea water (3.5% NaCl)	20	5.25	43
Glycerol (4% in H_2O)	20	8.77	30
	25	8.58 ± 0.34	29
		8.84	30
	30	9.0	43
		9.4	5
		9.08	30
Methanol	20	9.42	19
	30	9.64	19
Ethanol	0	10.42	19
	20	10.52	19
	40	10.60	19
n-Propanol	0	10.47	19
	20	10.69	19
	40	10.73	19
n-Butanol	0	10.71	19
	20	10.69	19
	40	10.75	19
Benzyl alcohol	30	10.19	19
	50	9.97	19
Ethylene glycol	25	9.88 ± 0.40	29
	26	9.6	31
	30	9.7	43
		9.93	5
		9.88 ± 0.035	32
Acetone	20	9.23	19
	40	9.51	19
Benzene	20	9.0	43
		8.4	44
	25	6.5	3
	40	8.5	3
Chlorobenzene	30	9.33	19
Diethylamine	30	10.30	19
Ethyl formate	30	9.8	43
Heptane	30	10.0	43
	40	10.05	45
Hexane	25	9.81 ± 0.39	29
	30	9.9	43
	40	10.39	45
Methyl acetate	30	9.7	43
Cyclohexane	30	10.1	43
Nitrobenzene	30	9.9	43
1,2-DHCP	30	11.8	43

TABLE 1 (*Continued*)

Substance	T (°C)	B/A	Reference
Carbon bisulfide	10	6.4	3
	25	6.2	3
	40	6.1	3
Chloroform	25	8.2	3
Carbon tetrachloride	10	8.1	3
	25	8.7	3
		7.85 ± 0.31	29
	40	9.3	3
Toluene	20	5.6	3
	25	7.9	3
	30	8.929	46
Pentane	30	9.87	45
Octane	40	9.75	45
Aqueous *t*-butanol			
(60% by volume)	20	11.5	47
Bismuth	318	7.1	43
Indium	160	4.6	43
Mercury	30	7.8	43
Methyl iodide	30	8.2	43
Potassium	100	2.9	43
Sodium	110	2.7	43
Sulfur	121	9.5	43
Tin	240	4.4	43
Monatomic gas	20	0.67	43
Diatomic gas	20	0.40	43
Liquid argon	−183.16	5.67	48
Liquid nitrogen	−195.76	6.6	24
Liquid helium	−271.38	4.5	24
Saturated marine sediments	20	19	49
	20	11.78	50

TABLE 2 **Published B/A Values of Biologic Materials**

Substance	T (°C)	B/A	Reference
Isotonic saline	20	5.540 ± 0.032	23
	30	5.559 ± 0.018	23
Bovine serum albumin			
20 g/100 ml H_2O	25	6.23 ± 0.25	29
38.8 g/100 ml H_2O	30	6.68	5
Dextrose (25%)	30	5.96	5
		6.11 ± 0.4	32
Dextrose (30%)	26	5.9	31
Dextran T150 (24%)	30	6.05	5
Dextran T2000 (26%)	30	6.03	5
Sucrose (30%)	25	5.50	31
D-Glucose anhydrous	25	5.85	51
Bovine liver	23	7.5–8.0	5
	30	7.23–8.9	5
	30	6.88	52
Bovine brain	30	7.6	5
Bovine heart	30	6.8–7.4	5
Bovine milk	26	5.1	53
Bovine whole blood	26	5.5	53

TABLE 2 *(Continued)*

Substance	T (°C)	B/A	Reference
Chicken fat	30	11.270 ± 0.090	23
Porcine liver	26	6.9	31
Porcine heart	26	6.8	31
Porcine kidney	26	6.3	31
Porcine spleen	26	6.3	31
Porcine brain	26	6.7–7.0	51
Porcine muscle	26	6.5–6.6	51
	30	7.5–8.1	5
Porcine tongue	26	6.8	31
Porcine fat	26	9.5–10.9	51
	30	10.9–11.3	5
Porcine whole blood	26	5.8	53
Human liver	30	6.54	30
Human breast fat	22	9.206	30
	30	9.909	30
	37	9.633	30
Human multiple myeloma	22	5.603	30
	30	5.796	30
	37	6.178	30
Corn oil	20	10.666 ± 0.074	23
	30	10.574 ± 0.026	23
Castor oil	20	11.270 ± 0.044	23
	30	11.006 ± 0.051	23
Olive oil	20	11.136 ± 0.042	23
	30	11.066 ± 0.641	23
Peanut oil	20	10.911 ± 0.065	23
	30	10.680 ± 0.038	23
Safflower oil	20	11.610 ± 0.102	23
	30	11.161 ± 0.083	23
Cod liver oil	20	10.958 ± 0.022	23
	30	10.867 ± 0.029	23
Tung oil	20	11.278 ± 0.031	23
	30	11.064 ± 0.041	23
Lamp oil	20	11.156 ± 0.053	23
	30	10.918 ± 0.104	23
Mineral oil	20	11.331 ± 0.020	23
	30	11.497 ± 0.038	23
Pump oil	20	11.791 ± 0.026	23
	30	11.451 ± 0.018	23
Silicone oil	20	11.381 ± 0.052	23
	30	11.461 ± 0.017	23

REFERENCES

1. R. T. Beyer and S. V. Letcher, *Physical Ultrasonics*, Pure & Appl. Physics Ser., Academic, New York, 1969.

2. L. Bjørnø, Nonlinear acoustics, in R. W. B. Stephens and H. G. Leventhall (Eds.), *Acoustics and Vibration Progress*, Vol. 2, Chapman & Hall, London, 1976.

3. R. T. Beyer, "Parameter of Nonlinearity in Fluids," *J. Acoust. Soc. Am.*, Vol. 32, 1960, pp. 719–721.

4. I. Rudnick, "On the Attenuation of Finite Amplitude Waves in a Liquid," *J. Acoust. Soc. Am.*, Vol. 30, 1958, pp. 564–567.

5. W. K. Law, L. A. Frizzell, and F. Dunn, "Determination of the Nonlinearity Parameter *B/A* of Biological Media," *Ultrasound Med. Biol.*, Vol. 11, No. 2, 1985, pp. 307–318.

6. Y. A. Kobelev and L. A. Ostrovsky, "Nonlinear Acoustic Phenomena Due to Bubble Drift in a Gas–Liquid Mixture," *J. Acoust. Soc. Am.*, Vol. 85, No. 2, 1989, pp. 621–629.

7. J. Wu and Z. Zhu, "Measurements of the Effective Nonlinearity Parameter B/A of Water Containing Trapped Cylindrical Bubbles," *J. Acoust. Soc. Am.*, Vol. 89, No. 6, 1991, pp. 2634–2639.

8. H. Endo, "Prediction of the Nonlinearity Parameter of a Liquid from the Percus-Yevick Equation," *J. Acoust. Soc. Am.*, Vol. 83, No. 6, 1988, pp. 2043–2046.

9. B. Hartmann, "Potential Energy Effects on the Sound Speed in Liquids," *J. Acoust. Soc. Am.*, Vol. 65, No. 6, 1979, pp. 1392–1396.

10. K. Yoshizumi, T. Sato, and N. Ichida, "A Physiochemical Evaluation of the Nonlinear Parameter B/A for Media Predominantly Composed of Water," *J. Acoust. Soc. Am.*, Vol. 82, No. 1, 1987, pp. 302–305.

11. C. M. Sehgal, B. Porter, and J. F. Greenleaf, "Relationship between Acoustic Nonlinearity and the Bound and Unbound States of Water," *IEEE Ultrason. Symp.*, Vol. 2, 1985, pp. 883–886.

12. A. P. Sarvazyan, D. P. Kharakoz, and P. Hemmes, "Ultrasonic Investigation of the pH-Dependent Solute–Solvent Interactions in Aqueous Solutions of Amino Acids and Proteins," *J. Phys. Chem.*, Vol. 83, No. 13, 1979, pp. 1796–1799.

13. J. Zhang, M. S. Kuhlenschmidt, and F. Dunn, "Influences of Structural Factors of Biological Media on the Acoustic Nonlinearity Parameter B/A," *J. Acoust. Soc. Am.*, Vol. 89, No. 1, 1991, pp. 80–91.

14. M. S. Cramer, and R. Sen, "Shock Formation in Fluids Having Embedded Regions of Negative Nonlinearity," *Phys. Fluids*, Vol. 29, No. 7, 1986, pp. 2181–2191.

15. D. T. Blackstock, "Nonlinear Acoustics (Theoretical)," *AIP Handbook of Physics*, McGraw-Hill, New York, 1972.

16. M. F. Hamilton and D. T. Blackstock, "On the Coefficient of Nonlinearity β in Nonlinear Acoustics," *J. Acoust. Soc. Am.*, Vol. 83, No. 1, 1988, pp. 74–77.

17. A. P. Sarvazyan, T. V. Chalikian, and F. Dunn, "Acoustic Nonlinearity Parameter B/A of Aqueous Solutions of Some Amino Acids and Proteins," *J. Acoust. Soc. Am.*, Vol. 88, No. 3, 1990, pp. 1555–1561.

18. Y. A. Basin and V. M. Kryachko, "Experimental Study of the Propagation of Compression and Expansion Pulses in a Sound Beam in a Highly Nonlinear Medium," *Sov. Phys. Acoust.*, Vol. 31, No. 4, 1985, pp. 255–257.

19. A. B. Coppens, R. T. Beyer, M. B. Seiden, J. Donohue, F. Guepin, R. H. Hodson, and C. Townsend, "Parameter of Nonlinearity in Fluids. II," *J. Acoust. Soc. Am.*, Vol. 38, 1965, pp. 797–804. In Equation A4 there is an omission of the squared factor of $(\partial c / \partial p)$.

20. L. Bjørnø and K. Black, "Higher-order Acoustic Nonlinearity Parameters of Fluids," in U. Nigul and J. Engelbrecht (Eds.), *Nonlinear Deformation Waves*, Springer, Berlin, 1983, pp. 355–361.

21. K. P. Thakur, "Non-linearity Acoustic Parameter in Higher Alkanes," *Acustica*, Vol. 39, 1978, pp. 270–272.

22. E. C. Everbach, Z. Zhu, P. Jiang, B. T. Chu, and R. E. Apfel, "A Corrected Mixture Law for B/A," *J. Acoust. Soc. Am.*, Vol. 89, No. 1, 1991, pp. 446–447.

23. E. C. Everbach, Tissue Composition Determination via Measurement of the Acoustic Nonlinearity Parameter B/A, Ph.D. Dissertation, Yale University, New Haven, CT (1989), p. 66.

24. H. A. Kashkooli, P. J. Dolan, Jr., and C. W. Smith, "Measurement of the Acoustic Nonlinearity Parameter in Water, Methanol, Liquid Nitrogen, and Liquid Helium-II by Two Different Methods: A Comparison," *J. Acoust. Soc. Am.*, Vol. 82, No. 6, 1987, pp. 2086–2089.

25. See, for example, W. N. Cobb, "Measurement of the Acoustic Nonlinearity Parameter of Biological Media, Ph.D. Dissertation, Yale University, New Haven, CT (1982).

26. See, for example, W. K. Law, Measurement of the Nonlinearity Parameter B/A in Biological Materials Using the Finite Amplitude and Thermodynamic Method, Ph.D. Dissertation, University of Illinois at Urbana-Champaign (1984).

27. C. Kammoun, J. Emery, and P. Alias, Determination of the Acoustic Parameter of Nonlinearity in Liquids at Very Low Pressure, in the Seventh Intern. Symp. on Nonlinear Acoust., Virginia Polytechnics Inst. and State Univ., 1976, pp. 146–149.

28. J. Emery, S. Gasse, C. Dugué, "Coefficient de nonlinéarité acoustique dans les melanges eau-methanol et eau-ethanol," *J. Phys.*, Vol. 11, No. 40, 1979, pp. 231–234.

29. Z. Zhu, M. S. Roos, W. N. Cobb, and K. Jensen, "Determination of the Acoustic nonlinearity Parameter B/A from Phase Measurements," *J. Acoust. Soc. Am.*, Vol. 74, No. 5, 1983, pp. 1518–1521.

30. C. M. Sehgal, R. C. Bahn, and J. F. Greenleaf, "Measurement of the Acoustic Nonlinearity Parameter B/A in Human Tissues by a Thermodynamic Method," *J. Acoust. Soc. Am.*, Vol. 76, No. 4, 1984, pp. 1023–1029.

31. X. Gong, Z. Zhu, T. Shi, and J. Huang, "Determination of the Acoustic Nonlinearity Parameter in Biological Media Using FAIS and ITD Methods," *J. Acoust. Soc. Am.*, Vol. 86, No. 1, 1989, pp. 1–5.

32. J. Zhang and F. Dunn, "A Small Volume Thermodynamic System for B/A Measurement," *J. Acoust. Soc. Am.*, Vol. 89, No. 1, 1991, pp. 73–79.

33. E. C. Everbach and R. E. Apfel, "An Interferometric Technique for B/A Measurement," *J. Acoust. Soc. Am.*, Vol. 98, No. 6, 1995, pp. 3428–3438.

34. R. E. Apfel, "Prediction of Tissue Composition from Ultrasonic Measurements and Mixture Rules," *J. Acoust. Soc. Am.*, Vol. 79, No. 1, 1986, pp. 148–152.

35. C. M. Sehgal, G. M. Brown, R. C. Bahn, and J. F. Greenleaf, "Measurement and Use of Acoustic Nonlinearity and Sound Speed to Estimate Composition of Excised Livers," *Ultrasound in Med. & Biol.*, Vol. 12, No. 11, 1986, pp. 865–874.

36. L. Adler and E. A. Hiedmann, "Determination of the

Nonlinearity Parameter B/A for Water and m-xylene," *J. Acoust. Soc. Am.*, Vol. 34, No. 4, 1962, pp. 410–412.

37. P. J. Westervelt, "Parametric Acoustic Array," *J. Acoust. Soc. Am.*, Vol. 35, 1963, pp. 535–537.

38. Y. Nakagawa, M. Nakagawa, M. Yoneyama, and M. Kikuchi, "Nonlinear Parameter Imaging Computed Tomography by Parametric Array," *Proc. IEEE Ultrasonics Symp.*, IEEE, New York, 1985, pp. 673–676.

39. D. T. Blackstock, "Finite-Amplitude Motion of a Piston in a Shallow, Fluid-filled Cavity," *J. Acoust. Soc. Am.*, Vol. 34, No. 6, 1962, pp. 792–802.

40. F. Eggers and T. Funck, "Ultrasonic Measurements with Milliliter Liquid Samples in the 0.5–100 MHz Range," *Rev. Sci. Instrum.*, Vol. 44, 1973, pp. 969–978.

41. A. P. Sarvazyan, "Development of Methods of Precise Ultrasonic Measurements in Small Volumes of Liquids," *Ultrasonics*, Vol. 20, 1982, pp. 151–154.

42. H. M. Merklinger, High Intensity Effects in the Nonlinear Acoustic Parametric End-Fire Array, Ph.D. Thesis, Appendix A, University of Birmingham, Birmingham, England (1971).

43. R. T. Beyer, *Nonlinear Acoustics*, Naval Ships Systems Command, Washington, D.C., 1974, Table 3–1.

44. O. Nomoto, "Nonlinear Parameter of the 'Rao Liquid'," *J. Phys. Soc. Jpn.*, Vol. 21, 1966, pp. 569–571.

45. K. L. Narayana and K. M. Swamy, "Acoustic Nonlinear Parameter (B/A) in n-pentane," *Acustica*, Vol. 49, 1981, pp. 336–339.

46. S. K. Kor and U. S. Tandon, "Scattering of Sound by Sound From Beyer's (B/A) Parameters," *Acustica*, Vol. 28, 1973, pp. 129–130.

47. C. M. Sehgal, B. R. Porter, and J. F. Greenleaf, "Ultrasonic Nonlinear Parameters and Sound Speed of Alcohol–Water Mixtures," *J. Acoust. Soc. Am.*, Vol. 79, No. 2, 1986, pp. 566–570.

48. K. M. Swamy, K. L. Narayana, and P. S. Swamy, "A Study of (B/A) in Liquefied Gases as a Function of Temperature and Pressure from Ultrasonic Velocity Measurements," *Acustica*, Vol. 28, 1973, pp. 129–130.

49. J. M. Hovem, "The Nonlinearity Parameter of Saturated Marine Sediments," *J. Acoust. Soc. Am.*, Vol. 66, No. 5, 1979, pp. 1463–1467.

50. L. Bjørnø, "Finite-Amplitude Wave Propagation through Water-saturated Marine Sediments," Acustica, Vol. 38, No. 4, 1977, pp. 196–200.

51. Z. Zhu, X. Gong, and J. Huang, Measurement of the Acoustic Nonlinearity Parameter B/A in Biological Media by Improved Thermodynamic Method, Proc. China-Japan Joint Conf. on Ultrasonics, May 11–14, Nanjing, 1987.

52. W. K. Law, L. A. Frizzell, and F. Dunn, "Comparison of Thermodynamic and Finite Amplitude Methods of B/A Measurement in Biological Materials," *J. Acoust. Soc. Am.*, Vol. 74, No. 4, 1983, pp. 1295–1297.

53. X. Gong, R. Feng, C. Zhu, and T. Shi, "Ultrasonic Investigation of the Nonlinearity Parameter B/A in Biological Media," *J. Acoust. Soc. Am.*, Vol. 76, No. 3, 1984, pp. 949–950.

21

FINITE-AMPLITUDE WAVES IN SOLIDS

MACK A. BREAZEALE

1   INTRODUCTION

The theory describing propagation of a finite-amplitude sound wave in air has been given.[1] An experimental proof of the validity of the approach also has been given.[2] The application of the theory to liquids has been made,[3] and the nonlinear equation for wave propagation in fluids has the same form as that for an ideal gas. Different thermodynamic quantities appear in the nonlinear equations, however. The same basic approach as used for fluids[4] has been used for solids.[5] The propagation of ultrasonic waves of finite amplitude in a crystal of cubic symmetry can be described by a nonlinear differential equation similar to that used for fluids. An approximation makes the equations identical. Using this approximation, one can show how nonlinear distortion of an initially sinusoidal ultrasonic wave takes place and how nonlinear distortion can be measured with the second-harmonic generation (SHG) technique. Subsequently it has been realized that the nonlinearity parameter is the controlling factor in nonlinear distortion of sound in air, fluids, and solids. We therefore describe the SHG technique and show how it is used to determine the non-linearity parameter of crystalline solids.[6] Description of the evaluation of nonlinearity parameters of liquids and gases will be left to other authors, as will the evaluation of other third-order elastic constants by stress–velocity techniques.

2   THEORY OF SOLIDS

The propagation of a finite-amplitude ultrasonic wave in the principal directions [100], [110], or [111] in a cubic

Encyclopedia of Acoustics, edited by Malcolm J. Crocker
ISBN 0-471-80465-7 © 1997 John Wiley & Sons, Inc.

solid or in any direction in an isotropic solid is described by an equation of the form

$$\rho_0 \frac{\partial^2 \xi}{\partial t^2} = K_2 \frac{\partial^2 \xi}{\partial a^2} + (3K_2 + K_3) \frac{\partial^2 \xi}{\partial a^2} \frac{\partial \xi}{\partial a}, \quad (1)$$

where ξ is the particle displacement and a is the distance measured in the direction of propagation. The coefficients K_2 and K_3 are to be interpreted as listed in Table 1, depending on the direction of propagation in the solid. For any direction in an isotropic solid one can use the K_2 and K_3 listed for the [100] direction.

3   THEORY OF FLUIDS

In an isentropic fluid the equation of state can be written in the form

$$P - P_0 = A \left[\frac{\rho - \rho_0}{\rho_0} \right] + \frac{B}{2!} \left[\frac{\rho - \rho_0^2}{\rho_0} \right]^2 + \cdots, \quad (2)$$

where

$$A = \rho_0 \left[\frac{\partial P}{\partial \rho} \right]_{P = P_0} = \rho_0 c^2 \quad (3)$$

and

$$B = \rho_0^2 \left[\frac{\partial^2 P}{\partial \rho^2} \right]_{P = P_0}, \quad (4)$$

TABLE 1 K_2 and K_3 for [100], [110], and [111] Directions in Cubic Lattice

Direction	K_2	K_3
[100]	C_{11}	C_{111}
[110]	$\frac{1}{2}(C_{11} + C_{12} + 2C_{44})$	$\frac{1}{4}(C_{111} + 3C_{112} + 12C_{166})$
[111]	$\frac{1}{3}(C_{11} + 2C_{12} + 4C_{44})$	$\frac{1}{9}(C_{111} + 6C_{112} + 12C_{144} + 24C_{166} + 2C_{123} + 16C_{456})$

Note: The coefficients C_{ij} are the ordinary elastic constants; the C_{ijk} are the third-order elastic constants.

where P is the pressure, ρ is the density, and c is the sound velocity. The subscript refers to reference values of the quantities. The quantity B/A has become an important measure of the nonlinearity of the fluid. For gases one can use an ideal-gas equation of state in the form

$$P = P_0 \left(\frac{\rho}{\rho_0} \right)^\gamma, \tag{5}$$

where γ is the ratio of specific heats:

$$\gamma = \frac{C_p}{C_v}. \tag{6}$$

Substituting Eq. (5) into the general form Eq. (2) leads to the fact that, for an ideal gas,

$$\frac{B}{A} = \gamma - 1. \tag{7}$$

The nonlinear equation describing the propagation of a finite-amplitude wave in a fluid can be written in the form[5]

$$\frac{\partial^2 \xi}{\partial t^2} = c_0^2 \frac{\partial^2 \xi / \partial a^2}{(1 + \partial \xi / \partial a)^{B/A + 2}}, \tag{8}$$

where c_0 is the velocity of sound measured with small amplitudes. The nonlinear equation for describing the propagation of a finite-amplitude wave in an ideal gas can be written in a similar form:

$$\frac{\partial^2 \xi}{\partial t^2} = c_0^2 \frac{\partial^2 \xi / \partial a^2}{(1 + \partial \xi / \partial a)^{\gamma + 1}}, \tag{9}$$

where one recognizes the fact that $\gamma + 1$ plays the same role in an ideal gas as $B/A + 2$ does for any fluid. By expanding the denominator on the right of these equations in a binomial series, one finds that either Eq. (8) or Eq. (9) can be written in the same form as Eq. (1). These results can be summarized as in Table 2. Proper use of Table 2 allows one to make the appropriate substitutions such that Eq. (1) can be used to describe the propagation of finite-amplitude waves in solids, liquids, or gases.

4 SOLUTION OF THE NONLINEAR EQUATION

The solution of the nonlinear equation (1) appropriate to measurement is made by assuming that the wave is initially sinusoidal: At $a = 0$, $\xi = A_1 \sin(-\omega t)$. With this assumption, one can obtain a perturbation solution in the form

$$\xi = A_1 \sin(ka - \omega t) + \beta \frac{A_1^2 k^2 a}{4} \cos 2(ka - \omega t) + \cdots, \tag{10}$$

TABLE 2 Parameters Used in Describing Waves of Finite Amplitude in Gases, Liquids, and Solids

Wave	Velocity Squared, c_0^2	Nonlinearity Parameter, 2β	Discontinuity Distance, L
In ideal gas	$\dfrac{\gamma P_0}{\rho_0}$	$\gamma + 1$	$\dfrac{2c_0^2}{(\gamma + 1)\omega^2 A_1}$
In liquid	$\dfrac{A}{\rho_0}$	$\dfrac{B}{A} + 2$	$\dfrac{2c_0^2}{(B/A + 2)\omega^2 A_1}$
In solid (cubic or isotropic)	$\dfrac{K_2}{\rho_0}$	$-\left(\dfrac{K_3}{K_2} + 3 \right)$	$\dfrac{2c_0^2}{-(K_3/K_2 + 3)\omega^2 A_1}$

where β is the nonlinearity parameter used by others in this book. The negative ratio of the coefficient of the nonlinear term to the linear term in Eq. (1),

$$2\beta = -\left(3 + \frac{K_3}{K_2}\right) \tag{11}$$

is the quantity to be determined from measurement. From 2β one can evaluate K_3, since $K_2 = \rho c_0^2$ is known. From values of K_3 in the three principal directions one can use Table 1 to determine combinations of third-order elastic constants from measured results.

Equation (10) shows that an initially sinusoidal ultrasonic wave generates a second harmonic as it propagates. To measure the nonlinearity parameter 2β, one needs to measure the absolute value of the fundamental displacement amplitude A_1 and that of the second harmonic,

$$A_2 = \beta \frac{A_1^2 k^2 a}{4}. \tag{12}$$

Since the propagation constant

$$k = \frac{2\pi}{\lambda} = \frac{\omega}{c_0} \tag{13}$$

is known and the sample length a can be measured, one can write

$$2\beta = \frac{8A_2 c_0^2}{A_1^2 \omega^2 a} \tag{14}$$

and determine 2β directly. It is common to plot A_2 as a function of A_1^2 and to evaluate the slope to determine 2β.

5 EXPERIMENTAL CONSIDERATIONS

5.1 Room Temperature Measurements

To determine the nonlinearity parameters, one needs only to measure the sample length a, the angular frequency ω from which one can calculate $k = \omega/c_0$, and the slope A_2/A_1^2. The problem is in measurement of A_2/A_1^2. This requires absolute measurement of amplitudes. At a fundamental frequency of 30 MHz the amplitude A_1 is of the order of 1 Å, and A_2 may be 1% of A_1. A very sensitive device is required.

Figure 1 is a block diagram of a system for measuring A_1 and A_2. A radio frequency (RF) tone burst of approximately 30 MHz is applied to an X-cut quartz transducer bonded to one end of the sample. The ultrasonic wave that propagates through the sample is detected at the other end by a capacitive receiver. The sample assembly is shown in Fig. 2. The capacitive receiver is a 1.016-cm-

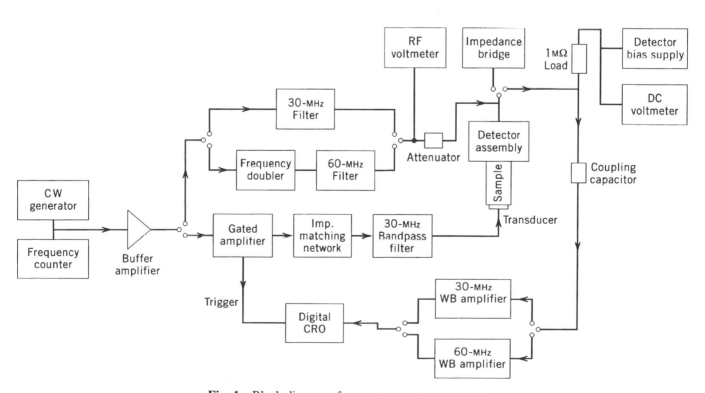

Fig. 1 Block diagram of room temperature apparatus.

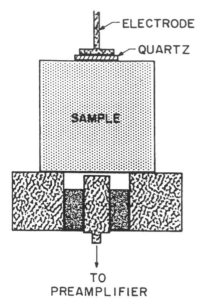

Fig. 2 Sample assembly.

diameter electrode placed approximately 5 μm from the conducting sample (both surfaces being optically flat), and a direct current (DC) bias on the order of 150 V is applied across the gap through a large resistor (approximately 1 MΩ). A substitutional signal giving the same output as the ultrasonic signal is introduced across the capacitive receiver in such a way that the current in the substitutional signal can be measured. This current i can be related to the amplitude A_1 of the ultrasonic wave by

$$i = 2A_1 V_b \omega \, \frac{\chi}{s},$$

where V_b is the DC bias on the receiver, ω is the angular frequency, χ is the capacitance of the receiver, and s is the spacing of the capacitive receiver plates.

The signal from the capacitive receiver is taken to either a 30- or a 60-MHz wide-bandpass amplifier. This amplified signal is detected and taken to the digital oscilloscope. The digital oscilloscope is used to select a portion of the first received echo and measure its amplitude.

After the fundamental and second-harmonic signals have been measured by the digital oscilloscope, a continuous-wave substitutional signal is introduced at the capacitive receiver. The 60-MHz signal is derived by doubling the 30-MHz signal with a ring bridge mixer. Both the 30- and the 60-MHz substitutional signals are filtered to ensure spectral purity. These two signals are adjusted to give the same output at the digital oscilloscope as the ultrasonic signals and are measured with an RF voltmeter. From these measurements and a knowledge of the circuit impedances, the current of the substitutional signal can be calculated that exactly matches the ultrasonic signal to be measured.

5.2 Samples

For precise measurements it is necessary to lap samples optically flat and parallel. Optical tolerance is usually one fringe over the sample surface and parallelism of the surfaces should be within 12 s of arc. Nonconductive samples must be coated with copper, silver, or gold to a thickness of approximately 1000 Å for electrical conductivity. Optimum sample size is 2.5 cm diameter and 2.5 cm long, the cylinder axis being one of the principal directions [100], [110], or [111]. The smallest practical sample size is a cube 0.5 cm on a side. The size of the ultrasonic transducer and the capacitive receiver are determined by the sample size. For smaller transducers it is necessary to correct the measurements for the effect of diffraction.[7]

5.3 Second-Order Elastic Constants

The second-order elastic constants are determined from measurement of velocity. Many sets of data, both at room temperature and lower temperatures, are readily available for comparison. Velocity measurements can be measured to an accuracy of 10^{-4}–10^{-2}, depending upon the sample. Relative measurements can be made to an accuracy of 10^{-6} with a stable frequency source. Such velocity measurements are described in Chapter 55.

5.4 Cryogenic Apparatus

The electronic apparatus for measurements at low temperatures is essentially the same as that for room temperature, with one exception. It is not necessary to calibrate at each temperature since relative measurements are adequate. A reference signal, then, is not necessary.

The sample holder is encased in a stainless steel can, as shown in Fig. 3. The can shown is surrounded by a second can, and the space between them is evacuated to provide an insulating jacket around the inner can shown. Radiation loss is reduced by polishing the cans to a high sheen. The cans are supported by three cupro-nickel tubes. Two of the tubes have a smaller cupro-nickel tube centered inside to provide a 50-Ω coaxial transmission line. The tubes simultaneously can be used as vacuum lines. The third tube houses leads for the temperature sensor and the heater and also is used to evacuate the inner can to a known pressure.

The entire apparatus can be suspended inside a standard helium research dewar. By sealing the dewar and pumping on the coolant, temperatures below the ambient pressure boiling point of the coolant (either liquid nitrogen or liquid helium) may be obtained.

The temperature in the inner can is controlled by an electric resistance heater connected to a commercial temperature controller. It can be varied continuously from

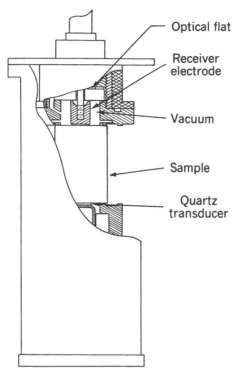

Fig. 3   Sample assembly for cryogenic measurements.

approximately 3 K to a few degrees above room temperature.

5.5   Cryogenic Nonlinearity Measurements

Only relative measurements need to be taken with the cryogenic apparatus. At each temperature one first adjusts the drive signal to the quartz transducer so that the fundamental ultrasonic wave received at the capacitive receiver has the same known value. The second harmonic then is measured by a slide-back technique: the

DC bias voltage on the capacitive receiver is adjusted so that the electrical signal coming from the capacitive receiver is the same at the new temperature as it was at the previous temperature. The second-harmonic amplitude A_2 of the ultrasonic wave at the two different absolute temperatures (T_1 and T_2) are then related by

$$A_2(T_2) = \frac{A_2(T_1)V_b(T_1)}{V_b(T_2)}.$$

Use of this equation allows one to use experimental data to plot the temperature dependence of the second harmonic and ultimately to plot the nonlinearity parameter as a function of temperature. If desirable, one can interpret data on different crystalline orientations in terms of combinations of third-order elastic constants.

6   EXPERIMENTAL RESULTS

Many experimental results are available in the literature. The results for cubic crystals are summarized by Breazeale and Philip[6]; however, it may be useful to make a comparison of the magnitudes of the nonlinearity parameters for different substances. A few common gases are listed in Table 3. The nonlinearity parameters,

TABLE 3   Values of γ and 2β for Gases (15°C)

Gas	γ	$2\beta = \gamma + 1$
He	1.66	2.66
Ar	1.67	2.67
O_2	1.40	2.40
N_2	1.40	2.40
CO_2	1.30	2.30
$H_2O(200°C)$	1.31	2.31
CH_4	1.31	2.31

TABLE 4   Values of B/A and 2β for Fluids at Atmospheric Pressure

Liquid	Temperature (°C)	B/A	$2\beta = B/A + 2$
Water, distilled	0	4.16	6.16
	20	4.96	6.96
	40	5.38	7.38
	60	5.67	7.67
	80	5.96	7.96
	100	6.11	8.11
Acetone	30	9.44	11.44
Benzene	30	9.03	11.03
Benzyl alcohol	30	10.19	12.19
CCl_4	30	11.54	13.54

TABLE 5 **Comparison of Structure, Bonding, and Acoustic Nonlinearity Parameters along the [100] Direction of Cubic Crystals**

Structure	Bonding Type	2β Range
Zincblende	Covalent	1.8–3.2
Flourite	Ionic	3.4–4.6
FCC	Metallic	4.0–7.0
FCC (inert gas)	Van der Waals	5.8–7.0
BCC	Metallic	5.0–8.8
NaCl	Ionic	13.5–15.4

Abbreviations: FCC, face-centered-cubic; BCC, body-centered-cubic.

obtained simply by evaluating $\beta = \gamma + 1$, cover the range between 2.3 and 2.7. For comparison, Table 4 lists the nonlinearity parameters for common liquids. In liquids the nonlinearity parameters range between 6 and 14. In

Table 5 are listed nonlinearity parameters for the [100] direction in cubic crystals having different structure and bonding. The solids listed cover the range of β between 2 and 15, a range that includes that of both gases and liquids. The relatively wide range of nonlinearity parameters of solids is one reason for interest in measuring this important physical property.

6.1 Evaluation of the Nonlinearity Parameters of Solids

Evaluation of specific nonlinearity parameters involves measurement of the amplitudes of the fundamental and the second harmonic at room temperature, then relative measurements at different temperatures. Recent data[8] on NaCl will serve as an example. The received second harmonic at room temperature as a function of the square of the fundamental is plotted in Fig. 4. The different

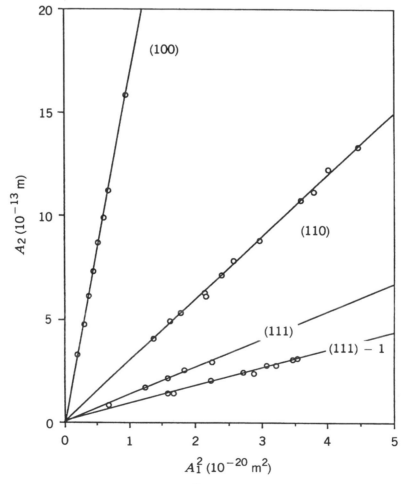

Fig. 4 Room temperature values of A_2 vs. A_1^2 for NaCl. (Two different samples gave different curves for the [111] direction.)

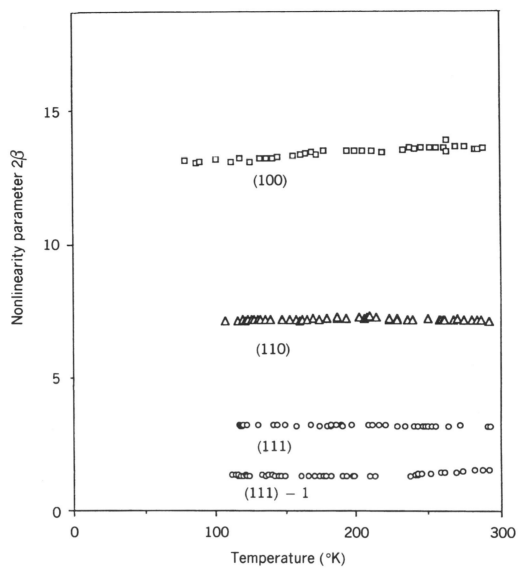

Fig. 5 Temperature dependence of the nonlinearity parameters of NaCl for different propagation directions.

slopes show that the nonlinearity parameter in the [100] direction is greater than that in other directions. Furthermore, in this case the nonlinearity parameter in the [111] direction was different for the two samples measured. By evaluating the slopes in Fig. 4, one can evaluate the nonlinearity parameters at room temperature; then relative measurement of the second harmonic at different temperatures allows one to evaluate the nonlinearity parameter as a function of temperature. A plot of the results for NaCl as a function of temperature is given in Fig. 5. Although the nonlinearity parameters typically are not strong functions of temperature, it is seldom that the linear behavior shown in Fig. 5 is observed for other cubic solids. A collection of data on a few other cubic and isotropic solids is shown in Figs. 6a–c. Liquid helium was used for the low-temperature measurements to show that there is not a wide variation in nonlinearity parameters with temperature. The range of nonlinearity parameters typically is between 0 and 16 for solids, although to date there is no specific reason to assume that nonlinearity parameters outside this range will not be observed.

Further data analysis is possible. In particular, third-order elastic constants calculated from these data and interpreted in terms of a model of the solid state is possible; however, such analysis is beyond the scope of the present work.

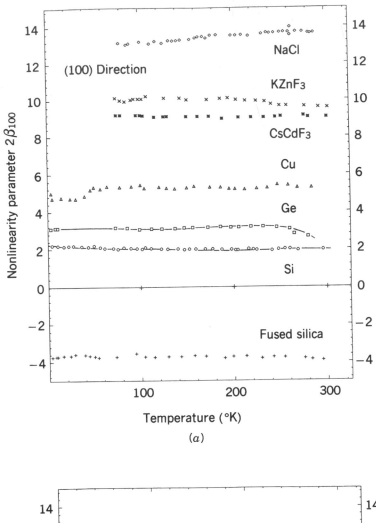

(a)

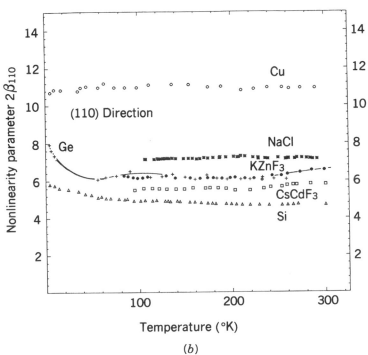

(b)

Fig. 6   Temperature dependence of the nonlinearity parameters of different cubic crystals for (a) [100]; (b) [110]; and (c) [111] directions.

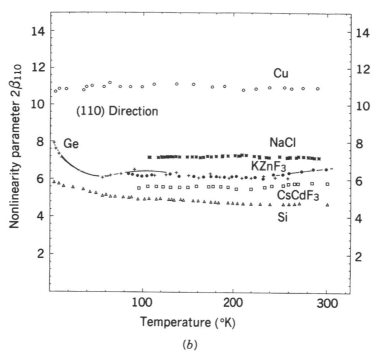

Fig. 6   (*Continued*)

REFERENCES

1. S. Earnshaw, "On the Mathematical Theory of Sound," *Phil. Trans. Roy. Soc. (Lond.)*, Vol. 150, 1860, p. 133.

2. A. L. Thuras, R. T. Jenkins, and H. T. O'Neil, "Extraneous Frequencies Generated in Air Carrying Intense Sound Wave," *J. Acoust. Soc. Am.*, Vol. 6, 1935, p. 173.

3. W. Keck and R. T. Beyer, *Phys. Fluids*, Vol. 3, 1960, p. 346.

4. R. T. Beyer, "Nonlinear Acoustics," in W. P. Mason (Ed.), *Physical Acoustics*, Vol. IIB, Academic, New York, 1965, pp. 231–264.

5. M. A. Breazeale and J. Ford, "Ultrasonic Studies of the Nonlinear Behavior of Solids," *J. Appl. Phys.*, Vol. 36, 1965, pp. 3486–3490.

6. M. A. Breazeale and J. Philip, "Determination of Third-Order Elastic Constants from Ultrasonic Harmonic Generation Measurements," in W. P. Mason and R. N. Thurston (Eds.), *Physical Acoustics*, Vol. XVII, Academic, New York, 1984, pp. 1–60.

7. B. D. Blackburn and M. A. Breazeale, "Nonlinear Distortion of Ultrasonic Waves in Small Crystalline Samples," *J. Acoust. Soc. Am.*, Vol. 76, 1984, pp. 1755–1760.

8. W. Jiang and M. A. Breazeale, "Temperature Variation of Elastic Nonlinearity of NaCl," *J. Appl. Phys.*, Vol. 68, 1990, pp. 5472–5477.

22

NONLINEAR STANDING WAVES IN CAVITIES

ALAN B. COPPENS AND ANTHONY A. ATCHLEY

1 INTRODUCTION

Finite-amplitude acoustic phenomena in resonant cavities have been of practical interest since the 1930s, when they were studied in connection with Panzer tank mufflers.[1] They continue to be of interest today, particularly in the areas of thermoacoustic heat engines and acoustic compressors.[2–4] The problem has been addressed by a number of authors using a number of different solution techniques, including Riemann invariant analysis, the method of characteristics, and finite-difference simulations.[5–16]

A rigorous theoretical model for nonlinear standing waves in even the simplest of geometries is quite involved. In any real situation thermal and viscous losses at the walls of the cavity must be considered in addition to mainstream losses. This requires in general a three-dimensional solution of the nonlinear wave equation that includes a description of the motion of the fluid within the boundary layer.

A more pragmatic approach is to adopt reasonable approximations that simplify the description of the complete acoustic field and allow relatively straightforward modeling of the additional contributions resulting from the boundary layers and from the nonideal natures of real cavities. Physically plausible approximations will be introduced where needed without detailed justification. The result will be a reasonably good description of the behavior of nonlinear standing waves in rigid-walled cavities of simple geometries for strengths less than those leading to the formation of shock waves.

The presence of the walls introduces phase speed dispersion and additional thermal and viscous loss terms, all depending on noninteger powers of the frequency. The dispersion will perturb the resonance frequencies of the system, thereby influencing the harmonic spectrum of the standing wave significantly.

The losses at the walls depend on the nature of the standing wave. For example, the particle velocity of a one-dimensional (plane) standing wave in a tube is perpendicular to the two ends of the tube; there is no shearing at those surfaces, but there still are thermal conductivity losses. The absorption coefficients resulting from the thermal losses at these ends must be added to the other absorption coefficients for this standing wave to obtain the total Q for the wave. In addition to these corrections, however, the nonideal nature of any real system (walls that are nonparallel, are not ideally rigid, etc.) introduces other perturbations that can be of similar importance. Consequently, while the Q's can be predicted reasonably well as long as the cavity is tightly constructed with smooth and clean walls, it is more accurate to measure the Q's of the standing waves in the cavity at strengths for which linear acoustics is applicable.

The phase speed dispersion could similarly be predicted for each standing wave, but in even well-designed cavities such predictions are usually insufficiently accurate because of irregularities in construction. As a consequence, for accurate prediction of the properties of the nonlinear standing wave, it is usually better to measure the actual resonance frequencies and cavity dimensions and obtain therefrom the effective phase speed c_p for each standing wave.

1.1 Response of a Linear System to a Distributed Volume Source

A generalized practical linear Helmholtz equation that describes the behavior of a standing wave in a cavity in

Encyclopedia of Acoustics, edited by Malcolm J. Crocker
ISBN 0-471-80465-7 © 1997 John Wiley & Sons, Inc.

the vicinity of resonance can be written in terms of the effective phase speed c_p and the measured quality factor Q appropriate for the closest mode:

$$\left[\left(c_p^2 + \frac{jc^2}{Q} \right) \nabla^2 + \omega^2 \right] \left[\frac{\mathbf{p}}{\rho_0 c^2} \right] = 0. \qquad (1)$$

In this equation ρ_0 is the equilibrium density of the fluid, $\mathbf{p}$ the complex pressure fluctuation about the equilibrium hydrostatic pressure, and c the free-field speed of sound given by $\sqrt{B/\rho_0}$, with B the adiabatic bulk modulus. Standing-wave solutions of the nonlinear wave equation can be obtained through a perturbation expansion. Each successive solution is found by treating the nonlinear term (comprised of the previous solution) as a volume source driving the linear wave equation. The method can be clarified by first treating a linear case: Introduce an appropriate ad hoc volume source as an inhomogeneous term in the linear wave equation and solve. For a mono-frequency three-dimensional standing wave in a rigid-walled cavity in Cartesian geometry the relevant inhomogeneous equation is

$$\left[\left(c_p^2 + \frac{jc^2}{Q} \right) \nabla^2 + \omega^2 \right] \left[\frac{\mathbf{p}}{\rho_0 c^2} \right]$$

$$= -\frac{\partial^2}{\partial t^2} \cos(k_x x) \cos(k_y y) \cos(k_z z) e^{j(\omega t + \varphi)}. \quad (2)$$

The forcing function is a standing wave with spatial dependence $\cos(k_x x) \cos(k_y y) \cos(k_z z)$ and frequency $\omega/2\pi$ not necessarily equal to the resonance value $\omega_r/2\pi$ of the standing wave. The phase speed is given by $c_p = \omega_r/\sqrt{k_x^2 + k_y^2 + k_z^2}$. The solution is the sum of steady-state (particular) and transient (homogeneous) terms. The transient term is an oscillating standing wave that decays with time as $\exp(-t/\tau_r)$, where $\tau_r = 2Q_r/\omega_r$ is the relaxation time. What remains after sufficient time is the steady-state solution. The real acoustic pressure $p = \text{Re}\{\mathbf{p}\}$ in the steady state is given as

$$\frac{p}{\rho_0 c^2} = -Q_r \cos \theta_r \cos(k_x x) \cos(k_y y) \cos(k_z z)$$

$$\cdot \sin(\omega t + \varphi + \theta_r), \qquad (3)$$

$$\tan \theta_r = Q_r \left[\left(\frac{\omega_r}{\omega} \right)^2 - 1 \right] \doteq -2Q_r \frac{\Delta\omega}{\omega_r}, \qquad (4)$$

where $\Delta\omega = \omega - \omega_r$. If the driving frequency is close to

the resonance value ($\Delta\omega \ll \omega_r$), then

$$Q_r \cos \theta_r \doteq \frac{1}{\sqrt{1/Q_r^2 + (2 \Delta\omega/\omega_r)^2}}. \qquad (5)$$

The amplitude $Q_r \cos \theta_r$ and phase angle θ_r depend on the shift in frequency away from resonance, exactly as they do for the standing wave driven by a velocity source at $x = 0$.

2 NONLINEAR PROBLEM

A conventional form of the nonlinear acoustic wave equation with shear viscosity η is

$$\left[c^2 \nabla^2 + \frac{4\eta/3}{\rho_0} \frac{\partial}{\partial t} \nabla^2 - \frac{\partial^2}{\partial t^2} \right] \left[\frac{p}{\rho_0 c^2} \right]$$

$$= -\frac{\partial^2}{\partial t^2} \left[\frac{\gamma + 1}{2} \left(\frac{p}{\rho_0 c^2} \right)^2 \right], \qquad (6)$$

where u is the acoustic particle velocity and γ is the ratio of specific heats. In the derivation of this equation, the nonlinear terms on the right-hand side and the absorptive term on the left-hand side have each been simplified by the use of lossless linear acoustic relationships. (These terms are considerably smaller than either of the other two and can therefore be approximated in this way without introducing any significant error in the solutions.) The same approximations can also be made when inserting an expression into the nonlinear term of Eq. (6) and when modifying the left-hand side to account for losses other than shear viscosity. If the solution to Eq. (6) is represented as a series in frequency, then the right-hand side can be written as a sum over frequency, with each term simplified into its linear, lossless equivalent. Each such term can then be expanded into complex form, and each will have the same form as the right-hand side of Eq. (2). The left-hand side of Eq. (6) can also be generalized to include all mainstream and wall losses, expanded into complex form, and then replaced by the left-hand side of Eq. (2) for each frequency of interest. Thus, under the assumption that the solution will be expressed in terms of the fundamental frequency and its harmonics, the right-hand side is calculated using lossless linear acoustic relationships, written as a sum of monofrequency components, and then extended to complex form. Equation (6) can then be expressed as a summation over all harmonics of the driver frequency:

$$\sum_n \left\{ \left[\left(c_n^2 + \frac{jc^2}{Q_n} \right) \nabla^2 + (n\omega)^2 \right] \left[\frac{\mathbf{p}(n\omega)}{\rho_0 c^2} \right] \right\}$$

$$= -\sum_n \frac{\partial^2}{\partial t^2} \left[\frac{\gamma+1}{2} \left(\frac{p}{\rho_0 c^2} \right)^2 \right]_{n\omega, \, complex} \tag{7}$$

As will be seen, in general, there may be an infinite number of terms associated with each n. In Eq. (7) c_n is the phase speed for the standing wave of frequency nf. The quality factor Q_n of the nth mode is given by $1/Q_n = 1/Q_b + 1/Q_w$, where $Q_b = 2\alpha_b/k$ and $Q_w = 2\alpha_w/k$ and α_b and α_w are the absorption coefficients for mainstream and wall losses, respectively. For one-dimensional standing waves in narrow tubes, $\alpha_w = (1/ac)(\eta_e \omega/2\rho_0)^{1/2}$, where a is the effective radius of a tube with circular cross section (or the cross-sectional area divided by half the perimeter for a tube of arbitrary cross-sectional shape) and the effective viscosity is defined as $\eta_e = \eta[1+(\gamma-1)\sqrt{\kappa/C_p\eta}]^{1/2}$, with κ the thermal conductivity of the fluid and C_p the heat capacity at constant pressure. Each separate term on the right-hand side is a component of the nonlinear forcing term extended to complex form. Each complex pressure $\mathbf{p}(n\omega)$ on the left-hand side is the match to each of the forcing terms. Obtaining the nonlinear standing-wave solution of Eq. (7) amounts to a systematic accumulation of the individual frequency components, which both generate the volume source and form the solution to it.

2.1 Perturbation Solution Method for Standing Waves

Equation (7) will first be solved using a perturbation approach. As a simple example, consider the one-dimensional case of plane-wave modes in a rigid-walled tube of length L. The method will be generalized later to three dimensions. The first perturbation solution p_1 solves the homogeneous linear wave equation (1) for a standing wave driven at frequency $f = \omega/2\pi$ close to the measured resonance frequency $f_1 = \omega_1/2\pi$ of the wave. If the driver at $x = 0$ has known particle displacement $\hat{x}X\cos(\omega t - \theta_1)$, then the resulting standing wave $p_1 = p_{11}$ is given as

$$\frac{p_{11}(x,t)}{\rho_0 c^2} = 2\,\frac{X}{L}\,Q_1 \cos\theta_1 \cos(k_1 x)\sin(\omega t), \tag{8}$$

where k_1 is determined from the dimensions, c_1 is determined from ω_1/k_1, and terms of higher order in α/k have been neglected. It is convenient to introduce a Mach number $M = 2(X/L)Q_1 \cos\theta_1$ so that the first approximate solution can be written as

$$\frac{p_{11}}{\rho_0 c^2} \doteq M \cos(k_1 x)\sin(\omega t). \tag{9}$$

Recall that the phase angle θ_1 is determined from Eqs. (4) or (5):

$$\tan\theta_1 = Q_1 \left[\left(\frac{\omega_1}{\omega} \right)^2 - 1 \right] \doteq -2Q_1 \frac{\Delta\omega}{\omega_1}, \tag{10}$$

$$Q_1 \cos\theta_1 \doteq \frac{1}{\sqrt{1/Q_1^2 + (2\,\Delta\omega/\omega_1)^2}}. \tag{11}$$

The next approximation $p_1 + p_2$ is substituted into Eq. (7), and only terms of lowest order in M are retained. The right-hand side consists of terms of the forms $\cos(2k_1 x)\cos(2\omega t)$, $\cos(2\omega t)$, and $\cos(2k_1 x)$ and constants. The frequency $2f$ will be very close to the resonance frequency f_2 of the first overtone, and the standing wave of frequency $2f$ near $2f_1$ should therefore have a propagation constant of $2k_1$. Note that f_2 will not be exactly $2f_1$ because of the dispersive effects from the wall losses and the geometric realities of the tube. Only the first of the forms listed above leads to a significant solution for p_2, since it corresponds to a standing wave being driven close to resonance; the others can be neglected because each corresponds to a standing wave being driven far away from its resonance. Use of linear, lossless relations to simplify the retained nonlinear term and then extension to complex form results in

$$\left[\left(c_2^2 + \frac{jc^2}{Q_2} \right) \nabla^2 + (2\omega)^2 \right] \left[\frac{\mathbf{p}_2}{\rho_0 c^2} \right]$$

$$= M\,\frac{\gamma+1}{2}\,\frac{M}{4}\,\frac{\partial^2}{\partial t^2}\cos(2k_1 x)e^{j(2\omega t)}. \tag{12}$$

Then, again following Section 1.1, the steady-state solution for $p_2 = p_{22}$ is found to be

$$\frac{p_{22}}{\rho_0 c^2} = M\,\frac{\gamma+1}{2}\,\frac{M}{4}\,Q_2 \cos\theta_2 \cos(2k_1 x)$$

$$\cdot \sin(2\omega t + \theta_2), \tag{13}$$

$$\tan\theta_2 = Q_2 \left[\left(\frac{\omega_2}{2\omega} \right)^2 - 1 \right]. \tag{14}$$

The term $\cos\theta_2$ gives the dependence on frequency of the amplitude of the nonlinearly generated standing wave at the second harmonic of the driven fundamental. The relative amplitude of p_{22} depends on how closely tuned $2f$ is to the resonance frequency f_2. Maximum relative amplitude occurs when $2f = f_2$. Note, however, that the amplitude of p_{22} also depends quadratically on the Mach number.

The third approximation comes from substituting $p_1 + p_2 + p_3$ into Eq. (7) to obtain an equation for p_3 in terms of p_1 and p_2. Retaining leading terms in M leaves p_3 on the left side and $p_1 + p_2$ on the right side. The products $(p_1 + p_2)^2$ and $(u_1 + u_2)^2$ involve several terms, but only those of forms $\cos(3k_1x)\cos(3\omega t+\theta_2)$ and $\cos(k_1x)\cos(\omega t+\theta_2)$ lead to significant solutions. The term p_3 is thus the sum of two terms, $p_3 = p_{33} + p_{31}$, where p_{33} has frequency $3f$ and p_{31} has frequency f. Each, expressed in complex form, satisfies a separate inhomogeneous equation,

$$\left[\left(c_3^2 + \frac{jc^2}{Q_3}\right)\nabla^2 + (3\omega)^2\right]\left[\frac{\mathbf{p}_{33}}{\rho_0 c^2}\right]$$

$$= 2M\left[\frac{\gamma+1}{2}\frac{M}{4}\right]^2 Q_2\cos\theta_2\frac{\partial^2}{\partial t^2}$$

$$\cdot\cos(3k_1x)e^{j(3\omega t+\theta_2)} \qquad (15)$$

and

$$\left[\left(c_1^2 + \frac{jc^2}{Q_1}\right)\nabla^2 + (\omega)^2\right]\left[\frac{\mathbf{p}_{31}}{\rho_0 c^2}\right]$$

$$= -2M\left[\frac{\gamma+1}{2}\frac{M}{4}\right]^2 Q_2\cos\theta_2\frac{\partial^2}{\partial t^2}$$

$$\cdot\cos(k_1x)e^{j(\omega t+\theta_2)}. \qquad (16)$$

Solutions are

$$\frac{p_{33}}{\rho_0 c^2} = 2M\left[\frac{\gamma+1}{2}\frac{M}{4}\right]^2 Q_2 Q_3\cos\theta_2\cos\theta_3\cos(3k_1x)$$

$$\cdot\sin(3\omega t+\theta_2+\theta_3), \qquad (17)$$

$$\tan\theta_3 = Q_3\left[\left(\frac{\omega_3}{3\omega}\right)^2 - 1\right] \qquad (18)$$

and

$$\frac{p_{31}}{\rho_0 c^2} = -2M\left[\frac{\gamma+1}{2}\frac{M}{4}\right]^2 Q_1 Q_2\cos\theta_1\cos\theta_2\cos(k_1x)$$

$$\cdot\sin(\omega t+\theta_1+\theta_2), \qquad (19)$$

$$\tan\theta_1 = Q_1\left[\left(\frac{\omega_1}{\omega}\right)^2 - 1\right]. \qquad (20)$$

Thus, p_3 contains third harmonic p_{33} and a correction p_{31} to the first harmonic. This latter term reduces the amplitude of the fundamental to compensate for the energy that has been stored in the second harmonic. The amplitude of the third harmonic depends strongly on the differences $2f - f_2$ and $3f - f_3$ as well as the quality factors Q_2 and Q_3.

If p_{11} were driven exactly at resonance and the resonance frequencies of the standing waves p_{22} and p_{33} were exactly twice and thrice that of the fundamental, then the angles θ_1, θ_2, and θ_3 would all be zero and the cosines of these angles would be unity. The terms p_1, p_2, and p_{33} would all be maximized and aligned for greatest effect, and p_{31} would be exactly out of phase with p_{11}. The correction p_{31} to the fundamental p_{11} would be to reduce it as strongly as possible, representing maximum energy loss from the fundamental into the second harmonic. The phases of the first three harmonics would act to steepen the positive-going portion of the pressure wave, maximally at the axis crossing of p_{11}.

Introducing the next approximation $p_1 + p_2 + p_3 + p_4$ yields standing waves p_{44} at the fourth harmonic and p_{42} at the second harmonic. The fourth harmonic is generated by the interaction of p_{11} with p_{33} and also p_{22} with itself. The wave at the second harmonic arises from the interaction of p_{11} and p_{31} and represents the resultant loss of energy from the second harmonic into the third.

2.2 Extension to Two- and Three-Dimensional Waves

While the cases described so far have been one dimensional, generalization to two- and three-dimensional standing waves follows the same mathematical steps used in developing the one-dimensional nonlinear standing waves. The products of one-, two-, or three-dimensional standing waves contain waves of the same dimensionality, but these terms are multiplied by a factor N that depends on the number of spatial dimensions present in the standing waves. A wave $\cos(kx)$ leads to a volume source $\cos^2(kx)$ which yields a term $\frac{1}{2}\cos(2kx)$ containing the factor $N = \frac{1}{2}$. A two-dimensional wave $\cos(k_x x)\cos(k_y y)$ leads to a volume source $\frac{1}{4}\cos(2k_x x)\cos(2k_y y)$ with $N = \frac{1}{4}$, and a three-dimensional wave yields $N = \frac{1}{8}$. Thus, the number m of spatial dimensions involved in the standing wave alters the amplitude of the resultant volume source by $N = (\frac{1}{2})^m$.

The various spatial forms $\cos(mk_x x)$ and $\cos(nk_x x)$ in the waves multiply together to give volume sources having cosines of arguments $(m \pm n)k_x$, and similarly for k_y and k_z. Examination reveals that only volume sources with frequencies near the resonance frequencies of their standing waves result in forced standing waves of significant amplitude. The resultant standing wave $p(x, y, z, t)$ comprises a *family* whose members constitute a harmonic sequence based on the frequency of the driven member: If the driven standing wave has the form $\cos(k_x x)\cos(k_y y)\cos(k_z z)\exp(j\omega t)$, where $\omega \approx c\sqrt{k_x^2 + k_y^2 + k_z^2}$, then the family of standing waves has members with forms $\cos(nk_x x)\cos(nk_y y)\cos(nk_z z)\cdot \exp(jn\omega t)$ for $n = 1, 2, 3, \ldots$.

If there are near degeneracies between family and non–family members, the simple model discussed here yields no excitation of a non–family member. It is observed experimentally that there can be some small coupling of energy into such degeneracies if the cavities are of poor geometry. The coupling probably has its roots in the corrections to the eigenfunctions arising from the perturbations in geometry. If the cavity is irregular, then the actual standing wave for a particular resonance will consist of a summation over the eigenfunctions of the geometrically regular cavity. The standing wave will be that for the ideally shaped cavity plus additional standing waves whose amplitudes and phases will be controlled by a separate expansion parameter measuring the dimensional irregularity. These corrections are usually relatively small for cavities of fairly regular form.

2.3 Useful Descriptive Parameters

The above perturbation calculations reveal useful descriptive parameters. Each successive perturbation solution contributes terms of higher products of the Mach number, the coefficient of nonlinearity $\beta = (\gamma + 1)/2$, and the quality factors Q_n.

The *strength parameter* is $M\beta Q_1$ [equal to half the Goldberg number $M\beta/(\alpha_1/k_1)$]. This is a convenient measure of the value of the nonlinear strength to absorptive loss. It describes the ability of the standing wave to overcome its absorption losses and accumulate nonlinear distortion. Each harmonic in the perturbation solution involves a factor $M\beta Q_n$, but since the Q's for the lower harmonics are usually fairly similar (related, at least approximately, by $\sqrt{\omega}$, according to the theoretical models for wall effects), it is sufficient to use $M\beta Q_1$ as a measure of the capacity of the standing wave to distort.

The *frequency parameter* is defined by $2Q_1\,\Delta\omega/\omega \doteq -\tan\theta_1$ (where $\Delta\omega = \omega - \omega_1$) and measures the deviation of the driving frequency from the resonance frequency of the fundamental standing wave in terms of the bandwidth of its resonance.

Nonlinear distortion would be maximized if the resonance frequencies of the relevant standing waves were exactly equal to the harmonics of the driving frequency

f. In real tubes this alignment is only approximate; the resonance frequencies are not harmonics, and the nonlinear excitation of standing waves at the harmonics of the fundamental is reduced. The frequency dependence of the amplitude of a particular term in the perturbation expansion depends on the products of the cosines of the phase angles θ_n, which in turn describe the deviation of the driving frequency of the volume source from the resonance frequency of the standing wave with matching spatial behavior. These considerations can be described by a collection of response functions

$$S_n = M\beta Q_n \cos\theta_n. \tag{21}$$

The first three perturbation solutions (Section 2.1) can now be written as

$$\frac{p_{11}}{\rho_0 c^2} = M\cos(k_1 x)\sin(\omega t), \tag{22}$$

$$\frac{p_{22}}{\rho_0 c^2} = \frac{1}{2}M(NS_2)\cos(2k_1 x)\sin(2\omega t + \theta_2), \tag{23}$$

$$\frac{p_{33}}{\rho_0 c^2} = \frac{1}{2}M(NS_2)(NS_3)\cos(3k_1 x)$$
$$\cdot \sin(3\omega t + \theta_2 + \theta_3), \tag{24}$$

$$\frac{p_{31}}{\rho_0 c^2} = -\frac{1}{2}M(NS_1)(NS_2)\cos(k_1 x)$$
$$\cdot \sin(\omega t + \theta_1 + \theta_2). \tag{25}$$

For this one-dimensional wave, $N = \frac{1}{2}$. The amplitude of each perturbation term depends on how great an overlap there is between the product of response functions appearing in each term of the volume source and the response function of the resultant driven standing wave.

There are additional parameters that help to describe the acoustic behavior of the cavity:

The fractional deviation from consonance is described by

$$h_n = \frac{\omega_n - n\omega_1}{n\omega_1}. \tag{26}$$

The magnitudes and signs of the values of h_n for a family of nonlinearly excited standing waves allow evaluation of how closely aligned are the resonance frequencies of the standing waves with the harmonics of the resonance frequency of the fundamental. The amount of overlap of the various response functions can be quantified by comparing each with the frequency-shifted response function of the fundamental through a parameter H_n,

$$H_n = 2h_n \frac{Q_n Q_1}{Q_n + Q_1}, \tag{27}$$

which is the ratio of the frequency interval $f_n - nf_1$ to the averaged bandwidths of the two response functions. The more dissimilar the values of the H_n's are, the more "mistuned" the cavity is and the weaker the nonlinear distortion will tend to be.

For a carefully designed cavity (good geometry and tightly fitted joints) with a fundamental resonance around 0.5–2 kHz, measured Q_n's will normally be around 200–1000, h_n's within about $\pm 10^{-3}$, and H_n's within ± 0.5 for the lowest several overtones. These representative values show that even though the overtones may be nearly harmonics of the fundamental, differing by no more than 1 Hz or so from consonance, the high Q's can cause the shifted response functions to overlap only slightly. Consequently, the appearance of the nonlinear standing wave can be significantly affected by slight changes in the driver frequency.

2.4   Extension to Larger Strength Parameters (Fourier Expansion)

For small strength parameters, successively higher perturbation solutions will reduce in amplitude relatively quickly. Each higher order perturbation solution introduces terms of one power greater in the strength parameter than the previous solution. Of these, some yield the first contributions to the next harmonic above that from the previous perturbation solution, and the others consist of higher order corrections to lower harmonics. If the strength parameter is small, these higher terms diminish quite rapidly in amplitude so that (1) the harmonic spectrum of the standing wave rolls off very rapidly and (2) the leading (lowest order) terms of each harmonic are appreciably greater than subsequent terms. Only the first few perturbation orders must be worked out, and calculation of the (weakly) distorted waveform is relatively quick.

For greater strength parameters, however, the relative importance of higher harmonics increases, and more orders must be included to describe accurately the nonlinear standing wave. Also, the contributions of higher order solutions to lower harmonics do not converge as rapidly as for lower strength parameters. The result is that accurate calculation of the amplitude and phase of the nth-harmonic standing wave depends on summing a large number of terms of the power series in the strength parameter. In this case it is appropriate to reformulate the solution based on a redefined set of p_n's:

The previous discussion has shown that the standing wave represents a Fourier series in frequency and that the individual terms are sums of those standing waves with resonance frequencies close to the associated harmonic component of the fundamental driving frequency. This means that an appropriate form for the standing wave is

$$\frac{p}{\rho_0 c^2} = \sum_n \frac{p_n}{\rho_0 c^2}, \tag{28}$$

where now each p_n is a *single* term

$$\frac{p_n}{\rho_0 c^2} = M_1 R_n \cos(nk_x x) \cos(nk_y y) \cos(nk_z z)$$
$$\cdot \sin(n\omega t + \phi_n). \tag{29}$$

The nth term has frequency nf and contains all contributions of that frequency from all perturbation corrections. (For example, now $p_1 = p_{11} + p_{31} + p_{51} + \cdots$.) Each ϕ_n represents the collective phase angle when all perturbation terms of frequency nf have been combined. The Mach number M_1 is redefined here and in what follows to describe the amplitude of the (entire) fundamental component of the nonlinear standing wave so that $R_1 = 1$. (This differs from what was done above, where M was the Mach number of the first solution p_{11}.) For simplicity, time is chosen so that $\phi_1 = 0$.

If Eqs. (28) and (29) are substituted into Eq. (7) and all terms arising from products that are not nearly resonant are discarded, the result is a collection of coupled transcendental equations to be solved for the allowed values of R_n and ϕ_n for all $n > 1$,

$$R_n e^{j(\phi_n - \theta_n)} = \frac{1}{2} N S_n \left(\sum_{j=1}^{n-1} R_j R_{n-j} e^{j(\phi_j + \phi_{n-j})} \right.$$
$$\left. -2 \sum_{j=1}^{\infty} R_{n+j} R_j e^{j(\phi_{n+j} - \phi_j)} \right), \tag{30}$$

with N as defined above and now $S_n = M_1 \beta Q_n \cos \theta_n$. The first summation on the right gives the generation of the nth harmonic from all lower harmonics and the migration of acoustic energy upward from lower to higher frequencies. The second summation expresses the depletion of the nth harmonic as a consequence of the passing of energy upward into higher harmonics.

2.5   Qualitative Shapes of the Nonlinear Standing Wave

The thermoviscous effects at the cavity walls introduce a first-order dispersion, which lowers the phase speed with decreasing frequency. Under the assumption that the wall losses alone describe the Q_n's and c_n's of the cavity, then α_n will behave as $\sqrt{\omega}$ and the phase speed will increase toward an asymptotic limit of c as frequency increases. This means that the higher resonance frequencies will run increasingly sharp with respect to the fundamental. If the driving frequency excites the fundamental standing wave at its resonance frequency, then the nth (nonlinearly excited) standing wave will be driven at a frequency nf below resonance f_n and will have positive phase angle θ_n. Examination of Eqs. (22)–(25) shows that the positive-going axis crossings of the higher harmonics will occur at times earlier than those for lower harmonics. The higher harmonics lead the lower ones, so that the region of steepest positive slope should occur at a time somewhat before the (positive-going) axis crossing of the fundamental. If the driving frequency is now increased, the phase angles will decrease and eventually become negative. The effect will be stronger for higher harmonics, so that the higher harmonics will begin to lag the lower and the region of greatest slope should occur at a later position on the fundamental. Thus, for weak shocks, as frequency increases from below the resonance of the fundamental to above, the shock front should migrate across the positive-going axis crossing of the fundamental, occurring at later times.

As an aid in visualizing the physical behavior of the pressure wave in the cavity, assume that a one-dimensional finite-amplitude standing wave is excited in a tube with fundamental frequency close to the lowest eigenfrequency of the cavity. The pressure waveform is

$$\frac{p}{\rho_0 c^2} = M_1 \sum_n R_n \cos(nkx) \, \sin(n\omega t + \phi_n), \qquad (31)$$

where $R_1 = 1$ and $\phi_1 = 0$. The other R_n and ϕ_n are found from solution of Eq. (30). In the idealized cavity, the Q's are related by $Q_n = Q_1 \sqrt{n}$ and the phases θ_n are given by

$$\tan \theta_n \sim (\sqrt{n} - 1) - 2Q_1 \sqrt{n} \, \frac{\Delta\omega}{\omega_1} \qquad (32)$$

[obtained from Eq. (4)].

If the frequency parameter $2Q_1 \, \Delta\omega/\omega_1$ is set equal to 1, then $\theta_n = -\pi/4$ for all n and each standing wave is excited at a frequency corresponding to the upper half power point on its response function. For this idealized case, an approximate solution can be obtained for a fairly high value of the strength parameter. This allows graphing representative waveforms that have clearly visible nonlinear distortion. If it is assumed that the phase angles ϕ_n tend toward $-(n-1)\pi/4$, then the simple form $R_n = An^{-b}$ for $n > 1$ (with $R_1 = 1$) comes close to satisfying Eq. (30) for $A \sim 0.6$ and $b \sim 1.6$. (This corresponds to a strength parameter of about 1.2 or a Goldberg number ~ 0.6.) Figure 1 shows the pressure wave for this approximate solution as a function of time at various locations in the tube. The progression of the pressure front as it splits and reflects back and forth between the ends of the tube can be seen clearly. Had the tube been driven at the second harmonic, there would have been two such pairs moving in opposite directions, superimposing at both ends of the tube and also in the middle. The sequence of sketches presented in Fig. 1 would then describe the sequence of events in one-half of the tube, and the behavior in the second half can be obtained by symmetry.

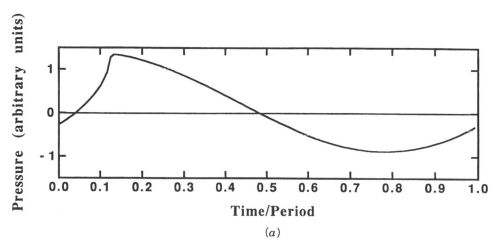

(a)

Fig. 1 Time dependence of an approximate nonlinear pressure waveform at various locations within an idealized tube with wall losses: $x/L =$ (a) 0, (b) $\frac{1}{8}$, (c) $\frac{1}{4}$, (d) $\frac{3}{8}$, (e) $\frac{1}{2}$, (f) $\frac{5}{8}$, (g) $\frac{3}{4}$, (h) $\frac{7}{8}$, (i) 1. Frequency parameter ~ 1, strength parameter ~ 1.2.

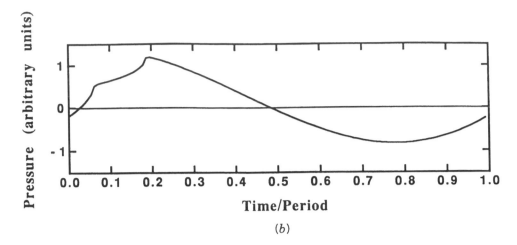

(b)

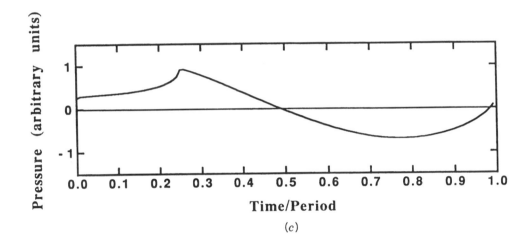

(c)

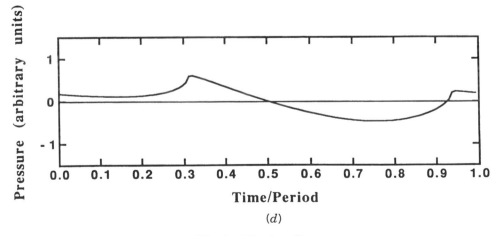

(d)

Fig. 1 (*Continued*)

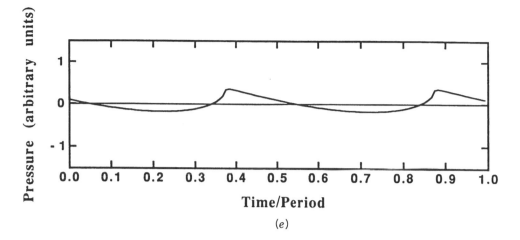

(e)

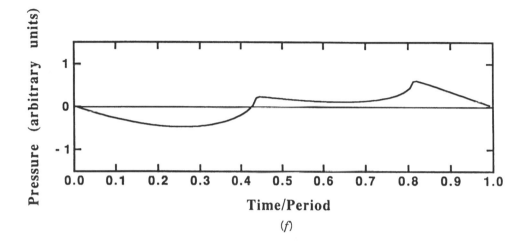

(f)

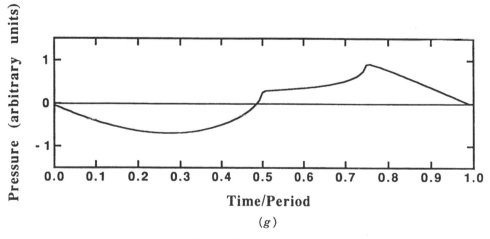

(g)

Fig. 1 (*Continued*)

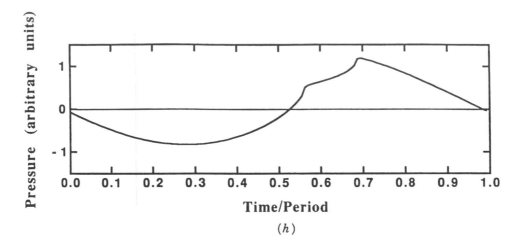

(h)

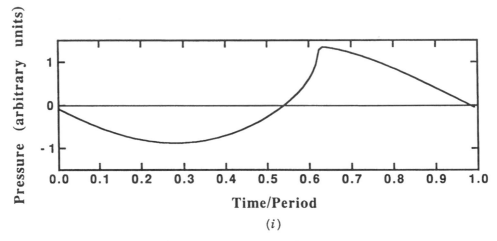

(i)

Fig. 1 (*Continued*)

REFERENCES

1. E. Lettau, "Messungen und Gasschwingungen Grosser Amplitude in Rohrleitungen," *Deut. Kraftfahrtforschung*, Vol. 39, 1939, pp. 1–17.

2. A. A. Atchley, H. E. Bass, and T. J. Hofler, "Development of Nonlinear Waves in a Thermoacoustic Prime Mover," in M. F. Hamilton and D. T. Blackstock (Eds.), *Frontiers of Nonlinear Acoustics 12th ISNA*, Elsevier Applied Science, New York, 1990, pp. 603–608.

3. D. F. Gaitan and A. A. Atchley, "Finite Amplitude Standing Waves in Harmonic and Anharmonic Tubes," *J. Acoust. Soc. Am.*, Vol. 93, 1993, pp. 2489–2495.

4. T. S. Lucas, "U.S. Patent Number 5,020,997: Standing Wave Compressor," *J. Acoust. Soc. Am.*, Vol. 90, 1991, p. 2875.

5. R. Betchov, "Nonlinear Oscillations of a Column of Gas," *Phys. Fluids*, Vol. 1, 1958, pp. 205–212.

6. R. A. Saenger and G. E. Hudson, "Periodic Shock Waves in Resonating Gas Columns," *J. Acoust. Soc. Am.*, Vol. 32, 1960, pp. 961–970.

7. W. Chester, "Resonant Oscillations in Closed Tubes," *J. Fluid Mech.*, Vol. 18, 1964, pp. 44–64.

8. A. B. Coppens and J. V. Sanders, "Finite-Amplitude Standing Waves in Rigid-Walled Tubes," *J. Acoust. Soc. Am.*, Vol. 43, 1968, pp. 516–529.

9. S. Tempkin, "Nonlinear Gas Oscillations in a Resonant Tube," *Phys. Fluids*, Vol. 11, 1968, pp. 960–963.

10. D. B. Cruikshank, Jr., "Experimental Investigation of Finite-Amplitude Acoustic Oscillations in a Closed Tube," *J. Acoust. Soc. Am.*, Vol. 52, 1971, pp. 1024–1036.

11. J. Jimenez, "Nonlinear Gas Oscillations in Pipes. Part 1. Theory," *J. Fluid Mech.*, Vol. 59, 1973, pp. 23–46.

12. B. R. Seymour and M. P. Mortell, "Resonant Acoustic Oscillations with Damping: Small Rate Theory," *J. Fluid Mech.*, Vol. 58, 1973, pp. 353–373.

13. B. Sturtevant, "Nonlinear Gas Oscillations in Pipes. Part 2. Experiment," *J. Fluid Mech.*, Vol. 63, 1974, pp. 97–120.

14. A. B. Coppens and J. V. Sanders, "Finite-Amplitude Standing Waves Within Real Cavities," *J. Acoust. Soc. Am.*, Vol. 58, 1975, pp. 1133–1140.

15. J. Keller, "A Note on Nonlinear Acoustic Resonances in Rectangular Cavities," *J. Fluid Mech.*, Vol. 87, 1978, pp. 299–303.

16. V. Sparrow, "Finite Amplitude Standing Wave Calculations Including Weak Shocks," in H. Hobaek (Ed.), *Advances in Nonlinear Acoustics*, World Scientific, River Edge, NJ, 1993, pp. 173–178.

23

NONLINEAR EFFECTS IN SOUND BEAMS

Mark F. Hamilton

1 INTRODUCTION

Many applications of intense sound, such as in sonar, medicine, and nondestructive evaluation, involve radiation from directional sources. Linear effects of diffraction determine length scales corresponding to near fields, focal lengths, and beamwidths. Depending on how these diffraction lengths compare with length scales corresponding to absorption and shock formation, the dominant nonlinear effects can be substantially different. This chapter describes these competing effects in detail, beginning with analytic solutions for harmonic generation in weakly nonlinear Gaussian beams, both unfocused and focused. Asymptotic and numerical solutions are used to illustrate shock formation, nonlinear side lobe generation, and self-demodulation of pulses in sound beams radiated by circular pistons. Hallmarks of nonlinear effects in sound beams include asymmetric waveform distortion and the increase in beamwidth and side lobe levels that accompanies shock formation along the axis of the beam. Another, in the case of bifrequency radiation, is the generation of a difference-frequency beam with high directivity and low side lobe levels.

2 PARABOLIC WAVE EQUATION

The combined effects of diffraction, absorption, and nonlinearity in *directional sound beams* (i.e., radiated from sources with dimensions that are large compared with the characteristic wavelength) are taken into account by the Khokhlov–Zabolotskaya–Kuznetsov (KZK) parabolic wave equation[1]:

Encyclopedia of Acoustics, edited by Malcolm J. Crocker
ISBN 0-471-80465-7 © 1997 John Wiley & Sons, Inc.

$$\frac{\partial^2 p}{\partial z \, \partial \tau} = \frac{c_0}{2} \left(\frac{\partial^2 p}{\partial r^2} + \frac{1}{r} \frac{\partial p}{\partial r} \right) + \frac{\delta}{2c_0^3} \frac{\partial^3 p}{\partial \tau^3}$$
$$+ \frac{\beta}{2\rho_0 c_0^3} \frac{\partial^2 p^2}{\partial \tau^2} . \tag{1}$$

Here p is the sound pressure, z the coordinate along the axis of the sound beam (assumed to be axisymmetric and propagating in the +z-direction), $\tau = t - z/c_0$ a retarded time, c_0 the small-signal sound speed, r the distance from the z-axis, δ the diffusivity of sound,[2] which accounts for absorption due to both viscosity and heat conduction, β the coefficient of nonlinearity, and ρ_0 the ambient density. Equation (1) provides an accurate description of the sound field at distances beyond several source radii and in regions close to the z-axis (e.g., within about $20°$ from the z-axis in the far field). The KZK equation may thus be used to model most regions of interest in directional sound beams. To the order of accuracy inherent in the KZK equation, it is consistent to use the plane-wave impedance relation $p = \rho_0 c_0 u$, where u is the z-component of the particle velocity vector. Consequently, source conditions expressed in terms of velocity are easily transformed into source conditions for the pressure. In the absence of nonlinearity ($\beta = 0$), Eq. (1) models the small-signal propagation of a directional beam in a thermoviscous fluid. Alternatively, when the first term on the right-hand side is omitted, Eq. (1) reduces to the Burgers equation [Eq. (16) of Chapter 18].

3 QUASILINEAR SOLUTIONS FOR HARMONIC RADIATION

Quasilinear solutions are based on the assumption that linear theory provides a good description of the harmonic components radiated directly by the source. The linear pressure field p_1 is obtained by solving Eq. (1) with $\beta = 0$. The secondary pressure field p_2 may then be interpreted as radiation from a volume distribution of *virtual sources* with strengths proportional to p_1^2. Solutions for the secondary field are obtained by substituting the linear solution p_1 into the nonlinear term in Eq. (1) and solving the resulting linear, inhomogeneous differential equation for p_2. Quasilinear solutions ($p = p_1 + p_2$) obtained in this way are valid provided the nonlinear interactions are relatively weak, that is, for $|p_2| \ll |p_1|$ throughout the beam.

Consider a bifrequency source in the plane $z = 0$ that radiates simultaneously at the two angular frequencies ω_a and ω_b. The linear solution for the primary wave field p_1 may be written

$$p_1(r,z,\tau) = \frac{1}{2j}[q_{1a}(r,z)e^{j\omega_a\tau} + q_{1b}(r,z)e^{j\omega_b\tau}] + \text{c.c.}, \quad (2)$$

where q_{1a} and q_{1b} are the complex-pressure amplitudes of the individual primary beams, $|q_{1a}|$ and $|q_{1b}|$ are the corresponding peak physical pressures, and c.c. designates complex conjugates of the preceding terms. An equivalent statement of Eq. (2) is $p_1 = \text{Im}\{q_{1a}e^{j\omega_a\tau} + q_{1b}e^{j\omega_b\tau}\}$. The functions q_{1a} and q_{1b} are slowly varying in z because the rapid variations on the wavelength scale have been factored out. When q_{1a} and q_{1b} are real, Eq. (2) reduces to $p_1 = q_{1a}\sin\omega_a t + q_{1b}\sin\omega_b t$. The solutions for q_{1a} and q_{1b} are given by[3]

$$q_1(r,z) = \left(\frac{jk}{z}\right)e^{-\alpha_1 z - jkr^2/2z}\int_0^\infty q_1(r',0)J_0\left(\frac{krr'}{z}\right)$$

$$\cdot\, e^{-jkr'^2/2z}r'\,dr', \quad (3)$$

where the subscripts a and b have been suppressed, J_0 is the Bessel function of the first kind of order zero, $q_1(r,0)$ is a source function, $k = \omega/c_0$ is the wavenumber, and $\alpha_1 = \delta\omega^2/2c_0^3$ is the thermoviscous attenuation coefficient. Since Eq. (3) is expressed in the frequency domain, it is not restricted to thermoviscous fluids. Simply replace the attenuation coefficients with suitable empirical relations or numerical values for the fluid under consideration. The same may be done in Eqs. (5) and (6) below.

The secondary wave field p_2 is composed of four components, the pressures at the second harmonics of the source frequencies ($2\omega_a$ and $2\omega_b$), the sum frequency ($\omega_+ = \omega_a + \omega_b$), and the difference frequency ($\omega_- = $

$\omega_a - \omega_b$, where $\omega_a > \omega_b$ is assumed):

$$p_2(r,z,\tau) = \frac{1}{2j}[q_{2a}(r,z)e^{j2\omega_a\tau} + q_{2b}(r,z)e^{j2\omega_b\tau}$$

$$+ q_+(r,z)e^{j\omega_+\tau} + q_-(r,z)e^{j\omega_-\tau}] + \text{c.c.} \quad (4)$$

Quasilinear solutions for the individual spectral components are obtained by substituting the known solution for the primary field [Eqs. (2) and (3)] into the nonlinear term in Eq. (1), substituting the unknown expression for the secondary field [Eq. (4)] into the linear terms, and solving the resulting system of inhomogeneous wave equations for the complex pressures q_{2a}, q_{2b}, q_+, and q_-.[4–6] We assume that no secondary-frequency components are radiated directly by the source, that is, $p_2(r,0,t) = 0$. The solution for the second-harmonic pressures is then

$$q_2(r,z) = \frac{j\beta k^2 e^{-\alpha_2 z}}{\rho_0 c_0^2}\int_0^z\int_0^\infty \frac{q_1^2(r',z')}{z-z'}J_0\left(\frac{2krr'}{z-z'}\right)$$

$$\cdot\, e^{\alpha_2 z' - jk(r^2+r'^2)/(z-z')}r'\,dr'\,dz'. \quad (5)$$

Equation (5) applies to either q_{2a} or q_{2b} once the appropriate subscript, a or b, is affixed to q_1, k, and α_2. The solution for the sum- and difference-frequency pressures is

$$q_\pm(r,z) = \pm\frac{j\beta k_\pm^2 e^{-\alpha_\pm z}}{2\rho_0 c_0^2}$$

$$\cdot\int_0^z\int_0^\infty \frac{q_{1a}(r',z')q_{1b}^{(*)}(r',z')}{z-z'}J_0\left(\frac{k_\pm rr'}{z-z'}\right)$$

$$\cdot\, e^{\alpha_\pm z' - jk_\pm(r^2+r'^2)/2(z-z')}r'\,dr'\,dz', \quad (6)$$

where the upper sign of $\pm$ is used for the sum frequency, the lower sign for the difference frequency. The notation $q_{1b}^{(*)}$ indicates that the complex conjugate of q_{1b} must be used when evaluating the difference-frequency component, but not when evaluating the sum-frequency component.

4 GAUSSIAN SOURCES

Source distributions with Gaussian amplitude profiles generate sound fields for which the integrals in Section 3 can be solved analytically for a variety of cases. Let the source functions be given by

$$q_{1a}(r,0) = p_{ga} \exp\left[-\left(\frac{r}{\epsilon_a}\right)^2\right], \qquad (7a)$$

$$q_{1b}(r,0) = p_{gb} \exp\left[-\left(\frac{r}{\epsilon_b}\right)^2\right], \qquad (7b)$$

where p_g and ϵ (with the appropriate subscripts) are the peak pressure and characteristic radius, respectively, of each Gaussian source.

4.1 Primary Beams

The linear solution obtained by substituting either of Eqs. (7) into Eq. (3) is

$$q_1(r,z) = \frac{p_g e^{-\alpha_1 z}}{1 - jz/z_0} \exp\left(-\frac{(r/\epsilon)^2}{1 - jz/z_0}\right), \qquad (8)$$

where the subscripts a and b are again suppressed. The quantity $z_0 = k\epsilon^2/2$ (which represents either $z_{0a} = k_a\epsilon_a^2/2$ or $z_{0b} = k_b\epsilon_b^2/2$) is referred to as the *Rayleigh distance*, and it marks the transition between the *near-field* and *far-field* regions of the primary beams. For $z \ll z_0$, Eq. (8) describes a collimated plane wave having a transverse amplitude distribution that matches that of the source. For $z \gg z_0$, Eq. (8) describes spherical waves having directivity

$$D_1(\theta) = \exp[-(\tfrac{1}{2}k\epsilon)^2 \tan^2\theta] \qquad (9)$$

in terms of the angle $\theta = \arctan(r/z)$ with respect to the z-axis. The pressure amplitude decays as $z^{-1}e^{-\alpha_1 z}$ in the far field.

4.2 Second-Harmonic Generation

The quasilinear solution for the second-harmonic pressure is obtained by substituting Eq. (8) into Eq. (5):

$$q_2(r,z) = \frac{jP_g e^{-\alpha_2 z + j(2\alpha_1 - \alpha_2)z_0}}{1 - jz/z_0} \exp\left(-\frac{2(r/\epsilon)^2}{1 - jz/z_0}\right)$$

$$\cdot \{E_1[j(2\alpha_1 - \alpha_2)z_0]$$

$$- E_1[j(2\alpha_1 - \alpha_2)(z_0 - jz)]\}, \qquad (10)$$

where $P_g = \beta p_g^2 k^2 \epsilon^2/4\rho_0 c_0^2$ has dimensions of pressure and $E_1(\sigma) = \int_\sigma^\infty u^{-1}e^{-u}\,du$ is the exponential integral,

which can be evaluated with standard mathematical software packages and reduces to simple asymptotic forms for large and small arguments. Comparison of the exponentials in Eqs. (8) and (10) reveals $q_2(r) \propto q_1^2(r)$, and the second-harmonic beamwidth is narrower at all ranges z, by a factor of $1/\sqrt{2}$, than the width of the primary beam. For $\alpha_2 > 2\alpha_1$ (as is the case for thermoviscous fluids, for which $\alpha_2 = 4\alpha_1$) and in the far field $[z \gg \max\{z_0, (\alpha_2 - 2\alpha_1)^{-1}\}]$, Eq. (10) describes spherical waves that decay as $z^{-2}e^{-2\alpha_1 z}$ and have directivity $D_2(\theta) = D_1^2(\theta)$.

With no absorption ($\alpha_1 = \alpha_2 = 0$), Eq. (10) reduces to

$$q_2(r,z) = \frac{jP_g \ln(1 - jz/z_0)}{1 - jz/z_0} \exp\left(-\frac{2(r/\epsilon)^2}{1 - jz/z_0}\right), \qquad (11)$$

which in the far field describes spherical waves that decay as $z^{-1}\ln(z/z_0)$, slightly slower than z^{-1} on account of energy transfer from the primary waves. Equation (11) reveals that the asymptotic decay rate $z^{-1}\ln(z/z_0)$ occurs not simply for $z/z_0 \gg 1$, which defines the far-field region in linear acoustics. Instead, the much stronger condition $\ln(z/z_0) \gg 1$ must be satisfied[5,6] [note that $\ln(-jz/z_0) = \ln(z/z_0) - j\pi/2$]. In general (not just for Gaussian beams), as long as effects of absorption are negligible, the far-field region in nonlinear acoustics may begin at distances that are orders of magnitude greater than in linear acoustics. Nonlinear near-field effects are particularly noticeable in harmonic beam patterns produced by piston sources (see Section 5.3).

4.3 Sum- and Difference-Frequency Generation

The general form of the quasilinear solution for the sum- and difference-frequency sound,[7] which is obtained from Eq. (6), requires numerical integration:

$$q_\pm(r,z) = \pm k_\pm P_\pm \exp\left(-\alpha_\pm z - \frac{k_\pm^2 r^2}{2f_\pm}\right)$$

$$\cdot \int_0^z \exp\left(-\alpha_T^\pm z' - \frac{k_\pm^2 k_a k_b(z_{0a} \mp z_{0b})^2 r^2}{2f_\pm(g_\pm \mp jf_\pm k_\pm z')}\right)$$

$$\cdot \frac{dz'}{g_\pm \mp jf_\pm k_\pm z'}, \qquad (12)$$

where $P_\pm = \beta k_\pm^2 z_{0a} z_{0b} p_{ga} p_{gb}/2\rho_0 c_0^2$, $\alpha_T^\pm = \alpha_a + \alpha_b - \alpha_\pm$, $f_\pm(z) = k_a z_{0a} + k_b z_{0b} - jk_\pm z$, and $g_\pm(z) = k_\pm^2 z_{0a} z_{0b} - j(k_b z_{0a} + k_a z_{0b})k_\pm z$. On axis, Eq. (12) reduces to

$$q_\pm(0,z) = \left(\frac{jP_\pm}{f_\pm}\right) \exp\left(-\alpha_\pm z \pm \frac{j\alpha_T^\pm g_\pm}{k_\pm f_\pm}\right)$$

$$\cdot \left\{E_1\left[\pm\frac{j\alpha_T^\pm g_\pm}{k_\pm f_\pm}\right]\right.$$

$$\left.- E_1\left[\pm\frac{j\alpha_T^\pm g_\pm}{k_\pm f_\pm}\left(1 \mp \frac{jf_\pm k_\pm z}{g_\pm}\right)\right]\right\}, \quad (13)$$

$$= \left(\frac{jP_\pm}{f_\pm}\right)\ln\left(1 \mp \frac{jf_\pm k_\pm z}{g_\pm}\right),$$

$$\alpha_a = \alpha_b = \alpha_\pm = 0. \quad (14)$$

When absorption is negligible, Eq. (12) reduces to

$$q_\pm(r,z) = \left(\frac{jP_\pm}{f_\pm}\right)\exp\left(-\frac{k_\pm^2 r^2}{2f_\pm}\right)$$

$$\cdot \left\{E_1\left[\frac{k_\pm^2 k_a k_b (z_{0a} \mp z_{0b})^2 r^2}{2f_\pm(g_\pm \mp jf_\pm k_\pm z)}\right]\right.$$

$$\left.- E_1\left[\frac{k_\pm^2 k_a k_b(z_{0a} \mp z_{0b})^2 r^2}{2f_\pm g_\pm}\right]\right\}. \quad (15)$$

Different asymptotic properties[8] are obtained for the sum- and difference-frequency fields according to the sign of the combined attenuation coefficient $\alpha_T^\pm$. Specifically, the relations $\alpha_T^+ < 0$ and $\alpha_T^- > 0$ apply to thermoviscous fluids (because $\alpha_\omega \propto \omega^2$), in which case the sum-frequency pressure decays in the far field as $z^{-2}e^{-(\alpha_a+\alpha_b)z}$ with directivity $D_+(\theta) = D_{1a}(\theta)D_{1b}(\theta)$.

The nonlinear interaction region where the difference-frequency sound is generated is often referred to as a *parametric array*.[9] An interesting and widely applied result is obtained with $\alpha_T^- > 0$, and for the case of equal source radii ($\epsilon_a = \epsilon_b = \epsilon$), neighboring primary frequencies ($\omega_a \approx \omega_b \gg \omega_-$) and strong absorption ($\alpha_T^- z_0 > 1$). The difference-frequency pressure then decays as $z^{-1}e^{-\alpha_- z}$ in the far field ($\alpha_T^- z \gg 1$), and the directivity is given by $D_-(\theta) = D_A(\theta)D_W(\theta)$, where[8]

$$D_A(\theta) = \exp\left[-\left(\frac{k_-\epsilon}{2\sqrt{2}}\right)^2 \tan^2\theta\right], \quad (16a)$$

$$D_W(\theta) = \frac{1}{\sqrt{1+(k_-/2\alpha_T^-)^2\tan^4\theta}}. \quad (16b)$$

For small $k_-\epsilon$, as is often the case for $\omega_a \approx \omega_b \gg \omega_-$, the *aperture factor*[10] $D_A(\theta)$ is a relatively weak function of

θ in comparison with the *Westervelt directivity*[9] $D_W(\theta)$, and the directivity of the difference-frequency radiation is therefore determined primarily by the latter. [The form of $D_W(\theta)$ given in Eq. (16b) is the parabolic (i.e., small-angle) approximation of the classical result obtained by Westervelt.[9]] Note that $D_W(\theta)$ does not depend on the source radius ϵ, only on the ratio k_-/α_T^-, and no side lobes are predicted. Consequently, the nonlinearly generated difference-frequency sound forms a more narrow beam than if it were radiated directly by the source of the primary beams. For sources other than Gaussian, the same far-field properties are obtained but with $D_A(\theta)$ replaced by the directivity function corresponding to direct, linear radiation at frequency ω_- by a source with amplitude distribution $q_{1a}(r,0)q_{1b}^*(r,0)$. See Chapter 53 for more extensive discussion of parametric arrays.

4.4 Focused Sources

When sources a and b are focused at distances d_a and d_b, respectively, Eqs. (7) are replaced by (with the subscripts a and b suppressed)

$$q_1(r,0) = p_g \exp\left[-\left(\frac{r}{\epsilon}\right)^2 + \frac{jkr^2}{2d}\right], \quad (17)$$

which reduces to Eqs. (7) when the sources are focused at infinity ($d = \infty$). Equations (7) are transformed into Eq. (17) if, in Eqs. (7), ϵ is replaced by $\tilde{\epsilon} = \epsilon(1-jG)^{-1/2}$, where $G = z_0/d$. Likewise, the effects of focusing on the radiated sound may be included simply by replacing ϵ by $\tilde{\epsilon}$ everywhere throughout the expressions for q_1, q_2, and q_+. In the expressions for q_-, however, ϵ_a is replaced by $\tilde{\epsilon}_a$ but ϵ_b is replaced by $\tilde{\epsilon}_b^*$. For example, Eq. (8) (with $\alpha_1 = 0$) and Eq. (11) become, respectively,

$$q_1(r,z) = \frac{p_g}{1-(1+jG^{-1})z/d}$$

$$\cdot \exp\left(-\frac{(1-jG)(r/\epsilon)^2}{1-(1+jG^{-1})z/d}\right), \quad (18)$$

$$q_2(r,z) = \frac{jP_g}{1-jG}\frac{\ln[1-(1+jG^{-1})z/d]}{1-(1+jG^{-1})z/d}$$

$$\cdot \exp\left(-\frac{2(1-jG)(r/\epsilon)^2}{1-(1+jG^{-1})z/d}\right). \quad (19)$$

Since $|q_1(0,d)| = Gp_g$, G is referred to as the linear *focusing gain*. The maximum value of $|q_1(0,z)|$ is located at $z = d(1+G^{-2})^{-1}$, and it approaches the geometric focus at $z = d$ with increasing G.

4.5 Approximation for Circular Piston Sources

Gaussian beam formulas can be used to approximate the field produced by a plane circular piston at locations beyond the near field and close to the axis (i.e., within the main radiation lobe). For a circular piston of radius a and with effective peak pressure $p_0 = \rho_0 c_0 u_0$, where u_0 is the velocity amplitude of the piston, Eq. (3) yields an axial pressure amplitude $(ka^2/2z)p_0 e^{-\alpha_1 z}$ and directivity $D_1(\theta) = 2J_1(ka \tan \theta)/ka \tan \theta$ for the far field. Comparison of these expressions with Eqs. (8) and (9) leads to the circular piston source transformation[11] $p_g = 2p_0$ and $\epsilon = a/\sqrt{2}$. Application of the transformation to Eq. (8) produces

$$q_1(r,z) = \frac{2p_0 e^{-\alpha_1 z}}{1 - j4z/ka^2} \exp\left(-\frac{2(r/a)^2}{1 - j4z/ka^2}\right). \quad (20)$$

Application of the same transformation to q_2 and $q_\pm$ yields good predictions for the secondary pressures beyond the near fields of the primary beams and close to the z-axis, particularly for weak attenuation (see Fig. A1 of Ref. 11).

5 PISTON SOURCES

Let the source condition be given by

$$p(r,0,t) = \begin{cases} p_0 E(t) \sin \omega_0 t, & r \leq a, \\ 0, & r > a, \end{cases} \quad (21)$$

where the envelope function $E(t)$ modulates the carrier signal of frequency ω_0. Equation (21) is used to model the vibration of a circular piston having characteristic velocity amplitude $u_0 = p_0/\rho_0 c_0$. For $E(t) = 1$, that is, a monofrequency source, evaluation of Eq. (3) for the magnitude of the axial pressure field yields $|q_1(0,z)| = 2p_0 e^{-\alpha_0 z}|\sin(z_0/2z)|$, where $z_0 = k_0 a^2/2$, $k_0 = \omega_0/c_0$, and α_0 is the attenuation coefficient at frequency ω_0. The axial solution is in good agreement with the exact solution of the wave equation for a baffled circular piston (i.e., without the parabolic approximation; see Chapter 9) at distances $z/a \gtrsim (k_0 a)^{1/3}$. The corresponding far-field directivity function was given in Section 4.5.

Equations (5) and (6) are not available in closed form for either monofrequency or bifrequency piston sources, although asymptotic expressions have been obtained for the far field.[5] The main far-field properties are similar to those discussed in Section 4 for Gaussian beams. The quasilinear integral solutions have been evaluated

numerically,[4–6] but sophisticated computational techniques are required because of the highly oscillatory nature of the integrands. An alternative approach is to solve the fully nonlinear KZK equation with finite-difference methods, either in the frequency domain[12] or in the time domain.[13]

5.1 Self-demodulation

Now let $E(t)$ vary slowly in comparison with $\sin \omega_0 t$. For strong absorption ($\alpha_0 z_0 > 1$), an approximate quasilinear solution ($p = p_1 + p_2$) of Eq. (1) for the axial pressure in the beam is given by[14]

$$p(0,z,\tau) = p_0 \left[f(\tau) - f\left(\tau - \frac{a^2}{2c_0 z}\right) \right.$$

$$\left. + \frac{\beta p_0 a^2}{16 \rho_0 c_0^4 \alpha_0 z} \frac{d^2 E^2}{d\tau^2} \right] * \mathcal{D}(z,\tau), \quad (22)$$

where $\mathcal{D}(z,t) = (c_0^3/2\pi \delta z)^{1/2} \exp(-c_0^3 t^2/2\delta z)$ is the dissipation function for a thermoviscous fluid, the aster-

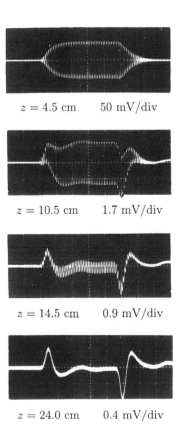

$z = 4.5$ cm   50 mV/div

$z = 10.5$ cm   1.7 mV/div

$z = 14.5$ cm   0.9 mV/div

$z = 24.0$ cm   0.4 mV/div

Fig. 1 Measured axial pressure waveforms that demonstrate the self-demodulation of a 10-MHz pulse in carbon tetrachloride.[16]

isk designates convolution with respect to time, and $f(t) = E(t) \sin \omega_0 t$. The high-frequency component (ω_0) is absorbed within the near field, and what remains in the far field is a distorted replica of the envelope $E(t)$. For $z \ll \alpha_E^{-1}$, where α_E is the attenuation coefficient at the characteristic frequency of the envelope, the far-field axial waveform is given simply by the Berktay result[15]

$$p(0, z, \tau) = \frac{\beta p_0^2 a^2}{16 \rho_0 c_0^4 \alpha_0 z} \frac{d^2 E^2}{d\tau^2}, \qquad \alpha_0^{-1} \ll z \ll \alpha_E^{-1}.$$

(23)

Experimental verification[16] of the *self-demodulation* phenomenon is shown in Fig. 1 for a 10-MHz pulse in carbon tetrachloride. At $z = 4.5$ cm the waveform is unaffected by nonlinearity. The self-demodulation process becomes visible near $z = 10.5$ cm, the high-frequency component is strongly attenuated in comparison with the envelope at $z = 14.5$ cm, and only the distorted (i.e., squared and twice differentiated) envelope remains at $z = 24.0$ cm. Equation (22) is in excellent agreement with measurements of this type.[14]

5.2 Waveform Distortion and Shock Formation

Predictions of waveform distortion and shock formation cannot be made on the basis of quasilinear solutions, and numerical computations are required. Shown in the left column of Fig. 2 are axial pressure waveforms, with the corresponding frequency spectra $S(\omega)$ in the right column (normalized by the peak spectral magnitude S_0 at $z = 0$), computed for a thermoviscous fluid with a finite-difference solution[13] of Eq. (1). The source waveform is a Gaussian tone burst with envelope $E(t) = \exp[-(\omega_0 t/3\pi)^2]$. The absorption is relatively weak ($\alpha_0 z_0 = 0.1$) and the source amplitude is relatively high [$p_0 = 4\rho_0 c_0^2/\beta(k_0 a)^2$, for which a plane wave would form a shock at distance $\frac{1}{2} z_0$ in a lossless fluid]. By the end of the (small-signal) near field ($z/z_0 = 1$), the combined effects of nonlinearity and diffraction on the waveform have produced sharpening of the positive cycles, rounding of the negative cycles, and the development of shock fronts. In addition, the peak positive pressures are approximately twice the peak negative pressures. The waveform distortion in the near field causes energy to be shifted primarily upward in the frequency spectrum. Farther away from the source, absorption filters out the nonlinearly generated high-frequency components and the shock fronts disappear ($z/z_0 = 10$), the relative importance of the low-frequency spectrum increases ($z/z_0 = 30$), and a self-demodulated, low-frequency signal is all that remains at $z/z_0 = 100$.

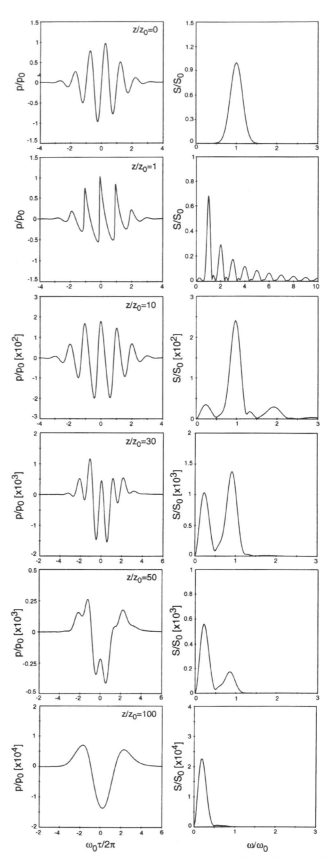

Fig. 2 Computed axial pressure waveforms (left column) and corresponding frequency spectra (right column) for a Gaussian tone burst radiated by a circular piston.[13]

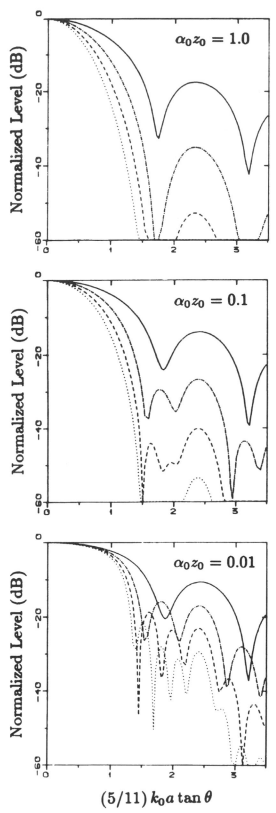

Fig. 3 Computed beam patterns at $z = 10z_0$ for the fundamental component $n = 1$ (—), second-harmonic component $n = 2$ ($-\cdot-$), third-harmonic component $n = 3$ ($--$), and fourth harmonic component $n = 4$ ($\cdots$) due to radiation of sound from a monofrequency circular piston.[17]

5.3 Harmonic Beam Patterns

In Fig. 3 are shown numerical calculations[17] of normalized beam patterns at range $z = 10z_0$ for $E(t) = 1$, that is, for a monofrequency source. The source pressure is $p_0 = 3\rho_0 c_0^2 / \beta (k_0 a)^2$, for which the plane-wave shock formation distance is $\frac{2}{3}z_0$. Three levels of absorption are considered: $\alpha_0 z_0 = 1.0$, 0.1, and 0.01. In the case of strong absorption ($\alpha_0 z_0 = 1.0$), the far-field structure is fully established at $z = 10z_0$, and the directivity function $D_n(\theta)$ for the nth-harmonic component is given very nearly by $D_n(\theta) = D_1^n(\theta)$. With increasing n, the harmonic components become more directional and exhibit greater side lobe suppression. For weak absorption ($\alpha_0 z_0 = 0.01$), the nth-harmonic beam pattern possesses n times as many side lobes as are present in the beam pattern at the fundamental frequency. Similar phenomona also occur in focused sound beams.[18] The additional side lobes are due to *near-field effects*, and they decay faster with range than do side lobes in the fundamental beam pattern.[6] However, the decay rates of different side lobes may differ by only a factor of order $\ln(z/z_0)$, and consequently the nonlinear near-field effects may be significant out to hundreds of Rayleigh distances. Finally, it can be seen that decreasing the effect of absorption (or similarly, increasing the source level and therefore the relative effect of nonlinearity) produces flattening of the main lobes, which causes the relative levels of the side lobes to increase (compare results for $\alpha_0 z_0 = 1.0$ and $\alpha_0 z_0 = 0.01$). Ultimately, nonlinear effects in the main lobe can become sufficiently strong that *acoustic saturation*[19] occurs, and a further increase in source level produces no increase in the axial pressure at a given distance.

Acknowledgment

The Office of Naval Reearch is gratefully acknowledged for supporting much of the work that produced the results described in this chapter.

REFERENCES

1. N. S. Bakhvalov, Ya. M. Zhileikin, and E. A. Zabolotskaya, *Nonlinear Theory of Sound Beams*, American Institute of Physics, New York, 1987.

2. J. Lighthill, *Waves in Fluids*, Cambridge University Press, New York, 1980.

3. G. S. Garrett, J. Naze Tjøtta, and S. Tjøtta, "Nearfield of a Large Acoustic Transducer. Part I: Linear Radiation," *J. Acoust. Soc. Am.*, Vol. 72, 1982, pp. 1056–1061.

4. G. S. Garrett, J. Naze Tjøtta, and S. Tjøtta, "Nearfield of a Large Acoustic Transducer. Part II: Parametric Radiation," *J. Acoust. Soc. Am.*, Vol. 74, 1983, pp. 1013–1020.

5. G. S. Garrett, J. Naze Tjøtta, and S. Tjøtta, "Nearfield of a Large Acoustic Transducer. Part III: General Results," *J. Acoust. Soc. Am.*, Vol. 75, 1984, pp. 769–779.

6. J. Berntsen, J. Naze Tjøtta, and S. Tjøtta, "Nearfield of a Large Acoustic Transducer. Part IV: Second Harmonic and Sum Frequency Radiation," *J. Acoust. Soc. Am.*, Vol. 75, 1984, pp. 1383–1391.

7. C. M. Darvennes and M. F. Hamilton, "Scattering of Sound by Sound from Two Gaussian Beams," *J. Acoust. Soc. Am.*, Vol. 87, 1990, pp. 1955–1964.

8. C. M. Darvennes, M. F. Hamilton, J. Naze Tjøtta, and S. Tjøtta, "Effects of Absorption on the Nonlinear Interaction of Sound Beams," *J. Acoust. Soc. Am.*, Vol. 89, 1991, pp. 1028–1036.

9. P. J. Westervelt, "Parametric Acoustic Array," *J. Acoust. Soc. Am.*, Vol. 35, 1963, pp. 535–537.

10. J. Naze Tjøtta and S. Tjøtta, "Effects of Finite Aperture in a Parametric Acoustic Array," *J. Acoust. Soc. Am.*, Vol. 68, 1980, pp. 970–972.

11. M. F. Hamilton and F. H. Fenlon, "Parametric Acoustic Array Formation in Dispersive Fluids," *J. Acoust. Soc. Am.*, Vol. 76, 1984, pp. 1474–1492.

12. J. Naze Tjøtta, S. Tjøtta, and E. Vefring, "Propagation and Interaction of Two Collinear Finite Amplitude Sound Beams," *J. Acoust. Soc. Am.*, Vol. 88, 1990, pp. 2859–2870.

13. Y.-S. Lee and M. F. Hamilton, "Time-Domain Modeling of Pulsed Finite-Amplitude Sound Beams," *J. Acoust. Soc. Am.*, Vol. 97, 1995, pp. 906–917.

14. M. A. Averkiou, Y.-S. Lee, and M. F. Hamilton, "Self-Demodulation of Amplitude and Frequency Modulated Pulses in a Thermoviscous Fluid," *J. Acoust. Soc. Am.*, Vol. 94, 1993, pp. 2876–2883.

15. H. O. Berktay, "Possible Exploitation of Non-linear Acoustics in Underwater Transmitting Applications," *J. Sound Vib.*, Vol. 2, 1965, pp. 435–461.

16. M. B. Moffett, P. J. Westervelt, and R. T. Beyer, "Large-Amplitude Pulse Propagation—A Transient Effect," *J. Acoust. Soc. Am.*, Vol. 47, 1970, pp. 1473–1474.

17. M. F. Hamilton, J. Naze Tjøtta, and S. Tjøtta, "Nonlinear Effects in the Farfield of a Directive Sound Source," *J. Acoust. Soc. Am.*, Vol. 89, 1985, pp. 202–216.

18. M. A. Averkiou and M. F. Hamilton, "Measurements of Harmonic Generation in a Focused Finite-Amplitude Sound Beam," *J. Acoust. Soc. Am.*, Vol. 98, 1995, pp. 3439–3442.

19. J. A. Shooter, T. G. Muir, and D. T. Blackstock, "Acoustic Saturation of Spherical Waves in Water," *J. Acoust. Soc. Am.*, Vol. 55, 1974, pp. 54–62.

24

NONLINEAR LUMPED ELEMENTS

WILLIAM E. ZORUMSKI

1 INTRODUCTION

Acoustics is usually based on the assumption of small motions. This is a good approximation for sound in the tolerable range of hearing, but there are many instances, such as jet engine fans and internal combustion engine exhausts, where intense sound waves must be considered. The amplitudes of these waves are sufficiently large so that nonlinear effects are significant. Noise control is an area where this occurs often. The sound must be controlled because it is intense, and this leads to the consideration of nonlinear effects in the control process. Lumped elements are an approximation, as discussed in Chapter 13, used when the size of the element is small in comparison to a typical acoustic wavelength. This chapter will consider some representative causes of nonlinearity in lumped elements and then focus on some details of nonlinear resistance and reactance as is exhibited by some porous materials and perforated plates used in noise control. It will be shown how the effect of nonlinearity is to cause an interaction between sound at different frequencies, which is a considerable complication in understanding, analysis, measurement, and design.

2 SOURCES OF NONLINEARITY

There are many sources of nonlinearity, but the physical example of the interaction of a sound wave in air with a perforated plate illustrates most of the nonlinear effects. These effects are observed in irregular materials such as woven composites and fibermetals also. Consider a plate, as shown in Fig. 1, that has N holes of diameter d per unit face area of the plate and is mounted a fixed

Encyclopedia of Acoustics, edited by Malcolm J. Crocker
ISBN 0-471-80465-7 © 1997 John Wiley & Sons, Inc.

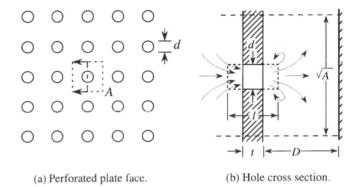

(a) Perforated plate face.　　(b) Hole cross section.

Fig. 1　Unsteady flow through a perforated plate.

distance D in from a rigid wall. The external area associated with a single hole is then $A = 1/N$, as denoted by the dashed square around the hole. Looking at the fine detail of a cross section through the plate, as shown on the right, we imagine air flow through the hole caused by a sound wave. The air stream must contract from a channel of cross section A and squeeze through the hole with area A_h. Energy is dissipated in this flow by viscous effects within the thickness t and by shed vortices as the flow exits the hole. As air flows into the space behind the plate, the pressure increases behind the plate, and as air flows out, the pressure decreases.

Now construct a harmonic oscillator model for this interaction. A "slug" of air of length l within the hole forms an effective mass $m = \rho A_h l$, but this mass is not constant. As velocity through the hole increases, pressure decreases due to the Bernoulli effect and density decreases. Also, a jet forms on the outflow side so that the length of the slug increases. Consequently, the inertial force is not simply proportional to acceleration $\ddot{x}$ but may depend as well on velocity $\dot{x}$ and displacement x. The damping force may be simply proportional to velocity $\dot{x}$ when velocity is small but can increase quickly when the exiting jet begins to shed vortices. The effective spring for the oscillator is the air between the plate and the wall, but the air pressure, assuming adiabatic compression and expansion, increases faster than the density so that the spring force is nonlinear also. The equation for the motion of the oscillator shown in Fig. 2 is

$$m\ddot{x} + c\dot{x} + kx = f \tag{1}$$

where $f = pA_h$ is the exciting force due to the pressure in front of the plate. This would be a standard equation, but the mass m, damping coefficient c, and spring coefficient k are not constants, as depicted by the straight dashed lines in Fig. 2, but instead are the slopes of the solid curves, which may all depend on x, $\dot{x}$, and $\ddot{x}$. These effects are all horribly complicated, even for the simple example of a perforated plate. A comprehensive discussion of the impedance of perforates is given by Melling.[1]

An alternative to modeling the flow details within the porous material is to adopt a phenomenologic approach, relying on measurements of certain assumed functions, to incorporate nonlinear effects into acoustic interactions. One such approach is given in the following section.

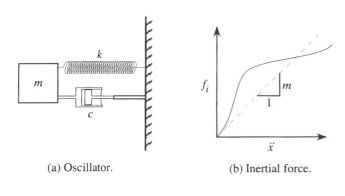

(a) Oscillator.   (b) Inertial force.

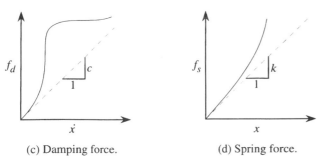

(c) Damping force.   (d) Spring force.

Fig. 2   Harmonic oscillator model for perforated plate.

3   NONLINEAR MATERIALS

3.1   Spectral Impedance

Impedance is the parameter used in linear acoustics to relate acoustic pressure to acoustic velocity. An impedance equation is derived readily from Eq. (1) by assuming complex harmonic motions of the form $e^{i\omega t}$, where ω is circular frequency and t is time. Let the velocity $\dot{x} = v$ so that acceleration is $\ddot{x} = i\omega v$ and displacement is $x = -iv/\omega$. Equation (1) is then

$$
\begin{aligned}
p &= \left[\frac{c}{A_h} + i\left(\omega\,\frac{m}{A_h} - \frac{1}{\omega}\,\frac{k}{A_h} \right) \right] v \\
&= [R + iX]v \\
&= Zv.
\end{aligned}
\tag{2}
$$

Equation (2) is still the equation of motion of a harmonic oscillator, but expressed in terms of variables common to acoustics. Resistance R is clearly proportional to the damping coefficient of the oscillator. Reactance X has two parts: The first, proportional to mass, increases with frequency, while the second, proportional to the spring constant, decreases. Acoustic impedance Z is the phased sum of the two parameters, $Z = R + iX$, and is the constant of proportionality relating acoustic pressure and velocity. More precisely, it is the constant of proportionality for complex pressure and velocity at the same frequency, and hence the qualifying adjective *spectral* is introduced to make this distinction from what follows.

3.2   Temporal Impedance

Temporal impedance is the phenomenologic operator introduced by Zorumski and Parrott[2] to relate instantaneous acoustic pressure and velocity when both are arbitrary functions of time:

$$Z_t = R_t + X_t\,\frac{\partial}{\partial t}. \tag{3}$$

Any attempt to derive this expression reduces eventually to so much hand waving. It is true if it works and only because it works. It is a postulate. The construction of the operator has X_t cast in the role of a masslike effect. An operator could be introduced with a springlike effect by adding an integral, with respect to time, to the right-hand side of Eq. (3). This term would complete the analogy with the spectral impedance operator. The more limited form shown in Eq. (3) was used in the instance where a porous plate or thin sheet of porous material was to be represented by an equation relating pressure difference across the sheet to velocity through the sheet:

$$\Delta p = Z_t v. \qquad (4)$$

Here, Δp would be the pressure difference between the front and back of the plate or sheet. Since the backing space is not included, there is less need for an integral in the operator.

3.3 Measurement of Nonlinear Impedance

Resistance The postulate represented by Eqs. (3) and (4) has important and immediate consequences for the measurement of temporal impedance. Imagine a sample of material placed in a tube, as shown in Fig. 3. The instantaneous pressure drop across the material is Δp, and the velocity through the material is v. Before measurements may be made, some assumptions must be made about the nature of R_t and X_t. If they are constants, the problem is linear. Nonlinearity is represented by the dependence of R_t and X_t on (perhaps) particle displacement, velocity, or acceleration. Whatever the dependence, it must be valid for all possible time functions $\Delta p(t)$ and $v(t)$, and one of these possibilities is the case of steady flow. In this case the temporal impedance is identical to R_t, and this term can be at most a function of the velocity v. That is, R_t is the steady-flow resistance. This function is generated experimentally by adjusting the flow v through the material, measuring the pressure difference for each flow rate, and plotting $\Delta p/v$ versus v, as shown in Fig. 3.

Reactance Measurement of reactance is more challenging because it necessarily involves a dynamic test. The method of Ref. 2 was to expose the test specimen to a periodic (but not necessarily harmonic) signal and plot $\Delta p(t)$ versus $v(t)$, as shown in Fig. 4. Since the sig-

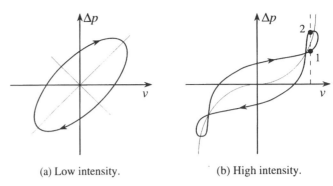

(a) Low intensity. (b) High intensity.

Fig. 4 Procedure for measuring temporal impedance.

nal is periodic, the plot must form a closed trajectory with time being a parameter representing position along the trajectory. The trajectory on the left of Fig. 4 is a typical result of a low-intensity test where the sample is behaving linearly. The observed trajectory is an inclined ellipse. The slope of the major axis is proportional to the resistance, and the ratio of the minor axis to the major axis is proportional to the reactance. When time increases in a clockwise sense moving along the trajectory, reactance is positive. On the right of Fig. 4 is a typical trajectory formed by a high-intensity test where the sample exhibits nonlinear behavior. Note the loops that have formed at the "ends" of the trajectory. Time is increasing while moving in a counterclockwise sense around these loops, which is an indication of negative reactance during this portion of the cycle.

Both resistance and reactance may be derived from a test of this sort if it is assumed that each is a function of velocity only. Simply select two times t_1 and t_2 on the trajectory where the velocities are equal by drawing a vertical line as shown on the right of Fig. 4. Accelerations $a = \dot{v}$ are not equal at these different times so that Eqs. (3) and (4) produce two simultaneous equations:

$$R_t[v] + a_1 X_t[v] = \Delta p(t_1), \qquad R_t[v] + a_2 X_t[v] = \Delta p(t_2).$$

$$(5)$$

Solving these equations gives both resistance and reactance for the selected velocity. Repeating the process for a sequence of velocities generates both functions.

It was verified in Ref. 2 that dynamically measured resistance was equal to flow resistance for fundamental test frequencies up to 4000 Hz and sound pressure levels up to 157 dB. This result supported the fundamental assumption of temporal impedance, that resistance is a function of velocity only, at least for that one material sample.

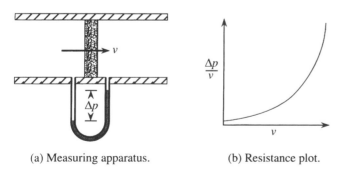

(a) Measuring apparatus. (b) Resistance plot.

Fig. 3 Flow resistance measurement.

4   EFFECTS OF NONLINEARITY

There are two cases where exact solutions are known[3] for the interaction of acoustic waves with a nonlinear lumped element. In the first case, an incident wave is transmitted through the element into a nonreflecting region behind the element. In the second case, the incident wave is periodic, and the space behind the element is one-quarter wavelength (of the fundamental) deep and terminates at a perfectly reflecting, or hard, wall. In both cases, a linear wave equation is assumed for the waves, but the lumped element is nonlinear. The waves thus give pressure and velocity in the forms

$$p(x, t) = f(x - t) + g(x + t), \qquad v(x, t) = f(x - t) - g(x + t).$$

$$(6)$$

All variables in Eq. (6) represent suitable nondimensional groups.

4.1   Nonreflecting Termination

Let the nonlinear element be placed at $x = 0$ and use subscripts 1 to denote waves where $x < 0$ and 2 to denote $x > 0$. Further, assume that the reactance is negligible so that the element is represented by

$$\Delta p(t) = R[v(t)]v(t), \qquad x = 0, \qquad (7)$$

where v is the velocity through the element. The nonreflecting condition gives $g_2(x + t) = 0$. Expressions for velocity through and pressure drop across the element are then

$$v(t) = f_1(-t) - g_1(t) = f_2(-t),$$

$$\Delta p(t) = f_1(-t) + g_1(t) - f_2(-t). \qquad (8)$$

Given the incident wave f_1, there results a single equation for the velocity $v(t)$:

$$(R[v(t)] + 2)v(t) - 2f_1(-t) = 0. \qquad (9)$$

Since the resistance must be positive, the function on the left of Eq. (9) is a monotonically increasing function of $v(t)$ and has a single real-valued solution. The solution for the case where $R[v] = 1 + v^2$ and $f_1(x - t) = \cos 2\pi(x - t)$ is

$$v(t) = -\left(\cos 2\pi t + \sqrt{1 + \cos^2 2\pi t} \right)^{-1/3}$$

$$+ \left(\cos 2\pi t + \sqrt{1 + \cos^2 2\pi t} \right). \qquad (10)$$

Clearly, a simple harmonic wave will be transmitted through a nonlinear element as a spectrum of harmonics of the incident wave.

4.2   Quarter-Wave Backing

Now consider the interaction of waves with the element at $x = 0$ but with a hard wall at $x = \frac{1}{4}$. Assume periodic waves such that $f(x - t + n) = f(x - t)$ and $g(x + t + n) = g(x + t)$ for all integers n. The condition of perfect reflection at $x = \frac{1}{4}$ gives $v(t + \frac{1}{2}) = -v(t)$, and if $R[v]$ is an even function, the following algebraic equation for the velocity through the element results:

$$(R[v(t)] + 1)v(t) + [f_1(\tfrac{1}{2} - t) - f_1(-t)] = 0. \qquad (11)$$

Using the same resistance and incident wave as before, the velocity through the element is

$$v(t) = -\frac{2}{3}\left(\cos 2\pi t + \sqrt{\frac{8}{27} + \cos^2 2\pi t} \right)^{-1/3}$$

$$+ \left(\cos 2\pi t + \sqrt{\frac{8}{27} + \cos^2 2\pi t} \right). \qquad (12)$$

5   CONCLUDING REMARKS

Lumped nonlinear acoustic elements cause interactions of waves with different frequencies. The effect of the element is influenced by the complete spectrum of the flow velocity through the element. The description of this complex process is simplified by using the time domain. In the time domain, acoustic impedance is replaced by a temporal impedance operator. The acoustic resistance that appears in this operator is identical to the steady-flow resistance for a broad range of frequencies so that the flow resistance is the primary descriptor of the nonlinear behavior of the element. The nature of the temporal reactance is not so clear. While it may be measured, it seems to depend on more than just the instantaneous flow velocity. Kuntz[4] has developed models of these functions for porous materials. This reference is recommended for the study of distributed nonlinear acoustic elements.

Grazing flow over porous materials has an important effect on their impedance. Rao and Munjal[5] have evaluated these effects for perforates. Further information on these effects and on the application of lumped-parameter methods to the analysis and design of internal engine mufflers is given by Sullivan and Crocker.[6–9]

REFERENCES

1. T. H. Melling, "The Acoustic Impedance of Perforates at Medium and High Sound Pressure Levels" *J. Sound Vib.*, Vol. 29, No. 1, 1973, pp. 1–65.

2. W. E. Zorumski and T. L. Parrott, "Nonlinear Acoustic Theory for Rigid Porous Materials," NASA TN D-6196, June 1971.

3. W. E. Zorumski, "Acoustic Scattering by a Porous Elliptic Cylinder with Nonlinear Resistance," Ph.D. Dissertation, Virginia Polytechnic Institute, March 1970.

4. H. L. Kuntz, II, "High Intensity Sound in Air Saturated Fibrous Bulk Porous Materials," ARL-TR-82-54, Applied Research Laboratories, University of Texas at Austin, September 1982.

5. K. N. Rao and M. L. Munjal, "Experimental Evaluation of Impedance of Perforates with Grazing Flows," *J. Sound Vib.*, Vol. 108, No. 2, July, 1986, pp. 283–295.

6. J. W. Sullivan and M. J. Crocker, "Analysis of Concentric-Tube Resonators Having Unpartitioned Cavities," *J. Acoust. Soc. Am.*, Vol. 64, No. 1, 1978, pp. 207–215.

7. J. W. Sullivan, "A method for Modelling Perforated Tube Muffler Components. I. Theory," *J. Acoust. Soc. Am.*, Vol. 66, No. 3, 1979, pp. 772–778.

8. J. W. Sullivan, "A Method for Modelling Perforated Tube Muffler Components. II. Application," *J. Acoust. Soc. Am.*, Vol. 66, No. 3, 1979, pp. 779–788.

9. J. W. Sullivan, "Some Gas Flow and Acoustic Pressure Measurements Inside a Concentric-Tube Resonator," *J. Acoust. Soc. Am.*, Vol. 76, No. 2, 1984, pp. 479–484.

25

CAVITATION

WERNER LAUTERBORN

1 INTRODUCTION

Cavitation, the rupture of liquids including the effects connected with it, may be classified according to the scheme given in Fig. 1. Cavitation is brought about by either tension in the liquid or a deposit of energy. The main technical areas where tension in a liquid plays a role are hydrodynamics and acoustics. Hydrodynamic cavitation occurs with ship propellers, turbines, pumps, and hydrofoils, devices that all introduce a pressure reduction in the liquid via Bernoulli pressure forces and eventually generate strong tension. Acoustic cavitation occurs with underwater sound projectors and vibrating containers as, for example, used for cleaning. In sufficiently strong sound fields the tension generated during the pressure reduction phase of the sound wave may produce cavitation. Energy deposit can be brought about in various ways. When light of high intensity is propagating through a liquid, for example, laser light into it, cavities may form by heating or dielectric breakdown. When, instead of photons, elementary particles are propagating (e.g., protons), cavities may form along their paths (bubble chamber). Other forms are local heat addition (boiling) and strong (static) electric fields (underwater spark). Acoustic cavitation has been emphasized in Fig. 1 because it is the topic of the present chapter. A selection of useful books and survey articles in descending order of their date of issue is given in Refs. 1–14.

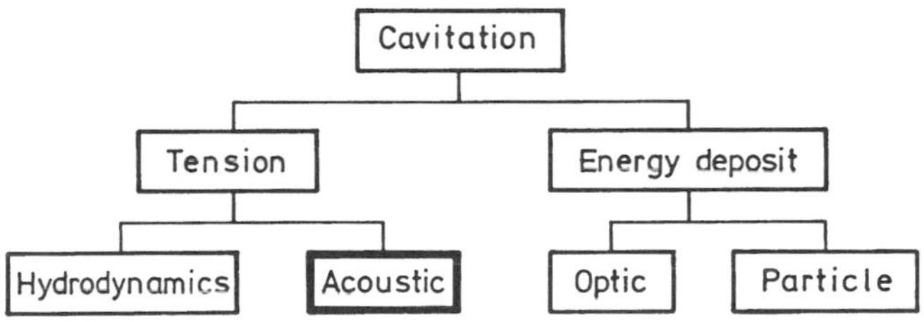

2 DEVICES

There exists a variety of devices to produce acoustic cavitation. The "Mason horn" (Fig. 2) is mainly used for

Encyclopedia of Acoustics, edited by Malcolm J. Crocker
ISBN 0-471-80465-7 © 1997 John Wiley & Sons, Inc.

Fig. 1 Classification scheme for the different types of cavitation.

cavitation erosion tests (besides in drilling). The probe to be tested forms the tip of an amplitude transformer (the horn) dipped into the liquid or is put at some distance opposite to the tip to get rid of the large accelerations encountered on the tip. The transducer driving the horn is usually made up of a sandwich of nickel plates (making use of magnetostriction) or plates of piezoelectric mate-

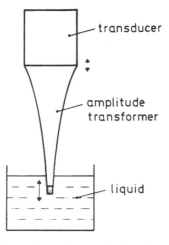

Fig. 2 Mason horn to produce cavitation for erosion studies.

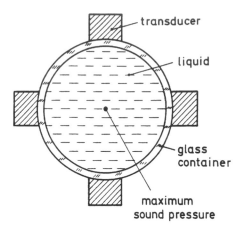

Fig. 3 Spherical or cylindrical container filled with liquid for cavitation threshold, bubble oscillation, and sonoluminescence studies.

rials (BaTiO$_4$ or PZT ceramics). Figure 3 shows a device to investigate cavitation thresholds, bubble oscillations, and sonoluminescence.[15, 16] A glass container, a sphere or cylinder, is driven by several transducers to generate a standing-wave pattern in the liquid at the fundamental (or some upper resonance) of the whole system. The fundamental resonance has the advantage that only one maximum of the sound pressure (and of the tension) occurs in the liquid. Also bubbles placed in the liquid can be "levitated," thus allowing extended study of their vibrations. Figure 4 shows another simple device to study cavitation thresholds, the dynamics of cavitation bubble clouds, and cavitation noise emission (acoustic chaos[17, 18]). It consists of a hollow cylinder of piezoelectric material sub-

merged in the liquid to be cavitated. When the system is driven sinusoidally to excite the fundamental resonance, maximum sound pressure and tension occurs at the center of the cylinder.

3 CAVITATION THRESHOLD

Sound waves in liquids need a certain intensity for cavitation to appear. Cavitation inception thus is a threshold process. It is commonly agreed that nuclei must mediate the inception, that is, tiny bubbles being stabilized by some mechanism against dissolution. There is the crevice model, where gas is trapped in conical pits of solid impurities,[19] and the skin model, with organic or surface-active molecules occupying the bubble wall.[20] When these nuclei encounter a sound field, they are set into oscillation and may grow by a process called rectified diffusion.[21] Rectified diffusion itself is a threshold

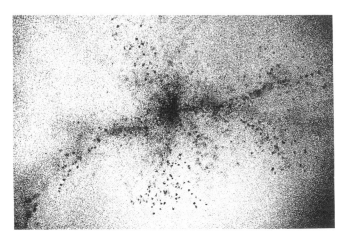

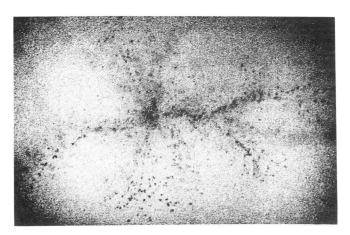

Fig. 5 Cavitation bubble cloud obtained inside a cylindrical transducer. Two planes from a holographic image. (Courtesy of F. Bader.)

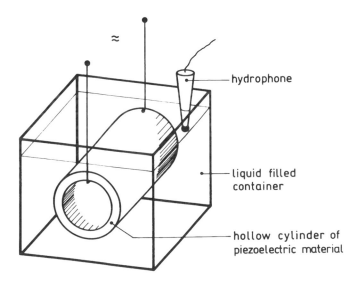

Fig. 4 Cylindrical transducer submerged in liquid for cavitation noise measurements (acoustic chaos).

process. It is the net effect of mass diffusion across the bubble wall due to varying bubble wall area and concentration gradients during oscillation. When a nucleus starts to grow, it eventually reaches a size where the oscillations turn into large excursions of the bubble radius, with strong and fast bubble collapse shattering the bubble into tiny fractions. This then marks the cavitation threshold, and acoustic cavitation sets in.

4 OBSERVED BUBBLE DYNAMICS AND PATTERN FORMATION

Substantial noise emission occurs upon acoustic cavitation. It is brought about by the action of the cavities or bubbles generated and set into oscillation in the sound field. Usually many bubbles are involved, forming a complex dynamic pattern (Fig. 5). The individu-

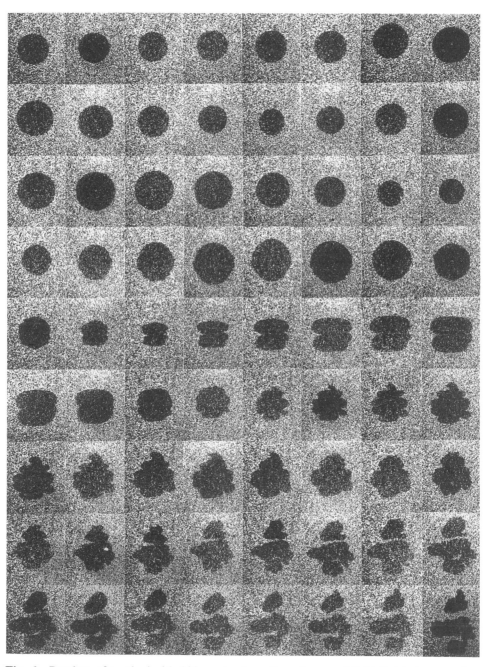

Fig. 6 Breakup of a spherical bubble under the action of a sound field that induces surface oscillations. Reconstructed images from a series of holograms taken at a rate of 66,700 holograms per second. Individual picture size is 2.4 × 2 mm. (Courtesy of W. Hentschel.)

ally moving and oscillating bubbles arrange themselves along a branchlike, filamentary structure, the filaments being called streamers. The pattern is steadily rearranging but is stable in gross appearance over hundreds of cycles of the driving sound field and resembles a Lichtenberg figure.

The processes leading to this type of structured bubble ensemble are very complex due to an interplay between competing mechanisms based on instabilities and cooperation. There are at least two time scales involved, a fast one connected with bubble oscillation and a slow one connected with bubble migration and spatial pattern formation. The spherical shape of a bubble is due to the surface tension of the liquid. This shape becomes unstable at higher oscillation amplitudes whereby surface waves[22, 23] are set up on the bubble wall, eventually leading to a breakup of the bubble (Fig. 6). This shattering yields a fast proliferation of small bubbles that is stopped when the bubbles become too small to be excited to large amplitudes.

In addition to the appearance of surface waves destroying a bubble, there is a related mechanism operating in aspherical geometries surrounding a cavitation bubble. The collapse of a bubble, that is, its fast approach to minimum size, may proceed with the formation of a liquid jet piercing the bubble.[5, 24, 25] The jet leads to a long protrusion sticking out of the bubble and to the formation of a vortex ring decaying into a host of microbubbles.

There is also the competing opposite phenomenon of coalescence in which, upon expansion, bubbles come into contact with each other. These effects surely are not sufficient to describe the pattern formation, as some attracting and guiding forces are needed. These are the Bjerknes forces,[26] which are attracting when two bubbles are oscillating in phase, and the radiation pressure forces, which drive the bubbles toward the pressure antinodes in a standing acoustic field when they are smaller than resonant size. Moreover, the bubbles themselves radiate sound and are thus interacting additionally. A first approach to account for the observed pattern formation in acoustic cavitation is given elsewhere.[27]

5 SINGLE-BUBBLE DYNAMICS

Cavitation bubble clouds are difficult to investigate both experimentally and theoretically. Therefore much effort is spent on single-bubble dynamics. Laser-produced bubbles are used to investigate the involved dynamics of single bubbles near boundaries with jet formation (Fig. 7), shock wave radiation, and vortex ring formation.[25, 28] Bubble collapse is accompanied by strong shock wave emission[28, 29] (Fig. 8). It comes about

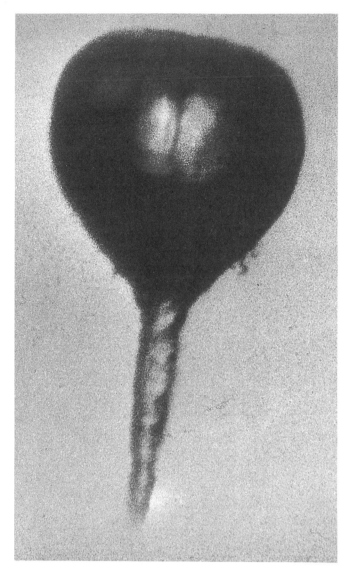

Fig. 7 Jet formation in the case of a bubble collapsing in the neighborhood of a solid wall (below the lower margin of the picture). The jet pierces the bubble from above and leads to the long protrusion.

through the sudden stop of the inflowing liquid when the bubble approaches a tiny fraction of its initial volume with a correspondingly high internal pressure. Simultaneously, high temperatures are reached inside the bubble, as indicated by the emission of a short flash of light (sonoluminescence).[15, 16]

Theory is not yet developed sufficiently to cope with the experimental dynamic system completely. Just the single spherical bubble can be treated fairly well.[11, 30] The simplest bubble model is called the Rayleigh–Plesset model:

$$\rho R \ddot{R} + \tfrac{3}{2}\rho \dot{R}^2 = P_i - P_e, \qquad (1)$$

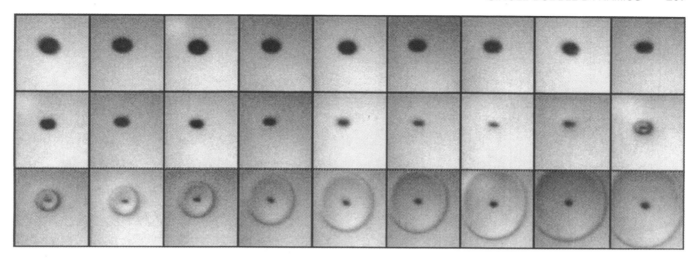

Fig. 8 High-speed photographic series of a collapsing, laser-produced bubble in water taken at 20.8 million frames per second. Picture size is 1.5 × 1.8 mm. (Courtesy of C. D. Ohl.)

where R is the bubble radius, an overdot means differentiation with respect to time, ρ is the density of the liquid, and P_i and P_e are internal (in the bubble) pressure and external (in the liquid including the surface to the bubble) pressure, respectively. The difference in pressure drives the bubble motion. The form of the inertial terms on the left-hand side is the result of the condensation of the spherical three-dimensional geometry to one dimension in the differential equation. Both P_i and P_e are functions of the radius R and time t, when gas and vapor fill the bubble, and surface tension σ, (kinematic) liquid viscosity μ, and a sound field are taken into account. Then the Rayleigh–Plesset model takes the form[31]

$$\rho R\ddot{R} + \frac{3}{2}\rho \dot{R}^2 = P_{gn}\left(\frac{R_n}{R}\right)^{3\kappa} + P_v - P_{\text{stat}}$$
$$- \frac{2\sigma}{R} - \frac{4\mu}{R}\dot{R} - P(t), \qquad (2)$$

with

$$P_{gn} = \frac{2\sigma}{R_n} + P_{\text{stat}} - P_v, \qquad (3)$$

the equilibrium gas pressure inside the bubble at equilibrium radius R_n, and

$$P(t) = -P_A \sin \omega t, \qquad (4)$$

the acoustic pressure of angular frequency ω and pressure amplitude P_A.

It is assumed that P_v, the vapor pressure in the bubble,

remains constant during the motion. In the case of water at 20°C, P_v will be small and will change the value of P_{stat}, the static pressure in the liquid at infinity corrected by the hydrostatic pressure at the location of the bubble, only slightly. The gas in the bubble is assumed to be compressed according to a polytropic gas law with the polytropic exponent κ. Usually κ is assumed constant during the motion.

In the bubble model of Eq. (2) viscosity is the only damping mechanism. A model taking the damping by sound radiation into account is the Gilmore model[12, 13, 32]:

$$\left(1 - \frac{\dot{R}}{C}\right)R\ddot{R} + \frac{3}{2}\left(1 - \frac{1}{3}\frac{\dot{R}}{C}\right)\dot{R}^2$$
$$= \left(1 + \frac{\dot{R}}{C}\right)H + \left(1 - \frac{\dot{R}}{C}\right)\frac{R}{C}\dot{H}, \qquad (5)$$

where C is the liquid sound velocity at the bubble interface and H the enthalpy evaluated at the bubble wall:

$$H = \int_{P_\infty}^{P(R)} \frac{dp}{\rho}. \qquad (6)$$

Even more advanced bubble models have been developed to better incorporate a time-varying pressure as given by an external sound field.[33] These models have been investigated for the response of a bubble in an acoustic field.[34, 35] A behavior typical for nonlinear oscillators is found.[36] Period-doubling cascades to chaos (aperiodic oscillations with broadband spectra) abound in

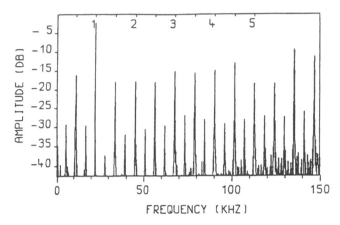

Fig. 9 Power spectrum of acoustic cavitation noise after second period doubling. One-fourth of the driving frequency (23.56 kHz) and its harmonics are present.[39]

all models, providing a clue to why the first subharmonic in the spectrum of the cavitation noise is so pronounced.

6 CAVITATION NOISE MEASUREMENTS

Measurements of the noise emission have been done with subsequent application of the new methods from nonlinear dynamics to the sound output from the liquid. These are phase space analysis, dimension analysis, and Lyapunov analysis.[37] A period-doubling route to chaos has been found.[38] Figure 9 gives a power spectrum of the noise output after two period doublings have taken place.[39] After a cascade of period doublings a chaotic noise attractor appears (Fig. 10). It is possible to determine the dimension of the noise attractors. Surprisingly small fractal dimensions between 2 and 3 are found. The appearance of fractal attractors as well as period-doubling sequences suggests that the system of oscillating bubbles in an acoustic field is a chaotic system. The definition of a chaotic system is that at least one of the Lyapunov exponents of the system making up the Lyapunov spectrum should be positive. A Lyapunov exponent is a measure of how fast two neighboring trajectories in phase space separate. The calculation of the Lyapunov spectrum for a (chaotic) noise attractor indeed yields a positive Lyapunov exponent.[40] Thus acoustic cavitation noise has been proven to be a chaotic system with only a small number of (nonlinear) degrees of freedom. According to the fractal dimension, only three variables should be sufficient to describe the dynamics of the system. This finding suggests that a high degree of cooperation must take place among the bubbles. The highly structured bubble ensemble confirms this view.

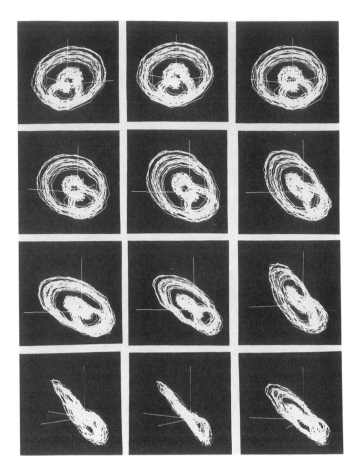

Fig. 10 Chaotic noise attractor as obtained by embedding the noise data into a three-dimensional phase space. (Courtesy of J. Holzfuss.)

7 CAVITATION EFFECTS

The violent motion of cavities in a sound field not only is the source of the peculiar noise emission discussed but also gives rise to other phenomena, notably the destructive action[41] on all kinds of material from soft tissue to hard steel. This action appears quite natural in view of the fact that high pressures and temperatures are reached in collapsing bubbles. This aggressive action may be put to good use for cleaning, for example, for removing grinding material from lenses. In the majority of cases, however, the destructive action is unwanted and may lead to failure of the parts suffering cavitation by progressive removal of material. This process is called cavitation erosion and was first observed with ship propellers. With acoustic cavitation it has been used for accelerated cavitation damage tests. Aluminum bronze and titanium proved to be quite cavitation resistant.[13,14] Figure 11 shows a damage pit from one single cavitation bubble collapsing on an aluminum specimen. The dam-

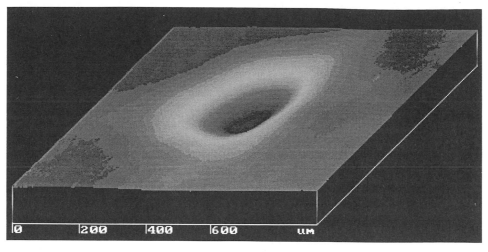

Fig. 11 Damage pit from a single laser-produced bubble collapsing in the neighborhood of a solid boundary (aluminum specimen). Pit depth 1.0 μm, pit diameter 560 μm. (Courtesy of A. Philipp.)

aging process is very involved and of a statistical nature even under highly controlled conditions. It seems that only a bubble collapsing in contact with the boundary leads to pit formation.

It has not yet been possible to study the dynamics of pit formation and its relation to bubble dynamics. There are several candidates presumably being involved; these are shock waves, the liquid jet, and the high pressure and temperature in the bubble that act on the surface when the bubble touches it in the final collapse phase.

Sound waves, in particular those in the megahertz range, are used in medicine for diagnosis and therapy. As cavitation is connected with high pressures and temperatures and the possibility of damage, it normally has to be avoided for reasons of safety. Therefore quite a number of experiments have been conducted to learn about the damage potential to biologic material (tissue, blood cells, etc.). It has been found that, whenever gas bubbles are present, the damage potential of ultrasonic waves is high.[43] Even a few 10 mW/cm^2 peak intensities suffice to induce damage. Damage may be reduced by pulsed ultrasound with pulse lengths of a few microseconds only. Unless the duty cycle is also low, however, damage may nevertheless occur, and even in one acoustic cycle preexisting cavities may be set into violent motion if they fit the appropriate conditions.

8 APOLOGY

Some basic topics of acoustic cavitation have been reviewed, essentially those pertaining to the process itself. Space limitations have not allowed us to include the plethora of effects of ultrasound in liquids that have connections to cavitation, as there are lithotripsy, chemical effects,[44] sonoluminescence, biological effects, and so on. Also, the author apologizes to all those having contributed to acoustic cavitation but that could not be mentioned here because of space constraints.

REFERENCES

1. J. R. Blake, J. M. Boulton-Stone, and N. H. Thomas (Eds.), *Bubble Dynamics and Interface Phenomena*, Kluwer, Dordrecht, 1994.

1a. T. G. Leighton, *The Acoustic Bubble*, Academic Press, London, 1994.

2. L. M. Lyamshev, "Radiation Acoustics," *Colloque de Physique C2*, Vol. 51, 1990, pp. C2-1 to C2-7.

3. F. R. Young, *Cavitation*, McGraw-Hill, London, 1989.

4. A. A. Atchley and L. A. Crum, "Acoustic Cavitation and Bubble Dynamics," in K. S. Suslick (Ed.), *Ultrasound: Its Chemical, Physical and Biological Effects*, VCH Publishers, New York, 1988, pp. 1–64.

5. J. R. Blake and D. C. Gibson, "Cavitation Bubbles Near boundaries," *Ann. Rev. Fluid Mech.*, Vol. 19, 1987, pp. 99–123.

6. L. A. Crum, "Acoustic Cavitation," in B. R. Avoy (Ed.), *Proceedings of the 1982 Ultrasonics Symposium*, IEEE Press, New York, 1983, pp. 1–12.

7. L. van Wijngaarden (Ed.), *Mechanics and Physics of Bubbles in Liquids*, Martinus Nijhoff Publishers, The Hague, 1982.

8. R. E. Apfel, "Acoustic Cavitation," in P. D. Edmonds (Ed.), *Methods of Experimental Physics*, Vol. 19, Academic Press, New York, 1981, pp. 355–411.

9. W. Lauterborn (Ed.), *Cavitation and Inhomogeneities in Underwater Acoustics*, Springer, Berlin, 1980.

10. E. A. Neppiras, "Acoustic Cavitation," *Phys. Rep.*, Vol. 61, 1980, pp. 159–251.

11. M. S. Plesset and A. Prosperetti, "Bubble Dynamics and Cavitation," *Ann. Rev. Fluid Mech.*, Vol. 9, 1977, pp. 145–185.

12. L. D. Rozenberg (Ed.), *High Intensity Ultrasonic Fields*, Plenum, New York, 1971.

13. R. T. Knapp, J. W. Daily, and F. G. Hammitt, *Cavitation*, McGraw-Hill, London, 1970.

14. H. G. Flynn, "Physics of Acoustic Cavitation in Liquids," in W. P. Mason (Ed.), *Physical Acoustics*, Vol. 1, Part B, Academic, New York, 1964, pp. 57–112.

15. L. A. Crum, "Sonoluminescence," *Physics Today*, September 1994, pp. 22–29.

16. A. J. Walton and G. T. Reynolds, "Sonoluminescence," *Adv. Phys.*, Vol. 33, 1984, pp. 595–660.

17. W. Lauterborn and J. Holzfuss, "Acoustic Chaos," *Int. J. Bifurcation Chaos*, Vol. 1, 1991, pp. 13–26.

18. W. Lauterborn, J. Holzfuss, and A. Billo, "Chaotic Behavior in Acoustic Cavitation," in M. Levy, S. C. Schneider, and B. R. Avoy (Eds.), *Proceedings of the 1994 Ultrasonics Symposium*, IEEE Press, New York 1994, pp. 801–810.

19. L. A. Crum, "Nucleation and Stabilization of Microbubbles in Liquids," *Appl. Sci. Res.*, Vol. 38, 1982, pp. 101–115.

20. D. E. Yount, E. W. Gillary, and D. C. Hoffmann, "A Microscopic Investigation of Bubble Formation Nuclei," *J. Acoust. Soc. Am.*, Vol. 76, 1984, pp. 1511–1521.

21. L. A. Crum, "Rectified Diffusion," *Ultrasonics*, Vol. 22, 1984, pp. 215–223.

22. M. S. Plesset and T. P. Mitchell, "On the Stability of the Spherical Shape of a Vapor Cavity in a Liquid," *Q. Appl. Math.*, Vol. 13, 1956, pp. 419–430.

23. H. W. Strube, "Numerische Untersuchungen zur Stabilität nichtsphärisch schwingender Blasen," *Acustica*, Vol. 25, 1971, pp. 289–330.

24. T. B. Benjamin and A. T. Ellis, "The Collapse of Cavitation Bubbles and the Pressure Thereby Produced Against Solid Boundaries," *Phil. Trans. Roy. Soc. Lond.*, Vol. A260, 1966, pp. 221–240.

25. A. Vogel, W. Lauterborn, and R. Timm, "Optical and Acoustic Investigations of the Dynamics of Laser-Produced Cavitation Bubbles Near a Solid Boundary," *J. Fluid Mech.*, Vol. 206, 1989, pp. 299–338.

26. L. A. Crum, "Bjerknes Forces on Bubbles in a Stationary Sound Field," *J. Acoust. Soc. Am.*, Vol. 57, 1975, pp. 1363–1370.

27. I. Akhatov, U. Parlitz, and W. Lauterborn, "Pattern For-mation in Acoustic Cavitation," *J. Acoust. Soc. Am.*, Vol. 96, 1994, pp. 3627–3635.

28. C. D. Ohl, A. Phillipp, and W. Lauterborn, "Cavitation Bubble Collapse Studied at 20 Million Frames per Second," *Ann. Phys.*, Vol. 4, 1995, pp. 26–34.

29. R. Hickling and M. S. Plesset, "Collapse and Rebound of a Spherical Bubble in Water," *Phys. Fluids*, Vol. 7, 1964, pp. 7–14.

30. A. Prosperetti, "Physics of Acoustic Cavitation," in D. Sette (Ed.), *Frontiers in Physical Acoustics*, North-Holland, Amsterdam, 1986, pp. 145–188.

31. B. E. Noltingk and E. A. Neppiras, "Cavitation Produced by Ultrasonics," *Proc. Phys. Soc. Lond.*, Vol. B63, 1950, pp. 674–685.

32. F. R. Gilmore, "The Growth or Collapse of a Spherical Bubble in a Viscous Compressible Liquid," California Institute of Technology Report No 26-4, Pasadena, CA, 1952, pp. 1–40.

33. J. B. Keller and M. Miksis, "Bubble Oscillations of Large Amplitude," *J. Acoust. Soc. Am.*, Vol. 68, 1980, pp. 628–633.

34. W. Lauterborn, "Numerical Investigation of Nonlinear Oscillations of Gas Bubbles in Liquids," *J. Acoust. Soc. Am.*, Vol. 59, 1976, pp. 283–293.

35. U. Parlitz, V. English, C. Scheffczyk, and W. Lauterborn, "Bifurcation Structure of Bubble Oscillators," *J. Acoust. Soc. Am.*, Vol. 88, 1990, pp. 1061–1077.

36. C. Scheffczyk, U. Parlitz, T. Kurz, W. Knop, and W. Lauterborn, "Comparison of Bifurcation Structures of Driven Dissipative Nonlinear Oscillators," *Phys. Rev. A*, Vol. 43, 1991, pp. 6495–6501.

37. W. Lauterborn and U. Parlitz, "Methods of Chaos Physics and Their Application to Acoustics," *J. Acoust. Soc. Am.*, Vol. 84, 1988, pp. 1975–1993.

38. W. Lauterborn and E. Cramer, "Subharmonic Route to Chaos Observed in Acoustics," *Phys. Rev. Lett.*, Vol. 47, 1981, pp. 1445–1448.

39. W. Lauterborn, "Acoustic Turbulence," in D. Sette (Ed.), *Frontiers in Physical Acoustics*, North-Holland, Amsterdam, 1986, pp. 124–144.

40. J. Holzfuss and W. Lauterborn, "Liapunov Exponents from Time Series of Acoustic Chaos," *Phys. Rev. A*, Vol. 39, 1989, pp. 2146–2152.

41. Y. Tomita and A. Shima, "Mechanisms of Impulsive Pressure Generation and Damage Pit Formation by Bubble Collapse," *J. Fluid Mech.*, Vol. 169, 1986, pp. 535–564.

42. I. S. Pearsall, *Cavitation*, Mills and Boon, London, 1972.

43. D. L. Miller, "Gas Body Activation," *Ultrasonics*, Vol. 22, 1984, pp. 261–271.

44. K. S. Suslick, "Sonochemistry," *Science*, Vol. 247, 1990, pp. 1439–1445.

26

SONOCHEMISTRY AND SONOLUMINESCENCE

Kenneth S. Suslick and Lawrence A. Crum

1 INTRODUCTION

High-energy chemical reactions occur during the ultrasonic irradiation of liquids.[1-6] The chemical effects of ultrasound, however, do not come from a direct interaction with molecular species. The velocity of sounds in liquids is typically about 1500 m/s; ultrasound spans the frequencies of roughly 15 kHz to 1 GHz, with associated acoustic wavelengths of $10-10^{-4}$ cm. These are not molecular dimensions. No direct coupling of the acoustic field with chemical species on a molecular level can account for sonochemistry or sonoluminescence. Instead, these phenomena derive principally from acoustic cavitation: the formation, growth, and implosive collapse of bubbles in a liquid. Cavitation serves as a means of concentrating the diffuse energy of sound. Bubble collapse induced by cavitation produces intense local heating, high pressures, and very short lifetimes. In clouds of cavitating bubbles, these hot spots[7,8] have equivalent temperatures of roughly 5000 K, pressures of about 2000 atm, and heating and cooling rates above 10^9 K/s.

Related phenomena occur with cavitation in liquid–solid systems. Near an extended solid surface, cavity collapse is nonspherical and drives high-speed jets of liquid into the surface.[9] This process can produce newly exposed, highly heated surfaces. Furthermore, during ultrasonic irradiation of liquid–powder slurries, cavitation and the shock waves it creates can accelerate solid particles to high velocities.[10] The resultant interparticle collisions are capable of inducing dramatic changes in surface morphology, composition, and reactivity.[11]

The chemical effects of ultrasound are diverse and include dramatic improvements in both stoichiometric and catalytic reactions.[12-15] In some cases, ultrasonic irradiation can increase reactivity by nearly a million-fold.[16] The chemical effects of ultrasound fall into three areas: homogeneous sonochemistry of liquids, heterogeneous sonochemistry of liquid–liquid or liquid–solid systems, and sonocatalysis (which overlaps the first two). Chemical reactions are not generally seen in the ultrasonic irradiation of solids or solid–gas systems.

Sonoluminescence may be considered a special case of homogeneous sonochemistry; however, recent discoveries in this field have heightened interest in the phenomenon in and by itself.[17,18] New data on the duration of the sonoluminescence flash suggest that under the conditions of single-bubble sonoluminescence (SBSL), a shock wave may be created within the collapsing bubble with the capacity to generate enormous temperatures and pressures within the gas.

Acoustic cavitation provides the potential for creating exciting new physical and chemical conditions in otherwise cold liquids and results in an enormous concentration of energy. If one considers the energy density in an acoustic field that produces cavitation and that in the collapsed cavitation bubble, there is an amplification factor of over 11 orders of magnitude. The enormous local temperatures and pressures so created result in phenomena such as sonochemistry and sonoluminescence and provide a unique means for fundamental studies of chemistry and physics under extreme conditions.

2 MECHANISTIC ORIGINS OF SONOCHEMISTRY AND SONOLUMINESCENCE

2.1 Hot-Spot Formation during Cavitation

The most important acoustic process for sonochemistry and sonoluminescence is cavitation. Compression of a

Encyclopedia of Acoustics, edited by Malcolm J. Crocker
ISBN 0-471-80465-7 © 1997 John Wiley & Sons, Inc.

gas generates heat. When the compression of cavities occurs in irradiated liquids, it is more rapid than thermal transport, which generates a short-lived, localized hot spot. There is a general consensus that this hot spot is the source of homogeneous sonochemistry. Rayleigh's early descriptions of a mathematical model for the collapse of cavities in incompressible liquids predicted enormous local temperatures and pressures.[19] Ten years later, Richards and Loomis reported the first chemical and biologic effects of ultrasound.[20] Alternative mechanisms involving electrical microdischarge have been occasionally proposed[21,22] but remain a minority viewpoint.

2.2 Two-Site Model of Sonochemical Reactivity

The transient nature of the cavitation event precludes conventional measurement of the conditions generated during bubble collapse. Chemical reactions themselves, however, can be used to probe reaction conditions. The effective temperature realized by the collapse of clouds of cavitating bubbles can be determined by the use of competing unimolecular reactions whose rate dependencies on temperature have already been measured. This technique of *comparative-rate chemical thermometry* was used by Suslick, Hammerton, and Cline to first determine the effective temperature reached during cavity collapse.[7] The sonochemical ligand substitutions of volatile metal carbonyls were used as these comparative-rate probes [Eq. (1), where the arrow with three small closing parentheses on top represents ultrasonic irradiation of a solution and L represents a substituting ligand]:

$$M(CO)_x \xrightarrow{)))} M(CO)_{x-n} + n\text{-}CO \xrightarrow{L} M(CO)_{x-n}(L)_n,$$

$$\text{where } M = Fe, Cr, Mo, W. \quad (1)$$

These kinetic studies revealed that there were in fact *two* sonochemical reaction sites: the first (and dominant site) is the bubble's interior gas phase while the second is an *initially* liquid phase. The latter corresponds either to heating of a shell of liquid around the collapsing bubble or to droplets of liquid ejected into the hot spot by surface wave distortions of the collapsing bubble.

The effective local temperatures in both sites were determined. By combining the relative sonochemical reaction rates for Eq. (1) with the known temperature behavior of these reactions, the conditions present during cavity collapse could then be calculated. The effective temperature of these hot spots was measured at ≈ 5200 K in the gas-phase reaction zone and ≈ 1900 K in the initially liquid zone.[7] Of course, the comparative-rate data represent only a composite temperature: During the

collapse, the temperature has a highly dynamic profile as well as a spatial temperature gradient. This two-site model has been confirmed with other reactions,[23,24] and alternative measurements of local temperatures by sonoluminescence are also consistent,[8] as discussed later.

2.3 Microjet Formation during Cavitation at Liquid–Solid Interfaces

Cavitation near extended liquid–solid interfaces is very different from cavitation in pure liquids. There are two proposed mechanisms for the effects of cavitation near surfaces: microjet impact and shock wave damage. Whenever a cavitation bubble is produced near a boundary, the asymmetry of the liquid particle motion during cavity collapse induces a deformation in the cavity.[9] The potential energy of the expanded bubble is converted into kinetic energy of a liquid jet that extends through the bubble's interior and penetrates the opposite bubble wall. Because most of the available energy is transferred to the accelerating jet, rather than the bubble wall itself, this jet can reach velocities of hundreds of meters per second. Because of the induced asymmetry, the jet often impacts the local boundary and can deposit enormous energy densities at the site of impact. Such energy concentration can result in severe damage to the boundary surface. Figure 1 shows a photograph of a jet developed in a collapsing cavity; Fig. 2 shows a micrograph of an impact site. The second mechanism of cavitation-induced surface damage invokes shock waves created by cavity col-

Fig. 1 Photograph of liquid jet produced during collapse of a cavitation bubble. The width of the bubble is about 1 mm.

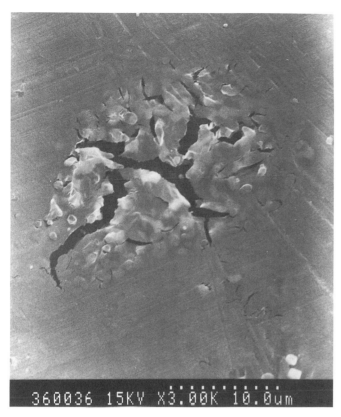

Fig. 2 Photomicrograph of a region of an aluminum foil (about 50 μm in thickness) exposed to collapsing cavitation bubbles produced by extracorporeal shock wave lithotriptor.

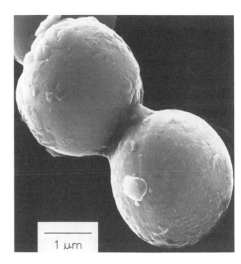

Fig. 3 Scanning electron micrograph of Zn powder after ultrasonic irradiation for 30 min at 288 K in decane under Ar at 20 kHz and $\approx$50 W/cm^2. (Reproduced from Ref. 10 with permission.)

lapse in the liquid. The impingement of microjets and shock waves on the surface creates the localized erosion responsible for ultrasonic cleaning and many of the sonochemical effects on heterogeneous reactions. The erosion of metals by cavitation generates newly exposed, highly heated surfaces. Further details of jet and shock wave production and associated effects are presented elsewhere.[25–27]

Distortions of bubble collapse require a surface several times larger than the resonance bubble size. Thus, for solid particles smaller than $\approx$200 μm, damage associated with jet formation cannot occur with ultrasonic frequencies of $\approx$20 kHz. In these cases, however, the shock waves created by homogeneous cavitation can create high-velocity interparticle collisions.[10,11] Suslick and co-workers have found that the turbulent flow and shock waves produced by intense ultrasound can drive metal particles together at sufficiently high speeds to induce effective melting in direct collisions (Fig. 3) and the abrasion of surface crystallites in glancing impacts (Fig. 4). A series of transition-metal powders were used to probe the maximum temperatures and speeds reached during interparticle collisions. Using the irradiation of Cr, Mo, and W powders in decane at 20 kHz and 50

W/cm^2, agglomeration and essentially a localized melting occur for the first two metals but not the third. On the basis of the melting points of these metals, the effective transient temperature reached at the point of impact during interparticle collisions is roughly 3000°C. From the volume of the melted region of impact, the amount of energy generated during collision was determined. From this, a lower estimate of the velocity of impact is roughly one-half the speed of sound.[10] These are precisely the effects expected on suspended particulates from cavitation-induced shock waves in the liquid.

3 SONOCHEMISTRY

High-intensity ultrasonic probes (50–500 W/cm^2) are the most reliable and effective source for laboratory-scale sonochemistry and are commercially available from several sources. Lower acoustic intensities can often be used

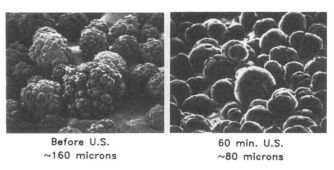

Before U.S.
~160 microns

60 min. U.S.
~80 microns

Fig. 4 Effect of ultrasonic irradiation on surface morphology of Ni powder. (Reproduced from Ref. 16 with permission.)

in liquid–solid heterogeneous systems, because of the reduced liquid tensile strength at the liquid–solid interface. For such reactions, a common ultrasonic cleaning bath will therefore often suffice. The low intensity available in these devices ($\approx$1 W/cm^2), however, can prove limiting. In addition, the standing-wave patterns in ultrasonic cleaners require accurate positioning of the reaction vessel. On the other hand, ultrasonic cleaning baths are easily accessible, relatively inexpensive, and usable on a moderately large scale. Finally, for larger scale irradiation, flow reactors with high ultrasonic intensities are commercially available in modular units of $\approx$20 kW.

3.1 Comparison of Cavitation Conditions to Other Forms of Chemistry

Chemistry is the study of the interaction of energy and matter. Chemical reactions require energy, in one form or another, to proceed: Chemistry stops as the temperature approaches absolute zero. One has only limited control, however, over the nature of this interaction. In large part, the properties of a specific energy source determines the course of a chemical reaction. Ultrasonic irradiation differs from traditional energy sources (such as heat, light, or ionizing radiation) in duration, pressure, and energy per molecule. The immense local temperatures and pressures and the extraordinary heating and cooling rates generated by cavitation bubble collapse mean that ultrasound provides an unusual mechanism for generating high-energy chemistry. Similar to photochemistry, very large amounts of energy are introduced in a short period of time, but it is thermal, not electronic, excitation. As in flash pyrolysis, high thermal temperatures are reached, but the duration is very much shorter (by more than 10^4) and the temperatures are even higher (by 5- to 10-fold). Similar to shock-tube chemistry or multiphoton infrared laser photolysis, cavitation heating is very short lived but occurs within condensed phases. Furthermore, sonochemistry has a high-pressure component, which suggests that one might be able to produce on a microscopic scale the same macroscopic conditions of high temperature–pressure "bomb" reactions or explosive shock wave synthesis in solids.

Control of sonochemical reactions is subject to the same limitation that any thermal process has: The Boltzmann energy distribution means that the energy per individual molecule will vary widely. One does have easy control, however, over the intensity of heating generated by acoustic cavitation through the use of various physical parameters (including thermal conductivity of dissolved gases, solvent vapor pressure inside the bubble, and ambient pressure). In contrast, frequency appears to be less important, at least within the range where cavitation can occur (a few hertz to a few megahertz), although there have been few detailed studies of its role.

3.2 Homogeneous Sonochemistry: Bond Breaking and Radical Formation

The chemical effects of ultrasound on aqueous solutions have been studied for many years. The primary products are H_2 and H_2O_2; other high-energy intermediates have been suggested, including HO_2, $H\cdot$, $OH\cdot$, and perhaps $e_{(aq)}^-$. The elegant work of Riesz and collaborators used electron paramagnetic resonance with chemical spin-traps to demonstrate definitively the generation of $H\cdot$ and $OH\cdot$ during ultrasonic irradiation, even with clinical sources of ultrasound.[28] The extensive work in Henglein's laboratory involving aqueous sonochemistry of dissolved gases has established clear analogies to combustion processes.[23,24] As one would expect, the sonolysis of water, which produces both strong reductants and oxidants, is capable of causing secondary oxidation and reduction reactions, as often observed by Margulis and co-workers.[29]

In contrast, the ultrasonic irradiation of organic liquids has been less studied. Suslick and co-workers established that virtually all organic liquids will generate free radicals upon ultrasonic irradiation, as long as the total vapor pressure is low enough to allow effective bubble collapse.[30] The sonolysis of simple hydrocarbons (e.g., n-alkanes) creates the same kinds of products associated with very high temperature pyrolysis. Most of these products (H_2, CH_4, and the smaller l-alkenes) derive from a well-understood radical chain mechanism.

The sonochemistry of solutes dissolved in organic liquids also remains largely unexplored. The sonochemistry of metal carbonyl compounds is an exception.[31] Detailed studies of these systems led to important mechanistic understandings of the nature of sonochemistry. A variety of unusual reactivity patterns have been observed during ultrasonic irradiation, including multiple ligand dissociation, novel metal cluster formation, and the initiation of homogeneous catalysis (discussed later) at low ambient temperature, with rate enhancements greater than 100,000-fold.

Of special interest is the recent development of sonochemistry as a synthetic tool for the creation of unusual materials.[32] As one example is the recent discovery of a simple sonochemical synthesis of amorphous and nanostructured materials, including transition metals, alloys, carbides, and oxides.[33] A second example is the sonochemical preparation of protein microspheres,[34] which have applications for medical diagnostic imaging, drug delivery, and blood substitutes.

3.3 Heterogeneous Sonochemistry: Reactions of Solids with Liquids

The use of high-intensity ultrasound to enhance the reactivity of metals as stoichiometric reagents has become a routine synthetic technique for many heterogeneous organic and organometallic reactions,[12–15] especially those involving reactive metals, such as Mg, Li, or Zn. This development originated from the early work of Renaud and the more recent breakthroughs of Luche.[12] The effects are quite general and apply to reactive inorganic salts and to main-group reagents as well.[35] Less work has been done with unreactive metals (e.g., V, Nb, Mo, W), but results here are promising as well.[11] Rate enhancements of more than 10-fold are common, yields are often substantially improved, and byproducts are avoided. A few simple examples of synthetic applications of heterogeneous sonochemistry are shown in Eqs. (2)–(7), taken from the work of Ando, Boudjouk, Luche, Mason, and Suslick, among others:

$$C_6H_5Br + Li$$

$$\xrightarrow{)))} C_6H_5Li + LiBr, \tag{2}$$

$$RBr + Li + R'_2NCHO$$

$$\xrightarrow[2.\,H_2O]{1.\,)))} RCHO + R'_2NH, \tag{3}$$

$$2o - C_6H_4(NO_2)I + Cu$$

$$\xrightarrow{)))} o\text{-}(O_2N)H_4C_6{-\!\!-}C_6H_4(NO_2) + 2CuI, \tag{4}$$

$$RR'HC{-\!\!-}OH + KMnO_{4(s)}$$

$$\xrightarrow{)))} RR'C{=\!\!=}O, \tag{5}$$

$$C_6H_5CH_2Br + KCN$$

$$\xrightarrow[Al_2O_3]{)))} C_6H_5CH_2CN, \tag{6}$$

$$MCl_5 + Na + CO$$

$$\xrightarrow{)))} M(CO)_6^- \quad (M = V, Nb, Ta). \tag{7}$$

The mechanism of the sonochemical rate enhancements in both stoichiometric and catalytic reactions of metals is associated with dramatic changes in morphology of both large extended surfaces and powders. As discussed earlier, these changes originate from microjet impact on large surfaces and high-velocity interparticle collisions in slurries. Surface composition studies by Auger electron spectroscopy and sputtered neutral mass spectrometry reveal that ultrasonic irradiation effectively removes surface oxide and other contaminating coatings.[11] The removal of such passivating coatings can dramatically improve reaction rates. The reactivity of clean metal surfaces also appears to be responsible for the greater tendency for heterogeneous sonochemical reactions to involve single electron transfer rather than acid–base chemistry.[36]

Green, Suslick, and co-workers examined another application of sonochemistry to difficult heterogeneous systems: the process of molecular intercalation.[37] The adsorption of organic or inorganic compounds as guest molecules between the atomic sheets of layered inorganic solid hosts permits the systematic change of optical, electronic, and catalytic properties for a variety of technologic applications (e.g., lithium batteries, hydrodesulfurization catalysts, and solid lubricants). The kinetics of intercalation, however, are generally extremely slow, and syntheses usually require high temperatures and very long reaction times. High-intensity ultrasound dramatically increases the rates of intercalation (by as much as 200-fold) of a wide range of compounds into various layered inorganic solids (such as ZrS_2, V_2O_5, TaS_2, MoS_2, and MoO_3). Scanning electron microscopy of the layered solids coupled to chemical kinetics studies demonstrated that the origin of the observed rate enhancements comes from particle fragmentation (which dramatically increases surface areas) and to a lesser extent from surface damage. The ability of high-intensity ultrasound to rapidly form uniform dispersions of micrometer-sized powders of brittle materials is often responsible for the activation of heterogeneous reagents, especially nonmetals.

3.4 Sonocatalysis

Catalytic reactions are of enormous importance in both laboratory and industrial applications. Catalysts are generally divided into two types. If the catalyst is a molecular or ionic species dissolved in a liquid, then the system is *homogeneous*; if the catalyst is a solid, with the reactants either in a percolating liquid or gas, then it is *heterogeneous*. In both cases, it is often a difficult problem either to activate the catalyst or to keep it active.

Ultrasound has potentially important applications in both homogeneous and heterogeneous catalytic systems. The inherent advantages of sonocatalysis include (1) the use of low ambient temperatures to preserve thermally sensitive substrates and to enhance selectivity, (2) the ability to generate high-energy species difficult to obtain from photolysis or simple pyrolysis, and (3) the mimicry of high-temperature and high-pressure conditions on a microscopic scale.

Homogeneous catalysis of various reactions often

uses organometallic compounds. The starting organometallic compound, however, is often catalytically inactive until the loss of metal-bonded ligands (such as carbon monoxide) from the metal. Having demonstrated that ultrasound can induce ligand dissociation, the initiation of homogeneous catalysis by ultrasound becomes practical. A variety of metal carbonyls upon sonication will catalyze the isomerization of l-alkenes to the internal alkenes,[31] through reversible hydrogen atom abstraction, with rate enhancements of as much as 10^5 over thermal controls.

Heterogeneous catalysis is generally more industrially important than homogeneous systems. For example, virtually all of the petroleum industry is based on a series of catalytic transformations. Heterogeneous catalysts often require rare and expensive metals. The use of ultrasound offers some hope of activating less reactive, but also less costly, metals. Such effects can occur in three distinct stages: (1) during the formation of supported catalysts, (2) activation of preformed catalysts, or (3) enhancement of catalytic behavior during a catalytic reaction. Some early investigations of the effects of ultrasound on heterogeneous catalysis can be found in the Soviet literature.[38] In this early work, increases in turnover rates were usually observed upon ultrasonic irradiation but were rarely more than 10-fold. In the cases of modest rate increases, it appears likely that the cause is increased effective surface area; this is especially important in the case of catalysts supported on brittle solids.[39] More impressive accelerations, however, have included hydrogenations and hydrosilations by Ni powder, Raney Ni, and Pd or Pt on carbon.[13] For example, Casadonte and Suslick discovered that hydrogenation of alkenes by Ni powder is enormously enhanced ($>10^5$-fold) by ultrasonic irradiation.[16] This dramatic increase in catalytic activity is due to the formation of uncontaminated metal surfaces from interparticle collisions caused by cavitation-induced shock waves.

4 SONOLUMINESCENCE

Ultrasonic irradiation of liquids can also produce light. This phenomenon, known as sonoluminescence, was first observed from water in 1934 by Frenzel and Schultes.[40] As with sonochemistry, sonoluminescence derives from acoustic cavitation. Although sonoluminescence from aqueous solutions has been studied in some detail, only recently has significant work on sonoluminescence from nonaqueous liquids been reported.

4.1 Types of Sonoluminescence

It is now generally thought that there are two separate forms of sonoluminescence: multiple-bubble sonoluminescence (MBSL) and single-bubble sonoluminescence (SBSL). When an acoustic field of sufficient intensity is propagated through a liquid, placing it under dynamic tensile stress, microscopic preexisting inhomogeneities act as nucleation sites for liquid rupture. This cavitation inception process results in many separate and individual cavitation events that would be distributed broadly throughout the acoustic field, especially if the liquid had a sufficiently large number of nuclei. Since most liquids such as water, have many thousands of potential nucleation sites per milliliter, the "cavitation field" generated by a propagating (or standing) acoustic wave typically consists of many bubbles, and is distributed over an extended region of space. If this cavitation is sufficiently intense to produce sonoluminescence, then we call this phenomenon multiple-bubble sonoluminescence.[17,41]

When an acoustic standing wave is excited within a liquid, pressure nodes and antinodes are generated. If a gas bubble is inserted into this standing wave, it will experience an acoustic force that will tend to force the bubble either to the node or antinode, depending upon whether it is respectively larger or smaller than its resonance size. Under the appropriate conditions, this acoustic force can balance the buoyancy force and the bubble is said to be "acoustically levitated." Such a bubble is typically quite small, compared to a wavelength (e.g., at 20 kHz, the resonance size is approximately 150 μm), and thus the dynamic characteristics of this bubble can often be examined in considerable detail, both from a theoretical and an experimental perspective.

It was recently discovered that under rather specialized but easily obtainable conditions a single, stable, oscillating gas bubble can be forced into such large-amplitude pulsations that it produces sonoluminescence emissions each (and every) acoustic cycle.[42,43] This phenomenon is called single-bubble sonoluminescence and has received considerable recent attention.[17,18,44,45]

4.2 Applications of MBSL to Measurements of Cavitation Thresholds

When the acoustic pressure amplitude of a propagating acoustic wave is relatively large (greater than ≈ 0.5 MPa) local inhomogeneities in the liquid often give rise to the explosive growth of a nucleation site to a cavity of macroscopic dimensions, primarily filled with vapor. Such a cavity is inherently unstable, and its subsequent collapse can result in an enormous concentration of energy. This violent cavitation event has been termed *transient* or *inertial cavitation* because the collapse of the cavity is primarily dominated by inertial forces.[46] A normal consequence of this rapid growth and violent collapse is that the cavitation bubble itself is destroyed. Although gas-filled residues from the collapse may give rise to reinitiation of the process, this type of cavitation is thought to be a temporally discrete phenomenon. When one examines the light emissions from transient inertial cavitation, one can see single isolated events associated (presumably) with individual collapses of imploding cavities.

Because acoustic cavitation is often associated with large energy concentrations and thus potentially damaging mechanical effects, its presence is often undesirable (e.g., in the use of diagnostic ultrasound for prenatal examinations). Since light can be detected at very low levels and with very high time resolution, sonoluminescence can be used as a sensitive indicator of violent acoustic cavitation.[47,48] This capability permits threshold determination of cavitation inception for acoustic pulses of short time duration, as often is the case for medical ultrasound.

The widespread use of medical ultrasound has made possible enormous advances in the noninvasive examination of internal organs and conditions. With these diagnostic devices, the acoustic pulses used to create images are often less than a microsecond in length and possess duty cycles on the order of 1 : 1000. Since the acoustic pressure amplitudes generated by these devices are relatively large, it has been possible to use sonoluminescence as a detection criterion for acoustic cavitation generated by microsecond-length pulses of ultrasound. These studies have indicated that cavitation can be generated by acoustic pulses similar to those used in diagnostic ultrasound instruments and that continued studies need to be undertaken to evaluate the potential risks of these devices.[49]

4.3 Origin of MBSL Emissions: Chemiluminescence

The spectrum of MBSL in water consists of a peak at 310 nm and a broad continuum throughout the visible region. An intensive study of aqueous MBSL was conducted by Verrall and Sehgal.[50] The emission at 310 nm is from excited-state OH·, but the continuum is difficult to interpret. The MBSL from aqueous and alcohol solutions of many metal salts have been reported and are characterized by emission from metal atom excited states.[51]

Flint and Suslick reported the first MBSL spectra of organic liquids.[52] With various hydrocarbons, the observed emission is from excited states of C_2 ($d^3\Pi_g$–$a^3\Pi_u$, the Swan lines), the same emission seen in flames. Furthermore, the ultrasonic irradiation of alkanes in the presence of N_2 (or NH_3 or amines) gives emission from CN excited states, but not from N_2 excited states. Emission from N_2 excited states would have been expected if the MBSL originated from microdischarge, whereas CN emission is typically observed from thermal sources. When oxygen is present, emission from excited states of CO_2, CH·, and OH· is observed, again similar to flame emission.

For both aqueous and nonaqueous liquids, MBSL is caused by chemical reactions of high-energy species formed during cavitation by bubble collapse. Its principal source is most probably not blackbody radiation or electrical discharge. The MBSL is a form of chemiluminescence.

4.4 Origin of SBSL Emissions: Imploding Shock Waves

It is known that the spectra of MBSL and SBSL are measurably different. For example, an aqueous solution of NaCl shows evidence of excited states of both OH· and Na in the MBSL spectrum; however, the SBSL spectrum of an identical solution shows no evidence of either of these peaks.[53] Similarly, the MBSL spectrum falls off at low wavelengths, while the SBSL spectrum continues to rise, at least for bubbles containing most noble gases.[54]

Perhaps the most intriguing aspect of SBSL is the extremely short duration of the sonoluminescence flash. Putterman and his colleagues, using the fastest photomultiplier tube (PMT) available, determined that this duration must be at least as short as 50 ps, perhaps even lower.[55] Moran et al., using a streak camera, presented evidence of a 12-ps flash duration.[56] Because these measurements test the limitations of current technology, an accurate assessment of this pulse duration cannot be reliably given at this time.

As described earlier, the most likely explanation for the origin of sonoluminescence is the hot-spot theory, in which the potential energy given the bubble as it expands to maximum size is concentrated into a heated gas core as the bubble implodes. To understand the origin of sonoluminescence emissions, it is necessary first to understand something about the bubble dynamics that describe the bubble's motion. The equation that describes the oscillations of a gas bubble driven by an acoustic field is known generally by the name *Rayleigh–Plesset*, one form of which, called the *Gilmore equation*, can be expressed as a second-order nonlinear differential equation given as

$$R\left(1 - \frac{U}{C}\right)\frac{d^2R}{dt^2} + \frac{3}{2}\left(1 - \frac{U}{3C}\right)\left(\frac{dR}{dt}\right)^2$$
$$- \left(1 + \frac{U}{C}\right)H - \frac{R}{C}\left(1 - \frac{U}{C}\right)\frac{dH}{dt} = 0. \quad (8)$$

The radius and velocity of the bubble wall are given by R and U, respectively. The values for H, the enthalpy at the bubble wall, and C, the local sound speed, may be expressed as follows using the Tait equation of state for the liquid:

$$H = \frac{n}{n-1} \frac{A^{1/n}}{\rho_0} \left\{ (P(R) + B)^{(n-1)/n} \right.$$

$$\left. - [P_\infty(t) + B]^{(n-1)/n} \right\}, \qquad (9)$$

$$C = [c_0^2 + (n-1)H]. \qquad (10)$$

The linear speed of sound in the liquid is c_0. The constants A, B, and n should be set to the appropriate values for water. Any acoustic forcing function is included in the pressure at infinity, $P_\infty(t)$. The pressure at the bubble wall, $P(R)$, is given by

$$P(R) = \left(P_0 + \frac{2\sigma}{R} \right) \left(\frac{R_0}{R} \right)^{3\gamma} - \frac{2\sigma}{R} - \frac{4\mu U}{R}, \qquad (11)$$

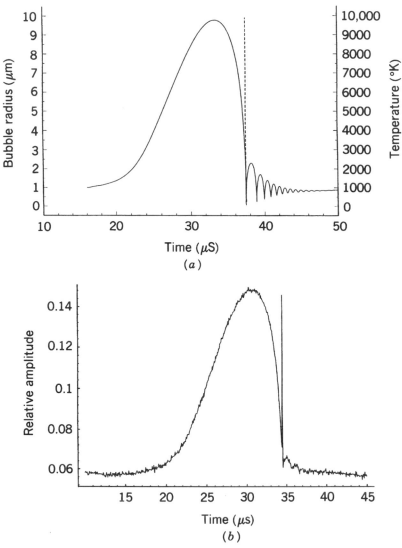

Fig. 5 (a) Theoretical response of a single gas bubble when driven under conditions of SBSL (solid line). Pressure amplitude 0.136 MPa, equilibrium radius 4.5 μm, and driving frequency 26.5 kHz. Bubble expands to several times its initial radius and then implosively collapses. Broken line: calculation of the interior bubble temperature assuming an adiabatic collapse of the cavity contents. (Courtesy of John Allen.) (b) Measured response of a gas bubble to conditions similar to those shown. Experimentally determined bubble radius reconstructed from the square root of the scattered laser light intensity. Intensity spike near bubble collapse is not due to the laser but results from the sonoluminescence emissions. With some minor adjustments, the theoretical and experimental curves for the bubble radius can be made to coincide. (Courtesy of Tom Matula.)

where the initial radius of the bubble at time zero is R_0. The ambient pressure of the liquid is P_0, the surface tension σ, the shear viscosity μ, and the polytropic exponent γ. The latter is set to 1.4 assuming the bubble behaves as an adiabatic system.

We can use the Gilmore equation to compute the behavior of a bubble undergoing SBSL for conditions similar to those in which this phenomenon is observed experimentally. Figure 5a shows an example of these computations. The solid line is the radius–time curve; the dashed line is the computed temperature in the interior of the bubble based upon the assumptions of this model.

It is possible to test experimentally certain aspects of these models. For example, using a light-scattering technique, various researchers have obtained measurements of the radius–time curve, simultaneous with the optical emissions,[41,57] as shown in Fig. 5b. Both the laser light scattered intensity and the SBSL can be acquired with a single photomultiplier tube. The SBSL emission is seen as the sharp spike in the figure, appearing at the final stages of bubble collapse. Note that these emissions occur at the point of minimum bubble size, as predicted by the hot-spot theory and that the general shape of the theoretical curve is reproduced. More quantitative assessment of these data has been made.[57,58]

The computations of single-bubble cavitation suggest that the temperature of the gas within the bubble would remain at elevated temperatures for times on the order of tens of nanoseconds; however, there is strong evidence that the pulse duration of the SBSL flash is at least three orders of magnitude shorter than this value. The most plausible explanation for this short flash interval and some of the observed spectra (see below) is that an imploding shock wave is created within the gas bubble during the final stages of collapse. If this shock wave does indeed exist, exciting possibilities can be inferred about the temperatures that could be attained within the bubble and the physics that might result. Indeed, speculations on the possibilities of inertial confinement fusion have been made.[59–61]

4.5 Spectroscopic Probes of Cavitation Conditions

Determination of the temperatures reached in a cavitating bubble has remained a difficult experimental problem. As a spectroscopic probe of the cavitation event, MBSL provides a solution. High-resolution MBSL spectra from silicone oil have been reported and analyzed.[8] The observed emission comes from excited state C_2 and has been modeled with synthetic spectra as a function of rotational and vibrational temperatures, as shown in

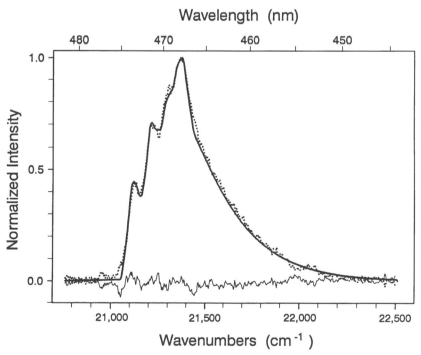

Fig. 6 Sonoluminescence of excited state C_2. Emission from the $\Delta\nu = +1$ manifold of the $d^3\Pi_g-a^3\Pi_u$ transition (Swan band) of C_2. ($\cdots\cdots$) Observed sonoluminescence from silicone oil (polydimethylsiloxane, Dow 200 series, 50 cSt viscosity) under a continuous Ar sparge at 0°C. (——) Best fit synthetic spectrum, with $T_v = T_r = 4900$ K. (——) Difference spectrum.

Fig. 6. From comparison of synthetic to observed spectra, the effective cavitation temperature is 5050 ± 150 K. The excellence of the match between the observed MBSL and the synthetic spectra provides definitive proof that the sonoluminescence event is a thermal, chemiluminescence process. The agreement between this spectroscopic determination of the cavitation temperature and that made by comparative-rate thermometry of sonochemical reactions[7] is surprisingly close.

The interpretation of the spectroscopy of SBSL is much more unclear. Some very interesting effects are observed when the gas contents of the bubble are changed.[55] These results are shown in Fig. 7. Note that doping a nitrogen bubble with small quantities of noble gases drastically affects the emission intensity. Furthermore, the spectra show practically no evidence of OH emissions and, when He and Ar bubbles are considered, continue to increase in intensity for smaller and smaller wavelengths. These spectra suggest that temperatures considerably in excess of 5000 K may exist within the bubble and lend some support to the concept of an imploding shock wave. Several other alternative explanations for SBSL have been presented, and there exists considerable theoretical activity in this particular aspect of SBSL.[62–66]

Acknowledgments

We acknowledge the support of the Office of Naval Research (L. A. C.), the National Institutes of Health through grant numbers DK43881 (L. A. C.) and HL25934 (K. S. S.), and the National Science Foundation through grant numbers CHE-9420758 and DMR-89-20538 (K. S. S.) and PHY-9311108 (L. A. C.).

REFERENCES

1. K. S. Suslick (Ed.), *Ultrasound: Its Chemical, Physical, and Biological Effects*, VCH Publishers, New York, 1988.

2. K. S. Suslick, *Science*, Vol. 247, 1990, p. 1439.

3. T. J. Mason (Ed.), *Advances in Sonochemistry*, Vols. 1–3, JAI Press, New York, 1990, 1991, 1993.

4. T. J. Mason and J. P. Lorimer, *Sonochemistry: Theory, Applications and Uses of Ultrasound in Chemistry*, Ellis Horword, Chichester, United Kingdom, 1988.

5. G. J. Price (Ed.), *Current Trends in Sonochemistry*, Royal Society of Chemistry, Cambridge, 1992.

6. O. V. Abramov, *Ultrasound in Liquid and Solid Metals*, CRC Press, Boca Raton, FL, 1994.

7. K. S. Suslick, D. A. Hammerton, and R. E. Cline, Jr., *J. Am. Chem. Soc.*, Vol. 108, 1986, p. 5641.

8. E. B. Flint and K. S. Suslick, *Science*, Vol. 253, 1991, p. 1397.

9. T. G. Leighton, *The Acoustic Bubble*, Academic, London, 1994, pp. 531–551.

10. S. J. Doktycz and K. S. Suslick, *Science*, Vol. 247, 1990, pp. 1067.

11. K. S. Suslick and S. J. Doktycz, *Adv. Sonochem.*, Vol. 1, 1990, pp. 197–230.

12. C. Einhorn, J. Einhorn, and J.-L. Luche, *Synthesis*, Vol. 1989, 1989, p. 787.

13. P. Boudjouk, *Comments Inorg. Chem.*, Vol. 9, 1990, p. 123.

14. J. M. Pestman, J. B. F. N. Engberts, and F. de Jong, *Recl. Trav. Chim. Pays-Bas*, Vol. 113, 1994, p. 533.

15. K. S. Suslick, "Sonochemistry of Transition Metal Compounds," in R. B. King (Ed.), *Encyclopedia of Inorganic Chemistry*, Vol. 7, 1996, Wiley, New York, pp. 3890–3905.

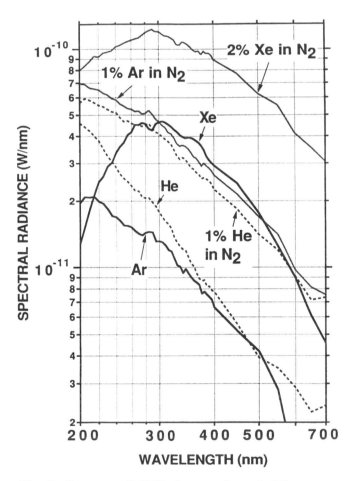

Fig. 7 Spectrum of SBSL for a variety of different gas mixtures in water. Pressure head 150 Torr. In contrast to spectra for MBSL, curves are relatively smooth and in some cases show a steady increase in intensity as one progresses to smaller wavelengths. Data have been corrected for the adsorption of water and quartz and for the quantum efficiency of the photodetector. (Reprinted with permission from Ref. 54. Copyright 1994 American Association for the Advancement of Science. Courtesy of Bob Hiller.)

16. K. S. Suslick and D. J. Casadonte, *J. Am. Chem. Soc.*, Vol. 109, 1987, p. 3459.

17. L. A. Crum, *Physics Today*, Vol. 47, 1994, p. 22.

18. S. J. Putterman, *Sci. Am.*, February 1995, p. 46.

19. Lord Rayleigh, *Philos. Mag.*, Vol. 34, 1917, p. 94.

20. W. T. Richards and A. L. Loomis, *J. Am. Chem. Soc.*, Vol. 49, 1927, p. 3086.

21. M. A. Margulis, *Ultrasonics*, Vol. 30, 1992, p. 152.

22. T. Lepoint and F. Mullie, *Ultrasonics Sonochem.*, Vol. 1, 1994, p. S13.

23. A. Henglein, *Ultrasonics*, Vol. 25, 1985, p. 6.

24. A. Henglein, *Adv. Sonochem.*, Vol. 3, 1993, p. 17.

25. C. M. Preece and I. L. Hansson, *Adv. Mech. Phys. Surf.*, Vol. 1, 1981, p. 199.

26. W. Lauterborn and A. Vogel, *Ann. Rev. Fluid Mech.*, Vol. 16, 1984, p. 223.

27. J. R. Blake and D. C. Gibson, *Ann. Rev. Fluid Mech.*, Vol. 19, 1987, p. 99.

28. P. Riesz, *Adv. Sonochem.*, Vol. 2, 1991, p. 23.

29. M. A. Margulis and N. A. Maximenko, *Adv. Sonochem.*, Vol. 2, 1991, p. 253.

30. K. S. Suslick, J. W. Gawienowski, P. F. Schubert, and H. H. Wang, *J. Phys. Chem.*, Vol. 87, 1983, p. 2299.

31. K. S. Suslick, J. W. Goodale, H. H. Wang, and P. F. Schubert, *J. Am. Chem. Soc.*, Vol. 105, 1983, p. 5781.

32. K. S. Suslick, *MRS Bulletin*, Vol. 20, 1995, pp. 29–34.

33. K. S. Suslick, S. B. Choe, A. A. Cichowlas, and M. W. Grinstaff, *Nature*, Vol. 353, 1991, p. 414.

34. K. S. Suslick and M. W. Grinstaff, *J. Am. Chem. Soc.*, Vol. 112, 1990, p. 7807.

35. T. Ando and T. Kimura, *Adv. Sonochem.*, Vol. 2, 1991, p. 211.

36. J.-L. Luche *Ultrasonics Sonochem.*, Vol. 1, 1994, p. S111.

37. K. Chatakondu, M. L. H. Green, M. E. Thompson, and K. S. Suslick, *J. Chem. Soc. Chem. Commun.*, 1987, p. 900.

38. A. N. Mal'tsev, *Zh. Fiz. Khim.*, Vol. 50, 1976, p. 1641.

39. B. H. Han and P. Boudjouk, *Organometallics*, Vol. 2, 1983, p. 769.

40. H. Frenzel and H. Schultes, *Z. Phys. Chem.*, Vol. 27b, 1934, p. 421.

41. L. A. Crum, *J. Acoust. Soc. Am.*, Vol. 95, 1994, p. 559.

42. D. F. Gaitan and L. A. Crum, in M. Hamilton and D. T. Blackstock (Eds.), *Frontiers of Nonlinear Acoustics*, 12th ISNA, Elsevier Applied Science, New York, 1990, pp. 459–463.

43. D. F. Gaitan, L. A. Crum, R. A. Roy, and C. C. Church, *J. Acoust. Soc. Am.*, Vol. 91, 1992, p. 3166.

44. B. P. Barber and S. J. Putterman, *Nature*, Vol. 352, 1991, p. 318.

45. L. A. Crum and R. A. Roy, *Science*, Vol. 266, 1994, p. 233.

46. H. G. Flynn, "Physics of Acoustic Cavitation in Liquids," in W. P. Mason (Ed.), *Physical Acoustics*, Vol. IB, Academic, New York, 1964, p. 157.

47. L. A. Crum and J. B. Fowlkes, *Nature*, Vol. 319, 1986, p. 52.

48. L. A. Crum and D. F. Gaitan, *Proc. Int. Soc. Opt. Eng.*, Vol. 1161, 1989, p. 125.

49. L. A. Crum, R. A. Roy, M. A. Dinno, C. C. Church, R. E. Apfel, C. K. Holland, and S. I. Madanshetty, *J. Acoust. Soc. Am.*, Vol. 91, 1992, p. 1113.

50. R. E. Verrall and C. Sehgal, in K. S. Suslick (Ed.) *Ultrasound: Its Chemical, Physical, and Biological Effects*, VCH Publishers, New York, 1988, pp. 227–287.

51. E. B. Flint and K. S. Suslick, *J. Phys. Chem.*, Vol. 95, 1991, p. 1484.

52. E. B. Flint and K. S. Suslick, *J. Am. Chem. Soc.*, Vol. 111, 1989, p. 6987.

53. T. J. Matula, R. A. Roy, P. D. Mourad, W. B. McNamara III, and K. S. Suslick, submitted.

54. R. Hiller, K. Weninger, S. J. Putterman, and B. P. Barber, *Science*, Vol. 266, 1994, p. 248.

55. B. P. Barber, R. Hiller, K. Arisaka, H. Fetterman, and S. J. Putterman, *J. Acoust. Soc. Am.*, Vol. 91, 1992, p. 3061.

56. M. J. Moran et al., "Direct Observations of Single Sonoluminescence Pulses," *UCRL-JC-118486* (Preprint), Lawrence Livermore National Lab, October 1994.

57. B. P. Barber and S. J. Putterman, *Phys. Rev. Lett.*, Vol. 69, 1992, p. 3839.

58. R. Lofstedt, B. P. Barber, and S. J. Putterman, *Phys. Fluids*, Vol. A5, 1993, p. 2911.

59. B. P. Barber, C. C. Wu, R. Lofstedt, P. H. Roberts, and S. J. Putterman, *Phys. Rev. Lett.*, Vol. 72, 1994, p. 1380.

60. C. C. Wu and P. H. Roberts, *Phys. Rev. Lett.*, Vol. 70, 1993, p. 3424.

61. W. C. Moss, D. B. Clarke, J. W. White, and D. A. Young, *Phys. Fluids*, Vol. 6, 1994, p. 2979.

62. L. Frommhold and A. A. Atchley, *Phys. Rev. Lett.*, Vol. 73, 1994, p. 2883.

63. R. G. Holt, D. F. Gaitan, A. A. Atchley, and J. Holzfuss, *Phys. Rev. Lett.*, Vol. 72, 1994, p. 1376.

64. C. C. Wu and P. H. Roberts, *Proc. R. Soc. Lond.* A, Vol. 445, 1994, p. 323.

65. J. Schwinger, *Proc. Natl. Acad. Sci.*, Vol. 89, 1992, p. 4091.

66. V. Kamath, A. Prosperetti, and F. N. Egolfopoulos, *J. Acoust. Soc. Am.*, Vol. 94, 1993, p. 248.

PART III

AEROACOUSTICS AND ATMOSPHERIC SOUND

27

INTRODUCTION

J<small>AMES</small> L<small>IGHTHILL</small>

1 BROAD OVERVIEW

1.1 Brief Survey of Part III

Part III, consisting of this introductory chapter and the six chapters following it, deals with sound in the air surrounding our planet. This nonhomogeneous fluid, usually in nonuniform motion, is the medium for much of the sound propagation that is of greatest human importance. Such propagation interacts with solid boundaries, including natural topography and man-made structures, while being strongly influenced also by distributions of wind and of atmospheric composition.

Sources of atmospheric sound long included a wide variety of natural and man-made phenomena at ground level, along with meteorologic processes ranging from lightning flashes to interactions between winds and structures. Moreover, in the twentieth century, exploitation of the air as a medium for transport introduced new sources of atmospheric sound, including aeroengine noise, airframe noise, and supersonic booms. Study of such sources forms a major part of aeroacoustics.

In addition to sound at frequencies to which the human ear is sensitive, atmospheric acoustics is also concerned with sources of sound at much lower frequencies (infrasound) and with its propagation, often over very long distances, through the atmosphere. By contrast, sound at ultrasonic frequencies achieves only modest penetration, being subject to a much higher level of atmospheric absorption; which, however, may be a source of interesting air motion known as acoustic streaming.

This chapter serves as a general introduction to Part III. First, some fundamental ideas that, historically, have

helped to give physical understanding of the field are introduced. Then their power in this regard is illustrated through applications to aeroengine noise, airframe noise, and supersonic booms, to the propagation of sound through nonuniform winds, and to certain aspects of acoustic streaming. Next a more detailed modern account of aeroengine noise, with special emphasis on jet noise and on techniques for its suppression, is given in Chapter 28. Some subtle features of the interactions between fluid motion and sound are outlined in Chapter 29, and Chapter 30 describes the streaming effects generated by both standing and traveling sound waves. General accounts, both of blast waves generated by explosions or by lightning flashes and of booms originating in the flight of supersonic aircraft, are given in Chapter 31.

A wide-ranging survey of propagation of sound in the atmosphere follows in Chapter 32. This takes into account the different mechanisms of attenuation in the air itself and at the ground and some effects of these acting in combination with refraction due to nonuniformities of wind velocity and of atmospheric composition. It outlines diffraction and scattering processes, including scattering by atmospheric turbulence, and ends with an account of acoustic methods (echosound) for probing the atmosphere. Part III is then concluded with a comprehensive description (Chapter 33) of atmospheric infrasound, which includes accounts of sources, propagation, measurement methods, and perception by the human body.

So Part III relates mainly to interactions of sound with air, including transmission through the atmosphere and both generation of sound by, and propagation of sound in, airflows (e.g., man-made flows—around aircraft or air machinery—or natural winds) as affected by the air's boundaries and atmospheric composition, with (conversely) generation of airflows by sound (acoustic streaming).

Encyclopedia of Acoustics, edited by Malcolm J. Crocker
ISBN 0-471-80465-7 © 1997 John Wiley & Sons, Inc.

From linear theory in Part I we use the wave equation's physical basis (Chapter 2), the short-wavelength ray acoustics approximation (Chapter 3), acoustic attenuation (Chapters 6 and 8), waveguides (Chapter 8), and multipole sources with the long-wavelength compact-source approximation (Chapter 9). From nonlinear acoustics in Part II we use the physics of waveform shearing and shock formation (Chapter 19).

Historically, the key to understanding how airflows interact with sound has been a comparison between the dynamic equations for airflows and the wave equation approximations of the elementary theory of sound. Although a feature common to both theories is mass conservation (very briefly, mass flux is a vector field whose divergence gives the local rate of decrease of mass density), nonetheless fundamental differences emerge in their treatment of momentum. The following critique of these differences offers a basis for understanding the interactions.

1.2 Role of the Momentum Equation

The techniques presented in Part III are centered on the momentum equation for air. Differences between the momentum equation and a wave equation approximation include the following:

1. The *linear effects* of gravity acting on stratified air. These effects allow independent propagation of "internal" gravity waves and of sound, except at wavelengths of many kilometers, when the atmosphere becomes a waveguide for global propagation of interactive acoustic gravity waves[1] (Chapter 32).

2. More important are *nonlinear effects* of the momentum flux $\rho u_i u_j$; that is, the flux, or the rate of transport across a unit area, of any ρu_i momentum component by any u_j velocity component. This term, neglected in linear acoustics, acts as a stress (i.e., force per unit area, since the rate of change of momentum is the force):

   (i) An airflow's momentum flux $\rho u_i u_j$ generates sound as would a distribution of (time-varying) imposed stresses; thus not only do forces between the airflow and its boundary radiate sound as distributed dipoles, but also stresses (acting on fluid elements with equal and opposite dipole-like forces) radiate as distributed quadrupoles.[2,3]

   (ii) The *mean momentum flux* $\langle \rho u_i u_j \rangle$ in any sound waves propagating through a sheared flow (with shear $\partial V_i / \partial x_j$) is a stress on that flow[1,4]; the consequent *energy exchange*

(from sound to flow when positive, vice versa when negative) is given as

$$\langle \rho u_i u_j \rangle \frac{\partial V_i}{\partial x_j}. \tag{1}$$

   (iii) Even without any preexisting flow, energy flux *attenuation* in a sound wave allows streaming to be generated by unbalanced stresses due to a corresponding attenuation in acoustic momentum flux; essentially, then, as acoustic energy flux is dissipated into heat, any associated acoustic momentum flux is transformed into a mean motion.[1,5]

3. Another (less crucial) momentum equation/wave equation difference is the *nonlinear deviation of pressure excess $p - p_0$* from a constant multiple, $c_0^2(\rho - \rho_0)$ of density excess:

   (i) For sound generation by airflows, this adds an isotropic term to the quadrupole strength per unit volume:

$$T_{ij} = \rho u_i u_j + [(p - p_0) - c_0^2(\rho - \rho_0)]\delta_{ij}. \tag{2}$$

   The last term is considered important mainly for flows at above-ambient temperatures.[2,6]

   (ii) For propagation of sound with energy density E through flows of air with adiabatic index γ, the mean deviation is about $\frac{1}{2}(\gamma - 1)E$, and the *total radiation stress*[7]

$$\langle \rho u_i u_j \rangle + \frac{1}{2}(\gamma - 1)E\delta_{ij} \tag{3}$$

   adds an isotropic pressure excess to the mean momentum flux [although the energy exchange (1) is unchanged in typical cases with $\partial V_i / \partial x_i$ essentially zero].

[Note that the special case of sound waves interacting on themselves (see Part II) may be interpreted as a combined operation of the *self-convection effect* (difference 2 above) and the (smaller) *sound speed deviation* (difference 3).]

2 COMPACT-SOURCE REGIONS

2.1 Sound Generation by Low-Mach-Number Airflows

The main nondimensional parameters governing airflows of characteristic speed U and length-scale L are the Mach number $M = U/c$ (where, in aeroacoustics, c is taken as the sound speed in the atmosphere into which sound radiates) and the Reynolds number $R = UL/\nu$ (where ν is the kinematic viscosity). Low-Mach-number airflows are compact sources of sound, with frequencies either narrow banded at moderate R or broadbanded at high R when, respectively, flow instabilities lead to regular flow oscillations or extremely irregular turbulence. Since, in either case, a typical frequency ω scales as U/L (Strouhal scaling), the compactness condition that $\omega L/c$ be small (Chapter 9) is satisfied if $M = U/c$ is small.[3]

A solid body that, because of flow instability, is subjected to a fluctuating aerodynamic force F scaling as $\rho U^2 L^2$ (at frequencies scaling as U/L) radiates as an *acoustic dipole* of strength F, with (Chapter 9) mean radiated power $\langle \dot{F}^2 \rangle / 12\pi\rho c^3$.

This acoustic power scales as $\rho U^6 L^2/c^3$ (a sixth-power dependence on flow speed). Therefore *acoustic efficiency*, defined as the ratio of acoustic power to a rate of delivery (scaling as $\rho U^3 L^2$) of energy to the flow, scales as $(U/c)^3 = M^3$. [Exceptions to compactness include bodies of high aspect ratio; thus, a long wire in a wind (where the scale L determining frequency is its diameter) radiates as a lengthwise distribution of dipoles.]

Away from any solid body a compact flow (oscillating or turbulent, with frequencies scaling as U/L) leads to quadrupole radiation (see difference 2, part (i) above) with total quadrupole strength scaling as $\rho U^2 L^3$. Acoustic power then scales (Chapter 9) as $\rho U^8 L^2/c^5$: an eighth-power dependence[2,3] on flow speed. In this case acoustic efficiency (see above) scales as $(U/c)^5 = M^5$. Such quadrupole radiation, though often important, may become negligible near a solid body when dipole radiation due to fluctuating body force (with its sixth-power dependence) is also present.[8,9]

For bodies that are not necessarily compact there exists a more refined calculation, using Green's functions for internally bounded space rather than for free space, that leads in general to the same conclusion: Quadrupole radiation with its eighth-power dependence is negligible alongside the sixth-power dependence of dipole radiation due to fluctuating body forces. Important exceptions to this rule include sharp-edged bodies, where features of the relevant Green's function imply a fifth-power dependence on flow speed of acoustic radiation from turbulence.[10–13]

2.2 Sound Generation by Turbulence at Not So Low Mach Numbers

The chaotic character of turbulent-flow fields implies that velocity fluctuations at points P and Q, although they are well correlated when P and Q are very close, become almost uncorrelated when P and Q are not close to one another.

Statisticians define the correlation coefficient C for the velocities $\mathbf{u}_P$ and $\mathbf{u}_Q$ as $C = \langle \mathbf{v}_P \cdot \mathbf{v}_Q \rangle / \langle \mathbf{v}_P^2 \rangle^{1/2} \langle \mathbf{v}_Q^2 \rangle^{1/2}$ in terms of the deviations $\mathbf{v}_P = \mathbf{u}_P - \langle \mathbf{u}_P \rangle$ and $\mathbf{v}_Q = \mathbf{u}_Q - \langle \mathbf{u}_Q \rangle$ from their means. When two uncorrelated quantities are combined, their mean-square deviations are summed:

$$\langle \mathbf{v}_P + \mathbf{v}_Q \rangle^2 = \langle \mathbf{v}_P^2 \rangle + 2\langle \mathbf{v}_P \cdot \mathbf{v}_Q \rangle + \langle \mathbf{v}_Q^2 \rangle$$
$$= \langle \mathbf{v}_P^2 \rangle + \langle \mathbf{v}_Q^2 \rangle \quad \text{if } C = 0.$$

Theories of turbulence define a correlation length l, with $\mathbf{u}_P$ and $\mathbf{u}_Q$ either well correlated or uncorrelated (C close to 1 or 0) when PQ is substantially $<l$ or $>l$. Roughly speaking, different regions of size l ("eddies") generate sound independently, and the mean-square radiated noise is the sum of the mean square outputs from all the regions.[14]

Typical frequencies in the turbulence are of order $\omega = v/l$, where v is a typical root-mean-square (rms) velocity deviation $\langle v^2 \rangle^{1/2}$, so that for each region the compactness condition $\omega l/c$ small is satisfied if v/c is small. Compactness, then, requires only that an rms velocity deviation v (rather than a characteristic mean velocity U) be small compared with c, which is less of a restriction on $M = U/c$ and can be satisfied at not-so-low Mach number.[6]

3 DOPPLER EFFECT

How is the radiation from such "eddies" modified by the fact that they are being *convected* at not-so-low Mach number? The term *Doppler effect*, covering all aspects of how the movement of sources of sound alters their radiation patterns, comprises (i) frequency changes,[1] (ii) volume changes,[2,3] and (iii) compactness changes.[6,15]

3.1 Frequency Changes

When a source of sound at frequency ω approaches an observer at velocity w, then in a single period $T = 2\pi/\omega$ sound emitted at the beginning travels a distance cT, while at the end of the period sound is being emitted from a source that is closer by a distance wT. The wavelength λ (distance between crests) is reduced to

$$\lambda = cT - wT = \frac{2\pi(c-w)}{\omega}, \qquad (4)$$

and the frequency heard by the observer (2π divided by the time λ/c between arrival of crests) is increased to the Doppler-shifted value[1]

$$\omega_r = \frac{\omega}{1 - (w/c)} \quad \text{(relative frequency)} \qquad (5)$$

that results from relative motion between source and observer. For an observer located on a line making an angle θ with a source's direction of motion at speed V, the source's velocity of approach toward the observer is $w = V \cos \theta$ and the relative frequency becomes

$$\omega_r = \frac{\omega}{1 - (V/c) \cos \theta} \qquad (6)$$

and is augmented or diminished, when θ is an acute or obtuse angle. Such Doppler shifts in frequency are familiar everyday experiences (Chapter 32).

3.2 Volume Changes

When an observer is approached at velocity w by a source whose dimension (in the direction of the observer) is l, sounds arriving simultaneously from the source's far or near sides have been emitted earlier or later by a time difference τ (say). In the time t for sound from the far side to reach the observer, after traveling a distance ct, the relative distance of the near side in the direction of the observer was increased from l to $l + w\tau$ before it emitted sound that then traveled a distance $c(t - \tau)$. Both sounds arrive simultaneously if

$$ct = l + w\tau + c(t - \tau), \qquad (7a)$$

giving

$$\tau = \frac{l}{c - w} \quad \text{and} \quad l + w\tau = \frac{l}{1 - (w/c)} = \frac{l\omega_r}{\omega}. \qquad (7b)$$

The source's effective volume during emission is increased, then, by the Doppler factor ω_r/ω (since the dimension in the direction of the observer is so increased whereas other dimensions are unaltered).[2,3]

If turbulent "eddies" are effectively being convected, relative to the air into which they are radiating, at velocity V, then Eq. (6) gives, for radiation at angle θ, the Doppler factor ω_r/ω, which modifies both the frequencies at which they radiate and the effective volume occupied by a radiating eddy.

But Eq. (2) specifies the quadrupole strength T_{ij} per unit volume for such an eddy. Without convection the pattern of acoustic intensity around a compact eddy of volume l^3 and quadrupole strength $l^3 T_{ij}$ would be

$$\frac{\langle (l^3 \ddot{T}_{ij} x_i x_j r^{-2})^2 \rangle}{16 \pi^2 r^2 \rho_0 c^5} \qquad (8)$$

(Chapter 9); and since different eddies of volume l^3 radiate independently, we can simply sum mean squares in the corresponding expressions for their far-field intensities. This gives

$$\frac{l^3 \langle (\ddot{T}_{ij} x_i x_j r^{-2})^2 \rangle}{16 \pi^2 r^2 \rho_0 c^5} \qquad (9)$$

as the intensity pattern radiated by a unit volume of turbulence. The Doppler effect modifies this, when the compactness condition is satisfied, by five factors ω_r/ω (one for the change in source volume l^3 and four for the frequency change as it affects the mean square of a multiple of the second time derivative of T_{ij}), and this intensity modification brings about an important preference for forward emission[15] by a factor

$$\left[1 - \left(\frac{V}{c} \right) \cos \theta \right]^{-5}. \qquad (10)$$

3.3 Compactness Changes

As V/c increases, however, the Doppler effect tends to degrade the compactness of aeroacoustic sources in relation to forward emission. Not only does $\omega l/c$ increase in proportion to Mach number, but an even greater value is taken by $\omega_r l/c$, a ratio that must be small if convected sources are to be compact. A restriction on the extent (10) of intensity enhancement for forward emission as V/c increases is placed by these tendencies.[6,15,16]

Indeed, they can develop to a point where the compact-source approximation (Chapter 9) may appropriately be replaced by the ray acoustics approximation (Chapter 3). Thus, for supersonic source convection ($V/c > 1$), the relative frequency (6) becomes infinite in the Mach direction

$$\theta = \cos^{-1} \frac{c}{V}, \qquad (11)$$

and radiation from the source proceeds along rays emitted at this angle.[17]

Explanatory note: The source's velocity of approach w toward an observer positioned at an angle (11) to its direction of motion is the sound speed c; thus, not only is the generated wavelength (4) reduced indefinitely (the ray acoustics limit) but, essentially, different parts of a signal are observed simultaneously: the condition of stationary phase satisfied on rays (Chapter 3).

Further note: The influences placing a limit on the signal propagated along rays may include the duration δ of well-correlated emission from turbulent eddies and, in addition, nonlinear effects (see Section 4.2).

3.4 Uniformly Valid Doppler Effect Approximations

The correlation length l for turbulence was presented in Section 2.2; a correlation duration δ can be characterized by the requirement that moving eddies have, respectively, well-correlated or uncorrelated velocities at times differing by substantially $<\delta$ or $>\delta$. Combined use of correlation length l and duration δ affords an approximation to the radiation pattern from convected eddies that has some value at all Mach numbers, spanning the areas of applicability of the compact-source and ray acoustics approximations.

Figure 1 uses space–time diagrams in which the space coordinate (abscissa) is distance in the direction of the observer. Figure 1a for unconvected eddies approximates the region of good correlation as an ellipse with axes l (in the space direction) and δ (in the time direction). Figure 1b shows such a region for convected eddies whose velocity of approach toward the observer is w; thus, it is Fig. 1a sheared by distance w per unit time.

Signals from far points F and near points N, in either case, reach the observer simultaneously, as do signals from other points on the line FN, if this line slopes by distance c (the sound speed) per unit time:

(i) Compact-source case with w/c small: The space component of FN in Fig. 1b is $l[1 - (w/c)]^{-1}$, just as in Eq. (7) for normal Doppler effects (neglecting finite δ).

(ii) Ray acoustics case with $w/c = 1$: The space component of FN is $c\delta$.

(iii) Intermediate case with w/c "moderately" <1: The space component of FN is l multiplied by an enhancement factor

$$\left[\left(1 - \frac{w}{c}\right)^2 + \left(\frac{l}{c\delta}\right)^2\right]^{-1/2}, \quad (12)$$

which represents the effective augmentation of source volume due to convection.[15]

Enhancement factor (12) is applied not only to the volume term l^3 in the quadrupole field (9) but also twice to each of the pair of twice-differentiated terms inside the mean square, essentially because time differentiations in quadrupole fields arise (Chapter 9) from differences in the time of emission by different parts of the quadrupole source region (and the time component of FN in Fig. 1b is simply the space component divided by c). As before, then, five separate factors (12) enhance the intensity field, and with w replaced by $V \cos \theta$, expression (10) for the overall intensity modification factor is replaced by

$$\left\{\left[1 - \left(\frac{V}{c}\right)\cos\theta\right]^2 + \left(\frac{l}{c\delta}\right)^2\right\}^{-5/2}. \quad (13)$$

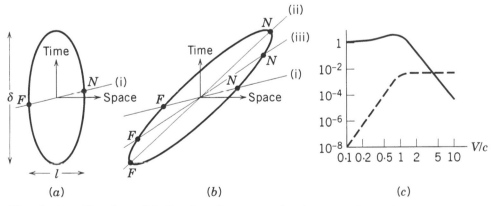

Fig. 1 A uniformly valid Doppler-effect approximation. (*a*) Space-time diagram for unconvected "eddies" of correlation length l and duration δ. (*b*) Case of "eddies" convected towards observer at velocity w; being Fig. 1a sheared by a distance w per unit time. Here, lines sloping by a distance c per unit time represent emissions received simultaneously by observer. Case (i): w/c small. Case (ii): $w/c = 1$. Case (iii): intermediate value of w/c. (*c*) The solid line is the average modification factor (13). The dashed line is the acoustic efficiency obtained by applying this factor to a low-Mach-number "quadrupole" efficiency of (say) 10^{-3} $(V/c)^5$.

This modification factor affords us an improved description of the influence of the Doppler effect not only on the preference for forward emission but also on the overall acoustic power output from convected turbulence.[6,15] For example, Fig. 1c gives a (plain line) log-log plot of the average (spherical mean) of the factor (13) as a function of V/c on the reasonable assumption that $l = 0.6V\delta$. As V/c increases, this average modification factor rises a little at first but falls drastically as $5(V/c)^{-5}$ for $V/c \gg 1$.

Now low-Mach-number turbulence away from solid boundaries (Section 2.1) should radiate sound with an acoustic efficiency scaling as $(U/c)^5$, where U is a characteristic velocity in the flow. With V taken as that characteristic velocity (although in a jet a typical velocity V of eddy convection would be between 0.5 and 0.6 times the jet exit speed), the modification of (say) an acoustic efficiency of $10^{-3}(V/c)^5$ for low Mach number by the average modification factor would cause acoustic efficiency to follow the broken-line curve in Fig. 1c, tending asymptotically to a constant value 0.005 (aeroacoustic saturation) at high Mach number. Such a tendency is often observed for sound radiation from "properly expanded" supersonic jets (see below).

4 INTRODUCTION TO AIRCRAFT NOISE

4.1 Aeroengine and Airframe Noise

How are aeroacoustic principles applied to practical problems such as the reduction of aircraft noise?[18–20]

In any analysis of the generation of sound by airflows, we may first need to ask whether the geometry of the problem has features that tend to promote resonance. For example, a long wire in a wind (Section 2.1) generates most sound when vortex-shedding frequencies are fairly close (and so can "lock on") to the wire's lowest natural frequency of vibration; giving good correlation of side forces, and so also of dipole strengths, all along the wire.

Again, a jet emerging from a thin slit may interact with a downstream edge (parallel to the slit) in a resonant way,[21,22] with very small directional disturbances at the jet orifice amplified by flow instability as they move downstream to the edge, where they produce angle-of-attack variations. Dipole fields associated with the resulting side forces can at particular frequencies renew the directional disturbances at the orifice with the right phase to produce a resonant oscillation. Some musical wind instruments utilize such jet-edge resonances, reinforced by coincidence with standing-wave resonances (Chapter 7) in an adjacent pipe. However, in the absence of such resonances (leading to enhanced acoustic generation at fairly well defined frequencies), airflows tend to generate acoustic "noise" whose reaction on the flow instability phenomena themselves is negligible.

Resonances analogous to the above that need to be avoided in aircraft design include the following:

(a) Panel flutter, generated at a characteristic frequency as an unstable vibration of a structural panel in the presence of an adjacent airflow.[23]

(b) Screeching of supersonic jets from nozzles which, instead of being "properly expanded" so that an essentially parallel jet emerges, produce a jet in an initially nonparallel form followed by shock waves in the well-known recurrent "diamond" shock cell pattern. The first of these, replacing the edge in the above description, can, through a similar feedback of disturbances to the jet orifice, generate a powerful resonant oscillation.[22,24,25]

Undesirable resonances may also be associated with aeroengine combustion processes.[26] We now turn to the aircraft noise of a broadbanded character that remains even when resonances have been avoided.

Aeroengine jet noise proper[6] (i.e., the part unrelated to any interaction of jet turbulence with solid boundaries) tends to follow a broad trend similar to that in Fig. 1, where, however, because the eddy convection velocity V is between 0.5 and 0.6 times the jet exit speed U, the acoustic efficiency makes a transition between a value of around $10^{-4}M^5$ in order-of-magnitude terms for subsonic values of $M = U/c$ and an asymptotically constant value of 10^{-2} or a little less for M exceeding about 2.

The above tendency for $M < 1$ implies that noise emission from jet engines may be greatly diminished if a given engine power can be achieved with a substantially lower jet exit speed, requiring of course a correspondingly larger jet diameter L. Furthermore, with acoustic power output scaling as $\rho U^8 L^2/c^5$ (Section 2.1) and jet thrust as $\rho U^2 L^2$, noise emission for a given thrust can be greatly reduced if U can be decreased and L increased by comparable factors.

Trends (along these lines) in aeroengine design toward large turbofan engines with higher and higher bypass ratios, generating very wide jets at relatively modest mean Mach numbers, have massively contributed to jet noise suppression (while also reducing fuel consumption). On the other hand, such successes in suppressing jet noise proper (originally, the main component of noise from jet aircraft) led to needs for focusing attention upon parallel reductions of other aircraft noise sources[27]:

(a) Those associated with the interaction of jet turbulence with solid boundaries, where sharp-edged boundaries (Section 2.1) pose a particular threat

(b) Fan noise emerging from the front of the engine and turbine noise emerging from the rear

(c) Airframe noise, including acoustic radiation from boundary layer turbulence and from interaction of that turbulence with aerodynamic surfaces for control purposes or lift enhancement

Some key areas of modern research on aeroengine and airframe noise are as follows:

For jet noise, techniques that relate acoustic output to vorticity distributions,[28,29] and to any coherent structures,[30,31] in jet turbulence and that take into account (cf. Section 5.4) propagation through the sheared flow in a wide jet[32,33]

For noise from fans and propellers, mathematically sophisticated ways of reliably estimating the extent of cancellation of dipole radiation from different parts of a rotating-blade system (alongside a good independent estimate of quadrupole radiation)[34]

For airframe noise, a recognition[35,36] that massive cancellations act to minimize noise radiation from boundary layer turbulence on a flat surface of uniform compliance; and, therefore, that avoidance of sharp nonuniformities in airframe skin compliance may promote noise reduction

4.2 Supersonic Booms

In addition to aeroengine and airframe noise, any aircraft flying at a supersonic speed V emits a concentrated "boom"-like noise along rays in the Mach direction (11). We sketch the theory of supersonic booms with the atmosphere approximated as isothermal (so that the undisturbed sound speed takes a constant value c even though the undisturbed density ρ varies with altitude), a case permitting a simple extension of the nonlinear analysis (Chapter 19) of waveform shearing and shock formation.[37] Then the rays continue as straight lines at the Mach angle for reasons summarized in the explanatory note below expression (11). (Actually, the slight refraction of rays by temperature stratification in the atmosphere, when taken into account in a generalized version of the theory, produces only somewhat minor modifications of the results.)

As such straight rays stretch out from a straight flight path along cone-shaped surfaces with semi-angle (11), any narrow tube of rays has its cross-sectional area A increasing in proportion to distance r along the tube.[1,38] In linear theory (Chapter 3), acoustic energy flux $u^2 \rho c A$ is propagated unchanged along such a ray tube [so that $u(\rho r)^{1/2}$ is unchanged], where u is air velocity along the ray tube. In nonlinear theory, $u(\rho r)^{1/2}$ is propagated unchanged but at a signal speed altered to

$$c + \frac{\gamma + 1}{2} u \qquad (14)$$

by self-convection and excess-wave-speed effects (Chapter 19).

This property can be described[1] by

$$\left[\left\{ \frac{1}{c} - \frac{\gamma + 1}{2} \frac{u}{c^2} \right\} \frac{\partial}{\partial t} + \frac{\partial}{\partial r} \right] u(\rho r)^{1/2} = 0, \qquad (15)$$

where the quantity in braces is the altered value of the reciprocal of the signal speed (14). Now a simple transformation of variables,

$$x_1 = r - ct, \qquad t_1 = \int_0^r (\rho r)^{-1/2} \, dr,$$

$$u_1 = \frac{\gamma + 1}{2} \frac{u}{c} (\rho r)^{1/2}, \qquad (16)$$

converts Eq. (15) into the familiar form (Chapter 19)

$$\frac{\partial u_1}{\partial t_1} + u_1 \frac{\partial u_1}{\partial x_1} = 0, \qquad (17)$$

which describes the waveform shearing at a uniform rate associated with shock formation and propagation in nonlinear plane-wave acoustics.

From the physically relevant solutions of Eq. (17), namely, those with area-conserving discontinuities (representing shocks), the famous N-wave solution is the one produced by an initial signal (such as an aircraft's passage through the air) that is first compressive and then expansive. The rules governing N-wave solutions of Eq. (17) (Chapter 19) are that the discontinuity Δu_1 at each shock falls off as $t_1^{-1/2}$ while the space (change Δx_1 in x_1) between shocks increases as $t_1^{1/2}$. These rules for the transformed variables (16) have the following consequences for the true physical variables: At a large distance r from the flight path the velocity change Δu at each shock and the time interval Δt between the two shocks vary as

$$\Delta u \approx \left[(\rho r) \int_0^r (\rho r)^{-1/2} \, dr \right]^{-1/2} \qquad \text{and}$$

$$\Delta t \approx \left[\int_0^r (\rho r)^{-1/2} \, dr \right]^{1/2}. \qquad (18)$$

On horizontal rays (at the level where the aircraft is flying), ρ is independent of r and Eqs. (18) take the greatly simplified form

$$\Delta u \approx r^{-3/4} \quad \text{and} \quad \Delta t \approx r^{1/4}, \tag{19}$$

appropriate to conical N-waves in a homogeneous atmosphere. Actually, the rules (19) apply also to the propagation of cylindrical blast waves generated by an exploding wire, since, here also, ray tube areas increase in proportion to r. (For spherical blast waves, see Chapter 31.)

On downward-pointing rays in an isothermal atmosphere ρ increases exponentially in such a way that the time interval Δt between shocks approaches the constant value obtained in Eqs. (18) by making the integral's upper limit infinite.[1,38] On the other hand, the shock strength (proportional to the velocity change Δu) includes the factor $(\rho r)^{-1/2}$, where the large increase in ρ from the flight path to the ground (as well as in r) enormously attenuates the supersonic boom. Below Concorde cruising at Mach 2, for example, an observer on the ground hears two clear shocks with an interval of around 0.5 s between them and yet with strength $\Delta p/p$ of only about 0.001.

5 PROPAGATION OF SOUND THROUGH STEADY MEAN FLOWS

5.1 Adaptations of Ray Acoustics

Useful information on sound propagation through steady mean flows[4,39] can be obtained by adaptations of the ray acoustics approximation. We sketch these here before, first, applying them (in Section 5.3) to propagation through sheared stratified winds and, second, giving indications of how effects of such parallel mean flows are modified at wavelengths too large for the applicability of ray acoustics.

Sound propagation through a steady airflow represents an autonomous mechanical system, one governed by laws that do not change with time. Then small disturbances can be Fourier analyzed in the knowledge that propagation of signals with different frequencies ω must proceed without exchange of energy between them.

Such disturbances of frequency ω involve pressure changes in the form $P \cos \alpha$, where P varies with position and the phase α is a function of position and time, satisfying

$$\frac{\partial \alpha}{\partial t} = \omega \quad \text{(frequency) and}$$

$$-\frac{\partial \alpha}{\partial x_i} = k_i \quad \text{(wavenumber)}, \tag{20}$$

a vector with its direction normal to crests and its magnitude 2π divided by a local wavelength.

In ray theory for any wave system,[1,4] we assume that the wavelength is small enough (compared with distances over which the medium—and its motion, if any—change significantly) for a well-defined relationship

$$\omega = \Omega(k_i, x_i) \tag{21}$$

to link frequency with wavenumber at each position. Equations (20) and (21) require that

$$-\frac{\partial k_j}{\partial t} = \frac{\partial^2 \alpha}{\partial x_j \, \partial t} = \frac{\partial \omega}{\partial x_j} = \frac{\partial \Omega}{\partial k_i} \left(-\frac{\partial^2 \alpha}{\partial x_i \, \partial x_j} \right) + \frac{\partial \Omega}{\partial x_j}$$

$$= \frac{\partial \Omega}{\partial k_i} \frac{\partial k_j}{\partial x_i} + \frac{\partial \Omega}{\partial x_j}, \tag{22}$$

yielding the basic law (in Hamiltonian form) for any wave system: On rays satisfying

$$\frac{dx_i}{dt} = \frac{\partial \Omega}{\partial k_i} \tag{23a}$$

wavenumbers vary as

$$\frac{dk_j}{dt} = -\frac{\partial \Omega}{\partial x_j}. \tag{23b}$$

These are equations easy to solve numerically for given initial position and wavenumber. However, the variations (23) of wavenumber ("refraction") produce no change of frequency along rays:

$$\frac{d\omega}{dt} = \frac{\partial \Omega}{\partial k_i} \frac{dk_i}{dt} + \frac{\partial \Omega}{\partial x_i} \frac{dx_i}{dt}$$

$$= \frac{\partial \Omega}{\partial k_i} \left(-\frac{\partial \Omega}{\partial x_i} \right) + \frac{\partial \Omega}{\partial x_i} \frac{\partial \Omega}{\partial k_i} = 0, \tag{24}$$

so that rays are paths of propagation of the excess energy, at each frequency, associated with the waves' presence.

For sound waves we write k as the magnitude of the wavenumber vector, expecting that at any point the value of the relative frequency ω_r in a frame of reference moving at the local steady-flow velocity u_{fi} will be $c_f k$ (the local sound speed times k); this implies[1,4] that

$$\omega_r = \frac{\partial \alpha}{\partial t} + u_{fi} \frac{\partial \alpha}{\partial x_i} = \omega - u_{fi} k_i, \tag{25a}$$

giving

$$\omega = \omega_r + u_{fi}k_i = c_f k + u_{fi}k_i \qquad (25b)$$

as the acoustic form of relationship (21). [*Note:* Rule (25) for relative frequency agrees with the Doppler rule (6), since the velocity of a source of frequency ω relative to stationary fluid into which it radiates is minus the velocity of the fluid relative to a frame in which the acoustic frequency is ω.]

Use of form (25b) of relationship (21) in the basic law (23) tells us that

$$\frac{dk_j}{dt} = -k\frac{\partial c_f}{\partial x_j} - k_i\frac{\partial u_{fi}}{\partial x_j} \qquad (26a)$$

on rays with

$$\frac{dx_i}{dt} = c_f\frac{k_i}{k} + u_{fi}, \qquad (26b)$$

where the last terms in these equations represent adaptations of ray acoustics associated with the mean flow. For example, the velocity of propagation along rays is the vector sum of the mean-flow velocity u_{fi} with a wave velocity of magnitude c_f and direction normal to crests.

5.2 Energy Exchange between Sound Waves and Mean Flow

The excess energy (say, E per unit volume) associated with the presence of sound waves is propagated along such rays; in particular, if attenuation of sound energy is negligible, then

Flux of excess energy along a ray tube = constant.

$$(27)$$

Note that this excess energy density E is by no means identical with the sound waves' energy density,

$$E_r = \langle \tfrac{1}{2}\rho_f u_{si}u_{si}\rangle + \langle \tfrac{1}{2}c_f^2\rho_f^{-1}\rho_s^2\rangle = c_f^2\rho_f^{-1}\langle\rho_s^2\rangle, \qquad (28)$$

where the subscript s identifies changes due to the sound waves and the equality of the kinetic and potential energies makes E_r simply twice the latter in a frame of reference moving at the local flow velocity [compare definition (25) of ω_r]. The kinetic-energy part of the excess energy density E is

$$\langle \tfrac{1}{2}(\rho_f + \rho_s)(u_{fi} + u_{si})^2\rangle - \tfrac{1}{2}\rho_f u_{fi}u_{fi}, \qquad (29)$$

which includes an extra term,

$$\langle \rho_s u_{fi}u_{si}\rangle = \left\langle \rho_s u_{fi}\frac{c_f}{\rho_f}\rho_s\frac{k_i}{k}\right\rangle = E_r\frac{u_{fi}k_i}{c_f k}$$

$$= E_r\left(\frac{\omega}{\omega_r} - 1\right), \qquad (30)$$

and E is the sum of expressions (28) and (30), giving

$$E = E_r\frac{\omega}{\omega_r}, \qquad (31a)$$

or, equivalently,

$$E_r = E\frac{\omega_r}{\omega} \quad \text{or} \quad \frac{E_r}{\omega_r} = \frac{E}{\omega}. \qquad (31b)$$

The quantity E/ω, called the *action density* in Hamiltonian mechanics, is identical in both frames of reference, and Eqs. (24) and (27) tell us that its flux along a ray tube is constant.[1,4]

However, Eq. (31) shows too that energy is exchanged between (i) the acoustic motions relative to the mean flow and (ii) the mean flow itself. For example, where sound waves of frequency ω enter a region of opposing flow (or leave a region where the mean flow is along their direction of propagation), the ratio ω_r/ω increases and so therefore does E_r/E: The sound waves gain energy at the expense of the mean flow.

The rate of exchange of energy takes the value (1) from Section 1. This is readily seen from the laws governing motion in an accelerating frame of reference, which is subject to an

Inertial force = $-$(mass)$\times$(acceleration of frame). (32)

If at each point of space we use a local frame of reference moving with velocity u_{fi}, then fluid in that frame has velocity u_{si} but is subject to an additional force (32), where, per unit volume, mass is ρ_f and the frame's acceleration takes the form

$$u_{sj}\frac{\partial u_{fi}}{\partial x_j}$$

Giving force $-\rho_f u_{sj}\dfrac{\partial u_{fi}}{\partial x_j}$

Doing work $-\rho_f\langle u_{si}u_{sj}\rangle\dfrac{\partial u_{fi}}{\partial x_j}$ (33)

per unit time on the local relative motions. This rate of energy exchange (33) proves to be consistent with the fact that it is the flux, not of E_r but of action E_r/ω_r, that is conserved along ray tubes.

Energy can be extracted from a mean flow, then, not only by turbulence but also by sound waves; in both cases, the rate of extraction takes the same form (33) in terms of perturbation velocities u_{si}. It represents the effect (Section 1) of that

$$\text{Mean momentum flux} = \rho_f \langle u_{si} u_{sj} \rangle \qquad (34)$$

or Reynolds stress[40] with which either the sound waves or the turbulent motions act upon the mean flow. For sound waves, by Eq. (28) for E_r and by the substitution

$$u_{si} = \frac{c_f}{\rho_f} \rho_s \frac{k_i}{k}$$

we arrive at

$$\text{Mean momentum flux} = E_r \frac{k_i k_j}{k^2}, \qquad (35)$$

so that the Reynolds stress is a uniaxial stress in the direction of the wavenumber vector having magnitude E_r. Strictly speaking, the complete

$$\text{Radiation stress} = E_r \left(\frac{k_i k_j}{k^2} + \frac{\gamma - 1}{2} \delta_{ij} \right) \qquad (36)$$

for sound waves includes not only the momentum flux (35) but also the waves' mean pressure excess,

$$\frac{1}{2} \left(\frac{\partial^2 p}{\partial \rho^2} \right)_{\rho = \rho_f} \langle \rho_s^2 \rangle = \frac{\gamma - 1}{2} \frac{c_f^2}{\rho_f} \langle \rho_s^2 \rangle = \frac{\gamma - 1}{2} E_r, \qquad (37)$$

acting equally in all directions[7]; however (Section 1), this isotropic component produces no energy exchange with solenoidal mean flows.

5.3 Propagation through Sheared Stratified Winds

The extremely general ray acoustics treatment outlined above for sound propagation through fluids in motion has far-reaching applications (in environmental and, also, in engineering acoustics), which, however, are illustrated below only by cases of propagation through parallel flows, with stratification of velocity as well as

of temperature.[1,4] The x_1-direction is taken as that of the mean-flow velocity $V(x_3)$, which, together with the sound speed $c(x_3)$, depends only on the coordinate x_3. Thus, V replaces u_{f1} in the general theory while c replaces c_f (and, for atmospheric propagation, x_3 is altitude). [*Note:* The analysis sketched here is readily extended to cases of winds veering with altitude, where u_{f2} as well as u_{f1} is nonzero.]

Either the basic law (23) or its ray acoustics form (26) provides, in general, "refraction" information in the form of three equations for change of wavenumber, while the single, far simpler, Eq. (24) is a consequence of, but by no means equivalent to, those three. By contrast, in the particular case when u_{f1} and c_f are independent of x_1 and x_2, Eq. (24) and, additionally, Eqs. (26) for $j = 1, 2$ give three simple results along rays that may be shown fully equivalent to the basic law:

$$\omega = \text{const}, \qquad k_1 = \text{const}, \qquad k_2 = \text{const}. \qquad (38)$$

If now we write the wavenumber (a vector normal to crests) as

$$(k_1, k_2, k_3) = (\kappa \cos \psi, \kappa \sin \psi, \kappa \cot \theta), \qquad (39)$$

so that κ is its constant horizontal resultant, ψ its constant azimuthal angle to the wind direction, and θ its variable angle to the vertical, and use Eq. (25) in the form

$$\omega = c(x_3)k + V(x_3)k_1 = c(x_3)\kappa \csc \theta + V(x_3)\kappa \cos \psi, \qquad (40)$$

we obtain an important extension to Snell's law from the classical case ($V = 0$) when the denominator is a constant:

$$\sin \theta = \frac{c(x_3)}{\omega \kappa^{-1} - V(x_3) \cos \psi}. \qquad (41)$$

This extended law (41) tells us how θ varies with x_3 along any ray, whose path we can then trace using Eq. (26) in the form

$$\frac{dx_1}{dt} = c(x_3) \cos \psi \sin \theta + V(x_3),$$

$$\frac{dx_2}{dt} = c(x_3) \sin \psi \sin \theta,$$

$$\frac{dx_3}{dt} = c(x_3) \cos \theta \qquad (42)$$

by simply integrating dx_1/dx_3 and dx_2/dx_3 with respect to x_3.

It follows that a ray tube covers the same horizontal area at each altitude, so that conservation of the flux of wave action E_r/ω_r along it implies that the vertical component

$$\frac{E_r}{\omega_r}\frac{dx_3}{dt} = E_r\kappa^{-1}\sin\theta\cos\theta \qquad (43)$$

of wave action flux is constant along rays, from which, with Eq. (28), sound amplitudes are readily derived.

Wind shear is able to reproduce all the main types of ray bending (Chapter 3) associated with temperature stratification, often to an enhanced extent. Roughly, the downward curvature of near-horizontal rays in reciprocal kilometres comes to

$$3\{V'(x_3)\cos\psi + c'(x_3)\}, \qquad (44)$$

where the velocity gradients are in reciprocal seconds and the factor 3 (km)$^{-1}$s outside the braces is an approximate reciprocal of the sound speed.[1,4]

When (44) is negative, curvature is upward; its magnitude with zero wind is at most 0.018 km^{-1} (because temperature lapse rate in stable atmospheres cannot exceed 10°C/km, giving $c' = 0.006$ s^{-1}) but with strong wind shear can take much bigger values for upwind propagation ($\psi = \pi$). In either case Fig. 2a shows how the lowest ray emitted by a source "lifts off" from the ground, leaving below it a zone of silence (on ray theory; actually, a zone where amplitudes decrease exponentially with distance below that ray).

When (44) is positive, curvature is downward, as found with zero wind in temperature inversion conditions (e.g., over a calm, cold lake) and even more with strong wind shear for downwind propagation ($\psi = 0$). Figure 2b shows how this leads to signal enhancement through multiple-path communication.

In summary, then, the very familiar augmentation of sound levels downwind, and diminution upwind, of a

source represent effects of the wind's shear (increase with altitude).

5.4 Wider Aspects of Parallel-Flow Acoustics

The propagation of sound through parallel flows at wavelengths too great for the applicability of ray acoustics can be analyzed by a second-order ordinary differential equation. Thus, a typical Fourier component of the sound pressure field takes the form

$$p_s(x_3)e^{i(\omega t - k_1 x_1 - k_2 x_2)}$$

with

$$\rho\frac{d}{dx_3}\left[\frac{1}{\rho(\omega - Vk_1)^2}\frac{dp_s}{dx_3}\right] + \left[\frac{1}{c^2} - \frac{k_1^2 + k_2^2}{(\omega - Vk_1)^2}\right]p_s = 0. \qquad (45)$$

Equation (45) can be used to improve on ray acoustics:

(a) Near caustics (envelopes of rays), where it allows a uniformly valid representation of amplitude in terms of the famous Airy function, giving "beats" between superimposed waves on one side of the caustic and exponential decay on the other[1]

(b) At larger wavelengths by abandoning ray theory altogether in favor of extensive numerical solutions of Eq. (45)

(c) To obtain waveguide modes for sound propagation in a two-dimensional duct (between parallel planes)[41–43]

On the other hand, in the case of a three-dimensional duct carrying parallel flow $V(x_2, x_3)$ in the x_1-direction, Eq. (45) is converted to a partial differential equation

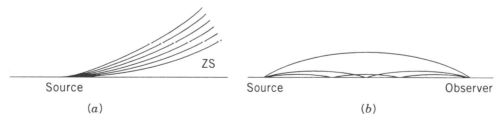

Fig. 2 Effects of ray curvature (44) on propagation from a source on horizontal ground. (a) Rays of given upward curvature (due to temperature lapse or upwind propagation) can leave a zone of silence (ZS) below the ray emitted horizontally. (b) Rays of given downward curvature (due to temperature inversion or downwind propagation) can enhance received signals through multiple-path communication.

(the first term being supplemented by another with d/dx_2 replacing d/dx_3, while k_2 is deleted) which is used for the following:

(d) To obtain waveguide modes in such ducts

(e) In calculations of propagation of sound through the wide jets, modeled as parallel flows, typical of modern aeroengines (Section 4.1)

(f) With aeroacoustic source terms included, in certain enterprising attempts at modeling jet noise generation and emission[32,33]

6   ACOUSTIC STREAMING

6.1   Streaming as a Result of Acoustic Attenuation

Sound waves act on the air with a Reynolds stress (34) even when mean flow is absent (so that subscript f becomes subscript zero). The j-component of force acting on a unit volume of air is then

$$F_j = -\frac{\partial}{\partial x_i} \langle \rho_0 u_{si} u_{sj} \rangle, \qquad (46)$$

namely, the force generating acoustic streaming.[5]

However, the force given in Eq. (46) could not produce streaming for unattenuated sound waves; indeed, the linearized equations (Chapter 2) can be used to show that

if: $p^M = \langle \frac{1}{2} c_0^2 \rho_0^{-1} \rho_s^2 - \frac{1}{2}\rho_0 u_{si} u_{si} \rangle$

then: $F_j - \frac{\partial p^M}{\partial x_j} = \left\langle \frac{\partial}{\partial t}(\rho_s u_{sj}) \right\rangle,$ (47)

which is necessarily zero (as the mean value of the rate of change of a bounded quantity). Accordingly, the fluid must remain at rest, responding merely by setting up the distribution p^M of mean pressure whose gradient can balance the force. [*Note:* In the ray acoustic approximation (28), p^M is itself zero, but the above argument does not need to use this approximation.]

Attenuation of sound waves takes place as follows:

(a) In the bulk of the fluid through the action of viscosity, thermal conductivity, and lags in attaining thermodynamic equilibrium (Chapter 6)

(b) Near solid walls by viscous attenuation in Stokes boundary layers (Chapter 8)

All these effects produce forces [Eq. (46)] that act to generate acoustic streaming. It is important to note, furthermore, that even the forces due solely to viscous attenuation, being opposed just by the fluid's own viscous resistance, generate mean motions that do not disappear as the viscosity μ tends to zero.[44–47]

6.2   Jets Generated by Attenuated Acoustic Beams

Attenuation of type (a) produces a streaming motion u_{fj} satisfying

$$\rho u_{fi} \frac{\partial u_{fj}}{\partial x_i} = F_j - \frac{\partial p}{\partial x_j} + \mu \nabla^2 u_{fj}. \qquad (48)$$

Substantial streaming motions can be calculated from this equation only with the left-hand side included,[48] although in the pre-1966 literature it was misleadingly regarded as "a fourth-order term" and so ignored, thus limiting all the theories to uninteresting cases when the streaming Reynolds number would be of order 1 or less.

We can use streaming generated by acoustic beams to illustrate the above principles. If acoustic energy is attenuated at a rate β per unit length, then a source at the origin that beams acoustic power P along the x_1-axis transmits the following distributions:

Power $Pe^{-\beta x_1}$ (49a)

and therefore

Energy per unit length $c^{-1} Pe^{-\beta x_1}$,   (49b)

which is necessarily the integral over the beam's cross section of energy density as well as of the uniaxial Reynolds stress (35). It follows that the force per unit volume (46) integrated over a cross section produces[5]

Force $c^{-1} P\beta e^{-\beta x_1}$ per unit length in the x_1-direction.

$$(50)$$

At high ultrasonic frequencies the force distribution (50) is rather concentrated, the distance of its center of application from the origin being just β^{-1} (which at 1 MHz, e.g., is 24 mm in air). Effectively, the beam applies at this center a total force $c^{-1}P$ [integral of distribution (50)].

The type[1,5] of streaming motion generated by this concentrated force $c^{-1}P$ depends critically on the value of $\rho c^{-1}P\mu^{-2}$: a sort of Reynolds number squared, which is about $10^7 P$ in atmospheric air (with P in watts). Streaming of the low-Reynolds-number "stokeslet" type predicted (for a concentrated force) by Eq. (48) with the left-hand side suppressed is a good approximation only for $P < 10^{-6}$ W.

For a source of 10^{-4} W power, by contrast, the force $c^{-1}P$ generates quite a narrow laminar jet with momentum transport $c^{-1}P$, and at powers exceeding 3×10^{-4} W this jet has become turbulent, spreading conically with a semiangle of about $15°$ and continuing to transport momentum at the rate $c^{-1}P$. Such turbulent jets generated by sound are strikingly reciprocal to a classical aeroacoustic theme!

At lower frequencies an acoustic beam of substantial power delivers a turbulent jet with a somewhat more variable angle of spread but one that at each point x_1 carries momentum transport

$$c^{-1}P(1 - e^{-\beta x_1}) \qquad (51)$$

generated by the total force (50) acting up to that point. This momentum transport in the jet represents the source's original rate of momentum delivery minus the acoustic beam's own remaining momentum transport (49). In summary, as acoustic power is dissipated into heat, the associated acoustic momentum transport is converted into a mean motion (which, at higher Reynolds numbers, is turbulent).[5]

6.3 Streaming around Bodies Generated by Boundary Layer Attenuation

Sound waves of frequency ω well below high ultrasonic frequencies have their attenuation concentrated, if solid bodies are present (Chapter 8), in thin Stokes boundary layers attached to each body. Then the streaming generated near a particular point on a body surface is rather simply expressed by using local coordinates with that point as origin, with the z-axis normal to the body and the x-axis in the direction of the inviscid flow just outside the boundary layer—the exterior flow. The Stokes boundary layer for an exterior flow

$$(U(x,y), V(x,y))e^{i\omega t} \qquad (52a)$$

has interior flow

$$(U(x,y), V(x,y))e^{i\omega t}[1 - e^{-z\sqrt{(i\omega \rho/\mu)}}]. \qquad (52b)$$

[Note that our choice of coordinates makes $V(0,0) = 0$ and that expressions (52) become identical outside the layer.] The streaming motion[1,5] is calculated from the equation

$$F_j^{int} - F_j^{ext} + \mu \frac{\partial^2 u_{fj}}{\partial z^2} = 0, \qquad (53)$$

with certain differences from Eq. (48) explained as follows:

(a) The first term is the force (46) generating streaming within the boundary layer.

(b) We are free, however, to subtract the second, since (see Section 6.1) it can produce no streaming, and conveniently, the difference is zero outside the layer.

(c) Gradients in the z-direction are so steep that the third term dominates the viscous force and, indeed, in such a boundary layer, dominates also the left-hand side of Eq. (48).

The solution of Eq. (53), which vanishes at $z = 0$ and has zero gradient at the edge of the layer, is obtained by two integrations, and its exterior value is

$$u_{fj}^{ext} = \mu^{-1} \int_0^\infty (F_j^{int} - F_j^{ext})z \ dz, \qquad (54)$$

where integration extends in practice, not to "infinity," but to the edge of the layer within which the integrand is nonzero. Expression (54) for the exterior streaming is yet again (see Section 6.1) independent of the viscosity μ, since Eq. (52) makes $z \ dz$ of order $\mu/\rho\omega$, and it is easily evaluated.

At $x = y = 0$ (in the coordinates specified earlier) the exterior streaming (54) has components

$$-U\frac{3 \ \partial U/\partial x + 2 \ \partial V/\partial y}{4\omega} \qquad (x\text{-component}),$$

$$-U\frac{\partial V/\partial x}{4\omega} \qquad (y\text{-component}) \qquad (55)$$

with zero z-component. This is a generalized form of the century-old Rayleigh law of streaming (which covers cases when V is identically zero).

For the complete streaming pattern, expressions (55) are, effectively, boundary values for its tangential component at the body surface (because the Stokes boundary layer is so thin). Therefore, any simple solver for the steady-flow Navier–Stokes equations with specified tangential velocities on the boundary allows the pattern to be determined. It is important to remember that the inertia terms in the Navier–Stokes equations must *not* be neglected unless the Reynolds number R_s based on the streaming velocity (55) is of order 1 or less, when, however, the corresponding streaming motions would (as in Section 6.2) be uninterestingly small.

In the other extreme case when R_s is rather large (at least 10^3) the streaming motion remains quite close to the body[48] within a steady boundary layer whose dimension (relative to that of the body) is of order $R_s^{-1/2}$. This layer is by no means as thin as the Stokes boundary layer, but it does confine very considerably the acoustic streaming motion. Equations (55) direct this motion toward one of the exterior flow's stagnation points, whence the steady-boundary-layer flow emerges as a jet—yet another jet generated by sound.[49,50]

Acknowledgments

I am warmly grateful to D. G. Crighton and N. Riley for invaluable advice on the text as well as to the Leverhulme Trust for generous support.

REFERENCES

1. J. Lighthill, *Waves in Fluids*, Cambridge University Press, 1978.

2. M. J. Lighthill, "On Sound Generated Aerodynamically. I. General Theory," *Proc. Roy. Soc. A*, Vol. 211, 1952, pp. 564–587.

3. M. J. Lighthill, "Sound Generated Aerodynamically. The Bakerian Lecture," *Proc. Roy. Soc. A*, Vol. 267, 1962, pp. 147–182.

4. J. Lighthill, "The Propagation of Sound through Moving Fluids," *J. Sound Vib.*, vol. 24, 1972, pp. 471–492.

5. J. Lighthill, "Acoustic Streaming," *J. Sound Vib.*, Vol. 61, 1978, pp. 391–418.

6. M. J. Lighthill, "Jet Noise. The Wright Brothers Lecture," *Am. Inst. Aeronaut. Astronaut. J.*, Vol. 1, 1963, pp. 1507–1517.

7. F. P. Bretherton and C. J. R. Garrett, "Wavetrains in Inhomogeneous Moving Media," *Proc. Roy. Soc. A*, Vol. 302, 1968, pp. 529–554.

8. N. Curle, "The Influence of Solid Boundaries on Aerodynamic Sound," *Proc. Roy. Soc. A*, Vol. 231, 1955, pp. 505–514.

9. J. E. Ffowcs Williams and D. L. Hawkins, "Sound Generation by Turbulence and Surfaces in Arbitrary Motion," *Phil. Trans. Roy. Soc. A*, Vol. 264, 1969, pp. 321–342.

10. J. E. Ffowcs Williams and L. H. Hall, "Aerodynamic Sound Generation by Turbulent Flow in the Vicinity of a Scattering Half Plane," *J. Fluid Mech.*, Vol. 40, 1970, pp. 657–670.

11. D. G. Crighton and F. G. Leppington, "Scattering of Aerodynamic Noise by a Semi-infinite Compliant Plate," *J. Fluid Mech.*, Vol. 43, 1970, pp. 721–736.

12. D. G. Crighton and F. G. Leppington, "On the Scattering of Aerodynamic Noise," *J. Fluid Mech.*, Vol. 46, 1971, pp. 577–597.

13. D. G. Crighton, "Acoustics as a Branch of Fluid Mechanics," *J. Fluid Mech.*, Vol. 106, 1981, pp. 261–298.

14. M. J. Lighthill, "On Sound Generated Aerodynamically. II. Turbulence as a Source of Sound," *Proc. Roy. Soc. A*, Vol. 222, 1954, pp. 1–32.

15. J. E. Ffowcs Williams, "The Noise from Turbulence Convected at High Speed," *Phil. Trans. Roy. Soc. A*, Vol. 255, 1963, pp. 469–503.

16. A. P. Dowling, J. E. Ffowcs Williams, and M. E. Goldstein, "Sound Production in a Moving Stream," *Phil. Trans. Roy. Soc. A*, Vol. 288, 1978, pp. 321–349.

17. J. E. Ffowcs Williams and G. Maidanik, "The Mach Wave Field Radiated by Supersonic Turbulent Shear Flows," *J. Fluid Mech.*, Vol. 21, 1965, pp. 641–657.

18. D. G. Crighton, "Basic Principles of Aerodynamic Noise Generation," *Prog. Aerospace Sci.*, Vol. 16, 1975, pp. 31–96.

19. M. E. Goldstein, *Aeroacoustics*, McGraw-Hill, New York, 1976.

20. M. E. Goldstein, "Aeroacoustics of Turbulent Shear Flows," *Ann. Rev. Fluid Mech.*, Vol. 16, 1984, pp. 263–285.

21. N. Curle, "The Mechanics of Edge-Tones," *Proc. Roy. Soc. A*, Vol. 216, 1953, pp. 412–424.

22. A. Powell, "On Edge-tones and Associated Phenomena," *Acustica*, Vol. 3, 1953, pp. 233–243.

23. E. H. Dowell, *Aeroelasticity of Plates and Shells*, Noordhoff, Leiden, 1975.

24. A. Powell, "The Noise of Choked Jets," *J. Acoust. Soc. Am.*, Vol. 25, 1953, pp. 385–389.

25. M. S. Howe and J. E. Ffowcs Williams, "On the Noise Generated by an Imperfectly Expanded Supersonic Jet," *Phil. Trans. Roy. Soc. A*, Vol. 289, 1978, pp. 271–314.

26. S. M. Candel and T. J. Poinsot, "Interactions between Acoustics and Combustions," *Proceed. Instit. Acoust.*, Vol. 10, 1988, pp. 103–153.

27. D. G. Crighton, "The Excess Noise Field of Subsonic Jets," *J. Fluid Mech.*, Vol. 56, 1972, pp. 683–694.

28. Powell, A. "Theory of Vortex Sound," *J. Acoust. Soc. Am.*, Vol. 36, 1964, pp. 177–195.

29. W. Möhring, "On Vortex Sound at Low Mach Number," *J. Fluid Mech.*, Vol. 85, 1978, pp. 685–691.

30. H. S. Ribner, "The Generation of Sound by Turbulent Jets," *Adv. Appl. Mech.*, Vol. 8, 1964, pp. 103–182.

31. J. E. Ffowcs Williams and A. J. Kempton, "The Noise from the Large-scale Structure of a Jet," *J. Fluid Mech.*, Vol. 84, 1978, pp. 673–694.

32. O. M. Phillips, "On the Generation of Sound by Supersonic Turbulent Shear Layers," *J. Fluid Mech.*, Vol. 9, 1960, pp. 1–28.

33. R. Mani, "The Influence of Jet Flow on Jet Noise, Parts 1 and 2," *J. Fluid Mech.*, Vol. 73, 1976, pp. 753–793.

34. A. B. Parry and D. G. Crighton, "Asymptotic Theory of Propeller Noise," *Am. Inst. Aeronaut. Astronau. J.*, Part I in Vol. 27, 1989, pp. 1184–1190 and Part II in Vol. 29, 1991, pp. 2031–2037.

35. A. Powell, "Aerodynamic Noise and the Plane Boundary," *J. Acoust. Soc. Am.*, Vol. 32, 1960, pp. 982–990.

36. D. G. Crighton, "Long Range Acoustic Scattering by Surface Inhomogeneities Beneath a Turbulent Boundary Layer," *J. Vibration, Stress and Reliability*, Vol. 106, 1984, pp. 376–382.

37. G. B. Whitham, "On the Propagation of Shock Waves through Regions of Non-uniform Area or Flow," *J. Fluid Mech.*, Vol. 4, 1958, pp. 337–360.

38. M. J. Lighthill, "Viscosity Effects in Sound Waves of Finite Amplitude," in G. K. Batchelor and R. M. Davies (Eds.), *Surveys in Mechanics*, Cambridge University Press, 1956, pp. 250–351.

39. D. I. Blokhintsev, *Acoustics of a Nonhomogeneous Moving Medium*, Moscow: Gostekhizdat, 1945. Also available in English translation as *National Advisory Committee for Aeronautics Technical Memorandum*, No. 1399, Washington, DC, 1956.

40. O. Reynolds, "On the Dynamical Theory of Incompressible Viscous Fluids and the Determination of the Criterion," *Phil. Trans. Roy. Soc. A*, Vol. 186, 1895, pp. 123–164.

41. D. C. Pridmore-Brown, "Sound Propagation in a Fluid Flowing through an Attenuating Duct," *J. Fluid Mech.*, Vol. 4, 1958, pp. 393–406.

42. P. Mungur and G. M. L. Gladwell, "Acoustic Wave Propagation in a Sheared Fluid Contained in a Duct," *J. Sound Vib.*, Vol. 9, 1969, pp. 28–48.

43. P. N. Shankar, "On Acoustic Refraction by Duct Shear Layers," *J. Fluid Mech.*, Vol. 47, 1971, pp. 81–91.

44. Lord Rayleigh, *The Theory of Sound*, 2nd ed., Vol. II, Macmillan, London, 1896.

45. W. L. Nyborg, "Acoustic Streaming Due to Attenuated Plane Waves," *J. Acoust. Soc. Am.*, Vol. 25, 1953, pp. 68–75.

46. P. J. Westervelt, "The Theory of Steady Rotational Flow Generated by a Sound Field," *J. Acoust. Soc. Am.*, Vol. 5, 1953, pp. 60–67.

47. W. L. Nyborg, "Acoustic Streaming," in W. P. Mason (Ed.), *Physical Acoustics, Principles and Methods*, Academic, New York, 1965.

48. J. T. Stuart, "Double Boundary Layers in Oscillatory Viscous Flow," *J. Fluid Mech.*, Vol. 24, 1966, pp. 673–687.

49. N. Riley, "Streaming from a Cylinder Due to an Acoustic Source," *J. Fluid Mech.*, Vol. 180, 187, pp. 319–326.

50. N. Amin and N. Riley, "Streaming from a Sphere Due to a Pulsating Acoustic Source," *J. Fluid Mech.*, Vol. 210, 1990, pp. 459–473.

28

AERODYNAMIC AND JET NOISE

ALAN POWELL

1 INTRODUCTION

The subject of aerodynamically generated noise became of considerable interest in about 1950 as the result of the appearance of the aircraft jet engine, soon recognized as a remarkably powerful sound source likely to cause environmental problems. Besides interest in measurements and empirical noise reduction techniques, attention focused on the *mechanism* by which the sound was generated, whereas the few precursor studies had mostly focused on the frequency of aerodynamic sound, as of the very familiar aeolian tone of the whistling of telephone wires in the wind.

This aeolian tone typifies aerodynamic sound in the presence of a *fixed* surface, since sound is still radiated even if the wire does not move at all. Instead it reacts to a fluctuating aerodynamic lift force and therefore is to be associated with sound radiation of *dipole* character (see Ref. 1 for an extended account).).

The aeolian tone is often periodic, a nearly pure tone, due to the regular formation of vortices of alternating sign in its wake, forming what is often call a Kàrmàn vortex street. This situation is totally different to that of an organ pipe or whistle, both with rigid surfaces, where the side-to-side oscillations of a jet excites a resonator. The mass of air in the resonator fluctuates as it resonates, and therefore the rate at which the net air flow enters the atmosphere fluctuates sympathetically and oppositely, these pulsations generating sound of *monopole* (simple-source) character. Such systems are not discussed here (but see Ref. 2).

Encyclopedia of Acoustics, edited by Malcolm J. Crocker
ISBN 0-471-80465-7 © 1997 John Wiley & Sons, Inc.

Sound may be generated even in the complex absence of a fixed surface. Jet noise typifies this situation, the turbulence of the jet generating noise long after it has left the jet nozzle. There is no surface to react to a net aerodynamic force, and so the dipole character is absent. Instead, it is of the next-order higher source type, namely the *quadrupole* (see Chapter 9, Section 5). Here one may visualize a spherical eddy. Its volume does not fluctuate if the flow speed is low enough, so there is no monopole sound. It cannot vibrate in position, for there is no surface to react to the force that would be needed to cause the momentum to fluctuate, and so there is no dipole radiation. But there is the possibility that the surface of the sphere vibrates like the four quarters of an orange skin, adjacent quarters moving in and out oppositely, there being no volume or momentum change. Such distortions can in fact be viewed as the source of aerodynamic sound in the absence of solid surfaces, though it is not the only physical interpretation.[3]

The following discussion is limited to the latter, that is, to the aerodynamic noise of free flows (no solid surfaces present). After consideration of the fundamentals of aerodynamic sound generation of very low speed flows, the more complex situations of subsonic and supersonic jets are briefly discussed.

There is now a very extensive literature covering practically all aspects of the subject. Generally the references are limited to a few key, historically significant or recent ones. For fuller treatments and more extensive bibliographies, covering closely related aspects of aeroacoustics, the reader may refer to some fairly recent reviews,[4–11] some books,[1,2,12] and a conference proceedings.[13] There is also a recent collection of comprehensive monographs covering aeronautic aeroacoustics in general.[14]

2   AERODYNAMIC NOISE SOURCES

In this section, the size of the source region is assumed to be "compact," that is, very small compared to the acoustic wavelength, and the flow Mach number is assumed to be very small, $M \ll 1$.

2.1   Lighthill Theory

Lighthill[15] manipulated the exact equations of continuity with zero source strength,

$$\frac{\partial \rho}{\partial t} + \frac{\partial \rho v_j}{\partial y_j} = 0,$$

and of momentum with zero externally applied force,

$$\frac{\partial \rho v_i}{\partial t} + \frac{\partial \rho v_i v_j}{\partial y_j} + \frac{1}{\rho}\frac{\partial p_{ij}}{\partial y_j} = 0,$$

so as to obtain the inhomogeneous wave equation (see Chapter 9, Section 2.1)

$$\frac{\partial^2 \rho}{\partial y_i^2} - \frac{1}{c_0^2}\frac{\partial^2 \rho}{\partial t^2} = -\frac{1}{c_0^2}\frac{\partial^2 T_{ij}}{\partial y_i \partial y_j},$$
$$\text{where } T_{ij} = \rho v_i v_j + p_{ij} - c_0^2 \rho \delta_{ij}, \qquad (1)$$

in which the standard acoustic form appears on the left side and everything else is put on the right side. Here c_0 is the sound speed in the stationary uniform acoustic medium surrounding the volume V_0 of unsteady sound-generating fluid flow, v_i is the velocity in the y_i-direction, ρ is the fluid density, and p_{ij} is the stress tensor (the force per unit area in the i-direction on the surface element with inward normal in the j-direction). The suffices i and j take the directions 1, 2, and 3 in turn but are to be summed if either one is repeated in a single term (e.g., there are three momentum equations, one for each of $i = 1, 2, 3$; but $\partial \rho v_i / \partial y_j \equiv \partial \rho v_1 / \partial y_1 + \partial \rho v_2 / \partial y_2 + \partial \rho v_3 / \partial y_3$).

This is interpreted as *Lighthill's acoustic analogy*: The *stationary* uniform acoustic medium is now taken to be *everywhere*, including the volume corresponding to the flow region V_0 with a *monopole* source distribution, given by the right side, embedded in it. The source strength is zero outside volume V_0 of the sound-generating flow. The double space differential reflects that the monopole and dipole strengths are both zero (corresponding to the assumed zero source strength and zero externally applied force), so the result is a quadrupole source field of strength T_{ij} per unit volume, (see Chap-

ter 9, Sections 4 and 5, where, however, T_{ij} is the total *point* source strength).

The equation is exact. As the density ρ appears on both sides, the solution is formally exact if the source strength T_{ij} is taken to be known *exactly*. Usually the source strength is approximated by use of the fluctuations in a similar incompressible flow; then terms representing the interaction of the sound with the flow itself (e.g., propagation and refraction of the sound through the moving fluid) are omitted. It is also commonly assumed that the extent of the source region is small compared to the acoustic wavelength λ of interest, so that the source arriving at the observation point $\mathbf{x}$ can be assumed to have been emitted simultaneously from throughout the source region, a great mathematical simplification. Then the sound pressure in the far field at distance x from the source region reduces to a point quadrupole:

$$p(\mathbf{x}, t) = \frac{1}{4\pi x c_0^2}\frac{\partial^2}{\partial t^2}\int_{V_0} T_{xx}\, dV(\mathbf{y})^*. \qquad (2)$$

This corresponds to Eq. (40) in Chapter 9, where now the asterisk means that the retarded time $(t - x/c_0)$ is to be taken for the integral over the compact source volume V_0, and $(1/c_0^2)\, \partial t^2/\partial t^2$ replaces $-k^2$, so that all frequencies are represented. Here the $T_{xx} \equiv T_{ij}x_i x_j/x^2$ is summed over all the longitudinal and lateral quadrupole terms for which $i = j$ and $i \neq j$, respectively.

The effect of viscous stresses can be neglected at other than small Reynolds number.[15] Specifically, the effect of dissipation by viscous forces results in a monopole-like contraction as the flow loses kinetic energy (imagine the lessening pressure drop in the center of a decaying vortex) and an expansion due to the corresponding heating.[11] The net effect is rather small, being of the order of 1/(Reynolds number) compared to that due to the vorticity movement per se.

For adiabatic fluctuations of an unheated flow, $\delta p = c_0^2 \delta \rho$.

Then taking the density to have the unperturbed value, everywhere, $T_{ij} = \rho_0 v_i v_j$ and $T_{xx} = \rho_0 v_x^2$, where v_x is the velocity component in the $\mathbf{x}$-direction.

If the sound pressures at three orthogonally related equidistant points $\mathbf{x}$, $\mathbf{y}$, and $\mathbf{z}$ are summed, then the integrals sum to twice the kinetic energy of the flow, which may be taken to be constant. Only lateral quadrupoles satisfy the condition of the sound pressure at such three far-field points always summing to zero. This *three-sound-pressures theorem*[11] indicates that, under the stated conditions, the assembly of quadrupole sources must be reducible to lateral ones alone.

Whereas a monopole source can be considered due to a fluctuation in mass $m = \rho_0 Q$ and a dipole due to a fluctuating force, that is, changes in momentum $mv = \rho_0 Qv$, these quadrupoles are due to fluctuations in momentum flux mv^2. A spherical surface in the fluid undergoes no vibratory expansion, or overall displacement, but only the distortions of the lateral quadrupole. These are similar to the surface of the four quarters of an orange, adjacent quarters moving oppositely in or out.

The sound intensity $I = p^2/\rho c$, with $q = \partial^2 (v_r^2)/\partial t^2$, is given by

$$I(\mathbf{x}) = \frac{\rho_0}{16\pi^2 x^2 c_0^5} \left\langle \int_{V_0} q(y_1)\, dV(y_1) \right.$$

$$\left. \times \int_{V_0} q(y_2)\, dV(y_2) \right\rangle, \qquad (3)$$

where $\langle \cdots \rangle$ indicates that the time average of the quantity is to be taken. Now

$$\langle \cdots \rangle = \int\int \langle q(y_1)q(y_2)\rangle\, dV(y_1)\, dV(y_2).$$

If homogeneous turbulence is considered, this time average depends only on the difference between y_2 and y_1, so put $y_2 = y_1 + y'$ and define the *correlation coefficient* R by

$$\langle q^2 \rangle R(y') = \langle q(y_1)q(y_1 + y')\rangle. \qquad (4)$$

Compactness of the whole source region, assumed earlier for convenience, may be relaxed somewhat, since now it is only necessary that $y' \ll \lambda$ when $R(y') \to 0$. Then

$$\langle \cdots \rangle = \langle q^2 \rangle \int dV(y_1) \int R(y')\, dV(y') = \langle q^2 \rangle V_0 \mathcal{V},$$

where the *correlation volume* $\mathcal{V}$ is defined by the second integral.

This $\mathcal{V}$ may be considered to be the definition of the volume of a *noise-producing eddy* of strength $\langle q^2 \rangle$ per unit volume. There are $V_0/\mathcal{V}$ such eddies in the noise-generating volume V_0. Then

$$I(\mathbf{x}) = \frac{\rho_0 \langle q^2 \rangle \mathcal{V} V_0}{16\pi^2 x^2 c_0^5}. \qquad (5)$$

2.2 Vortex Theory

It is shown in classical (incompressible) hydrodynamics that the fluid flow induced by a vortex ring (of circulation Γ and area A_i) is exactly the same as that induced by a sheet of dipoles (of strength Γ per unit area of the same area A_i).[11] The total dipole strength is ΓA_i and the momentum in both cases is $m_i = \rho_0 \Gamma A_i$; see Fig. 1. In this theory[16] the aeolian tone, of dipole nature, is generated by the fluctuating force $dF_i/dt = d^2 m_i/dt^2$ exerted by a cylinder on the surrounding fluid when the vortex ring formed by the "bound" circulation about the cylinder and the cast-off "starting" vortex are stretched, or rather, the rate of change of stretching, $d^2 A_i/dt^2$, that can be taken as the downstream acceleration of the vortex as it breaks loose from the cylinder. The line of action of the dipole is through the vortex and perpendicular both to it and to its acceleration (but may be considered to act at the cylinder with negligible quadrupole error). The maxima of the dipole sound field lie along the line of action, approximately normal to the flow direction. The far-field sound pressure is therefore given by

$$p(\mathbf{x}, t) = -\frac{\rho_0}{4\pi x c_0} \frac{\partial}{\partial t} \int_b L_x\, db^*, \qquad \mathbf{L} = \Gamma \times u, \qquad (6)$$

where the product of the cross-stream span b of the vortex and its downstream velocity u is $bu - dA_i/dt$. The simple similarity argument indicates that the sound power is proportional to the flow velocity to the sixth power.

In a free flow, no net force occurs between the flow and the surrounding fluid, so the force due to a given accelerating vortex *must* be exactly balanced by an equal and opposite action elsewhere in the flow. As each one acts as a dipole, together they must constitute a quadruple with strength proportional to the distance between them, the orientation determining the relative lateral and longitudinal components; see Chapter 9, Section 5.3, and

Fig. 1 Left: Streamlines of incompressible flow field due to vortex pair. Center: Streamlines due to a sheet of dipoles. Right: Streamlines due to bound vortex about cylinder and accelerating cast-off vortex (mean flow omitted).

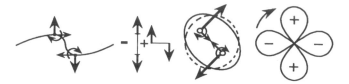

Fig. 2 Left: Horizontal acceleration of vorticity results in vertical dipoles, which form a longitudinal and a lateral quadrupole. Center: Spinning vortices accelerate toward their center, resulting in a rotating lateral quadrupole. Right: Rotating lateral quadrupole field due to spinning vortices.

Fig. 2. The far-field sound pressure is given by

$$p(\mathbf{x}, t) = -\frac{\rho_0}{4\pi x c_0^2} \int \frac{\partial^2}{\partial t^2} y_x \mathcal{L}_x \, dV^*, \qquad \mathcal{L} = \zeta \times \boldsymbol{u}. \tag{7}$$

The three-sound-pressure theorem and, for the sound power, the eighth-power law naturally apply.

Although their physical interpretations are very different, the vortex theory and Lighthill's are closely connected: since $\partial v_i v_j/\partial y_j = v_i \, \partial v_j/\partial y_j + v_j \, \partial v_i/\partial y_j$ (the first term vanishes because of continuity), which then equals $v_j(\partial v_j/\partial y_i - \partial v_i/\partial y_j) + \frac{1}{2}\partial v^2/\partial y_i$, that is, the velocity times the vorticity ζ (vortex strength per unit area), with the last term ultimately vanishing when integrated over the source volume.

Further developments include a recasting of the result into a source term depending on $\partial \zeta^3/\partial t^3$ rather than on L (possibly attractive for numerical methods) and the inclusion of the effect of entropy (temperature) variation in the flow.[11,17]

Apart from the useful physical interpretation, the vortex theory is very convenient when the vorticity can be specified, as for idealized vortex flows, for example, the spinning vortex pair (Fig. 2), vortex rings colliding head-on with each other, and simulated jet flows[13,18] when the vorticity can be viewed as inducing the local incompressible (hydrodynamic) flow field as well as the acoustic field.

2.3 Dilatation Theory

The dilatational theory is an alternative acoustic analogy that has a simple but unfortunately deceptive physical appeal. In an incompressible aerodynamic flow there are pressure fluctuations p_{inc}, and it may be imagined that if a small degree of compressibility is admitted then at some point where the pressure is positive, say, there would be a corresponding compression of the fluid. This compression then acts like a local monopole. For a free flow the

pattern of positive and negative pressure volumes yields a quadrupole form, and simple similarity leads to the eighth-power law.

The theory may be developed *formally* for a free flow,[6] finding p_{inc} from the equation $\nabla^2 p_{\text{inc}} = \partial^2 T_{ij}/\partial y_i \, \partial y_j$, for example, so that

$$p(\mathbf{x}, t) = -\frac{1}{4\pi x c_0^2} \int \frac{\partial^2 p_{\text{inc}}}{\partial t^2} \, dV^*. \tag{8}$$

Unfortunately one cannot distinguish the pressure in the original T_{ij} source region from that in the surrounding resultant ("incompressible") near field, so as p_{inc} falls off only as r^{-3}, the region of integration must extend out to *at least* a wavelength: The source region is *essentially* noncompact.[11] While this spoils the simple appeal of the physical argument, the method is mathematically correct for some simple test cases[6] but has not been used to develop jet noise theory.

2.4 Alternatives to Acoustic Analogies

Several alternatives to the acoustic analogy methods have been, or are being developed. Space permits only a brief mention of these newer important aspects of aeroacoustics.

Contiguous Method The imaginary rotating ellipsoid-like surface formed by the streamlines about the spinning vortices in Fig. 2 can be considered to drive the *contiguous* acoustic field external to them. In fact, if the pressure $p_a(\mathbf{a})$ on a surface of radius a can be estimated for some *incompressible* (hydrodynamic) flow contained within, then the pressure $p(\mathbf{x})$ in the acoustic field follows immediately; for the far field

$$p(\mathbf{x}) = \frac{1}{3} \frac{a^3}{x c_0^2} \frac{\partial^2 p_a^*}{\partial t^2}. \tag{9}$$

This simple formulation[11] [readily derived from Eq. (42) in Chapter 9] obviously depends on some knowledge of the source type, specifically lateral quadrupole for free flows.

Matched Asymptotic Expansions The preceding method has a mathematical counterpart in the method of asymptotic expansions (MAE), in which the *inner* hydrodynamic flow field, say of a vortex pair, is matched analytically to the *outer* external acoustic fields.[4] This method is more general and rigorous provided that corresponding care is taken, the inner flow not necessarily being incompressible or compact and the source type not needing to be known initially. An important aeroacoustic application to the instability waves of a jet is addressed later for supersonic jets.[19]

Computational Aeroacoustics   The advent of very fast computers is making possible numerical solutions of the fundamental equations of motion of noise-producing flows, that is, direct numerical simulation without the introduction of "models" of any sort. For example, the results for the two-dimensional choked jet (neglecting viscosity and heat transfer) bear a satisfying likeness to photographic and other evidence.[20] Such methods may be used to confirm various models, for example, the vortex theory, that are very much more simple to use. In practice, though, it is more likely that the numerical model will be evaluated against a simple mode and, once validated, extended to conditions that the simpler models plainly cannot address. For example, for a pair of spinning vortices, direct computation yields results that agree precisely with the results of vortex theory (an acoustic model) at very small Mach number and extends the results to the compressible region (higher Mach numbers) that the acoustic model formulations are ill-equipped to handle.[11] Also possible are numerical simulation experiments to test simple physical hypotheses, as in the case of the edge tone, where the point dipole of theory can be inserted in the jet flow to represent the action of the edge[8] (see later).

3   JET NOISE

In this section some applications to the noise of turbulent jets are briefly considered. While the jet exhaust velocity may be supersonic, the convection velocity of the eddies in the flow is initially taken to be subsonic relative to the ambient sound speed.

3.1   Simple Similarity

Simple scaling based on Lighthill's theory can give useful indications, and its limitations gives some hint about important omitted effects.

By *simple similarity* is meant taking a basic equation, such as Eq. (5) or (7), and assuming that the turbulent velocities v_i are proportional to some characteristic velocity U (e.g., jet exit velocity) and the $\partial/\partial t$ are like frequency and so proportional to U/D and both V and V_0 are proportional to D^3, where D is a convenient characteristic dimension (e.g., jet nozzle diameter). Then it follows that $p(\mathbf{x}) \sim \rho_0 U^4 D / x c_0^2$ in any given direction and the corresponding intensity at a point is $I(\mathbf{x}) \sim \rho_0 U^8 D^2 / x^2 c_0^5$. Lighthill's *eighth-power law*[15] for

the sound power W follows:

$$W = \frac{K \rho_0 U^8 D^2}{c_0^5},   \tag{10}$$

where K is a constant and the factor $\rho_0 U^8 D^2 / c_0^5$ is often called *Lighthill's parameter*. As the kinetic power of a jet, say, is proportional to $\frac{1}{2}\rho_0 U^2 \cdot U D^2$ (ignoring the density differences), the fraction of the power converted to sound energy is the *noise-generating efficiency* η:

$$\eta \sim M^5,   \tag{11}$$

where $M = U/c_0$ is the Mach number of the flow referenced to the *ambient* speed of sound. No provision has been made for convection of the sources or for flow interaction effects.

The earliest measurements of jet noise showed that the intensity and noise power varied very closely with the eighth power of the jet exit velocity. In fact, it is now generally accepted that

$$P \approx \frac{(3\text{–}4) \times 10^{-5} \cdot \rho_0 U^8 D^2}{c_0^5},   \tag{12}$$

and in terms of the mechanical power P_M of the jet,

$$P \approx 10^{-4} M^5 P_M.   \tag{13}$$

These apply quite well to unheated air jets (with low initial turbulence and noise entering the nozzle) from low speeds (say $U \approx 50$ ms^{-1}) to slightly above choking ($U \approx 310$ ms^{-1}) and to the jets of turbojets near full power ($U \approx 610$ ms^{-1}).[21,22] Note that it is the ambient density that occurs in the former equation, not that of the jet mixing region.

From the theoretical point of view, the interesting implication is that the sum effect of omitted considerations—such as source convection and refraction—are either not important or balance each other out, so far as *noise power* is concerned.

As this simply derived eighth-power law agrees rather well with experimental results, one may take the *empirical* approach that the similarity method can be applied to the most important noise-generating regions of a turbulent jet. There are two regions of the jet in which the turbulence is *self-preserving*, that is, follows a similarity relationship: first, the nearly two-dimensional thin shear layer in region A not too far from the nozzle exit and adjacent to the nonturbulent potential core (see Fig. 3) and, second, the axially symmetric region B relatively

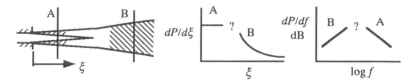

Fig. 3 Left: Similarity regions of a turbulent jet. Center: Axial source strength distribution. Right: Corresponding spectral shape.

far downstream. These are characterized by the lateral dimensions of the turbulent regions, growing very nearly proportionally to the distance ξ from the nozzle, and by the total shear velocity that equals the jet exit velocity U in region A but falls inversely with ξ in region B. Then the noise power produced per unit length is constant in region A but falls very rapidly, as ξ^{-7}, in region B, with an indeterminate transition between them.[5,21] The relative level for the two regions is indeterminate (see Fig. 3).

Early theoretical ideas suggested that high shear gradients "amplify" the turbulent sound generation; but according to similarlity considerations, high shear rates and the high frequency adjacent to the jet exit are just offset by the smallness of the eddy size and depth of the shear layer. If the unknown constants for the two regions are taken to be the same and the regions are assumed to join each other without a transitional region between them, then three-fourths of the noise is generated in region A.

Measurements in the very near field and the use of an acoustic focusing mirror do indicate that relatively little sound is generated in region B but that in region A it is weighted toward the transition region.[13]

The associated noise power spectrum $dP/df = dP/d\xi \cdot d\xi/df$, assuming similarity of local spectrum shapes, first increases as f^2 and then decreases as f^{-2}, the relative levels again being undefined[21] (see Fig. 3). The peak of the spectrum naturally follows some fixed value of Strouhal number fD/U. This result is independent of the spectral shape of the local source (except that it must have slopes steeper than the resultant ones).[5]

Experiments showed that the average spectra of noise power of unheated air jets, turbojets, and rockets do have a *shape* fairly close to that just discussed. However, these spectra mostly collapse onto the common shape using the parameter fD/c_j, where c_j is the speed of sound in the jet, *not* fD/U, as shown in Fig. 4. That the peak frequency does not scale proportionally to jet velocity was evident from the earliest measurements.[5]

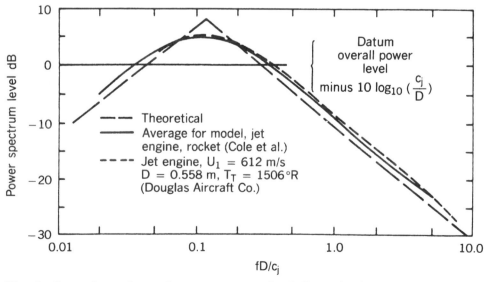

Fig. 4 Comparison of sound power spectrum level from simple similarity and from measurements of unheated air jets, jet engines and rockets. Note that the abcissa scale is fD/c_j, not fD/U. (From A. Powell, *Noise Control Eng. J.*, Vol. 8, 1977, pp. 69–80, 108–119. Copyright Institute of Noise Control Engineering of the USA, used with permission.)

3.2 Turbulent Self-Noise and Shear Noise

For jet noise, two classes of quadrupoles may be identified. First, *self-noise* is that due to the turbulence, assumed to act as if it were isotropic and locally homogeneous. These quadrupoles are lateral ones, oriented at random and therefore producing, on the average, a uniform directionality.[6]

The second is that due to the turbulence being sheared. In Lighthill's original theory the quantity $\partial/\partial t \cdot \rho_0 v_i v_j = p e_{ij}$, p being the local pressure and e_{ij} the local rate of strain tensor,[15] was interpreted as producing lateral quadrupole noise generators, with the maxima at 45° to the jet axis.

A more recent alternative[6] is to take from Eq. (1) the quadrupole source terms $2\rho_0 \partial U_1/\partial y_2 \cdot \partial v_2/\partial y_1$, where U_1 is the mean velocity sheared in the direction y_2 (or, equivalent after integration, putting $v_1 = U_1 + v_1'$ in T_{ij}). This results in longitudinal quadrupoles with the maxima along the jet axis, having a $\cos^4 \theta$ variation for their intensity. Assuming the turbulence to be isotropic and superimposed on the sheared flow, this *shear noise* power is estimated to be about equal to that of the self-noise.[6]

3.3 Convection and Refraction

The considerations leading to the eighth-power law, based on Eq. (5), have a major shortcoming: The observed strong downstream bias of jet noise (see Fig. 5) is completely absent. This bias has been attributed to the movement of the noise-producing eddies at about 0.6 of the jet exit velocity relative to the external medium into which they radiate. It may be incorporated by the method discussed in Chapter 9, Section 11, or by the use of moving axes[15] applied to the sources in the acous-

tic model. For quadrupoles, convection at Mach number $M_{con} = U_{con}/c_0$ introduces the directionality factor $(1 - M_{con} \cos \theta)^{-6}$ for the sound intensity, where θ is the angle measured from the downstream axis; however, the effective volume of the noise-producing eddies is reduced,[22] so the net effect becomes $(1 - M_{con} \cos \theta)^{-5}$. Thus there is a considerable amplification in the downstream direction, falling to none at right angles to the jet and with a reduction in the upstream direction. A corresponding factor applies to supersonic convection, $M_{con} \cos \theta - 1)^{-5}$. These are both obviously unsatisfactory as $M_{con} \cos \theta \rightarrow 1$ when it might be considered that the degeneration of constituent monopoles into a quadrupole can no longer take place, and a proper combination is the factor

$$[(1 - M_{con} \cos \theta)^2 + \alpha^2 M_{con}^2]^{-5/2}, \qquad (14)$$

where α depends on the length and time scale of the turbulence.[6,22] Plausible values of α result in the sound power in the region of $M_{con} = 1$ more modestly exceeding the extrapolation of the eighth-power law.

Thus the gross bias of the jet noise toward the downstream direction can be explained. Decreasing turbulence levels with jet velocity were suggested to partly compensate for the rise above the eighth-power law,[6,21] supported by some measurements of noise intensity at $\theta = 90°$ varying more slowly than with the eighth power of velocity.

However, there are two shortcomings. First, the necessary Doppler effect on frequency, $(1 - M_{con} \cos \theta)^{-1}$, worsens the aforementioned disparity with the observed spectra of noise power, and second, this directionality has the maximum in the jet direction instead of the observed sharp minimum.

Refraction The sound power of a source is not much affected by (say rigid) boundaries more than a quarter of a wavelength or so from it, that is, beyond some *region of influence*. It may be presumed that the same is true if there is a shear of the medium there. Thus the effects of convection will be limited to the differential velocities occurring within the region of influence, while mean velocity changes outside this region will cause *refraction*. For a given jet velocity, the shorter wavelengths of the local characteristic spectrum will scale with the local shear layer thickness in region A of Fig. 3, and so the convection and refraction effects will both tend to be constant. For increasing jet velocity, the position of the source of a given high frequency (relative to the local characteristic frequency) will move downstream and suffer constant convection effects but increasing refraction, away from the jet in the downstream

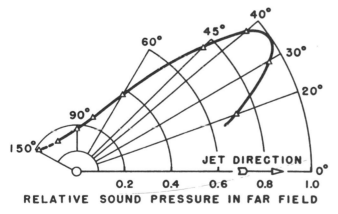

Fig. 5 Directivity of the sound pressure for a jet engine on a linear scale. (From A. Powell, "On the Generation of Noise by Turbulent Jets," ASME Paper 59-AV-53, 1958, with permission.)

quadrant but toward the jet in the upstream quadrant. The longest (local) wavelengths will undergo full convection effects and little refraction and the shortest little convection but strong refraction.[5] Experiments with a sound source introduced into a jet show the anticipated minimum along the downstream axis, while a very cold jet shows a local maximum there, the lower sound speed causing inward refraction.[6] Such considerations provide a plausible qualitative explanation of the directional characteristics of turbulent jets at subsonic Mach numbers, in particular, the fact that the peak frequency of the noise spectra in the important intense downstream direction varies very little with jet speed.

Flow Interaction The important effects of convection and refraction cannot be simply separated for moderate acoustic wavelengths in realistic flows that have flow shear; together they typify *flow interaction* of the mean flow with the acoustic radiation within it. The formal solution to Lighthill's equation is exact and includes all such effects, provided that the source terms are known *exactly*. The common use of incompressible approximations clearly eliminates all interaction between the acoustic perturbations and the mean flow, such as that due to refraction.

The alternative is to incorporate mean-flow interaction effects in the wave operator on the left side of the equation, so that the right (source) side can still be approximated in a fairly simple manner. This is a convective wave equation, with the convective velocity varying in a shear flow. Although such equations are difficult to solve, very significant progress has been made, providing much detailed insight.[14]

From the point of view of general fundamental theory, attention is drawn to developments describing flows in terms of mean, turbulent, acoustic, and thermal components.[23,24]

Turbulent Jet Noise Reduction The challenge in reducing the high noise levels of the jets of propulsive engines is to do so with the minimum impact on the thrust. By far the most progress has come about through "simply" reducing the jet velocity, maintaining the thrust constant, for then Eq. (10) gives $W \sim (\text{thrust})U^6/c_0^5$. The earliest emphasis was on altering the nozzle shape to variations on the "cooky-cutter" forms, some rather extreme, or of using multiple smaller nozzles. Some of these produced up to 10 dB noise reductions for a modest thrust loss. More recent by-pass and fan-jet engine designs involve much lower jet velocities, large streamlined center bodies, and annular jets that may be subdivided in various ways.[14]

4 SUPERSONIC JETS AND ROCKETS

While the noise of subsonic turbulent jets can be considered to be due to turbulent mixing at subsonic convection velocities, for supersonic jets the convection velocity is supersonic, and this brings about a fundamental change in the source mechanism. Moreover, there are shock waves present in the flow, and these introduce an entirely different source mechanism.

4.1 Shock-Free Jet Noise

Supersonic Convection The eighth-power law must ultimately give way to a power no greater than 3, or else the acoustic power would exceed the jet kinetic power. Actually, at the very high speeds of rockets the noise power does tend to the third power of the velocity,[5,21] the sound power then being about 0.5% of the jet power. The sound power for very large rockets (up to 10^{10} W) also levels out to about the same 0.5% of the mechanical power.[25] When the convection Mach number exceeds unity, the convection amplification factor discussed earlier, $(M_{\text{con}} \cos \theta - 1)^{-5}$, when combined with the U^8, does in fact tend to the U^3 that has been observed.[5,21] Of course, in the acoustic analogy, the source strength has to be assumed given, so to some extent such agreement might be fortuitous.

Instability Wave Radiation Close to the nozzle exit of supersonic jets, a pattern of almost straight sound waves of very high frequency, making an acute angle with the jet direction, has been long observed in schlieren and shadowgraph photographs, usually attributed simply to Mach waves generated by supersonic eddies of short lifetime. A more recent explanation is that the shear layer there is unstable and the instability waves radiate sound as their amplitude increases and then decreases. The acoustic analogy may be employed,[26] while alternatively the exact solution to the equations of motion (within plausible simplifications) also reveals these nearly plane sound waves, but propagating at a phase velocity somewhat less than the sound speed in the rather extensive near field.[27]

Most of the noise from a supersonic jet emanates from further downstream than for a subsonic jet and appears largely attributable to the large-scale eddy structures associated with instability waves there.[26–28]

4.2 Shock-Generated Noise

Vortex–Shock Interaction In the elementary one-dimensional case of sound transmission across a boundary, the incident wave gives rise to a transmitted and a reflected wave, these two waves being necessary to satisfy the two boundary conditions at the interface, namely the continuity of velocity and pressure. All the fluctuations occur in a single perturbation mode, namely that of acoustic waves. If the interface is a shock wave, with the fluid on the incident wave side approaching it at supersonic velocity, then no reflected wave can propagate upstream. In a sense, it cannot leave the shock wave interface and so causes fluctuations in its strength and therefore in the entropy change across it, these entropy changes being carried downstream at the subsonic flow velocity. Thus the two waves now necessary to satisfy the boundary conditions are a transmitted sound wave and a convected entropy (temperature) wave.[8] The sound wave is amplified, perhaps considerably, by this nonlinear interaction. The resultant perturbations are now of two modes: acoustic and entropy.

Similarly, an entropy wave convected into a shock wave results in the two modes of a convected entropy wave and a sound wave propagating downstream, the latter being relatively intense in acoustic measure.

In three dimensions, three perturbation modes occur, the additional one being vorticity. Thus a vortex swept into a shock wave results in a changed vortex being convected away from the shock wave, an entropy wave, and a sound wave.[29,30] In the same way the interaction of turbulence with a shock wave results the generation of sound.

Screech There are many situations in which *flow resonance* occurs. The flow—either a shear flow or of a jet—is inherently unstable over a certain frequency range, or more strictly, over a range of disturbance wavelengths in the flow. It is self-excited into resonance-like sound-generating oscillations in the region of maximum instability, all in the total absence of any physical resonator (such as an organ pipe).[9] The classic case is that of the edge tone, in which a laminar slit jet blows onto a wedge, the periodic interaction there resulting in the jet being disturbed as it leaves the orifice. The induced sinuosity of the jet grows very rapidly as it is convected toward the edge, taking energy from the mean jet flow, some of which is radiated as sound of dipole character associated with the fluctuating fluid force on the edge.

Two criteria, concerning the phase and loop gain, respectively, must be met if such oscillations are to be maintained.[8] First, one might think that there must be an integer number N of wavelengths in the feedback loop. But because of possible noncanceling phase changes at both the orifice and the edge, a noninteger fraction p must be introduced, so the possible oscillation frequencies must really be proportional to $N + p$. Second, in the limit cycle, the amplification in the jet—a factor of possi-

bly hundreds or thousands—must result in the loop gain being unity. As the orifice-to-edge distance is increased, the sinuosities in the jet become longer and consequently less unstable. At some point the oscillation flips to a shorter, more unstable wavelength for a new limit cycle with the next greater integer value of N; the process may be repeated several times. Thus steady falls in frequency are interrupted by sudden jumps upward.

Supersonic jets are characterized by the formation of a stationary "cell" structure of nearly periodically varying pressure, with each cell terminating in shock waves across the jet, the nearly equal spacing between them increasing with jet pressure ratio. The characteristic "screech" noise of such jets is due to a feedback mechanism analogous to that of the classical edge tone, except that the acoustic source is attributed to the interaction of the jet sinuosities with the periodically spaced shock waves, the latter being virtually fixed in space. Together the acoustic sources—assumed to be monopoles—form a *stationary* phased array that is *not* compact, so unlike the simple dipole of the edge tone, the directionality of the sound is markedly frequency dependent, though of course the frequency still has to be such that the jet is excited in the region of maximum instability.[8,10,32]

As a simple example, assume that the sound is emitted equally from three shock waves spaced a distance s apart, then the sound pressure at angle β to the jet downstream direction is proportional to

$$\frac{1}{3} + \frac{2}{3} \cos\left(\frac{2\pi s}{\Lambda}\left(1 - M_{\text{con}}\cos\beta\right)\right), \qquad (15)$$

where M_{con} is the convection Mach number and $\Lambda = M_{\text{con}}\lambda = M_{\text{con}}c/f$ is the wavelength of the jet instability.[32] This accounts for the relatively high sound levels passing upstream, past the nozzle exit. Moreover, if it is postulated that the feedback is maximized in the limit cycle, given that the oscillations are to be in the region of maximum jet instability, then the sound waves from all the sources arrive all in phase with each other at the nozzle; that is, there is constructive interference. With $\cos\beta = -1$, the resonant frequency is then given by the simple formula[8,10]

$$f = \frac{c}{s}\frac{M_{\text{con}}}{1 + M_{\text{con}}}. \qquad (16)$$

This expression is independent of the number and strength of the assumed sources and of the distance of the source array from the nozzle. The same formula results for a continuously distributed source with its strength being proportional to the disturbance wave amplitude and the local pressure in the shock cells (the fundamental wavelength of which is the shock spacing).[10] It also applies to supersonic jets from convergent–divergent nozzles.[10] Thus, for screech the gain criterion fixes the frequency, whereas it was the phase criterion for the edge tone. The dipole of the edge tone is located at a definite point, that is, in the vicinity of the edge of the wedge, but for screech the distance to the effective source center is determined primarily by the limit cycle of the unstable waves in the jet, that is, by both the rate at which the instability grows and how rapidly it dissipates the shock structure.

The directionality of the screech tone can be estimated by formulas akin to Eq. (15). The agreement with experiment is very satisfactory, especially considering the simplicity of the formula.[33]

In the edge tone, frequency jumps occur as the nozzle–edge distance increases. For screech the shock wave spacing become greater as the pressure ratio increases: The instability mode for the screech of choked jets (from convergent nozzles) of high aspect ratio ("two dimensional") is (almost) always sinuous, and the acoustic wavelength is closely proportional to that spacing. But for round nozzles there is a change in the instability mode, taking the following form as the pressure ratio increases: first, one or two axially symmetric (varicose, toroidal); second, sinuous (flapping); third, a strong and very stable helical, (spinning); and sometimes fourth, the sinuous one reappears.[34] The corresponding jumps in frequency are all upward except for the last one.

For supersonic jets (from convergent–divergent nozzles) at Mach 1.41, the modes appear to be just axially symmetric (torroidal) and helical.[10]

Screech Reduction As screech involves feedback, the reduction of any of the factors in the gain criterion of the feedback loop may be hypothesized to reduce the amplitude of the limit cycle.[35] The principal means are as follows: First, the sound pressure perturbing the jet is lessened by removal of a flat face (thick lip) of the nozzle[33,35] its replacement by a sound absorbant face,[35] or shielding the jet at the exit.[35] Similarly a reflector at a variable axial distance upstream of the nozzle will accentuate the feedback at some distances and attenuate it at others.[36,37] Second, the growth of the initially miniscule instability waves in the jet may be disrupted by a rough edge to the nozzle[35] or the addition of cambered vanes protruding radially into the jet at the exit[35] or of small circumferential tabs pressing the jet boundary inward.[10] Third, the acoustic source vanishes when the shock cells are absent, as resulting from ventilating the nozzle of its excess pressure[35] or by the use of a shock-free divergent nozzle at its design pressure.[38]

Broadband Shock Noise Besides the discrete frequency instability modes of screech, any other large-scale structures (eddies) interacting with the shock waves will naturally result in sound emission. The everpresent more random eddy structure can be visualized as made up of a broadband of frequencies and spacewise Fourier components, each of which will radiate with the directionality given by the principles of Eq. (15) or refinements thereon.[10] Thus, in complete contrast to the directionality of turbulent mixing noise, the lowest frequencies will be emphasized by a tendency to constructive interference in the upstream direction and the highest frequencies in the downstream direction, according to the maxima of Eq. (15). Recasting Eq. (15) the angle of the directional maximum for a given frequency can be expected to be given by

$$\beta = \cos^{-1}\left(\frac{1}{M_{\text{con}}} - \frac{c}{sf} \right). \qquad (17)$$

Screech evidently may be viewed as a special case when feedback occurs, requiring that the maximum is close to the upstream direction.[10]

A detailed analysis of the interaction of the constituent instability waves with the nearly periodic structure of the jet results in good agreement with measurements of the noise field.[10] Measurements of the intensity of the broadband noise from choked jets (convergent nozzle) show a proportionality of the rms value to the shock strength for which there is a theoretical basis[10]; the same relationship holds for the supersonic jets from convergent–divergent nozzles.[10]

REFERENCES

1. W. K. Blake, *Mechanics of Flow-Induced Sound and Vibration*, Academic, 1986.

2. N. H. Fletcher and T. H. Rossing, *The Physics of Musical Instruments*, Pt. IV: *Wind Instruments*, Springer-Verlag, New York, 1990, pp. 345–491.

3. A. Powell, "Why Do Vortices Generate Sound?" *Trans. ASME*, Vol. 17, 1995, pp. 252–260.

4. D. G. Crighton, "Basic Principles of Aerodynamic Noise Generation," *Prog. Aerospace Sci.*, Vol. 16, 1975, pp. 31–96.

5. A. Powell, "Flow Noise: A Perspective on Some Aspects of Flow Noise, and of Jet Noise in Particular," *Noise Control Eng.*, Vol. 8, 1977, pp. 69–80, 108–119.

6. H. S. Ribner, "Perspectives on Jet Noise," American Institute of Aeronautics and Astronautics, Paper No. AIAA-81-0428R, 1981.

7. J. M. Seiner, "Advances in High Speed Jet Aeroacoustics," American Institute of Aeronautics and Astronautics, Paper No. AIAA-84-2275, 1984.

8. A. Powell, "Some Aspects of Aeroacoustics: from Rayleigh until Today," *J. Vibr. Acoust.*, Vol. 112, 1990, pp. 145–159.

9. D. Rockwell, "Oscillations of Impinging Shear Layers," *AIAA J.*, Vol. 21, No. 5, 1983, pp. 645–664.

10. C. K. W. Tam, "Supersonic Jet Noise," *Ann. Rev. Fluid Mech.*, Vol. 27, 1995, pp. 17–43.

11. A. Powell, "Why Do Vortices Generate Sound?" *Trans. ASME*, Vol. 117, 1995, pp. 252–260.

12. M. E. Goldstein, *Aeroacoustics*, McGraw-Hill, New York, 1976.

13. E.-A. Muller (Ed.), *Mechanics of Sound Generation in Flows*, Springer-Verlag, Berlin, 1979.

14. H. H. Hubbard, (Ed.) *Aerodynamics of Flight Vehicles: Theory and Practice*, Vol. 1: *Noise Sources*, Vol. 2: *Noise Control*. Acoustical Society of America, New York, 1994.

15. M. J. Lighthill, "On Sound Generated Aerodynamically, I General Theory," *Proc. Roy. Soc.*, Vol. A 211, 1952, pp. 564–587.

16. A. Powell, "Theory of Vortex Sound," *J. Acoust. Soc. Am.*, Vol. 36, 1964, pp. 177–195.

17. M. J. Howe, "Contributions to the Theory of Aerodynamic Sound, with Application to Excess Jet Noise and the Theory of the Flute," *J. Fluid Mech.*, Vol. 71, No. 4, 1975, pp. 625–673.

18. T. Kambe, "Acoustic Emissions by Vortex Motions," *J. Fluid Mech.*, Vol. 173, 1986, pp. 643–666.

19. C. K. W. Tam and D. E. Burton, "Sound Generation by Instability Waves of Supersonic Flows," *J. Fluid Mech.*, Vol. 138, No. 7, 1984, pp. 273–295.

20. Y. Umeda, R. Ishii, T. Matsuda, A. Yasuda, K. Sawada, and E. Shima, "Instability of Astrophysical Jets. II. Numerical Simulation of Two-Dimensional Choked Under-Expanded Slab Jets," *Prog. Theor. Phys.*, Vol. 4, No. 6, 1990, pp. 856–866.

21. A. Powell, "On the Generation of Noise by Turbulent Jets," ASME Paper 59-AV-53, 1958.

22. J. E. Ffowcs Williams, "The Noise from Turbulence Convected at High Speed," *Phil. Trans. Roy. Soc. (Lond.)*, Vol. A255, 1963, pp. 469–503.

23. P. E. Doak, "Momentum Potential Theory of Energy Flux Caused by Momentum Fluctuations," *J. Sound. Vib.*, Vol. 131, 1989, pp. 67–90.

24. P. L. Jenvey, "The Sound Power from Turbulence; A Theory of the Exchange of Energy between the Acoustic and Non-Acoustic Fields," *J. Sound. Vib.*, Vol. 131, 1989, pp. 37–66.

25. S. H. Guest, "Acoustic Efficiency Trends for High Thrust Boosters," NASA Tech. Note D-1999, 1964.

26. J. E. Ffowcs Williams and A. J. Kempton," The Noise from the Large-Scale Structure of a Jet," *J. Fluid Mech.*, Vol. 84, No. 4, 1978, pp. 673–694.

27. C. K. W. Tam, "Directional Acoustic Radiation from a Supersonic Jet Generated by Shear Layer Instability," *J. Fluid Mech.*, Vol. 46, No. 4, 1974, pp. 757–786.

28. J. M. Seiner, D. K. McLaughlin, and C. H. Liu, "Supersonic Jet Noise Generated by Large-Scale Instabilities," NASA Tech. Paper 2072, 1982.

29. H. S. Ribner, "Cylindrical Sound Waves Generated by Shock-Vortex Interaction," *AIAA J.*, Vol. 23, No. 11, 1985, pp. 1708–1715.

30. K. R. Meadows, A. Kumar, and M. Y. Hussain, "Computational Study on the Interaction between a Vortex and a Shock Wave," *AIAA J.*, Vol. 29, No. 2, 1991, pp. 171–179.

31. H. S. Ribner, "Spectra of Noise and Amplified Turbulence Emanating from Shock Turbulence Interaction," *AIAA J.*, Vol. 25, No. 3, 1987, pp. 436–442.

32. A. Powell, "On the Noise Emanating from a Two-Dimensional Jet Above the Critical Pressure," *Aero. Quart.*, Vol. 4, 1953, pp. 103–122.

33. T. D. Norum, "Screech Suppression in Supersonic Jets," *AIAA J.*, Vol. 21, No. 2, 1983, pp. 283–303.

34. A. Powell, Y. Umeda, and R. Ishii, "Observations of the Oscillating Modes of Choked Circular Jets," *J. Acoust. Soc. A.*, Vol. 92, 1992, pp. 2823–2836.

35. A. Powell, "The Reduction of Choked Jet Noise," *Proc. Phys. Soc. Series B*, Vol. 67, 1954, pp. 313–329.

36. T. D. Norum, "Control of Jet Shock Associated Noise by a Reflector," American Institute of Aeronautics and Astronautics, AIAA Paper No. 84-2279, 1984.

37. R. T. Nagel, J. W. Denham, and A. G. Papathanasiou, "Supersonic Jet Screech Tone Cancellation," *AIAA J.*, Vol. 21, No. 11, 1983, pp. 1541–1545.

38. C. K. W. Tam and H. K. Tanna, "Shock Associated Noise of Supersonic Jets from Convergent-Divergent Nozzles," *J. Sound. Vib.*, Vol. 81, No. 3, 1982, pp. 337–358.

29

INTERACTION OF FLUID MOTION AND SOUND

J. E. FFOWCS WILLIAMS

1 INTRODUCTION

Fluid motion and sound are not necessarily different things, and the cases in which they can be usefully separated and their interaction examined are rather special. If the waves are sufficiently short, their progress can be followed and their characteristics recognized to be essentially those of simple one-dimensional waves. The situation is different if the propagation properties vary more abruptly and the acoustic wavelength is not small. For example, sound is both reflected from and transmitted through a plane vortex sheet. The situation is complicated by the fact that the sheet is unstable; sound interacts best with flow when the flow is unstable. General flow/acoustic interaction phenomena are understood only in a qualitative way, and exact statements that bear on the problem are very rare. That is why this chapter is devoted to statements relating flow and sound and emphasizing the sources of sound in fast flow.

2 THE PHYSICAL PROBLEM

There is little risk of confusion in defining the small-amplitude vibrational field of a compressible homogeneous material as sound. Bulk motion convects the entire wave field, any particular wave crest moving at the speed of sound augmented by material drift. Inhomogeneous convective motion deforms the geometric arrangement of both the wave field and the material elements through which the waves propagate. Materials with low resistance to those staining motions are fluids, and the nontrivial cases of *fluid motions* all involve a substantial and

Encyclopedia of Acoustics, edited by Malcolm J. Crocker
ISBN 0-471-80465-7 © 1997 John Wiley & Sons, Inc.

continuous rearrangement of a particle's shape. Again it might be reasonable to define the small-amplitude vibration of this distorting material as sound, but to do so without recognizing that it might be quite unlike the vibrational field of a uniform material is to risk confusing the picture by oversimplification.

The simplest cases are those in which "packets" of unidirectional sound waves travel in virtually "uniform" material. If the waves are sufficiently short compared with the size of the uniform zone, their progress can be followed and their characteristics recognized to be identical with those of a simple one-dimensional wave; waves travel through matter in a direction normal to wavefronts, energy flowing with them. The existence of sound imparts the material with vibrational and elastic energy, and bulk motion augments both the wave speed and the rate at which energy flows. The packet of waves, identifiable by the crest structure spanning across the wave's path, moves through the material by radiation and additionally drifts with the flow. Gradual changes in flow conditions, those that cause the "port" and "starboard" sides of the packet to move at different speeds, will rotate the packet and bend the path it follows. These are the principles underlying the method of ray tracing, one of the most potent techniques for describing the progress made by sound propagating through slowly varying flows and through material of slowly varying composition. Short-wave sound moves along rays bent by gradients in either flow velocity or sound speed. The wave activity tends to grow where rays converge; speed of sound gradients do not destroy the sound's energy-conserving properties but wind gradients do. The general theory of ray tracing is clearly described by Lighthill,[1] who builds on the earlier foundations laid down by Brekovskikh[2] and Blokhintzev.[3] This is the best way of thinking about how sound travels through the stratified ocean or is refracted by wind gradients. Though the underlying theory exploits the fact that the propagation properties change very slowly, the effect of those changes need not be small. The wind can refract sound to create zones of effective shadow; it can also guide sound waves to concentrate and focus them. These effects are routinely accounted for in engineering practice, with the ray-tracing calculations being performed with the aid of widely available computational codes.

But the situation is different if the propagation properties vary more abruptly and the acoustic wavelength is not small compared to the scale of material change. Of course, ray theory may still provide a useful guide even though it has lost its formal validity. It is certainly interesting to note the qualitative similarity that exists between the refractive properties measured in a jet shear layer and those predicted by ray theory (though wave acoustics are much better).[4,5] It is also tempting to explain the apparent inability of sound to traverse high-speed boundary layers in terms of ray trajectories, even though the flow profile is not thick enough to justify that explanation.[6]

New effects enter with abrupt changes in flow, Miles[7] having been the first to show how sound was both reflected back from and transmitted through a plane vortex sheet, the acoustic energy changing because of work done by the flow during its interaction with sound. The situation is complicated by the fact that the sheet is unstable; local acoustic sources induce exponentially growing disturbances that soon render the linear calculations meaningless.[8] Despite that complication, linear perturbation methods have proved extremely impressive in describing how sound escapes from a pipe, interacting with and modifying the form of the early jet enclosed in the vortex sheet continuation of the pipe wall.[9] Acoustic seeding of emergent jet flow can provoke jets to modify their noise-producing behavior; the broadband turbulence-induced noise can be changed by harmonic jet excitation, but a proper explanation to that mechanism must await more inroads into the understanding of turbulence.[10,11]

Turbulence has known wave-scattering properties,[12] the modeling of which is one of the principal applications of new computational methods.[13,14] Ray techniques have pointed to the tendency for rays to cross, giving multiple paths and multiple arrival times for the receipt of signals. The atmospheric propagation of sound displays these features in a characteristic spasmodic fading of distant signals, with the waveform of the signals being grossly distorted because of its interaction with atmospheric turbulence. The boom from a high-flying supersonic aircraft evolves in a homogeneous atmosphere into a characteristic N-wave, but in a typical turbulent atmosphere both the starting and trailing shocks are commonly spread out over a couple of metres, with distinct individual spikes being formed by coalescing rays. Crow[15] has explained the origins of these changes to be in the unsteady Reynolds stresses that give a scattering source strength proportional to the product of the velocities in the sound and turbulence field.

Vorticity is an element of flow that is foreign to sound, but there is no doubt that unsteady vortex fields create sound as a by-product.[16] Vorticity is created in regions of high velocity gradient, at the lip of a jet nozzle for instance. Flow usually separates from a sharp edge rather than flowing tightly around it as would a potential flow, which sound is. But the usually noisy shedding of vorticity into a downstream vortical wake can, remarkably, act to reduce the scattered field. Howe[17] demonstrated that the sound caused by the vortex acted to oppose what would otherwise be scattered from the sharp edge. Another example of vorticity as a sound-absorbing feature was discovered by Howe[18] in his study of the acoustic properties of rigid but porous screens. The potential interaction of sound with rigid screens is conservative; they convert the energy of the incident sound into that of sound transmitted through and reflected from the screen. But flow through the screen tends to separate at the sharp edges of the perforations, and if the right balance of steady through-flow and porosity is selected, the incident sound can be completely absorbed in the flow,[19] an absorption dependent on the secondary field created by the unsteady vortex shedding.

The interaction between a line vortex and a sharp-fronted pressure wave has long been a source of fascination to aerodynamicists and acousticians, mainly because the flow can be realized in a simple shock tube[20] and because it might provide a tractable example of the otherwise far-too-complicated turbulence–acoustic interaction problem. Though little mystery now remains in that example,[21] the general interaction of sound with flow of nonnegligible Mach number has fallen into the hands of computational aeroacousticians, and their field is still in its infancy. What is sound and what is flow is hard to tell, with the subject's repertoire of worked examples corroborated by independent theoretical or experimental checks yet to be established. Brentner's[22] early calculation of the flow created by impulsive boundary motion and the evident interchangeability revealed by that calculation of flow and acoustic elements provide a fascinating preview of what might yet be to come.

Sound interacts best with flow if the flow is unstable or the sound is especially strong. In whistles, organ pipes, and other forms of wind-driven resonators the energizing feature of the flow is in the unsteady vortex elements shed into an unstable shear layer to grow as they travel toward an edge. There they are scattered into sound, a sound that reacts back to provoke the shedding of another vortex element, to travel and grow and repeat the cycle. The interaction is especially strong when the vortex-shedding frequency coincides with the resonance frequency.[23] The screech of an imperfectly expanded supersonic jet flow results from cyclic backscattering of instability waves from their interaction with compressive waves trapped in the jet.[24]

All these flow–acoustic interaction phenomena are understood in a qualitative way, but exact statements that bear on the problem are very rare. Remarkable in its ability to relate flow and sound is the acoustic analogy produced by Lighthill[25]; his is an exact statement that anchors the flow–acoustic interaction subject and yields definite relationships between flow and sound, though arguably they have only limited predictive force. At this stage in the subject where the most important acoustic-coupled flows are at nonnegligible Mach number, definite results can be invaluable, so little is certain from the inevitably noisy experimental data and intuitive theoretical models. That is the reason why the remaining part of this chapter is devoted to statements relating flow and sound, particularly the sources of sound in fast flow. They may be useful in drawing out effects that remain hidden from view until finite-Mach-number terms are emphasized.

3 SOUND SOURCES IN MOVING FLUID

Sound propagates through uniform still fluid according to the wave equation $\partial^2\phi/\partial t^2 - c^2 \nabla^2\phi = 0$. If the fluid moves uniformly, it will convect the sound field with it and a Gallilean transformation equates the two cases.

We can regard $\partial^2\phi/\partial t^2 - c^2 \nabla^2\phi$ as the source of sound; we call it q. Given q and a boundary condition, the sound field ϕ is determined. Should the source q not be specified independently of ϕ, any "solution" giving ϕ in terms of q would still be only a secondary equation for ϕ. Then, ϕ is not a sound field; it is something else, similar to sound only inasmuch as q is independent of ϕ.

When the fluid is moving and the sound is better specified relative to a moving reference frame, there are two distinct phenomena affecting the level and structure of the sound field. First, there is the Doppler effect, by which successive wave crests are closer together when laid down by a following source. Second, the waves can be stronger because they can spend longer within the source, accumulating more from the source all the time they remain in touch. A moving point source at $\mathbf{y} = \mathbf{y}_s(\tau)$ is modeled by

$$q(\mathbf{y}, \tau) = Q(\tau)\,\delta[\mathbf{y} - \mathbf{y}_s(\tau)] \qquad (1)$$

and generates the outwardly propagating sound field

$$\phi(\mathbf{x}, t) = \frac{Q(t - r^*/c)}{4\pi c^2 r^* |1 - M_r|^*}, \qquad (2)$$

$r = |\mathbf{x} - \mathbf{y}|$ signifying the distance separating the source from the observer and cM_r the speed at which the source approaches. The asterisk implies values at the time when the source emitted the sound destined to arrive at $(\mathbf{x}, t)$:

$$\tau^* = t - \frac{r^*}{c}, \qquad r^* = |\mathbf{x} - \mathbf{y}_s(\tau)^*|. \qquad (3)$$

The source moves with velocity $d\mathbf{y}_s/d\tau$, so that it approaches the observer with speed

$$cM_r = \frac{d\mathbf{y}_s}{d\tau} \cdot \frac{(\mathbf{x} - \mathbf{y}_s)}{|\mathbf{x} - \mathbf{y}_s|}.$$

The source time τ and reception time t are connected for a particular source and a particular observation position $\mathbf{x}$. The sound reaching $\mathbf{x}$ at time t was emitted from the source at $\mathbf{y}_s(\tau)$ at time $\tau = \tau^*$, the emission time scale being stretched when the source approaches:

$$\frac{d\tau^*}{dt} = 1 - \frac{1}{c}\frac{dr^*}{dt} = 1 - \frac{d}{dt}\frac{|\mathbf{x} - \mathbf{y}_s(\tau^*)|}{c}$$

$$= 1 + \frac{(\mathbf{x} - \mathbf{y}_s)}{|\mathbf{x} - \mathbf{y}_s|} \cdot \frac{d\mathbf{y}_s}{d\tau}\frac{1}{c}\frac{d\tau^*}{dt},$$

$$\frac{d\tau^*}{dt} = \frac{1}{(1 - M_r)^*}. \qquad (4)$$

The perturbations of a real fluid caused by unsteady vorticity, by inhomogeneous heating, by the motion of a flow structure, or by foreign bodies moving or deforming in flow, by anything at all in fact, can all be represented through Lighthill's acoustic analogy as being pure sound in a uniform medium at rest, the sources of that sound being prescribed through the analogy. The analogy is useful if the perturbations are clearly recognizable as sound over an extensive region, the sound sources being outside that region and specified independently. That is not generally the case, most fluid motion being different from sound. The analogy is then an exact but somewhat sterile prescription of what sources would be needed to drive a nonexistent acoustic medium to support the density perturbations of the real world. The analogy is designed for the sound radiated by powerful machinery into uniform flow, and for that it is uniquely useful, giving insight, approximation techniques, and an exact formal methodology; accounting for the interaction of fluid motion and sound is reduced to a matter of extracting useful results from the formal method, and the main difficulty of the subject is to avoid the confusion likely to follow the inevitable simplifying approximation.

Sound in a source-free region V can be calculated by Kirchhoff's theorem in terms of boundary conditions on S, a fixed control surface enclosing V with its unit normal $\mathbf{n}$ leading into V:

$$\phi(\mathbf{x}, t) = -\frac{1}{4\pi} \frac{\partial}{\partial x_i} \int_S \frac{\phi^* n_i}{r} \, dS(\mathbf{y})$$
$$- \frac{1}{4\pi} \int_S \frac{1}{r} \left[\frac{\partial \phi}{\partial n}\right]^* dS(y), \qquad (5)$$

the boundary values being integrated over all boundary positions $\mathbf{y}$ at the retarded time $\tau^* = t - r/c$. A difficulty of the general problem is illustrated by using Eq. (5) to estimate the field at $(\mathbf{x}, t)$ in terms of approximations to the boundary conditions. For example, when the boundary is hard and $\partial \phi/\partial n = 0$, the field is given unambiguously by the first of the two integrals, the dipole term. Even when the boundary is not hard, the field is still *determined* once the dipole term is known, so is the boundary value $\partial \phi/\partial n$, which cannot be arbitrarily specified on S. Approximating either of the two surface integrals introduces errors that are not easy to quantify.

A generalization of Eq. (5) to give the field in terms of conditions specified on a moving surface $S(\boldsymbol{\eta})$ that encloses the space surrounding the observation point has been given in Ref. 26:

$$\phi(\mathbf{x}, t) = -\frac{1}{4\pi} \frac{\partial}{\partial x_t} \int_{S(\boldsymbol{\eta})} \left[\frac{\phi n_i A}{r|1 - M_r|}\right]^* dS(\boldsymbol{\eta})$$
$$- \frac{1}{4\pi c^2} \frac{\partial}{\partial t} \int_{S(\boldsymbol{\eta})} \left[\frac{\phi y_n A}{r|1 - M_r|}\right]^* dS(\boldsymbol{\eta})$$
$$- \frac{1}{4\pi} \int_{S(\boldsymbol{\eta})} \left[\left(\frac{\partial \phi}{\partial n} - \frac{v_n}{c^2} \frac{\partial \phi}{\partial \tau}\right) \frac{A}{r|1 - M_r|}\right]^*$$
$$\cdot dS(\boldsymbol{\eta}). \qquad (6)$$

The surface element in the moving reference frame (fixed $\boldsymbol{\eta}$) moves with velocity $\mathbf{v}$ and passes the point $\mathbf{y} = \boldsymbol{\eta}$ at a fixed reference time, the unit area of the moving surface $S(\boldsymbol{\eta})$ corresponding to an area A of the instantaneous reference time surface $S(\mathbf{y})$.

Equations (5) and (6) are two exact equations, one giving the field in terms of conditions on a moving surface and the other in terms of conditions on a fixed control surface. They could, for example, both be used to specify sound fields generated by a motionless source outside both the fixed and moving boundary surfaces. The two equations appear completely different if casually regarded as the field of moving boundary sources without recognizing that the formal problem requires that the boundary values be the very special values determined from the solution itself. The sometimes singular values of the Doppler factor $|1 - M_r|^{-1}$ in the high-Mach-number form of each term of Eq. (6) need not imply that the field shows any trace of this high-Mach-number effect; the Doppler factors arise because the control boundary surface is moving and need not imply the motion of any real source.

Both Eqs. (5) and (6) are true for any interior sound field generated by exterior sources, and it is obviously a hazardous matter to seek in the equation particular features of the sound without the knowledge of all the boundary conditions. The exact equations of motion are similar in this respect but different enough in general to allow some surprising features. Lighthill's analogy by which the real density fluctuations are regarded as linear waves driven by specific but wave-dependent sources takes a strikingly simple form,

$$\frac{\partial^2(\rho - \rho_0)}{\partial t^2} - c^2 \, \nabla^2(\rho - \rho_0) = \frac{\partial^2 T_{ij}}{\partial x_i \, \partial x_j},$$
$$T_{ij} = \rho u_i u_j + p_{ij} - c^2(\rho - \rho_0)\delta_{ij}. \qquad (7)$$

Here, T_{ij} is Lighthill's stress tensor, ρ the mass density, u the fluid velocity, and p_{ij} the compressive viscous stress tensor. The constants ρ_0 and c represent the mean density and the speed of sound. Equation (7) is a statement of the Navier–Stokes equations manipulated to appear as an inhomogeneous wave equation. The quadrupole strength density T_{ij} is needed to drive an acoustic medium to mimic exactly the density perturbation of the real fluid. Reference 26 gives an integral form of this equation that gives the density field in terms of T_{ij} and conditions on a moving boundary, an equation that has become the starting point for many studies of the sound generated by bodies moving at high speed.

If a surface S marks the fluid particles enclosing a foreign body moving and deforming arbitrarily in flow, flow that is surrounded externally by weakly disturbed homogeneous material through which sound radiates out to infinity, then the first Ffowcs Williams–Hawkings representation of Eq. (7) is appropriate.

$$4\pi c^2(\rho - \rho_0)(\mathbf{x}, t) = \frac{\partial^2}{\partial x_i \, \partial x_j} \int_V \left[\frac{T_{ij}J}{r|1 - M_r|} \right]^* d^3\boldsymbol{\eta}$$

$$- \frac{\partial}{\partial x_i} \int_S \left[\frac{p_{ij}n_jA}{r|1 - M_r|} \right]^* dS(\boldsymbol{\eta})$$

$$+ \frac{\partial}{\partial t} \int_S \left[\frac{\rho_0 v_n A}{r|1 - M_r|} \right]^* dS(\boldsymbol{\eta}). \tag{8}$$

This is an exact result for the fully nonlinear equations and is surprisingly simple when compared with its linear form, which is effectively Eq. (6). At low Mach number the volume quadrupole term is usually negligible compared with the dipole and monopole surface integrals and tends often to be ignored.

At first sight the sound of propeller blades, in which the last (monopole) term is specified by kinematic conditions and the dipole term is known once the blade loading is determined, can be calculated by a straightforward evaluation of this equation. But in practice the blade loading is not easy to specify exactly even when the blade flow is steady and laminar, the turbulent surface loading is very problematic, and the influence of the volume quadrupole is unclear. It takes great skill to build this prescription of the sound field into a blade noise prediction scheme.[27]

The simplest case for illustrating the effect of flow on a boundary-induced source is when the motion is linear and T_{ij} is negligible. The relative flow at speed U is then very nearly uniform and parallel to the boundary surface, and the normal velocity v_n is equal to $\partial\xi/\partial t + (\partial\xi/\partial y_1)$, ξ being the surface elevation above the y_1-axis that is parallel to the uniform mean flow. When the boundary encloses a finite-sized body, it must be thin for this linearization to apply. In the particular case when the boundary surface is plane, Eq. (8) simplifies because S may then be taken as the $y_3 = 0$ surface; $A = 1$ and the monopole and dipole terms are equal [see Ref. 28 and Eq. (19) in Chapter 86]:

$$(\rho - \rho_0)(\mathbf{x}, t) = \frac{\rho_0}{2\pi c^2} \frac{\partial}{\partial t} \int_{\eta_3 = 0} \left[\frac{v_n}{r|1 - M_r|} \right]^*$$
$$\cdot \, dS(\boldsymbol{\eta}) \tag{9a}$$

$$= -\frac{1}{2\pi c^2} \frac{\partial}{\partial x_3} \int_{\eta_3 = 0} \left[\frac{p}{r|1 - M_r|} \right]^*$$
$$\cdot \, dS(\boldsymbol{\eta}). \tag{9b}$$

These are convenient forms for evaluating the sound

of flow over a slightly deformed plane surface and are simplest when only a small section of the surface is displaced from its nominal position. Reference 29 gives the far field of such a planar compact source by evaluating Eq. (9a) and shows that

$$(\rho - \rho_0)(\mathbf{x}, t) = \frac{\rho_0 \ddot{Q}(\tau)^*}{2\pi r_0|1 - M_{r_0}|^3}, \qquad \text{where}$$

$$\ddot{Q}(\tau)^* = \int_{s_0} [\ddot{\xi}] \, dS(\boldsymbol{\eta}), \tag{10}$$

and r_0 is the distance of the observer at $\mathbf{x}$ from the vibrating part of the surface at the time when it launched its sound to the observer; the source region was then approaching at speed cM_{r_0}. The effect of motion on this source is to modify the field strength at a large distance from the small wave *emission* zone by three powers of the Doppler factor and to alter the time scale of the sound from that of the source by the Doppler factor. The Doppler amplification is at first sight surprising, the three powers being a characteristic of quadrupole sources,[25] and warns that it is not always a trivial matter to deduce the influence of flow from exact representations of the field. It comes from Eq. (9a) only after recognizing that the part of v_n given by $U(\partial\xi/\partial y_1)$ tends to integrate to zero.

A second case in which the effect of low-speed motion can be worked out exactly is due to Dowling.[30] She considered a compact pulsating sphere of radius $a(\tau)$ moving at low Mach number through infinite homogeneous inviscid fluid at rest. She worked out the effect of slow flow by deriving an expression for the distant sound accurate to first order in M_{r_0}, the Mach number at which the sphere was approaching the distant observer at the time it radiated the observed sound. The volume quadrupoles are negligible because of their low-Mach-number inefficiency, but a dipole arises as the sphere experiences an unsteady drag $2\pi\rho a^2 \dot{a} U$.[31] The linear element of the dipole term in Eq. (8) therefore contains a term arising from the component of drag directed toward the observer, $2\pi\rho a^2 \dot{a} c M_{r_0}$, making

$$- \frac{\partial}{\partial x_i} \int_S \left[\frac{p_{ij}n_jA}{r|1 - M_r|} \right] dS(\boldsymbol{\eta})$$

$$= \frac{1}{r_0} \frac{\partial}{\partial t} \left[2\pi\rho_0 a^2 \dot{a} M_{r_0} \right]^* \tag{11}$$

to first order in M_{r_0}.

The monopole term of Eq. (8) can be evaluated once it is recognized that the contact between the fluid and surface must be continuous[32]:

$$\frac{\partial}{\partial t} \int_s \left[\frac{p_0 v_n A}{r|1 - M_r|} \right] dS(\boldsymbol{\eta}) = \frac{\rho_0}{r_0} \frac{\partial}{\partial t} \left[\frac{4\pi\rho_0 a^2 \dot{a}}{(1 - M_{r_0})^2} \right]^*.$$

(12)

The sum of Eqs. (11) and (12) give the first-order effects of source motion to be

$$4\pi c^2 (\rho - \rho_0)(\mathbf{x}, t) = \frac{4\pi\rho_0 a^2}{r_0} \frac{\partial}{\partial t} \left[\frac{\dot{a}(\tau)^*}{(1 - M_{r_0})^{5/2}} \right]^*,$$

(13)

so that, bearing in mind the Doppler time contraction [Eq. (4)],

$$c^2 (\rho - \rho_0)(\mathbf{x}, t) = \frac{\rho_0 a^2 \ddot{a}(\tau)^*}{r_0 (1 - M_{r_0})^{7/2}}.$$

(14)

This $3\frac{1}{2}$ power of the Doppler factor amplification due to real source motion is more complicated than the effects of flow on hypothetical sources; flow affects both the coupling of a source to the sound and *also* the strength of the source. The small pulsating sphere, often regarded as the physical embodiment of a simple point source, is evidently quite different from it when flow is considered, Eqs. (14) and (2) representing their respective fields. The modification of sources by flow is just as important as the effect of flow on the radiation efficiency.

The distinction between sound and flow is difficult to maintain at high speed. Equation (8), though exact, is hard to interpret when T_{ij} depends on the sound; very few cases exist for evaluating the explicit form and importance of extensively distributed quadrupole terms. If the acoustic analogy is used to calculate known shock waves generated by impulsively started plane boundaries or attached to wedges in supersonic flow,[33] it is seen that the quadrupoles mainly account for the self-convective aspects of nonlinear waves and are weak provided the boundary-induced velocity is much smaller than the sound speed; when that is not so, the sources representing the waves in the exact analogy have a highly unphysical form; for example, they can arrive at the observer before their sound is heard. It seems likely that the general use of the analogy in high-speed flow is usefully confined to flows that are only weakly disturbed by boundary motions and where T_{ij} is negligible outside strictly limited source regions. Other fields, where a nonvanishing T_{ij} represents all nonlinear effects over extensive regions of space, are easier to approach through

a modification of Lighthill's method. One such modification extends the analogy to sound generated near and interacting with the interface between two extensive regions of uniform fluid in relative motion, the sources of waves being confined to the proximity of the interface.

The interface matters a lot, even when it is only linearly disturbed by sound. Sound passing from one region to the other is partly reflected and partly refracted and exchanges energy with the moving fluid. The irradiated plane vortex sheet is one of the simplest idealizations of sound negotiating inhomogeneous flow, but even that is a far from straightforward problem. For a start the vortex sheet is highly unstable and if disturbed by a nearby source will soon degenerate as instability waves grow and destroy the otherwise calm state. This would be avoided if the source were anticipated by an exactly opposite instability wave to that produced by the source so that the two combine, by linear superposition, to cancel one another at large time. This noncausal response is the only way that the vortex sheet modeling is consistent with linear theory and is an essential part of the generalization of Lighthill's acoustic analogy to account for the presence of vortex layers.[34] The sound generated at the turbulent interface of moving streams is identical to the sound that would be generated by Lighthillian quadrupoles moving with the fluid but reflected in a hypothetical backing surface, which behaves exactly as if it were a linearly disturbed vortex sheet in otherwise uniform flow. Beyond the surface it is as if the moving quadrupoles' sound passed through a linearly disturbed vortex sheet into the uniformly moving fluid.

Instabilities are the cause of turbulence, but sound provoked the instability, to make turbulence, to make sound, and so on. In this view of a statistically steady turbulent flow finiteness of the linear response is a more important aspect to preserve in an analogy than is causality, which is patently inappropriate. This will be so whatever scheme is used to synthesize the actual sound by superposing a set of linear perturbations onto an unstable primary flow.

Even the planar shear layer displays considerable modifying influence on the sound of turbulence, Ref. 35 predicting that no sound can travel parallel to the layer and that the angular directivity of the sound is arranged in beams and shadows with tens of decibels contrast occurring within a few degrees variation of the angle at which sound leaves a highly supersonic flow. These aspects are not inconsistent with experimental jet noise studies, the vortex sheet modeling of the sound's interaction with jet flow bringing the acoustic analogy into far better accord with experiment than is possible when that interaction is ignored.[36]

Early jet noise modeling accounted for the convective effect of flow on the source but not its subsequent interaction with sound. Source motion through a still atmosphere amplifies the radiated sound and directs it preferentially ahead of the source. But if the source radiates its sound into uniform surrounding fluid that moves at the same speed, there is no such amplification, and the wrong impression is given by concentrating on the convective terms in the fixed-atmosphere form of the analogy. Jet noise displays both effects to varying degrees, the low-frequency noise being amplified by convection while the moving jet flow shields the high-frequency elements. Short waves are refracted away from the jet.[37] Ray theory is more appropriate to the short-wave case but is not easy to accommodate in the acoustic analogy. Equation (8) does give the sound once the flow is known, but T_{ij} is unlikely to be adequately specific other than for compact sources. There are hardly any examples simple enough to model exactly, and one should not expect more to come from the modeling than the useful qualitive deductions that can be made on dimensional grounds.

The sound is also given precisely in the vortex sheet extension of the acoustic analogy. In that, the solution of Lighthill's quadrupole-driven wave equation is applied in two stages. First, it is applied in the moving flow in a convected reference frame with the mean density and sound speed equal to those of the moving primary flow. Second, the Lighthill equation is applied in the fixed rest frame, with the density and sound speed of the surrounding fluid. The convective form is solved in the first region and the ordinary form in the other, the solution proceeding via a Green's function, which is the response of the homogeneous problem to an impulsive point source in the two regions connecting across a vortex sheet with the usual continuity of pressure and material position. This modified analogy[34] points to the field of a compact jet being closely related to that of a plane turbulent mixing layer, the density perturbation differing only because of a speed-sensitive directionality factor D_{ij} in the general expression

$$\rho'(\mathbf{x}, t) = \frac{D_{ij}}{4\pi x c_0^4} \int \left[\frac{D}{Dt} \left\{ \frac{1}{1 - N_r} \frac{D}{Dt} \right. \right.$$
$$\left. \left. \cdot \left(\frac{\rho^* T_{ij}}{\rho(1 - N_r)} \right) \right\} \frac{d^3 \mathbf{\eta}}{1 - N_r} \right], \quad (15)$$

where the square brackets indicate retarded time, N_r is the Mach number at which the mean flow approaches the observer at $\mathbf{x}$, ρ is the density at the emission time, and ρ^* is the density at the reference time when $\mathbf{\eta} = \mathbf{y}$. Mani's proof that the interaction of longitudinal quadrupoles with a jet flow results in an amplification term $(1 - M_r)^{-2}(1 - N_r)^{-3}$ is contained in this formula, M_r being based on the source convection speed

and N_r on the jet speed. This is a good illustration of how regions of moving fluid affect both the generation and refraction of sound.

The interaction of sound with flow is known once the Reynolds stress terms in T_{ij} are known. That interaction is usually weak because of the inefficiency of compact quadrupoles but could be very strong should the flow contain material inhomogeneities. The isotropic element in T_{ij} contains the term $p - c^2 \rho$, the unsteady parts of which depend on the difference between changes in $c^2 \rho$ and change in p, a vanishing difference in a perfect acoustic medium. But inhomogeneous material has density gradients unrelated to pressure, and the acoustic balance fails, leaving a strong isotropic source in unsteady flow. In fact, Morfey[38] showed that this effect produces a linear source term even in isentropic fluid, the Lighthill equation then being effectively

$$\frac{\partial^2 \rho}{\partial t^2} - c^2 \nabla^2 \rho = \frac{\partial}{\partial x_i} \left((\rho^* - \rho_0) \frac{\partial u_i}{\partial t} \right). \quad (16)$$

Each particle of mean density ρ^* accelerating in unsteady flow is being acted on by a force different to what would be needed to produce that acceleration in sound, $\rho_0(\partial u_i / \partial t)$, and that difference acts as a dipole source, fundamentally more efficient than the quadrupoles that account for the interactions between sound and homogeneous fluid. This effect is an important contributor to the noise of propulsive jets, which are inevitably hotter and lighter than their environment. The noise of many of those jets is actually dominated by this effect at subsonic speeds, increasing in proportion to M^6 rather than conforming with the familiar eighth-power law.

REFERENCES

1. M. J. Lighthill, *Waves in Fluids*, Cambridge University Press, Cambridge, 1978.

2. L. M. Brekovskikh, *Waves in Layered Media*, Academic, New York, 1960.

3. D. I. Blokhintzev, "Acoustics of a Nonhomogeneous Moving Medium," Tech. Mem. No. 1399, National Advisory Communittee on Aeronautics, Washington, DC, 1956.

4. L. K. Schubert, "Numerical Study of Sound Refraction by a Jet Flow I. Ray Acoustics," *J. Acoust. Soc. Am.*, Vol. 51, 1972, pp. 439–446.

5. L. K. Schubert, "Numerical Study of Sound Refraction by a Jet Flow II. Wave Acoustics," *J. Acoust. Soc. Am.*, Vol. 51, 1972, pp. 447–463.

6. D. B. Hanson, "Shielding of Prop-fan Cabin Noise by the Fuselage Boundary Layer," *J. Sound Vib.*, Vol. 92, 1984, pp. 591–598.

7. J. W. Miles, "On the Reflection of Sound at an Interface of Relative Motion," *J. Acoust. Soc. Am.*, Vol. 29, 1957, pp. 226–228.

8. J. D. Morgan, "The Interaction of Sound with a Semi-infinite Vortex Sheet," *Quart. J. Mech. Appl. Math.*, Vol. 27, 1974, pp. 465–487.

9. R. M. Munt, "The Interaction of Sound with a Subsonic Jet Issuing from a Semi-infinite Cylindrical Pipe," *J. Fluid Mech.*, Vol. 83, 1977, pp. 609–640.

10. C. J. Moore, "The Role of Shear-Layer Instability Waves in Jet Exhaust Noise," *J. Fluid Mech.*, Vol. 80, 1977, pp. 321–367.

11. D. Bechert and E. Pfizenmaier, "On the Amplification of Broadband Jet Noise by a Pure Tone Excitation," *J. Sound Vib.*, Vol. 43, 1975, pp. 581–587.

12. V. I. Tatarski, *Wave Propagation in a Turbulent Medium*, McGraw-Hill, New York, 1961.

13. C. R. Truman and M. J. Lee, "Effects of Organised Turbulence Structures on the Phase Distortion in a Coherent Optical Beam Propagating through a Turbulent Shear Flow," *Phys. Fluids A*, Vol. 5, No. 2, 1990, pp. 851–857.

14. P. H. Blanc-Benon, D. Juvé, and G. Comte-Bellot, "Occurrence of Caustics for High-Frequency Acoustic Waves Propagating through Turbulent Fields," *Theoret. Comput. Fluid Dynamics*, Vol. 2, 1991, pp. 271–278.

15. S. C. Crow, "Distortion of Sonic Bangs by Atmospheric Turbulence," *J. Fluid Mech.*, Vol. 37, 1969, pp. 529–563.

16. W. Möhring, "On Vortex Sound at Low Mach Number," *J. Fluid Mech.*, Vol. 85, 1978, pp. 685–691.

17. M. S. Howe, "The Influence of Vortex Shedding on the Generation of Sound by Convected Turbulence," *J. Fluid Mech.*, Vol. 76, 1976, pp. 711–740.

18. M. S. Howe, "The Influence of Vortex Shedding on the Diffraction of Sound by a Perforated Screen," *J. Fluid Mech.*, Vol. 97, 1980, pp. 641–653.

19. A. P. Dowling and I. J. Hughes, "The Absorption of Sound by Perforated Linings," *J. Fluid Mech.*, Vol. 218, 1990, pp. 299–335.

20. M. A. Hollingsworth and E. J. Richard, "A Schlieren Study of the Interaction between a Vortex and a Shock Wave in a Shock Tube," Aeronautical Research Council (report) 17,985, FM 2323, 1955, AD 140 845.

21. H. S. Ribner, "Cylindrical Sound Wave Generated by Shock-Vortex Interaction," *AIAA J.*, Vol. 23, 1985, pp. 2708–1715.

22. K. S. Brentner, "Direct Numerical Calculation of Acoustics: Solution Evaluation through Energy Analysis," *J. Fluid Mech.*, Vol. 254, 1993, pp. 267–281.

23. A. Powell, "On the Edgetone," *J. Acoust. Soc. Am.*, Vol. 33, 1961, pp. 395–409.

24. A. Powell, "The Noise of Choked Jets," *J. Acoust. Soc. Am.*, Vol. 25, 1953, pp. 385–389.

25. M. J. Lighthill, "On Sound Generated Aerodynamically, I General Theory," *Proc. Roy. Soc.*, Vol. 221A, 1952, pp. 564–587.

26. J. E. Ffowcs Williams and D. L. Hawkings, "Sound Generation by Turbulence and Surfaces in Arbitrary Motion," *Phil. Trans. Roy. Soc.*, Vol. A264, 1969, pp. 312–342.

27. F. Farasat and K. S. Brentner, "The Uses and Abuses of the Acoustic Analogy in Helicopter Rotor Noise Prediction," *J. Am. Helicopter Soc.*, Vol. 33, No. 1, 1988, pp. 29–36.

28. A. P. Dowling and J. E. Ffowcs Williams, *Sound and Sources of Sound*, Ellis-Horwood, Chichester, UK, 1983.

29. J. E. Ffowcs Williams and D. J. Lovely, "Sound Radiation into Uniformly Flowing Fluid by Compact Surface Vibration," *J. Fluid Mech.*, Vol. 71, 1978, pp. 689–700.

30. A. P. Dowling, "Convective Amplification of Real Simple Sources," *J. Fluid Mech.*, Vol. 74, 1976, pp. 529–546.

31. Milne-Thomson, *Theoretical Hydrodynamics*, 5th ed., Macmillan, New York, 1968.

32. D. G. Crighton, A. P. Dowling, J. E. Ffowcs Williams, M. Heckl, and F. G. Leppington, *Modern Methods in Analytical Acoustics Lecture Notes*, Springer-Verlag, 1992.

33. J. E. Ffowcs Williams, "On the Role of Quadrupole Source Terms Generated by Moving Bodies," AIAA 5th Aeroacoustics Conference, Paper 79-0567, 1979.

34. A. P. Dowling, J. E. Ffowcs Williams, and M. E. Goldstein, "Sound Production in a Moving Stream," *Phil. Trans. Roy. Soc.*, Vol. 288, No. 1353, 1978, pp. 321–349.

35. J. E. Ffowcs Williams, "Sound Production at the Edge of a Steady Flow," *J. Fluid Mech.*, Vol. 66, No. 4, 1974, pp. 791–816.

36. R. Mani, "The Influence of Jet Flow on Jet Noise Parts 1 and 2," *J. Fluid Mech.*, Vol. 73, No. 4, 1976, pp. 753–793.

37. P. A. Lush, "Measurement of Subsonic Jet Noise and Comparison with Theory," *J. Fluid Mech.*, Vol. 46, 1971, pp. 477–501.

38. C. L. Morfey, "Amplification of Aerodynamic Noise by Convected Flow Inhomogeneities," *J. Sound Vib.*, Vol. 31, 1973, pp. 391–397.

30

ACOUSTIC STREAMING

N. RILEY

1 INTRODUCTION

The term *acoustic streaming* has come into general use to describe the mean, or time-independent, motion that is induced in a fluid flow dominated by its fluctuating components. Such streaming is not, therefore, exclusively associated with sound waves but is also to be found, for example, in the flow induced by waves on the free surface of a liquid.[1] In another context, acoustic streaming in liquids may play a crucial role in the hearing process due to time-independent motion induced in the cochlear fluid within the inner ear.[2,3]

A theoretical analysis of a streaming phenomenon was first given by Rayleigh.[4,5] He considered, in particular, the streaming induced by standing waves between plane walls. Streaming of another kind, often referred to as the "quartz wind," was observed when ultrasonics came into general use. Such streaming may be generated by any source that projects a high-intensity beam of sound into a body of fluid. It was originally associated with quartz oscillators in liquids, but subsequent observations were made in air by Walker and Allen.[6] The two types of streaming referred to above have a common origin, namely *attenuation*. In the first case, which we refer to as type (a), acoustic energy dissipation occurs within the Stokes boundary layer adjacent to a solid boundary, whereas in the second, type (b), it occurs within the main body of fluid. The force that drives the motion owes its origin, directly or indirectly and in general, to the Reynolds stresses developed in the flow, as outlined in Chapter 27.

In the next section, both types of streaming from a theoretical standpoint are considered, from which can emerge a clear physical understanding of the phenomenon. Applications areas are discussed in a final section.

2 THEORY

2.1 Type (a)

To appreciate the origins of a steady, or time-independent, component of streaming within a fluid, consider first a situation in which there is a body force $\overline{\mathbf{F}} = F_0\mathbf{F} = F_0(F_1, F_2, 0)$ acting on an incompressible fluid, with $\nabla \wedge \mathbf{F} = (0, 0, \Phi) \neq 0$. Here, $\mathbf{F}$ is assumed to fluctuate with frequency ω and to have a time average $\langle\mathbf{F}\rangle \equiv 0$. For this two-dimensional flow situation quantities are nondimensionalized with a velocity $U_0 = F_0/\rho\omega$, typical length a, and time ω^{-1} so that the equation for the vorticity $\nabla \wedge \mathbf{v} = (0, 0, \zeta)$ becomes (cf. Batchelor[7])

$$\frac{\partial \zeta}{\partial t} + \frac{1}{S}(\mathbf{v} \cdot \nabla)\zeta = \frac{1}{RS}\nabla^2\zeta + \Phi, \tag{1}$$

where $S = \omega a/U_0$ is the Strouhal number and $R = U_0 a/\nu$ a Reynolds number. For $S \gg 1$ a perturbation solution is developed as

$$\mathbf{v}(\mathbf{x}, t) = \mathbf{v}_0^{(u)}(\mathbf{x}, t) + \mathbf{v}_0^{(s)}(\mathbf{x}) + S^{-1}\mathbf{v}_1(\mathbf{x}, t) + \cdots$$

$$\zeta(\mathbf{x}, t) = \zeta_0^{(u)}(\mathbf{x}, t) + \zeta_0^{(s)}(\mathbf{x}) + S^{-1}\zeta_1(\mathbf{x}, t) + \cdots, \tag{2}$$

which anticipates, via the terms with superscript (s), a nontrivial time-averaged flow, with all time-dependent quantities having the property that their time averages, $\langle\cdot\rangle$, vanish.

Encyclopedia of Acoustics, edited by Malcolm J. Crocker
ISBN 0-471-80465-7 © 1997 John Wiley & Sons, Inc.

Substitution of Eqs. (2) into Eq. (1) yields, at leading order, $\zeta_0^{(u)} = \Phi^t$, where a superscript t denotes integration with respect to time. At $O(S^{-1})$

$$\frac{\partial \zeta_1}{\partial t} + \{(\mathbf{v}_0^{(u)} + \mathbf{v}_0^{(s)}) \cdot \nabla\}(\zeta_0^{(u)} + \zeta_0^{(s)}) = \frac{1}{R} \nabla^2(\zeta_0^{(u)} + \zeta_0^{(s)}),$$

whose time average yields

$$\frac{1}{R} \nabla^2 \zeta_0^{(s)} - (\mathbf{v}_0^{(s)} \cdot \nabla)\zeta_0^{(s)} = \langle(\mathbf{v}_0^{(u)} \cdot \nabla)\Phi^t\rangle \qquad (3)$$

as the equation for the steady component of flow. This equation is simply Helmholtz's equation for two-dimensional steady flow with a distributed time-independent source term. Flows governed by Eq. (3) are not common, although an interesting example involving a fluctuating force field, known as *g*-jitter, in an otherwise gravity-free environment, has been given by Amin.[8] If the body force is conservative, $\nabla \wedge \mathbf{F} = 0$ and $\Phi \equiv 0$, so that $\zeta_0^{(s)}$ has only a trivial solution, unless the boundary conditions are inhomogeneous. If the unsteady flow is initiated by a sound wave, frequency ω, and wavenumber k, with velocity amplitude U_0 and $ka \ll 1$, then a steady streaming will be induced within the Stokes layer at a body surface that will persist at its edge, and in dimensionless form be $O(S^{-1})$. This deduction is from Rayleigh's law of streaming [Chapter 27, Eq. (55)]. Accordingly, Eq. (3) will have a nontrivial solution $O(S^{-1})$. Writing $\mathbf{v}_0^{(s)} = S^{-1}\mathbf{v}_1^{(s)}$, $\zeta_0^{(s)} = S^{-1}\zeta_1^{(s)}$, Eq. (3) becomes

$$\frac{1}{R_s} \nabla^2 \zeta_1^{(s)} - (\mathbf{v}_1^{(s)} \cdot \nabla)\zeta_1^{(s)} = 0, \qquad (4)$$

which suggests that the flow outside the Stokes layer is governed by the Navier–Stokes equations with Reynolds number $R_s = R/S$. Early work on this type of streaming was essentially carried out for $R_s \ll 1$, and an excellent review is given by Nyborg.[9] Only in the 1960s was the crucial role of the streaming Reynolds number R_s recognized.[10]

Unfortunately, Eq. (4) is incomplete. Rather more subtle arguments have to be used to determine the equation that governs the steady streaming outside the Stokes layer in such a sound field. With $\Phi = 0$ and $R = SR_s$, Eq. (1) is the governing equation. Replacing Eq. (2) is

$$\mathbf{v}(\mathbf{x}, t) = \sum_{n=0}^{\infty} S^{-n} \mathbf{v}_n(\mathbf{x}, t), \qquad \zeta(\mathbf{x}, t) = \sum_{n=0}^{\infty} S^{-n} \zeta_n(\mathbf{x}, t),$$

$$(5)$$

where now, in particular, $\mathbf{v}_1(\mathbf{x}, t) = \mathbf{v}_1^{(u)}(\mathbf{x}, t) + \mathbf{v}_1^{(s)}(\mathbf{x})$ and $\zeta_1(\mathbf{x}, t) = \zeta_1^{(u)}(\mathbf{x}, t) + \zeta_1^{(s)}(\mathbf{x})$. Substituting Eq. (3) into Eq. (1) gives at $O(1)$, $\zeta_0(\mathbf{x}, t) \equiv 0$ so that $\mathbf{v}_0(\mathbf{x}, t)$ is an unsteady potential flow field determined by the far-field conditions and body shape. At $O(S^{-1})$, $\zeta_1^{(u)} \equiv 0$, so that, again, $\mathbf{v}_1^{(u)}$ is a potential flow field. Terms $O(S^{-2})$ now give

$$\zeta_2 = -(\mathbf{v}_0^t \cdot \nabla)\zeta_1^{(s)}, \qquad (6)$$

while, at $O(S^{-3})$,

$$\frac{1}{R_s} \nabla^2 \zeta_1^{(s)} - (\mathbf{v}_1 \cdot \nabla)\zeta_1^{(s)} = \frac{\partial \zeta_3}{\partial t} + (\mathbf{v}_0 \cdot \nabla)\zeta_2, \qquad (7)$$

so that, time averaging,

$$\frac{1}{R_s} \nabla^2 \zeta_1^{(s)} - (\mathbf{v}_1^{(s)} \cdot \nabla)\zeta_1^{(s)} = \langle(\mathbf{v}_0 \cdot \nabla)\zeta_2\rangle. \qquad (8)$$

With ζ_2 as in Eq. (6), considerable manipulation yields

$$\langle(\mathbf{v}_0 \cdot \nabla)\zeta_2\rangle = \{\langle(\mathbf{v}_0^t \cdot \nabla)\mathbf{v}_0\rangle\} \cdot \nabla \zeta_1^{(s)} = (\mathbf{v}_d \cdot \nabla)\zeta_1^{(s)},$$

where $\mathbf{v}_d$ is the Stokes drift velocity, so that Eq. (8) becomes

$$\frac{1}{R_s} \nabla^2 \zeta_1^{(s)} - (\mathbf{v}_l^{(s)} \cdot \nabla)\zeta_1^{(s)} = 0, \qquad (9)$$

where $\mathbf{v}_l^{(s)} = \mathbf{v}_1^{(s)} + \mathbf{v}_d$, which may be compared with Eq. (4). The difference is that in Eq. (9) the mean vorticity $\zeta_1^{(s)}$ is convected, in general, with the Lagrangian mean velocity $\mathbf{v}_l^{(s)}$ rather than the Eulerian mean velocity $\mathbf{v}_1^{(s)}$. The boundary condition for Eq. (9) is, once more, the slip velocity provided by Rayleigh's law of streaming at the edge of the thin Stokes layer.

Two simple examples illustrate the above. Rayleigh's problem for standing waves between plane walls, a distance $2a$ apart, has

$$\mathbf{v}_0(\mathbf{x}, t) = (\sin akx \cos t, 0, 0),$$

which gives a streaming velocity at the edge of each Stokes layer, according to Rayleigh's law of streaming,

$$u_e = -\frac{3}{8} \frac{ak}{S} \sin 2akx.$$

With this and the assumption $R_s \ll 1$, $ak \ll 1$, the solution of Eq. (9) gives

$$\mathbf{v}_1^{(s)} = \{\tfrac{3}{16}ak(1 - 3y^2)\sin 2akx, -\tfrac{3}{8}a^2k^2(y - y^3)$$

$$\cdot \cos 2akx, 0\},$$

where y is measured from the symmetry plane. The implied recirculating flow pattern is shown in Fig. 1. This problem, in a different context, has been studied by Haddon and Riley[11] for $ka = O(1)$ and arbitrary R_s. As R_s becomes large, the recirculating regions of flow shown in Fig. 1, viscous dominated for small R_s, become, essentially, regions of inviscid flow, in which the vorticity is uniform and surrounded by viscous boundary layers that are thin compared with the plate spacing but much thicker than the Stokes layers.

As a second example, consider an acoustic line source located at a distance $al(l > 1)$ from the center of a circular cylinder radius a. With $ka \ll 1$ Riley[12] has shown that, according to Rayleigh's law, the streaming velocity at the edge of the Stokes layer at the cylinder surface is given as

$$u_e = \frac{3}{2S}\left\{\frac{4l^3\sin\theta - l^2(l^2 + 1)\sin 2\theta}{(l^2 + 1 - 2l\cos\theta)^2}\right\}, \qquad (10)$$

where θ is a polar angle measured from the line joining the center of the cylinder to the source. Of greatest interest is the steady streaming for $R_s \gg 1$. In that case the dominant streaming is confined to viscous boundary layers, thickness $O(aR_s^{-1/2})$, which are flung off as jets along the axis of symmetry. These boundary layers are, again, much thicker than the Stokes layers. This situation is illustrated in Fig. 2 for $l = 1.5$. The jet directed to the source has strength six times its opposite number in this case; the figure also shows the streamlines of entrainment into the boundary layer as the jet fluid is replenished. The case of a point source close to a sphere has also been considered by Amin and Riley[13] with qualitatively similar results. If $l \rightarrow \infty$, and the source strength is appropriately increased, we recover the case of plane sound waves traveling at right angles to the axis of the circular cylinder for which, in the far field, $\mathbf{v}_0 \sim (\cos t, 0, 0)$. In that case the jets of Fig. 2 assume equal strength.[10,14] An experiment analogous to this situation has been carried

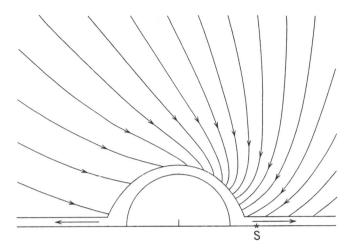

Fig. 2 Streaming for $R_s \gg 1$. An acoustic source at S induces a steady streaming on the cylinder in a boundary layer outside the Stokes layer. This results in a jet, stronger on the source than on the lee side. The inviscid flow outside replenishes the flow within the boundary layer.

out by Davidson and Riley,[15] which confirms both qualitatively and quantitatively the main predictions of the theory.

In both of the above examples the Stokes drift velocity $\mathbf{v}_d \equiv \langle(\mathbf{v}_0^l \cdot \nabla)\mathbf{v}_0\rangle \equiv 0$. An example in which $\mathbf{v}_d \neq 0$, so that outside the Stokes layer vorticity is convected by the Lagrangian mean velocity as in Eq. (9), has been considered by Riley.[16,17] In that example plane waves, out of phase by $\tfrac{1}{2}\pi$, travel in mutually perpendicular directions at right angles to a cylinder so that, in the far field, $v_0 \sim \{\cos t, \cos(t + \tfrac{1}{2}\pi), 0\}$. The streaming motion now assumes a different character. Rayleigh's law of streaming predicts a unidirectional velocity of slip for the outer flow at the edge of the Stokes layer. For a circular cylinder it is uniform and is accommodated by a potential vortex for all values of R_s. For elliptic cylinders, if $R_s \gg 1$ the streaming motion is again dominated by potential vortex flow except close to the boundary where a viscous layer of thickness $O(aR_s^{-1/2})$ adjusts the potential flow to the slip velocity. For these flows no jets penetrate from the boundary into the main body of fluid, and the streaming motion now exerts a net torque on the cylinder.

2.2 Type (b)

Acoustic streaming of type (a) owes its origin to dissipation of acoustic energy in the thin Stokes layer adjacent to a solid boundary. This is made manifest as a slip velocity at the edge of the layer, independent of viscosity, according to Rayleigh's law of streaming. By contrast, streaming of type (b) is associated with acoustic energy dissipation within a body of fluid penetrated by an ultra-high frequency beam. In such a one-dimensional

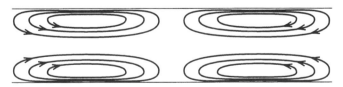

Fig. 1 Steady streaming flow due to standing waves in a channel. Fluid is directed to the channel interior at the nodes. The lateral extent shown is $\lambda/2a$.

beam, with ω^{-1} as a scale for time, c/ω for length, and the unattenuated sound wave velocity amplitude U_0 for velocity, the dimensionless equation for the speed u of the fluid is (cf. Ref. 5)

$$\frac{\partial^2 u}{\partial t^2} - \frac{\partial^2 u}{\partial x^2} - \frac{4}{3}\delta \frac{\partial^3 u}{\partial x^2 \partial t} = -\varepsilon \frac{\partial}{\partial t}\left(u\,\frac{\partial u}{\partial x}\right), \quad (11)$$

where $\delta = \nu\omega/c^2 \ll 1$ includes only the viscous contribution to the diffusivity of sound, ignoring contributions from heat conduction and irreversibility due to delays in attaining thermodynamic equilibrium (cf. Ref. 18), and $\varepsilon = U_0/c \ll 1$. Although ε, δ are both small, and in some circumstances comparably small, the term $O(\delta)$ is properly included with the other linear terms in Eq. (11) if attenuation is to be correctly incorporated. In other words changes that take place on a length scale $c/\omega\delta$ are also important at leading order. The first-order solution of Eq. (11) is

$$u_0 = e^{-(2/3)\delta x}\cos(x-t)$$

if the beam originates at $x = 0$. The net force per unit volume contributed by the Reynolds stresses, in $x > 0$, is

$$F = -\rho_0 U_0^2\,\frac{\omega}{c}\left\langle u_0\,\frac{\partial u_0}{\partial x}\right\rangle = \tfrac{1}{3}\rho_0 U_0^2\,\frac{\delta\omega}{c}\,e^{-(4/3)\delta x}. \tag{12}$$

If a beam of radius r_i is projected along a circular pipe of radius a and length L such that $r_i/a < 1$, $a/L \ll 1$ and, in addition, $\delta\omega L/c \ll 1$, then the force (12) is constant. Fully developed acoustic streaming flow in the pipe in which the beam force is resisted by viscosity gives, for the dimensionless axial velocity,

$$\frac{4\nu}{\varepsilon\delta\omega a^2}\,u^{(s)} = \begin{cases} \left(\dfrac{r_i}{a}\right)^2 - 2\left(\dfrac{r_i}{a}\right)^2\log\left(\dfrac{r_i}{a}\right) - \left(\dfrac{r}{a}\right)^2, \\[4pt] \qquad r < r_i, \\[10pt] -2\left(\dfrac{r_i}{a}\right)^2\log\left(\dfrac{r}{a}\right), \\[4pt] \qquad r > r_i. \end{cases}$$

The velocity profiles for two values of r_i/a are shown in Fig. 3. But more interesting acoustic streaming flows

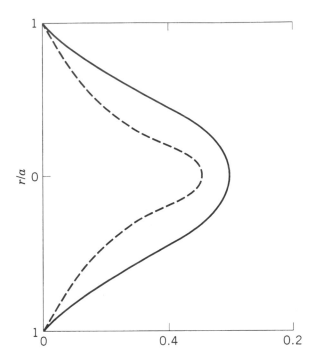

Fig. 3 Fully developed streaming flow in a circular tube due to an ultrasonic beam: (——) $u^{(s)}$ when $r_i/a = 0.5$; (-----) $3u^{(s)}$ when $r_i/a = 0.2$.

can be realized when there is no pipelike constraint. The principles for such flows have been clearly set out by Lighthill.[18,19]

For a narrow ultrasonic beam of cross-sectional area S_b the force per unit volume (12) may be integrated across the beam to give the total force acting, per unit length of the beam, F_l; a further integration along its length gives the total force acting, F_t:

$$F_l = \frac{1}{3}\,\rho_0 U_0^2 S_b\,\frac{\delta\omega}{c}\,e^{-(4/3)\delta x}, \tag{12a}$$

$$F_t = \frac{1}{4}\,\rho_0 U_0^2 S_b. \tag{12b}$$

Lighthill[18,19] shows that $F_t = P/c$, where P is the power emitted by the acoustic source in a narrow beam, from which $U_0 = 2(P/\rho_0 c S_b)^{1/2}$. If the total force (12b) produces acoustic streaming with typical velocity U_s in a system of length l, then $U_s = U_0 S_b/l^2 = 2(PS_b/\rho_0 c l^4)^{1/2}$, to give a streaming Reynolds number $R_s = U_s l/\nu = 2(PS_b/\mu\nu c l^2)^{1/2} = O(QS_b)^{1/2}$ for an acoustic source of Q microwatts, assuming typical values for μ, ν, and c in air, with $l = 0.1$ m. For a beam of 1 μW and area $S_b = 2\times 10^{-4}$m^2, $R_s = O(10^{-2})$, which is in the slow-flow regime, wherein the applied force is resisted by viscous stresses only. For $S_b^{1/2} \ll l$ and $c^3/\omega^2\nu \ll l$, correspond-

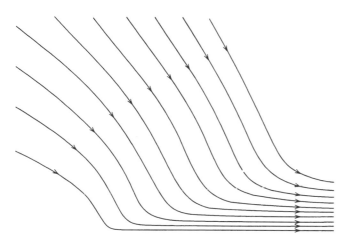

Fig. 4 Streamlines for an axisymmetric jet due to a point force for which $F_t/\rho\nu^2 = 3282$.

ing to frequencies in excess of 1 MHz, the applied force is essentially a point force and may be represented as $F_t\delta(\mathbf{x})$ to give a streaming velocity field[18,19] with velocity scale $F_t/8\pi\mu l$,

$$\mathbf{v}^{(s)} = \left(\frac{r^2 + x^2}{r^3}, \frac{xy}{r^3}, \frac{xz}{r^3} \right),$$

which is, of course, the stokeslet velocity field. For sources of greater power the streaming Reynolds number R_s increases, and inertia terms in the governing equations may not be neglected. The acoustic streaming flow for the point force then assumes an axisymmetric jetlike character. Such jet flows have been studied by Squire.[20] In Fig. 4 we show the streamline pattern for $F_t/\rho\nu^2 = 3282$ calculated by Squire. This jet results from an acoustic power source $Q = 335$ μW, and for a beam of cross section $S_b = 2 \times 10^{-4}$m^2, $R_s = O(1)$. As the acoustic power and streaming Reynolds number increase, the induced jet flow becomes turbulent. Lighthill[19] presents a theory for these turbulent jets. In this theory the point force approximation is abandoned, and the exponential variation of the force (12) is accommodated.

At all but the lowest streaming Reynolds numbers R_s, when the force due to acoustic energy dissipation is resisted by viscous stresses alone, acoustic streaming of both type (a) and type (b) is, in general, characterized by jetlike flows. Jets and acoustic energy are readily associated, but usually in a manner opposite to that outlined in this chapter.

3 AREAS OF APPLICATION

Areas in which the ideas outlined in Section 2 find application may be divided into two distinct categories. In the first are devices that take advantage of acoustic streaming phenomena. By contrast, the second and larger category includes situations in which acoustic streaming is an inevitable consequence of some other acoustic process.

In the first category advantage of type (b) streaming was used to measure *absorption coefficients* of common liquids.[21] An ultrasonic beam was directed along an initially straight, liquid-filled tube turned at right angles and absorbed. The unidirectional streaming flow was diverted to a side tube for its return. Absorption coefficients were inferred from measurements of the flow in the side tube. With such coefficients known, *transducer calibration* may be achieved by suitably measuring the acoustic streaming velocity in a beam to determine the velocity amplitude U_0. Nature may have turned acoustic streaming to advantage in the *hearing process*. Motion of the ossicles produces waves in the cochlear fluid. Lighthill[2,3] suggests that streaming associated with these waves may provide the stimulus for the inner hair cells to transform acoustic signals to neural activity. This streaming has a peak at the "characteristic place" for a particular frequency. Perhaps conversely, the design of *woodwind instruments* seeks to minimize the amount of streaming in tone-hole regions. This is believed, for the low-register tones, to induce an instability of the vibration sustained by the reed generation mechanism.[22]

In the second category acoustic streaming arises in diverse areas. The growth rate of bubbles in a sound field, by the process of *rectified diffusion*, is known to be influenced by the presence of acoustic streaming.[23,24] This is confirmed by the studies of Davidson,[25] who shows that the transport of gas across the interface of a spherical volume-preserving bubble due to plane sound waves can be greatly enhanced by acoustic streaming, particularly at high streaming Reynolds numbers R_s. Large values of R_s are difficult to achieve in this *microstreaming* situation since the streaming velocities are smaller by a factor $(\nu/\omega a^2)^{1/2}$ for a fluid–fluid interface, compared with a fluid–solid interface,[26] due, of course, to the smaller dissipation rates in the former case. This type of microstreaming was first observed by Nyborg[9] as a by-product in a study of *biological processes*. Examples of such processes in which a shear mechanism due to acoustic streaming, as can be brought about by ultrasonic diagnostics, may affect aggregates of cells in suspension are discussed by Rooney[27] and Williams.[28] If the shear results in cell disaggregation, a further consequence of the streaming could be removal of the actively secreted coat of mucopolysaccharide. This in turn could alter such diverse functions as antigen–antibody reactions and the resistance by fetal cells to infection by certain virus particles. There is also interest in the role of acoustic streaming, developed in a purely oscillatory flow field, in problems dealing with *biophysical fluid* transport.[29,30] *Acoustic levitation* and *acoustic positioning*, in particular the latter, where acoustic forces can be used to manipulate materials in the low-gravity environment of space where containerless melting, undercooling, and freezing of materials can be carried out, provide a technologic example in which acoustic streaming is influential. For example, acoustic streaming induced by two perpendicular plane waves out of phase by $\frac{1}{2}\pi$ will exert a torque on the material. Such cases are considered by Busse and Wang,[31] Lee and Wang,[32] and for large streaming Reynolds numbers Riley.[17,33] In a solidification process the hot liquid material will not maintain a fixed shape in the applied acoustic field. Oscillations of it will induce a streaming of type (a) within it, as described by Trinh and Wang[34] and Wang.[35] Such internal flows within a melt may greatly influence the kinetics of crystallization as well as the macrostructure of the resulting solid. With large temperature differences present in these materials-processing situations, the effect of *heat transfer* by acoustic streaming cannot be ignored. Experimental and theoretical studies of this mode of heat transfer have been made by Leung and Wang[36] and by Davidson,[37] who demonstrates analytically a significant enhancement of heat transfer for large values of the streaming Reynolds number R_s.

REFERENCES

1. M. S. Longuet-Higgins, "Mass Transport in Water Waves," *Phil. Trans. Roy. Soc. A*, Vol. 245, 1953, pp. 535–581.

2. M. J. Lighthill, "Biomechanics of Hearing Sensitivity," *Trans ASME: J. Vib. Acoust.*, Vol. 113, 1991, pp. 1–13.

3. M. J. Lighthill, "Acoustic Streaming in the Ear Itself," *J. Fluid Mech.*, Vol. 239, 1992, pp. 551–606.

4. Lord Rayleigh, "On the Circulation of Air Observed in Kundt's Tubes, and on Some Allied Acoustical Problems," *Phil. Trans. Roy. Soc. A*, Vol. 175, 1883, pp. 1–21.

5. Lord Rayleigh, *The Theory of Sound*, 2nd ed., Vol. II, MacMillan, 1896.

6. J. P. Walker and C. H. Allen, "Sonic Wind and Static Pressure in Intense Sound Fields," *J. Acoust. Soc. Am.*, Vol. 22, 1950, p. 680A.

7. G. K. Batchelor, *An Introduction to Fluid Mechanics*, Cambridge University Press, U.K., 1967.

8. N. Amin, "The Effect of g-jitter on Heat Transfer," *Proc. Roy. Soc. A*, Vol. 419, 1989, pp. 151–172.

9. W. L. Nyborg, "Acoustic Streaming," in W. Mason (Ed.), *Physical Acoustics IIB*, Academic, New York, 1965, Chap. 11.

10. J. T. Stuart, "Double Boundary Layers in Oscillatory Viscous Flows," *J. Fluid Mech.*, Vol. 24, 1966, pp. 673–687.

11. E. W. Haddon and N. Riley, "A Note on the Mean Circulation in Standing Waves," *Wave Motion*, Vol. 5, 1983, pp. 43–48.

12. N. Riley, "Streaming from a Cylinder Due to an Acoustic Source," *J. Fluid Mech.*, Vol. 180, 1987, pp. 319–326.

13. N. Amin and N. Riley, "Streaming from a Sphere Due to a Pulsating Source," *J. Fluid Mech.*, Vol. 210, 1990, pp. 459–473.

14. N. Riley, "Oscillating Viscous Flows," *Mathematika*, Vol. 12, 1965, pp. 161–175.

15. B. J. Davidson and N. Riley, "Jets Induced by Oscillatory Motion," *J. Fluid Mech.*, Vol. 53, 1972, pp. 287–303.

16. N. Riley, "Stirring of a Viscous Fluid," *Z. Angew. Math. Phys.*, Vol. 22, 1971, pp. 645–653.

17. N. Riley, "Circular Oscillations of a Cylinder in a Viscous Fluid," *Z. Angew. Math. Phys.*, Vol. 29, 1978, pp. 439–449.

18. M. J. Lighthill, *Waves in Fluids*, Cambridge University Press, U.K., 1978.

19. M. J. Lighthill, "Acoustic Streaming," *J. Sound Vib.*, Vol. 61, 1978, pp. 391–418.

20. H. B. Squire, "The Round Laminar Jet," *Q. J. Mech. Appl. Math.*, Vol. 4, 1951, pp. 321–329.

21. J. E. Piercy and J. Lamb, "Acoustic Streaming in Liquids," *Proc. Roy. Soc. A*, Vol. 226, 1954, pp. 43–50.

22. D. H. Keefe, "Acoustic Streaming, Dimensional Analysis of Nonlinearities, and Tone-Hole Mutual Interactions in Woodwinds," *J. Acoust. Soc. Am.*, Vol. 73, 1983, pp. 1804–1820.

23. R. K. Gould, "Rectified Diffusion in the Presence of, and Absence of, Acoustic Streaming," *J. Acoust. Soc. Am.*, Vol. 56, 1974, pp. 1740–1746.

24. C. C. Church, "A Method to Account for Acoustic Microstreaming when Predicting Bubble Growth Rates Produced by Rectified Diffusion," *J. Acoust. Soc. Am.*, Vol. 84, 1988, pp. 1758–1764.

25. B. J. Davidson, "Mass Transfer Due to Cavitation Microstreaming," *J. Sound Vib.*, Vol. 17, 1971, pp. 261–270.

26. B. J. Davidson and N. Riley, "Cavitation Microstreaming," *J. Sound Vib.*, Vol. 15, 1971, pp. 217–233.

27. J. Rooney, "Shear as a Mechanism for Sonically Induced Biological Effects," *J. Acoust. Soc. Am.*, Vol. 52, 1972, pp. 1718–1724.

28. A. R. Williams, "Disorganization and Disruption of Mammalian and Amoeboid Cells by Acoustic Microstreaming," *J. Acoust. Soc. Am.*, Vol. 52, 1972, pp. 688–693.

29. T. W. Secomb, "Flow in a Channel with Pulsating Walls," *J. Fluid Mech.*, Vol. 88, 1978, pp. 273–288.

30. C. Thompson, "Acoustic Streaming in a Waveguide with Slowly Varying Height," *J. Acoust. Soc. Am.*, Vol. 75, 1984, pp. 97–107.

31. F. H. Busse and T. G. Wang, "Torque Generated by Orthogonal Acoustic Waves. Theory," *J. Acoust. Soc. Am.*, Vol. 69, 1981, pp. 1634–1638.

32. C. P. Lee and T. G. Wang, "Acoustic Radiation Forces on a Heated Sphere Including Effects of Heat Transfer and

Acoustic Streaming," *J. Acoust. Soc. Am.*, Vol. 83, 1988, pp. 1324–1331.

33. N. Riley, "Acoustic Streaming About a Cylinder in Orthogonal Beams," *J. Fluid Mech.*, Vol. 242, 1992, pp. 387–394.

34. E. H. Trinh and T. G. Wang, "Large-Amplitude Free and Driven Drop-Shape Oscillations: Experimental Observations," *J. Fluid Mech.*, Vol. 122, 1982, pp. 315–338.

35. T. G. Wang, "Equilibrium Shapes of Rotating Spheroids and Drop Shape Oscillations," *Adv. Appl. Mech.*, Vol. 26, 1988, pp. 1–62.

36. E. W. Leung and T. G. Wang, "Force on a Heated Sphere in a Horizontal Plane Acoustic Standing Wave Field," *J. Acoust. Soc. Am.*, Vol. 77, 1985, pp. 1686–1691.

37. B. J. Davidson, "Heat Transfer from a Vibrating Cylinder," *Int. J. Heat Mass Transfer*, Vol. 16, 1973, pp. 1703–1727.

31

SHOCK WAVES, BLAST WAVES, AND SONIC BOOMS

Richard Raspet

1 INTRODUCTION

Many sounds in the atmosphere originate from strong impulsive sources. Blast waves from explosions, thunder from lightning, sonic booms from aircraft, ballistic waves from projectiles, and N-waves from spark sources all involve large-amplitude waves. This chapter serves as a guide to strong- and weak-shock theory as applied to impulses.

First the basic equations governing shock propagation and decay are examined. Next, the application of these equations to the prediction of source waveforms and decay with distance is described. Plots of the waveforms and calculated pressure decay with range and scaling laws to relate these plots to arbitrary sources are provided. These relations can be applied to spherical and cylindrical explosions, sparks, and exploding wires. Next, nonlinear propagation and weak-shock theory are reviewed, and their application to far-field predictions is described. More detailed information on finite-wave and weak-shock propagation is contained in Chapter 19.

2 BASIC EQUATIONS

The basic equations necessary to describe shock wave propagation express conservation of mass, Newton's second law for fluids, conservation of energy, and an equation of state relating pressure to density.[1] In regions away from shocks, the Eulerian form of these equations are

$$\frac{\partial \rho}{\partial t} + \boldsymbol{\nabla} \cdot (\rho \boldsymbol{u}) = 0, \tag{1}$$

Encyclopedia of Acoustics, edited by Malcolm J. Crocker
ISBN 0-471-80465-7 © 1997 John Wiley & Sons, Inc.

$$\rho \left(\frac{\partial \boldsymbol{u}}{\partial t} + \boldsymbol{u} \cdot \boldsymbol{\nabla} \boldsymbol{u} \right) = -\boldsymbol{\nabla} p, \tag{2}$$

$$\rho \, \frac{\partial}{\partial t} (\tfrac{1}{2} u^2 + e) + \boldsymbol{u} \cdot \boldsymbol{\nabla} (\tfrac{1}{2} u^2 + e) = 0, \tag{3}$$

and

$$p = p(\rho, T), \tag{4}$$

where ρ is the fluid density, $\mathbf{u}$ is the particle velocity, T is the temperature, p is the pressure, and e is the internal energy per unit mass. These equations may be linearized in terms of the fluctuation densities and pressure to form the basic acoustic equations.

If Eqs. (1)–(4) are solved for wave propagation without any attenuation mechanism, they will result in multivalued pressure waveforms. In reality, attenuation mechanisms such as viscosity and heat conduction become important for large pressure gradients and lead to small but finite rise times of the pressure waveform.

For sufficiently strong waves, discontinuities form and the differential equations expressed in Eqs. (1)–(3) do not hold. The governing equations across the shocks are the Rankine–Hugoniot equations. These may be expressed as

$$[\rho(u - u_{\text{sh}})]_+ = [\rho(u - u_{\text{sh}})], \tag{5}$$

$$[\rho u(u - u_{\text{sh}}) + p]_+ = [\rho u(u - u_{\text{sh}}) + p]_-, \tag{6}$$

and

$$[\rho(\tfrac{1}{2} u^2 + e)(u - u_{\text{sh}}) + pu]_+ = [\rho(\tfrac{1}{2} u^2 + e)(u - u_{\text{sh}}) + pu]_-, \tag{7}$$

where e is the internal energy per unit mass, the subscript minus signifies the variable just behind the shock, and the subscript plus signifies the variable just in front of the shock. Equations (5), (6), and (7) express conservation of mass, Newton's second law, and work–energy conservation, respectively.

In many cases the air or gas surrounding an explosion or energy release can be considered a polytropic gas, that is, a gas with a constant ratio of specific heats.[2] The shock relations for a polytropic gas are written conveniently in terms of shock strength $z = (p_- - p_+)/p_+$ and Mach number of the shock relative to the flow ahead, $M = (u_{sh} - u_+)/c_+$. The shock relations are then given by

$$\frac{u_- - u_+}{c_+} = \frac{z}{\gamma\{1 + [(\gamma + 1)/2\gamma]z\}^{1/2}}, \qquad (8)$$

$$\frac{\rho_-}{\rho_+} = \frac{1 + [(\gamma + 1)/2\gamma]z}{1 + [(\gamma - 1)/2\gamma]z}, \qquad (9)$$

and

$$\frac{c_-}{c_+} = \left(\frac{1 + z\{1 + [(\gamma - 1)/2\gamma]z\}}{1 + [(\gamma + 1)/2\gamma]z}\right)^{1/2}. \qquad (10)$$

We examine shock theory as applied to explosion waves in the next section.

3 EXPLOSION WAVES

3.1 Spherical Explosions

The simplest model of an explosion is a point blast. Exact solutions for this problem have been developed by Von Neumann,[3] Taylor,[4] and Sedov.[5] The explosion is modeled as an energy release E concentrated at a point in space with ambient density ρ_0. Dimensional analysis can be used to derive relations between the position of the shock, $r(t)$, and the overpressure in terms of E, ρ_0, and t:

$$r(t) = k\left(\frac{E}{\rho_0}\right)^{1/5} t^{2/5} \qquad (11)$$

and

$$p = \frac{8}{25}\frac{k^2\rho_0}{\gamma + 1}\left(\frac{E}{\rho_0}\right)\frac{2}{5}t^{-6/5}, \qquad (12)$$

or, equivalently,

$$p = \frac{8}{25}\frac{k^5}{\gamma + 1}Er^{-3}. \qquad (13)$$

The decay of a strong spherical shock as r^{-3} is a general characteristic of the strong-shock regime.

The dimensionless constant k is determined by requiring conservation of energy and is a function only of the ratio of specific heats. We require that

$$E = \int_0^{r(t)}\left(\frac{p}{\gamma - 1} + \frac{\rho u^2}{2}\right)4\pi r^2\, dr \qquad (14)$$

be conserved. This has been evaluated numerically by Taylor[4] and analytically by Sedov[5] and Von Neumann.[3] For $\gamma = 1.4$, at standard pressure, Eq. (13) becomes

$$p = \frac{0.155E}{r^3}. \qquad (15)$$

Sedov presents a number of similarity solutions to more complicated problems, but the simple point explosion serves as a basis to understanding the numerical solutions for strong-shock development and for understanding scaling laws.

It is useful to express the variables in terms of dimensionless variables as used in similarity solutions. The development of explosion waves from realistic explosions or differing initial conditions is not expected to obey the scaling laws exactly since realistic initial conditions introduce a characteristic length. For strong enough sources, however, the influence of the initial geometry is reduced as the radius of the shock increases. The variables of overpressure, radius, and time can all be expressed in terms of dimensionless variables in terms of the ambient density and the total energy release[6]:

$$\frac{p}{p_0} \quad \text{(scaled pressure)}, \qquad (16)$$

$$\frac{c_0 p_0^{1/3} t}{E^{1/3}} \quad \text{(scaled time)}, \qquad (17)$$

$$\frac{p_0^{1/3} r}{E^{1/3}} \quad \text{(scaled distance)}, \qquad (18)$$

where p_0 is the ambient pressure, c_0 is the speed of sound in air, E is the energy release, r is the radial distance, p is the shock pressure, and t is the time from the explosion. The model of a point energy release works well for nuclear explosions. For chemical explosions, the difference in the equation of state of the explosion products and the finite initial distribution of kinetic energy lead to lower initial shock pressures and higher efficiencies in the conversion of energy to the blast wave.

Brode numerically integrated the gas dynamic equations for a variety of spherical explosions with different initial conditions and equations of state. In his 1955 paper[7] he demonstrated that the blast wave from an isothermal sphere of 199 atm initial pressure "assumes the general shape and values of the point source solution (to within 10 percent) after the shock wave has engulfed a mass of air 10 times the initial mass of the sphere."

In a later paper, Brode[8] calculated the blast wave from the detonation of a sphere of TNT with a loading density of 1.5 g/cm³. He used realistic equations of state for the TNT and for air. Figure 1 displays Brode's results for TNT in air compared to the point source in air and in an ideal gas. Also included are calculations from Berry, Butler, and Holt that used an ideal gas with $\gamma = 3$ for the explosion products and an ideal gas with $\gamma = 1.2$ for the surrounding gas and calculations by Wecken for RDX-TNT in an ideal gas. The similarity of the curves beyond a reduced radius of 0.2 shows that the blast wave from a strong explosion is only weakly dependent on initial conditions and that the scaling laws are a good approximation for relating different yield explosions. The shock decay follows a $1/r^3$ dependence down to about 10 atm. Brode[7] also suggests fits to the decay curve for the point source at lower pressures. At longer distances weak-shock theory holds and the decay approaches $1/r$. ANSI standard S2.20[9] approximates the long-range decay of explosion waves to be $r^{-1.1}$. This power law is an approximate fit to theoretical results prepared by the Air Force Weapons Laboratory.

The principal difference in initial conditions is in the effective energy yield. Nuclear explosions approximate a point release of energy and leave a higher percentage of energy near the origin. The efficiencies of various explosives in terms of their weight are listed in Table 1 (taken from Ref. 9). Data from different types of explosions can be related by converting to the TNT equivalent weight.

Spherical blast waveforms for noise calculations can be developed from Brode's calculations. Figure 2 presents the pressure versus radius at fixed time based upon Brode's calculation for a TNT explosion. Extrapolations using finite-wave analysis and weak-shock theory may be used for lower amplitude waves. Figure 3 compares a measured waveform from a blasting cap to Brode's waveform extrapolated to larger distance with a finite-wave propagation algorithm.[10,11]

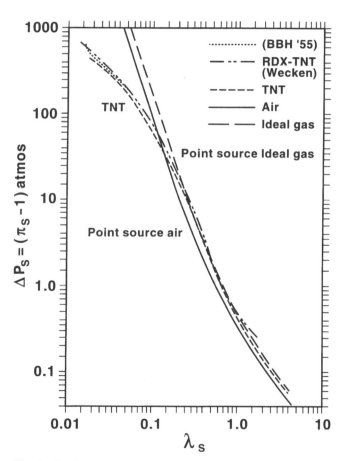

Fig. 1 Peak overpressure ($\Delta p_s = \pi_s - 1$) in atmospheres versus shock radius ($\lambda_s = R_s/\alpha$, $\alpha^3 = W_{\text{tot}}/p_0$) with comparisons between TNT and point source blasts in both real air and an ideal gas ($\gamma = 1.4$). The dash–double-dot curve is from the results of Wecken for 50% RDX, 50% TNT. The short dotted curve is from the Berry, Butler, and Holt calculations for PENT. (From H. L. Brode, "Blast Wave from a Spherical Charge," *Phys. Fluids*, Vol. 2, 1959, pp. 217–229.)

TABLE 1 Equivalent Explosive Weight Approximations Based on TNT

Explosive	Equivalent Weight
TNT	1.00
Tritonal	1.07
Composition B	1.11
HBX-1	1.17
HBX-3	1.14
TNETB	1.36
Composition C-4	1.37
H-6	1.38
Pentolite	1.42
PETN	1.27
Nitroglycerine	1.23
RDX-Cyclonite	1.17
Nitromethane	1.00
Ammonium nitrate	0.84
Black powder	0.46
Nuclear explosives	0.79

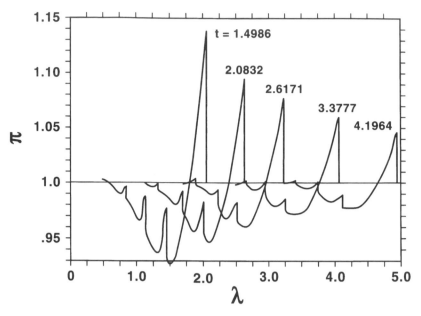

Fig. 2 Pressure ($\pi = p/p_0$) versus radius at indicated times for a TNT blast. (From H. L. Brode, "Blast Wave from a Spherical Charge," *Phys. Fluids*, Vol. 2, 1959, pp. 217–229.)

Measured and empirical waveforms can also be used in calculations. Figure 4 is a waveform for 0.57 kg of C-4 plastic explosive measured at 30 m over an acoustically hard surface.[12] The calibration constant is 3.17×10^5 Pa/V. Reed[13] has used the following waveform in blast noise attenuation calculations and it is a reasonable fit to measured waves:

$$p(t) = \Delta p \left(1 - \frac{t}{t_+}\right)\left(1 - \frac{t}{\tau}\right)\left[1 - \left(\frac{t}{\tau}\right)^2\right],$$

$$0 < t < \tau. \tag{19}$$

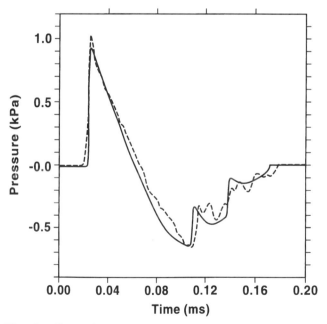

Fig. 3 Comparison of the measured waveform with the output of the finite-wave program. (From R. Raspet, J. Ezell, and S. V. Coggeshall, "Diffraction of an Explosive Transient," *J. Acoust. Soc. Am.*, Vol. 79, 1985, pp. 1326–1334.)

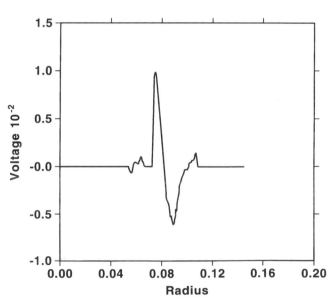

Fig. 4 Measured explosion pulse shape at 30 m for 0.57 kg of C-4 plastic explosive. (From R. Raspet, H. E. Bass, and J. Ezell, "Effect of Finite Ground Impedance on the Propagation of Acoustic Pulses," *J. Acoust. Soc. Am.*, Vol. 74, 1983, pp. 267–274.)

For a 1.0-kiloton nuclear explosion $R_0 = 2.7$ km, $t_+ = 0.375$ s, $\tau = 1.375$ s, and $\Delta p = 2.55$ kPa. The long negative duration of this pulse is typical of larger explosions. The waveforms illustrated may be scaled to different charge weights and explosive types by Eqs. (16)–(18) and Table 1. Charge weight or explosive mass may be substituted for the energy when scaling between different explosives.

3.2 Cylindrical Explosions

Cylindrical explosions and line source energy releases can be analyzed using appropriate scaling laws and solutions. The development of the wave from an instantaneous energy release along a line in an ideal gas has been solved by Lin[14] in connection with the problem of shock generation by meteors or missiles. Plooster has evaluated the development and decay of cylindrical sources with more realistic initial energy distributions and atmospheres.[15] Plooster applied his procedure to the prediction of the shock wave from long sparks[16] and from lightning.[17]

Plooster[15] and Lin[14] have used a scaled radius that involves an additional constant determined by the ratio of the specific heats of the fluid:

$$\frac{p}{p_0} \quad \text{(scaled pressure)}, \qquad (20)$$

$$\frac{r}{r_0} \quad \text{(scaled distance)}, \qquad (21)$$

$$\frac{c_0 t}{r_0}, \quad \text{(scaled time)}. \qquad (22)$$

where $r_0 = (E_0/b\gamma p_0)^{1/2}$ and b is a constant dependent on γ, the ratio of the specific heats. For $\gamma = 1.4$, $b = 3.94$.

Figure 5 displays Plooster's result for A, the line source in an ideal gas; B, an isothermal cylinder, constant density, ideal gas; C, an isothermal cylinder, constant density, real gas; D, an isothermal cylinder, low density, ideal gas; and E, an isothermal cylinder, high density, ideal gas. As in the spherical case, the solutions differ mainly at short distances and in the efficiency of the energy release coupling into the explosion waves. Beyond a reduced radius of 0.09, the decay rates are similar. Figure 6 displays the pressure wave calculated by Plooster at scaled distances from 1.44 to 5.95.

Similar calculations for exploding wires were first carried out by Rouse[18] and by Sakurai.[19] Rouse found that the line source model produced good estimates of the shock strength and decay from exploding wires because the errors introduced by the presence of the copper vapor partially canceled the errors in the equation of state for air.

A strong-shock estimate of the decay of cylindrical waves from an instantaneous line source in air characterized by $\gamma = 1.4$ at standard pressure is given by

$$p = \frac{0.216E}{r^2}. \qquad (23)$$

Decay proportional to r^{-2} is a general property of cylindrical strong shocks.

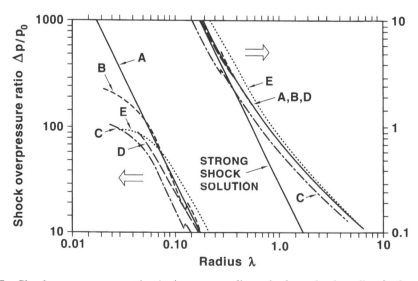

Fig. 5 Shock overpressure ratio $\Delta p/p_0$ versus dimensionless shock radius λ for initial conditions described in text. (From M. N. Plooster, "Shock Waves from Line Sources. Numerical Solutions and Experimental Measurements," *Phys. Fluids*, Vol. 13, 1970, pp. 2665–2675.)

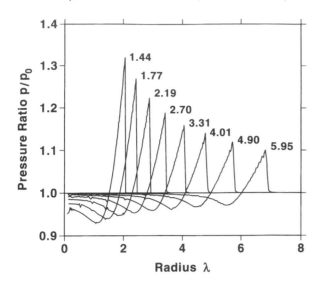

Fig. 6 Pressure ratio p/p_0 versus radius λ for a line source of energy in an ideal gas at indicated times after energy release. (From M. N. Plooster, "Shock Waves from Line Sources. Numerical Solutions and Experimental Measurements," *Phys. Fluids*, Vol. 13, 1970, pp. 2665–2675.)

4 NONLINEAR WAVE PROPAGATION OF PLANE, CYLINDRICAL, AND SPHERICAL WAVES

The development and propagation of sonic booms, ballistic waves from projectiles, and N-waves from weak sparks can usually be treated with nonlinear acoustic equations and with weak-shock theory.[1,2] The propagation of explosion waves can be treated in the same manner once the overpressure has decayed into the weak-shock region.

4.1 Nonlinear Propagation

If second-order terms are retained in the acoustic equations, the largest effect is the dependence of the sound velocity on amplitude, which may be expressed as

$$c \approx c_0 + \frac{\beta p'}{\rho_0 c_0} \qquad (24)$$

for plane waves, where c_0 is the small-amplitude velocity. Here, $\beta = 1 + B/2A$, where

$$A = \left(\frac{\rho \, \partial p}{\partial \rho} \right)_0 \quad \text{and} \quad B = \left(\rho^2 \frac{\partial^2 p}{\partial \rho^2} \right)_0 .$$

For an ideal gas, $\beta = (\gamma + 1)/2$, where γ is the ratio of specific heats.

The dependence of sound speed on amplitude leads to the distortion of the waveform as it propagates. For plane waves, this distortion can be expressed parametrically by

$$p(x, t) = g(\psi), \qquad (25)$$

$$t = \psi + \frac{x}{c_0} - \frac{x}{c_0^2} \frac{\beta g(\psi)}{\rho_0 c_0}, \qquad (26)$$

where $g(t)$ is $p(0, t)$ and t is ψ when $x = 0$.

The distortion is illustrated in Fig. 2 of Chapter 19 for a single-cycle wave. The slope of a waveform dp/dx will become infinite at time

$$t = \frac{\rho_0 c_0^2}{\beta (dp/dt)_{\max}}, \qquad (27)$$

where $(dp/dt)_{\max}$ is the largest negative slope of the initial waveform. At this time weak-shock theory must be employed to determine the subsequent propagation of the wave.

The nonlinear equations may be generalized to different geometries and to a refractive atmosphere through the definition of the age variable. In this construction it is assumed that nonlinear effects do not appreciably affect geometric acoustic ray directions and ray tube areas. Again, the principal effect is the distortion of the waveform. The strength of the wave and its distortion are inversely proportional to the square root of the ray tube area. The geometric decay of nonlinear waves in cylindrical and spherical geometries may be derived by combining the linear geometric decay with the change in travel times due to nonlinear effects.

The linear decay of a wave is expressed by

$$p = \left[\frac{A(r_0)}{A(r)} \right]^{1/2} g[t - \tau(r)], \qquad (28)$$

where $A(r)$ is the ray tube area and $\tau(r)$ is the linear ray travel time from r_0 to r, and

$$c = c_0 + \frac{\beta p}{\rho_0 c_0} = \left(\frac{d\tau}{dr} \right)^{-1} + \left(\frac{d\mathcal{A}}{dr} \right) \left(\frac{d\tau}{dr} \right)^{-2} g(\psi), \qquad (29)$$

where the age variable $\mathcal{A}(r)$ is given by

$$A = \int_{r_0}^{r} \beta \left[\frac{A(r_0)}{A(r)} \right]^{1/2} \frac{\hat{e}_{\text{ray}} \cdot \hat{n}}{\rho_0 c_0 (c + \mathbf{v} \cdot \hat{n})^2} \, dr$$

$$= \frac{\beta}{\rho_0 c_0^3} \int_{r_0}^{r} \left[\frac{A(r_0)}{A(r)} \right] \, dr, \tag{30}$$

where $\mathbf{v}$ is the wind velocity, $\hat{n}$ is the unit vector in the direction normal to the wavefront, and $\hat{e}_{\text{ray}}$ is the unit vector in the ray direction. The first integral in Eq. (30) is the general form for an inhomogeneous atmosphere with winds; the second applies only to a homogeneous quiescent atmosphere. If no shock forms, then $g(\psi)$ must be constant for an observer moving at the nonlinear velocity and the distortion can be determined from the parametric equation

$$t = \psi + \tau(r) - g(\psi) \mathcal{A}(l). \tag{31}$$

This equation with Eq. (28) can be used to specify the distortion of cylindrical and spherical waves in a homogeneous atmosphere:

Cylindrical:

$$p = \left[\frac{r_0}{r} \right]^{1/2} g(\psi), \tag{32}$$

$$t = \psi + \frac{r - r_0}{c_0} - 2g(\psi) \frac{\beta}{\rho_0 c_0^3} (\sqrt{r} - \sqrt{r_0})$$

$$\cdot \sqrt{r_0}. \tag{33}$$

Spherical:

$$p = \frac{r_0}{r} g(\psi), \tag{34}$$

$$t = \psi + \frac{r - r_0}{c_0} - g(\psi) \frac{\beta}{\rho_0 c_0^3} r_0 \ln\left(\frac{r}{r_0} \right). \tag{35}$$

The distance required for an initial waveform to form a shock may also be calculated:

Cylindrical:

$$r = r_0 \left(1 + \frac{\rho_0 c_0^3 / \beta r_0}{(dp/dt)_{\text{max}}} \right), \tag{36}$$

Spherical:

$$r = r_0 \exp\left(\frac{\rho_0 c_0^3 / \beta r_0}{(dp/dt)_{\text{max}}} \right). \tag{37}$$

Weak-shock theory must be used once a shock forms.

4.2 Weak-Shock Theory

Nonlinear theory predicts the formation of multivalued waveforms. Since density and pressure are physical quantities, such waveforms are not physically possible. If the Rankine–Hugoniot equations are expanded for small density and pressure variations ($\Delta \rho / \rho < 0.1$), it can be shown that the entropy change across the shock is third order in the pressure fluctuation and the difference terms may be treated as isentropic derivatives yielding

$$u_{\text{sh}} - u_{\text{av}} = \pm c_{\text{av}} \tag{38}$$

and

$$\Delta u = \pm \frac{\Delta p}{\rho_{\text{av}} c_{\text{av}}}. \tag{39}$$

If we let $g(\psi_-)$ and $g(\psi_+)$ be the acoustic pressure behind and in front of the shock,

$$u_{\text{sh}} = c_0 + \frac{1}{2} \beta \frac{g(\psi_+) + g(\psi_-)}{\rho_0 c_0}, \tag{40}$$

where $g(\psi)$ and ψ are defined in Eqs. (25) and (26). Comparison of this equation and Eq. (24) shows that the shock velocity is slower than the second-order acoustic velocity behind the shock and faster than the second-order velocity in front of the shock. The shock "takes in" points on the waveform in front of and behind the shock.

The position of the shock and behavior of the wave after a shock forms can be determined from the multivalued waveform by the equal-areas rule.[20] The equal-areas rule specifies that the area between the shock position and the multivalued predicted waveform must be zero:

$$A(t) = \int_{\psi_-}^{\psi_+} [x(\psi, t) - x_{\text{sh}}(t)] \frac{dg(\psi)}{d\psi} \, d\psi. \tag{41}$$

See Fig. 4 of Chapter 19. This construction along with weak-shock theory in general is valid only until the shock becomes weak enough for dissipative effects to disperse the shock.

If weak-shock theory were uniformly valid, all impulses with a positive-overpressure front lobe and negative-overpressure rear lobe would form an N-shaped wave, that is, a front shock with positive overpressure p_{max} decaying linearly to a rear shock with negative pressure $-p_{\text{max}}$. Sonic booms from aircraft, ballistic waves from bullets, and strong sparks often form N-waves.

Spherically spreading explosions and weak sparks decay too rapidly for the rear shock to form from the lower pressure and longer duration rear lobe of the pulse. The overpressure decay of the front lobe, however, can usually be treated by weak-shock decay of a triangular pulse.

We may use the equal-area rule to determine the decay and lengthening of an N-wave or a triangular pulse. The analysis of the corrected travel time along a path can be adapted for the analysis of N-waves by noting that the shock velocity is $c_0 + \beta p_{av}/\rho_0 c_0$.

An N-wave with initial overpressure $p(r_0)$ and duration $2T_0$ at r_0 will have duration $2T(r)$ and overpressure $p(r)$ at position r given by

$$T(r) = T_0 \left[1 + \frac{p(r_0)}{T_0} \mathcal{A}(r) \right]^{1/2}, \qquad (42)$$

$$p(r) = \left[\frac{\mathcal{A}(r_0)}{\mathcal{A}(r)} \right]^{1/2} p(r_0) \left[1 + \frac{p(r_0)}{T_0} \mathcal{A}(r) \right]^{-1/2}. \qquad (43)$$

The conserved quantity in the propagation of an N-wave is $P(r)T(r)[A(r)/(A(r_0)]^{1/2}$. The area under the N-wave is preserved as if the geometric spreading is taken into account. Specific forms are easily derived for plane, cylindrically, and spherically spreading waves:

Plane waves:

$$p(x) = p(x_0) \left[1 + \frac{\beta}{\rho_0 c_0^3} \frac{p(x_0)}{T_0} x \right]^{-1/2}, \qquad (44)$$

$$T(x) = T_0 \left[1 + \frac{\beta}{\rho_0 c_0^3} \frac{p(x_0)}{T_0} x \right]^{1/2}. \qquad (45)$$

Cylindrical waves:

$$p(r) = p(r_0) \sqrt{\frac{r_0}{r}}$$
$$\cdot \left[1 + \frac{2\beta}{\rho_0 c_0^3} \frac{p(r_0)}{T_0} \sqrt{r_0}(\sqrt{r} - \sqrt{r_0}) \right]^{-1/2}, \qquad (46)$$

$$T(r) = T_0 \left[1 + \frac{2\beta}{\rho_0 c_0^3} \frac{p(r_0)}{T_0} \sqrt{r_0}(\sqrt{r} - \sqrt{r_0}) \right]^{1/2}. \qquad (47)$$

Spherical waves:

$$p(r) = \frac{r_0}{r} p(r_0) \left[1 + \frac{\beta}{\rho_0 c_0^3} \frac{p(r_0)}{T_0} r_0 \ln \frac{r}{r_0} \right]^{-1/2}, \qquad (48)$$

$$T(r) = \frac{r_0}{r} T_0 \left[1 + \frac{\beta}{\rho_0 c_0^3} \frac{p(r_0)}{T_0} r_0 \ln \frac{r}{r_0} \right]^{1/2}. \qquad (49)$$

Although weak-shock theory does not explicitly include any absorption mechanisms, these are included in the assumption that discontinuities are formed and that the Rankine–Hugoniot jump conditions hold.

The energy loss inherent in weak-shock propagation can be demonstrated using the N-wave propagation equations. The energy carried by the N-wave across a surface at distance r such that kr is large is given by

$$E(r) = A(r) \int_{-\infty}^{\infty} \frac{p^2 \, dt}{2\rho c} = \frac{A(r)}{3\rho c} p(r)^2 T(r)$$
$$= E(r_0) \left[1 + \frac{p(r_0)}{T_0} \mathcal{A}(r) \right]^{-1/2}. \qquad (50)$$

The energy loss across the shock is not sensitive to the dissipative mechanism that produces the gradient. Only if the shock is dispersed are the details of the physical mechanism important.

The relationship between the duration and pressure of a spherically spreading N-wave [Eqs. (48) and (49)] has been exploited for the calibration of high-pressure wideband capacitor microphones.[21,22] Wright has achieved calibration accurate to ±1.0 dB using 0.5–1.0 cm long sparks with 0.01–0.10 J energy discharge.

5 SOURCE MODELING OF BALLISTIC WAVES AND SONIC BOOMS

The shock wave from projectiles and sonic booms at moderate Mach numbers can be predicted using the nonlinearization process described in the previous section. The generated shocks are sufficiently weak at low Mach numbers that the waveform can be calculated by the nonlinear distortion of the wave predicted from linear theory.

An example of such a process is the prediction of the ballistic wave from supersonic thin cylindrical objects moving at velocity V.[1] The displacement of air by the body moving supersonically along the x-axis is equivalent to a distribution of monopole volume sources.

The result of this displacement is a wave with conical constant-phase surfaces. The ray paths are straight lines normal to the Mach cone with apex angle

$$\sin \theta_m = \frac{c_0}{V}. \tag{51}$$

Along the rays the linear solution can be put in the standard form for cylindrical propagation in terms of a reference distance s_0:

$$p = \left(\frac{s_0}{s}\right)^{1/2} g[t - \tau(l)], \tag{52}$$

where

$$\tau(l) = \frac{s}{c_0} + \frac{x_0}{V} \quad \text{and} \quad g(t) = \frac{\rho_0 V^2 \sqrt{m} F_w(Vt)}{\sqrt{m^2 - 1}\sqrt{2s_0}}.$$

Here, F_w is the Whitham F-function

$$F_w(\xi) = \frac{1}{2\pi} \frac{d^2}{d\xi^2} \int_0^\infty \frac{A(\xi - \eta)}{\sqrt{\eta}} d\eta. \tag{53}$$

The nonlinear distortion of the wave can be calculated using Eqs. (28) and (30). The final waveform is calculated using the equal-areas rule.

The discussion above is restricted to steady supersonic motion of an axisymmetric body in a homogeneous quiescent atmosphere. Whitham[23] describes the extension of the F-function to include the effects of lift on a thin wing. Hayes[24,25] has incorporated the effect of an inhomogeneous atmosphere and winds on sonic booms into a computer program to predict sonic booms on the ground. If the ray tube area goes to zero, focusing occurs and diffraction corrections are necessary to predict the waveform. Recent work has applied improved numerical methods and computational fluid dynamics for the prediction of sonic booms for complicated aircraft shapes at high Mach numbers.

6 SPECIAL CASE: N-WAVE PROPAGATION FROM BALLISTIC WAVES AND SONIC BOOMS

If the attenuation mechanisms are weak so that the weak shocks are not dispersed, every waveform of finite duration and a positive pressure front lobe will asymptotically become an N-wave whose maximum pressure at cylindrical distance r is given as

$$p_{max} = \frac{\rho_0 c_0^2 (M^2 - 1)^{1/2} S_{max}^{1/2} K}{2^{1/4} \beta^{1/2} r^{3/4} L^{1/4}}, \tag{54}$$

where S_{max} is the maximum cross section of the body, L the length of the body, and

$$K^2 = \frac{\sqrt{L}}{S_{max}} \max \int_{-\infty}^{\xi} F_w(\xi) \, d\xi, \tag{55}$$

where K depends only on the shape of the body.

The positive phase duration of the N-wave may be approximated as

$$T = \frac{2^{3/4} \sqrt{\beta} M S_{max}^{1/2} K r^{1/4}}{L^{1/4} c_0 (M^2 - 1)^{3/8}}. \tag{56}$$

For these expressions to be valid, the weak shock must have formed an N-wave. Whether this occurs or not depends on the initial strength of the shock, the initial duration, and the spreading law. The predicted duration must be longer than the linear duration L/V for the N-wave to occur. Equations (54) and (56) may still be applied to a positive-pressure triangular pulse of a shock, even when the rear pulse has not formed a shock due to a longer initial duration.

DuMond, Cohen, Panofsky, and Deeds[26] present extensive measurements and weak-shock predictions for ballistic waves from a variety of small arms. The Army report, "Acoustical Considerations for a Silent Weapon's System: A Feasibility Study," contains a good review of the theory as well as measurements of the ballistic wave from a number of weapons.[27]

The shock wave from high-energy sparks and the ballistic wave from small arms typically form clean N-waves since they are relatively strong sources with small initial duration. Sonic booms from fighter aircraft usually form N-waves in spite of their long initial duration since the wave propagates a long distance while maintaining the weak-shock conditions. The sonic boom from large aircraft do not reach their asymptotic form by the time they propagate to the ground due to their large initial duration.

7 SOURCE MODELING OF FINITE-LENGTH SPARKS AND LIGHTNING

The strength and duration of a shock wave from a spherically symmetric or cylindrically symmetric source can be used to approximate the energy released. Detailed numerical calculations can be used to infer information about energy relations in the source region.[16]

For finite-length sparks, the shock development is primarily cylindrical close to the spark but becomes spherical at long distances from the spark. The study of thunder recorded at a distance as a diagnostic of energy released in lightning requires assumptions as to the nature of the propagation at long distances.

Wright and Medendorp[28] developed a model for a finite-length spark using a superposition of spherically propagating N-waves. They had good success reproducing the angular dependence of the waveshape on the propagation angle. They adjusted the duration of the spherical N-waves to best reproduce the measured data. The success of this linear model for shock propagation is due to the weak sparks used in the study (0.01–0.10 J).

Ribner and Roy[29] applied the Wright and Medendorp model for finite-length sparks to simulate thunder. They summed spherically spreading Wright–Medendorp waves emitted from zig-zag segments of a simulated lightning channel. These summations reproduce thunder that sounds realistic.

Few[30,31] modeled the shock development as cylindrical out to the relaxation length, then as spherical weak decay. The relaxation lengths that relate the spherical and cylindrical decay are defined as

$$r_0 = \left(\frac{E_t}{\frac{4\pi}{3} p_0} \right)^{1/3} \quad \text{(spherical)} \qquad (57)$$

and

$$r_0 = \left(\frac{E_l}{\pi p_0} \right)^{1/2} \quad \text{(cylindrical)}, \qquad (58)$$

where E_t is the total energy release and E_l is the energy release per unit length. Few argues that the tortuous nature of the lightning channel leads to spherical decay at ranges beyond one relaxation length and that the duration of the pulses are given by $2.6r_0$.

Plooster[17] and Bass[32] have modeled the propagation of thunder from lightning as finite-wave propagation from a cylindrical source. Since a cylindrical wave undergoes less decay with distance and greater finite-wave distortion, this approach leads to smaller estimates of the energy per unit length of the lightning. Few's estimates are usually on the order of 10^5 J/m while Bass and Plooster's estimates using cylindrical wave properties are two orders of magnitude smaller.

REFERENCES

1. A. D. Pierce, *Acoustics, an Introduction to Its Physical Principles and Applications*, Acoustical Society of America, Woodbury, NY, 1989.

2. G. B. Whitham, *Linear and Nonlinear Waves*, Wiley-Interscience, New York, 1974.

3. J. Von Neumann, "The Point Source Solution," in A. H. Taub (Ed.), *The Collected Works of John Von Neumann*, Vol. VI, Pergamon, New York, 1963, pp. 219–237.

4. G. I. Taylor, "The Formation of a Blast Wave by a Very Intense Explosion: I. Theoretical Discussion," *Proc. Roy Soc. A*, Vol. 201, 1950, pp. 159–174.

5. L. I. Sedov, *Similarity and Dimensional Methods in Mechanics*, Academic, New York, 1961.

6. W. E. Baker, *Explosions in Air*, University of Texas Press, Austin, TX, 1973.

7. H. L. Brode, "Numerical Solutions of Spherical Blast Waves," *J. Appl. Phys.*, Vol. 26, 1955, pp. 766–775.

8. H. L. Brode, "Blast Wave from a Spherical Charge," *Phys. Fluids*, Vol. 2, 1959, pp. 217–229.

9. "Estimating Air Blast Characteristics for Single Point Atmospheric Propagation and Effects," ANSI S2.20–1983, American National Standards Institute.

10. R. Raspet, J. Ezell, and S. V. Coggeshall, "Diffraction of an Explosive Transient," *J. Acoust. Soc. Am.*, Vol. 79, 1985, pp. 1326–1334.

11. H. E. Bass, J. Ezell, and R. Raspet, "Effect of Vibrational Relaxation on the Atmospheric Attenuation and Rise Time of Explosion Waves," *J. Acoust. Soc. Am.*, Vol. 74, 1983, pp. 1514–1517.

12. R. Raspet, H. E. Bass, and J. Ezell, "Effect on Finite Ground Impedance on the Propagation of Acoustic Pulses," *J. Acoust. Soc. Am.*, Vol. 74, 1983, pp. 267–274.

13. J. W. Reed, "Atmospheric Attenuation of Explosion Waves," *J. Acoust. Soc. Am.*, Vol. 61, 1977, pp. 39–47.

14. S-C. Lin, "Cylindrical Shock Waves Produced by Instantaneous Energy Release," *J. Appl. Phys.*, Vol. 25, 1954, pp. 54–57.

15. M. N. Plooster, "Shock Waves from Line Sources. Numerical Solutions and Experimental Measurements," *Phys. Fluids*, Vol. 13, 1970, pp. 2665–2675.

16. M. N. Plooster, "Numerical Simulation of Spark Discharges in Air," *Phys. Fluids*, Vol. 14, 1971, pp. 2111–2123.

17. M. N. Plooster, "Numerical Model of the Return Strokes of the Lightning Discharge," *Phys. Fluids*, Vol. 14, 1971, pp. 2124–2133.

18. C. A. Rouse, "Theoretical Analysis of the Hydrodynamic Flow in Exploding Wire Phenomena," in W. G. Chace and H. K. Moore (Eds.), *Exploding Wires*, Plenum, New York, 1959.

19. A. Sakurai, "On the Propagation of Cylindrical Shock Waves," in W. G. Chace and H. K. Moore (Eds.), *Exploding Wires*, Plenum, New York, 1959.

20. L. D. Landau, "On Shock Waves at Large Distances from the Place of Their Origin," *U.S.S.R. J. Phys.*, Vol. 9, 1945, pp. 496–503.

21. W. M. Wright, "Propagation in Air of N-Waves Produced by Sparks," *J. Acoust. Soc. Am.*, Vol. 73, 1983, pp. 1948–1955.

22. B. A. Davy and D. T. Blackstock, "Measurement of the Refraction and Diffraction of a Short N Wave by a Gas Filled Soap Bubble," *J. Acoust. Soc. Am.*, Vol. 49, 1971, pp. 732–737.

23. G. B. Whitham, "On the Propagation of Weak Shock Waves," *J. Fluid Mech.*, Vol. 1, 1956, pp. 291–318.

24. W. D. Hayes, R. C. Haefeli, and H. E. Kulsrud, "Sonic Boom Propagation in a Stratified Atmosphere with Computer Program," NASA CR-1299, 1969.

25. W. D. Hayes and H. L. Runyan, "Sonic Boom Propagation through a Stratified Atmosphere," *J. Acoust. Soc. Am.*, Vol. 51, 1972, pp. 695–701.

26. W. M. Dumond, E. R. Cohen, W. K. H. Panofsky, and E. Deeds, "A Determination of the Wave Forms and Laws of Propagation and Dissipation of Ballistic Shock Waves," *J. Acoust. Soc. Am.*, Vol. 18, 1946, pp. 97–118.

27. G. R. Garinther and J. B. Moreland, "Acoustical Consideration for a Silent Weapon System: A Feasibility Study," Human Engineering Laboratories, Technical Memorandum 10-66, Defense Documentation Center, Cameron Station, Alexandria, VA, AD521 727/H, 1966.

28. W. M. Wright and N. W. Medendorp, "Acoustic Radiation for a Finite Line Source with N-Wave Excitation," *J. Acoust. Soc. Am.*, Vol. 43, 1968, pp. 966–971.

29. H. S. Ribner and D. Roy, "Acoustics of Thunder: A Quasilinear Model for Tortuous Lightning," *J. Acoust. Soc. Am.*, Vol. 72, 1982, pp. 1911–1925.

30. A. A. Few, "Power Spectrum of Thunder," *J. Geophys. Res.*, Vol. 74, 1969, pp. 6926–6934.

31. A. A. Few, "Acoustic Radiation from Lightning," in H. Volland (Ed.), *CRC Handbook of Atmospherics*, CRC Press, Boca Raton, FL 1982, Vol. 2, pp. 257–290.

32. H. E. Bass, "The Propagation of Thunder Through the Atmosphere," *J. Acoust. Soc. Am.*, Vol. 67, 1980, pp. 1959–1966.

32

ATMOSPHERIC SOUND PROPAGATION

LOUIS C. SUTHERLAND AND GILLES A. DAIGLE

1 INTRODUCTION

This chapter summarizes the current state of knowledge in outdoor sound propagation, including geometric spreading and atmospheric absorption, propagation over ground, reflection and diffraction by obstacles, and refraction by a stationary but nonhomogeneous atmosphere. The important influence of turbulence on sound propagation in the atmosphere and the related topic of acoustic sounding of the atmosphere are treated at the end of the chapter. Propagation losses due to finite-amplitude effects are treated in Part II.

Recent reviews[1-3] reflect the great strides made in this field since the late 1960s, primarily in the area of atmospheric absorption, ground and terrain effects, numerical evaluation of propagation in nonhomogeneous media, and the development in 1968[1] of acoustic sounders (sodar) to probe the atmosphere acoustically.

2 SPREADING LOSSES IN OUTDOOR SOUND PROPAGATION

For this chapter, it will be convenient to describe the total attenuation A_T, in decibels, due to sound propagation as the sound level at a source minus the level at a receiver:

$$A_T = L_{ps} - L_{pr} = 20 \log \left[\frac{p(s)}{p(r)} \right], \qquad (1)$$

where L_{ps} is the sound pressure level with a root-mean-square (rms) sound pressure $p(s)$ at a distance s near the

source and L_{pr} is the corresponding sound pressure level with an rms sound pressure $p(r)$ at the receiver a distance r from the source.

This attenuation, normally a positive quantity, will be expressed as the sum of three nominally independent terms:

$$A_T = A_s + A_a + A_e, \qquad (2)$$

where A_s is the attenuation due to geometric spreading, A_a is the attenuation due to atmospheric absorption, and A_e is the excess attenuation due to all other effects including attenuation A_g by the ground in a homogeneous atmosphere, refraction by a nonhomogeneous atmosphere, attenuation by diffraction and reflection by a barrier, and scattering or diffraction effects due to turbulence.

The "excess attenuation" term A_e is not subdivided into the individual elements suggested by its description since they may not all be present or may not necessarily act independently.

A general expression for the spreading loss A_s, in decibels, between any two positions at distances r_1, r_2 from an acoustic source can be given in the form

$$A_s = 20g \log \left(\frac{r_2}{r_1} \right), \qquad (3)$$

where r_2, r_1 are the distances between the acoustic center of the source and the farthest (i.e., receiver) and closest (i.e., source) positions, respectively, and $g = 0$ for plane-wave propagation such as within a uniform pipe (i.e., no spreading loss), $g = \frac{1}{2}$ for cylindrical propagation from a line source, and $g = 1$ for spherical wave propagation from a point source.

Encyclopedia of Acoustics, edited by Malcolm J. Crocker
ISBN 0-471-80465-7 © 1997 John Wiley & Sons, Inc.

The latter two conditions correspond to the commonly specified condition of 3 and 6 dB loss (respectively) per doubling of distance from the source.

Finite arrays of stationary, incoherent sources are commonly utilized to model community noise levels from large sources such as highways, industrial plants, or broad distributions of ambient noise sources. The spreading losses for these cases can be conveniently evaluated using linear or planar arrays of such incoherent sources in the form of finite line, circular, or rectangular source arrays[4] or infinite source arrays to model the horizontal[5] and vertical[6] distribution of ambient noise in a community. Prediction models have also been developed, empirically and experimentally, to describe sound propagation in urban, built-up areas from motor vehicles[7] and V/STOL aircraft such as helicopters.[8]

As discussed in Section 5, major deviations from these simple spreading loss models occur outdoors when nonuniformity of the atmosphere is considered. For example, downwind propagation from sources of very low-frequency (e.g., <25 Hz) energy, such as wind turbines,[9] can, under some conditions, exhibit a spreading loss corresponding to a line, instead of a point, source.

3 ATTENUATION OF OUTDOOR SOUND BY ATMOSPHERIC ABSORPTION

A sound wave traveling through air free of any particles is attenuated due to atmospheric absorption caused by (1) classical (heat conduction and shear viscosity) losses and (2) molecular relaxation losses associated with an exchange between molecular translational and molecular rotational or vibration energy. These loss components vary with temperature and atmospheric pressure and, for molecular vibrational relaxation, with humidity content.

The theoretical foundation[2] for atmospheric absorption by Knudsen and Kneser has been refined by recent experimental data to define two key humidity- and temperature-dependent parameters: the vibrational relaxation frequencies for oxygen and nitrogen. This combination of theory and experimentally based algorithms now provides a firm basis for standard expressions to accurately predict the attenuation coefficient for atmospheric absorption of pure-tone sounds.[10]

The attenuation A_a, in decibels, over a path length r, in meters, due only to atmospheric absorption can be expressed by

$$A_a = -20 \log\left[\frac{p(r)}{p(0)}\right] = -20 \log[\exp(-\alpha r)] = ar, \quad (4)$$

where $p(r)$ is the sound pressure after traveling the distance r, $p(0)$ is the initial sound pressure at $r = 0$, α is the attenuation coefficient in Nepers per meter, and a is the attenuation coefficient in decibels per meter (= 8.686α).

While spreading losses are nominally independent of frequency and weather, atmospheric absorption losses are strongly dependent on these parameters. At long propagation distances and for high frequencies, atmospheric absorption is usually much greater than spreading losses for sound propagation outdoors.

The attenuation coefficient for pure tones is shown in Fig. 1 as a function of frequency, with relative humidity as a parameter, for a temperature of 20°C and a pressure of 1 standard atmosphere. Values for other atmospheric pressures can be obtained from this figure by employing the pressure scaling indicated by the abscissa and ordinate scales and the pressure-scaled relative humidity.[11] Values for this pure-tone attenuation coefficient, given in Table 1 for a limited range of frequencies, temperatures, and relative humidities, are readily computed for other conditions (i.e., <330 K and <2 atm) from the standard algorithms.[10]

The effective atmospheric attenuation of a constant-percentage band of a broadband noise is normally less than for a pure-tone sound due to the finite bandwidth and slope of the filter skirts. For a one-third octave band of noise with an exact midband frequency f_m, the atmospheric attenuation $A_{ab}(f_m)$ in decibels is closely approximated by the following expression. It is a theoretically based, empirically refined function of the total pure-tone atmospheric attenuation $A_{at}(f_m)$ at this same midband frequency. The expression is invalid if $A_{at}(f_m)$ is more than about 50 dB[12]:

$$A_{ab}(f_m) = A_{at}(f_m)[1 + 0.00533[1 - 0.2303A_{at}(f_m)]]^{1.6}. \quad (5)$$

Some atmospheric attenuation also occurs in fog,[3] in dust in the air,[13] and at frequencies below about 10 Hz, from absorption due to electromagnetic radiation of moist air molecules.[14]

4 ATTENUATION OF OUTDOOR SOUND PROPAGATION OVER THE GROUND

Sound propagation in a still, uniform atmosphere over a path near the ground surface, such as illustrated in Fig. 2, is changed by the interaction between the direct source-to-receiver sound over path r_1 and sound received from the ground surface. The latter consists primarily of ground-reflected sound over the path r_2. For a spheri-

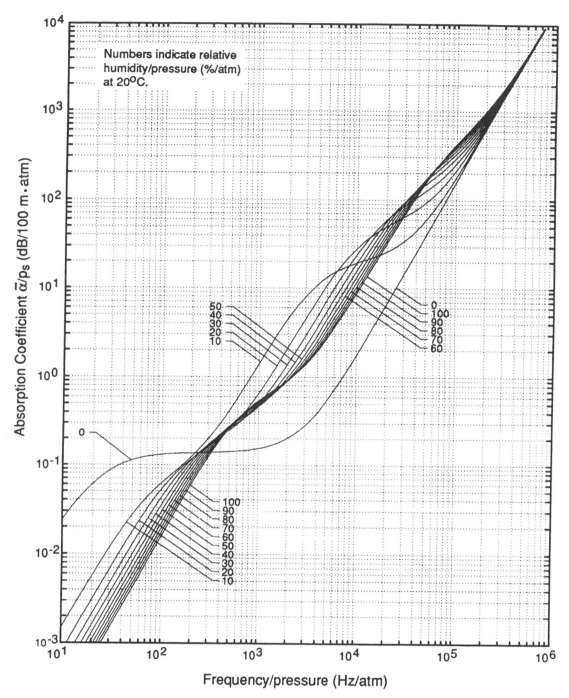

Fig. 1 Attenuation coefficient for atmospheric absorption as a function of frequency and relative humidity. All parameters are scaled by atmospheric pressure so the chart may be used for any pressure within linear limits of the perfect-gas law. (From Ref. 11 with permission.)

cal wavefront, this is augmented by a ground wave and, under some circumstances, by a surface wave. This section treats (1) acoustic impedance of the ground and (2) propagation of spherical sound waves over such ground.

4.1 Acoustic Impedance of Ground Surfaces

The specific acoustic impedance at a ground surface for normal incidence Z_2 is the complex ratio of the acoustic pressure at the surface and the resulting nor-

TABLE 1 **Atmospheric Attenuation Coefficient a (dB/km) at Selected Preferred Frequencies**[a]

Temperature	Relative Humidity (%)	62.5 Hz	125 Hz	250 Hz	500 Hz	1000 Hz	2000 Hz	4000 Hz	8000 Hz
30.0°C	10	0.362	0.958	1.82	3.40	8.67	28.5	96.0	260
(86 °F)	20	0.212	0.725	1.87	3.41	6.00	14.5	47.1	165
	30	0.147	0.543	1.68	3.67	6.15	11.8	32.7	113
	50	0.091	0.351	1.25	3.57	7.03	11.7	24.5	73.1
	70	0.065	0.256	0.963	3.14	7.41	12.7	23.1	59.3
	90	0.051	0.202	0.775	2.71	7.32	13.8	23.5	53.3
20.0°C	10	0.370	0.775	1.58	4.25	14.1	45.3	109	175
(68 °F)	20	0.260	0.712	1.39	2.60	6.53	21.5	74.1	215
	30	0.192	0.615	1.42	2.52	5.01	14.1	48.5	166
	50	0.123	0.445	1.32	2.73	4.66	9.86	29.4	104
	70	0.090	0.339	1.13	2.80	4.98	9.02	22.9	76.6
	90	0.071	0.272	0.966	2.71	5.30	9.06	20.2	62.6
10.0°C	10	0.342	0.788	2.29	7.52	21.6	42.3	57.3	69.4
(50 °F)	20	0.271	0.579	1.20	3.27	11.0	36.2	91.5	154
	30	0.225	0.551	1.05	2.28	6.77	23.5	76.6	187
	50	0.160	0.486	1.05	1.90	4.26	13.2	46.7	155
	70	0.122	0.411	1.04	1.93	3.66	9.66	32.8	117
	90	0.097	0.348	0.996	2.00	3.54	8.14	25.7	92.4
0.0°C	10	0.424	1.30	4.00	9.25	14.0	16.6	19.0	26.4
(32 °F)	20	0.256	0.614	1.85	6.16	17.7	34.6	47.0	58.1
	30	0.219	0.469	1.17	3.73	12.7	36.0	69.0	95.2
	50	0.181	0.411	0.821	2.08	6.83	23.8	71.0	147
	70	0.151	0.390	0.763	1.61	4.64	16.1	55.5	153
	90	0.127	0.367	0.760	1.45	3.66	12.1	43.2	138

[a] Values actually computed at the exact midband frequency $f_m = 1000 \times 10^{0.3n}$, where the integer n varies from -4 to $+3$.

mal component of particle velocity into the ground. For a semi-infinite media, this specific acoustic impedance for normal incidence is the same as the characteristic impedance Z_c throughout the medium and is expressed, here, in terms of its normalized value $Z_2/\rho_0 c_0$, where $\rho_0 c_0$ is the characteristic impedance of air. In this chapter, this normalized value will be referred to as simply the surface impedance Z_s.

The characteristic impedance for a medium with a resistance to steady flow is a convenient first approximation to the surface impedance of a semi-infinite rigid-frame porous medium, a simple ground surface model. For such a medium with a density and speed of sound in air of ρ_0 and c_0 and flow resistivity σ in the medium at a frequency f, the characteristic impedance Z_c and acoustic propagation constant $k_b = \beta + i\alpha$ are given by[15]

$$Z_c = \rho_0 c_0 \left[1 + i \left(\frac{\sigma}{2\pi f \rho_0} \right) \right]^{1/2} \tag{6a}$$

and

$$k_b = \frac{2\pi f}{c} \left[1 + i \left(\frac{\sigma}{2\pi f \rho_0} \right) \right]^{1/2}. \tag{6b}$$

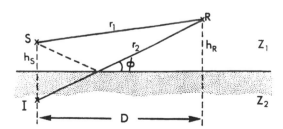

Fig. 2 Geometry of sound propagation over the ground illustrating the source (S), image source (I), and the direct (r_1) and reflected (r_2) path lengths to the receiver (R).

Transforming the complex roots, the characteristic impedance, normalized by the characteristic impedance of air, is

$$\frac{Z_c}{\rho_0 c_0} = [\tfrac{1}{2}(A+1)]^{1/2} + i[\tfrac{1}{2}(A-1)]^{1/2}, \qquad (7)$$

where $A = [1 + (2\pi f \rho_0/\sigma)^{-2}]^{1/2}$.

For time dependence in the form $\exp(-i2\pi f t)$, the reactive part of the impedance is positive, corresponding to a stiffness or spring reactance.[16] This "acoustic spring" represents the compressible air in ground surface pores.

A dimensionless frequency $f\rho_0/\sigma$ appears as the key scaling parameter in this approximation for the characteristic impedance and propagation constant for porous ground. While the general form of Eq. (7) is found in more complex models for propagation through porous media,[15] other parameters, as indicated later, besides flow resistivity are required to more accurately define surface impedance for many ground surfaces. However, flow resistivity retains its dominant influence on the frequency variation of impedance.

In a benchmark study, Delany and Bazley[17] found that the measured characteristic impedance Z_c and acoustic propagation constant k_b of a wide range of absorbent, porous materials could be defined in terms of the same dimensionless frequency $f\rho_0/\sigma$ found in Eq. (7). Using a standard value for ρ_0 of 1.205 kg/m³, they developed the following expressions to define the measured characteristic impedance and propagation constant of these materials in terms of the ratio of frequency f, in hertz, to flow resistivity σ, in Pa $\cdot$ s/m² (MKS Rayls/m):

$$\frac{Z_c}{\rho_0 c_0} = \left[1 + 0.0511\left(\frac{f}{\sigma}\right)^{-0.75}\right]$$
$$+ i\left[0.0768\left(\frac{f}{\sigma}\right)^{-0.73}\right], \qquad (8a)$$

$$k_b = \frac{2\pi f}{c}\left\{\left[1 + 0.0858\left(\frac{f}{\sigma}\right)^{-0.70}\right]\right.$$
$$\left. + i\left[0.175\left(\frac{f}{\sigma}\right)^{-0.59}\right]\right\}. \qquad (8b)$$

Although their experimental data fit these empirical expressions well for f/σ from 0.01 to 1.0 m³/kg, Delany and Bazley cautioned that outside this range other power law relationships may be required. Nevertheless, Delany and Bazley as well as Chessell[18] subsequently found that Eq. (8a) provides a reasonable first approximation to surface impedance in their separate evaluations of ground effects on propagation of aircraft noise over grassy ter-

rains. For such terrains, f/σ varies from approximately 10^{-4} to 0.1 m³/kg.

Delany and Bazley assumed that the flow resistivity σ of their materials was actually an effective value σ_e equal to the "DC" value σ multiplied by the porosity Ω. The latter is typically about 0.4–0.7 for grass-covered ground surfaces.[19]

An alternative model by Miki[20] essentially eliminates one problem with the Delany–Bazley model, which predicts, at low frequencies, the physically unrealizable condition of a negative-resistance component for a thin, hard-backed layer for most ground surfaces.[21]

Subject to these limitations, the surface impedance for a semi-infinite ground can be estimated with Eq. (8a) given suitable values for the effective flow resistivity. Table 2 lists published values from a number of sources (e.g., Refs. 22–25) for empirically determined effective flow resistivity and measured or computed values of porosity for a full range of ground surfaces. Effective flow resistivities that best fit ground attenuation data are less than directly measured values for grassy surfaces, but the opposite is true for relatively homogeneous ground surfaces such as sand, silty soil, and snow. This supports the need for more complex models for surface impedance for the general case.[26]

Two examples of these more accurate models for the surface impedance of the ground are provided in the following.

Four-Parameter Model This general model has been employed in a benchmark paper presenting a comparison of sound propagation computation programs (see Ref. 51). With this model, the characteristic impedance Z_c (and hence the surface impedance for a homogeneous, locally reacting media) and acoustic propagation constant k_b can be expressed as

$$\frac{Z_c}{\rho_0 c_0} = \frac{q/\Omega}{B^{1/2}C^{1/2}} \quad \text{and} \quad k_b = \frac{2\pi f q}{c_0}\frac{C^{1/2}}{B^{1/2}}, \qquad (9)$$

where

$$B = \left[1 - \frac{1}{\epsilon\sqrt{i}}T(2\epsilon\sqrt{i})\right],$$

$$C = \left[1 + \frac{\gamma-1}{\epsilon\sqrt{iN_{\text{pr}}}}T(2\epsilon\sqrt{iN_{\text{pr}}})\right]$$

and Ω is porosity, the ratio of air to total volume; $q^2 = \Omega^{-n'}$ a tortuosity parameter that accounts for the total

TABLE 2 Flow Resistivity and Porosity Data for Ground Surfaces[22–25]

| Types of Ground | Effective Flow Resistivity,[a] kPa · s/m² | | Porosity |
	Range	Average	
Upper limit[b]	2.5×10^5–25×10^5	800,000	
Concrete, painted	200,000	200,000	
Concrete, depends on finish	30,000–100,000	65,000	
Asphalt, old, sealed with dust	25,000–30,000	27,000	
Quarry dust, hard packed	5,000–20,000	12,500	
Asphalt, new, varies with particle size	5,000–15,000	10,000	
Dirt, exposed, rain-packed	4,000–8,000	6,000	
Dirt, old road, filled mesh	2,000–4,000	3,000	
Limestone chips, $\frac{1}{2}$–1-in. mesh	1,500–4,000	2,750	
Dirt, roadside with <4 in. rocks	300–800	550	
Sand, various types	40–906	317	0.35–0.47
Soil, various types	106–450	200	0.36–0.55
Grass lawn or grass field	125–300	200	0.48–0.55
Clay, dry (wheeled/unwheeled)	92–168	130	0.47–0.55
Grass field, 16.5% moisture content	75	75	
Forest floor (Pine/Hemlock)	20–80	50	
Grass field, 11.9% moisture content	41	41	
Snow, various types	1.3–50	29	0.56–0.76

[a]Effective resistivity value inferred from fitting measured ground attenuation data to predicted values based on Delany–Bazley impedance model and local reaction ground attenuation model (Ref. 22).
[b]Upper limit of effective flow resistivity due to thermal and viscous boundary layer, = $(1.16 \times 10^5)f^{1/3}$ (kPa · s/m²) (an equivalent value, according to resistance portion of the Delany–Bazley impedance model; Ref. 22).

deviation of pore axes from a normal to the surface; $\epsilon = \sqrt{\pi/\Omega}\,(q/s_p)[f\rho_0/\sigma]^{1/2}$; $T(x) = J_1(x)/J_0(x)$ the ratio of Bessel functions of the first kind and order 1 and 0, respectively; n' a grain shape factor, typically taken to be 0.5; s_p a pore shape factor, typically 0.25 (these latter two parameters account for friction at the pore walls); N_{pr} the Prandtl number for air (0.71 at 15°C); and $\gamma = 1.4$ the ratio of specific heats for air.

Various approximations to Eq. (9) exist[26] for a restricted range of the same frequency parameter f/σ found in ϵ above and in Eq. (8).

Thin-Layer Model For a thin homogeneous layer (e.g., grassy sod) of thickness d over a rigid semi-infinite backing, the surface impedance Z_s is proportional to the characteristic impedance Z_c of the layer medium and is given by[15]

$$Z_s = Z_c \coth(-ik_b \cdot d) \qquad (10)$$

where k_b is the acoustic propagation constant for the layer media.

Values for Z_c and k_b can be taken from Eq. (8) or (9).

In addition to applying prediction models, surface impedances for ground surfaces have also been obtained by (1) measurements with impedance tubes directed into the ground;[2] (2) calculation from standing-wave ratios,[2] Fourier spectra,[27] or phase gradients of interference patterns of ground-reflected sound;[25] or (3) trial-and-error calculation adjusting impedance parameters until predicted short-range ground attenuation matches measurements.[22–24] This latter semiempirical approach is considered a very practical way to determine surface impedance.

Published data on surface impedance for grassy surfaces obtained by the direct measurement methods are compared in Fig. 3 with predicted values for the characteristic impedance based on the Delany and Bazley[17] model. Predicted impedance values using Eq. (8a) for effective flow resistivities of 40 and 400 kPa · s/m² provide an approximate upper and lower bound for these measured surface impedance data from the different site–investigator data sets[27–31], identified in the figure.

4.2 Boundary Conditions at the Ground

Either local or extended reaction[15] conditions are normally assumed at a ground surface boundary. For a local reaction boundary, (1) the speed of sound in the ground

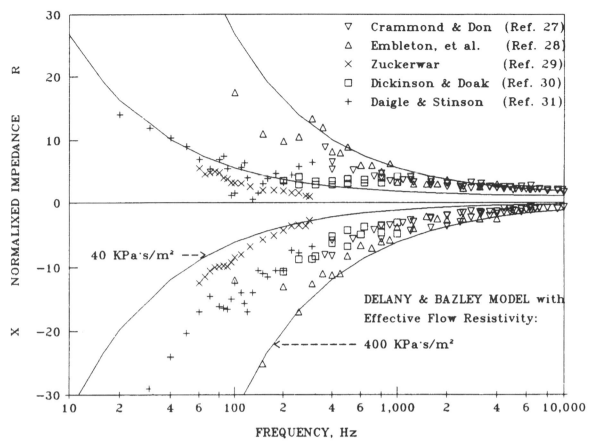

Fig. 3 Normal impedance measurements for grass surfaces compared to values predicted by the Delany–Bazley model. (Data from Refs. 27–31.)

surface is much less than the speed of sound in air, (2) the surface impedance is essentially independent of the incidence angle of any impinging sound wave, and (3) the dilatational sound waves transmitted into the ground travel normal to the surface. For an extended reaction boundary, the opposie is true and the refraction angle of the sound wave transmitted into the ground depends upon the incidence angle of the incident sound wave. While a locally reacting boundary model is usually suitable for predicting propagation over ground,[18] an extended reaction model may be more realistic when the acoustic surface properties vary significantly with depth or correspond to a medium with a relatively low flow resistance[32] or the "ground" is a body of water that has a much higher sound speed than air.

A third more complex impedance model treats the ground as an elastic medium capable of transmitting both acoustic dilatational and seismic shear waves from an acoustic wave impinging on the surface.[33] However, this complexity is normally not required for evaluation of sound propagation over the ground since the seismic impedance is 20–60 times greater than the acoustic impedance of the ground.

4.3 Attenuation of Spherical Acoustic Waves over Ground

According to Fig. 2, the sound pressure at the observer height h_r above the ground from a point source at a height h_s above the ground will be the sum of the direct sound pressure $p(r_1)[= P_d \exp(-kr_1)/r_1]$ from source to receiver over path r_1 and the indirect or ground-reflected sound pressure $p(r_2) = P_r \exp(-kr_2)/r_2$ nominally over the path r_2. A time variation $\exp(-i2\pi f t)$ is understood for both, and k is the wavenumber $2\pi f/c$. The indirect sound can be considered as originating from an image source located below the ground the same distance h_s as the actual source is above the ground. The ground attenuation A_g in decibels at the observer due solely to the presence of the ground surface can be expressed by[18]

$$A_g = 10 \log \left[\frac{|P_d \exp(-kr_1)/r_1 + P_r \exp(-kr_2)/r_2|^2}{|P_d \exp(-kr_1)/r_1|^2} \right]$$

$$= 10 \log \left[1 + \frac{r_1^2}{r_2^2} |Q_s|^2 + 2 \frac{r_1}{r_2} |Q_s| C_c \right], \qquad (11)$$

where $Q_s = \overline{[p(r_2)/p(r_1)]} = |Q_s|e^{i\theta}$ is the spherical wave reflection factor with a magnitude $|Q_s|$ and phase angle θ, $C_c = \overline{p(r_2)p(r_1)}/[\overline{p^2(r_2)p^2(r_1)}]^{1/2}$ is the correlation coefficient between the direct and ground "reflection," and the overbar denotes a time average.

For a wide-band random noise source with an approximately constant sound pressure spectrum level over a one-third octave frequency band with a bandwidth δf and midband frequency f_m, the correlation coefficient C_c is given by[18]

$$C_c = \frac{\sin[\mu\,\delta r\,f_m/c_0]}{[\mu\,\delta r\,f_m/c_0]}\cos\left[\tau\delta r\,\frac{f_m}{c_0}+\theta\right], \quad (12)$$

where δr is the path length difference $(r_2 - r_1)$ between the direct and reflected wavefronts moving with the sound speed c_0. The phase shift in the latter, $[\tau\delta r(f_m/c_0) + \theta]$ is due to the path length difference δr and the phase angle θ of the spherical wave reflection factor. The parameter $\mu = \pi(\delta f)/f_m$ is 0.725 for a one-third octave band filter and 0 for a pure-tone source and $\tau = 2\pi[1 + (\delta f/2f_m)^2]^{1/2}$ is approximately 2π for a one-third octave band source and 2π for a pure tone.

Atmospheric turbulence can add an additional fluctuating random phase shift term to the argument of the cosine term in Eq. (12). For this condition, with large differences in path lengths, C_c can approach zero.

For a very hard ground, if the path difference δr is much less than the wavelength c_0/f, C_c approaches 1.0 and the excess ground attenuation A_g approaches -6 dB, equivalent to pressure doubling at a hard surface. If δr is much greater than a wavelength, C_c approaches 0 so the excess ground attenuation approaches -3 dB, equivalent to a simple energy summation of the two uncorrelated direct and reflected signals. As the incidence angle ϕ shown in Fig. 2 approaches 90°, the total phase shift between the direct and reflected signals [the argument of the cosine term in Eq. (12)] becomes dominated by the path length difference and is nearly independent of the surface impedance.

A "ground dip," or minimum, in excess ground attenuation occurs when the total phase shift between the direct and reflected signals (i.e., the argument of the cosine term in C_c) is equal to an odd multiple of π. As the incidence angle ϕ becomes very small, the path length difference δ_r also becomes small and the total phase shift is due almost entirely to the phase angle θ in the spherical wave reflection factor. The latter is dominated by the surface impedance, which therefore controls the frequency of the ground dip. This behavior is put to practical use to measure, indirectly, the surface impedance of the ground, given a model for the total ground attenuation pattern as

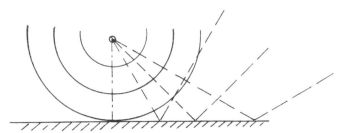

Fig. 4 Sketch illustrating varying incidence angle of spherical wavefront impinging on a plane surface.

a function of surface impedance parameters for a given source–receiver geometry.[22,34]

As shown in Fig. 4 by the straight sound ray paths from a point source in a uniform atmosphere, there is not a single incidence angle for all parts of the spherical wavefront striking the ground plane. The Weyl–Van der Pol solution provides a widely used approach to this classical problem of matching the boundary conditions for a spherical wavefront impinging on a plane, finite-impedance surface.[2,35] In this case, the spherical reflection factor Q_s is expressed as the sum of a plane-wave reflection factor R_p plus a boundary correction term $(1 - R_p)F(w)$. The added component provides the necessary correction to the reflected plane-wave component to match the boundary conditions of the spherical wavefront to the ground plane surface.[2] This component, called a ground wave, is governed by the same inverse-square spreading loss as the reflected wave but, its magnitude is proportional to the unreflected fraction $(1 - R_p)$ and, at large distances, has an additional inverse-square law loss of its own. Thus, the spherical reflection factor Q_s is given as

$$Q_s = R_p + (1 - R_p)F(w). \quad (13)$$

The first application of this approach by Rudnick built on previous studies of the analogous electromagnetic field over a conducting surface.[2] The many subsequent solutions to the problem differ, for example, in how the spherical wavefront is synthesized, in the use of an extended or local reaction impedance model for the ground, or in the details of the complex contour integration involved.[2,35,36]

The plane-wave reflection coefficient term R_p is given by

$$R_p = \frac{Z_s\sin\phi - 1}{Z_s\sin\phi + 1}, \quad (14)$$

where Z_s is the surface impedance defined earlier and ϕ is the incidence angle shown in Fig. 2.

The quantity $F(w)$, called the boundary loss factor by Rudnick,[2] is actually the first term of a more accurate asymptotic series solution[35] and is defined in terms of a complex complimentary error function erfc($-iZ$) by

$$F(w) = 1 + i[\pi w]^{1/2} e^{-w} \,\text{erfc}(-iw^{1/2}), \qquad (15)$$

where w is a numerical distance given, for a locally reacting medium, by[18]

$$w = i \, \frac{kr_2}{2} \, \frac{[\sin(\phi) + 1/Z_s]^2}{1 + (1/Z_s)\sin(\phi)}. \qquad (16)$$

The denominator, in this expression $1 + (1/Z_s)\sin\phi$, is usually assumed equal to 1. This numerical distance, often expressed as $w^{1/2}$, can be considered a nondimensional distance r_2 in wavelengths modified by a function of the incidence angle ϕ and surface impedance Z_s. [Note that the positive square root of w is used in the argument $-iw^{1/2}$ for erfc($-iw^{1/2}$)]. Algorithms defined by Chien and Soroka[37] for computation of $F(w)$ are available in a more convenient form.[38] For an extended reaction surface, $F(w)$ must be multiplied by an additional correction term.[39]

The ground wave embodied in the boundary loss factor $F(w)$ is augmented by a surface wave that exists, as illustrated in Fig. 5, only between the flat ground and a conical surface (with an apex at the "image source" position) whose vertical projection defines the maximum value of the incidence angle for which a surface wave can occur. This angle is defined in terms of the following function of the surface impedance $|Z_s|e^{i\beta}$:

$$\sin(\phi) < \frac{[\text{Im}(Z_s) - \text{Re}(Z_s)]}{|Z_s|^2}. \qquad (17)$$

For a grazing incidence wave with $\phi = 0$, this general condition requires only that the imaginary part, Im(Z_s),

of the surface impedance exceed the real part, Re(Z_s). This criterion also applies for the existence of surface waves for propagation of near-grazing incidence plane waves traveling over an impedance ground.[40]

Over a range of distances (10–1000 m), frequencies (10–10,000 Hz), and flow resistivities (σ_e = 25–25,000 kPa $\cdot$ s/m^2) of practical interest, the magnitude of the numerical distance w will vary from less than 0.01 to greater than 100. Figure 6 shows the marked change in $20 \log |F(w)|$ over this range of $|w|$ with the surface impedance phase β as a parameter. For this plot, the values of $|w|$ were computed for the source and receiver on the surface of a local reaction ground.

The increase in the boundary loss factor $F(w)$ for $\beta > 45°$ is due to the addition of the surface wave component. For realistic values of β ($< 90°$), as $|w|$ becomes larger than about 10, this added surface wave decays exponentially along the ground to a value below the residual ground wave and the latter decreases by an additional 6 dB per doubling of $|w|$.

For very small values of $|w|$, $F(w)$ approaches 1.00, and the spherical reflection coefficient Q_s also approaches 1.0, so that the total sound pressure at a receiver is simply the direct signal plus a perfectly reflected signal modified in phase only by the path length difference δr and in magnitude by the different spreading loss (r_1/r_2). For nearly horizontal propagation, these quantities approach 0 and 1.0, respectively, resulting in the basic 6 dB pressure doubling as for a rigid plane. For $|w| \approx 1$, the ground wave and, when present, the surface wave component, will make a substantial difference in the total ground attenuation.

Surface wave propagation has been clearly demonstrated in laboratory studies and by a unique outdoor experiment with impulse sounds propagated over snow.[41] The important physical characteristics of surface waves observed in this study, confirming theory, are shown in Fig. 7 by the broad, delayed pulse following the initial impulse sound made up of the direct, reflected, and ground wave. This shows that the phase velocity of the

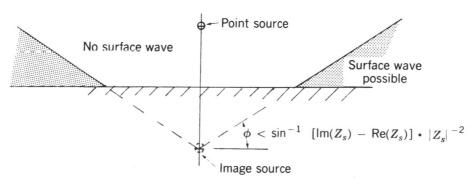

Fig. 5 Region where surface wave can exist. (Based on Ref. 40 with permission.)

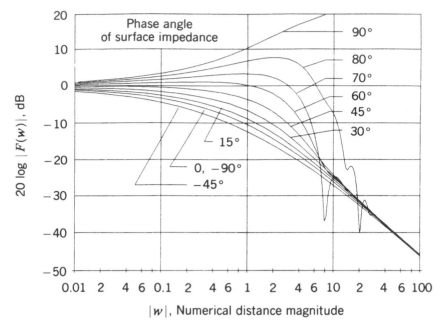

Fig. 6 Absolute magnitude of boundary loss factor $F(w)$ as a function of frequency with phase angle β of surface impedance as parameter. Source and receiver on surface of normally reacting ground.

surface wave is less than the speed of sound in air and the amplitude of the surface wave decreases exponentially[40] vertically and horizontally above the surface.

The separate wave components for propagation over the ground are illustrated by the classic long-range sound

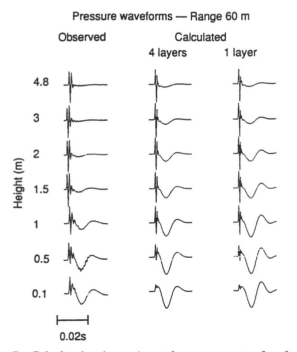

Fig. 7 Calculated and experimental measurements of surface waves over snow. (From Ref. 41.)

propagation data obtained by Parkin and Scholes.[42] Figure 8, from a previous review[2] of these data, breaks down the predicted contributions to the total ground attenuation by the direct (D) wave, the reflected (R) component, the ground (G) wave, and the surface (S) wave for three different distances representative of the range covered by the Parkin and Scholes data.

For a distance of 31.2 m from the source, there is no contribution by a surface (S) wave, and for frequencies greater than about 1 kHz, the minima are for path length differences that are approximately even multiples of one-half wavelength, indicating that the surface impedance is relatively low (e.g., soft surface) for this configuration and frequency range.

For a range of 125 m, there is a broad minimum (i.e., ground dip), just as for the shorter range, centered at about 500 Hz, which is also characteristic of the propagation for short distances over soft ground. It is the result of cancellation between direct and reflected waves caused primarily by the phase change on reflection.

The curves in Fig. 8 indicate that the ground dip broadens and deepens with increasing distance until most of the audible frequency range is included in this region of destructive interference (i.e., an acoustic "shadow"). The measurements are in reasonable agreement with theoretical predictions for 125 m using impedances obtained from a similar "soft" site.[2]

At 500 m, a further broadening of this ground dip or shadow zone extends to both higher and lower frequencies, which leaves most of the energy in the frequency

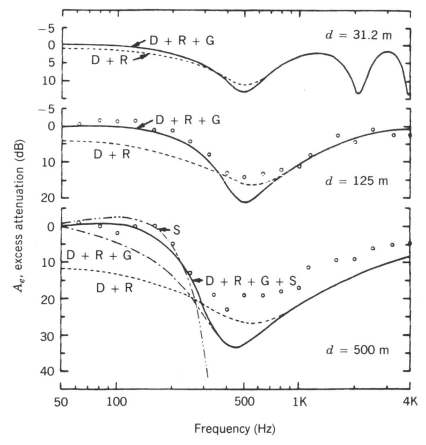

Fig. 8 Ground attenuation measured over several distances compared to theoretical prediction showing contributions by direct (D), plane-wave reflected (R), ground wave (G), and surface wave (S) components. (From Ref. 2 with permission.)

range 50–200 Hz in the surface wave. At high frequencies, the measured excess attenuation is consistently less than predicted due to turbulence effects discussed in Section 7.

4.4 Practical Examples of Ground Attenuation

To illustrate ground attenuation for a range of practical cases, values have been evaluated with Eqs. (8) and (10), assuming a homogeneous, locally reacting ground surface, for the following representative source heights h_s a receiver height h_r of 1.2 m and typical source–receiver distances encountered in the analysis of environmental sound:

Source	Source Height h_s, m	Receiver Height h_r, m
Tire noise	0.01	1.2
Passenger car engine exhaust noise	0.3	1.2
Truck exhaust noise (cab over engine) or aircraft engine exhaust (typical)	3.0	1.2

The resulting values for the ground attenuation are shown in Figs. 9a,b for distances from 15 to 1500 m for the first two of these three source heights for a typical grass surface with a flow resistivity of 250 kPa · s/m², approximately the same as for Fig. 8. The overall trend in Fig. 9 is similar to that of Fig. 8 but covers a wider range of source–receiver geometry. The ground attenuation for the higher source height of 3 m exhibited a pattern similar to that for the 0.3 m height but with less attenuation for the ground dip and a more complex pattern of constructive and destructive interference for the shortest distance of 15 m.

Significant departures from the curves in Fig. 9 can occur at high frequencies and long distances due to turbulence (see Section 7) when the excess attenuation due to interference between direct and reflected waves would otherwise be greater than about 20–25 dB, an approximate upper bound for total excess attenuation outdoors in a real atmosphere.

A limited examination of the effect of changing flow resistance is presented in Figs. 10a,b for a source height of 0.3 m and a range of 150 m. In Fig. 10a the ground is assumed to be homogeneous while in Fig. 10b a

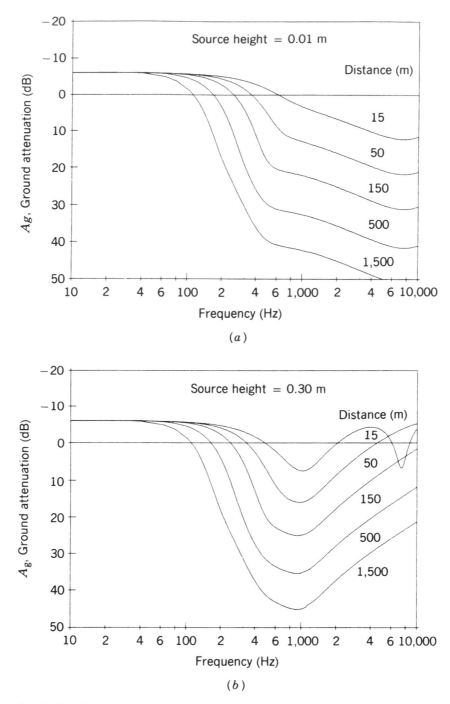

Fig. 9 Predicted ground attenuation for average grass surface using Delany–Bazley impedance model for receiver height h_r of 1.2 m, distances from 15 to 1500 m for source heights h_s of (*a*) 0.01 m (i.e., tire noise) and (*b*) 0.3 m (e.g., passenger car exhaust noise).

0.05-m-thick layer over a hard backing is assumed. For both parts, the range of values for σ_e (25–25,000 kPa $\cdot$ s/m^2) covers ground surfaces from soft snow to old, sealed asphalt. Figure 10 shows that the ground attenuation changes markedly as this key impedance parameter changes, showing, as expected, a maximum ground dip

for snow and little ground attenuation for the hardest surface.

Figure 10*b* shows a significant effect of surface waves at low frequencies for the lowest flow resistivities due to the greater reactive component of surface impedance for this hard-backed layer model for the ground.

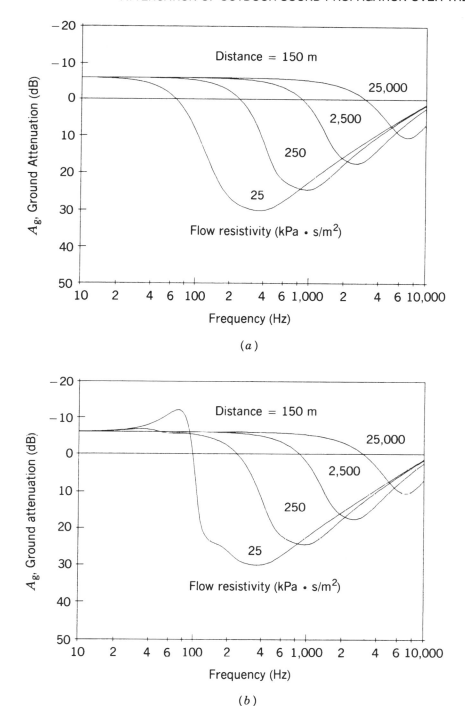

Fig. 10 Predicted ground attenuation for different surfaces using Delany–Bazley impedance model with effective resistivity σ_e of 25–25,000 kPa · s/m^2. Receiver height h_r = 1.2 m, distance 150 m and, source height h_s = 0.3 m. (*a*) Ground is semi-infinite; (*b*) 0.05 m thick layer over a hard backing.

When the surface impedance changes abruptly, such as at the edge of a roadway, analytical methods[43] are available as well as an approximate approach involving weighting the excess attenuation values according to the distance traveled over each surface.[25]

4.5 Attenuation through Foliage and Trees

For sound attenuation through foliage and trees, the main effect at low frequencies is to enhance ground attenuation, the roots making the ground more porous.[44] At high frequencies, where the dimensions of leaves become comparable with the wavelength, there is also a significant attenuation caused by scattering.[45]

In a forest, the vertical gradients of wind and temperature are reduced at elevations up to approximately the height of the trees, thus reducing the effects of refraction from such gradients as considered in Section 5.

4.6 Attenuation through Built-up Areas

As within forests, the effect of refraction on the propagation of sound in city streets is small, and for the same reason, the obstruction to the flow of air causes the viscous and thermal boundary layer to approach the height of buildings. The ground attenuation effect in the city is also much smaller than in the country, partly because of paving, but also because the buildings produce an interference pattern much different from that encountered over flat terrain. However, prediction of the propagation of noise from airports, freeways, and so on, out into the suburbs is based largely on empirical methods not well found on analytical models or extensive experimental data.[46]

5 REFRACTION IN OUTDOOR SOUND PROPAGATION

The preceding sections have discussed propagation in a still uniform atmosphere where sound follows straight ray paths. Although this is a reasonable assumption at shorter ranges, the simple description of the direct and ground-reflected path shown in Fig. 2 is no longer valid at longer ranges. Under most weather conditions both the temperature and the wind velocity vary with height above the ground. The speed of sound relative to the ground is a function of temperature and wind velocity, and hence it also varies with height, causing the sound waves to propagate along curved paths.[47]

During the day solar radiation heats the earth's surface, resulting in warmer air near the ground. This condition, called a temperature lapse, is most pronounced on sunny days but can also exist under overcast skies. A temperature lapse is the common daytime condition during most of the year and ray paths curve upward. After sunset there is often radiation cooling of the ground, which produces cooler air near the surface and forms a temperature inversion. Within the temperature inversion, the temperature increases with height and ray paths curve downward.

When there is wind, its speed decreases with decreasing height because of drag on the moving air at the ground. Therefore, the speed of sound relative to the ground increases with height during downwind propa-

gation, and ray paths curve downward. For propagation upwind the sound speed decreases with height, and ray paths curve upward. There is no refraction in the vertical direction produced by wind when the sound propagates directly crosswind.

5.1 Downward Refraction

Downward-curving sound rays (see Chapters 1 and 3) are shown in Fig. 11. The effect of the downward-curving rays is to increase the grazing angle ϕ in Fig. 2 and hence modify the ground attenuation shown in Figs. 8 and 9. Further, under specific conditions that depend on source and receiver heights, horizontal range, and the strength of the refraction, additional ray paths are possible that involve one or more reflections at the ground. These additional rays can nullify the ground attenuation.

It is possible to modify[48] the Weyl–Van der Pol solution to account for refraction. However, the general solution[47] for the pressure $p(r)$ in Eq. (1) at a receiver a distance r from the source is expressed as a Hankel transform:

$$p(r) = \int_{\infty}^{\infty} H_0^1(Kr)P(K,z)K\ dK, \qquad (18)$$

where H_0^1 is the Henkel function of the first kind and of order 0. K is the horizontal wavenumber, and z is the vertical distance above the ground. In general, it is not possible to obtain closed-form solutions to Eq. (18), and numerical algorithms are required to compute Eq. (18). These will be discussed later, in Section 5.3. However, in the case of simple sound speed profiles, a residue series solution[49] to Eq. (18) can be obtained. The form of profile that is most convenient for physical interpretation and mathematical computation is one where the

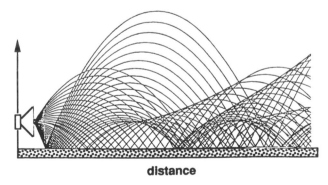

distance

Fig. 11 Sound rays in the atmosphere during downward refraction. At the longer distance, additional ray paths are seen that involve more than one reflection at the ground. There are also regions of focusing called caustics.

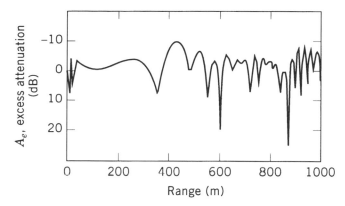

Fig. 12 Excess attenuation predicted for a frequency of 500 Hz for propagation during downward refraction. The ground impedance is characterized using the Delany–Bazley model [Eq. (8a)] with an effective flow resistivity σ_e of 200 kPa · s/m². The sound speed profile is given by $c(z) = 340(1 + 5.88 \times 10^{-4}Z)$ (m/s). Source height 2 m; receiver height 5 m.

sound velocity increases linearly with height z above the ground:

$$c(z) = c_0(1 + az). \qquad (19)$$

In Eq. (19), a is the coefficient of increase in velocity with height. The sound rays between source and receiver are circular arcs. We note that a linear variation with height is a good approximation for many cases, although it is not necessarily achieved in practice.

An example of the excess attenuation A_e, in decibels, obtained during downward refraction at 500 Hz is shown by the curve in Fig. 12. The excess attenuation

is obtained using one of the numerical algorithms (see Section 5.3) to compute Eq. (18) and then using Eqs. (1)–(4). On average, the excess attenuation is constant with increasing distance. In simple terms, the attenuation due to the ground is nullified by the energy provided by the additional rays shown in Fig. 11. The sound pressure levels therefore increase to the levels predicted by geometric spreading and atmospheric absorption alone, but in general not above such levels. Increases above such levels are due to focusing of the various rays and are inevitably accompanied by decreases caused by defocusing elsewhere in the sound field, which leads to the various dips observed in Fig. 12.

5.2 Upward Refraction

When the sound speed decreases with height, the sound rays are bent upward, away from the ground. For realistic sound speed profiles, there is a shadow boundary formed by sound rays from various regions of the ensonified sound field and beyond which there is an acoustic shadow region. If the relation between sound speed and height is linear, as in Eq. (19), with a being negative, the rays are arcs of circle and the shadow boundary is a single limiting ray that just grazes the ground. A limiting ray with radius of curvature R is shown in Fig. 13 for a source located at a height h_s. Beyond the limiting ray no direct sound energy can penetrate causing the acoustic shadow region.

Equation (18) also applies in the case of upward refraction as the general solution for the pressure, and the numerical algorithms discussed later can be used for computation. There is also a residue series solution[50] to

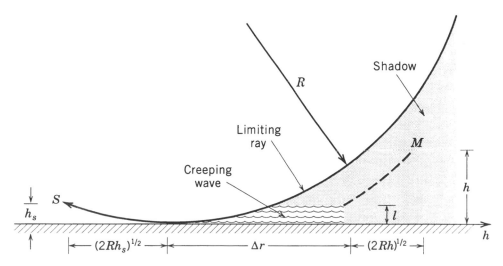

Fig. 13 Sketch illustrating the limiting ray during upward refraction in the case where the sound speed profile is a linear function of height. The limiting ray is an arc of a circle with radius of curvature R.

Eq. (18) in the case of a linear sound speed profile. The residue series provides a physical interpretation of the diffraction within the shadow region. When the receiver is deep within the shadow, but not close to the ground, the series can be approximated by the first term only and, further, the various functions in the solution can be approximated by their asymptotic behavior. The result is

$$
p(r) = \left(\frac{1}{2k_0 r l^2} \right)^{1/2} \frac{1}{K} \left(\frac{l}{h_s} \right)^{1/4} \left(\frac{l}{z} \right)^{1/4} \exp(-\alpha \, \Delta r)
$$
$$
\cdot \exp\left(i\omega \left[\tau(h_s) + \frac{\Delta r}{v} + \tau(z) \right] \right), \qquad (20)
$$

where $k_0 = \omega/c(0)$, $l = (R/2k_0^2)^{1/3}$, and τ corresponds to travel times. The function K essentially incorporates the impedance of the ground and the function α describes the attenuation of the wave along the portion of the path Δr with reduced phase velocity $v < c$. Because of the attenuation and reduced phase velocity, the term *creeping wave* is often used to describe the part of the sound field along the path Δr. The quantity l can then be identified as the thickness of the creeping wave.

Equation (20) suggests (see Fig. 13) that the energy received at M initially leaves the source and travels along the limiting ray to the ground with a travel time $\tau(h_s)$ along a distance $(2Rh_s)^{1/2}$. Then it propagates via the creeping wave in the air along the surface a distance Δr with reduced phase velocity v and weak attenuation. At the appropriate distance, the energy is then shed from the creeping wave and travels to M at a height $z = h$ along the ordinary geometric acoustic rays with travel time $\tau(h)$ along a distance $(2Rh)^{1/2}$.

Unfortunately, the simple picture provided by Eq. (20) has limited success in predicting the excess attenuation during propagation in an upward-refracting atmosphere. Beyond a few hundred meters Eq. (20) predicts large attenuation that is not supported by experimental data.[2,50] Atmospheric turbulence contributes significantly to the excess attenuation, and numerical methods that incorporate the effects of turbulence must be used to calculate the acoustic pressure (see Section 7).

5.3 Numerical Methods

Numerical techniques originally developed for applications in underwater acoustics have been adapted in recent years for atmospheric predictions in the presence of refraction. The numerical techniques can be broadly classified as variations of the fast-field program (FFP) or forms of the parabolic equation (PE). A review of the various FFP and PE techniques in common use, includ-

ing a comprehensive list of references, can be found in Ref. 51.

Fast-Field Program Models Fast-field programs permit the prediction of sound pressure in a refracting atmosphere at an arbitrary receiver on or above a flat continuous ground from a point source somewhere above the ground. The sound speed can be specified from Eq. (19) or as an arbitrary function of height above the ground. The basis of the FFP method is to work numerically from exact integral representations of the sound field within the layered atmosphere in terms of coefficients that may be determined from the ground impedance. The method gets its name from the discrete Fourier transform used to evaluate these integrals.

The starting point is the Hankel transform in Eq. (18), from which it is straightforward to show that $P(K, z)$ satifies

$$
\frac{d^2 P}{dz^2} + [k^2(z) - K^2]P = -2\delta(z - h_s). \qquad (21)
$$

The solution for $P(K, z)$ in Eq. (21) is found and the total field at frequency f is calculated at any range r by carrying out the inverse transform. The indefinite integral is replaced by a finite sum using discrete Fourier transforms. If the maximum value of wavenumber in the sum is $K_{\max}$ and N discrete values of K are introduced, then the wavenumber intervals are given by $\Delta K = K_{\max}/(N - 1)$ and correspond to range intervals $\Delta r = 2\pi/N \, \Delta k$, so, for example,

$$
p(r_m) = 2(1 - i)\sqrt{\frac{\pi}{r_m}} \, \Delta K \sum_{n=0}^{N-1} P(K_n)\sqrt{K_n} \, e^{2i\pi nm/N},
$$
$$
(22)
$$

where $K_n = n \, \Delta K$ and $r_m = m \, \Delta r$ (or $r_0 + m \, \Delta r$, where r_0 is the desired starting range).

An implementation[51] of Eq. (22) was used to calculate the excess attenuation shown in Fig. 12 for the case of downward refraction. In the case of upward refraction, Eq. (22) agrees with the residue series solution described by Eq. (20) (see Section 7).

Parabolic Equation Models This technique assumes that wave motion for a particular problem is always directed away from the source or that there is very little backscattering. Making the change of variable $U = pr^{1/2}$, the far-field assumption ($kr \gg 1$) leads to the Helmholtz wave equation for the field U in two dimensions (r, z):

$$\frac{\partial^2 U}{\partial r^2} + \frac{\partial^2 U}{\partial z^2} + k^2 U = 0. \qquad (23)$$

We define the operator $Q = \partial^2/\partial z^2 + k^2$, and if k is independent of range, Eq. (23) can be written as

$$\left(\frac{\partial}{\partial r} + i\sqrt{Q}\right)\left(\frac{\partial}{\partial r} - i\sqrt{Q}\right) U = 0. \qquad (24)$$

The factors within the parentheses represent propagation of incoming and outgoing waves, respectively, if a time dependence $\exp(-i2\pi f t)$ is assumed. Considering only the outgoing wave, Eq. (24) reduces to

$$\frac{\partial U}{\partial r} = i\sqrt{Q}U. \qquad (25)$$

Most implementations of the PE method can be traced back to Eq. (25).

It is convenient to define a new, more slowly varying wave u, where $u = U \exp(-ik_0 r)$. The equation for u is given by

$$\frac{\partial u}{\partial r} = i(\sqrt{Q} - k_0)u, \qquad (26)$$

and the formal operator solution to Eq. (26) for advancing the field is

$$u(r + \Delta r) = e^{i\Delta r(\sqrt{Q} - k_0)}u(r). \qquad (27)$$

That is, the field at a distance $r + \Delta r$ is calculated by multiplying the field previously calculated at r by the operator in Eq. (27). For example, the implementations[51] of Eq. (27) yield the same acoustics pressure, and hence the same excess attenuation in decibels, as Eq. (22).

6 DIFFRACTION IN OUTDOOR SOUND PROPAGATION

Far from any boundaries or above a simple infinite plane boundary, a sound field propagates in a relatively simple way, and as seen in the previous sections, this simplicity can be exploited by describing the propagation in terms of ray paths. However, if a large solid body blocks the sound field, the ray theory of sound propagation predicts a shadow region behind the body with sharply defined boundaries, so in principle, on one side of the boundary there is a sound field and close by on the other side of the boundary there is essentially silence. This does

not happen in practice; as the waves propagate, sound "leaks" across this sharp boundary. Diffraction effects are most clearly evident in the vicinity of solid boundaries or along geometric ray boundaries such as the limiting ray discussed in Fig. 13. A more complete discussion of diffraction can be found in Ref. 47.

Acoustic diffraction occurs in conjunction with a wide range of solid bodies: Some such as thin solid barriers are erected alongside highways or are carefully located to shield residential communities from ground operations of aircraft; others such as buildings are often built for other purposes but fortuitously provide some beneficial shielding; yet others such as undulating ground or low hills occur naturally and provide shielding at much larger distances and bring forth other manifestations of diffraction such as the creeping waves referred to earlier.

The simplest and most widely used procedure for determining the reduction of sound pressure level due to diffraction around the edge of a barrier is described in terms of a Fresnel number.[47] This is simply the minimum increase in distance that the sound must travel around the edge of the barrier to go from source to receiver (Fig. 14) divided by the half-wavelength $\lambda/2$ at the frequency of interest. Thus, Fresnel number N is given as

$$N = \frac{2}{\lambda}(d_1 + d_2 - d). \qquad (28)$$

The attenuation by diffraction A_d, in decibels, is then given as a function of Fresnel number by

$$A_d = 10\log(20N). \qquad (29)$$

Equation (29) yields the curve in Fig. 14. This curve is obtained[52] from diffraction theory assuming a thin knife-edge barrier and no ground and then empirically allows for the presence of the ground by reducing the loss of sound level by about 2 dB. This prediction curve is not exact because the empirical correction does not account for the frequency dependence (here, the Fresnel number dependence) of the ground reflection interference in a specific configuration of source, barrier, and receiver heights and separation distances. In order to obtain a more exact prediction of the sound field behind the barrier, the complex interference spectrum resulting from the sum of four paths shown in Fig. 15 must be calculated. Nonetheless, the curve in Fig. 14 is correct to about ± 5 dB in most cases and is the mean curve through the interference spectrum that would be measured, and can be predicted, in any specific circumstances.[53]

Rigorously, the attenuation provided by a barrier above a natural ground surface replaces the ground attenuation present before the construction of the barrier.

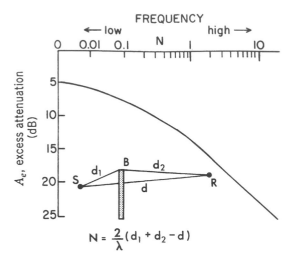

Fig. 14 Sketch defining the Fresnel number N in terms of the path difference and wavelength λ. The curve is the barrier attenuation as a function of Fresnel number.

Therefore, barrier performance is best measured by its insertion loss $IL_{barrier}$, in decibles, defined as the difference in sound pressure level before and after the barrier is constructed:

$$IL_{barrier} = L_p(\text{before}) - L_p(\text{after}). \qquad (30)$$

Detailed studies of traffic noise reduction by barriers show average insertion loss of 5–8 dB but rarely exceeding 10 dB. In general, the insertion loss of a barrier is limited by the effects of atmospheric turbulence to about 15–25 dB. As discussed earlier (and see Section 7), turbulence scatters sound energy that penetrates the shadow behind the barrier, thus resulting in an upper limit to the insertion loss by a barrier.[54]

When barriers are used specifically to attenuate sound, it is good practice to locate them, when possible, as closely as possible to either the source or the receiver. A barrier of given height then results in a large value of the diffraction angle θ and a greater path difference ($d_1 + d_2 - d$). At distances between source and receiver greater than a few hundred meters, it is difficult to provide man-made barriers large enough to provide any noticeable attenuation. Naturally occurring topographic features such as hills can often function as barriers, blocking the line of sight between source and receiver.

The use of absorbing materials can increase barrier performance,[55] and the general principles can be obtained from simplified calculations.[56] The calculations demonstrate that the effectiveness of absorbing material increases as either the source or receiver is moved closer to the barrier and decreases as the frequency decreases. Field studies also show that barrier effectiveness is reduced when the propagation over the barrier is downwind. The rays from the source reach the receiver behind the barrier by curving over the top of the barrier edge, thus providing direct ensonification. However, this does not mean that the barrier has completely lost its effectiveness. In this case, the "before" levels, L_p(before), in the calculation of the insertion loss correspond to the ray-tracing picture in Fig. 11. The multiple rays close to the ground responsible for the before levels in Eq. (30) are effectively blocked by the barrier. Model experiments and theory[57] suggest that the barrier still provides positive insertion loss during downwind propagation.

7 ATMOSPHERIC TURBULENCE

The atmosphere is an unsteady medium with random variations in temperature, wind velocity, pressure, and density. In practice, only the temperature and wind velocity variations significantly affect acoustic waves over a short time period. During the daytime these inhomogeneities are normally much larger than is generally appreciated. Fluctuations in temperature of 5°C that last several seconds are common and 10°C fluctuations not uncommon. The wind velocity fluctuates in a similar manner and has a standard deviation about its mean value that is commonly one-third of the average value. When sound waves propagate through the atmosphere, these random fluctuations scatter the sound energy. The total field is then the sum, in amplitude and phase, of these scattered waves and the direct line-of-sight wave, resulting in random fluctuations in amplitude and phase. The acoustic fluctuations are in some respects analogous to more familiar optical phenomena such as the twinkling of light from a star.

Turbulence can be visualized as a continuous spectrum of eddies.[58] For example, since the average horizon-

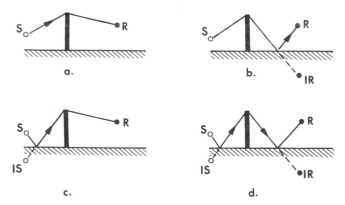

Fig. 15 Four paths contributing to the sound field behind a barrier above ground.

tal wind velocity varies as a function of height (see Section 5), this variation creates turbulent motion or eddies of a size called the outer scale of turbulence. The size of the outer scale L_0, is typically of the order of metres. In the range of eddy size smaller than L_0, the kinetic energy is transferred to eddies of smaller size. This energy transfer can be visualized as a process of eddy fragmentation where large-scale eddies cascade into eddies of ever-decreasing size. The fluid motion is almost completely random and irregular, and its features can be described in statistical terms. This range of eddy sizes is called the inertial range. As the eddy size becomes smaller, virtually all the energy is dissipated into heat and almost no energy is left for eddies of size smaller than l_0. This size l_0 is called the inner scale of turbulence and is typically of the order of 1 mm. The three characteristic ranges of eddy sizes of the turbulent atmosphere are illustrated in Fig. 16. The points are an example of the power spectral density of the time-varying signal recorded by an anemometer of the wind velocity fluctuations. A similar power spectral density is obtained from the time-varying signal recorded by a fast-response thermometer.

The scattering of sound by turbulence produces fluctuations in the sound pressure level. The fluctuations initially increase with increasing distance of propagation, sound frequency, and strength of turbulence but quickly reach a limiting value.[59] This saturation minimizes the nuisance of coping with fluctuating levels during noise measurements from relatively distant sources. For example, when the noise from aircraft propagates under clearly line-of-sight conditions over distances of a few kilometers, the measured sound pressure levels fluctuate about their mean value with a standard deviation of no more than 6 dB.

Another effect of atmospheric turbulence that has traditionally been considered important is the direct attenuation of sound. In a highly directed beam, the turbulence attenuates the beam by scattering energy out of it. However, for a spherically expanding wave this attenuation is negligible. In a simpleminded way, the energy scattered out from the line of sight is replaced by energy scattered back to the receiver from adjacent regions. This implies that the energy level of the rms sound pressure in an unsteady medium is the same as the level in the absence of turbulence. The only mechanism by which turbulence could provide attenuation in a spherical wave field is backscattering. However, it seems that the attenuation provided by backscattering is much smaller than the attenuation due to molecular absorption. This is an important result for the use of the PE models.

An important effect of atmospheric turbulence is the degradation of the ground attenuation and the reduction of the deep shadows produced behind barriers or during propagation in upward refraction conditions. As dis-

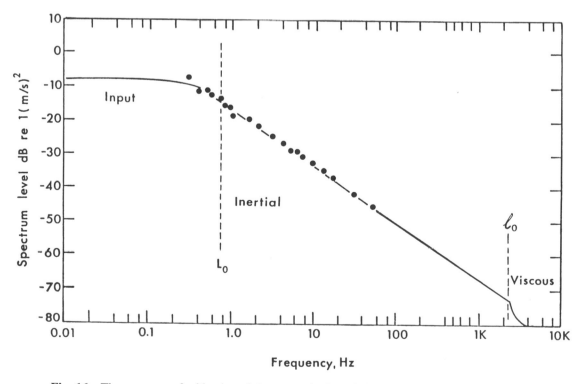

Fig. 16 Three ranges of eddy size of the atmospheric turbulence spectrum. The points are the result of a fast Fourier transform analysis of a wind velocity recording.

cussed in Section 4.3, the effects of atmospheric turbulence on the attenuation of sound over ground can be incorporated into the theory by adding an additional fluctuating random term in Eq. (12). The resulting modified expressions[60] provide an explanation for the discrepancy observed in Fig. 8 between the solid curve and the experimental points. The scattering of sound behind a noise barrier can be calculated [54] by using the standard cross section for scattering by atmospheric turbulence (see Section 8). In order to include the effects of turbulence in the case of propagation in upward refraction conditions, the numerical models discussed in Section 5.3 can be extended.[61,62]

For example, in the PE model, the wavenumber is separated into a part representing the variation of the sound speed with height z above the ground and a part that represents a small random perturbation $\mu \ll 1$,

$$k(r, z) = k_0 \left[\frac{c_0}{c(z)} + \mu(r, z) \right]. \tag{31}$$

The operator for advancing the field in Eq. (27) can then be approximated by the product

$$e^{i\Delta r(\sqrt{Q} - k_0)} = e^{i\langle \mu \rangle k_d \Delta r} e^{i\Delta r(\sqrt{Q_d} - k_0)}, \tag{32}$$

where $Q_d = \partial^2/\partial z^2 + k_d^2$ and $k_d = k_0 c_0/c(z)$. For computational purposes, the stochastic part of the index of refraction is assumed to have a simple correlation function approximated by a Gaussian distribution

$$\langle \mu_1 \mu_2 \rangle = \langle \mu^2 \rangle \exp\left(\frac{-r^2}{L^2} \right). \tag{33}$$

The Fourier transform of the correlation function in Eq. (33) is then an approximation to the power spectral density shown in Fig. 16. For small-scale turbulence[63] near the ground, $L \approx 1$–7 m and $\langle \mu^2 \rangle \sim 10^{-6}$.

The curves in Fig. 17 were calculated[64] from the PE using the operator in Eq. (32) for a frequency of 500 Hz during upward refraction for $\langle \mu^2 \rangle$ values of 0.5, 2, and 10×10^{-6} (lower, middle, and upper curves, respectively). Beyond 200 m, the calculation shows that the excess attenuation is between 20 and 30 dB, in agreement with results found in many field data sets.[2] The pictures in Fig. 18 illustrate more visually the effects of turbulence on the excess attenuation shown in Fig. 17. The top of Fig. 18 is a contour plot of sound levels when the atmosphere is assumed nonturbulent, that is, $\langle \mu^2 \rangle = 0$. The same contour plot for $\langle \mu^2 \rangle = 2 \times 10^{-6}$ is shown at the bottom of Fig. 18. Sound energy (white areas) scattered into the shadow region is clearly observed.

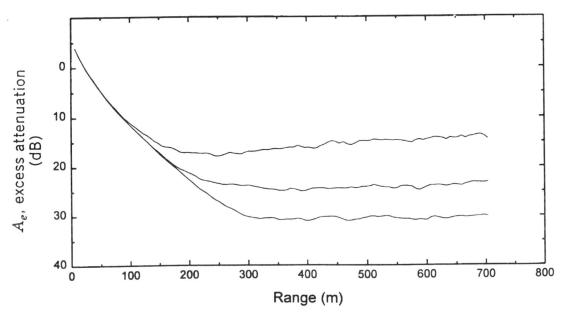

Fig. 17 Excess attenuation predicted for a frequency of 500 Hz for propagation during upward refraction in a turbulent atmosphere. The ground impedance is characterized using the Delany–Bazley model [Eq. (8a)], with an effective flow resistivity of 300 kPa · s/m². The sound speed profile is given by $c(z) = c(0) - 0.5 \ln(z)$ (m/s) and the turbulence strengths were $\langle \mu^2 \rangle = 0.5, 2, 10 \times 10^{-6}$ (lower, middle, and upper curves, respectively). Source and receiver heights are 0.3 and 0.1 m, respectively.

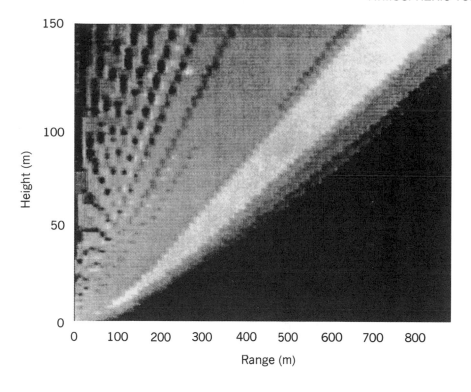

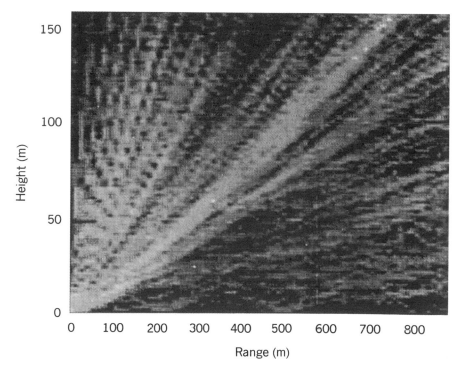

Fig. 18 Contour plots of sound levels in the presence of upward refraction. The white areas are regions of higher sound levels. (Top) No turbulence; (Bottom) in the presence of turbulence.

8   ACOUSTIC SOUNDING OF THE ATMOSPHERE

Meteorologists divide the structure of the atmosphere into a succession of layers, each having different characteristics. Beginning at the surface, the troposphere represents a region of, mainly, decreasing temperature with increasing altitude. The lowest part of the troposphere, the atmospheric boundary layer, which varies in thickness from 1 or 2 km during sunny daytime to a few hundred meters at night, departs from the general temperature decrease. An increasing temperature in the boundary layer is called an inversion and for, example, acts as a lid that prevents the dispersion of atmospheric pollutants. The stratosphere lies just above the troposphere at an altitude of between 10 and 20 km. The temperature in the upper level of the stratosphere increases with height. In the next layer, the mesosphere, at an altitude of about 50 km, temperatures again decrease with height, and the molecular composition of the atmosphere begins to change.

The invention of the echosonde (also called the sodar) added a powerful tool for atmospheric research and for applications to forecasting, especially in the atmospheric boundary layer. The use of the echosonde has become so widespread that results derived with it have become an ubiquitous part of the environmental impact studies required in many countries around the world. In its more complex form—the Doppler echosonde—the instrument can provide measurements of mean winds, sea breezes, and flows in and out of valleys. The monostatic echosonde relies on backscatter (scattering angle

$\theta_s = \pi$) directly back along the ensonification path from a field of turbulence-driven temperature fluctuations along that path. The bistatic echosonde ($\theta_s < \pi$) uses a separated transmitter and receiver, with a signal that results from both temperature and velocity fluctuations. Figure 19 illustrates a bistatic configuration.

The equations of fluid mechanics, together with two assumptions valid in acoustic remote sensing (but invalid in aeroacoustics), that (1) that the turbulence remains a nonzero vorticity and incompressible field that produces no sound while (2) the acoustic wave remains a vorticity-free longitudinal compression field that produces no changes in the turbulence, lead to the governing acoustic remote-sensing equation. If it is further assumed that the turbulence "eddy" sizes lie within the inertial range (see Fig. 16), a good estimate of the scattering cross section (in reciprocal metres) as a function of scattering angle θ_s is obtained,[1]

$$\sigma_s = 1.52 k_0^{1/3} \cos^2 \theta_s \left[0.13 C_n^2 + \left(\frac{C_v^2}{4c_0^2} \right) \cos^2 \left(\frac{\theta_s}{2} \right) \right]$$

$$\cdot \left[2 \sin \left(\frac{\theta_s}{2} \right) \right]^{-11/3} . \tag{34}$$

In Eq. (34), k_0 is the initial wavenumber, C_v^2 the structure parameter for the velocity fluctuations, and C_n^2 the structure parameter for refractive index fluctuations. There is no scattering at right angles, and Eq. (34) predicts the strong forward nature of the scattering by turbulence. In fact, for very small scattering angles Eq. (34) very quickly exceeds unity and becomes infinite at 0°. For this reason, although scattering can be invoked to explain the sound pressure level fluctuations discussed earlier, it is inappropriate to use Eq. (34) to quantify the fluctuations. Instead, other techniques such as perturbation methods are invoked. Equation (34) can be used, though, to calculate the scattered sound energy behind a barrier.[54]

For direct backscatter ($\theta_s = 180°$) velocity fluctuations do not contribute to the scattering because of the $\cos(\theta_s/2)$ multiplying the term containing the C_v^2. Therefore the intensity display for a monostatic echosonde system transmitter and receiver collocated reflects the temperature structure only. By monitoring the backscatter intensity as a function of time from transmission, the relative strength of temperature fluctuations as a function of altitude can be estimated.

The echosonde equation, like its radar and sonar counterparts, relates the received acoustic power P_r to the transmitted power P_t and frequency f and other atmospheric and system design variables. We define L as

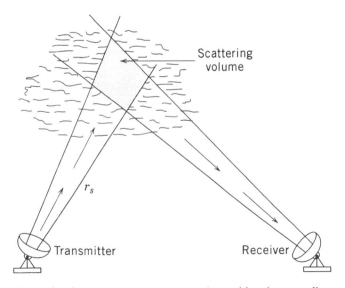

Fig. 19   Scattering volume used in a bistatic sounding measurement.

the propagation attenuation, S the pulse stretching factor ($S \sim 1$), $\alpha(\theta_s)$ a dimensionless aspect factor that accounts for the shape of the scattering volume, l_p the pulse length, and g the antenna directivity gain factor. Then the echosonde equation is

$$P_t = P_t LS\alpha(\theta_s) \frac{l_p}{2} \frac{gA_r}{r_s^2} \sigma_s(\theta_s, f), \qquad (35)$$

where A_r is the area of the receiver antenna aperture and r_s is the distance from the receiving antenna to the scattering volume. Equation (35) represents the key link between theory and experiments and provides the means for interpreting observation.

REFERENCES

1. E. H. Brown and F. F. Hall, Jr. "Advances in Atmospheric Acoustics," *Rev. Geophys. Space Phys.*, Vol. 16, 1978, pp. 47–110.

2. J. E. Piercy, T. F. W. Embleton, and L. C. Sutherland, "Review of Sound Propagation in the Atmosphere," *J. Acoust. Soc. Am.*, Vol. 61, 1977, pp. 1403–1418.

3. M. E. Delany, "Sound Propagation in the Atmosphere, A Historical Review," *Proc. Inst. Acoust.*, Vol. 1, 1978, pp. 32–72.

4. R. B. Tatge, "Noise Radiation by Plane Arrays of Incoherent Sources," *J. Acoust. Soc. Am.*, Vol. 52, 1972, pp. 732–736.

5. E. A. G. Shaw and N. Olson, "Theory of Steady-State Urban Noise for an Ideal Homogeneous City," *J. Acoust. Soc. Am.*, Vol. 51, 1972, pp. 1781–1793.

6. L. C. Sutherland, "Ambient Noise Level Above Plane with Continuous Distribution of Random Sources," *J. Acoust. Soc. Am.*, Vol. 57, 1975, pp. 1540–1542.

7. B. H. Sharp and P. R. Donavan, "Motor Vehicle Noise," in C. M. Harris (Ed.), *Handbook of Noise Control*, McGraw-Hill, New York, 1979, pp. 32–16.

8. P. R. Donavan and R. H. Lyon, "Model Study on the Propagation from V/STOL Aircraft into Urban Environs," *J. Acoust. Soc. Am.*," Vol. 55, 1974, p. 485(A).

9. W. E. Zorumski and W. L. Willshire, "The Acoustic Field of a Point Source in a Uniform Boundary Layer Over an Impedance Plane," Paper before AIAA Aeroacoustics Conference, July, 1986, Paper No. AIAA-86-1923.

10. American National Standards Institute, "Method for the Calculation of the Absorption of Sound by the Atmosphere," ANSI S1.26-1995 (Revision of ANSI S1.26-1978).

11. H. E. Bass, L. C. Sutherland, A. J. Zuckerwar, D. T. Blackstock, and D. M. Hester, "Atmospheric Absorption of Sound: Further Developments," *J. Acoust. Soc. Am.*, Vol. 97, 1995, pp. 680–683.

12. P. D. Joppa, L. C. Sutherland, and A. J. Zuckerwar, "A New Approach to the Effect of Bandpass Filters on Sound Absorption Calculations," *J. Acoust. Soc. Am.*, Vol. 88, S1, 1990, p. S73.

13. D. C. Henley and G. B. Hoidale, "Attenuation and Dispersion of Acoustic Energy by Dust," *J. Acoust. Soc. Am.*, Vol. 54, 1973, pp. 437–445.

14. J. B. Calvert, J. W. Coffman, and C. W. Querfeld, "Radiative Absorption of Sound by Water Vapour in the Atmosphere," *J. Acoust. Soc. Am.*, Vol. 39, 1966, pp. 532–536.

15. P. M. Morse and K. U. Ingard, *Theoretical Acoustics*, McGraw-Hill, New York, 1968.

16. G. A. Daigle, T. F. W. Embleton, and J. E. Piercy, "Some Comments on the Literature of Propagation Near Boundaries of Finite Acoustical Impedance," *J. Acoust. Soc. Am.*, Vol. 66, 1979, pp. 918–919.

17. M. E. Delany and E. N. Bazley, "Acoustical Properties of Fibrous Absorbent Materials," *Appl. Acoust.*, Vol. 3, 1970, pp. 105–116.

18. C. I. Chessell, "Propagation of Noise Along a Finite Impedance Boundary," *J. Acoust. Soc. Am.*, Vol. 62, 1977, pp. 825–834.

19. M. J. M. Martens, L. A. M. van der Heijden, H. H. J. Walthaus, and W. J. J. M. van Rens, "Classification of Soils Based on Acoustic Impedance, Air Flow Resistivity, and Other Physical Soil Properties," *J. Acoust. Soc. Am.*, Vol. 78, 1985, pp. 970–980.

20. Y. Miki, "Acoustical Properties of Porous Materials—Modifications of Delany–Bazley Models," *J. Acoust. Soc. Jpn. (E)*, Vol. 1, 1990, pp. 19–23.

21. L. C. Sutherland, "Review of Ground Impedance for Grass Surfaces—Delany and Bazley Revisited," *Proc. Sixth Int. Symp. on Long Range Sound Prop.*, Ottawa, Canada, 1994, pp. 460–479.

22. T. F. W. Embleton, J. E. Piercy, and G. A. Daigle, "Effective Flow Resistivity of Ground Surfaces Determined by Acoustical Measurements," *J. Acoust. Soc. Am.*, Vol. 74, 1983, pp. 1239–1244.

23. L. N. Bolen and H. E. Bass, "Effects of Ground Cover on the Propagation of Sound through the Atmosphere," *J. Acoust. Soc. Am.*, Vol. 69, 1981, pp. 950–954.

24. H. M. Hess, K. Attenborough, and N. W. Heap, "Ground Characterization by Short Range Propagation Measurements," *J. Acoust. Soc. Am.*, Vol. 87, 1990, pp. 1975–1986.

25. T. F. W. Embleton and G. A. Daigle, "Atmospheric Propagation," Chap. 12, in H. H. Hubbard (Ed.), NASA Ref. Pub. 1258, Vol. 2, WRDC Tech. Report 90–3052, Aug. 1991.

26. K. Attenborough, "Ground parameter Information for Propagation Modeling," *J. Acoust. Soc. Am.*, Vol. 91, 1992, pp. 418–427.

27. A. J. Crammond and C. G. Don, "Effects of Moisture Content on Soil Impedance," *J. Acoust. Soc. Am.*, Vol. 82, 1987, pp. 293–301.

28. T. F. W. Embleton, J. E. Piercy, and N. Olson, "Outdoor

Propagation Over Ground of Finite Impedance," *J. Acoust. Soc. Am.*, Vol. 59, 1976, pp. 267–277.

29. A. Zuckerwar, "Acoustical Ground Impedance Meter," *J. Acoust. Soc. Am.*, Vol. 73, 1983, pp. 2180–2186.

30. P. J. Dickinson and P. E. Doak, "Measurements of the Normal Acoustic Impedance of Ground Surfaces," *J. Sound Vib.*, Vol. 13, 1970, pp. 309–322.

31. G. A. Daigle and M. R. Stinson, "Impedance of Grass-Covered Ground at Low Frequencies Measured Using a Phase Difference Technique," *J. Acoust. Soc. Am.*, Vol. 81, 1987, pp. 62–68.

32. K. Attenborough, "Review of Ground Effects on Outdoor Sound Propagation from Continuous Broadband Sources," *Appl. Acoust.*, Vol. 24, 1988, pp. 289–319.

33. J. M. Sabatier, H. E. Bass, L. N. Bolen, and K. Attenborough, "Acoustically Induced Seismic Waves," *J. Acoust. Soc. Am.*, Vol. 80, 1986, pp. 646–649.

34. J. M. Sabatier, R. Raspet, and C. K. Fredericksen, "An Improved Procedure for the Determination of Ground Parameters Using Level Difference Measurements," *J. Acoust. Soc. Am.*, Vol. 94, 1993, pp. 396–399.

35. M. A. Nobile and S. I. Hayek, "Acoustic Propagation over an Impedance Plane," *J. Acoust. Soc. Am.*, Vol. 78, 1985, pp. 1325–1336.

36. Y. L. Li, M. J. White, and M. H. Hwang, "Green's Functions for Wave Propagation Above an Impedance Plane," *J. Acoust. Soc. Am.*, Vol. 96, 1994, pp. 2485–2490.

37. C. F. Chien and W. W. Soroka, "A Note on the Calculation of Sound Propagation Along an Impedance Surface," *J. Sound Vib.*, Vol. 69, 1980, pp. 340–343.

38. R. K. Pirinchieva, "Model Study of Sound Propagation Over Ground of Finite Impedance," *J. Acoust. Soc. Am.*, Vol. 90, 1990, pp. 2678–2682; see also Erratum in *J. Acoust. Soc. Am.*, Vol. 94, 1993, p. 1722.

39. K. B. Rasmussen, "Sound Propagation Over Grass Covered Ground," *J. Sound Vib.*, Vol. 78, 1981, pp. 247–255.

40. G. L. McAninch and M. K. Myers, "Propagation of Quasiplane Waves Along an Impedance Boundary," *Proc. 26th AIAA Aerospace Sciences. Mtg.*, Reno, NE, 1988, Paper No. AIAA-88-0179.

41. D. G. Albert, "Observations of Acoustic Surface Waves Propagating Above a Snow Cover," *Proc. Fifth Int. Symp. Long Range Sound Prop.*, 1992, pp. 10–16.

42. P. H. Parkin and W. E. Scholes, "The Horizontal Propagation of Sound from a Jet Engine Close to the Ground at Hatfield," *J. Sound Vib.*, Vol. 2, 1965, pp. 353–374.

43. G. A. Daigle, J. Nicolas, and J.-L. Berry, "Propagation of Noise Above Ground Having an Impedance Discontinuity," *J. Acoust. Soc. Am.*, Vol. 77, 1985, pp. 127–138.

44. D. Aylor, "Noise Reduction by Vegetation and Ground," *J. Acoust. Soc. Am.*, Vol. 51, 1972, pp. 201–209.

45. T. F. W. Embleton, "Sound Propagation in Homogeneous Deciduous and Evergreen Woods," *J. Acoust. Soc. Am.*, Vol. 35, 1963, pp. 1119–1125.

46. R. H. Lyon, "Role of Multiple Reflections and Reverberations in Urban Noise Propagation," *J. Acoust. Soc. Am.*, Vol. 55, 1974, pp. 493–503.

47. A. D. Pierce, *Acoustics—An Introduction to Its Physical Principles and Applications*, McGraw-Hill, New York, 1981.

48. K. M. Li, "A High-Frequency Approximation of Sound Propagation in a Stratified Moving Atmosphere Above a Porous Ground Surface," *J. Acoust. Soc. Am.*, Vol. 95, 1994, pp. 1840–1852.

49. R. Raspet, G. E. Baird, and W. Wu, "Normal Mode Solution for Low Frequency Sound Propagation in a Downward Refracting Atmosphere Above a Complex Impedance Plane," *J. Acoust. Soc. Am.*, Vol. 91, 1992, pp. 1341–1352.

50. A. Berry and G. A. Daigle, "Controlled Experiments on the Diffraction of Sound by a Curved Surface," *J. Acoust. Soc. Am.*, Vol. 83, 1988, pp. 2047–2058.

51. K. Attenborough, S. Taherzadeh, H. E. Bass, X. Di, R. Raspet, G. R. Becker, A. Güdesen, A. Chrestman, G. A. Daigle, A. L'Espérance, Y. Gabillet, K. Gilbert, Y. L. Li, M. J. White, P. Naz, J. M. Noble, and H. A. J. M. Van Hoof, "Benchmark Cases for Outdoor Sound Propagation Models," *J. Acoust. Soc. Am.*, Vol. 97, 1995, pp. 173–191.

52. Z. Maekawa, "Noise Reduction by Screens," *Appl. Acoust.*, Vol. 1, 1968, pp. 157–173.

53. T. Isei, T. F. W. Embleton, and J. E. Piercy, "Noise Reduction by Barriers on Finite Impedance Ground," *J. Acoust. Soc. Am.*, Vol. 67, 1980, pp. 46–58.

54. G. A. Daigle, "Diffraction of Sound by a Noise Barrier in the Presence of Atmospheric Turbulence," *J. Acoust. Soc. Am.*, Vol. 71, 1982, pp. 847–854.

55. S. I. Hayek, "Mathematical Modeling of Absorbent Highway Noise Barriers," *J. Acoust. Soc. Am.*, Vol. 31, 1990, pp. 77–100.

56. A. L'Espérance, J. Nicolas, and G. A. Daigle, "Insertion Loss of Absorbent Barriers on Ground," *J. Acoust. Soc. Am.*, Vol. 86, 1989, pp. 1060–1064.

57. Y. Gabillet, H. Schroeder, G. A. Daigle, and A. L'Espérance, "Application of the Gaussian Beam Approach to Sound Propagation in the Atmosphere," *J. Acoust. Soc. Am.*, Vol. 93, 1993, pp. 3105–3116.

58. V. I. Tatarskii, *Wave Propagation in a Turbulent Atmosphere*, McGraw-Hill, New York, 1961.

59. G. A. Daigle, J. E. Piercy, and T. F. W. Embleton, "Line-of-Sight Propagation through Atmospheric Turbulence near the Ground," *J. Acoust. Soc. Am.*, Vol. 74, 1983, pp. 1505–1513.

60. G. A. Daigle, "Effects of Atmospheric Turbulence on the Interference of Sound Waves Above a Finite Impedance Boundary," *J. Acoust. Soc. Am.*, Vol. 79, 1986, pp. 613–627.

61. K. E. Gilbert, R. Raspet, and X. Di, "Calculation of Turbulence Effects in an Upward Refracting Atmosphere," *J. Acoust. Soc. Am.*, Vol. 87, 1990, pp. 2428–2437.

62. R. Raspet and W. Wu, "Calculation of Average Turbulence Effects on Sound Propagation Based on the Fast Field Program Formulation," *J. Acoust. Soc. Am.*, Vol. 97, 1995, pp. 147–153.

63. M. A. Johnson, R. Raspet, and M. T. Bobak, "A Turbulence Model for Sound Propagation from an Elevated Source Above Level Ground," *J. Acoust. Soc. Am.*, Vol. 81, 1987, pp. 638–649.

64. X. Di and G. A. Daigle, "Prediction of Noise Propagation During Upward Refraction Above Ground," *Proc. INTER-NOISE 94*, Yokohama, Japan, 1994, pp. 563–566.

33

INFRASOUND

THOMAS B. GABRIELSON

1 INTRODUCTION

Ever present but generally inaudible, infrasound forms a continuum of acoustic radiation below the useful frequency range of human hearing. In contrast to the more familiar audible sound, natural infrasound often reflects global phenomena propagating over thousands of kilometres.

Infrasound covers that part of the acoustic spectrum below 20 Hz. Above about 0.003 Hz (300-s period), the restoring force for acoustic oscillations results from compressibility of the air; below this frequency in gravitationally stable air masses, transverse oscillations are possible in which buoyancy provides the restoring force. Waves having both compressional and transverse components are known as acoustic-gravity waves. At even lower frequencies (periods of hours), the transverse component overwhelms the compressional component and the waves are known as gravity waves.

Infrasound is used to monitor natural events such as volcanic eruptions, earthquakes, severe storms (including microbursts), oceanic disturbances ("microbaroms"), avalanches, meteors, and motion of the auroras; to monitor man-made events such as explosions or supersonic aircraft and rocket flight; and to probe the middle and upper atmosphere.

2 SOURCES OF INFRASOUND

Naturally occuring infrasound is produced continually by wind interaction with mountain ranges (mountain-

Encyclopedia of Acoustics, edited by Malcolm J. Crocker
ISBN 0-471-80465-7 © 1997 John Wiley & Sons, Inc.

associated infrasound) and intermittently from auroral motion, meteors, avalanches, storms, volcanoes, and earthquakes (Table 1). One of the most persistent source mechanisms is nonlinear interaction of ocean waves.[1] Particularly intense when storms are active at sea, these signals are known as microbaroms if the transmission path is through the atmosphere and microseisms if the transmission path is through the solid earth. The lack of correlation between microbaroms and microseisms is a reflection of the relatively rapid temporal changes in the atmospheric transmission path compared to the seismic transmission path.

Highly energetic infrasound is more commonly produced from man-made sources but does occur naturally on occasion. Two hours of infrasonic arrivals were recorded in Antarctica from the 1982 explosive eruption of El Chichon,[2] and infrasound from the 1980 eruption of Mt. St. Helens was recorded in China.[3] At least three man-made sources lead to long-distance propagation of infrasound: large explosions, rocket launches and reentry, and supersonic aircraft flight.

In any infrasonic measurement, there is a background of noise. Noise resulting from local nonacoustic pressure fluctuations (caused principally by wind at the sensor) can be substantially reduced by sensor construction (see Section 4) and electronic filtering. Acoustic noise generated by turbulence remote from the sensor is not so easily reduced. The common feature of this background is a power spectrum that decreases with increasing frequency. Power laws ranging from f^{-1} to f^{-3} (3–9 dB/octave) are evident in various data; however, f^{-2} (6 dB/octave) seems to be representative of this noise from about 0.1 to 10 Hz.[4] Theoretically, acoustic radiation from shear turbulence would give this same power law as would nonacoustic contributions from eddies and waves.

TABLE 1   Natural Sources of Infrasonic Acoustic Waves

Source	Period (s)	Typical Levels	Characteristics	Mechanism	Altitude
Microbaroms	2–8	0.01–1 Pa	Diurnal, semidiurnal, seasonal variations	Nonlinear interaction of ocean waves	Sea level
Aurora	10–100	0.1–0.5 Pa	Not observed from 1200–1600 local time	Supersonic motion of auroral arc, joule heating or electromagnetic force	100+ km
Meteors	0.2–18	0.05–1 Pa at 200–1200 km	Often observe both stratospheric and thermospheric arrivals	Explosive interaction with the atmosphere	25 km and above
Avalanches	0.5–2	0.02–0.05 Pa at 100 km	Dominant spectral peak	Monochromatic structure may result from periodic leading-edge roll of avalanche	Surface
Volcanic eruptions	>100	15 Pa at 10,000 km from Mt. St. Helens	Long duration (hours) at long distances, multiple arrivals	High-energy explosive compression of atmosphere	Surface to 10+ km
Strong earthquakes	8–30	0.1–2 Pa at thousands of kilometres for large events	In addition to airborne arrivals, locally generated seismic-acoustic arrivals	Ground motion at epicenter and from intermediate regions, ground motion from seismic waves at receiving station	Surface
Mountain-associated waves	10–50	0.1–3 Pa	No diurnal variation, long duration, −10.5 dB/octave power spectrum	Associated with wake turbulence from wind flow over mountain ridges	Surface to several kilometres
Severe storms	0.1–50	0.05–0.3 Pa at 30–800 km	Not correlated with lightning, −3 to −6 dB/octave power spectrum from 2–16 Hz	Mechanism uncertain: vigorous convection, tornadoes, fronts?	Surface to tropopause

3   PROPAGATION CHANNELS

Infrasound propagates to extremely large distances in the atmosphere for two primary reasons: The absorption of infrasound is small, and strong refracting channels are formed naturally in the atmosphere. A secondary contributor is the high reflectivity of the ground at infrasonic frequencies. While its actual behavior has not been completely determined, below 20 Hz the ground is often adequately treated as a perfectly rigid reflector.

Atmospheric refracting channels are produced by vertical gradients in temperature and wind speed. Temperature in the "average" atmosphere first decreases with altitude (troposphere) to about 11 km and then is constant (tropopause) to about 20 km. Absorption of solar radiation by ozone produces a temperature increase (stratosphere) to about 45 km and a region of roughly constant temperature (stratopause) to about 55 km. Above that, the temperature decreases (mesosphere) to 85 km or so. The temperature then increases again (thermosphere) from solar heating of molecular oxygen. The sound speed profile based on the *1976 Standard Atmosphere*[5] temperature profile is shown in Fig. 1.

For ground-based observations of infrasound, the two principal propagation channels are the stratospheric duct, formed between the ground and the stratopause, and the thermospheric duct, formed between the ground and 100 km or so in the thermosphere. In addition, downwind propagation channels in the planetary boundary layer can form from vertical wind shear, and very strong, low-altitude channels are produced by strong temperature inversions common in the Arctic or over snow-covered surfaces in winter (Table 2). Vertical gradients in wind speed associated with the jet stream can also produce ducting.

High-altitude wind influences the stratospheric and thermospheric channels. There is a strong seasonal reversal of wind over a large region of altitude (tens of kilometres) in the vicinity of the stratopause.[6] In winter at midlatitudes, there is a strong west-to-east flow of up to 100 m/s at 70 km altitude; in summer, the dominant flow is east to west at up to 60 m/s and 50 km altitude. Since, without wind, the sound speed in the stratopause is approximately equal to that at the ground, the effectiveness of the stratospheric channel is strongly influenced by these winds.[7] For propagation paths from east to west, the winter wind reduces the effective sound speed in the stratopause, which allows much of the energy to escape into the thermospheric duct; the summer wind strengthens the stratospheric channel. For propagation from west to east, the channel is stronger in winter than summer.

Through the fall, winter, and spring, there is a strong semidiurnal tidal variation in the thermospheric wind[8] with an amplitude of up to 100 m/s above 80 km. This

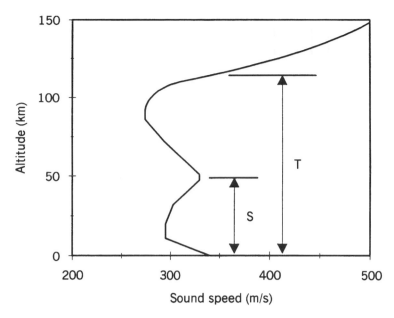

Fig. 1 Vertical sound speed profile for the 1976 Standard Atmosphere. The two primary propagation channels are the stratospheric channel (S) and the thermospheric channel (T).

introduces a semidiurnal fluctuation in infrasonic signals propagating through the thermosphere; this fluctuation can be used to identify this path.

Absorption of infrasound is so low in the troposphere and stratosphere that signals, even of a few hertz, can propagate to long distances. At these altitudes, the absorption can be determined using the procedures in Chapter 32. In the lower thermosphere, the pressure is low enough that the vibrational relaxation frequencies of oxygen and nitrogen are below the band of acoustic infrasonics. Over most of the band and at these altitudes, the ratio of frequency to ambient pressure is below 10 Hz/Pa, so that rotational relaxation need not be considered. Lacking detailed measurements of absorption at these low pressures, low frequencies, and very low humidity, the classical absorption coefficient is an adequate first estimate. For more details about the physics of absorption under these conditions, see Chapter 56.

The pressure at 80 km is about 1 Pa (or 10^{-5} atm). Assuming that the dry-air limits for relaxation frequency can be scaled by pressure, the oxygen and nitrogen relaxation frequencies should be 10^{-4}–10^{-5} Hz—below the

acoustic region. For a bulk viscosity of 0.6 times the normal viscosity[9] η and a Prandtl number of 0.7, the classical absorption coefficient at the pressure p_s is

$$\alpha_{cl} = \frac{5\pi^2 \eta f^2}{\gamma p_s c}, \tag{1}$$

where γ is the ratio of specific heats and c is the sound speed. At 80 km altitude, this is $1.5 \times 10^{-6} f^2$ nepers per metre, or $13 f^2$ decibels per 1000 km. This would attenuate energy strongly above 1 Hz over the long paths typical of thermospheric propagation, but energy below a few tenths of hertz would propagate to long distances. This is consistent with the observed spectral content of infrasound refracted by the thermosphere.

Microbaroms (3–6-s periods) can be used to monitor conditions in the upper atmosphere by also monitoring microseisms to track changes in source strength.[8] Microbaroms recorded on the East coast show marked semidiurnal variation in the fall, winter, and early spring, indicating propagation through the thermospheric duct.

TABLE 2 Natural Propagation Channels

Channel	Altitude	Cycle Range	Frequency (Hz)
Thermosphere	100+ km	200–1000 km	<0.5
Stratosphere	50 km	10–300 km	<10
Planetary boundary layer	Several kilometres	1–10 km	
Arctic inversion	10–100 m	20–1000 m	

For propagation from the ocean to the coast (east to west), the dominant prevailing upper atmosphere wind reduces the effectiveness of the stratospheric duct during the winter. In the summer, when the stratospheric wind has reversed, propagation from east to west is enhanced in the stratospheric channel and the semidiurnal variation is not observed.

Systematic observations of long-range infrasound from transatlantic supersonic transport (SST) flights also reflect some of the features of high-altitude propagation channels.[10] In this case, the source (12–17 km altitude) is in the tropopause, a sound speed minimum. If the winds in the stratopause are in the same direction as the propagation and are of sufficient speed, energy refracted by the stratospheric channel reaches the ground. In this case, there are often two arrivals, one from stratospheric refraction and another from thermospheric refraction. Otherwise, the ground arrival is the thermospheric path. Most of the energy in the stratospheric arrival is between 0.5 and 2 Hz, while the thermospheric arrival is concentrated between 0.13 and 0.3 Hz.

At sufficiently low frequency, the effects of gravity become important in the propagation equation. The buoyant restoring force for vertical displacement becomes comparable to the restoring force from adiabatic compressibility. The resulting waves, which have both transverse and longitudinal components, are known as acoustic-gravity waves; their periods range from about 300 s to several hours.[11,12] For low enough frequency, the compressibility effects are negligible compared to buoyancy and the waves are called gravity waves.

Acoustic-gravity (and gravity) waves are supported only in those regions of the atmosphere that are convectively stable. That is, if a portion of air is displaced vertically and the buoyant force opposes that motion, then the region is stable. The natural frequency of the resulting free oscillation of such a displaced volume of air is the Brunt–Väisälä frequency. If air is treated as an ideal gas, this frequency is

$$f_{BV}^2 = -\frac{g}{4\pi^2 T}\left[\frac{\rho g T(\gamma-1)}{p_s\gamma} + \frac{dT}{dz}\right] \qquad (2)$$

or, for $g = 9.8$ m/s^2, temperature T in kelvin, and height z in metres,

$$f_{BV}^2 = \frac{0.0025}{T}\left[1 + 100\frac{dT}{dz}\right]. \qquad (3)$$

If f_{BV}^2 is negative, the region is unstable for vertical displacements; a parcel of air displaced upward would continue to move upward. If f_{BV}^2 is positive, then the region is stable; f_{BV} is an upper limit for acoustic-gravity waves. The *1976 Standard Atmosphere* definition has no regions of instability, but in reality, the troposphere often has temperature gradients sufficiently negative for instability. Based on the *Standard Atmosphere*, the Brunt–Väisälä period in the tropopause, stratosphere, and stratopause is about 300 s. Long-range propagation of acoustic-gravity waves near this period are often considered diagnostic of source location in the tropopause or higher. Large storm systems that reach the tropopause can generate acoustic-gravity waves; the explosive eruption of El Chichon in 1982 penetrated the tropopause, and acoustic-gravity waves were received as far away as the antipodal point in Australia.[2]

Since the standard atmospheric sound speed profile supports refractive propagation to altitudes in excess of 100 km, not only are extremely long range paths possible, so also are extremely large amplitudes. Ignoring for the moment temperature differences, the density of the atmosphere decreases exponentially with altitude. Since the energy density in an acoustic wave is proportional to ρv^2, where v is the acoustic particle velocity, conservation of energy suggests that the velocity, and therefore the displacement, should increase exponentially (as $\rho^{-1/2}$). Consequently, displacement amplitudes of hundreds of metres (with a concomitant decrease in acoustic pressure) may be produced in the thermosphere.[12]

Not all infrasound is propagated entirely through the atmosphere. Strong earthquakes generate surface waves in the solid earth that have frequencies in the infrasound region. These waves travel with seismic speeds (kilometres per second) and so the coincidence angle for radiation into the atmosphere is almost vertical. Consequently, this component of the radiation is only detected acoustically when and where the seismic wave is present. The effective propagation speed from the epicenter is the seismic speed, and the acoustic amplitude is directly related to the amplitude of the ground motion. Strong infrasonic signals from the Alaskan earthquake of 1964 were received near Washington, DC, by this seismic-acoustic path.[13] About 5000 km from the epicenter, the Rayleigh wave produced ground motion with a 25-s period and 4 cm amplitude; this generated a local acoustic wave with an amplitude of 2 Pa. Other components of radiation, generated for example at the earthquake's epicenter or from regions with significant changes in terrain (mountain ranges), can also be detected under favorable propagation conditions in the atmosphere.

4 INSTRUMENTATION

Most systems for detecting infrasound below 0.1 Hz consist of a capacitor or moving-coil microphone element (or a pressure transducer) mounted inside a chamber. The chamber is divided into a front volume and a back volume with the microphone element mounted in the dividing wall. A capillary leak (the inlet) into the front volume determines the high-frequency cutoff, and another capillary leak between the front and back volumes determines the low-frequency cutoff.[14] In practice, the chamber volumes may be filled with steel wool and buried to reduce sensitivity to temperature fluctuations. The inlet to the chamber is either exposed directly to the air or connected to a noise-reducing pipe. Above 1 IIz, capacitor microphone elements with only a back volume can be used in large wind screens. The back volume and capillary (back vent) roll the response off at lower frequencies. Some standard low-frequency capacitor microphones can be modified by reducing the back-vent aperture to increase the vent resistance, thereby lowering the low-frequency rolloff. In any infrasonic measurement system, it is particularly important to reduce the low-frequency response below the band of interest since the overall background spectrum goes roughly as f^{-2}. By doing this acoustically, the dynamic range requirement on the electronics is reduced.

Most infrasonic systems must be designed to reduce response to wind-generated local pressure fluctuations. Large wind screens are of some value, but these fluctuations can be reduced greatly with pipe arrays.[15,16] By effectively spreading the sampling region of a single sensor over a region much larger than the correlation distance of these pressure fluctuations but smaller than the acoustic wavelength of the intended signal, the signal-to-noise ratio can be improved substantially. This is achieved by connecting the sensor to one or more long pipes with capillary tubes spaced along the pipe to serve as sampling ports.

Based on observed correlation distances for nonacoustic pressure fluctuations, the minimum sampling port spacing is on the order of tens of centimetres for frequencies above 1 Hz, while below 0.01 Hz spacings of more than 5 m can be used. The acoustic wavelength sets an upper limit on the pipe length (and, therefore, the number of sampling ports). If the maximum dimension is greater than about 0.1 wavelength, the apparatus will start to exhibit noticeable directionality. This limits pipe array lengths to 30 or 40 m at 1 IIz, but at 0.01 Hz pipes of several kilometres may still be effective. Ideally, the signals should add coherently over all the inlets and the nonacoustic fluctuations should add incoherently. For N inlets, the signal power would then be proportional to N^2 and the noise power to N. Therefore, the theoretical gain in signal-to-noise power ratio would be N. In some circumstances, it may be more advantageous to distribute the sampling ports over an area with multiple pipes instead of over a line with a single pipe.

Frequently, several sensor systems are placed in a spatial array, and the outputs are processed for horizontal direction of arrival, horizontal phase speed ("trace" speed), and correlation. In this manner, further gain can be achieved against interfering signals and the origin of the desired signals can be determined. For example, a phase speed of the same order as the local sound speed indicates a nearly horizontal arrival as would be the case for a path through the stratospheric channel (since the sound speed in the stratopause is about equal to the sound speed at the ground). A phase speed of about twice the normal sound speed is often indicative of a thermospheric path. Phase speeds an order of magnitude higher than the local sound speed are observed for acoustic waves generated by local passage of seismic waves.

Suitable measures for infrasonic acoustic signals include frequency spectrum, time wave shape, spatial correlation, horizontal phase speed, and horizontal and vertical arrival direction. Local temperature and wind speed and direction must also be measured over the array site in order to relate measured phase speed to vertical direction of arrival. To locate unknown sources, the sound speed profile and the wind speed and direction profiles over the suspected propagation path must be known. For measurements undertaken in environments in which nonacoustic turbulence-generated signals are likely to interfere, the local wind speed and temperature profiles should be measured so that the turbulent stability of the flow can be determined.

This short introduction to infrasound can scarcely begin to cover the richness of the subject. Often ignored because it is not perceived, a continuous and highly variable background of infrasonic acoustic energy is present everywhere on earth and under the sea. In addition to the source mechanisms mentioned above, some machinery and transportation systems generate infrasound, which can excite very low frequency resonances in buildings and structures, which can in turn be a source of annoyance and fatigue to the occupants. A number of large mammals produce infrasonic sounds with sufficient variety and connection to behavior patterns that strongly suggest communication, perhaps over relatively long distances. Even a casual reading of the references will reveal the broad scope of existence and application of infrasound.

REFERENCES

1. E. Posmentier, "A Theory of Microbaroms," *Geophys. J. R. Astron. Soc.*, Vol. 13, 1967, pp. 487–501.

2. L. McClelland, T. Simkin, M. Summers, E. Nielsen, and T. Stein (Eds.), *Global Volcanism 1975–1985*, Prentice Hall, Englewood Cliffs, NJ, 1989, pp. 467–472.

3. L. Wen-Yi, "Detection and Analytical Results of the Infrasonic Signal from the Eruption of St. Helens Volcano," *J. Low Freq. Noise Vibr.*, Vol. 4, 1985, pp. 98–103.

4. E. Gossard, "Spectra of Atmospheric Scalars," *J. Geophys. Res.*, Vol. 65, 1960, pp. 3339–3351.

5. *U.S. Standard Atmosphere*, U.S. Government Printing Office, Washington, DC, 1976.

6. E. Batten, "Wind Systems in the Mesosphere and Lower Thermosphere," *J. Meteorol.*, Vol. 18, 1961, pp. 283–291.

7. T. Georges and W. Beasley, "Refraction of Infrasound by Upper-Atmosphere Winds," *J. Acoust. Soc. Am.*, Vol. 61, 1977, pp. 28–34.

8. W. Donn and D. Rind, "Microbaroms and the Temperature and Wind of the Upper Atmosphere," *J. Atmos. Sci.*, Vol. 29, 1972, pp. 156–172.

9. A. Pierce, *Acoustics*, Acoustical Society of America, New York, 1989, p. 553.

10. W. Donn, "Exploring the Atmosphere with Sonic Booms," *Am. Scientist*, Vol. 66, 1978, pp. 724–733.

11. T. Beer, *Atmospheric Waves*, Wiley, New York, 1974, pp. 29–34.

12. I. Tolstoy, *Wave Propagation*, McGraw-Hill, New York, 1973, pp. 63–66.

13. J. Young and G. Greene, "Anomalous Infrasound Generated by the Alaskan Earthquake of 28 March 1964," *J. Acoust. Soc. Am.*, Vol. 71, 1982, pp. 334–339.

14. J. Macdonald, E. Douze, and E. Herrin, "The Structure of Atmospheric Turbulence and Its Application on the Design of Pipe Arrays," *Geophys. J. R. Astron. Soc.*, Vol. 26, 1971, pp. 99–110.

15. F. Daniels, "Noise-Reducing Line Microphone for Frequencies Below 1 cps," *J. Acoust. Soc. Am.*, Vol. 31, 1959, pp. 529–531.

16. R. Burridge, "The Acoustics of Pipe Arrays," *Geophys. J. R. Astron. Soc.*, Vol. 26, 1971, pp. 53–69.

PART IV

UNDERWATER SOUND

34

INTRODUCTION

Ira Dyer

In this introduction, I broadly survey the subject of underwater sound. My purpose is to help those wishing to glimpse its various facets, to better understand its at times confusing nomenclature, and to sense what is in the chapters of Part IV. I will set the stage by illuminating its application areas, describing its professional names and general outlooks, and discussing a few trends and scientific challenges that may help explain the formulation of some of its major topics. Many will find it appropriate to go directly to Chapters 35–53, for they contain the central technical ideas of the subject.

Underwater sound involves a large range of distances and a wide spread of frequencies. Although no firm limits exist, present practice entails distances from about 1 m to 20,000 km and frequencies from about 1 Hz to 1 MHz. (Since the speed of sound in seawater is about 1.5 km/s, the latter correspond to wavelengths as large as 1.5 km and as small as 1.5 mm.) Within this broad sweep of parameters, a rich mixture of physical ideas, analytical and numerical methods, data, and empirical approaches reside, which unfold in Chapters 35–53.

Sound waves are the principal means of long-distance wireless signaling in the ocean. While electromagnetic waves are carried by wires or fibers at the ocean bottom with high reliability, useful bandwidth, and low cost, in wireless use such waves cannot overcome the conductivity of seawater for the distances needed. On the other hand, sound waves propagating in seawater can provide long-distance signal links, although they do so with somewhat restricted bandwidths, imperfect reliability, and higher costs than we might like. But because sound can work, a wide range of applications are now

Encyclopedia of Acoustics, edited by Malcolm J. Crocker
ISBN 0-471-80465-7 © 1997 John Wiley & Sons, Inc.

to be found, from monitoring global climate change to detecting sunken ships on the bottom.

It might be tempting to think that sound waves in the ocean are direct analogs of electromagnetic waves in the air, since both can propagate in their respective media reasonably well. But acoustic wavelengths are relatively enormous (i.e., the sound propagation speed relatively tiny), so that the technologies are, with some exceptions, substantially different. For example, a side-scan sonar image of a familiar object on the sea bottom often requires an expert for its unequivocal identification, while a photographic image of the same object in air can be reliably identified without special training. The side-scan sonar image is simply a stack of *temporal* signals scattered by the object and taken in sequence as the sonar is towed by. The photographic image is, in contrast, a *spatial* display of signals scattered simultaneously by the object to a fixed receiver. While physically related, the two technologies, and their match to a human observer, are far different! One difference that often is crucial is the size–wavelength ratio for detecting and recognizing an object: The ratio is many orders of magnitude larger for the visible band of the electromagnetic spectrum, compared to the ratio for typical acoustic waves in seawater.

1 SOUND IN THE SEA

At the dawn of the sonar age, roughly from 1910 to 1940, sonars were high-frequency, short-range devices for which the classical three-dimensional inverse-square spreading law sufficed. (By short range I mean an observation distance comparable to or less than the ocean depth, which might be, as examples, less than 200 m on the continental shelf or less than 5000 m in the deep ocean.) With the growth of interest in longer ranges, the applicable spreading law changed from the three-dimensional model to a two-dimensional one; containment of sound at long ranges within the ocean's horizontal boundaries made a two-dimensional model essential. Since the ocean's gradients of ambient pressure, temperature, and salinity are larger in the vertical compared to the horizontal direction (Chapter 35), acousticians adopted range-independent water column approximations. These, then, when coupled to range-independent models for the bottom geology, completed the two-dimensional idea. A spreading law of inverse range, rather than inverse-*square* range, clearly means that larger sound power can be received in two-dimensional propagation and, therefore, that many applications at long ranges are feasible. In the main, this is the applications context today, as is reflected in most of the subsequent chapters of this Part.

Deterministic approaches to propagation prediction are relatively simple for short-range three-dimensional propagation. While more complicated for long-range two-dimensional cases, deterministic methods are nonetheless at hand and provide essential predictive and interpretive capabilities (see Chapter 36). Complexity in such methods arises precisely because the two-dimensional containment of sound produces multiple overlapped sound paths (multiple modes) with differing phases or time delays.

Statistical rather than deterministic methods that describe two-dimensional propagation are also in use.[1–6] These, in essence, treat the multipath phases as random, that is, the multipaths as incoherent. Statistical methods are easier to apply, implicitly average over the more troublesome complexities, agree well with the broad trend of measurements, but submerge details of the propagation process induced by coherence among the multipaths. To expand the latter point, the ocean is a dynamic medium with spatial and temporal fluctuation scales that readily perturb the phases of long-range sound paths. When these perturbations are not known or little understood, or when broad predictive/interpretive trends suffice, statistical methods are viable alternatives to the deterministic ones. Indeed, although powerful deterministic methods were already available, I estimate that up until about 1975, statistical methods were favored by practitioners.

About 1975, practitioners' choices for long-range two-dimensional propagation modeling began to shift strongly to determinism, and that trend continues to this day. Many factors account for the shift, the main one being that research and operational sonars increasingly include high-resolution capabilities. Large arrays and optimum array processing algorithms in such high-resolution systems refine the ocean's angular domain

being measured or interrogated. Large bandwidths in such sonars enable coherent processing of signals in short well-resolved time windows and reduce the range-wise extent of the measurement/interrogation volume. With use of high-resolution sonars, the ocean thus can be viewed more deftly through a few or even one path, rather than through many overlapping paths. Consequently, sound is now a tool for studying detailed ocean processes or for more precisely detecting and tracking targets in the ocean. The limits of sonar in pursuit of such high-resolution applications are not yet reached, promising broader scientific challenges and further important applications for many years to come.

2 SUBFIELDS OF UNDERWATER SOUND

2.1 Ocean Acoustical Metrology

An early application of sound in the ocean (about 1925) was in depth measurement, a use that continues to this day, although often with much higher resolution of bottom topography than that needed for surface ship navigation (its original objective). Depth sounders are suggestive of a class of uses that can be called *ocean acoustical metrology*, in which one observes changes in sound waves as affected by one or more ocean properties (depth, sea height, water temperature, current, etc.). The observation is then inverted to reveal the ocean property. In many cases other sensors could do the same, but sound can measure remotely (as in depth), synoptically (as in eddy tomography), and rapidly (as in sea ice fracture), capabilities that are increasingly important for large-area monitoring, long-time sampling, or fast diagnosing.

Oceanographers, marine geologists/geophysicists, and marine biologists need acoustic measurement tools to understand basic oceanic processes (see Chapter 50), and many have adopted the name *acoustical oceanography* for this form of ocean acoustical metrology. Offshore petroleum engineers, marine soil engineers, naval architects, fishery resource managers, and ocean environmental engineers likewise need such tools (as in Chapters 44, 50, and 51). Because the problems of these user groups are different in scope and outlook, distinctions are likely to evolve, but for now the user groups are seen as one within the name acoustical oceanography.

Whatever distinctions might evolve, the disciplines in acoustical oceanography have invariably become joint ones, for example oceanography *and* acoustics. The implications of the interdisciplinary nature of acoustical oceanography are important and far-reaching: in the conduct of research and in the development of measurement systems, intersections of one or more relevant disciplines with acoustics are firmly established and highly productive. An observer of this subfield, or a newcomer to it, would better appreciate its activities with this in mind. But it must be said that in education such inter-disciplinary intersections are at best still at the formative stages.

2.2 Ocean Acoustics

The design, development, and underlying science of acoustical systems for use in the oceans generally falls in a class of activities called *ocean acoustics*. Probably only those professionals close to acoustical oceanography or ocean acoustics care about the distinctions in the names, but the objectives of the two subfields are different. In the latter, one wants to detect an adversary's submarine, to navigate an underwater search vehicle with bottom-fixed beacons, to find harvestable schools of fish, and to receive and decipher mammalian vocalizations as just a few examples. In such applications the focus is on the task the system is intended to perform, as affected by the relevant ocean properties, and not on measurement of those properties per se (as in acoustical oceanography). There is, however, little or no difference between the fundamentals of the two subfields as they relate to ocean science, since issues such as propagation, scattering, and the like need to be addressed in each. Thus the chapters in Part IV are properly silent on the distinction between ocean acoustics and acoustical oceanography, although a sensitive observer will at times detect a leaning to engineering as well as to ocean science in ocean acoustics.

The bases of ocean acoustics are those interdisciplinary ones at the base of acoustical oceanography, almost always with additional intersections between acoustics and certain engineering sciences, such as signal processing, instrumentation, and system design and optimization. In its fullest form, therefore, ocean acoustics could be called "sonar engineering," but this name is less commonly used. Bonds in ocean acoustics at the various intersections of acoustics with its other disciplines are, for the most part, quite strong and well developed. This strength cuts across the avenues of research, development, design, and education.

I believe that two large challenges confront the science of ocean acoustics: (1) more detailed knowledge is required of the ocean related to the increasing resolution of sonars and (2) more powerful predictive/interpretive tools are needed where the two-dimensional environmental assumptions cannot be justified. Scattering from a rough ocean bottom, as an example under (1), begs for detailed knowledge of rock outcrops within *small* bottom footprints, each of which may be a sample from a spatially *nonstationary* random process. And, as an example under (2), propagation on the continental shelf requires acoustic models that can efficiently deal with anisotropic heterogeneities in the water column or the bottom.

Systems used in acoustical oceanography and ocean acoustics are of two kinds: *passive sonars* and *active sonars*; these are described in Chapter 49. A sonar that receives signals created by breaking surface gravity waves (to measure wind stress) is, for example, a passive sonar; one that radiates an acoustic signal *and* receives that signal scattered from an autonomous underwater vehicle (to track it) is an example of an active sonar. The fundamentals of passive and active sonars have much in common, and when they do, the following chapters need not distinguish between them. But the fundamentals can also differ. For an active sonar one needs to address interfering or unwanted acoustic scatter. This is scatter of the radiated acoustic signal from oceanic features, also called *reverberation*, rather than that from the target, and such cases are treated separately.

A major difference between passive and active sonars is in their signal processing design, implementation, and display. But there are also large areas of commonality, for example in array configuration and signal handling. All are topics of extraordinary interest, challenge, and importance and are covered in Chapter 49. These are to be coupled with other engineering topics such as transducers (which connect a sonar's shipboard subsystems to the ocean) and with topics in the acoustics of the oceans to form an overall view of ocean acoustics.

2.3 Marine Structural Acoustics

Another subfield of underwater sound is *marine structural acoustics*, which is covered in Chapters 41–43 and (indirectly) 46. This subfield deals with elastic waves in floating or submerged structures and their coupling to acoustic waves in the water. Two factors account for its difficulty and importance: First, structures of practical interest are usually quite large and geometrically complicated, and therefore can contain a variety of elastic wave types that interact at structural discontinuities to produce a total elastic wave field of high complexity. Second, since structure/water densities and wave speeds are not too dissimilar, this complicated structural wave field is strongly coupled to the acoustic field in the water. Thus, for example, sound scattered from a ship's hull is different from the incident signal, not only in its signal amplitude, but also in its time (or frequency) characteristics.

The underlying disciplines of marine structural acoustics are structural dynamics *and* acoustics, an intersection that is reasonably well established in research and development and nearly so in education. In theory and in practice, however, marine structural acoustics has yet to develop substantive intersections with basic disciplines in the strength of structures, perhaps because issues in the latter are largely static or quasi-static. The net result is that a marine structure is, with few exceptions, considered separately from the two perspectives, and thus not jointly optimized for strength and acoustics. I believe the challenge of joining marine structural acoustics with the strength-related disciplines is not only important but also achievable.

Other challenges in marine structural acoustics are also important. Structures of interest to the underwater sound community have size–wavelength ratios of about 1–10^3. The lower half of this regime (1–30) is one in which details of the acoustic–elastic wave coupling is crucial in high-resolution applications. But exact analytical or numerical methods cannot now account for the immense complexity in a practical structure. While the ultimate challenge for the lower half of the size regime is to develop such methods, an intermediate challenge is to develop hybridized partial models that can robustly predict/interpret acoustic–elastic coupling for practical structures. This is on the horizon and realistically achievable. Chapters 41 and 42 provide approaches to partial models that are not yet, however, in a hybrid amalgam.

In the upper half of the size–wavelength regime (30–10^3), acoustic–elastic coupling is dominated by the shape and material properties of the structure's outer skin. Recent progress in this size regime for scattering of underwater sound waves from elastic bodies is covered in Chapter 43. The methods described there are generally applicable to acoustic–elastic coupling, for example in radiation from structures as well as in scattering.

2.4 Hydroacoustics

Finally, *hydroacoustics* (also known as *hydrodynamic noise*) is a subfield of underwater sound that addresses noise from fluctuating forces on marine propellers, from turbulent boundary layers acting on moving vehicles, from vortices shed by flow over cavities, in fact from any portion of an unsteady-flow field. Chapter 45 covers this subfield. When the unsteadiness is strong, the noise can be detected at large distances. (Example: Propeller noise of a merchant ship traveling at high speed can be detected across an ocean!) But even when relatively weak, hydrodynamic noise can interfere with nearby active or passive sonars; such locally important noise is commonly termed *self-noise*. (Sidelight: Sonar domes or outer decouplers between sonar transducers and a flow field are used to reduce noise from turbulent boundary layers and, if properly configured, do not degrade acoustic transmission!)

Fluid dynamics *and* acoustics are the intersecting disciplines underlying hydroacoustics. In some important areas, marine structural acoustics is closely tied to hydroacoustics (e.g., in understanding and design of the aforementioned sonar domes); in such cases, exquisite atten-tion needs to be given to the wavenumber content of both the acoustic and the fluctuating hydrodynamic fields as they may match/mismatch to the wavenumber properties of the structure.

The greatest scientific challenges in hydroacoustics mirror those in fluid dynamics, at least for flows at low Mach and high Reynolds numbers. For example, the wavenumber content of anisotropic turbulence, up to and beyond the Kolmogorov scale, could well be important in propeller noise but is little understood.

3 MAJOR SCIENCE/ENGINEERING TOPICS OF UNDERWATER SOUND

As may be surmised from my emphasis on it in the previous sections, *propagation* commands a large share of attention among underwater sound professionals. It includes absorption of sound in sea water caused by chemical relaxation, notably the Epsom salt and the boric acid reactions, as described in Chapter 35. While present in quite small concentrations, these constituents are significant in setting the maximum practical range sound waves can reach at a set frequency (or the maximum practical frequency for a set range). Specifically, maximum range and maximum frequency are inversely related via chemical relaxation (or other loss mechanisms), so that long-range sonars operate at low frequencies, and vice versa.

Propagation also includes the effects of geometric spreading of sound, which, as discussed in Section 1 is, for short ranges, the classical three-dimensional inverse-square spreading law. For short ranges, one can usually disregard wave refraction, but not so for long ranges. Instead wave refraction, caused by spatial sound speed gradients (which originate from spatial gradients of temperature, salinity, and density, as detailed in Chapter 35), not only changes the spreading law but also modifies vital details of the multipaths (see Chapter 36). Specifically, to reach long ranges, the refracted sound in most cases interacts effectively with only *one* of the ocean's top and bottom interfaces and in a few cases with *neither* interface. Refraction in a range-independent ocean thus forms two-dimensional propagation at long ranges, but the containing horizontal interfaces need not be the physical top or bottom. With use of ray language (Chapter 36), refraction bends the sound rays away from one or both physical interfaces and does so periodically in range to form a *duct* or *channel* that contains the sound in two dimensions. Equivalently, with use of mode language (also Chapter 36), those modes whose amplitudes are small near one or both physical interfaces form the two-dimensional duct.

While sound will not transmit effectively from water to air through the top interface, because of the large contrast in density and bulk modulus, it can do so effectively from water to the geological medium below the bottom interface. When a ray path strikes the physical bottom, or equivalently, when a mode's amplitude near the bottom is not small, the propagation parameters given in Chapter 37 need to be invoked. In this process, the propagation medium is greatly expanded via good coupling, since it then includes the geological medium, whose thickness is typically of the same order as or much thicker than the seawater itself. *Ocean seismic acoustics* is the name sometimes applied to these cases, to emphasize the physical importance of coupling between the two media.

Because they are usually rough, *scattering* can occur at the physical ocean interfaces provided, of course, that rays (modes) strike one or both of them. In propagation our concern is with *forward scatter*, for example, with that portion of the redirected sound field at the interface pointing around the forward propagation direction. Chapter 38 describes the main ideas and results in forward scatter. For active sonars, with the receiving sonar at or near the radiating sonar, our conern also is with *backscatter* at an interface, for this determines the interface *reverberation* (defined as backscatter reaching the receiving sonar).

Reverberation (see Chapter 40) can interfere with the signal and thereby limit sonar performance. For a given interface, the characteristics of forward scatter and backscatter are not in general the same, and consequently in the chapters just referred to they are covered separately.

Scattering also occurs in the ocean volume, with causes as diverse as fish schools or turbulence patches. Other scattering sources are found in the geological volume, such as shellfish debris buried in sediment. As in the interface case, volume scattering is divided into forward and back categories. Chapters 40 and 44 should be accessed for details, but the user of this book will surely note that less practical knowledge is available for volume than for interface scattering, a consequence of its lesser importance in the operation of most sonar systems.

Sometimes a given scattering mechanism degrades or interferes with a signal and sometimes it is the signal. Scattering from the rough underside of pack ice, for example, degrades active sonar tracking of an underwater vehicle in the Arctic Ocean or provides signals for the ice-keel avoidance sonar mounted on the vehicle.

In the context of active sonar, *target strength* deals with man-made structures (Chapters 42 and 43) and fish and fish schools (Chapter 44) as sources of scatter. A sonar using scatter in the back direction (simply, "backscatter") is known as a *monostatic active* one. *Bistatic active* sonars that use an arbitrary scatter angle are technically feasible, although more costly because of the need for an additional sonar platform. (*Mono* in monostatic refers to the single location of the radiating and receiving subsystems of the active sonar *bi* in bistatic to the two separate horizontal locations. *Multistatic active* sonars, with multiple horizontal locations, are also technically feasible. The origin of *static* carries no essential meaning as these names are used today.)

Another topic relating to man-made structures is *ship noise* (Chapter 46). Beyond the hydrodynamic and self-noise introduced previously, ships or ocean platforms typically contain noisy machinery. Machinery can be significant sources of self-noise and also of noise radiated to large ranges. As examples, machinery noise radiated by submarines, if uncontrolled, can lead to easy detection at long ranges, or machinery aboard offshore petroleum platforms is often of concern because of possible adverse effects on nearby sea life.

A ubiquitous feature of signals observed in the ocean is their *fluctuations* (see Chapter 39). Unsteadiness in source properties is not a factor; in most cases fluctuations are dominated by effects of ocean scattering and/or multipaths. (As described previously in the Introduction, multipaths are readily developed by refraction and concomitant trapping of sound within the ocean's top and bottom boundaries.) Fluctuations affect a sonar's ability to relay information, and thus their causes and characteristics are important in underwater sound. Whether the application entails a long-range sonar for detecting and tracking a target or a much shorter range sonar for *telemetry* of broadband data (see Chapter 51 on telemetry), fluctuations set performance bounds via time spreads, frequency spreads, short-term amplitude variations, and the like.

Performance also depends upon signal level relative to noise. Beyond backscatter (i.e., reverberation) for an active sonar, *ambient noise* in the ocean must be considered for both active and passive sonars as the ultimate limit to performance. Chapter 48 covers ambient noise. Such noise is created by various mechanisms (breaking surface waves, rain, ships distributed over a large area, etc.). As variable as the weather, ambient noise has its own spatial and temporal fluctuation scales that, besides its mean level, affect sonar performance.

Directional receivers discriminate against (spatially filter) ambient noise and therefore improve performance. In the main, directionality is achieved with use of an *array* of elemental receivers. To be useful, an array's size should be larger than about one wavelength; the larger it is compared to wavelength, the more directional it is. Simply put, an array filters noise by receiving it through the array's weakly responding side lobes, while the desired signal is received through the strongly responding main directional lobe. Chapter 49 should be read for details on receiving arrays.

Arrays are also used in sonars to radiate sound. (Again, see Chapter 49.) As in reception, an array's size compared to wavelength is important. A large array directs most of the sound power, through its main lobe, to the desired angular region, leaving only a small fraction of the total power to be radiated in undesired directions via the weak side lobes. In radiation, therefore, an array optimizes the use of sonar power.

Both receiving and transmitting arrays need to be steered, which is done by phase (or time) modulation among individual transducers forming an array. Early in sonar history, mechanical motion of the entire array accomplished steering, but except for a few cases, this older technology has been supplanted by electrical control of the signals among the array elements; see Chapter 49.

Parametric arrays (described in Chapter 53) make use of seawater nonlinearity in response to very high amplitude sound. Size for such an array is determined by the absorption length of high-amplitude sound at its two somewhat different "pump" frequencies, set much larger than the signal frequency (the difference in the pump frequencies). Thus the acoustical size of a parametric array, set by its absorption length, is considerably larger than its actual size, an advantage one must pay for with increased source power.

Sound waves in an ideal fluid are described by three scalar thermodynamic variables (acoustic perturbations in pressure, temperature, and density) and by one vector kinematic variable (the three components of particle velocity, say). Basic fluid equations connect these six variables so that the complete sound field can be uniquely described by one thermodynamic variable plus the one kinematic (vector) variable. Far from a discontinuity (sound source, scatterer, boundary), the list of variables required to describe the field can be reduced to one. For underwater sound, as in air acoustics, it is sound pressure. *Transducers* (Chapter 52) typically used in underwater sound produce an electrical output proportional to the sound pressure acting on it, and vice versa, and thereby connect the electrical domain of the sonar with the acoustical domain of the ocean. That sound pressure is the variable of underwater sound has no fundamental significance; its use reflects the simplicity of a scalar quantity and the accident of technological evolution centered on sound pressure transducers.

Close to a sound field discontinuity, more than one scalar variable is needed. This need arises mainly in research or testing contexts, and when it does, the sensors used are usually pressure and vector velocity transducers or vector intensity transducers.

Rather than electrical, some sources transduce chemical, hydaulic, or pneumatic energy to sound pressure (Chapter 47). Compared with electrically driven sources, the latter can produce very high source strengths. Thus, these are often used by petroleum exploration geophysicists to overcome high losses suffered by waves propagating in the sea bottom or by global climate monitoring scientists to sense sound signals propagated via very long (transoceanic) paths to give but two of many application examples.

The references that follow are particular to the development and fruition of statistical methods for prediction and interpretation of sound waves that propagate to long ranges in two-dimensional ocean ducts. These are included here because such methods still have practical use and are not treated elsewhere in Part IV.

Ample references to all other topics are included in Chapters 35–53.

REFERENCES

1. L. M. Brekhovskikh, *Waves in Layered Media*, 1st ed. (trans. D. Lieberman), Academic Press, New York, 1960, pp. 415–426.

2. L. M. Brekhovskikh, "The Average Field in an Underwater Sound Channel," *Sov. Phys.–Acoust.*, Vol. 11, 1965, pp. 126–134.

3. P. W. Smith, Jr., "Averaged Impulse Response of a Shallow-Water Channel," *J. Acoust. Soc. Am.*, Vol. 50, 1971, pp. 332–336.

4. D. E. Weston, "Intensity-Range Relations in Oceanographic Acoustics," *J. Sound Vib.*, Vol. 18, 1971, pp. 271–287.

5. P. W. Smith, Jr., "Spatial Coherence in Multipath or Multimodal Channels," *J. Acoust. Soc. Am.*, Vol. 60, 1976, pp. 305–310.

6. L. M. Brekhovskikh and Yu. Lysanov, Fundamentals of Ocean Acoustics, Springer-Verlag, New York 1982, pp. 104–108.

35

ESSENTIAL OCEANOGRAPHY

F. H. FISHER AND P. F. WORCESTER

1 INTRODUCTION

Propagation of sound in the ocean is determined by how the speed of sound varies along its path. The decay in intensity of sound with increasing range is governed by geometric spreading losses and sound absorption due to two chemicals in the ocean. Intensity losses due to geometric spreading and interaction with boundaries are beyond the scope of this chapter.

The ocean is a complex chemical solution of many salts that modify the sound speed and sound absorption of its solvent water. Sound speed increases with temperature, pressure, and concentration of the salts. Sodium chloride, the principal salt, has the greatest effect on sound speed; magnesium sulfate and boric acid are important ones for sound absorption. The ratio of the concentration of salts in open-ocean seawater to the sodium chloride concentration is essentially constant. The total concentration of seawater salts is called salinity. A salinity $S = 35$ corresponds to about a 3.5% concentration of salts, the nominal average value for open-ocean water, away from rivers, ice, and heavy rainfall regions. (In older literature salinity was expressed in parts per thousand with the symbol ‰[1]). Drake et al.[2] point out that more than three-quarters of the ocean has a salinity within 1 of the median, which is 34.69.

Vertical and horizontal variations of sound speed in the ocean determine the paths by which sound propagates and the resulting, often dramatic, effects on intensity caused by refraction. Chemical sound absorption decreases the intensity of the sound energy with increasing range, in addition to the losses due to geometric

Encyclopedia of Acoustics, edited by Malcolm J. Crocker
ISBN 0-471-80465-7 © 1997 John Wiley & Sons, Inc.

spreading. At frequencies above a few kilohertz, scattering due to thermal microstructure can cause attenuation beyond that due to absorption.

Equations for sound speed and absorption are given along with tables of representative values for seawater of salinity 35.

2 SOUND SPEED

Sound speed varies greatly with latitude. Considerable variability in sound speed occurs due to changing seasons, which affect near-surface ocean temperatures. Internal waves and warm-core or cold-core eddies cause significant fluctuations in sound speed in the upper waters of the ocean.

2.1 Mean Sound Speed Distribution

The vertical and horizontal distribution of sound speed in the ocean determines the paths by which sound propagates. (Ocean currents also affect acoustic propagation but are normally much less important.) The vertical sound speed gradient is almost always much greater than the horizontal gradient. Figure 1 shows a series of vertical sound speed profiles extending from 60° S to 55° N along 150° W in the Pacific Ocean. These were computed from annual average profiles of temperature and salinity taken from a climatological data base constructed by Levitus.[3] Sound speed fluctuations due to internal waves, mesoscale eddies, and other small-scale oceanographic variability have been suppressed by a combination of time-averaging and spatial smoothing. Other climatologies are available.[4]

A striking feature of the profiles is the sound speed

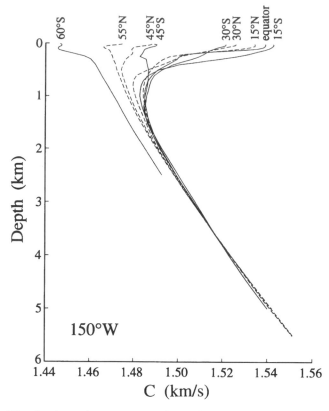

Fig. 1 Annual average sound speed profiles from 60° S to 55° N along 150° W in the Pacific Ocean. Profiles north (south) of the equator are dashed (solid). The profiles were constructed from temperature and salinity data extracted from Levitus[3] using the Del Grosso sound speed equation.

minimum present at about 1000 m depth in low and mid-latitudes, known as the SOFAR axis (for Sound Fixing And Ranging). Above the axis sound speed increases toward the surface with increasing temperature in the main thermocline. Below the axis temperature is nearly constant and sound speed increases with increasing pressure. Since sound is refracted toward regions of low sound speed, acoustic energy can travel long distances in the waveguide formed by the sound speed minimum without interacting with the lossy surface and bottom. At high latitudes, however, temperature varies only weakly as a function of depth, leading to profiles in which sound speed increases more or less monotonically with increasing pressure.

Not shown in Fig. 1 is the seasonal variability of the upper few hundred metres. In most areas a shallow thermocline forms each summer and disappears each winter. The Levitus[3] data base includes seasonal and monthly averages as well as the annual averages shown here. Acoustic propagation is sensitive to the vertical gradient of the sound speed profile, so it is important to use an accurate profile to make acoustical predictions.

While two-dimensional propagation codes are commonly used, horizontal gradients can be important in some situations, including very long range propagation and propagation in the vicinity of strong oceanic fronts and through eddies. Figure 2 shows the depth of the sound channel axis and the axial sound speed for the entire world, constructed from selected data obtained over many years (Ref. 5; see also Refs. 6 and 7). The shoaling of the SOFAR axis toward the poles is the most dramatic feature. Strong horizontal gradients are evident in regions with major currents, such as the Gulf Stream, and in regions with permanent major fronts.

2.2 Sound Speed Measurements

Vertical sound speed profiles can be obtained using any of a variety of instruments that measure temperature and salinity as a function of depth for subsequent conversion to sound speed or using sound velocimeters that directly measure sound speed as a function of depth using acoustical techniques. Synoptic maps of the sound speed distribution over large areas can be obtained using the technique of ocean acoustic tomography, although with significantly lower vertical resolution than can be obtained from point measurements.

The most accurate temperature and salinity profiles are obtained with instruments that record electrical conductivity, temperature, and pressure while being lowered on a wire from shipboard. The instruments are generically called CTDs (for Conductivity–Temperature–Depth). Salinity, depth, sound speed, and other quantities of interest are then computed from the basic measurements. Fofonoff and Millard[8] give algorithms for computing a variety of the fundamental properties of seawater from temperature–conductivity–pressure data. (As noted below, more recent work indicates that the sound speed equation due to DelGrosso[9] is preferable to the one used by Fofonoff and Millard.) Since stopping a ship to make a CTD cast is time consuming, less accurate, expendable instruments that can be used while a ship is underway have been developed. The simplest of these is the eXpendable BathyThermograph (XBT), which consists of a streamlined probe containing a thermistor. The probe is connected to a shipboard recorder by a fine wire that unspools from canisters in the probe and on shipboard and breaks when the probe reaches its maximum depth. The fall rate of the probe is calibrated so that the time since launch can be converted to depth. To compute sound speed, the temperature provided by the XBT must be augmented with salinity information. In many parts of the ocean, particularly below the surface layers, temperature and salinity are highly correlated, so that an archival T–S (Temperature–Salinity) relation can be used to obtain a salinity estimate from

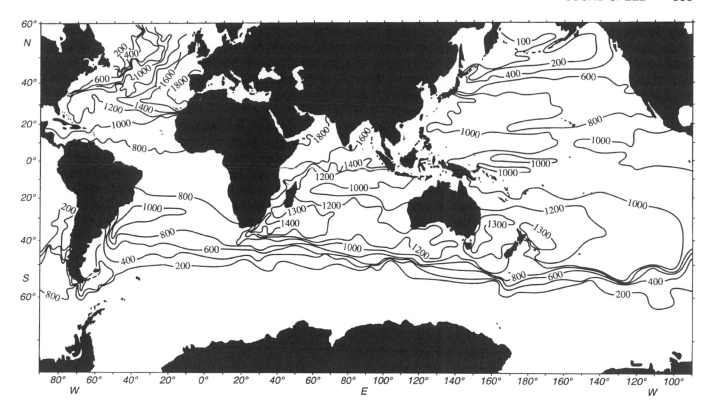

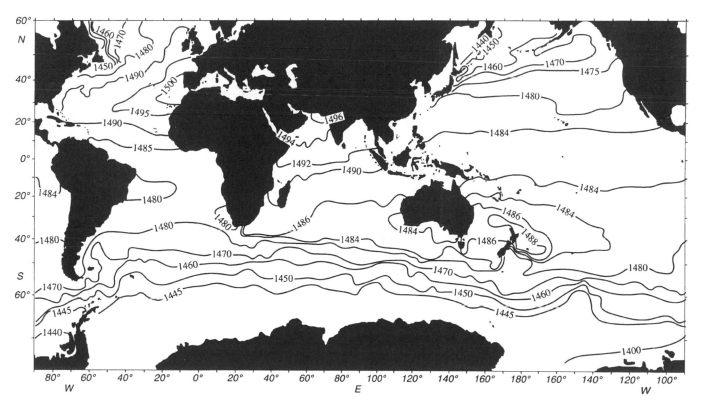

Fig. 2 Sound channel axis depth in metres (top), and axial sound speed in metres per second (bottom). The axial depth and sound speed are modified from Munk and Forbes.[5] (Reprinted, by permission, from Ref. 5.)

the temperature data. For regions where the relation between temperature and salinity is more variable or when more accurate profiles are required, eXpendable Conductivity–Temperature–Depth (XCTD) probes are available (at significantly greater cost). Aircraft-deployable versions of the XBT and XCTD also exist.

Rather than measuring temperature and salinity to compute sound speed, sound velocimeters directly measure the travel times of acoustic pulses transmitted over a known path length of order 10 cm to obtain sound speed.[10,11] Sound velocimeters are ultimately calibrated using baths of seawater and pure water. Sound velocimeters are available both as units that can be lowered from shipboard (typically as an auxiliary sensor on a CTD) and as expendable units (XSVs).

Sound velocimeters operate at very high frequencies, in the megahertz region. Because of relaxation sound absorption processes, discussed below, there are small dispersive effects that lower the sound speed near and below each relaxation frequency from that measured at a few megahertz with a sound velocimeter. These dispersions amount to less than 0.1 m/s, less than one part in 15,000. At atmospheric pressure and 25°C, there could be as much as a 5 cm/s higher sound speed measured at 4 MHz than that applicable to sound propagation below 1 kHz.[12]

Ocean acoustic tomography is a technique for remotely sensing the sound speed field over large areas by precisely measuring the travel times of pulses transmitted between a number of sources and receivers.[13,14] Mathematical inverse methods are then used to infer the three-dimensional sound speed structure in the region. The inferred structure is a smoothed version of the actual structure, rather than the detailed vertical profiles obtained from point measurements. The technique is in a sense a large-scale sound velocimeter, although the complexity of acoustic propagation requires inverse methods to interpret quantitatively the measured travel times in terms of the three-dimensional sound speed field. While the technique is not yet routine, tomographic instruments are commercially available.

2.3 Sound Speed Equation

Recent field data[15–19] indicate that the sound speed equation due to Del Grosso[9] is preferred over that given in Fofonoff and Millard[8] based on the work of Chen and Millero.[20] Dushaw et al.[18] summarize the history of laboratory measurements of sound speed and the construction of sound speed equations from these measurements. Del Grosso's equation is

$$C_{STP} = C_{000} + \Delta C_T + \Delta C_S + \Delta C_P + \Delta C_{STP},$$

where

$$C_{000} = 1402.392,$$

$$\Delta C_T = 0.501109398873 \times 10^1 T$$
$$- 0.550946843172 \times 10^{-1} T^2$$
$$+ 0.221535969240 \times 10^{-3} T^3,$$

$$\Delta C_S = 0.132952290781 \times 10^1 S$$
$$+ 0.128955756844 \times 10^{-3} S^2,$$

$$\Delta C_P = 0.156059257041 \times 10^0 P$$
$$+ 0.244998688441 \times 10^{-4} P^2$$
$$- 0.883392332513 \times 10^{-8} P^3,$$

$$\Delta C_{STP} = -0.127562783426 \times 10^{-1} TS$$
$$+ 0.635191613389 \times 10^{-2} TP$$
$$+ 0.265484716608 \times 10^{-7} T^2 P^2$$
$$- 0.159349479045 \times 10^{-5} TP^2$$
$$+ 0.522116437235 \times 10^{-9} TP^3$$
$$- 0.438031096213 \times 10^{-6} T^3 P$$
$$- 0.161674495909 \times 10^{-8} S^2 P^2$$
$$+ 0.968403156410 \times 10^{-4} T^2 S$$
$$+ 0.485639620015 \times 10^{-5} TS^2 P$$
$$- 0.340597039004 \times 10^{-3} TSP,$$

where C is in metres per second, T is in degrees Celsius, S is in parts per thousand, and P is in kilograms per square centimetre gauge. [Del Grosso's paper does not indicate which International Temperature Scale (ITS) was used, but ITS-68 was the standard at the time his work was done. His equation implies accuracies of millimetres per second, which require that temperature be known to $O(1 \text{ m°C})$. Standard ITS-68 and ITS-90, the current standard, differ by as much as 0.005°C.] Pressure can be converted to depth using the algorithms in Fofonoff and Millard.[8] Values of sound speed are given in Table 1 as a function of depth, pressure, and temperature for $S = 35$. (Millero and Li[21] provide a correction to the Chen and Millero equation to make it consistent with the Del Grosso[9] equation but with a wider range of applicability.)

3 SOUND ABSORPTION

A brief description of the history of sound absorption in both pure water and seawater might be aptly summed up as a progressive series of discoveries of excess or "anomalous" absorptions. Sound absorption in pure water is itself considered anomalous since it is

**TABLE 1   Sound Speed as Function of Depth D and Temperature T
for Seawater of Salinity 35**

D (m)	Sound Speed (m/s)				
	$T = 0°C$	$T = 5°C$	$T = 10°C$	$T = 20°$	$T = 30°C$
0	1449.08	1470.64	1489.78	1521.47	1545.47
100	1450.69	1472.27	1491.42	1523.11	1547.07
200	1452.30	1473.90	1493.07	1524.76	1548.67
300	1453.92	1475.53	1494.71	1526.41	1550.27
400	1455.55	1477.17	1496.36	1528.06	1551.88
500	1457.18	1478.81	1498.01	1529.72	1553.48
1000	1465.40	1487.07	1506.31	1537.99	1561.51
1500	1473.75	1495.43	1514.67	1546.30	1569.55
2000	1482.22	1503.87	1523.09	1554.63	1577.63
2500	1490.79	1512.39	1531.56	1563.00	1585.74
3000	1499.48	1520.98	1540.09	1571.39	1593.89
3500	1508.26	1529.65	1548.67	1579.83	1602.09
4000	1517.13	1538.38	1557.31	1588.30	1610.34
4500	1526.08	1547.17	1565.98	1596.80	1618.64
5000	1535.11	1556.02	1574.70	1605.35	1627.01
5500	1544.22	1564.93	1583.46	1613.94	1635.46
6000	1553.38	1573.87	1592.26	1622.57	1643.98

Source: According to Ref. 9.

three times greater than expected on the basis of classical consideration of losses due to shear viscosity absorption mechanisms. Ocean measurements of sound propagation losses in excess of those expected on the basis of geometric spreading and absorption due to pure water led to the discovery of "anomalous" absorption mechanisms in the ocean.

3.1   Brief History of Anomalous Absorption

Early measurements at frequencies above 5 kHz at modest ranges led to the discovery of the first excess absorption phenomenon in the ocean.[22] Since pure water exhibited no known relaxations, chemicals in seawater were the likely source of the excess absorption. Liebermann[23] showed how pressure- and temperature-dependent chemical relaxation processes could produce sound absorption. By the late 1940s[24] and early 1950s,[25] it was found to be due to a small concentration of magnesium sulfate (~0.02 M), which increases sound absorption below 100 kHz by a factor of 30 above that due to water. The effect arises from a pressure-dependent relaxation with a substantial volume change between ion pairs. The multistate ion pair model developed by Eigen and Tamm[26,27] explains the observed large decrease in absorption with increasing pressure.[28] As longer range and lower frequency propagation experiments were performed, another anomalous absorption effect was found.[29] At frequencies below 1 kHz, a pressure-dependent boric acid structural relaxation process increases sound absorption by up to a factor of 5–10 times above that due to magnesium sulfate at atmospheric pressure, depending on the acidity (pH) of the ocean. This is due to a very small concentration (~0.0004 M) of boric acid, which has a structural relaxation with a very large volume change leading to the large absorption observed.[12,30]

Figure 3[31] illustrates the current state of knowledge with respect to the principal sound absorption mechanisms in the ocean. Mellen et al.[32] have published a detailed summary of two decades of research on attenuation of sound in the ocean.

3.2   Sound Absorption Equation

Comprehensive summaries have been published of what is known about sound absorption in all oceans by Francois, Garrison, and their collaborators.[31,33,34] They included extensive experimental in situ data as well as laboratory measurements and summarized the contributions of boric acid, magnesium sulfate, and pure water for ocean water of pH 8. Their results are expressed by the equations

Total absorption = boric acid contribution

+ $MgSO_4$ contribution

+ pure-water contribution,

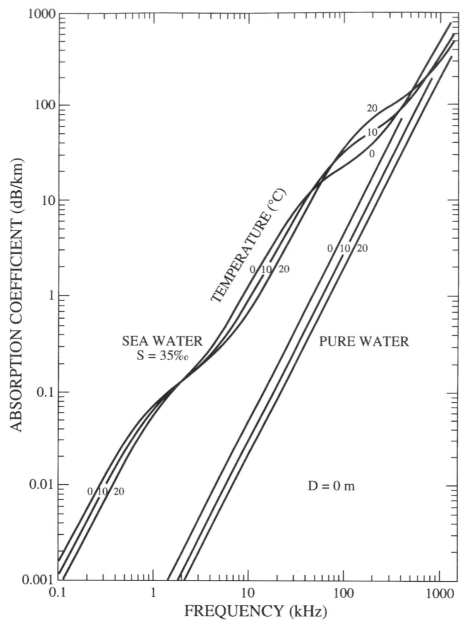

Fig. 3 Sound absorption in decibels per kilometre as a function of frequency at three temperatures at atmospheric pressure (zero depth) for S = 35, according to Francois and Garrison.[31] (Reprinted, by permission, from Ref. 31.)

$$a = \frac{A_1 P_1 f_1 f^2}{f^2 + f_1^2} + \frac{A_2 P_2 f_2 f^2}{f^2 + f_2^2} + A_3 P_3 f^2 \quad \text{dB/km}$$

for frequency f in kilohertz.

For the *boric acid contribution*

$$A_1 = \frac{8.86}{c} \times 10^{(0.78\text{pH}-5)} \quad \text{dB/(km·kHz)},$$

$$P_1 = 1,$$

$$f_1 = 2.8 \left(\frac{S}{35} \right)^{0.5} \times 10^{(4-1245/\theta)} \quad \text{kHz},$$

where c is the sound speed (in metres per second), given approximately by

$$c = 1412 + 3.21T + 1.19S + 0.0167D,$$

T is the temperature (in degrees Celsius), $\theta = 273 + T$, S is the salinity (in parts per thousand), and D is the depth (in metres).

For the *MgSO₄ contribution*

$$A_2 = 21.44 \frac{S}{c} (1 + 0.025T) \quad \text{dB/(km·kHz)},$$

$$P_2 = 1 - 1.37 \times 10^{-4}D + 6.2 \times 10^{-9}D^2,$$

$$f_2 = \frac{8.17 \times 10^{(8 - 1990/\theta)}}{1 + 0.0018(S - 35)} \quad \text{kHz.}$$

For the *pure-water contribution*

for $T \leq 20°C$,

$$A_3 = 4.937 \times 10^{-4} - 2.59 \times 10^{-5}T + 9.11 \times 10^{-7}T^2$$
$$- 1.50 \times 10^{-8}T^3 \quad \text{dB/(km·kHz)}^2$$

for $T > 20°C$,

$$A_3 = 3.964 \times 10^{-4} - 1.146 \times 10^{-5}T + 1.45 \times 10^{-7}T^2$$
$$- 6.5 \times 10^{-10}T^3 \quad \text{dB/(km·kHz)}^2$$

$$P_3 = 1 - 3.83 \times 10^{-5}D + 4.9 \times 10^{-10}D^2.$$

From these equations, sample values of sound absorption as a function of temperature and frequency for seawater of salinity $S = 35$ and pH 8 are shown in Table 2.

Since pH in the ocean varies with geographic region, the coefficient A_1 for boric acid also varies. This is shown in Fig. 4 for the depth of the sound channel axis.[35] While Lovett's formula for A differs slightly from Francois and Garrison's, the results are nearly identical.

4 FUTURE PROBLEMS

Absolute calibration of sound velocimeters as a function of pressure and temperature for use in the ocean is an important problem that needs to be addressed, initially to an accuracy of 0.1 m/s and ultimately to 1 cm/s, a goal that now appears to be possible.[36] As propagation experiments at sea become more sophisticated, it is necessary to have environmental data such as the sound speed profile to higher absolute accuracies, at least to 0.1 m/s. For geodesy research, measuring displacement changes along faults at the bottom of the ocean requires accuracies of 1 cm in distance measurements at ranges of 1 km,[36] which in turn requires knowledge of sound speed to 1 cm/s.

Knowledge of chemical sound absorption as a function of pressure and temperature is important in doing quantitative backscattering bathymetric research, in which the geophysical properties of the ocean bottom can be inferred from the strength of the return signals. The temperature dependence of the sound absorption of magnesium sulfate in seawater has not been measured at deep ocean depths. Two other chemical relaxations, minor compared to the ones discussed here, have yet to be incorporated into the absorption equations; one is due to calcium sulfate and the other is a complex one involving the carbonate system. Interaction or coupling between the various relaxation absorption processes poses difficult and as yet untreated theoretical and experimental challenges.

TABLE 2 Sound Absorption as Function of Frequency and Temperature at Salinity 35 and pH 8

Frequency (kHz)	Sound Absorption (dB/km)				
	−1.8°C	0°C	10°C	20°C	30°C
0.4	0.020	0.019	0.015	0.011	0.008
0.6	0.036	0.035	0.029	0.022	0.017
0.8	0.051	0.051	0.045	0.036	0.028
1	0.064	0.064	0.060	0.052	0.042
2	0.122	0.122	0.123	0.122	0.115
4	0.291	0.282	0.245	0.232	0.231
6	0.558	0.532	0.420	0.360	0.335
8	0.918	0.872	0.660	0.526	0.457
10	1.36	1.29	0.963	0.737	0.606
20	4.40	4.29	3.35	2.44	1.80
40	10.8	11.2	11.0	8.75	6.42
60	15.4	16.4	19.7	17.7	13.7
80	18.8	20.2	27.3	27.9	23.0
100	21.8	23.4	33.6	38.1	33.7
200	39.8	40.6	54.4	77.1	91.1
400	105	100	96.1	124	170
600	214	199	159	173	225
800	366	338	247	236	280
1000	562	515	359	316	344

Source: According to Ref. 31.

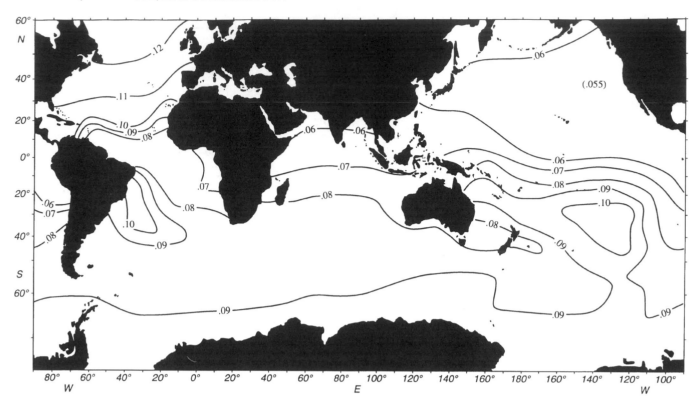

Fig. 4   The coefficient A of the boric acid term for sound absorption as a function of location as modified from Lovett.[35] (Reprinted, by permission, from Ref. 35.)

REFERENCES

1. H. U. Sverdrup, M. W. Johnson, and R. H. Fleming, *The Oceans*, Prentice-Hall, Englewood Cliffs, N.J., 1942, pp. 50–56.

2. C. L. Drake, J. Imbrie, J. A. Knauss, and K. Turekian, "Physical Properties of Seawater," in C. L. Drake et al. (Eds.), *Oceanography*, Holt, Rinehart and Winston, New York, 1978, pp. 52–67.

3. S. Levitus, "Climatological Atlas of the World Ocean," NOAA Professional Paper 13, 1982.

4. W. J. Teague, M. J. Carron, and P. J. Hogan, "A Comparison between the Generalized Digital Environmental Model and Levitus Climatologies," *J. Geophys. Res.*, Vol. 95, 1990, pp. 7167–7183.

5. W. H. Munk and A. M. G. Forbes, "Global Ocean Warming: An Acoustic Measure?" *J. Phys. Oceanogr.*, Vol. 19, 1989, pp. 1765–1778.

6. J. Northrup and J. G. Colborn, "Sofar Channel Axial Speed and Depth in the Atlantic Ocean," *J. Geophys. Res.*, Vol. 79, 1974, pp. 5633–5641.

7. R. H. Johnson and R. A. Norris, "Geographic Variation of Sofar Speed and Axis Depth in the Pacific Ocean," *J. Geophys. Res.*, Vol. 73, 1968, pp. 4695–4700.

8. N. P. Fofonoff and R. C. Millard, Jr., "Algorithms for Computation of Fundamental Properties of Seawater," *UNESCO Tech. Pap. Mar. Sci.*, Vol. 44, 1983, pp. 1–53.

9. V. A. Del Grosso, "New Equation for the Speed of Sound in Natural Waters (with Comparisons to Other Equations)," *J. Acoust. Soc. Am.*, Vol. 56, 1974, pp. 1084–1091.

10. M. Greenspan and C. Tschiegg, "Sing-Around Ultrasonic Velocimeter for Liquids," *Rev. Sci. Instrum.*, Vol. 28, 1957, pp. 897–901.

11. K. V. Mackenzie, "A Decade of Experience with Velocimeters," *J. Acoust. Soc. Am.*, Vol. 50, 1971, pp. 1321–1333.

12. V. P. Simmons, "Investigation of the 1 kHz Sound Absorption Anomaly in Sea Water," Ph.D. Dissertation, University of California, San Diego, 1975.

13. W. H. Munk and C. Wunsch, "Ocean Acoustic Tomography: A Scheme for Large Scale Monitoring," *Deep-Sea Res.*, Vol. 26, 1979, pp. 123–161.

14. W. H. Munk, P. F. Worcester, and C. Wunsch, *Ocean Acoustic Tomography*, Cambridge University Press, Cambridge, England, 1995.

15. P. F. Worcester, B. D. Dushaw, and B. M. Howe, "Gyre-Scale Reciprocal Acoustic Transmissions," in J. Potter and A. Warn-Varnas (Eds.), *Ocean Variability & Acoustic Propagation*, Kluwer Academic Publishers, Dordrecht, The Netherlands, 1991, pp. 119–134.

16. J. L. Spiesberger and K. Metzger, "New Estimates of Sound Speed in Water," *J. Acoust. Soc. Am.*, Vol. 89, 1991, pp. 1697–1700.

17. J. L. Spiesberger and K. Metzger, "A New Algorithm for Sound Speed in Seawater," *J. Acoust. Soc. Am.*, Vol. 89, 1991, pp. 2677–2688.

18. B. D. Dushaw, P. F. Worcester, B. D. Cornuelle, and B. M. Howe, "On Equations for the Speed of Sound in Seawater," *J. Acoust. Soc. Am.*, Vol. 93, 1993, pp. 255–275.

19. J. L. Spiesberger, "Is Del Grosso's Sound-Speed Algorithm Correct?" *J. Acoust. Soc. Am.*, Vol. 93, 1993, pp. 2235–2237.

20. C. Chen and F. J. Millero, "Speed of Sound in Seawater at High Pressures," *J. Acoust. Soc. Am.*, Vol. 62, 1977, pp. 1129–1135.

21. F. J. Millero and Xu Li, Comments on "On Equations for the Speed of Sound in Seawater," [*J. Acoust. Soc. Am.*, 93 1993, pp. 255–275]; *J. Acoust. Soc. Am.*, Vol. 95, 1994, pp. 2757–2759.

22. E. B. Stephenson, "Absorption Coefficients of Supersonic Sound in Open Sea Water," U.S. Naval Research Laboratory Report S-1549, 1939.

23. L. N. Liebermann, "Sound Absorption in Chemically Active Media," *Phys. Rev.*, Vol. 76, 1949, pp. 1520–1524.

24. R. W. Leonard, P. C. Combs, and L. R. Skidmore, "The Attenuation of Sound in Synthetic Sea Water," *J. Acoust. Soc. Am.*, Vol. 21, 1949, p. 63.

25. D. A. Bies, *J. Chem. Phys.*, Vol. 23, 1955, pp. 428–434.

26. M. Eigen and K. Tamm, "Sound Absorption in Electrolyte Solutions Due to Chemical Relaxation," a translation of the original paper [*Z. Elektrochem.*, Vol. 66, 1962, pp. 107–121] with an introduction by K. Tamm, Univ. of Calif., San Diego, SIO Reference 68–41, 1968.

27. K. Tamm, "Acoustic Relaxation in Electrolyte Solutions," in *Proceedings of the International School of Physics "Enrico Fermi," Course XXVII*, D. Dette (Ed.), Academic Press, New York, 1963, pp. 175–222.

28. F. H. Fisher, "Ultrasonic Absorption in $MgSO_4$ Solutions as a Function of Pressure and Dielectric Constant," *J. Acoust. Soc. Am.*, Vol. 38, 1965, pp. 805–812.

29. W. H. Thorp, "Analytic Description of the Low-Frequency Attenuation Coefficient," *J. Acoust. Soc. Am.*, Vol. 42, 1967, pp. 270–271.

30. F. H. Fisher and V. P. Simmons, "Sound Absorption in Sea Water," *J. Acoust. Soc. Am.*, Vol. 62, 1977, pp. 558-564.

31. R. E. Francois and G. R. Garrison, "Sound Absorption Based on Ocean Measurements: Part II: Boric Acid Contribution and Equation for Total Absorption," *J. Acoust. Soc. Am.*, Vol. 72, 1982, pp. 1879–1890.

32. R. H. Mellen, P. M. Scheifele, and D. G. Browning, *Global Model for Sound Absorption in Sea Water*, NUSC Scientific and Engineering Studies, New London, CT, 1987.

33. R. E. Francois and G. R. Garrison, "Sound Absorption Based on Ocean Measurements: Part I: Pure Water and Magnesium Sulfate Contributions," *J. Acoust. Soc. Am.*, Vol. 72, 1982, pp. 896–907.

34. G. R. Garrison, R. E. Francois, E. W. Early, and T. Wen, "Sound Absorption Measurements at 10–650 kHz in Arctic Waters," *J. Acoust. Soc. Am.*, Vol. 73, 1983, pp. 492–501.

35. J. R. Lovett, "Geographic Variation of Low Frequency Sound Absorption in the Atlantic, Indian and Pacific Oceans," *J. Acoust. Soc. Am.*, Vol. 67, 1980, pp. 338–340.

36. F. N. Spiess, "Suboceanic Geodetic Measurements," *IEEE Trans. Geosci. Remote Sensing*, Vol. GE-23, 1985, pp. 502–510.

36

PROPAGATION OF SOUND IN THE OCEAN

WILLIAM A. KUPERMAN

1 INTRODUCTION

The ocean is an acoustic waveguide bounded above by the air–sea interface and below by a viscoelastic layered structure, the latter commonly called the ocean bottom. The physical oceanographic parameters, as ultimately represented by the ocean sound speed structure, make up the index of refraction of the water column waveguide. The combination of water column and bottom properties leads to a set of generic sound propagation paths descriptive of most propagation phenomena in the ocean. This chapter first reviews the qualitative properties of the various propagation paths. Then the ocean acoustic wave equations with the appropriate coefficients and boundary conditions are presented. Sound propagation models are essentially algorithms for solving the equations; the word *model* is used because the technique used for realistic ocean scenarios typically is implemented as a computer model (program). Many models exist because it is not computationally efficient to use a single algorithm for all frequencies and ocean environments (e.g., water depth, variable bathymetry). Acoustic measurements at sea normally show significant fluctuations that are not predicted by the deterministic class of models considered here; an output from these models represents an average prediction. The models discussed are only as realistic as the input environment, and they do not include temporal ocean variability. A basic set of models are reviewed and results obtained from these models are presented to further elucidate the physics of sound propagation in the sea. Throughout this chapter, the depth coordinate z is positive in the downward direction. The appendix reviews some relevant units.

Encyclopedia of Acoustics, edited by Malcolm J. Crocker
ISBN 0-471-80465-7 © 1997 John Wiley & Sons, Inc.

2 QUALITATIVE DESCRIPTION OF OCEAN SOUND PROPAGATION PATHS

Chapter 35 summarizes essential details of physical oceanography vis à vis acoustics. Here we summarize aspects of oceanography that impact propagation paths before we go into more detail on the propagation itself.

2.1 Selective Review of the Ocean Environment

Figure 1 illustrates a typical set of sound speed profiles indicating greatest variability near the surface as a function of season and time of day. In a warmer season (or warmer part of the day), the temperature increases near the surface and hence the sound speed decreases with depth. In nonpolar regions, the oceanographic properties of the water near the surface result from mixing activity originating from the air–sea interface. This near-surface mixed layer has a constant temperature (except in calm, warm surface conditions as described above). In this isothermal mixed layer, the sound speed profile can increase with depth due to the pressure gradient effect. This is the "surface duct" region.

Below the mixed layer is the thermocline, in which the temperature decreases with depth and the sound speed decreases with depth. Below the thermocline, the temperature is constant (about 4°C, a thermodynamic property of salt water at high pressure) and the sound speed increases because of increasing pressure. Therefore, there exists a depth between the deep isothermal region and the mixed layer with a minimum in sound speed; this depth is often referred to as the axis of the deep sound channel. However, in polar regions, the water is coldest near the surface, and hence the minimum sound speed is at the ocean–air (or ice) interface, as indicated in Fig. 1. In continental shelf regions (shallow water) with

SOUND SPEED (METERS/SEC)

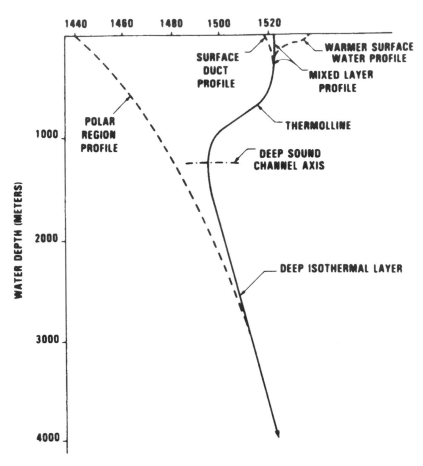

Fig. 1 Generic sound speed profiles.

water depths on the order of a few hundred metres, only the upper region of the sound speed profile in Fig. 1, which is dependent on season and time of day, affects sound propagation in the water column.

Figure 2 is a contour display of the sound speed structure of the North and South Atlantic,[1] with the deep sound channel indicated by the heavy dashed line. Note the geographic (and climatic) variability of the upper ocean sound speed structure and the stability of this structure in the deep isothermal layer. For example, as explained above, the axis of the deep sound channel becomes shallower toward both poles, eventually going to the surface.

2.2 Sound Propagation Paths in the Ocean

Figure 3 is a schematic of the basic types of propagation in the ocean resulting from the sound speed profiles (indicated by the dashed lines) discussed in the last section. These sound paths can be understood from a simplified statement of Snell's law: Sound bends locally toward regions of low sound speed (or sound is "trapped" in regions of low sound speed). Paths A and B correspond to surface duct propagation where the minimum sound speed is at the ocean surface (or at the bottom of the ice cover for the Arctic case). Path C, depicted by a ray leaving a deeper source at a shallow horizontal angle, propagates in the deep sound channel, whose axis is at the shown sound speed minimum. This local minimum tends to become more shallow toward polar latitudes converging to the Arctic surface minimum. Hence, for midlatitudes, sound in the deep channel can propagate long distances without interacting with lossy boundaries; propagation via this path has been observed over distances of thousands of kilometres. Also, from the above description of the geographical variation of the acoustic environment combined with Snell's law, we can expect that shallow sources coupling into the water column at polar latitudes will tend to propagate more horizontally around an axis that becomes deeper toward the midlatitudes. Path D, which is at slightly steeper angles than those associated with path C, is convergence zone propagation, a

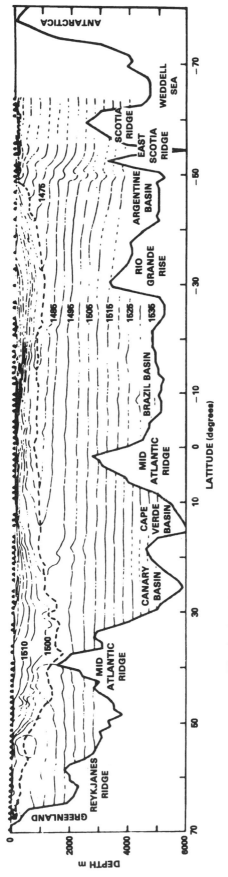

Fig. 2 Sound speed contours at 5-m/s intervals taken from North and South Atlantic along 30.50° W. Dashed line indicates axis of deep sound channel (from Ref. 1).

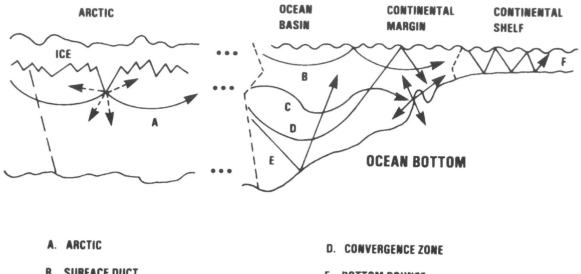

ARCTIC   OCEAN   CONTINENTAL   CONTINENTAL
                         BASIN   MARGIN   SHELF

A. ARCTIC   D. CONVERGENCE ZONE

B. SURFACE DUCT   E. BOTTOM BOUNCE

C. DEEP SOUND CHANNEL   F. SHALLOW WATER

Fig. 3   Schematic representation of various types of sound propagation in the ocean.

spatially periodic (~35–65 km) refocusing phenomenon producing zones of high intensity near the surface due to the upward refracting nature of the deep sound speed profile. Referring back to Fig. 1, there may be a depth in the deep isothermal layer at which the sound speed is the same as it is at the surface. This depth is called the *critical depth* and, in effect, is the lower limit of the deep sound channel. A *positive* critical depth specifies that the environment supports long-distance propagation without bottom interaction, whereas a *negative* one implies that the bottom ocean boundary *is* the lower boundary of the deep sound channel. The bottom bounce path *E*, which interacts with the ocean bottom, is also a periodic phenomenon but with a shorter cycle distance and a shorter total propagation distance because of losses when sound is reflected from the ocean bottom. Finally, the right-hand side of Fig. 3 depicts propagation in a shallow-water region such as a continental shelf. Here sound is channeled in a waveguide bounded above by the ocean surface and below by the ocean bottom.

The modeling of sound propagation in the ocean is further complicated because the environment varies laterally (range dependence), and all environmental effects on sound propagation are dependent on acoustic frequency in a rather complicated way that often makes the ray-type schematic of Fig. 3 misleading, particularly at low frequencies. Finally, a quantitative understanding of acoustic loss mechanisms in the ocean is required for modeling sound propagation. These losses are, aside from geometric spreading, volume attenuation (Chapters 35 and 38), bottom loss (i.e., a smooth water–bottom

interface is not a perfect reflector), and surface, volume (including fish), and bottom scattering loss (Chapters 38–40 and 44). Here we will only review those aspects of bottom loss that impact propagation structure.

2.3   Bottom Loss

Ocean bottom sediments are often modeled as fluids since the rigidity (and hence the shear speed) of the sediment is usually considerably less than that of a solid such as rock. In the latter case, which applies to the "ocean basement" or the case where there is no sediment overlying the basement, the medium must be modeled as an elastic solid, which means it supports both compressional and shear waves.

Reflectivity, the amplitude ratio of reflected and incident plane waves at an interface separating two media, is an important measure of the effect of the bottom on sound propagation. For an interface between two fluid semi-infinite half-spaces with density ρ_i and sound speed c_i, $i = 1, 2$, as shown in Fig. 4a [assuming a harmonic time dependence of $\exp(-i\omega t)$], the reflectivity is given by

$$\mathcal{R}(\theta) = \frac{\rho_2 k_{1z} - \rho_1 k_{2z}}{\rho_2 k_{1z} + \rho_1 k_{2z}}, \qquad (1)$$

with

$$k_{iz} = \frac{\omega}{c_i} \sin \theta_i \equiv k_i \sin \theta_i, \qquad i = 1, 2. \qquad (2)$$

(a)

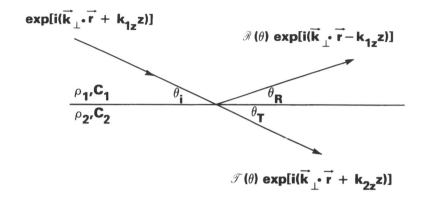

(b)

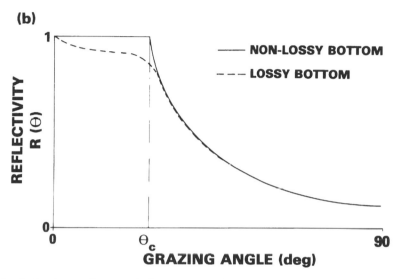

Fig. 4 Reflection and transmission process. Grazing angles are defined relative to the horizontal.

The incident and transmitted grazing angles are related by Snell's law,

$$k_\perp = k_1 \cos \theta_1 = k_2 \cos \theta_2, \qquad (3)$$

where the incident grazing angle θ_1 is also equal to the angle of the reflected plane wave. The function $\mathcal{R}(\theta)$ is also referred to as the *Rayleigh reflection coefficient* and has unit magnitude (total internal reflection) when the numerator and denominator of Eq. (1) are complex conjugates. This occurs when k_{2z} is purely imaginary, and using Snell's law to determine θ_2 in terms of the incident grazing angle, we obtain the *critical grazing angle* below which there is perfect reflection,

$$\cos \theta_c = \frac{c_1}{c_2}, \qquad (4)$$

so that a critical angle can exist only when the speed in

the second medium is higher than that of the first. Using Eq. (2), Eq. (1) can be rewritten as

$$\mathcal{R}(\theta) = \frac{\rho_2 c_2 / \sin \theta_2 - \rho_1 c_1 / \sin \theta_1}{\rho_2 c_2 / \sin \theta_2 + \rho_1 c_1 / \sin \theta_1} \equiv \frac{Z_2 - Z_1}{Z_2 + Z_1}, \qquad (5)$$

where the right-hand side is in the form of *impedances* $Z_i(\theta_i) = \rho_i c_i / \sin \theta_i$, which are the ratios of the pressure to the vertical particle velocity at the interface in the ith medium. Written in this form, more complicated reflection coefficients become intuitively plausible. Consider the case in Fig. 4a where the second medium is elastic and thus supports shear as well as compressional waves with sounds speeds c_{2s} and c_{2p}, respectively. The Rayleigh reflection coefficient is then given by

$$\mathcal{R}(\theta) = \frac{Z_{2,\text{tot}} - Z_1}{Z_{2,\text{tot}} + Z_1}, \qquad (6)$$

with the total impedance of the second medium being

$$Z_{2,\text{tot}} \equiv Z_{2s} \sin^2 2\theta_{2s} + Z_{2p} \cos^2 2\theta_{2p}. \qquad (7)$$

Snell's law for this case is

$$k_1 \cos \theta_1 = k_{2s} \cos \theta_{2s} = k_{2p} \cos \theta_{2p}. \qquad (8)$$

In lossy media, attenuation can be included in the reflectivity formula by taking the sound speed as complex so that the wavenumbers are subsequently also complex, $k_i \rightarrow k_i + \alpha_i$.

Figure 4b depicts a simple bottom *loss* curve derived from the Rayleigh reflection coefficient formula where both the densities and sound speed of the second medium are larger than those in the first medium, with unit reflectivity indicating perfect reflection. For loss in decibels, 0 dB is perfect reflecting, 6 dB loss is an amplitude factor of $\frac{1}{2}$, 12 dB loss is of $\frac{1}{4}$, and so on. For a lossless bottom, severe loss occurs above the critical angle in the water column due to transmission into the bottom. For the lossy (more realistic) bottoms, only partial reflection occurs at all angles. With paths involving many bottom bounces (shallow-water propagation), bottom losses as small as a few-tenths of a decibel per bounce accumulate and become significant because the propagation path may involve many tens of bounces. Further information on the acoustic properties of the ocean bottom are given in Chapter 37.

Path E in Fig. 3, the bottom-bounce path, often involves paths that correspond to angles near or above the critical angle; therefore, after a few bounces, the sound level is highly attenuated. On the other hand, for shallow angles, many bounces are possible. Hence, in shallow water, path F, most of the energy that propagates is close to the horizontal and this type of propagation is most analogous to waveguide propagation. In fact, as shown in Fig. 5, there exists a small cone from which energy propagates long distances (θ_c is typically 10°–20°). Energy outside the cone is referred to as the near field (or continuous spectrum), which eventually escapes the waveguide. The trapped field originating from within the cone is referred to as the normal-mode field (or discrete spectrum) because there are a set of angles corresponding to discrete paths that constructively interfere and make up the normal (natural) modes of the shallow-water environment.

3 SOUND PROPAGATION MODELS

Sound propagation in the ocean is mathematically described by the wave equation, whose parameters and boundary conditions are descriptive of the ocean envi-

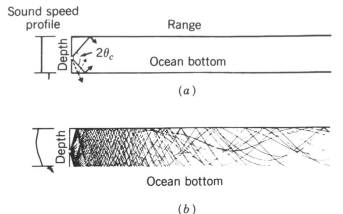

Fig. 5 Ocean waveguide propagation: (*a*) long-distance propagation occurs within a cone of $2\theta_c$; (*b*) same as (*a*) but with nonisovelocity water column showing refraction.

ronment. There are essentially four types of models (computer solutions to the wave equation) to describe sound propagation in the sea: ray theory, the spectral method or fast-field program (FFP), the normal mode (NM), and the parabolic equation (PE). All of these models allow for the fact that the ocean environment varies with depth. A model that also takes into account horizontal variations in the environment (i.e., sloping bottom or spatially variable oceanography) is termed range dependent. For high frequencies (a few kilohertz or above), ray theory is the most practical. The other three model types are more applicable and usable at lower frequencies (below a kilohertz). The hierarchy of underwater acoustics models is shown in schematic form in Fig. 6. The models discussed here are essentially two-dimensional models since the index of refraction has much stronger dependence on depth than on horizontal distance. Nevertheless, bottom topography and strong ocean features can cause horizontal refraction (out of the range depth plane). Ray models are most easily extendable to include this added complexity; though fully three dimensional models are beginning to emerge, the latter are extremely computationally intensive. A compromise that often works for "weak" three-dimensional problems is the $N \times 2D$ approximation, which consists of combining a set of two dimensional solutions along radials to produce a three-dimensional solution.[2]

3.1 Wave Equation and Boundary Conditions

The wave equation is typically written and solved in terms of pressure, displacement, or velocity potentials. For a velocity potential φ, the wave equation in cylindrical coordinates with the range coordinates denoted by $\mathbf{r} = (x, y)$ and the depth coordinate denoted by z (taken

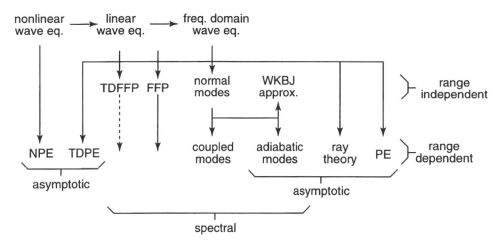

Fig. 6 Hierarchy of underwater acoustics models (TD, time domain).

positive downward) for a source-free region is

$$\nabla^2 \varphi(\mathbf{r}, z, t) - \frac{1}{c^2} \frac{\partial^2 \varphi(\mathbf{r}, z, t)}{\partial t^2} = 0, \qquad (9)$$

where c is the sound speed in the wave-propagating medium. With respect to the velocity potential, the velocity $\mathbf{v}$ and pressure p are given by

$$\mathbf{v} = \nabla \varphi, \qquad p = -\rho \frac{\partial \varphi}{\partial t}, \qquad (10)$$

where ρ is the density of the medium. The wave equation is most often solved in the frequency domain; that is, a frequency dependence of $\exp(-i\omega t)$ is assumed to obtain the Helmholtz equation ($K \equiv \omega/c$)

$$\nabla^2 \varphi(\mathbf{r}, z) + K^2 \varphi(\mathbf{r}, z) = 0. \qquad (11)$$

In underwater acoustics both fluid and elastic (shear-supporting) media are of interest. In elastic media the field can be expressed in terms of three scalar displacement potentials $\boldsymbol{\Phi}(\mathbf{r}, z) \equiv \{\phi(\mathbf{r}, z), \psi(\mathbf{r}, z), \Lambda(\mathbf{r}, z)\}$, corresponding to compressional (P), vertically polarized (SV), and horizontally polarized (SH) shear waves, respectively. In the limiting case of a fluid medium, where shear waves do not exist, $\boldsymbol{\Phi}(\mathbf{r}, z)$ represents the compressional potential $\phi(\mathbf{r}, z)$. Most underwater acoustic applications involve only compressional sources that only excite the P and SV potentials, eliminating the SH potential $\Lambda(\mathbf{r}, z)$. The displacement potentials satisfy the Helmholtz equation with the appropriate compressional or shear sound speeds c_p or c_s, respectively,

$$c_p = \left[\frac{\lambda + 2\mu}{\rho} \right]^{1/2}, \qquad c_s = \left[\frac{\mu}{\rho} \right]^{1/2}, \qquad (12)$$

where λ and μ are the Lamé constants.

The most common plane interface boundary conditions encountered in underwater acoustics are described below: For the ocean surface there is the pressure release condition where the pressure (normal stress) vanishes; for the appropriate solution of the Helmholtz equation, this condition is

$$p = 0, \qquad \varphi = 0 \quad \text{or} \quad \phi = 0. \qquad (13)$$

The interface between the water column (layer 1) and an ocean bottom sediment (layer 2) is often characterized as a fluid–fluid interface. The continuity of pressure and vertical particle velocity at the interface yields the following boundary conditions in terms of pressure:

$$p_1 = p_2, \qquad \frac{1}{\rho_1} \frac{\partial p_1}{\partial z} = \frac{1}{\rho_2} \frac{\partial p_2}{\partial z}, \qquad (14)$$

or velocity potential:

$$\rho_1 \varphi_1 = \rho_2 \varphi_2, \qquad \frac{\partial \varphi_1}{\partial z} = \frac{\partial \varphi_2}{\partial z}. \qquad (15)$$

These boundary conditions applied to the plane-wave fields in Fig. 4a yield the Rayleigh reflection coefficient given by Eq. (1).

For an interface separating two solid layers, the boundary conditions are continuity of vertical displacement $w(\mathbf{r}, z_i)$, tangential displacements $\mathbf{u}(\mathbf{r}, z_i) = (u_x, u_y)$, normal stress $n = \sigma_{zz}$, and tangential stresses $\mathbf{t} = $

$(\sigma_{xx}, \sigma_{yz})$. In cyclindrical coordinates with azimuthal symmetry, the radial and vertical components of displacements (in homogeneous media) u and w, respectively, are

$$u(r, z) = \frac{\partial \phi}{\partial r} + \frac{\partial^2 \psi}{\partial z^2}, \quad (16)$$

$$w(r, z) = \frac{\partial \phi}{\partial z} - \frac{1}{r} \frac{\partial}{\partial r} r \frac{\partial \psi}{\partial r}, \quad (17)$$

and the normal and tangential stresses are

$$\sigma_{zz}(r, z) = (\lambda + 2\mu) \frac{\partial w}{\partial z} + \lambda \frac{\partial u}{\partial r}, \quad (18)$$

$$\sigma_{rz}(r, z) = \mu \left(\frac{\partial u}{\partial z} + \frac{\partial w}{\partial r} \right). \quad (19)$$

Continuity of these quantities at the interface between two solids are the boundary conditions. For a fluid–solid interface, the rigidity μ vanishes in the fluid layer and the tangential stress in the solid layer vanishes at the boundary. If at least one of the media is elastic, these boundary conditions permit the existence of interface or surface waves (see Chapter 37) such as Rayleigh waves at the interface between a solid and vacuum, Scholte waves at a fluid–solid interface, and Stoneley waves at a solid–solid interface. These waves are normally only excited when the source is acoustically close, in terms of wavelengths, to the interface.

The Helmholtz equation for an acoustic field from a point source with angular frequency ω is

$$\nabla^2 G(\mathbf{r}, z) + K^2(\mathbf{r}, z) G(\mathbf{r}, z) = -\delta^2(\mathbf{r} - \mathbf{r}_s) \delta(z - z_s),$$

$$K^2(\mathbf{r}, z) = \frac{\omega^2}{c^2(\mathbf{r}, z)}, \quad (20)$$

where the subscript s denotes the source coordinates. The range-dependent environment manifests itself as the coefficient $K^2(\mathbf{r}, z)$ of the partial differential equation for the appropriate sound speed profile. The range-dependent bottom type and topography appear as boundary conditions on scalars and tangential and normal quantities, as discussed above. The acoustic field from a point source $G(\mathbf{r})$ is obtained either by solving the boundary value problem of Eq. (20) (spectral method or normal modes) or by approximating Eq. (11) by an initial-value problem (ray theory, parabolic equation).

3.2 Ray Theory

Ray theory is a geometric, high-frequency approximate solution to Eq. (9) of the form

$$G(\mathbf{R}) = A(\mathbf{R}) \exp[iS(\mathbf{R})], \quad (21)$$

where the exponential term allows for rapid variations as a function of range and $A(\mathbf{R})$ is a more slowly varying "envelope" that incorporates both geometric spreading and loss mechanisms. The geometric approximation is that the amplitude varies slowly with range [i.e., $(1/A)\nabla^2 A \ll K^2$] so that Eq. (20) yields the eikonal equation

$$(\nabla S)^2 = K^2. \quad (22)$$

The ray trajectories are perpendicular to surfaces of constant phase (wavefronts) S and may be expressed mathematically as

$$\frac{d}{dl} \left(K \frac{d\mathbf{R}}{dl} \right) = \nabla K, \quad (23)$$

where l is the arc length along the direction of the ray and $\mathbf{R}$ is the displacement vector. The direction of average flux (energy) follows that of the trajectories, and the amplitude of the field at any point can be obtained from the density of rays.

The ray theory method is computationally rapid and extends to range-dependent problems. Furthermore, the ray traces give a physical picture of the acoustic paths. It is helpful in describing how noise redistributes itself when propagating long distances over paths that include shallow and deep environments and/or midlatitude to polar regions. The disadvantage of conventional ray theory is that it does not include diffraction and such effects that describe the low-frequency dependence (*degree of trapping*) of ducted propagation.

3.3 Wavenumber Representation or Spectral Solution

The wave equation can be solved efficiently with spectral methods when the ocean environment does not vary with range. The term *fast-field program (FFP)* had been used because the spectral methods became practical with the advent of the fast Fourier transform (FFT). Assume a solution of Eq. (20) of the form

$$G(\mathbf{r}, z) = \frac{1}{2\pi} \int_{-\infty}^{\infty} d^2 k \, g(\mathbf{k}, z, z_s) \exp[i\mathbf{k} \cdot (\mathbf{r} - \mathbf{r}_s)], \quad (24)$$

which then leads to the equation for the depth-dependent Green's function $g(\mathbf{k}, z, z_s)$,

$$\frac{d^2g}{dz^2} + [K^2(z) - k^2]g = \frac{1}{2\pi}\,\delta(z - z_s). \qquad (25)$$

Furthermore, we assume azimuthal symmetry, $kr > 2\pi$ and $\mathbf{r}_s = 0$, so that Eq. (24) reduces to

$$G(r,z) = \frac{\exp(-i\pi/4)}{(2\pi r)^{1/2}} \int_{-\infty}^{\infty} dk\,(k)^{1/2}g(k,z,z_s)\exp(ikr). \qquad (26)$$

We now convert the above integral to an FFT form by setting $k_m = k_0 + m\,\Delta k$; $r_n = r_0 + n\,\Delta r$, where $n, m = 0, 1, \ldots, N - 1$, with the additional condition $\Delta r\,\Delta k = 2\pi/N$, where N is an integral power of 2:

$$G(r_n,z) = \frac{\Delta k\,\exp[i(k_0 r_n - \pi/4)]}{(2\pi r)^{1/2}}$$
$$\cdot \sum_{m=0}^{N-1} X_m \exp\left(\frac{2\pi i m n}{N}\right),$$
$$X_m = (k_m)^{1/2}g(k_m,z,z_s)\exp(imr_0\,\Delta k). \qquad (27)$$

Although the method was initially labeled "fast field," it is fairly slow because of the time required to calculate the Green's functions [solve Eq. (27)]. However, it has advantages when one wishes to calculate the "near-field" region or to include shear wave effects in elastic media[3]; it is also often used as a benchmark for other less exact techniques. Recently, a range-dependent spectral solution technique has been developed.[4]

3.4 Normal-Mode Model

Rather than solve Eq. (25) for each g for the complete set of k's (typically thousands of times), one can utilize a normal-mode expansion of the form

$$g(\mathbf{k},z) = \sum_n a_n(\mathbf{k})u_n(z), \qquad (28)$$

where the quantities u_n are eigenfunctions of the eigenvalue problem

$$\frac{d^2u_n}{dz^2} + [K^2(z) - k_n^2]u_n(z) = 0. \qquad (29)$$

The eigenfunctions u_n are zero at $z = 0$, satisfy the local boundary conditions descriptive of the ocean bottom properties, and satisfy a radiation condition for $z \rightarrow \infty$.

They form an orthonormal set in a Hilbert space with weighting function $\rho(z)$, the local density. The range of discrete eigenvalues corresponding to the poles in the integrand of Eq. (26) is given by the condition

$$\min[K(z)] < k_n < \max[K(z)]. \qquad (30)$$

These discrete eigenvalues correspond to discrete angles within the critical angle cone in Fig. 5 such that specific waves constructively interfere. The eigenvalues k_n typically have a small imaginary part α_n, which serves as the modal attenuation representative of all the losses in the ocean environment (see Ref. 2 for the formulation of normal-mode attenuation coefficients). Solving Eq. (20) using the normal-mode expansion given by Eq. (28) yields (for the source at the origin)

$$G(r,z) = \frac{i}{4}\,\rho(z_s)\sum_n u_n(z_s)u_n(z)H_0^1(k_n r) \qquad (31)$$

The asymptotic form of the Hankel function can be used in the above equation to obtain the well-known normal-mode representation of a cylindrical (axis is depth) waveguide:

$$G(r,z) = \frac{i\rho(z_s)}{(8\pi r)^{1/2}}\exp\left(\frac{-i\pi}{4}\right)\sum_n \frac{u_n(z_s)u_n(z)}{k_n^{1/2}}$$
$$\cdot \exp(ik_n r). \qquad (32)$$

Equation (32) is a far-field solution of the wave equation and neglects the continuous spectrum $[k_n < \min[K(z)]$ of inequality (30)] of modes. For purposes of illustrating the various portions of the acoustic field, we note that k_n is a horizontal wavenumber so that a "ray angle" associated with a mode with respect to the horizontal can be taken to be $\theta = \cos^{-1}[k_n/K(z)]$. For a simple waveguide the maximum sound speed is the bottom sound speed corresponding to $\min[K(z)]$. At this value of $K(z)$, we have, from Snell's law, $\theta = \theta_c$, the bottom critical angle. In effect, if we look at a ray picture of the modes, the continuous portion of the mode spectrum corresponds to rays with grazing angles greater than the bottom critical angle of Fig. 4b and therefore outside the cone of Fig. 5. This portion undergoes severe loss. Hence, we note that the continuous spectrum is the near (vertical) field and the discrete spectrum is the (more horizontal, profile-dependent) far field (falling within the cone in Fig. 5).

The advantages of the NM procedure are that the solution is available for all source and receiver configurations once the eigenvalue problem is solved; it is easily extended to moderately range dependent conditions using the adiabatic approximation; it can be applied (with more effort) to extremely range dependent environments using coupled-mode theory. However, it does not include a full representation of the near field.

3.5 Adiabatic Mode Theory

All of the range-independent normal-mode "machinery" developed for environmental ocean acoustic modeling applications can be adapted to mildly range dependent conditions using adiabatic mode theory. The underlying assumption is that individual propagating normal modes adapt (but do not scatter or "couple" into each other) to the local environment. The coefficients of the mode expansion, a_n in Eq. (28), now become mild functions of range, that is, $a_n(\mathbf{k}) \rightarrow a_n(\mathbf{k}, \mathbf{r})$. This modifies Eq. (32) as follows:

$$G(\mathbf{r}, z) = \frac{i\rho(z_s)}{(8\pi r)^{1/2}} \exp\left(\frac{-i\pi}{4}\right) \sum_n \frac{u_n(z_s)u_n(z)}{\overline{k_n}^{1/2}}$$

$$\cdot \exp(i\overline{k_n}r), \tag{33}$$

where the range-averaged wavenumber (eigenvalue) is

$$\overline{k_n} = \frac{1}{r} \int_0^r k_n(r')\, dr' \tag{34}$$

and the $k_n(r')$ are obtained at each range segment from the eigenvalue problem (29) evaluated at the environment at that particular range along the path. The quantities u_n and v_n are the sets of modes at the source and the field positions, respectively.

Simply stated, the adiabatic mode theory leads to a description of sound propagation such that the acoustic field is a function of the modal structure at both the source and the receiver and some average propagation conditions between the two. Thus, for example, when sound emanates from a shallow region where only two discrete modes exist and propagates into a deeper region with the same bottom (same critical angle), the two modes from the shallow region adapt to the form of the first two modes in the deep region. However, the deep region can support many more modes; intuitively, we therefore expect the resulting two modes in the deep region will take up a smaller more horizontal part of the cone of Fig. 5 than they take up in the shallow region. This means that sound rays going from shallow to deep tend to become more horizontal which is consistent with a ray picture of downslope propagation. Finally, fully coupled mode theory for range-dependent environ-

ments has been developed[5] but requires extremely intensive computation.

3.6 Parabolic Equation Model

The PE method was introduced into ocean acoustics and made viable with the development of the "split-step" algorithm, which utilized FFTs at each range step.[6] Subsequent numerical developments greatly expanded the applicability of parabolic equation.

Standard PE–Split Step Algorithm The PE method is presently the most practical and encompassing wave-theoretic range-dependent propagation model. In its simplest form, it is a far-field narrow-angle ($\sim \pm 20°$ with respect to the horizontal, adequate for most underwater propagation problems) approximation to the wave equation. Assuming azimuthal symmetry about a source, we express the solution of Eq. (21) in cylindrical coordinates in a source-free region in the form

$$G(r, z) = \psi(r, z) \cdot J(r), \tag{35}$$

and we define $K^2(r, z) \equiv K_0^2 n^2$, n therefore being an *index of refraction* c_0/c, where c_0 is a reference sound speed. Substituting Eq. (35) into Eq. (11) in a source-free region and taking K_0^2 as the separation constant, J and ψ satisfy the equations

$$\frac{d^2J}{dr^2} + \frac{1}{r}\frac{dJ}{dr} + K_0^2 J = 0, \tag{36}$$

$$\frac{\partial^2\psi}{\partial r^2} + \frac{\partial^2\psi}{\partial z^2} + \left(\frac{1}{r} + \frac{2}{J}\frac{\partial J}{\partial r}\right) + \left(\frac{\partial\psi}{\partial r}\right)$$

$$+ K_0^2 n^2 \psi - K_0^2 \psi = 0. \tag{37}$$

Equation (36) is a Bessel equation, and we take the outgoing solution, a Hankel function, $H_0^1(K_0 r)$, in its asymptotic form and substitute it into Eq. (37), together with the "paraxial" (narrow-angle) approximation

$$\frac{\partial^2\psi}{\partial r^2} \ll 2K_0 \frac{\partial\psi}{\partial r} \tag{38}$$

to obtain the parabolic equation (in r)

$$\frac{\partial^2\psi}{\partial z^2} + 2iK_0 \frac{\partial\psi}{\partial r} + K_0^2(n^2 - 1)\psi - 0, \tag{39}$$

where we note that n is a function of range and depth. We use a marching solution to solve the parabolic equation. There has been an assortment of numerical solutions, but the one that still remains the standard is the so-called split-step algorithm.[6]

We take n to be a constant; the error this introduces can be made arbitrarily small by the appropriate numerical gridding. The Fourier transform of ψ can then be written as

$$\chi(r,s) = \frac{1}{2\pi} \int_{-\infty}^{\infty} \psi(r,z) \exp(-isz)dz, \qquad (40)$$

which together with Eq. (39) gives

$$-s^2\chi + 2iK_0 \frac{\partial \chi}{\partial r} + K_0^2(n^2 - 1)\chi = 0. \qquad (41)$$

The solution of Eq. (41) is simply

$$\chi(r,s) = \chi(r_0,s) \exp\left[-\frac{K_0^2(n^2-1) - s^2}{2iK_0}(r - r_0)\right], \qquad (42)$$

with specified initial condition at r_0. The inverse transform gives the field as a function of depth,

$$\psi(r,z) = \int_{-\infty}^{\infty} \chi(r_0,s) \exp\left[\frac{iK_0}{2}(n^2-1)\Delta r\right]$$
$$\cdot \exp\left[-\frac{i\Delta r}{2K_0} s^2\right] \exp(isz)\, ds, \qquad (43)$$

where $\Delta r = r - r_0$. Introducing the symbol $\mathcal{F}$ for the Fourier transform operation from the z-domain [as performed in Eq. (40)] and $\mathcal{F}^{-1}$ as the inverse transform, Eq. (4) can be summarized by the range-stepping algorithm

$$\psi(r + \Delta r, z) =$$
$$\exp\left[\frac{iK_0}{2}(n^2-1)\Delta r\right]$$
$$\cdot \mathcal{F}^{-1}\left[\left(\exp\left(-\frac{i\Delta r}{2K_0} s^2\right)\right) \mathcal{F}[\psi(r,z)]\right], \qquad (44)$$

which is often referred to as the split-step marching solution to the PE. The Fourier transforms are performed using FFTs. Equation (44) is the solution for n constant, but the error introduced when n (profile or bathymetry)

varies with range and depth can be made arbitrarily small by increasing the transform size and decreasing the range step size. It is possible to modify the split-step algorithm to increase its accuracy with respect to higher angle propagation.[7]

Generalized or Higher Order PE Methods

Methods of solving the parabolic equation, including extensions to higher angle propagation, elastic media, and direct time-domain solutions including nonlinear effects, have recently appeared (see Refs. 8 and 9 for additional references). In particular, accurate high-angle solutions are important when the evironment supports acoustic paths that become more vertical, such as when the bottom has a very high speed and, hence, a large critical angle with respect to the horizontal. In addition, for elastic propagation, the compressional and shear waves span a wide-angle interval. Finally, Fourier synthesis for pulse modeling requires high accuracy in phase, and the high-angle PEs are more accurate in phase, even at the low angles.

Equation (39) with the second-order range derivative that was neglected because of inequality (38) can be written in operator notation as

$$[P^2 + 2iK_0P + K_0^2(Q^2 - 1)]\psi = 0, \qquad (45)$$

where

$$P \equiv \frac{\partial}{\partial r}, \qquad Q \equiv \sqrt{n^2 + \frac{1}{K_0^2} \frac{\partial^2}{\partial z^2}}. \qquad (46)$$

Factoring Eq. (45) assuming weak range dependence and retaining only the factor associated with outgoing propagation yields a one-way equation

$$P\psi = iK_0(Q - 1)\psi, \qquad (47)$$

which is a generalization of the parabolic equation beyond the narrow-angle approximation associated with inequality (38). If we define $Q = \sqrt{1 + q}$ and expand Q in a Taylor series as a function of q, the standard PE method is recovered by $Q \approx 1 + 0.5q$. The wide-angle PE to arbitrary accuracy in angle, phase, and so on, can be obtained from a Padé series representation of the Q operator,[8]

$$Q \equiv \sqrt{1 + q} = 1 + \sum_{j=1}^{n} \frac{a_{j,n}q}{1 + b_{j,n}q} + O(q^{2n+1}), \qquad (48)$$

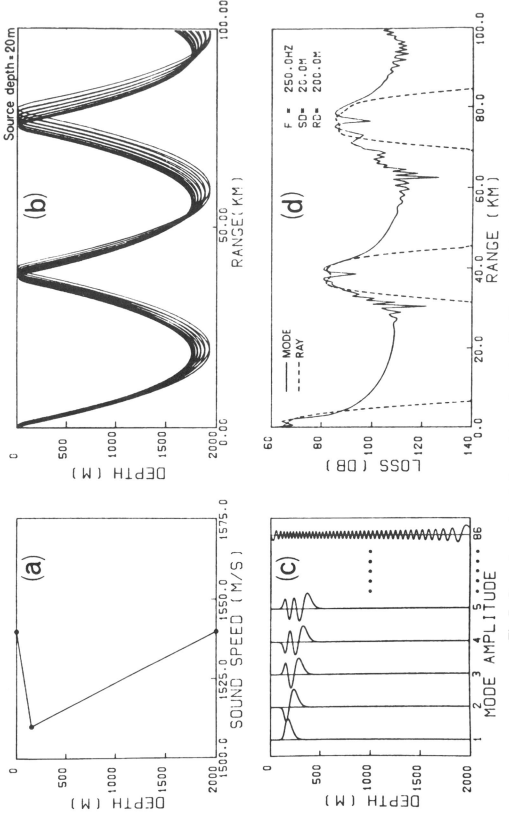

Fig. 7 Ray and normal-mode theory: (*a*) sound speed profile; (*b*) ray trace; (*c*) normal modes; (*d*) propagation calculations.

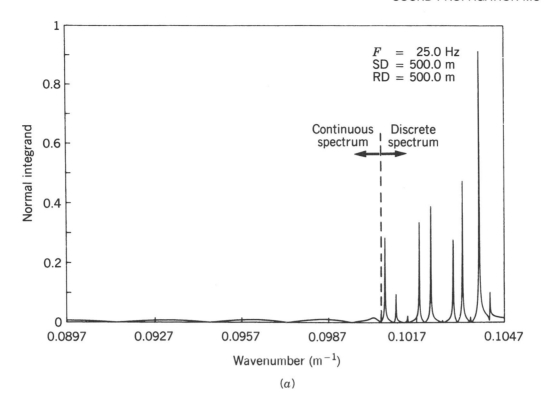

F = 25.0 Hz
SD = 500.0 m
RD = 500.0 m

Continuous spectrum   Discrete spectrum

(a)

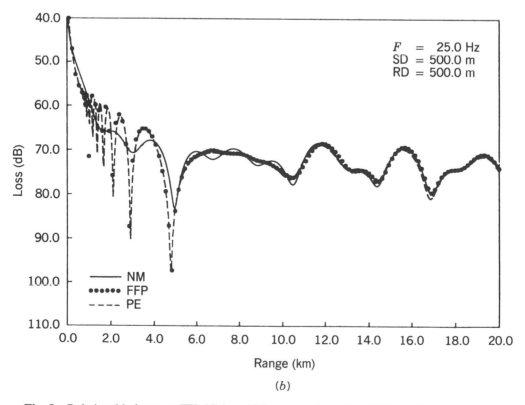

F = 25.0 Hz
SD = 500.0 m
RD = 500.0 m

——— NM
●●●●● FFP
----- PE

(b)

Fig. 8 Relationship between FFP, NM, and PE computations: (a) FFP Green's function from Eq. (25); (b) NM, FFP, and PE propagation results showing some agreement in near field complete agreement in far field.

where n is the number of terms in the Padé expansion and

$$a_{j,n} = \frac{2}{2n+1}\,\sin^2\left(\frac{j\pi}{2n+1}\right),$$

$$b_{j,n} = \cos^2\left(\frac{j\pi}{2n+1}\right). \tag{49}$$

The solution of Eq. (47) using Eq. (49) has been implemented using finite-difference techniques for fluid and elastic media.[8] A split-step Padé algorithm[10] has recently been developed that greatly enhances the numerical efficiency of this method.

4   QUANTITATIVE DESCRIPTION OF PROPAGATION

All of the models described above attempt to describe reality and to solve in one way or another the Helmholtz equation. They therefore should be consistent, and there is much insight to be gained from understanding this consistency. The models ultimately compute propagation loss, which is taken as the decibel ratio (see Appendix) of the pressure at the field point to a reference pressure, typically 1 m from the source.

Figure 7 shows convergence zone type propagation for a simplified profile. The ray trace in Fig. 7b shows the cyclic focusing discussed in Section 1.2. The same profile used to calculate normal modes is shown in Fig. 7c, which when summed according to Eq. (32) exhibit the same cyclic pattern as the ray picture. Figure 7d shows both the normal-mode (wave theory) and ray theory result. Ray theory exhibits sharply bounded shadow regions, as expected, whereas the normal-mode theory, which includes diffraction, shows that the acoustic field does exist in the shadow regions and the convergence zones have structure.

Normal-mode models sum the discrete modes, which roughly correspond to angles of propagation within the cone of Fig. 5. The spectral method can include the

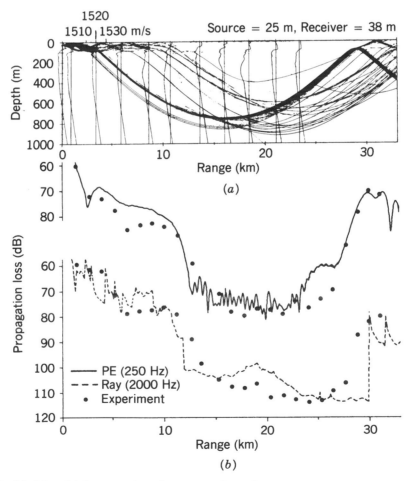

Fig. 9   Model and data comparison for a range-dependent case: (a) profiles and ray trace for a case of a surface duct disappearing; (b) 250 Hz PE and 2 kHz ray trace comparisons with data.

full field, discrete plus continuous, the latter corresponding to larger angles. The discussion below Eq. (32) defines these angles in terms of horizontal wavenumbers, and eigenvalues of the normal-mode problem are a discrete set of horizontal wavenumbers. Hence the integrand (Green's function) of the spectral method has peaks at the eigenvalues associated with the normal modes. These peeks are shown on the right of Fig. 8a. The smoother portion of the spectrum is the continuous part corresponding to the larger angles. Therefore, the consistency we expect between the normal mode and the spectral method and the physics of Fig. 5 is that the continuous portion of the spectral solution decays rapidly with range so that there should be complete agreement at long ranges between normal-mode and spectral solutions. The Lloyd's mirror effect, a near-field effect, should also be exhibited in the spectral solution but not the normal-mode solution. These aspects are apparent in Fig. 7b. The PE solution appears in Fig. 7b and is in good agreement with the other solutions, but with some phase error associated with the average wavenumber that must be chosen in the split-step method. The PE solution, which contains part of the continuous spectrum including the Lloyd mirror beams, is more accurate than the normal-mode solution at short range; more recent PE results[8] can be made arbitrarily accurate in the forward direction.

Range-dependent results[2] are shown in Fig. 9. A ray trace, a ray trace field result, a PE result, and data are plotted together for a range-dependent sound speed profile environment. The models agree with the data in general, with the exception that the ray results predict too sharp a leading edge of the convergence zone.

Upslope propagation is modeled with the PE in Fig. 10. As the field propagates upslope, sound is dumped into the bottom in what appears to be discrete beams.[11] The flat region has three modes, and each is cut off successively as sound propagates into shallower water. The ray picture also has a consistent explanation of this phenomenon. The rays for each mode become steeper as they propagate upslope. When the ray angle exceeds the critical angle, the sound is significantly transmitted into the bottom. The locations where this takes place for each of the modes is identified by the three arrows.

As a final example of how physical insight can be derived from models, we present a range-independent normal-mode[12] study of the optimum frequency of propagation in a shallow-water environment with a summer profile, as indicated in Fig. 11a, with the source (S) and receiver (R) also indicated. Frequency-versus-range contours of propagation loss obtained from a wideband experiment (analyzed in one-third-octave bands) and from an incoherent (no cross terms) sum of modes

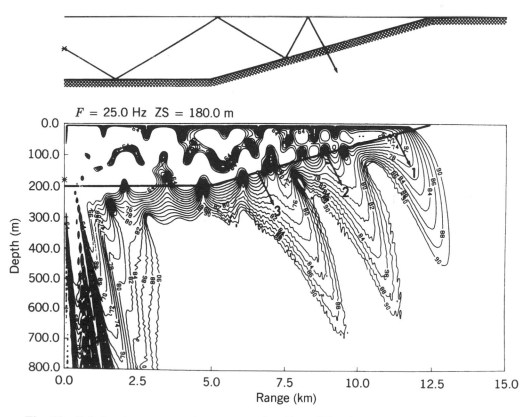

$F = 25.0$ Hz $ZS = 180.0$ m

Fig. 10 Relation between up-slope propagation (from PE calculation) showing individual mode cutoff and energy dumping in the bottom and a corresponding ray schematic.

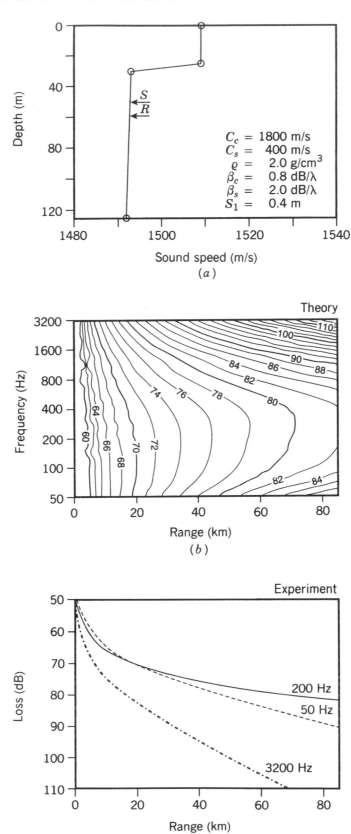

Fig. 11 (*a*) Shallow-water environment with summer and winter (isovelocity) profiles; (*b*) frequency vs. depth propagation loss contours; (*c*) third-octave propagation loss curves.

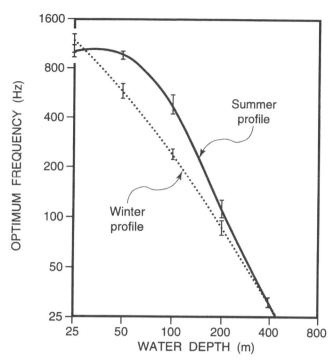

Fig. 12 Optimum frequency is strongly dependent on the depth of the ocean waveguide.

is shown in Figs. 11*b* and 11*c*. One obtains the conventional type of propagation loss curves by making a horizontal cut through the contour plots, as shown in Fig. 11*d*. We note here, as an aside, that in shallow-water environments propagation loss obtained by incoherently summing the modes is approximately equal to one-third-octave frequency averaging, which has the effect of averaging away model interference. The frequency versus-range contours reveal an optimum frequency in the 200–400-Hz region. This can be seen by observing the 80-dB contour, which goes out to long ranges in the region, whereas other frequencies, at say a range of 70 km, have much higher losses.

The underlying physics of the optimum frequency is further elaborated in Fig. 12. For the sound speed profiles in Fig. 11*a*, the optimum frequency is shown to be a strong function of water depth. As a matter of fact, for the source receiver configuration under study it is actually a function of the dominant duct, which would be whole water column for an isovelocity winter profile and the duct below the thermocline in the summer. Next, vis à vis the optimum frequency, after water depth in importance is bottom type, which also has a dominant effect on the actual propagation loss levels.

APPENDIX: UNITS

The decibel is the dominant unit in underwater acoustics and denotes a ratio of intensities (not pressures) expressed in terms of a logarithmic (base 10) scale. Two intensities I_1 and I_2 have a ratio I_1/I_2 in decibels of $10 \log I_1/I_2$ decibels. Absolute intensities can therefore be expressed by using a reference intensity. The presently accepted reference intensity is based on a reference pressure of one micropascal: the intensity of a plane wave having an rms pressure equal to 10^{-5} dyn/cm^2. Therefore, taking 1 μPa as I_2, a sound wave having an intensity, of, say, one million times that of a plane wave of rms pressure 1 μPa has a level of $10 \log(10^6/1) \equiv 60$ dB re 1 μPa. Pressure (p) ratios are expressed in dB re 1 μPa by taking $20 \log p_1/p_2$, where it is understood that the reference originates from the intensity of a plane wave of pressure equal to 1 μPa.

The average intensity I of a plane wave with rms pressure p in a medium of density ρ and sound speed c is $I = p^2/\rho c$. In seawater, ρc is 1.5×10^5 g/cm · s, so that a plane wave of rms pressure 1 dyn/cm^2 has an intensity of 0.67×10^{-12} W/cm^2. Substituting the value of a micropascal for the rms pressure in the plane-wave intensity expression, we find that a plane-wave pressure of 1 μPa corresponds to an intensity of 0.67×10^{-22} W/cm^2 (i.e., 0 dB re 1 μPa).

REFERENCES

1. J. Northrup and J. G. Colborn, "Sofar Channel Axial Sound Speed and Depth in the Atlantic Ocean," *J. Geophys. Res.*, Vol. 79, 1974, p. 5633.

2. F. B. Jensen, W. A. Kuperman, M. B. Porter, and H. Schmidt, *Computational Ocean Acoustics*, AIP Press, Woodbury, NY 1994.

3. H. Schmidt and F. B. Jensen, "A Full Wave Solution for Propagation in Multilayered Viscoelastic Media with Application to Gaussian Beam Reflection at Fluid-Solid Interfaces," *J. Acoust. Soc. Am.*, Vol. 77, 1985, p. 813.

4. H. Schmidt, W. Seong, and J. T. Goh, "Spectral Superelement Approach to Range-Dependent Ocean Acoustic Modeling," *J. Acoust. Soc. Am.*, Vol. 98, 1995, p. 465.

5. R. B. Evans, "A Coupled Mode Solution for Acoustic Propagation in a Waveguide with Stepwise Depth Variations of a Penetrable Bottom," *J. Acoust. Soc. Am.*, Vol. 74, 1983, p. 188.

6. F. D. Tappert, "The Parabolic Approximation Method," in J. B. Keller and J. S. Papadakis, (Eds.), *Wave Propagation and Underwater Acoustics*, Springer Verlag, Berlin, 1977.

7. D. J. Thomson and N. R. Chapman, "A Wide-Angle Split-Step Algorithm for the Parabolic Equation," *J. Acoust. Soc. Am.*, Vol. 74, 1983, p. 1848.

8. M. D. Collins, "Higher-Order Padé Approximations for Accurate and Stable Elastic Parabolic Equations with Applications to Interface Wave Propagation," *J. Acoust. Soc. Am.*, Vol. 89, 1991, p. 1050.

9. B. E. McDonald and W. A. Kuperman, "Time Domain

Formulation for Pulse Propagation Including Nonlinear Behavior at a Caustic," *J. Acoust. Soc. Am.*, Vol. 81, 1987, p. 1406.

10. M. D. Collins "A Split-Step Padé Solution for the Parabolic Equation Method," *J. Acoust. Soc. Am.*, Vol. 93, 1993, p. 1736.

11. F. B. Jensen and W. A. Kuperman, "Sound Propagation in a Wedge Shaped Ocean with a Penetrable Bottom," *J. Acoust. Soc. Am.*, Vol. 67, 1980, p. 1564.

12. F. B. Jensen and W. A. Kuperman, "Optimum Frequency of Propagation in Shallow Water Environments," *J. Acoust. Soc. Am.*, Vol. 73, 1983, p. 813.

37

PROPAGATION IN MARINE SEDIMENTS

LeRoy M. Dorman

1 INTRODUCTION

The propagation of sound in marine sediments can be important in ocean acoustics at frequencies below 50 Hz and is, moreover, an important tool for study of seafloor geology. Propagation is dependent on the physical properties of the sediments, which are much more variable than the properties of the overlying water. The most important physical properties are compressional velocity, shear velocity, density, anisotropy, and attenuation. These physical properties are influenced by the source of the sedimentary materials and also by the environment. These dependencies allow a degree of predictability. Although the variability is high, there are some systematic regional variations (which control source materials, mechanical winnowing, and chemical alteration) that are useful in making predictions for regions of no data. This chapter contains a discussion of in situ measurement techniques (sources, propagation types, receivers, and analytical methods), representative samples of in situ determinations of most of the important physical properties in several different geological regions, as well as a discussion of predictability.

2 REGIONAL VARIABILITY AND PREDICTABILITY

The range of water depths we must treat is shown by the frequency distribution of seafloor depths that shows peaks near zero and near 3800 m. The shallow depths are found at the ocean (or continental) margins and the

Encyclopedia of Acoustics, edited by Malcolm J. Crocker
ISBN 0-471-80465-7 © 1997 John Wiley & Sons, Inc.

deeper peak represents the deep-sea basins. The shallowest part (depth < 100–400 m) of the continental margin is called the shelf and has a slope of less than (typically half) a degree. The deeper portion of the margin is called the continental slope and has a depth gradient of $1°–3°$. The surficial sediments of the shelf and slope are similar in nature.

The sediments that form the margins of continents are, in large part, similar to the sediments that cover the crystalline continental basement. They are, of course, always water saturated, the condition we find below the water table on continents. At the mouths of rivers, sediments derived from the continents are now being deposited. The sediment delivered by rivers is derived from the erosion of continental rock and is called lithogenous (rock generated). These rock-derived sediments form more than 70% of the volume of seafloor sediments.[1] Wide continental shelves, such as those on the east coast of North America, on the west coast of Northern Europe, and in the Gulf of Mexico, form effective filters, trapping the largest and most swiftly falling particles. The margins of the oceans are typically much more productive biologically than the basins, so the continental (terrigenous) sediments are mixed with those of shallow marine origin. Submarine canyons are often found at the mouths of great rivers. These canyons form channels for the transport of turbidity currents (rapid movements of water-saturated sediments), which form deltas or fans on the floor of the deep-ocean basins.

In the deep oceans, the conditions we find are distinctly different from those on the continents or their margins. Continental sources of solids (except for fine wind-blown components) are confined to the edges, so most of the sediment reaching the central portions of the ocean basins is detritus from the plants and animals living above. The volume of these biogenous sediments is about 30% of the total, although about 55% of the area of the seafloor is dominantly biogenic.[2] Chemically, the biogenous sediments may be carbonate based (calcareous) or silicate based (siliceous).

The sedimentation rate for calcareous sediments is typically an order of magnitude higher than that for siliceous sediments. The distribution of the two types on the sea floor is controlled primarily by the solubility of the carbonates, which increases with pressure and with acidity (dissolved CO_2) and by the accumulation rate of carbonates. The depth at which CO_2 dissolution exceeds accumulation is called the calcite compensation depth (CCD). Below the CCD the sediments that survive in volume will be siliceous oozes and wind-transported clays. In the equatorial regions, where the biological productivity is high, oozes dominate; elsewhere clays will prevail. Thus in most areas deeper than the CCD (3800–5000 m) the sediments are predominantly fine-grained silicates. As we increase the CO_2 content of the atmosphere the CCD will rise.

The circulation of deep water in the ocean basins appears to be the most important factor controlling the CCD. In the North Atlantic the bottom water is the North Atlantic Deep Water, which is young and has not yet absorbed much CO_2. This condition lowers the CCD to ~ 5.5 km. The deeper parts of the Pacific Basin are Filled with Antarctic Bottom Water, which is older and more acidic (since it is CO_2 rich). This chemical aggressiveness raises the CCD there to the ~ 4-km range. The CCD is locally depressed in the equatorial regions by the high supply of carbonates.

3 PHYSICAL PROPERTIES

To a first approximation, the seafloor is an elastic solid that allows the propagation of shear disturbances as well as the propagation of compressional or sound waves. Also to a first approximation, whose accuracy depends on the particular site, the seafloor is laterally homogeneous and vertically heterogeneous. That is, the partial derivatives of physical properties in the vertical direction are much larger than the corresponding derivatives taken in the horizontal directions. The relative sizes of the variations are, however, significantly greater in the seafloor than in the ocean itself, and this increases the complexity of propagation effects.

When the wavelength is short compared to the scale of variation of physical properties, the compressional and shear waves decouple and can be treated separately. At the "high" frequencies at which the compressional and shear wave propagation decouple, ray theory is a useful tool. In an elastic solid, the physical observables directly affecting wave propagation are the compressional and

shear velocities (and their anisotropy), the compressional and shear attenuation, and the density. The best-known and understood properties are the compressional velocity and the density.

3.1 Compressional Velocity

The compressional velocity (or simply the sound velocity) is the simplest acoustic property to measure. In an isotropic elastic medium, the propagation velocity α is related to the elastic moduli and density by

$$\alpha^2 = \frac{\lambda + 2\mu}{\rho} = \frac{\kappa + (4/3)\mu}{\rho},$$

where λ and μ are the Lamé parameters, κ is the incompressibility or bulk modulus, and ρ is the density.

Numerous laboratory measurements of sound velocity, along with the basic measurements of sedimentology (density, porosity, and grain size), have led to the development of simple velocity predictors based on grain size and sediment type that are accurate to 1% or better given mean grain size and 3% or better given only the sediment type.[3,4] These laboratory measurements are typically made at frequencies of a few hundred kilohertz.

The sound velocity at the seafloor varies from 20% above the velocity of water (for coarse sands) to a few percent below it (for clays and oozes).[5] Figure 1 shows examples of velocities measured in deep-sea fans and on the North American continental margin.

The depth dependence of the physical properties of seafloor sediments is primarily controlled by compaction, which first squeezes out the water and then compresses and deforms particle grains. Both the density and elastic moduli increase, but the stiffnesses increase faster, producing a positive gradient (increase in velocity with depth) in the range 0.5–2 s^{-1}. The gradient decreases with depth but linear approximation is adequate for the upper few hundred meters.

The common occurrence of a surficial velocity less than that of water, if only a few percent, creates for compressional waves an acoustic waveguide that can affect propagation significantly.[10]

Figure 2 shows profiles that extend deeper than those shown in Fig. 1. At depths of 10–20 km, compaction and metamorphosis increase the velocities of sedimentary rocks to values indistinguishable from the 6–7 km/s of crystalline crustal rocks.

3.2 Shear Velocity

The shear velocity in shallow marine sediments is much lower and is, in general, much more variable than the

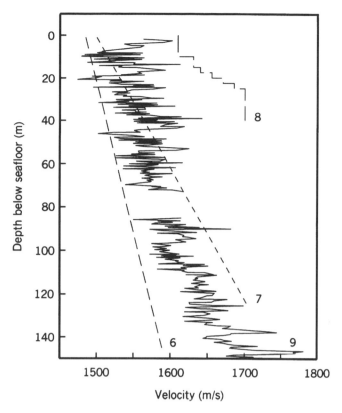

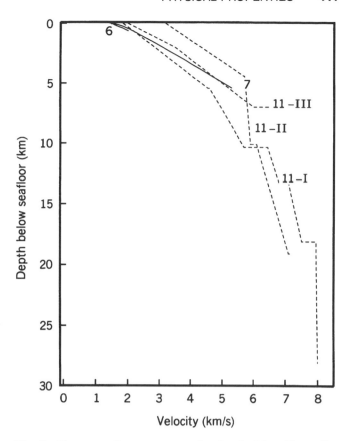

Fig. 1 Sediment compressional velocities, the upper 140 m, displayed according to sediment type. The long dashed line represents clays, in this case from the Monterey[6] and Bengal Fans,[7] and the irregular dashes represent shallow-water silts and sands.[8] The irregular solid line is sample and logging data from the carbonate sediments of the Ontong Java Plateau.[9] The number by each line is the reference number of the data source. For the fans, the surficial velocity is slightly below the water velocity.

compressional velocity (see Fig. 3). The shear velocity, in terms of the elastic constants, is simply

$$\beta^2 = \frac{\mu}{\rho},$$

where μ is the shear modulus. There are relatively few measurements of shear velocity in situ, and the predictive relations with sedimentary parameters are less well developed. The surficial shear velocities from in situ measurements tend to be lower than those from samples. The sedimentary compressional velocity increases gradually from that of water (1450–1550 m/s), and the surficial velocity ratio (sediment sound velocity divided by water velocity) varies by ±10%. The shear velocity of water is zero, so the transition even to the slow shear velocities of the seafloor sediments is abrupt.

The shear velocity structure always contains a strong waveguide, since surficial seafloor shear velocities are

Fig. 2 Deeper sediment compressional velocities. The solid lines are from deep-sea fans[6,7] and the Carolina Trough (eastern U.S. continental margin.)[11]

virtually always lower than 1.5 km/s. This has a strong effect on the noise structure.[13,17,18]

The few in situ observations of near-surface shear velocity in deep-water sediments shown in Fig. 3 show a distinct difference between the calcareous (depth $Z <$ CCD) and pelagic clay sediments ($Z >$ CCD), with the shear velocity in the calcareous sediments rising more rapidly. We can attribute this to the different grain sizes and shapes of the sediments. The calcareous sediments are composed of sand-sized particles having sharp edges and corners while the clays consist of finer platelets. Only a small increase in lithostatic pressure is required for the carbonates to touch and to start the process of cementation leading to the formation of limestone. The clay particles, in contrast, require greater lithostatic pressure to create enough contact area to develop shear strength. This pattern shows up also in compressional wave velocities[5] but is much more pronounced in shear waves.

3.3 Anisotropy

The process of sedimentary compaction leads to anisotropy in seismic velocity. Many particles, especially clays

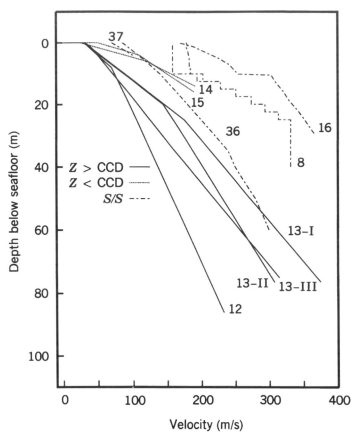

Fig. 3 Sediment shear velocities measured in situ. The solid lines are data from the Eastern Pacific,[12,13] where the sediments are clays and are below the CCD. The short dashes are data from the Madiera Abyssal Plain[14] (a calcareous clay) and from the Somalia Basin in the Indian Ocean[15] (a calcareous ooze). Note that the profiles from the deep-sea sites form two distinct groups. The long–short dashes represent shallow-water silts and stands.[8,16]

and silts, are platelike in shape and composed of minerals that are anisotropic in crystal structure. When these are randomly dispersed in a fluid and stirred, their orientation is random and the sound velocity in the resultant fluid is isotropic. The process of compaction under gravity, however, leads to a preferred orientation of the flatter particles, with the platelets tending to be horizontal. This produces seismic anisotropy with symmetry around the vertical axis, called transverse isotropy.[19] Additionally, laminations of materials of differing seismic velocity will produce apparent anisotropy, again with transverse symmetry.[20] The compressional wave anisotropy observed in continental sedimentary rocks ranges up to 40%[21] and up to 13% in carbonates.[22]

Berge and others recently analyzed shallow-water interface waves and found that transverse isotropy in the upper layers was required by data from a silty clay bottom.[8] The splitting of the upper part of the line marked 3 in Fig. 3 shows the velocities for horizontally

polarized (faster) and vertically polarized (slower) shear waves.

3.4 Attenuation

The propagation of waves in seafloor sediments can be affected greatly by loss due to anelastic attenuation. The Boltzmann treatment of viscoelastic media starts with material creep and relaxation functions and leads to an elegant and compact representation of the anelastic effects, which cause the elastic moduli, which are real for perfectly elastic media, to become complex ($m_0 + im_1$) (Ref. 23, Chapter 9).

More specific to modeling propagation in sediments whose pores are filled with fluid is the model of Biot (see Ref. 16). This analysis gives rise to a "slow wave" or P-wave of the second kind, which can be thought of as propagating primarily in the infilling fluid.

Both these treatments produce complex-valued elastic

moduli. In linear viscoelastic theory these are constants while in the Biot case the moduli depend on frequency. When complex moduli are used in the wave equation, the propagating-wave solutions become waves that decay as they propagate. The presence of attenuation can produce frequency dependence of propagation velocity and this is, in fact, observed.

In practice, attenuation is measured by observing loss along paths of varying length or by direct waveform modeling. The amplitude decay of a plane wave with distance due to attenuation is $\Psi = e^{i(\omega t - kx)}e^{-\alpha^{(e)}x}$, where t is time, k is wavenumber, x is distance, and $\alpha^{(e)}$ is the imaginary component of wavenumber, or the *attenuation*. The units of $\alpha^{(e)}$ are nepers per meter. This often converted to α', loss in decibels per metre, by multiplying by 8.686, or to $\alpha^{(f)}$ in decibels per metre by kilohertz. The logarithmic decrement δ is $\alpha^{(e)}\lambda$. (See Ref. 24 for a discussion). In most materials, losses per cycle are nearly independent of frequency (rather than being nearly constant per path length). This produces an apparent frequency dependence of the attenuation coefficient since wavelength is dependent on frequency. Because of this, the dimensionless measures, such as α^λ, decibels per wavelength, or $1/Q$ are used more frequently now. The simplest dimensionless quantity is the quality factor Q, defined by its reciprocal, as in engineering usage, is related to the terms defined above[25]:

$$
\frac{1}{Q(\omega)} = -\frac{\Delta E}{2\pi E} = \frac{\delta}{\pi} = \frac{m_0}{m_1} = \frac{\alpha_e V}{\pi f}
$$
$$
= \frac{\alpha^{(f)} V}{27,278} = \frac{\alpha^{(\lambda)}}{27.28}.
$$

Here V is wave velocity.

Although dependence of Q on frequency ω is indicated in this equation (and required by causality), the frequency dependence of energy loss per cycle has been difficult to determine and now manifests itself as a slightly higher loss at 10 kHz compared to 100 Hz.

For compressional waves, $1000/Q$ varies between 6 and 30 in surficial shalllow-water sediments.[26,27] There are systematic dependences on grain size and porosity. Deeper than a few metres, $1000/Q$ drops to 5 or so.[28]

For shear waves, losses are higher, with $1000/Q$ in the range 6–190. Below the upper few metres of the sedimentary section, the attenuation drops an order of magnitude, to $1000/Q$ of 5.[12,16]

3.5 Density

The density of seafloor materials is higher than the density of water and is lowest for oozes and higher for pelagic clays and highest for calcareous and terrigenous sediments.[5]

4 LABORATORY MEASUREMENTS

The most common acoustic measurements available are those made in the laboratory on cores. Laboratory measurements have the advantage of ease of control of conditions but require care in sampling, especially in the case of the soft surficial samples, which are difficult to obtain and are easily damaged.

4.1 Sampling

The first step toward the measurement of the physical properties of sediments is sampling. The unconsolidated nature of the soft sediments of the upper seafloor make sampling a difficult process, and undisturbed samples of the upper few metres are elusive. The sampling tools most commonly used at sea are gravity-driven devices, which rely upon gravity-generated kinetic energy to drive a container into the seafloor. The kinetic energy is limited by the weight handling of available oceanographic winches and by the viscosity of seawater, which restricts terminal velocity of falling weights or winch speed of 6–10 m/s for non-free-falling weights. A wide variety of devices for sampling are discussed and illustrated by Lee[29] and by Mudroch and MacKnight.[30]

Underwater rotary drilling from fixed platforms or dynamically positioned ships can return samples from depths of many kilometres subbottom. The Deep Sea Drilling Program [now the Ocean Drilling Program (ODP)] has obtained hundreds of cores from the deep-ocean basins as well as the continental margins. Drilling by the oil industry is economically driven, so coring (which shows drilling progress) is done selectively. The ODP, in contrast, is scientifically driven and, by policy, cores continuously and has amassed a large core library.

4.2 Measurements

The samples available are usually cores of a few centimetres in size so the appropriate frequencies for sampling velocities are ultrasonic, usually a few hundred kilohertz. Seismic velocities are, in most cases, very weakly frequency dependent, so, with some caution, extrapolation to lower frequencies is warranted.[31]

Ultrasonic measurements are most valuable when made under temperature and pressure conditions like those in situ. The pressure vessels required must incorporate compressional (and shear wave) generators and detectors. Velocity measurements are made in both vertical and horizontal directions, when possible, to test for departures from isotropy.

The velocity measurements derive from distance and travel time observations, both of which are simple to do. Measurements of anelasticity require more care, especially for samples that are poorly consolidated.[16]

5   FIELD MEASUREMENTS

Conventional marine seismic measurements fall into two main categories: reflection, in which the geometry of wave propagation is predominantly vertical and whose sensitivity is primarily to variations of acoustic impedance ($\alpha\rho$), and refraction, in which the geometry is predominantly horizontal and whose sensitivity is primarily to the velocity and its depth gradient. The reflection geometry is the primary one for the petroleum exploration industry and is treated at greater length in Chapter 69. A third category, interface wave experiments, which are used for determination of the shear velocities of the upper few tens of metres, are specialized efforts to eliminate the unavoidable disturbance of the delicate upper part of the sediments by the sampling process.

In the basic refraction experiment, airgun or explosive sources are fired along one or more lines radiating from a receiver. The general increase of velocity with depth causes the turning point for each ray to occur at only one depth, allowing the problem of extraction of the depth-versus-velocity function to be cast as a linear inverse problem.[7,32]

5.1   Interface Waves

At a boundary between a solid and a liquid, there exists a class of guided waves that are similar in character to the Rayleigh waves that propagate with energy bound to the earth's surface. The physical phenomenon that binds the propagation of these waves to the earth's surface is the fact that, when the phase velocity is less than that of the lowest propagation velocity in a medium, the wave function becomes inhomogeneous, showing a spatial dependence that is a real exponential (growing or decaying) rather than the imaginary exponential (sinusoidal) behavior characterizing freely propagating waves. These waves are the extension of Rayleigh propagation to phase velocities less than that of water. When the phase velocity is less than that of water, the waves cannot propagate freely in the water column but are bound to the seafloor. These waves are named after Stoneley, who described propagation at a boundary between two elastic media, or Scholte, who studied more extensively the particular case in which one medium is a liquid. The shear velocity at the seafloor is very low, rising rapidly with depth beneath the water layer. This depth dependence gives rise to dispersion. Analysis of the dispersion of these waves yields the depth dependence of the physical properties of the seafloor, primarily the shear velocity.

The phase velocity of the Scholte wave is low, less than the phase velocity of water. In a laterally homogeneous seafloor, sound propagating in water will not excite interface wave modes. The addition of scatterers, however, will cause conversion from freely traveling waves to trapped waves.

5.2   In Situ Ultrasonic Measurements

Richardson[33] used a diver-operated acoustic and sampling device to measure compressional and shear velocities in place while the seafloor was being sampled.

5.3   Sources

The source energy required for satisfactory field observations depends on the distance the signal must travel (because of geometric spreading and anelastic losses) and on the frequency (since the anelastic losses are fairly constant per cycle). Very short range work from submarines[34,35] can be done with electric blasting caps. Shallow-water measurements have used a shotgun shell,[16] small explosive charges,[36,37] and a sled-mounted airgun.[8] Interface wave studies require sources on the seafloor for effective excitation and midwater explosions are sometimes used to remove near-source reverberation from the source signature.[6,7] Studies in deep water can require sources of 2–25 kg of high explosives, since the bubble frequency of the explosions is high (150 Hz for 24 kg at 3800 m depth) and the frequency range of interest is 1–5 Hz.[13,15,18,38] A large weight has been used to generate Scholte waves in deep water.[14] Refraction experiments penetrating tens of kilometres in depth utilize explosions up to ~50 kg of explosive for a single shot.[11]

5.4   Detectors

The simplest detectors are hydrophones, but seismometers allow recording of multiple components of ground motion, especially valuable in anisotropy studies.[8] Arrays of sensors allow use of multichannel signal processing methods to separate signals traveling at different velocities.[18] The simplest recording technique is to monitor the data on land or aboard ship.[15,16] In deep-water experiments or long-range ones, autonomous instruments, containing their own recorders and time control, are required.[11,18,39,40]

6 SUMMARY

This chapter has discussed wave propagation in marine sediments, with emphasis on shallow in situ measurements and their relationship to the geological environment. The deep-water shear velocity observations fall into two distinct groups, depending on the water depth with respect to the CCD.

Acknowledgments

This work was supported by the Office of Naval Research under contract N00014-90-J-1275. I thank E. Hamilton and P. Berge for their valuable comments.

REFERENCES

1. E. Seibold and W. H. Berger, *The Sea Floor*, Springer Verlag, Berlin, 1993.

2. E. Seibold and W. H. Berger, *The Sea Floor*, Springer Verlag, Berlin, 1982.

3. E. L. Hamilton and R. T. Bachman, "Sound Velocity and Related Properties of Marine Sediments," *J. Acoust. Soc. Am.*, Vol. 72, 1982, pp. 1891–1904.

4. A. Nur, H. Yin, and D. Marion, "Wave Velocities in Sediments," in J. Hoven, R. Stoll, and M. Richardson, (Eds.), *Shear Waves in Marine Sediments*, Kluwer, Amsterdam, 1991, pp. 131–140.

5. E. L. Hamilton, "Geoacoustic Modeling of the Sea Floor," *J. Acoust. Soc. Am.*, Vol. 68, 1980, pp. 1313–1340.

6. R. K. Brienzo, "Velocity and Attenuation Profiles in the Monterey Deep-Sea Fan," *J. Acoust. Soc. Am.*, Vol. 92, 1992, pp. 2109–2125.

7. L. M. Dorman and R. S. Jacobson, "Linear Inversion of Body Wave Data, Part I: Velocity Structure from Travel-Times and Ranges," *Geophysics*, Vol. 46, 1981, pp. 138–150.

8. P. A. Berge, S. Mallick, G. J. Fryer, N. Barstow, J. A. Carter, G. H. Sutton, and J. I. Ewing, "In situ Measurement of Transverse Isotropy in Shallow-Water Marine Sediments," *Geophys J. Int.*, Vol. 104, 1991, pp. 241–254.

9. W. H. Berger, L. W. Kroenke, L. A. Mayer, and Shiboard Scientific Party, *Ontong Java Plateau, Leg 130: Synopsis of Major Drilling Results*, Vol. 130, in Proc. ODP, Init. Repts., L. W. Kroenke, W. H. Berger, T. R. Janacek, et al. (Eds.), Ocean Drilling Program, College Station, TX, 1991, pp. 497–537.

10. E. L. Hamilton, "Sound Channels in Surficial Marine Sediments," *J. Acoust. Soc. Am.*, Vol. 48, 1970, pp. 1296–1298.

11. A. M. Tréhu, A. Ballard, L. M. Dorman, J. F. Getrust, K. D. Klitogord, and A. Schreiner, "Structure of the Lower Crust Beneath the Carolina Through, U. S. Continental Margin," *J. Geophys. Res.*, Vol. 94, 1989, p. 10585–10600.

12. G. Nolet and L. M. Dorman, "Waveform Modeling of Scholte Waves," *Geophys. J. Intl.*, Vol. 125, 1996, pp. 385–396.

13. A. E. Schreiner, L. M. Dorman, and L. D. Bibee, "Shear Wave Velocity Structure from Interface Waves at Two Deep Water Sites in the Pacific Ocean," in J. Hovem, R. Stoll, and M. Richardson (Eds.), *Shear Waves in Marine Sediments*, Kluwer, La Spezia, Italy, 1991, pp. 231–238.

14. R. B. Whitmarsh and R. C. Lilwall, "A New Method for the Determination of In-situ Shear-wave Velocity in Deep-sea Sediments," in *Proceedings of Conference "Oceanology International 1982,"* Spearhead, Kingston-upon-Thames, 1982.

15. D. Davies, "Dispersed Stoneley Waves on the Ocean Bottom," *Bull Seismol. Soc. Am.*, Vol. 55, 1965, pp. 903–918.

16. R. D. Stoll, *Sediment Acoustics*, Springer Verlag, Berlin, 1989.

17. L. M. Dorman, A. E. Schreiner, and L. D. Bibee, "The Effects of Shear Structure on Sea Floor Noise," in J. Hovem, R. Stoll, and M. Richardson (Eds.), *Shear Waves in Marine Sediments*, Kluwer, Dordrecht, 1991, pp. 239–245.

18. A. E. Schreiner and L. M. Dorman, "Coherence Lengths of Seafloor Noise: Effects of Seafloor Structure," *J. Acoust. Soc. Am.*, Vol. 88, 1990, pp. 1503–1514.

19. J. E. White, *Seismic Waves: Radiation. Transmission and Attenuation*, McGraw-Hill, New York, 1965.

20. G. E. Backus, "Long-Wave Elastic Anisotropy Produced by Horizontal Layering," *J. Geophys. Res.*, Vol. 67, 1962, pp. 4427–4440.

21. L. F. Uhrig and F. A. van Melle, "Velocity Anisotropy in Stratified Media," *Geophysics*, Vol. 20, 1955, pp. 774–779.

22. R. L. Carlson, C. H. Schaftenaar, and R. P. Moore, "Causes of Compressional-wave Anisotropy in Carbonate-bearing, Deep-sea Sediments," *Geophysics*, Vol. 49, 1984, pp. 525–532.

23. J. A. Hudson, *The Excitation and Propagation of Elastic Waves*, Cambridge University Press, Cambridge, England 1980.

24. F. B. Jensen, W. A. Kuperman, M. B. Porter, and H. Schmidt, *Computational Ocean Acoustics*, AIP Press, Woodbury, NY, 1993, p. 612.

25. R. J. O'Connell and B. Budianski, "Measures of Dissipation in Viscoelastic Media," *Geophys. Res. Lett.* Vol. 5, 1978, pp. 5–8.

26. E. L. Hamilton, "Compressional-Wave Attenuation in Marine Sediments," *Geophysics*, Vol. 37, 1972, pp. 620–646.

27. E. L. Hamilton, "Attenuation of Shear Waves in Marine Sediments," *J. Acoust. Soc. Am.*, Vol. 60, 1976, pp. 334–338.

28. R. S. Jacobson, "An Investigation into the Fundamental Relationships between Attenuation, Phase Dispersion, and Frequency Using Seismic Refraction Profiles over

Sedimentary Structures," *Geophysics*, Vol. 52, 1987, pp. 72–87.

29. H. J. Lee, "State of the Art: Laboratory Determination of the Strength of Marine Soils," Vol. 883, in *Strength Testing of Marine Sediments: Laboratory and In-Situ Measurements*, American Society for Testing and Materials, Philadelphia, 1985, pp. 181–250.

30. A. Mudroch and D. MacKnight, *CRC Handbook of Techniques for Aquatic Sediments Sampling*, CRC Press, Boca Raton, FL, 1991.

31. W. R. Bryant, R. H. Bennett, and C. E. Katherman, "Shear Strength, Consolidation, Porosity, and Permeability of Oceanic Sediments," Vol. 7, in *The Oceanic Lithosphere*, C. Emiliani (Ed.), 1981, pp. 1555–1616.

32. K. Aki and P. G. Richards, *Quantitative Seismology: Theory and Methods*, W. H. Freeman, San Francisco, 1980.

33. M. D. Richardson, "Spatial Variability of Surficial Shallow Water Sediment Geoacoustic Properties," in T. Akal and J. M. Berkson *Ocean Seismo-Acoustics: Low-Frequency Underwater Acoustics*, Plenum, New York, 1986, pp. 527–536.

34. H. P. Bucker, J. A. Whitney, and D. L. Keir, "Use of Stoneley Waves to Determine the Shear Velocity in Ocean Sediments," *J. Acoust. Soc. Am.*, Vol. 36, 1964, pp. 139–143.

35. E. L. Hamilton, H. P. Bucker, D. L. Keir, and J.

A. Whitney, "Velocities of Compressional and Shear Waves in Marine Sediments Determined In Situ from a Research Submarine," *J. Geophys. Res.*, Vol. 75, 1970, pp. 4039–4049.

36. F. B. Jensen and H. Schmidt, "Shear Properties of Ocean Sediments Determined from Numerical Modeling of Scholte Wave Data," in T. Akal and J. M. Berkson, (Eds.), *Ocean Seismo-Acoustics: Low-Frequency Underwater Acoustics*, Plenum, New York, 1986, pp. 683–692.

37. M. Snoek, "Interface-Wave Propagation Studies: An Example of Seismo-acoustic Propagation in Non-homogeneous Materials," SACLANTCEN MEMORANDUM serial no. SM-229, SACLANT Undersea Research Centre, 1990.

38. A. W. Sauter, L. M. Dorman, and A. E. Schreiner, "A Study of Sea Floor Structure Using Ocean Bottom Shots," in T. Akal and J. M. Berkson (Eds.), *Ocean Seismo-acoustics, Low Frequency Underwater Acoustics*, Plenum, New York, 1986, pp. 673–681.

39. R. D. Moore, L. M. Dorman, C.-Y. Huang, and D. L. Berliner, "An Ocean Bottom, Microprocessor Based Seismometer," *Mar. Geophys. Res.*, Vol. 4, 1981, pp. 451–477.

40. A. W. Sauter, J. Hallinan, R. Currier, T. Barash, B. Wooding, A. Schultz, and LeRoy M. Dorman, "A New Ocean Bottom Seismometer," Proceedings of Conference: Marine Instrumentation '90, 1990, pp. 99–140.

38

ATTENUATION BY FORWARD SCATTERING: MEASUREMENTS AND MODELING

ROBERT H. MELLEN, HANS G. SCHNEIDER, AND DAVID G. BROWNING

1 INTRODUCTION

Propagation loss is taken here as $PL = 10 \log_{10}(I_0/I_R)$ (in decibels), where I_R is acoustic intensity at range $R(m)$ and I_0 is the effective intensity at 1 m from the source. The components are taken as geometric spreading loss and attenuation.[1]

Attenuation is generally frequency dependent, and the associated losses tend to increase linearly with range. Absorption is the principal mechanism in seawater. However, other mechanisms are often involved, and the objective is to identify and quantify them.

Propagation experiments are usually carried out in uniform sound channels, which are formed by downward refraction within the thermocline and upward refraction by the pressure gradient at greater depths (see Chapter 35). With source and receiver near the axis, the average spreading loss at longer ranges becomes cylindrical and excess losses due to surface and bottom effects are minimized.

For sound channel conditions, a simple method of estimating attenuation is to subtract cylindrical spreading loss from measured propagation loss versus range and fit the results by linear regression. For such applications, this method has been found to compare favorably with computer-based propagation models.

Explosives are generally used as wide-band sources.

Results of an experiment in the sound channel in the Mediterranean Sea are shown in Fig. 1. Both source and receiver were at 200 m depth, close to the chan-

nel axis. Spreading loss becomes cylindrical for $R > 1$ km and the ordinate is taken empirically as $PL^* = PL - 10 \log_{10}[R(km)] - 50$ (dB), where PL is the measured propagation loss. Attenuation coefficients at selected frequencies (in decibels per kilometre) are obtained from PL^* versus range by linear regression.

Absorption involves conversion of acoustic energy into heat and is usually the dominant attenuation component in seawater, at least at high frequencies. Ionic relaxation is the principal mechanism in seawater (see Chapter 35). Losses in excess of absorption, with a systematic frequency dependence, mean that other mechanisms are involved. Analysis of such excess losses will therefore depend critically on the accuracy of modeling the absorption.

2 ABSORPTION MODEL

The NUSC Global Model for Sound Absorption in Sea Water[2] is used in the calculations here. This model is based on field measurements supported by extensive laboratory measurements of the pH-dependent relaxations involved. The three-component absorption formula is

$$A_a = A_1(\text{MgSO}_4) + A_2(\text{B(OH)}_3) + A_3(\text{MgCO}_3),$$

$$A_n = \frac{(S/35)C_n F_n F^2}{F^2 + F_n^2},$$

$$C_1 = 0.5 \times 10^{-Z(km)/20}, \qquad F_1 = 50 \times 10^{T/60},$$

$$C_2 = 0.1 \times 10^{(pH-8)}, \qquad F_2 = 0.9 \times 10^{T/70},$$

$$C_3 = 0.03 \times 10^{(pH-8)}, \qquad F_3 = 4.5 \times 10^{T/30}. \qquad (1)$$

Encyclopedia of Acoustics, edited by Malcolm J. Crocker
ISBN 0-471-80465-7 © 1997 John Wiley & Sons, Inc.

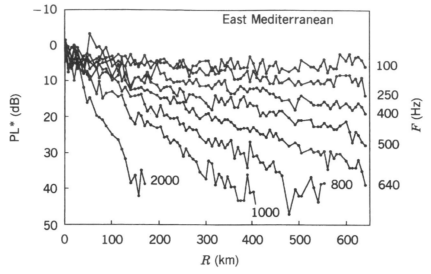

Fig. 1 Mediterranean propagation loss corrected for cylindrical spreading (see text).

The terms in Eq. (1) are absorption coefficent A_a (in decibels per kilometre), frequency F (in kilohertz), relaxation frequency F_n (in kilohertz), temperature T (in degrees Celsius), salinity S (in parts per thousand), and depth Z (in kilometres). Pure-water absorption is negligible below 100 kHz and is omitted for simplicity. The mean values for the World Ocean, $S = 35$ ppt and pH 8.0, are used as reference values. The pH range is roughly 7.7–8.3, corresponding to an absorption range of nearly a factor of 4 at lower frequencies. Formula error is estimated to be within ±15% providing that the salinity range is limited to roughly 30–40 ppt and the pH error is less than ±0.05 units. Typical profiles of pH versus depth in selected regions are shown in Fig. 2.

3 SOUND CHANNEL EXPERIMENTS

Figure 3 compares the absorption model predictions and data from sound channel experiments in the Mediterranean, North Atlantic, and North Pacific.[2] The axis depths for both the Atlantic and Pacific cases are roughly 1 km. Axial values of pH and temperature have been used for calculations. Good agreement between model and data indicates no significant excess attenuation in all three cases.

Other experiments in the Pacific Ocean showing marked excess attenuation at lower frequencies have been reviewed by Kibblewhite and Hampton.[3] Analyses of Tasman Sea data for two tracks are shown in Fig. 4.

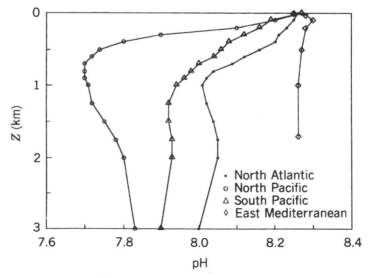

Fig. 2 Profiles of pH vs. depth.

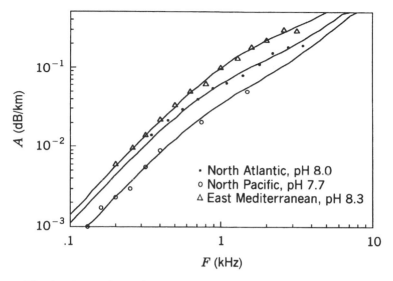

Fig. 3 Comparison of attenuation data with predicted absorption.

Since the excess loss is independent of frequency, the likely mechanism is scattering of energy out of the channel by temperature inhomogeneities having scales that are large compared to the acoustic wavelengths.

The reliability of the very small excess losses reported in many deep-ocean experiments has frequently been questioned. Support for the scatter hypothesis is offered by experiments in shallower water, where the channels are weaker and the losses consequently much greater.

Results of an experiment in the shallow waters of Hudson Bay[2] are shown in Fig. 5. In this case, the excess loss is also independent of frequency and much greater than probable experimental error.

Results of an experiment in the Baltic Sea[4] are shown in Fig. 6. In this case, absorption is very small because of the low salinity (minor modifications of the relaxation parameters in the absorption formula were required here). The excess loss is again quite independent of frequency, large compared to absorption, and well above the probable experimental error.

In other Baltic Sea experiments, concurrent measurements were made with a second receiver outside the sound channel.[5] Propagation data at 4 kHz, shown in Fig. 7, have been corrected for cylindrical spreading, and the dashed lines are deterministic predictions for the average sound speed profile. Analysis showed that the increased

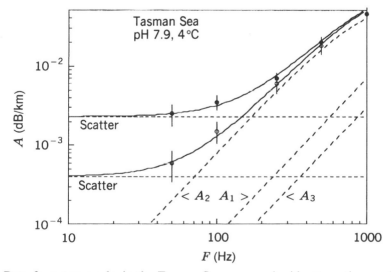

Fig. 4 Data from two tracks in the Tasman Sea compared with attenuation model (solid curves) assuming scatter loss components independent of frequency. Individual components are shown by dashed lines.

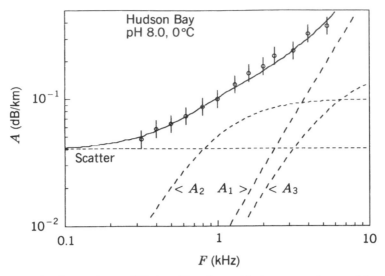

Fig. 5 Comparison of Hudson Bay data with the attenuation model.

levels outside the channel account for the excess loss within the channel, giving further support to the scatter hypothesis.

4 RAY DIFFUSION THEORY

Frequency-independent scatter loss means that ray theory is appropriate. Random migration of rays out of the channel and into the bottom is therefore treated as a small-angle diffusion process governed by the equation[6]

$$\frac{\partial W}{\partial R} = D \frac{\partial^2 W}{\partial \theta^2} , \qquad (2)$$

where θ is the grazing angle, W is the angular density of rays, R is the range, and D is the diffusion constant. A simple model is a channel with axis near the surface and a constant sound speed gradient. The scatterers are taken as isotropic and the bottom as infinitely absorbing. For Gaussian statistics, $D = \pi^{1/2}\mu^2/\rho$, where $\mu^2 = \langle(\delta C/C)^2\rangle$ is the variance of the refractive index, C is the depth-dependent mean sound speed, δC is its variability, and ρ is the correlation length. The long-range asymptotic solution for D independent of depth and range is approximately[7]

$$W \approx a_0 \cos\left(\frac{\pi\theta}{2\theta_0}\right) \exp\left[-DR\left(\frac{\pi}{2\theta_0}\right)^2\right] , \qquad (3)$$

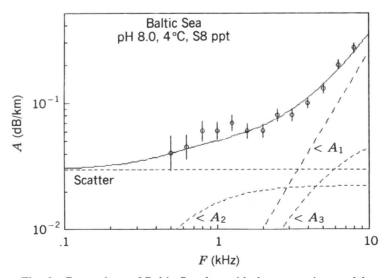

Fig. 6 Comparison of Baltic Sea data with the attenuation model.

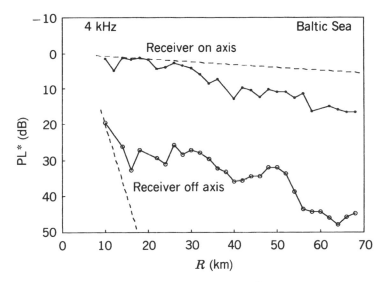

Fig. 7 Baltic Sea sound channel propagation loss PL^* at 4 kHz with source on the axis. Points: receiver on axis; circles: receiver outside the channel. The dashed lines are deterministic model calculations including absorption.

where constant a_0 depends on initial conditions and θ_0 is the limiting (bottom-grazing) ray angle. For small angles $\theta_0^2 \approx 2\Delta C/C_0$, where C_0 is the axial sound speed and ΔC is the difference between bottom and axial values of C. The attenuation coefficient is then

$$A_s \approx \frac{\pi^2 D c_0}{16 \Delta C} \qquad \text{(Np/unit distance).} \qquad (4)$$

For the isothermal gradient $g \approx 0.018$ s^{-1}, $\Delta C \approx g\Delta Z$, where ΔZ is the distance from the axis to the bottom. Unit distance is approximated as the skip distance of the limiting ray $R_0 = 2C_0 g^{-1} \sin \theta_0$.

In the Hudson Bay experiment, conditions were quite isothermal below the thermocline and the deep-water approximation is therefore appropriate. Correlation analysis of the temperature variations over the water column gave $\mu^2 \approx 1 \times 10^{-7}$ and $\rho \approx 15$ m. For the experimental conditions $\Delta Z \approx 100$ m and $D \approx 1 \times 10^{-8}$ m^{-1}, Eq. (4) yields $A_s \approx 5 \times 10^{-2}$ dB/km, compared to the measured value $A_s \approx 4 \times 10^{-2}$ dB/km.

Baltic conditions below the axis were far from isothermal. The measured value $\Delta C \approx 20$ m/s corresponds to an effective $\Delta Z \approx 1000$ m compared to the actual value $\Delta Z \approx 60$ m. For the same value of D, Eq. (4) gives $A_2 \approx 5 \times 10^{-3}$ dB/km, compared to the measured value $A_s \approx 3 \times 10^{-2}$ dB/km. The discrepancy is believed to be due to the unusually disturbed oceanographic conditions, which apparently increase the diffusion constant nearly an order of magnitude. The higher value is also consistent with the results in Fig. 7.

For sound channels in deep water with $\Delta Z \approx 3$ km and $D \approx 1 \times 10^{-8}$ m^{-1}, Eq. (4) gives $A_s \approx 2 \times 10^{-3}$ dB/km, which is comparable to the higher values observed in deep sound chanel experiments.[3]

Effects of finite bottom loss have been investigated and found to be a secondary consideration.[8] The main question is whether taking the scatterers as isotropic and averaging over the water column is a good approximation.

In the internal wave model,[9] temperature variations in the thermocline decrease exponentially with depth. Correlation scales are also depth dependent and non-isotropic, the horizontal length ρ_h being up to the order of kilometres and much greater than the vertical length ρ_v. In this case, the diffusion factor can be approximated as $D(z) \approx a\mu^2 \rho_h/\rho_v^2$, where a is of the order unity and all the parameters are uniquely defined by the buoyancy–frequency profile and latitude. Stratification effects in the deep sound channel were examined by averaging diffusion over a single cycle of the periodic ray paths, and the calculated value, $A_s \approx 5 \times 10^{-4}$ dB/km, falls in the lower range of experimental values.[10]

5 SURFACE DUCTS

Ducted propagation also occurs in a mixed surface layer, the propagation mode involving upward refraction and periodic surface reflection. Potential loss mechanisms are scatter by surface waves and entrained air. These smaller scale multiple-scatter processes can be treated as frequency-dependent diffusion. Bubble oscillations are

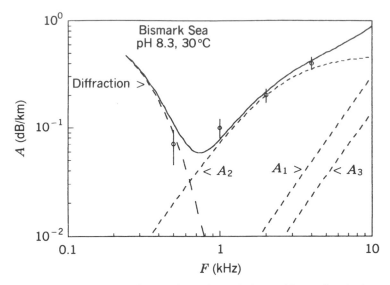

Fig. 8 Comparison of Bismark Sea surface channel data with predicted absorption plus diffraction loss.

dissipative and can cause absorption as well. Diffraction is another important loss mechanism is surface ducts.[11] In this case, the loss decreases exponentially with frequency, resulting in an effective low-frequency cutoff that depends on the layer depth.

Results of an experiment in the Bismark Sea[12] are shown in Fig. 8. For the duct depth 90 m, diffraction loss is negligible above about 500 Hz and no excess attenuation is evident in this region.

Results of an experiment in the Caribbean Sea[13] are shown in Fig. 9. For the duct depth 60 m, diffraction loss is negligible above about 1 kHz. The lower solid curve is

predicted attenuation based on the deterministic model. The data points for sea states 1+ and 2+ indicate apparent excess attenuation due to some other mechanism. Reasonable fit is achieved by adding a loss component with Rayleigh scatter (F^4) dependence below about 2 kHz, as shown by the dashed curve.

6 DISCUSSION

The excess losses observed in many sound channel experiments appear to be consistent with ray diffusion

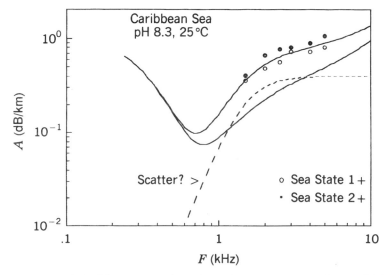

Fig. 9 Comparison of Caribbean Sea surface channel data with attenuation model. The lower solid curve includes absorption and diffraction losses. The dashed curve is added in the upper solid curve.

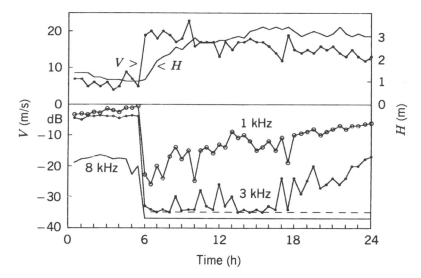

Fig. 10 Time series of wind speed V and wave height H (upper curves) and relative levels (in decibels) at 1, 3, and 8 kHz for 10 km range in the North Sea (lower curves). The dashed and solid horizontal lines represent the threshold levels for 3 and 8 kHz, respectively.

theory based on internal scatter by temperature inhomogeneities with scales that are large compared with the acoustic wavelengths involved. This hypothesis is further supported by the shallow-water experiment of Fig. 7, wherein the increase in levels outside the channel accounts for the losses within the channel. However, attempts to account for the wide regional variability of deep-water results have met with only limited success. Specific water masses and boundaries have been identified in some cases,[3] but the oceanographic data needed to determine the scattering parameters are lacking.

In most surface duct experiments by the range variation method, absorption and diffraction account for total attenuation. Lack of evidence for surface loss is probably due to low sea states and the masking effect of high absorption. However, transmission experiments between fixed sites using narrow-band sources have shown temporal variations. Experiments in the Bristol Channel by Weston and Ching[14] showed an excess loss proportional to $F^{3/2}V^4$, where V is wind speed. This is qualitatively consistent with small-perturbation theory for surface wave scattering,[11] which has been reviewed by Mellen.[15] Christian et al.[16] have also examined an empirical surface loss model based on an analysis of AMOS data from the 1950s using an earlier absorption model. They showed that reasonable quantitative agreement with theory can be achieved using an absorption correction based on the current absorption model.

Experiments in the North Sea reported by Wille et al.[17] indicate that bubble effects become dominant at higher frequencies and wind speeds. As shown in Fig. 10, these losses increase with wind speed far more rapidly than the wave height, and the process is therefore

best parameterized by wind speed. Novarini and Bruno[18] have proposed an approximate formula for loss in surface ducts based on single-bubble resonance theory and an empirical model for the bubble distribution versus depth:

$$A_b \approx 6.83 \times 10^{-5} F^{1.35} V^{3.26} D^{-S(V)}$$
$$\text{(dB/limiting ray cycle)}, \qquad (5)$$

where F is frequency (in kilohertz), D is duct depth (in metres), V is wind speed (in knots), and $S(V) = 1.1V^{-0.7}$.

A review of the experimental evidence by Wille[19] indicates that effects of wind-induced loss in shallow water falls into three frequency regimes: (1) low frequencies are practically unaffected; (2) surface scattering is the dominant mechanism for acoustic wavelengths comparable to the wave height; and (3) higher frequencies also suffer additional scatter loss and/or absorption by bubbles, which becomes the dominant mechanism for $V > 10$ m/s. Experiments similar to that of Fig. 7 would allow discrimination between scattering and absorption.

Experiments in the Bristol Channel reported by Weston[20] also indicate that aggregations of fish can be an important mechanism for attenuation in shallow water, although the relative importance of absorption and scatter is uncertain.

The mechanism responsible for the excess loss in the Caribbean Sea surface duct shown in Fig. 9 remains unresolved. Wind speeds were apparently too low for either wave or bubble effects and internal scattering seems unlikely because of the homogeneity of a mixed surface layer. Attenuation by schools of fish or other marine life is a possible mechanism here also.

REFERENCES

1. R. J. Urick, *Principles of Underwater Sound*, McGraw-Hill, New York, 1983.

2. *Global Model for Sound Absorption in Sea Water*, NUSC Scientific and Engineering Studies, Naval Underwater Systems Center, 1987.

3. A. C. Kibblewhite and L. D. Hampton, "A Review of Deep Ocean Sound Attenuation at Very Low Frequencies," *J. Acoust. Soc. Am.*, Vol. 67, 1980, pp. 147–157.

4. H. G. Schneider, R. Thiele, and P. C. Wille, "Measurement of Sound Absorption in Low Salinity Water of the Baltic Sea," *J. Acoust. Soc. Am.*, Vol. 77, 1985, pp. 1409–1412.

5. H. G. Schneider, "Average Sound Intensities in Randomly Varying Sound-Speed Structures," in J. Potter and A. Warn-Varnas (Eds.), *Ocean Variability and Acoustic Propagation*, Kluwer Academic Publishers, Dordrecht, The Netherlands, 1991, pp. 283–292.

6. L. A. Chernov, *Wave Propagation in a Random Medium*, McGraw-Hill, New York, 1975.

7. R. H. Mellen, D. G. Browning, and J. R. Ross, "Attenuation in Randomly Inhomogeneous Sound Channels," *J. Acoust. Soc. Am.*, Vol. 56, 1974, pp. 80–82.

8. R. H. Mellen and H. G. Schneider, "Diffusion Loss in Refractive Sound Channels: Bottom-Loss Effects," *J. Acoust. Soc. Am.*, Vol. 62, 1977, pp. 1038–1041.

9. C. Garrett and W. H. Munk, "Space-Time Scales of Internal Waves," *J. Geophys. Res.*, Vol. 80, 1975, pp. 291–297.

10. R. H. Mellen, D. G. Browning, and L. Goodman, "Diffusion Loss in a Stratified Sound Channel," *J. Acoust. Soc. Am.*, Vol. 60, 1976, pp. 1053–1055.

11. L. Brekhovskikh and Y. Lysanov, *Fundamentals of Ocean Acoustics*, Springer Verlag, New York, 1982, pp. 160–181.

12. R. H. Mellen and D. G. Browning, "Attenuation in Surface Ducts," *J. Acoust. Soc. Am.*, Vol. 63, 1978, pp. 1624–1626.

13. H. R. Baker, A. G. Pieper and C. W. Searfoss, "Measurements of Sound Transmission Loss at Low Frequencies 1.5–5 KC," Naval Res. Lab. Rep. 4225, Sept. 1953.

14. D. E. Weston and P. A. Ching, "Wind Effects in Shallow-Water Acoustic Transmission," *J. Acoust. Soc. Am.*, Vol. 86, 1989, pp. 1530–1555.

15. R. H. Mellen, "On Underwater Sound Scattering by Surface Waves," *IEEE J. Oceanic Eng.*, Vol. 14, 1990, pp. 245–247.

16. R. J. Christian, D. G. Browning, and D. G. Williams, "Revisions to an Empirical Surface Loss Model Using a Correction for pH-Dependent Attenuation," NUWC-NPT Technical Document 10,715, Sept. 1994.

17. P. Wille, L. Geyer, and E. Schunk, "Measurements of Wind Dependent Acoustic Transmission Loss in Shallow Water under Breaking Wave Conditions," in H. Merklinger (Ed.), *Progress In Underwater Acoustics*, Plenum Press, New York, 1987, pp. 501–508.

18. J. C. Novarini and D. R. Bruno, "Effects of the Sub-Surface Bubble Layer on Sound Propagation," *J. Acoust. Soc. Am.*, Vol. 72, 1982, pp. 510–514.

19. P. C. Wille, "Acoustic Properties of Sea Water," in J. Sundermann (Ed.), *Landolt-Börnstein, Zahlenwerte und Funktionen. New Series, Part V (Geophysics)*, Vol. 3a, Springer Verlag, Berlin, 1986.

20. D. E. Weston, "Scattering from Inhomogeneities," in L. Bjorno (Ed.), *Underwater Acoustics and Signal Processing*, D. Reidel Publishing, Dordrecht, The Netherlands, 1981, pp. 99–122.

39

VOLUME SCATTERING IN UNDERWATER ACOUSTIC PROPAGATION

Shimshon Frankenthal

1 INTRODUCTORY BACKGROUND

1.1 Introduction

Fluctuations in acoustic signals propagating underwater are caused by the stochastic variability of the index of refraction (or sound speed) in the ocean and by reflection from its randomly irregular boundaries. The objective of their study is to establish the relation between the statistics of the environmental fluctuations and the statistics of the fluctuations of propagating signals. The motivation is twofold: to be able to assess the impact of fluctuations on underwater communication and detection systems and to study the information provided by these fluctuations about the environmental processes that produce them.

In principle, the subject is part and parcel of the broad theory of wave propagation in random media, which includes, among others, studies of optical propagation through the turbulent atmosphere and the passage of radio waves through space or across the ionosphere.[1–3] Nonetheless, the ocean environment imposes peculiarities of its own, which have motivated the development of specific techniques. A major component of ocean variability—the internal waves—exhibits a strong vertical–horizontal anisotropy and a strong vertical inhomogeneity; also, its spatial and temporal characteristics are not "frozen" but are related by a hydrodynamic dispersion relation.[4,5] Moreover, acoustic propagation in the ocean is dominated by the depth profile of the sound speed and the attendant refraction of the acoustic waves. This is responsible for the long-range waveguidelike propagation in the acoustic channel that forms about the sound speed minimum, which is unaffected by surface or bottom interactions. Indeed, some of the major difficulties are encountered when one attempts to deal with stochastic phenomena in the presence of refraction, with its attendant caustics and shadow zones. Propagation in shallow water is also affected by surface and bottom scattering; these effects will not be considered here.

Considerations such as the above, as well as the signal frequencies, determine the choice of either the mode format or the ray format to account for the acoustic propagation.[6] Each of these, in turn, entails a different format to account for the stochastic effects. It would be difficult to encapsulate the variety of regimes and choices in a single neat summary, and ineffective to catalogue this variety in full. Moreover, the results are rarely reducible to formulas or simple recipes; rather, they are embodied in computer codes, "solvers," or simulators. This chapter is therefore structured as a guide to the fundamental theoretical approaches for dealing with the stochastic aspects of underwater acoustic propagation, with suitable reference to detailed sources for specific results or computational methods and the caveats attendant on their use. Our approach is to view the ocean as a (random) system whose response to a basic excitation (e.g., a monochromatic point source) is provided by the solution of the wave equation, reviewed in Section 1.2. The subsequent two sections review the relevant signal statistics and medium statistics. Sections 2 and 3 then expose how these elements are incorporated in the mode- and ray-oriented formulations, respectively. Section 4 contains a brief review of the numerical techniques that have evolved to deal with the problem and a comparison between the theoretical predictions and some of the relevant experimental observations.

Encyclopedia of Acoustics, edited by Malcolm J. Crocker
ISBN 0-471-80465-7 © 1997 John Wiley & Sons, Inc.

1.2 Wave Propagation

The propagation of pressure signals $\tilde{p}$ is governed by the *wave equation*

$$\left(\nabla^2 - \frac{1}{c^2}\frac{\partial^2}{\partial t^2}\right)\tilde{p} = 0, \qquad c = c_0(1 + \bar{\mu} + \mu), \qquad (1)$$

where $\bar{\mu}$ and μ measure, respectively, the average and fluctuating components of the variation of sound speed c about its reference level c_0. We shall assume both $\bar{\mu} \ll 1$ and $\mu \ll 1$.

The following successive substitutions adapt this equation to (a) *monochromatic signals* at the frequency ω (and the corresponding wavenumber k) and (b) *"directed" propagation*, which is confined to a narrow cone about the range direction z ($\mathbf{u}$ denotes transverse position)

$$\tilde{p} = \hat{p}e^{-i\omega t}, \qquad (2a)$$

$$\hat{p} = pe^{ikz}, \qquad (2b)$$

where

$$k = \frac{\omega}{c_0}. \qquad (2c)$$

They produce, successively, the *Helmholtz wave equation* and its *parabolic approximation*[7]

$$[\nabla^2 + k^2(1 - 2\bar{\mu} - 2\mu)]\hat{p} = 0, \qquad (3a)$$

$$\frac{\partial p}{\partial z} = \left(\frac{i}{2k}\frac{\partial^2}{\partial \mathbf{u}^2} - ik\bar{\mu} - ik\mu\right)p. \qquad (3b)$$

The parabolic equation (3b) neglects the term $\partial^2 p/\partial z^2$ on the grounds that p varies slowly with range on the scale of a wavelength. This equation ascribes the range variation of p to three "causes": *diffraction, refraction* ($\bar{\mu}$), and *scattering* due to the random refraction μ. A formal solution of this equation is the path integral,[4,8] which sums the phase contributions of *all* the paths $\mathbf{u}_p(z')$ that connect a point source to the observation point $(z, \mathbf{u})$,

$$p(z, \mathbf{u}) = N \int d(\text{paths})e^{ik(\bar{S}+S_\mu)},$$

$$\bar{S} = \int_0^z dz'\left[\frac{1}{2}\left(\frac{\partial \mathbf{u}_p}{\partial z'}\right)^2 - \bar{\mu}(y)\right],$$

$$S_\mu = -\int_0^z dz_p\, \mu[z', \mathbf{u}_p(z'), t]. \qquad (4)$$

Here, $\bar{S}$ and S_μ are the average and fluctuating components of the acoustic length and the normalizations factor N is usually adjusted to 1 in the absence of fluctuations.

Asymptotic (high-k) evaluations of integrals such as (4) center on *paths of minimal acoustic length*, that is, *rays*, which provide a skeletal description of the propagation. Within the narrow-cone approximation of the parabolic equation, both Snell's law and $\delta\bar{S} = 0$ give[9–11]

$$\partial_{z'}\mathbf{u} = \boldsymbol{\sigma}, \qquad \mathbf{u}(0) = \mathbf{u}_s; \qquad (5a)$$

$$\partial_{z'}\boldsymbol{\sigma} = -\bar{\mu}'(\mathbf{u}), \qquad \boldsymbol{\sigma}(0) = \boldsymbol{\sigma}_s; \qquad (5b)$$

$$\bar{\mu}' \equiv \partial_\mathbf{u}\bar{\mu}. \qquad (5c)$$

This system tracks the range evolution of the transverse position $\mathbf{u}$ and slope (or direction) $\boldsymbol{\sigma}$ of the deterministic ray that emerges from the plane $z' = 0$ at $\mathbf{u}_s$ with the slope $\boldsymbol{\sigma}_s$. In a depth-stratified ocean, the gradient $\bar{\mu}'$ has only a y-component.

The matrix pairs $(\partial\mathbf{u}/\partial\boldsymbol{\sigma}_s, \partial\boldsymbol{\sigma}/\partial\boldsymbol{\sigma}_s)$ and $(\partial\mathbf{u}/\partial\mathbf{u}_s, \partial\boldsymbol{\sigma}/\partial\mathbf{u}_s)$ are the sensitivities of the ray position and slope at z' to its initial slope and position. To track them, denote derivatives of $\mathbf{u}$ by $\bar{\bar{\xi}}_{1,2}$ and derivatives of $\boldsymbol{\sigma}$ by $\bar{\bar{\eta}}_{1,2}$ and solve the linear systems [derivatives of Eq. (5); see Refs. 4 and 12]

$$\partial_{z'}\bar{\bar{\xi}}_i = \bar{\bar{\eta}}_i, \qquad \bar{\bar{\xi}}_1(0) = 0, \qquad \bar{\bar{\xi}}_2(0) = I_2, \qquad (6a)$$

$$\partial_{z'}\bar{\bar{\eta}}_i = -\bar{\mu}''(\mathbf{u})\bar{\bar{\xi}}_i, \qquad \bar{\bar{\eta}}_1(0) = I_2, \qquad \bar{\bar{\eta}}_2(0) = 0, \qquad (6b)$$

where I_2 is a unit matrix, and the matrix $\bar{\mu}'' \equiv \partial_\mathbf{u}\bar{\mu}'$ consists of the transverse derivatives of the components of the vector $\bar{\mu}'$ (in a depth-stratified ocean, only the yy-component of $\bar{\bar{\mu}}''$ is nonzero). Since $\bar{\bar{\eta}}_1$ tracks the differential spreading of rays that emerge from $\mathbf{u}_s$ in various directions about $\boldsymbol{\sigma}_s$, the determinant $|\bar{\bar{\xi}}_1|$ tracks the area of a differential ray tube about the ray (5). With a point source at $(0, \mathbf{u}_s)$, the pressure at a point $(z, \mathbf{u})$ is approximately

$$p(z, \mathbf{u}) = \left(\frac{1}{4\pi}\right)|\bar{\bar{\xi}}_1|^{-1/2}\exp(ik\bar{S}), \qquad (7)$$

where the phase is determined by the acoustic path length $\bar{S}$ along the ray tracked by Eq. (5) and the amplitude by the conservation of the radiant energy flux.

The *Fresnel zone* is defined about a point $(z_r, \mathbf{u}_r)$ on a ray whose endpoints are $(0, \mathbf{u}_s)$ and $(z, \mathbf{u})$. It contains the points $P_r(z_r + \Delta z_r, \mathbf{u}_r + \Delta\mathbf{u}_r)$ such that the total acoustic length along the discontinuous path that consists of two ray segments that connect P_r and the endpoints exceeds the acoustic length of the original ray by less than a half-wavelength (π/k). The radiation that propagates along the paths that constitute the boundary of the Fresnel zone interferes destructively at $(z, \mathbf{u})$ with the radiation along the original ray. In a depth-stratified ocean (planar rays), the trace of this zone in a transverse plane z_r is an ellipse whose axes are

$$\Delta u_i = \left[\frac{2\pi}{k} g_i(z_r, z_r) \right]^{1/2} \quad \text{with} \quad i = x, y, \quad (8a)$$

where g_i are the Green's functions associated with the ray spread equations (6):

$$\left(\frac{\partial^2}{\partial z'^2} + \overline{\overline{\mu}}''_{ii} \right) g_i = \delta(z' - z_r)$$

$$\text{with } g_i(z' = 0) = g_i(z' = z) = 0. \quad (8b)$$

1.3 Statistics of Signal Waves

Many of the statistics of interest are related to moments of the pressure p, defined as ensemble averages $(\langle \cdot \rangle)$ over all the realizations of the random ocean. Besides the average pressure

$$\Gamma_0(z, \mathbf{u}, t) = \langle p(z, \mathbf{u}, t) \rangle, \quad (9)$$

one can define higher coherence moments in a range plane z compactly by using the label arguments j (and l) to identify combinations of the transverse position $\mathbf{u}$, signal wavenumber k, and time t at which the pressure (and its conjugate) are sampled:

$$\Gamma_{nm} = \left\langle \prod_{j=1}^{n} p(z, j) \prod_{l=1}^{m} p^*(z, l) \right\rangle. \quad (10)$$

The *mutual coherence function* Γ_{11} is best treated using average and separation variables

$$\Gamma_{11}(z, \mathbf{u}, \mathbf{q}, k_s, k_d, t, \tau)$$
$$= \langle p(z, \mathbf{u}_1, k_1, t_1) p^*(z, \mathbf{u}_2, k_2, t_2) \rangle, \quad (11a)$$

$$\mathbf{u}_{1,2} = \mathbf{u} \pm \tfrac{1}{2} \mathbf{q}, \qquad k_{1,2} = k_s \pm \tfrac{1}{2} k_d,$$
$$t_{1,2} = t \pm \tfrac{1}{2} \tau. \quad (11b)$$

The Fourier transforms of Γ_{11} with respect to the separation variables $(\mathbf{q} \to \boldsymbol{\sigma}, k_d \to t_d, \tau \to \omega_D)$ are also meaningful. The effects of scattering are manifest in the characteristic quantities that determine the dependence of Γ_{11} (and its transforms) on these separation variables (and their conjugate counterparts).

In monochromatic propagation, the lengths $L_{qx,y}$ associated with the $\mathbf{q}$-dependence of Γ_{11} roughly determine how far one can separate two receivers about $(z, \mathbf{u})$ before scattering "destroys" the correlation between their signals. With $\mathbf{q} = 0, \Gamma_{11}$ is the average intensity at $(z, \mathbf{u})$,

$$\langle I(z, \mathbf{u}) \rangle = \Gamma_{11}(z, \mathbf{u}, 0). \quad (12)$$

The transform $\mathbf{q} \to \boldsymbol{\sigma}$ produces the directional distribution of the signal at $(z, \mathbf{u})$, and the conjugate characteristic quantities $(k L_{qx,y})^{-1}$ determine its directional spread.

For signals with a finite bandwidth, bichromatic correlation (of signals at different frequencies or wavenumbers k_1 and k_2) is important. Interest in this entity is readily motivated by the expression for the intensity of a time-dependent signal $\tilde{p}(t)$ with a spectrum $\hat{p}(\omega)$,

$$I(t) = \frac{1}{4\pi^2} \int_{-\infty}^{\infty} d\omega_d \, e^{-i\omega_d t} \int_{-\infty}^{\infty} d\omega_s \, \hat{p}(\omega_1) \hat{p}^*(\omega_2)$$

$$\text{with } \omega_{1,2} = \omega_s \pm \tfrac{1}{2} \omega_d. \quad (13)$$

Here, the intensity spectrum is the inner integral, whose integrand relates directly to the bichromatic coherence. Scattering will significantly decorrelate two pure tones if their wavenumber separation exceeds the *coherence bandwidth* k_c, which characterizes the k_d-dependence of Γ_{11}. The conjugate *multipath spread time* $t_p = (c_0 k_c)^{-1}$ determines the broadening of a pulse that was initially sharp.

Finally, the time-dependent properties of the medium are manifest in the (t, τ)-dependence of Γ_{11}. Here, the *coherence time* τ_c, which is associated with the τ-dependence of Γ_{11}, determines how long a pure tone will remain coherent, and its conjugate *Doppler-bandwidth* $\omega_{Dc} = \tau_c^{-1}$ is the broadening of a spectral line that was initially sharp.

The *monochromatic fourth moment* is customarily treated using the average position $\mathbf{u}$ and the three separation variables $\mathbf{q}, \mathbf{s},$ and $\mathbf{p}$:

$$\Gamma_{22}(z, \mathbf{u}_1, \mathbf{u}_2, \mathbf{u}_3, \mathbf{u}_4) = \langle p(z, \mathbf{u}_1) p^*(z, \mathbf{u}_2) p^*(z, \mathbf{u}_3) p(z, \mathbf{u}_4) \rangle$$
$$= \Gamma_{22}(z, \mathbf{u}, \mathbf{q}, \mathbf{s}, \mathbf{p}), \quad (14a)$$
$$\mathbf{u}_{1,2} = \mathbf{u} \pm \tfrac{1}{2} \mathbf{s} \pm \left(\tfrac{1}{2} \mathbf{p} + \tfrac{1}{4} \mathbf{q} \right),$$
$$\mathbf{u}_{3,4} = \mathbf{u} \pm \tfrac{1}{2} \mathbf{s} \pm \left(\tfrac{1}{2} \mathbf{p} + \tfrac{1}{4} \mathbf{q} \right). \quad (14b)$$

Setting all the separation coordinates to zero yields the mean-square intensity and the scintillation index associated with it,

$$\langle I^2(z, \mathbf{u}) \rangle = \Gamma_{22}(z, \mathbf{u}, 0, 0, 0), \quad (14c)$$

$$\sigma_I^2 \equiv \frac{\langle I^2 \rangle - \langle I \rangle^2}{\langle I \rangle^2}, \qquad (14d)$$

whereas setting $\mathbf{q}$ and either $\mathbf{s}$ or $\mathbf{p}$ to zero yields the intensity correlation

$$\langle B_I(z, \mathbf{u}, \mathbf{s}) \rangle = \langle I(z, \mathbf{u} + \mathbf{s}/2) I(z, \mathbf{u} - \mathbf{s}/2) \rangle$$
$$= \Gamma_{22}(z, \mathbf{u}, \mathbf{s}, 0, 0). \qquad (14e)$$

1.4 Statistics of the Medium Refractivity

General Definitions At any given time, we may view the random refractivity as a space-dependent random process $\mu(\mathbf{r})$. Such a description may be adequate, for example, when the random medium may be considered "frozen" over the brief duration of a given acoustic transmission. Successive transmissions may then be assumed to experience different realizations of $\mu(\mathbf{r})$. For such a process, the correlation function is the ensemble average

$$B_\mu(\mathbf{r}_1, \mathbf{r}_2) \equiv \langle \mu(\mathbf{r}_1) \mu(\mathbf{r}_2) \rangle = B_\mu(\boldsymbol{\rho}, \mathbf{r})$$
$$\text{with } \mathbf{r}_{1,2} = \mathbf{r} \pm \tfrac{1}{2} \boldsymbol{\rho}. \qquad (15)$$

If the process is statistically homogeneous, then B_μ depends only on $\boldsymbol{\rho} = \mathbf{r}_1 - \mathbf{r}_2$. Also, B_μ and the variance $\langle \mu^2 \rangle$ can be related to the spectrum $\psi_\mu(\mathbf{K})$ of the process; thus

$$B_\mu(\mathbf{r}_1, \mathbf{r}_2) = B_\mu(\boldsymbol{\rho}) = \int d\mathbf{K}\, e^{i\mathbf{K} \cdot \boldsymbol{\rho}} \psi_\mu(\mathbf{K}), \qquad (16a)$$

$$\langle \mu^2 \rangle = B_\mu(0) = \int d\mathbf{K}\, \psi_\mu(\mathbf{K}). \qquad (16b)$$

One may also define a structure function for this process and relate it to the spectrum,

$$D_\mu(\boldsymbol{\rho}) \equiv \langle |\mu(\mathbf{r}_1 + \boldsymbol{\rho}) - \mu(\boldsymbol{\rho})|^2 \rangle$$
$$= 2 \int d\mathbf{K}\, (1 - \cos \mathbf{K} \cdot \boldsymbol{\rho}) \psi_\mu(\mathbf{K}). \qquad (17)$$

For a real homogeneous process, this is equivalent to setting

$$D_\mu(\boldsymbol{\rho}) = 2[B_\mu(0) - B_\mu(\boldsymbol{\rho})] \quad \text{or} \quad B_\mu(\boldsymbol{\rho})$$
$$= \tfrac{1}{2} [D_\mu(\boldsymbol{\rho}) - D_\mu(\boldsymbol{\rho} \to \infty)]$$
$$\text{with } D_\mu(\boldsymbol{\rho} \to \infty) = 2\langle \mu^2 \rangle. \qquad (18)$$

Several scales l_i (with corresponding wavenumbers $K_i^* = 1/l_i$) may be associated with such a process: They characterize the $\boldsymbol{\rho}$-dependence of B_μ (or the $\mathbf{K}$-dependence of ψ_μ) in various regimes. The largest characteristic scale is often termed the "outer" scale: it characterizes the manner in which B_μ approaches 0 and D_μ approaches $2\langle \mu^2 \rangle$ as the separation $\boldsymbol{\rho}$ approaches infinity (and thus how the spectrum ψ_μ approaches a finite limit as K approaches zero).

The structure function D_μ serves as a viable descriptor of nonhomogeneous processes as well, provided that their first increments constitute homogeneous processes. Under such circumstances, the spectrum ψ_μ no longer approaches a finite value as $\mathbf{K}$ approaches 0, so that a variance does not exist, and D_μ does not approach a finite limit as $\boldsymbol{\rho}$ approaches infinity. Although the relations (16) and (18) no longer apply in this situation, Eq. (17) still does. A typical example would be a process whose spectrum is a power law K^{-p} with $p < 5$. Both conceptually and practically, the simplest way to deal with such a process is by considering a homogeneous process whose outer scale becomes infinite: It is an idealization motivated by the desire to make asymptotic statements that are independent of the outer scale (e.g., when the latter is very large and highly variable).

The frequently invoked Markov approximation[1,2,13] is formally embodied in the statement

$$B_\mu(\boldsymbol{\rho}) = B_\mu(\mathbf{q}, \rho_z) = \delta(\rho_z) A(\mathbf{q})$$

where $\boldsymbol{\rho} = (\mathbf{q}, \rho_z)$, $\qquad (19a)$

$$A(\mathbf{q}) = \int_{-\infty}^{\infty} d\rho_z\, B_\mu(\mathbf{q}, \rho_z) = 2\pi \int d\boldsymbol{\kappa}\, e^{i\boldsymbol{\kappa} \cdot \mathbf{q}} \psi_\mu(\boldsymbol{\kappa})$$

where $\mathbf{K} = (\boldsymbol{\kappa}, K_z)$, $\qquad (19b)$

which asserts that the process is completely uncorrelated in the range direction. This serves to idealize many practical situations in which the range resolution of interest far exceeds the correlation length in that direction. The function $A(\mathbf{q})$, which depends on the separation $\mathbf{q}$ in planes transverse to the range direction, then contains all the statistical information that is needed: It might be termed the "transverse" correlation. Indeed, by analogy to D_μ, one then defines a transverse structure function

$$D_A(\mathbf{q}) = 2[A(0) - A(\mathbf{q})]. \qquad (20)$$

One must, however, bear in mind that both A and D_A have the dimension of length, which arises from the integration in Eq. (19b); this is the so-called integral scale l_p,[1] which "lumps" the details of the dependence of the correlation function on the separation ρ_z.

To describe a space–time random process $\mu(\mathbf{r}, t)$, the average and separation time arguments (t, τ) defined as in Eq. (11b) must be incorporated in B_μ. The correlation of a homogeneous and stationary process is then related to the space–time spectral density $\psi_\mu(\mathbf{K}, \Omega)$ by

$$B_\mu(\boldsymbol{\rho}, \tau) = \iint d\Omega \, d\mathbf{K} \, e^{i(\mathbf{K} \cdot \boldsymbol{\rho} - \Omega \tau)} \psi_{s\mu}(\mathbf{K}, \Omega). \quad (21)$$

Ocean Variability

Ocean variability due to internal waves[4,5,14] is indeed space and time dependent. However, in adapting the statistical constructs above to the treatment of this variability, the following points must be borne in mind:

1. The time variability is stationary and therefore introduces τ in B_μ and its conjugate Ω in $\psi_{s\mu}$. However, the hydrodynamics of the internal waves imposes a dispersion relation[4,5] $\Omega = f_{\text{dis}}(\mathbf{K})$ between the frequency Ω and the spatial wavenumber $\mathbf{K}$. Formally, the spectral density $\psi_{s\mu}$ must therefore contain a delta function $\delta[\Omega - f_{\text{dis}}(\mathbf{K})]$, whose coefficient may be treated as a spatial spectrum $\psi_\mu(\mathbf{K})$ (alternatively, either component of $\mathbf{K}$ may be replaced by Ω consistently with the dispersion relation).

2. The space variability may be termed "locally" homogeneous and anisotropic. The first attribute implies that its statistics (e.g., correlation lengths and variance) vary slowly with depth on a "global" scale that far exceeds the local vertical correlation length. The anisotropy is manifest in a large disparity between the local correlation lengths associated, respectively, with the horizontal variability (which is isotropic) and the vertical variability.

3. The spectrum involves the depth-dependent Brunt–Wisala buoyant frequency $\omega_b(y)$ and the latitude-dependent inertial frequency ω_i associated with Earth's rotation, as well as the surface variance $\langle \mu_0^2 \rangle$ of the refractivity fluctuations ($\sim 2.5 \times 10^{-7}$), the scale L_0 of the sound channel (~ 1 km), and a characteristic mode number $j_*(=3)$. We have[4]

$$\psi_\mu(\mathbf{K}, y) = \langle \mu_0^2 \rangle \overline{\omega}_b^3 \overline{\omega}_i \, \frac{2}{\pi^3} \, \frac{K_H}{(K_H^2 + \overline{\omega}_i^2 K_V^2)^2} \, \frac{|K_V| K_V^*}{K_V^2 + K_V^{*2}},$$

$$0 < K_H < (1 - \overline{\omega}_i^2)^{1/2} K_V, \quad (22)$$

$$\overline{\omega}_b(y) = \frac{\omega_b(y)}{\omega_b(0)}, \quad (22a)$$

$$\overline{\omega}_i(y) = \frac{\omega_i}{\omega_b(y)}, \quad (22b)$$

$$K_V^* = j_* \, \frac{\pi}{L_0} \, \overline{\omega}_b(y), \quad (22c)$$

and

$$\langle \mu^2(y) \rangle = \langle \mu_0^2 \rangle \overline{\omega}_b^3 W(\overline{\omega}_i) \quad (22d)$$

where

$$W = \frac{2}{\pi} \left(\tan^{-1} \alpha - \frac{\alpha}{1 + \alpha^2} \right) \quad \text{and}$$

$$\alpha = \overline{\omega}_i^{-1} (1 - \overline{\omega}_i^2)^{1/2}.$$

The spectrum (22) is confined to a cylindrical cone about the vertical axis K_V in the wavenumber space $\mathbf{K}(K_H, K_V)$; the depth dependences of the refractivity variance $\langle \mu^2 \rangle$ in Eq. (22d) is governed by the buoyant frequency [in Eq. (22d), $W \sim 1$ when $\overline{\omega}_i \ll 1$]; so is also the vertical cutoff K_V^* in Eq. (22c); the horizontal cutoff in Eq. (22) is much smaller because $\overline{\omega}_i$ is small.

4. The spectrum (22) implies that the transverse correlation A and its associated structure function D_A depend on the depth y as well as on $\mathbf{q}$. This is easy to rationalize when one considers that this depth dependence is very slow on the scale of the vertical correlation length.

Caveats

Two cautionary notes must be sounded before leaving this section. First, the internal wave spectrum above is based on extensive measurements and was used to interpret acoustic observations in at least one major experiment, namely AFAR.[4] However, environmental measurements taken during the MATE experiment suggest the presence of a "contaminant" spectrum, termed Finestructure, due to ocean processes that do not obey the dispersion relation mentioned above. The modifications are discussed in Ref. 15, and their application in the interpretation of acoustic MATE observations are discussed in Ref. 16.

Second, the Markov approximation, which rationalizes the uses of A and D_α above, effectively "eliminates" the correlation length l_z from consideration. With this approximation, narrow angular scattering in the ith transverse direction is assured only when the correlation length l_i in that direction satisfies $k l_i \gg 1$. However, it was shown[17–19] that even when this condition is violated the effect of angular scatter are limited provided that $k l_z \gg 1$. Thus, in anisotropic media, the Markov approximation must be used with great caution.

2 MODAL TREATMENT OF VOLUME SCATTERING

2.1 Background and General Formulation

The modal description of radiative propagation is very well established, especially in waveguidelike environments.[6,20] It is therefore natural to couch the study of the effects of random variability in medium properties in the same format. Exhaustive treatments of certain aspects of this approach are available.[21,22] Perhaps the most widespread tool has been the coupled-power equations[21–25] (coupled fluctuation equations were also suggested[21–23] whose main application is the study of the redistribution of energy among modes and radiative losses from the waveguide structure. To complement the above, we emphasize here the development of mathematical models for the monochromatic spatial coherence, which is of particular significance to the issue of array performance.

To incorporate scattering within the modal format, it is convenient to proceed from the Helmholtz wave equation (3a) and introduce the parabolic approximation later. The essence of the ideas is best exposed in the context of a range-invariant environment; we shall later outline the extension to a range-dependent environment.

The mode format expresses the solution of the Helmholtz equation (3a) as follows:

$$\hat{p}(x, y, z) = \sum_i \hat{p}_i(\mathbf{r}_H) Y_i(y) \quad \text{where} \quad \mathbf{r}_H = (x, z),$$

$$(23)$$

$$\left[\frac{d^2}{dy^2} + k^2[1 + 2\overline{\mu}(y)] - \beta_i^2 \right] Y_i(y) = 0, \qquad (23a)$$

$$\left[\frac{\partial^2}{\partial \mathbf{r}_H^2} + \beta_i^2 \right] \hat{p}_i(\mathbf{r}_H) = -\sum_j \mu_{ij}(\mathbf{r}_H) \hat{p}_j(\mathbf{r}_H), \qquad (23b)$$

$$\mu_{ij}(\mathbf{r}_H) = 2 \int_0^H dy \, k^2 \mu(\mathbf{r}_H, y) Y_i(y) Y_j(y). \qquad (23c)$$

The modal eigenfunctions Y_i account for the average refractivity $\overline{\mu}$: They satisfy Eq. (23a) as well as the conditions imposed at the horizontal boundaries of the model ocean. The eigenvalues β_i of Eq. (23a) define the horizontal propagation constants in the modal amplitude equation (23b). This equation is obtained by inserting Eq. (23a) in the wave equation and using the orthonormal property of the eigenfunctions. The modal amplitudes also satisfy the source conditions and the Sommerfeld

radiation condition at large distances. The random variability of the refractive index, without which the modes would be uncoupled, enters via its depth integral $\mu_{ij}(\mathbf{r}_H)$.

When the eigenvalue spectrum of (23a) possesses both a discrete component (modes for $i < N$) and a continuous λ-dependent component, the "improper" modal function $Y(y, \lambda)$ is determined by an equation such as (23a) with λ replacing β_i^2. All summations terminate at N; the remainder of each sum over the indices i or j is replaced by an integral over the continuous variables λ or λ' that replace them. An equation such as (23b), with the replacements indicated above, governs the amplitude function $\hat{p}(\mathbf{r}_H, \lambda)$ of the improper modal function $Y(y, \lambda)$; and functions $\mu_j(\mathbf{r}_H, \lambda), \mu_i(\mathbf{r}_H, \lambda')$, and $\mu(\lambda, \lambda')$, defined by analogy to Eq. (23c), describe scattering between the modes and the continuous spectrum and within that spectrum.[21]

Equations (23b) for the modal amplitudes are the basis of mode-coupling theory. We shall use them to compute the overall two-point coherence $\hat{\Gamma}$ of the propagating field. This can be done expeditiously using the parabolic counterpart of Eq. (23b):

$$\hat{p}_i = p_i e^{i\beta_i z}, \qquad (24a)$$

$$i2\beta_i \frac{\partial p_i}{\partial z} + \frac{\partial^2 p_i}{\partial x^2} = -\sum \mu_{ij}(\mathbf{r}) p_j(\mathbf{r}) e^{i(\beta_j - \beta_i)z},$$

$$(24b)$$

where the term $\partial^2 p/\partial z^2$ is omitted, as in the derivation of Eq. (3b). The overall coherence $\hat{\Gamma}$, defined as in Eq. (11) for the pressure $\hat{p}$ in Eq. (23), assumes the form

$$\hat{\Gamma}_{11}(x_1, y_1, x_2, y_2, z) = \langle \hat{p}(x_1, y_1, z) \hat{p}^*(x_2, y_2, z) \rangle$$

$$= \sum_i \sum_j \Gamma_{ij}(x_1, x_2, z) Y_i(y) Y_j(y)$$

$$\cdot e^{i(\beta_i - \beta_j)z}, \qquad (25a)$$

$$\Gamma_{ij}(x_1, x_2, z) = \langle p_i(x_1, z) p_j^*(x_2, z) \rangle. \qquad (25b)$$

This introduces a matrix of modal coherences Γ_{ij} whose equations can be derived from the paraxial approximation (24b) to the coupled-mode equations. To derive the general form, multiply Eq. (24b) for p_i by p_j^*, write the conjugate of (24b) for p_j^* and multiply it by p_i, and ensemble average their sum. The result assumes the form

$$\frac{\partial}{\partial z} \Gamma_{ij}(x_1, x_2, z) = \frac{i}{2} \left(\frac{1}{\beta_i} \frac{\partial^2}{\partial x_1^2} - \frac{1}{\beta_j} \frac{\partial^2}{\partial x_2^2} \right)$$

$$\cdot \Gamma_{ij}(x_1, x_2, z) + Sct, \qquad (26)$$

and the problem is to perform the ensemble average that produces the scattering term Sct. This can be accomplished by invoking the Markov approximation (19) and employing functional analysis techniques[2] (for the case at hand, this is done in Ref. 23) or by using the technique discussed in connection with the moment equations in Section 2.2.[14] (For yet another procedure, see the treatment of cylindrical spreading in Ref. 26). Either technique is subject to the caveats

$$\left(\frac{1}{k}, l_H, l_V\right) \ll \Delta z \ll k^2 l_H, \qquad (27a)$$

$$1 \ll k l_H, \qquad (27b)$$

where Δz is the smallest range resolution element of interest. As may well be expected, the scattering term Sct involves statistics of the μ_{ij}'s of Eq. (23c), namely

$$\sigma_{ijkm}(\mathbf{r}'_H, \mathbf{r}''_H) = \langle\mu_{ij}(\mathbf{r}'_H)\mu_{km}(\mathbf{r}''_H)\rangle = \int_0^H dy' \int_0^H dy''$$

$$\cdot k^4 B_\mu(\mathbf{r}', \mathbf{r}'') Y_i(y') Y_j(y') Y_k(y'') Y_m(y''), \qquad (28a)$$

where $B_\mu(\mathbf{r}', \mathbf{r}'')$ is the correlation (15) of the refractivity fluctuations that involves the vertical and horizontal correlation lengths l_V and l_H. When the μ_{ij}'s are statistically homogeneous and isotropic in $\mathbf{r}_H$, the σ's depend only on the horizontal separations

$$\sigma_{ijkm}(\mathbf{r}'_H, \mathbf{r}''_H) = \sigma_{ijkm}(\rho_x, \rho_z), \qquad \rho_x = x' - x'',$$

$$\rho_z = z' - z'' \qquad (28b)$$

The evaluation of Sct produces distinct equations for the range evolution of the (self-) modal ($i = j$) and the cross-modal ($i \neq j$) coherences. The modal coherences Γ_{ii} are coupled only to one another and obey

$$\left[\frac{\partial}{\partial z} - \frac{i}{2\beta_i}\left(\frac{\partial^2}{\partial x_1^2} - \frac{\partial^2}{\partial x_2^2}\right) + S_{ii}\right]$$

$$\cdot \Gamma_{ii} + \sum_{k \neq i} S_{ik}\Gamma_{kk} = 0, \qquad (29)$$

$$S_{ii}(\rho_x) = \frac{1}{2\beta_i^2} \int_0^\infty d\rho_z [\sigma_{iiii}(0, \rho_z) - \sigma_{iiii}(\rho_x, \rho_z)]$$

$$+ \frac{1}{2\beta_i} \sum_{k \neq i} \frac{1}{\beta_k} \int_0^\infty d\rho_z \cos(\beta_k - \beta_i)$$

$$\cdot \rho_z \sigma_{ikki}(0, \rho_z), \qquad (29a)$$

$$S_{ik}(\rho_x) = -\frac{1}{2\beta_i^2} \int_0^\infty d\rho_z \cos(\beta_k - \beta_i)\rho_z \sigma_{ikik}(\rho_x, \rho_z). \qquad (29b)$$

The cross-modal coherence functions Γ_{ij}, which are completely uncoupled, obey

$$\left[\frac{\partial}{\partial z} - \frac{i}{2}\left(\frac{1}{\beta_i}\frac{\partial^2}{\partial x_1^2} - \frac{1}{\beta_j}\frac{\partial^2}{\partial x_2^2}\right) + C_{ij}\right]\Gamma_{ij} = 0, \qquad (30)$$

$$C_{ij}(x_1 \quad x_2) = \frac{1}{4}\left[\frac{1}{\beta_i^2}\int_0^\infty d\rho_z \sigma_{iiii}(0, \rho_z) + \frac{1}{\beta_j^2}\int_0^\infty\right.$$

$$\cdot d\rho_z \sigma_{jjjj}(0, \rho_z) - \frac{2}{\beta_i\beta_j}\int_0^\infty$$

$$\left. \cdot d\rho_z \sigma_{iijj}(x_1 - x_2, \rho_z)\right]$$

$$+ \frac{1}{4}\left[\sum_{k \neq i} \frac{1}{\beta_i\beta_k}\int_0^\infty d\rho_z e^{i(\beta_k - \beta_i)\rho_z}\right.$$

$$\cdot \sigma_{ikki}(0, \rho_z) + \sum_{k \neq j} \frac{1}{\beta_j\beta_k}\int_0^\infty$$

$$\left. \cdot d\rho_z e^{-i(\beta_k - \beta_j)\rho_z}\sigma_{jkkj}(0, \rho_z)\right]. \qquad (30a)$$

The original equations (24b) and (25a) suggest that double sums might be present; the absence of such sums and the decoupling between the various coherences are a consequence of certain "selection rules" that stem from the oscillatory phase terms in Eq. (25a).

In Eqs. (29) and (30), we have isolated the terms that involve sums over $k \neq i$ and $k \neq j$, whose integrands contain factors that oscillate with ρ_z at frequencies $\beta_k - \beta_{i,j}$. Since l_H governs the ρ_z-dependence of the various σ's, these integrals are negligible if

$$(\beta_k - \beta_i)l_H \gg 1 \quad \text{for all} \quad \beta\text{'s.} \tag{31}$$

Although the β's are of the order of the wavenumber k and we have asummed $kl_H \gg 1$ in Eq. (4b), the condition stated in Eq. (31) ensues only if *all* the relative modal separations are not too small. When Eq. (31) is satisfied, we may ignore the scattering coefficient S_{ik} in Eq. (29b) and the second terms in the scattering coefficients S_{ii} in Eq. (29a) and C_{ij} in Eq. (30a).

2.2 Modal Coherence Functions Γ_{ii} and Power Coupling

In the presence of scattering, the modal coherence functions via the coefficients S_{ik} defined in Eq. (29b). However, because the diffraction term in Eq. (29) involves only β_i, the equations simplify readily if we replace x_1 and x_2 by their average and difference counterparts x and ρ_x:

$$\left[\frac{\partial}{\partial z} - \frac{i}{\beta_i} \frac{\partial^2}{\partial x \, \partial \rho_x} + S_{ii}(\rho_x) \right] \Gamma_{ii}(z, x, \rho_x)$$

$$+ \sum_{k \neq i} S_{ik}(\rho_x) \Gamma_{kk}(z, x, \rho_x) = 0. \tag{32}$$

In general, these equations must be solved numerically. However, in certain cases, the diffractive operator $-i \, \partial^2 / \partial x \, \partial \rho_x$ is absent, and Eq. (32) becomes a set of ordinary differential equations in z with constant (ρ_x-dependent) coefficients. This is relevant when the source coherence is x-independent or if one is interested only in the x-integrated modal coherence or in the integrated modal power flux across any z-plane. The solutions[25] are linear combinations of the form $\gamma_n(\rho_x) \exp[-z/l_n(\rho_x)]$, where the characteristic decay lengths l_n are roots of the characteristic equations associated with Eq. (32).

Even when the source coherence depends on the x-coordinate, an analytic solution of the modal coherence equations is still possible, provided that the condition stated in Eq. (31) is satisfied. Equations (32) are then decoupled because $S_{ik} = 0$ and can be solved by well-known techniques (see Section 2.2). Their Green's functions are then

$$G_{ii}(x, \rho_x, z; x', \rho_x') = \frac{\beta_i}{2\pi z} \exp\left[i \frac{\beta_i}{z} (x - x')(\rho_x - \rho_x') \right.$$

$$\left. - z \int_0^1 d\xi \, S_{ii}[\rho_x' + (\rho_x - \rho_x')\xi] \right], \tag{33a}$$

and when the source coherence is x independent, they produce the very simple solution

$$\Gamma_{ii}(\rho_x, z) = \Gamma_{ii}(\rho_x, 0) \, \exp[-z S_{ii}(\rho_x)]. \tag{33b}$$

In Eq. (33b), S_{ii} is indeed just the first term of Eq. (29a), since the second term in that equation becomes negligible together with S_{ik} when Eq. (31) is satisfied.

2.3 Cross-Modal Coherence Functions

The cross-modal coherence functions contribute to the overall coherence (25) yet are often ignored or neglected in modal treatments of underwater propagation. Note that there are N^2 such modes, so that, even though their phases $(\beta_i - \beta_j)z$ are quasi-random at sufficiently large ranges, their total contribution may still be of the order of the contribution of the N modal coherence functions. It is therefore important to establish when their neglect can be justified (if at all) as a consequence of the effects of scattering.

To do so, we must establish bounds on the scales that dominate the range dependence of the coherence. These arise from rewriting the coefficient C_{ij} in Eq. (30a); thus

$$C_{ij}(\rho_x) = C_0 - C_1(\rho_x) = C_0 - C_1(0) + C_1(0)$$

$$\cdot \left[1 - \frac{C_1(\rho_x)}{C_1(0)} \right] = \frac{1}{l_x} + \frac{1}{l_D} D_\sigma\left(\frac{\rho_x}{l_H} \right), \tag{34}$$

and identifying the structure function D_σ and the variance relevant scales

$$D_\sigma\left(\frac{\rho_x}{l_H} \right) = 1 - \frac{C_1(\rho_x)}{C_1(0)}, \tag{34a}$$

$$\frac{1}{l_D} = C_1(0), \tag{34b}$$

$$\frac{1}{l_x} = C_0 - C_1(0), \tag{34c}$$

$$\frac{1}{l_0} = \frac{1}{l_x} + \frac{1}{l_D}. \tag{34d}$$

Here, C_0 is the component of C_{ij}, where the σ's are taken at $\rho_x = 0$ and $C_1(\rho_x)$ is the rest.

While the length scale l_0 of Eq. (34d) is clearly shorter than l_D of Eq. (34b) and l_x of Eq. (34c), these three scales are generally of the same order, unless $C_1(0)$ is either much smaller than C_0 (and then $l_0 \sim l_x \ll l_D$) or very close to it (and then $l_0 \sim l_D \ll l_x$).

The structure function D_σ in Eq. (34a) varies from 0 (when $\rho_x = 0$) to 1 (as $\rho_x \to \infty$) on the scale l_H. It provides two additional parameters that will be needed subsequently:

$$\gamma = l_H \sqrt{\frac{2\beta_i \beta_j}{|\beta_i - \beta_j| l_D}}, \tag{35a}$$

$$\alpha = \left(\frac{\gamma^p}{D_0} \right)^{1/(p+2)} \tag{35b}$$

with

$$D_\sigma(\xi) \to D_0 \xi^p \quad \text{as} \quad \xi \to 0. \tag{35c}$$

Here, γ is a dimensionless scale for D_σ that depends on mode separation and α reflects the parameters D_0 and p, which characterize the assumed power law behavior Eq. (35c) of D_σ when $\rho_x \ll l_H$. The bounds on the z-dependence of the Γ_{ij} correspond to the asymptotic limits $\gamma \to \infty$ and $\gamma \to 0$. All else being equal, the former limit corresponds to neighboring modes at very high frequencies and the latter limit to distant modes.

When $\gamma \ll 1$, D_σ may be replaced by its asymptotic limit 1. The coefficient C_{ij} in Eq. (34) is then C_0, and Eq. (30) shows that Γ_{ij} decays exponentially with z on the scale l_0 of Eq. (34d).

When $\gamma \gg 1$, D_σ obeys Eq. (35c), and one can show that the z-dependence of Γ_{ij} is governed by the shorter of the scales l_x of Eq. (34c) and $\alpha^2 l_D$ of Eqs. (34b) and (35b). If $C_1(0)$ is not too close to C_0 [see the discussion following Eq. (34)], l_D and l_x are of the same order, and since $\alpha^2 \gg 1$ [see Eq. (35b)], l_x is the shorter, and therefore dominant, scale. The exception occurs if C_0 is very close to $C_1(0)$. Reference to Eq. (30a) reveals that this might happen, for example, if (1) the summation terms are negligible and (2) the ρ_z-integrals of the various σ's in the remaining coefficient are all equal (this takes place in a channel that is much shallower than the vertical correlation length l_V). In this case, we find that $l_x \propto (\beta_i - \beta_j)^{-2}$ and therefore becomes very large precisely as γ becomes large. The dominant scale of the range behavior of the coherence is then $\alpha^2 l_D$ rather than l_x.

2.4 Extension to Range-Dependent Waveguides

When the average sound speed profile depends on the horizontal position $\mathbf{r}_H$, it is necessary to allow the eigen-

functions Y_i of Eq. (23a) to depend on $\mathbf{r}_H$,[27,28] with attendant but surmountable complications in the right-hand member of Eq. (23b). Indeed, Eq. (23b) has been adapted[29,30] to handle a randomly varying surface height in this manner. However, a full treatment of the coherence by this procedure is yet to be produced.

3 RAY-ORIENTED TREATMENT OF VOLUME SCATTERING

There are two formalisms for incorporating the effects of scattering in a ray-oriented treatment of acoustic propagation. They are based, respectively, on the path integral solution (4) of the wave equation and on the solution of the equations that govern the various moments defined in Section 1.3. The two approaches are interrelated.[31,32] Essentially, both approaches rely on the localization of the propagation along the ray trajectories and employ spread tracing and asymptotic techniques to determine the major stochastic effects that originate near these rays.

3.1 Path Integrals

The path integral formalism[4,8,33] employs Eq. (4) to express the moments of the field directly as multiple-order path integrals. Ensemble averaging is applied to the random phase factor, which involves a sum S_T of (presumably Gaussian) S_μ's and therefore produces $\exp(-0.5 S_T^2)$. An asymptotic evaluation of the resultant integrals is possible when the dominant contributions are localized about rays. Since the integrals are normalized to produce 1 in the absence of fluctuations, the results of the asymptotic evaluations are multiplicative factors that account for the effect of scattering on the ambient field. These results are expressed in terms of three quantities: the strength parameter Φ^2, which in some cases provides a measure of the phase variance; the structure function D_ϕ of the receiver phase; and the diffraction parameter Λ, a fluctuation-weighted measure of the ratio of the Fresnel radius to the vertical correlation length l_V along the ray, which indicates the degree to which the fluctuations in a bundle of rays that interfere at the receiver are correlated. The diffraction parameter is defined and the other two quantities approximated by ray integrals, which involve the variance $\langle \mu^2 \rangle$ of the refractivity fluctuations and their correlation length L_p along the ray (which also depends on the ray slope), the phase correlation f between two points displaced vertically about the ray, and the Green's function g of the ray spread equation (6):

$$\Phi^2 \equiv k^2 \int_0^z dz' \langle \mu^2 \rangle L_p, \tag{36a}$$

$$D_\phi(1,2) \equiv \left\langle \left| k_1 \int_1 \mu \, ds - k_2 \int_2 \mu \, ds \right|^2 \right\rangle$$

$$\sim 2k^2 \int_0^z dz' \langle \mu^2 \rangle L_p f, \tag{36b}$$

$$\Lambda \equiv \frac{k^2}{\Phi^2} \int_0^z \frac{dz' \langle \mu^2 \rangle L_p |g|}{k l_V^2}, \tag{36c}$$

$$\Phi_s = \max[1, \Lambda^{-1}], \tag{36d}$$

$$\Phi_p = \Lambda^{-\max[1, p/4]}, \qquad \Lambda < 1. \tag{36e}$$

The parameters Φ^2 and Λ provide a basis—the transition lines Φ_s and Φ_p of Eqs. (36d) and (36e)—for classifying propagation into the unsaturated, partly saturated, and saturated regimes. In the unsaturated regime ($\Phi < \Phi_p$ and $\Phi < \Phi_s$), where the fluctuations are primarily in the phase and are either weak or well correlated, the statistics can be worked out using perturbation solutions of the wave equation (the Rytov or the supereikonal approximations; see Refs. 2, 4, and 34). In the saturated regime $\Phi_s < \Phi$, the cumulative decorrelating effect of scattering and diffraction is manifest in signal amplitude fluctuations, and the perturbation treatment is invalid. The partial-saturation regime occurs in "scaleless" media with a power law spectrum $\psi_\mu \sim K^{-p}$, where the uncorrelated effects of the small-scale μ-fluctuations can coexist with the still-correlated effects of the large-scale μ-fluctuations. The asymptotic evaluations of the path integrals are generally feasible in the latter two regimes: They produce the asymptotic statistics and corrections thereto.

However, for the *mutual coherence function* Γ_{11}, path integral evaluations are available[35] for points that are far from caustics. In the *monochromatic* case, $\Gamma_{11} = 0.5D_\phi$. The phase structure function D_ϕ can be approximated [see Eqs. (34)–(39) in Ref. 35] by simple powers of the separations in time or in transverse coordinates, provided that these separations are sufficiently small relative to certain characteristic values whose evaluation is discussed below. In the *bichromatic* case and the related *pulse propagation* problem [see Eq. (13)] we have [see Ref. 35 and Eqs. (66)–(68)]

$$\Gamma_{11} = \exp\left[i\omega_d \tau_1 - 0.5\left(\frac{\omega_d}{\omega_c}\right)^2 \right],$$

$$\omega_c = \left(\frac{\Phi}{\omega} + \tau_0\right)^{-1}. \tag{37}$$

Using Eq. (13), this reveals that scattering produces a delay τ_1 and a broadening $1/\omega_c$ of intensity pulses, where ω_c is the coherence bandwidth.

References 36 and 37 present computations of the parameters listed in Eq. (36) for the internal-wave spectrum and provide simplified approximations for them. These approximations are also used in the computation of the characteristic separations mentioned above and of τ_1 and τ_0. Unfortunately, even these approximations, and the caveats attendant on their use, are difficult to encapsulate here. The interested reader is urged to consult the above references directly for the proper use of the results.

3.2 Moment Equations

The moment equations are derived directly from the parabolic wave equation (3b). The zeroth-moment equation (the average field) is obtained by ensemble averaging that equation directly. To illustrate the derivation of the higher order moment equations, consider the equation for the monochromatic second moment, which can be derived in the same manner as Eq. (26). The major conceptual problem in the derivation is the formulation of the ensemble average of the scattering term, which is a sum of products of the random μ and several random pressures. There are two formal procedures for accomplishing this. One approach relies on the Markov approximation (19) of the refractivity correlation function and on the use of functional derivatives.[1,2] The other approach uses a perturbation solution of the wave equation in a narrow range slice to express the moment at $z + \Delta z$ in terms of its value at z and converts this expression to a differential equation.[38,39] The latter process requires that Δz be sufficiently small to allow treating scattering and diffraction as independent perturbations yet remain large compared to the wavelength and the correlation scales, so as to warrant treating the pressure at z and the (forward-scattering) fluctuations μ within Δz as statistically independent; see Eq. (27). In either case, the end result is that the scattering term is proportional to the moment in question, with a coefficient that reflects the statistics of the medium.

For the arbitrary correlation moment Γ_{nm} of Eq. (10), the equation is (after Ref. 40)

$$\frac{\partial \Gamma_{nm}}{\partial z} = (D_{nm} - R_{nm} - S_{nm})\Gamma_{nm}, \tag{38}$$

where the diffraction, refraction, and scattering terms are, respectively,

$$D_{nm} = \frac{i}{2}\left[\sum_{j=1}^{j=n} \frac{1}{k_j} \frac{\partial^2}{\partial \mathbf{u}_j^2} - \sum_{j=1}^{j=m} \frac{1}{k_j'} \frac{\partial^2}{\partial \mathbf{u}_j'^2} \right], \tag{38a}$$

$$R_{nm} = i \left[\sum_{j=1}^{j=n} k_j \overline{\mu}(\mathbf{u}_j) - \sum_{j=1}^{j=m} k_j' \overline{\mu}(\mathbf{u}_j') \right], \qquad (38b)$$

$$2S_{nm} = \sum_{j=1}^{n} \sum_{l=1}^{n} k_j k_l A(\mathbf{u}_j - \mathbf{u}_l) - 2 \sum_{j=1}^{n} \sum_{l=1}^{m} k_j k_l' A$$

$$\cdot (\mathbf{u}_j - \mathbf{u}_l') + \sum_{j=1}^{m} \sum_{l=1}^{m} k_j' k_l' A(\mathbf{u}_j' - \mathbf{u}_l'). \qquad (38c)$$

Excepting the most elementary cases (first and second moments, with no refraction), the practical solutions of these moment equations are all approximate. Various techniques[41,42] have been used to obtain limiting solutions for the fourth moment; the method of successive scatters[43] even produces an exact formal solution in the form of a sum of multiple convolutions. A systematic approach, pursued in Refs. 44–54, combines Fourier transforms, multiscale embedding, and asymptotic techniques to transform the moment equations (38) into first-order partial differential equations. These lead to approximate solutions, in terms of approximate Green's functions or propagators, that produce a given coherence moment in terms of its prescribed distribution in a source plane or aperture. Rays and spreading ray tubes emerge as characteristics of the first-order partial differential equation, and scattering enters as an integral of the appropriate S_{nm} coefficient of Eq. (38c) over a tube surrounding the main ray. The Green's functions appear to span all the propagation regimes; they converge to known results at both the perturbation limit and the saturation limit mentioned above. The multiscale embedding technique also permits the incorporation of caustic corrections.

The moments $\Gamma_0 (= \langle p \rangle,$ the average field), Γ_{11} (field correlation and average intensity), and Γ_{22} (intensity correlation and scintillation index) have been treated for both monochromatic and bichromatic excitations. The following sequence of example reviews the known solutions and illustrates the mathematical techniques involved.

The equation for the average field and its point source solutions are

$$\left[\frac{\partial}{\partial z} - \frac{i}{2k} \frac{\partial^2}{\partial \mathbf{u}^2} + ik\overline{\mu}(\mathbf{u}) + \frac{k^2}{2} (A(0, \mathbf{u})) \right]$$

$$\cdot \langle p(z, \mathbf{u}) \rangle = 0, \qquad (39)$$

$$\langle p(z, \mathbf{u}) \rangle = p^0(z, \mathbf{u}) \exp\left(-\frac{k^2}{2} zA(0) \right) \quad \text{with}$$

$$p^0 = \frac{1}{2\pi z} \exp\left[ik \frac{u^2}{2z} - i \frac{\pi}{2} \right], \qquad (39a)$$

$$\langle p(z, \mathbf{u}) \rangle = p^0(z, \mathbf{u}) \exp(-\tfrac{1}{2} \Phi^2) \quad \text{where}$$

$$\Phi^2 = \int_0^z dz' A[0, \mathbf{u}(z')], \qquad (39b)$$

where p^0 is the field that prevails in the absence of sound speed fluctuations: The exact solution (39a) obtains in a nonrefractive ($\overline{\mu} = 0$) uniform ($\partial_{\mathbf{u}} A = 0$) scattering medium[2]; the approximate solution (39b), with p^0 given by Eq. (7), obtains in a general medium, with Φ^2 expressed as an integral of A along the ray (5).

The equation for the coherence of a monochromatic field is best treated in terms of the sum and difference coordinates (11b), which recasts it in the form

$$\left\{ \frac{\partial}{\partial z} - \frac{i}{k} \frac{\partial}{\partial \mathbf{u}} \cdot \frac{\partial}{\partial \mathbf{q}} + ik \left[\overline{\mu}\left(\mathbf{u} + \frac{\mathbf{q}}{2} \right) - \overline{\mu}\left(\mathbf{u} - \frac{\mathbf{q}}{2} \right) \right] \right.$$

$$\left. + \frac{k^2}{2} D_A(\mathbf{q}, \mathbf{u}) \right\} \Gamma_{11}(z, \mathbf{u}, \mathbf{q}) = 0. \qquad (40)$$

In nonrefracting, statistically homogeneous media, the exact solution is[2]

$$\Gamma_{11}(z, \mathbf{u}, \mathbf{q}) = \iint d\mathbf{u}_s \, d\mathbf{q}_s \, \Gamma_s(\mathbf{u}_s, \mathbf{q}_s) G_{TS}(\mathbf{u}, \mathbf{q}, \mathbf{u}_s, \mathbf{q}_s) \quad \text{with}$$

$$\Gamma_s(\mathbf{u}, \mathbf{q}) = \Gamma_{11}(0, \mathbf{u}, \mathbf{q}), \qquad (41a)$$

$$G_{TS} = \frac{k^2}{4\pi^2 z^2} \exp\left[i \frac{k}{z} (\mathbf{u} - \mathbf{u}_s) \cdot (\mathbf{q} - \mathbf{q}_s) \right]$$

$$\cdot \exp\left(-\frac{k^2}{2} \int_0^z dz' D_A\left[\mathbf{q} - (\mathbf{q}_s) \frac{z'}{z} \right] \right). \qquad (41b)$$

The first two factors of the Green's function G_{TS} account for diffraction. The third factor accounts for the scattering in the presence of diffraction.

Away from caustics, it suffices to replace the refractive term in Eq. (40) by the first-order term $k\overline{\mu}'(\mathbf{u}) \cdot \mathbf{q}$ of its series expansion in $\mathbf{q}$ [this "quadratic" approximation is of course exact when $\overline{\mu} \propto (y - y_{\min})^2$]. To clarify the role of the rays, we first examine the limit where fluctuations are absent. Fourier transforming the $\mathbf{q}$-dependence then introduces the directional spectrum of the intensity,[9,12]

$$\hat{\Gamma}(z, \mathbf{u}, \boldsymbol{\sigma}) = \int_{-\infty}^{\infty} d\mathbf{q} \, e^{-(ik\mathbf{q} \cdot \boldsymbol{\sigma})} \Gamma(z, \mathbf{u}, \mathbf{q}), \qquad (42a)$$

which is governed by the first-order partial differential equation

$$\left[\frac{\partial}{\partial z} + \boldsymbol{\sigma} \cdot \frac{\partial}{\partial \mathbf{u}} - \overline{\boldsymbol{\mu}}'(\mathbf{u}) \cdot \frac{\partial}{\partial \mathbf{q}} \right] \hat{\Gamma} = 0$$

$$\text{with} \quad \hat{\Gamma}(0, \mathbf{u}, \boldsymbol{\sigma}) = \hat{\Gamma}_s(\mathbf{u}, \boldsymbol{\sigma}) \qquad (42b)$$

and remains constant along its characteristics, namely the rays (5):

$$\hat{\Gamma}(z, \mathbf{u}, \boldsymbol{\sigma}) = \hat{\Gamma}_s(\mathbf{u}_s, \boldsymbol{\sigma}_s). \qquad (42c)$$

Here, $\mathbf{u}_s(z, \mathbf{u}, \boldsymbol{\sigma})$ and $\boldsymbol{\sigma}_s(z, \mathbf{u}, \boldsymbol{\sigma})$ are the position and slope, at the source plane $z = 0$, of the ray that crosses the range plane z at $\mathbf{u}$ with a slope $\boldsymbol{\sigma}$; they are obtained by integrating Eq. (5) "backward" from the point $(z, \mathbf{u}, \boldsymbol{\sigma})$ to the source point $(0, \mathbf{u}_s, \boldsymbol{\sigma}_s)$. Transforming Eq. (6c) to obtain the coherence and then replacing $\boldsymbol{\sigma}$ by $\mathbf{u}_s$ as integration variable, we find

$$\Gamma(z, \mathbf{u}, \mathbf{q}) = \frac{k^2}{4\pi^2} \int d\boldsymbol{\sigma} \, e^{-ik\mathbf{q} \cdot \boldsymbol{\sigma}}$$

$$\cdot \int d\mathbf{q}_s \, e^{-ik\mathbf{q}_s \cdot \boldsymbol{\sigma}_s} \Gamma_s(\mathbf{u}_s, \mathbf{q}_s), \qquad (42d)$$

$$\Gamma(z, \mathbf{u}, \mathbf{q}) = \iint d\mathbf{u}_s \, d\mathbf{q}_s \Gamma_s(\mathbf{u}_s, \mathbf{q}_s) G_{TR} \quad \text{where}$$

$$G_{TR} = \frac{k^2}{4\pi^2} \frac{1}{|\partial_{\boldsymbol{\sigma}} \mathbf{u}_s|} \exp[-ik(\mathbf{q} \cdot \boldsymbol{\sigma} - \mathbf{q}_s \cdot \boldsymbol{\sigma}_s)]. \qquad (42e)$$

This expresses the coherence in terms of its prescribed distribution Γ_s at the source plane and a (refractive) Green's function that depends entirely on the geometry of the rays. In the expression for G_{TR}, $\boldsymbol{\sigma}$ and $\boldsymbol{\sigma}_s$ are, respectively, the initial and final slopes of the ray that connects the points $(0, \mathbf{u}_s)$ and $(z, \mathbf{u})$, expressed as functions of $z, \mathbf{u}$, and $\mathbf{u}_s$; the determinant $|\partial_{\boldsymbol{\sigma}} \mathbf{u}_s| = |\partial \mathbf{u}_s / \partial \boldsymbol{\sigma}|$

of Eq. (6) (reciprocity guarantees $|\partial_{\boldsymbol{\sigma}} \mathbf{u}_s| = |\partial_{\boldsymbol{\sigma}_s} \mathbf{u}|$) measures the size of a spreading flux tube about the ray; in nonrefractive media it reduces to the z^{-2} factor in Eq. (41b).

The results of Eqs. (41) and (42) do not combine directly to provide the Green's function when both scattering and refraction are present. Dealing with the complete problem requires the use of a two-scale embedding procedure[44,45] that relies on the fact that the dependence of the coherence on $\mathbf{q}$ involves two widely disparate scales whose ratio $\epsilon = (kl_{\mathbf{q}})^{-1}$ is small. To take advantage of this disparity: (a) apply a Fourier transform $\mathbf{u} \rightarrow \boldsymbol{\eta}$ to (4); (b) embed the result in a higher dimensional space that also includes $\overline{\boldsymbol{\eta}} = \epsilon\boldsymbol{\eta}$ and $\overline{\mathbf{q}} = \epsilon\mathbf{q}$; and (c) apply Fourier transforms $\boldsymbol{\eta} \rightarrow -\mathbf{v}$ and $\mathbf{q} \rightarrow \boldsymbol{\sigma}$ and retain the leading-order term in the result, which is a first-order partial differential equation for the multiply transformed coherence $\overline{\overline{\Gamma}}$:

$$\left[\frac{\partial}{\partial z} + \boldsymbol{\sigma} \cdot \frac{\partial}{\partial \mathbf{v}} + \overline{\boldsymbol{\eta}} \cdot \frac{\partial}{\partial \overline{\mathbf{q}}} - \overline{\boldsymbol{\mu}}'(\mathbf{v}) \cdot \frac{\partial}{\partial \boldsymbol{\sigma}} - \overline{\overline{\boldsymbol{\mu}}}''(\mathbf{v})\overline{\mathbf{q}} \right.$$

$$\left. \cdot \frac{\partial}{\partial \overline{\boldsymbol{\eta}}} + \frac{k^2}{2} D_A(\overline{\mathbf{q}}, \mathbf{v}) \right] \overline{\overline{\Gamma}} = 0. \qquad (43)$$

Besides the $(\mathbf{v}, \boldsymbol{\sigma})$ rays (5), this also has the $(\overline{\mathbf{q}}, \overline{\boldsymbol{\eta}})$ spread characteristics (6) that affect the first argument of the structure function D_A: $\overline{\overline{\Gamma}}$ decays exponentially with the integral Q of D_A along these characteristics. Retracing the transforms and embedding process, we obtain a solution such as (42e) with the Green's function (after Ref. 45)

$$G_{TRS} = \frac{k^4}{(2\pi)^4} \int d\boldsymbol{\eta} \, d\mathbf{v} \, \frac{1}{|\partial_{\boldsymbol{\sigma}} \mathbf{u}_s|} \exp\{ik[\boldsymbol{\eta} \cdot (\mathbf{u} - \mathbf{v})$$

$$+ \boldsymbol{\sigma} \cdot \mathbf{q} - \boldsymbol{\sigma}_s \cdot \mathbf{q}_s]\} \exp -Q, \qquad (43a)$$

$$Q = \frac{k^2}{2} \int_0^z dz' \, D_A[\overline{\overline{\xi}}_1(z', \mathbf{v}, \mathbf{u}_s)\mathbf{q}$$

$$+ \overline{\overline{\xi}}_2(z', \mathbf{v}, \mathbf{u}_s)\boldsymbol{\eta}, \mathbf{v}'(z', \mathbf{v}, \mathbf{u}_s)], \qquad (43b)$$

where $\boldsymbol{\sigma}_s$ and $\boldsymbol{\sigma}$ have the interpretation following Eq. (42d), $\mathbf{v}'$ is the transverse position along the ray, $\overline{\overline{\xi}}_1$ is a solution of the ray spread equations (6) with the initial conditions $\overline{\overline{\xi}} = I_2$ (unit matrix) and $\overline{\overline{\eta}} = 0$, while $\overline{\overline{\xi}}_2$ satisfies the same system with the initial conditions interchanged. Approximate though it is, this function reduces to the Green's functions in Eq. (42d) or (41b) when $D_A = 0$ (no scattering) or $\overline{\mu} = 0$ (no refraction), respectively. Also, at sufficiently high wavenumbers k, the integral in

Eq. (43a) may be evaluated asymptotically (the dominant contributions arise from the regions where $\boldsymbol{\eta} \sim 0$ and $\mathbf{v} \sim \mathbf{u}$); it reduces to

$$G_{\Gamma RS\infty} = G_{\Gamma R} \, \exp[-Q_\infty],$$

$$Q_\infty = \frac{k^2}{2} \int_0^z dz' \, D_A[\bar{\bar{\xi}}_1(z', \mathbf{u}, \mathbf{u}_s)\mathbf{q}], \quad (43c)$$

where $G_{\Gamma R}$ is the diffractive–refractive Green's function (42d). This coincides with the corresponding expression obtained using path integrals.[4,35] Since the relation between the respective asymptotic arguments implied by Eq. (43) and by the path integrals has not yet been clarified, it is uncertain whether Eq. (43) is indeed "more general" than its asymptotic limit (43c).

To get the behavior near caustics, it is necessary to add the third-derivative term in our approximation for the refractive term in Eq. (40). In the absence of scattering, a two-scale embedding suffices to deal with this and produces the familiar Airy function corrections.[44] A three-scale embedding is required to study scattering near the caustic transition.[46]

The treatment of bichromatic correlation also requires the use of the two-scale embedding.[47] The Green's function again resembles its path integral counterpart[4]

$$G_{\Gamma RSB} = \exp[-\tfrac{1}{2}(k_1 - k_2)^2 \Phi^2 G_{\Gamma SB} \quad (44)$$

except that, like its path integral counterpart, $G_{\Gamma SB}$ is a relatively complicated function.

The embedding technique also resolves the equation for the monochromatic and bichromatic fourth moment in a homogeneous medium[48–51] and in a refractive scattering medium.[52] In broad outline, the procedure has also been extended to situations where the main direction of propagation follows a curved ray as well as to the treatment of all moments.[53,54]

4 NUMERICAL TECHNIQUES AND EXPERIMENTAL OBSERVATIONS

4.1 Numerical Techniques

Direct numerical procedures for solving the moment equations rely on the parabolic character of the moment equations (38): Given the source distribution of any moment in the plane $z = 0$, its transverse derivative can be computed and used to evaluate the range derivative of that moment, so that the entire distribution can be incrementally "marched" to the next range plane. Such techniques have evolved from the original *split-step* algorithm[55] to rather sophisticated *moment equation solvers*.[56] Conceptually, the process is equivalent to the assignment of the ensemble-averaged scattering effects in a given range increment to the single-range plane. This plane is treated as a "phase screen" with prescribed ensemble-averaged stochastic properties, which impulsively changes the phase distribution of the incident signal.

With increasing computational capability, it has become possible to simulate the effect of a random environment. This is accomplished by solving Eq. (3) numerically for the propagation of the basic wavefunction through each of many individual realizations of the random medium, selected at random to fit a prescribed statistical distribution, and performing the statistical processing on the outcome of the solutions and their moments. Early efforts, which used phase screens to simulate propagation in two-dimensional medium models,[57,58] were later extended to three-dimensional media.[59] This approach was recently implemented using direct solutions of the parabolic wave equation for individual medium realizations[60] and used to check the analytical predictions of the path integral expressions for pulse broadening; see Eq. (37). Even though this approach is computer intensive, it is capable of handling realistic propagation scenarios outside the asymptotic regimes that can be treated analytically. It also produces dramatic and useful visualizations of features averaged over by the statistical treatment.

4.2 Comparisons with Experimental Observations

The path integral formulation and computations relating thereto have been employed almost exclusively in the analysis of the early AFAR experiment[4] and the December 1978 propagation experiment.[61] The theoretical predictions for the second-order coherence statistics are consistent with all the AFAR observations and also with the pulse propagation observations in the December 1978 experiment. However, they failed to agree with the CW observations of the latter experiment. The path integral results were also employed in the early analysis of the Cobb Seamount experiment in 1971.[62] However, later phases of that analysis, as well as the analysis of the ensuing MATE experiment,[16,63,64] relied increasingly on the solution of the moment equations for the prediction of the intensity statistics (or the fourth moment). These solutions employed a combination of successive scatter, multiscale embedding, and path integral solutions of the moment equations and neglected the effect of the acoustic channel because the MATE experiment was carried out in a region where the channel sound speed profiles were linear. The theoretical predictions were fairly consistent with measurements of the intensity autocorrelation and scintillation index. However, they still fail to predict correctly the observed spectra of the two-frequency correlation of the intensity.

The more recent Slice89 experiment,[65] which measured the effects of scattering on pulse propagation at a 1000-km range, has been compared extensively with computed predictions based on numerical simulation[60] that employed a model Garrett–Munk spectrum. Most of the observed effects are also present in the simulation results. However, there are still quantitative discrepancies, which are presently attributed to the incompleteness and/or inapplicability of the spectral model to the experimental propagation scenario.

4.3 Concluding Comments

The above summarizes the main concepts needed to understand the issues addressed in the study of volume scattering in the ocean acoustic environment, and the major theoretical avenues for dealing with these issues. As indicated in the Introduction, work in this area has not yet produced simplified recipes that can be encapsulated in standard formulas, graphs, or tables. At best, it has produced computational algorithms that must be implemented with great care.

The two ray-oriented formulations are strongly interrelated; they are indeed identical in content, though markedly different in implementation. These approaches have considerable intuitive appeal, which is nonetheless countermanded by complications that arise from the strong anisotropy of the scattering medium. These complications are largely circumvented by the modal formulation. The latter was also deemed computationally intensive, especially at higher depths and frequencies where many modes are present. However, with improvement in numerical capabilities, the severity of this handicap may well be diminishing.

It may appear that, with the advent of improved numerical techniques and the ever-increasing capacity for performing numerical computations, the significance of analytical treatments may be waning. Nonetheless, theoretical treatments are still needed, at very least, to establish benchmarks for checking computational algorithms. Moreover, even the most sophisticated simulation algorithms still rely basically on techniques for solving the wave equation, and these are still undergoing development. Finally, there are still issues—such as wide-angle scattering and backscattering—that await formulations that are conductive to efficient computation. Hopefully, future editions of this handbook will reflect progress along these lines.

Acknowledgments

The author expresses his gratitude for the hospitality of ComSERC at Howard University, Washington, DC, where he was visiting when parts of this chapter were written, and for the support extended him by AFOSR under contract No. F49620-94-1-03039 during that period.

REFERENCES

1. V. I. Tatarskii, *The Effect of the Turbulent Atmosphere on Wave Propagation*, Israel Program for Scientific Translation, Jerusalem, U.S. Department of Commerce, NTIS, Springfield, VA, TT68-50464, 1971.

2. A. Ishimaru, *Wave Propagation and Scattering in Random Media*, Academic Press, New York, 1978.

3. A. M. Prokhorov, F. V. Bunkin, K. S. Gochelashvili, and V. I. Shishov, "Laser Irradiance Propagation through Turbulent Media," *Proc. IEEE*, Vol. 63, 1975, pp. 790–811.

4. S. M. Flatte, R. Dashen, W. H. Munk, K. M. Watson, and F. Zachariasen, *Sound Transmission through a Fluctuating Ocean*, Cambridge University Press, New York, 1979.

5. C. Garrett and W. H. Munk, "Spacetime Scales of Internal Waves: A Progress Report," *J. Geophys. Res.*, Vol. 80, 1975, pp. 291–297.

6. C. B. Officer, *Introduction to the Theory of Sound Transmission*, McGraw-Hill, New York, 1958.

7. M. Leontovich and V. Fok, "Solution of the Problem of Propagation of Electromagnetic Waves along the Earth's Surface by the Parabolic Equation Method," *Zh. Exsp. Teor. Fiz.*, Vol. 16, 1946, pp. 557–573.

8. R. Dashen, "Path Integrals for Waves in Random Media," *J. Math. Phys.*, Vol. 20, 1979, pp. 894–920.

9. J. J. McCoy and M. J. Beran, "Propagation of Beamed Signals through Inhomogeneous Media," *J. Acoust. Soc. Am.*, Vol. 59, 1976, pp. 1142–1149.

10. A. M. Whitman and M. J. Beran, "Numerical Calculation of the Intensity Distribution in Sound Channels Using Coherence Theory," *J. Acoust. Soc. Am.*, Vol. 63, 1978, pp. 1727–1732.

11. S. Frankenthal, "Beam Intensity Calculations with Uncertain Sound-Speed Profiles, *J. Acoust. Soc. Am.*, Vol. 76, 1984, pp. 198–204.

12. S. Frankenthal, "Scintillations of Partially-Coherent Signals in Non-scattering Channels," *J. Acoust. Soc. Am.*, Vol. 81, 1987, pp. 1399–1405.

13. S. Frankenthal, "Close-Range Scintillations in Anisotropic Scattering Media," *J. Acoust. Soc. Am.*, Vol. 77, 1985, pp. 1395–1402.

14. Y. J. F. DeSaubies, "Analytical Representations of Internal Wave Spectra," *J. Phys. Oceanogr.*, 1976, Vol. 4, pp. 976–981.

15. M. D. Levine and J. D. Irish, "A Statistical Description of Fine Structure in the Presence of Internal Waves," *J. Phys. Oceanogr.*, Vol. 11, 1981, pp. 676–691.

16. T. E. Ewart and S. A. Reynolds, "The Mid-Ocean Acoustic Transmission Experiment, MATE," *J. Acoust. Soc. Am.*, Vol. 75, 1983, pp. 785–802.

17. J. J. McCoy and M. J. Beran, "Directional Spectral Spreading in Randomly Inhomogeneous Media," *J. Acoust. Soc. Am.*, Vol. 66, 1979, pp. 1468–1481.

18. M. J. Beran and J. J. McCoy, "Propagation through an Anisotropic Random Medium," *J. Math. Phys.*, Vol. 15, 1974, pp. 1901–1912.

19. M. J. Beran and J. J. McCoy, "Propagation through an Anisotropic Random Medium. An Integrodifferential Formulation," *J. Math. Phys.*, Vol. 17, No. 7, 1976, pp. 1186–1189.

20. L. M. Brekhovskikh, *Waves in Layered Media*, Academic Press, New York, 1980.

21. W. Kohler and G. C. Papanicolau, "Wave Propagation in a Randomly Inhomogeneous Ocean," in J. B. Keller and J. S. Papadakis (Eds.), *Wave Propagation and Underwater Acoustics*, Springer Verlag, New York, 1977, pp. 153–223.

22. D. Marcuse, *Theory of Dielectric Optical Waveguides*, Academic Press, New York, 1974.

23. L. B. Dozier and F. B. Tappert, "Statistics of Normal-Mode Amplitudes in a Random Ocean. I. Theory and II. Computations," *J. Acoust. Soc. Am.*, Vol. 63, 1978, pp. 353–365 and Vol. 64, 1978, pp. 533–547.

24. S. T. McDaniel, "Mode Coupling and the Environmental Sensitivity of Shallow-Water Propagation Loss Predictions," *J. Acoust. Soc. Am.*, Vol. 82, 1987, pp. 217–223.

25. G. R. Sutton and J. J. McCoy, "Spatial Coherence of Acoustic Signals in Randomly Inhomogeneous Waveguides," *J. Math. Phys.*, Vol. 18, 1977, pp. 1052–1057.

26. C. Penland, "Acoustic Normal Mode Propagation through a Three-Dimensional Internal Wave Field," *J. Acoust. Soc. Am.*, Vol. 78, No. 4, 1985, pp. 1356–1365.

27. A. D. Pierce, "Extension of the Method of Normal Modes to Sound-Propagation in an Almost-Stratified Medium," *J. Acoust. Soc. Am.*, Vol. 37, 1965, pp. 19–27.

28. D. M. Milder, "Ray and Wave Invariants for SOFAR Channel Propagation," *J. Acoust. Soc. Am.*, Vol. 46, 1969, pp. 1259–1263.

29. C. Allan Boyles, *Acoustic Waveguides*, Wiley, New York, 1984.

30. S. T. McDaniel, "Coupled Power Equations for Cylindrically Spreading Waves," *J. Acoust. Soc. Am.*, Vol. 60, 1976, pp. 1285–1289.

31. B. J. Uscinski, C. Macaskill, and M. Spivak, "Path Integrals for Wave Intensity Fluctuations in Random Media," *J. Sound Vibration*, Vol. 106, No. 3, 1986, pp. 509–527.

32. J. L. Codona, D. B. Creamer, S. M. Flatte, R. G. Frehlich, and F. S. Heyney, "Moment-Equation and Path-Integral Techniques for Wave Propagation in Random Media," *J. Math. Phys.*, Vol. 27, No. 1, 1986, pp. 171–177.

33. S. M. Flatte, "Wave Propagation in Random Media: Contributions from Ocean Acoustics," *Proc. IEEE*, Vol. 71, 1983, pp. 1267–1294.

34. W. H. Munk and F. Zachariasen, "Sound Propagation through a stratified Fluctuating Ocean—Theory and Observations," *J. Acoust. Soc. Am.*, Vol. 59, 1976, pp. 818–838.

35. R. Dashen, S. M. Flatte, and S. A. Reynolds, "Path Integral Treatment of Acoustic Mutual Coherence Functions for Rays in a Soundchannel," *J. Acoust. Soc. Am.*, Vol. 77, 1985, pp. 1716–1722.

36. R. Esswein and S. M. Flatte, "Calculation of the Phase Structure Function Density from Oceanic Internal Waves," *J. Acoust. Soc. Am.*, Vol. 70, 1981, pp. 1387–1396, 1981.

37. R. Esswein and S. M. Flatte, "Calculation of the Strength and Diffraction Parameters in for Oceanic Sound Transmission," *J. Acoust. Soc. Am.*, Vol. 67, 1980, pp. 1523–1531.

38. M. Beran and T. L. Ho, "Propagation of the Fourth-Order Coherence Function in a Random Medium (a Non-perturbative Formulation," *J. Opt. Soc. Am.*, Vol. 59, 1969, pp. 1134–1138.

39. B. J. Uscinski, *The Elements of Wave Propagation in a Random Medium*, McGraw-Hill, New York, 1977.

40. L. C. Lee, "Wave Propagation in a Random Medium: A Complete Set of the Moment Equations with Different Wavenumbers," *J. Math. Phys.*, Vol. 15, 1974, pp. 1431–1435.

41. V. I. Shishov, "Strong Fluctuations of the Intensity of a Plane Wave Propagating in a Random Medium," *Sov. Phys. JETP*, Vol. 34, 1972, pp. 744–748.

42. R. L. Fante, "Inner-Scale Size Effect on the Scintillation of Light in a Turbulent Atmosphere," *J. Opt. Soc. Am.*, Vol. 73, 1983, pp. 277–281.

43. B. J. Uscinski, C. Macaskill, and T. Ewart, "Intensity Fluctuations I—theory," *J. Acoust. Soc. Am.*, Vol. 74, 1983, pp. 1474–1483.

44. S. Frankenthal, M. J. Beran, and A. M. Whitman, "Caustic Corrections Using Coherence Theory," *J. Acoust. Soc. Am.*, Vol. 71, 1982, pp. 348–358.

45. M. J. Beran, A. M. Whitman, and S. Frankenthal, "Scattering Calculations Using the Characteristic Rays of the Coherence Function," *J. Acoust. Soc. Am.*, Vol. 71, 1982, pp. 1124–1130.

46. R. Mazar and M. J. Beran, "Intensity Corrections in a Random Medium in the Neighborhood of a Caustic," *J. Acoust. Soc. Am.*, Vol. 72, 1982, pp. 1269–1275.

47. S. Frankenthal, "The mutual Coherence Function in a Scattering Channel—A Two-Scale Solution," *J. Acoust. Soc. Am.*, Vol. 85, 1989, pp. 104–113.

48. C. Macaskill, "An Improved Solution of the Fourth-

Moment Equation for Intensity Fluctuations," *Proc. R. Soc. London, Ser. A*, Vol. 386, 1983, pp. 461–474.

49. S. Frankenthal, M. J. Beran, and A. M. Whitman, "Two-Scale Solutions for Intensity Fluctuations in Strong Scattering," *J. Opt. Soc. Am. A*, Vol. 1, 1984, pp. 585–597.

50. A. M. Whitman and M. J. Beran, "Two-Scale Solution for Atmospheric Scintillations," *J. Opt. Soc. Am. A*, Vol. 2, 1985, pp. 2133–2143.

51. R. Mazar, J. Gozani, and M. Tur, "Two-Scale Solution for the Intensity Fluctuations in Two-Frequency Wave Propagation in a Random Medium," *J. Opt. Soc. Am. A*, Vol. 2, 1985, pp. 2152–2160.

52. M. J. Beran and R. Mazar, "Intensity Fluctuations in a Quadratic Channel," *J. Acoust. Soc. Am.*, Vol. 82, 1987, pp. 588–592.

53. R. Mazar and L. B. Felsen, "High Frequency Coherence Functions Propagated along Ray Paths in the Inhomogeneous Background of a Weakly Random Medium. I. Formulation and Evaluation of the Second Moment," *J. Acoust. Soc. Am.*, Vol. 81, 1987, pp. 925–937; "II. Higher moments," *J. Acoust. Soc. Am.*, Vol. 82, 1987, pp. 593–600.

54. R. Mazar and L. B. Felsen, "Stochastic Geometrical Theory of Diffraction," *J. Acoust. Soc. Am.*, Vol. 86, 1989, pp. 2292–2308.

55. F. D. Tappert, "Parabolic Equation Method in Underwater Acoustics," *J. Acoust. Soc. Am.*, Vol. 55, 1973, p. 534.

56. C. Macaskill and T. E. Ewart, "Numerical Solution of the Fourth-Moment Equation for Acoustic Intensity Correlation and Comparison with MATE," *J. Acoust. Soc. Am.*, Vol. 99, 1996, pp. 1419–1429.

57. C. L. Rino and J. Owen, "Numerical Simulations of Intensity Scintillations Using the Power-Law Phase-Screen Model," *Radio Sci.*, Vol. 15, p. 41, 1980.

58. C. Macaskill and T. E. Ewart, "Computer Simulation of Two-Dimensional Random Wave Propagation," *IMA J. Appl. Math.*, Vol. 33, 1984, p. 1.

59. J. M. Martin and S. M. Flatte, "Intensity Images and Statistics from Numerical Simulation in 3-D Random Media," *Appl. Optics*, Vol. 27, No. 11, 1988, pp. 2111–2126.

60. J. A. Colosi, S. M. Flatte, and C. Bracher, "Internal-Wave Effects on 1000-km Oceanic Pulse Propagation: Simulation and Comparison with Experiment," *J. Acoust. Soc. Am.*, Vol. 96, No. 1, 1984, pp. 452–468.

61. P. F. Worceser, G. O. Williams, and S. M. Flatte, "Fluctuations of Resolved Acoustic Multipaths at Short Ranges in the Ocean," *J. Acoust. Soc. Am.*, Vol. 70, No. 3, 1981, pp. 825–840.

62. T. E. Ewart, "Acoustic Fluctuations in the Open Ocean—A Measurement Using a Fixed Refracted Path," *J. Acoust. Soc. Am.*, Vol. 60, 1976, pp. 46–59.

63. T. E. Ewart, C. Macaskill, and B. J. Uscinski, "Intensity Fluctuations. Part II: Comparison with the Cobb Seamount Experiment," *J. Acoust. Soc. Am.*, Vol. 74, 1983, pp. 1484–1499.

64. T. E. Ewart, C. Macaskill, and B. J. Uscinski, "The MATE Acoustic Frequency Cross-Correlation of Intensity," *J. Acoust. Soc. Am.*, Vol. 77, No. 5, 1985, pp. 1732–1741.

65. T. F. Duda, S. M. Flatte, J. A. Colosi, B. D. Cornuelle, J. A. Hildebrand, W. S. Hodgkiss Jr., A. P. Worcester, B. M. Howe, J. A. Mercer, and R. C. Spindel, "Measured Wave-Front Fluctuations in 1000-km Pulse Propagation in the Pacific Ocean," *J. Acoust. Soc. Am.*, Vol. 92, No. 2, 1992, pp. 939–955.

40

BACKSCATTERING FROM ROUGH SURFACES AND INHOMOGENEOUS VOLUMES

R. P. CHAPMAN, O. Z. BLUY, AND P. C. HINES

1 INTRODUCTION

When a sound pulse is transmitted through the ocean, energy is scattered back toward the source by inhomogeneities in the volume of the ocean and its boundaries. The intensity of this backscattered sound picked up by a receiver is referred to as *reverberation*, a term adopted from room acoustics that describes the same phenomenon. It is a generally decaying signal that, depending on factors such as the frequency and strength of the transmitted pulse and the background noise level, persists for times ranging from fractions of seconds to minutes. Reverberation has a significant impact in a number of areas: In fisheries research it is used to determine the depth, size, and number of marine organisms at a given location; in physical oceanography it is used to measure current and to investigate the characteristics of bubble clouds; in geophysics it is effective in producing detailed maps of the bottom and in delineating mineral deposits; in active sonar it is the background against which submarines, torpedoes, and mines are to be detected.

2 SCATTERING THEORY

Scattering may occur from the water–air interface, the water–bottom interface, or inhomogeneities within the water column or the ocean bottom. In this section the acoustic scattering models most frequently employed in underwater acoustics research are outlined.

Encyclopedia of Acoustics, edited by Malcolm J. Crocker
ISBN 0-471-80465-7 © 1997 John Wiley & Sons, Inc.

2.1 Scattering from Periodic Surfaces

One of the earliest attempts to model acoustic scattering from a rough interface mathematically was by J. W. Strutt[1] (Lord Rayleigh) at the end of the nineteenth century. His work concentrated on scattering from interfaces with periodic undulations. Due to this periodicity, he assumed that the scattered pressure wave propagated in discrete modes making angles θ_i with the vertical such that

$$\sin(\theta_i) - \sin(\theta) + i \; \frac{\lambda}{\Lambda}, \qquad i = 0, \pm 1, \pm 2, \dots, \quad (1)$$

where θ is the angle of the incident wave measured from the vertical, λ is the acoustic wavelength, and Λ is the wavelength of the surface. Although the validity of the assumptions employed by Rayleigh has been widely debated for many years (Refs. 2–4 and the references contained therein), it is generally accepted that his model agrees reasonably well with experiment so long as the interface relief is not too rough. More recently, the definition of what constitutes not too rough has become the more interesting question. For example, Millar[3] states that the Rayleigh hypothesis should only be valid for surface slopes less than 0.448, whereas Jackson et al.[4] have shown the hypothesis to be valid for a sinusoidal surface with a slope of 0.628.

2.2 Scattering from a Randomly Rough Interface

Rayleigh's scattering model deals with scattering from a periodic interface. However, rough interfaces encountered in underwater acoustics problems are usually random. The majority of the mathematical models devoted to scattering from a randomly rough interface fall into two categories: the small roughness perturbation approximation (commonly referred to as the Rayleigh–Rice method) and the Helmholtz integral with Kirchhoff boundary condition.

The small roughness perturbation approximation was developed by Steven Rice[5] to describe the scattering of electromagnetic waves from a perfectly conducting, randomly rough interface that is "almost but not quite flat." He based the development on the same boundary conditions used by Rayleigh for the periodic surface, with the result that it suffers the same limitations. However, as discussed above, the extent of the limitations is not entirely clear. The validity of Rice's expression for the scattered field "should be" further limited by his approximations of the constants in his expression for the total electric field. These constants, correct to order $f^2(x, y)$ [where the surface roughness is described by $z = f(x, y)$] are determined by assuming that

$$kf(x, y) \ll 1 \quad \text{and} \quad f_x, f_y \ll 1, \tag{2}$$

where k is the wavenumber of the incident radiation and f_x, f_y are the gradients of $f(x, y)$ in the x- and y-directions. These are the small wave height and small surface slope approximations, respectively. We employ the phrase "should be" because Jackson et al.[4] present evidence that the validity of the small-roughness perturbation approximation extends beyond the constraints of Eq. (2). They base their argument on the equivalence (at least to fifth order) of terms obtained employing the small-roughness perturbation approximation with those obtained from the extinction theorem, a much more rigorous perturbation method.* Marsh[7] applied the Rayleigh–Rice method to scattering of acoustic waves from a one-dimensional pressure-release interface. Kuo extended Marsh's model to include non-pressure-release interfaces[8] and the effect of variations in the bottom surface density.[9]

Following an excellent tutorial on the small-roughness perturbation approximation, Ogilvy[10] presents the following expression for the backscattered intensity from a pressure-release surface (Dirichlet boundary condition):

$$I = \frac{4k^4 \sin^4 \phi}{r^2} A_m P(kA, kB), \tag{3}$$

where ϕ is the angle the incoming wave makes with the horizontal (hereafter denoted the grazing angle), k the acoustic wavenumber, A_m the area of the mean plane of the scattering surface, and $P(kA, kB)$ the surface power spectrum, the arguments of which are the x- and y-components of the change in wave vector. If the correlation length of the surface is much shorter than the horizontal component of the wavelength, then $P(kA, kB) \simeq 1$ and the backscattered intensity is proportional to k^4 and $\sin^4 \phi$. The reader will also find derivations of the small-roughness perturbation approximation contained in Refs. 11–13.

Even if the limitations on the small-perturbation approximation are less restrictive than those contained in Eq. (2), the method still applies only to rough interfaces with small gradients and small surface heights. For interfaces that are not smooth enough, the Helmholtz integral with Kirchhoff boundary condition may provide a better solution. It is generally assumed[14–16] that this latter approach is valid if the radius of curvature of the interface is large compared to the acoustic wavelength.* The Helmholtz integral equates the scattered field to the sum of the contributions from an infinite number of point sources (or Huygen's wavelets) located on the boundary. The integral is given by

$$p_s = \frac{1}{4\pi} \iint_S \frac{\partial p_s}{\partial n} \left(\frac{e^{ikr}}{r} \right) - p_s \frac{\partial}{\partial n} \left(\frac{e^{ikr}}{r} \right) dS, \tag{4}$$

where p_s is the scattered pressure and n is the inward-pointing surface normal. As can be seen from the integral, the scattered field is given in terms of the scattered pressure on the boundary and its derivative with respect to the surface normal. The Kirchhoff approximation (also known as the tangent plane equation) assumes that the pressure at any point on the scattering surface is equivalent to the pressure that would be present on a plane tangent to the surface at that point.

Eckart[18] used the Helmholtz integral with the Kirchhoff approximation to examine the case of acoustic waves originating underwater, incident on a randomly rough pressure-release surface for which we have the

---

*Note that Thorsos and Jackson[6] have concluded that for a surface with a Gaussian roughness spectrum and Gaussian surface statistics, if the correlation length of the interface roughness becomes "too long or too short," the approximation becomes inaccurate independent of the limitation defined by Eq. (2).

*Thorsos[17] has concluded that for a surface with a Gaussian roughness spectrum the key parameter in determining the validity of the approximation is the ratio of surface correlation length to acoustic wavelength, not the ratio of the radius of curvature to acoustic wavelength.

boundary condition $p_s + p_0 = 0$, where p_0 is the incident wave and $\partial p_s/\partial_n = \partial p_0/\partial n$. In this way he equated the scattered pressure at any point (in the half-space bounded by the ocean surface) to an integral containing pressure terms that involve only the incident field. To allow solution of the resulting integral, Eckert observed that for an interface with gentle undulations one could write $\partial/\partial n \simeq \partial/\partial z$, where z is the coordinate measured in the vertical direction. This is a small-slope approximation. Beckmann and Spizzichino[13] improved on Eckert's application of the Kirchhoff approximation by equating the scattered field and the normal derivative of the scattered field to the values one would obtain on a plane tangent to the surface at any point.

Following the derivation of the Helmholtz integral with the Kirchhoff approximation, Ogilvy[10] presents expressions for the scattered field for several limiting cases of roughness as well as several forms of the surface correlation function. The expression for a very rough, pressure-release, isotropic, Gaussian surface (for which the Helmholtz integral with the Kirchhoff approximation is frequency employed) is given by

$$I = \frac{\lambda_0^2}{16\pi\sigma^2 \sin^4 \phi r^2} A_m \exp\left(-\frac{\lambda_0^2 \cot^2 \phi}{4\sigma^2}\right), \quad (5)$$

where ϕ is the grazing angle, σ^2 is the surface variance, λ_0 is the correlation length, r is the distance from the scattering patch to the source-receiver, and A_m is the area of the mean plane of the scattering surface. Thus, for a very rough Gaussian surface the backscattered intensity will be independent of frequency. The reader is also directed to Refs. 11–13 and 19 for derivation of the Helmholtz integral with the Kirchhoff approximation.

Many surfaces encountered in underwater acoustics do not exhibit the smoothly varying, random nature required by either the small-roughness perturbation approximation or the Helmholtz integral with the Kirchhoff approximation. For example, rocks sprinkled over a flat ocean floor would introduce discontinuities and/or slopes of infinite gradient. In such instances, the boss theories of Biot[20] or Twersky[21] may be applicable. As summarized by Ogilvy,[10] the Biot method "determines an effective boundary condition for the smooth plane on which the bosses are situated, and uses this to determine the scattered field. The (Twersky) approach uses a Green's function method to determine the effective surface reflection coefficient, by summing over all the bosses." The main disadvantage of the latter two theories is that solutions have only been obtained for the Dirichlet and Neumann boundary conditions.

There are few instances in which the perturbation

approximation, the boss model, or the Helmholtz integral with the Kirchhoff approximation holds completely. The composite roughness method was developed to treat surfaces with more than one roughness scale. Use of the method assumes that the surface can be decomposed into a large-scale surface over which the latter approximation is valid and a small-scale surface over which the perturbation approximation[16, 22–24] or the boss model[25] holds. Unfortunately, space limitations restrict discussion of the composite roughness method, and the interested reader is referred to the references.

The topic of underwater acoustic backscattering would not be complete without mentioning Lambert's law.[26] It was originally applied to optical scattering and is written

$$I = I_i \mu \sin \phi_i \sin \phi_f, \quad (6)$$

where I_i is the incident intensity, μ is a constant for a given surface, and ϕ_i is the incident grazing angle and ϕ_f the scattered grazing angle. For backscatter $\phi_i = \phi_f$. As stated by MacKenzie,[27] who first adopted the law (more appropriately termed Lambert's rule) for acoustic scattering, the physical basis of the law is that "an illuminated matte surface looks almost equally bright when viewed from any angle." The advantages of Lambert's law are its mathematical simplicity and its ability to provide a reasonable description of some of the (frequency-independent) acoustic backscatter data. Lambert scattering has been widely adopted by modelers because of its frequent validity and ease of use; it is particularly useful in those situations where sound arrives by one path and returns by another. Its disadvantages are that it cannot account for any fine structure in the grazing angle dependence and it provides no physical understanding of the scattering problem.

2.3 Self-Shadowing Effects

If one deals with a sufficiently rough surface or interest lies in the region of shallow grazing angles, one must account for the self-shadowing that occurs when part of a surface acts to shield another part from the incident radiation.

Beckmann[28] accounted for shadowing effects by introducing a shadowing function S, the probability that a point on the surface $z = f(x)$ will be illuminated by the incident radiation. The probability that a point on the surface is shadowed depends on both its height and its slope. These two parameters are related, but Beckmann treated them as independent to simplify the ensuing mathematics. Wagner[29] followed Beckmann's definition of the shadowing function but maintained the interdependence of the height and surface slope. Thorsos[17] employs Wagner's scattering function, but when dealing with short (surface) correlation lengths, he employs an effective surface slope given by $s' = 1/\sqrt{kL}$, where L is the surface correlation length and k is the acoustic wavenumber.

2.4 Volume Scattering from the Surface, Bottom, and Water Column

Thus far, only the scattering that results from an impenetrable interface roughness has been considered. We now outline circumstances in which volume inhomogeneities within the water column and/or the bottom contribute to scattering.

If an acoustic wave penetrates the water–bottom interface and the bottom is inhomogeneous, scattering from the inhomogeneities will occur. In 1963, Stockhausen[30] and Nolle et al.[31] independently developed scattering models in which the incident wave penetrates the interface and is scattered from within the bottom. Stockhausen's model[30] assumes that the bottom is composed of solid spherical particles that act to scatter the acoustic energy. Nolle et al.[31] consider an average scattering amplitude factor per unit volume and a term δ to account for local deviations in the scattering strength of individual scattering centers. More recently, Hines[32] employed a perturbation technique developed by Chernov[33] to derive a mathematical model of acoustic scattering from a smooth sediment bottom in which fluctuations in sediment porosity result in scattering. The model allows for penetration of the incident wave into the bottom at subcritical grazing angles and retransmission of scattered spherical waves through the interface.

A contribution to volume scattering is also associated with the ocean surface. In sufficiently rough seas, a layer of air bubbles will be trapped in the first few meters of water beneath the air–water interface. The bubbles are generated primarily by breaking waves and may last tens of minutes after the breaking event. Data obtained by Farmer and Vagle[34] depict vertical bubble plumes varying in depth from several centimetres to several metres. Based on this, Henyey,[35] McDonald,[36] and Prosperetti et al.[37] have each developed models for scattering from these bubble plumes.

Finally, other inhomogeneities within the water column also contribute to volume scattering, the dominant inhomogeneity being marine life. Particularly effective, at their resonant frequencies, are organisms that use gas bladders to control their buoyancy. While the physics of scattering from fish swim bladders is fairly well understood, the mathematics is not included here, since a detailed treatment of the subject is contained in Chapter 44. However, backscatter data from marine organisms is contained in Section 3.6.

2.5 Numerical Models

The increase in computer speed over the past decade has made it possible to obtain exact or nearly exact numerical solutions to the scattering problem for several geometries. Thorsos and Jackson[38] use integral equation methods to obtain solutions for scattering from a one-dimensional surface with a Gaussian roughness spectrum that obeys the Dirichlet boundary condition. The paper is particularly useful because it compares the results to those obtained using traditional approximations such as the perturbation method and the Helmholtz integral with the Kirchhoff approximation.

An alternative to the smoothly varying, random surface is the facet ensemble or wedge assembly method employed by Medwin and Novarini,[39] for example. They suggest that since the ocean surface is wedge-like rather than pointlike, it should be modeled by a series of fine, contiguous wedges. They use the normal coordinate theory of Biot and Tolstoy[40,41] to obtain the scattered pressure from each wedge of a computer-generated Neumann–Pierson sea surface and sum the returns (accounting for the proper time delay) from each wedge comprising the surface. Further applications of the method can be found in Clay and Kinney[42] and the reference contained therein.

Collins and Evans[43] derive a method to include single scattering in the range-dependent parabolic equation (PE) and then obtain estimates of the reflected field for several environments. Stephen[44] employs the finite-difference solution to the two-way wave equation to develop a "numerical scattering chamber." He accounts for inhomogeneities and surface roughness by representing each grid point in the chamber by its physical parameters (compressional and shear wave speeds, density, etc.). Schmidt[45] obtains numerical solutions for scattering in a three-dimensional environment by employing a hybrid, wavenumber integration model.

3 EXPERIMENT

To compare measurements made with different geometries, equipments, environmental backgrounds, or acoustic frequencies, reverberation levels are normally converted to scattering strengths. In this section, scattering strengths are defined and equations relating them to reverberation levels are presented. The section concludes with a summary of the principal characteristics reported for surface, underice, volume, and bottom reverberation.

3.1 Scattering Strength

The theory of reverberation in the ocean was first developed during World War II by scientists at the University of California, Division of War Research.[46] With some refinements their basic approach has been adopted by most researchers in the field. Using the nomenclature developed by Urick[26,47] to improve the visualization of the problem, a scattering strength of a unit volume of the ocean or of a unit area of its boundaries is defined that, combined with the acoustic characteristics of the sonar and the geometry of the experiment, determines the reverberation level. The scattering strengths are defined in Fig. 1. On the left in the figure, I_I is the intensity of a plane wave incident on a scattering volume, ΔV, and I_s is the scattered intensity at distance r from the scattering volume. When r is a unit distance and ΔV is a unit volume, S_V is defined as the scattering strength of a unit volume. The scattering strength S_S at grazing angle θ of a unit area of an ocean boundary is defined in a similar manner on the right in the figure. These definitions are sufficiently general to cater to situations such as those encountered in bistatic sonar, where the incident and scattered angles are different. Very often, in practice, experimental geometries are selected so that these angles are coincident, in which case the scattering strengths are referred to as backscattering strengths.

3.2 Reverberation Equations

Equations relating scattering strengths to reverberation levels are derived using the above definitions and the following assumptions: The reverberation intensity, which is a function of the time elapsed since the emission of a short-duration pulse, is the sum of the intensities received from the individual scatterers; multiple scattering can be neglected; sound travels to the scatterer and back via the same path; and the duration of the pulse is negligible compared to the time taken to reach the scatterers. For colocated sources and receivers in an infinite, nonrefracting medium with uniform scattering, the volume reverberation equation for a continuous wave (cw) pulse is

$$RL = SL - 2H + 10\log\psi + 20\log r + 10\log\frac{c\tau}{2} + S_v,$$
(7)

where RL is the root-mean-square (rms) pressure level of the reverberation in decibels relative to 1 μPa, SL is the transmitted pulse strength at 1 m in decibels relative to 1 μPa and H is the transmission loss in decibels between the sound source and the scatterers. The next three terms define the scattering volume, where r is the range in metres, c is the sound velocity in metres per second, and τ is the pulse length in seconds. The term ψ is the ideal solid angle beamwidth of the source–receiver combination, which can be envisioned as the "opening angle of the ideal beam pattern having a flat response within ψ and none beyond, which for reverberation is equivalent to the actual two-way beam pattern."[26] For sources and receivers in a semi-infinite medium with a scattering boundary, the surface or bottom reverberation equation for a cw pulse is

$$RL = SL - 2H + 10\log\phi + 10\log r + 10\log\frac{c\tau}{2} + S_s,$$
(8)

where ϕ is the ideal plane angle beamwidth of the source–receiver combination. Values of ψ and ϕ for various transducers are presented in Ref. 26.

Explosives and omnidirectional hydrophones are often used for investigating sound scattering over a wide range of frequencies, the equations being reformulated in terms of the energy density of the explosive pulse.[48–51] Under conditions where refraction and attenuation can be neglected, the reverberation in a narrow band from layered scattering in the volume of the ocean can be described by the equation[51]

$$RL = E + 10\log\int_{d_1}^{d_2} M_z\, dz - 30\log t + K,$$
(9)

where RL is the rms pressure level of the reverberation in the frequency band considered in decibels relative to 1 μPa, E is the energy per unit area level in the same frequency band measured at 100 m from the source in decibels relative to 1 μPa$^2 \cdot$ s, and t is the time in seconds. In this formulation the volume scattering coefficient M, as defined in Fig. 1, depends on the depth z. When the

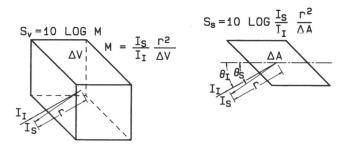

Fig. 1 Definition of volume and surface scattering strengths.

integration extends over the whole water depth D, the quantity $10 \log \int_0^D M_z \, dz$ is referred to as the column strength. The constant K depends on the experimental conditions; for an infinite ocean, it is -10; for a semi-infinite ocean where surface-reflected paths are involved, it is -4. Equation (9) also applies to scattering from the sea surface or bottom,[50] with $K = -10$ and the column strength replaced by S_S.

The above equations suffice for many practical situations. However, variants appear in the literature, as the trend has been to extend reverberation investigations to situations that violate some of the basic assumptions used in the derivations. Currently there is interest in hybrid path situations where sound arrives at the scatterer by a number of paths, is scattered at different angles, and arrives back at the receiver by multiple paths.[52]

3.3 Experimental Considerations

Reverberation may arise from a combination of scatterers, as indicated in Figs. 2 and 3. It generally decays with time but may increase dramatically after a new source of scatterers is encountered. For the situation illustrated, of omnidirectional sound sources and receivers located relatively close to the sea surface in the deep ocean, the following sequence occurs. The initial reverberation is due to deep scattering layers (DSLs). This volume reverberation decay is interrupted, first by the specular return from the sea surface followed by surface scattering and then by the specular (fathometer) return from the sea bottom followed by scattering from the interface and sub-

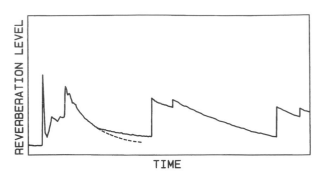

Fig. 3 Reverberation level versus time curve for the experimental arrangement shown in Fig. 2.

bottom structure. Generally there are additional bottom reverberation peaks at the time intervals required for the sound to make additional excursions between the surface and bottom. A considerable amount of information can be obtained from such plots of reverberation level versus time: the column strength; surface scattering strength as a function of grazing angle down to some minimum angle that depends on the relative strengths of the surface and volume scattering; and bottom scattering strengths down to grazing angles of about $30°$, at which point they are swamped by the second fathometer returns.

The simple arrangement shown has been used to obtain much of the scattering strength information presented in this section. Most of the rest was measured with a variety of directional cw-pulsed equipment such as the tiltable transducers used to investigate surface and bottom reverberation.

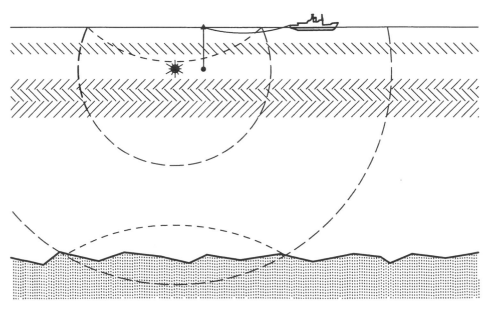

Fig. 2 Typical experimental arrangement for measuring scattering strengths with a point charge sound source and an omnidirectional receiving hydrophone.

3.4 Surface Reverberation

Measurements made in the early 1940s using a 24-kHz hull-mounted sonar showed that, at small grazing angles, surface scattering strengths were independent of grazing angle and increased with increasing sea state.[46] From this, it was inferred that the scattering arose from a layer of bubbles generated by breaking waves and trapped below the sea surface. In the mid-1950s, Urick and Hoover[53] investigated scattering strengths at 60 kHz in the open ocean using a tiltable transducer suspended from a surface buoy. The results, shown on the left in Fig. 4, indicate a dependence on wind speed and grazing angle consistent with three dominant scattering mechanisms: from 0° to 30°, scattering from entrapped air bubbles; from 30° to 70°, scattering from the rough sea surface; and at steeper angles, specular reflection from wave facets. Similar measurements made by Garrison et al.[54] in sheltered waters are shown on the right in the figure. Their measurements differ from those in the open ocean[53] by continuing to drop off at the smaller grazing angles. Presumably this arises because fewer bubbles are generated for a given wind speed in confined waters.

In the early 1960s, a number of investigations of surface backscattering at lower frequencies as a function of wind speed, grazing angle, and frequency were carried out in the open ocean using explosives as sound sources and omnidirectional hydrophones as receivers. Results obtained by Chapman and Harris[50] and Chapman Scott[55] in the 0.4–0.8-kHz and 3.2–6.4-kHz bands are shown in Fig. 5. There are evident similarities between these data and those observed at higher frequencies. At grazing angles in excess of 60°, scattering is independent of frequency and decays more steeply with grazing angle

the lower the wind speed. The results are consistent with reflection from wave facets and have been modeled[55] using Eckart's theory[18] and Cox and Munk's[56] wave facet slopes. At smaller grazing angles, scattering strengths fall off less rapidly with decreasing grazing angle and become dependent on frequency. The smaller the frequency, wind speed, and grazing angle, the more rapid is the decrease in scattering strength with decreasing grazing angle. These effects are described by the empirical equation developed by Chapman and Harris[50] from data covering the frequency range from 0.4 to 6.4 kHz and grazing angles less than 40°:

$$S_s = 3.3\beta \log(\theta/30) - 42.4 \log \beta + 2.6, \qquad (10)$$

where β, the slope of the reverberation level versus the grazing angle curve in decibels per angle doubled, is $158(vf^{1/3})^{-0.58}$, where v is the wind speed in knots and f is the frequency in hertz.

Scattering theory[23,57,58] provides a capability to model backscattering from the rough sea surface, and agreement with experiment is obtained at low wind speeds. However, predicted scattering strengths at high wind speeds are well below those obtained experimentally.[35,36,59,60] This is shown in Fig. 6, where scattering strength is plotted as a function of wind speed for a frequency of 3.5 kHz and a grazing angle of 15°. This figure by McDaniel[60] shows results from a summary of a number of data sets, the scattering strengths generated by Eq. (10), and the calculation of the scattering of the rough sea surface based on two-component scattering theory. Although theory and experiment agree at low

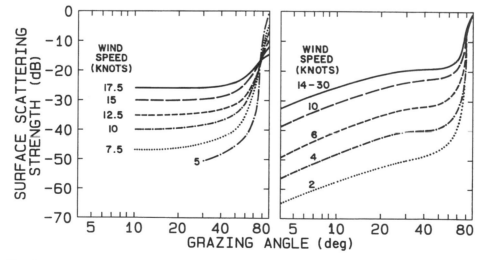

Fig. 4 Surface scattering strength at 60 kHz as a function of wind speed and grazing angle. [Reprinted, by permission, from Ref. 53 (left) and Ref. 54 (right).]

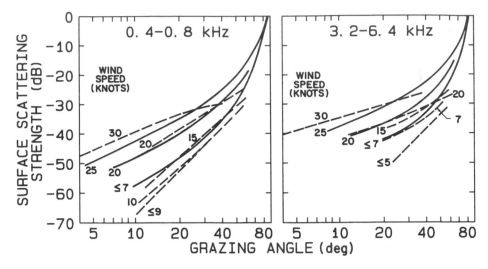

Fig. 5 Low-frequency surface scattering strengths measured in the open ocean by Chapman and Harris[50] (-----) and Chapman and Scott[55] (———).

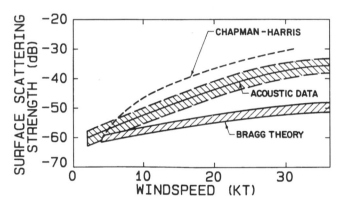

Fig. 6 Surface scattering strengths, at a frequency of 3.5 kHz and a grazing angle of 15°, versus wind speed. (Reprinted, by permission, from S. T. McDaniel, *J. Acoust. Soc. Am.*, Suppl. 1, Vol. 84, 1988.)

wind speeds, they diverge by 15–20 dB at the higher wind speeds. To explain this discrepancy, Henyey,[35] McDonald,[36] and Prosperetti et al.[37] modeled the reverberation as scattering from plumes of microbubbles. Microbubbles are the small bubbles that remain after the large bubbles generated by breaking waves have risen to the surface. These bubbles are of long duration and can be carried to depths in excess of 10 m by downwelling currents, especially Langmuir circulation.[34] For frequencies exceeding several kilohertz, the scattering from these plumes is expected to be dominated by resonant scattering of the microbubbles. At lower frequencies, it is the compressibility of the plume, rather than the characteristics of the individual microbubble, that is responsible for the scattering. Scattering strengths at 24 knots computed by Henyey[35] for both the bubble plumes

and the rough sea surface are shown plotted as a function of frequency in Fig. 7, along with the experimental observations of Refs. 50 and 55. The modeled results provide a satisfactory fit to the experimental data. Scattering from both the plumes and rough surface falls off with decreasing frequency, the former more rapidly than the latter. Thus at 24 knots, it is predicted that scattering from surface roughness would dominate at frequencies below 125 Hz. This is consistent with the conclusion of Ogden and Erskine,[59] who examined a more extensive data base of surface backscattering strengths mea-

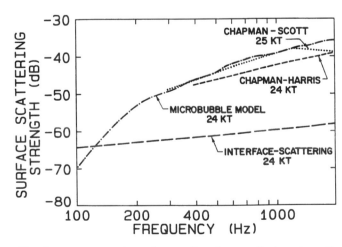

Fig. 7 Frequency dependence of surface scattering strength at a grazing angle of 10°. Henyey's[35] modeled results for microbubble plumes (· – – ·) and interface scattering (————). Experimental results from Chapman and Harris[50] (-----) and Chapman and Scott[55] (·····). (Reprinted, by permission, from Ref. 35.)

sured with explosive sources. They found that, for grazing angles less than 40° and wind speeds below 14 knots, rough surface scattering described the results down to 250 Hz. At lower frequencies scattering from rough surfaces was consistent with the observations even at much higher wind speeds.

Although bubble plume scattering accounts for many of the observed characteristics of surface backscattering at high wind speeds, the story is not yet complete and there is considerable modeling activity going on. McDaniel,[61] based on an analysis of surface reverberation fluctuations at 3 kHz and higher, has proposed an alternative mechanism. She concludes that resonant scattering from a thin layer of microbubbles generated just below the sea surface by breaking waves better describes the observed fluctuations. As existing bubble plume scattering models predict increased scattering with frequency, it was further concluded that their predicted contribution to backscatter at somewhat lower frequencies is questionable.

3.5   Under-Ice Reverberation

Several investigators[62–66] have used explosive sound sources to examine the dependence of sea-ice reverberation on ice condition, frequency, and grazing angle; Hayward and Yang[66] used a vertical array as a receiver; the others used omnidirectional hydrophones. Not unexpectedly, ice roughness has a strong impact on scattering strength.

The roughest ice conditions are found in the Arctic ice pack in spring. A combination of first-year and multiyear ice results in relatively smooth stretches of ice interrupted by large ridgelike structures, extending to depths of a few tens of metres. Summer melting results in some reduction in roughness, which is later restored by the freezing, breaking, crushing, and refreezing caused by the low temperatures and high winds of winter.

Scattering strengths measured under the spring Arctic ice pack by Milne[62] and Brown and Milne[63] are shown in Fig. 8. For the range of grazing angles displayed (5–20°) scattering strengths decrease with decreasing grazing angle and frequency at frequencies below about a kilohertz. At higher frequencies they tend to become independent of both frequency and grazing angle.

Scattering strengths for frequencies slightly above 1 kHz are shown in Fig. 9 for a number of ice conditions. The greatest scattering is from the Arctic ice pack in spring.[63] The next highest scattering strengths are those from well-rotted Arctic ice in summer, which are similar to scattering strengths from the open sea surface at a wind speed of 30 knots.[64] Greene and Stokes[67] were able to obtain good agreement with the Arctic data, for both spring and summer conditions, using a model that attributed the scattering to a distribution of triangular wedges covered with small-scale roughness due to ice rubble. The calculations were based on diffractive scattering from the small-scale roughness using the one-dimensional roughness spectrum published by Mellen.[68]

Also shown in the figure are two sets of measurements[65] made in the Gulf of St. Lawrence, an area that is ice free most of the year and hence lacking in multiyear ice. The scattering strengths from the roughest condition encountered, a floe of heavily broken and refrozen ice, produced scattering strengths comparable to those reported for the

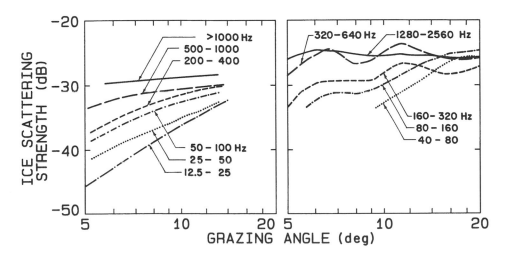

Fig. 8 Scattering strengths measured under the Arctic ice pack in spring. Note that the scattering strength levels on the left have been reduced by 17 dB to account for the source strength correction discussed in Ref. 63. [Reprinted, by permission, from Ref. 62 (left) and Ref. 63 (right).]

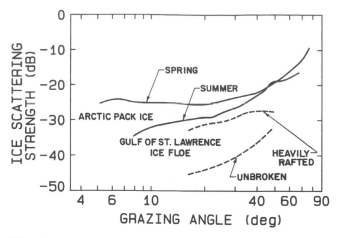

Fig. 9 Scattering strengths under the Arctic ice pack at 1.28–2.56 kHz (——)[63,64] and in the Gulf of St Lawrence at 1.6–3.2 kHz (-----).[65]

summer Arctic ice pack. The scattering strengths for the smoothest ice condition encountered were taken from an ice floe showing little if any rafting. These examples indicate that the smoother the ice, the lower the scattering strength and the steeper the decay with decreasing grazing angle.

3.6 Volume Reverberation

The first reported measurements of volume reverberation, made off San Diego in 1942, indicated that the principal scatterers were confined to a horizontal layer that experienced a pronounced diurnal migration in depth.[46] Because of this behavior, it was inferred that the scatterers were marine organisms. Initially, the layer was named the ECR layer after its discoverers, Eyring, Christiansen, and Raitt.[69] Later, it became known as the deep scattering layer, or DSL. It is now known that in any ocean area there may well be several scattering layers not all of which undergo diurnal migrations.

Below acoustic frequencies of about 15 kHz, the principal scatterers are fish that use swim bladders to control their buoyancy.[70] At higher frequencies scattering from zooplankton dominates. The geographic extent of a given species, and hence its contribution to the reverberation background, tends to be limited by oceanographic boundaries such as those shown in Fig.10[71,72] (Also included are the sites used later in this section to illustrate particular features of volume reverberation.) Because the boundaries run primarily east–west, reverberation changes more strikingly with latitude than with longitude. Generally, sound scattering decreases as location shifts from the equator, has a minimum near 30°

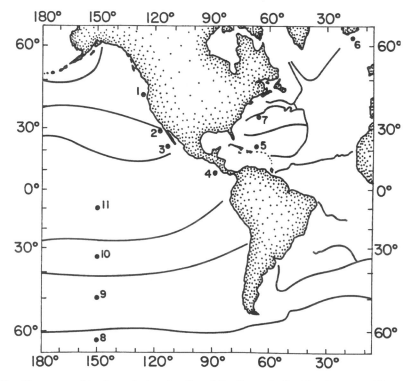

Fig. 10 Oceanographic boundaries in the Atlantic and Pacific Oceans. The principal boundaries are from Ref. 72. Those in the north west Atlantic are from Ref. 71. The location of volume reverberation sites referred to in the text are indicated.

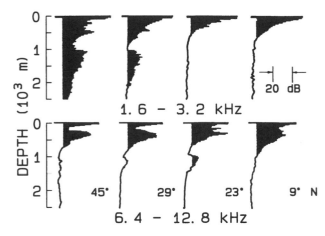

Fig. 11 Reverberation level versus depth profiles along the west coast of North America.[73] Left-to-right sites 1–4 of Fig. 10.

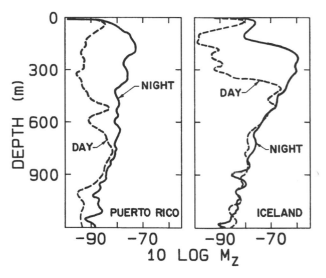

Fig. 12 Volume-backscattering strength at 5 kHz versus depth profiles for typical sites north of Puerto Rico (site 5) and south of Iceland (site 6).[73]

latitude, increases towards the polar front in both the Arctic and Antarctic, and then decreases at higher latitudes. The major changes in reverberation conditions generally take place close to the oceanographic boundaries. Between boundaries, the reverberation levels, volume scattering strength-versus-depth profiles, and column strength spectra often maintain recognizable features over hundreds of kilometres. This is illustrated in Fig. 11 with traces of reverberation level versus depth for two frequency bands at four sites along the west coast of North America.[73] The sound source was a surface-fired charge and the receiver a downward-looking wideband conical array. The shaded area indicates a common reference level for the two frequencies. The traces show the consistency in reverberation conditions that can occur over long distances and the striking changes that can occur in crossing oceanographic boundaries. Similar results have been obtained by Shevtsov et al.[74], who conducted three meridional transits of the North Pacific using 12- and 30-kHz echo sounders. The large differences in the reverberation characteristics found in both investigations at frequencies an octave or two apart are indicative of a process dominated by resonant scatterers.

The migration of resonant scatterers toward the sea surface at night can also cause very large changes in scattering between daytime and nightime conditions, as ilustrated in Fig. 12. The scattering strengths were measured at two sites in the North Atlantic, one representative of conditions between Puerto Rico and Bermuda and the other of conditions between Newfoundland and Iceland.[73] The figure also illustrates the generalization that the bulk of the diurnal changes occur at the shallower depths, with the deeper layers showing little if any effect.

Because of the dependence on frequency, column

strength spectra are useful for looking at gross changes in scattering with location and time of day. Column strengths can be obtained with simple and easily deployed equipment such as point charges and omnidirectional hydrophones and are thus well suited for wide-area surveys, both by ship[75] and aircraft.[76] Akal et al.[77] recently demonstrated an elegant surveying technique employing deep-fired charges and the upward-looking end-fire beam of a vertical line array that provides both column strength spectra and remarkably detailed plots of backscattering strength versus depth. Examples of daytime and nighttime column strength spectra for a site in the Sargasso Sea north of Bermuda are shown in Fig. 13. There is an increase in scattering at night all across the frequency band. The peaks in the daytime spectra are due to resonant scattering from scatterers of effective radius 0.54, 0.38, and 0.18 cm at depths of 800, 650, and 460 m, respectively. The pronounced peak in the nighttime spectra at 3.5 kHz is attributed to the migration of the 0.18-cm scatterers to near the sea surface at night.

The pattern referred to earlier, on the dependence of sound scattering on latitude, has been observed in both the Atlantic and Pacific Oceans but is best illustrated in the South Pacific, where the oceanographic boundaries run most closely east–west. Representative column strength spectra from the four principal oceanographic regions of the South Pacific are shown in Fig. 14.[75] Typical of polar regions, scattering in the Antarctic is low and experiences little day–night change. In the warmer areas there is a pronounced increase in scattering at night. Scattering increases substantially on crossing the polar front, experiences a minimum in the subtropical region (near 30°), and again increases toward the equator.

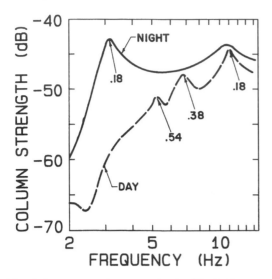

Fig. 13 Column strength spectra in the North Sargasso Sea (site 7). The effective radius of the gas bladder of the dominant scatterers responsible for the resonant peaks is indicated. (Reprinted, by permission, from Ref. 51.)

3.7 Bottom Reverberation

The first reported measurements of backscattering strengths of the ocean bottom were made during World War II, primarily at 24 kHz and small grazing angles.[46] Since then, numerous experiments have been carried out to extend the range of frequencies, grazing angles, and bottom characteristics. Considerable variability, attributed to such factors as bottom material, volume inhomogeneity, and interface roughness, has been observed in scattering strengths measured under apparently similar conditions. This has led to the development of increasingly sophisticated experiments and models to cope with the complexity.

Shallow Water   It is customary to treat reverberation in shallow water separately from that in deep water. This arose because of observed differences in scattering strengths, attributed in part to the less rugged topography, but primarily to the rich variety of sedimentary conditions found on continental shelves.[78] In an effort to avoid

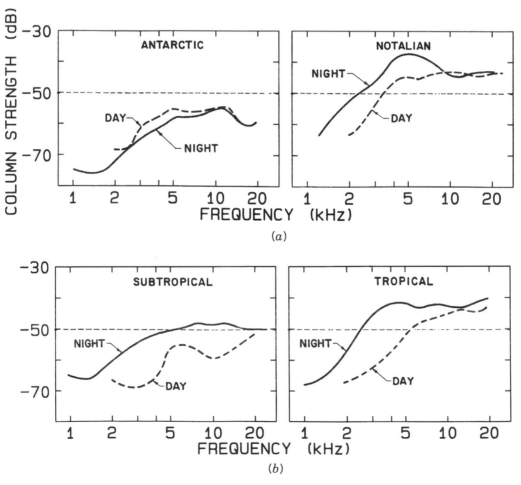

Fig. 14   Column strength spectra for the main oceanographic areas of the South Pacific (sites 8–11).[75]

TABLE 1   Shallow-Water Bottom Backscattering Experiments

Authors	Year	Bottom Type	Frequency (kHz)	Grazing Angle (deg)
MacKinney and Anderson[79]	1964	Mud, sand, gravel	12.5–290	1–90
Wong and Chesterman[80]	1968	Rock sand, silt, clay	48	0.4–8
Boehme et al.[81]	1985	Sand	30–95	2–10
Jackson et al.[82]	1986	Sand, gravel, silt	20–85	5–90
Stanic et al.[83]	1988	Sand	20–180	5–30
Stanic et al.[84]	1989	Shell-covered area	20–180	5–30
Gensane[85]	1989	Sand, clay, gravel	8–40	4–90
Jackson and Briggs[86]	1992	Sand, silt	15–45	5–90

contamination of the data by surface reverberation, a particularly serious problem when making bottom backscattering measurements at small grazing angles in shallow water, experiments are normally conducted at close range using high-frequency, directional sound sources and short cw pulses. A representative list of experiments including bottom types, frequencies, and grazing angles is given in Table 1.

Measurements made on these and other experiments[26,47,79–85] indicate that, over a wide range of bottom conditions and grazing angles, scattering strengths are either independent of or only weakly dependent on frequency. Bunchuk and Zhitkovskii[78] point out that, based on theoretical considerations, this is readily explainable at large, but not at small, grazing angles. A number of observers have reported on the dependence of backscattering strength on frequency at small grazing angles (below the critical angle).[79–85] Frequently no dependence is found; however, modest increases in slope, up to a maximum of about 4.8 dB/octave,[79,81] have been reported for sand-covered bottoms. Small negative slopes have also been observed, Stannic et al.[84] reporting −0.75 dB/octave for a rough coarse shell covered bottom. The results for other experimenters and bottom conditions generally fall well between these extremes.

Bottom scattering is often associated with bottom material, with scattering strength decreasing from rock through sand and silt to clay. This is illustrated in Fig. 15 based on a 1968 report by Wong and Chesterman[80] on measurements made at 48 kHz and a survey of the literature. This dependence on bottom type suggests that the particulate nature of the bottom is a significant factor in the scattering process.[79] However, as scattering strengths within each general sediment type can vary by 10–20 dB,[83] rather more than the differences between the various classes, other factors must also be considered.

Early experiments[79,80] indicated that scattering strength

tends to become independent of grazing angle at small grazing angles, as shown in the figure. This has now been called into question,[81,82] as later experiments working in similar environments have been unable to duplicate this effect; at grazing angles below the critical angle they measure fall-offs that typically follow or exceed those of Lambert's law down to the smallest angles observed.[81–85]

Gensane,[85] who compared slopes of scattering strength versus grazing angle over the complete range of grazing angles for sand, clay, and gravel bottoms, reported three scattering regimes. Below 10°, scattering strength slopes often exceeded that of Lambert's law; between 20° and 60°, they were generally flatter; and above 60°, they rose very rapidly. These results are consistent with those of other observers and are well described by three theoretical mechanisms: at near normal incidence, roughness scattering described by the Kirchhoff approximation; at intermediate angles, sediment volume scattering; and at angles below the sediment critical angle, where the

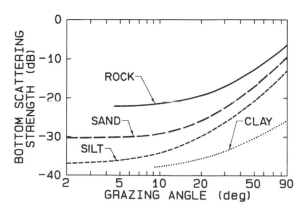

Fig. 15 Dependence of scattering strength on bottom type.[80] (Produced by Urick in Ref. 26. Copyright 1983. Reproduced by permission of McGraw-Hill.)

acoustic wave scarcely penetrates the bottom, interface roughness scattering described by the Rayleigh–Rice approximation.

Comprehensive models, which include the various scattering mechanisms, have been developed by Jackson et al.[22] and Mourad and Jackson[87] to account for the scattering strengths observed in shallow water over the frequency range 10–100 kHz. Two examples from Jackson and Briggs[86] showing experimental and modeled results for two sedimentary conditions are shown in Fig. 16. Figure 16a indicates that, for a fine sand bottom with ripples, roughness scattering dominates at both large and small grazing angles; in between, from approximately 30° to 50°, the contributions from roughness and sedimentary volume scattering are nearly equal. This is in marked contrast to the situation shown in Fig. 16b where, for a fine sediment mixed with coarse material, sediment volume scattering dominates at all except the steepest grazing angles.

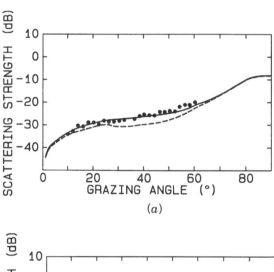

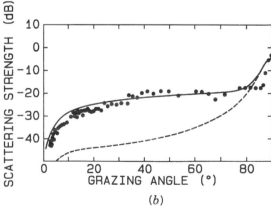

Fig. 16 Comparison of backscattering model and data for (a) fine sand with well-defined ripples, 25 kHz, and (b) fine sediment mixed with coarse material, 20 kHz. The dashed curve shows the model prediction for roughness scattering and the solid curve shows the total including sediment volume scattering. (Reprinted by permission from Ref. 86.)

Deep Water In deep water, measurements of bottom backscattering strength are usually made either from near the sea surface or close to the sea bottom. Ranges to the scatterers tend to be greater and the measurement frequencies lower than those employed in shallow water.

The first measurements of bottom scattering strength in the deep ocean that covered essentially the complete range of grazing angles were made by Patterson[88] at 2.5 kHz using an omnidirectional sound source and receiver operating close to the ocean bottom. The results follow the characteristic shape of scattering strength versus grazing angle noted earlier for shallow water, a rapid drop-off from normal incidence to 60°, a flattening down to 20°, and a rapid decrease at smaller grazing angles. The data also showed a marked peak in the vicinity of the critical angle of 30°, similar to that obtained by Jackson et al.,[22] using Kuo's[8] model for scattering from a rough interface. Patterson developed an empirical model to describe the observed backscattering strengths:

$$S_f = 10 \log[A(B + C)], \qquad (11)$$

where A is the Rayleigh reflection coefficient, B is a constant times "some sort of random probability density function," and C is a constant times the sine of the grazing angle raised to some power.

In the early 1960s, a number of investigations[89–92] were carried out to measure the dependence of bottom backscattering strength in the deep ocean on bottom condition, frequency, and grazing angle. Results in the 3.2–6.4-kHz band for two extensive surveys[91,92] in the North Atlantic are shown in Fig. 17. The measurements were made with explosives and omnidirectional hydrophones deployed near the sea surface, thus restricting grazing angles to those greater than 30°. Zhitkovskii and his associates[91,93] carried out an extensive theoretical effort to account for observed dependences of backscattering on bottom characteristics, frequency, and grazing angle. It was concluded that, for ridges, where the irregularities are large compared to the wavelength, scattering is independent of frequency and follows Lambert's law. For the smoother areas scattering could be described adequately by the two-scale roughness model of Kur'yanov.[23] Large-scale roughness dominates the scattering process at steep grazing angles and the scattering is independent of frequency. At smaller grazing angles, roughness that is small compared to a wavelength is important and there is a strong frequency dependence of the scattering strength, up to the theoretical limit of 12 dB/octave.

Increasingly, sound sources and receivers are deployed close to the bottom to permit measurements over essentially the complete range of grazing angles. Although the initial measurements of frequency dependence were

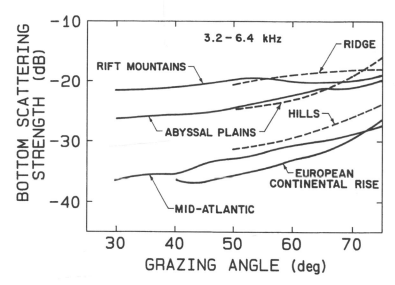

Fig. 17 Scattering strengths at a number of locations in the deep North Atlantic by Zhitkovskii and Volovova[91] (-----) and Chapman[92] (———). (Copyright 1967. Reproduced courtesy of Plenum Publishing Corporation.)

carried out with explosives,[93–95] the time taken to raise the equipment to the surface for reloading has precluded their use in area surveys. Since then a number of investigators have overcome this problem by employing cw projectors operating at several frequencies.[93,96,97] Greater variations have been reported in deep than in shallow water on the dependence of scattering strength at small grazing angles on frequency. Hines and Barry,[96] who made measurements at frequencies from 0.8 to 2.4 kHz for a variety of abyssal plain sedimentary conditions, observed slopes ranging from 0 to 9 dB per octave. Zhitkowksii,[93] operating at frequencies between 2 and 16 kHz in abyssal hills, observed significant differences at sites only 200–300 m apart. The presence of ferromanganese nodules increased local scattering by 10–15 dB and caused the scattering strength to increase at 12 dB/octave, as expected for scatterers that are small compared to the wavelength. The concept was extended by Weydert,[97] who showed that by using several cw frequencies above 30 kHz it is possible to estimate the fraction of the sea floor covered by nodules and that by employing lower frequencies their size distribution can be determined.

Mourad and Jackson[98] have extended their modeling to lower frequencies—from 1 kHz down to 100 Hz—and have compared the results with measurements made in the North Atlantic on the Sohm Abyssal Plain and on the Bermuda and Continental Rises.[94,96,99] They conclude that scattering from within the sediments is the dominant deep-water scattering mechanism over the frequency range considered of 100 Hz to 10 kHz. For slow sediments (sound speed ratio less than unity), such as those found on the Continental and Bermuda Rises, dominance of volume over interface roughness scattering is predicted for grazing angles of 1°–5° to about 75°. A decrease in scattering strength of 8–10 dB is indicated as frequency increases from 100 Hz to 1 kHz. At higher frequencies backscattering tends to become independent of frequency. For hard, fast (sound speed ratio greater than unity) sediments such as those found on the Sohm Abyssal Plain, they find that volume scattering dominates from near the critical angle up to about 70°. The bottom backscattering decreases as frequency increases consistent with dominance by sediment volume scattering. Below the critical angle, the volume scattering process dominates at frequencies between 100 Hz and 1 kHz. Backscattering strength increases with increasing frequency consistent with the eventual dominance of total backscattering strength by surface roughness at high frequencies.

REFERENCES

1. J. W. Strutt, *The Theory of Sound*, Dover, New York, 1945.
2. W. C. Meecham, "Variational Method for the Calculation of the Distribution of Energy Reflected from a Periodic Surface," *J. Appl. Phys.*, Vol. 27, 1956, pp. 361–367.
3. R. F. Millar, "On the Rayleigh Assumption in Scattering by a Periodic Surface II," *Proc. Cambridge Philos. Soc.*, Vol. 69, 1971, pp. 217–225.
4. D. R. Jackson, D. P. Winebrenner, and A. Ishimaru, "A Comparison of Perturbation Theories for Rough Surface Scattering," *J. Acoust. Soc. Am.*, Vol. 83, 1988, pp. 961–969.

5. S. O. Rice, "Reflection of Electromagnetic Waves from Slightly Rough Surfaces," *Comm. Pure and Appl. Math.*, Vol. 4, 1951, pp. 351–378.

6. E. I. Thorsos and D. R. Jackson, "The Validity of the Perturbation Approximation for Rough Surface Scattering Using a Gaussian Roughness Spectrum," *J. Acoust. Soc. Am.*, Vol. 86, 1989, pp. 261–277.

7. H. W. Marsh, "Exact Solution of Wave Scattering by Irregular Surfaces," *J. Acoust. Soc. Am.*, Vol. 33, 1961, pp. 330–333.

8. E. Y. T. Kuo, "Wave Scattering and Transmission at Irregular Surfaces," *J. Acoust. Soc. Am.*, Vol. 36, 1964, pp. 2135–2142.

9. E. Y. T. Kuo, "The Effects of Boundary Surface Inhomogeneities on Acoustic Scattering. I. Theory," *J. Acoust. Soc. Am.*, Vol. 81, 1987, pp. 1762–1766.

10. J. A. Ogilvy, *Theory of Wave Scattering from Random Rough Surfaces*, Adam Hilger, Bristol, UK, 1991.

11. F. G. Bass and I. M. Fuks, *Wave Scattering from Statistically Rough Surfaces*, Pergamon Press, New York, 1979.

12. L. Brekhovskikh and Yu Lysanov, *Fundamentals of Ocean Acoustics*, Springer Verlag, Berlin, 1982.

13. P. Beckmann and A. Spizzichino, *The Scattering of Electromagnetic Waves from Rough Surfaces*, Macmillan, New York, 1963.

14. D. Mintzer, "Discussion of the Paper by C. Eckart on Sea Surface Scattering," *J. Acoust. Soc. Am.*, Vol. 25, 1953, 1015(L).

15. W. C. Meecham, "On the Use of the Kirchhoff Approximation for the Solution of Reflection Problems," *J. Rational Mech.*, Vol. 5, 1956, pp. 323–334.

16. S. T. McDaniel and A. D. Gorman, "An Examination of the Composite Roughness Scattering Model," *J. Acoust. Soc. Am.*, Vol. 73, 1983, pp. 1476–1486.

17. E. I. Thorsos, "The Validity of the Kirchhoff Approximation for Rough Surface Scattering Using a Gaussian Roughness Spectrum," *J. Acoust. Soc. Am.*, Vol. 83, 1988, pp. 78–92.

18. C. Eckart, "The Scattering of Sound from the Sea Surface," *J. Acoust. Soc. Am.*, Vol. 25, 1953, pp. 566–570.

19. C. S. Clay and H. Medwin, *Acoustical Oceanography: Principles and Applications*, Wiley, New York, 1977.

20. M. A. Biot, "On the Reflection of Acoustic Waves on a Rough Surface," *J. Acoust. Soc. Am.*, Vol. 30, 1958, pp. 479–480.

21. V. Twersky, "On the Scattering and Reflection of Sound by Rough Surfaces," *J. Acoust. Soc. Am.*, Vol. 29, 1957, pp. 209–215.

22. D. R. Jackson, D. P. Winebrenner, and A. Ishimaru, "Application of the Composite Roughness Model to High-Frequency Bottom Backscattering," *J. Acoust. Soc. Am.*, Vol. 79, 1986, pp. 1410–1422.

23. B. F. Kur'yanov, "The Scattering of Sound at a Rough Surface with Two Types of Irregularity," *Sov. Phys. Acoust.*, Vol. 8, 1963, pp. 252–257.

24. W. Bachmann, "A Theoretical Model for the Backscattering Strength of a Composite-Roughness Sea Surface," *J. Acoust. Soc. Am.*, Vol. 54, 1973, pp. 712–716.

25. A. Purcell, "Bistatic Acoustic Scattering at the Seabed," Defence Research Establishment Atlantic Contractor Report, DREA CR/92/412, 1992.

26. R. J. Urick, *Principles of Underwater Sound*, 3rd ed., McGraw-Hill, New York, 1983.

27. K. V. Mackenzie, "Bottom Reverberation for 530- and 1030-cps Sound in Deep Water," *J. Acoust. Soc. Am.*, Vol. 33, 1961, pp. 1498–1504.

28. P. Beckmann, "Shadowing of Random Rough Surfaces," *Trans. IEEE Antennas Propagation*, Vol. 13, 1965, pp. 284–288.

29. R. J. Wagner, "Shadowing of Randomly Rough Surfaces," *J. Acoust. Soc. Am.*, Vol. 41, 1967, pp. 138–147.

30. J. H. Stockhausen, "Scattering from the Volume of an Inhomogeneous Half-Space," Naval Research Establishment, Dartmouth N.S., Canada, Report 63/9, 1963.

31. A. W. Nolle, W. A. Hoyer, J. F. Mifsud, W. R. Runyan, and M. B. Ward, "Acoustic Properties of Water-Filled Sands," *J. Acoust. Soc. Am.*, Vol. 35, 1963, pp. 1394–1408.

32. P. C. Hines, "Theoretical Model of Acoustic Backscatter from a Smooth Seabed," *J. Acoust. Soc. Am.*, Vol. 88, 1990, pp. 324–334.

33. L. A. Chernov, *Wave Propagation in a Random Medium*, English trans. McGraw-Hill, New York, 1975.

34. D. M. Farmer and S. Vagle, "Waveguide Propagation of Ambient Sound in the Ocean-Surface Bubble Layer," *J. Acoust. Soc. Am.*, Vol. 86, 1989, pp. 1897–1908.

35. F. S. Henyey, "Acoustic Scattering from Ocean Microbubble Plumes in the 100 Hz to 2 kHz Region," *J. Acoust. Soc. Am.*, Vol. 90, 1991, pp. 399–405.

36. B. E. McDonald, "Echoes from Vertically Striated Subresonant Bubble Clouds: A Model for Ocean Surface Reverberation," *J. Acoust. Soc. Am.*, Vol. 89, 1991, pp. 617–622.

37. A. Prosperetti, N. Q. Lu, and H. S. Kim, "Active and Passive Acoustic Behaviour of Bubble Clouds at the Ocean's Surface," *J. Acoust. Soc. Am.*, Vol. 93, 1993, pp. 3117–3127.

38. E. I. Thorsos and D. R. Jackson, "Studies of Scattering Theory Using Numerical Methods," *Waves in Random Media*, Vol. 3, 1991, pp. S165–S190.

39. H. Medwin and J. C. Novarini, "Backscattering Strength and the Range Dependence of Sound Scattered from the Ocean Surface," *J. Acoust. Soc. Am.*, Vol. 69, 1981, pp. 108–111.

40. M. A. Biot and I. Tolstoy, "Formulation of Wave Propagation in Infinite Media by Normal Coordinates with an Application to Diffraction," *J. Acoust. Soc. Am.*, Vol. 29, 1957, pp. 381–391.

41. I. Tolstoy, *Wave Propagation*, McGraw-Hill, New York, 1973, Chapter 8.

42. C. S. Clay and W. A. Kinney, "Numerical Computations of Time-Domain Diffractions from Wedges and Reflec-

tions from Facets," *J. Acoust. Soc. Am.*, Vol. 83, 1988, pp. 2126–2133.

43. M. D. Collins and R. B. Evans, "A Two-Way Parabolic Equation for Acoustic Backscattering in the Ocean," *J. Acoust. Soc. Am.*, Vol. 91, 1991, pp. 1357–1368.

44. R. A. Stephen, "A Numerical Scattering Chamber for Studying Reverberation in the Seafloor," in D. D. Ellis, J. R. Preston, and H. G. Urban (Eds.), *Ocean Reverberation*, Kluwer Academic Publishers, Dordrecht, Netherlands, 1993, pp. 227–232.

45. H. Schmidt, "Numerical Modeling of Three-dimensional Reverberation from Bottom Facets," in D. D. Ellis, J. R. Preston, and H. G. Urban (Eds.), *Ocean Reverberation*, Kluwer Academic Publishers, Dordrecht, Netherlands, 1993, pp. 105–112.

46. "Physics of Sound in the Sea," Natl. Defense Res. Comm., Div 6, Sum. Technical Report 8, 1946.

47. R. J. Urick, "The Backscattering of Sound from a Harbour Bottom," *J. Acoust. Soc. Am.*, Vol. 26, 1954, pp. 231–235.

48. R. J. Urick, "Generalized Form of the Sonar Equations," *J. Acoust. Soc. Am.*, Vol. 34, 1962, pp. 547–554.

49. S. Machlup and J. B. Hersey, "Analysis of Sound-Scattering Observations from Non-Uniform Distributions of Scatterers in the Ocean," *Deep-Sea Res.*, Vol. 3, 1955, pp. 1–22.

50. R. P. Chapman and J. H. Harris, "Surface Backscattering Strengths Measured with Explosive Sound Sources," *J. Acoust. Soc. Am.*, Vol. 34, 1962, pp. 1592–1597.

51. R. P. Chapman and J. R. Marshall, "Reverberation from Deep Scattering Layers in the Western North Atlantic," *J. Acoust. Soc. Am.*, Vol. 40, 1966, pp. 405–411.

52. D. D. Ellis and J. B. Franklin, "The Impact of Hybrid Ray Paths, Bottom Loss and Facet Reflection on Ocean Bottom Reverberation," in H. M. Merklinger (Ed.), *Progress in Underwater Acoustics*, Plenum Press, New York, 1986, pp. 75–84.

53. R. J. Urick and R. M. Hoover, "Backscattering of Sound from the Sea Surface: Its Measurement, Causes, and Application to the Prediction of Reverberation Levels," *J. Acoust. Soc. Am.*, Vol. 28, 1956, pp. 1038–1042.

54. G. R. Garrison, S. R. Murphy, and D. S. Potter, "Measurement of the Backscattering of Underwater Sound from the Sea Surface," *J. Acoust. Soc. Am.*, Vol. 32, 1960, pp. 104–111.

55. R. P. Chapman and H. D. Scott, "Surface Backscattering Strengths Measured over an Extended Range of Frequencies and Grazing Angles," *J. Acoust. Soc. Am.*, Vol. 36, 1964, pp. 1735–1737.

56. C. Cox and W. Munk, "Measurement of the Roughness of the Sea Surface from Photographs of the Sun's Glitter," *J. Opt. Soc. Am.*, Vol. 44, 1954, pp. 838–850.

57. E. I. Thorsos, "Acoustic Scattering from a Pierson-Moskowitz Sea Surface," *J. Acoust. Soc. Am.*, Vol. 88, 1990, pp. 335–349.

58. R. Dashen, F. Henyey, and D. Wurmser, "Calculations of Acoustic Scattering from the Ocean Surface," *J. Acoust. Soc. Am.*, Vol. 88, 1990, pp. 310–323.

59. P. M. Ogden and F. T. Erskine, "An Empirical Prediction Algorithm for Low-Frequency Acoustic Surface Scattering Strengths," NRL/FR/5160-92-9377, April 28, 1992.

60. S. T. McDaniel, "High-Frequency Sea Surface Scattering: Recent Progress," *J. Acoust. Soc. Am.*, Suppl. 1, Vol. 84, 1988, pp. S121.

61. S. T. McDaniel, "Sea-Surface Reverberation Fluctuations," *J. Acoust. Soc. Am.*, Vol. 94, 1993, pp. 1551–1559.

62. A. R. Milne, "Underwater Backscattering Strengths of Arctic Pack Ice," *J. Acoust. Soc. Am.*, Vol. 36, 1964, pp. 1551–1556.

63. J. R. Brown and A. R. Milne, "Reverberation under Arctic Sea-Ice," *J. Acoust. Soc. Am.*, Vol. 42, 1967, pp. 78–82.

64. J. R. Brown, "Reverberation under Arctic Ice," *J. Acoust. Soc. Am.*, Vol. 36, 1964, pp. 601–603.

65. R. P. Chapman and H. D. Scott, "Backscattering Strengths of Sea Ice," *J. Acoust. Soc. Am.*, Vol. 39, 1966, pp. 1191–1193.

66. J. T. Hayward and T. C. Yang, "Low-Frequency Arctic Reverberation. I: Measurement of Under-ice Backscattering Strengths from Short-range Direct-path Returns," *J. Acoust. Soc. Am.*, Vol. 93, 1993, pp. 2517–2523.

67. R. R. Greene and A. P. Stokes, "A Model of Acoustic Backscatter from Arctic Sea Ice," *J. Acoust. Soc. Am.*, Vol. 78, 1985, pp.1699–1701.

68. R. H. Mellen, "Underwater Acoustic Scattering from Arctic Ice," *J. Acoust. Soc. Am.*, Vol. 40, 1966, pp. 1200–1202.

69. C. F. Eyring, R. J. Christensen, and R. W. Raitt, "Reverberation in the Sea," *J. Acoust. Soc. Am.*, Vol. 20, 1948, pp. 462–475.

70. J. B. Hersey and R. H. Backus, "New Evidence that Migrating Gas Bubbles, Probably the Swim Bladders of Fish, Are Largely Responsible for Scattering Layers on the Continental Rise North of New England," *Deep-Sea Res.*, Vol. 7, 1954, pp. 190–191.

71. R. H. Backus, J. E. Craddock, R. L. Haedrich, and D. L. Shores, "The Distribution of Mesopelagic Fishes in the Equatorial and Western North Atlantic Ocean," in G. B. Farquhar (Ed.) *Proceedings of International Symposium on Biological Sound Scattering in the Ocean*, Maury Center Report 005, Dept. of the Navy, Wash., D.C., 1970, pp. 20–40.

72. G. Dietrich, *General Oceanography*, Wiley, New York, 1963.

73. R. P. Chapman, O. Z. Bluy, and R. H. Adlington, "Geographic Variations in the Acoustic Characteristics of Deep Scattering Layers," in G. B. Farguhar (Ed.), *Proceeding of an International Symposium on Biological Sound Scattering in the Ocean*, Maury Center Report 005, Dept. of the Navy, Wash., D.C., 1970, pp. 306–317.

74. V. P. Shevtsov, A. S. Salamatin, and V. I. Yusupov, "Large-Scale Structure of the Three-Dimensional Sound Scatter-

ing Field in the Pacific Ocean," *Oceanology*, Vol. 28, 1988, pp. 294–299.

75. R. P. Chapman, O. Z. Bluy, R. H. Adlington, and A. E. Robison, "Deep Scattering Layer Spectra in the Atlantic and Pacific Oceans and Adjacent Seas," *J. Acoust. Soc. Am.*, Vol. 56, 1974, pp. 1722–1734.

76. E. E. Davis, "Quasi-Synoptic Measurements of Volume Reverberation in the Western North Atlantic," in G. B. Farquhar (Ed.), *Proceeding of an International Symposium on Biological Sound Scattering in the Ocean*, Maury Center Report 005, Dept. of the Navy, Wash., D.C., 1970, pp. 294–305.

77. T. Akal, R. K. Dullea, G. Guido, and J. H. Stockhausen, "Low-Frequency Volume Reverberation Measurements," *J. Acoust. Soc. Am.*, Vol. 93, 1993, pp. 2535–2548.

78. A. V. Bunchuk and Yu. Yu. Zhitkovkii, "Sound Scattering by the Ocean Bottom in Shallow-Water Regions," *Sov. Phys. Acoust.*, Vol. 26, 1980, pp. 363–370.

79. C. M. McKinney and C. D. Anderson, "Measurement of Backscattering of Sound from the Ocean Bottom," *J. Acoust. Soc. Am.*, Vol. 36, 1964, pp. 158–163.

80. H. K. Wong and W. D. Chesterman, "Bottom Backscattering Near Grazing Incidence in Shallow Water," *J. Acoust. Soc. Am.*, Vol. 44, 1968, pp. 1713–1718.

81. H. Boehme, N. P. Chotiros, L. D. Rolleigh, S. P. Pitt, A. L. Garcia, T. G. Goldsberry, and R. A. Lamb, "Acoustic Backscattering at Low Grazing Angles from the Ocean Bottom. Part I. Bottom Backscattering Strength," *J. Acoust. Soc. Am.*, Vol. 77, 1985, pp. 962–974.

82. D. R. Jackson, A. M. Baird, J. J. Crisp, and P. A. G. Thomson, "High-Frequency Bottom Backscatter Measurements in Shallow Water," *J. Acoust. Soc. Am.*, Vol. 80, 1986, pp. 1188–1199.

83. S. Stannic, K. B. Briggs, P. Fleisher, R. I. Ray, and W. B. Sawyer, "Shallow Water High-Frequency Bottom Scattering off Panama City, Florida," *J. Acoust. Soc. Am.*, Vol. 83, 1988, pp. 2134–2144.

84. S. Stannic, K. B. Briggs, P. Fleisher, W. B. Sawyer, and R. I. Ray, "High-Frequency Acoustic Backscattering from a Coarse Shell Ocean Bottom," *J. Acoust. Soc. Am.*, Vol. 85, 1989, pp. 125–136.

85. M. Gensane, "A Statistical Study of Acoustic Signals Backscattered from the Sea Bottom," *IEEE J. Ocean. Eng.*, Vol. 14, 1989, pp. 84–93.

86. D. R. Jackson and K. B. Briggs, "High-Frequency Bottom Backscattering: Roughness versus Sediment Volume Scattering," *J. Acoust. Soc. Am.*, Vol. 92, 1992, pp. 962–977.

87. P. D. Mourad and D. R. Jackson, "High Frequency Sonar Equation Models for Bottom Backscatter and Forward Loss," in *Proceedings of Oceans '89*, IEEE Press, New York, 1989, pp. 1168–1175.

88. R. B. Patterson, "Backscatter of Sound from a Rough Boundary," *J. Acoust. Soc. Am.*, Vol. 35, 1963, pp. 2010–2013.

89. R. J. Urick and D. S. Saling, "Backscattering of Explosive Sound from the Deep Sea Bed," *J. Acoust. Soc. Am.*, Vol. 34, 1962, pp. 1721–1724.

90. A. W. Burstein and J. J. Keane, "Backscattering of Explosive Sound from Ocean Bottoms," *J. Acoust. Soc. Am.*, Vol. 36, 1964, pp. 1596–1597.

91. Yu. Yu. Zhitkovskii and L. A. Volovova, "Sound Scattering from the Ocean Bottom," *Congrès International d'Acoustique*, Liège, Paper E67, 1965.

92. R. P. Chapman, "Sound Scattering in the Ocean," in V. M. Albers (Ed.), *Underwater Acoust.*, Vol. 2, Plenum Press, New York, 1967, pp. 161–183.

93. Yu. Yu. Zhitkovskii, "The Sea Bottom Backscattering of Sound (The History and Modern State)," in H. M. Merklinger (Ed.), *Progress in Underwater Acoustics*, Plenum Press, New York, 1986, pp. 15–23.

94. H. M. Merklinger, "Bottom Reverberation Measured with Explosive Charges Fired Deep in the Ocean," *J. Acoust. Soc. Am.*, Vol. 44, 1968, pp. 508–513.

95. J. P. Buckley and R. J. Urick, "Backscattering from the Deep Sea Bed at Small Grazing Angles," *J. Acoust. Soc. Am.*, Vol. 44, 1968, pp. 648–650.

96. P. C. Hines and P. J. Barry, "Measurements of Acoustic Backscatter from the Sohm Abyssal Plain," *J. Acoust. Soc. Am.*, Vol. 92, 1992, pp. 315–323.

97. M. M. P. Weydert, "Measurements of the Acoustic Backscatter of Selected Areas of the Deep Seafloor and Some Implications for the Assessment of Manganese Nodule Resources," *J. Acoust. Soc. Am.*, Vol. 88, 1990, pp. 350–366.

98. P. D. Mourad and D. R. Jackson, "A Model/Data Comparison for Low-frequency Bottom Backscatter," *J. Acoust. Soc. Am.*, Vol. 94, 1993, pp. 344–358.

99. A. E. Robison, "Bottom Reverberation in the North Atlantic," Defence Research Establishment Atlantic, Dartmouth N.S., Technical Memorandum 75/B, 1975.

41

SOUND RADIATION FROM MARINE STRUCTURES

DAVID FEIT

1 INTRODUCTION

Marine structures radiate sound as a result of the time-dependent pressure fluctuations communicated to the surrounding water medium by the vibratory motions of their hull envelopes. These motions are in response to unsteady forces and moments generated within the hull by the many machines necessary to the ship's operation. The effects of the forces are transmitted to the hull plating via the structures supporting the machinery.

Ross (Ref. 1, pp. 326–347) characterizes the internal sources as being primarily due to (1) mechanical imbalances, (2) electromagnetic force fluctuations, and (3) friction between moving parts and impact sounds such as gear tooth impacts or piston slap. The present treatment is not concerned with the sources but examines the dynamic response of the hull and the sound radiation arising from the vibrations of the wetted hull surfaces due to the sources.

For present purposes water is assumed to be nonviscous and slightly compressible. Because of the inviscid nature of water, it is only the normal component of the hull-plating vibratory response that imparts its motion to the contiguous fluid particles, and this in turn generates unsteady pressure fluctuations in the water. These pressure fluctuations not only radiate sound to distant observers but also react back on the structure in the form of *radiation loading*. This latter effect significantly complicates the problem, altering the vibrations by effectively adding mass to the structure, thus reducing the frequency and magnitude of vibration. Furthermore, the fluid motions carry energy away from the structure, hav-

ing the effect of an additional resistive force, acting to reduce the magnitude of vibration and dissipate the vibratory energy.

In addition to internal machinery sources, a ship's propeller also generates time-varying forces and moments as it rotates through the nonuniform flow that exists in the wake of the ship. These are transmitted to the hull via the propeller shaft and thrust bearing. At the same time, the propeller generates a rotating pressure field that acts as an unsteady force on nearby appendages or hull surfaces, creating vibrations and its accompanying radiation of sound.

In this chapter we assume that the unsteady forces and/or moments are prescribed and concern ourselves only with the vibrations and sound generated by the structure in response to these forces.

2 VIBRATION AND RADIATION FROM SHIP HULLS

Ship hulls are very complex structures. Therefore, to gain an understanding of the mechanisms involved in ship sound radiation, the structures are highly idealized. This allows us to arrive at approximate explicit results to describe the fundamental phenomena that are involved, rather than give an exact one-to-one correspondence with realistic structural configurations. To effect the latter would require a detailed discussion of computational structural acoustics as applied to ship structures. The general subject of computational methods in acoustics is treated more thoroughly in Chapters 11, 14, and 15.

Besides the assumptions made with respect to ship structures, we deal with the radiation problem in as

Encyclopedia of Acoustics, edited by Malcolm J. Crocker
ISBN 0-471-80465-7 © 1997 John Wiley & Sons, Inc.

simple a way as possible. To achieve this, we assume that the structures representing the ships are completely submerged in an infinite body of water. Therefore, for surface ships the radiation results must be modified to account for the free surface. This may be accomplished by adding to the source distributions, determined by the ship's vibratory motions, image source distributions of equal and opposite strength located at image points with respect to the free surface. Such an assumption yields an overall dipole-like directivity pattern in the vertical plane for a surface ship far-field pressure.

Ship hulls are complicated structures, and the frequency range over which they vibrate and radiate sound is extremely wide. It is neither recommended nor possible to postulate a single all-encompassing model that can be used to investigate the problem over such a wide range. It is useful to divide the frequency range into three parts designated as low frequency (LF), mid frequency (MF), and high frequency (HF). Such a division is described by Ross (Ref. 1, pp. 100–102), among others.

Within each of these ranges we shall postulate appropriate modeling procedures that have proven to be useful. In practice, modeling of the structure used for estimating the hull vibrational response is more complicated than that required for estimation of the radiated sound. This is due to the wide variability and uniqueness of design for each type of ship structure. Here we use very simplified and idealized models for illustrative purposes only.

2.1 Low-Frequency Range

In the LF range, which extends from about 1 Hz up to the frequency at which an acoustic wavelength is on the order of half the ship's length (e.g., for a ship that is 250 ft long, this corresponds to 40 Hz), the hull vibrates as a nonuniform elastic beam. Because of the low frequencies involved and the stiffness of the structure, a localized excitation causes the entire structure to vibrate, and the vibration patterns are said to be *global*, that is, they extend over the entire length of the body.

In calculating the vibration response in this range the water is assumed to be incompressible. The water strongly affects the motion by the addition of *added mass*. This is the inertia effect of the surrounding medium. Sound radiation can then be estimated by assuming that the vibrating body is a distribution of acoustic volume sources whose strength is determined by the product of the normal acceleration and the surface area.

When considered as a rigid body, the hull has six degrees of freedom that are superimposed on its steady forward speed. The orientation of a ship relative to a typical set of coordinate axes is shown in Fig. 1. In this pre-

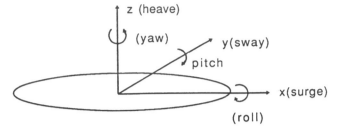

Fig. 1 Axes showing directions and names of a ship's rigid body motions (both linear and angular).

sentation we deal with the vibrations and sound radiation of ships that are completely submerged, and we therefore illustrate the ship as a prolate spheroid. We call the rigid-body translational motions in the x-, y-, and z-directions *surge, sway*, and *heave*, respectively, while the angular motions about the same axes are called *roll, pitch*, and *yaw*.

The ship's elastic vibrations in which a cross section vibrates uniformly, back and forth, in the surge direction are referred to as *longitudinal* vibrations, while vibrations of the section in the two directions transverse to the longitudinal axis are known as *whipping, bending*, or *flexural* vibrations. The latter can occur in either the vertical or horizontal planes. As the hull stretches and shortens in longitudinal vibration, there are radial motions of the hull plating induced by the Poisson ratio effect. The motion normal to the structure's surface is the only vibratory component that is coupled to the water medium for a nonviscous fluid.

In modeling for vibration and sound radiation in the LF range, the ship structures are assumed to be beams, freely suspended in the water medium, with longitudinally varying cross-sectional properties. To handle the longitudinal variation, a *lumped-parameter* multi-degree-of-freedom model of the beams can be used in determining the ship's LF vibration response. For illustrative purposes only, we shall consider the axial variation of stiffness, both longitudinal and transverse bending, to be negligible. This allows us to introduce analytic expressions rather than numerical solutions for the vibration distributions.

2.2 Midfrequency Range

The MF range covers those frequencies between the LF and the HF range. Here the response of the hull to a localized force extends approximately over a compartment length. In this range the acoustic wavelength varies from being somewhat larger than a compartment length to a fraction of the circumference of a ship's cross section. As an example, we consider a submarine hull to be adequately modeled as a cylindrical shell separated into compartments by bulkheads. Thus for a submarine having a diameter of 30 ft and a length of 250 ft, the MF range would go from about 40 Hz to approximately 200 Hz where the cylinder's circumference is about four acoustic wavelengths.

Here we typically model the hull compartment as a finite-length elastic cylindrical shell terminated structurally by bulkheads, or end closures. Resonances of the shell section play a significant part in the sound radiation phenomena, and the entire length of the compartment is assumed to participate in the vibration and sound radiation. The resonance frequencies are sensitive to the boundary conditions assumed at the ends of the cylindrical segments as well as the dynamics of the structural systems internal to the hull.

To make the acoustics problem more amenable to analysis, the remaining portion of the hull is modeled as a rigid cylindrical baffle extending to infinity in both fore and aft directions. Any more refined investigation would necessitate an analysis that makes use of numerical methods.

2.3 High-Frequency Range

In this range the vibrating forces excite only a small portion of the hull. The lower boundary of this range is sometimes taken as the ring resonance frequency $f_r = c_p/2\pi a$, where c_p is the hull plating compressional wave velocity and a is the compartment cylinder radius. The effects of curvature of the hull plating and the importance of resonances diminish. Here details of the hull plating and the stiffening structures take on added significance to the radiation process. Analytical methods have been most successful in this range, and it is from such analyses that much of our intuitive knowledge about the radiation of elastic structures arises.

3 VIBRATION AND RADIATION IN LOW-FREQUENCY RANGE

3.1 Longitudinal Vibrations

We treat the hull as an equivalent beam in the LF range. The motions of a beamlike ship include longitudinal (when dealing with longitudinal motions beams are sometimes referred to as rods), flexural, and torsional vibrations. For the cases to be illustrated here, the torsional vibrations of the ship are of no significance to the acoustic radiation problem. For this treatment the internal mass distributions are assumed to be rigidly attached at discrete points along the length of the hull and for simplicity do not add any bending or longitudinal stiffness. They do, however, add distributed mass to the beam model. The internal mass is assumed to be distributed in

such a way that there is no mass coupling of longitudinal and flexural vibrations.

With these assumptions, the equation of motion (assuming harmonic time variation of the form $e^{-i\omega t}$ for all time-dependent variables) that determines the longitudinal vibrations is given by

$$\frac{d}{dx}\left(EA\frac{dU}{dx}\right) + (\rho_s A + M)\omega^2 U = -F(x). \quad (1)$$

Here $U(x)$ is the longitudinal component of displacement, E is Young's modulus, $A(x)$ is the cross-sectional area, ρ_s is the density of the hull material, $M(x)$ is the internal mass distribution per unit length, and $F(x)$ is the externally applied longitudinal force distribution. The displacement $U(x)$ is subject to the free–free boundary conditions at the ends, given by

$$\frac{dU(0)}{dx} = \frac{dU(L)}{dx} = 0. \quad (1a)$$

Assuming that the cross-sectional area and the internal mass distribution are uniform, Eq. (1) can be solved in terms of the *eigenmodes* of the system $\cos(n\pi x/L), n = 0, 1, 2, \ldots, \infty$. For a compact force of magnitude F_0 applied at the longitudinal location $x = x_0$, the solution is

$$U(x) = \frac{F_0}{M_T L}\left[\frac{1}{\omega^2} + 2\sum_{n-1}^{\infty}\frac{\cos(n\pi x/L)\cos(n\pi x_0/L)}{\omega^2 - \omega_n^2}\right],$$

$$(2)$$

where $\omega_n = C_B\sqrt{M_P/M_T}(n\pi/L)$ is the nth *resonance* frequency of the hull vibration and M_P/M_T represents the ratio of pressure hull mass to the total mass of the ship.

Examination of Eq. (2) reveals that if the excitation frequency corresponds to any of the resonance frequencies, the solution becomes unbounded. This situation is circumvented by ascribing an ad hoc measure of damping to the structure. We achieve this by assuming that the Young's modulus is complex according to the relation $E^* = E(1 - i\eta_s)$, where η_s is the structural loss factor. The first term on the right-hand side of Eq. (3) represents the rigid-body surging motion of the body, while the terms appearing in the summation are the elastic vibration mode contributions. At each of the resonance frequencies of the hull the vibration displacement is inversely proportional to the structural loss factor.

In Fig. 2 we show the real and imaginary parts of the normalized response for two different excitation frequen-

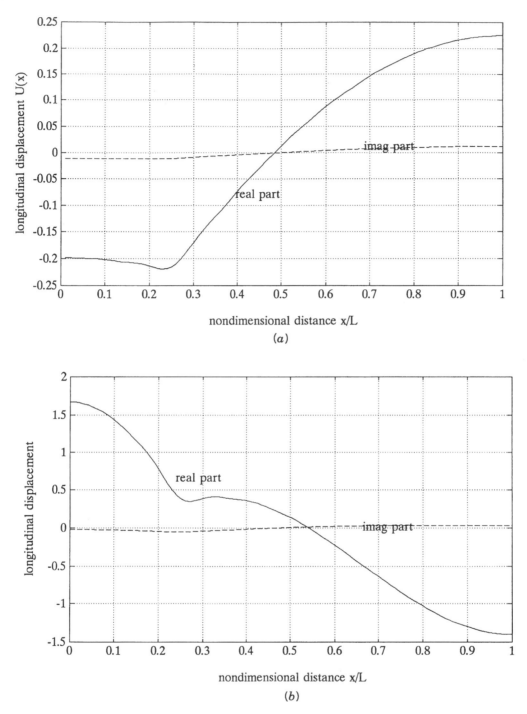

Fig. 2   Forced response: (*a*) nondimensional frequency = 0.5 and (*b*) nondimensional frequency = 1.5.

cies, one at a frequency below the first elastic mode and the other midway between the first and second resonance frequencies of the beam. The apparent discontinuity in the slope of the response at $x = L/4$ reflects the fact that the applied load is represented as a point load.

The far-field radiated pressure can be calculated in this LF range by using the response function (2) and cal-

culating the equivalent volumetric source distribution on the surface of the cylinder. The pressure radiated by an idealized point source located at the origin of a Cartesian coordinate system is given by

$$p(R, \theta) = \frac{\rho \ddot{Q}}{4\pi R} \, e^{ikR}, \tag{3}$$

where $k = \omega/c$ is the acoustic wavenumber, c is the speed of sound in the acoustic medium, and $\ddot{Q}$ is the volumetric source strength. The parameter $R = \sqrt{x^2 + y^2 + z^2}$ measures the distance between the observation point and the source location.

The radiating sources are, in this case, the two end faces of the cylinder, with displacements $U(0)$, $U(L)$ and the cylindrical surface, which has a radial component of displacement due to Poisson's ratio effect given by $W(x) = -\nu a \, dU/dx$. The expression for the far-field pressure in terms of the hull response function, assuming a circular cross section of radius a, is then given by

$$p(R, \theta) = -\frac{\rho(\omega a)^2}{4\pi R_0} \, e^{ikR_0} \left[U(L) e^{-i(kL/2)\cos\theta} \right.$$

$$- U(0) e^{i(kL/2)\cos\theta}$$

$$\left. -2\nu \int \frac{dU(x)}{dx} \, e^{-ikx\cos\theta} \, dx \right], \quad (4)$$

where θ is the polar angle made by the observation point position vector and the centerline of the axis along which it vibrates. Here, R_0 is the distance to the origin. The first two terms inside the brackets on the right-hand side of Eq. (4) are the end-face contributions while the term under the integral represents the radiation from the cylindrical surface.

The far-field directivity patterns for the two cases illustrated in Fig. 2 are shown in Figs. 3 and 4. The higher frequency case shown in Fig. 4 reveals a more

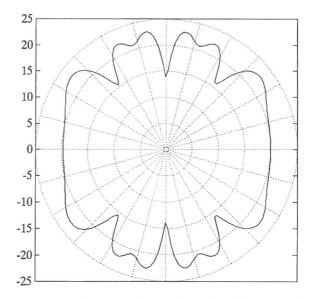

Fig. 4 Directivity pattern for nondimensional frequency = 1.5 (see Fig. 2*b*).

highly directive pattern. This is to be expected because the driving frequency is three times as high as that used in the first case, making the beam length three times longer in terms of acoustic wavelengths. In each pattern the levels are given in decibels normalized to the pressure at $\theta = 0$.

3.2 Flexural Vibrations

For forces that are applied in the vertical or horizontal direction, the beam representing the ship structure vibrates in flexure. The equation of motion (again assuming harmonic time variation) for such motions $V(x)$, using the theory of beam flexural vibrations, is given by

$$\frac{d^2}{dx^2} \left(EI \, \frac{d^2V}{dx^2} \right) - m(x)\omega^2 V = F_V, \quad (5)$$

where I is the beam's cross-sectional moment of inertia and $m(x)$ is the total mass distribution per unit length, including both the beam's cross-sectional mass and its distributed internal mass.

A beam that is free at both ends must satisfy the condition that $d^2V/dx^2 = d^3V/dx^3 = 0$ at the ends $x = 0, L$. These conditions admit the two rigid-body modes, which in the vertical plane are a heaving and pitching motion and in the horizontal plane would be swaying and yawing motion. For this case as well as any other case of flexural vibrations, except that of a uniform beam simply supported at both ends, the non-rigid-body solutions to Eq. (5) are functions of eigenvalues that cannot be

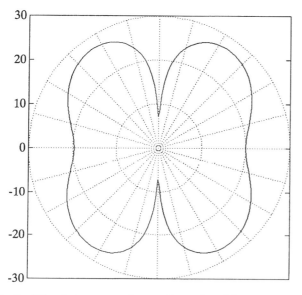

Fig. 3 Directivity pattern for nondimensional frequency = 0.5 (see Fig. 2*a*).

explicitly obtained. To solve such problems, especially in the realistic situation where the beam properties are a function of position along the length of the ship, numerical solutions are used.

For present purposes we assume that the flexural response has been determined by one of the numerical approaches. An approximate expression for the far-field radiated pressure assumes the radiation is from a distribution of dipoles oriented in the direction of the response along the beam axis. The result, first derived by Junger[2] and also by Chertock,[3] is

$$p(R, \theta, \phi) = \frac{i\rho\omega^2 k}{2\pi R} \sin\theta$$

$$\cos\phi \int_0^L A(x)V(x)e^{-ikx\cos\theta}\,dx, \quad (5a)$$

where $A(x)$ is the cross-sectional area distribution, the angle ϕ is the angle between the plane of vibrations and the plane determined by the observation point position vector and the centerline of the ship, and θ is the angle in the vibration plane measured from the position vector projection to the beam axis perpendicular.

4 VIBRATION AND RADIATION IN MIDFREQUENCY REGION

In the MF range we assume that the ship's radiation can be modeled as if the radiation emanates from a limited region of the ship. Typically, the ship structure has a number of strong discontinuities such as bulkheads separating it into compartments. It then becomes appropriate in this MF region to model only that portion of the hull between major bulkheads.

An adequate representation of a ship's compartment in this frequency range is a ring-stiffened cylindrical shell simply supported by rigid and motionless end caps. The latter simulate the compartment bulkheads. The ring stiffeners are accounted for by assuming an equivalent orthotropic shell with a different bending stiffness in the circumferential and longitudinal directions. A cylindrical shell of length L_c, the compartment length, radius a, and thickness h is shown in Fig. 5.

The sound radiated by a shell, or for that matter by any structural model, depends only on the velocity component normal to the surface. We therefore focus our attention on w, the radial displacement (the radial velocity $\dot{w}$ in the case of periodic excitation is simply related to the displacement by the relation $\dot{w} = -i\omega w$, where ω is the circular frequency corresponding to the excitation frequency).

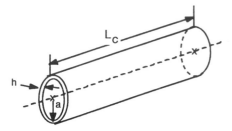

Fig. 5 Cylindrical shell.

The shell's motion is described as a superposition of responses in various modes. Each mode's response is determined by the generalized parameters such as modal mass, stiffness, and resistance. In this fashion, the radial response of the shell is written as a modal sum of the form

$$w(z, \phi) = \sum_{m=1}^{\infty} \sum_{n=1}^{\infty} W_{mn} \cos n\phi \sin k_m z, \quad (6)$$

where W_{mn} is the amplitude of the mode having $2n$ (circumferential mode number) nodes in the circumferential direction and $k_m = m\pi/L_c$ is the axial wavenumber, where m is the number of half wavelengths within the length L_c of the shell.

A highly exaggerated picture of the radial displacement around the shell circumference at a particular cross section is shown in Fig. 6 for $n = 1, 2, 3$. The mode $n = 1$ corresponds to the flexural vibration of a beam having a thin cylindrical shell cross section. The higher modes $n = 2, 3, \ldots$ are more characteristic of shells and are referred to as lobar modes. The $n = 0$ mode is axisymmetric (no ϕ-dependence) and is known as the breathing mode of the shell.

The modal amplitudes W_{mn} are determined by the generalized modal parameters, which are here explicitly listed:

F_{mn} Modal force, or component of force that acts on a particular mode, a measure of how well the force distribution matches the displacement distribution in a given mode

M_{mn} Modal generalized mass

ω_{mn} Modal resonance frequency

R_{mn} Total modal resistance, including the acoustic radiation resistance $R_{mn}^{(a)}$ and a structural modal resistance $R_{mn}^{(s)}$, related to the intrinsic loss factor discussed earlier

A detailed derivation of the above parameters, especially the modal resistance expressions, is beyond the

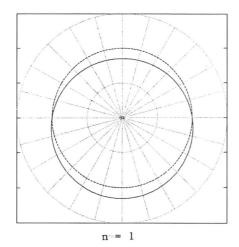

n = 1

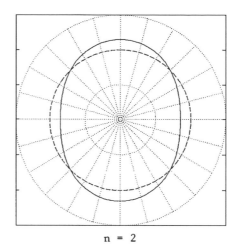

n = 2

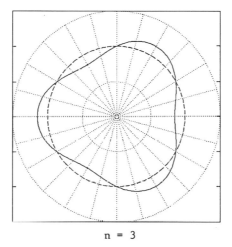

n = 3

Fig. 6 Cylindrical shell model used in the midfrequency range.

scope of this presentation but can be found in Junger and Feit.[4] Here approximations to those parameters that are necessary for an estimation of the vibrational response are presented. We now consider the case of a cylindrical shell section of length L_c driven radially by a concentrated force of magnitude F at the midpoint of the shell section, $z = L_c/2, \phi = 0$; the modal amplitudes of the velocity response $\dot{W}_{mn}$ are given by

$$\dot{W}_{m0} = \frac{F}{\pi a L_c} \{\sin(m\pi/2)\}/R_{m0} - i\omega\rho_s h[1 - (\omega_0/\omega)^2]$$

$$\cdot (1 + \rho L_c/\rho_s h_m m\pi), \tag{7a}$$

$$\dot{W}_{mn} = \frac{2F}{\pi a L_c} \{\sin(m\pi/2)\}/R_{mn} - i\omega\rho_s h(1 - (\omega_{mn}/\omega)^2)$$

$$\cdot (1 + 1/n^2 + \rho/\rho_s h_m k_s). \tag{7b}$$

These expressions are in the form characteristic of the response of single-degree-of-freedom systems. The modal resonance frequencies can be approximated using the formulas

$$\omega_{m0} = \frac{c_p}{a} \frac{1}{(1 + \rho L_c/\rho_s h_m m\pi)^{1/2}}, \tag{8a}$$

$$\omega_{m1} = \frac{c_p}{a} \{[(1 - \nu^2)(m\pi/k_s L_c)^4$$

$$+ (h^2/12a^2)(k_s a)^4]/(1 + \rho/2\rho_s k_s h_m)\}^{1/2}, \tag{8b}$$

$$\omega_{mn} = \frac{c_p}{a} \{[(1 - \nu^2)(m\pi/k_s L_c)^4 + [(m\pi/L)^2$$

$$+ \sqrt{D_c/D_a}(n/a)^2]^2(h^3 a^2/12 h_m)]/$$

$$[1 + \rho/\rho_s k_s h_m(n^2/n^2 + 1)]\}^{1/2}. \tag{8c}$$

The new quantities introduced in the above formulas, in addition to ν, Poisson's ratio (for steel $\nu = 0.29$), and ρ_s, the density of shell material, are $h_m = M_0/\rho_s A$, the effective shell thickness derived from the total shell mass M_0, the shell surface area A, the helical wavenumber given by the relation $k_s = [(m\pi/L_c)^2 + (n/a)^2]^{1/2}$, and D_c/D_a, the ratio of the circumferential bending stiffness to the axial bending stiffness calculated as for a flat orthotropic plate. In calculating the resonance frequencies of the modes given in Eq. (7), the added mass effects of the fluid loading have also been approximated using expressions derived in Junger and Feit.[4]

Some of the modal responses are critically dependent on the radiation resistance portion of the total modal resistance. To estimate this factor, we must consider the acoustic radiation properties of the modes. The modal radiation resistance is a measure of how efficiently the spatial characteristics of a particular mode give rise to a radiated sound pressure. The critical parameter of the vibrational pattern in this regard is the scale of the vibration pattern, or structural wavelength, relative to the acoustic wavelength at the frequency in question. The scale of the structural vibration is conveniently given by the axial wavelength $2L_c/m$ for any axial mode of order m and the circumferential wavelength $2\pi a/n$.

Using the results and experiences of previous studies, the modes can be organized into various types:

1. *Surface modes* are those whose helical wavelength is greater than the acoustic wavelength at a given frequency. The radiation resistance of such a mode is approximately

$$\frac{R_{mn}^{(a)}}{\rho c} \approx 1. \tag{9}$$

2. *Edge modes* are modes whose axial wavelength is less than the acoustic wavelength while the circumferential wavelength is larger than the acoustic wavelength. An estimate of the radiation resistance of such a mode is

$$\frac{R_{mn}^{(a)}}{\rho c} = 2\left(\frac{\omega L_c}{c}\right)\frac{1}{m^2\pi^2}, \qquad \frac{\omega L_c}{c} \gg 1, \tag{10a}$$

$$\frac{R_{mn}^{(a)}}{\rho c} = 4\left(\frac{\omega L_c}{c}\right)\frac{1}{m^2\pi^2}, \qquad \frac{\omega L_c}{c} \ll 1. \tag{10b}$$

3. *End modes* are a special case that arises for $n = 0$ modes when $\omega a/c < 1$, and an estimate of its radiation resistance is

$$\frac{R_{m0}^{(a)}}{\rho c} \approx 2\left(\frac{\omega L_c}{c}\right)\left(\frac{\omega a}{c}\right)\frac{1}{m^2\pi^2},$$

$$\frac{\omega L_c}{c} \gg 1, \tag{11a}$$

$$\frac{R_{m0}^{(a)}}{\rho c} = 4\left(\frac{\omega L_c}{c}\right)\left(\frac{\omega a}{c}\right)\frac{1}{m^2\pi^2},$$

$$\frac{\omega L_c}{c} \ll 1. \tag{11b}$$

4. For the remaining modes not satisfying the above conditions, the acoustic resistance is taken as zero.

The structural damping part of the modal resistance is given by

$$R_{m0}^{(s)} \approx \eta_s \rho_s h_m \left(\frac{\omega_{m0}}{\omega}\right)^2 \left(1 + \frac{\rho L_c}{\rho_s h_m m\pi}\right), \tag{12a}$$

$$R_{mn}^{(s)} \approx \eta_s \rho_s h_m \left(\frac{\omega_{mn}}{\omega}\right)^2 \left(1 + \frac{1}{n^2} + \frac{\rho}{\rho_s k_s h_m}\right),$$

$$n \geq 1. \tag{12b}$$

The response of the shell and the near-field sound pressure are dominated by the resonant modes that radiate a negligible amount of sound power to the far field. Because they radiate little sound power, these modes are controlled by the structural damping, that is, $R_{mn}^{(s)} \gg R_{mn}^{(a)}$. When we combine Eqs. (7) and (12), we see that the modal amplitude and, therefore, the structural response can be effectively controlled by increased structural damping. Increased damping reduces the resonant response of the shell and hence the vibration level of the shell itself as well as its near-field sound pressures. It does not, however, significantly reduce the total radiated sound power that is dominated by a large number of nonresonant, efficiently radiating surface modes and a lesser number of resonant edge modes.

5 VIBRATION AND RADIATION IN HIGH-FREQUENCY REGION

As the excitation frequency increases, we reach the regime in which the effects of the radiating surface curvature become negligible. In this range circumferential stresses due to curvature become much less important than the out-of-plane bending stresses. We therefore assume that the behavior of the actual structure can be approximated by that corresponding to a flat elastic plate. The vibratory response of a plate is most easily evaluated using a structural theory rather than an elasticity theory when the thickness of the plate is small relative to a flexural wavelength. The flexural wavelength in inches of a steel plate is given by the formula $\lambda_f = 612\sqrt{h/f}$, where f is in hertz and h is in inches.

The parameters that determine the vibrational response of the plate are its mass density per unit area m and its bending stiffness $D = Eh^3/[12(1 - \nu^2)]$. In most applications the plates are stiffened by frames. If the distance between the frames is small compared to a flexural wavelength, then the stiffened plate can be conveniently idealized as an orthotropic plate with different bending stiffnesses for each of two orthogonal directions. We write D_x as the bending stiffness in the stiffest direction (for example, in the previous section we assumed the cylindrical shell was framed circumferentially and D_x would correspond to the circumferential bending stiffness) and D_y as the bending stiffness in the least stiff direction (for the previous case this corresponds to the axial direction, i.e., the direction perpendicular to the frames).

As in the previous discussions the motion of the plate is affected by the presence of the water. Depending on the excitation frequency, the water adds either additional mass to the plate or a resistive component dissipating the motion of the plate. We consider such a plate to be of infinite extent in all directions. Whether it be *isotropic* (plate of uniform thickness with no variation of bending stiffness in any direction) or *orthotropic* (bending stiffness different for different directions because of the framing), the vibrational response to an oscillating force consists of two parts. One part is an outwardly propagating wave that spreads out in all directions, analogous to the outwardly spreading wave caused by a stone dropped onto the quiet surface of a lake. This is referred to as the *propagating field*. The other part of the response consists of a pistonlike motion that is confined to the vicinity of the excitation point and is referred to as the *near field*.

The speed at which the wave crests of the propagating field spread out varies with frequency and the properties of the plate. This propagation speed c_f is called the flexural wave speed and for a steel plate is given by $c_f = 612\sqrt{fh}$ inches per second. From this formula we note that the flexural wave speed is proportional to the square root of the frequency. As the frequency of excitation increases, we find that there is a frequency at which the flexural wave speed coincides with the acoustic wave speed. This frequency is called the *coincidence frequency*.

If one were to calculate the flexural wavelength at the coincidence frequency or higher, we would find that it is equal or greater than the acoustic wavelength at this same frequency. This gives rise to an enhanced radiation

of sound in a specific direction known as the *coincidence direction*. The coincidence frequency for a steel plate in water is related to the plate thickness through the equation $f_c = 9300/h$, where frequency is given in hertz and h is measured in inches.

Consider an orthotropic plate of infinite extent lying in the (x, y) plane, driven by a normal point force $F_0 e^{-i\omega t}$ applied at the origin. The bending stiffness in the x-direction is D_x, and D_y is that in the y-direction. The radiated far-field pressure is then given by

$$p(R, \theta, \phi) = \frac{-ikF_0 e^{ikR}}{2\pi R} \; [\cos \theta]/1 - i(\omega m/\rho c)$$
$$\cdot \cos \theta [1 - (\omega/\omega_c)^2 \sin^4 \theta (\cos^2 \phi$$
$$+ \sqrt{D_y/D_x} \sin^2 \phi)^2], \qquad (13)$$

where θ is the polar angle between the perpendicular to the plate and the observation point vector and ϕ is the circumferential angle between the x-axis and the projection of that same vector on the plane of the plate.

For very low frequencies such that $\omega m/\rho c \ll 1$, the pressure field is that of a force acting directly on an acoustic half-space,

$$p(R, \theta, \phi) = -\frac{-ikF_0 e^{ikR}}{2\pi R} \cos \theta. \qquad (14)$$

From this expression we see that the plate presence has no effect on the radiated pressure.

At somewhat higher frequencies but still in the frequency range less than coincidence, the radiated pressure becomes

$$p(R, \theta, \phi) = \frac{-ikF_0 e^{ikR}}{2\pi R} \frac{\cos \theta}{1 - i(\omega m/\rho c)\cos \theta}. \qquad (15)$$

Here it is only the mass per unit area of the plate that is significant, while the stiffness of the plate plays no role. The stiffness terms do not play a significant role until the frequency approaches coincidence and above.

In Fig. 7 we plot the polar angle dependence of the far-field radiated pressure for a number of frequencies. In each curve the pressures are plotted in decibels relative to the on-axis pressure, that is, on the normal to the plate. For illustrative purposes we have assumed that the plate is of uniform thickness, having no stiffeners and therefore isotropic. This accounts for the fact that the pressure is independent of the circumferential angle.

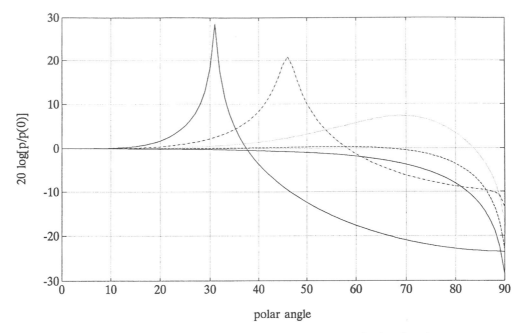

Fig. 7   Far-field directivity: polar angle plotted against far-field radiated pressure.

We see that in the frequency range above coincidence the pressure maximum occurs at a specific angle, the *coincidence angle*, which is off the normal and is given by the relation $\theta_c = \sin^{-1} \sqrt{f_c/f}$. As the frequency increases, the coincidence angle approaches the normal. In this extremely high frequency range, a higher order plate theory such as that of Timoshenko-Mindlin (Ref. 4, pp. 214–215) would be more appropriate. Had such a theory been used in the calculations, the coincidence angle would asymptotically approach a limiting angle $\theta_c = \sin^{-1}(c/c_R)$ where c_R is the Rayleigh wave speed for the plate material.

REFERENCES

1. D. Ross, *Mechanics of Underwater Sound*, Pergamon Press, New York, 1976, pp. 326–347.

2. M. C. Junger, "Sound Radiation by Resonances of Free-Free Beams," *J. Acoust. Soc. Am.*, Vol. 52, No. 1 (Part 2), pp. 332–334.

3. G. Chertock, "Sound Radiated by Low-Frequency Vibrations of Slender Bodies," Vol. 57, No. 5, May 1975, pp. 1007–1016.

4. M. C. Junger and D. Feit, *Sound, Structures and Their Interation*, 2nd ed., MIT Press, Cambridge, MA, 1986.

42

TRANSIENT AND STEADY-STATE SCATTERING AND DIFFRACTION FROM UNDERWATER TARGETS

LOUIS R. DRAGONETTE AND CHARLES F. GAUMOND

1 INTRODUCTION

Much of the understanding of the physics of underwater acoustic scattering has been derived from investigations of the scattering of plane waves by simple shapes, especially elastic spheres and spherical shells and the infinite-elastic cylinder and cylindrical shell. The spherical and cylindrical shapes lend themselves to exact solutions and thus to the very accurate calibration and verification of the experimental and numerical approaches required to deal with the more complicated problems of interest. The term *canonical* in this chapter will be limited to those two simple shapes.

Three formulations of the canonical problems are of particular interest.

The normal-mode series solution, or Rayleigh series, is the description best suited to accurate computation of scattering by the canonical shapes as a function of frequency.[1-12] The relative ease and accuracy of this technique over a broad bandwidth permits, among other uses, the simulation of transient responses[4,13] and the calibration of sophisticated transient and steady-state measurement techniques.

The "creeping-wave" formulation, a contour integral representation of the scattering problem,[14-19] led to the circumferential wave descriptions of scattering. This description has been especially valuable in the understanding of the contributions of diffraction (Franz waves), the "leaky-Rayleigh-surface" wave, and guided "leaky Lamb" waves and had a major influence in research on active classification approaches.[20,21]

Encyclopedia of Acoustics, edited by Malcolm J. Crocker
ISBN 0-471-80465-7 © 1997 John Wiley & Sons, Inc.

The formalization of a resonance theory of acoustic scattering,[22] analogous to nuclear reaction theory,[23] grew out of comparisons between the Rayleigh series and contour integral formulations. (See Chapter 5, Section 3.) Families of free-body resonances of an aluminum cylinder were identified with particular elastic surface waves.[24] A physically meaningful separation of the Rayleigh series solution into a geometric, rigid background term and a term containing the elastic resonance behavior was accomplished[22] and demonstrated.[25-27] The introduction of this resonance formalism[22] has been, subsequently, the source of a prolific area of published theoretical scattering research,[28-31] only a small sampling of which will be indicated here.

Scattering is quantitatively described in terms of the form function, which is related to the impulse response of a target, and target strength, which is the metric used in the active sonar equations. The techniques used to empirically determine these metrics will be outlined.

2 FORM FUNCTION

The form function is a dimensionless measure of the complex amplitude A of the pressure field scattered from a target. The form function requires the scattered pressure p_s to be normalized by the incident pressure p_0, the range r between the target and receiver, and a characteristic length dimension a of the target. The frequency is expressed by the dimensionless parameter $ka = 2\pi a/\lambda$, where λ is the wavelength of the incident sound in the surrounding water. The normalization factor for a given shape is, in general, chosen so that the magnitude of the backscattered form function is unity for perfectly rigid or soft boundary conditions in the high-frequency limit.

The scattering examples considered here will involve incident plane pressure waves, $p_0 = P_0 e^{-ikz}$, with harmonic time dependence $e^{+i\omega t}$; the scattered pressure field p_s approaches either $A(\Omega)e^{-ikr}/r$ in three dimensions or $A(\theta)e^{-ikr}/\sqrt{r}$ in two dimensions for the distances $r \gg a^2/\lambda$ large enough to be in the far-field or Fraunhofer zone.[32]

For a spherically shaped target the characteristic length dimension is the radius a and the far-field form function is defined as

$$f(\theta, ka) = \frac{2r}{a}\frac{p_s}{P_0}. \tag{1a}$$

The case most discussed in the literature is backscattering, $\theta = \pi$. The form function $f(\pi, ka)$ is then generally presented as a function of ka only; that is, when the angular dependence is subdued, backscattering $\theta = \pi$ is understood. Thus the backscattered form function is written as

$$f(ka) = \frac{2r}{a}\frac{p_r}{P_0}, \qquad \theta = \pi. \tag{1b}$$

For the infinite cylinder, the characteristic length dimension is again the radius a, but the far-field spreading of the scattered pressure has a $1/\sqrt{r}$ dependence. When the incident wave vector is normal to the cylinder axis, the far-field form function for backscattering is given as

$$f(ka) = \sqrt{\frac{2r}{a}}\frac{p_s}{P_0}. \tag{2}$$

3 NORMAL-MODE SOLUTIONS

Normal-mode series solutions for the far-field function of spherical[1–5,7,8,10–12] and cylindrical[1,2,6,9,12] shapes with a variety of boundary conditions are found in the literature. Consideration of these shapes, as evidenced by this large but not complete list, are predominate in the literature since these shapes lend themselves to solution by a separation-of-variables technique. The form function depends, of course, on the boundary conditions as well as the target shape. Results and interpretation for the case of solid elastic spheres and infinite cylinders are given below. The perfectly "rigid" (Neumann) and "soft" (Dirichlet) boundary conditions are considered here as special cases of the solid elastic target.

3.1 Spheres and Infinite Cylinders

For spherical geometries the far-field form function is given in Eq. (3) of Chapter 5 as

$$\sin \eta_n e^{i\eta_n} = \frac{j_n(ka)L_n - (ka)j_n'(ka)}{h_n^{(2)}(ka)L_n - (ka)h_n^{(2)\prime}(ka)}. \tag{3}$$

The functions j_n and h_n are spherical Bessel functions of the first and third kind, respectively, with primes denoting derivatives with respect to the argument. The factor L_n is a function of frequency and of the material properties of the sphere. For a rigid sphere $L_n = 0$, and for a pressure-release sphere $L_n = \infty$. The a_n are given by $j_n(ka)/h_n^{(2)}(ka)$ for the case of Dirichlet boundary conditions and $j_n'(ka)/h_n^{(2)\prime}(ka)$ for Neumann boundary conditions. The L_n for the solid elastic sphere with and without absorbtion are found in Ref. 10; for a spherical shell the L_n elements are found in Ref. 4. (Note that the complex conjugate of L_n is given in Ref. 4 where $e^{-i\omega t}$ is assumed.)

Normal-mode-series solutions in the infinite cylindrical geometry are found in many references[1,2,6,9,12] (Note that Refs. 1, 2, 9, and 12 use $e^{+i\omega t}$ while Ref. 6 uses $e^{-i\omega t}$.) In this geometry the far-field form functions have the form

$$f(ka) = \sqrt{\frac{2}{\pi ka}} \sum_{n=0}^{\infty} \epsilon_n C_n \cos(n\phi), \tag{4}$$

where ϵ_n is the Neumann factor ($\epsilon_n = 2, n = 0; \epsilon_n = 1, n > 0$) and C_n are given as

$$C_n = \frac{J_n(ka)L_n - ka J_n'(ka)}{H_n^{(2)}(ka)L_n - ka H_n^{(2)\prime}(ka)}, \tag{5}$$

with L_n given as quotients of determinants B_n and D_n:

$$L_n = \frac{\rho}{\rho_1}\frac{B_n}{D_n}. \tag{6}$$

The rank and elements of the determinants depend on the boundary conditions; ρ and ρ_1 are the densities of water and the material, respectively. Flax[9] gives the matrix elements required to determine L_n for the case of a cylindrical shell comprised of two absorbing layers with a fluid-filled interior. Instructions are given in the same reference for the reduction of this broad solution to the simpler cases of the solid elastic cylinder and the air- and water-filled cylindrical shells using the appropriate subset of these matrix elements. The solutions of the Neumann and Dirichlet problems reduce, as in the case of the sphere, to simplified expressions for the L_n. These are

$$L_n = 0 \quad \text{and} \quad C_n = \frac{J_n'(ka)}{H_n^{(2)'}(ka)}$$

for the rigid cylinder and

$$L_n = \infty \quad \text{and} \quad C_n = \frac{J_n(ka)}{H_n^{(2)}(ka)}$$

for the soft cylinder.

The form functions for rigid and elastic spheres and infinite cylinders are given in Figs. 1a and 1b, respectively. The form functions contain contributions from both the geometric and elastic properties of the scatterer. Figure 1a shows $|f(ka)|$ for a rigid sphere and spheres of aluminum, tungsten carbide, and brass. The form function of the rigid sphere is determined by two physical mechanisms—specular reflection from the surface of the sphere and diffraction of energy around the sphere. These are sketched in Fig. 1 of Chapter 5. The diffracted component can be described mathematically as a dispersive surface wave traveling wholly in the surrounding fluid at a speed slower than the compressional wave speed in the fluid. Peaks in the form function of a rigid sphere (and cylinder) occur at ka values for which the specular reflection and the creeping wave add in phase; nulls when the waves destructively interfere. The initial peak occurs near the same ka value (ka equals 1.15 for the sphere and 0.85 for the cylinder) for both the rigid and for each metal scatterer, thus providing a shape clue, that is, one that is independent of material. All solid elastic spheres and cylinders fabricated from materials whose densities and shear speeds are greater than the density and the sound speed in water produce form functions that at low ka are similar to the rigid-body form function and may be said to deviate from a "rigid background" at mid and high ka values. The form function curves for liquid[33] and plastic[10] spheres have unique properties. The amplitude of the initial peaks in Figs. 1a and 1b are related to the densities of the material and approach $|f| = 1$ as the density becomes larger.

A more significant material property can be determined by the ka value at which the form function curves first deviate significantly from the rigid case. As predicted by Faran,[1] Vogt[11] demonstrated for elastic-solid spheres that substantial variations occur above a critical ka value that depends on the resonance frequencies of the free sphere. Figure 2, computed by Vogt,[11] shows the amplitudes of the resonance terms plotted versus ka for three metals. The modes are labeled by the integers n, l. The n's, which correspond to each term in the normal-mode series, can take on values from 0 to ∞; the l's represent the overtones beginning with the fundamental $l = 1$. Just as the creeping-wave interaction with the specular provides unique shape clues, the results given in Figs. 1 and 2 show that unique information about the material composition may be obtained directly from the form function. The lowest observable mode for an elastic sphere is the (2, 1) spheroidal mode. Vogt[11] demonstrated that the shear speeds of the materials considered in Fig. 2 are inversely related to the ka position of the (2, 1) mode, that is, for two materials

$$\frac{c_{s1}}{(ka)_1} \sim \frac{c_{s2}}{(ka)_2}, \tag{7}$$

where c_{s1} and c_{s2} are the shear velocities of materials 1 and 2 and $(ka)_m$ is the position of the (2, 1) mode for material m. This expression gives resonable accuracy for all homogeneous materials whose density and shear speeds are greater than the density and sound speed in water. Similar elastic phenomena are exhibited in cylinders.[22, 25]

4   CREEPING-WAVE FORMULATION

Uberall and collaborators,[14–19] at the Catholic University, rewrote the Rayleigh series descriptions of the scattering by spheres and cylinders as contour integrals using the Watson transformation. This is discussed in Chapter 5. However, the growth of computer technology makes the Rayleigh series more computationally tractable for generating form functions of spheres and cylinders. The significant use of the Watson transform is the physical interpretation that follows. Residues of contributing poles (ν_l) are interpreted as attenuated circumferential waves having a phase velocity.[24]

$$c_l(ka) = \frac{(ka)c}{\Re \nu_l}. \tag{8}$$

Franz poles were investigated first[14]; this work led to the creeping-wave description of the diffraction around a rigid cylinder or sphere. The term *creeping-wave series* is now commonly used as a descriptor the Watson transform analysis, even though in elastic bodies important poles give rise to elastic circumferential waves with speeds generally much faster than the speed of

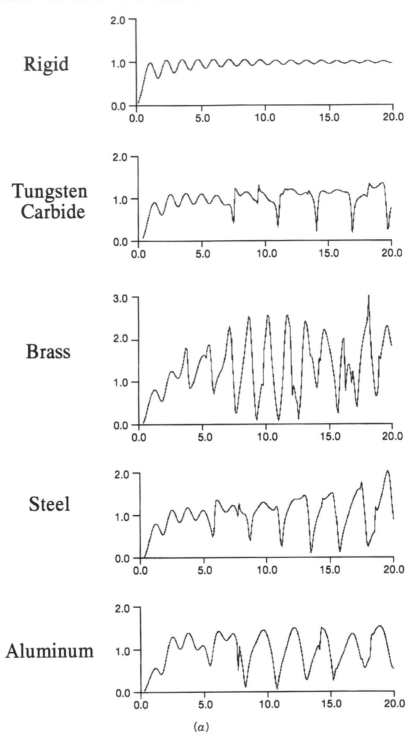

(a)

Fig. 1 (a) Form function for rigid, tungsten carbide, brass, steel, and aluminum spheres. (b) Form function for rigid, tungsten carbide, brass, steel, and aluminum cylinders. Computations require compressional speed, shear speed, and density as the independent variables. For the four materials these are: tungsten carbide (6860 m/s, 4185 m/s, 13,800 kg/m^3); brass (4700 m/s, 2110 m/s, 8600 kg/m^3); steel (5950 m/s, 3240 m/s, 7700 kg/m^3); aluminum (6376 m/s, 3120 m/s, 2710 kg/m^3).

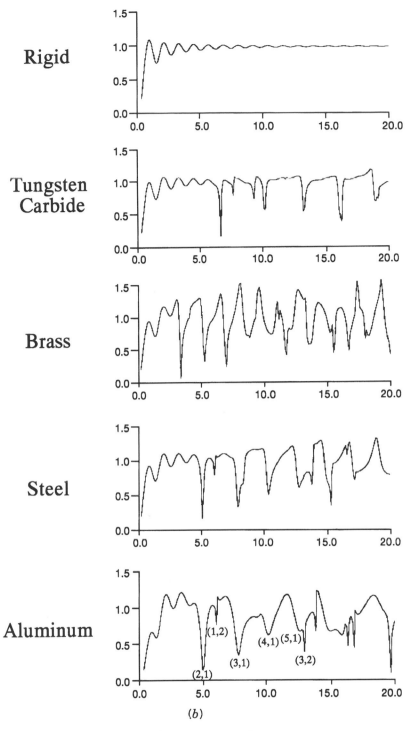

Rigid

Tungsten
Carbide

Brass

Steel

Aluminum

(b)

Fig. 1 (Continued)

sound in water. Poles corresponding to Rayleigh-like and Stonely-like waves (the Rayleigh and Stonely designations are defined for an infinite half-space) have been investigated[15]; whispering gallery waves,[19] which hug the inside surface, and general refraction mechanisms[18] have also been studied.

The physical relationship between the modal resonances observed in the form functions computed by the Rayleigh series and the circumferential wave descriptions obtained from the creeping-wave formulation have been established.[24–27] If we consider the form function for an aluminum cylinder given in Fig. 1b, the labeled

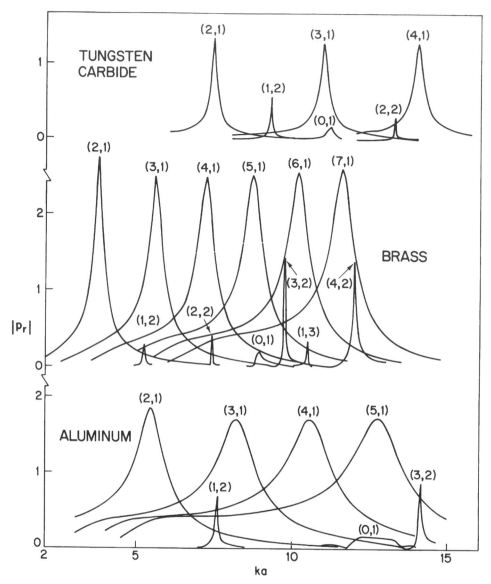

Fig. 2 Modulus of resonant terms $|f_n|$ as a function of ka for tungsten carbide, aluminum, and brass spheres.

resonance effects in $|f(ka)|$ occur near ka values that satisfy the *in vacuo* resonance condition.[24] Each term n in the normal-mode series is the nth multipole with $n = 0$ the breathing mode, $n = 1$ the dipole, $n = 2$ the quadrupole, and so on. These motions can be resolved into a pair of standing waves $e^{\pm in\phi}e^{+i\omega t}$ circumnavigating the target in opposite directions with a modal phase velocity

$$c_n(ka) = \frac{(ka)c}{n}. \qquad (9)$$

A comparison of Eqs. (8) and (9) gives the physical relationship between the circumferential waves discussed in the Catholic University series of papers and the modal resonances.

When $\Re(\nu_t) = n$, Eqs. (8) and (9) become identical in form so that the modal velocity and the circumferential velocity become equal. For the case of the aluminum cylinder response given in Fig. 1b, the R_1 (Rayleigh-type) wave labeled by Doolittle[15] is related to the modes labeled $(n, 1)$ in Fig. 1b, the R_2 wave of Doolittle to the modes labeled $(n, 2)$, and so on. At the ka position of the $(2, 1)$ resonance the circumference of the cylinder is exactly two wavelengths of the R_1 wave, the circumference is three wavelengths long at the position of the $(3, 1)$ resonance, and so on. A particular circumferential wave R_l is related to all of the resonances labeled with the eigenvalue l.

5 RESONANCE THEORY OF SCATTERING FORMULATION

This topic is discussed in detail in Chapter 5. Briefly, clear resonance behavior is observed in the form function computations considered in Figs. 1a and 1b, but the exact Rayleigh series expressions [e.g., Eqs. (3) and (4)] do not explicitly show resonance terms. A resonance term would contain a denominator of the form $1|(k - k_n + i\Gamma_n)$, where k_n is the resonance wavenumber and Γ_n is the resonance width. The formalism of the classical resonance theory of nuclear reactions[23] was applied to the solid elastic infinite-cylinder and spherical geometries,[22] and a meaningful separation of the Rayleigh series into the form

$$f(ka) = \sum_{n=0}^{\infty} f_n(ka)$$

$$= \sum_{n=0}^{\infty} f_n^R(ka) + \sum_{n=0}^{\infty} f_n^E(ka) \qquad (10)$$

was demonstrated. In Eq. (10) each partial-wave term $f_n(ka)$ in the exact normal-mode series solution has been separated into a term $f_n^R(ka)$ from the rigid-body solution and a resonance term $f_n^E(ka)$ in the form $1/(k - k_n + i\Gamma_n)$. For descriptions and examples see Chapter 5. The utility of the resonance formalism in describing the physics of acoustic scattering has been demonstrated. Uberall and collaborators have applied the formalism to numerous problems with various background conditions.[28–31] References 28–31 are a small sample of the existing literature on this subject.

6 SHELLS

The three techniques of computation and analysis considered above have been applied to the shell problem.[4,6,16,29] A consideration of the properties of Lamb waves[34] on a flat plate provides a useful analysis tool for the high ka scattering by a curved shell. Figure 3 shows a computation of the phase velocity versus frequency–thickness product (fd) for the first asymmetric and first symmetric Lamb waves for a thin aluminum plate. The ordinate shows the phase velocity V_p; a surface wave can be generated by an incident sound wave in water at an angle of incidence $\sin \theta_i = c/V_p$. The symmetric wave can be generated by phase matching at any fd value represented on the plot; the asymmetric wave can only be generated above a critical fd value at which V_p is equal to the sound speed in water. This critical value is indicated by the arrow in Fig. 3. The Lamb wave curves presented are defined for a plate in vacuo. There are significant modifications caused by fluid loading[35,36] and curvature.[37] For an infinite cylin-

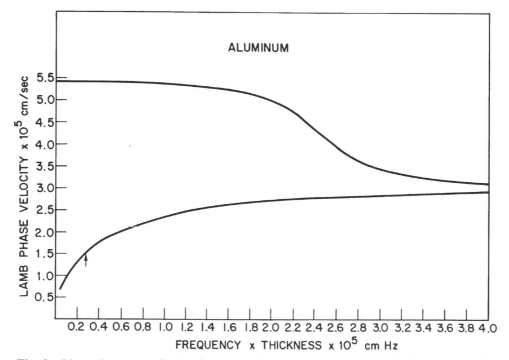

Fig. 3 Dispersion curves for the fundamental symmetric and asymmetric Lamb waves for an infinite aluminum plate in a vacuum.

der this simple model predicts that the high ka scattering below the coincidence frequency for asymmetric waves would consist of an interference between specular reflection and a symmetric leaky Lamb wave that follows the circumference of the cylinder. The backscattering from a steel shell with $d/a = 0.015$ (here d is the thickness) is given in Fig. 4, which plots the form function magnitude versus ka. Below $ka \sim 60$, the form function is well modeled by the interference of a specularly reflected wave and a symmetric Lamb wave. Above $ka \sim 60$ the coincidence frequency, the form function includes effects from the asymmetric wave.[36] The symmetric Lamb wave generated by phase matching is in phase with the incident acoustic wave. The specular reflection is out of phase with the incident wave at the lower ka values, and as frequency is increased, it becomes in phase. Thus the impedance boundary condition for a thin shell is frequency dependent. The shell presents a soft background at low ka and presents a rigid background in the high-frequency limit. The steady-state, low-frequency nulls in the form function occur at the ka values for which the cylinder circumference is an integral number of wavelengths of the symmetric Lamb wave. At these ka values the elastic response is a maximum due to constructive interference caused by the multiple circumnavigations of the Lamb wave. Nulls in the form function occur in the frequency band over which the specular response is 180° out of phase with the incident wave (soft background); peaks occur at high ka when the shell presents a rigid background and the specular is in phase. The absolute amplitudes of the interaction decrease with frequency because of the damping of the Lamb wave by reradiation into the water as it circumnavigates the cylinder. As frequency increases, the same circumferential path represents a larger number of wavelengths traversed. For other shapes similar plate wave generation angles and paths can be determined to provide a qualitative description of the form of the expected response.

7 TARGET STRENGTH

Target strength is a descriptor of the scattering amplitude widely used in sonar[38]; it is defined as the logarithm of the ratio of the reflected and incident sound intensities weighted by the target-to-receiver range r:

$$\text{TS} = 10 \log\left(\frac{I_s}{I_i}\right) + 20 \log r. \qquad (11a)$$

Since the steady-state intensity is $p^2(f)/\rho c$, this equation can be expressed as

$$\text{TS} = 20 \log\left(\frac{p_r}{p_0}\right) + 20 \log r, \qquad (11b)$$

where p_r and p_0 are steady-state pressures. Note that the energy spectral density defined as $\tilde{p}(f) = \int pe^{-i2\pi ft}\,dt$ may be used instead of the steady-state pressure. Target strength is given as decibels relative to 1 m^2, since the unit of r currently in common use in the U.S. Navy is the metre. Previously the yard was used, which leads to a definition that is smaller by 0.78 dB. The use of standard units allows the use of TS in the sonar equation.[38] Table 1 gives the standard units of applicable terms in the sonar equation in the time domain and frequency domain. Using the terminology of the sonar equation, steady-state target strength can be expressed as

$$\text{TS} = \text{EL} - \text{SL} + \text{TL}_1 - \text{TL}_2, \qquad (11c)$$

where $20 \log(|\tilde{p}_0(f)|) = \text{SL} - \text{TL}_1$ and the other terms are defined in Table 1. Other definitions of TS that depend on the transient, incident pressure signal are integrated target strength (ITS) and peak target strength (PTS):

$$\text{ITS} = 10 \log \frac{\int |p_s(t)|^2\,dt}{\int |p_i(t)|^2\,dt} + 20 \log r, \qquad (11d)$$

$$\text{PTS} = 10 \log \frac{\max |p_s(t)|^2}{\max |p_i(t)|^2} + 20 \log r. \qquad (11e)$$

The target strength definitions implicitly assume that the receiver is located in the far field of the target where the effects of range can be compensated by the factor r. This "projection of the scattered amplitude back to a distance of 1 m from the target center" then removes the range dependence in the definition of TS.

Target strength can be found from expressions of form function. The backscattered TS of a spherically shaped scatterer can be obtained from Eqs. (1b) and (11b) as

$$\text{TS} = 20 \log(a/2) + 20 \log |f(ka)| \qquad (12)$$

and for the infinite cylinder at normal incidence as

$$\text{TS} = 10 \log(ar/2) + 20 \log |f(ka)|. \qquad (13)$$

An extensive collection of formulas for high-frequency limits for many shapes can be found in Urick.[38] These formulas assume that the targets are rigid; thus $|f(ka)| \approx 1$. An expression that approximates the target strength of a finite-length elastic cylinder at normal incidence is

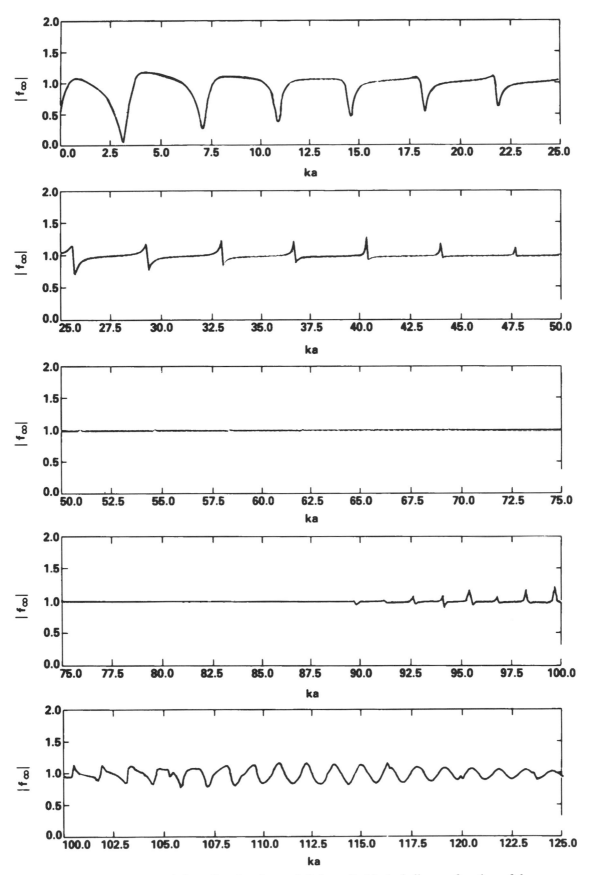

Fig. 4 Backscattered form function for an infinite cylindrical shell as a function of ka. The material is steel; the shell has an inner-to-outer-radius ratio b/a of 0.985. Indicated in parentheses on the ka scale is the equivalent fd value for a plate of the same thickness. The units of the fd values are $fd \times 10^{-5}$ cm·Hz.

TABLE 1   Standard Units

Quantity	Abbreviation	Expression	Unit
Pressure		$p(t)$	μPa
Signal level (time)	EL	$20\log|p(t)|$	dB re μPa
Signal level (frequency)	EL	$20\log|\tilde{p}(f)|^a$	dB re μPa $\cdot$ s
Source level (time)	SL	$20\log|A(t)|^b$	dB re μPa $\cdot$ m
Source level (frequency)	SL	$20\log|\tilde{A}(f)|^{a,b}$	dB re μPa $\cdot$ m $\cdot$ s
Transmission loss (time)	TL	$20\log r^b$	dB re m
Transmission loss (time)	TL	$20\log|g(r,r',t-t')|^c$	dB re m/ s
Transmission loss (frequency)	TL	$20\log|\tilde{g}(r,r',f)|^d$	dB re m/ s
Noise level (time)	NL	$10\log\left(\dfrac{1}{T}\displaystyle\int_0^T |n(t)|^2 dt\right)$	dB re μPa/Hz$^{1/2}$
Noise level (frequency)	NL	$10\log\left(\dfrac{1}{T}\left|\displaystyle\int_0^T n(t)e^{-i2\pi ft}dt\right|^2\right)$	dB re μPa/Hz$^{1/2}$
Form function (frequency)		$f(\theta,f)$	Dimensionless
Target strength (frequency)	TS	$10\log\dfrac{a}{2}|(f\theta,f)|$	dB re m
Target strength (time)	TS	$10\log\dfrac{a}{2}|(\tilde{f}\theta,t)|$	dB re m/s

^{a}Sometimes 10 log 2 is added to these frequency-domain definitions in order to include the contributions of the negative frequencies to the energy density of the signal.
^{b}Pressure is related to source level as $p(t) = A(t)/r$ or $\tilde{p}(f) = \tilde{A}(f)/r$ for an infinite, homogeneous, and isotropic medium.
^{c}A time-invariant medium is assumed. In this case the pressure at time t and position r is related to a source at time t' and position r' by a temporal convolution with $g(r,r',t-t')$.
^{d}A time-invariant medium is assumed and $\tilde{p}(f) = \tilde{A}(f)\tilde{g}(r,r',f)$.

derived from Eq. (11b) in this chapter and Section 10.6 in Junger and Feit[32] as

$$\text{TS} = 10\log\left(\frac{al^2}{2\lambda}\right) + 20\log|f(ka)|, \qquad (14)$$

where l is the length of the cylinder and λ is the wavelength of the incident sound wave. Note that the approximation uses the far-field expression for the form function of an infinite cylinder. Equation (14) is approximate since it does not include three-dimensional scattering effects such as the reflection and reradiation of axially traveling waves at the cylinder ends.

Table 2 gives several high-frequency, rigid-body target strengths that can be properly combined to form a more complicated scatterer. An extensive list is found in Urick.[38]

When specular scattering is the dominant echo mechanism, the bistatic TS may be estimated from the monostatic TS by a simple rotation of the monostatic beam pattern. Figure 5a shows the monostatic beam pattern computed for a rigid hemispherically end-capped cylinder from formulas given in Table 2. The pattern in Fig. 5a may be interpreted in two ways. The source/receiver can be fixed and the target rotated over 360° from the starting position shown, or the target can be considered fixed and the source/receiver rotated over the full 360°.

The bistatic beam pattern for the same target, with source and receiver separated by an angle θ_r, is estimated by rotating the monostatic pattern by the bistatic angle θ_b, defined as $\theta_b = \frac{1}{2}\theta_r$. In the example given in Fig. 5b, the source and receiver are separated by 60° and the monostatic beam pattern displayed in Fig. 5a is rotated through 30° to give the approximation to the bistatic pattern. The interpretation of Fig. 5b is similar to that of Fig. 5a; namely, the target may be considered fixed in the position shown while the source and receiver are rotated through 360°, maintaining a 60° separation, or the source and receiver are fixed and the target rotated through 360°. The source, receiver, and target positions giving the maximum response are indicated in Fig. 5b. The rotation technique for bistatic approximation will fail when elastic mechanisms are dominant and also when the receiver-to-source separation is much larger than 90°.

8 MEASUREMENT METHODS

A form function or target strength experiment requires measurement of the parameters given in Eq. (1) or (2),

TABLE 2 Target Strength of Various Shapes

Shape	Incident Angle	TS
Sphere	Any	$10\log\dfrac{a^2}{4}$
General convex shape	Any	$10\log\dfrac{r_1 r_2}{4}$
Infinite cylinder	Normal	$10\log\dfrac{ar}{4}$
Finite cylinder	θ_i	$10\log\dfrac{al^2}{2\lambda}\left(\dfrac{\sin\beta}{\beta}\right)^2\cos^2\theta_i$
Finite plate with area A and reflection coefficient α_r	Normal	$10\log\left(\dfrac{A}{\lambda}\right)^2 + 10\log\alpha^r$

Note: The symbols r_1 and r_2 are the radii of curvature at the point of specular reflection and $\beta = kl\sin\theta_i$.

namely the scattered pressure $\tilde{p}_s$, the incident pressure $\tilde{p}_0$, and the range r as functions of frequency. Obtaining $\tilde{p}_s$ and $\tilde{p}_0$ over a band of frequencies can be accomplished either by capturing a time-limited signal and Fourier transforming it or by selecting the steady-state portion of an echo generated by a narrow-band, continuous-wave pulse at each frequency. The scattered pressure should be measured at a range r sufficiently far from the scatterer

so that an acceptably small phase error kl^2/r is incurred across the scatterer of lateral length l. A phase error of less than $\pi/8$ radians (45°) requires $r \geq 2l^2/\lambda$. The measurement of range r to sufficient accuracy is straightforward in a laboratory setting and needs no discussion. The measurement of range, or more properly transmission loss from the target to the receiver, in a natural environment can be quite problematical and is outside

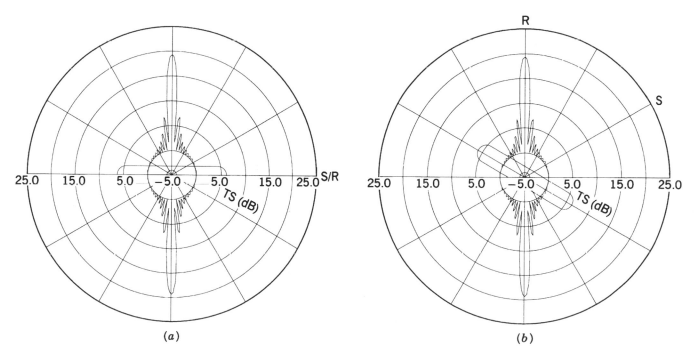

(a) (b)

Fig. 5 (a) Monostatic beam pattern for a rigid, hemispherically end-capped cylinder as computed from the formula in the Table 2. (b) An approximation of the bistatic beam pattern that would be obtained with a source-to-receiver separation of 60°. The approximation is made by rotating the monostatic pattern through the bistatic angle of 30°.

the scope of this section. In controlled environments, pressure measurements can be made either directly or indirectly. The dynamic range between p_0 and p_s is usually sufficiently large to preclude simultaneous measurement with the same hydrophone. Often the measurement of p_0 is obtained by removal of the target and replacement with a desensitized hydrophone at the same position. Alternatively, the incident pressure can be obtained by measuring a reference pulse p_r with a desensitized hydrophone at a range r_{sr} from the source. Assuming spherical spreading from the source, the incident pressure p_0 can be inferred from p_r as

$$|\tilde{p}_0| = |\tilde{p}_r| \frac{r_{sr}}{r_{st}}, \tag{15}$$

where r_{st} is the distance from the source to the target. In terms of measured quantities the target strength is

$$\mathrm{TS} = 20\log\frac{|\tilde{p}_s|}{|\tilde{p}_r|} + 20\log\frac{r_{st}r}{r_{sr}}. \tag{16}$$

The dynamic range between p_s and p_r is still large enough to preclude using the same hydrophone system to measure both easily. One approach is to use a fast switch to increase the sensitivity of the hydrophone after the time when p_r is present at the hydrophone and before the arrival of p_s. This approach does not require a calibrated receiver. Another approach is to use two hydrophones of differing sensitivity to measure p_s and p_r. In this case each hydrophone must be adequately calibrated.

For controlled laboratory measurements the sphere calibration technique is often the easiest method for calibration of scattering measurements.[26] For $ka \gg 1$, the target strength of a rigid sphere is given by $20\log(a/2)$. The target strength of a test object can in principle be obtained by comparison of the transient response of a rigid sphere, p_{sph}, with that of the test object to the same incident signal. If the sphere identically replaces the target, a simple modification of Eq. (16) gives

$$\mathrm{TS} = 20\log\frac{|\tilde{p}_s(f)|}{|\tilde{p}_{\mathrm{sph}}(f)|} + 20\log\frac{a}{2} + 20\log\frac{r_{st}r}{r_{sr}}. \tag{17}$$

In the sphere and test object are measured at different ranges, a simple modification to the range term is required. As indicated by the form functions discussed previously with reference to Fig. 1a, no sphere will act as a rigid reflector in water; however, for an impulsive incident signal there is a significant frequency band over which the elastic and geometric responses of a sphere can be temporally isolated.[26] Figure 6a shows the echo backscattered by a solid tungsten carbide sphere that was illuminated by a broadband signal whose useful band-

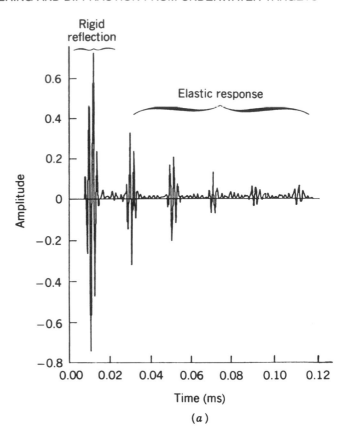

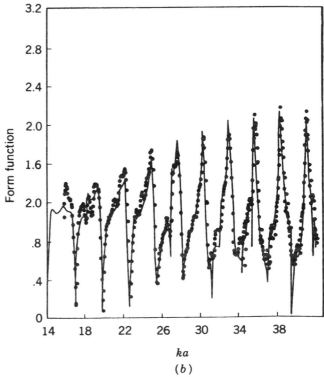

Fig. 6. (a) Backscattered echo from a tungsten carbide sphere, showing a clear separation of specular and elastic contributions. (b) Comparison of theory and experiment. The experiment was processed by using the rigid reflection identified in (a) as a calibration standard.

width covers the frequency range between $14 < ka < 40$. The form function of the sphere computed by a normal-mode series is given as the solid curve in Fig. 6b. In Fig. 6a the specular and elastic responses of the sphere are observed to be effectively isolated in the transient echo. The time-windowed Fourier transform of the initial specular echo is taken to produce $\tilde{p}_{sph}(f)$ in Eq. (17), and the Fourier transform of the entire response gives $\tilde{p}_s(f)$. The form function obtained using $\tilde{p}_{sph}(f)$ and $\tilde{p}_s(f)$ is given as the points in Fig. 6b.

High-frequency calibration methods that involve the use of liquid-filled spheres are found in Refs. 39 and 40. Descriptions of the techniques used for precise lake measurements are found in the work performed by the ARL University of Texas; Refs. 41 and 42 are recommended introductions to that body of work.

REFERENCES

1. J. J. Faran, "Sound Scattering by Solid Cylinders and Spheres," Office of Naval Research Technical Memorandum No. 22, DTIC No. AD827027, March 15, 1951.

2. J. J. Faran, "Sound Scattering by Solid Cylinders and Spheres," *J. Acoust. Soc. Am.*, Vol. 23, 1951, p. 405.

3. R. Hickling, "Analysis of Echoes from a Solid Elastic Sphere in Water," *J. Acoust. Soc. Am.*, Vol. 34, 1962, p. 1582.

4. R. Hickling, "Analysis of Echoes from a Hollow Metallic Sphere in Water," *J. Acoust. Soc. Am.*, Vol. 36, 1964, p. 1124.

5. R. R. Goodman and R. Stern, "Reflection and Transmission of Sound by Elastic Spherical Shells," *J. Acoust. Soc. Am.*, Vol. 34, 1962, p. 338.

6. R. D. Doolittle and H. Uberall, "Sound Scattering by Elastic Cylindrical Shells," *J. Acoust. Soc. Am.*, Vol. 39, 1966, p. 272.

7. Werner G. Neubauer, Richard H. Vogt, and Louis R. Dragonette, "Acoustic Reflection from Elastic Spheres, I. Steady-State Signals," *J. Acoust. Soc. Am.*, Vol. 55, 1974, p. 1123.

8. Louis R. Dragonette, Richard H. Vogt, Lawrence Flax, and Werner G. Neubauer, "Acoustic Reflections from Elastic Spheres and Rigid Spheres and Spheroids. II. Transient Analysis," *J. Acoust. Soc. Am.*, Vol. 55, 1974, p. 1130.

9. L. Flax and W. G. Neubauer, "Acoustic Reflection from Layered Elastic Absorptive Cylinders," *J. Acoust. Soc. Am.*, Vol. 61, 1977, p. 307.

10. Richard H. Vogt, Lawrence Flax, Louis R. Dragonette, and Werner G. Neubauer, "Monostatic Reflection of a Plane Wave from an Absorbing Sphere," *J. Acoust. Soc. Am.*, Vol. 57, 1975, p. 558.

11. R. Vogt and W. G. Neubauer, "Relationship between Acoustic Reflection and Vibrational Modes of Elastic Spheres," *J. Acoust. Soc. Am.*, Vol. 60, 1976, p. 15.

12. L. Flax, "High ka Scattering of Elastic Cylinders and Spheres," *J. Acoust. Soc. Am.*, Vol. 62, 1982, p. 1502.

13. A. J. Rudgers, "Acoustic Pulses Scattered by a Rigid Sphere Immersed in a Fluid," *J. Acoust. Soc. Am.*, Vol. 45, 1969, p. 900.

14. H. Uberall, R. D. Doolittle, and J. V. McNicholas, "Use of Sound Pulses for a Study of Circumferential Waves," *J. Acoust. Soc. Am.*, Vol. 39, 1966, p. 564.

15. R. D. Doolittle, H. Uberall, and P. Ugincius, "Sound Scattering by Elastic Cylinders," *J. Acoust. Soc. Am.*, Vol. 43, 1968, p. 1.

16. P. Ugincius and H. Uberall, "Creeping-Wave Analysis of Acoustic Scattering by Elastic Cylindrical Shells," *J. Acoust. Soc. Am.*, Vol. 43, 1968, p. 1025.

17. W. G. Neubauer, P. Ugincius, and H. Uberall, "Theory of Creeping Waves in Acoustics and Their Experimental Demonstration," *Zeitschrift Naturforschung*, Vol. 24, 1969, p. 693.

18. D. Brill, H. Uberall et al., "Acoustic Waves Transmitted through Solid Elastic Cylidners," *J. Acoust. Soc. Am.*, Vol. 50, 1970, p. 921.

19. J. W. Dickey, G. V. Frisk, and H. Uberall, "Whispering Gallery Wave Modes on Elastic Cylinders," *J. Acoust. Soc. Am.*, Vol. 59, 1976, p. 1339.

20. N. C. Yen, Louis R. Dragonette, and Susan K. Numrich, "Time-Frequency Analysis of Acoustic Scattering from Elastic Objects," *J. Acoust. Soc. Am.*, Vol. 87, 1990, p. 2359.

21. S. K. Numrich, L. R. Dragonette, and L. Flax, "Classification of Submerged Targets," in V. K. Varadan and V. V. Varadan (Eds.), *Elastic Wave Scattering and Propagation*, Ann Arbor Press, 1982, MI, p. 149–175.

22. L. Flax, L. R. Dragonette, and H. Uberall, "Theory of Elastic Resonance Excitation by Sound Scattering," *J. Acoust. Soc. Am.*, Vol. 63, 1978, p. 723.

23. G. Breit and E. P. Wigner, "Capture of Slow Neutrons," *Phys. Rev.*, Vol. 49, 1936, p. 519.

24. H. Uberall, L. R. Dragonette, and L. Flax, "Relation between Creeping Waves and Normal Modes of Vibration of a Curved Body," *J. Acoust. Soc. Am.*, Vol. 61, 1977, p. 711.

25. L. R. Dragonette, "Influence of the Rayleigh Surface Wave on the Backscattering by Submerged Aluminum Cylinders," *J. Acoust. Soc. Am.*, Vol. 65, 1979, p. 1570.

26. Louis R. Dragonette, S. K. Numrich, and Laurence J. Frank, "Calibration Technique for Acoustic Scattering Measurements," *J. Acoust. Soc. Am.*, Vol. 69, 1981, p. 1186.

27. J. W. Dickey and H. Uberall, "Surface Wave Resonances in Sound Scattering from Elastic Cylinders," *J. Acoust. Soc. Am.*, Vol. 63, 1978, p. 319.

28. L. Flax, G. C. Gaunaurd, and H. Uberall, "Theory of Resonance Scattering," in W. P. Mason and R. N. Thurston (Eds.), *Physical Acoustics*, Vol. 15, Academic Press, New York, 1981.

29. J. D. Murphy, J. George, A. Nagl, and H. Uberall, "Isola-

tion of the Resonant Component in Acoustic Scattering from Fluid-Loaded Elastic Spherical Shells," *J. Acoust. Soc. Am.*, Vol. 65, 1979, p. 368.

30. H. Uberall, Y. J. Stoyanov, A. Nagl, M. F. Werby, S. H. Brown, J. W. Dickey, S. K. Numrich, and J. M. D'Archangelo, "Resonance Spectra of Elongated Elastic-Objects," *J. Acoust. Soc. Am.*, Vol. 81, 1987, p. 312.

31. M. F. Werby, H. Uberall, A. Nagl, S. H. Brown, and J. W. Dickey, "Bistatic Scattering and Identification of the Resonances of Elastic Spheroids," *J. Acoust. Soc. Am.*, Vol. 84, 1988, p. 1425.

32. M. C. Junger and D. Feit, *Sound, Structures and Their Interaction*, 2nd ed., MIT Press, Cambridge, MA, 1986.

33. C. M. Davis, et al., "Acoustic Scattering from Silicon Rubber Cylinders and Spheres," *J. Acoust. Soc. Am.*, Vol. 63, 1978, p. 1694.

34. H. Lamb, "On Waves in an Elastic Plate," *Proc. Roy. Soc.*, Vol. A93, 1917, p. 114.

35. J. F. M. Scott, "The Free Modes of Propagation of an Infinite Fluid-Loaded Thin Cylindrical Shell," *J. Sound Vib.*, Vol. 125, 1988, p. 241.

36. M. Talmant and J. Ripoche, "Study of the Pseudo-Lamb Wave S_0 Generated in Thin Cylindrical Shells Insonified by Short Ultrasonic Pulses in Water," in H. M. Merklinger (Ed.), *Progress in Underwater Acoustics*, Plenum, New York, 1987, pp. 137–144.

37. P. L. Marston, "Phase Velocity of Lamb Waves on a Spherical Shell: Approximate Dependence on Curvature from Kinematics," *J. Acoust. Soc. Am.*, Vol. 85, 1989, p. 2663.

38. R. J. Urick, *Principles of Underwater Sound*, 3rd ed. McGraw-Hill, New York, 1983, p. 291.

39. B. M. Marks and E. E. Mikeska, "Reflections from Focused Liquid-Filled Spherical Reflectors," *J. Acoust. Soc. Am.*, Vol. 59, 1976, p. 813.

40. D. L. Folds and C. D. Loggins, "Target Strength of Liquid-Filled Spheres," *J. Acoust. Soc. Am.*, Vol. 73, 1983, p. 1147.

41. G. R. Barnard and C. M. McKinney, "Scattering of Acoustic Energy by Solid and Air-Filled Cylinders in Water," *J. Acoust. Soc. Am.*, Vol. 33, 1961, p. 226.

42. C. W. Horton and M. V. Mechler, "Circumferential Waves in a Thin-Walled Air-Filled Cylinder in a Water Medium," *J. Acoust. Soc. Am.*, Vol. 51, 1971, p. 295.

43

QUANTITATIVE RAY METHODS FOR SCATTERING

Philip L. Marston

1 INTRODUCTION

Rays have long been used to understand optical imaging systems and natural phenomena. Subsequent to the derivation of wave equations for light and sound, the mathematical basis of ray descriptions of wave fields was established. Some of the basis is reviewed elsewhere in this volume (Chapters 3–5 and 42). Pierce[1] gives an introduction to ray acoustics and summarizes the history of ray methods. The purpose of the present chapter is to illustrate how ray concepts and geometric constructions can be applied to simple and complicated high-frequency scattering situations associated either with reflection from underwater targets or with the coupling of sound with internal motion. Caustics can be produced where elementary ray methods fail to properly describe wave fields so that corrections associated with diffraction are important. Examples included illustrate the analysis of scattered wave fields near caustics.

The chapter is organized as follows. Section 2 summarizes general considerations and procedures for scatterers having smooth but possibly complicated shapes. Much of this discussion is also relevant to the reflection of sound in air from solid objects and to ultrasonic imaging and nondestructive evaluation. Examples are considered in Sections 3 and 4, where the mechanical response of the scatterer gives rise to large backscattering amplitudes. While the emphasis there is on spherical or cylindrical targets, the analysis of surface wave contributions could be extended to other smooth shapes. References are made to review publications and to derivations of ray theory and numerical comparisons with exact the-

ory for the applications considered. For the time-harmonic signals considered, the amplitude is proportional to $\exp(-i\omega t)$, where ω is the radian frequency. Scattering of transients and tone bursts are also mentioned. The medium surrounding the scatterer has a uniform phase velocity c. It is assumed that the wavelength in the surrounding fluid is somewhat smaller than the principal radii of the scatterer's surface except in the discussion of diffraction by sharp edges. Some asymptotic corrections are noted to illustrate the significance of frequency on the accuracy of simple ray formulations.

2 CURVED REFLECTORS, CAUSTICS, PHYSICAL OPTICS APPROXIMATIONS, AND EDGE RAYS

2.1 Reflected Ray Tubes and Associated Scatterng Amplitudes

The elementary ray approach to calculating scattering amplitudes involves the construction of rays reflected from the scatterer. The angles of incidence and reflection are equal for each ray, and incident and reflected rays lie in the plane containing the local surface normal. This is illustrated in Fig. 1, where $\hat{k}$ denotes the direction of the incident ray, $\hat{s}_3$ is the normal at the point of reflection, and $\hat{n}$ is the direction of the reflected ray. Associated with the reflected ray is a reflected wavefront that is normal to the ray. To approximate the amplitude of the reflected wave field, it is necessary to describe the spreading of the reflected wavefront with propagation. A convenient way to analyze this spreading is to consider the paths of rays adjacent to (and displaced infinitesimally from) the specific ray under consideration. These displaced rays form the boundary of a ray tube that contains the ray of

Encyclopedia of Acoustics, edited by Malcolm J. Crocker
ISBN 0-471-80465-7 © 1997 John Wiley & Sons, Inc.

interest. The distance to the source of the incident wave affects the spreading of the outgoing ray tube. Figure 1 illustrates the results of such a construction for a source located at S that is a finite distance d from the specular point P. As summarized below, a general coordinate-free description of the outgoing ray tube may be constructed if the local curvatures of the reflecting surface are known.[2,3]

In general, at each point on a smooth surface, there will be two principal curvatures and two orthogonal principal directions that are perpendicular to the local surface normal.[4] At the specular point P in Fig. 1, these curvatures are denoted by positive numbers $(a_1)^{-1}$ and $(a_2)^{-1}$ for a convex surface, and the corresponding principal directions are $\hat{s}_1$ and $\hat{s}_2$; a_1 and a_2 are the principal radii of the reflecting surface Σ and may be calculated from an equation for Σ as described in Ref. 2. Wavefronts may also be characterized by principal radii and directions. Let r_1 and r_2 denote the principal radii of the reflected wavefront at P. The corresponding wavefront curvatures are given by[2,3]

$$\kappa_l = \frac{1}{r_l} = \frac{1}{d} + \frac{1}{a_0 \cos \beta}$$

$$\pm \left(\frac{1}{a_0^2 \cos^2 \beta} - \frac{4}{a_1 a_2} \right)^{1/2},$$

$$l = 1, 2, \tag{1}$$

$$\frac{1}{a_0} = \frac{\sin^2 \theta_1}{a_2} + \frac{\sin^2 \theta_2}{a_1}, \tag{2a}$$

$$\cos \theta_l = \hat{k} \cdot \hat{s}_l, \tag{2b}$$

$$\cos \beta = -\hat{k} \cdot \hat{s}_3, \tag{2c}$$

where β is the angle of incidence and $\theta_{1,2}$ denotes the angle between $\hat{k}$ and $\hat{s}_{1,2}$. As indicated by the virtual foci in Fig. 1, the wavefront spreads from different centers of curvature that are located distances r_1 and r_2 behind P. After propagating a distance q from P to a general observation point Q, the principal radii of the outgoing wavefront become $r_1 + q$ and $r_2 + q$.

The amplitude of the reflected wave at Q may now be calculated. The essential idea is that the power per unit area in the ray tube varies inversely with the area of the tube so that the pressure amplitude is proportional to $(\text{area})^{-1/2}$. By inspection of Fig. 1, the area of the ray tube varies as $(r_1 + q)(r_2 + q)/r_1 r_2$. Therefore, if $p(q = 0)$ denotes the pressure amplitude of the reflected wavefront at P, the amplitude of the scattered wave is

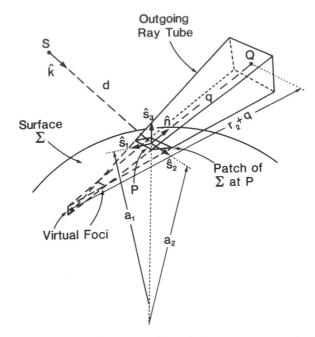

Fig. 1 Geometric construction of the pressure amplitude reflected from the surface to a point Q. The ray tube associated with the incoming ray from the source at S is not shown. The principal radii r_1 and r_2 for the reflected wavefront are related to the surface radii a_1 and a_2 by Eqs. (1) and (2).

$$p(q) = p(0) \left| \frac{r_1 r_2}{(q + r_1)(q + r_2)} \right|^{1/2} e^{ikq}, \tag{3}$$

where $k = \omega/c$. The directions of the reflected and incident rays are related by $\hat{n} = \hat{k} + 2\hat{s}_3 \cos \beta$. It remains to express $p(0)$ in terms of the amplitude of the wave p_{inc} that is incident on P from the source at S. The simplest assumption is to relate these by a local coefficient R so that in the illuminated region $p(0) = Rp_{\text{inc}}$. Modeling the scatterer as a fixed rigid reflector gives $R = 1$. This is known as the Kirchhoff approximation for a rigid surface[5] and is often useful if the material impedance of the scatterer greatly exceeds that of the surrounding fluid. In other situations, it can be acceptable to approximate R by the reflection coefficient for a flat interface with the angle of incidence β at the specular point under consideration.[2] The approximation of Eq. (3) with $p(0) = Rp_{\text{inc}}$ breaks down near certain critical angles of incidence where diffractive corrections become important.[2,3,6,7]

In Section 2.2, Eqs. (1)–(3) are applied to far-field scattering. Prior to considering those examples, it is appropriate to comment on general situations that may arise in the reflection of sound from curved surfaces. If either or both of the r_l are negative, the reflected wavefront converges to a caustic,[2,8] where $r_1 + q$ or $r_2 + q$ vanishes, as discussed in Section 2.3. This may occur if the surface is concave or if the incident wavefront is converging. A converging spherical wave is allowed for in Eq. (1) by taking $d < 0$. More generally, the principal curvatures of the incident wavefront need not be the same. For that case r_1 and r_2 are derivable from discussions given in Refs. 1, 3, and 8. While the discussion above has centered on the curvatures of reflected wavefronts, a related method of analysis given in Refs. 3 and 8 allows the principal curvatures of refracted wavefronts to be calculated. The method, known as generalized ray tracing, has applications to the design of optical instrumentation as well as to scattering.[2,3]

2.2 Specular Contributions to Far-Field Scattering

The results of Eqs. (1)–(3) take on a simple form if the distances d and q greatly exceed the principal radii of the reflector a_1 and a_2. For any surface, the product of the principal curvatures is known as the Gaussian curvature.[4] For the situation under consideration the Gaussian curvature of the reflected wavefront at P reduces to

$$\kappa_g = \kappa_1 \kappa_2 = 4(a_1 a_2)^{-1} = 4\kappa_g^{\Sigma}, \qquad (4)$$

where κ_g^{Σ} is the Gaussian curvature at P of the reflector. The scattering angle θ is $\pi - 2\beta$. Assuming there is only a single specular ray to the distant observation point of interest, Eq. (3) reduces for a rigid object to[2]

$$p = \frac{p_{\text{inc}}}{r} \tilde{f} e^{ikr}, \qquad (5a)$$

$$|\tilde{f}|^2 = \frac{d\sigma}{d\Omega} = (4\kappa_g^{\Sigma})^{-1}, \qquad (5b)$$

where $\tilde{f}$ is a complex dimensional form function and $d\sigma/d\Omega$ is a differential cross section giving the ratio of the area of the incident ray tube to the solid angle of the outgoing ray tube. In Eq. (5), r denotes the distance to the observer from some reference point O that is usually taken to lie within or on the boundary of the scatterer. The incident wave is phase referenced at O and the phase of $\tilde{f}$ depends on the scattering angle and may be calculated geometrically. The essential result in Eq. (5), known in acoustics since World War II, is the dependence of $d\sigma/d\Omega$ on κ_g^{Σ}. This dependence is not related to the angle of incidence β, a result that simplifies appli-

cations to inverse problems.[3] Even for rigid objects, however, Eq. (5b) gives an incomplete description of the scattering since it neglects ordinary forward diffraction, contributions from creeping waves, and $O(ka_j)^{-1}$ asymptotic corrections. Nevertheless, at high frequencies the specular reflection can be the major source of backscattering, and the relationship given in Eq. (5) can be useful for estimating the differential cross section of a general curved surface.[2,9]

An elementary application of Eq. (5) is the reflection of sound from a sphere of radius a as diagrammed in Fig. 2a. It is conventional[2,10] to express the far-field pressure for scattering by spheres as

$$p(r, \theta) = \frac{p_{\text{inc}} f a}{2r} e^{ikr} \qquad (6)$$

in terms of a complex dimensionless form function $f(\theta)$ given exactly by a partial-wave series. Here $r \gg ka^2$ is the distance from the center C of the sphere and p_{inc} is the incident amplitude at C in the absence of a sphere. Equation (5) reduces to $d\sigma/d\Omega = a^2/4$ for specular

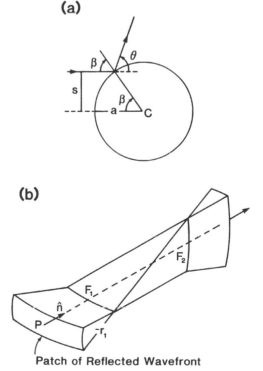

Fig. 2 (a) Reflection from a sphere or a circular cylinder of radius a. The incident ray is displaced a distance s from the axis. (b) Ray tube for the situation where both of the principal radii of the reflected wavefront are negative. Focal lines F_1 and F_2 show where the area of the tube vanishes.

reflection from a rigid sphere so that the specular contribution to f, call it f_S, has $|f_S| = 1$ for all scattering angles θ. Consequently the normalization is such that the exact $|f|$ is the amplitude relative to reflection from a rigid sphere of the same size calculated geometrically. To facilitate subsequent discussion, it is convenient to introduce an effective reflectivity R_e. A calculation of the path length for reflection gives the phase of f_S with the result[2]

$$f_S = R_e e^{-i2ka\sin(\theta/2)}, \tag{7a}$$

$$R_e = 1. \tag{7b}$$

For backscattering the leading asymptotic correction to f_S may be incorporated by taking[11] $R_e = 1 - 3i/2ka$. The correction affects the phase more than the magnitude and is important when ka is not large. Recently, Nussenzveig[7] derived a uniform approximation to scattering by impenetrable spheres that is applicable to all angles for ka down to ≈ 1. While that approximation has a ray interpretation, it involves integrals of special functions.

It is instructive to consider the modifications required to describe far-field scattering by an infinite circular cylinder of radius a. For simplicity, attention is restricted to the situation of a plane wave incident perpendicular to the cylinder's axis. Equation (6), relating the dimensionless form function f and the scattering, is replaced by[2,10]

$$p(r, \theta) = \left(\frac{a}{2r} \right)^{1/2} p_{\text{inc}} f(\theta) e^{ikr}, \tag{8}$$

where r is the distance from the cylinder's axis. Consider now the ray calculation of the contribution to f due to specular reflection from a rigid cylinder. Figure 2a remains applicable to this case, but in the evaluation of Eq. (1), one of the principal radii of the surface, say a_2, is infinite. For plane-wave incidence (corresponding to $d = \infty$), one of the principal curvatures of the reflected wavefront, κ_2, vanishes while the other is given by $\kappa_1 = r_1^{-1} = 2a^{-1}(\cos \beta)^{-1} = 2a^{-1}[\cos(\theta/2)]^{-1}$. Insertion of these limits into Eq. (3) and use of the Kirchhoff approximation gives a specular contribution to f of $f_S = [\sin(\theta/2)]^{1/2} \exp[-i2ka\sin(\theta/2)]$. The phase follows from the path length as in Eq. (7). This result, which agrees with an asymptotic analysis by Debye translated in Ref. 3, differs from the result for spheres in Eq. (7) in that now, while $|f_S| = 1$ for backscattering, $|f_S|$ vanishes as the scattering angle $\theta \to 0$ and the wavefront curvature κ_1 diverges. (In the sphere case κ_1 also diverges but the product $\kappa_1\kappa_2$ remains independent of θ.) Eval-

uation of the exact $|f(\theta)|$ for rigid cylinders confirms that when ka is not small, there is a contribution to f that varies with θ roughly in proportion to $[\sin(\theta/2)]^{1/2}$. As is the case for impenetrable spheres, f contains forward diffractive and associated creeping-wave contributions that become dominant if θ is close to or less than $(ka)^{-1}$ radians.

In experiments, cylinders are never infinite in length, and the incident wave may be curved or not perpendicular to the cylinder's axis. The ray analysis may be modified for other source geometries by application of Eqs. (1)–(3) while a finite length L introduces diffractive corrections noted in Ref. 9 and (including the case of tilted cylinders) Refs. 2 and 12. For a finite cylinder, the amplitude varies as r^{-1} instead of $r^{-1/2}$ in the far field where $r \gg k(L^2 + 4a^2)/8$. Even though Eq. (8) may be of limited applicability to experiments, the scattering by infinite circular cylinder remains of theoretical importance.

2.3 Near-Field and Far-Field Caustics and Reflection from Concave Surfaces

A surface that is locally concave has one or both of the principal radii a_1 and a_2 as negative in Eqs. (1) and (2). For sufficiently large source distances d, one or both of the curvatures κ_l of the reflected wavefront are negative. At some distance from the specular point, the denominator in Eq. (3) vanishes and the predicted amplitude unphysically diverges. The vanishing denominator locates the caustics of the reflected wavefront, and near such caustics there are essential diffractive corrections to the wave field. Figure 2b illustrates the ray tube for an example in which both of the ρ_l are negative so that there are distinct caustics formed along foci F_1 and F_2. Sliding the specular point P to other points on the reflecting surface will ordinarily cause F_1 and F_2 to shift and to each trace out a caustic surface. The surface may be at least locally two-sheeted when both r_l are negative or it may be single sheeted if only one of the r_l is negative for rays reflected to the region of interest. Caustics lacking axial symmetry can be classified using catastrophe optics and wave fields near caustics approximated in terms of canonical diffraction integrals such as Airy and Pearcey functions.[2,13] It is noteworthy that Eq. (3) may be used to approximate the amplitude subsequent to propagation through a caustic provided that a phase shift of $-\pi/2$ is introduced for each focal curve F_l crossed.[1–3]

While the discussion above has emphasized the location of the caustic as the center of curvature of an outgoing wavefront, the shapes of associated wavefronts may be obtained from the shape of the caustic. For two-dimensional situations, outgoing wavefronts are obtained as involutes of the caustic. An involute curve is traced out by unwinding a taut thread tangent to the caustic.[2,4]

Especially relevant to scattering problems are situations where either or both of the κ_l vanish along some curve on the outgoing wavefront. The caustic surface for propagation from such points is shifted to be infinitely distant, that is, to the far field of the reflector. Such caustics are characterized by scattering directions and are known as directional cuastics.[2] The outgoing wavefront is locally flat on the curve where the Gaussian curvature $\kappa_g = \kappa_1 \kappa_2$ vanishes. When the wave incident on the reflector is flat, Eq. (4) is applicable and such points are associated with the vanishing of the Gaussian curvature of the reflector. Diffractive corrections are calculated as described in Sections 2.4 and 2.5. If κ_g vanishes over a region of the outgoing wavefront, as in the example of the finite-length circular cylinder mentioned earlier, the width and height of the region strongly affect the far-field scattering.[2]

2.4 Physical Optics Approximation of Wave Fields

Consider the situation illustrated in Fig. 3 where sound from a source at (u_s, v_s, z_s) is reflected from a curved surface. As an alternative to the ray tube analysis given

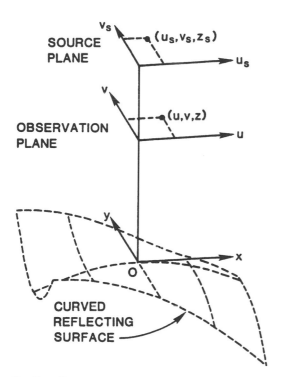

Fig. 3 Coordinate system used in the physical optics approximation of the amplitude in the observation plane. The displacement of the surface relative to the xy plane is specified by $h(x, y)$. The specific surface shown produces a cusp caustic in the observation plane, as illustrated in Fig. 4. (Reproduced, by permission, from Ref. 15, p. 2870.)

in Eq. (3), the wave field may be approximated by an integral as summarized below. For simplicity contributions to the wave field diffracted by the edges of the reflector will be neglected so that the reflector is taken to be infinitely wide. Suppose that the complex amplitude of the reflected wave were exactly known in a plane, to be referred to as the "exit plane" in subsequent discussion; then the evolution of the wave field with propagation may be computed exactly from the Rayleigh–Sommerfeld propagation integral[2]

$$
p(u, v, z) = -\frac{1}{2\pi} \iint_{-\infty}^{\infty} dx\, dy\, p(x, y, z = 0)
$$

$$
\cdot \frac{\partial}{\partial z}\left(\frac{e^{ikr'}}{r'}\right), \qquad z \geq 0, \qquad (9)
$$

where $r' = [(u - x)^2 + (v - y)^2 + z^2]^{1/2}$ is the distance of the observation point at (u, v, z) from a point in the exit plane at $(x, y, 0)$. A derivation of this relationship is reviewed in Ref. 14 in the context of near-field acoustic holography. Figure 3 illustrates an example where the xy exit plane is taken to be adjacent to the reflector. In a physical optics approximation (POA), $p(x, y, 0)$ is approximated geometrically from properties of the incident wave and the local reflectivity R of the surface. (The reader is cautioned that the detailed formulation of physical optics approaches may vary, as evident from discussions in Refs. 2, 5, and 12.) Evaluation of Eq. (9) from the stationary phase approximation[2,13] yields a ray representation for the reflected wave field that is equivalent to that given by Eq. (3), with possible caustic phase shifts as discussed in Section 2.3. For wave fields containing caustics, the advantage of the POA is that the integral in Eq. (9) may be locally approximated by a canonical diffraction integral so that the amplitude diverges only in the limit $k \to \infty$. Assuming that the slope of the reflecting surface is small in the physically relevant region, $p(x, y, 0)$ may be well approximated by $A(x, y) \exp[ik W(x, y)]$, where $W(x, y)$ is the displacement of an outgoing wavefront (or equiphase surface) from the exit plane and the amplitude factor A is taken to have a constant phase.[2]

To facilitate the aforementioned simplification of Eq. (9), it is often convenient to assume that the source and observation points lie sufficiently close to the z-axis that a paraxial approximation of the form $r = (u^2 + v^2 + z^2)^{1/2} \approx z[1 + (u^2 + v^2)/2z^2]$ is applicable in calculations of the phase of the integrand. This is also known as a Frensel approximation of the phase. The derivative in Eq. (9) may also be simplified for paraxial observation points having $kr \gg 1$. These approximations give

$$p(u,v,z) = \frac{-ik}{2\pi r} \int\!\!\int_{-\infty}^{\infty} A(x,y) e^{ik\phi(x,y;u,v,z)} \, dx \, dy, \quad (10)$$

$$\phi \approx z - W(x,y) + \frac{z(U^2 + V^2)}{2} - (xU + yV)$$
$$+ \frac{x^2 + y^2}{2z}, \quad (11)$$

where $U = u/z$ and $V = v/z$. The phase of A is selected such that $|W(x,y)| \ll z$ for the relevant region of the exit plane.

As an example, consider the reflection of sound from a gradually sloped surface specified by the height $h(x,y)$ relative to the exit plane. For (x,y) where $h > 0, A(x,y)\exp(-ikW)$ approximates an upward-going wave in an unbounded fluid that propagates to give the reflected wave in the physical region $z > h$. If the source of the incident wave lies close to the z-axis, paraxial approximation of the distance from S to the reflector gives

$$W(x,y) \approx 2h(x,y) + \frac{xu_s + yv_s}{z_s} - \frac{x^2 + y^2}{2z_s}, \quad (12)$$

and $A(x,y) \approx p_{\text{inc}}(x = 0, y = 0, z = 0) \, R(x,y)$. Terms proportional to $x^2 + y^2$ in Eqs. (11) and (12) may be omitted in the Fraunhofer or far-field approximation of the phase. Far-field reflection from a rigid sphere, when approximated by Eqs. (10)–(12), reduces to a result consistent with f_S from Eq. (7) for θ near π when the stationary phase approximation is used to evaluate Eq. (10). The asymptotic phase correction noted below Eq. (7) is also relevant to the PAO.

2.5 Diffraction Catastrophes

For caustics classified by catastrophe theory mentioned in Section 2.3, the associated wave fields are referred to as diffraction catastrophes.[2,13] A simple example of such a wave field is given by taking $h(x,y) = h_1 x^2 + h_2 y^2 x + h_3 y^2$, where h_j are constants and $h_2 \neq 0$. It is appropriate to introduce the Pearcey function[2,13]

$$\text{Pe}(w_2, w_1) = \int_{-\infty}^{\infty} \exp[i(s^4 + w_2 s^2 + w_1 s)] \, ds \quad (13)$$

as the canonical diffraction integral of cusp caustics. Letting the reflection coefficient R be constant, it has been shown[2,15] that the two-dimensional integral in Eq. (10) is proportional to $k^{1/4}\text{Pe}(w_2, w_1)$, with w_2 and w_1 proportional to $k^{1/2}(u - u_{\text{cp}})$ and $k^{3/4}(v - v_{\text{cp}})$, respectively,

where u_{cp} and v_{cp} locate the cusp point of a caustic surface that intersects the uv observation plane in Fig. 3. To verify this analysis, a metallic surface was fabricated having the form of $h(x,y)$ noted above. The surface was placed in water and subjected to ultrasonic tone bursts and the wave field was scanned with a small hydrophone. Figure 4 shows the observed wave field pattern and contours of the pattern calculated from the physical optics approximation.[15] The agreement is satisfactory considering the surface parameters h_j were determined by fitting the function $h(x,y)$ to a measured surface shape not exactly described by the assumed form of $h(x,y)$. The sampled time interval is chosen so that only the reflected wave field is viewed. The wave field is strongest in the region close to, but displaced slightly from, the transverse cusp caustic formed by the intersection of the caustic surface with the observation plane.

The caustic curve in Fig. 4 separates a region where there are three specular points for reflections to (u,v,z) from a region where there is only one.[2,15] It is calculated from the distance function ϕ in Eq. (11). The transition results from a merging of rays on the caustic surface and is generic to other caustics described by catastrophe theory. Associated with a merging of reflected rays is a merging of the propagation distance from (u_s, v_s, z_s) to (u,v,z). This has been observed as a merging of echoes when a transient source is used instead of tone bursts.[2,15]

The example of the cusp diffraction catastrophe considered here displays features representative of those predicted for other catastrophes.[2,13] The reader is cautioned, however, that the paraxial approximation of the distance function ϕ in Eq. (11) does not properly describe the caustic and associated wave field for all catastrophes.

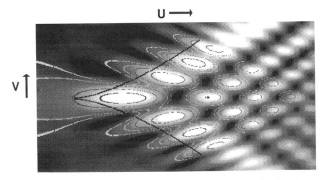

Fig. 4 Wave field recorded by scanning a hydrophone through a transverse cusp diffraction catastrophe. Bright regions correspond to high-pressure amplitudes of the 800-kHz reflected wave. Predicted equiamplitude contours of the Pearcey function pattern are shown. The predicted caustic is the dark cusp curve. The displayed field has a width of 15 cm and height of 8 cm and the distance from the curved reflector was 67 cm. (Reproduced, by permission, from Ref. 15, p. 2874.)

For reasons reviewed in Ref. 2, it may be necessary to include additional terms in the expansions of ϕ for certain classes of caustics except for far-field observers and sources. The effect of the finite width or aperture of a reflector may be accounted for by a convolution with the aperture diffraction pattern.[2]

2.6 Edge Rays and Geometric Theory of Diffraction

Suppose that a scatterer contains sharp edges. Then it is possible to separate the wave field into direct and edge ray contributions. Consider first the example of a straight rigid wedge or a rigid half-plane. The Sommerfeld solution to the wave field (see, e.g., Ref. 1) may be expressed in terms of Fresnel integrals. Keller, in publications reprinted in Ref. 3, showed that an asymptotic expression of the Fresnel integral may be used to identify an edge ray contribution to the wave field and formulated an approach to approximating such wave fields for curved edges for which exact solutions are not available. That original method, or the geometric theory of diffraction (GTD) representation of wave fields, failed at shadow boundaries. Subsequently, uniform representations were formulated that are applicable near the shadow boundary, and these are discussed in Refs. 2, 3, 7, and 12.

2.7 Fresnel Volumes and Heuristic Criteria for Elementary Geometric Methods

Kravtsov[16] has reviewed the idea of identifying a Fresnel volume associated with each ray from a source to an observer and shows how that identification can be useful for judging the applicability of elementary ray methods such as GTD and those discussed in Section 2.1. Let $\phi(S, Q)$ denote the path length along a ray from a source at S to an observer at Q. Consider paths that deviate from $\phi(S, Q)$ by $\lambda/2$, where λ is the wavelength. The boundary of the Fresnel volume is taken to be the surface generated by those points on the deviant paths that are most distant from the true ray. For simple propagation through a uniform media from S to Q, the Fresnel volume is a prolate spheroid. The definition may be generalized[7,3,16] to include propagation from an initial wavefront with unequal curvatures to Q. Elementary geometric results such as Eq. (3) and the method of stationary phase require that when there is more than one ray to Q, an appreciable fraction of the Fresnel volumes of the different rays do not overlap. This is not the situation when Q is close to (or on) a caustic surface since an appreciable overlap of their Fresnel volumes is associated with a merging of rays. Consequently, the divergent geometric amplitudes are replaced by canonical diffraction integrals, as illustrated in Section 2.5. The criterion is also applicable to understanding the failure of the elementary formulation of GTD near a shadow boundary. As the boundary is approached, there is an appreciable overlap of the Fresnel volumes of the direct and edge rays.[2,3,16]

The intersection of the Fresnel volume with its associated initial wavefront defines the first Fresnel zone on that wavefront for propagation to the specified observer at Q. The stationary phase approximation of double integrals, Eq. (11), for ray amplitudes may be viewed as the contribution from the patch of the initial wavefront given by the Fresnel zone.[3]

3 LIQUID-FILLED SPHERICAL REFLECTORS AND GLORY SCATTERING

The geometric and physical optics methods summarized in Section 2 may be used to understand an enhancement in the backscattering from penetrable fluid[17] and elastic[2] spheres. Associated with rays that are refracted and internally reflected so as to leave the sphere parallel to but offset from the backscattering axis is an axial caustic extending to infinity. An analogous caustic in light scattered from drops is known as the glory.[2] Figure 5a shows the relevant ray for a situation of practical significance: liquid-filled spherical reflectors used for sonar calibration and tracking.[18] For simplicity, the shell that confines the fluid is taken to be sufficiently thin that any effect on the rays is neglected in the analysis summarized below from Refs. 2 and 17. The outgoing wavefront is toroidal such that its Gaussian curvature vanishes on a circle having the same radius $b_n = a \sin \theta_n$ as the impact parameter for the class (designated by the index n) of ray considered. The two-chord glory ray shown, designated by $n = 2$, exists provided the acoustical refractive index $N = c/c_i$ is between $\sqrt{2}$ and 2. Here c_i denotes the speed of sound in the inner liquid. Ray methods are used to approximate the amplitude of the outgoing wave in an exit plane that just touches the sphere at the optic axis. The curvature $(r_n)^{-1}$ of the wavefront is negative for the value of N used in Fig. 5a so that the toroidal wave converges through an external focal circle. The Fraunhofer approximation and the axial symmetry simplify Eq. (10) so that the magnitude of the contribution to the form function f becomes

$$|f_n(\gamma = \pi - \theta)|$$

$$\approx \frac{2kQ_n^{-1/2}}{\pi} \left| \int_0^\infty sB(s)J_0(ks \sin \gamma)e^{-ikW} \, ds \right|,$$

$$\gamma \ll 1 \text{ rad}, \qquad (14)$$

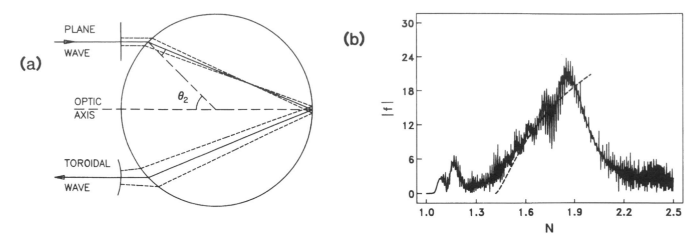

Fig. 5 Ray analysis relevant to thin-walled liquid-filled spherical reflectors. (*a*) Ray tube for the glory ray when the acoustical refractive index $N = 1.85$. The wavefront curvature has been exaggerated for clarity. (*b*) Exact $|f|$ for backscattering by a neutrally buoyant liquid sphere with $ka = 100$ plotted as a function of N. The dashed curve is the contribution $|f_2|$ from the ray in (*a*). The assumed quadratic form of the outgoing wavefront W breaks down as $N \to 2$ as W becomes proportional to s^4. For that case $|f_2|$ is calculated in Ref. 2.

where s is the distance from the optic axis, the toroidal wavefront shape is $W(s) = -(s - b_n)^2/2r_n$, $Q_n = |r_n/(r_n - a)|$ is a ratio of the outgoing ray tube width to that of the incident width, $B(s) = T_1 R^{n-1} T_2$ is a product of reflection and transmission factors for the ray at s, and γ is the backscattering angle. If $kb_n \gg 1$, the integral may be approximated giving the simple result[22] $|f_n| \approx (ka)^{1/2} 2^{2/3} \pi^{1/2} (b_n/a) Q_n^{-1/2} |B(s = b_n) J_0(kb_n \sin \gamma)|$. The comparison of $|f_2(\gamma = 0)|$ with the exact $|f|$ from the partial-wave series shown in Fig. 5*b* gives a satisfactory agreement over the range of N where the analysis is applicable. The $(ka)^{1/2}(b_n/a)$ enhancement factor and the dependence on the Bessel function J_0 are characteristic of axial focusing. In other examples,[2] reflections along the optic axis were significant.

4 HOLLOW SPHERICAL AND CYLINDRICAL SHELLS AND RESONANCES

Consider now backscattering by hollow metallic shells. As illustrated in Fig. 6, the coupling of sound with waves guided by spherical shells gives rise to backward-directed toroidal wavefronts. The construction of the local and far-field outgoing wave amplitudes is more complicated than for Fig. 5 since the interaction of the acoustic field with the guided wave must be analyzed. The associated coupling coefficient G_l is approximated by the Watson transformation (Chapters 5 and 42) or by other methods.[19-26] Here l is an index that specifies the type of guided wave. The contributions to the scattering are found to be governed by surface wave parame-

ters and by the shape and perimeter of the surface. Consider first the case of a leaky wave that by definition has a phase velocity c_l (along the outer surface of the shell) that is somewhat larger than c. The analysis of the leakage of sound to the surrounding fluid in Ref. 6 for plane surfaces is discussed for curved surfaces in Ref. 24. [The notion of Fresnel volume (Section 2.7) was used to describe the widths of regions for the launching and detachment of leaky waves and for anticipating effects of shell truncations.] In addition to the clockwise ray shown in Fig. 6, there will be an equal contribution to the backscattering from a counterclockwise ray not shown. In the discussion below, f_{ml} is the resulting combined contribution to the form function in Eq. (6) from the mth complete circumnavigation. For $m = 0$, the wave has gone only between points B and B'. For steady-state scattering, the contributions from repeated circumnavigations are superposed, giving a total contribution to f of[19-23]

$$f_l = \sum_{m=0}^{\infty} f_{ml}, \qquad (15a)$$

$$\eta_l = 2ka \left[\frac{c}{c_l} (\pi - \theta_l) - \cos \theta_l \right] - \frac{(j+1)\pi}{4}, \qquad (15b)$$

$$f_{ml} = -G_l \exp(i\eta_l) \exp[-2(\pi - \theta_l)\beta_l] Z^m, \qquad (16a)$$

$$Z = -j \exp(-2\pi\beta_l) \exp\left(i2\pi ka \frac{c}{c_l} \right), \qquad (16b)$$

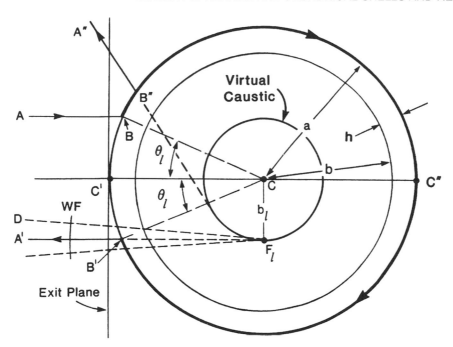

Fig. 6 Ray diagram for leaky Lamb wave contributions to backscattering from empty spherical shells. The wave is excited at B and radiates back from B'. WF is an evolute of the virtual caustic and designates a backward directed toroidal wavefront since the figure may be rotated around the CC' axis. The figure is also applicable to right circular cylinders where WF becomes a cylindrical wavefront. For solid spheres or cylinders, the leaky wave can be a generalization of a Rayleigh wave. (Reproduced, by permission, from Ref. 22, p. 2546.)

where the index $j = 1$ for spheres, $\theta_l = \arcsin(c/c_l)$, and β_l is the radiation damping of the guided wave in nepers per radian. For the case of a solid sphere, Eqs. (15) and (16) were derived directly from the Watson transformation, which also gave an expression for G_l in terms of the material parameters.[19,20] The ray diagram is also applicable to a cylinder with its axis perpendicular to the incident wave. The contribution to the scattered pressure is given through Eq. (8), and unlike the sphere, there are no polar caustics at C' and C'', so that j becomes -1. Approximations applicable to leaky waves requiring $\beta_l \ll kac/c_l$ nepers per radian express G_l in terms of the surface wave parameters on spheres and cylinders[20–27]:

$$[G_l]_{\text{sp}} \approx 8\pi\beta_l \frac{c}{c_l}, \tag{17a}$$

$$[G_l]_{\text{cy}} \approx \left[\frac{8\pi\beta_l}{(\pi ka)^{1/2}} \right] \exp \frac{i\pi}{4}. \tag{17b}$$

A phase correction $-2\tan^{-1}(x_N/ka \cos\theta_l)$ becomes important for thin shells depending on the value of the dimensionless "null frequency" $x_N = \rho a/\rho_E h$, where ρ_E and h are the density and thickness of the shell. Larger β_l more strongly couple the guided wave with sound.

The series in Eq. (15a) is a geometric series that may be analytically summed.[19–23] At resonances, the f_{ml} all have the same phase and $|f_l|$ is enhanced. A ray approximation to f is given by summing a specular contribution f_S with contributions from existing leaky waves in the ka region of interest. Such a synthesis has been verified by comparison with the exact $|f|$ from the partial-wave series for solid spheres[19,20] and spherical[22] and cylindrical[27] shells. From the Watson transformation, the required surface wave parameters are given by finding the roots in ν of $D_\nu(ka) = 0$, where D is the denominator of the applicable partial-wave series with the integer partial-wave index replaced by a complex number ν. The numerical procedure and examples for thick[21,22] and thin[3,28] shells are published. The synthesis for scattering by thick shells has been found to be applicable for ka as small as 7. Contributions are included in the f_S for shells from the reverberation of longitudinal waves within the shell[22]; they shift the phase of the effective reflection coefficient R_e in Eq. (7) by $-2\tan^{-1}(x_N/ka)$.[25] Additional support for the leaky ray model comes from measurements of the backscattering amplitude of short tone bursts[19,21] and pressure impulses[29] where different circumnavigations and ray types are separated in time. The impulse response also contains a specular feature indicative of the mass-per-area $\rho_E h$ of the shell.[29]

At sufficiently high frequencies there are large backscattering enhancements not describable by the diagram in Fig. 6 for objects as simple as an empty spherical shell. These have been observed and modeled with modified ray diagrams for the coincidence frequency region[3,28,29] and a process associated with a negative group velocity leaky wave.[30] For metallic objects in water, the ka regions of the former and latter enhancements are typically near a/h and $\pi ac_L/ch$, respectively, where c_L is the longitudinal wave speed in the elastic material. They are expected to be relevant only to high-frequency sonar systems.

Acknowledgment

Preparation of this chapter was supported by the U.S. Office of Naval Research.

REFERENCES

1. A. D. Pierce, *Acoustics, An Introduction to Its Physical Principles and Applications*, Acoustical Society of America, New York, 1989, Chaps. 7 and 8.

2. P. L. Marston, "Geometrical and Catastrophe Optics Methods in Scattering," in *Physical Acoustics*, Vol. 21, Academic, Boston, 1992, pp. 1–234.

3. P. L. Marston (Ed.), *Selected Papers on Geometrical Aspects of Scattering*, SPIE, Bellingham, WA, 1994.

4. D. J. Struik, *Lectures on Classical Differential Geometry*, Dover, New York, 1988.

5. J. A. Ogilvy, *Theory of Wave Scattering from Random Rough Surfaces*, Hilger, Bristol, 1991.

6. H. L. Bertoni and T. Tamir, *Appl. Phys.*, Vol. 2, 1973, pp. 157–172.

7. H. M. Nussenzveig, *Diffraction Effects in Semiclassical Scattering*, University Press, Cambridge, U.K., 1992.

8. O. N. Stravrodis, *Optics of Rays, Wavefronts and Caustics*, Academic, New York, 1972.

9. R. J. Urick, *Principles of Underwater Sound*, 2nd ed., McGraw-Hill, New York, 1975.

10. L. Flax, G. C. Gaunaurd, and H. Überall, "Theory of Resonance Scattering," in *Physical Acoustics*, Vol. 10, Academic, New York, 1981, pp. 191–294.

11. J. J. Boweman, T. B. A. Senior, and P. L. E. Uslenghi (Eds.), *Electromagnetic and Acoustic Scattering by Simple Shapes*, Hemisphere, New York, 1987.

12. P. Y. Ufimtsev, "Theory of Acoustical Edge Waves," *J. Acoust. Soc. Am.*, Vol. 86, 1989, pp. 463–474.

13. M. V. Berry and C. Upstill, "Catastrophe Optics: Morphologies of Caustics and Their Diffraction Patterns," in *Progress in Optics*, Vol. 18, Elsevier, New York, 1980, pp. 257–346.

14. J. D. Maynard, E. G. Williams, and Y. Lee, *J. Acoust. Soc. Am.*, Vol. 78, 1985, pp. 1395–1413.

15. C. K. Frederickson and P. L. Marston, *J. Acoust. Soc. Am.*, Vol. 92, 1992, pp. 2869–2877; Vol. 95, 1994, pp. 650–660.

16. Y. A. Kravtsov, "Rays and Caustics as Physical Objects," in *Progress in Optics*, Vol. 26, Elsevier, New York, 1988, pp. 227–348.

17. P. L. Marston and D. S. Langley, *J. Acoust. Soc. Am.*, Vol. 73, 1983, pp. 1464–1475; (E) Vol. 75, 1985, p. 1128.

18. D. L. Folds and C. D. Loggins, *J. Acoust. Soc. Am.*, Vol. 73, 1983, pp. 1147–1151.

19. K. L. Williams and P. L. Marston, *J. Acoust. Soc. Am.*, Vol. 78, 1985, pp. 1093–1102; Vol. 79, 1986, pp. 1702–1708.

20. P. L. Marston, *J. Acoust. Soc. Am.*, Vol. 83, 1988, pp. 25–37.

21. S. G. Kargl and P. L. Marston, *J. Acoust. Soc. Am.*, Vol. 85, 1989, pp. 1014–1028; (E) Vol. 89, 1991, p. 2426.

22. S. G. Kargl and P. L. Marston, *J. Acoust. Soc. Am.*, Vol. 88, 1990, pp. 1103–1113; Vol. 89, 1991, pp. 2545–2558.

23. J. M. Ho and L. B. Felsen, *J. Acoust. Soc. Am.*, Vol. 88, 1990, pp. 2389–2414.

24. P. L. Marston, *J. Acoust. Soc. Am.*, Vol. 96, 1994, pp. 1893–1898; Vol. 97, 1995, pp. 34–41.

25. D. A. Rebinsky and A. N. Norris, *J. Acoust. Soc. Am.*, Vol. 98, 1995, pp. 2368–2371.

26. P. L. Marston, *Wave Motion*, Vol. 22, 1995, pp. 65–74.

27. N. H. Sun and P. L. Marston, *J. Acoust. Soc. Am.*, Vol. 91, 1992, pp. 1398–1402.

28. G. Kaduchak and P. L. Marston, *J. Acoust. Soc. Am.*, Vol. 93, 1993, pp. 224–230.

29. G. Kaduchak, C. S. Kwiatkowski, and P. L. Marston, *J. Acoust. Soc. Am.*, Vol. 97, 1995, pp. 2699–2708.

30. G. Kaduchak, D. H. Hughes, and P. L. Marston, *J. Acoust. Soc. Am.*, Vol. 96, 1994, pp. 3704–3714.

44

TARGET STRENGTH OF FISH

KENNETH G. FOOTE

1 INTRODUCTION

The target strength (TS) or backscattering cross section σ of a fish is a physical measure of the potential of the animal to produce an echo. It has been actively investigated for over 30 years[1] because of its importance in acoustic surveys of fish stocks[2] and other remote measurements of fish. In surveying by echo counting, the sampling volume depends on TS.[3] In surveying by echo integration, the measured area or column backscattering coefficient s_A is converted to an area fish density by dividing s_A by σ.[4] Knowledge of target strength is also important for modeling reverberation in sonar performance studies and for evaluating the sampling efficiency of trawls, among other applications.

In this chapter, target strength is defined and the physics of scattering by fish is described. Methods of determining target strength are summarized. Several equations are presented together with critical comments on their applicability. Some open problems are mentioned.

2 DEFINITIONS

The backscattering cross section σ of a finite target is quite often defined with respect to several idealized conditions: (1) the incident field is monochromatic, or single-frequency in spectral composition; (2) the incident field is weak, so that associated propagation and scattering processes are linear; (3) the target is in the transducer farfield; (4) the target subtends a small angle relative to

the transducer and hence is uniformly insonified; (5) the backscattered field can be measured far from the target; and (6) the measuring instrument or receiver has a flat frequency response function. If a field quantity such as pressure or velocity potential is denoted by ψ, then

$$\sigma = \lim_{r \to \infty} \frac{4\pi r^2 |\psi_{\text{bsc}}|^2}{|\psi_{\text{inc}}|^2}, \tag{1}$$

where r is the measurement range relative to the target.[5]

The first and sixth assumptions are routinely violated, requiring a significant generalization of Eq. (1). To incorporate the effect of finite signal duration, the squared amplitudes are replaced by measures of energy. To incorporate the effect of receiver bandwidth, the measures are referred to the receiver. Thus,

$$\sigma = \lim_{r \to \infty} \frac{4\pi r^2 \epsilon_{\text{bsc}}}{\epsilon_{\text{inc}}}, \tag{2}$$

where ϵ_{bsc} denotes the energy in the received echo and ϵ_{inc} denotes the energy in the incident field as it would be received by the transducer when oriented toward the source and placed at the target location. This definition is tantamount to that used in calibration work.[6]

Because of the large range in values of σ that are typically encountered in scattering phenomena, it is often convenient to employ a logarithmic measure. This is the target strength,

$$\text{TS} = 10 \log \frac{\sigma}{4\pi}, \tag{3}$$

where the common, or base-10, logarithm is understood

Encyclopedia of Acoustics, edited by Malcolm J. Crocker
ISBN 0-471-80465-7 © 1997 John Wiley & Sons, Inc.

and σ is expressed in SI units. The reference implied in this definition is that of the idealized perfectly reflecting sphere of 2-m radius, for which TS = 0 dB. An alternative, equivalent definition is TS = $10 \log[\sigma/(4\pi r_0^2)]$, where r_0 is a reference distance, typically 1 m.

The several equations thus far concern scattering by a single target in a fixed state, including orientation and biological condition in the case of fish. Quite often it is necessary to study an ensemble of scatterers or scattering states. In this case, the mean backscattering cross section $\overline{\sigma}$ may be the most applicable quantity. It is derived from the value $\sigma(\alpha)$ characterizing the target in state α by integrating over the probability density function $f(\alpha)$:

$$\overline{\alpha} = \int_{-\infty}^{\infty} \sigma(\alpha)f(\alpha)\,d\alpha. \qquad (4)$$

The generalization to the multivariate case is immediate. The distribution of σ is also frequently the sought quantity.

If the same quantity were to be expressed logarithmically, then the averaging would still be done in the intensity or σ-domain before logarithmic conversion of the result. For example, the quantity $10 \log(\overline{\sigma}/4\pi)$ is often called the average or mean target strength and denoted $\overline{\text{TS}}$, for which it is a convenient abbreviation. It is used in the same way in this chapter.

3   PHYSICS

Insofar as certain physical properties of fish differ from those of the immersion medium, echo formation is inevitable. In general, the density and speed of longitudinal sound waves in fish flesh[7] are different from the corresponding properties of the immersion medium. Excitation of transverse waves in the fish body also produces echoes.

Some fishes bear gas-filled bladders called swimbladders (Fig. 1). When present, the swimbladder generally dominates scattering, because of the large contrast in compressibility, or density, between the bladder and surrounding fish flesh and water. At ultrasonic frequencies well above resonance, the swimbladder contributes significantly to the total echo energy, hence σ.[8] At or near resonance, the swimbladder clearly dominates the fish echo.[9]

Swimbladdered fish are either physostomes or physoclists, as the swimbladder is open, with external duct, or closed, without the same. Examples of physostomes are clupeoids, such as herring (*Clupea harengus*). Examples of physoclists are gadoids, such as cod (*Gadus morhua*) and walleye pollock (*Theragra chalcogramma*),

Fig. 1 Side view of a Norwegian spring-spawning herring (*Clupea harengus*), female, length 370 mm, weight 399 g, with exposed swimbladder. (Adapted from an original drawing by H. T. Kinacigil with permission.)

and many mesopelagic fish of the families Gonostomatidae, Myctophidae, and Sternoptychidae.

Nonswimbladdered fish [e.g., mackerel (*Scomber scombrus*)] produce rather weak echoes compared with their bladdered counterparts. The echoes may, however, still be sufficient for routine acoustic detection.

The target strength of fish also depends on fish size. This is commonly characterized by a measure of fish length, such as total length, or distance from the tip of the snout to the end of the tail fin. Fish size is also characterized by mass, but it is usual to refer target strength to fish length and, if required, to fish mass by means of the length–mass relationship.

At very low frequencies, Rayleigh scattering dominates, and σ is proportional to the square of fish volume and the fourth power of frequency. At or near swimbladder resonance, for bladdered fish, σ varies with fish size and frequency as, for example, a damped air bubble[10] or undamped cylindrical gas inclusion[11] would. At higher frequencies, fish scattering may be regarded as arising from a number of distinct scattering processes. The combined effect of these explains, very roughly, an increase in σ with fish size as the square of fish length, or geometric cross section of the fish, and approximate constancy of σ with respect to frequency. More insight can be gained into high-frequency scattering by one of two approaches. If the number of scattering processes is low, as is the case with many swimbladdered fishes, the scattering may be modeled as a deterministic process. If the number is high, the scattering may be modeled as a stochastic process.

At frequencies where the fish length is at least several times the wavelength, σ is a sensitive function of fish orientation. Since fish properties are often studied or applied in bulk, statistical measures of fish target strength may be referred to distribution functions of fish orientation, which also characterize behavior.[12,13]

4   METHODS OF DETERMINATION

4.1   Measurements in Resonance Region

Measurement of target strength at low frequencies near or at the frequency of swimbladder resonance requires special techniques to overcome at least two practical difficulties. These are generation of a suitably wideband waveform for resonance excitation and separation of transmitted and echo waveforms.

Continuous-Wave Source

McCartney and Stubbs[14] have measured target strength across the region of swimbladder resonance with a continuous-wave source, with frequency swept from 20 Hz to 20 kHz. Two ring hydrophones receive both transmitted and echo waveforms. One hydrophone is placed around the small, cylindrical net cage that holds a single fish specimen, while the second hydrophone is suspended 1 m away and at equal range from the source to measure the source level. Measurements repeated both with and without the fish present allow compensation to be made for the net cage and measurement geometry. Løvik and Hovem[15] have employed the same technique.

Impulse Source

Holliday[16] has used seal control bombs consisting of 2.5 g of flash powder to generate a low-frequency wide-band signal. Comparison of the measured power spectral density of the explosive shot with that of the echo enables the frequency-dependent scattering function to be extracted. This has been done for aggregations and schools of a number of species. Since the square of the magnitude of the scattering or acoustic transfer function is proportional to σ, the data can also be expressed directly in terms of target strength.

4.2 Measurements at Ultrasonic Frequencies

Nearly all measurements of target strength are performed at the same frequencies as are used in acoustic surveys of fish abundance. These frequencies are generally ultrasonic because of the need to detect and sometimes resolve single fish at ranges up to several hundred meters. Typical parameters characterizing this process are transmit power of 0.1–5 kW over 0.1–3 ms with transducer beamwidth of 2°–10°, as measured between opposite −3-dB levels. Signal waveforms are typically pulsed sinusoids, with carrier or center frequency in the range 30–500 kHz. Transducer sizes seldom exceed 50 cm in maximum linear dimension and are generally less than about 35 cm on a side.

Measurement methods at ultrasonic frequencies are divided into three groups.[17] In the first two groups, the target fish is measured in situ, as the fish is found in nature. Such measurements are further categorized as being indirect or direct, according to the method of removal of the effect of transducer directivity. In the third measurement group, the target fish is observed ex situ, or outside of its natural environment, in an attempt to achieve greater control over the fish than is possible when the measurement is performed in situ.

Indirect in Situ Methods

Three quite different methods are described. The first two depend on acoustic resolution of individual fish, while the third may apply at arbitrary densities.

The echo energy ϵ, or proportional quantity such as the echo intensity for sufficiently long transmit pulses, due to a single fish may be expressed as the product $gb^2\sigma$, where g is a gain factor that may contain equipment constants or other scaling factors and b^2 is the product of transmit and receive beam patterns. The problem to be addressed is determination of the probability density function (pdf) of σ from an ensemble of measurements of ϵ. Solutions have been specified in each of three domains, whose basis equations are the following:

$$\epsilon = gb^2\sigma \quad \text{(intensity domain)}, \tag{5a}$$
$$\epsilon^{1/2} = (gb^2\sigma)^{1/2} \quad \text{(amplitude domain)}, \tag{5b}$$
$$\log \epsilon = \log(g\sigma) + \log b^2 \quad \text{(logarithmic domain)} \tag{5c}$$

In the intensity domain the pdf f of each of the variables in Eq. (5a) are connected through the integral equation

$$f_E(\epsilon) = \int_0^1 f_A\left(\frac{\epsilon}{\beta}\right) f_B(\beta)\, \frac{d\beta}{\beta}, \tag{6}$$

where E denotes the variable ϵ, A denotes $g\sigma$, and B denotes b^2. A similar equation applies in the amplitude domain. In the logarithmic domain, the several pdf's are connected by the integral equation

$$f_Z(z) = \int_{-\infty}^0 f_X(z-y) f_Y(y)\, dy, \tag{7}$$

where Z denotes the variable $\log \epsilon$, X denotes $\log(g\sigma)$, and Y denotes $\log b^2$.

Nonparametric solutions to the integral equations make no assumptions about the pdf of σ, $\sigma^{1/2}$, or $\log \sigma$. The simple recursive solution to Eq. (7), when discretized, is known to be unstable. To avoid instabilities, therefore, Degnbol et al.[18] added the constraint that the resulting pdf f_X be nonnegative. Ehrenberg[19] and Robinson[20] have each solved Eq. (6) by approximating the pdf of σ by one or several polynomials. Clay solved the same equation by deconvolution using the Laplace Z-transform.[21]

Parametric solutions do assume definite forms for the pdf of σ, $\sigma^{1/2}$, or $\log \sigma$. Early parametric solutions assumed that $\sigma^{1/2}$ is Rayleigh distributed. These were found to apply only to large fish sizes relative to the wavelength.[22] Clay and Heist[23] subsequently employed

Rice's two-parameter pdf. The parameters, called the concentrated and distributed scattering components σ_c and σ_d, respectively, are related through their ratio $\gamma = \sigma_c/\sigma_d$. When γ is small, the Rayleigh distribution obtains; when γ is large, the Gaussian distribution results. These two limiting cases apply respectively to large and small fish.

The statistical methods described here work best or only for a narrow size range of scatterers and high signal-to-noise ratio in the fundamental measurement.

Single-Target Echo Integration. In another case where fish are acoustically resolved as single targets, it is possible to measure the area fish density by each of two techniques. The echo counting technique determines the area fish density ρ_A directly, while the echo integration technique measures the area backscattering coefficient s_A.[4] Since s_A is just the product of ρ_A and the mean value $\bar{\sigma}$, $\bar{\sigma} = s_A/\rho_A$, which is Ona and Hansen's solution.[24]

Multiple-Target Echo Integration. In the case where fish targets cannot be individually resolved, it may be possible to estimate $\bar{\sigma}$ by means of echo integration accompanied by a nonacoustic means of enumeration. Two such methods are based on postsurveying seining[25] and counting of migrating fish at a weir.[26]

Direct in Situ Methods Two powerful methods for determining fish target strength in situ are based on multi-element transducers, with determination of target position in the beam. Joint acoustic and photographic observation constitutes a third direct method,[27] but this is only mentioned.

Dual-Beam Method. According to the dual-beam method developed by Ehrenberg,[28] a circular array of elements transmits a narrow beam but receives echoes with each of two beams, a similarly narrow beam formed by all of the elements and a relatively broad beam due to a small central core of elements. The echo energy or proportional echo intensity for the same fish as received on the narrow (N) and wide (W) beams is given as

$$\epsilon_N = g_N b_N^2 \sigma \tag{8a}$$

and

$$\epsilon_W = g_W b_N b_W \sigma. \tag{8b}$$

If $b_W \doteq 1$, then, to a good approximation,

$$\sigma = \frac{g_N \epsilon_W^2}{g_W^2 \epsilon_N}. \tag{9}$$

Thus, given prior knowledge of g_N and g_W and simultaneous measurements of ϵ_N and ϵ_W, σ can be immediately determined. The assumption $b_W = 1$ can be relaxed.[29]

Split-Beam Method. Use of a transducer divided into quadrants allows the formation of quadrant beams, which can be paired to detect alongships and athwartships phase differences.[30,31] These allow the target position to be determined, which specifies the value of b^2 in Eq. (5a). Simultaneous measurement of ϵ allows immediate solution for σ.

Ex Situ Methods To avoid the greatest hazard of in situ measurement, which is not knowing precisely what the target is, ex situ techniques have been developed and pursued since the beginning of fish target strength studies. These are summarized here in two separate categories.

Tethered Single-Fish Measurement. Killed, stunned, anesthetized, and sometimes swimming fish have been suspended in a known position of a transducer beam by tethering. Since b^2 is known, measurement of ϵ in Eq. (5a) allows σ to be determined. Midttun gives 17 references to such measurements.[1] A significant aim of these has been description of the orientation dependence of target strength. Combination of such measurements with information on orientation[12] may specify the effective σ or TS for application to fish in the wild.

Caged Fish. Measurement of swimming fish confined to a cage has been a popular method for determining target strength. This may depend on the spatial distribution of fish in the cage if the transducer beam is narrow, but the echo integration method can, with care, be applied in any case and σ determined as in the above multiple-target echo integration technique. Edwards and Armstrong[32] describe an extensive series of caged-fish measurements. Other variants on the basic method are found in the bibliography to Ref. 33.

4.3 Model Calculations

Model calculations have been performed for both swimbladder-bearing and nonswimbladdered fish. Four models are briefly described.

Resonant Frequency Model The basic model for swimbladdered fish at or near resonance has been given by Andreeva.[34] Accordingly,

$$\sigma = \frac{4\pi a^2}{(\omega_0^2/\omega^2 - 1)^2 + (\omega_0 \delta/\omega)^2}, \quad (10)$$

where a is the equivalent spherical radius of the swimbladder, ω_0 is the angular resonant frequency, and δ is the damping factor at resonance. Determination of ω_0 was originally performed for small spherical air bubbles. This has been extended to the ellipsoidal shape by Weston.[35] The model has been further developed by Love.[10]

Prolate Spheroidal Model   Swimbladdered fish have been modeled at high frequencies by Furusawa,[36] assuming the equivalence of the fish to an ideal, soft ellipsoid. Furusawa has in the same work modeled nonswimbladdered fish as a homogeneous liquid spheroid with similar mean physical properties. The attraction of a spheroidal model is that the wave equation separates and can be directly solved. By considering the effect of gas in the interior of a prolate spheroid, Stanton has shown that a resonant-like behavior is obtained,[37] but with low Q and without the need for a large damping factor in Eq. (10).

Kirchhoff Model   Swimbladder-bearing fishes have also been modeled by assuming the equivalence of the fish and exact swimbladder shape as an ideal pressure-release surface.[38,39] At sufficiently high frequencies, well above resonance, the incident pressure field on the swimbladder surface S can be expressed as the field that would exist in the absence of diffraction; hence

$$\sigma = 4\pi \left| \lambda^{-1} \oint_S \exp(2i\mathbf{k} \cdot \mathbf{r}) \mathcal{H}(\hat{k} \cdot \hat{n}) \hat{k} \cdot \hat{n} \, dS \right|^2, \quad (11)$$

where λ is the wavelength, $\mathbf{k}$ is the wavevector, $\mathbf{r}$ is the position vector of the surface element of infinitesimal area dS, $\hat{n}$ is the unit normal to dS at $\mathbf{r}$, and $\mathcal{H}(x)$ is the Heaviside step function, with values $1, \frac{1}{2}, 0$ for $x > 0$, $x = 0$, $x < 0$, respectively.

Stochastic Model   Clay and Heist[23] have modeled a swimbladdered fish by two line arrays of point scatterers, one representing the swimbladder and the other representing the body. The swimbladder array is concentrated and considered rigid, while the distributed body array may flex in simulating swimming. The orientation of the two arrays relative to the transducer is also allowed to vary with movement. Scattering simulations performed for a range of orientations and degrees of body flexure show consistency with the Ricean distribution, given

association of the unobservable concentrated and distributed scattering components through the parameter γ, already defined in Section 4.2. This parameter depends on both fish behavior and morphology.

5 EMPIRICAL EQUATIONS AND OTHER DATA FOR APPLICATION

There is an enormous diversity of data on fish target strength. For reasons of uncertainty over data quality, the data presented here are deliberately limited to several cases.

For echo sounding applications, average and maximum values of σ or TS for fish in dorsal aspect are useful. The following equations for $\overline{\text{TS}}$ are based on in situ measurements of $\overline{\sigma}$ at 38 kHz (Fig. 2)[40]:

$$\overline{\text{TS}} = 20 \log l - 67.5 \quad \text{(physoclists)}, \quad (12a)$$

$$\overline{\text{TS}} = 20 \log l - 71.9 \quad \text{(physostomes)}, \quad (12b)$$

where l is the total fish length in centimeters. The nominal length ranges spanned by the basic data are 10–100 cm for the physoclists and 10–30 cm for the physostomes. Maximum values of the dorsal aspect TS may be derived from corresponding mean values by adding 7.1 dB, that is, $\text{TS}_{\text{max}} = \overline{\text{TS}} + 7.1$. For nonswimbladdered fish, σ may very roughly be assumed to be 5–10% of the corresponding value of σ for physoclists.[8] Mean and maximum values of TS given here may be assumed to apply at least approximately over the frequency range 38–120 kHz.

For the arbitrary aspect and fish species, without regard to the presence or absence of a swimbladder, Love[41] has expressed target strength through the standard relation

$$\frac{\sigma}{\lambda^2} = a \left(\frac{l}{\lambda} \right)^b, \quad (13)$$

where the parameters a and b are tabulated by aspect. The parameter b is close to 2 in nearly all cases. Interestingly, Love's equation for σ averaged over all aspects, namely $\sigma = 0.015 l^2$, is tantamount to the following:

$$\text{TS} = 20 \log l - 69.2. \quad (14)$$

6 OPEN PROBLEMS

Despite a long history of work on fish target strength, surprisingly much remains to be learned. Particular out-

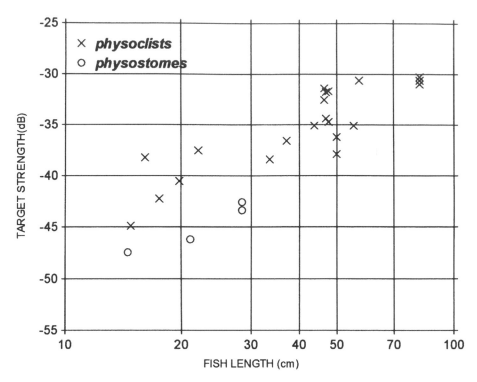

Fig. 2   Scatter diagram of mean value of target strength and mean fish length for physoclists and physostomes as measured in situ at 38 kHz.

standing problems include the dependence of target strength on biological state, depth, depth history, and frequency. There is strong evidence for an inverse relationship of TS and fat content for herring, for instance, which may be attributed to the buoyancy function of the swimbladder.[42] Such knowledge needs to be refined and applied to other species. The effect of depth excursion, including rates of depth change on fish with *rete mirabile*, or the network of fine blood vessels that facilitate gas exchange with the swimbladder, hence controlling its state of inflation, is a completely open problem. The problem of the frequency dependence is also open, although present findings are that this must be rather weak, at least over the relative l/λ range of 1–100.[41] For fish exceeding about 35 cm in length, there may be no systematic frequency dependence at ultrasonic frequencies.[43]

Even the length dependence has not been fully specified. The most uncertain size regime is that of small fish, with lengths less than about 15 cm, for which in situ measurement is especially difficult.

The connection between TS and fish behavior seems to be well established, but there are few data on the orientation distribution of fish.[44] The connection between backscattering and extinction cross sections is beginning to be elucidated.[45,46]

Acknowledgments

Reviews by or discussions with J. Barger, C. S. Clay, and T. K. Stanton were useful in revising the manuscript. H. T. Kinacigil is thanked for preparing the original drawing from which Fig. 1 was derived, and M. Ostrowski is thanked for preparing Fig. 2.

REFERENCES

1. L. Midttun, "Fish and Other Organisms as Acoustic Targets," *Rapp. P.-v. Réun. Cons. int. Explor. Mer*, Vol. 184, 1984, pp. 25–33.

2. D. N. MacLennan, "Acoustical Measurement of Fish Abundance," *J. Acoust. Soc. Am.*, Vol. 87, 1990, pp. 1–15.

3. I. L. Kalikhman and W. D. Tesler, "The Effective Parameters of the Real Acoustic Beam," *FAO Fish. Rep.*, Vol. 300, 1983, pp. 9–17.

4. H. P. Knudsen, "The Bergen Echo Integrator: an Introduction," *J. Cons. int. Explor. Mer*, Vol. 47, 1990, pp. 167–174.

5. J. J. Bowman, T. B. A. Senior, and P. L. E. Uslenghi, "Introduction," in J. J. Bowman, T. B. A. Senior, and P. L. E. Uslenghi (Eds.), *Electromagnetic and Acoustic Scattering by Simple Shapes*, North-Holland, Amsterdam, 1969, pp. 1–86.

6. K. G. Foote, "Optimizing Copper Spheres for Precision Calibration of Hydroacoustic Equipment," *J. Acoust. Soc. Am.*, Vol. 71, 1982, pp. 742–747.

7. K. Shibata, "Experimental Measurement of Target Strength of Fish," in H. Kristjonsson (Ed.), *Modern Fishing Gear of the World*, Vol. 3, Fishing News (Books), London, 1971, pp. 104–108.

8. K. G. Foote, "Importance of the Swimbladder in Acoustic Scattering by Fish: A Comparison of Gadoid and Mackerel Target Strengths," *J. Acoust. Soc. Am.*, Vol. 67, 1980, pp. 2084–2089.

9. W. E. Batzler and G. V. Pickwell, "Resonant Acoustic Scattering from Gas-Bladder Fishes," in G. B. Farquhar (Ed.), *Proceedings of an International Symposium on Biological Sound Scattering in the Ocean*, Department of the Navy, Washington, D.C., 1970, pp. 168–179.

10. R. H. Love, "Resonant Acoustic Scattering by Swimbladder-Bearing Fish," *J. Acoust. Soc. Am.*, Vol. 64, 1978, pp. 571–580.

11. C. S. Clay, "Low-Resolution Acoustic Scattering Models: Fluid-Filled Cylinders of Fish with Swim Bladders," *J. Acoust. Soc. Am.*, Vol. 89, 1991, pp. 2168–2179.

12. O. Nakken and K. Olsen, "Target Strength Measurements of Fish," *Rapp. P.-v. Réun. Cons. int. Explor. Mer*, Vol. 170, 1977, pp. 52–69.

13. K. G. Foote, "Effects of Fish Behaviour on Echo Energy: The Need for Measurements of Orientation Distributions," *J. Cons. int. Explor. Mer*, Vol. 39, 1980, pp. 193–201.

14. B. S. McCartney and A. R. Stubbs, "Measurements of the Acoustic Target Strengths of Fish in Dorsal Aspect, Including Swimbladder Resonance," *J. Sound Vib.*, Vol. 15, 1971, pp. 397–420.

15. A. Løvik and J. M. Hovem, "An Experimental Investigation of Swimbladder Resonance in Fishes," *J. Acoust. Soc. Am.*, Vol. 66, 1979, pp. 850–854.

16. D. V. Holliday, "Resonance Structure in Echoes from Schooled Pelagic Fish," *J. Acoust. Soc. Am.*, Vol. 51, 1972, pp. 1322–1332.

17. K. G. Foote, "Summary of Methods for Determining Fish Target Strength at Ultrasonic Frequencies," *ICES J. mar. Sci.*, Vol. 48, 1991, pp. 211–217.

18. P. Degnbol, H. Lassen, and K.-J. Staehr, "*In situ* Determination of Target Strength of Herring and Sprat at 38 and 120 kHz," *Dana*, Vol. 5, 1985, pp. 45–54.

19. J. E. Ehrenberg, "A Method of Extracting the Fish Target Strength Distribution from Acoustic Echoes," *Proc. IEEE Conf. Eng. Ocean Environ.*, Vol. 1, 1972, pp. 61–64.

20. B. J. Robinson, "An *in situ* Technique to Determine Fish Target Strength with Results for Blue Whiting (*Micromesistius poutassou* Risso)," *J. Cons. int. Explor. Mer*, Vol. 40, 1982, pp. 153–160.

21. C. S. Clay, "Deconvolution of the Fish Scattering PDF from the Echo PDF for a Single Transducer Sonar," *J. Acoust. Soc. Am.*, Vol. 73, 1983, pp. 1989–1994.

22. J. E. Ehrenberg, T. J. Carlson, J. J. Traynor, and N. J. Williamson, "Indirect Measurement of the Mean Acoustic Backscattering Cross Section of Fish," *J. Acoust. Soc. Am.*, Vol. 69, 1981, pp. 955–962.

23. C. S. Clay and B. G. Heist, "Acoustic Scattering by Fish—Acoustic Models and a Two Parameter Fit," *J. Acoust. Soc. Am.*, Vol. 75, 1984, pp. 1077–1083.

24. E. Ona and K. Hansen, "*In situ* Target Strength Observations on Haddock," *Coun. Meet. int. Coun. Explor. Sea* (B:39), 1986.

25. O. Hagstrøm and I. Røttingen, "Measurements of the Density Coefficient and Average Target Strength of Herring Using Purse Seine," *Coun. Meet. int. Coun. Explor. Sea*, (B:33), 1982.

26. T. J. Mulligan and R. Kieser, "Comparison of Acoustic Population Estimates of Salmon in a Lake with a Weir Count," *Can. J. Fish. Aquat. Sci.*, Vol. 43, 1986, pp. 1373–1385.

27. L. V. Long and T. Aoyama, "Photographic and Acoustic Techniques Applied to a Study of the Side Aspect Target Strength of Free-Swimming Fish," *Bull. Jpn. Soc. Sci. Fish.*, Vol. 51, 1985, pp. 1051–1055.

28. J. E. Ehrenberg, "Two Applications for a Dual-Beam Transducer in Hydroacoustic Fish Assessment Systems," *Proc. IEEE Conf. Eng. Ocean Environ.*, Vol. 1, 1974, pp. 152–154.

29. R. Kieser, "An Extension of the Dual Beam Algorithm," *Coun. Meet. int. Coun. Explor. Sea*, (B:23), 1988.

30. J. E. Ehrenberg, "A Comparative Analysis of *in situ* Methods for Directly Measuring the Acoustic Target Strength of Individual Fish," *IEEE J. Ocean Eng.*, Vol. OE-4, 1979, pp. 141–152.

31. K. G. Foote, F. H. Kristensen, and H. Solli, "Trial of a New, Split-Beam Echo Sounder," *Coun. Meet. Int. Coun. Explor. Sea*, (B:21), 1984.

32. J. I. Edwards and F. Armstrong, "Measurement of the Target Strength of Live Herring and Mackerel," *FAO Fish. Rep.*, Vol. 300, 1983, pp. 69–77.

33. K. G. Foote, "A Critique of Goddard and Welsby's Paper 'The Acoustic Target Strength of Live Fish'," *J. Cons. int. Explor. Mer*, Vol. 42, 1986, pp. 212–220.

34. I. B. Andreeva, "Scattering of Sound by Air Bladders of Fish in Deep Sound Scattering Ocean Layers," *Sov. Phys. Acoust.*, Vol. 10, 1964, pp. 17–20.

35. D. E. Weston, "Sound Propagation in the Presence of Bladder Fish," in V. M. Albers (Ed.), *Underwater Acoustics*, Vol. 2, Plenum, New York, 1967, pp. 55–88.

36. M. Furusawa, "Prolate Spheroidal Models for Predicting General Trends of Fish Target Strength," *J. Acoust. Soc. Jpn. (E)*, Vol. 9, 1988, pp. 13–24.

37. T. K. Stanton, "Simple Approximate Formulas for Backscattering of Sound by Spherical and Elongated Objects," *J. Acoust. Soc. Am.*, Vol. 86, 1989, pp. 1499–1510.

38. K. G. Foote, "Rather-High-Frequency Sound Scattering by

Swimbladdered Fish," *J. Acoust. Soc. Am.*, Vol. 78, 1985, pp. 688–700.

39. K. G. Foote and J. J. Traynor, "Comparison of Walleye Pollock Target Strength Estimates Determined from *in situ* Measurements and Calculations Based on Swimbladder Form," *J. Acoust. Soc. Am.*, Vol. 83, 1988, pp. 9–17.

40. K. G. Foote, "Fish Target Strengths for Use in Echo Integrator Surveys," *J. Acoust. Soc. Am.*, Vol. 82, 1987, pp. 981–987.

41. R. H. Love, "Target Strength of an Individual Fish at Any Aspect," *J. Acoust. Soc. Am.*, Vol. 62, 1977, pp. 1397–1403.

42. E. Ona, "Physiological Factors Causing Natural Variations in Acoustic Target Strength of Fish," *J. Mar. Biol. Assoc. U.K.*, Vol. 70, 1990, pp. 107–127.

43. R. W. G. Haslett, "The Target Strengths of Fish," *J. Sound Vib.*, Vol. 9, 1969, pp. 181–191.

44. K. G. Foote and E. Ona, "Tilt Angles of Schooling Penned Saithe," *J. Cons. int. Explor. Mer*, Vol. 43, 1987, pp. 118–121.

45. K. G. Foote, "Correcting Acoustic Measurements of Scatterer Density for Extinction," *J. Acoust. Soc. Am.*, Vol. 88, 1990, pp. 1543–1546.

46. K. G. Foote, E. Ona, and R. Toresen, "Determining the Extinction Cross Section of Aggregating Fish," *J. Acoust. Soc. Am.*, Vol. 91, 1992, pp. 1983–1989.

45

FUNDAMENTAL UNDERWATER NOISE SOURCES

WILLIAM BLAKE

1 INTRODUCTION

In this chapter we will examine the essential features of sound from underwater acoustic sources that typically emerge from hydrodynamic cavitation, bubbly flow, single-phase flow past elastic surfaces and bodies, and elastic surfaces that are driven by flow. Emphasis will be on fundamentals; interested readers will find more complete discussion in Refs. 1–3. The theme of this chapter is to bring out essential features of underwater noise source mechanisms in a unified way. The development of the fundamentals of these source types began in the 1950s, and extends to today. We shall survey the generic types of sources, examine the general acoustics of flow-driven surfaces, and finally summarize the characteristics of flow excitation forces.

Underwater acoustics is of obvious importance to underwater vehicles and ships, but it is also of importance to the noises radiated by piping systems and pumps in processing industries. Underwater acoustic sources also have competed with acoustic signals in mobile active acoustic devices used in Doppler positioning and oceanography as well as in towed streamers and fairings used in off-shore oil exploration.

2 PROPERTIES OF ELEMENTARY UNDERWATER SOUND SOURCES

2.1 Canonical Sources

The features of flow-induced sound from underwater sources differ somewhat from the features of aerody-

namically generated sound. This is a consequence of the low Mach numbers of single-phase hydrodynamic flows, which make these canonical sources very inefficient. In practical instances, enhancements in the acoustic radiation efficiencies of hydrodynamic flow sources require the involvement of resonators, scattering sites in the form of spatially local discontinuities in the surface impedances of structures, bubbles, or cavitation. Bubbly flows and cavitation are among the most efficient of hydrodynamic sources because these are monopoles and have counterparts in subsonic combustion flow aeroacoustics. Underwater sources in the absence of cavitation are principally dipole and are dominated by vibration of contiguous surfaces as well as by localized flow dipoles. These sources also have aeroacoustic counterparts, although contribution from flow-induced vibration is often of lesser importance in aeroacoustics. Quadrupole noise is typically not of such importance in single-phase underwater acoustics as it is in aeroacoustics because of typically low Mach numbers of hydroacoustics. The presence of bubbles can, however, elevate the relevance of quadrupole sources. In the case of noncavitating flow, this different hierarchy of sources is determined by the generally long acoustic wavelengths of the sound in low-Mach-number liquid flow compared with the characteristic dimensions of the sources and the fact that fluid phase impedances may often be comparable to those of typical elastic boundary surfaces.

All hydroacoustic systems of sound sources may be regarded as a superposition of spatial distributions of three sources: monopoles, dipoles, and quadrupoles. These elementary classifications apply to both mechanically driven and flow-driven bodies. In the case of a flow-generated source system, a monopole source may be modeled as an unsteady mass injection $\dot{q}$, dipoles by a spatial distribution of force divergence $\nabla \mathbf{f}$, and

Encyclopedia of Acoustics, edited by Malcolm J. Crocker
ISBN 0-471-80465-7 © 1997 John Wiley & Sons, Inc.

quadrupoles by a spatial distribution of Reynolds stress $\partial^2 T_{ij}/\partial x_i\,\partial x_j$. The acoustic wave equation that expresses the field of these sources is

$$\nabla^2 p(\mathbf{x},t) - c_0^2 p(\mathbf{x},t) = -q(\mathbf{x},t) + \nabla \mathbf{f}(\mathbf{x},t) - \frac{\partial^2 T_{ij}(\mathbf{x},t)}{\partial x_i\,\partial x_j}.$$

(1)

The source terms in this expression are per unit volume and they include all the stress distributions acting on the fluid, all the localized forces, and mass injection sites in the fluid. The effect of bounding surfaces that are adjacent to the fluid is simply to reflect and scatter the acoustic field pressures from these sources. These effects may be properly dealt with mathematically in the solution of the appropriate Green's function for the geometry involved.

In the case of mechanically driven plates and shells, the dipole classification applies to the mechanism of sound radiation from localized forces. Such forces may arise both from a mechanical drive (such as a machine attached to a plate) as well as from reaction forces that exist at stiffeners and attached masses on plates that are subjected to flow excitation. This latter mechanism results from local reactions at impedance discontinuities that can convert (scatter) plate vibration at one wavenumber into sound radiation to all directions. Without these wavenumber conversion (scattering) sites, the primary plate vibration field may only radiate sound weakly or into only preferred directions.

One physical system that allows us to conceptualize a source region such as that existing in the case of flow past bluff bodies is shown in Fig. 1. A blunt body is illustrated in a flow with downstream vortex shedding and associated unsteady lift fluctuations generated back on the shedding body. Due to low ambient static pressures, cavitation is shown to be developed at the low-pressure regions of the body surface and, in the more remote wake, a region of turbulent stresses is generated. The coexisting combination of unsteady cavitation, lift, and turbulent Reynolds stresses provides the required dynamics for sound generation by monopole, dipole, and quadrupole sources, respectively. Although the illustration is idealized, many real underwater flow sources are complex in the multiplicity of excitations and source types. In general, body vibration can make this picture even more complex by introducing additional dipoles.

Solutions to Eq. (1) for an unbounded fluid that is also free of reflecting boundaries apply to these sources as they are confined to specific regions. For the monopole sources

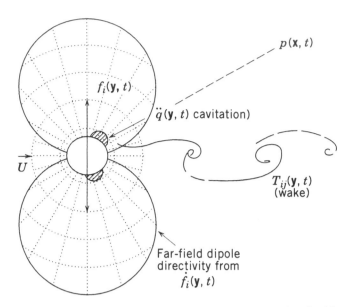

Fig. 1 Notional diagram of acoustic sources associated with flow–body interaction.

$$p(\mathbf{x},t) = \frac{1}{4\pi} \iiint_v \frac{\ddot{q}(\mathbf{y},t - r/c_0)}{r}\, dV(\mathbf{y}),$$

(2a)

which, in the case of a nonconvecting volume source, reduces to

$$p(\bar{x},t) = \frac{\rho \ddot{Q}(\mathbf{y}_0, t - r/c_0)}{4\pi r},$$

(2b)

where $\ddot{Q}$ represents the net volumetric acceleration in the case of acoustically compact monopole regions. Acoustically compact regions are those for which the characteristic spatial dimensions of the sources are small compared with both the acoustic range $r = |\mathbf{x} - \mathbf{y}|$ from the sources to the field observation point and an acoustic wavelength over the entire frequency range of interest. This statement invokes the notion of a characteristic frequency of the source dynamics in order to specify this wavelength. We continue to develop these notions in the context of Fig. 1. For the dipole sources

$$p(\mathbf{x},t) = \iiint_v \frac{(\partial f_i/\partial y_i)(\mathbf{y}, t - r/c_0)}{4\pi r}\, dV(\mathbf{y}),$$

(3a)

which reduces to

$$p(x,t) = \frac{1}{4\pi c_0}\left(\frac{x_i}{r}\right)\frac{\partial F_i}{\partial t}\left(y_0, t - \frac{r}{c_0}\right)$$

(3b)

for compact dipole source regions, where F_i represents the resultant force on the fluid and x_i/r is the direction cosine ($\cos \phi$) from the force vector to the observer.

Finally, the quadrupole source field yields

$$p(x,t) = \iiint_v \frac{\partial^2 T_{ij}}{\partial y_i \, \partial y_j} \frac{1}{4\pi r} \, dV(\mathbf{y}), \qquad (4a)$$

which reduces to

$$p(x,t) = \frac{1}{4\pi c_0^2} \frac{x_i x_j}{r^3} \iiint_v \frac{\partial^2 T_{ij}}{\partial t^2} \cdot \left(\mathbf{y}, t - \frac{r}{c_0}\right) \, dV(\mathbf{y}) \qquad (4b)$$

for a compact (i,j) quadrupole source region. It is to be emphasized that Eqs. (2b), (3b), and (4b) apply to far-field acoustic pressures. The acoustic far field is loosely defined as r much larger than the largest spatial dimension of the source field and $k_0 r \gg 1$.

2.2 Speed Dependencies of Canonical Flow Sources

We still assume that the sources are compact, nonconvecting, and situated near a coordinate $\mathbf{y}_0$, but we now assume that the time variation of the sources is simple harmonic with a frequency ω_s. Source convection may be safely ignored in underwater acoustics since the convection Mach number is negligible with respect to unity. This allows the retardation $t - r/c_0$, where $r = |\mathbf{x} - \mathbf{y}_0|$, to be replaced by a simple phase representation. Then for monopoles

$$p(x,t) = -\rho \omega_s^2 Q \frac{e^{i(k_0 r - \omega_s t)}}{4\pi r}, \qquad (2c)$$

where $Q e^{-i\omega_s t}$ represents the time-varying net volumetric acceleration. The dipoles give the field

$$p(x,t) = -i\left(\frac{\omega_s}{c_0}\right) F \cos \phi \left(\frac{e^{i(k_0 r - \omega_s t)}}{4\pi r}\right), \qquad (3c)$$

where F represents the amplitude of the net harmonic force on the fluid and ϕ is the angle from the force vector to the observer position. A similar expression could be derived for quadrupoles, but it has limited practical utility.

Figure 2 illustrates these directional features for a force dipole induced by a propeller operating in nonuniform flow as measured in an anechoic wind tunnel.[4] Typically, the flow-generated forces have a resultant direction, arising from the orientation of lift or drag, thrust, or side-force fluctuations. Equation (3c) and Fig. 2 show that the sound field is directed along the axes of each of the resultant thrust and forces acting on the fluid by the propeller.

The quadrupole field is more amorphous, showing minimal directional preference because the random-turbulence quadrupole sources all contribute to the sound with a multitude of directivities, with mean convection being responsible for a slight directionality. This has been discussed elsewhere in this handbook in connection with aeroacoustic sources. Accordingly, quadrupole sources will receive little further consideration in this chapter.

These relationships, though approximate and strictly valid for acoustically compact sources, have valuable application to underwater acoustics. Typically due to the relatively low Mach numbers of underwater flows and the characteristic frequencies involved, the source regions are compact, so that Eqs. (2b), (2c), (3b), and (3c) actually become exact for certain underwater applications and have great utility in estimation and extrapolation.

It is a characteristic of flow-generated sources that the characteristic frequency is determined by the flow velocity U and a typical body dimension L (say, diameter in the case of a circular cylinder or boundary layer thickness in the case of extended flow surfaces) such that a dimensionless frequency (called a Strouhal number)

$$S_t = \frac{\omega_s L}{U} \qquad (5)$$

may be defined. For dynamically similar flows, the Strouhal number is a constant and then necessarily determined by fluid dynamic considerations. Thus the wavenumber ω_s/c_0 appearing in Eqs. (3c) and (4c) may be written as

$$\frac{\omega_s}{c_0} = S_t \left(\frac{U}{c_0}\right) L^{-1}. \qquad (6)$$

Substitution of Eq. (6) into Eqs. (2b) and (2c) yields the important dimensional relationships for the mean-square radiated sound pressure of the monopole

$$\overline{p^2}(\mathbf{r}) = \frac{\rho^2 U^4}{16\pi^2} \frac{L^2}{r^2} [S_t^4 \tilde{Q}^2]. \qquad (7)$$

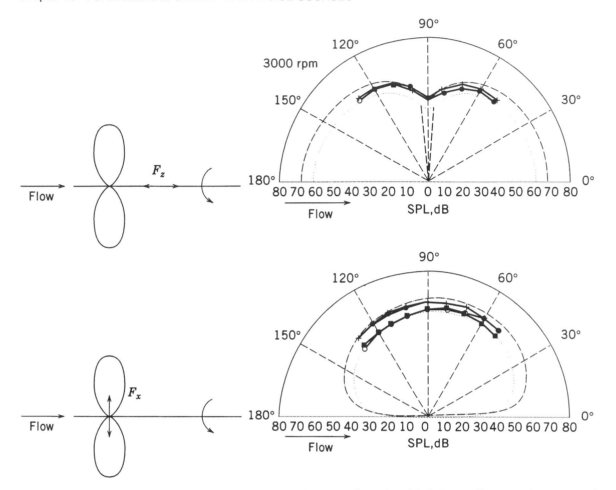

Fig. 2 Measured and predicted dipole sound patterns for a four-bladed propeller operating in three- and four-cycle wakes. Experiment: (○) 0° half-plane; (●) 90° half-plane; (×) 180° half-plane. Theory: (- - - -) low; (- · · -) high.

The term in square brackets is dimensionless, involving a dimensionless volume quantity $\tilde{Q}$, so that the leading terms carry all the dimensionality of the sound. No directivity functions appear in Eq. (7) because monopole sound (e.g., cavitation noise) is omnidirectional. For the dipole sound we have

$$\overline{p^2}(\mathbf{r}) = \frac{\rho^2 U^4}{16\pi^2} \frac{L^2}{r^2} \cos^2 \phi \left(\frac{U}{c_0}\right)^2 S_t^2 C_F^2, \qquad (8)$$

where C_F represents a dimensionless force coefficient

$$C_F = \frac{F}{\rho U^2 L^2}.$$

Again, S_t and C_F are characteristics of the fluid dynamics and, particularly at low Mach numbers, they are strictly

hydrodynamic parameters. The $\cos^2 \phi$ factor orients the directionality of the sound field with the direction of the resultant force. This directionality of the sound as well as the presence of the Mach number are direct consequences of the gradient of the forces that appears in Eqs. (1) and (3a). Equations (7) and (8) form the bases of many practical hydroacoustic scaling rules.

For the quadrupole sound, Eq. (4b) yields the familiar eighth-power rule for velocity dependence,

$$\overline{p^2} = \frac{\rho^2 U^4}{16\pi^2} \left(\frac{U}{c_0}\right)^4 \frac{L^2}{r^2} \tau(\theta, \phi) S_t^4, \qquad (9)$$

where $\tau(\theta, \phi)$ represents a dimensionless resultant quadrupole strength term. This dimensionless term carries with it all the resultant directivity features of the quadrupole source field, as is discussed in Chapter 28.

These expressions show the hierarchy of increasing order of Mach number dependence and directivity of the sources as the geometric order of the sources increases. Thus, for the low Mach numbers of underwater acoustics, quadrupole sound is irrelevant in comparison to other dipoles and monopoles when they also exist. Accounting for the existence of monopoles would appear to be simple since these are typically due to volumetric pulsations of bubbles suspended in the flow and convected through a pressure field. In the linear acoustics limit (i.e., low Mach number of bubble wall motion), the sound is determined by the net value of $Q(\mathbf{y}, \omega_s)$. However, as we shall discuss in Section 3 on cavitating flows, such sources are hydrodynamically very complicated, and generalized descriptions of cavitation noise are only available by empirical means.

Flow dipoles may be generated by the resultant flow-induced forces on finite-extent blunt bodies and propellers, as illustrated in Fig. 2, as well as by turbulent flow over continuous but acoustically compact elastic surfaces with discontinuities such as stiffeners and trailing edges. In this case of localized unsteady flow sources, a commonly used expression for sound power radiated from a compact flow dipole is determined from Eq. (3c) by squaring and integrating over the polar angle to give the power spectral density

$$P_{\text{rad}}(\omega) = \frac{\omega^2 \overline{F^2}(\omega)}{12\pi\rho c_0^3}. \tag{10}$$

In the other extreme, as discussed in Section 4.3, flat rigid surfaces that support spatially homogeneous turbulent boundary layers generate quadrupole radiation because such surfaces serve only as reflectors. For rigid extended surfaces to generate dipole sound, they must have edges, curvature, bumps, or roughness since these elements represent wave conversion sites that scatter the relatively high flow-convected wavenumbers of turbulence (ω/U) to the lower wavenumbers of sound (ω/c_0). Such wavenumber conversion mechanisms will also be discussed in Section 4.3. All real surfaces are finite, curved, and stiffened elastic, so that such surfaces, when excited by nominally homogeneous turbulent boundary layers, emit dipole sound.

Underwater quadrupole sound as generated by jet flow is dominated by so-called lip dipoles and other possible coexisting sources so that quadrupoles per se generated by jet flows are also typically irrelevant. Lip dipoles are due to the interaction of the turbulence in the jet efflux grazing and interaction with the lip of the jet nozzle (which acts analogously to a trailing edge). These are dominant in low-Mach-number turbulent jet flows.

2.3 Low-Frequency Sound from Localized (Point) Forces on a Plate in Fluid

In this section we examine the necessary ingredient for the scattering of flow-driven vibration to sound at surface structural discontinuities. As described above, when a plate of mass impedances ($-i\omega m_s$) contains localized attachments and stiffeners, a vibration incident on these stiffeners generates localized forces at these sites. When the incident vibration field is excited by flow, it is comprised of the length and time scales associated with the flow-induced surface pressure. Due to the interaction forces at the discontinuities of structure, the sound is locally radiated, and it may differ significantly in intensity and directionality from the sound that would result from the incident primary wave field on the plate without the stiffeners. The canonical radiation mechanism for this phenomenon is dipole and may be regarded as due to a point force applied to a plate that is immersed in fluid. A point force applied to a plate of area density m_s with dense fluid on one side and vacuum on the other, F, generates a sound pressure[5] that is given by the familiar dipole relationship

$$p(r, \phi, t) = k_0 F f(\phi) \frac{e^{i(k_0 r - \omega x)}}{2\pi r}, \tag{11}$$

where

$$f(\phi) = \left[\frac{\rho c_0 / \cos\phi}{\rho c_0 / \cos\phi + (-im_s\omega)} \right] (-i\cos\phi), \tag{12a}$$

$$f(\phi) = \begin{cases} \dfrac{\beta\cos\phi}{\cos\phi + i\beta}, & \tag{12b} \\[2ex] \dfrac{Z_f}{Z_s + Z_f} (-i\cos\phi), & \tag{12c} \end{cases}$$

where Z_f is the acoustic impedance of the fluid adjacent to the plate and Z_s is the inertial impedance of the plate. The factor

$$\beta = \frac{\rho c_0}{m_s\omega} \tag{13}$$

is a fluid-loading factor that expresses the ratio of fluid impedance to characteristic plate impedance ($i\omega m_s$) and the angle is measured from the direction of the force. Figure 3 illustrates the dependence of this dipole on plate inertia and observation angle. We see that at low frequencies (large β) the directivity pattern of the sound is that of an isolated dipole and the dipole strength is double

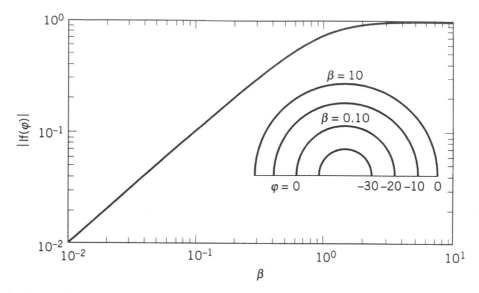

Fig. 3 Radiated sound pressure from a point force acting on an effectively infinite elastic plate with fluid on one side.

that of the original dipole due to reflection at the $y = 0$ surface. At higher frequencies (small β) the directivity appears omnidirectional (except near $\phi = \pm\pi/2$, where the fluid impedance exceeds the plate impedance near grazing incidence).

This equation applies remarkably well to practical underwater structures that are large enough that the radiated sound field is dominated by the drive point rather than by reflections from boundaries or effects of finite size of the structure. It shows that the sound is essentially that of a point dipole applied to the fluid whose strength is modified by the impedance of the plate, particularly at large β (low enough frequency that $m_s\omega$ is smaller than ρc_0).

It is well to remark that the intensity and directionality of the sound resulting from the primary plate vibration field is determined by the structural acoustics of the structure. Below the acoustic coincidence frequency of the plate, the above equation often accounts for the dominant contribution of the radiated sound from the point-driven structure. Analogous relationships apply for line force and moment-driven structures. We shall return to the acoustics of flow-driven surfaces in Section 4.

3 SOUND FROM CAVITATING FLOWS

Although the mechanism of sound generation from cavitation is fundamentally quite simple (i.e., monopole), hydrodynamic cavitation noise is an extremely complex subject about which little detail is known. A classical reference on hydrodynamic cavitation is the book by Knapp et al.[6] The adjective *hydrodynamic* is invoked

here in order to distinguish flow-induced cavitation from spark-induced cavitation. The latter form of cavitation is used in the laboratory to study specific aspects of single-bubble-cavity dynamics. There has not been a substantive development in theoretical modeling of cavitation noise for engineering application since the work of Fitzpatrick and Strasberg in the 1950s.[7] Most follow-up approaches have been empirical and based on dimensional analysis of single-bubble analogies.

Theoretical work on cavity dynamics has focussed on the dynamics of single bubbles and bubble clusters (e.g., Ref. 8; see also Ref. 2 for an extensive reference list). These researches have examined the influences of liquid phase compressibility, entrapped gas in bubbles, gas–liquid thermodynamics, and bubble interactions. Even for idealized bubble populations, the bubble mechanics are nonlinear and highly complex.[8,9] The resulting mathematics is accordingly involved. With regard to mathematical modeling of ship propeller cavitation, Baiter[10] has examined low frequency sound and effects of modulation.

Practical description of hydrodynamic cavitation noise is made difficult because of the large number of flow parameters that can influence the acoustically relevant details of cavity dynamics. There is unfortunately little connection between cavitation noise theory and observed cavitation noise of ship propellers and turbines. Experimentally, it has been difficult to control (or even measure) some of the relevant parameters. These include Reynolds number, the concentration of dissolved and free gas in the liquid, the gas/liquid phase dynamic transfer properties of the liquid, particularly over short time scales, and distributions of cavitation nucleation sites. On the other end of the spectrum, analytical work has been largely devoted to single spherical bubbles and dynamics of bubble clouds. Controlled experimental studies of cavitation noise are also difficult. Some of the most recent are given in Refs. 11–15.

Notwithstanding these complexities, attempts have been made to develop scaling and prediction rules that have engineering use. These start with Eq. (2a) for linear acoustic monopole sources. Notions of single-bubble cavitation noise assist in the establishment of dimensional length and time scales for the monopole source; the resulting scaling then depends on the ambient hydrostatic pressure, the size of the body, and the vapor pressure of the fluid. Several rules for scaling and predicting cavitation noise from propellers have thus emerged over the past 20 years.[6,7,11–21]

Cavitation in propellers and turbomachine rotors occurs in the tip vortex or the gap between the end wall and the blade tip and on the blade surfaces. The blade surface cavitation takes two major forms: bubbly cavities occurring individually or in clusters on or near the blade surface on its suction side (or back) and sheet cavities that extend along the chord from the line of minimum pressure at the leading edge. Sheet cavitation can occur on either side of the rotor, depending on its load distribution or its operating point. The common names given to these forms of cavitation are tip vortex, gap, back bubble (BUBBLE), leading-edge pressure surface (LEPS), and leading-edge suction surface (LESS). Figure 4 illustrates the surface pressure characteristics on hydrofoils and turbomachinery blades, P_s which cause cavitation environments. Cavitation in the shear layers of free jets occurs as individual bubbles and bubble clusters which are convected with the turbulent vortices.

The principal parameter that determines the occur-

rence of cavitation and cavitation noise is the cavitation index σ, defined as[6]

$$\sigma = \frac{P - P_v}{(1/2)\rho U^2},\qquad(14)$$

where P is the ambient hydrostatic pressure, P_v is the vapor pressure and ρ the density of the fluid, and U is the relative velocity between the blade and the fluid. This dimensionless number is the single most relevant parameter in determining the practical scaling of cavitation onset and the occurrence of cavitation noise. The condition for cavitation is that $\sigma < \sigma_i$, where σ_i is the cavitation inception index for the blade. The cavitation inception index is a function of the hydrodynamic design of the blades, the disturbance field in which the blades must operate, a variety of viscous effects, and the amount of free and dissolved gas in the fluid. It is thus roughly predictable in the design stage of the propeller and turbine rotor, but it must eventually be measured on either a model or a prototype design. Simple dimensional analysis leads to a description of the spectrum of the sound pressure in proportional frequency bands ($\Delta f \propto f$, in this case we will consider one-third-octave bands) of the form that follows immediately from Eq. (7) by replacing U^2 by P and L by propeller or pump diameter D, that is,

$$\overline{p^2} \propto P^2 \left(\frac{D}{r}\right)^2\qquad(15)$$

with frequency scaled as

$$f \propto D^{-1}\sqrt{\frac{P}{\rho}}.\qquad(16)$$

These relationships assume that strict geometric dynamic similitude for the cavitation occurs.[1,2,21] Due to several possible scale effects, strict similitude may not occur, particularly to the extent that $\sigma_i)_{model} = \sigma_i)_{full}$. This lack of similitude may result in differing speeds for a similar extent of cavitation at model and full scale. For example, the inception of tip vortex cavitation has a Reynolds number scale effect that delays cavitation inception on models, compared with full size. Thus, an attempt has been made to include the inception index in the semiempirical rule given below. This approach also borrows some notions of single-bubble cavitation dynamics and yields the form

$$\overline{p^2}(r, f, \Delta f) = p_m^2 G\left(\frac{f}{f_m}\right).\qquad(17)$$

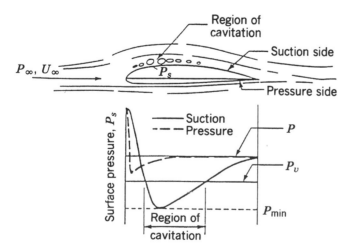

Fig. 4 Illustration of a cavitating hydrofoil, its surface pressure distribution, and its region of cavitation.

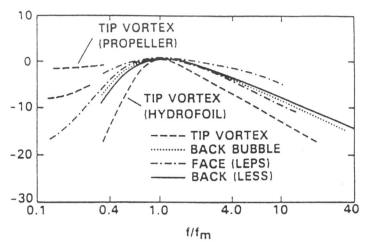

Fig. 5 Measured spectral forms for various types of hydrodynamic cavitation.

Figure 5, from Ref. 19, shows the characteristic spectrum forms for the four forms of cavitation. The tip vortex cavitation appears to have a variable level at low frequencies. This may occur because of the large-scale unsteadiness of propeller blade cavitation due to ship wake effects. Figure 6 shows an example of the collapse of the data for the maximum spectrum level, in this case the tip vortex cavitation, which has a roughly 10-dB spread. As an example of the utility of the cavitation noise prediction formulas for ship propellers, Fig. 6 shows a prediction for the twin propellers of the research vehicle *R/V Athena*. Here it is assumed that the dominant form of cavitation is the leading-edge suction side (LESS) type. The LESS type is the most common form of propeller surface cavitation and is noisier than tip vortex cavitation noise when it occurs. An example for axial flow turbomachinery is a little less reliable. Figure 7

shows measured and predicted noise levels for the drive pump of a water tunnel.[17] The measured sound levels are expressed as sound power due to the semireverberant nature of water tunnel enclosures.

4 GENERAL FEATURES OF COMPLEX HYDROSTRUCTURAL ACOUSTIC ELASTIC SYSTEMS

4.1 Description of the System

We now return the broad subject of the hydroacoustics of flow-excited structures in the absence of cavitation. We devote our attention to the role of structural surfaces in the conversion of flow energy to sound energy. The physical complexities of the hydroacoustic sources for

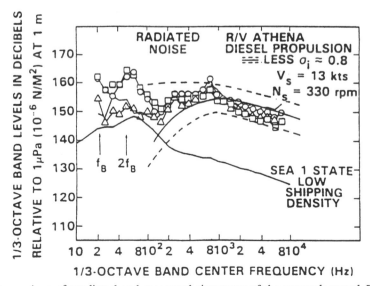

Fig. 6 Comparison of predicted and measured signatures of the research vessel *RV Athena*.

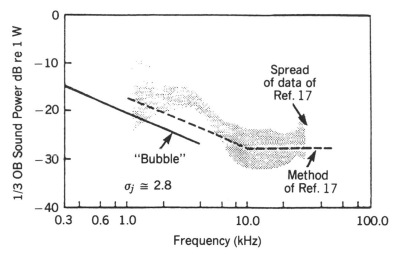

Fig. 7 Comparison of predicted and inferred sound power levels for impeller cavitation measured within the MIT water tunnel.

flow over real elastic surfaces can be viewed as a coupled system in which the flow–surface interaction leads to unsteady surface forces and pressure fluctuations in the fluid as well as to induced surface vibration. The fluid-borne pressures and forces and the structure-borne vibration combine to yield a sound field that is a superposition of both contributions. Thus the effect of surfaces in hydroacoustic source mechanisms is both hydrodynamic and structural acoustic. Hydrodynamic motion develops sources in the fluid and on the surface, while structural vibration induced by flow forcing introduces new surface sources. These new sources are due to impedance discontinuities occurring at localized structural stiffening and constraining boundaries as well as to surface finiteness and curvature.

Figure 8 illustrates these general interactions as a coupled hydrostructural acoustic system. The discipline of classical structural acoustics, the subject of Chapter 41, considers the coupling of vibration and sound as identified in the lowest three blocks in the figure. With the exception of cavitation, the discipline of aeroacoustics of rigid surfaces typically considers the upper portion of the figure. Thus, the science of underwater sound generation involves a combined discipline of hydrostructural acoustics in which are studied the sound, structure, and flow interactions. The characteristics of the sound generated by this system are determined by the relative acoustic length and time scales of the acoustic field, the hydrodynamic motion, and the flow-induced surface vibration. The sound field characteristics include the magnitude, the speed dependence, and the directivity.

Domination of the flow–surface interaction by both potential and viscous effects develops the flow unsteadiness, which establishes the spatial and temporal scales of the flow sources. Flows over bluff bodies thus often involve the development of viscous-dominated shear flows that accompany strong surface pressure gradients on the body (and thus body forces) and generate vortex street wakes. Thus trailing-edge flows, propeller unsteady forces, unsteady body forces of vortex street wakes, and turbulent boundary layer pressure fluctuations are all consequences of these flows. Cavitation occurs on surfaces when the surface pressure locally dips below a critical pressure (nearly the vapor pressure of the fluid). Inflow unsteadiness may cause the cavitation extent to become strongly time dependent if it results from variation of the surface pressure, as through angle-of-attack variations of lifting surfaces.

In the case of hydrodynamically generated sources due to flow–surface interaction, it is the nature of unsteady fluid mechanics that for certain flows there may be a coupling between the development of disturbances in the fluid and the flow-induced vibration of the surface. When surface vibration occurs, this motion can couple with the unsteady viscous-dominated shear flow at the surface and alter (typically strengthen) the generation of vortex sound sources (and forces). Such coupled flow–structure sources typically occur over a speed and frequency range for which the fundamental shear flow frequency becomes coincident with a frequency of a surface vibration mode and the amplitude of motion becomes comparable to a viscous disturbance scale of the shear flow. The essential behavior is a feedback mechanism that accounts for propeller singing, hole tones, and cavity tones; characteristics of many of these are discussed in Section 6. Damping and stiffening of the surface structure can significantly minimize the likelihood of these tones by adding dissipation (loop delay) and by detuning the characteristic frequencies of the structure and flow.

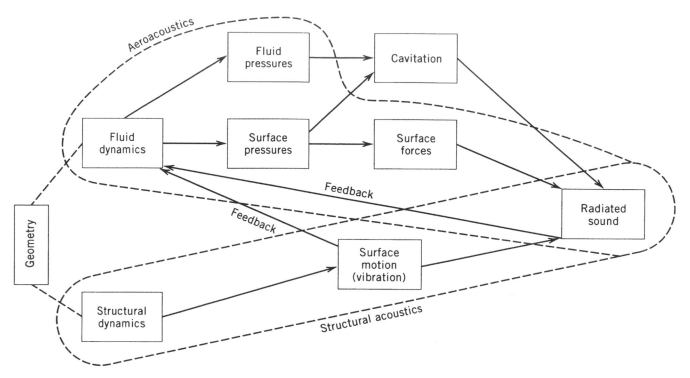

Fig. 8 Interfaces of various classes of mechanisms in a hydrostructural acoustic system.

Even when there is no flow–surface feedback mechanism, surface motion typically plays a significant role in hydroacoustics source mechanisms. This is because the motion of the surface can contribute to the sources. This is done in three fundamental ways: the surface becomes a finite-impedance reflector of the sources so that the radiated sound is simply altered by a reflection coefficient; the finite spatial extent of the surface or structural discontinuities or curvature of the surface may cause localized surface forces that contribute to the radiated field as dipoles; and trailing-edge flows of extended surfaces can generate additional wake sources (by dipole mechanisms).

The mathematical representation of the characteristics of the source complex are described with Helmholtz's integral equation, which gives the superposition of direct sound radiation and the contributions from surfaces. Thus

$$p(\mathbf{x}, \omega) = \iiint_{v} \sigma(\mathbf{y}, \omega) g(\mathbf{x}, \mathbf{y}, \omega) \, dV(\mathbf{y})$$

$$+ \iint_{s} \nabla_n p(\mathbf{y}, \omega) g(\mathbf{x}, \mathbf{y}, \omega) \, dS(\mathbf{y})$$

$$- \iint p(\mathbf{y}, \omega) \nabla_n(\mathbf{x}, \mathbf{y}, \omega) \, dS(\mathbf{y}), \quad (18)$$

where $\sigma(\mathbf{y}, \omega)$ represents the volume density of sources and $g(\mathbf{x}, \mathbf{y}, \omega)$ is the free-space Green function. On the surface, the normal gradient ∇_n of the pressure is related to the normal surface velocity $u_n(\mathbf{y}, \omega)$ by

$$-\nabla_n p(\mathbf{y}, \omega) = \rho \dot{u}_n(\mathbf{y}, \omega). \quad (19)$$

The volume integral includes the effects of sources within a region of turbulent flow, and the two surface integrals include all effects of the surface motion $u_n(\mathbf{y}, \omega)$ and normal stresses $p(\mathbf{y}, \omega)$ on the surface–fluid interface. The effects of surfaces thus also may extend to the modification of the source density term $\sigma(\mathbf{y}, \omega)$ due to the generation of new turbulent sources. For a given turbulent volume source density, the surface integrals provide all the surface stresses, surface wave scattering, and geometric reflection and diffraction. For the remainder of this section we will assume that the fluid phase is devoid of bubble and cavitation motion.

The next two sections will examine the consequences of the limits of low- and high-frequency behavior of hydroacoustic sources that are made complex by the coexistence of turbulent sources and vibration response of adjacent surfaces. These sections are based on some of the earliest and fundamental work in flow-induced noise modeling.[22–24] Since then, there have been many extensions.[25–27]

We shall examine the behavior of the hydroacoustic system in two frequency limits in order to clearly see the parametric ranges for which important physical mechanisms are dominant. Understanding of these generalities is important in identifying generic control strategies for mitigating underwater sound. We shall see that the sound from flow-induced vibration is determined by the so-called blocked-flow forces on the rigid surface and the relative impedances of the fluid and surface. This finding is common to both acoustically compact and acoustically extended surfaces.

4.2 Sound from Flow around Acoustically Compact Elastic Bodies: Low-Frequency Limits

In the low-frequency limit, the radiating surface is compact compared with an acoustic wavelength. Figure 9 illustrates the case of a bluff body, shown here as a blunt lifting surface of chord C and span L in a uniform flow of mean velocity U. The surface is incompressible and is free to vibrate under the influence of flow-exciting pressure fluctuations on the upper and lower surfaces, designated as p^+ and p^-, where the superscripts plus and minus denote quantities evaluated on the "upper" and "lower" surfaces, respectively. We assume that the time dependence can be suppressed so that all disturbances may be considered in the frequency domain as Fourier transforms. The Helmholtz equation, Eq. (18), for this situation then gives the far-field sound pressure as a superposition of three contributions that involve the volume integration of the wake quadrupoles, surface pressure dipoles, and surface vibration dipoles. Values of the integrands are determined by the differentials between quantities on the upper and lower surfaces. Flow-induced vibration is constrained to be an oscillatory rigid-body translation so that the velocities on the upper and lower surfaces are equal and opposite,

$$u^+ = -u^- = u_n(\mathbf{y}_s, \omega), \qquad (20)$$

because of the assumption of incompressibility of the surface. The compactness of the surface applies to low frequencies for which

$$\frac{\omega C}{c_0} \ll 1. \qquad (21)$$

The compactness of the source allows simple expansions of the Green function with position in the source zone relative to the observer. In these expressions, the free-space Green function is

$$g(\mathbf{x}, \mathbf{y}, \omega) = \frac{e^{i(k_0 r - \omega t)}}{4\pi r}, \qquad (22)$$

where $r = |\mathbf{x} - \mathbf{y}|$, so a simplified expression for the far-field pressure is

$$
\begin{aligned}
p(\mathbf{x}, \omega) = {} & \frac{1}{4\pi} \iiint_v \frac{\sigma}{r} e^{ik_0 r} \, dV(\mathbf{y}) \\
& + \iint_s [-i\omega u_n(\mathbf{y}, \omega)](g^+ \quad g^-) \, dS(\mathbf{y}) - \cdots \\
& - \iint_s \nabla_n g(\mathbf{x}, \mathbf{y}_s, \omega) \\
& \cdot [(p^+(\mathbf{y}, \omega) - p^-(\mathbf{y}, \omega)] \, dS(\mathbf{y}), \qquad (23)
\end{aligned}
$$

where the Green functions on the upper and lower surfaces are given by

$$g^+ = g(\mathbf{x}, y^+, \omega), \qquad g^- = g(\mathbf{x}, y^-, \omega) \qquad (24)$$

and

$$g = g(\mathbf{x}, \mathbf{y}_s, \omega),$$

respectively, and $\mathbf{y}_s$ is the coordinate of the median surface of the body with coordinate $\mathbf{y}_0$. Under these stipulations, Eq. (18) reduces to (see Refs. 2 and 28)

$$
\begin{aligned}
p(\mathbf{x}, \omega) \cong {} & \iiint_v \sigma(\mathbf{y}, \omega) g(\mathbf{x}, \mathbf{y}_s) \, dV(\mathbf{y}) \\
& + ik_0 \cos \phi L \, g(\mathbf{x}, \mathbf{y}_s, \omega), \qquad (25)
\end{aligned}
$$

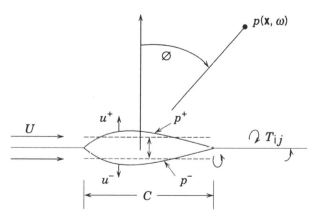

Fig. 9 Conceptual diagram of an acoustically compact elastic body interacting with flow.

where $\mathcal{L}$ denotes the spectrum of unsteady lift *on the surface* including the effect of a contribution to the pressure jump that is due to the fluid reaction to the motion of the surface, as discussed below.

The first term of this expression is the direct quadrupole sound emitted by the wake turbulence; it is of order $U/c_0 \ll 1$ relative to the remaining terms. The second term is the contribution from surface pressure dipoles.

Since the fluid reacts to surface motion, these dipoles are affected by the response of the surface. Equation (26) gives an approximate equation of motion of the surface,

$$\overline{Z_s \bar{u}_n} = \overline{(p^+ - p^-)} - \overline{\Delta p_{fl}} = (\mathcal{L}_h(\omega) + \frac{\mathcal{L}_h(\omega)}{CL}, \qquad (26)$$

where (Δp_{fl}) is the differential of the reaction pressure of the fluid on the vibrating surface, $\mathcal{L}_h(\omega)$ is the unsteady lift that would act on the surface if it was rigid, and Z_s represents the impedance of the surface to the differential pressure. In the acoustically compact limit, the pressure differential $\mathcal{L}_{fl}(\omega)$ is essentially reactive and determines the added mass.[29] Thus, approximately, for the inertial loading on the compact body

$$\frac{-\mathcal{L}_{fl}(\omega)}{CL} = -i\omega m_a \overline{U}_n. \qquad (27)$$

Upon substitution of these expressions into Eq. (25), we can isolate the dominant parametric behavior of the flow-excited vibrating body, particularly as it relates to the influence of body motion due to the relative impedances of the fluid and the body. Thus

$p(\mathbf{x}, \omega)$

$$= \begin{cases} ik_0 \cos \phi \mathcal{L}_h(\omega) \left(\dfrac{e^{ik_0 r}}{4\pi r} \right) \left\{ 1 - \dfrac{Z_f}{Z_s + Z_f} \right\}, & (28a) \\[2em] ik_0 \cos \phi \mathcal{L}_h(\omega) \left(\dfrac{e^{ik_0 r}}{4\pi r} \right) \left\{ \dfrac{Z_f}{Z_s + Z_f} \right\}, & (28b) \end{cases}$$

where we have let $Z_f = -i\omega m_a$. The leading term is recognized as the basic dipole sound from the flow–body interaction if the body is not free to vibrate; as given by Eq. (3c), note that L_h is the negative of the force applied to the fluid. The second term in the curly brackets of Eq. (28a) gives an adjustment due to the contribution of surface dipoles caused by the forced motion of the surface. This term is recognized as the leading term of $f(\phi)$ in Eq. (12c). This contribution is determined by the radiation efficiency of the surface motion as expressed by the relative impedances of the body and fluid since

these govern the level of vibration that responds to the unsteady lift on the body $L_h(\omega)$. The effect of surface motion may destructively interfere with the direct dipole contribution if $Z_f \approx Z_s$, i.e. the sound from a compact, unrestrained rigid neutrally buoyant body vanishes. The second term in curly brackets we shall see is also analogous to a term appearing later in Eq. (36). For the many cases for which the excitation forces are due to turbulent flow, we must consider the sound in the statistical sense of a sound pressure spectral density. In this case the sound pressure spectrum is

$$\Phi_{pp}(\mathbf{x}, \omega) = \left[\frac{k_0^2 \langle \mathcal{L}_h^2(\omega) \rangle}{(4\pi r)^2} \right] \left| \frac{Z_s}{Z_s + Z_f} \right|^2. \qquad (28c)$$

4.3   Sound from Flow over Extended Surfaces: High-Frequency Limits

We will now consider the case of sound and vibration that is caused by turbulent flow past bodies extended in the flow direction, such as boundary layers on hulls (analogous to boundary layers on aircraft fuselages), internal flow in long pipes, and external flow along pipes. At high enough frequencies that the flow bounding surface is larger in the flow direction than an acoustic wavelength and all controlling length scales of the flow and flexural vibration wavelengths are also small, we may approximate the out-of-plane vibration of the surface by a flow-induced vibration of a flat surface. We will consider that the driving flow is turbulent and spatially homogeneous in the plane of the surface; this is a non-restrictive provision that will greatly simplify analysis and permit ready conclusions. These surfaces will, in general, contain discontinuities such as stiffeners and other attached impedances; these features will determine the characteristics of the resulting sound and vibration fields. Under the driving of flow-induced surface pressures associated with the boundary layer, the plate will deform in flexure.

Figure 10 illustrates the model problem of this section, which was first solved by Ffowcs Williams,[22,23] with many extensions in the ensuing years.[25–27] This example problem will enable us to identify the controlling physical parameters that are relevant to the flow-induced vibration sound of extended elastic surfaces. Figure 10a shows the surface as a homogeneous one without stiffeners or impedance discontinuities; Fig. 10b shows the surface with such discontinuities. Both problems will be examined in order to show how such discontinuities become important to the sound field.

The mathematical treatment of these types of flows is

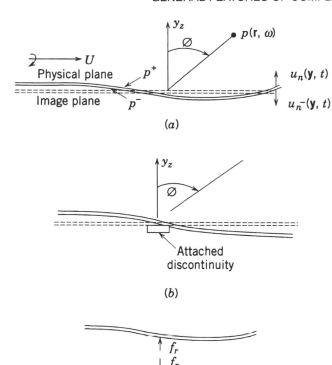

Fig. 10 Conceptual diagrams of flow-excited panels: (*a*) homogeneous flow and elastic surface; (*b*) homogeneous flow and impedance discontinuity; (*c*) free-body diagram of (*b*).

facilitated by considering the spatial Fourier transforms of pressure and flow-induced response. Thus, for example, we express the flexural surface velocity as

$$u(\mathbf{x}, t) = \iiint_{-\infty}^{\infty} U(\mathbf{k}, \omega) e^{i(k\mathbf{x} - \omega x)} d^2\mathbf{k}\, d\omega, \qquad (29)$$

where the magnitude squared of the transformed variable $U(\mathbf{k}, \omega)$ is the wavenumber of the frequency spectral density of the panel velocity. Similarly, we define a pressure transform

$$p(\mathbf{x}, t) = \iiint_{-\infty}^{\infty} P(\mathbf{k}, \omega) e^{i(k\mathbf{x} - \omega x)} d^2\mathbf{k}\, d\omega \qquad (30)$$

and a pressure spectral density that is the magnitude squared of $p(\mathbf{k}, \omega)$. The problem is solved by a method of images applied to the Helmholtz integral equation in order to eliminate one of the surface integrals. Following a method also used by Powell,[24] an image system is assumed in the lower half plane ($y_2 < 0$) (see also the chapters on statistical methods in acoustics).

We shall define the turbulent flow sources to be invariant upon reflection from the upper half plane to the lower half plane and occupy a thin region δ adjacent to the surface such that $k_0\delta \ll 1$. By invoking an impedance model for the surface velocity $U_n(\mathbf{k}, \omega)$, the transform of the surface pressure is

$$P(\mathbf{k}, \omega) = Z_s(\mathbf{k}, \omega) U_n(\mathbf{k}, \omega), \qquad (31)$$

where $Z_s(\mathbf{k}, \omega)$ is the impedance of the fluid-loaded surface that includes both the structural impedances and the additional impedance loading of the fluid, that is,

$$Z_s(\mathbf{k}, \omega) = Z_m(\mathbf{k}, \omega) + Z_f(\mathbf{k}, \omega), \qquad (32)$$

where m and f denote material and fluid properties, respectively, of the surface. Equation (31) is the analogy of Eqs. (26) and (27).

If the surface is perfectly rigid, the analysis gives only a perfect reflection (Powell's reflection principle,[24]) and the familiar doubling of the surface pressure by the rigid wall. We denote this pressure the so-called blocked pressure $P_{\text{blocked}}(\mathbf{k}, \omega)$, and it is completely determined by the quadrupole sources in the turbulence above the plate. When the wall is elastic and spatially homogeneous, as currently modeled, the surface pressure becomes

$$P(\mathbf{k}, \omega) = P_{\text{blocked}}(\mathbf{k}, \omega) \left[\frac{Z_s(\mathbf{k}, \omega)}{Z_m(\mathbf{k}, \omega)} \right], \qquad (33)$$

and the surface pressure spectral density on the elastic surface is related to the blocked surface pressure spectral density by

$$\Phi(\mathbf{k}, \omega) = \left| \frac{Z_m(\mathbf{k}, \omega)}{Z_m(\mathbf{k}, \omega) + Z_f(\mathbf{k}, \omega)} \right|^2 \Phi_{\text{blocked}}(\mathbf{k}, \omega). \quad (34)$$

This surface pressure spectrum is still determined solely by the turbulent *quadrupole* field above the plate. Thus, the effect of a spatially homogeneous elastic plate is to reflect the sources; finite surface impedance modifies this reflection, but it does not introduce any additional sources.

The radiated sound pressure is given by an inverse transform

$$p_{\text{rad}}(\mathbf{x}, t) = \iiint_{-\infty}^{\infty} P(x_2 = 0, \mathbf{k}, \omega)$$

$$\cdot \exp[i(\sqrt{k_0^2 - k^2}\, x_2 + \mathbf{k}_{1,3} \cdot \mathbf{x}_{1,3}) - i\omega t]$$

$$\cdot d^2\mathbf{k}\, d\omega. \qquad (35)$$

Since the radical in the integrand becomes imaginary for $|k| > k_0$, pressure components at wavenumbers greater than acoustic are evanescent with distance above the plate. Thus, only surface pressure at wavenumbers that are less than or equal to k_0 are capable of contributing to the sound field. Since the dominant contribution of surface pressures is contained at convected wavenumbers

$$k_c = \frac{\omega}{U} \tag{36}$$

and since the acoustically relevant radiating pressures are determined by wavenumbers less than acoustic, that is,

$$k_c < k_0 = \frac{\omega}{c_0}, \tag{37}$$

the radiating pressures occupy only a fraction

$$\frac{k_c}{k_0} = \frac{U}{c_0} \tag{38}$$

of the total effective wavenumber spectral range of the pressure spectrum. Since the Mach numbers of underwater vehicles and flows are typically small, this frac-tion accounts for the weak radiation from the quadrupole sound that results from flow over geometrically and structurally homogeneous flat extended surfaces.

Fully analytical predictions that are based on the above relationships and methods are essentially limited by our knowledge of the spectrum of the blocked wall pressure, $\Phi_{\text{blocked}}(\mathbf{k}, \omega)$. The features of the wavenumber frequency spectrum of the blocked wall pressure are well known only for certain boundary layer flow types and only in a range of wavenumbers near ω/U. Figure 11 illustrates a typical spectrum that has been nondimensionalized on the mean wall shear stress τ_w and the boundary layer displacement thickness δ^*. The dominating behavior near the convective wavenumber ω/U is clear. The remainder of the spectrum is reliably known only down to wavenumbers of order $\sim 0.03 \, \omega/U$.

When the surface has structural inhomogeneities, as depicted in Fig. 10b, then the sound field may be remarkably enhanced due to the creation of additional dipoles at the spatially localized stiffeners as discussed above. To see how the stiffener or attachment affects the vibration and to derive a simple solution that gives the relative importance of scattered to incident vibration sound, we consider the case of a single-point attachment to a plate that has a vertical translational input impedance Z_r, as illustrated in the free-body diagram in Fig. 10c. The attachment is situated at the origin of the plate coordinate system and has a velocity u_r and causes a reaction

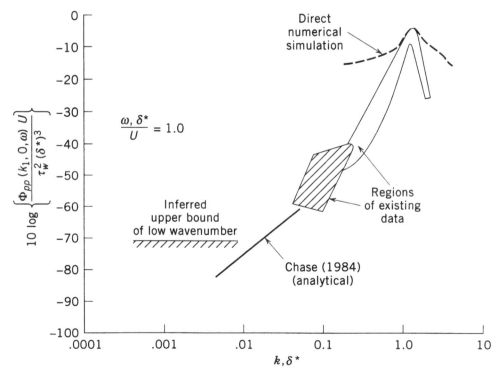

Fig. 11 Wavenumber spectrum of turbulent boundary layer wall pressure at fixed.

force F_r on the plate. We shall consider the attachment at the plate to be a point in order to draw parallels with previous discussion in this chapter. This is not a limitation to the general conclusions, which also apply to other geometries of structural discontinuities. Incorporating this force on the fluid-loaded plate yields, instead of Eq. (31) (see, e.g., Ref. 23),

$$Z_s(\mathbf{k}, \omega) U_n(\mathbf{k}, \omega) = P(\mathbf{k}, \omega) + \frac{F_r}{(2\pi)^2}, \quad (39)$$

where the impedance $Z_s(\mathbf{k}, \omega)$ is given by Eq. (35).

The consequences of the discontinuity is to introduce a dipole of strength F_r to the sound field that would have existed without the presence of the discontinuity.

The effect of the discontinuity compared with the homogeneous plate, we apply the imaging approach to remove the explicit reference to the plate velocity to obtain is given in terms of the blocked pressure spectrum and the impedance variables:

$$P(\mathbf{k}, \omega) = P_{\text{blocked}}(\mathbf{k}, \omega) \left[\frac{Z_m(\mathbf{k}, \omega)}{Z_m(\mathbf{k}, \omega) + Z_f(\mathbf{k}, \omega)} \right]$$

$$+ \left\{ 1 - \frac{Z_m(\mathbf{k}, \omega) - Z_f(\mathbf{k}, \omega)}{Z_m(\mathbf{k}, \omega) + Z_f(\mathbf{k}, \omega)} \right\} \frac{F_r}{(2\pi)^2} k_0$$

$$\cdot \cos \phi \, G(x_2, \mathbf{k}, \omega), \quad (40)$$

where $G(x_2, \mathbf{k}, \omega)$ is the two-dimensional Fourier transform of the free-space Green function

$$G(x_2, \mathbf{k}, \omega) = \frac{i}{2} \frac{e^{i\sqrt{k_0^2 - k^2}|x_2|}}{\sqrt{k_0^2 - k^2}}. \quad (41)$$

The first term has already appeared [Eq. (33)] and the new term in curly brackets is identical to that given in Eqs. (11)–(13) for the impedance relationship of the point-driven fluid-loaded plate. Here, the driving force applied to the plate is F_r. To see this identity more clearly, we use inverse Fourier transformation of only the last term to recover the expression for the point dipole applied to the fluid,

$$p_{\substack{\text{rad} \\ \text{dipole}}}(\mathbf{r}, \omega) = ik_0 \left[\frac{Z_m - Z_f}{Z_m + Z_f} \right] F_r \cos \phi \, \frac{e^{i|k_0 r|}}{4\pi r} \quad (42a)$$

or

$$p_{\substack{\text{rad} \\ \text{dipole}}}(\mathbf{r}, \omega) = ik_0 \left[\frac{Z_f}{Z_m + Z_f} \right] F_r \cos \phi \, \frac{e^{i|k_0 r|}}{4\pi r}, \quad (42b)$$

where Z_m and Z_f represent values of impedance evaluated at wavenumbers $k = k_0 \sin \phi$, where ϕ is the angle between the observer vector and the plate normal. The force that is induced by the discontinuity depends on the relationship between the input impedance of the discontinuity and the point input impedance of the fluid-loaded plate. The point force on the plate (which is negative of that applied to the fluid) is

$$F_r = \frac{Z_r Z_{P_0}}{Z_r + Z_{P_0}} \iint \frac{P_{\text{blocked}}(\mathbf{k}, \omega)}{Z_m(\mathbf{k}, \omega) + Z_f(\mathbf{k}, \omega)} \, d^2\mathbf{k}, \quad (43)$$

where Z_{P_0} is the point impedance of the fluid-loaded plate. The term in Eqs. brackets in Eq. (42b) will be recognized as analogous to the bracketed term of Eqs. (12a) and (28b) in expressing the essential dipole nature of the localized effects of real surface impedance and geometry nonuniformities in fluid loaded elastic surfaces.

5   GENERALIZATIONS

The above three sections form the essential result of this section, and this result has general implications to the flow-forced vibration and sound of structures; see Fig. 8. The effect of the extended surface alone on the flow-induced sound from the turbulent sources is as a specular reflector; the sound field will be established by the identical wave vector spectral content in the vibration as the sources that produce the pressure field on the rigid surface; however, it is weighted by a function that expresses the effect of the plate impedance in altering the acoustic reflection coefficient. This is expressed by Eq. (34) for the extended surface and by Eq. (28c) for the compact surface. The surface which is acoustically compact in one dimension but not in the other dimension is just a one-dimensional extended surface in this context. The presence of structural discontinuities will introduce localized forces that cause nonspecular acoustic radiation. This nonspecular radiation results because the discontinuities convert the wave vectors of the vibrations that are the same as wave vectors of the pressure field into contributions at lower and higher wavenumbers by generating vibrations of the surface at these localized surface distortions. With the addition of the vibration, lower wavenumbers occur that are also less than or equal to the acoustic wavenumber, so that the radiated sound is enhanced. The magnitude of the interaction forces will be determined by the input impedances of the plate and attached structure; the acoustic directivity will be determined by the spatial geometry of the attachment (i.e., the dimensionality, the size and orientation, the type of constraint in translation or rotation). Regardless of detail, the overall effect will be to produce dominating dipole sound. These effects are the dipole source contribution adjustments that appear in Eq. (40) and are evaluated in Eqs. (11) and (42).

If there are distributions of multiple discontinuities (as rib arrays and the like), the above physics and solution approach still applies; however, the interactions between discontinuities through the plate and the adjacent fluid field make the solution procedure more complex due to the acoustical reinforcements and interferences associated with source arrays that are made coherent by the basic structure. The resulting sound field will still depend on the magnitude of the interaction forces between the plate and the discontinuities, but it will also depend on the phase and spatial separation scale of the array of discontinuities. In these cases, fluid–strength path interactions may introduce flow-excited surface resonances, and the array behavior could introduce preferred far-field acoustic directionality.

Since the dipole contribution to the flow-induced noise is likely to be dominant in most realistic situations, and since these situations are generally rather complex, it has been found useful to consider the effects of flow-induced plate vibration and sound in a product form that invokes the notion of the structural radiation efficiency σ_{rad}. This is done particularly for engineering analysis of one-third-octave-band levels. Thus, if the flow-induced mean-square vibration velocity of the ribbed plate averaged over its area is $\overline{V^2}(\omega)$, then a general expression for the acoustic power radiated is developed by methods given in Chapter 78 on statistical methods in acoustics and Refs. 2, 30, and 31. The acoustic power is

$$P_{\mathrm{rad}}(\omega) = \rho_0 c_0 A \sigma_{\mathrm{rad}} \overline{V^2}(\omega) \qquad (44)$$

and the mean-square surface velocity is

$$\overline{V^2}(\omega) = \sum \int\int |Z_s(\mathbf{k}, \omega)|^{-2} \Phi_{\mathrm{blocked}}(\mathbf{k}, \omega)\, d\mathbf{k}, \quad (45)$$

where the summation symbol represents summation over all modes of the flow-driven structure and the integration is over all wavenumbers such that $|k| < k_0$.

6 SURVEY OF FORCING CHARACTERISTICS OF SOME IMPORTANT FLOW-INDUCED SOUNDS

The previous sections discussed the essential acoustical features and radiation mechanism for various classes of flow sources and flow-driven surfaces. In this section, we briefly discuss the characteristics of various classes of flow-induced surface excitation forces and pressures for noncavitating flows. Cavitating flows have already been discussed in Section 3 as well as elsewhere in this volume. Specifically, in this section, we will discuss the behaviors of forcing function spectral quantities $\langle \mathcal{L}_h^2(\omega) \rangle$ and $\Phi_{\mathrm{blocked}}(k, \omega)$ that appear in the preceding sections.

Thus, these flow-induced surface force distributions may do the following:

- Occupy one or two space dimensions [e.g., $\mathcal{L}_h(\omega)$ or $\Phi_{\mathrm{blocked}}(\mathbf{k}, \omega)$]
- Be convected over the surface in a mean flow direction or spatially localized by separated flow [e.g., $\Phi_{\mathrm{blocked}}(\mathbf{k}, \omega)$ of a turbulent boundary layer]
- Be broadband in frequency [e.g., $\Phi_{\mathrm{blocked}}(\mathbf{k}, \omega)$ of a turbulent boundary layer] or be nearly pure tone [e.g., $\mathcal{L}_h(\omega)$ for vortex shedding from a bluff body]

The discussion of specific magnitudes and length–time scales for the relevant forcing function is beyond the scope of this chapter. They are fully surveyed and discussed in Refs. 1 and 2. We shall only describe the general features of these characteristics. The following is a summary of various major classes of forcing function by surface type.

6.1 Dipole Forcing Functions on Lifting Surfaces and Blunt Bodies

These are roughly one dimensional at the low Mach numbers of hydroacoustics since they are typically aligned with leading and trailing edges or the axis of a vortex-shedding body. Trailing-edge forcing may occupy a narrow frequency band when orderly vortex shedding occurs at bluff (blunt) trailing edges just as behind circular cylinders. Trailing-edge flow noise and forcing originates because of two classes of flows: turbulent flow across the edge of the surface and in the viscous wake of the trailing edge. Leading-edge forcing is caused by interaction of the lifting surface with incoming gusts. The gust occurs because of relative motion between the surface and its incoming mean flow. If the inflow contains a time mean distortion (as a pipe flow upstream of a

TABLE 1 Important Source Types and Their Parameters

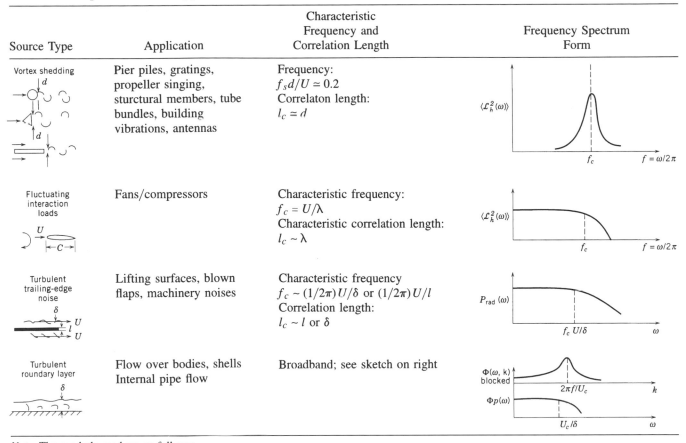

Source Type	Application	Characteristic Frequency and Correlation Length	Frequency Spectrum Form
Vortex shedding	Pier piles, gratings, propeller singing, sturctural members, tube bundles, building vibrations, antennas	Frequency: $f_s d/U \simeq 0.2$ Correlaton length: $l_c \simeq d$	
Fluctuating interaction loads	Fans/compressors	Characteristic frequency: $f_c = U/\lambda$ Characteristic correlation length: $l_c \sim \lambda$	
Turbulent trailing-edge noise	Lifting surfaces, blown flaps, machinery noises	Characteristic frequency $f_c \sim (1/2\pi)U/\delta$ or $(1/2\pi)U/l$ Correlation length: $l_c \sim l$ or δ	
Turbulent roundary layer	Flow over bodies, shells Internal pipe flow	Broadband; see sketch on right	

Note: The symbols used are as follows:

f – frequency, $\omega/2\pi$

d = diameter

U = approach velocity

δ = boundary layer thickness

C = chord

l_c = correlation length

λ = characteristic eddy (gust) scale

l = thickness of blunt ($l > \delta$) trailingedge

$P_{\text{rad}}(f)$ = sound power in proportional frequency band $Df \propto f$

$\Phi(k, \omega)$ = wavenumber frequency spectral density of wall pressure

pump rotor), the interaction force on the rotor that passes through this flow distortion will be periodic. This periodicity is in accordance with the repetitive passage of the blades through the flow distortion. If the inflow is also turbulent, then the interaction forces will be broadband in frequency. Table 1 summarizes each of these three generic forcing function characteristics.

6.2 Spatially Distributed Turbulent Boundary Layer on Extended Surfaces

This class of excitation is distributed over a large area and is typically convected at a velocity U_c, where U_c is a substantial percentage of the mean advance velocity of the surface. This convection gives a spatial bias to the forcing so that the excitation is concentrated at spatial scales or order $U_c/2\pi f$ at each frequency. As previously described and illustrated in Fig. 11, the wave vector spectrum extends to low values of wave vector, and this is due to the nonfrozen characteristics of the turbulence. Thus in Table 1 we depict the surface pressure field as broadband both in frequency and wavenumber when the integral over all wavenumbers,

$$\Phi_p(\omega) = \int\int_{-\infty}^{\infty} \Phi_p(\mathbf{k}, \omega)\, d^2\mathbf{k}, \tag{46}$$

expresses the overall frequency content of the boundary layer pressure. If flow separation occurs due to a localized geometric discontinuity on the surface, then the sound-producing character of surface pressure will more closely resemble that of an edge flow, as described above because of the stationary and localized (i.e., nonconvected) nature of the flow separation.

6.3 Flow Tones due to Feedback Mechanisms

In the cases of flow tones, all sources shown in Table 2 are dipoles, and the data derived for the table are essentially of aeroacoustic origin; see also Blake and Powell.[32] Little is known about the acoustic amplitudes because they depend on the details of each feedback mechanism that sustains the tone, particularly in underwater application. In aeroacoustic application, feedback is by a fluid path; in underwater acoustics the path may also be through the vibrating structure. The frequencies of the tones are often established by the matching of fundamental dispersion characteristics of allowed modes of flow instabilities and the resonance frequencies or geometry of the structure that is attached to the flow. Thus the frequencies shown are likely to have application to both airborne and waterborne flows.

The behavior of flow oscillator tones that occur with coupled flow elastic phenomena manifest tones that occur with $n = 1, 2, 3, \ldots,$ resonance frequencies and have the frequency–speed relationship with jumps as illustrated in Fig. 12. In the case of edge, hole, and cavity tones a pronounced velocity gradient is produced by the jet or grazing flow across the opening. Such flows with strong velocity profiles are unstable and are very sound and vibration sensitive. These sensitive disturbances in the field have a characteristic wavelength in the flow direction, say $\lambda_c = U_c/f$, where U_c is the flow convection velocity of the sensitive disturbance and f is frequency. The sounds of such flows often exhibit a series of stages, or orders of tones. A downstream body situated a distance L away on which the flow impinges sends back a disturbance to the upstream point. When $\lambda_c = \alpha L/n$, where α is a factor that depends on the par-

TABLE 2 Types of Flow Tones

Source Type	Application	Characteristic or Fundamental Frequency
D-1 Edge tone	Laminar rectangular jet impinging on sharp edge, musical instruments	Tonal: $\dfrac{f_s L}{U_j} \cong \dfrac{1}{2}\left(n + \dfrac{1}{4}\right)$ $n = 1, 2, \ldots, 4 < L < 20$ s = jet width
D-2 Hole tone	Staged throttling devices	Tonal: $0.011\sqrt{\mathcal{R}_d} > \dfrac{fb}{U_L} > 0.5$ $\mathcal{R}_d < 3000,$ $\dfrac{L}{d} = 2, 2.5, 3$ d = jet diameter
D-3 Cavity tone	Cavities, bleed ports	Laminar boundary layer: $\dfrac{fb}{U} \cong 0.022\dfrac{b}{\theta}$ Turbulent boundary layer $\dfrac{fb}{U} = 0.33\left(n - \dfrac{1}{4}\right)$ $n = 1, 2, \ldots$ b = opening width
D-4 Gaps	Flaps on lifting surfaces	$\dfrac{fw}{U} \sim (0.6, \ldots, 0.7)\left(n - \dfrac{1}{2}\right)$ $n = 1, 2, \ldots$ $\dfrac{fw}{U} \sim (0.5, \ldots, 0.6)\left(n - \dfrac{1}{4}\right)$ $n = 1, 2, \ldots$

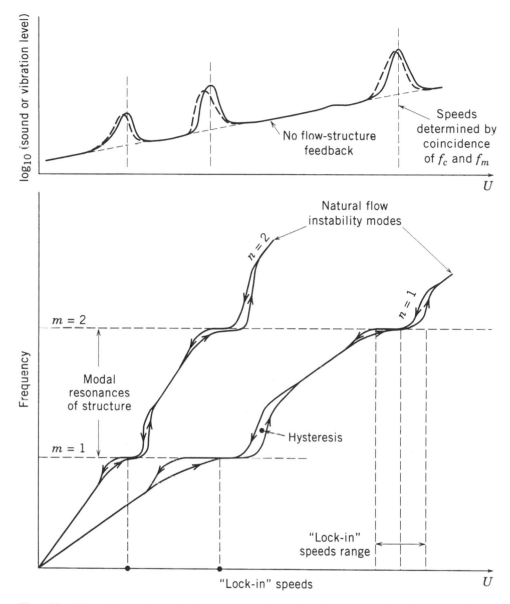

Fig. 12 Illustration of speed dependence of typical flow tones involving nonlinear flow–structure interaction. At "lock-in" very large amplitudes occur.

ticular flow geometry and n is the order of the tone, then the reinforced tone frequencies are

$$f_n = \frac{n U_c}{\alpha L}. \qquad (47)$$

The flow presents multiple modes, each mode being roughly established by the number of half waves that "fit" between the leading and trailing edges of the opening. The exact end points of each stage usually depend on whether the mean flow velocity increases or decreases and the mobility characteristics of the structure. Each stage persists as long as the phase and disturbance

growth may be sustained by the coupled flow acoustic (elastic) system.

In the case of singing hydrofoils and propellers, only a single fluid mode is typical and the tone is sustained by coupled modal vibration and vortex shedding. In this case, the structure has a series of resonance frequencies, say f_m, for which the fluid disturbance has a characteristic frequency, say f_c, which varies as, say, $f_c d/U = $ const when d is a transverse dimension of the body that generates the flow.

Thus the self-sustained frequency is always near both a resonance frequency of a flexural mode of the structure and an intrinsic instability frequency of the flow. Such instability frequencies are described by a Strouhal number Eq. (5), where L is a characteristic dimension of the flow. Self-sustained vibration and sound at "lock-in" has a very much larger amplitude than at other speeds.

REFERENCES

1. D. Ross, *Mechanics of Underwater Noise*, Permagon Press, 1984.

2. W. K. Blake, "*Mechanics of Flow-Induced Sound and Vibration*," 2 Vols., Academic Press, New York, 1986.

3. D. G. Crighton, A. P. Dowling, J. E. Ffowcs Williams, M. Heckl, and F. G. Leppington, "*Modern Methods in Analytical Acoustics*," Springer Verlag, 1992.

4. S. Subramanian and T. J. Mueller, "An Experimental Study of Propeller Noise Due to Cyclic Flow Distortions," *J. Sound Vibration*, Vol. 183, 1995, pp. 907–923.

5. G. Maidanik and E. M. Kerwin, "Influence of Fluid Loading on the Radiation from Infinite Plates below the Critical Frequency," *J. Acoust. Soc. Am.*, Vol. 40, 1966, pp. 1034–1038.

6. R. T. Knapp, J. Daily, and F. G. Hammitt, *Cavitation*, McGraw-Hill, New York, 1970.

7. H. Fitzpatrick and M. Strasberg, "Hydrodynamic Sources of Sound," *Proceedings 1st Symposium of Naval Hydrodynamics*, Washington, D.C., 1956, pp. 241–280.

8. G. L. Chanine and C. R. Sirham, "Collapse of a Simulated Multibubble System," ASME Cavitation and Multiphase Flow Forum, Albuquerque, NM, 1985.

9. G. VanWijngaarden, "One Dimensional Flow of Liquids Containing Small Gas Bubbles," *Am. Rev. Fluid Mech.*, Vol. 4, 1970, pp. 369–396.

10. H. J. Baiter, "On Different Notions of Cavitation Noise and What They Imply," ASME International Symposium on Cavitation and Multiphase Flow Noise, Anaheim, CA, 1986.

11. W. K. Blake and M. M. Sevik, "Recent Developments in Cavitation Noise Research," ASME Symposium on Cavitation Noise, Phoenix, AZ, 1982.

12. W. K. Blake, "Propeller Cavitation Noise: The Problems of Scaling and Prediction," ASME Symposium on Cavitation and Multiphase Flow Noise, Anaheim, CA, 1986.

13. W. K. Blake, M. J. Wolpert, and F. E. Geib, "Cavitation Noise and Inception as Influenced by Boundary Layer Development on a Hydrofoil," *J. Mech.*, Vol. 80, 1974, pp. 617–640.

14. S. Ceccio and C. E. Brennan, "Observations of the Dynamics and Acoustics of Travelling Bubble Cavitation," *J. Fluid Mech.*, Vol. 233, 1991, pp. 633–660.

15. V. Arakeri and V. Shanmuganathan, "On the Evidence for the Effect of Bubble Interference in Cavitation Noise," *J. Fluid Mech.*, Vol. 159, 1985, pp. 131–150.

16. G. Bark, "Development of Distortions in Sheet Cavitation on Hydrofoils," ASME Jets and Cavities-International Symposium, Miami Beach, FL, 1985.

17. P. Abbot, D. S. Greeley, and N. A. Brown, "Water Tunnel Pump Cavitation Noise Investigations," ASME International Symposium on Cavitation Research Facilities and Techniques, Boston, MA, 1987.

18. N. A. Brown, "Cavitation Noise Problems and Solutions," in *International Symposium on Shipboard Acoustics*, Elsevier, 1977.

19. W. K. Blake, H. Hemingway, and T. C. Mathews, "Two Phase Flow Noise," Noise-Con 88, Purdue University, June 1988.

20. Y. L. Levkovskii, "Modelling of Cavitation Noise," *Sea Prop.-Acoust.* (Engl. Trans.), Vol. 13, 1968, pp. 337–339.

21. M. Strasberg, "Propeller Cavitation Noise after 25 Years of Study," *Proc. ASME Sym. Noise Fluids*, Atlanta, GA, 1977.

22. J. E. Ffowcs Williams, "Sound Radiation from Turbulent Boundary Layers Formed on Compliant Surfaces," *J. Fluid Mech.*, Vol. 22, 1965, pp. 347–358.

23. J. E. Ffowcs Williams, "The Influence of Simple Supports on the Radiation from Turbulent Flow Near a Plane Compliant Surface," *J. Fluid Mech.*, Vol. 26, 1966, pp. 641–649.

24. Powell, A., "Aerodynamic Noise and the Plane Boundary," *J. Acoust. Soc. Am.*, Vol. 32, 1960, pp. 982–990.

25. D. G. Crighton and J. E. Oswell, "Fluid Loading with Mean Flow. I. Response of an Elastic Plate to Localized Excitation," *Phil. Trans. R. Soc. Lon.*, Vol. A335, 1991, pp. 559–592.

26. D. G. Crighton and D. Innes, "The Modes and Forced Response of Elastic Structures under Heavy Fluid Loading," *Phil. Trans. R. Soc. Lon.*, Vol. A312, 1984, pp. 291–341.

27. M. S. Howe, "Sound Produced by an Aerodynamic Source Adjacent to a Partly Coated Finite Elastic Plate," *Phil. Trans. R. Soc. Lon.*, Vol. A436, 1992, pp. 351–372.

28. R. Martinez, "Thin Shape Breakdown (TSB) of the Helmholtz Integral Equation," *J. Acoust. Soc. Am.*, Vol. 90, 1991, pp. 2728–2738.

29. W. K. Blake, D. Noll, R. Martinez, and Y. T. Lee, "Dynamics of Fluid-Coupled Neighboring Substructures at Low Frequencies," *Int. J. Comp. Str.*, 1997 to be published.

30. R. H. Lyon, *Statistical Energy Analysis of Dynamical Systems*, MIT Press, Cambridge, MA, 1975.

31. F. Fahey, *Sound and Structural Vibration*, Academic Press, New York, 1985.

32. W. K. Blake and A. Powell, "The Development of Contemporary Views of Flow-Tone Generation," in *Recent Advances in Aeroacoustics*, A. Krothapalli and C. A. Smith (Eds.), Springer Verlag, New York, 1986.

46

SHIP AND PLATFORM NOISE, PROPELLER NOISE

ROBERT D. COLLIER

1 INTRODUCTION

Ship noise is a major part of the field of underwater acoustics. Ship noise reduction and control is an important factor in the performance of underwater acoustic systems and in the habitability of the vessel for the crew and passengers. A singularly important text is *Mechanics of Underwater Noise*,[1] which provides detailed information on both mechanical and hydrodynamic noise sources and radiation. Chapter 41 summarizes the physics of sound radiation from ship structures and illustrates the basic mechanisms with simple mathematical models. The effects of fluid loading on acoustic radiation of plates and cylindrical structures are described in Chapter 11. In addition, detailed descriptions of hydroacoustic noise sources, including propellers, cavitation, vortex shedding, and turbulent boundary layer flow-induced noise, are covered in Chapter 45. These chapters provide valuable information directly related to the sources and characteristics of ship noise.

One of the principal objectives of this chapter is to provide engineering procedures for estimating ship machinery source levels and structural vibration transmission losses to arrive at hull vibration levels. The subsequent calculations for acoustic radiation are highly dependent on the details of the ship design and operational factors. Empirical data are presented to provide engineering guidance on radiated noise levels and, in particular, procedures for estimating propeller radiated noise are given, including estimates of cavitation noise. This chapter also discusses the underwater noise effects on ship sensors, that is, platform or self-noise, which has a significant impact on the performance of sonar sys-

tems (Chapter 49). Ship noise as it relates to human factors is a critical issue in ship design and operations. This chapter provides design guidance on criteria for environmental noise in interior ship compartments and discusses some of the material issues relating to noise control in the marine environment. Detailed information on noise control in interior spaces may be found in Parts VIII and IX.

2 RADIATED NOISE

2.1 General Characteristics

In naval operations the noise radiated by a ship is a dominant source of information, that is, signal, for underwater sonar systems. Radiated noise from ships can be an important contributor to ocean ambient noise, as discussed in Chapter 48, and as a factor in oceanographic research and geophysical exploration and cruise ships operating in environmentally sensitive areas. Engineering estimates for noise predictions of specific classes of ship machinery are given in this chapter based on weight, power, and foundation types. The effectiveness of vibration isolation systems, including examples of two-stage systems with rafts, is presented in Chapter 71. The important role of structural transmission mechanisms and their interaction with fluid systems is dealt with in Chapter 11.

The four principal groups of radiated noise sources are (1) machinery vibration caused by propulsion machinery and ships' services and auxiliary machinery, including steam, water, and hydraulic piping systems; (2) propellers, jets, and other forms of in-water propulsion; (3) acoustic noise within compartments below the waterline; and (4) hydroacoustic noise generated external to the hull by flow interaction with appendages, cavities, and other

Encyclopedia of Acoustics, edited by Malcolm J. Crocker
ISBN 0-471-80465-7 © 1997 John Wiley & Sons, Inc.

TABLE 1   Representative Ship-Radiated Noise Source Levels at 1 yd

Ship Class	Source Noise Levels (dB re 1 μPa 1 Hz Band)						
	0.1 kHz	0.3 kHz	1.0 kHz	3.0 kHz	5.0 kHz	10.0 kHz	25.0 kHz
Freighter, 10 knots	152	142	131	121	117	111	103
Passenger, 15 knots	162	152	141	131	127	121	113
Battleship, 20 knots	176	166	155	145	141	135	127
Cruiser, 20 knots	169	159	148	138	134	128	120
Destroyer, 20 knots	163	153	142	132	128	122	114

discontinuities. A summary of the these noise sources and their radiated noise characteristics is discussed in Ref. 2, from which Table 1 gives representative source levels as a function of frequency for a range of surface ships. The data for ships operating between 10 and 20 knots illustrate the dominance of low-frequency noise.

The speed/power dependence of the radiated noise of surface ships is further illustrated by the measurements plotted in Fig. 1.[1] The 9-knot noise spectrum is governed by machinery sources while the significant increase in low-frequency noise as speed is increased is due to both propulsion machinery and the inception and development of propeller cavitation noise. The latter source of radiated noise is dealt with in Section 4. Ross[1] provides estimation formulas for broadband source levels as a function of size or displacement tonnage and speed or power. For example,

$$L_s = 134 + 60 \log \frac{U_a}{10\,\text{knots}} + 9 \log \text{DT} \qquad (1)$$

where U_a is ship speed in knots and DT is displacement tonnage. Ross states that this formula is applicable for frequencies above 100 Hz and ships weighing under 30,000 tons. The acoustic efficiencies of ships have been found to range from 0.3 to 5 W of acoustic power for ship mechanical propulsion power of 1 MW; Ross suggests acoustic conversion efficiencies of 1×10^{-6} for machinery sources and 1.5×10^{-6} for cavitating propellers.

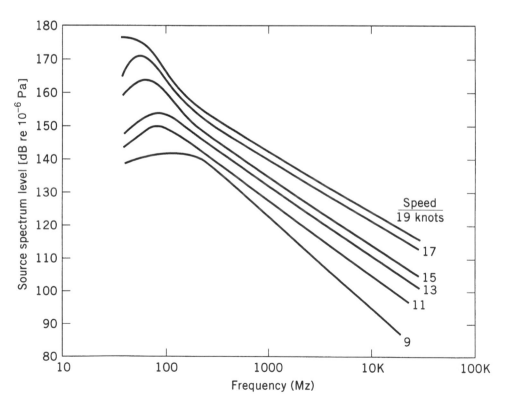

Fig. 1   Radiated noise of passenger ship *Astrid*, U.S. Office of Scientific R&D, published 1960.[1]

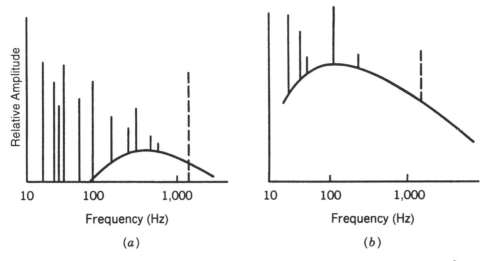

Fig. 2 Diagrammatic spectra of submarine noise at (*a*) low and (*b*) high speeds.[2]

2.2 Noise Spectra

Noise spectra are generally classified in two types: (a) broadband noise having a continuous spectrum such as that associated with cavitation and (b) tonal noise containing discrete frequency or line components related to machinery, gears, and modulation of broadband noise. In addition to steady-state noise, ship noise is also characterized by transient and intermittent noise caused by impacts, loose equipment, or unsteady flow that have

particular spectral properties. Ship noise is generally a combination of continuous and tonal noise covering the audio spectrum and is usually concentrated in the low-frcquency region. Figure 2 shows a diagrammatic comparison of two radiated noise spectra in which auxiliary machinery tonal noise governs the low-speed condition and propulsion system speed-related frequencies are superimposed on broadband propeller cavitation noise at high speeds. Figure 3 gives an overview of the frequency range of the major sources of radiated noise and

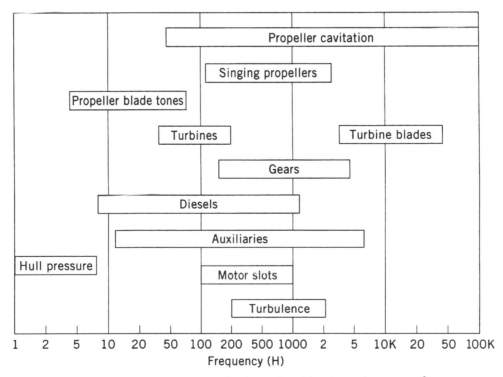

Fig. 3 Frequency ranges of noise radiated by ship noise sources.[3]

shows that the majority of machinery sources produce noise from 10 to 1 kHz and significant sources such as cavitation and turbine lines may extend into the 10-kHz region and above.[3]

2.3 Machinery Noise

The principal mechanisms that generate vibratory forces involve mechanical unbalance, electromagnetic force fluctuations, impact, friction, and pressure fluctuations. The classes of machinery that produce noise may be categorized according to their functions, such as (a) propulsion machinery (diesel engines, steam turbines, gas turbines, main motors, reduction gears, etc.) and (b) auxiliary machinery (pumps, compressors, generators, air conditioning equipment, hydraulic control systems, etc.).

Figure 4 is a schematic view of the machinery components of a diesel–electric propulsion system and their associated noise sources, which are described as follows:

1. Piston slap, which is a dominant noise mechanism of diesel engines and is caused by the impact of a piston against the cylinder wall, results in a spectrum made up of a large family of harmonically spaced tonals.
2. Mechanical imbalances of the generator and auxiliary machinery result in fluctuating forces and moments that are proportional to the square of the angular speed. Since the force is proportional to the angular speed. Since the force is proportional to

vibration velocity, the radiated power increases as the fourth power of rotational speed.

3. Electromagnetic force fluctuations of the main drive motor are related to changes in the flux density, which are a function of the number of poles and result in low-frequency line spectra.
4. Reduction gear noise is dominated by gear tooth impacts and results in tones at multiples of the tooth contact frequency. Helical gears are significantly quieter than spur gears.
5. Propeller noise, which is discussed in Section 3, consists of two major components: (a) direct radiation from the propeller blades and (b) low-frequency hull vibration modes induced by hydrodynamic fluctuating forces acting on the blades and transmitted through the propeller shaft and thrust bearings to the hull. The hull response is thus related to the shaft rpm and the number of propeller blades. The modes of hull vibration and resulting sound radiation are discussed in Chapter 41.

2.4 Machinery Vibration Levels

The prediction of radiated noise from machinery sources is based on the traditional noise model, which involves source levels, transmission path dynamics including vibration isolation and foundation transfer functions, and hull vibration and radiation. The role of foundation struc-

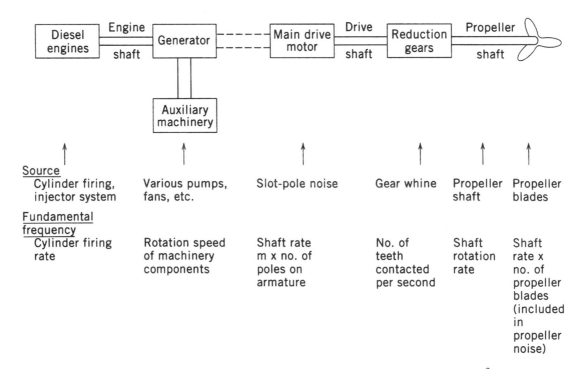

Fig. 4 Machinery components and noise sources on a diesel–electric ship.[2]

TABLE 2 Machinery Source Levels: Octave Band Baseline Vibration Levels

Machinery Type	Baseline Vibration Levels, L_{VB} (dB re 10^{-6} cm/s)										
	Center Frequency										
	16 Hz	31.5 Hz	63 Hz	125 Hz	250 Hz	500 Hz	1 kHz	2 kHz	4 kHz	8 kHz	16 kHz
Gas turbines											
Small high speed	80	85	95	100	100	90	110	100	100	95	95
Intermediate	110	104	106	103	98	93	86	80	73	67	57
Electric motors											
AC	102	96	90	84	78	72	66	60	54	48	42
DC	88	88	83	78	76	73	68	58	53	48	43

tural mobilities is a critical part of this system and directly determines the vibration levels of the source machinery, as discussed in Chapter 71. The purpose of this section is to provide engineering estimates for machinery vibration levels, based on an extensive data base, as an input to structural design and noise radiation calculations developed for SNAME.[4]

The source level algorithms are provided for engineering guidance. Reliable measured data should be used when available. Also, the parametric dependencies of the prediction algorithms can be used to scale from measured data for similar classes of machinery. It should be noted that the source vibration levels assume that the machine is mounted on low-frequency vibration isolation mounts. Thus the effect of rigid mounting is to decrease machinery source vibration levels due to the increased constraining effect of the foundation relative to that of resilient mounts.

Table 2 gives source vibration velocity levels in decibels relative to 10^{-6} cm/s as a function of octave bands for two types of gas turbines and for representative classes of electric motors.

Table 3 gives formulas for overall machinery source levels expressed as baseline vibration velocity in decibels relative to 10^{-6} cm/s. The machinery parameters include w = gross weight in kilograms, kW = rated power in kilowatts, and actual and rated rpm. The types of machinery include diesels, generators, pumps, and reduction gears. For each type, adjustments to the overall levels are given in Table 4 to obtain estimates of vibration levels for the designated octave bands. Alternative prediction procedures may be found in other handbooks.[5–7]

2.5 Structural Vibration Transfer Functions

The effects of the transmission of vibratory energy (structure-borne noise) along and through different structural paths are expressed as transfer functions that relate the input (machinery) and output (foundation or hull) vibration levels. In this section an empirical procedure is given for estimating transfer functions for different transmission paths. The U.S. Navy has published several design handbooks to guide structural designers.[8–11] It should be noted that state-of-the-art PC-based noise models can provide greatly refined and more accurate transfer function estimates for specified structural configurations, and the accompanying estimates should only be used for guidance. Furthermore, several vibration trans-

TABLE 3 Machinery Vibration Overall Source Levels

Machinery Class	Overall Baseline Vibration Source Level, L_{VB} (dB re 10^{-6} cm/s)
Diesel	$-20\log(w) + 20\log(\text{kW}) + 30\log\left(\dfrac{\text{rpm}}{\text{rpm}_0}\right) + 136$
Reduction gears	$64 + 10\log(\text{kW})$
Generator	$53 + 10\log(\text{kW}) + 7\log(\text{rpm})$
Pumps	
Nonhydraulic	$65 + 10\log(\text{kW})$
Hydraulic	$63 + 10\log(\text{kW})$

Note: w = gross weight (kg), kW = rated power, rpm = given rotational speed, rpm_0 = rated rotational speed. See Table 4 for octave band adjustments.

TABLE 4   Machinery Vibration Source Levels[a]: Octave Band Adjustments to Overall Baseline Vibration Levels (Table 3)

Machinery Class	Vibration Source Levels										
	Center Frequency										
	16 Hz	31.5 Hz	63 Hz	125 Hz	250 Hz	500 Hz	1 kHz	2 kHz	4 kHz	8 kHz	16 kHz
Diesel	0	−3	−4	−4	5	−6	−6	−10	−18	−29	−44
Reduct. Gears	0	−2	1	−11	−12	−3	1	−5	−16	−32	−38
Generator	0	3	8	5	−1	−5	−10	−15	−21	−27	−35
Pumps											
Nonhydraulic	10	10	12	19	11	9	4	−6	−8	−15	−25
Hydraulic	10	20	27	32	33	36	30	25	20	5	−10

[a]See Table 3.

mission paths must generally be considered for ship noise predictions.

Vibration Isolation Mountings   The transfer function or transmission loss is expressed as the log ratio of vibration velocity above the mounts at the attachment to the machine subbase to that on the foundation structure below the mounting system. In this design guide, sources of vibration are divided into three weight classes: class I, less than 450 kg; class II, 450–4500 kg; class III, over 4500 kg.

Table 5 gives representative transmission loss values for four types of mounting configurations: hard mounted (no vibration isolation), distributed isolation material (DIM), single-stage low-frequency isolation mounts, and

TABLE 5   Representative Transmission Loss Values versus Octave Band Center Frequency for Ship Machine Mounting Arrangements (dB)

Machinery Weight Class	Transmission Loss, dB								
	Center Frequency								
	31.5 Hz	63 Hz	125 Hz	250 Hz	500 Hz	1000 Hz	2000 Hz	4000 Hz	8000 Hz
Hard Mounted									
I	13	10	8	5	3	2	1	0	0
II	9	7	5	3	1	0	0	0	0
III	5	3	2	1	0	0	0	0	0
Distributed Isolation Material									
I	0	1	5	9	12	15	15	15	15
II	0	0	1	3	5	8	9	10	10
III	0	0	1	2	3	3	4	5	8
Low Frequency Isolation Mounts									
I	20	25	30	30	30	30	30	30	30
II	12	16	20	23	25	25	25	25	25
III	5	6	8	12	15	18	20	20	20
Two-Stage Mounting System									
I	25	33	40	45	50	50	50	50	50
II	22	30	35	40	45	48	50	50	50
III	20	25	30	30	35	45	50	50	50

Note: Machinery weight classes: class I, under 450 kg; class II; 450–4500 kg; class III; over 4500 kg. Values based on relatively rigid high-impedance foundation structures.

two-stage isolation with intermediate rafts. The estimates are presented as average values for the designated octave bands. For the hard-mounted case lightweight machines can be expected to have a modest loss at low frequencies, while all classes of machinery have little or no loss above 250 Hz. The DIM installations are effective for lightweight machines above 250 Hz, but their performance decreases as the machinery weight increases. The losses of low-frequency mounts closely follow theoretical predictions for lightweight machines, that is, 20–30 dB transmission loss over the given frequency range. However, as the weight of a machine increases relative to that of the foundation structure, the degree of isolation decreases, reflecting the overall impedance relationships. Two-stage mounting systems have been implemented extensively in ship designs and have proved to be highly effective. Similar types of estimates of transfer functions for representative machinery foundation structures may be used to arrive at hull vibration levels.

2.6 Hull Vibration-Radiated Noise

The relationships between hull structural vibration levels and radiated noise are discussed in Chapters 11 and 41. Assuming that the radiation efficiency σ_r is known, a first-order estimate of the sound radiated from hull plating can be calculated by

$$L_s = L_v + 10 \log \sigma_r + 10 \log A_p + 10 \log n + 41 \quad (2)$$

where L_s is the equivalent source level at 1 yd, L_v is the space average vibration velocity level of the hull plating, A_p is the area of a single hull plate in inches squared, and n is the number of radiating panels.

2.7 Hull Vibration Transmission

Hull vibration (calculated or measured) in the area of machinery foundation attachment locations (e.g., deep frames) is transmitted through the steel hull structure. Thus, larger areas of the hull may contribute to radiated noise. Also, the transmission of machinery excited hull vibration into sonar array structures can contribute to sonar self-noise (see Section 4). A useful guideline for broadband transmission loss in typical damped ship structures is 0.5–0.8 dB/ft for cases of free-layer damping and 1.7 dB/ft for structures, including wetted hulls, with constrained layer damping. Structural details are needed to establish frequency dependence.

2.8 Hull Grazing Sound Transmission Loss

Table 6 provides frequency-dependent expressions of propagation losses for sound traveling in the water along

TABLE 6 Hull Grazing Transmission Loss versus Frequency (dB)

Octave-Band Center Frequency (Hz)	Transmission Loss at Distance r (dB re r_0 = 1 yd)
250	$10 \log r/r_0$
500	$13 \log r/r_0$
1000	$17 \log r/r_0$
2000	$23 \log r/r_0$
4000	$27 \log r/r_0$
8000	$27 \log r/r_0$

the length of the hull, that is, grazing sound. The hull is considered to be an air-backed baffle and the grazing sound transmission loss applies to a 1-yd source level for far-field radiation. The theoretical basis for these estimates involves energy transmission through both the hull structure and the water path.

3 PROPELLER NOISE

3.1 General Characteristics

Propulsion propellers constitute a major source of ship noise and, similar to fans, aircraft turbo-props, helicopters, and other devices with rotating blades operating in nonuniform flow fields, are the subject of continuing noise control efforts. The interaction of blade forms and hydrodynamic flows and the resulting dynamic response of the blades and the associated acoustic radiation are complex phenomena that depend on a wide range of design and operational variables that do not lend themselves to simple models and noise predictions. Hydrodynamic noise sources are discussed in Chapter 45; in particular, Section 3 in that chapter provides the theoretical background for cavitation inception, development, and associated noise.

The guidelines for estimating propeller noise included in this chapter are based on the original work of Ross[1] and the more recent engineering analyses presented in the report of the Nordic cooperative project.[6] Marine propeller noise reduction has benefited from parallel aerodynamic acoustic studies, which are the subject of other chapters in the handbook. For example, significant reductions in both propeller and fan noise have been achieved through smoothing and control of inlet flows and the design of skewed blades.

3.2 Propeller Noncavitating Noise

There are three types of noncavitating propeller noise: (a) mechanical blade tonals related to propeller shaft speed

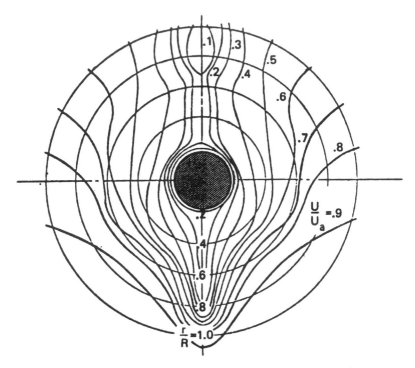

Fig. 5 Wake diagram for a single-screw merchant ship.[1]

and the number of blades; (b) propeller broadband noise related to blade vibratory response to turbulence ingestation and trailing-edge vortices; and (c) propeller singing due to coincidence of vortex shedding and blade resonant frequencies. Blade tonals and harmonics result from oscillating components of forces or propeller thrust variations caused by circumferential variations of the wake inflow velocity. Figure 5 illustrates, by use of equivelocity contours, the velocity variations in the plane of a single-propeller merchant ship. The flow speed varies from 10 to 90% of the forward speed of the propeller. These velocity differences cause large variations of the angle of attack and associated lift forces, which lead to significant fluctuations in thrust and torque during each revolution of the shaft and, in turn, to high-level, low-frequency hull vibration. Thus, the most important design consideration is the relationship between the harmonic structure of the wake and the number and blade form of the propellers.

The primary propeller design factors include diameter, shaft rpm, number of blades, expanded area ratio, blade load distribution, skew distribution, blade tip–hull clearance, and the spatial and temporal characteristics of the inflow field.

3.3 Propeller Cavitation Noise

As stated above, a description of cavitation is provided in Chapter 45. There are four types of propeller cavita-

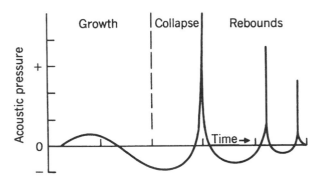

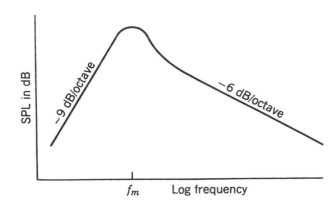

Fig. 6 Cavitation pressure pulses from collapsing cavity and idealized spectrum.[1]

tion: driving face, suction face, tip vortex, and hub vortex. The blade tip speed governs cavitation inception, as shown in Eq. 14 of that chapter. Broadband cavitation noise results from the growth and collapse of a sheet of bubbles occupying a volume on the individual blades. Figure 6 illustrates this process for a single cavitation bubble with the resultant idealized spectrum. The general noise spectrum of blade cavitation is shown in Figure 7 and has four principal spectral regions:

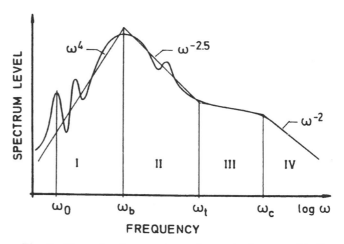

Fig. 7 General noise spectrum of a cavitating propeller.[6]

I. Low frequency: contains blade frequency ω and harmonics; mean power level increases as ω^4.

II. Midfrequency: starts at bubble frequency ω_b; mean power level increases as $\omega^{-5/2}$.

III. Intermediate frequency: transition region between regions II and IV.

IV. High frequency: starts at bubble frequency ω_c; mean power level decreases as ω^{-2}.

In regions I and II the fluctuations of the sheet cavitation volumes may be represented by a large bubble that acts as an acoustic monopole. In region IV, the power is caused by cavity collapse or by shock wave generation by nonlinear wave propagation. Region III contains a mixture of regions II and IV. Figure 8 is an example of

a comparison between predicted and full-scale measured source levels for a 32,000-ton vessel.[6]

4 PLATFORM AND SONAR SELF-NOISE

4.1 General Characteristics

Conceptually, self-noise is that noise in a sonar system attributable to the presence of the platform, as illustrated

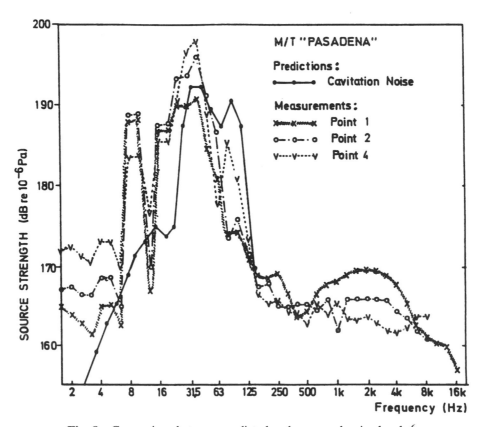

Fig. 8 Comparison between predicted and measured noise levels.[6]

in Figure 9. In practice, sonar self-noise measurements always include the contribution of the ambient noise. The self-noise term in the sonar equation set forth in Chapter 49 is also all inclusive. Thus, sonar self-noise is really the total noise level of a sonar when there is no target present.

Two kinds of self-noise measurements are made: "platform-noise" measurements are self-noise measurements made with omnidirectional hydrophones; "sonar self-noise" measurements are made at the output of the designated array. Platform noise is L_N in the sonar equation, while sonar self-noise L_e is the equivalent of $L_N - N_{DI}$, where N_{DI} is the measured array gain. It is important to measure the sonar self-noise directly because the array signal-to-noise gain depends upon the spatial properties of the noise field. The directivity index, which is based on isotropic noise, is a first-order approximation of the ability of an array to discriminate against noise, but it usually does not equal the actual array gain. In fact, coherent noise sources can appear as target signals on sonar displays.

In treating sonar self-noise, it is convenient to consider six dominant noise sources:

1. Ambient ocean noise
2. Local machinery sources
3. Remote machinery sources
4. Propeller noise

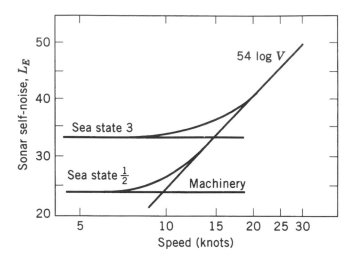

Fig. 10 Speed-dependent self-noise as a function of sea state, relative sonar noise level vs. speed, knots.[1]

5. Local flow noise
6. Local cavitation and/or bubble sweepdown

Ambient noise is discussed in Chapter 47. It generally dominates only at slow speeds. Figure 10 illustrates how at low sea states (e.g., sea state $\frac{1}{2}$) ambient noise may control sonar self-noise only at low ship speeds before other speed-dependent noise sources begin to dominate. At sea state 3, ambient noise controls sonar self-noise up to higher ship speeds.

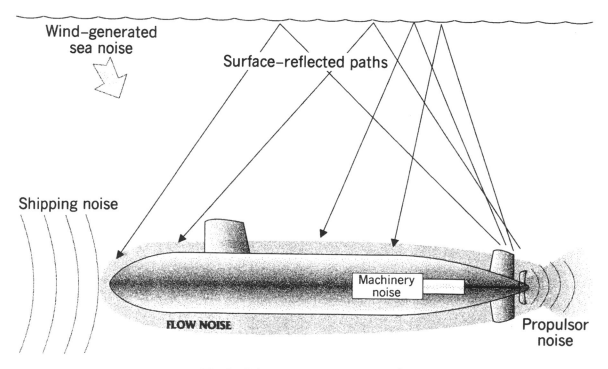

Fig. 9 Primary sources of self-noise.[3]

At the other extreme, *local cavitation* on the sonar dome is generally a problem only at the highest speeds of surface ships. In moderate to high sea states, bow wave splash is a significant noise source that extends along the line of the breaking bow wave. It is highly dependent on ship motion with respect to the seaway. Machinery can contribute noise into the sonar arrays by vibrational paths and these sources may be significant over a wide speed range. Propeller noise may be transmitted to sonar arrays by hull vibration, hull grazing (see Table 6), and by surface reflected paths. Each type of noise source has individual speed and power dependencies that contribute in varying degrees to the overall sonar self-noise level as a function of speed.

At higher speeds *local flow noise* is a dominant noise source. The mechanisms whereby turbulent boundary layer pressure fluctuations excite flush-mounted hydrophones, sonar domes, and structures local to the sonar are discussed in Chapter 45. The wavenumber of the convective turbulent excitation is much larger than that of the acoustic signal of the same frequency. It is thus possible to design hydrophones and arrays to discriminate against flow noise while maximizing signal gain.

The importance of remote machinery and propellers as contributors to self-noise is apparent when close correlations between radiated noise and sonar self-noise are recognized. The two principal paths are surface boundary reflection or scattering and hull vibration. Hull vibration is particularly important for low-frequency noise while the acoustic path surface ship sonar self-noise is often higher in shallow water than in deep water. Generally speaking, the reduction of radiated noise can also be important to improving sonar performance.

4.2 Dome Design

The term *dome* refers to a vaulted structure. It probably originated when rounded projections were first installed to protect protruding hydrophones. Today the term encompasses any structure housing arrays or hydrophones, whatever the shape, and sometimes even describes the supporting structure and array. Domes may now comprise the whole front portion of a submarine or the bulbous bow of a surface ship. They are either conformal domes (i.e., conforming to the general shape of the ship) or appendage domes (i.e., protruding into the water flow around the ship, as illustrated in Fig. 11).

The structural and acoustic factors involved in dome design and ship installations are dealt with in Ref. 12 and illustrated in Fig. 12. The acoustic factors in dome design, which are numbered in Fig. 12, are (1) transmission through material, (2) compressional and flexural coincidence angles, (3) flow excitation, (4) refraction, (5) internal reflection, (6) structure-borne noise, (7) water-borne noise, (8) reverberation, (9) cavitation, (10) bubbles, and (11) fouling.

When designing and installing sonar domes, the

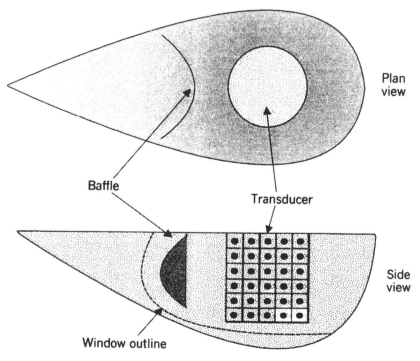

Fig. 11 Sonar dome schematic.

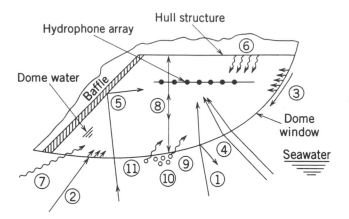

Fig. 12 Acoustic considerations in dome design.

emphasis is on protecting the tranducers, minimizing ship impact, and enhancing signal-to-noise ratio. The interposition of the dome, or window, reduces the signal level and, therefore, low-transmission-loss structures and materials are mandatory. At low frequencies, where large arrays and domes are required, the wavelengths are large and attenuation through the material is low. The thicker windows required for mechanical strength do not therefore cause unacceptable attenuation losses at these frequencies. Thus, design becomes a compromise between acoustic, hydrodynamic and structural requirements and the location, size, operating frequency, and shape of the array.

The dome or window should be made of tough, hard-to-damage, material because hydroelastic excitation or flutter around a damaged portion can cause noise. It can also be excited by the flow of water past its surface and radiated to the hydrophone if it is constructed of an elastic material (e.g., steel or fiberglass). A highly damped, low-modulus material such as rubber reduces flow noise problems. But in every case, it is important to maintain both the fairness and smoothness of the dome and the adjacent hull surfaces. Step discontinuities must be avoided.

The fluid inside the dome may become warmer than the surrounding sea and, thus, may not have the same speed of sound as the external water. This results in refractions and focusing, which leads to degradation in beam former performance and gives erroneous bear-

ings. Thus dome design should provide for flushing and replacement of the internal dome water.

The dome is usually equipped with baffles aft of the array to reduce the effect of ship machinery and propeller noise on the sonar array. However, these baffles can also mask targets in the stern sectors. Flat or curved surfaces inside the dome can reflect incoming signals, and non-reflecting surface materials may be needed to reduce the possibility of spurious signals appearing in the array output. Table 7 gives values for diffraction losses as a function of frequency for typical water and air-backed sonar baffles.

4.3 Flow-Induced Noise

Turbulence is generated along the hull of a ship as it passes through the water. The fluctuating pressures associated with a turbulent boundary layer radiate noise directly into the water (flow noise) and excites vibrations in ship structures because the mechanism of direct radiation of flow noise into the water is inefficient. Flow noise is usually not a significant source of ship noise. However, flow-induced noise transmitted through excitation and radiation of sonar domes and adjacent hull structures becomes a significant source of sonar self-noise at higher speeds. Chapters 11 and 45 provide more detailed information on hydroacoustic noise and structure–fluid interactions.

The estimates of platform self-noise given below are sound pressure levels at individual element locations within sonar domes. For flow-induced noise, platform noise levels inside sonar domes are estimated by adding the values given below to the baseline level:

$$L_N = 34 + 45\log(V) - 20\log(h) + 10\log(A) \quad (3)$$

where V is ship speed in knots, h is the thickness of the sonar dome in centimetres, and A is the surface area of the sonar dome in square metres.

Octave Band Adjustments to L_N (dB)

Center Frequency, Hz	31.5	63	125	250	500	1000	2000	4000	8000
Flow Noise	39	28	19	16	13	10	7	4	1

TABLE 7   Hull-Mounted Sonar Array Baffle Diffraction versus Frequency Loss (dB)

Type of Barrier/Baffle	Diffraction Loss (dB)					
	250 Hz	500 Hz	1000 Hz	2000 Hz	4000 Hz	8000 Hz
Water backed	0	0	0	5	5	5
Air backed	8	11	14	17	20	23

TABLE 8   Noise Level Criteria for Interior Compartments for New Construction, U.S. Navy Ships

Noise Category[a]	Compartment Noise Level Criteria (dB re 20 μPa)								
	31.5 Hz	63 Hz	125 Hz	250 Hz	500 Hz	1000 Hz	2000 Hz	4000 Hz	8000 Hz
A-12	66	63	60	57	54	53	48	45	42
A-3	75	72	69	66	63	60	57	54	51
B	75	72	69	66	63	60	57	54	51
C	72	69	66	63	60	57	54	51	48
D	91	88	85	82	79	76	73	70	67
E	82	79	74	73	70	67	64	61	58

[a]Noise categories: (1) A-12, command and control centers, equivalent to large quiet offices to allow normal conversation at 6 ft; (2) A-3 and B, sonar rooms and crew's living quarters; (3) C, crew's birthing areas; (4) D, machinery rooms and working spaces; (5) D-1, engine rooms requiring ear protection; (6) E, compartments and spaces subjected to acative sonar transmissions.

5   INTERIOR COMPARTMENT NOISE

5.1   General Characteristics

The airborne noise levels in ships constitute a major area of ship acoustics. The high airborne noise levels of propulsion machinery and reduction gears in confined machinery spaces constitute a serious hearing and communication problem. Living and working spaces must meet habitability criteria for the long-term health and comfort of ship's personnel. References 4, 5, and 7 provide applicable design guidelines.

The latest available (1991) Coast Guard and Navy airborne noise criteria are given by the noise category designations defined in Table 8:

- Category A: Spaces where direct speech communication must be understood with minimal error and without need for repetition. Acceptable noise level is based on a talker–listener distance of either 3 or 12 ft. Category A-3 applies when extreme talker-to-listener distance is less than 6 ft. Category A-12 applies when the extreme talker-to-listener distance is 6 ft or greater.
- Category B: Spaces where comfort of personnel in their quarters is the primary consideration and communication considerations secondary.
- Category C: Spaces where it is essential to maintain especially quiet conditions.
- Category D: High-noise-level areas where voice communication is not important, car protection is not provided, and prevention of hearing loss is the primary consideration.
- Category E: High-noise-level areas where voice communication is at short distances and there is high vocal effort and where amplified speech and telephones are normally available.

TABLE 9   Machinery Noise: Sound Power Source Levels (dB re 10^{-12} W) Octave Band Adjustments

Machinery Class	Sound Power Source Levels (dB re 10^{-12} W)								
	31.5 Hz	63 Hz	125 Hz	250 Hz	500 Hz	1000 Hz	2000 Hz	4000 Hz	8000 Hz
	Equation (4)								
Diesel engines									
Intake	21	21	27	28	26	24	20	13	4
Exhaust	44	40	46	42	34	30	24	14	6
Casing	4	6	15	18	17	15	11	4	0
	Equation (5)								
Gas turbines									
Exhaust	22	22	22	22	22	20	16	14	4

Source: From Ref. 13.

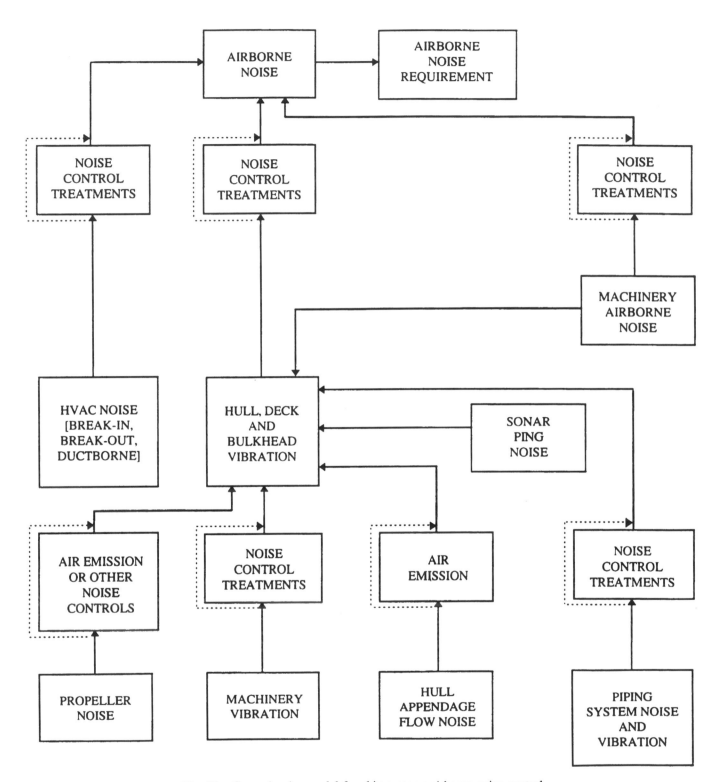

Fig. 13 General noise model for ship systems airborne noise control.

Associated with each noise category are sound pressure levels in octave bands and/or requirements on the speech interference level (SIL), where SIL is defined as the arithmetic average of the sound pressure levels in the 500-, 1000-, and 2000-Hz octave bands. For design purposes it is permissible to assign to each of these bands the value of the SIL requirement.

Additional noise categories are sometimes defined for spaces where noise levels are higher than the levels of noise category D and personnel are instructed to wear ear protectors, e.g., D-1 engine rooms.

5.2 Airborne Noise Sources

The acoustic source levels of representative machinery noise sources in ships are expressed as sound power in octave bands with adjustments given in Table 9[13]:

1. Diesel engines—intake, exhaust and casing radiation, baseline:

$$L_{WB} = 58 + 10\log(kW)\,dB\,re\,10^{-12}\,W \qquad (4)$$

2. Gas turbines, intermediate—exhaust:

$$L_{WB} = 74 + 10\log(kW)\,dB\,re\,10^{-12}\,W \qquad (5)$$

5.3 Acoustic Transmission Paths

Noise control measures in ships are similar to those in other architectural acoustic applications. An extensive and broad base of materials technology has been developed and applied by the U.S. Navy and industry to meet surface ship and submarine habitability and environmental requirements. This technology is largely available to nonmilitary ship designers. A general noise model is given in Fig. 13. In addition, Statistical Energy Analysis noise models are available to support alternative designs for noise control treatment combinations.

Examples of available noise control technology are as follows:

1. Internal high-transmission-loss "sandwich"-type composite treatments (e.g., fiberglass blanket/loaded vinyl septum/fiberglass blanket) or tuned

TABLE 10 Transmission Loss for Representative Ship Structure and Materials

Material No.	Panel Description, Type and Thickness	125 Hz	250 Hz	500 Hz	1000 Hz	2000 Hz	4000 Hz	STC
1	Bare aluminum bulkhead, $\frac{1}{4}$ in.	20	25	30	35	30	39	30
2	One layer, 2 in. MIL-A-23054	18	27	37	47	48	57	39
3	One layer, 2 in. MIL-A-23054 + 1 lb/ft² lead vinyl	14	37	51	56	59	69	38
4	Two layers, 2 in. MIL-A-23054 + 1 lb/ft² lead vinyl	23	44	57	65	68	75	38
5	One layer, 1 in. Mylar-faced fiberglass	17	23	33	44	45	59	35
6	One layer, 1 in. Mylar-faced fiberglass and W.R. 1 lb/ft² lead	21	34	49	60	63	75	40
7	Two layers, 1 in. Mylar-faced fiberglass and W.R. 1 lb/ft² lead	20	33	51	61	61	72	39
8	One layer, 2 in. Mylar-faced fiberglass	18	25	38	48	48	67	38
9	One layer, 2 in. Mylar-faced fiberglass and F.G.R. 1 lb/ft² LD	12	36	49	57	60	72	36
10	Two layers, 2 in. Mylar-faced fiberglass and F.G.R. 1 lb/ft² LD	17	38	54	57	60	78	41
11	Two layers, 2 in. Mylar-faced fiberglass (no septum)	16	33	47	56	55	73	40
12	One layer, 2 in. fiberglass with 14 oz/ft² lead face septum	14	35	45	57	58	67	38
13	Two layers, 2 in. fiberglass with 14 oz/ft² lead septum	17	38	52	58	55	65	41
14	Quilted blanket, two layers 1 in. fiberglass with lead septum	14	28	40	47	48	53	37

Note: Tests in conducted accordance with ASTM E90.

TABLE 11   Coefficients of Sound Absorption for Representative Shipboard Materials

Material Type	Thickness (in.)	Coefficient of Sound Absorption					
		125 Hz	250 Hz	500 Hz	1000 Hz	2000 Hz	4000 Hz
Board	1.0	0.07	0.25	0.70	0.90	0.75	.70
2.5 ± 0.5	1.5						
lb/ft.^{2}		0.15	0.45	0.90	0.90	0.80	0.75
Navy II	2.0	0.25	0.70	0.90	0.85	0.75	0.75
	0.5	0.04	0.10	0.20	0.40	0.45	0.55
	1.0	0.06	0.20	0.45	0.65	0.65	0.65
	2.0	0.15	0.40	0.75	0.75	0.75	0.70
	3.0	0.20	0.60	0.90	0.80	0.80	0.75
	4.0	0.25	0.65	0.95	0.85	0.85	0.80
Navy III	2.0	0.43	0.96	1.0	1.0	0.70	0.35

double-wall constructions for reduction of airborne noise transmission through the hull or through partitions between compartments

2. Internal sound absorption materials for installation on bulkheads and overheads of ship compartments to reduce individual compartment noise levels

3. Vibration damping materials for application to internal ship structures to reduce transmission of structure-borne noise to interior compartments

Transmission loss data on a representative sample of commercial products that meet U.S. Navy requirements are given in Table 10.[14]

5.4   Acoustic Absorption Materials

There are two classes of acoustic materials for sound absorption applications in (1) "clean" spaces and (2) machinery and equipment spaces and a third class to supplement structural transmission loss of decks, bulkheads, and interior joiner work.

Type I treatments are simply designed to reduce the reverberant sound field by increasing the sound absorption characteristics of interior bulkheads and overheads. Typically, 2-in.-thick fiberglass blankets with perforated facings are used for this purpose. Such treatments typically provide on the order of 5 dB (A) noise reduction.

Type II treatments are similar in function to type I treatments but generally consist of a 2-in. layer of fibrous glass material faced with an impervious fabric for protection and to reduce the risk of degradation in absorption characteristics due to oil and water contamination. However, the facing material generally degrades the absorption performance of the overall treatment.

Type III treatments are designed to supplement the transmission loss afforded by baseline structural decks and bulkheads and interior joiner work. Typically, these treatments incorporate relatively high density fiberglass installed as two separate blankets separated by a thin septum of lead-loaded or barium-sulfate-loaded vinyl. The outer, exposed layer of fibrous glass material can have either a perforated facing to maximize sound absorption or a thin facing impervious to oil and water.

Table 11 provides data on ship acoustic materials.[15]

Acknowledgments

The author wishes to acknowledge the contributions of many colleagues to the database which underlies the information presented in this chapter. In particular, the selection and summarizing of this material has benefitted from discussions with Dr. Donald Ross, author of the text *Mechanics of Underwater Noise*, and Daniel Nelson, Bolt, Beranek and Newman, Inc., who originated a major part of the noise and vibration performance predictions for shipboard machinery systems.

REFERENCES

1. D. Ross, *Mechanics of Underwater Noise*, Peninsula Publishing, Los Altos, CA, 1987.

2. R. J. Urick, *Principles of Underwater Sound for Supervisors*, McGraw-Hill, New York, 1967.

3. D. Ross and R. D. Collier, *Mechanics of Underwater Noise*, Course Notes, Applied Technology Institute, Columbia, MD, 1985–94.

4. R. W. Fischer, C. B. Burroughs, and D. L. Nelson, "Design Guide for Shipboard Airborne Noise Control," *SNAME Tech. Res. Bull.*, 1983, pp. 3–37.

5. *Noise Control Handbook*, British Ship Research Association, Naval Architecture Department, February 1982.

6. S. Nilsson and N. P. Tyvand (Eds.), *Noise Sources in Ships: I Propellers, II Diesel Engines*, Nordic Cooperative

Project: Structure Borne Sound in Ships from Propellers and Diesel Engines, Nordforsk, Norway, 1981.

7. *Handbook for Shipboard Airborne Noise Control*, Bolt Beranek and Newman, Inc., Technical Publication 073-0100, U.S. Coast Guard and U.S. Naval Ship Engineering Center, February 1974.

8. *Design Handbook. Resilient Mounts*, U.S. Naval Sea Systems Command, NAVSEA 0900-LP-089-5010, 1977.

9. *Design Handbook. Distributed Isolation Material*, U.S. Naval Sea Systems Command, NAVSEA S9078-AA-HBK-010-DIM, 1982.

10. *Design Handbook. Piper Hangers*; U.S. Naval Sea Systems Command, NAVSEA S9073-A2 HBK-010, undated.

11. *Design Handbook of Vibration Damping*, Mare Island Naval Shipyard, Report No. 11-77, U.S. Naval Sea Systems Command, January 1979.

12. *Design Handbook for Sonar Installations*, U.S. Naval Underwater Systems Center, TD 6059, undated.

13. D. L. Nelson, *Main Propulsion Gas Turbine Exhaust Noise*, Bolt Beranek and Newman, Inc., TM-339, February 1977.

14. R. D. Collier, *Noise Control Materials and Application*, Bolt Beranek and Newman, Inc., Report No. 6637, 1987.

15. *Design Data Sheet. Ship Damping and Special Acoustic Materials*, Bolt Beranek and Newman, Inc., DDS-636-1, U.S. Naval Sea Systems Command, September 1980.

47

UNDERWATER EXPLOSIVE SOUND SOURCES

VALERY K. KEDRINSKII

1 INTRODUCTION

High-rate changes in the physical and chemical states of matter accompanied by abrupt heat release and increase in pressure and temperature are usually defined as an *explosion*. Chemical matter capable of such transformations is called an *explosive*. Processes proceeding for parts per million of a second with velocities of the order of 1 km/s and higher are called *explosive processes*. Radiation sources or systems based on such processes are called *explosive sound sources*. They generate shock and compression waves, that is, high-pressure zones with high energy density.

Such waves may be generated not only by explosives but also by shock tubes, which in the simplest case comprise two sections separated by a diaphragm: one section filled with high-pressure gas (or a vacuum subsection equipped with a piston accelerated by a working gas) and the other containing liquid under normal pressure. A quick opening of the diaphragm or the impact of the piston on it generates a shock wave propagating through the liquid. Such sources are called the *explosive-type sources*.

Some examples of explosive sound sources are condensed and gaseous explosives, electric charges, and original analogies of hydrodynamic shock tubes such as airguns, waterguns, watershocks, and powerful laser beams focused into a liquid. The main advantage of these sources over traditional ones is a high level of transformation of energy, generated by the explosive and the explosive-type sources, into an acoustic one in the form of a high-power wide-band signal that is noise resistant and capable of propagating over hundreds of kilometres.

Encyclopedia of Acoustics, edited by Malcolm J. Crocker
ISBN 0-471-80465-7 © 1997 John Wiley & Sons, Inc.

In this chapter the different types of sources and their characteristics, parameters of detonation as well as shock waves in near and far fields of standard explosive sources, and spectral characteristics will be considered, as well as nontraditional sources especially designed for solving the problems of directivity and control of radiation energy.

2 MAIN CHARACTERISTICS OF EXPLOSIVE SOUND SOURCES

The explosive process can be characterized by two stages. First, prior to radiation, high energy is stored inside the bounded volume. For example, in the case of explosives and electric charges, this energy is released at the detonation front due to chemical transformation (~ 1 kcal/g matter) or repumping of electric energy, stored in the capacitor bank, into the discharge channel. The radiation process itself may be defined as the second stage, although there is no exact time boundary, since it may start before the storage stage is over. Start with estimating the energy state of the above sources to analyze their possibilities as radiators.

2.1 Explosive Detonation Parameters

Combustion and detonation are the two main types of explosive chemical transformations (Table 1). The first, as a rule, takes place in gaseous mixtures. Detonation is typical for high-energy explosives. *Detonation* means the process of shock wave propagation through an explosive, after which there is a stable zone of a chemical reaction accompanied by heat release. The rate of this process, D, is a constant for a given explosive, which, along with the known density of the explosive, ρ_0, explosion heat Q, heat capacity of the reaction products, C_V, and isentropic index γ, defines the main parameters of the prod-

TABLE 1 Parameters of Detonation in Gaseous Mixtures and High Explosives

Explosive	ρ_0 (g/cm^3)	D (m/s)	p_* (kbar)	T_* (K)
2H + O$_2$	1.17×10^{-3}	2630	0.038	3960
CH$_4$ + 2O$_2$	1.17×10^{-3}	2220	0.027	4080
2C$_2$H$_2$ + 5O$_2$	1.17×10^{-3}	3090	0.051	5570
TNT	1.62	7050	215	2350
RDX (Cyclotrimethylene trinitramine)	1.8	8600	360	3750
PETN (Pentaerythritol tetranitrate)	1.77	8400	340	4150
TETRYL (Tetranitroaniline)	1.7	7850	265	2940

Source: Ref. 1.

cuts (pressure p_*, density ρ_*, and temperature T_*):

$$p_* = \frac{\rho_0 D^2}{\gamma + 1}, \quad \rho_* = \frac{\gamma + 1}{\gamma} \rho_0, \quad T_* = \frac{2\gamma}{\gamma + 1} \frac{Q}{C_V}.$$

$$(1)$$

($\gamma \cong 3$) and gaseous mixtures ($\gamma \cong 1.25$) differ by several orders of magnitude. The orders of magnitude of the same parameters at nuclear underwater explosions are estimated as 10^7 kbar and 10^7 K. In most cases, a physical model of instantaneous detonation is applicable to explosives. According to this model, detonation of the whole charge occurs instantaneously inside the closed volume. Pressure $\bar{p}_*$ and temperature $\bar{T}_*$ are the same throughout the volume, density $\bar{\rho}_*$ being equal to the initial density of the charge, ρ_0. The value of $\bar{p}_*$ is found from the condition that the explosive heat fully defines the internal energy of the explosive products:

$$\bar{p}_* = (\gamma - 1)\rho_0 Q = P_*/2. \quad (2)$$

2.2 Electric Explosion

A schematic of hydroacoustic signal generation by this method is shown in Fig. 1. The capacitor bank with capacitance C (in microfarads) is charged up to a high voltage U (in kilovolts). The circuit comprises the electric one (2) with the gaseous gap (G), the working gap (W) of length l submerged in the water, and the induc-

tance L allowing control of the discharge regimes in circuit 2. As a rule, the gap G is ionized by an external high-voltage pulse and the capacitor bank is closed in the load, being in the form of a column of dielectric and weakly conducting liquid between the electrodes of the gap W. In this rather complicated process, two main stages are distinguished: (i) a predischarge stage in which a conducting channel is created in the dielectric due to dissociation and ionization of molecules and (ii) breakdown, at which the characteristics of the explosive gap W as a high-energy density source are formed. When the charge is aperiodic (i.e., the oscillation process in circuit 2 is not observed), the hydrodynamic and physical characteristics of the discharge in the liquid are similar to those typical of an explosive process. The energy stored in the capacitor bank is determined from

$$E_0 = CU^2/2 \quad \text{J}. \quad (3)$$

Experience shows that the total coefficient of energy transfer (into the explosive cavity, η_c, and into the shock wave, η_*), $\eta = \eta_c + \eta_*$, depends on the capacitance C and length l of the discharge gap W, the value of E_0 being fixed. However, the value of η_* as the efficiency parameter is used more often (Table 2).

To estimate η, it is necessary to take into account the values of internal energy remaining in the explosive products after shock wave generation, that is, to determine a value of η_c. It may be done on the pulsation period T_b of the explosive cavity. In the experiments with $C = 0.1\mu$F, $U = 30$ kV, and $l = 4$ cm, this period is equal to approximately 5–6 ms. The internal energy of the cavity (cylindrical in form) found from the theoretical estimate,

$$E^* \simeq \pi \rho l^3 \left(\frac{p_0 T_b}{\rho l} \right)^2,$$

is about $0.1E_0$ at the ambient pressure $p_0 = 1$ atm, that is, $\eta_c \simeq 10\%$. Thus $\eta \simeq 40\%$, that is, twice as low as

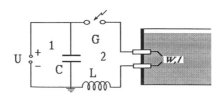

Fig. 1 Principal scheme of the experimental arrangement meant for spark discharge study in water.

TABLE 2 Discharge Contour, Time τ_d of Discharge, and Its Efficiency η_* at $E_0 \simeq 45$ J

U (kV)	C (μF)	l (cm)	τ_d (μs)	η_*(%)	Type of Discharge
30	0.1	5	7.5	32	Aperiodic
21.2	0.2	2.5	17.5	25	Aperiodic
15	0.4	1.2	25	13	Periodic
10	0.9	0.5	140	3	Periodic

Source: Ref. 2.

in for cylindrical explosives. An energy of ~ 1 kJ, close to the heat released by a 1-g explosive, may in principle be stored in the capacitor bank due to the increase in the capacitance C and voltage U; however, as is seen from Table 2, with increasing C, η_* sharply decreases. To stabilize the electric discharge, a thin manganin wire can be mounted in the gap W (the method of shock wave generation by exploding wires).

3 HYDRODYNAMIC SOURCES OF EXPLOSIVE TYPE

Pneumatic systems called airguns, waterguns, and watershocks are explosive-type sources. The principle of operation of these systems is storage of compressed gas energy and the transfer of this energy directly into the liquid in the form of an explosive cavity (airgun) or transformation of it into a potential energy of the surrounding liquid (watergun). It can also be analogous to a version of hydrodynamic shock tubes (watershock).

3.1 Airguns

An airgun (Fig. 2) in the simplest case comprises two elements: a pressure chamber of constant volume (p.c.) and a source, being a gaseous cavity (g.c.), in the surrounding liquid. The compressed air is supplied from the pressure chamber through the system of ports equipped with the controlling piston (P) into the surrounding liquid. Ideally, the losses in the port and the remaining portion of gas in the pressure chamber are not taken into account, and the internal energy of compressed gas is converted into the internal energy of the gas in the surrounding liquid. The formed cavity is assumed to be spherical or cylindrical and the gas inside it to obey adiabatic law. The dynamics of such cavity of radius R and volume V may be described by the law of conservation of energy:

$$\frac{p_{01}V_{01}}{\gamma - 1} - \frac{p_1 V_1}{\gamma - 1} = \frac{pV}{\gamma - 1} + p_0 V + \frac{\rho \dot{V}^2}{8\pi R} + E_{\text{rad}}. \quad (4)$$

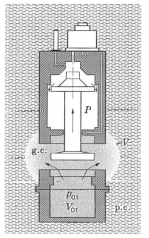

Fig. 2 Principal diagram of PAR airgun function (Luskin Bernard, Bolt Associates).

Here the initial energy of the compressed gas in the pressure chamber, $E_0 = p_{01}V_{01}/(\gamma - 1)$, is spent on producing the internal gas energy in the cavity, $pV/(\gamma - 1)$. The potential energy $p_0 V$ is spent on producing the kinetic energy of the liquid, $\rho \dot{V}^2/8\pi R$, surrounding the cavity and the acoustic wave energy E_{rad}.

The values of E_0 for airguns approximate the analogous values for explosives: When $V_1 = 1.2 \times 10^3$ in.[3] and $p_{01} = 345$ bars, $E_0 = 1.7$ MJ. This is equivalent to 0.4 kg of TNT; however, the efficiency of energy transfer into the source is less due to a relatively low explosive cavity formation velocity; for explosives, this is the detonation velocity. In the case under consideration, this is the velocity of gas mass transfer from the pressure chamber.[3]

$$\frac{dm}{dt} = k_e A \sqrt{2\rho(p_u - p_d)}, \quad (5)$$

where k_e is the coefficient of losses, which depends on the peculiarities of the structure, A is the port area, and p_u and p_d are the upstream and downstream pressures, respectively, in the orifice. Such systems are often used in seismic investigations to design high-power pulsed sources of low-frequency sound in the frequency range 10–200 Hz. Their efficiency is likely to be dependent on depth.

3.2 Watershocks

Figure 3 illustrates the principles of operation of hydrodynamic shock tubes that result in a powerful sound source, the so-called watershock.[4] The piston, placed inside the pneumatic cylinder, is accelerated in a vacuum under the action of a supply pressure and stores a kinetic energy, which, through a transmitting piston, is partially transformed into the energy of a shock wave generated via the acceleration of a liquid column in the horn.

At an abrupt stop, on the surface of the transmitting piston there appear a cavitation zone and rarefaction waves that suppress possible oscillations and define the profile of a resulting wave. At the horn exist, the shock wave profile is triangular with a subsequent rarefaction phase. The wave amplitude at the beginning of the horn is estimated by the simple expression $p_* = \rho C u_p$, where u_p is the velocity of the transmitting piston at the moment of impact on the liquid column. Pressure variations in the wave as the latter propagates over a cross-sectional area of the horn are associated with the decrease in energy flux density.

3.3 Waterguns

These pulsed pneumatic systems, intended for the generation of powerful single pulses by flux cumulation, comprise the main chamber, partially filled with seawater, and compressed air. The latter ejects water through the special ports with the help of shuttles. After the shuttle has stopped, the liquid moves by inertia and separates from the system, and a vapor cavity with potential energy $p_0 V_{max}$ and a very low pressure is formed in the surrounding space. The subsequent collapse of this cavity, being almost empty, generates an implosion pulse in the surrounding space.

4 WAVE FIELD AND SPECTRAL CHARACTERISTICS OF EXPLOSIVE AND EXPLOSIVE-TYPE SOURCES

For most of the above-mentioned sources, the structure of the wave pattern and the relation with the source dynamics are analogous to an underwater explosion (Fig. 4). After the detonation wave has refracted at the charge–water interface, a shock wave is formed in the liquid, and the explosive products form a gaseous cavity with high internal pressure. The cavity makes decaying elastic radial oscillations that are the source of acoustic waves (pulsations) in a sea. The number of these is restricted, as a rule, by the instability effect of the collapsing cavity form.

4.1 Parameters of Underwater Explosion

The general peculiarities of the structure and parameters of the wave generated by such sources are well known.[5–9] Presented in Fig. 4 are the designations and the main relationships for relatively short distances.[6] Here W is the TNT explosive weight, in pounds, $Z_a = h + 33$ is the full "depth," h and r are the explosion depth and distance in feet, pressure is measured in psi, and time is measured in seconds. The shock wave has the form of an exponent near the front with amplitude p_m and positive phase dura-

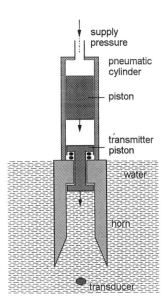

Fig. 3 Experimental arrangement of watershock.[4]

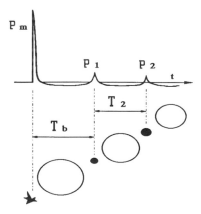

Fig. 4 Sketch of water form of an underwater explosion and its main parameters[6]: $p_m = 2.08 \times 10^4 (W^{1/3}/r)^{1.13}$, $\tau_0 = 0.555 W^{1/3}/Z_a^{5/6}$, $p_1 = 3300 W^{1/3} r^{-1}$, $\tau_1 = 1.1 W^{1/3}/Z_a^{5/6}$, $p_2 = 0.22 p_1$, $T_b = 4.34 W^{1/3}/Z_a^{5/6}$, $T_3 = 7.4 W^{1/3}/Z_a^{5/6}$.

tion τ_0. Pulsations are in the form of double exponents with positive phase durations τ_1, τ_2 and amplitudes p_1, p_2, respectively. The wave pressure profile is characterized by the negative phase 7:

$$\Delta t = 3.1 \frac{W^{1/3}}{Z_a^{5/6}} \quad \text{s.}$$

4.2 Characteristic Energy Spectrum of Explosive Sources

The energy spectrum of an explosive source is shown in Fig. 5. It consists of three components.[8] The first defines the spectral characteristic of the shock wave (SW) and describes a high-frequency part of the spectrum where the energy level falls off with frequency at the rate of 6 dB/octave ($\sim f^{-2}$). The second defines a spectral composition of first and second pulsations. The third component is associated with the determination of a low-frequency interval of the spectrum (IF) via combination of pulses I_0, I_1, and I_2 of the shock wave and pulsations. Its characteristic times are T_b and T_2, and the duration of the rarefaction phase between I_0 and I_2 is $T_3 \simeq T_b + T_2$. Figure 5 presents also the theoretical asymptote (TA), summation result (SF), and bubble pulse frequency (BP, 15 Hz).

For a fundamental pulsation frequency of the cavity filled with the explosive products,

$$f_b = \frac{1}{T_b} = 0.23 Z_a^{5/6} W^{-1/3}, \qquad (6)$$

the spectrum (Fig. 6)[10] has a sharp maximum that decreases as the explosion depth increases[11]:

$$10 \log \left(\frac{E_{max}}{W^{4/3}} \right) = -6.2 \log Z_a + 33.5, \qquad (7)$$

and in the region $f < f_b$ it is described by the relation

$$\frac{E}{E_{max}} = \left(\frac{f}{f_b} \right)^{2.67}. \qquad (8)$$

The experimental data for relatively small charges (less than 10 lb) at large depths (2.2×10^4 ft and less) recorded above the charge near a free surface demonstrate the variations in energy distribution between a shock wave and pulsations with depth. This is due to transformations of a wave profile and pulsations. The explosive detonation depth becomes one of the most significant param-

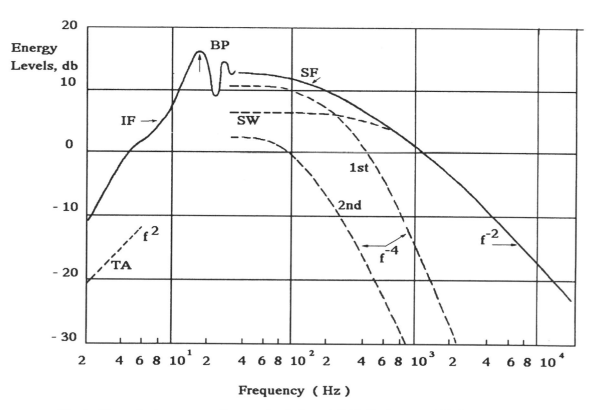

Fig. 5 Theoretical spectrum of energy levels for a 1-lb TNT charge at 120 ft, as measured 100 yd away.[8]

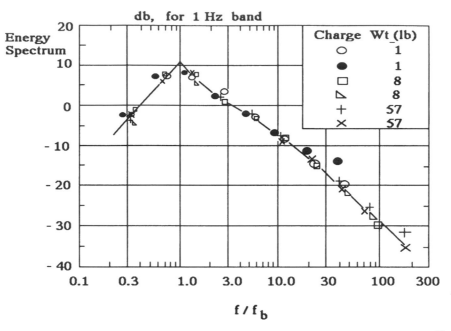

Fig. 6 Energy flux spectrum at large burst depth of 4500 ft versus reduced frequency.[10]

eters. It is determined from the characteristic period of the spectrum, T_s[6]:

$$h = 5.82W^{0.4}(T_s + \theta_e)^{-1.2} - 33, \qquad (9)$$

which is scalloped in form and remains the same when propagating over large distances, even if the signal is distorted. Here $\theta_e = 2.3$ ms is the empirical constant obtained from the data recorded at distances of 460–570 km and 1150–1300 km from the explosion of a 1.8-lb charge (W) at depths of 60 ft, the receiver being at a depth of about 3 km.

It should be noted that, for the above-mentioned reasons, the spectrum transformations have to be estimated with great care, since upon explosion of large charges at large depths, cavity migration may take place. The point is that the kinetic energy of the radial pulsations of the cavity is converted into an energy of vertical motion upon the cavity collapse, thereby cavity minimum size and, consequently, pulsation amplitude may sharply vary.

4.3 Peculiarities of Pneumatic Systems

As stated above, a watergun[12] operates by creating a low-pressure vapor cavity that, when collapsing under ambient pressure p_0, radiates a single-implosion pulse (Fig. 7). The radiation process is adequate here to the phase of the first pulse of the explosion cavity, and its characteristics are likely to be used to estimate the watergun operation.

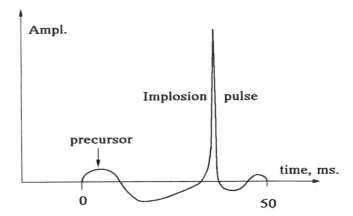

Fig. 7 Ideal near-field signature of watergun.[12]

The profile of the wave generated by an airgun is like that illustrated in Fig. 4 for explosives, but its amplitude is significantly less and the pulsations decay less. In the frames of a spherical or cylindrical model their frequency, being the fundamental one in the spectrum for the airgun, is defined by the relations[13]

$$f_s = 0.027p_0^{0.93}E_0^{-0.27},$$

$$f_{cyl} = 2.86p_0^{0.93}\left(\frac{E_0}{1}\right)^{-1/2}. \qquad (10)$$

On the basis of Eq. (4), the wave pressure may be estimated by

TABLE 3   Efficiency of Explosive Sound Sources in Low-Frequency Range

N	Type of Radiator Energy	Transformation Efficiency (%)			
		Total	0–100 Hz	20–50 Hz	100–1000 Hz
1	TNT, 1.5 g	24	0.3	0.09	2.77
2	TNT, 44 g	24	1.07	0.317	5.94
3	TNT, 255 g	24	1.96	0.595	10.1
4	TNT, 1274 g	24	3.24	1.03	12.8
5	Combustion of gaseous mixture, 45 kcal	1.95	1.2	0.415	0.656
6	Detonation of gaseous mixture, 39 kcal	1.45	0.7	0.233	0.659
7	Airgun, 35 kcal (different rates of port opening)	0.75– –1.5	0.75– –1.47	0.31– –0.61	0.03

$$p_r(t) = \frac{\rho}{4\pi r}\, \ddot{V}. \qquad (11)$$

It has been confirmed experimentally that airgun efficiency is maximized at low frequencies (~10 Hz) and sharply decreases as the explosion depth increases.

4.4 Comparative Characteristics of Spectra

An interesting effect may be well illustrated by experimental data on the low-frequency range of the energy spectrum of acoustic signals emitted by different sources[14] (a stoichiometric mixture of propane butane with oxygen, a 61-in.[3] airgun chamber, and 2000 psi pressure were used in the experiments).

As is seen from Table 3, the most amount of energy is released in the band up to 1 kHz; however, for airguns it is already released in the range 0–100 Hz. With almost the same energies of radiators, from 35 to 45 kcal (cases 2 and 5–7), the efficiency of transformation into acoustic energy over the range 20–50 Hz (underlined numbers) is of the same order.

The use of low-energy explosives as an acoustical source is less effective compared to high ones, due to the low detonation velocities. Their spectrum features are closer to the gas mixtures.

5 ARRAY SYSTEMS

The aim of airgun or watergun arrays is to cumulate wave energy emitted in a given direction.[15,16] The system comprises n individual sources (Fig. 8) with fixed distances d between them. A maximum energy E_{max} of the signal generated by the system depends on the array parameters and wavelength. The configuration of sources and time intervals between detonations of the airguns are responsible for the radiation direction, an index for which is defined by the relation[13]

$$D_1 = 10\log\frac{E_{max}}{nE_0}.$$

The problem of synchronization in firing each airgun in the array is, certainly, of great importance. If all the sources are initiated simultaneously, the emitted energy

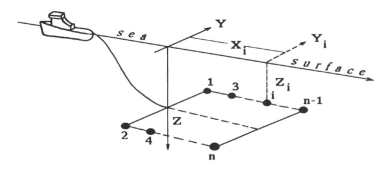

Fig. 8 Sketch of experiment with n-airgun array (A. M. Ziolkovski, 1986).

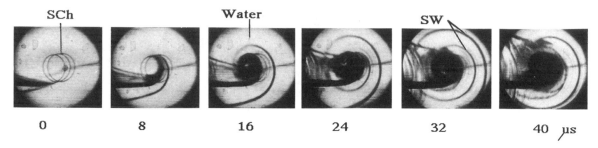

Fig. 9 Formation of shock wave sequence at underwater explosion of spatial spiral charge.

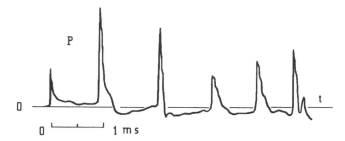

Fig. 10 Waveform of hydroacoustic signal from plane spiral charge; $r = 20$ m, time scale 1 ms.

(18.6 ± 2.3 kJ for seven sources) is much higher than in the case of individual firing (7.5 ∓ 0.9 kJ). The total efficiency of the airgun system is somewhat less than 2%. To describe the behavior of arbitrary arrays of airguns, numerical methods have been successfully used.[3]

It is known that the requirements of guarantee of directivity and low-frequency radiation are a very difficult problem for explosive sources which are often applied at the attempts to solve a wide spectrum of problems of geophysics, navigation, and propagation in the ocean. Nevertheless, a special distribution of array type or geometry of a charge proves to be appropriate for this purpose.

Some approaches are based on the well-known representations of classical acoustics. Among them is a linear cord charge or a chain of spherical charges,[8] specially arranged, for example, in line, where each is initiated individually and successively with a given time interval between initiations. These intervals are settled, for example, at the known length and the detonation velocity of the detonation cord connecting the charges coming from the given fundamental frequency and spectrum maximum.

Continuous systems of the type of plane or spatial spiral charges designed of high explosive cord are of particular interest.[19] Their advantages over other types are controllable frequency characteristics of the spectrum and a wave shape that is "prescribed" by the parameters of a spiral. In such a system the detonation front moves along the spiral coils rotating relative to its axis and provides the cyclic radiation of the shock waves in the given direction. Figure 9 illustrates streak-camera records of the formation of a succession of shock waves (SW) upon detonation of the spatial spiral charge (SCh). The signal shape may be either a prescribed succession of shock waves or a single positive signal with modulated amplitude. In any case its duration is defined by the cord length l and the detonation velocity D: $\tau = l/D$.

A typical pressure profile in the form of a succession of shock waves, recorded at a 20 m distance by a pressure gauge placed on the axis of a plane spiral 1 m in external radius and with a 0.1-m step, is displayed in Fig. 10. The parameters of this signal may be estimated in a simple way: The shock amplitudes are determined from the data for spherical charges with an equivalent explosive weight for each coil, the distance between shocks being defined by the coil length and detonation velocity. The effect is maximized when an axial component of detonation velocity in a spatial spiral is equal to the velocity of wave propagation through a liquid.

REFERENCES

1. F. Baum, L. Orlenko, K. Stanyukovich, V. Chelyshev, B. Shekhter, *Physics of Explosion*, 2nd ed., Nauka, Moscow, 1975, pp. 74–116, 259–298.

2. N. Roy and D. Frolov, "Generation of Sound by Spark Discharges in Water," *Proceedings of 3d International Congress on Acoustics*, Elsevier, Stuttgart, 1959, pp. 321–325.

3. R. Laws, L. Hatton, and G. Parkes, "Energy Interaction in Marine Airgun Arrays," E.A.E.G., Paper No. 1756, 18 Oct. 1989, pp. 1-37.

4. A. Laake and G. Meier, "Sound Generation by the Watershock," *Proceedings of ICA-12*, Toronto, Canada, 1986, pp. I–3.

5. R. H. Cole, *Underwater Explosions*, reprinted Dover, New York, 1965, Chap. 7.

6. S. Mitchell, N. Bedford, and M. Weinstein, "Determination of Source Depth from the Spectra of Small Explosions Observed at Long Ranges, *J. Acoust. Soc. Am.*, Vol. 60, No. 4, Oct. 1976, pp. 825–828.

7. M. Blaik and E. Christian, "Near-Surface Measurements of Deep Explosives, Part I–II," *J. Acoust. Soc. Am.*, Vol. 38, No. 1, 1965, pp. 50–62.

8. D. Weston, Underwater Explosions as Acoustic Sources," *Proceedings of the Physical Society*, Vol. 76, Part 2, No. 488, 1960, pp. 233–249.

9. A. Arons, "Underwater Explosion Shock Wave Parameters at Large Distances from the Charge," *JASA*, Vol. 26, No. 3, 1954, pp. 343–346.

10. E. Christian, "Source Levels for Deep Underwater Explosions," *J. Acoust. Soc. Am.*, Vol. 42, No. 4, 1967, pp. 905–907.

11. A. Kibblewhite and R. Denham, "Measurements of Acoustic Energy from Underwater Explosions," *J. Acoust. Soc. Am.*, Vol. 48, No. 1, Pr. 2, 1970, pp. 346–351.

12. E. Tree, R. Lugg, and Y. Brummitt, "Why Waterguns? *Geophys. Prospect.*, Vol. 34, 1986, pp. 302–329.

13. J. Barger and W. Hamblen, "The Airgun Impulsive Underwater Transducer," *J. Acoust. Soc. Am.*, Vol. 68, No. 4, Oct. 1980, pp. 1038–1045.

14. M. Balashkand, "Comparison of Acoustic Efficiencies of Some Explosion Sound Sources in Water," *Dokl. Akad. Nauk SSSR*, Vol. 194, No. 6, 1970, pp. 1309–1312.

15. R. Bailey and P. Garces "On the Theory of Airgun Bubble Interaction," *Geophysics*, Vol. 53, No. 2, Feb. 1988, pp. 192–200.

16. R. Laws, G. Parkers, and L. Hattou, "Energy Interaction. The Long-Wave Interaction of Seismic Waves," *Geophys. Prospect.*, Vol. 36, 1988, pp. 333–348.

17. A. M. Ziolkovski, "The Scaling of Airgun Arrays, Including Depth Dependence and Interactions," *Geophys. Prospect.* Vol. 34, 1986, pp. 343–408.

18. E. Lavrentyev and O. Kuzyan, *Explosions in a Sea*, Sudostroienie, Leningrad, 1977, pp. 60–64.

19. V. K. Kedrinskii, "Peculiarities of Shock Wave Structure at Underwater Explosions of Spiral Charges," *Prikladnaia Mekhanika i Tekhnicheskaia fizika*, No. 5, 1980, pp. 57–59 (*J. Appl. Mech. Tech. Phys.* in translation).

48

OCEAN AMBIENT NOISE

Ira Dyer

1 INTRODUCTION AND DEFINITIONS

Ocean ambient noise is sound in the ocean, unwanted generally because it interferes with operation of sonar systems or other underwater sound devices. Except with use of frequency filters and/or receiving arrays, ambient noise typically cannot be controlled by system operators.

Ambient noise mechanisms are as diverse as sea surface agitation, mammalian vocalization, or distant shipping. It is usual to consider such noise as an oceanic property (in that it can be related, for the three examples cited, to breaking waves under action of wind stress, to food chains and water column parameters such as temperature, or to commercial ships that inject noise into the deep sound channel). Also since ambient noise in most cases propagates from its source to a sonar, the noise can be affected by sound propagation properties of the ocean.

Noise observed omnidirectionally is designated by its mean-square pressure $p_N^2(f)$, dependent upon frequency f. In general, the noise has a *spectral density* $S_n(f)$, so that the omnidirectional observation at f is

$$p_N^2(f) = \int_0^\infty S_n(f_0)F^2(f_0;f)df_0$$

$$\approx \int_{f-b/2}^{f+b/2} S_n(f_0)\,df_0 \approx S_n(f)b, \qquad (1)$$

where F^2 is the mean-square response of the receiver's frequency filter (with normalization $F^2 = 1$ at its maximum) and b is the receiver bandwidth that, for the latter

parts of Eq. (1), is assumed small enough to consider the noise spectral shape approximately constant within it and the filter response outside it to be essentially zero. Sonar bandwidth clearly is a spectral filter against the usually continuous and wide-frequency distribution of the noise. Note that p_N^2 is weighted by the electrical frequency filter and thus is considered an *effective* acoustical quantity.

Ambient noise is not spatially isotropic, and it is therefore necessary to consider its *spatial* or *directional* density. Hence spectral density is composed of

$$S_n(f) = \int_{4\pi} S(f,\Omega)d\Omega, \qquad (2)$$

where the integrand is the spatial density at f and varies with the solid angle Ω. (In spherical coordinates, $d\Omega = \cos\theta\,d\theta\,d\phi$, with θ the vertical angle measured from the horizontal and ϕ the azimuthal angle.)

In log measure, the noise level L_N is given as

$$L_N = 10\log p_N^2(f) \qquad \text{dB re 1 } \mu\text{Pa}, \qquad (3)$$

and the spectrum level L_n is

$$L_n = 10\log S_n(f) \qquad \text{dB re 1 } \mu\text{Pa and 1 Hz.} \qquad (4)$$

Also

$$L_N = L_n + 10\log b \qquad \text{dB re 1 } \mu\text{PA} \qquad (5)$$

in which b is in hertz.

When noise is observed with a receiving array, the performance of the array as a spatial filter against the generally wide spatial distribution of the noise is mea-

Encyclopedia of Acoustics, edited by Malcolm J. Crocker
ISBN 0-471-80465-7 © 1997 John Wiley & Sons, Inc.

sured by the array gain AG:

$$
AG = 10 \log \left[\frac{S_n(f)}{\int_{4\pi} S(f,\Omega)B^2(f,\Omega)\,d\Omega} \right] \quad \text{dB.} \quad (6)
$$

Here, B^2 is the array's mean-square beam pattern[1,2] (with normalization $B^2 = 1$ at the maximum of the array's main lobe) and is determined by the amplitude and phase response of the array's transducers, as distributed in the array. The equivalent (postarray) noise level L_{N_e}, or equivalent spectrum level L_{n_e}, is Eq. (3) or (4), respectively, less Eq. (6). (Alternatively, AG can be defined in terms of the noise spatial covariance[2]; both formulations will be used subsequently in this chapter.) The denominator of the argument of Eq. (6),

$$
S_{n_e}(f) = \int_{4\pi} S(f,\Omega)B^2(f,\Omega)\,d\Omega, \quad (7)
$$

is the effective (postarray) noise spectral density that, when compared with the in-water (prearray) density $S_n(f)$, determines AG.[*] It is clear that one needs both the spatial density $S(f,\Omega)$ and the mean-square beam pattern $B^2(f,\Omega)$ to determine spatial filtering of the noise by a receiving array. Good spatial filtering ultimately requires the array side lobes to have low sensitivity in the direction of the noise maximum. Note that if the array main lobe is directed toward the noise maximum, little or no spatial filtering occurs.

By comparison with Eq. (1), B^2 in the spatial domain is the analog of F^2 in the frequency domain. An approximation to Eq. (6) [or Eq. (7)] analogous to that of Eq. (1) can be reached with use of the array's main-lobe beamwidth Ω_e. But the side lobes of an array typically are much more energetic than the out-of-band response of a frequency filter. Thus the spatial maximum of the noise, when received via the side lobes, can be quite important in setting the total noise entering the sonar. An approximation such as in Eq. (1) is, therefore, not as useful, and the integral in Eq. (6) [or Eq. (7)] usually has to be evaluated.

Finally, ambient noise may be considered, in the time domain, as a statistically random process. In addition to frequency spectrum and spatial directionality, sonar per-

formance depends directly on the staistical properties of ambient noise.[2] Its *probability density*, in general, is well established.[2] But its statistical stationarity, also crucial in determining performance[2] is, in general, less so. Beyond ocean conditions, stationarity is affected by sonar design and operational parameters (bandwidth b, beamwidth Ω_e, sonar speed, etc.), which cannot be addressed in general in a chapter on ambient noise. Nevertheless, this chapter provides some information on stationarity, intended largely to direct the reader to its fuller treatment.

2 SPECTRAL DENSITY ESTIMATES

Some years ago Wenz published a compendium of ambient noise spectral levels for the open ocean,[3] shown in Fig. 1. Remarkably it has withstood the test of time and serves as the best single survey of such noise. He distinguished between prevailing and intermittent (or local) noise mechanisms, the former ever-present spatially and temporally, the latter observed sporadically. The prevailing noises are considered the more important and are described in Sections 2.1–2.4 in terms of their spectral density $S_n(f)$.

2.1 Turbulence-Related Pseudosound

At the lowest frequencies, turbulence in or graininess of the seawater can induce pressure fluctuations at a hydrophone. This noise is strictly pseudosound rather than a result of propagating sound waves. Its spectral level will depend on the turbulence or grain wavenumber spectrum as well as on the hydrophone size, shape, and motion. No general estimate can readily be given, since these details vary greatly from application to application. But this prevailing noise mechanism is rarely of practical importance, at least because most sonars operate at $f \geq 10$ Hz, for which noise from other sources is almost always more important.

2.2 Distant Shipping Noise

Noise radiated by a surface ship is distributed broadly in frequency,[4] but when propagated to long ranges, it is shaped by the low-pass filter of the ocean[5] to a peak in the range $10 \leq f \leq 10^2$ Hz. Distant shipping in most cases is the dominant noise in this frequency range. Busy shipping areas or lanes have $O(10^3)$ ships within them at any one time,[6] so that ship traffic noise is sensed as a virtually continuous rather than a discrete spectrum.

Wenz's summary for distant traffic noise in the deep ocean is roughly appropriate to all sensor depths except those near the ocean bottom, in which as much as 25 dB

---

[*]Formulation of Eqs. (6) and (7) is based on the assumption of many incoherent noise components in bandwidth b and beamwidth Ω_e, a condition usually met in practice.

less can be observed below the critical depth[7] near the bottom. (At such depths, there are no strong long-range deep-channel paths, because sound speeds are higher there than those near the surface, where the noise is created.)

Distant shipping noise prevails over most of the ocean. Exceptions include the ice-covered Arctic Ocean and some seas sheltered from ocean traffic (provided their own traffic is very light).

2.3 Surface Agitation Noise

At still higher frequencies (about $10^2 \leq f \leq 10^4$ Hz) formation and oscillation of subsurface air bubbles created by breaking waves cause noise that dominates the spectrum. Only recently has this mechanism come under detailed study,[8–11] but it is highly probable that the essential ideas are now well in hand. Sea surface bubble noise is parameterized in Fig. 1 by wind force, and this remains

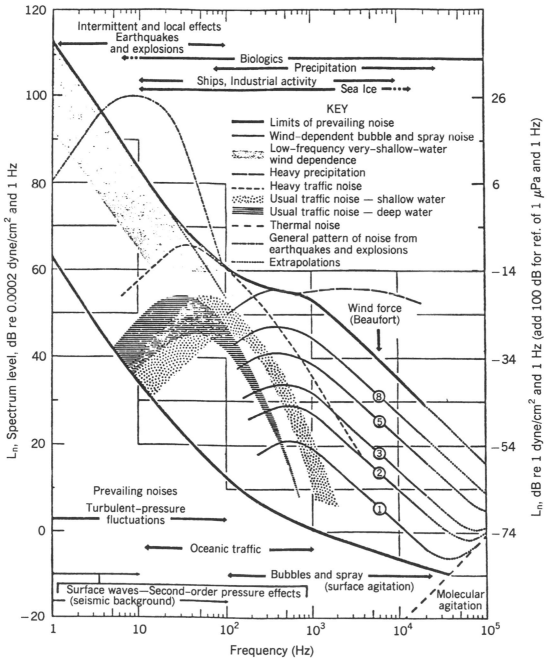

Fig. 1 Wenz's ambient noise spectra.[3] (Add 100 dB to the right-hand scale to obtain L_n in dB re 1 μPa and 1 Hz.)

the most practical way to describe noise generated by bubbles from breaking waves.[12,13] But it is also so that more fundamental parameters, such as wind stress and wave age, may ultimately succeed wind force for estimating the spectrum more precisely.

As stated, Wenz's summary of noise generated bubbles from breaking waves is for the deep open ocean. In shallow water no fully systematic estimates for bubble-caused noise are at hand, at least because shallow-water environments vary greatly in depth, sound speed profile, and bottom loss.[14] An estimate for a wide range of wind force, however, is in Fig. 2 and is useful for gaining at least a rough idea of such noise without detailed knowledge of the shallow-water environment.

As concluded in Section 2.2, noise from distant traffic typically dominates at low frequencies ($10 \leq f \leq 10^2$ Hz). But there are ocean areas that either are shielded or are remote from distant shipping. In such areas a wind-dependent noise component at low frequencies is observed instead;[12] see Fig. 3. The wind-related mechanism has not been firmly identified for $10 \leq f \leq 10^2$ Hz (but many possibilities have been suggested[13]).

2.4   Molecular Agitation Noise

At high frequencies ($f \geq 10^2$ kHz or so, dependent upon the wind agitation crossover), ambient noise is dominated by noise from molecular agitation. See Fig.

1. Dynamic forces from molecular momentum reversals at a pressure sensor cause this noise.[15] Molecules have kinetic energy set by the absolute temperature, but because absolute temperature varies only slightly even for the widest variations in ocean conditions, the single line in Fig. 1 can be considered broadly applicable and can be extrapolated as f^2 for $f \geq 10^2$ kHz. This noise is not strictly sound; its local molecular origin will be of importance later when its covariance properties are estimated.

2.5   Intermittent or Local Noise

In this chapter, I exclude details of several intermittent or local ambient noise sources, but they can be important and thus are included by reference. Earthquakes[16–18] and precipitation[19–21] have recently been studied as sources of noise and can dominate their respective spectral ranges for the duration of such events. Earthquakes cause noise below about 10 Hz, via induced motion of the bottom above the epicenter, which then creates noise in the sound channel by scattering from nearby rough boundaries. Because of its low frequencies, earthquake noise rarely is an issue in sonar performance. Precipitation noise, however, is usually observed for $f > 10$ kHz and can affect sonars operating in that frequency regime, of which there is a wide variety of designs for many different applications. Such noise is created by direct forcing

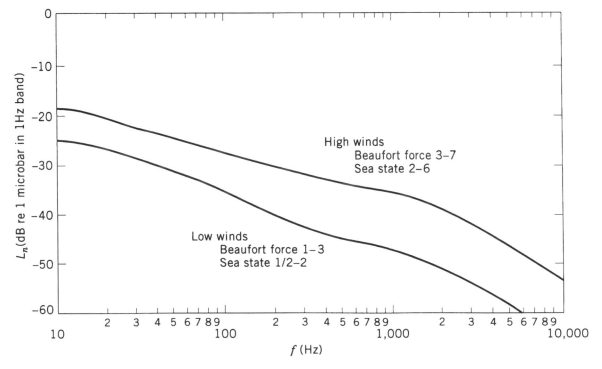

Fig. 2   Wind-driven sea surface noise in shallow water. (Add 100 dB to obtain L_n in dB re 1 μPa and 1 Hz.)

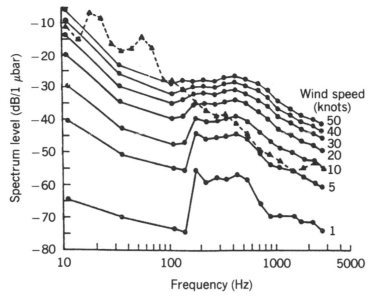

Fig. 3 Spectrum levels of non-wind-dependent (▲) and wind dependent (●) portions of ambient noise, extracted from measured levels.[11] (Add 100 dB to obtain L_n in dB re 1 μPa and 1 Hz.)

of the water surface as the precipitates strike and by wake cavity formation and oscillation as the precipitates drive down into the water.

Noise near and under the pack ice of the Arctic also has been recently studied. It is dominated by various ice fracture processes,[22–26] is pervasive in the sonar frequency range, and is always present to a varying degree. I think of it as local or at least special in character, however, and exclude its details.

Noise from marine life is present to varying degrees

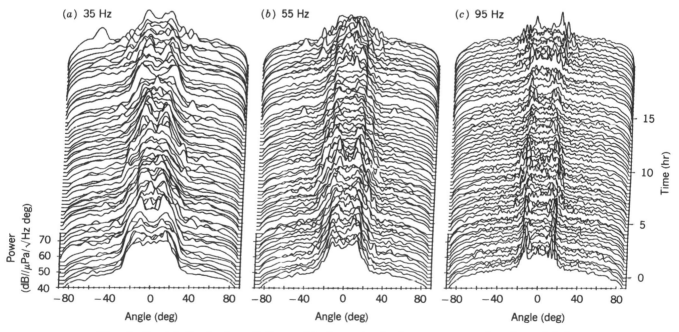

Fig. 4 Spatial distribution of distant shipping noise in deep water as a function of f, θ, and time.[28] Frequencies f are (a) 35, (b) 55, and (c) 95 Hz. Data were averaged in 0.5-Hz bandwidths. Time corresponds to wind increasing in speed from 2 to 12 m/s. The array was at a depth of 850 m, and the spatial density levels are in dB re 1 μPa, 1 Hz, and 1°.

in specific ocean areas of appropriate nutritional content. Marine life can create noise throughout the frequency range of interest, but its presence is temporally and spatially sporadic. While not all species create sound, enough do to provide even the occasional sonar operator with opportunities to marvel at its richness and to practice patience as its passage is awaited. Again the inclusion of such noise is by reference only.[27]

3 SPATIAL DENSITY ESTIMATES

3.1 Directionality of Distant Shipping Noise

In the deep ocean the distribution of distant shipping noise in vertical angle is closely symmetric about $\theta = 0$ (the horizontal) and has at most weak frequency and depth dependence. See Figs. 4 and 5.[28] Data in these figures show an angular "pedestal" standing 18 dB or so

above a broad "platform" in vertical angle. At the lower and higher frequencies the pedestal width θ values are $\approx \pm 20°$ and $\approx \pm 15°$, respectively. These widths correspond to the sound field contained within the deep sound channel. The pedestal is more rounded at the lower frequencies, more peaked at the edges for the higher frequencies, the difference likely due at the lower frequencies to more intense scattering associated with distant ocean boundaries.[29] The platform level is wind speed dependent (an effect consistent with Fig. 3), and at the low wind speeds corresponding to distant shipping dominance, it is roughly estimated to be below the pedestal peak by $-18 + 15 \log |\cos \theta|$ decibels.

Synoptic measurements of the distribution of distant shipping noise in horizontal (azimuthal) angle ϕ are not available, but within a large shipping area one can surmise it to be approximately uniform in 2π, and well away from a shipping lane, to be roughly uniform in π.

Systematic data for the spatial distribution of shipping

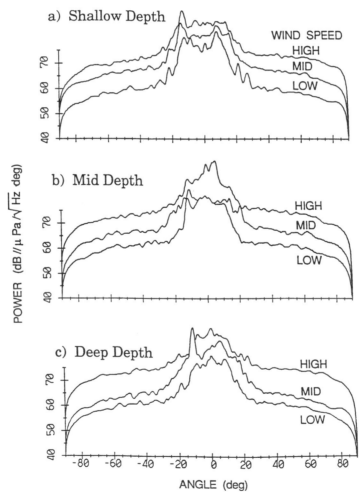

Fig. 5 Spatial distribution of distant shipping noise in deep water at 75 Hz versus depth (170, 850, and 2650 m) and wind speed (3, 7, and 11 m/s).[28]

noise in shallow water are also not available. Because propagation in shallow water is quite variable, each case would have to be estimated from knowledge of shipping activity and of sound speed profile and other ocean parameters. But it should be symmetric in θ, with a pedestal width determined by the low-order modes (which usually have little or no excess attenuation).

3.2 Directionality of Surface Bubble Noise

An estimate of the density in vertical angle θ for surface bubble noise is in Fig. 6, based on measurements for $0 \le \theta \le 90°$ (the up-looking quadrant).[30] The data were acquired near the deep bottom, apparently below the critical depth. The density maximizes at $\theta = 90°$ (vertically up), with a 3-dB-down width of about $35° \pm 5°$. Air bubble dynamics, the physical mechanism for surface noise, has dipolar radiation.[8] For a vertical dipole the directivity function is proportional to $\sin^2 \alpha$, where α is the vertical angle at the source measured from the horizontal. In the absence of refraction (and sea surface tilt), this transforms to a proportionality of $\sin \theta$ for the *spatial density* in solid angle Ω.[*] Thus the 3-dB-down width of the density would be expected to be $30°$, reasonably close to the observed value of about $35°$.

Similarly, in the nonrefracting limit, the density for a dipolar surface distribution near $\theta = 0°$ (measured at the receiver) would have a deep minimum as a consequence of the deep minimum near $\alpha = 0°$ (at the source). But refraction must be accounted for at such small angles. In cases of sound speed at depth larger than that at the surface, refraction has the effect that radiation is observed at depth only for $|\alpha| > 0°$, based on Snell's law.[31] This accounts for the absence of a deep minimum in Fig. 6. For example, an *excess* of 1% of sound speed at depth over the surface sound speed would give a trough of only -9 dB at $\theta = 0°$, rather than a deep minimum.

In cases of sound speed at observation depth smaller than that at the surface, the effect, while different from the foregoing one, is equally important. For example, a *decrement* of 1% of sound speed at depth would make the minimum centered on $\theta = 0°$ wider by about $16°$.

As shown in Fig. 6, there is evidence that a broad secondary maximum occurs around $\theta = -50°$ in the downward-looking quadrant, a result of noise scattered from the ocean bottom.[32] Of course a secondary maximum cannot be observed too close to the bottom but is likely to be observed at midwater or higher depths.

The frequency dependence of the density shown in Fig. 6 is typically weak. Noise in the upper quadrant can become somewhat more directive (3-dB-down angle

*This assumes a uniform surface distribution of dipoles, takes account of inverse-square spreading, and is for a spatial density in constant increments of Ω.

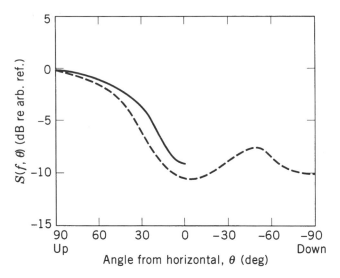

Fig. 6 Estimated spatial distribution of sea surface wind agitation noise in the deep ocean. [The arbitrary reference may be converted to an absolute one by setting the integral of $S(f, \Omega)$ from the figure to $S_n(f)$, as in Eq. (2).] The solid curve is based on data acquired deep in the water column for $200 < f < 1400$ Hz,[30] and the dash curve on data at a depth of 175 m for $f = 1000$ Hz.[31] For simplicity, both have been smoothed by the present author.

smaller than $35°$) at high frequencies ($f > 10$ kHz) and high wind speeds (> 15 m/s). This is so largely because a quasistatic bubble layer exists just below the sea surface, the quiescent debris of bubbles created by earlier breaking waves. The layer acts to increase the attenuation for high-frequency and low-angle-noise radiation paths.[33]

There is some indication from at-sea data that the spatial density is not azimuthally uniform, exhibiting small maxima for ϕ normal to the wave crests and shallow minima for ϕ parallel to them.[34]

Spatial density data in shallow water are not generally available, but measurements in a sloping basin of 500 m depth yield density estimates about the same as the foregoing.[35] The 3-dB-down angle, however, is somewhat broader (about $40°$ rather than $35°$). The additional width of about $5°$ is perhaps within the uncertainty of the deep-water case or perhaps caused by more intense scattering from the nearby sloping bottom.

3.3 Postarray Noise from Molecular Motion

Noise from molecular agitation cannot be described by a spatial density originating from propagating sound waves.[15] But forces from the molecular motion acting on the sensor are spatially and temporally uncorrelated, the molecular scales for all practical purposes giving rise to a delta function space–time correlation. Consequently the postarray spectral density is precisely

$$S_{n_e}(f) = S_n(f), \qquad (8)$$

where $S_n(f)$ is the quantity shown in Fig. 1 for molecular agitation. The postarray value is independent of array size and design, simply because the spectral density grows as $\sigma S_n(f)$, where σ is the total surface area of the sensor (or array) but the sensor (or array) has a gain of $1/\sigma$ against the noise, which is uncorrelated over it.[2]

4 ESTIMATES OF STATISTICAL STATIONARITY

4.1 Stationarity of Distant Shipping Noise

Distant shipping noise is at times statistically stationary, at times not. Its stationarity is affected by the bandwidth b and beamwidth Ω_e forming the noise time series, that is, by the number of distant ships at various ranges sensed simultaneously. Some examples help to provide an image of its wide variation. For omnidirectional data in a one-third-octave band centered on 63 Hz and for short observation periods, less than about 1 h, it is approximately statistically stationary. For longer periods, between about 4 and 24 h, it is nonstationary.[3] For very long observation periods, between about 1 and 10^2 days, it once again can be considered approximately stationary.[3] (One may surmise its return to a nonstationary process for periods longer than about 10^2 days but less than several years.) The major reason for these stationarity shifts is variability in long-range propagation, as affected by temporal fluctuation scales of the ocean.

Still other data show that distant shipping noise is stationary for observation periods less than about $\frac{1}{3}$ h when sensed through $b \approx 0.13$ Hz at 75 Hz, also omnidirectionally.[36] When sensed omnidirectionally in the horizontal but directionally in the vertical, in one-third-octave bands at several frequencies from 23 to 100 Hz, the noise is stationary for about 1 h.[37] These data indicate at least some of the sensitivity of random-process stationarity to observation conditions. Other factors are also important, such as wind-driven seas that can affect scattering in long-range propagation paths.

4.2 Stationarity of Surface Bubble Noise

Ambient noise due to bubbles can be taken as stationary, one might surmise, for observation periods less than the wind persistence time, typically about 4 h. Omnidirectional data, however, at several frequencies from 0.1 to 1.6 kHz in one-third-octave bands show it to be stationary for periods of at most $\frac{1}{20}$ h![38] Further, directional data in the octaves 0.8–1.6 and 1.6–3.2 kHz show stationarity for periods < 1 s in an upward-looking beam and < 10 s in a horizontally looking beam.[39,40]

Apparently, bubble creation by each breaking wave is temporally discrete (<1 s), so that an upward beam senses noise beyond it as a nonstationary eventlike process. A horizontal beam, in contrast, simultaneously senses a much larger ocean surface area and hence a superposition of many events; the typical wave period of about 10 s then sets the transition to nonstationarity.

Because its source is more local than that of distant shipping noise, one might also surmise that surface bubble noise does not get imprinted significantly with the fluctuation scales of long-range propagation paths.

4.3 Stationarity of Molecular Noise

Molecular noise in the ocean is a statistically stationary random process for indefinitely long observation periods.

REFERENCES

1. I. Dyer, *Fundamentals and Applications of Ocean Acoustics*, Cambridge University Press, New York, 1998 (in preparation).

2. J. E. Barger, "Sonar Systems," Chap. 49.

3. G. M. Wenz, "Acoustic Ambient Noise in the Ocean: Spectra and Sources," *J. Acoust. Soc. Am.*, Vol. 34, 1962, pp. 1936–1956.

4. R. J. Collier, "Ship Noise," Chap. 46.

5. F. H. Fisher and P. F. Worcester, "Essential Oceanography," Chap. 35.

6. I. Dyer, "Statistics of Distant Shipping Noise," *J. Acoust. Soc. Am.*, Vol. 53, 1973, pp. 564–570.

7. J. A. Shooter, T. E. Demary, and A. F. Wittenborn, "Depth Dependence of Noise Resulting from Ship Traffic and Wind," *J. Oceanic Eng.* Vol. 15, 1990, pp. 292–298.

8. H. Medwin and M. M. Beaky, "Bubble Sources of the Knudsen Sea Noise Spectra," *J. Acoust. Soc. Am.*, Vol. 86, 1989, pp. 1124–1130.

9. A. Prosperetti, "Bubble Related Ambient Noise in the Ocean," *J. Acoust. Soc. Am.*, Vol. 84, 1988, pp. 1042–1054.

10. G. E. Updegraff and V. C. Anderson, "Bubble Noise and Wavelet Spills Recorded 1 m below the Ocean Surface," *J. Acoust. Soc. Am.*, Vol. 89, 1991, pp. 2264–2279.

11. H. C. Pumphrey and J. E. Ffowcs Williams, "Bubbles as Sources of Ambient Noise," *J. Oceanic Eng.*, Vol. 15, 1990, pp. 268–274.

12. W. W. Crouch and P. J. Burt, "The Logarithmic Dependence of Surface-Generated Ambient Sea-Noise Spectrum Level on Wind Speed," *J. Acoust. Soc. Am.*, Vol. 51, 1972, pp. 1066–1072.

13. D. J. Kewley, D. G. Browning, and W. M. Carey, "Low-Frequency Wind-Generated Ambient Noise Source Levels," *J. Acoust. Soc. Am.*, Vol. 88, 1990, pp. 1894–1902.

14. F. Ingenito, and S. N. Wolf, "Site Dependence of Wind-Dominated Ambient Noise in Shallow Water," *J. Acoust. Soc. Am.*, Vol. 85, 1989, pp. 141–145.

15. R. H. Mellen, "The Thermal Noise Limit in the Detection of Underwater Acoustic Signals," *J. Acoust. Soc. Am.*, Vol. 24, 1952, pp. 478–480.

16. R. H. Johnson, R. A. Norris, and F. K. Duennebier, "Abyssally Generated T-Phases," in L. Knopoff, C. L. Drake, and P. J. Hart, (Eds.), *The Crust and Upper Mantle of the Pacific Area*, Geophys. Mono. No. 12, Am. Geophys. Union, Washington D.C., 1968, pp. 70–78.

17. R. E. Keenan and I. Dyer, "Noise from Arctic Ocean earthquakes," *J. Acoust. Soc. Am.*, Vol. 75, 1984, pp. 819–825.

18. R. E. Keenan and L. R. L. Merriam, "Arctic Abyssal T Phases: Coupling Seismic Energy to the Ocean Sound Channel via Under-ice Scattering," *J. Acoust. Soc. Am.*, Vol. 89, 1991, pp. 1128–1133.

19. J. A. Scringer, D. J. Evans, and W. Yee, "Underwater Noise Due to Rain: Open Ocean Measurements," *J. Acoust. Soc. Am.*, Vol. 85, 1989, pp. 726–731.

20. H. C. Pumphrey, L. A. Crum, and L. Bjorno, "Underwater Sound Produced by Individual Drop Impacts and Rainfall," *J. Acoust. Soc. Am.*, Vol. 85, 1989, pp. 1518–1526.

21. D. G. Browning, D. G. Williams, and V. Sadowski, "Revised Standard Precipitation Noise Curves for Sonar Performance Modeling," NUWC-NL Tech. Memo. No. 931116, Naval Undersea Warfare Center, Division Newport, New London Detachment, New London, CT, 1993.

22. N. C. Makris and I. Dyer, "Environmental Correlates of Pack Ice Noise," *J. Acoust. Soc. Am.*, Vol. 79, 1986, pp. 1434–1440.

23. T. C. Yang, C. W. Votaw, G. R. Giellis, and O. I. Diachok, "Acoustic Properties of Ice Edge Noise in the Greenland Sea," *J. Acoust. Soc. Am.*, Vol. 82, 1987, pp. 1034–1038.

24. J. K. Lewis and W. W. Denner, "Higher Frequency Ambient Noise in the Arctic Ocean," *J. Acoust. Soc. Am.*, Vol. 84, 1988, pp. 1444–14552.

25. Y. Xie, and D. M. Farmer, "Acoustical Radiation from Thermally Stressed Sea Ice," *J. Acoust. Soc. Am.*, Vol. 89, 1991, pp. 2215–2231.

26. N. C. Makris and I. Dyer, "Environmental Correlates of Arctic Ice-Edge Noise," *J. Acout. Soc. Am.*, Vol. 90, 1991, pp. 3288–3298.

27. R. J. Urick, *Ambient Noise in the Sea*, Peninsula Publishing, Los Altos, CA, 1986, pp. 7.1–7.19.

28. B. J. Sotirin and W. S. Hodgkiss, "Fine-Scale Measurements of the Vertical Ambient Noise Field," *J. Acoust. Soc. Am.*, Vol. 87, 1990, pp. 2052–2063.

29. W. M. Carey, R. B. Evans, J. A. Davis, and G. Botseas, "Deep-Ocean Vertical Noise Directionality," *J. Oceanic Eng.*, Vol. 15, 1990, pp. 324–334.

30. E. H. Axelrod, B. A. Schoomer, and W. A. Von Winkle, "Vertical Directionality of Ambient Noise in the Deep Ocean at a Site Near Bermuda," *J. Acoust. Soc. Am.*, Vol. 37, 1965, pp. 77–83.

31. W. A. Kuperman, "Propagation of Sound in the Ocean," Chap. 36.

32. B. A. Becken, "Sonar," in Ven Te Chow (Ed.), *Advances in Hydroscience*, Vol. 1, Academic Press, New York, 1964, p. 1.

33. D. M. Farmer and D. D. Lemon, "The Influence of Bubbles on Ambient Noise in the Ocean at High Wind Speeds," *J. Phys. Ocean.*, Vol. 14, 1984, pp. 1762–1777.

34. W. F. Hunter, "An Introduction to Acoustic Exploration," in R. W. B. Stephens (Ed.), *Underwater Acoustics*, Wiley-Interscience, London, 1970, pp. 91–127.

35. R. M. Kennedy and T. V. Goodnow, "Measuring the Vertical Directional Spectra Caused by Sea Surface Sound," *J. Oceanic Eng.*, Vol. 15, 1990, pp. 299–310.

36. W. J. Jobst and S. L. Adams, "Statistical Analysis of Ambient Noise," *J. Acoust. Soc. Am.*, Vol. 62, 1977, pp. 63–71.

37. V. C. Anderson, "Envelope Spectra for Signals and Noise in Vertically Directional Beams," *J. Acoust. Soc. Am.*, Vol. 65, 1979, pp. 1480–1487.

38. T. Arase and E. M. Arase, "Deep-Sea Ambient-Noise Statistics," *J. Acoust. Soc. Am.*, Vol. 44, 1968, pp. 1679–1684.

39. W. S. Hodgkiss and V. C. Anderson, "Detection of Sinusoids in Ocean Acoustic Background Noise," *J. Acoust. Soc. Am.*, Vol. 67, 1980, pp. 214–219.

40. V. C. Anderson, "Nonstationary and Nonuniform Oceanic Background in a High-Gain Acoustic Array," *J. Acoust. Soc. Am.*, Vol. 67, 1980, pp. 1170–1179.

49

SONAR SYSTEMS

JAMES E. BARGER

1 INTRODUCTION

The principal application of underwater acoustics is to sonar, the acoustical analog of radar. The variation among sonar systems is very great; long-range detection sonars may operate at frequencies lower than 50 Hz, while mine detection sonars may operate at frequencies higher than 50 kHz, a relative frequency range greater than three decades. The purpose of most sonar systems is to detect and localize a particular target: submarines, mines, fish, the ocean floor, surface ships. All of these systems have a common architecture, and this chapter deals with their common components. Another class of sonar systems are designed to measure some particular quantity: depth of the ocean, speed of a ship, or speed of ocean current. A final class seeks to image remote objects: side-scan sonars and sub-bottom profilers are examples.

The other chapters in this part deal with the individual physical processes that act on the performance of a sonar system. This chapter describes how the measures of these physical processes are combined to analyze the performance of sonar systems. The generalized active and passive sonar systems are separated into their components. First described are the functions of each component, then analytical performance models, and finally, system performance prediction equations.

Encyclopedia of Acoustics, edited by Malcolm J. Crocker
ISBN 0-471-80465-7 © 1997 John Wiley & Sons, Inc.

2 FUNCTIONS OF SONAR SYSTEM COMPONENTS

2.1 Generalized Active Sonar System

A schematic diagram of the generalized active sonar system is shown in Fig. 1, where the principal system components and their interconnections are shown. This section describes the function of each component in the order of signal flow.

The four system components shown in Fig. 1. that form the signal transmission portion represent the majority of sonar systems that use electroacoustic transducers as their sound source. These transducers are functionally linear, radiating sound pressure waveforms that are proportional to the electrical waveforms impressed upon them. Linear transducers described in Chapter 52 and nonlinear transducers that are functionally linear are described in Chapter 53. The waveform generator initiates the sonar cycle by producing a pulse that can be a combination of up to four different types of waveforms. The first type of waveform is a constant-amplitude sinusoidal pulse that is useful to detect moving targets, because the Doppler frequency shift impressed by the target upon the reflected pulse enables the background of reverberation from fixed objects to be filtered out, unmasking the target echo. Moreover, the range rate of the target can be measured accurately from the magnitude of the Doppler frequency shift. The second type of waveform is a frequency shift pulse having sudden discrete changes in otherwise constant sinusoidal frequency used both to obtain the benefits of narrow-band

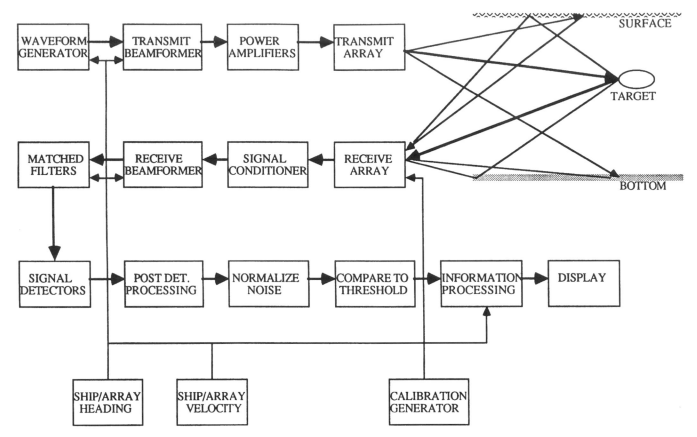

Fig. 1 Schematic diagram of an active sonar system showing the principal components and their interconnections.

pulses and to combat the propagation channel fading that inhibits sound propagation within bands at unpredictable frequencies. The third type of waveform is one having frequency modulation by one of several schemes. The increased bandwidth of these waveforms provides greater range resolution than from unmodulated narrowband pulses. The benefits are less reverberation power from extended scatterers and accurate target range measurement. The fourth type of waveform is pulse repitition of similar pulses. Pulse trains provide accurate measurement of target range rate and unmasking by reverberation and are the only type of waveform that can be used with systems that employ functionally nonlinear sound sources such as air guns, sparkers, explosives, and other impulsive sound sources. Impulsive sources are described in Chapter 47.

The transmit beamformer accepts waveforms from the generator and produces individual copies of them for each transducer in the transmit array. The function of the transmit beamformer and array together is to radiate beams of sound having power distributed narrowly in azimuth angle and vertical angle. There are two reasons for using relatively narrow transmit beams. First, many

systems are input power limited, so that sufficient power can be radiated into only one narrow sector at one time. Second, reverberation power coming from any azimuth can be minimized with respect to the target echo by radiating preferentially into vertical angles that reach the target at the expense of angles that reach either the surface or the bottom. If directive transmit beams are used, it is often necessary to stabilize them in both vertical and azimuth angle against ship or array motion. Furthermore, it is necessary to transmit sequentially toward successive azimuth angles to ensonify all potential targets. This rotational directional transmission (RDT) may require that successive pulse waveforms contain mutually exclusive frequencies so that the direct path signal to the receiver from the nth + 1 pulse does not blank the receiver during reception of the nth pulse.

A bank of power amplifiers drive the transducers in the transmit array. These amplifiers have small gain amplitude and phase errors so that the beamformed signal is not distorted in a significant way. Typically, gain errors need to be less than ±0.5 dB and phase errors need to be less that ±10° to ensure that transmitted side-lobe amplitudes remain at least 30 dB below the main lobe.

The transmit array itself is composed either of electroacoustic transducers (the subject of Chapters 52 and 53) or of impulse sources (which are the subject of Chapter 47). Transmit arrays are subjected to several special requirements that are not discussed in the transducer chapters. It is necessary to avoid cavitation that occurs either at the face of the transducers or in the sound near field in front of the array whenever the peak sound pressure approaches the local ambient pressure. It is also necessary to avoid beam pattern distortion caused by mutual coupling between transducers. Whenever a transducer's radiation impedance is significantly modified by sound radiated from all other transducers in the array, the resulting transducer amplitude and phase are changed. Finally, the transmit array is often housed in a streamlined dome to facilitate flow around it. The dome must be acoustically transparent while not inducing cavitation on its surface in the intense sound field.

Sound from the transmitted beam can find its way to the receiver array by several processes, as shown in Fig. 1. Considering first the desired path, sound propagates to the target and then scatters from it, and the echo propagates back. Details of sound propagation are considered in Chapter 36, and details of sound scattering from targets are considered in Chapters 42 and 43. There are four distinct features of sound propagation that are important to sonar systems. Sound attenuation, called transmission loss, is the loss of signal power in decibels between a point 1 m from the source and a distant point. Time spread is the time increase in length of a pulse propagating over the path. Frequency spread is the bandwidth increase of a pulse propagating over the path. Correlation time is the time interval over which the other propagation features remain essentially unchanged by the internal fluctuations within the ocean.

A portion of sound power that strikes the surface scatters back to the receiver as surface reverberation. The essential features of the surface backscattering of sound are covered in Chapters 39 and 40, where surface scattering strength is introduced as the measure of surface backscatter. Two different kinds of surface reverberation can occur. First, there is the continuous backscatter from the component of surface roughness that is uniformly distributed over the surface. Second, there occur at the surface when significant winds blow plumes of bubbles that are convected downward and act as discrete targets that can mimic submarines.

All sound paths produce continuous backscatter that originates at inhomogeneities of the ocean's mechanical properties (sound, speed, and density) or at marine animals. The backscattered sound is called volume reverberation, and its features are discussed in Chapter 39. Typically, volume reverberation is not important at frequencies below 1 kHz. The marine animals that cause the most intense backscatter are confined to a layer whose depth has diurnal migrations.

The final path of sound to the receiver shown in Fig. 1 represents bottom reverberation. The essential features of the bottom backscattering of sound are covered in Chapter 40, where the bottom scattering strength is introduced as the measure of bottom backscatter. Bottom backscatter has both continuous and discrete components, just as surface backscatter has. The discrete bottom targets that can mimic submarines are either concentrated bottom roughness or inhomogeneities of sub-bottom properties.

The receive array is often a spatially distributed array of individual hydrophones having the purpose of preferentially selecting incoming signals according to their arrival azimuth and vertical angles. In some systems, the receiving array function is performed by an acoustic lens or reflector. The subsequent signal processing equipment is separated into many parallel channels, called beams, each pointed at a different incoming azimuth and vertical angle. Two general purposes are served by the receiving array. First, the total noise in each beam is less than the total noise arriving at the receiving array because of the preferential array response in the direction of the beam. Second, the direction of the target is indicated by the direction of the beam that receives the target echo. The receive array takes many different forms in different applications. Towed line arrays are long, neutrally buoyant flexible hoses that contain hydrophones at regular intervals and are towed behind the sonar platform. Hull arrays are spherical, cylindrical, or conformal arrays attached below or on the platform hull and contained within a streamlined, transparent sonar dome or window. Variable-depth systems (VDSs) have arrays contained in tow bodies.

Each hydrophone in the receive array is connected to its own preamplifier and signal conditioning electronics. Generally, the preamplifier output is low-pass filtered at a frequency at or above the Nyquist sampling frequency and digitized. Many systems incorporate a test signal insertion point in series with each hydrophone. The array is calibrated by inserting the same signal simultaneously into each hydrophone. It is then convenient to measure the amplitude of all of the digitized signals and to multiply each one by a different constant that causes the resulting products for all hydrophones to be equal. This procedure corrects for all signal amplitude variations that may occur in the analog portion of each hydrophone signal path.

The receive beamformer accepts the conditioned signals from each hydrophone and combines them in many parallel channels to form a set of beams, usually about equal in number to the number of hydrophones in the array. Flush-mounted arrays that have flow over them are populated with many additional hydrophones to help average out the turbulent wall pressure fluctuations. Each beam is characterized by a maximum sensitivity to incoming signals on its own major lobe, or maximum response axis (MRA), with diminished sensitivity to incoming signals away from its MRA. Usually each MRA is defined by its relative azimuth and vertical angle, so that the MRA changes as the platform heading changes. To avoid this, stabilized beams are sometimes formed by considering the platform heading, so that the MRA is defined by its true heading. Forward platform motion induces frequency shifts in the echos coming from all targets except those that are exactly abeam. The beamformer can, by adjusting its clock rate, remove the Doppler shifts caused by platform motion by a process called own Doppler nullification (ODN).

The output of each beam is filtered in a way designed to maximize the signal-to-noise ratio at the filter output. The general name of this type of filter is *matched filter*. When the signal transmission sequence includes more that one kind of signal, there must be a matched filter for each signal type attached to each beam. When the transmitted signal is sensitive to Doppler distortion, there must be a bank of filters each of which is matched to a different resolvable Doppler shift.

Signal detectors follow the matched filters, because the filter outputs are zero-mean signals. Subsequent averaging to reduce the confounding effects of noise fluctuations would reduce the signals to zero, if not for the detectors. Any nonlinear process, such as half- or full-wave rectification or mth law exponentiation would do. The usual choice is square-law envelope detection.

Often the signal is spread out in time or frequency by the propagation path to and from the target. When this occurs, the matched filter performance is degraded somewhat, because the filter was matched to the transmitted waveform and not to the received waveform. The loss in matched filter performance can be substantially recovered by postdetection processing, which involves averaging the output of the detectors over the same frequency or time window that was introduced by the propagation paths. It is helpful to convert the data to its logarithm in certain cases of reverberation interference that contain a large number of false echoes that are louder than some target echoes.

The next step, the normalizer, would not be necessary if the noise could be characterized as a time-invariant random process. But it never can be, so that fluctuations in the noise can be confused with possible target echoes. Moreover, the reverberation component of interference dies out with increased range, so that it is not time invariant either. The normalizer essentially divides the output of each envelope detector by its mean value determined over a time longer that the signal duration but shorter that the characteristic time of noise power fluc-

tuation. This normalization can also be done in azimuth by dividing the signal in one beam by the average signal in many adjacent beams at the same time (range). Signal normalization by its mean in time and bearing eliminates the effect of long-term changes in background interference, so that the nonsignal outputs always have the same mean.

The normalizer output is continuously compared to a fixed output value called the threshold value for detection. Each time the normalizer output exceeds this threshold value, then a tentative detection (called a detection) is made and ascribed to the beam and Doppler filter to which that normalizer is attached. These detections are passed to the information processing subsystem for subsequent analysis.

The first purpose of the information processor, called classification, is to determine which of the detections are correct ones and which are false ones. Then the target position and its course made good are estimated by a process called localization and tracking. These functions rely upon several classes of logic and processing. One class is target connectivity. Successive detections are pieced together in tracks that are plausible ones in terms of target speed and maneuvers. Another class is echo parameter plausibility. The range and bearing extent of each detection is measured to see if it represents a target much larger that a submarine. Range rate (Doppler shift) is measured to see if it is consistent with the hypothetical track. Echo power level is compared with plausible echo power levels according to the estimated target range as a test that a submarine caused the echo. Sometimes various moments of the echo envelope are computed and compared to characteristic values for submarines. Localization parameters (range, relative bearing, and range rate) are estimated from the individual sonar detections. These values are used in algorithms called trackers to form geographical tracks of the target. Trackers begin new tracks when detections are inconsistent with previous tracks and prune tracks when no new detections are received for a long period that can extend them.

The final system block is the man–machine interface, or the display. Typically displays are formed on high-resolution cathode-ray tubes (CRTs). Multiple A-scan displays show signal-amplitude-versus-time plots of the outputs of several beam normalizers in a vertical stack. The common abscissa of each beam output is target range. If enough beams are included, the operator sees an intensity-modulated area in the azimuth, range plane. In some displays, the threshold crossing detections are overlayed on this display to call the operator's attention to the echo. A zoom feature allows any portion of the display to be greatly expanded to allow the individual time features of the echo to be seen and evaluated. A synoptic or geographical situation display is shown either in relative coordinates or geographical coordinates. In the first case, the sonar platform is shown at the center of polar coordinates range and true bearing. In the second case, the platform is shown at its geographical position. Detections and target tracks are shown on these displays, sometimes with bathymetric overlays to aid in the classification of bottom echoes.

2.2 Generalized Passive Sonar System

The generalized passive sonar system is a subset of the active system, because in the passive case the target itself radiates the signal that is to be detected by the sonar system. Figure 1 shows that there are three classes of sound propagation paths between the target and the receiver: direct, surface reflected, and bottom reflected. At ranges much greater than the water depth there is an ensemble of paths, most of which are reflected many times from both the bottom and the surface. As before, the ensemble of paths is characterized by propagation loss, called transmission loss. The time spread of the paths is of no direct concern, because target noise radiation is for the most part of a continuous nature. Frequency spread induced by the time-varying propagation path due to motion limits the narrowness of the processing bandwidth.

Receive arrays for passive systems do not differ in a fundamental way from those for active systems. Passive systems do not need to be colocated with a sound source but must be located in quiet places. Therefore, passive receive arrays are located on the ocean bottom, towed behind quiet ships, or mounted on hulls with elaborate sound attenuation systems between them and the hull. Since the hydrophones in the arrays do not need to serve the dual purpose of transmitting the sound signal, they are much smaller and lighter.

Signal conditioners and receive beamformers do not differ in a significant way from their active counterparts. Matched filters for passive systems do differ greatly, because the time-continuous noise emitted from targets is completely different than the pulse train sequences emitted from active sonar systems. There are two distinct types of passive sonar-matched filters. One type is designed to detect the discrete spectral line components in the target-radiated noise signature. The other type is designed to detect the continuous spectral background of the target-radiated noise signature. The former filter is used for narrow-band detection, while the latter filter is used for broadband detection.

Postdetection processing of both narrow-band and broadband filters consists of rather long time averaging. The longer the averaging, the weaker the signal that can be detected, within limits imposed by the noise and propagation path. In the active sonar case the averaging time

is set equal to the time spread of the propagation path, for to average longer would reduce the signal amplitude. In the passive case the averaging time is also to be no longer than the duration of the signal. This time is determined by how long the target remains at a relatively low transmission loss range.

Normalizing is of crucial importance in passive sonar, because the signals last for relatively long times and therefore noise fluctuations over long times must be removed without removing the signals themselves. The principal interference in passive sonar systems is the noise radiated from individual merchant or other nontarget surface shipping, for these noise sources mimic the ships that are to be detected.

3 ANALYTICAL REPRESENTATION OF SONAR SYSTEM ELEMENTS

This section describes the performance of sonar system elements.

3.1 Arrays and Beamformers

A transmit or receive array and its beamformer functions as a unit either to radiate a pattern of sound or to receive a pattern of sound. The fundamental relationships between array parameters and array performance are described in this section. The parameters of beam patterns that affect signal reception are beamwidth, sidelobe levels, and directivity index. The parameter of beam patterns that affects noise reception is array gain.

Beamforming and Array Beam Patterns The beam pattern of a transmitting array describes its directional distribution of radiated sound power. In a similar way, the beam pattern of a receiving array describes its directional sensitivity to incoming sound power. A transmitting array consists of N discrete radiators and a receiving array consists of N discrete receiving hydrophones. When the radiators are linear, reciprocal transducers, they are often used also as the receiving hydrophones. A transmitting beamformer forms a time-delayed and amplitude-adjusted replica of the transmitted waveform for each transducer in the array to produce a desired transmission beam pattern. The principal consideration for transmit beam patterns is to maximize sound power in target directions by transmitting as little power as possible in other directions. A receiving beamformer obtains the outputs of each hydrophone and combines them all to produce a desired receiving beam pattern. The principal consideration for receive beam patterns is to minimize sensitivity in directions away from the target direction, thereby increasing the target signal-to-noise ratio. The beam pattern equations in this section are described as receiving array beam patterns, but they are equally valid as transmitting array beam patterns.

The sound pressure distribution at the array aperture caused by an incoming plane sound wave having sound pressure amplitude P_0, frequency f, and wave vector $\mathbf{k}$ is

$$p(\mathbf{x}, t) = P_0 e^{i(\mathbf{k} \cdot \mathbf{x} - 2\pi f t)}, \tag{1}$$

where the direction of $\mathbf{k}$ coincides with the direction of wave propagation and its magnitude is equal to $|k| = 2\pi f/c$. Beam patterns are defined for monochromatic waves, so that the sound pressure distribution is represented simply by its spatial factor:

$$P(\mathbf{x}) = P_0 e^{i\mathbf{k} \cdot \mathbf{x}}. \tag{2}$$

This sound pressure distribution is sampled spatially by an array of identical hydrophones. The distribution of hydrophone sensitivities over the array aperture is

$$s(\mathbf{x}) = \sum_{n=1}^{N} \sigma_n f(\mathbf{x} - \mathbf{x}_n) e^{i\mathbf{k}_0 \cdot \mathbf{x}}, \tag{3}$$

where $\mathbf{x}_n$ are the center locations of the hydrophones, σ_n are the receiving sensitivities of the hydrophones, $f(\mathbf{x})$ is the spatial distribution of hydrophone sensitivity over its face, and $\mathbf{k}_0$ is the wave vector to which the array beamformer is steered. The beamformer function is to add up all of the hydrophone outputs to obtain the beamformer output signal Y. The beam pattern D is equal to the normalized output signal power, $D = 10 \log |Y(k)|^2/|Y(k_0)|^2$:

$$Y = \int P(\mathbf{x})s(\mathbf{x}) \, d\mathbf{x} = P_0 F(\mathbf{k} - \mathbf{k}_0) \sum_{n=1}^{N} \sigma_n e^{i(\mathbf{k} - \mathbf{k}_0) \cdot \mathbf{x}_n}. \tag{4}$$

The function $F(\mathbf{k})$ is the Fourier transform of the hydrophone sensitivity function $f(\mathbf{x})$. Equation (4) is a product of two functions: $F(\mathbf{k})$ is the beam pattern of an individual hydrophone and the summation term is the beam pattern of an array of point hydrophones located at the centers of the actual ones. This result is known as the product theorem, or pattern multiplication theorem for arrays.

The product theorem is an important result, because it can be used to obtain the beam pattern of a planar array of hydrophones at the centers of a rectangular grid, a very common array arrangement. If the columns of hydrophones are considered to be line arrays with the

beam pattern $F(\mathbf{k})$, and if the rows of hydrophones are considered to be line arrays with the beam pattern $G(\mathbf{k})$, then the beam pattern of the array is equal to $G(\mathbf{k})F(\mathbf{k})$. The beam patterns of linear point arrays are of fundamental importance.

The beam pattern of an unshaded linear point array of N hydrophones spaced d apart and steered in the direction θ_0 and measured from the array normal is given as

$$D(\theta) = \left(\frac{\sin NX}{N \sin X} \right)^2, \tag{5}$$

where

$$X = \frac{kd}{2} \sin(\theta - \theta_0).$$

The beam pattern is a periodic function of the sound arrival angle θ. The beam pattern has maximum values, or lobes, equal to 1 at values of $X = n\pi$. Only the beam pattern lobe centered at $X = 0$ is the desired lobe. Any other lobes that appear in the real sound angle range of $-90°$ to $+90°$ are unwanted "grating lobes." The condition for avoiding grating lobes for any steering angle is

$$d < \frac{\lambda}{2}. \tag{6}$$

The beamwidth Δ of the main lobe of the beam pattern is its angular width at the half-power points, or $\Delta = 2\theta'$, where $D(\theta') = 0.5$,

$$\Delta = \frac{\lambda}{L} \geq \frac{2}{N}. \tag{7}$$

the second equality holds at the grating lobe limit given by Eq. (6).

All other lobes in the range of sound angles θ besides the main lobe are called minor lobes, or sometimes side lobes. For the beam pattern described by Eq. (5), the minor lobes occur at values of X that satisfy the equation $\tan(NX) = N \tan(X)$. If the length of the array $L = d(N - 1)$ is many wavelengths, then the side-lobe arguments and levels are given in Table 1.

The near side-lobe levels for a linear unshaded array are not much less sensitive that the major lobe. Expected signal power levels from targets and interfering ships can range over values that differ by at least 30 dB. Therefore, a loud target in the direction of a side lobe can overwhelm a weaker target in the direction of the main lobe, causing it to be missed. The usual correction for this problem is to shade the amplitudes of the array elements to produce lower side-lobe levels.

TABLE 1   Side Lobes of Long Line Arrays

Side-Lobe Number	X	Side-Lobe Level (dB)
1	4.5	−13.3
2	7.7	−17.8
3	10.9	−20.8
4	14.1	−23.0
5	17.2	−24.7

TABLE 2   Amplitude Coefficients and Beamwidths of Four Chebyshev Arrays

Side-Lobe Level	A_0	A_1	A_2	Beamwidth (deg)
Unshaded	1	1	1	28
−20	4.32	3.35	2.33	33
−25	8.42	6.12	3.25	36
−30	15.98	10.92	4.27	38

Dolph–Chebyshev Arrays   The shaded linear array that provides equal-level side lobes and at the same time the narrowest possible beam width is based on the Chebyshev polynomials.[1] The number of elements N is related to the order m of Chebyshev polynomial by

$$N = 2m + 1. \tag{8}$$

The ratio of main-lobe to side-lobe signal sensitivities is R ($R = 10$ for −20-dB side lobes). Two design parameters are a and b;

$$a = \frac{e + 1}{1 - \cos[kd(1 + \sin \theta_0)]},$$

$$b = \frac{e \cos[kd(1 + \sin \theta_0)] + 1}{\cos[kd(1 + \sin \theta_0)]}, \tag{9}$$

where

$$e = \cosh\left(\frac{1}{m} \cosh^{-1} R\right).$$

The mth-order Chebyshev polynomial can be derived from the following recursion relationship:

$$T_{m+1} = 2x T_m(x) - T_{m-1}(x), \tag{10}$$

where $T_0(x) = 1$ and $T_1(x) = x$.

The argument x is defined by the transformation $x = a \cos \psi + b$. The array shading coefficients A_n are obtained from

$$T_m(x) = A_0 + 2 \sum_{n=1}^{(N-1)/2} A_n \cos n\psi. \tag{11}$$

The shading coefficients are determined by equating Eq. (11), term by term, in $\cos n\psi$. As an example, Bartberger[2] developed four six-element arrays, summa-rized in Table 2. Beam broadening is not great, and the side lobes are much reduced with respect to unshaded arrays, for which the first side-lobe level is only −13 dB. Typical arrays are shaded for −25- or −30-dB side lobes. For these arrays, an unwanted noise on a side lobe would have to be more than 30 dB louder than a weaker signal on the main lobe to mask it. Sometimes this difference is not great enough, and adaptive shading is used.

Side-lobe control can also be accomplished by spatial shading (as opposed to amplitude shading). The theory of aperiodic array design is described by Steinberg.[1] The objective of aperiodic array design is to obtain acceptable beam patterns with fewer array elements than are required to satisfy the grating lobe requirement [Eq. (6)].

Other shading schemes having particular uses are described by Stenberg[1] and Nielson.[3]

Adaptive Beamforming   The individual hydrophone gains are derived from within the beamformer, based on the nature of the hydrophone outputs, in adaptive beamforming. The most common goal of adaptive beamforming in sonar systems is called adaptive nulling. In this process, each beam is assigned an expected signal direction, and receiving sensitivity for that beam is maintained at a constant value in the assigned direction. When a strong signal is detected from another direction (outside the main lobe), a null sensitivity is generated in that direction. Several strong, out-of-main-lobe signals can be nulled at once. Several methods of adaptive nulling are described by Steinberg.[1]

Array Gain and Directivity Index   Array gain (AG) is defined as the increase, in decibels, of the signal-to-noise ratio caused by a receiving array and its beamformer. Array gain depends both upon the geometry of the array and the nature of the noise field around the array. The directivity index (DI) of a transmitting array is defined as the ratio, in decibels, of the signal power in the transmitting direction to the signal power averaged over all directions. The directivity index of an array is numerically equal to its array gain in the special case of a three-dimensional isotropic noise field.

The signal and noise amplitudes at the output of each array element are given by $s(t)$ and $n(t)$. The signal from the beamformed beam pointing at the target is $y_s(t)$, and the noise in the same beam is $y_n(t)$. There are N elements in the array, and the sensitivity of the ith element is σ_i. The output signal and noise powers are

$$y_s^2(t) = \left(\sum_{i=1}^{N} \sigma_i S_i(t) \right)^2,$$

$$\overline{y_n^2} = \sum_{i=1}^{N} \sigma_i n_i(t) \sum_{j=1}^{N} \sigma_j n_j(t). \qquad (12)$$

The correlation coefficient of the noise at elements i and j is defined as the covariance of the noise amplitude at the two elements, normalized by the noise power. The noise power is assumed to have the same value in all elements. Moreover, the noise covariance C is a function only of the distance between the two elements x_{ij} if the noise field is statistically homogeneous over the array face:

$$C_{ij}(x_{ij}) = \frac{\overline{n_i n_j}}{\overline{n^2}}. \qquad (13)$$

The array gain of an N-element unshaded array ($\sigma_i = 1$) is equal to $10 \log(g)$, where

$$g = \frac{\overline{y_s^2}/\overline{y_n^2}}{s^2/\overline{n^2}} = \frac{N^2}{N + 2\sum_{i=1}^{N} \sum_{j \neq i}^{N} C_{ij}(x_{ij})}. \qquad (14)$$

The gain ratio g can be written as the ratio of two terms: the ratio of beam signal power to element signal power and the ratio of beam noise power to element noise power. This leads to the connection between the logarithmic quantities array gain (AG), signal gain (SG), and noise gain (NG):

$$g = \frac{\overline{y_s^2}}{s^2} \cdot \frac{\overline{n^2}}{\overline{y_n^2}}, \qquad AG = SG - NG. \qquad (15)$$

A common way of calibrating the beamformer is in terms of the incident signal sound power. This sound power is, of course, independent of the measurement array, so that the signal power at any element output should be the same as the signal power at the output of the beam that points at the incident wave. Unless the beamformer has loss due to incorrect element summation and delay, the signal gain ratio is 1, and SG = 0 dB. With this convention, the array gain is numerically equal to the negative of the noise gain. The noise power $\overline{y_n^2}$ is less than the element noise power $\overline{n^2}$.

Array Gain and Directivity Index of Unshaded Line Arrays

In the important case of a line array of N elements spaced by distance a, the distance between the ith and jth elements is equal to $(j - i)a$. The double summation in Eq. (14) becomes a single summation over $m = j - i$:

$$g^{-1} = \frac{1}{N} + \frac{2}{N^2} \sum_{m=1}^{N-1} (N - m)C(am). \qquad (16)$$

When the noise field over the face of the array is uncorrelated from element to element, $C(x_{ij})$ is zero. From Eqs. (14) and (16), the array gain is in this case equal to $10 \log(N)$. The noise field correlation coefficients are tabulated in Table 3 for four different homogeneous noise fields.

The first tabulated noise field is composed of a superposition of plane sound waves of the same frequency having equal power in all propagation directions distributed throughout three dimensions. This is an idealized noise field that is seldom approximated in the ocean. This case corresponds to conditions for which the directivity index is defined, so that this case leads to the directivity index. The second noise field is a superposition of plane sound waves of the same frequency having equal

TABLE 3   Noise Correlation Coefficients for Four Noise Fields

Noise Field	Correlation Coefficient
Three-dimensional isotropic	$\sin(ka)/ka$
Two-dimensional isotropic, elements in propagation plane	$J_0(ka)$
Cosine directivity, elements parallel to null plane of cosine	$2J_1(ka)/ka$
Cosine directivity, elements parallel to peak of cosine	$[\cos(ka) + ka\sin(ka) - 1]/(ka)^2$

power in all propagation directions distributed over a single plane. This noise field is approximated in the ocean at low frequencies, where the noise is mostly generated by distant shipping and therefore arrives more or less horizontally from all azimuthal directions. The third noise field is a superposition of plane waves of the same frequency having power proportional to the cosine of their vertical angle. This noise field is approximated in the ocean at high frequencies where dipole noise generated at the surface by wind waves predominates. The correlation coefficient for the third tabulated case is for elements in a line parallel to the surface, so that it is appropriate for towed line arrays. The correlation coefficient for the fourth tabulated case is for elements in a line perpendicular to the surface, so that it is appropriate for vertical line arrays.

The array gain for the broadside beam of a 50-element line array is shown on Fig. 2 for each of the four different correlation coefficients listed in Table 3. All four cases asymptote to a value of $10 \log(N)$ for large values of the argument ka. This is generally true, because the noise correlation functions all tend to zero at separations a that are large with respect to a wavelength. The element spacing a is equal to one-half wavelength at $ka = \pi$. This is the largest spacing that will not produce aliasing lobes when the array is steered toward endfire, and arrays are often not used at higher frequencies. But larger values of array gain occur up to $ka = 2\pi$. Values of array gain are about 2 dB larger than the DI against the two-dimensional noise field (representing low-frequency

ambient noise). Values of array gain are about 1 dB smaller than the DI against the dipole noise field (representing high-frequency ambient noise). This illustrates why towed arrays provide higher gain at lower frequencies. The vertical line array achieves about 15 dB more gain than the others. This illustrates the beneficial effect of directing the main lobe in a direction free of the predominant noise.

The DI for the unshaded linear array can be written as $DI = 10 \log(2L/\lambda)$, expressing the fact that in the three-dimensional isotropic noise field the elements must be spaced at a distance of $\lambda/2$.

Directivity Index of Two-Dimensional Arrays

The DI (also the array gain in three-dimensional isotropic noise) of two-dimensional planar arrays is a function only of the area of the array A and the wavenumber k:

$$DI = 10 \log\left(\frac{k^2 A}{\pi}\right). \tag{17}$$

This equation is valid for an array of any shape, provided the smallest dimension is longer than about two wavelengths. This equation is also valid if the array is made up of N-point hydrophones if the spacing between adjacent hydrophones is less than one-half wavelength. For a uniformly spaced planar array, $A = Nsw$, where s and w are the hydrophone spacings in the length and width directions of the planar array.

Array Gain of Shaded Arrays

The array gain of an array in isotropic noise is not sensibly changed by shading. An example of this is given by Bartberger.[2] He tabulated the array gain against a three-dimensional isotropic noise field for a six-element line array having Chebyshev polynomial shading. The results are listed in Table 4. Severe shading reduces the DI by only 0.7 dB. If a noise field is reliably nonisotropic, array gain will be increased by shading for low side-lobe levels in the noisy directions.

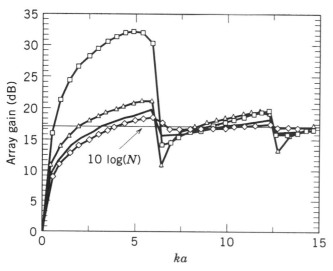

Fig. 2 Array gain for the broadside beam of 50-element line arrays as functions of normalized element spacing ka: (——) three-dimensional noise field; ($\triangle$) two-dimensional omnidirectional noise field; ($\diamond$) line parallel to a surface of dipole sources; ($\square$) line perpendicular to a surface of dipole sources.

TABLE 4   Array Gain of Six-Element Chebyshev Shaded Line Array

Array	Gain (dB)
Unshaded	7.78
Shaded, 15-dB side lobes	7.75
Shaded, 20-dB side lobes	7.53
Shaded, 25-dB side lobes	7.26
Shaded, 30-dB side lobes	7.02

3.2 Signal Processing in Active Sonar Systems

The signal processing chain is a part of Fig. 1. In this section each component of the chain is described, more or less in the order that the signal proceeds through the system.

Matched Filters The purpose of the matched filters is to maximize the signal-to-noise ratio (SNR) in each beamformer output beam. Each different signal waveform requires a unique filter to achieve this purpose, thus the name "matched." The typical active sonar signal bandwidth is not so broad that the noise power spectral density over the band cannot be assumed flat. Moreover, the noise amplitude probability density function (PDF) is assumed to be Gaussian. The desired filter is the one that produces the maximum value of SNR when a signal $x(t)$, together with additive white Gaussian noise $u(t)$ having a power spectral density N_0, is passed through it. The filter is linear and characterized by the impulse response $h(t)$. The output of the filter caused by the signal input is $s(t)$, and the output caused by the noise input is $n(t)$:

$$s(t) = \int_{-\infty}^{\infty} x(t')h(t' - t)\, dt',$$

$$n(t) = \int_{-\infty}^{\infty} u(t')h(t' - t)\, dt'. \tag{18}$$

The SNR is defined as the signal power at the filter output at time t_m that it reaches its maximum value divided by the noise power at the filter output:

$$\text{SNR} = \frac{s(t_m)^2}{n^2} = \frac{S(t_m)}{N}. \tag{19}$$

The derivation of the matched filter impulse response $h(t)$ is explained in many books, for example, Davenport and Root[4] and Levanon.[5] The impulse response of the desired filter is $h(t) = Kx^*(t_m - t)$, where K is a constant. This result is interpreted as the complex conjugate of the input signal $x(t)$ that has been inverted in time and delayed by the time at which the output signal reaches its maximum value. It is necessary for t_m to be greater than the time span T of the signal $x(t)$ itself, so that the value of $h(t)$ is zero for negative times (causality). The value of SNR at the time t_m of maximum signal output, as defined in Eq. (19), is proportional to the signal energy:

$$\text{SNR} = \frac{E}{N_0}, \quad \text{where } E = \int_{-\infty}^{\infty} x(t)^2\, dt. \tag{20}$$

The time duration τ of the matched filter output pulse can be deduced from the foregoing. Looking first at Eq. (20), we see the SNR is the ratio of two input energies. The filter gain factor K divides out, so that the ratio will apply also to the corresponding two output energies. Equating Eqs. (19) and (20) and denoting the signal energy by $S\tau$ and the noise power by N_0B lead to an important result,

$$\tau B = 1. \tag{21}$$

The time duration τ of the output signal pulse is called the pulse resolution time. The pulse time span T can be much larger than the resolution time τ. The pulse compression ratio c_r is given by

$$c_r = \frac{T}{\tau} = TB. \tag{22}$$

These important results mean that no parameter of the active sonar signal other than its energy and bandwidth has any bearing on the SNR or the range resolution at the output of the matched filter. It is shown in Section 3.3 that most of the system's detection performance depends only upon SNR and the bandwidth of the signal, so that only signal energy and bandwidth are prime system parameters.

Postfilter Envelope Detection Some form of nonlinear circuit is introduced following the matched filtering operation. The matched filters are linear and their inputs contain no DC component, so that their signal and noise outputs are both zero-mean time series. This is an inappropriate form for the signals to have for several reasons. First, the signal may have either positive or negative polarity. The difficulty of finding the signal in noise is increased if its larger portions can be either above or below the axis. Second, averaging or smoothing of the filter outputs is often helpful to supress noise fluctuations that tend to mimic targets. Averaging the signal in this zero-mean form will reduce its amplitude just as much as the noise with no net gain. Third, signals in this form display their entire waveforms. The only part of the waveform that is of interest in detection is its envelope. Therefore, each matched filter is followed immediately by an envelope detector.

There are several essentially equal methods of envelope detection, but the one most commonly used is an instantaneous square-law device followed by a low-pass filter. If the low-pass filter is "matched" to the signal at the squarer output, then it has a cutoff frequency equal to the bandwidth of the signal. It is equivalent to saying that the low-pass filter is an averager, which is matched to the squared signal when its averaging time equals the time span of the signal. It is common for the propagation path to be composed of several "multipaths." These multipaths are said to be irresolvable when the time delay between adjacent paths is less than the signal's resolution time. It is common in active sonar systems to "over-average' with the low-pass filter, so that the averaging time is set to approximate the sum of the pulse resolution time and the multipath time spread. Doing this causes no loss in the amplitude of the filtered signal, but it reduces the fluctuations in the noise. Whether the low-pass filter overaverages or not, it will also eliminate the noise spectral components that are generated by the square-law device at twice the passband frequency of the matched filter. Failure to filter out these high-frequency noise components would unnecessarily double the noise power.

Signals for Active Sonar Systems
The preceding section explained that sonar systems invariably use filters that are matched to the transmitted signals, for these filters maximize the peak SNR. It is intuitively obvious that maximum system detection performance is achieved by maximizing SNR. But there are other functions to be performed by the system. In particular, once the target is detected, it is necessary to measure accurately both the range to the target and its range rate. A different type of requirement is set by the maximum signal power output achievable from the transmitting array, so that some minimum pulse time span is required for the signal energy to produce a sufficiently large SNR.

Range to a target is determined by the time delay δ between the time the signal is transmitted and the time that its echo is received:

$$R = \frac{c\delta}{2}. \tag{23}$$

The effective sound speed c can be determined by calculating the travel time over the propagation path using methods described in Chapter 46. It is often sufficiently accurate to use a value of 0.8 nautical miles/s, which is found to match closely the effective sound speed of many different propagation paths in the ocean.

Measurement of target range can only be accomplished to an accuracy set by the time resolution of the pulse. The range resolution ΔR is given as

$$\Delta R = \frac{c\tau}{2} = \frac{c}{2B}. \tag{24}$$

Range rate is determined by measuring in some way

the time scaling impressed upon an echo by a moving target. If the transmitted signal is $x(t)$, the echo from a point target moving with respect to the platform with range rate $\dot{R}$ will be scaled in time as $x(t')$, where t' is related to t by a function of Mach number M, where $M = \dot{R}/c$:

$$t' = t\,\frac{1 + M}{1 - M}. \tag{25}$$

The Mach number is always small, for even a 90-knot torpedo closes range at a Mach number of only 0.03. This leads to the common approximation of Eq. (25):

$$t' = t(1 + 2M). \tag{26}$$

For narrow-band signals, the motion-induced time scaling causes a frequency shift, called the Doppler shift. Denoting the transmitted frequency by f_t, the echo frequency by f_e, and the Doppler shift frequency by f_d, the following equation relates the relative Doppler frequency to the Mach number:

$$f_e = f_t + f_d = f_t(1 - 2M). \tag{27}$$

The negative sign means that for receding targets, for which range rate is positive, the Doppler shift is negative due to stretching in time and the echo has a lower frequency than has the transmitted signal.

Measurement of target range rate can only be accomplished to an accuracy set by the frequency resolution of the pulse. The frequency resolution $\Delta f = \nu$ is a fundamental relationship

$$\Delta f = \nu = \frac{1}{T}, \tag{28}$$

where T is the pulse time span.

The capacity for a sonar signal to measure target range (travel time) and target range rate (Doppler shift) is illustrated by its ambiguity function. The ambiguity function of a signal is defined to be its own matched filter's response to its time-delayed and Doppler-shifted self. In other words, the ambiguity function of a signal is a two-dimensional map of its own filter's output on the time delay, Doppler shift plane. This map shows how sharp the signal response is to changes in target range and speed. The output of the matched filter is given by Eq. (18). The signal waveform is expressed in the complex form $x(t) = u(t)e^{i2\pi ft}$, where $u(t)$ is the complex modulation function whose magnitude $|u(t)|$ is the envelope of the real signal, and $2\pi f$ is the carrier frequency. The time-delayed and Doppler-shifted matched filter output is then[6]

$$\chi(\delta, f_d) = \int_{-\infty}^{\infty} u(t)u^*(t + \delta)e^{i2\pi f_d t}\, dt. \qquad (29)$$

The squared magnitude $|\chi(\delta, f_d)|^2$ is the ambiguity function.

The ambiguity function illustrates the pulse resolution time τ and resolution frequency ν by its central width in the δ, f_d coordinates. There are three general types of signal waveforms from the point of view of their ambiguity functions. First are single-frequency pulses, or "CW pulses," which are used to obtain a fine resolution of range rate or of echo frequency. The ambiguity function of such a pulse is shown on Fig. 3, both as a quasi three-dimensional plot and as a contour plot. The time span T of the pulse is two delay units. The figure shows that the pulse resolution τ of the ambiguity function is relatively large, being equal to the pulse time span at the half-amplitude points. The frequency resolution of the pulse is relatively small, being equal approximately to the inverse of the pulse time span. The pulse compression ratio c_r is unity, so the time–bandwidth product BT for a sinusoidal pulse is unity. The CW pulses can be amplitude shaded to reduce side lobes in the Doppler dimension of the ambiguity surface. Long CW pulses are used to separate high Doppler targets from low Doppler reverberation.

A second general type of waveform is designed to provide both fine range and Doppler frequency resolution at the expense of range–frequency ambiguity. This type is illustrated by a linear frequency modulated pulse, in which the frequency changes by $f(t) = kt$, where k is a constant. The ambiguity function of such a pulse is shown in Fig. 4, both as a quasi three-dimensional plot and as a contour plot. The time span of the pulse is $T = 1$ delay unit and $k = 10$, leading to a pulse bandwidth $B = 10$. The plot shows that the time delay width of the ambiguity function at zero Doppler (resolution time)

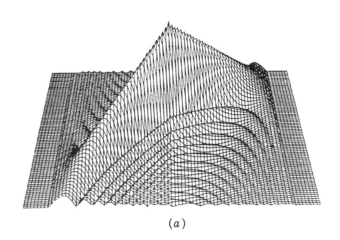

(a)

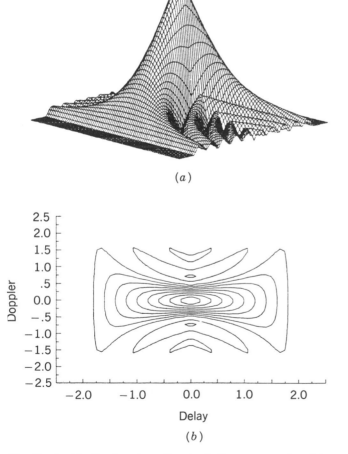

(a)

(b)

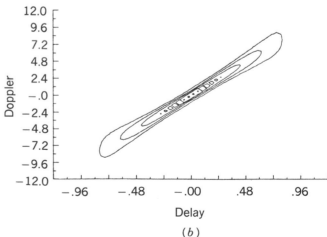

(b)

Fig. 3 Ambiguity function of a single-frequency pulse: (a) three-dimensional view; (b) contour plot. Pulse span $T = 2$. (From Ref. 5, Fig. 7.2.)

Fig. 4 Ambiguity function of a linear FM pulse ($t = 1$, $k = 10$): (a) three-dimensional view; (b) contour plot. (From Ref. 5, Fig. 7.5.)

is only about equal to 0.1. The product TB is 10, and the pulse compression is equal to 0.1, in agreement with Eq. (22). The figure also illustrates the resolution frequency ν. At zero delay, the width of the ambiguity function is unity, in agreement with Eq. (28). If the pulse time span had been 10 units and $k = 1$, then the pulse resolution time would still have been equal to $1/B = 0.1$ but for a compression factor of 100. The resolution frequency would have been reduced to 0.1. This type of pulse is very useful if T must be large to obtain the necessary signal energy from the transmitting array. Even though T is large, it is possible still to resolve small targets because the pulse resolution time τ is small. There is a problem of range–Doppler ambiguity with this waveform, as shown on Fig. 4. Targets having nonzero Doppler appear to be at the wrong range. This ambiguity may be advantageous in reducing the number of matched filters needed to detect targets.

The third general type of waveform is designed to avoid the range–Doppler ambiguity of the linear frequency-modulated pulse but to retain both its pulse compression and Doppler resolution features. These waveforms are coded to have thumb tack ambiguity functions, meaning a significant response occurs only for targets jointly having a narrow dimension in both Doppler and range coordinates. An example of this type of coded pulse are Costas signals. Costas signals are sequences of equal-time-span single-frequency tones whose frequencies skip in a way that time and frequency shift overlaps are minimized. The ambiguity function of a seven-tone Costas signal is shown in Fig. 5. If the seven different frequencies are ranked by frequency, the signal sequence shown is 4, 7, 1, 6, 5, 2, 3. The pulse resolution τ of the sequence is equal to the inverse sequence bandwidth, in this case the frequency difference between tones 1 and 7. The frequency resolution of the function is equal to the inverse time span of the sequence, or 7 times the span of each tone. The particular choice of frequency sequence is made to keep all of the ambiguity surface away from the central peak flat, so that no target ambiguity occurs.

3.3 Target Detection Decisions by Active Sonar Systems

The outputs of the envelope detectors contain target echoes superimposed onto a background of noise. The echoes occur at unknown time delays δ and contain unknown Doppler shift frequencies f_d. Their pulse resolution times τ are known. The detection decision of whether an echo is present is to be made for each time cell of length τ in which an echo could occur. In general, this decision is made by comparing the envelope detector output in each time interval to a threshold value.

The decision that a target is present is made when the detector output is larger than the threshold, and the target absent decision is made whenever the threshold is not exceeded. The value at which the threshold is set clearly has an important role in the validity of the detection decisions. If the detection threshold is set to a low value, even weak echoes will exceed it, and the probability of achieving detection is high. But the low threshold value will often be exceeded by large fluctuations of the noise, so the probability of a false detection (called a false alarm) will be high also. Conversely a high value of threshold will result in detecting only very large echoes, so that the probability of detection will be small. So also will the probability of false alarm be small. The proper adjustment of the threshold is a different matter for different statistical descriptions of both the signal echoes and the noise.

For each case of signal and noise statistics considered in this section, equations for both the probability of detection and the probability of false alarm are given as functions of SNR. The value SNR^* needed to achieve jointly a particular set of probabilities of detection and false alarm is called the detection index, and when expressed in decibels, it is called the detection threshold DT (note that Urick[7] defines detection threshold in terms of a "signal to noise ratio referred to a 1-Hz band of noise," rather than the dimensionless ratio used in this book):

$$DT = 10 \log \text{SNR}^*. \qquad (30)$$

We begin with single nonfluctuating pulse detection in stationary noise. But the requirements to measure range rate and to improve detection probability require that multiple pulses be used. Moreover, signal amplitudes fluctuate from pulse to pulse over real propagation paths. Several cases of increasing complexity and generality follow.

Single-Pulse Detection of Nonfluctuating Signals in Stationary Noise The simplest case is the detection of a single pulse in stationary noise having a Gaussian pdf. The probability of false alarm is determined by first calculating the pdf $p_n(r)$, which is the probability that the amplitude of the envelope detector output lies between r and $r + dr$, given that noise only is present. The probability of false alarm P_{FA} is determined by integration of $p_n(r)$ from the threshold value V_T to infinity. In other words, the probability of false alarm is equal to the probability that the noise envelope exceeds the threshold. The result is given by Levanon:[5]

$$P_{\text{FA}} = e^{-V_T^2/2N}, \qquad (31)$$

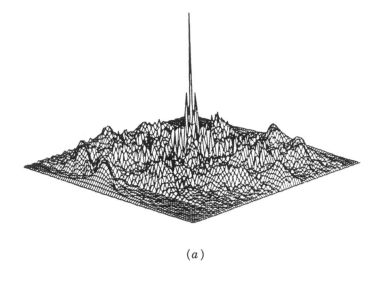

(a)

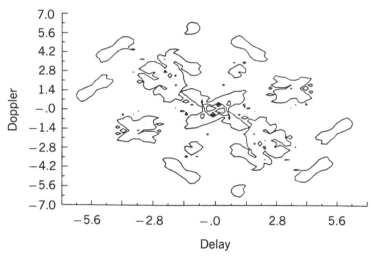

(b)

Fig. 5 Ambiguity function of a length 7 Costas signal-coding sequence 4, 7, 2, 6, 5, 2, 3: (a) three-dimensional view; (b) contour plot. (From Ref. 5, Fig. 8.3.)

where the noise power N is equal to $N_0 B$. The probability of detection is determined by first calculating the pdf $p_{s+n}(r)$, which is the probability that the amplitude of the envelope detector output lies between r and $r+dr$, given that the signal is present in the noise. The probability of detection P_D is equal to the probability that the signal plus noise envelope exceed the threshold. An approximation that is valid for values of SNR $= E/N_0$ that are

$$P_D = \frac{1}{2}\left[1 - \text{erf}\left(\frac{V_T}{\sqrt{2N}} - \sqrt{\text{SNR}}\right)\right]. \quad (32)$$

The threshold value V_T can be eliminated from Eqs. (31) and (32), yielding a threefold relationship between P_D, P_{FA}, and SNR. This relationship is widely pub-

lished (for example Fig. 12.6 in Urick[7]). The full curve, valid also for small SNR, can be found as Fig. 3.4 in Levanon.[5] An approximation due to Albersheim,[8] accurate for $10^{-7} < P_{\text{fa}} < 10^{-3}$ and $0.1 < P_D < 0.9$, is

$$\text{SNR} = A + 0.12AB + 1.7B, \quad (33)$$

where

$$A = \ln \frac{0.62}{P_{FA}}, \qquad B = \ln \frac{P_D}{1 - P_D}.$$

Detection of Multiple Nonfluctuating Pulses in Stationary Noise Single-pulse detection cannot achieve simultaneously large P_D and small P_{FA} without very large SNR, a luxury not often available. But the target will be in view for several consecutive pulses, and it is beneficial to defer the detection decision until after several pulses have been transmitted and received. There are two general ways to integrate consecutive pulses. One way is called coherent integration, a process that requires the propagation path to remain the same in all respects for the duration of the pulse ensemble. Due to motion, this typically does not happen in the ocean for usefully long time periods. Noncoherent integration requires only that the nominal sound speed of the propagation path remain the same over the pulse ensemble period, and this does obtain. In noncoherent integration, M pulses are combined after the envelope detectors by sampling the data stream simultaneously at the transmit pulse spacings and summing the samples obtained. The threefold relationship between P_{FA}, P_D and SNR is given by Levanon[5] in terms of the integral $\Phi(T)$:

$$\Phi(T) = \frac{1}{\sqrt{2\pi}} \int_T^\infty e^{x^2/2}dx = \frac{1}{2}\,\mathrm{erfc}\left(\frac{T}{\sqrt{2}}\right). \quad (34)$$

Since the inverse of the complementary error function given by Eq. (34) appears in many of the SNR equations to follow, it is plotted on Fig. 6:

$$\mathrm{SNR} = \frac{\Phi^{-1}(P_{FA}) - \Phi^{-1}(P_D)}{\sqrt{M}}. \quad (35)$$

There is an Albersheim approximation to this threefold relationship that is analogous to Eq. (33):[8]

$$10\log\mathrm{SNR} = -5\log M + \left[6.2 + \frac{4.54}{\sqrt{M + 0.44}}\right]$$
$$\cdot x\log(A + 0.12AB + 1.7B). \quad (36)$$

Comparison of Eq. (36) with Eq. (33) shows that the noncoherent summation of M pulses requires a lower SNR to achieve a given P_D and P_{FA}. The required SNR is reduced by approximately $\sqrt{M}$. Therefore, noncoherent summation of pulses is very beneficial.

Detection of Single Fluctuating Pulses in Stationary Noise

When considering pulse train to pulse train detection with fluctuations between trains, the simplest case is for a single pulse per train. The signal power of the echoes have a Rayleight pdf. For this case, the threefold relationship between P_D, P_{FA}, and SNR is particularly simple:

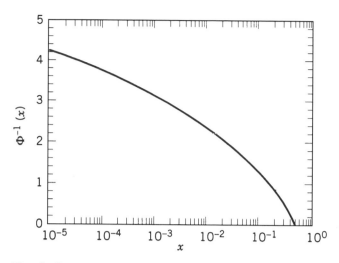

Fig. 6 Inverse complementary error function used in detection SNR equations.

$$P_D = P_{FA}e^{1/(1 + \mathrm{SNR})}. \quad (37)$$

If SNR = 10, then the value of P_D required for $P_{FA} = 10^{-4}$ is 0.43. This compares to a value of 0.60 from Fig. 3.4 in Levanon.[5] Therefore, the Rayleigh fluctuations in echo amplitude cause a 28% decrease in the probability of detection at $P_{FA} = 10^{-4}$.

Detection of Multiple Fluctuating Pulses in Stationary Noise

When the advantages of multiple pulse echo noncoherent integration are sought, it is usually found that the transmission loss over the propagation path changes between adjacent pulse trains and sometimes even within a pulse train. Four different cases are commonly differentiated. In Swerling cases I and III there is no amplitude change during the integration time, while in cases II and IV there is. In Swerling cases I and II the echo amplitudes are assigned a Raleigh pdf, while in Swerling cases III and IV the echo amplitudes are assigned a chi-square pdf. The Rayleigh pdf for echo amplitude represents physically the echoes from a large number of targets having similar target strength but randomly different ranges. This condition is met for large distributed targets such as surface or bottom reverberation. The chi-square pdf for echo amplitude represents physically the echoes from a dominant nonfluctuating target immersed in a sea of distributed targets. This condition is met for a point target surrounded in reverberation from an extended target at about the same range.

For the Swerling case I, the threefold relationship between P_D, P_{FA}, and SNR is given by Levanon.[5] The SNR is still defined as the ratio of signal energy to noise PSD (power spectral density, see Eq. 20), but in this case of fluctuating signals, the signal energy is taken as the average of the Raleigh-distributed signal power times the pulse width:

$$SNR = \frac{-\Phi^{-1}(P_{FA})}{\ln(P_D)\sqrt{M}}. \tag{38}$$

Comparison of Eq. (38) with Eq. (35) shows that the SNR needed for specified values of P_{FA} and P_D is reduced by the square root of the number of pulses noncoherently summed both for fluctuating and for nonfluctuating targets.

For the Swerling II case the signal power of each pulse is an uncorrelated random variable, having Rayleigh pdf, within the pulse train. The threefold relationship between P_D, P_{FA}, and SNR is given by Levanon:[5]

$$SNR = \frac{\Phi^{-1}(P_{FA}) - \Phi^{-1}(P_D)}{\Phi^{-1}(P_D) + \sqrt{M}}. \tag{39}$$

Consideration of the different Swerling cases I and II provides an important result. For values of the operating parameters $10^{-7} < P_{FA} < 10^{-3}$ and $0.1 < P_D < 0.9$ the SNR values required for detection are less for case II than for case I. The reason for this is that when no fluctuations take place within a pulse train, the whole train will be undetectable in a fade. When fluctuations do take place during the train, the fades are averaged out by the pulse train itself. This is an important consideration in designing active sonar pulse trains, because the fading time constant is a function of frequency, range, and range rate.

Swerling cases III and IV assume a chi-square distribution for the echo amplitudes. For the Swerling case II the signal power of each pulse is an uncorrelated random variable, having Rayleigh pdf, within the pulse train. Approximations to the threefold relationship between P_D, P_{FA}, and SNR is given by Levanon.[5] These relationships are accurate to a decibel for $M < 100$:

$$SNR = \frac{2}{MQ} [\sqrt{M}\Phi^{-1}(P_{FA}) + 2], \tag{40}$$

where

$$\ln(P_D) = \ln(1 + Q) - Q.$$

The threefold relationship between P_D, P_{FA}, and SNR for the Swerling case IV is also given by Levanon.[5] It

is found to be identical to Eq. (35), which was given for noncoherent integration of nonfluctuating targets.

Binary Integration of Pulse Trains Binary integration is another way to implement noncoherent integration over the pulse echo train. After the thresholding operation, the data stream becomes a sequence of 0's and 1's, one digit for each time interval (range bin) in which an echo could occur, with 1's representing only the time intervals when the threshold was crossed. If N pulses were transmitted, the binary integrator samples the data stream at N range bins separated from each other by the time intervals between the N pulses. A detection decision is made whenever $M < N$ occurrences of 1's are found within the N range bins. Clearly the choice of M is similar to selecting a secondary threshold. If M is large or equal to N, then the SNR of all N echoes must be large enough to cross the primary threshold. Likewise, the chance that noise could exceed the threshold in all N range bins simultaneously is small. But if M were much smaller than N, then the probability that M echoes would cross the threshold is larger and the probability that M noise threshold crossings would occur is also larger.

The probability of detection for M of N binary integration is given by Levanon in terms of the probability p that an echo in a range cell will exceed the threshold value.[5] The probability of false alarm is given in terms of the probability P_F that noise will exceed the threshold in one range cell:

$$P_D = \sum_{i=M}^{N} \binom{N}{i} p^i (1 - P)^{N-i},$$

$$P_{FA} = \binom{N-1}{M-1} P_F^M (1 - P_F)^{N-M+1}. \tag{41}$$

If the target range rate is nonzero, then the time spacing of the echoes will not be the same as the time spacing of the transmitted pulses. The time axis of the echoes will be scaled according to Eq. (26). In this event, no detections will ever occur because the binary integrator's range cells will not match more than one echo. The solution is to construct a family of binary integrators, each having its N range cells spaced by a differently scaled version of the transmission sequence. Each of these Doppler integrators is tuned to a different range rate. When detections occur, the target range is determined by the range cell having the detection, and the target range rate is determined by the binary integrator that made the detection.

When a region of intense backscatter, or clutter, is encountered, many range cells will contain echoes that cross the threshold. In this case, the *M*-of-*N* detectors will indicate detections at several contiguous range cells. Since a ship target could not return such a continuum of echoes, the fact of detection in contiguous range cells is sufficient to classify the target as clutter.

3.4 Signal Normalization for Constant False Alarm Rate

All the detection decision strategies described in the previous section assumed the noise to be stationary; that is, it can be characterized for all time by constant values of PSD and all moments of noise amplitude. This situation does not obtain in sonar practice. Ambient noise mean values fluctuate from time to time with a standard deviation of about 5 dB. Sonar self-noise can fluctuate by 10 dB or more as the platform speeds up or turns. Reverberation from continuous surfaces, such as the bottom, can easily increase the background noise by 10–20 dB. But the false alarm rate is extremely sensitive to changes in background noise. For the Gaussian noise, the probability of false alarm in a range cell is given by Eq. (31). If the desired P_{FA} is 10^{-4}, then the threshold must be set at a value equal to 4.29 times the rms noise amplitude. When the noise power increases by just 3 dB, then the P_{FA} increases by a factor of 100 to a value of 10^{-2} if the threshold is not changed.

There are many strategies for normalizing the outputs of the envelope detectors to achieve a constant false alarm rate (CFAR). The general idea is to form an estimate of the background noise in each range cell and then to divide the envelope amplitude in each cell by its own noise estimate. This normalized envelope will fluctuate, even in the absence of an echo, because the noise estimate for each cell will be only an estimate, not its true value. The detection threshold is set, often empirically, above the typical values achieved by the fluctuating normalized cell amplitudes, to achieve a specified false alarm rate. When the range cell contains an echo, its normalized envelope will cross the threshold if the SNR is large enough. The challenge is to avoid overestimating the noise, as will occur if false target or target echoes are present in the cells from which the noise estimates are made, for then the cell with the signal will be "normalized" too much and the detection will be missed. Equally important is to not underestimate the noise, for then the normalized outputs of cells containing only noise are increased, causing too many false alarms.

Cell-Averaging CFAR In a cell-averaging CFAR system the noise estimate is made from adjacent range cells, Doppler cells, azimuthal cells, or any combination of the three. The "split-window" normalizer uses two contiguous groups of $M/2$ cells on each side of the cell

under test, separated from the cell under test by "guard bands" that are a few cells wide. The guard bands are to ensure that the echo from a spread target does not spill over into the cells from which the noise estimate is made. An analysis of the cell averaging is given by Levanon[5] for the case of a single pulse whose echo fluctuates with a Rayleigh pdf. The cell under test has been normalized by the average of M cells, half on each side of the cell under test.

$$\text{SNR} = \frac{(P_D/P_{FA})^{1/M} - 1}{1 - P_D^{1/M}}. \tag{42}$$

In the limit as M approaches infinity, the noise estimate will be perfect, and the SNR′ necessary for detection is given by

$$\text{SNR}' = \frac{\log(P_{FA}/P_D)}{\log P_D}. \tag{43}$$

The normalization loss associated with the CFAR process is defined as the increase in SNR required for CFAR detection as compared to the SNR′ needed for unnormalized detection:

$$\text{Loss}_{\text{norm}} = \frac{\text{SNR}}{\text{SNR}'}. \tag{44}$$

As a typical example, a system may be set to operate at a P_D of 0.5 and at a P_{FA} of 10^{-4}, with $M = 50$ noise estimation cells. For this example, the SNR for normalized operation is 13.5, and the SNR′ for unnormalized operation is 12.3. The normalization loss is 1.1, or 0.4 dB.

There are two situations that defeat the proper operation of this simple cell-averaging CFAR. When a range-wise region of clutter spans part but not all of the noise estimation cells, then the noise estimate can be too large or too small, depending on whether the cell under test is in the clear. Also, when one or more strong echoes lie in some of the noise estimation cells, then the noise estimate will be too large. The result can cause either too many false alarms or missed detections. The common way to mitigate these effects is to eliminate from the noise estimate all cells that have unusual amplitudes.

The method of censoring suspected target or false target echoes in the averaging cells is called multiple-pass outlier removal. In this method, all cells are used to form the first-pass estimate of the average noise amplitude. All cells are censored that have amplitudes that exceed the average estimate by a factor proportional to the standard deviation of the cell amplitudes. The remaining cells are used to form the second-pass estimate of the average noise amplitude, and again any cells used in the average are censored as before. The average formed from the remaining cells is then used to normalize the cell under test.

If the cell amplitudes were statistically homogeneous, then the problems leading to cell censoring would not occur. Whenever the cell amplitudes are so inhomogeneous that the censoring technique does not work well, then an alternative method called clutter map CFAR is available.

Clutter Map CFAR Clutter map CFAR uses averages of past values from the cell under test to normalize its present value. In order to work, clutter map CFAR requires that the cell amplitudes be independent from pulse train to pulse train. This requirement is often met because of platform motion and propagation fluctuations induced by ocean motion. Clutter map CFAR is memory intensive, for it requires that 50 or more past scans be stored for every range, Doppler, and azimuth cell. A common way to reduce the memory requirement is to form a moving average by adding a portion of each new cell amplitude to the current average value. This method is called exponential cell smoothing. If a_n is the nth value of the cell amplitude and z_n is the nth cell average estimate, then

$$z_n = (1 - W)z_{n-1} + Wa_n \qquad (45)$$

The SNR relationship with P_D and W is plotted for $P_{FA} = 10^{-6}$ in Fig. 7. The $W = 0$ case represents no CFAR thresholding in a stationary noise field. The increased SNR required in a nonstationary field, dealt with by exponential cell smoothing, is the CFAR loss. When $W = 0.5$, the loss is 7 dB at $P_D = 0.5$.

3.5 Signal Processing by Passive Sonar Systems

The signals to be detected by passive systems are determined by the targets and are classified into two different types. The radiated-noise source PSDs of ships are described in Chapter 46. The first classification of signals to be detected is narrow-band spectral components, the so-called line components. The second classification of signals is the continuous-spectra component, the so-called broadband component. Different signal processing chains are needed for each type of signal.

Narrow-Band Passive Detection Processing The matched filter for a narrow-band signal is simply a filter with a passband that has the same bandwidth B as the signal. The SNR at the matched filter output is equal to the signal power S divided by the mean noise power N. It is convenient to write SNR in terms of the noise PSD:

$$\text{SNR} = \frac{S}{N_0 B}. \qquad (46)$$

The outputs of the matched filters (a bank of narrow-band filters) are envelope detected in the same way as active signals. The postdetection smoothing is likewise similar to active signal processing, in that the signal is averaged for a time T that is equal to the time span of the signal. In the passive case, T might seem to extend for as long as the target is in view, but this is not the case. Signal fading of the kind discussed for active signals is impressed by motion through the medium, and T is limited to the characteristic time of fluctuations in propagation loss. The postdetection smoothing reduces the noise variance by a factor that is a function of the smoothing time–bandwidth product BT. This factor is plotted in Fig.

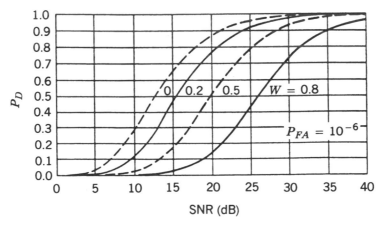

Fig. 7 Plot of P_D versus SNR for clutter map CFAR with exponential averaging weight W (Rayleigh fluctuating target). (From Ref. 5, Fig. 12.6.)

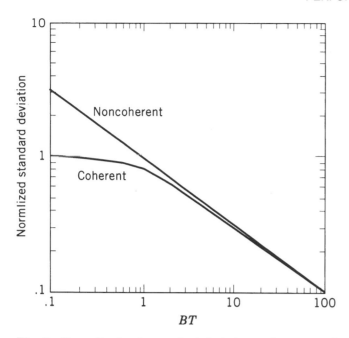

Fig. 8 Normalized noise standard deviation at the output of an envelope detector for white band-limited Gaussian noise versus bandwidth averaging–time product BT.

8, where it is seen to asymptote to $(BT)^{-0.5}$. Assuming that BT is large enough (larger than about 3), the required SNR^* at the square-law detector and smoother output is

$$SNR^* = \left(\frac{S}{N}\right)\sqrt{BT} = \left(\frac{S}{N_0}\right)\sqrt{\frac{T}{B}}. \qquad (47)$$

The narrow-band detection threshold is the signal power to noise PSD ratio, in decibels, necessary to achieve the desired SNR^*:

$$DT_0 = 10\log\left(\frac{S}{N_0}\right) = 10\log(SNR^*) + 5\log\left(\frac{B}{T}\right). \qquad (48)$$

The values of SNR necessary for specified detection and false alarm probabilities are given in Section 3.3. Equation (47) shows that DT is minimized (and therefore performance is maximized) by small bandwidths and large smoothing times as long as B is large enough to completely contain the signal.

Broadband Passive Detection Processing
The matched filter for passive broadband signals is the Eckart filter. The Eckart filter has the spectral shape of the signal, weighted by the noise PSD.[9] Since neither the signal nor the noise spectral shapes are usually known beforehand, the total bandwidth of the broadband filter is sim-

ply kept small, so that both the signal and noise are likely to be white within the filter band. In this case, the SNR^* at the output of the square-law detector and averager is, where again the asymptotic value of noise variance reduction is taken,

$$SNR^* = \left(\frac{S_0}{N_0}\right)^2 BT. \qquad (49)$$

The broadband detection threshold is the signal PSD to noise PSD level required for detection at the specified performance:

$$DT = 10\log\left(\frac{S_0}{N_0}\right) = 5\log\left(\frac{SNR^*}{BT}\right). \qquad (50)$$

Equation (50) shows that performance is maximized by large values of both bandwidth and smoothing time.

Normalization in Passive Detection
Noise interference in passive receivers is the same as it is in active ones, apart from reverberation. The normalization process that removes the effect of changes in time, frequency, and bearing of the mean noise values is described in Section 3.4. In narrow-band passive detection, the usual detection decision is based upon signal power changes over time and frequency. Normalization of the envelope-detected matched filter outputs across the frequency coordinate is called *noise spectral equalization* (NSE), which is very similar to the cell-averaging CFAR described in Section 3.4, but in NSE the noise estimate for the cell under test is made from adjacent frequency cells. Whenever loud signals (outliers) from other targets or from other spectral components from the same target occur in the noise-estimating cells, they are usually censored by a two-pass outlier replacement scheme.

The first noise estimate is made as an average of all cells in the noise-estimating windows. Next, each cell value in the noise-estimating windows is compared to this initial average. Each cell value that exceeds the initial average value by a factor α is replaced by the initial average. A second noise estimate is made from the new values in the noise-estimating window. Each cell value that exceeds the second average by the factor α is replaced by the second average. Finally, the cell under test value is divided by the second noise estimate. If this ratio exceeds the detection ratio, a detection is called.

4 PERFORMANCE PREDICTION FOR ACTIVE SONAR SYSTEMS

The principal tools for performance prediction, called the sonar equations, relate the mean values of the principal sonar system and environmental parameters. The sonar equations are written in terms of detection threshold in this section. In practice, the required value of DT is determined as a function of the SNR needed for specified P_D and P_{FA} values. The sonar equations are then solved for transmission loss (TL). The value of TL thus determined is named the system figure of merit (FOM). The FOM is clearly a function of nonsystem parameters such as TS and N_0, so that FOM must be quoted together with the target and noise environment used. When system performance at ranges less that the FOM range are considered, the difference in TL between the FOM range and the reduced range is called the signal excess.

4.1 Performance of Active Sonar Systems

The fundamental parameter that determines the performance of an active sonar system is its SNR, defined by Eq. (20) as the ratio of signal energy to noise PSD at the processor input. The detection threshold is the required value of this ratio, expressed in decibels, to achieve a specified detection and false alarm probability, as given by Eq. (30). The sonar equation appropriate for noise-dominated interference is derived from the definition of detection threshold. (Another form of the sonar equation, written to describe the ratio of signal power to noise PSD, is given by Urick[7] and Tucker and Gazey[10]):

$$\text{DT} = \text{ESL} - \text{TL}_1 + \text{TS} - \text{TL}_2 + \text{SG} - L - N_0 - \text{NG} \quad (51)$$

where DT is the detection threshold, the signal energy to noise power spectral density ratio, in decibels, needed for the specified detection and false alarm probability; ESL is the energy source level of the transmitted waveform, in dB re $\mu\text{Pa}^2 \cdot \text{s/m}$, directed at the propagation path leading to the target; TL_1 is the transmission loss from the transmitter to the target in decibels; TS is the target strength of the target in decibels; TL_2 is the transmission loss from the target to the receiver in decibels; SG is the signal gain of the receiving array, in decibels, in the beam directed at the propagation path from the target; L is the sum of all system processing losses in decibels; N_0 is the noise power spectral density at the receiving array in dB re $\mu\text{Pa}^2/\text{Hz}$; and NG is the noise gain of the receiving array in decibels. It is common to define the receiving array gain as AG = SG − NG in the noise-limited sonar equation.

A form of the sonar equation valid for distributed reverberation interference, applicable when the reverberation power in a receiving beam dominates the noise power, is

$$\text{DT} = \text{TL}_1' - \text{TL}_1 + \text{TS} - \text{SS} - 10 \log A_s + \text{TL}_2' - \text{TL}_2$$
$$+ \text{SG} - \text{SG}' - L + L', \quad (52)$$

where primed symbols represent the reverberation paths for the same quantities unprimed; SS is the surface or bottom scattering strength in dB re m^2, or the volume scattering strength in dB re m^3; and A_s is the scattering area in square metres, or the scattering volume in cubic metres. The values of signal gain and system loss can be different for the reverberation signal than for the target signal because the two "targets" are not colocated and do not have the same size. The scattering area for bottom- and surface-distributed targets is equal to $\frac{1}{2} r \Psi c t$, where r is the range to the scattering area, Ψ is the effective azimuthal beamwidth of the receiver beam, c is the speed of sound, and t is the effective pulse length after matched filtering. The scattering volume for volume-distributed targets is equal to $\frac{1}{2} r^2 \Psi \Theta c t$, where Θ is the vertical beamwidth of the receiver beam.

The sonar equations relate the mean values of the processes that comprise the signal chain. It is often necessary to estimate the variance of the SNR to determine the range of performance changes to expect. The symbol for the standard deviation of a process is S_a, where a designates the process

$$S_{\text{DT}}^2 = S_{\text{TL}_1}^2 + S_{\text{TS}}^2 - 2CS_{\text{TL}_1}S_{\text{TS}} + S_{\text{TL}_2}^2 + S_{\text{AG}}^2 - S_{N_0}^2. \quad (53)$$

If all of the individual processes in the echo signal chain were statistically independent, then the variance of the detection threshold values would equal the sum of the variances of all of the other processes. If, on the other hand, two of the processes are correlated with each other, then a term containing the correction coefficient C appears. In Eq. (53) the outgoing transmission loss TL_1 is correlated with target strength. This kind of correlation can occur whenever the target is large with respect to the spatial scale of amplitude fluctuations in the transmitted sound field. In this case, the target averages out some of the fluctuations in the sound field. No term is included in Eq. (53) for variations in ESL, because all pulse trains are transmitted alike. An equation for the expected variance in DT values for reverberation-limited operation can be developed from Eq. (52). Correlation can occur between the outgoing transmission loss and the scattering strength term, because the scattering area is usually larger that the area over which variations of transmission loss occur.

4.2 Performance of Passive Sonar Systems

The passive sonar equations yield the expected value of the detection threshold in terms of the mean target radiated-noise source level, the mean transmission loss, the mean array gain, and the mean background noise. Two different detection thresholds are defined for passive sonar. The detection threshold for narrow-band detection,

DT$_0$, is defined in Eq. (48) as the required decibel ratio of the signal power and noise PSD at the matched filter input. The detection threshold for broadband detection, DT, is defined by Eq. (50) as the required decibel ratio of the signal PSD and noise PSD at the matched filter input.

Target-radiated noise source levels, called target signatures, are the subject of Chapter 124. The broadband components of target signatures are characterized by their PSD, referred to 1 m, and denoted by SL in dB re μPa^2/Hz/m. The narrow-band components of target signatures are characterized by their power levels, referred to 1 m, and denoted by SL$_0$ in dB re μPa^2/m.

The broadband passive sonar equation is

$$DT = SL - TL - N_0 + AG - L. \tag{54}$$

The narrow-band passive sonar equation is

$$DT_0 = SL_0 - TL - N_0 + AG - L. \tag{55}$$

These equations relate mean values of the processes to yield the expected value of the detection thresholds. Since none of the processes are correlated with each other, the variances of the detection thresholds are given by the sum of the variances of each process:

$$S_{DT}^2 = S_{TL}^2 + S_{SL}^2 + S_{AG}^2 + S_{N_0}^2. \tag{56}$$

The variances of source levels, SL, are small for submarines but can be large for surface ships. Surface ship source levels vary periodically with the ship's encounter rate with sea waves.

REFERENCES

1. B. D. Steinberg, *Principles of Aperture and Array System Design*, Wiley, New York, 1976.

2. C. L. Bartberger, *Lecture Notes on Underwater Acoustics*, Defense Documentation Center, Alexandria, Virginia, 1965.

3. R. O. Neilsen, *Sonar Signal Processing*, Artech House, Boston, 1985.

4. W. B. Davenport and W. L. Root, *An Introduction to the Theory of Random Signals and Noise*, McGraw-Hill, New York, 1958.

5. N. Levanon, *Radar Principles*, Wiley, New York, 1988.

6. M. I. Skolnik, 2nd ed., *Introduction to Radar Systems*, McGraw-Hill, New York, 1980.

7. R. J. Urick, *Principles of Underwater Sound*, McGraw-Hill, New York, 1975.

8. W. J. Albersheim, "A Closed-Form Approximation to Robertson's Detection Characteristics," *Proc. IEEE*, Vol. 69, 1981.

9. W. S. Burdic, *Underwater Acoustic System Analysis*, Prentice-Hall, Englewood Cliffs, NJ, 1984.

10. D. G. Tucker and B. K. Gazey, *Applied Underwater Acoustics*, Pergamon Press, New York, 1966.

50

OCEANOGRAPHIC AND NAVIGATIONAL INSTRUMENTS

ROBERT C. SPINDEL

1 INTRODUCTION

A wide variety of fundamental ocean measurements are made with underwater sound. The characteristics of sound propagation are affected by the properties of seawater itself, its temperature, salinity, and chemical composition, by objects within the ocean volume, and by interaction with surface and bottom boundaries. Thus, information about these parameters is embedded in propagating sound waves and can be extracted by appropriate techniques. Acoustic devices are used to measure water depth, geologic properties of marine sediments, currents, turbulence, internal waves, mesoscale variability, bubbles created by breaking waves, rainfall, particulate matter in the water column, and fish and plankton density, distribution, and type. Acoustic systems are also used to measure ship speed through the water and over the bottom and to provide high-accuracy positioning and navigation.

2 GEOLOGIC MEASUREMENTS

2.1 Echo Sounders

The most widely used acoustic instrument is the so-called fathometer, echo or depth sounder, which in its most common form measures the round-trip travel time of an acoustic pulse emitted by a transducer at the ocean surface, reflected from the ocean bottom and received back at the same transducer. Travel time is converted to range by multiplying half the round-trip time by the speed of sound c. Most echo sounders are calibrated for nominal sound speeds $c = 1500$ m/s or 800

Encyclopedia of Acoustics, edited by Malcolm J. Crocker
ISBN 0-471-80465-7 © 1997 John Wiley & Sons, Inc.

fathoms/s (4800 ft/s). Accuracies achieved with no correction for departures from these nominal values are within a few percent of water depth. Where greater precision is required, corrections are made based on catalogued historical sound speed data or by measuring local sound speed directly. Carter has provided correction tables based on historical observations of worldwide sound speed profiles[1].

Factors affecting the performance of an echo sounder are its operating frequency, the characteristics of the emitted signal—its amplitude, duration, and modulation—the size of the transmitter and receiver aperture, and signal conditioning and processing algorithms. Depth accuracy is governed by the precision of the travel time measurement, σ_t, which is controlled by signal bandwidth BW and received signal-to-noise ratio SNR:

$$\sigma_t \propto \frac{1}{\text{BW}\sqrt{\text{SNR}}}. \tag{1}$$

Spatial resolution is governed by beamwidth ϕ, which is proportional to wavelength λ and inversely proportional to aperture size. For both a circular piston transducer of diameter d or continuous line array of length d, the full half-power beamwidth is approximately $\phi \approx \lambda/d$. The usual Rayleigh criterion for resolution implies that two features a distance R from the sonar will be resolved if their spatial separation exceeds $R\phi$. Since the earliest returned echo is the assumed bottom depth, wide beams that may span local depressions or elevations can result in erroneous readings.

Depth sounders intended for shallow, coastal waters operate in the 30–100-kHz range. Beams a few degrees wide can be produced with transducers only a few centimetres in diameter. The most common echo sounders for deep water operate near 12 kHz where a typical

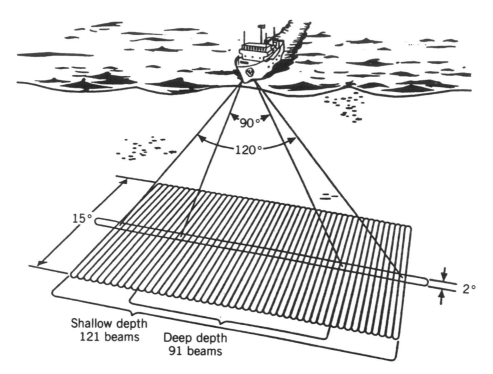

Fig. 1 Beam patterns of a typical multibeam, swath echo sounder. Swath width depends on water depth. Coverage in 3000 and 5000 m depth waters is about 120° and 90°, respectively. (Courtesy SeaBeam Instruments, Inc., East Walpole, MA.)

30-cm-diameter transducer produces an approximate 25° beam that resolves bottom features about 2 km apart in 5-km-deep waters. Advanced echo sounders use phased arrays with hull size apertures to transmit and receive simultaneously on many narrow beams, thereby providing high resolution and broad coverage. A typical multibeam swath echo sounder, as depicted in Fig. 1, has 1.5° beams, providing spatial resolution of about 150 m in 5-km-depth water. From 40 to 150 beams enable the system to sweep out a swath on the bottom up to 20 km wide. The beams are electronically stabilized to compensate for vessel pitch, heave, and roll. Automatic contour plotting algorithms convert measured travel times directly into bathymetric charts.

Precision echo sounders allow pulse duration to be varied from a fraction of a millisecond to tens of milliseconds. Shorter pulses provide better depth resolution but produce less intense echoes and therefore lower SNR. Pulse shapes are varied; the most common is rectangular. Some depth sounders transmit modulated signals, and returning echoes are processed by replica correlation. This results in SNR gain by deemphasizing noise that is uncorrelated with the signal.

2.2 Sub-bottom Profilers

Below about 3.5 kHz, energy penetrates readily to the bottom, where it is reflected from discontinuities and inhomogeneities in the sediments, thereby revealing sub-bottom stratigraphy and structures. Conventional piezo-electric and magnetic transducers are generally used at frequencies above 100 Hz; explosive materials, airguns, and electromechanical devices are used for lower frequencies. (See Chapters 37 and 47.)

2.3 Side-Scan Sonars

The side-scan sonar operates by transmitting a fan-shaped beam perpendicular to its direction of motion.[2] The beam, which is narrow in azimuth (the direction of motion) and wide in elevation, is produced by a linear array of transducers. Figure 2 is an example of the output of a typical side-scan sonar.

Reflections from the seafloor produce a shadowgraph of the swath swept out by the sonar. Some side-scan systems incorporate a second row of transducers parallel to the first to obtain differential phase or differential arrival time measurement of returning signals in order to estimate bottom relief. Such systems produce bathymetry as well as imagery. High-resolution side-scan sonars in the 100–500-kHz range are used for precision surveys for pipeline installation and detailed ocean bottom search. Their range is limited to several hundred metres due to the high acoustic attenuation at these frequencies, but the small acoustic wavelength allows resolution of 10 cm to 1 m with practical aperture sizes. Because of their limited

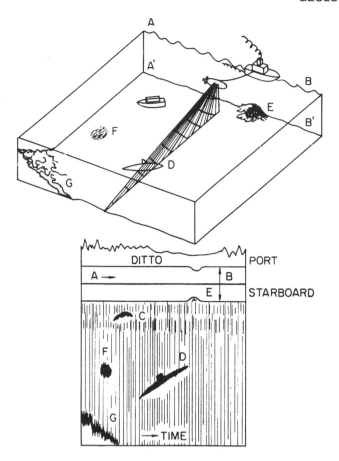

Fig. 2 Top drawings show how the side-scan sonar image is developed from a narrow beam along the direction of travel. (From H. Edgerton, *Sonar Images*, 1986, p. 15. Reprinted by permission of Prentice-Hall, Englewood Cliffs, NJ.) Bottom panel is a side-scan image of the *Breadalbane* on the seafloor in the High Arctic. (Courtesy Klein Associates, Inc., Salem, NH.)

range, in deep water they must be towed near the bottom. Low-frequency side-scan sonars operating at several kilohertz provide kilometer ranges and broad-area coverage but with less resolution because of the very large array needed to produce a narrow beam. They are used primarily for geologic surveying.

2.4 Synthetic Apertures for Echo Sounding

Synthetic aperture sonars, which operate on the same principle as synthetic aperture radars wherein an aperture of length L is synthesized from N subapertures of length $l = L/N$, have been tested but have not been widely used. The attractive feature of such systems is that synthetic apertures many wavelengths long can be formed with a smaller, real aperture system moving along a known (usually straight) path, thereby providing very narrow beams with consequently high spatial resolution mapping. Unfortunately, position of the subapertures must be known accurately, to within $\lambda/2$, to achieve full array gain, and this difficult requirement has limited the implementation of synthetic aperture sonars.

2.5 Seafloor Sediment Properties

Empirical geoacoustic models that relate acoustic properties of marine sediments, such as sound speed, impedance, and attenuation, to physical properties such as density, porosity, mean grain size, compressibility, and chemical composition are used to characterize sediments by acoustic measurements.[3] Signals reflected from the seafloor and sub-bottom are either fitted directly to empirical models to determine sediment type or inverted first to obtain impedance and attenuation coefficients, which are then related to sediment type. Wide-band sonars operating at 2–25 kHz have demonstrated most success.[4]

3 PHYSICAL OCEANOGRAPHY

3.1 Current Meters

Travel Time Instruments The difference in travel time, δt, of acoustic pulses traversing the same path r in opposite directions is proportional to the velocity of the current, u, along the path,

$$\delta t = \frac{2ru}{c^2},\qquad (2)$$

where c is the average speed of sound along the path. Current meters based on this principle are in wide use. Typically, $r \sim 10$ cm, so $\delta t \sim 10^{-9}$ s for $u = 1$ cm/ s.

Two orthogonal paths allow velocity to be resolved into vector components.

This technique also has been used to measure currents along $r \approx 1000$ km paths. (See Section 3.2.)

A related method for observing river and ocean flow depends on the advection by currents of particulate matter or turbulent cells that constitute acoustic impedance discontinuities. Scintillations in the forward scattered acoustic signal transmitted approximately orthogonal to the principal flow direction and received at two spatially separated points approximately parallel to the flow can be correlated, and the rate of advection can be calculated.[5]

Doppler Instruments Doppler shifts of acoustic pulses scattered from the ocean bottom or from scatterers within the water column are used to deduce water currents and ship speeds.[6] Acoustic Doppler current meters rely on scattering from particles or other acoustic impedance discontinuities in the water, such as bubbles or thermal microstructure that are swept along with the current. Instruments with multiple acoustic beams at various angles (usually orthogonal) allow computation of vector velocity. Estimates of the Doppler shift of short segments of the scattered return signal provides range resolution. The Doppler shift of a segment of the return signal t_0 seconds after transmission and τ seconds long gives an estimate of water velocity at range $ct_0/2$ with resolution $c\tau/2$. Typical Doppler current metres operate in the 50 kHz–200 kHz range, measure currents to centimetres per second accuracy, and when employed in the range-gated mode have ranges of several hundred metres and resolution of several metres.

Acoustic ship speed logs are described in Section 5.4.

3.2 Acoustic Tomography: Currents and Temperature

Ocean acoustic tomography is a technique for measuring the three-dimensional sound speed and current fields of an ocean volume. The travel time of acoustic signals transmitted between multiple points on the perimeter of a volume are related to the sound speed and current fields in the interior. Temperature change δT in degrees Celsius is related to sound speed change δc to first order; $\delta T \approx \delta c (m/s)/4.6$. Tomography consists of measuring the travel times and deducing the interior fields.[7]

Figure 3 shows a conceptual ocean tomography network sampling the ocean in both horizontal and vertical dimensions. Spatial resolution in the horizontal plane is determined by the number and placement of instruments. In deep-water applications, multipath propagation in the vertical plane arising as a result of the natural background ocean sound speed profile provides depth reso-

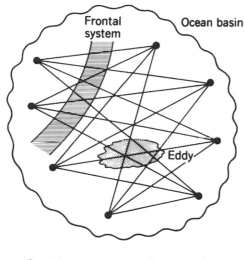

C (m/s)   Range (km)

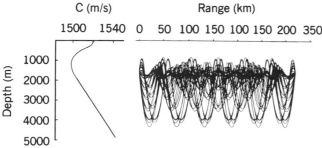

Fig. 3 In an ocean acoustic tomography system transceivers send and receive acoustic pulses along many paths, as shown in the upper panel plan view. Ocean features having different sound speed or current characteristics, such as eddies and frontal systems, alter the travel times of the pulses. The measured travel time changes are used to infer the sound speed and current fields interior to the tomographic array. The lower panel shows a typical natural background sound speed profile in the deep ocean (at left) and the multiple acoustic paths that exist between transceivers moored at 1200 m depth, 300 km apart. Acoustic energy directed downward is refracted upward, and vice versa, resulting in channeled, multipath propagation.

lution. Thus, a single instrument, usually at a depth near the axis of the deep sound channel, is sufficient at each measurement point. (See Chapter 36 for discussion of the deep sound channel.)

In the limit of geometric optics, the travel time along a path, i, in the presence of a current, $\overline{u}(\overline{x}, t)$, is given as

$$T_i(t) = \int_i \frac{ds}{c(\overline{x}, t) + \overline{u}(\overline{x}, t)}. \tag{3}$$

With a reference sound speed field $c_0(x)$ and a perturbation field $\delta c(x, t) \ll c_0(x)$, defined as

$$c(\overline{x}, t) = c_0(\overline{x}) + \delta c(\overline{x}, t) \tag{4}$$

and a reference travel time defined as

$$T_{0i} = \int_{0i} \frac{ds}{c_0(\overline{x})}, \tag{5}$$

the changes in travel time over the reference travel time, in opposite directions along the same path, due to the current and sound speed perturbation are

$$\delta T_i^+ = T_i^+ - T_{0i} = - \int_{0i} \frac{\delta c(\overline{x}, t) + \overline{u}(\overline{x}, t)}{c_0^2(\overline{x})} ds \tag{6}$$

and

$$\delta T_i^- = T_i^- - T_{0i} = - \int_{0i} \frac{\delta c(\overline{x}, t) - \overline{u}(\overline{x}, t)}{c_0^2(\overline{x})} ds, \tag{7}$$

respectively. Sums and differences of these two equations give

$$s_i = -2 \int_{0i} \frac{\delta c(\overline{x}, t)}{c_0^2(\overline{x})} ds \tag{8}$$

and

$$d_i = -2 \int_{0i} \frac{\overline{u}(\overline{x}, t)}{c_0^2(\overline{x})} ds. \tag{9}$$

Thus, the sum travel times are related to the sound speed perturbation; the difference travel times are related to the water velocity. The tomographic inverse problem determines the fields δc and u from measurement of the sum and difference travel times. The inversion is accomplished using standard linear inverse techniques in which the fields are parameterized with a finite number of discrete parameters derived from a model of the ocean.[8]

Tomography transceivers operating in the 200–500 Hz range have been used to measure ocean features over 1000-km basins.[9]

3.3 Inverted Echo Sounders

The round-trip travel times of acoustic pulses emitted by upward-looking echo sounders placed on moorings or on the ocean bottom are used to deduce changes in the depth of the main ocean thermocline, the region of rapidly decreasing temperature that separates warm surface waters from deep cold waters. The measurement is based on the fact that the integrated vertical sound speed is a function of vertical thermocline migration, increasing with decreasing thermocline depth.

Upward-looking sonars mounted on submarines and operating at high frequencies are used to monitor ice draft for submarine navigational purposes. Similar sonars operating near 300 kHz mounted on the seafloor or moorings are used to obtain time series of average ice thickness.

3.4 Wind and Rain

The impact of raindrops on the sea surface generates underwater ambient noise signatures in the frequency band 1–40 kHz depending on drop size, rainfall rate, and spatial density. Wind speed affects wave production and breaking, which in turn changes the background ambient noise.[10,11]. Thus, rainfall and wind speed can be deducted from ambient noise measurements (see Chapter 48).

3.5 Scatterometers

These are a class of high-frequency sonars used to measure a variety of phenomena based on backscattering from impedance discontinuities in the water column. They are used to study the intensity and distribution of material such as suspended particulate matter (pollutants, sediments), marine organisms, and bubbles.[12,13] These sonars operate in the range of several hundred kilohertz to several megahertz depending on the relative backscattering strength of the scatterers of interest. For a small nonresonant sphere of radius a, the backscattering cross section is

$$\frac{\sigma_s}{\pi a^2} = 4(ka)^4 \left[\left(\frac{e-1}{3e} \right)^2 + \frac{1}{3} \left(\frac{g^{-1}}{2g+1} \right)^2 \right],$$

$$ka \ll 1, \tag{10}$$

where e is the ratio of the elasticity of the sphere to water and g is the ratio of the density of the sphere to water. If the sphere is rigid, such as in the case of solid particulates, e, $g \gg 1$. Then, in the region $ka \ll 1$, known as the Rayleigh scattering region, scattering strength is proportional to f^4. Near $ka = 1$ there is a transition to a region of geometric scattering, where scattering strength no longer increases with frequency. If the scatterer is a gas bubble, such as in the case of fish swimbladders (see Chapter 44), or wave-generated bubbles, then e, $g \ll 1$, and scattering is large because the bracketed term dominates. At its resonant frequency f_r, where $a \approx \delta$, the scattering cross section of a bubble is larger still,

$$\frac{\sigma_s}{\pi a^2} = 4 \left(\frac{f_r}{\Delta f} \right)^2 = 4Q^2, \tag{11}$$

where Δf is the width of the resonance peak and Q varies from 10 to 100. Scatterer type and size are inferred from measurements of scattering strength and resonant frequency.[14]

4 BIOLOGICAL OCEANOGRAPHY

Sonars operating in the range of several kilohertz to 100 kHz are used routinely for fish finding and in some cases for estimating stock abundance (see Chapter 44). The fish swimbladder behaves roughly as a gas bubble. Bladder sizes, and therefore fish type and size, are differentiated by resonant frequency. At very high frequencies, several hundred kilohertz to several megahertz, scatterometers are used to study smaller species such as plankton, ctenophore. and some jellyfish.[13] (See Section 3.5.)

5 ACOUSTIC POSITIONING AND NAVIGATION

Acoustic systems are used for precise positioning and navigation of surface vessels and for underwater instruments and vehicles. They are used in oceanographic studies; geophysical prospecting; underwater exploration and surveying; subsea oil and gas production; vessel dynamic positioning; and test and performance evaluation of military systems such as sonars, torpedoes, and missile range and accuracy. These systems operate by measuring the time of arrival, phase, or Doppler shift of acoustic signals transmitted between a reference frame and the navigated point. In the case of missile tests, the origin of the underwater signature created by the impact is located by triangulation.

Acoustic positioning systems fall into three broad classes, long, short, and ultra-short baseline, distinguished by the separation between the reference elements. In a long-baseline system the reference elements are spaced about the same distance as the ranges from them to the navigated point, as shown in Fig. 4. Long baselines are commonly used when the reference elements are placed on the seafloor. In short-baseline systems the reference element spacing is much less than the ranges to the object. These are used in ship-mounted applications, where a long baseline, which gives greater precision, is not possible. Both the long- and short-baseline systems generally obtain ranges by measuring the transit time of acoustic pulses. The third class of system, the ultra-short-baseline system, has the reference element spacing $\sim\lambda$. Azimuth and depression angles to the navigated object are obtained by measuring phase differences of a pulse emitted by the object and received at the reference elements (usually three). Ultra-short-baseline systems are compact but not as accurate as those with longer baselines.

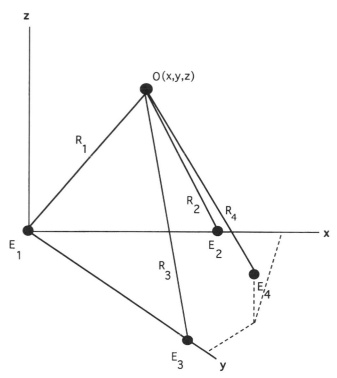

Fig. 4 Basic acoustic positioning system geometry. Four non-coplanar reference elements E_i locate the object $O(x, y, z)$ unambiguously. In most applications, three elements suffice, since it is known a priori where object lies with respect to reference element plane. Redundant ranges obtained from additional reference points beyond three or four can be used to reduce random errors due to system noise or fluctuations in acoustic travel times due to ocean inhomogeneities. Illustrated is a long-baseline system, where the ranges between the reference elements are roughly equal to the ranges to the navigated object. In a short-baseline system, distances between reference elements are much less than to the navigated object.

5.1 Travel Time Navigation

In a travel time system, the position of a point, $O(x, y, z)$, is computed from measurement of the ranges R_i obtained by determining the travel time of acoustic pulses transmitted between the reference elements and the point. Four non-coplanar reference elements locate $O(x, y, z)$ unambiguously; three locate it either above or below the plane defined by the three elements; and two locate it on a circle. All three implementations are used. When fewer than four elements are employed, the ambiguity is resolved by operational constraints. For example, if it is known that $O(x, y, z)$ is on the sea surface, three elements suffice; if it is known that $O(x, y, z)$ is both on the sea surface and on one side of the baseline defined by two reference elements, then two are sufficient.

To first order, the acoustic paths are assumed to be straight lines and the ranges are based on a nominal sound speed c or a computed average sound speed based on measurement of the local vertical sound speed profile $c(z)$. Thus, $R_i = ct_i$, where t_i are the measured travel times. More precise ranges are obtained by accounting for the actual path of the acoustic signal, which can depart significantly from a straight line under certain circumstances, for example, in deep water when the angle between the reference element plane and the range vector is small ($< 50°$).[15]

In one form of acoustic range employing travel time measurements, clocks in the reference elements and navigated point are synchronized. Signals are usually transmitted from $O(x, y, z)$ and received at the reference elements, but transmission in the opposite direction is also employed. In the latter case, signals from different reference elements are distinguished by modulation coding or frequency. A common variation replaces the reference elements with transponders, devices that respond to the reception of a signal by transmitting one of their own, thereby eliminating the need for synchronized clocks. The round-trip travel times are measured, and $R_i = ct_i/2$ is computed. Accuracy of pulse systems can be as high as a few centimetres, depending on the operating frequency and range. For ranges of kilometre order, frequencies in the 5–20-kHz range are used, and accuracies of less than a metre are easily obtained. For shorter range systems, greater acoustic attenuation can be tolerated, thus allowing the use of higher operating frequencies. These yield higher accuracies because shorter pulses with steeper rise times can be transmitted. Short-range systems operating in the several hundred kilohertz range give centimetre accuracy.

In short-baseline systems transmission is usually from $O(x, y, z)$ to a cluster of reference receivers. The reference element spacing is small compared to the range to $O(x, y, z)$, so small errors in travel time measurement result in relatively larger errors in position. Thus, overall accuracy is not as great as with a long baseline, but these systems have the advantage of being compact, and the reference elements are easily mounted on the hull of a vessel.

The ultra-short-baseline systems that measure signal phase rather than arrival time yield angles to the object; range is computed by triangulation. This is usually the least precise of the commonly used systems, but it is very compact and provides adequate positioning in many applications, particularly when short ranges are involved. To avoid ambiguities, receiving elements must be spaced less than $\lambda/2$. Some systems combine phase and travel time measurements, thus measuring both angle and range.[16]

5.2 Doppler Navigation

Acoustic navigation and positioning are also accomplished by measuring Doppler shifts induced by motion.

Doppler shift provides a direct measurement of velocity, and it also can be integrated to yield spatial translation. Either the reference elements, which are fixed in space, or the navigated point emit constant frequencies. In the former case, each reference element emits a different frequency f_i, and the Doppler-shifted receptions at the navigated object, f_{d_i}, are measured. In the latter case the navigated object emits a constant frequency, and the Doppler-shifted signals received by the fixed reference elements, f_{d_i}, are measured. Velocities along the paths from reference element to navigated point are

$$V_i = c \left(1 - \frac{f_{d_i}}{f_i} \right). \tag{12}$$

This technique is less common than transit time navigation, primarily because Doppler systems provide relative, rather than absolute, position. Also, because translation must be determined by integration, signal fades disrupt continuous tracking. However, under certain circumstances more precise relative positioning can be obtained than with a transit time system. Thus, the two systems are frequently combined.[17]

5.3 Reference Element Calibration

The position of the reference elements must be known. For short- and ultra-short-baseline systems element positions can often be determined easily at the time of manufacture or installation by direct measurement. For long-baseline systems, determining the position of the reference elements is more difficult, especially if the elements are placed on the inaccessible seafloor. In this case two methods are employed. In the first, a ship acoustically measures ranges to the bottom units at a sufficient number of points on the ocean surface to solve the set of resulting simultaneous equations that relate element positions, ship positions, and measured ranges. Six independent ship positions and ranges are sufficient to determine the relative position of three elements on the seafloor. The second method is self-calibrating. The reference elements transmit pulses between each other, thereby measuring their separations directly. This method is not always possible, especially in deep waters having negative sound speed gradients with consequent strong upward acoustic refraction that limits obtainable horizontal ranges. Optimized techniques for surveying reference element positions have been devised.[18]

5.4 Acoustic Speed Logs

Speed logs measure ship speed through the water. Two types are in general use. One, a high-frequency sonar in the 10–100-kHz range mounted on the hull of the vessel, relies on measuring the Doppler shift of signals scattered from particulate or biologic matter in the water. It is assumed that water motion, and hence scatterer motion, can be neglected, in comparison with ship speed. This assumption is not always valid, especially in high current regions such as the Gulf Stream, rivers, or tidal inlets where water currents can be as high as 2–4 m/s. Doppler-shifted returns of signals scattered from the ocean bottom are also used to estimate ship speed. Orthogonal beams allow forward and thwartship velocities to be computed.

The second type of log is the correlation log. An acoustic signal is projected vertically downward, and scattered signals from the bottom or from scatterers in the water column are received by transducers spaced several centimetres from the transmitter. The separation in time of the transmitted and received pulses, obtained by correlation, is a measure of ship speed. If the transmitter and received hydrophones are oriented along the bow to stern line, forward speed is measured; if oriented across the beam, thwartship speed is obtained. High frequencies (150–200 kHz) are used in shallow-water systems (300 m), and lower frequencies (10–25 kHz) are used in deep-bottom-tracking systems.[19]

REFERENCES

1. D. J. T. Carter, *Echo-Sounding Correction Tables*, 3rd ed., NP139, The Hydrographic Dept., Ministry of Defense, Taunton, England, 1980.

2. H. E. Edgerton, *Sonar Images*, Prentice-Hall, Englewood Cliffe, NJ, 1986.

3. R. D. Stoll, *Sediment Acoustics*, Springer Verlag, New York, 1989.

4. S. Panda, L. R. Leblanc, and S. G. Schock, "Sediment Classification Based on Impedance and Attenuation Estimation," *J. Acoust. Soc. Am.*, Vol. 95, No. 5, Pt. 1, 1994, pp. 3022–3035.

5. D. M. Farmer and G. B. Crawford, "Remote Sensing of Ocean Flows by Spatial Filtering of Acoustic Scintillations: Observations," *J. Acoust. Soc. Am.*, Vol. 90, No. 3, 1991, pp. 1582–1591.

6. G. F. Appell, T. N. Mero, R. Williams, and W. E. Woodward, "Remote Acoustic Doppler Sensing: Its Application to Environmental Measurements," in T. McGuinness and H. H. Shih (Eds.), *Current Practices and New Technology in Ocean Engineering*, Amer. Soc. Mech. Eng., New York, 1986.

7. W. Munk and C. Wunsch, "Ocean Acoustic Tomography: A Scheme for Large Scale Monitoring," *Deep-Sea Res.*, Vol. 26A, 1979, pp. 123–161.

8. B. Cornuelle, C. Wunsch, D. Behringer, T. Birdsall, M. Brown, R. Heinmiller, R. Knox, K. Metzger, W. Munk, J. Spiesberger, R. Spindel, D. Webb, and P. Worces-

ter, "Tomographic maps of the ocean mesoscale. Part 1: Pure Acoustics," *J. Phys. Oceanogr.*, Vol. 15, 1985, pp. 133–152, WHOI #5961.

9. P. F. Worcester, B. D. Cornuelle, and R. C. Spindel, "A Review of Ocean Acoustic Tomography: 1987–1990", in *U.S. National Report to the International Union of Geodosy and Geophysics (IUGG) 1987–1990, Contributions in Oceanography*, American Geophysical Union, 1991, pp. 557–570.

10. J. A. Nystuen, "Rainfall Measurements Using Underwater Ambient Noise," *J. Acoust. Soc. Am.*, Vol. 79, 1986, pp. 972–982.

11. J. A. Nystuen, C. C. McGlothin, and M. S. Cook, "The Underwater Sound Generated by Heavy Precipitation," *J. Acoust. Soc. Am.*, Vol. 93, 1993, pp. 3169–3177.

12. M. H. Orr and L. Baxter, "Dispersion of Particles after Disposal of Industrial and Sewage Wastes," in I. W. Duedall, B. H. Ketchum, P. K. Park, and O. R. Kester (Eds.), *Wastes in the Ocean*, Wiley, New York, 1983.

13. P. H. Wiebe, C. H. Greene, and T. K. Stanton, "Sound Scattering by Live Zooplankton and Micronekton: Empiri-cal Studies with a Dual Beam Acoustical System," *J. Acoust. Soc. Am.*, Vol. 88, No. 5, 1990, pp. 2346–2360.

14. C. S. Clay and H. Medwin, *Acoustical Oceanography: Principals and Applications*, Wiley, New York, 1977.

15. M. M. Hunt, W. M. Marquet, D. A. Moller, K. R. Peal, W. K. Smith, and R. C. Spindel, "An Acoustic Navigation System," Rep. 74-6, Woods Hole Oceanographic Institution, Woods Hole, MA, 1974.

16. M. J. Morgan, *Dynamic Positioning of Offshore Vessels*, PPC Books, Petroleum Publishing, Tulsa, OK, 1978.

17. R. C. Spindel, R. P. Porter, W. M. Marquet, and J. L. Durham, "A High-Resolution Pulse-Doppler Underwater Acoustic Navigation System," *IEEE J. Oceanic Eng.*, Vol. 1, No. 1, 1976, pp. 6–13.

18. A. G. Mourad, D. M. Fubara, A. T. Hopper, and G. Y. Ruck, "Geodetic Location of Acoustic Ocean-bottom Transponders from Surface Positions," *EOS, Trans. Amer. Geophysics Union*, Vol. 53, 1972, pp. 644–649.

19. B. Woodward, W. Forsythe, and S. K. Hole, "Estimating Backscattering Strength for a Correlation Log," *IEEE J. Oceanic Eng.*, Vol. 19, No. 3, 1994, pp. 476–483.

51

ACOUSTIC TELEMETRY

Josko A. Catipovic

1 INTRODUCTION

The underwater acoustic channel permits data telemetry at modest rates. It has been exploited for simple voice communication as well as sophisticated digital telemetry of television images and command signals for control of subsea wellheads, underwater vehicles, and other instrumentation. Its capacity is limited primarily by bandwidth constraints and acoustic ambient noise and also by unique characteristics of the undersea channel, such as ducted, multipath propagation in the deep ocean, surface and bottom reflection and scattering, especially in shallow water, and random variability arising from time-varying surface changes and changes in the transmission medium itself. The channel transfer function is therefore stochastic at certain scales and is spread in both time and frequency. In this chapter we discuss the limitations imposed by these channel constraints and describe the implementation and performance of typical underwater telemetry systems.

2 CHANNEL CAPACITY

Transmission capacity constrained by the Shannon limit states that error-free data transmission is possible if and only if the data rate is lower than some maximum data rate, termed the channel capacity C[1]:

$$C = W \log_2 \left(1 + \frac{P}{WN_0} \right), \tag{1}$$

Encyclopedia of Acoustics, edited by Malcolm J. Crocker
ISBN 0-471-80465-7 © 1997 John Wiley & Sons, Inc.

where P is the average received signal power, N_0 is the ambient noise level, and W is the bandwidth available for information transmission. Defining the signal-to-noise ratio (SNR) as

$$\text{SNR} = 10 \log_{10} \left(\frac{P}{WN_0} \right), \tag{2}$$

we note that for a SNR of 15 dB, a typical objective for relatively error free data transmission is $C/W \sim 5$ bits/s/Hz; that is, the maximum data rate is roughly 5 times the available bandwidth. A further consequence of Eq. (1) is the inverse relationship between received signal power and available bandwidth. As more bandwidth is available for transmission of each data bit, the power required to transmit that bit is reduced. Most systems operate either in the bandwidth-limited or energy-limited regimes, bandwidth-limited systems attempt to maximize the number of transmitted bits per second per hertz of available bandwidth and are able to operate at high SNR. An excellent example is telephone modems, some of which operate at 28,800 bps over the 2400-Hz telephone channels. Energy-limited systems are constrained by the amount of energy available for the transmission of a single data bit. An example of energy-constrained acoustic systems is deep ocean tomography transmitters, which typically send each waveform over many seconds and many hertz of bandwidth in an attempt to maximize energy efficiency. These energy-limited ocean basin transmitters attempt to approach the "infinite-bandwidth" channel capacity (in bits per second):

$$C = \frac{P}{N_0 \ln 2}. \tag{3}$$

The system designer is constrained by both power and bandwidth. Typically, for short deployments or applications where ample power is available, the telemetry system attempts to maximize bandwidth efficiency. Conversely, for long, energy-limited deployments, the number of transmitted data bits per joule of battery power typically dominates, and bandwidth efficiency is sacrificed.

2.1 Bandwidth and Ambient Noise Constraints

Acoustic channel bandwidth is limited by sound absorption which roughly varies quadratically with frequency. The reader is referred to Ref. 2 and Chapter 109 for an in-depth discussion of acoustic attenuation. A useful rule of thumb is to select the frequency resulting in a 10-dB attenuation loss at a desired transmission range. For example, this results in an upper frequency limit of ~50 kHz at 1 km, 12 kHz at 10 km, and 1.5 kHz at 100 km.

Although the entire band between 0 Hz and the attenuation limit is available for use, the limited bandwidth of practical transducers often restricts systems to a single octave. Most existing acoustic telemetry systems operate within the octave of bandwidth upper limited by the attenuation limit in order to maximize available bandwidth and because ocean ambient noise levels usually decrease with frequency.

Ambient noise levels are readily available for a variety of underwater environments.[2] Published noise levels are generally adequate for predicting ambient conditions at the receiver. However, these generally refer to average conditions, whereas a robust system is primarily affected by extremal or infrequent high-noise events. Near man-made structures such as offshore drilling sites, peak noise levels are particularly bothersome. In many cases, the receiver platform noise dominates SNR considerations, particularly if the receiver is on a ship or is being moved through the water. Decreasing receiver acoustic self-noise is possibly the most important operational aspect of undersea telemetry.

2.2 Multipath and Fluctuation Limits

The relatively slow sound speed in the ocean gives rise to a very long multipath when compared to other channels characterized by speed-of-light signal propagation. The SOFAR waveguide and many coastal and harbor environments have characteristic reverberation times from tens of milliseconds to several seconds, and time-variant long-delay multipath must be recognized by the system designer as a basic channel characteristic present in all but a few propagation geometries.

The SOFAR channel at ranges over 1 convergence zone yields a number of distinct arrivals. Typical multipath duration is ~1 s, with individual path root-mean-square (rms) fluctuation of approximately 10 ms. The fluctuation statistics reflect the underlying time-variable environmental process, such as the surface elevation spectrum or internal wave spectrum. For example, for a surface-reflecting ray, the rms travel time fluctuation is given by:

$$\langle (\Delta t)^2 \rangle \sim \frac{2\sigma_s \sqrt{M}}{c} \sin \theta, \tag{4}$$

where σ_s is the rms surface elevation, M is the number of surface reflections, and θ is the angle of incidence. The fluctuation spectrum is directly related to the surface elevation spectrum.[3] Surface-induced fluctuation is typically the most significant contribution to received signal dynamics in the deep ocean.

In shallow water, reflections from objects and channel boundaries dominate the multipath; the problem becomes geometry specific, and no generic solutions are available. It is important to note that the multipath itself is not a detrimental phenomenon. Rather, single-path temporal fluctuations and multipath time stability are primarily performance problems.

3 ACOUSTIC CHANNELS

The underwater acoustic channels are readily divided into four categories for purposes of data telemetry. Each category spans a range of propagation geometries with similar channel-induced performance limits. Each category has seen the development of distinct types of telemetry systems, ranging from simple "grafts" of telephone modems to elaborate joint channel–data estimators required for operation with highly dynamic multipaths.[4]

1. The simplest case is the near-vertical deep-water channel encountered, for example, when communicating between the ocean surface and ocean bottom instrumentation in the deep ocean. Transmission range is typically 3–10 km. Over this channel, often called the reliable acoustic path (RAP), the multipath is limited to discrete surface and bottom bounces and can be easily eliminated by appropriate transducer directionality or baffling. The direct path undergoes minimal fluctuation, and the received signal is well modeled as a combination of the transmitted signal and additive noise. Data rate is limited primarily by the available bandwidth and the ambient noise level. A number of communication systems developed for the radio frequency (RF) or telephone channel have been demonstrated successfully in deep water. The

15–30-kHz frequency band is typically utilized, yielding data rates of 10–50 kbits/s. Several systems for deepwater telemetry are currently commercially available.[3]

2. The very shallow water 2–10-km-range channel exhibits a complex and dynamic multipath structure, particularly in enclosed bodies such as harbors and bays. Significant acoustic interaction with the surface, midwater microstructure, and bottom produces a rapidly variant, extended multipath that forms the principal limitation to data telemetry in this environment. The upper frequency limit ranges from 10 to 100 kHz. To date, equalization techniques have been unable to track the multipath dynamics in this channel, although this is an area of active research. This has greatly compromised available data rates, and most current systems operate below 10 kbits/s, except at very short (<500-m) ranges.[3]

3. The continental shelf is amenable to telemetry at ranges up to ~100 km. Primary performance limitation in this case is the high spatial variability of the received acoustic pressure field. The spatial and frequency dependence of transmission loss makes it difficult to maintain a continuous telemetry channel, although data can be transmitted robustly for at least several hours per day. This channel also introduces complex bottom-interacting multipath, and multipath duration can be a performance constraint for higher data rates. Recent work in equalization of the continental shelf channel has resulted in 1–2 kbits/s data rate over ~80-km ranges.[4]

4. The long-range deep-water SOFAR channel supports telemetry at several convergence zones at data rates up to ~1 kbits/s. While significant multipath is present, it is relatively stable and hence can be equalized.[5] The primary performance limitation in this case is the ambient noise at the receiver; that is, this channel is essentially power limited, particularly since deep-ocean systems are difficult to deploy and long deployment times are required. Reducing noise levels at the receiver and placing transmitters and receivers deep in the SOFAR channel are key operational steps to long-range data telemetry. A prime example of ocean-basing scale telemetry systems are acoustic tomography transceivers, described in Chapter 124.

4 DATA TRANSMISSION SYSTEMS

This section describes acoustic telemetry systems in order of increased data rate and reliability. Possibly the first undersea communication system was the UQC-1 underwater telephone, which used analog modulation to transmit human voice. Soon after, the first Frequency Shift Keyed (FSK) systems were developed for simple control of undersea instrumentation.[6]

The FSK modulation consists of transmitting one of two possible frequency tones to signal the state of a single bit. The transmitter transmits at a frequency f_0 or f_1 for a duration Δt. Typically $\Delta t = 1/|f_1 - f_0|$. Note this technique requires a time–bandwidth product of 2 to transmit a single bit; that is, the required bandwidth in hertz is twice the data rate in bits/per second.

The time–frequency parameters are selected primarily with regard to multipath duration L. If $\Delta t \gg L$, the effect of channel multipath is negligible. However, the system may become Doppler sensitive if the tone frequency separation $1/\Delta t$ becomes smaller than a channel Doppler shift B. In practice, if $\Delta t < 10$ ms, the system becomes unacceptably sensitive to multipath in all but the vertical acoustic channel. Thus 100 bits/s has been a practical limitation to FSK telemetry.

Several techniques are used to increase FSK data rate. Frequency hopping the tones allows for a longer channel clearing time following each tone transmission, but at the expense of increased bandwidth per data bits/per second. While this method indeed increases data rate, it has been largely supplanted in favor of Multiple FSK, or MFSK, which implements a number of FSK systems in parallel. Its principal advantage is the ability to tailor the time–frequency cell to channel multipath and Doppler characteristics without compromising data rate, since the number of tones is adjustable to occupy the desired bandwidth. Typical tone spacings range from ~1 Hz for the long-range deep-water channel to ~80 Hz for the shallow-water channels where underwater vehicle motion and currents cause significant Doppler shifts.

The MFSK system found wide use in digital acoustic telemetry because it is well matched to typically encountered multipath durations and Doppler spreads while being insensitive to details of multipath fluctuation. A number of MFSK systems were constructed, and the method is widely used for underwater data transmission.[3]

4.1 MFSK Receivers

The MFSK receivers are essentially banks of narrowband filters that integrate and dump the received waveform.[7] The demodulation is typically performed with a fast Fourier transform (FFT) and the data decoding with a digital signal processing (DSP) chip. The FFT resolution and duration are matched to the tone frequency spacing and Δt. An external synchronization system is used to section data prior to FFT demodulation. Doppler corrections are typically performed by adjusting the digitizing frequency to match Doppler drift.[3]

For each FSK tone pair at the output of the demodulator FFT, the cell with larger received energy is declared as the received bit. In many underwater acoustic links, the received energy is χ^2 (chi-square) distributed; that is, the channel behaves as the Rayleigh fading channel. The received bit error probability for this case is[7]

$$P_{err} = \frac{1}{2 + SNR},\qquad(5)$$

where SNR is the received signal-to-noise ratio per bit. This results in generally mediocre performance. Error correction coding is used to improve bit error performance at the cost of increased bandwidth and receiver complexity. Error correction coding for fading channel is outside the scope of this discussion, but an excellent source is Ref. 7. At the cost of increasing bandwidth utilization to 4 Hz/bit/s, and a modest increase in receiver complexity, bit SNR required for acceptable system performance is typically reduced to 15–20 dB. Additional techniques such as interleaving, spatial diversity, automatic repeat request, and code combining offer additional robustness in extremal channel conditions.

In summary, incoherently detected MFSK is a robust modulation method for underwater telemetry. Robust performance can generally be reached at 0.25 bit/s/Hz at 20 dB SNR. However, this performance point is well below the performance bound determined by channel capacity considerations [Eq. (1)], which bounds performance at ~6.5 bits/s/Hz at 20 dB SNR.

4.2 Coherent Modulation

Phase-coherent signal modulation can approach channel capacity much closer than incoherently detected MFSK. This method encodes information into the phase, as well as amplitude, of the transmitted signal and is therefore more bandwidth efficient.[7] The transmitted signal can be expressed as

$$x(t) = a(t)e^{[j\omega t + \Phi(t)]}\qquad(6)$$

where $a(t)$ and $\Phi(t)$ are selected at each bit duration. Quadrature phase shift keying (QPSK) assigns four phase angles to each of four possible two-bit messages:

Data	00	01	10	11	
$\Phi(t)$	0	90	180	270	(degrees)
$a(t)$	1	1	1	1	

A more bandwidth-efficient quadrature amplitude modulation jointly modulates amplitude and phase:

Data	000	001	010	011	100	101	110	111
$\Phi(t)$	0	45	90	135	180	235	270	315
$a(t)$	$1+\sqrt{3}$	$\sqrt{2}$	$1+\sqrt{3}$	$\sqrt{2}$	$1+\sqrt{3}$	$\sqrt{2}$	$1+\sqrt{3}$	$\sqrt{2}$

A large number of modulation methods are useful in this context.[7] The receiver receives the waveform after it has propagated through a randomly time variant, multipath channel. Since received phase and amplitude are severely perturbed by the time-variant acoustic channel, phase-coherent receivers are required to track and equalize the channel multipath and phase fluctuations.[6] A phase-coherent receiver is typically based around a detrital channel estimator/equalizer. A possible receiver architecture is shown in Fig. 1. This receiver uses a feedforward linear equalizer to correct for channel fluctuations and remove some intersymbol interference (ISI). It is followed by a Doppler correction stage implemented with a digital phase-locked loop (PLL) and finally by a data decoder/feedback equalizer, which removes residual ISI and recovers the data stream.[4] Other receiver realizations have also been used.[5,7]

Performance of phase-coherent systems is typically limited by channel variability. The shallow water and continental shelf channels, in particular, have significant surface interactions and exhibit high levels of dynamics,

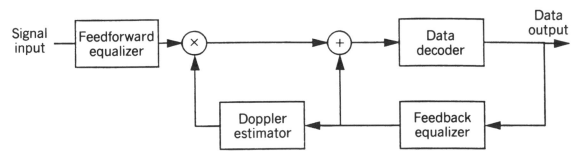

Fig. 1   Phase-coherent receiver structure.

TABLE 1 Phase-Coherent Telemetry Performance Milestones

Channel	Data Rate	Range (km)
Shallow water	20 kbits/s	8
Continental shelf	2000 bits/s	80
Deep water	1000 bits/s	200

as described by Eq. (4). The deep-ocean vertical channel is amenable to coherent telemetry. Receiver structures developed for RF and telephone channels operate reliably. A number of operational systems exist, with data rates from 5 bits/s to 30 kbits/s over the 5-km vertical channel. Recent improvements in equalizer performance have permitted coherent signaling over the other three types of underwater channels.[4] Some milestone results are summarized in Table 1.[4,5]

Phase-coherent acoustic data transmission is currently an active research topic, and further improvements in reliability, bandwidth, and power efficiency are likely. Performance improvement over incoherently demodulated MFSK telemetry is significant. For example, QAM bandwidth efficiency is 3 bits/s/Hz. It is decodable at 14 dB SNR over the long-range deep-water channel, offering a roughly order-of-magnitude improvement in bandwidth and power efficiency over MFSK.[5] However, phase-coherent methods are much more sensitive to environmental variability, and there is presently a large subset of acoustic channels where coherent methods do not perform well.

5 RELATIVE MERITS OF THE SIGNALING METHODS

It is a mistake to exclude FSK and MFSK signaling from consideration simply because phase-coherent methods outperform them on some occasions. For instance, FSK signalling is in wide use on a number of undersea acoustic products, such as transponders and releases. These systems are frequently built around a hard-limiting (clipper) detector or a phase-locked loop (PLL) tone detector. The receiver can be realized in simple hardware, and its performance is adequate for a large number of applications.[8] Its performance can be considerably enhanced by digitizing the acoustic signal and using an adaptive-threshold detector combined with a matched input filter and a forward error correction code.[9] Typically, data rates up to 80 bits/s can be achieved quite robustly at ranges up to 10 km, although higher data rate–range combinations have been reported.[9]

Increasing this data rate is easily done with a MFSK

system, where multiple tones are transmitted on adjacent bands to fill up the available bandwidth. Tone "chip" duration of 80 s^{-1} appears to yield an efficient trade-off between Doppler sensitivity and robustness to multipath-induced fading. Systems with data rates of 5–10 kbits/s at ~10-km ranges have been implemented.[3] There are also several military systems that operate at much longer ranges.

The MFSK systems are typically implemented around a DSP FFT implementation. Incoming data are corrected for Doppler with an adaptive digital sampler, blocked, and match-filtered with an FFT. The FFT outputs are used by the decision device to recover the data stream. In strongly fading channels, receiver spatial diversity offers a large performance increase at some additional complexity.[10]

Phase-coherent systems offer the highest performance, but as yet, their robustness has not been demonstrated in long-term deployments. A number of performance-limiting issues are being studied but have not been definitely resolved to data:

1. Long multipath duration increases the number of equalizer taps required for channel compensation. Since the taps are adaptively updated, increasing their number augments receiver self-noise, which eventually dominates the system SNR. In particular, long-range SOFAR propagation is characterized by multipath arrivals spread over several seconds, and phase-coherent techniques have difficulties. Overall multipath duration can be decreased by combining an equalizer with a beamformer or by using narrow transmitter beams to constrain the channel to a narrow beam.[5,11] Both techniques have shown considerable promise but have yet to be implemented in practical systems

2. Doppler tracking, in particular tracking and compensating for Doppler shifts of several frequency-spread multipath arrivals, has not been solved. There are several applicable frequency-tracking algorithms available, but they all appear to significantly increase receiver self-noise.

3. System robustness to interference, whether broadband impulsive sources such as snapping shrimp or strong tonal interference, is inadequate in current systems. Again, several solutions are available but have not been sufficiently tested in operation.

The above examples should alert the reader that phase-coherent systems are not the solution for many applications, although a great deal of work in this area is underway and they are likely to mature in the near future.

Meanwhile, the simple incoherent systems are likely to stay with us for the foreseeable future.

REFERENCES

1. R. Kennedy, *Fading Dispersive Communication Channels*, Wiley, New York; 1969. A. B. Baggeroer, "Acoustic Telemetry—An Overview," *IEEE J. Oceanic Eng.*, Vol. OE-9, October 1984, pp. 229–235.

2. R. J. Urick, *Principles of Underwater Sound for Engineers*, McGraw-Hill, New York, 1967.

3. J. Catipovic, "Performance Limitations in Underwater Acoustic Telemetry," *IEEE J. Oceanic Eng.*, Vol. 15, No. 3, July 1990, pp. 205–216.

4. M. Stojanovic, J. Catipovic, and J. Proakis: "Phase-Coherent Digital Communications for Underwater Acoustic Channels," *IEEE J. Oceanic Eng.*, Vol. 10, No. 1, 1994, pp. 100–111.

5. M. Stojanovic, J. Catipovic, and J. Proakis: "Adaptive Multichannel Combining and Equalization for Underwater Acoustic Communication," *J. Acoust. Soc. Am.*, Vol. 94, No. 3, pt. 1, September 1993, pp. 1621–1631.

6. *IEEE Journal of Oceanic Engineering*, Special Issue on Acoustic Telemetry, January 1991.

7. J. Proakis, *Digital Communications*, McGraw-Hill, New York, 1989.

8. P. Hearn: "Underwater Acoustic Telemetry," *IEEE Trans. on Comm. Tech.*, Vol. CT-14, December 1963, pp. 839–843.

9. R. S. Andrews and L. F. Turner, "On the Performance of Underwater Data Transmission Systems Using Amplitude Shift-Keying Techniques," *IEEE Trans. Sonics Ultrasonics*, Vol. SU-23, No. 1, January 1976, pp. 64–7.

10. J. Catipovic and L. Freitag, "Spatial Diversity Processing for Underwater Acoustic Telemetry," *IEEE J. Oceanic Eng.*, Vol. 16, No. 1, January 1991, pp. 86–87.

11. H. A. Quazi and W. L. Konrad, "Underwater Acoustic Communications," *IEEE Comm. Mag.*, 1982, pp. 24–30.

52

TRANSDUCERS

JOSEPH E. BLUE AND ARNIE LEE VAN BUREN

1 INTRODUCTION

A general understanding of sonar transducers is essential to underwater acousticians. When one thinks of sonar transducers, one thinks in terms of frequencies from around 10 Hz to 2 MHz. The lower end of this range is of particular interest in underwater acoustics. Thus this chapter concentrates on sonar transducers for lower frequencies. It includes a brief description of some of the more common transducing materials including their application to hydrophones. (Transducing principles for these materials are covered in Chapter 155.) A discussion of the types of low-frequency projectors is given next. This is followed by a description of transducer calibration including information on national primary and secondary standards. For more thorough discussions on sonar transducers, standards, and calibration, the reader is referred to Wilson,[1] Stansfield,[2] Kinsler and co-workers,[3] and Bobber.[4]

2 SONAR TRANSDUCERS

The term *sonar transducer* is used for underwater sound projectors, for hydrophones, and for reciprocal transducers that can be used as a projector or receiver of sound. The sound pressure level (SPL) from a projector in decibels relative to 1 μPa at 1 m is

$$SPL = S_v + 20 \log_{10} V, \qquad (1)$$

where S_v is the transmitting voltage response, and V is the root-mean-square (rms) input voltage to the projector.

Similarly,

$$SPL = S_I + 20 \log_{10} I, \qquad (2)$$

where S_I is the transmitting current response and I is the rms current. Hydrophone sensitivity is usually expressed in terms of free-field voltage sensitivity under open-circuit conditions or

$$M_H = 20 \log V_{oc} - 20 \log_{10} \frac{p}{p_0}, \qquad (3)$$

where V_{oc} is the open-circuit voltage from the hydrophone, p is the rms acoustic pressure, and p_0 is the reference pressure, which is usually taken to be 1 μPa.

If a transducer is reciprocal, then

$$\frac{m_H}{s_I} = J, \qquad (4)$$

where J is the spherical wave reciprocity coefficient, and m_H and s_I are related to M_H and S_I, respectively, through the relationships $M_H = 20 \log_{10} m_H$, and $S_I = 20 \log_{10} s_I$.

3 TRANSDUCER MATERIALS

The most common transduction materials used in sonar transducers are either piezoelectric, ferroelectric, or magnetostrictive. The term *piezoelectric* is often used for both true single-crystal materials such as Rochelle salt, quartz, tourmaline, ammonium dihydrogen phosphate, and lithium sulfate, which are naturally piezoelectric

Encyclopedia of Acoustics, edited by Malcolm J. Crocker
ISBN 0-471-80465-7 © 1997 John Wiley & Sons, Inc.

because of the asymmetry of their crystalline structure, and for ferroelectric polycrystalline ceramics, which must be poled in order that they have piezoelectric behavior. Piezoelectric materials have properties other than stability that affect their suitability for sonar transducer applications. These include piezoelectric constants, dielectric constant, resistivity, and anisotropy.

Two types of piezoelectric constants describe the relationships between electrical parameters, such as charge density, or electric field and mechanical parameters, such as stress or strain. Since the materials are anisotropic, the direction of the electrical and mechanical parameters are specified by subscripts. The g_{ij} constant is the quotient (strain)/(applied charge density) or (electric field)/(applied stress). The subscript i specifies the direction of the electric field and j the direction of induced strain or applied stress. The d_{ij} constant is the quotient (strain)/(applied electric field) or (charge density)/(applied stress). Subscripts 1, 2, and 3 pertain to the orthogonal axes and 4, 5, and 6 to shear motions about the 1, 2, and 3 axes, respectively. The g and d constants are related through the dielectric constant by[5]

$$d_{ij} = \epsilon \epsilon_0 g_{ij}, \qquad (5)$$

where ϵ_0 is the dielectric constant of free space and ϵ is the dielectric constant of the material relative to free space.

If one considers a very simple hydrophone constructed using a rectangular block of piezoelectric material of dimensions w, l, and t poled along the t-direction, the hydrophone sensitivity would be

$$m_H = (g_{31} + g_{32} + g_{33})t, \qquad (6)$$

where t is the thickness in metres. The combination $g_{31} + g_{32} + g_{33}$ is defined to be the hydrostatic constant g_h. It is applicable to a low-frequency acoustic pressure field acting on all sides of the material. Similarly the hydrostatic constant d_h is given by $d_{31} + d_{32} + d_{33}$. Table 1 compares various materials for hydrophones operating in the hydrostatic mode at low frequencies. The $g_h d_h$ constant is useful as a figure of merit when intrinsic hydrophone noise must be considered. For most materials, the quantity $g_{31} + g_{32}$ is opposite in sign and nearly equal in magnitude to g_{33} so that g_h is small compared to g_{33}. Materials such as lithium sulfate for which the magnitude of $g_{31} + g_{32}$ is small compared to that of g_{33} are called *volume expanders*. These materials do not require pressure-

TABLE 1 Comparisons of Low-Frequency Pressure Sensitivity of Various Piezoelectric Materials Operating in Hydrostatic Mode

Material	$g_h d_h$ (10^{-15} m^2/N)	Density (kg/m^3)	Sensitivity (dB re 1 V/μPa)	Thickness (cm)	Relative Permittivity, K_{33}^T
PZT-4	184	7,500	206	1.25	1,300
PbNb$_2$O$_4$	1,620	6,000	187	1.25	225
Li$_2$SO$_4$	1,990	2,060	187	0.64	10
SbSI (22°C)	15,730	5,000	191	0.76	1,360
PZT foam	13,600	3,200	180	1.25	190
PZT epoxy	1,290	2,800	195	0.34	54
PZT/rubber	17,300	2,800	—	—	170
Epoxy/PZT	16,500	5,400	—	—	410
PVDF (EMI)	1,260	1,800	203	0.06	14
PVDF/Nylon	973	1,800	210	0.03	11
PVDF (NBS)	250	1,800	209	0.06	11
Pennwalt copolymer	2,700	1,800	198	0.10	9
Ca-PT	1,450	6,700	202	0.25	200
1-3 Composites					
PT/epoxy	2,046	2,700	199	0.21	43
PZT/epoxy	527	2,950	209	0.3	840
0-3 PT/neoprene composites					
PR-304	1,045	5,300	200	0.2	40
PR-306	1,160	5,300	199	0.2	38
PR-307	4,884	5,900	194	0.2	45
3-3 Porous PZT	4,876	5,600	199	0.2	200
BST glass ceramic	830	6,300	201	0.1	12

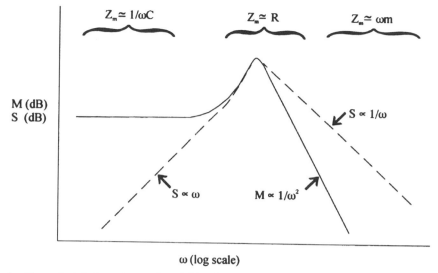

Fig. 1   Piezoelectric transducer; free-field voltage sensitivity M (solid line) and transmitting current response S (dashed line) as a function of angular frequency.

release material (Corprene, air-cell rubber, paper, etc.) on the edge faces to avoid a large reduction in sensitivity when these faces are exposed to the acoustic field. Since a material normally cannot be dynamically soft and statically hard, the use of pressure-release shielding limits the magnitude of hydrostatic pressure that can be applied. It also reduces the potential stability of the hydrophone with respect to temperature and hydrostatic pressure. Lithium sulfate is still used in its hydrostatic mode in standard transducers in the form of disks or plates because it is also extremely stable with time.

While hydrostatic mode hydrophones are desirable from the viewpoint of stability, they are not necessarily the best choice from a sensitivity viewpoint. Most hydrophones and sonar projectors are more complex than the example given above and use modes other than the hydrostatic. The most common modes used are the 33 and 31 modes with some type of "shielding" provided to isolate the effects of the other modes of vibration. Figures 1 and 2 show typical free-field voltage sensitivity and transmitting current response curves for simple piezoelectric and moving-coil or magnetostrictive hydrophones and projectors.

Ceramics that have electrical properties analogous to the magnetic properties of ferromagnetic materials are called ferroelectric ceramics. The ceramics came into general use for transducers in the early 1950s and have rapidly replaced crystals in many applications. High

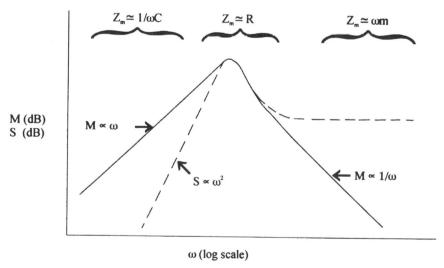

Fig. 2   Moving-coil and magnetostrictive transducer, free-field voltage sensitivity M (solid line) and transmitting current response S (dashed line) as a function of angular frequency.

dielectric and piezoelectric constants, variety of shapes available, low cost, and inherent ruggedness are reasons for the popularity of this material.

Of the several kinds of ceramic, barium titanate and lead zirconate titanate are the most popular. Some types of lead metaniobate have characteristics that make it peculiarly suited to broadband, high-pressure transducers, and it is incorporated in some transducers of this type.

The magnetostrictive and the electrostrictive effects are both second-order effects.[5] Applying magnetic field and electric fields to these types of materials, respectively, results in nonlinear strains. A bias field is maintained in these types of materials to provide a linear response region for the application of small varying fields. Some of the more useful magnetostrictive materials have been iron, cobalt, nickel, and their alloys. Recently metallic glass alloys have been developed with electromechanical coupling factors of 0.96. Also rare-earth–iron compounds have been developed by Levgold[6] and Clark et al.[7] with terbium, holmium, and iron. An electrostrictive material, lead-magnesium niobate, with high strain capability may show promise as a transducer material.

4 SONAR PROJECTOR TYPES

The design of sonar transducers must begin with a knowledge of the radiation of sound. Morse and Ingard[8]

give a good account of radiation in Chapter 7 of their book. Wilson[1] discusses transducer arrays and beam patterns and also briefly describes array element interactions. Of particular interest to underwater acousticians is the production of low-frequency sound. Some of the problems with low-frequency sonar projectors are discussed by Wilson.[1] Many of these problems occur because the dimensions of the projector are usually small relative to the wavelength λ of the sound produced.

As an example, consider a pulsating sphere of radius a where $\lambda \gg a$ or $ka \ll 1$, with $k = 2\pi/\lambda$ being the wavenumber. In this case, the acoustic pressure on the surface of the sphere is well approximated by

$$p(a) \simeq \rho c v_n (ka)^2 + j\rho c v_n ka, \qquad (7)$$

where v_n is the surface velocity and ρ and c are the density and sound speed in the surrounding medium. The acoustic impedance of the sphere Z is then given by

$$Z = R_r + jX_r \simeq \rho c (4\pi a^2)(ka)^2 + j\rho c (4\pi a^2)(ka). \qquad (8)$$

The simple approximations given here for the acoustic resistance R_r and the acoustic reactance X_r are accurate for ka less than about 0.2. Figure 3 shows the behavior of the acoustic resistance and reactance of the pulsating sphere for larger values of ka.

For $ka \ll 1$ the sound pressure radiated into the water, referenced back to 1 m, can be expressed in terms of the

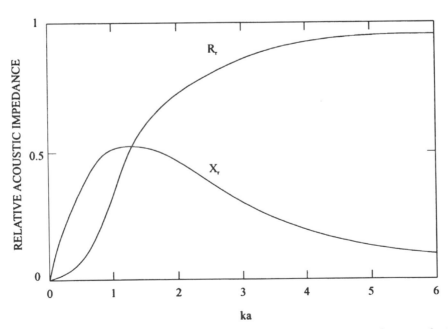

Fig. 3 Components of acoustic impedance for a spherical harmonic wave from a pulsating sphere of radius a.

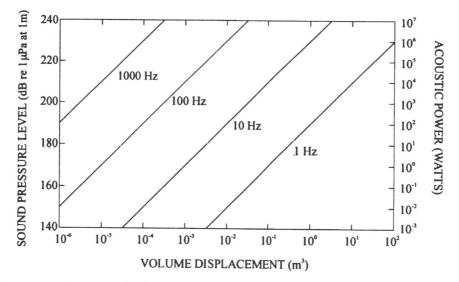

Fig. 4 Sound pressure level and acoustic power radiated from a harmonically pulsating sphere of radius a with $ka \ll 1$.

volume displacement Ξ of the sphere by

$$|p| \simeq \frac{\omega^2 \rho \Xi}{4\pi}. \qquad (9)$$

Figure 4 shows the sound pressure level and the corresponding radiated acoustic power $P = v_n^2 R_r = 4\pi |p|^2/\rho c$ for this case as a function of the volume displacement. Note that at 100 Hz, a volume displacement of 0.001 m^3 will radiate 10,000 W of acoustic power and produce a sound pressure level of 211 dB re 1 μPa at 1 m. However, as can be seen from Fig. 3 for $ka \ll 1$, the total input power necessary to supply the acoustic power is controlled primarily by the acoustic reactance since it dominates the much smaller acoustic resistance. Thus the basic problem is one of poor acoustic loading due to a very long acoustic wavelength in water at low frequencies. Simply stated, this means that the transfer of mechanical power at the surface of the radiator to the water under the form of radiated acoustic power is very inefficient. Since the overall efficiency is the product of the radiation and mechanical transduction efficiencies, it will also be low even if the mechanical transduction efficiency is high.

Because high-power, low-frequency sound production requires large volume displacements, the projector must meet the conflicting requirements of counteracting large hydrostatic forces and offer a pressure release mechanism to the interior of the vibrating surface. Virtually none of the pressure-release techniques used at high frequencies are practical at low frequencies.

The most commonly used pressure-release mecha-nism at low frequencies is compressed gas. If the interior of a transducer is filled with gas at the same pressure as the surrounding water, the transducer is obviously balanced against the forces due to hydrostatic pressure, and the large impedance mismatch provides an excellent pressure release. Compressed gas systems, however, are not without disadvantages. Since most low-frequency transducers have a large internal volume, applications requiring many depth changes may require a large high-pressure gas storage volume. Also, for applications at very great depths, high-pressure gas systems can become complicated and pose a reliability problem.

Transducers can be made essentially independent of operating depth by filling the enclosed volume with liquid. To do so, however, provides essentially no pressure release. For a given frequency and acoustic output, liquid-filled transducers will be larger, heavier, and require larger driving forces than will transducers using some other compensation mechanism.

As a compromise usable to moderate depths, sealed, air-filled, oval metal tubes can be inserted into the liquid-filled cavity to increase its compliance. This technique provides decreasing pressure-release capability as the depth is increased until the compliant tubes are collapsed by the hydrostatic pressure. Some transducer designs can be made to be simply self-supporting by filling the internal cavity with air at some predetermined pressure. The primary disadvantage, of course, is the severely limited depth capability.

Before discussing any of the specific transducer types, it should be noted that there is no general way to quantitatively compare different transduction mechanisms. The ratio of output power to total weight is frequently used, but it neglects the effects of bandwidth, transduction efficiency, and reliability. Since all of the available low-frequency transducers cannot be treated here, only those most likely to be encountered will be addressed. The most common transduction mechanism in use is, of course, piezoelectric ceramics. The most numerous sonar transducers in use are ceramic-driven longitudinal transducers, often called tonpilz transducers.[2] The tonpilz design is not feasible at low frequencies, however, because of the size that would be required to generate the required volume velocities. Low-frequency ceramic designs attempt to take advantage of vibrational modes not normally used at higher frequencies.

4.1 Ceramic Flexural Disc

The trilaminar configuration of the ceramic disc transducer shown in Fig. 5 lends itself reasonably well to the high-power, low-frequency application. In the trilaminar configuration, an inactive disc (normally steel or aluminum) is laminated between two ceramic disc composites; when the two ceramic discs are driven in opposition, a flexing motion is produced in the trilaminar structure. To keep the size of the ceramic within reasonable limits, the ceramic discs may be assembled in a mosaic instead of one piece.

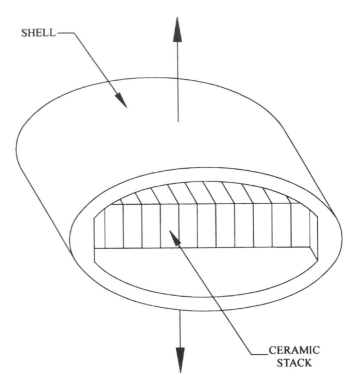

Fig. 6 Flextensional transducer.

4.2 Flextensional Transducers

In a common form, the flextensional transducer shown in Fig. 6 consists of an elliptically shaped housing, or shell, with a longitudinally vibrating ceramic stack mounted along its major axis. Unlike the flexural disc transducer, the housing (not the ceramic) forms the radiating surface. The ceramic stack is compressively prestressed by the shell to assure that it does not go into tension and fracture at high drive levels.

A single large shell may be used or several small shells may be stacked together in a line configuration. In either case, the open ends of the shell are sealed and the resulting internal volume may be either gas filled or oil filled and compliant tubes inserted. For relatively shallow depths, the transducer can be made self-supporting by filling the cavity with air at atmospheric pressure. To do so, however, means that the prestress on the ceramic and, therefore, the safe driving voltage decreases as a function of depth.

The flextensional transducer does offer a good power-to-weight ratio, but it is a resonant device and has a quality factor Q higher than most nonceramic designs. It can also be highly efficient. Its primary disadvantage compared to other low-frequency ceramic transducers is its difficult design, particularly for low resonance frequencies.

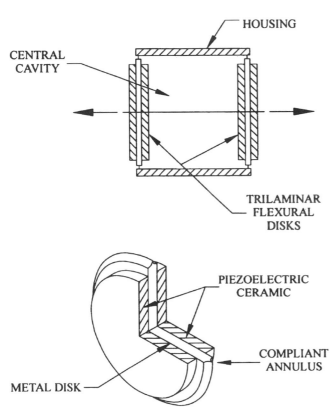

Fig. 5 Flexural disk transducer.

4.3 Ceramic Bender Bar Transducers

The ceramic bender bar transducer shown in Fig. 7 typically consists of multiple "bars" arranged in a "barrel stave" configuration around a cylindrical housing. Each bar consists of two segmented stacks of ceramic and is "hinged" at each end. When the stacks are driven in opposition, a bending motion is produced in the bars.

The barrel stave configuration of the transducer results in a central cavity that is normally oil filled to compensate for hydrostatic pressure. Compliant tubes are inserted into the cavity to increase its compliance and to provide the necessary pressure-release mechanism for radiation from the inner surfaces of the bars. Some very low frequency designs do, however, use compressed air as the pressure-release mechanism.

The transducer is capable of producing moderately high output power levels over a frequency range of an octave or so and at depths to several hundred metres; it does have the advantage of proven reliability. However, its design is such that it uses a very large amount of ceramic, thus it is heavy and expensive.

As in the case of the flexural disc, the input power is limited by the electric field and maximum stress that the ceramic can withstand.

4.4 Moving-Coil Transducer

The electrodynamic, or moving-coil, transducer shown in Fig. 8 is one of the oldest designs still in use and derives its driving force from the interaction between an alternating current moving in a conductor and a large magnetic field. In the most common configuration, the force

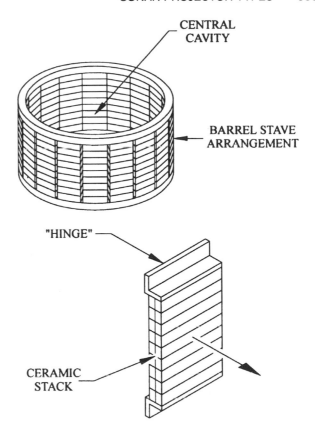

Fig. 7 Bender bar transducer.

is used to drive a rigid piston radiator. When applied to the requirements of low-frequency sources, the moving coil offers some distinct advantages. It can be, and usually is, designed to have a very low resonance frequency.

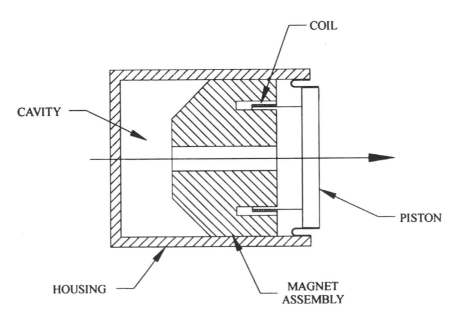

Fig. 8 Moving-coil transducer.

Being a typically large compliance system, it can accommodate large linear displacements. Also, wide operating bandwidths are relatively easily achieved.

It does, of course, also have several disadvantages. The moving coil is typically an inefficient transduction mechanism. It is a relatively small force device. As such it is capable of producing only moderate output power levels when used as a single surface radiator. The limitation upon the maximum input electrical power is determined by how well the heat generated in the coil can be dissipated. For higher acoustic output, the moving-coil transducer is normally used in arrays. It has an additional disadvantage. Being a large compliance system with a gas pressure-release mechanism, low-frequency moving-coil transducers typically exhibit large changes in performance as a function of depth.

4.5   Hydraulically Actuated Transducers

In a common form, as shown in Fig. 9, the hydraulically actuated transducer consists of two opposing flexural discs driven by a central hydraulic amplifier. A low-level electrical signal with the desired waveform is used to control the hydraulic amplifier while the hydraulic power is supplied by an electrically driven pump. The hydraulic

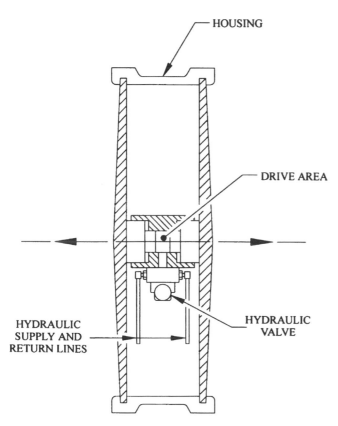

HOUSING

DRIVE AREA

HYDRAULIC
SUPPLY AND
RETURN LINES

HYDRAULIC
VALVE

Fig. 9   Hydraulically actuated transducer.

system can essentially by housed within the transducer module, eliminating the need for handling high-pressure hydraulic lines from the surface. The transducer is normally designed to be self-supporting to depths of several hundred metres and may be gas compensated to go deeper.

Hydraulic transduction seems ideally suited for low-frequency broad bandwidth applications because of the ability to produce very large mechanical forces from a relatively small package and yet allow for the required large linear displacements. A hydraulically actuated source typically produces moderately high acoustic output levels over bandwidths of two octaves or more. The power-to-weight ratio is comparable to or slightly less than that for some of the ceramic sources.

The primary disadvantage of hydraulic transduction is reliability; it is a relatively complex system and, as such, must be maintained to a greater extent than other transduction mechanisms.

4.6   Impulse-Type Sources

Restrictions on the use of impulse-type sources such as sparkers, air guns, and "water hammer" devices are primarily set by the requirements of the application. If a repeatable complex waveform from a continuous-wave (CW) source is not essential, the use of impulse sources should be considered since they are generally easier to design and operate. There is no general basis for comparing impulse source with other types.

In the discussion above on flexural disc, flextensional, and bender bar transducers, no mention was made of utilizing rare-earth–iron alloys as the transduction material in place of ceramic. These alloys can, of course, be used, and the potentially greater transduction energy density will provide certain advantages. However, the overall efficiency of the transducer will still be relatively small at low frequencies because of the dominance of the low radiation efficiency associated with a small acoustic size.

There are other transduction mechanisms, such as used in the tow-powered source and the thermoacoustic source, that have not been discussed here because they are still very much in the development phase. They are radically different from conventional mechanisms.

5   TRANSDUCER CALIBRATION

The American National Standards Institute (ANSI) and the Acoustical Society of America are jointly responsible for establishing procedures for the calibration of underwater acoustic transducers. ANSI S1.20-1988 covers this topic.[9] The Underwater Sound Reference Detachment (USRD) of the Naval Undersea Warfare Center (the USRD was part of the Naval Research Laboratory until July 1995) functions as the standardizing institution for underwater acoustic measurements. It maintains the national primary standards, special hydrophones that are designed to maintain stable sensitivity over large temperature and pressure ranges and over a long time. USRD uses two methods for absolute calibration of the primary standard hydrophones: (1) free-field reciprocity calibration and (2) coupler reciprocity calibration.

Free-field reciprocity calibration is used in open water or tanks down to about 100 Hz. Below 100 Hz lack of a suitable robust reciprocal transducer limits this method. Its use requires three transducers (a projector P, the hydrophone H to be calibrated, and a reciprocal transducer T that can be used as both a hydrophone and a projector). A series of three projector–hydrophone measurements are made using either P or T as a projector and H or T as a hydrophone. The measurements are made under free-field conditions with the hydrophone located in the far field of the projector. The three experimental setups are indicated in Fig. 10. The input current and output voltage values are complex, that is, they include both amplitude and phase. Extension of conventional three-transducer reciprocity calibration to include phase is given by Luker and Van Buren.[10] The receiving voltage sensitivity of the hydrophone H is obtained using

$$M_H - \left(\frac{4\pi e_{PH} e_{TH} d_1 d_3}{j\omega \rho e_{PT} i_T d_2} \exp[jk(d_1 + d_3 - d_2)] \right)^{1/2},$$

(10)

where ω is the angular frequency. Standard practice in the United States is to ignore the phase calibration and do magnitude calibration only. The distances d_1, d_2, and d_3 in this case are usually all the same. The difficulty in determining the phase of M_H using this method lies in accurately determining both the sound speed and the measurement distances d_1, d_2, and d_3. However, we can avoid this difficulty by positioning all three transducers

P, H, and T in a straight line with H located between P and T. This assures that $d_2 = d_1 + d_3$ to give

$$M_H = \left(\frac{4\pi e_{PH} e_{TH} d_1 d_3}{j\omega \rho e_{PT} i_T d_2} \right)^{1/2}.$$

(11)

The accuracy of the phase of M_H calculated using this expression is limited only by the accuracy of the phase measurements of the voltages and current.

Reciprocity couplers are small chambers used for absolute calibration of the low-frequency primary standard hydrophones. They operate over the range from about 10 Hz to several kilohertz and over the full range of ocean temperatures and pressures. This and other reciprocity calibration techniques are discussed by Bobber.[4]

USRD maintains a standard transducer loan program[11] where a calibrated transducer is issued to a customer for a specific time (usually 1 year). Loan standards are calibrated by secondary methods using primary standards. The hydrophones are calibrated by comparison of their output with that of the primary standard while the projectors are calibrated by directly measuring their output with the primary standard, either pulsed or continuous wave swept under good free-field conditions. Two separations are used to ensure that the transducers are responding properly to the pressure field. Failure of the transducer response to follow the inverse spreading law for spherical waves may imply improper far-field conditions or an inappropriate pressure-gradient sensitivity of the transducer, among other possibilities. Rigging must be carefully designed so as not to appreciably affect the calibration, and care must be taken to see that there are no bubbles adhering to the transducers.

Most transducers used in U.S. calibration facilities are USRD loan standards calibrated by secondary methods. Some facilities perform free-field reciprocity calibration on their own or USRD standards. However, most naval activities insist on maintaining traceability of its basic measurements to the USRD since the USRD maintains

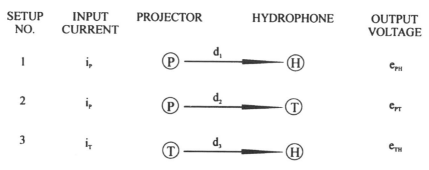

Fig. 10 Measurement setups for free-field reciprocity calibration.

traceability of its basic measurements used in setting underwater acoustic standards to the National Institute of Standards and Technology (NIST).

REFERENCES

1. O. B. Wilson, *Introduction to Theory and Design of Sonar Transducers*, Peninsula Publishing, Los Altos, CA, 1988.

2. D. Stansfield, *Underwater Electroacoustic Transducers*, Bath University Press, Bath, UK, 1990.

3. L. E. Kinsler, A. R. Frey, A. B. Coppens, and J. V. Sanders, *Fundamentals of Acoustics*, 3rd ed., Wiley, New York, 1982.

4. R. J. Bobber, *Underwater Electroacoustic Measurements*, Peninsula Publishing, Los Altos, CA, 1988.

5. D. A. Berlincourt, D. R. Curran, and H. Jaffe, "Piezoelectric and Piezomagnetic Materials and their Function in Transducers," in W. P. Mason (Ed.), *Physical Acoustics*, Vol. IA, Academic Press, New York, 1964.

6. S. Levgold, J. Alstad, and J. Rhyne, "Giant Magnetostriction in Dysprosium and Holmium Single Crystals," *Phys. Rev. Lett.*, Vol. 10, No. 12, 1963, pp. 509–511.

7. A. E. Clark, R. M. Bozorth, and B. F. DeSavage, "Anomalous Thermal Expansion and Magnetostriction of Single Crystals of Dysprosium," *Phys. Lett. (Netherlands)*, Vol. 5, No. 2, 1963, pp. 100–102.

8. P. M. Morse and K. U. Ingard, *Theoretical Acoustics*, McGraw-Hill, New York, 1968.

9. ANSI S1.20-1988, "Procedures for Calibration of Underwater Electroacoustic Transducers" (available from Standards Secretariat, Acoustical Society of America).

10. L. D. Luker and A. L. Van Buren, "Phase Calibration of Hydrophones," *J. Acoust. Soc. Am.*, Vol. 70, No. 2, 1981, pp. 516–519.

11. USRD Transducer Catalog, August 1994, Naval Undersea Warfare Center, Underwater Sound Reference Detachment, P.O. Box 568337, Orlando, FL 32856-8337.

53

NONLINEAR SOURCES AND RECEIVERS

MARK B. MOFFETT AND WILLIAM L. KONRAD

1 INTRODUCTION

Acoustic nonlinearities of seawater can be exploited to provide parametric acoustic sources and receivers having useful characteristics. Parametric sources allow the generation of very narrow beams having extremely low side-lobe levels from physically small projectors. These characteristics make feasible high-resolution sonars, especially where only small projectors can be deployed. For example, a parametric source has been designed and built to be carried in the mouth of a sea lion, providing the animal with an active sonar capability. Unlike a conventional endfire array of hydrophones, the parametric receiver does not require hardware between the pump source projector and the hydrophone. Therefore the parametric receiver should find application where a conventional array is not feasible. Because the beam pattern of the second harmonic generated during the propagation of a high-amplitude sound wave is narrower and has lower side-lobe levels than the fundamental, harmonic generation can be usefully applied to reverberation-limited applications. For example, low-frequency sonars might use the second harmonic with its characteristic narrower beam and low side lobes to improve performance under reverberation-limited conditions.

2 PARAMETRIC SOURCES

2.1 Concept

Parametric sources make use of the nonlinearity of a propagation medium (e.g., seawater) to generate secondary, difference-frequency signals from high-power,

high-frequency, directional primary beams. A (linear) directional projector driven at two primary frequencies $f_0 \pm \frac{1}{2}f$ produces a virtual endfire array of acoustic sources at the difference frequency f (see also Sections 4.3 and 5.2 in Chapter 23). The length of the endfire array is limited either by absorption of the primary frequencies or by saturation, that is, loss of primary power to the (nonlinear) generation of harmonics. Because the effective array length can be made extremely large, highly directional difference-frequency beams can be generated by a projector that is not large compared to the difference-frequency wavelength. Very low side-lobe levels usually accompany the parametric beam, due to to the exponential shading that results from primary absorption. Parametric generation is inefficient, however, so that high primary powers are required.

2.2 Design

Far-Field Source Level and Beamwidth The primary projector is assumed to be a planar array whose elements are resonant near the mean primary frequency f_0. Figures 1–3 can be used to determine the far-field difference-frequency source level from projectors having square or circular radiating surfaces for frequency downshift ratios f_0/f = 5, 10, 20.[1] The ordinate is the parametric gain

$$G = SL - SL_0, \qquad (1)$$

where SL is the root-mean-square (rms) source level at the difference frequency and SL_0 is the rms source level of one primary-frequency component. (The two primary-frequency components are assumed to be equal in amplitude.) The abscissa in Figs. 1–3 is the scaled (to 1 kHz) primary source level,

Encyclopedia of Acoustics, edited by Malcolm J. Crocker
ISBN 0-471-80465-7 © 1997 John Wiley & Sons, Inc.

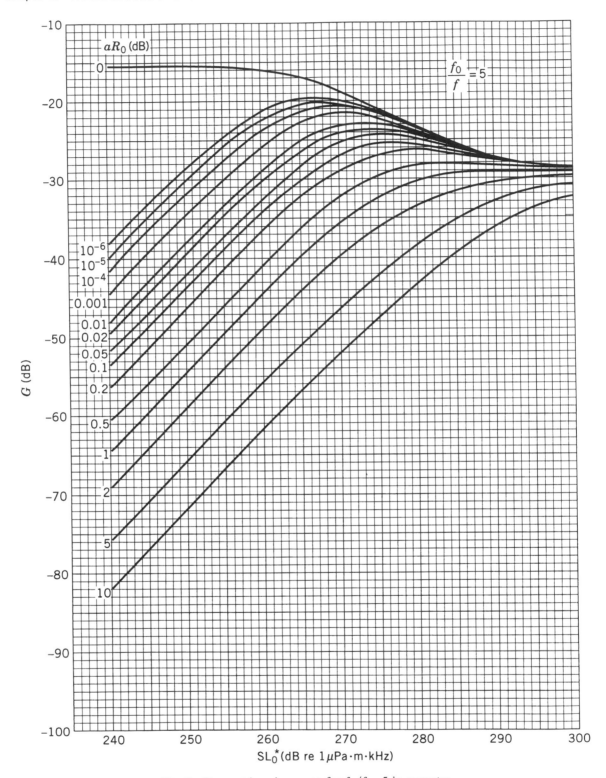

Fig. 1 Parametric gain curves for $f_0/f = 5$ in seawater.

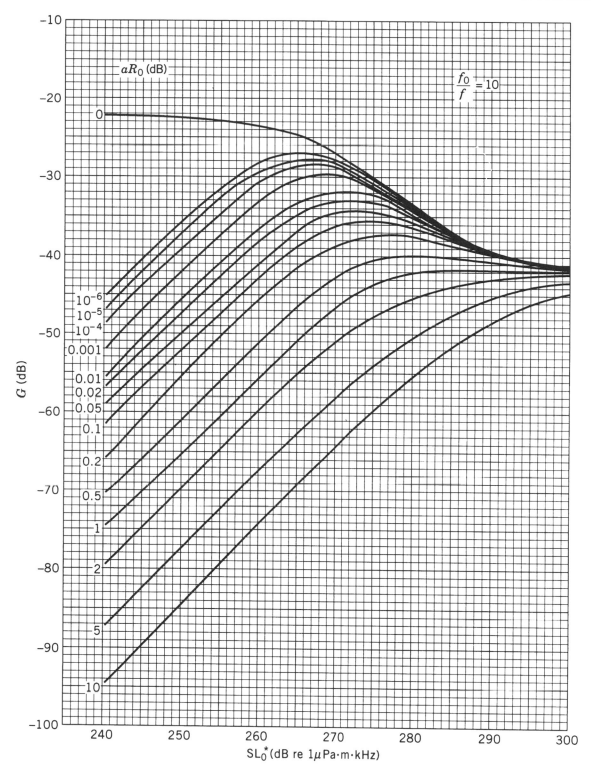

Fig. 2 Parametric gain curves for $f_0/f = 10$ in seawater.

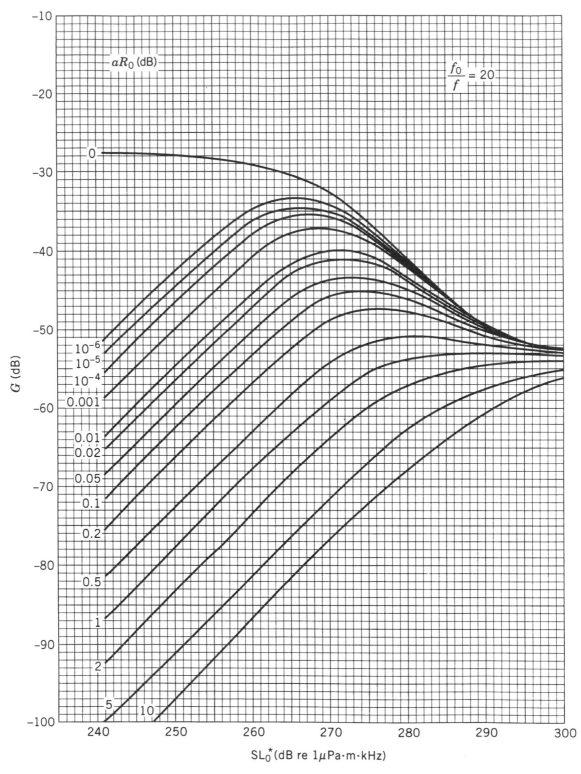

Fig. 3 Parametric gain curves for $f_0/f = 20$ in seawater.

$$SL_0^* = SL_0 + 20 \log f_0, \qquad (2)$$

where f_0, the mean primary frequency, is expressed in kilohertz. The parameter for the various curves of Figs. 1–3 is a aR_0, expressed in decibels, where a is the absorption coefficient (in decibels per metre) at f_0, and

$$R_0 = \frac{A f_0}{c}, \qquad (3)$$

where A is the projector area and c is the sound speed in the water medium. The Rayleigh length R_0 is a measure of the primary near-field collimation length, and so the parameter aR_0 is the amount of absorption loss (in decibels) occurring within the primary near field.

As an example, consider a 0.5×0.5 m square projector operated at a mean primary frequency f_0 of 100 kHz and a difference frequency f of 20 kHz. The downshift ratio f_0/f is 5, and so the curves of Fig. 1 obtain. The area A of the projector is $(0.5)^2 = 0.25$ m², and the Rayleigh length R_0 from Eq. (3) is approximately 17 m. The absorption coefficient a at 100 kHz is about 0.03 dB/m in seawater, and so the parameter aR_0 is approximately 0.5 dB. If the rms primary source level is, say, 237 dB re 1 μPa · m for each of the 90- and 110-kHz components of the primary waveform, the scaled source level, from Eq. (2), is 277 dB re 1 μPa · m · kHz. From Fig. 1, the parametric gain G is -29 dB. Then, from Eq. (1), the source level at 20 kHz is 208 dB re 1 μPa · m. With the same 237 dB re 1 μPa · m primary source level, the parametric gain G (from Figs. 2 and 3) would be -40 and -51 dB for f of 10 and 5 kHz, respectively.

There are four distinct operating regimes[1] for parametric sources, and they are all represented in Figs. 1–3. At values of the scaled source level SL_0^* above 280 dB re 1 μPa · m · kHz, the parametric gain curves begin to flatten, indicating that saturation, that is, harmonic distortion (see also Section 4 in Chapter 19 and Section 4 in Chapter 23), of the primary beam is limiting the effective length of the parametric array and, therefore, parametric gain G as well as broadening the beam. (Our 100-kHz, 237-dB re 1 μPa · m example approaches this limit.) At lower scaled source levels and at values of aR_0 exceeding 1 dB, the array is small signal and absorption limited within the projector near field, and Westervelt's analysis[2] applies. At very small values of aR_0, most of the parametric generation occurs within the projector far field. Both of these absorption-limited regimes, represented by the curves sloped at 45° in Figs. 1–3, can be treated with the analysis of Berktay and Leahy.[3] Finally, there is the regime represented by the uppermost curves of Figs. 1–3 for $aR_0 = 0$ dB. In this (impractical) case, because there is assumed to be no small-signal absorp-

tion whatsoever, the parametric source is saturation limited in the far field; these curves are included in Figs. 1–3 because they place an absolute limit on the parametric gain.

The far-field, half-power beamwidths of absorption-limited parametric sources of rectangular (or square) shape are obtainable from the curves[3] of Fig. 4. The ordinate of Fig. 4 is the ratio of the parametric difference-frequency beamwidth BW to the Westervelt[2] beamwidth,

$$BW_W = 4 \sin^{-1} \left[0.40 \left(\frac{\alpha c}{f} \right)^{1/2} \right], \qquad (4)$$

where α is the absorption coefficient in nepers per length, that is,

$$\alpha = \frac{a}{8.686}, \qquad (5)$$

and f is the difference frequency in hertz. The abscissa of Fig. 4 is the ratio of one primary beamwidth BW_1 to BW_W. (All beamwidths herein are defined as full-width, half-power, i.e., total, 3-dB beamwidths.) The parameter for the dashed curves of Fig. 4 is the ratio of the other primary beamwidth, BW_2, to BW_1. If the projector array is square ($BW_1 = BW_2$), the solid curve applies. For example, a projector of dimensions 0.2×1 m operating at a primary frequency of 100 kHz would have primary beamwidths of $BW_1 = 3.8°$ and $BW_2 = 0.8°$. If the difference frequency were 20 kHz, the Westervelt beamwidth, from Eq. (4), would be $BW_W = 1.5°$. From

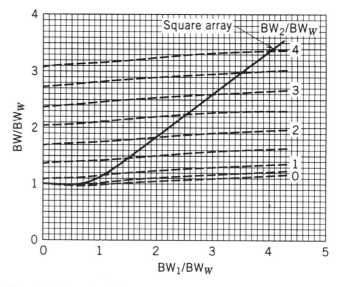

Fig. 4 Normalized beamwidths for absorption-limited parametric sources of rectangular aperture. (After Ref. 3.)

Fig. 4, the difference-frequency beamwidths can be read as $1.15BW_W = 1.7°$ and $2.05BW_W = 3.1°$. It should be noted that beamwidths calculated with the aid of Fig. 4 do not contain an aperture factor[4], that is, the directivity of the projector aperture at the difference frequency has not been accounted for. Thus, for aperture dimensions much larger than the difference-frequency wavelength, the beamwidths calculated from Fig. 4 will be too large. (In our 0.2×1 m example, the aperture-corrected beamwidths would be $3.1° \times 1.5°$ at 20 kHz.)

Effective Length

Because of the extent of the virtual source distribution that forms the parametric endfire array, calculations of the difference-frequency sound pressure level (SPL) and beamwidth (BW) within the source region are often required. The rms SPL on the axis of a circular-piston, absorption-limited parametric source may be determined from[5]

$$SPL = \Gamma + 2SL_0 + 20 \log \frac{f}{r} - 286.6, \qquad \text{dB re } 1\mu Pa, \quad (6)$$

where Γ is the ordinate of Fig. 5, SL_0 is expressed in decibels relative to $1\ \mu Pa \cdot m$ rms (per tone), f is in kilohertz, and the range r is in metres. The numeric 286.6 is $20 \log(1000\rho c^3/2^{1/2}\pi\beta)$, where ρ is the density of seawater and β its nonlinearity parameter. The beamwidth, BW, may be obtained[5] from the ratio BW/BW_0 of Fig. 6, where BW_0 is the circular-piston beamwidth at the primary frequency f_0.

As an example in the use of Figs. 5 and 6, consider a 0.24-m-diameter projector operated in seawater at $f_0 = 100$ kHz to generate a difference frequency $f = 20$ kHz. The Rayleigh length R_0 from Eq. (3) is 3.0 m. The absorption coefficient a is approximately 0.03 dB/m, and from Eq. (5), that is equivalent to $\alpha = 3.45 \times 10^{-3}$ Np/m. Thus the appropriate parameter for the curves in Figs. 5 and 6 is $2\alpha R_0 f_0/f = 0.10$. At a range r of 45 m, where the abscissa quantity $rf/R_0 f_0$ of Figs. 5 and 6 has the value $(45 \times 20)/(3.0 \times 100) = 3.0$, we read $\Gamma = 4.0$ dB from Fig. 5 and $BW/BW_0 = 0.94$ from 53.6. If the primary source level is 220 dB re $1\ \mu Pa \cdot m$, Eq. (6) yields SPL = 150.4 dB re $1\ \mu Pa$. The primary beamwidth is $3.8°$; therefore, the difference-frequency beamwidth BW is $3.6°$ at the 45-m range.

It is useful to define an effective length r_{eff} for the parametric array as the distance from the projector where the difference-frequency source level is 1 dB different from (it may be greater of less than) the far-field value. Figure 7 gives,[6] for absorption-limited parametric sources, the dimensionless effective length $2\alpha r_{eff}$ for circular and square projectors (solid curve) and for a $10:1$ aspect ratio rectangular projector (dashed curve). For small values of the abscissa, $2\alpha R_0 f_0/f$, most of the difference-frequency generation takes place in the primary far field, and the source level increases monotonically with range (see upper curves of Fig. 5). In that case, the source level at r_{eff} is 1 dB less than the far-field value. On the other hand, for large values of $2\alpha R_0 f_0/f$, most of the difference-frequency generation occurs in the primary near field and the source level has a maximum that is higher than the far-field value (see the lower curves of

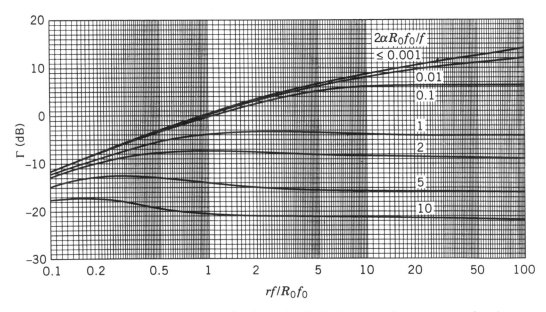

Fig. 5 Normalized parametric gain for absorption-limited parametric sources as a function of normalized range.

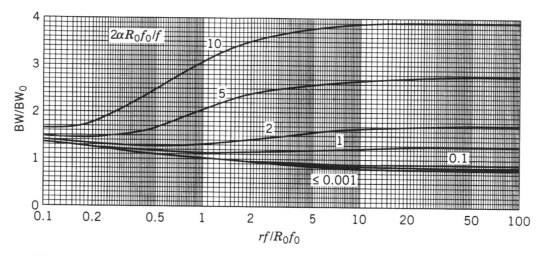

Fig. 6 Secondary-to-primary beamwidth ratio for absorption-limited parametric sources as a function of normalized range.

Fig. 5). In this case, the source level at r_{eff} is 1 dB greater than the far-field value. When $2\alpha R_0 f_0/f$ is near unity, there is a transition between these two kinds of behavior, and a discontinuous jump occurs in the curves of Fig. 7. When $2\alpha R_0 f_0/f$ is small, the effective length r_{eff} is on the order of $(4\alpha)^{-1}$, but when $2\alpha R_0 f_0/f$ becomes greater than unity, r_{eff} quickly increases to values approaching $5/\alpha$. It can also be seen in Fig. 7 that, except in the neighborhood of $2\alpha R_0 f_0/f = 1$, the rectangular projector effective length does not differ greatly from that for a square or circular projector. From Fig. 7, the effective length of the 0.24-m-diameter parametric source example just discussed is $0.7/2\alpha = 0.35 \times 8.686/0.03 \sim 100$ m.

2.3 Operation

Signal Generation and Projector Drive Methods

There are several means of generating the primary signal and feeding it to the parametric source projector. The

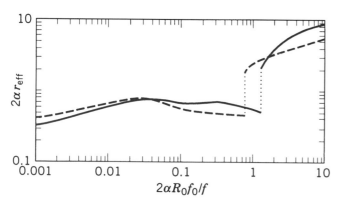

Fig. 7 Normalized effective length of absorption-limited parametric sources: (——) square or circular projectors; (---) rectangular projector, 10:1 aspect ratio.

simplest is shown in Fig. 8a, where the two primary frequencies $f_0 \pm \frac{1}{2}f$, are simply added and fed to a linear power amplifier that drives the projector. A second method, shown in Fig. 8b, is more convenient, because it allows the difference frequency f to be adjusted while keeping the carrier frequency fixed, that is, the upper and lower sidebands at $f_0 \pm \frac{1}{2}f$ remain equally spaced about the center carrier frequency f_0. Normally f_0 is adjusted to fall near the maximum transducer response. Both of the above generation methods require a linear power amplifier to avoid the generation of large amounts of difference frequency that, fed to projector, would result in direct radiation of the difference frequency.

Two other methods of projector feed allow the use of more efficient switching (class S) amplifiers.[7] Figure 8c shows separate primary frequencies fed to each power amplifier, the outputs of which are then summed, usually by connecting their output transformer secondaries in series. It should be noted that each of the amplifiers in Fig. 8c must handle the peak envelope power (i.e., four times the power in each primary-frequency component), because interaction of the two primary components produces back electromotive forces in the transformer primary windings. Another method, also permitting the use of switching amplifiers, is shown in Fig. 8d. Here, the projector transducer elements are separated into two groups (checkerboarded) with $f_0 - f/2$ fed into one group and $f_0 + \frac{1}{2}f$ fed to the other group. The two groups of elements must be fed in such a way as to avoid the radiation of large side lobes. The checkerboard technique has the advantage that each amplifier and transducer element is required to handle only the power of one primary component instead of the peak envelope power, as in Figs. 8a–c. However, this advantage is largely lost because the transducer array is effectively thinned (one-half of the

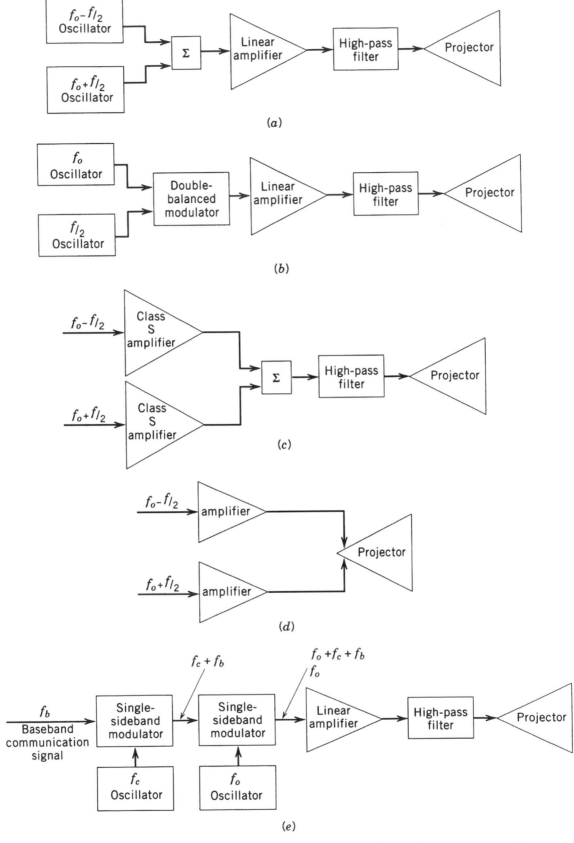

Fig. 8 Signal generation and projector drive methods: (*a*) low-level summing; (*b*) modulation; (*c*) high-level summing; (*d*) checkerboard; (*e*) communication system.

elements fed at each primary frequency). This thinning results in a loss of main-lobe source level of almost 3 dB because of higher side-lobe radiation.

The parametric source is capable of transmitting wideband communication signals.[8] Figure 8*e* illustrates one means of generating single-side-band signals for speech, simultaneous multifrequency, or video communication. Here a single-side-band modulator is fed by the carrier at f_c and the intelligence to be transmitted at the base band f_b. The resulting side band is mixed with a second (master) carrier frequency f_0 near the projector resonance. Two band shifts are used in order to avoid high downshift ratios (that would be on the order of f_0/f_b, resulting in very low parametric gains) and to allow flexibility in the choice of the difference-frequency band to be propagated. As an example, for underwater speech transmission, f_c could be 8 kHz with base-band speech (300–3000 Hz) fed to the first single side-band modulator. The resulting output (8.3–11 kHz) is then mixed in a second single-side-band modulator with f_0 = 170 kHz to produce a composite signal consisting of the master carrier f_0 at 170 kHz and the speech in the band 178.3–181.0 kHz. Medium interaction then generates the difference frequencies of 8.3–11 kHz, which are compatible with the conventional underwater (UQC) telephone. Similarly, video or other wide-band information can be transmitted by the parametric source by appropriate adjustment of the oscillator frequencies. In all cases the narrow, essentially side-lobeless, parametric beam minimizes multipath distortion (ghosts). It should be noted that signal distortion resulting from intermod- ulation between the side-band frequencies can be minimized by maintaining the master carrier at a higher level than the signal frequencies.[9] In the case of the speech system example, the carrier level is about 18 dB higher than the band level of the speech in the upper side band as fed to the projector. For more detail on these and other applications, see Refs. 8 and 9.

Beam Pattern Considerations

As indicated above, power fed to the projector at the difference frequency must be minimized to prevent direct radiation at that frequency.[9] Even though the efficiency of the projector at the difference frequency is usually quite low compared to the efficiency at the primary frequencies (where the projector is at or near resonance), small amounts of power fed to the transducer at the difference frequency can adversely affect the pattern because of the similarly low efficiency of parametric difference-frequency generation in the medium. The result of direct difference-frequency radiation is shown in the dotted portion of the beam pattern in Fig. 9. Here a large downshift ratio results in a particularly poor pattern (compared to the normal parametric beam pattern shown as the solid curve in Fig. 9). In a well-designed parametric source the side lobes can be 60 dB or more below the main lobe. Direct radiation of the difference frequency can be reduced by using a high-pass filter between the amplifier and projector. In most cases the use of parallel tuning of the (piezoceramic) transducer is sufficient to reduce direct radiation to acceptable levels. Excessive side-lobe levels at the primary frequencies can also adversely affect

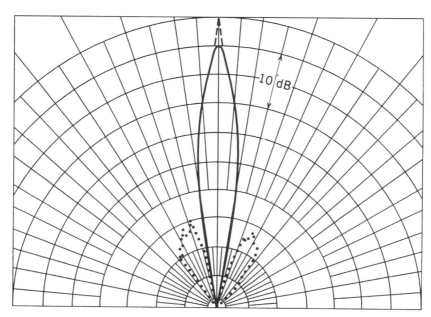

Fig. 9 Parametric source beam patterns: (———) normal pattern; (· · ·) side lobes caused by direct radiation; (- - -) excess response on axis caused by receiver nonlinearity.

the pattern, because they act like additional parametric sources.

Receiving System   The receiving hydrophone for measuring source levels and beam patterns from a parametric source is often located relatively close to the projector, that is, at ranges much less than r_{eff}. The result is that the primary-frequency components arrive at the hydrophone at levels that can be much higher than those at the difference frequency. If precautions are not taken, those primary-frequency levels will overload the receiving amplifier, generating unwanted components, including the difference frequencies, in the receiving system. The effect on the beam pattern is illustrated by the dashed curve in Fig. 9. To avoid this problem, it is necessary to attenuate the primaries prior to entering the active components of the amplifier. This can be done by inserting a low-pass filter ahead of any amplification. The filter can be RC or LC and should be designed to minimize loading the hydrophone.

3   PARAMETRIC RECEIVERS

The parametric acoustic receiving array, first suggested by Westervelt,[2] consists of a large-amplitude pump source at frequency f_0 directed at a receiving hydrophone, as shown in Fig. 10. When the signal (at frequency f) to be received arrives at the array, it interacts with the pump wave to produce sum- and difference-frequency components at $f_0 \pm f$ that are received at the hydrophone along with the high-level pump wave. After filtering, these side bands can be detected. Their levels will be proportional to the signal amplitude. The beam pattern $D(\theta)$ is identical to that of a continuous endfire array of length equal to the projector-to-hydrophone separation distance L, that is,[10]

$$D(\theta) = \frac{\sin[(2\pi f L/c)\sin^2(\theta/2)]}{(2\pi f L/c)\sin^2(\theta/2)}, \qquad (7)$$

where f is the signal frequency and θ is the angle between the pump–hydrophone axis and the incoming signal.

For a parametric receiver consisting of a piston pump projector and an identical piston hydrophone separated by the array length L, the sound pressure level of the side-band signals is[11]

$$SPL_{\pm} = SPL + SL_0 - EXDB - aL + 20\log f_0$$

$$- 10\log\left[1 + \left(\frac{R_0}{L}\right)^2\right] - 286.6 \qquad \text{dB re } 1\mu\text{Pa},$$

$$(8)$$

where SPL is the sound pressure level of the (unshifted) signal in decibels relative to 1 μPa, SL_0 is the pump rms source level in decibels relative to 1 μPa · m, f_0 is the pump frequency in kilohertz, aL is the absorption loss in decibels of the pump wave over the length L, and R_0 is the pump projector Rayleigh length, defined in Eq. (3). EXDB is the extra loss in decibels of the pump wave due to finite-amplitude distortion, a quantity that can be measured for the pump projector acting alone. (Thus, the sound pressure level of the pump wave averaged over the face of the hydrophone is approximated by

$$SL_0 - EXDB - aL - 10\log(R_0^2 + L^2) \qquad \text{dB re } 1\mu\text{Pa}, \qquad (9)$$

where R_0 and L are given in metres.) The pump frequency f_0 is normally much higher in frequency and

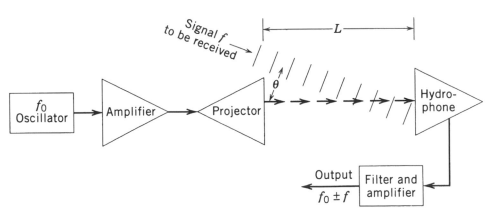

Fig. 10   Parametric receiver.

amplitude than those of the signal. Therefore careful processing and filtering must be done in the receiving system to separate out the parametric side bands from the extremely strong pump signal.[12]

4 HARMONIC GENERATION

As noted above, the propagation of a finite-amplitude sound wave is accompanied by the generation of harmonics (see also Section 5 in Chapter 19 and Section 4 in Chapter 23). The amplitude of the second-harmonic component is the largest. (For a discussion of higher harmonics, see Section 5.3 in Chapter 23.) In the far field of the projector and at ranges for which the fundamental loss ar exceeds a few decibels, the second harmonic can be assumed to be locally generated. Then, the second-harmonic pressure level is[13,14]

$$SPL_2 = 2(SL_1 - EXDB) - 20 \log r + 20 \log f_1 - 4ar$$
$$+ 20 \log[Ei(2\alpha r) - Ei(2\alpha R_1)] - 286.6$$
$$dB \, re 1\mu Pa, \tag{10}$$

where SL_1 is the rms source level at the fundamental frequency in decibels relative to 1 $\mu Pa \cdot m$, EXDB the (measured) extra loss in decibels due to harmonic generation, r the range in metres, f_1 the fundamental frequency in kilohertz, a the fundamental absorption coefficient in decibels per metre, Ei the exponential integral function,[15] $\alpha = a/8.686$, and R_1 the fundamental Rayleigh length [see Eq. (3)] for the projector. The quantity EXDB may be determined by measurement of the fundamental level,

$$SPL_1 = SL_1 - EXDB - 20 \log r - ar \qquad dB \, re 1\mu Pa, \tag{11}$$

at the range r. If the fundamental beam pattern $D_1(\theta, \phi)$ is likewise known, the second-harmonic beam pattern $D_2(\theta, \phi)$ can be estimated as[16]

$$D_2(\theta, \phi) = [D_1(\theta, \phi)]^2, \tag{12}$$

where θ and ϕ are the polar and azimuthal angles about the projector main response axis.

REFERENCES

1. M. B. Moffett and R. H. Mellen, "Model for Parametric Acoustic Sources," *J. Acoust. Soc. Am.*, Vol. 61, 1977, pp. 325–337.

2. P. J. Westervelt, "Parametric Acoustic Array," *J. Acoust. Soc. Am.*, Vol. 35, 1963, pp. 535–537.

3. H. O. Berktay and D. J. Leahy, "Farfield Performance of Parametric Transmitters," *J. Acoust. Soc. Am.*, Vol. 55, 1974, pp. 539–546.

4. M. B. Moffett and R. H. Mellen, "On Parametric Source Aperture Factors," *J. Acoust. Soc. Am.*, Vol. 60, 1976, pp. 581–583.

5. M. B. Moffett and R. H. Mellen, "Nearfield Characteristics of Parametric Acoustic Sources," *J. Acoust. Soc. Am.*, Vol. 69, 1981, pp. 404–409.

6. M. B. Moffett and R. H. Mellen, "Effective Lengths of Parametric Acoustic Sources," *J. Acoust. Soc. Am.*, Vol. 70, 1981, pp. 1424–1426. See also "Erratum," *J. Acoust. Soc. Am.*, Vol. 71, 1982, p. 1039.

7. W. L. Konrad and W. L. Clay, "Summing Amplifier," U.S. Patent 4,311,929, Jan. 19, 1982.

8. A. H. Quazi and W. L. Konrad, "Underwater Acoustic Communications," *IEEE Comm. Mag.*, March, 1982, pp. 24–30.

9. W. L. Konrad, "Design and Performance of Parametric Sonar Systems," Naval Underwater Systems Center Technical Report 5227, Sept. 24, 1975.

10. P. H. Rogers, A. L. Van Buren, A. O. Williams, Jr., and J. M. Barber, "Parametric Detection of Low-Frequency Acoustic Waves in the Nearfield of an Arbitrary Directional Pump Transducer," *J. Acoust. Soc. Am.*, Vol. 66, 1974, pp. 528–534.

11. M. B. Moffett, W. L. Konrad, and J. C. Lockwood, "A Saturated Parametric Acoustic Receiver," *J. Acoust. Soc. Am.*, Vol. 66, 1979, pp. 1842–1847.

12. D. F. Rohde, T. G. Goldsberry, W. S. Olsen, and C. R. Reeves, "Band Elimination Processor for an Experimental Parametric Acoustic Receiving Array," *J. Acoust. Soc. Am.*, Vol. 66, 1979, pp. 484–487.

13. M. B. Moffett, "Measurement of Fundamental and Second Harmonic Pressures in the Field of a Circular Piston Source," *J. Acoust. Soc. Am.*, Vol. 65, 1979, pp. 318–323.

14. M. H. Safar, "The Propagation of Spherical Acoustic Waves of Finite Amplitude in Fresh and Sea Water," *J. Sound Vib.*, Vol. 13, 1970, pp. 1–7.

15. M. Abramowitz and I. A. Stegun, *Handbook of Mathematical Functions*, Dover Books, New York, 1965, pp. 231–232.

16. J. C. Lockwood, T. G. Muir, and D. T. Blackstock, "Directive Harmonic Generation in the Radiation Field of a Circular Piston," *J. Acoust. Soc. Am.*, Vol. 53, 1973, pp. 1148–1153.